Advances in Chemistry

A Selection of C. N. R. Rao's Publications (1994 – 2003)

WORLD SCIENTIFIC SERIES IN 20TH CENTURY CHEMISTRY

Consulting Editors: D. H. R. Barton (Texas A&M University)
F. A. Cotton (Texas A&M University)
Y. T. Lee (Academia Sinica, Taiwan)
A. H. Zewail (California Institute of Technology)

Published:

Vol. 1: Molecular Structure and Statistical Thermodynamics
— Selected Papers of Kenneth S. Pitzer
by Kenneth S. Pitzer

Vol. 2: Modern Alchemy
— Selected Papers of Glenn T. Seaborg
by Glenn T. Seaborg

Vol. 3: Femtochemistry: Ultrafast Dynamics of the Chemical Bond
by Ahmed H. Zewail

Vol. 4: Solid State Chemistry
— Selected Papers of C. N. R. Rao
edited by S. K. Joshi and R. A. Mashelkar

Vol. 5: NMR in Structural Biology
— A Collection of Papers by Kurt Wüthrich
by Kurt Wüthrich

Vol. 6: Reason and Imagination: Reflections on Research in Organic Chemistry
— Selected Papers of Derek H. R. Barton
by Derek H. R. Barton

Vol. 7: Frontier Orbitals and Reaction Paths
— Selected Papers of Kenichi Fukui
edited by K. Fukui and H. Fujimoto

Vol. 8: Quantum Chemistry
— Classic Scientific Papers
translated and edited by Hinne Hettema

Vol. 9: The Oxidation of Oxygen and Related Chemistry
— Selected Papers of Neil Bartlett
edited with introductory essays by Neil Bartlett

Vol. 10: Linus Pauling
— Selected Scientific Papers
*edited by Barclay Kamp, Linda Pauling Kamb, Peter Jeffress Pauling,
Alexander Kamb and Linus Pauling, Jr.*

Vol. 11: Across Conventional Lines — Selected Papers of George A. Olah
edited by George A. Olah and G. K. Surya Prakash

World Scientific Series in 20th Century Chemistry – Vol. 12

ADVANCES IN CHEMISTRY

A Selection of C. N. R. Rao's Publications (1994 – 2003)

Editors

J. Gopalakrishnan

Indian Institute of Science

G. U. Kulkarni

Jawaharlal Nehru Centre for Advanced Scientific Research, India

World Scientific

New Jersey • London • Singapore • Hong Kong

Published by

World Scientific Publishing Co. Pte. Ltd.

5 Toh Tuck Link, Singapore 596224

USA office: Suite 202, 1060 Main Street, River Edge, NJ 07661

UK office: 57 Shelton Street, Covent Garden, London WC2H 9HE

British Library Cataloguing-in-Publication Data
A catalogue record for this book is available from the British Library.

The editors and publisher would like to thank the following publishers for their permission to reproduce the articles found in this volume:

The Royal Society of Chemistry (*J. Mater. Chem., Chem Commun., Chem. Soc. Rev., Dalton Trans., J. Chem. Soc. Dalton Trans.*); American Chemical Society (*J. Phys. Chem., Chem. Mater., J. Phys. B, Acc. Chem. Res., J. Phy. D, J. Am. Chem. Soc., Inorg. Chem.*); Blackwell Science Publishers (*A Chemistry for the 21st Century monograph, IUPAC, Ed. M.W. Roberts*; *A Chemistry for the 21st Century monograph, IUPAC, Ed. V.V. Boldyrev*); Elsevier Science Ltd., Oxford (*Solid State Commun., Chem. Phys. Lett., J. Mol. Struct., J. Mol. Struct. (Theochem), Micro. Meso. Mater., Solid State Sci., Mater. Res. Bull., Euro. J. Solid State Inorg. Chem.*); Wiley-VCH (Wiley) (*Chem. Euro. J., ChemPhysChem, Angew. Chem. Int. Ed.*); American Institute of Physics (*Phys. Rev.*); Institute of Physics Publishing (*J. Phys. Condens. Matter*); Taylor and Francis (*Mol. Phys.*).

ADVANCES IN CHEMISTRY
A Selection of C.N.R. Rao's Publications (1994–2003)

ISBN 981-238-599-1

Printed in Singapore by Mainland Press

Professor C.N.R. Rao

FOREWORD

Professor Chintamani Nagesa Ramachandra Rao (C.N.R. Rao) is a towering personality in Indian science in the post-Independence era, whose immense contributions span over five decades covering several facets of Chemical Sciences. Professor Rao's prolific research work delineated in more than 1200 publications and 36 books, has made an impact on a wide range of topics that include chemical spectroscopy, molecular structure, various aspects of the solid state — synthesis, structure and properties. To him, it concerns very little whether the system under investigation is organic, inorganic or a composite material, but it is the property or the phenomenon that interests him always. For example, he has worked with equal felicity on rutile-anatase, metal-insulator, plastic-crystalline, para-ferri-ferro magnetic transitions as well as on charge-ordering, phase separation, donor-acceptor complexes, superconductors, and last but not least, the fascinating field of fullerenes and nanomaterials.

Solid State and Materials Chemistry has been very close to his heart during the last two decades, where he has made significant contributions to the synthesis of novel materials, developing new routes to synthesis, copper oxide based superconductors, manganese oxide based magnetoresistive materials, among others. Particularly noteworthy is his unique work on porous and open-framework solids, where he has not only assembled a variety of crystalline architectures but also unravelled the fundamental principles governing the formation of structures of different dimensionalities. A natural extension has been into the fascinating world of the supramolecular chemistry, where for example, he has synthesized a novel organic zeolite and a hybrid solid containing layers of silver sheets separated by an organic spacer. He has established a world class research laboratory for the synthesis and investigation of nanomaterials in various forms of aggregation including clusters, tubes and wires; a special Y-junction nanotube showing rectification, doped gallium nitride nanowires and several nanoobjects based on transition metal chalcogenides are some of the important materials created in his laboratory.

It is with a profound sense of respect and gratitude to Professor C.N.R. Rao that we present to the scientific community, this monograph containing a selected collection of seventy of his research publications spanning the last one decade (1994–2003). It has not been an easy task to make this selection, especially because another set of equally important papers could have been easily collected from his 350 publications during this period. In making this selection, **an effort** has been made to bring out the flavour of Professor Rao's contribution to the entire gamut of Chemistry and Physics of Materials, in general. The collection of papers under different sections is preceded by Professor Rao's comments which embellish the contents with a personal touch. It must be added that such an elegant publication would not have been possible without the support from the Third World Academy of Sciences, Trieste, and the cooperation of the Publisher, World Scientific Publishing Co., Singapore especially, Dr. K.K. Phua, and Ms. Lakshmi Narayanan.

We do believe that the present volume will not only be treasured by his friends, close associates and admirers but also be cherished by young practitioners. For the two of us, it has indeed been a great privilege and honour to serve as Editors of this special monograph felicitating Professor C.N.R. Rao on his 70th birthday.

May God Almighty bless him with many more
years of research activity and highest scientific
laurels!

G.U. Kulkarni   J. Gopalakrishnan

Contents

I. Highlights of Materials Chemistry — 1
C.N.R. Rao

1. Novel Materials, Materials Design and Synthetic Strategies: Recent Advances and New Directions, *J. Mater. Chem.* **9**, 1 (1999). — 5

2. The Metal–Nonmetal Transition: A Global Perspective, (with P.P. Edwards and T.V. Ramakrishnan), *J. Phys. Chem.* **99**, 5228 (1995). — 19

3. Virtues of Marginally Metallic Oxides, *Chem. Commun.*, 2217 (1996). — 31

4. Phase Separation in Metal Oxides, (with P.V. Vanitha and A.K. Cheetham), *Chem. Eur. J.* **9**, 829 (2003). — 37

5. Science and Technology of Nanomaterials: Current Status and Future Prospects, (with A.K. Cheetham), *J. Mater. Chem.* **11**, 2887 (2001). — 45

II. Transition Metal Oxides (including Cuprate Superconductors) — 53
C.N.R. Rao

6. Chemical Strategies for the Synthesis of Metal Oxides, in *Reactivity of Solids: Past, Present and Future, A 'Chemistry for the 21st Century' monograph*, IUPAC, Ed. V.V. Boldyrev, Blackwell Science Publishers, Oxford, p. 237, 1996. — 55

7. Solid–Solid Interfaces in the Epitaxial Films of Complex Oxides Deposited by Chemical Methods, (with A.R. Raju and H.N. Aiyer), in *Interfacial Science, A 'Chemistry for the 21st Century' monograph*, IUPAC, Ed. M.W. Roberts, Blackwell Science Publishers, p. 1, 1997. — 70

8. Sub-micrometre Spherical Particles of TiO_2, ZrO_2 and PZT by Nebulized Spray Pyrolysis of Metal–Organic precursors, (with P. Murugavel, M. Kalaiselvam and A.R. Raju), *J. Mater. Chem.* **7**, 1433 (1997). — 89

9. A Study of Cubic Bismuth Oxides of the Type $Bi_{26-x}M_xO_{40-\delta}$ 95
   (M = Ti, Mn, Fe, Co, Ni or Pb) Related to γ-Bi_2O_3,
   (with N. Rangavittal and T.N. Guru Row),
   Eur. J. Solid State Inorg. Chem. **31**, 409 (1994).

10. Effect of Cation Size and Disorder on the Structure and   109
    Properties of the Rare Earth Cobaltates, $Ln_{0.5}A_{0.5}CoO_3$,
    (with P.V. Vanitha, A. Arulraj and P.N. Santosh),
    Chem. Mater. **12**, 1666 (2000).

11. Orbital Ordering as the Determinant for Ferromagnetism in   114
    Biferroic $BiMnO_3$, (with A.M. dos Santos, A.K. Cheetham,
    T. Atou, Y. Syona, Y. Yamaguchi, K. Ohoyama and H. Chiba),
    Phys. Rev. **B66**, 64425 (2002).

12. Structure–Property Relationships in Superconducting   118
    Cuprates, (with A.K. Ganguli),
    Chem. Soc. Rev., 1 (1995).

13. Stripes and Superconductivity in Cuprates — Is there a   125
    Connection? (with N. Kumar),
    ChemPhysChem **4**, 439 (2003).

III. Colossal Magnetoresistance, Charge Ordering and   131
    Related Aspects of Rare Earth Manganates
    C.N.R. Rao

14. Colossal Magnetoresistance, Charge Ordering and Other   133
    Novel Properties of Manganates and Related Materials,
    (with A.K. Raychaudhuri),
    in *Colossal Magnetoresistance, Charge Ordering and Related
    Properties of Manganese Oxides*, Eds. C.N.R. Rao and B. Raveau,
    World Scientific, Singapore, 1998, p. 1.

15. Charge, Spin and Orbital Ordering in the Perovskite   175
    Manganates, $Ln_{1-x}A_xMnO_3$ (Ln = Rare Earth, A = Ca or Sr),
    J. Phys. Chem. **B104**, 5877 (2000).

16. Electron–Hole Asymmetry in the Rare-Earth Manganates: A   188
    Comparative Study of the Hole- and Electron-Doped Materials,
    (with K. Vijaya Sarathy, P.V. Vanitha, R. Seshadri
    and A.K. Cheetham),
    Chem. Mater. **13**, 787 (2001).

17. Electric-Field-Induced Melting of the Randomly Pinned   197
Charged-Ordered States of Rare-Earth Manganates and
Associated Effects, (with A.R. Raju, V. Ponnambalam,
S. Parashar and N. Kumar),
Phys. Rev. **B61**, 594 (2000).

18. Occurrence of Re-entrant Ferromagnetic Transitions in   202
Rare-Earth Manganates on Cooling the Charge-Ordered States,
(with A.K. Kundu and P.V. Vanitha),
Solid State Commun. **125**, 41 (2003).

19. Electronic Phase Separation in the Rare-Earth Manganates   206
$(La_{1-x}Ln_x)_{0.7}Ca_{0.3}MnO_3$ (Ln = Nd, Gd and Y),
(with L. Sudheendra),
J. Phys. Condens. Matter **15**, 3029 (2003).

IV. Nanoparticles   **219**
C.N.R. Rao

20. Growth of Nanometric Gold Particles in Solution Phase   221
(with R. Seshadri, G.N. Subbanna, V. Vijayakrishnan,
G.U. Kulkarni and G. Ananthakrishna),
J. Phys. Chem. **99**, 5639 (1995).

21. Size-Dependent Chemistry: Properties of Nanocrystals,   227
(with G.U. Kulkarni, P.J. Thomas and P.P. Edwards),
Chem. Euro. J. **8**, 29 (2002).

22. Superlattices of Metal and Metal–Semiconductor Quantum   234
Dots Obtained by Layer-by-Layer Deposition of Nanoparticle
Arrays, (with K. Vijaya Sarathy, P.J. Thomas and G.U. Kulkarni),
J. Phys. Chem. **B103**, 399 (1999).

23. X-ray Photoelectron Spectroscopic Investigations of Cu–Ni,   237
Au–Ag, Ni–Pd and Cu–Pd Bimetallic Clusters,
(with K.R. Harikumar and S. Ghosh),
J. Phys. Chem. **A101**, 536 (1997).

24. Magic Nuclearity Giant Clusters of Metal Nanocrystals   242
Formed by Mesoscale Self-Assembly,
(with P.J. Thomas and G.U. Kulkarni),
J. Phys. Chem. **B105**, 2515 (2001).

25. Novel Effects of Metal Ion Chelation on the Properties of   245
Lipoic Acid-Capped Ag and Au Nanoparticles,
(with S. Berchmans and P.J. Thomas),
J. Phys. Chem. **B106**, 4647 (2002).

V. Nanotubes and Nanowires 251
C.N.R. Rao

26. Carbon Nanotubes from Organometallic Precursors, 253
(with A. Govindaraj),
Acc. Chem. Res. **35**, 998 (2002).

27. Production of Bundles of Aligned Carbon and 263
Carbon–Nitrogen Nanotubes by the Pyrolysis of Precursors
on Silica-Supported Iron and Cobalt Catalysts,
(with M. Nath, B.C. Satishkumar, A. Govindaraj and C.P. Vinod),
Chem. Phys. Lett. **322**, 333 (2000).

28. B–C–N, C–N and B–N Nanotubes Produced by the 271
Pyrolysis of Precursor Molecules over Co Catalysts,
(with R. Sen, B.C. Satishkumar, A. Govindaraj,
K.R. Harikumar, G. Raina, J.-P. Zhang and A.K. Cheetham),
Chem. Phys. Lett. **287**, 671 (1998).

29. Boron–Carbon Nanotubes from the Pyrolysis of C_2H_2–B_2H_6 277
Mixtures, (with B.C. Satishkumar, A. Govindaraj,
K.R. Harikumar, J.-P. Zhang and A. K. Cheetham),
Chem. Phys. Lett. **300**, 473 (1999).

30. Novel Experiments with Carbon Nanotubes: Opening, 282
Filling, Closing and Functionalizing Nanotubes,
(with B.C. Satishkumar, A. Govindaraj, J. Mofokeng and
G.N. Subbanna),
J. Phys. **B29**, 4925 (1996).

31. The Decoration of Carbon Nanotubes by Metal Nanoparticles, 292
(with B.C. Satishkumar, E.M. Vogl and A. Govindaraj),
J. Phys. **D29**, 3173 (1996).

32. A Study of Micropores in Single-Walled Carbon Nanotubes 296
by the Adsorption of Gases and Vapors,
(with M. Eswaramoorthy and R. Sen),
Chem. Phys. Lett. **304**, 207 (1999).

33. Pressure-Induced Phase Transformation and the Structural 300
Resilience of Single-Walled Carbon Nanotube Bundles,
(with S.M. Sharma, S. Karmakar, S.K. Sikka, P.V. Teredesai,
A.K. Sood and A. Govindaraj),
Phys. Rev. **B63**, 205417 (2001).

34. Hydrogen Storage in Carbon Nanotubes and Related 305
Materials, (with G. Gundiah, A. Govindaraj, N. Rajalakshmi
and K.S. Dhathathreyan),
J. Mater. Chem. **13**, 209 (2003).

35. Inorganic Nanotubes, (with M. Nath), 310
    Dalton Trans., 1 (2003).

36. Metal Nanowires and Intercalated Metal Layers in 334
    Single-Walled Carbon Nanotube Bundles,
    (with A. Govindaraj, B.C. Satishkumar and M. Nath),
    Chem. Mater. **12**, 202 (2000).

37. Synthesis of Metal Oxide Nanorods Using Carbon 338
    Nanotubes as Templates, (with B.C. Satishkumar,
    A. Govindaraj and M. Nath),
    J. Mater. Chem. **10**, 2115 (2000).

38. Nanowires, Nanobelts and Related Structures of Ga_2O_3, 343
    (with G. Gundiah and A. Govindaraj),
    Chem. Phys. Lett. **351**, 189 (2002).

39. Synthesis and Characterization of Silicon Carbide, Silicon 349
    Oxynitride and Silicon Nitride Nanowires,
    (with G. Gundiah, G.V. Madhav, A. Govindaraj and
    Md. M. Seikh),
    J. Mater. Chem. **12**, 1606 (2002).

40. Photoluminescence Spectra and Ferromagnetic Properties 355
    of GaMnN Nanowires, (with F.L. Deepak, P.V. Vanitha
    and A. Govindaraj),
    Chem. Phys. Lett. **374**, 314 (2003).

VI. Molecular Solids **361**
    C.N.R. Rao

41. Investigations of Diamond–Graphite Hybrids and Fullerenes 363
    with Seven-Membered Rings, (with R. Sen, R. Sumathy and
    B.C. Satishkumar),
    J. Mol. Struct. **436–437**, 11 (1997).

42. Polymerization and Pressure-Induced Amorphization of C_{60} 371
    and C_{70}, (with A. Govindaraj, H.N. Aiyer and R. Seshadri),
    J. Phys. Chem. **99**, 16814 (1995).

43. A Combined Experimental and Theoretical Study of the 374
    Charge-Transfer Compound between $C_{60}Br_8$ and
    Tetrathiafulvalene, (with A. Govindaraj, R. Sumathy
    and A.K. Sood),
    Mol. Phys. **89**, 267 (1996).

44. Hydrothermal Synthesis of Organic Channel Structures: 385
1:1 Hydrogen-Bonded Adducts of Melamine with Cyanuric
and Trithiocyanuric Acids, (with A. Ranganathan and
V.R. Pedireddi),
J. Am. Chem. Soc. **121**, 1752 (1999).

45. An Organic Channel Structure Formed by the   387
Supramolecular Assembly of Trithiocyanuric Acid
and 4,4'-bipyridyl, (with A. Ranganathan, V.R. Pedireddi
and S. Chatterjee),
J. Mater. Chem. **9**, 2407 (1999).

46. Self-Assembled Four-Membered Networks of Trimesic Acid   392
Forming Channel Structures, (with S. Chatterjee,
V.R. Pedireddi and A. Ranganathan),
J. Mol. Struct. **520**, 107 (2000).

47. A Novel Hybrid Layer Compound Containing Silver   401
Sheets and An Organic Spacer, (with A. Ranganathan,
V.R. Pedireddi and A.R. Raju),
Chem. Commun., 39 (2000).

48. Experimental and Theoretical Electronic Charge Densities in   403
Molecular Crystals, (with G.U. Kulkarni and R.S. Gopalan),
J. Mol. Struc. (Theochem) **500**, 339 (2000).

49. An Experimental Charge Density Study of the Effect of the   427
Noncentric Crystal Field on the Molecular Properties of
Organic NLO Materials, (with R.S. Gopalan and G.U. Kulkarni),
ChemPhysChem **1**, 127 (2000).

VII. Porous Solids   437
C.N.R. Rao

50. Evidence for Supramolecular Organization of Alkane and   439
Surfactant Molecules in the Process of Forming
Mesoporous Silica, (with N. Ulagappan),
Chem. Commun., 2759 (1996).

51. Phase Transformations in Mesoporous Zirconia,   441
(with Neeraj),
J. Mater. Chem. **8**, 1631 (1998).

52. Mesoporous Phases Based on SnO_2 and TiO_2,   445
(with N. Ulagappan),
Chem. Commun., 1685 (1996).

53. Mesoporous Aluminoborates, (with S. Ayyappan),   447
Chem. Commun., 575 (1997).

54. High Catalytic Efficiency of Transition Metal Complexes   449
Encapsulated in a Cubic Mesoporous Phase,
(with M. Eswaramoorthy and Neeraj),
Chem. Commun., 615 (1998).

55. Metal Chalcogenide–Organic Nanostructured Composites   451
from Self-Assembled Organic Amine Templates,
(with Neeraj),
J. Mater. Chem. **8**, 279 (1998).

56. Synthesis of Hexagonal Microporous Silica and   453
Aluminophosphate by Suparmolecular Templating of a
Short-Chain Amine, (with M. Eswaramoorthy and S. Neeraj),
Micro. Meso. Mater. **28**, 205 (1999).

57. Macroporous Oxide Materials with Three-Dimensionally   459
Interconnected Pores, (with G. Gundiah),
Solid State Sci. **2**, 877 (2000).

58. Macroporous Carbons Prepared by Templating Silica   465
Spheres, (with G. Gundiah and A. Govindaraj),
Mater. Res. Bull. **36**, 1751 (2001).

VIII. Open Framework Materials   **473**
    C.N.R. Rao

59. Aufbau Principle of Complex Open-Framework Structures   475
of Metal Phosphates with Different Dimensionalities,
(with S. Natarajan, A. Choudhury, S. Neeraj and A.A. Ayi),
Acc. Chem. Res. **34**, 80 (2001).

60. Understanding the Building-Up Process of Three   483
Dimensional Open-Framework Metal Phosphates:
Acid Degradation of the 3D Structures to Lower
Dimensional Structures, (with A. Choudhury),
Chem. Commun., 366 (2003).

61. A New Route for the Synthesis of Open-Framework   485
Metal Phosphates Using Organophosphates,
(with S. Neeraj, P.M. Forster and A.K. Cheetham),
Chem. Commun., 2716 (2001).

62. An Open-Framework Iron Phosphate with Large Voids   487
Exhibiting Spin-Crossover,
(with A. Choudhary and S. Natarajan),
Chem. Commun., 1305 (1999).

63. An Unusual Open-Framework Cobalt(II) Phosphate with   489
a Channel Structure that Exhibits Structural and Magnetic
Transitions, (with A. Choudhury, S. Neeraj and S. Natarajan),
Angew. Chem. Int. Ed. **39**, 3091 (2000).

64. Organically Templated Mixed-Valent Iron Sulfates Possessing   492
Kagomé and Other Types of Layered Networks,
(with G. Paul, A. Choudhury and E.V. Sampathkumaran),
Angew. Chem. Int. Ed. **41**, 4297 (2002).

65. Three-Dimensional Organically Templated Open-Framework   496
Transition Metal Selenites,
(with A. Choudhury and U. Kumar),
Angew. Chem. Int. Ed. **41**, 158 (2002).

66. Synthesis of a Hierarchy of Zinc Oxalate Structures from   500
Amine Oxalates, (with R. Vaidhyanathan and S. Natarajan),
J. Chem. Soc., Dalton Trans., 699 (2001).

67. Aliphatic Dicarboxylates with Three-Dimensional   508
Metal–Organic Frameworks Possessing Hydrophobic
Channels, (with R. Vaidhyanathan and S. Natarajan),
Dalton Trans., 1459 (2003).

68. Hybrid Open-Framework Iron Phosphate–Oxalates   514
Demonstrating a Dual Role of the Oxalate Unit,
(with A. Choudhury and S. Natarajan),
Chem. Euro. J. **6**, 1168 (2000).

69. Hybrid Inorganic–Organic Host–Guest Compounds: 522
Open-Framework Cadmium Oxalates Incorporating
Novel Extended Structures of Alkali Halides,
(with R. Vaidhyanathan and S. Natarajan),
Chem. Mater. **13**, 3524 (2001).

70. Sodalite Networks Formed by Metal Squarates,   532
(with S. Neeraj, M.L. Noy and A.K. Cheetham),
Solid State Sci. **4**, 1231 (2002).

Brief Biodata of Professor C.N.R. Rao   539

Scientific Colleagues   541

I. Highlights of Materials Chemistry

C.N.R. Rao

CSIR Centre of Excellence in Chemistry,
Chemistry & Physics of Materials Unit
Jawaharlal Nehru Centre for Advanced Scientific Research
Jakkur P.O., Bangalore-560 064, INDIA
cnrrao@jncasr.ac.in

MATERIALS CHEMISTRY as we understand it today is relatively of recent origin. A few decades ago, the subject generally included cement, steel, and a few other topics of an applied nature. In the last 25 years or so, the subject has emerged to be recognized as an important new direction in modern chemistry, having incorporated all the salient aspects of solid state chemistry as well.[1,2] While solid state chemistry may be considered to be a more formal academic subject representing the chemical counterpart of solid state physics, materials chemistry deals with structure, response, and function and has the ultimate purpose of developing novel materials or understanding structure-property relations and phenomena related to a wide range of materials. Structure and synthesis are integral parts of the subject, and they are fully utilized in the strategies for tailor-making materials with desired and controllable properties, be they electronic, magnetic, optical, dielectric, thermal, adsorptive, or catalytic. The materials can be organic, inorganic, or biological and can be in any condensed state of matter. Materials chemistry is thus a truly interdisciplinary subject incorporating the principles, systems, and practice of all branches of chemistry and based on sound foundations of physics. With this description, it becomes difficult to classify materials compared to earlier years when it was common to classify them as ceramics, metals, organics and so on. We now have organic metals, superconductors, and nonlinear materials as well as the ceramic counterparts. It is probably more convenient to classify materials based on properties or phenomena. For example, porous solids, superconductors, polymers, ferroics, and composites cover all types of chemical constituents. However, it is common to distinguish molecular solids from extended solids, as they represent two limiting descriptions.

Materials chemistry contains all the elements of modern chemistry. These include synthesis, structure, dynamics, and properties. In synthesis, one employs all possible methods and conditions from high-temperature and high-pressure techniques to mild solution methods (chimie douce or soft chemistry).[3] Chemical methods generally tend to be more delicate, often yielding novel as well as metastable products. Kinetic rather than thermodynamic control of reactions favors the formation of such structures. Supramolecular organization provides ways of designing novel materials.[4]

All available methods of diffraction, microscopy, and spectroscopy are used for structure elucidation in present-day materials chemistry.[1,2] For detailed structure determination, even powders suffice for the most part because of the advances in diffraction profile analysis. These advances in structural tools enable more meaningful correlations of structure with properties and phenomena. Catalysis is becoming more of a science partly because of our ability to unravel the structures and surfaces of catalysts. Phase transitions of all varieties[5] are being investigated more and more by chemists.

Materials chemistry is as old as chemistry itself. The early controversy over whether compounds were stoichiometric (e.g., NaCl) or nonstoichiometric (Fe_xS) may be considered to be the beginning of chemists' foray into the chemistry of solids. The problem was only solved in the late 1960s when it became clear that vacancies are not tolerated in high concentrations in oxides and other inorganic solids.

The study of nonstoichiometric compounds, decompositions of solids, and other solid state reactions occupied much of the attention of solid state chemists in the 1950s. There was also some effort during that period to understand the electrical and optical properties of organic solids such as anthracene. By the late 1960s, however, there was considerable work on complex inorganic materials, especially oxides and sulfides of various structures. One could even rationalize the electronic and magnetic properties of such materials based on their electronic and crystal structures. Oxide metals became known during this period, as did conducting polymers. Metal-insulator transitions in oxides such as V_2O_3 also came to the fore during this period. Transition metal oxides constituted a major part of early materials chemistry and continue to dominate the subject. It is not surprising that some of the major discoveries are also in the area of oxide materials. These include high-temperature superconductivity and colossal magnetoresistance. In this section, articles of general interest dealing with various important aspects of materials chemistry are included.

References

1. C.N.R. Rao and J. Gopalakrishnan, *New Directions in Solid State Chemistry*, Cambridge University Press, Second Edition, 1997.
2. C.N.R. Rao (Ed.), *Chemistry of Advanced Materials*, IUPAC 21st Century monograph, Blackwell, Oxford, 1993.

3. C.N.R. Rao, *Chemical Approaches to the Synthesis of Inorganic Materials*, John Wiley, New York, 1994.

4. W. Jones and C.N.R. Rao (Eds.), *Supramolecular Organization and Materials Design*, Cambridge University Press, 2002.

5. C.N.R. Rao and K.J. Rao, *Phase Transitions in Solids*, McGraw Hill, New York, 1978.

Novel materials, materials design and synthetic strategies: recent advances and new directions†

JOURNAL OF
Materials
Feature Article
CHEMISTRY

C. N. R. Rao‡

Solid State and Structural Chemistry Unit, Indian Institute of Science, Bangalore 560 012, India and Chemistry & Physics of Materials Unit, Jawaharlal Nehru Centre for Advanced Scientific Research, Jakkur P.O., Bangalore 560 064, India

Received 5th May 1998, Accepted 12th June 1998

There have been major advances in solid state and materials chemistry in the last two decades and the subject is growing rapidly. In this account, a few of the important aspects of materials chemistry of interest to the author are presented. Accordingly, transition metal oxides, which constitute the most fascinating class of inorganic materials, receive greater attention. Metal–insulator transitions in oxides, high temperature superconductivity in cuprates and colossal magnetoresistance in manganates are discussed at some length and the outstanding problems indicated. We then discuss certain other important classes of materials which include molecular materials, biomolecular materials and porous solids. Recent developments in synthetic strategies for inorganic materials are reviewed. Some results on metal nanoparticles and nanotubes are briefly presented. The overview, which is essentially intended to provide a flavour of the subject and show how it works, lists references to many crucial reviews in the recent literature.

I have been working in solid state and materials chemistry for over four decades. When I first got interested in the subject, it was not an integral part of main-stream chemistry. In the first book on solid state chemistry[1a] that I edited in 1970, it was stated in the preface, that 'Solid state chemistry has not become part of the formal training programmes in chemistry. Being one of the frontiers of chemistry, it has a tremendous future and undoubtedly demands the active involvement of many more chemists'. In 1986, in the first edition of the book *New Directions in Solid State Chemistry*,[1b] the preface states, 'At the present time, solid state chemistry is mainly concerned with the development of new methods of synthesis, new ways of identifying and characterizing materials and of describing their structure, and above all, with new strategies of tailor-making materials with desired and controllable properties, be they electronic, magnetic, dielectric, optical, absorptive or catalytic. It is heartening that the subject is increasingly coming to be recognized as an emerging area of chemical science'. In the last few years, solid state chemistry has given room to the broader area of materials chemistry. The subject has come of age and there are many reviews and books dealing with it.[2] I am delighted to have this opportunity to present an overview of the subject in this first materials chemistry discussion meeting. In so doing, I will try to present the highlights of some of the areas and refer to important reviews in the literature, but I cannot help dealing with those aspects which are of personal interest to me in somewhat greater detail. It is indeed impractical to cover every aspect of this vast subject in

Table 1 A description of materials chemistry

Constituent units	State[b]	Function	Advanced technology
atoms molecules[a] ions[a]	crystalline (molecular, ionic, polymeric, metallic *etc.*) non-crystalline (glasses) clusters and nanomaterials liquid crystalline	miniaturization selectivity and recognition transformation transduction transport energy storage	nanolithography microelectronics magnets sensors and transducers photonic devices energy devices porous solids and membranes micromachines

[a]Inorganic or organic. [b]In pure (monophasic) form or in the form of aggregates or composites.

an article. Furthermore, all research in solid state chemistry is not necessarily related to materials development. Materials chemistry, as distinct from solid state chemistry, deals with structure, response and function, and has an ultimate technological objective. I illustrate this aspect of materials chemistry in Table 1.

If one were to list the most important discoveries in solid state and materials science in the last decade, it would include high-temperature cuprate superconductors (1986), fullerenes and related materials (1990), mesoporous silica (1992) and colossal magnetoresistance (CMR) in manganates (1993). Three of these deal with metal oxides. I have been involved in research on metal oxides for many years and it has been exciting to witness the increasing importance gained by these materials. I shall deal with certain aspects of the chemistry of transition metal oxides at some length. This discussion is presented as a personal account and would also serve as a case-study. I will indicate the developments in molecular materials, porous solids and biomaterials and highlight out the role of chemical synthesis in inorganic systems. I will end the article with a discussion of some of my recent interests in nanotubes and metal nanoparticles.

Transition metal oxides

Transition metal oxides constitute the most fascinating class of materials, exhibiting a variety of structures and properties.[3] The metal–oxygen bond can vary anywhere between highly ionic to covalent or metallic. The unusual properties of transition metal oxides are clearly due to the unique nature of the outer d-electrons. The phenomenal range of electronic and magnetic properties exhibited by transition metal oxides is noteworthy. Thus, the electrical resistivity in oxide materials spans the extraordinary range of 10^{-10} to 10^{20} ohm cm. We have oxides with metallic properties (*e.g.* RuO_2, ReO_3, $LaNiO_3$) at one end of the range and oxides with highly insulating behaviour (*e.g.* $BaTiO_3$) at the other. There are also

†Basis of the presentation given at Materials Chemistry Discussion No. 1, 24–26 September 1998, ICMCB, University of Bordeaux, France.
‡Also at the Materials Research Laboratory, University of California, Santa Barbara, CA 93106-5050, USA.

oxides that traverse both these regimes with changes in temperature, pressure, or composition (*e.g.* V_2O_3, $La_{1-x}Sr_xVO_3$). Interesting electronic properties also arise from charge density waves (*e.g.* $K_{0.3}MoO_3$), charge-ordering (*e.g.* Fe_3O_4) and defect ordering (*e.g.* $Ca_2Mn_2O_5$, $Ca_2Fe_2O_5$). Oxides with diverse magnetic properties anywhere from ferromagnetism (*e.g.* CrO_2, $La_{0.5}Sr_{0.5}MnO_3$) to antiferromagnetism (*e.g.* NiO, $LaCrO_3$) are known. Many oxides possess switchable orientation states as in ferroelectric (*e.g.* $BaTiO_3$, $KNbO_3$) and ferroelastic [*e.g.* $Gd_2(MoO_4)_3$] materials. Then, there are a variety of oxide bronzes showing a gamut of properties.[4] Superconductivity in transition metal oxides has been known for some time, but the highest T_c reached was around 13 K; we now have oxides with T_cs in the region of 160 K. The discovery of high T_c superconductors[5] focused worldwide scientific attention on the chemistry of metal oxides and at the same time revealed the inadequacy of our understanding of these materials.

The unusual properties of transition metal oxides that distinguish them from the metallic elements and alloys, covalent semiconductors and ionic insulators are due to several factors. (a) Oxides of d-block transition elements have narrow electronic bands, because of the small overlap between the metal d and the oxygen p orbitals. The bandwidths are typically of the order of 1–2 eV (rather than 5–15 eV as in most metals). (b) Electron correlation effects play an important role, as expected because of the narrow electronic bands. The local electronic structure can be described in terms of atomic-like states [*e.g.* $Cu^+(d^{10})$, Cu^{2+} (d^9) and Cu^{3+} (d^8) for Cu in CuO] as in the Heitler–London limit. (c) The polarizability of oxygen is also of importance. The divalent oxide ion, O^{2-}, does not exactly describe the state of oxygen and configurations such as O^- have to be included, especially in the solid state. This gives rise to polaronic and bipolaronic effects. Species such as O^-, which are oxygen holes with a p^5 configuration instead of the filled p^6 configuration of O^{2-}, can be mobile and correlated. (d) Many transition metal oxides are not truly three-dimensional, but have low-dimensional features. For example, La_2CuO_4 and La_2NiO_4 with the K_2NiF_4 structure are two-dimensional compared to $LaCuO_3$ and $LaNiO_3$, which are three-dimensional perovskites. Because of the varied features of individual oxides, it has not been possible to establish satisfactory theoretical models for complex transition metal oxides. I started working on transition metal oxides in the late 1950s and my early preoccupation was with the structures, defects and phase transitions of oxides such as TiO_2, Pr_6O_{11} and Bi_2O_3. Soon I became interested in the electronic and magnetic properties of oxides, especially those with the perovskite structure. The perovskite structure dominates all classes of materials, whether they are dielectrics or superconductors.

Transition metal oxides that transform from the metallic state to an insulating or semiconducting state are of great interest. It has always amazed me how a simple oxide like V_2O_3 conducts like copper at ordinary temperatures and becomes like wood on cooling it to 150 K. Interest in these transitions has blossomed over the years and the subject has itself become a frontier area of investigation.[6,7] Typical transitions of this type in oxide materials are: (a) pressure-induced transitions, as in NiO, in which the pressure increases the wavefunction overlap between neighbors to induce a change from localized to itinerant behavior of electrons; (b) transitions as in Fe_3O_4 involving charge-ordering; (c) transitions in $LaCoO_3$ that are initially induced because of the different spin configurations of the transition metal ion; electron transfer between the two spin states initiates a process that eventually renders the oxide metallic around 1200 K; (d) transitions as in EuO arising from the disappearance of spin polarization band-splitting effects when the ferromagnetic Curie temperature is reached; (e) compositionally induced transitions, as in $La_{1-x}Sr_xCoO_3$ and $LaNi_{1-x}Mn_xO_3$, in which changes of band

structure in the vicinity of the Fermi level are brought about by a change in composition or are due to disorder-induced localization; (f) transitions in two-dimensional systems, such as La_2NiO_4, in which Ni—O—Ni interactions can only occur in the *ab* plane (unlike in the three-dimensional analogue in $LaNiO_3$) and (g) temperature-induced transitions in a large class of oxides such as Ti_2O_3, VO_2 and V_2O_3. Although I have only mentioned metal–insulator transitions in oxides, it should be noted that these transitions are found in a variety of other systems such as doped semiconductors, metal-ammonia solutions and expanded metals. Global aspects of metal–insulator transitions are truly fascinating.[6,7] In spite of considerable effort, we do not yet understand many aspects of these transitions, specially the 10-million fold jump in resistivity at the transition in V_2O_3.

The perovskite structure is ideal for 180° cation–anion–cation interactions and there are no cation–cation interactions. One can vary the covalency (or the interaction between the metal and oxygen orbitals) in this structure by varying the central cation and/or the transition metal ion and obtain a wide variation in properties. Interestingly, a high proportion of oxide materials of technological importance (*e.g.* ferroelectrics, superconductors, CMR materials) possess the perovskite structure. If we examine the electronic and magnetic properties of the family of perovskites of the type $LnMO_3$ (Ln, rare earth; M, first-row transition metal), we find a great variety. The titanates and nickelates possessing low-spin trivalent transition metal ions are metallic or nearly metallic. On the other hand, $LnCrO_3$, $LnMnO_3$ and $LnFeO_3$ are antiferromagnetic insulators. $LaCoO_3$ and other cobaltates are interesting, in that Co^{3+} can be in the low-spin (t_{2g}^6) or the high spin ($t_{2g}^4e_g^2$) state. While at low temperatures, $LaCoO_3$ is a diamagnetic insulator, with increasing temperature, the high-spin population increases eventually resulting in a phase transition due to the ordering of the two spin states. At high temperatures (*ca.* 1000 K), the material becomes metallic, resulting from the electron transfer between different spin/oxidation states of cobalt. Substituting for Ln in $LnCoO_3$ with a divalent ion such as Sr^{2+} or Ba^{2+}, as in $Ln_{1-x}Sr_xCoO_3$, progressively renders it metallic and ferromagnetic due to the fast electron transfer between the Co^{3+} and Co^{4+} states. We have investigated these cobaltates by a variety of methods including ^{57}Co Mössbauer spectroscopy, ferromagnetic resonance, photoemission spectroscopy and EXAFS. Electron transport properties of $LnCoO_3$ doped lightly with Sr^{2+} and other similar oxide systems can be understood in terms of variable range hopping as in disordered systems. The metal–nonmetal transition in $Ln_{1-x}Sr_xCoO_3$ [Fig. 1(*a*)] is a situation similar to impurity doping.[8] Another system showing metal-nonmetal transitions is $LaNi_{1-x}M_xO_3$ (M = Mn, Co, *etc.*), where $LaNiO_3$ ($x = 0.0$) is metallic [see Fig. 1(*b*) for the Mn system]. On increasing x, the material becomes an insulator.[8] Interestingly, in all such systems, the coefficient of resistivity changes sign across a

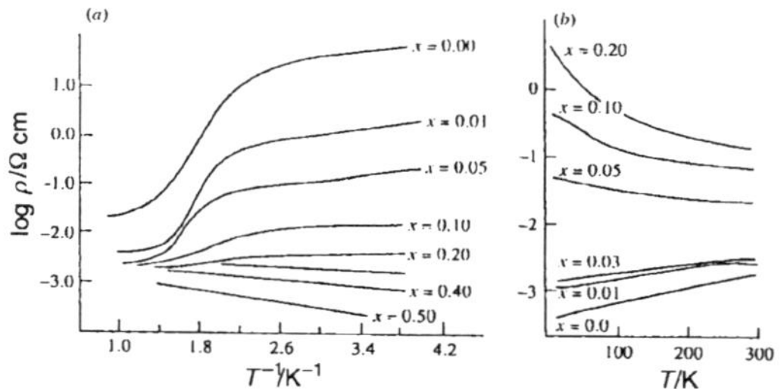

Fig. 1 Compositionally controlled metal–insulator transitions in (*a*) $La_{1-x}Sr_xCoO_3$ and (*b*) $LaNi_{1-x}Mn_xO_3$ [from Rao (reproduced with permission from ref. 8)].

universal value of resistivity, corresponding to Mott's minimum metallic conductivity.

Studies of the properties of quasi two-dimensional oxides with the K_2NiF_4 structure were initiated in this laboratory as early as 1970. It was found that Cu in Ln_2CuO_4 had no magnetic moment and that La_2CuO_4 was antiferromagnetic. Magnetic and electrical properties of La_2NiO_4 and its oxygen-excess nonstoichiometry were investigated. Solid solutions of La_2CuO_4 (T) and Nd_2CuO_4 (T') showed an interesting phase diagram. When we wrote an overview article[9] on the structure and properties of two-dimensional oxides with the K_2NiF_4 structure in 1984, little did we realize then that La_2CuO_4 was going to become famous because of its high-temperature superconductivity. The effect of dimensionality on the electronic and magnetic properties of layered nickel oxides (e.g. $LaNiO_3$, $La_4Ni_3O_{10}$, $La_2Ni_2O_7$, La_2NiO_4 with decreasing dimensionality from 3 to 2) was examined some time ago. Materials of this type have the general formula $(LaO)(LaMO_3)_n$ where the (LaO) is the rock salt layer and the $LaMO_3$ is the perovskite layer.[10] When M = Ni, the $n = \infty$ member ($LaNiO_3$) is metallic, but the metallicity decreases as n decreases. The $n = 1$ material has the K_2NiF_4 (two-dimensional) structure. One of the more fascinating families of such materials prepared by us was of the type $(SrO)(La_{1-x}Sr_xMnO_3)_n$. These materials have now become familiar because of the colossal magnetoresistance recently discovered in them.[11] In $La_{2-x}Sr_xNiO_4$, the electronic and magnetic properties depend crucially on x and on the oxygen stoichiometry.

Of the many other oxide systems we have investigated, I would like to briefly discuss oxides forming intergrowth structures. Several metal oxide systems exhibit chemically well defined recurrent intergrowth structures with large periodicities, rather than random solid solutions with variable composition. The ordered intergrowth structures themselves frequently show the presence of wrong sequences. High resolution electron microscopy (HREM) enables direct examination of the extent to which a particular ordered arrangement repeats itself, and the presence of different sequences of intergrowths, often of unit cell dimensions. Selected area electron diffraction, which forms an essential part of HREM, provides useful information (not generally provided by X-ray diffraction) regarding the presence of supercells due to the formation of intergrowth structures. Several systems forming ordered intergrowth structures have been discovered and they generally exhibit homology.[12]

Aurivillius described a family of oxides of the general formula $Bi_2A_{n-1}B_nO_{3n+3}$ where the perovskite slabs, $(A_{n-1}B_nO_{3n+1})^{2-}$, n octahedra thick, are interleaved by $(Bi_2O_2)^{2+}$ layers. Typical members of this family are Bi_2WO_6 ($n = 1$), $Bi_3Ti_{1.5}W_{0.5}O_9$ ($n = 2$), $Bi_4Ti_3CrO_{12}$ ($n = 3$) and $Bi_5Ti_3CrO_{15}$ ($n = 4$). These oxides form intergrowth structure of the general formula $Bi_4A_{m+n-2}B_{m+n}O_{3(m+n)+6}$ involving alternate stacking of two Aurivillius oxides with different n values (Fig. 2). The method of preparation simply involves heating a mixture of the component metal oxides around 1000 K. Ordered intergrowth structures with (m, n) values of (1,2), (2,3) and (3,4) have been fully characterized by X-ray diffraction and HREM. What is amazing is that such intergrowth structures with long-range order are indeed formed while either member (m and n) can exist as a stable entity. These materials seem to be truly representative of recurrent intergrowth. The periodicity found in recurrent intergrowth solids formed by the Aurivillius family of oxides is indeed impressive.[12] WO_3 forms tetragonal, hexagonal or perovskite-type bronzes by interaction with alkali and other metals. The family of intergrowth tungsten bronzes (ITB) involving the intergrowth of nWO_3 slabs and one to three strips of the hexagonal tungsten bronze (HTB), first described by Kihlborg, is of relevance to the discussion here. In the intergrowth

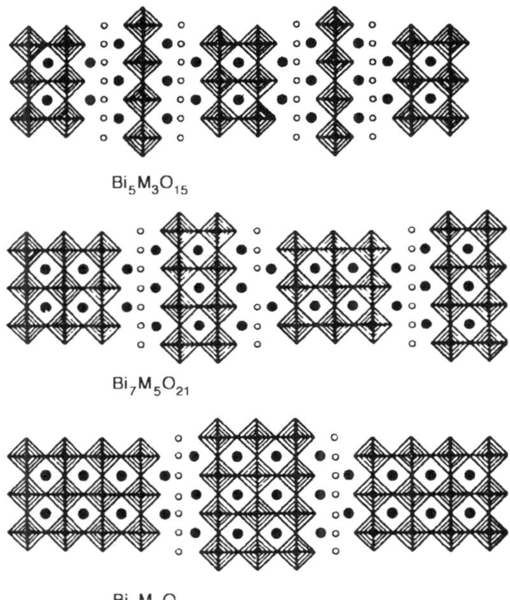

Fig. 2 Different types of intergrowth structures formed by the Aurivillius family of bismuth oxides.

tungsten bronzes of the general formula M_xWO_3, x is generally 0.1 or less and depends on whether the HTB strip is one or two tunnels wide. In the ITB phases of bismuth characterized by us some time ago, the HTB strips were always one tunnel wide. Displacement of adjacent tunnel rows due to the tilting of WO_3 octahedra often results in doubling of the long-period axis of the ITB. Evidence for the ordering of the intercalating bismuth atoms in the tunnels is found in terms of satellites around the superlattice spots in the electron diffraction patterns. The occurrence of recurrent intergrowth structures continues to fascinate chemists and crystallographers alike.

Superconducting cuprates

Late in 1986 the solid state community received the big news of the discovery of a high temperature oxide superconductor.[5] The material that was found to be superconducting with a T_c of ca. 30 K was based on the cuprate La_2CuO_4 (Fig. 3). In February 1987, we decided to look for an yttrium cuprate in the Y–Ba–Cu–O system as a candidate for high T_c superconductivity. Wu et al.[13] announced an yttrium cuprate with superconductivity above liquid nitrogen temperature around March 1987. Our strategy to prepare such a material was different. Wu et al. obtained a mixture of black and green compounds in their effort to make an yttrium cuprate doped with Ba. The black compound was superconducting ($T_c \approx 90$ K), but they did not know its composition or structure. Since we knew that Y_2CuO_4 could not be made, we started examining the $Y_{3-x}Ba_{3+x}Cu_6O_{14}$ system. We obtained 90 K superconductivity in the $x = 1$ composition corresponding to $YBa_2Cu_3O_7$. When we independently discovered 90 K superconductivity in March 1987, we knew the composition and the approximate structure.[14]

We worked on a variety of cuprates but unfortunately the area was so competitive and overcrowded that it became difficult to get due credit for much of the work. In spite of these limitations, we made several contributions to this area. These include, besides the independent synthesis and charac-

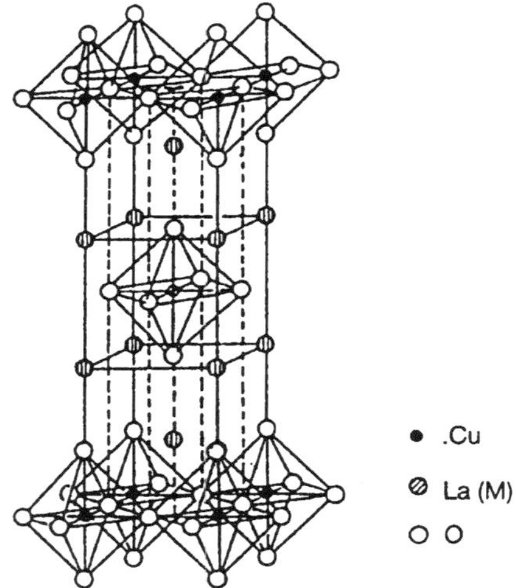

Fig. 3 Structure of La_2CuO_4 and $La_{2-x}A_xCuO_4$ (A = Sr or Ba). The CuO_6 octahedra are Jahn–Teller distorted.

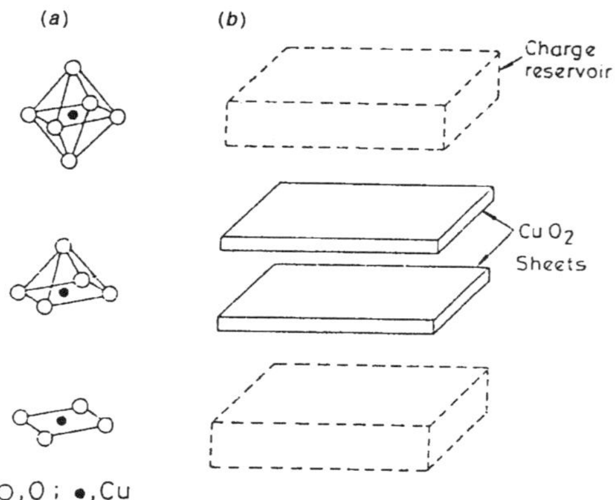

Fig. 4 (a) Cu–O polyhedra found in superconducting cuprates. (b) Schematic representation of the cuprates.

terization of $YBa_2Cu_3O_7$, the first observation of non-resonant microwave absorption in high T_c cuprates, metastability of $YBa_2Cu_3O_{7-\delta}$ (in the 60 K regime, $\delta = 0.3$–0.4) and its transformation to a novel type of $YBa_2Cu_4O_8$, the importance of Cu–O charge transfer energy in relation to the superconductivity of cuprates and the nature of charge carriers, optimization of hole concentration in cuprates and the relation between T_c and hole concentration. On the synthetic front, we made some contributions which we can consider to be Bangalore brand. For example, we could make a series of superconductors in the Tl–Ca–Sr–Cu–O system by substitution of Tl by Pb or Ca by a rare earth metal; a new family of superconductors of the type $TlSr_{n+1}Cu_nO_{2n+3}$ ($n = 1$ and $n = 2$) wherein Sr is partly substituted by a rare earth or Tl by Pb was discovered. Another noteworthy contribution was the synthesis of bismuth cuprate superconductors without superlattice modulation. This was significant since all the Bi cuprates showed such modulation. The modulation could be removed by suitable cation substitution (Pb/Bi) and the adjustment of oxygen stoichiometry. An interesting aspect was the synthesis and characterization of oxyanion derivatives of cuprate superconductors containing anions such as CO_3^{2-}, SO_4^{2-}, PO_4^{3-} and BO_3^{3-}, as integral parts of the oxide framework. Besides the mercury cuprates with the highest T_c to date, many interesting cuprates have been synthesized in the last 4–5 years.[15,16] These include copper oxycarbonates, copper oxyhalides, $Sr_2CuO_2F_{2+\delta}$ and ladder cuprates. Here, I must place on record the enormous contributions of Raveau and coworkers[17] to the synthesis and characterization of cuprate superonductors.

The several families of cuprate superconductors with holes as carriers all have the general features shown in Fig. 4. These features are common to the mercury cuprate showing the highest T_c as well. Electron superconductors of the type $Nd_{2-x}M_xCuO_4$ (M = Ce, Th) with the T' structure possess square-planar CuO_4 units instead of the elongated octahedra in the T structure. The hole superconductors generally have the square pyramidal units. Many of the cuprate families have an antiferromagnetic insulator member at one end of the composition. What is more interesting is that all the cuprates are at a metal–insulator boundary. Some of them undergo a metal–nonmetal transition as a function of composition (e.g. $Bi_2Ca_{1-x}Y_xSr_2Cu_2O_8$ and $TlY_{1-x}Ca_xSr_2Cu_2O_7$ with change in x). The essential feature of the cuprates is the presence of

CuO_2 sheets with or without apical oxygens. The mobile charge carriers in the cuprates are in the CuO_2 sheets. All the cuprates have charge reservoirs as exemplified by the Cu–O chains in the 123 and 124 cuprates and the TIO, BiO and HgO layers in the other cuprates. That the CuO_2 sheets are the seat of high-temperature superconductivity is demonstrated by the fact that intercalation of iodine between BiO layers in the bismuth cuprates does not affect the superconducting transition temperature while introduction of fluorite layers between the CuO_2 sheets adversely affects superconductivity. In the different series of cuprates with varying number of CuO_2 sheets studied hitherto, the T_c reaches a maximum when $n = 3$ except in single thallium layer cuprates where the maximum is at $n = 4$. The infinite layered cuprates, where the CuO_2 sheets are separated by alkaline earth and other cations, show T_c values in the 40–110 K range. Superconductivity in these materials appears to be due to the presence of Sr–O defect layers corresponding to the insertion of $Sr_3O_{2\pm x}$ blocks. We shall briefly present the common features of cuprate superconductors[18] and examine some of the outstanding problems.

The chemistry of the majority of the cuprate superconductors is governed by oxygen holes. (Rouxel[19] has carried out fine work with sulfur holes in metal sulfides). The excess positive charge in cuprates can be represented in terms of the formal valence of copper, which in the absence of holes will be $+2$ in the CuO_2 sheets. In hole superconducting cuprates, it is generally around $+2.2$. In electron superconductors, it would be less than $+2$ as expected. The actual concentration of holes, n_h, in the CuO_2 sheets in $La_{2-x}A_xCuO_4$, $YBa_2Cu_3O_7$ and Bi cuprates is readily determined by redox titrations. In the 123 cuprates, the concentration of mobile holes in the CuO_2 sheets can be delineated from that in the Cu–O chains. Determination of n_h in thallium cuprates poses some problems, but in single Tl–O layer cuprates, chemical methods have been developed to obtain reasonable estimates. Generally, T_c in a given family of cuprates reaches a maximum value at an optimal value of n_h as shown in Fig. 5; the maximum is around $n_h \approx 0.2$ in most cuprates. Notice that the points in the underdoped region in Fig. 5 fall close to a straight line, but deviations occur in the overdoped region. Single layer thallium cuprates also show this behaviour. In $Tl_{1-y}Pb_yY_{1-x}Ca_xSr_2Cu_2O_7$ where the substitution of Tl^{3+} by Pb^{4+} has an effect opposite to that due to the substitution of Y^{3+} by Ca^{2+}, the T_c becomes a maximum at an optimal value of $(x - y)$, which is a measure of the hole concentration. By suitably manipulating x and y, the T_c of this system can be increased from 85–90 K to 110 K.

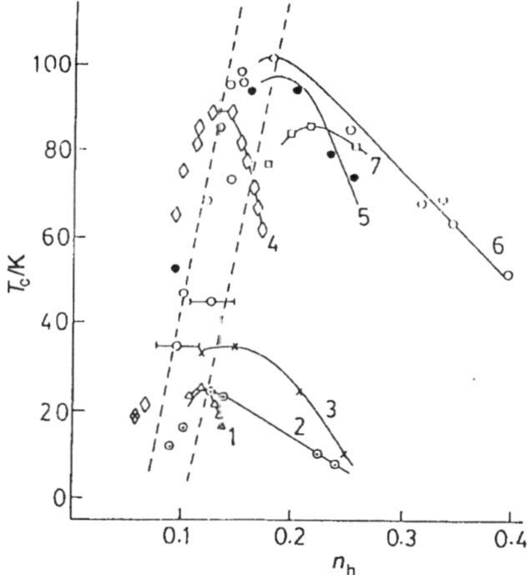

Fig. 5 Variation of T_c with the hole concentration, n_h: 1,2,6 and 7, Bi cuprates; 5, 123 cuprates; 3, $La_{2-x}Sr_xCuO_4$; 4, Tl cuprates (from Rao and Ganguli[18a]).

It is possible to obtain useful correlations between T_c and the Cu–O distances, both in-plane and apical. The Cu and oxygen Madelung energies also provide good correlations.[18a] The T_cs vary sensibly with the Cu–O charge-transfer energy as determined by core-level intensities. In spite of such relations with structural parameters, we are far from understanding the high T_c cuprates, particularly in the normal state.[18] The doped and undoped cuprates are both strongly correlated and we do not fully understand the strong correlation in these materials. Many puzzles remain. (i) For example, we do not fully understand the global composition–temperature phase diagram and the apparent lack of symmetry between hole and electron superconductors. (ii) The cuprates are unusual metals in-plane and insulators perpendicular to it; on cooling, they become three-dimensional superconductors. The anomalous temperature variation of resistivity in cuprates is noteworthy (Fig. 6), particularly the extraordinary linear variation of the ab plane metallic resistivity. (iii) The ac optical conductivity data do not conform the behavior expected of metals. (iv) The absence of splitting in the energy of states coupled by bilayer tunneling (as found from ARPES) is surprising. All the above observations cannot be reconciled with the picture of an anisotropic metal. (v) The linear resistivity (ab plane) with respect to temperature, with zero-intercept and nearly constant slope (independent of T_c) cannot be explained easily. The anisotropy is large and temperature dependent. The Hall resistivity varies inversely with temperature. (vi) Photoemission spectra suggest that the quasiparticles are not well defined in the normal state. (vii) Above all, there appears to be a characteristic low energy scale in cuprates. These pseudogap effects are indeed unique to cuprates. (viii) The superconducting state also has some unusual features, such as anisotropic order parameter, and well defined quasiparticles. Even a doped ladder cuprate has been rendered superconducting. Clearly, the status of our theoretical understanding of cuprate superconductors is far from satisfactory. It appears that cuprates constitute an experimentally overdetermined system. In the meantime, if one discovers a room-temperature superconductor, it will be an experimentalist's joy, but a theoretician's nightmare. It is noteworthy that the high T_c cuprates are finding applications in electronic devices, although their use in fabricating generators and magnets may take time.

Colossal magnetoresistance, charge ordering and related properties of rare earth manganates

Giant magnetoresistance was known to occur in bilayer and granular metallic materials. Perovskite manganates of the general formula $Ln_{1-x}A_xMnO_3$ (Ln = rare earth, A = alkaline earth) have created wide interest because they exhibit colossal magnetoresistance (CMR). These oxides become ferromagnetic at an optimal value of x (or Mn^{4+} content) and undergo an insulator–metal (I–M) transition around the ferromagnetic T_c. These properties are attributed to double exchange associated with electron hopping from Mn^{3+} to Mn^{4+}. The double-exchange which favours itinerant electron behavior is opposed by the Jahn–Teller distortion due to the presence of Mn^{3+}. The manganates show charge ordering especially when the average size of the A-site cations is small. Charge ordering also competes with double exchange and favours insulating behavior. Thus, the manganates exhibit a variety of properties associated with spin, charge as well as orbital ordering. In what follows, we discuss some important aspects of colossal magnetoresistance, charge ordering and related properties of the manganates.[8,11,20]

$LaMnO_3$ is an insulator with an orthorhombically distorted perovskite structure ($b>a>c\sqrt{2}$) due to Jahn–Teller distortion. Typically, as prepared samples of $LaMnO_3$ contain some Mn^{4+}. $LaMnO_3$ with a small proportion of Mn^{4+} ($<5\%$) becomes antiferromagnetically ordered at low temperatures ($T_N \approx 150$ K). When the La^{3+} in $LaMnO_3$ is progressively substituted by a divalent cation as in $La_{1-x}A_xMnO_3$ (A = Ca, Sr or Ba), it becomes ferromagnetic with a well defined Curie temperature, T_c, and metallic below T_c. The saturation moment is typically ca. 3.8 μ_B which is close to the theoretical estimate based on localized spin-only moments, suggesting that the carriers are spin-polarized. Below T_c, the manganates exhibit metal-like conductivity. Fig. 7 illustrates how the ferromagnetism and the I–M transition occur around T_c in a typical $La_{1-x}Ca_xMnO_3$ composition. The simultaneous observation of itinerant electron behavior and ferromagnetism in the manganates is explained by Zener's double-exchange mechanism. The basic process in this mechanism is the hopping of a d hole from Mn^{4+} (d^3, t_{2g}^3, $S=3/2$) to Mn^{3+} (d^4, $t_{2g}^3e_g^1$, $S=2$) via the oxygen, so that the Mn^{4+} and Mn^{3+} ions change places. The t_{2g}^3 electrons of the Mn^{3+} ion are localized on the Mn site giving rise to a local spin of 3/2, but the e_g state, which is hybridized with the oxygen 2p state, can be localized or itinerant. There is a strong Hund's rule interaction between the e_g and the t_{2g}^3 electrons. Goodenough pointed out that the ferromagnetism is governed not only by double exchange,

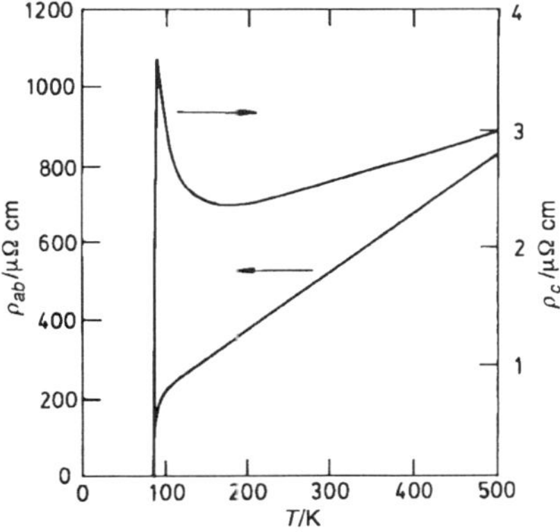

Fig. 6 Temperature variation of resistivity of $Bi_2Sr_2CaCu_2O_8$ single crystal along the ab plane and c-axis [data from Ong *et al.* as quoted by Ramakrishnan[18c] reproduced with permission from ref. 18(c)].

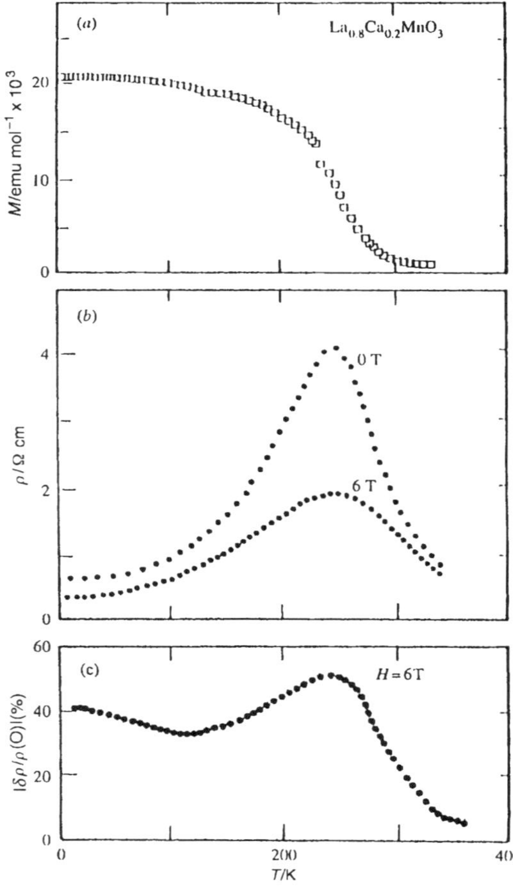

Fig. 7 Temperature variation of (*a*) the magnetization, (*b*) the resistivity (at 0 and 6 T) and (*c*) the magnetoresistance in $La_{0.8}Ca_{0.2}MnO_3$ [from Rao (reproduced with permission from ref. 8)].

but also by the nature of the superexchange interactions. It should be noted that the Mn^{3+}–O–Mn^{4+} superexchange interaction is ferromagnetic while both the Mn^{3+}–O–Mn^{3+} and Mn^{4+}–O–Mn^{4+} interactions are antiferromagnetic.

The orthorhombic distortion of $LaMnO_3$ decreases as La is progressively substituted by a divalent ion and until it becomes rhombohedral or pseudocubic. The Mn–O–Mn angle approaches $180°$ as the distortion decreases. When the size of the A-site cations is small, orthorhombic distortion is favoured. The key factor responsible for the distortion in $LaMnO_3$ is the Jahn–Teller distortion due to Mn^{3+} ions. Accordingly, $LaMnO_3$ has three Mn–O distances (1.91, 1.96 and 2.19 Å). Across the I–M transition, the Jahn–Teller distortion decreases in $Ln_{1-x}A_xMnO_3$, the distortion being greater in the insulating regime. The structural parameters, specifically the oxygen thermal parameters, show a significant change across the M–I transition. Clearly, Mn^{4+} ions decrease the Jahn–Teller distortion, increase the itinerancy of electrons through double exchange and increase the tolerance factor of the perovskite structure. Temperature–composition phase diagrams for $La_{1-x}A_xMnO_3$ have been worked out with increasing x or Mn^{4+} content. A composition of $x \approx 0.3$ is ideal to observe a good I–M transition and negative CMR. The magnitude of CMR is maximum at the I–M transition (Fig. 7) and scales

with the field-induced magnetization. CMR is found in cation-deficient $LaMnO_3$ containing Mn^{4+} as well, and the features of the magnetic and metal–insulator transitions are similar to those in Fig. 7.

CMR is found over a wide range of compositions of $La_{1-x}Ca_xMnO_3$ ($x<0.5$). However, this material shows significant lattice effects in particular for low x. EXAFS studies reveal significant changes in the local structure in the $x=0.3$ composition in the 80–300 K range which are attributed to the formation of small polarons due to the Jahn–Teller distortion when $T>T_c$. Lattice polaron formation and local structural distortion have important implications on the electron transport (high resistivity) in these materials. $La_{1-x}Sr_xMnO_3$ is somewhat different from $La_{1-x}Ca_xMnO_3$, the A-site ion size being larger. The ferromagnetic metallic regime in $La_{1-x}Sr_xMnO_3$ extends over larger values of x and the Curie temperatures are higher. Magnetization in the manganates decreases as T^2 and the resistivity increases as T^2. In $La_{1-x}Sr_xMnO_3$ ($x=0.17$), the local spin moments and charge carriers couple strongly to changes in the structure and the structure can be switched by the application of a magnetic field depending on the temperature. The magnetic field which causes this crossover from the orthorhombic to the rhombo-hedral structure is rather low (*ca.* 2 T), showing thereby that a very small energy difference exists between these two structures.

Hydrostatic pressure stabilizes the ferromagnetic metallic state in $La_{1-x}A_xMnO_3$ (*i.e.* increases T_c). A similar effect can be brought about by increasing the radius of the A-site cation. There is a direct relationship between T_c and the average radius of the A-site cation, $<r_A>$, the T_c increasing with $<r_A>$. A small $<r_A>$ gives rise to a distortion of the MnO_6 octahedra by bending the Mn–O–Mn bond and in turn causing the narrowing of the e_g bandwidth to a greater extent. Accordingly, yttrium substitution in $La_{1-x}A_xMnO_3$ decreases T_c, and enhances the magnitudes of resistivity as well as magnetoresistance. By plotting the T_c against $<r_A>$, one obtains a phase diagram separating the ferromagnetic metal and the paramagnetic insulator regimes. In Fig. 8, we show a phase diagram obtained in this manner. In the lower left-hand corner, we see the ferromagnetic insulator regime. CMR generally decreases with increase in $<r_A>$, just as the peak resistivity at the I–M transition.

The effect of particle size on the electron transport and magnetic properties of polycrystalline $La_{0.7}Ca_{0.3}MnO_3$ has been investigated. Although T_c decreases with decreasing particle size, the magnetoresistance (MR) is insensitive to particle size. The I–M transition becomes broader when the particle size is small, but the %MR is nearly constant over a

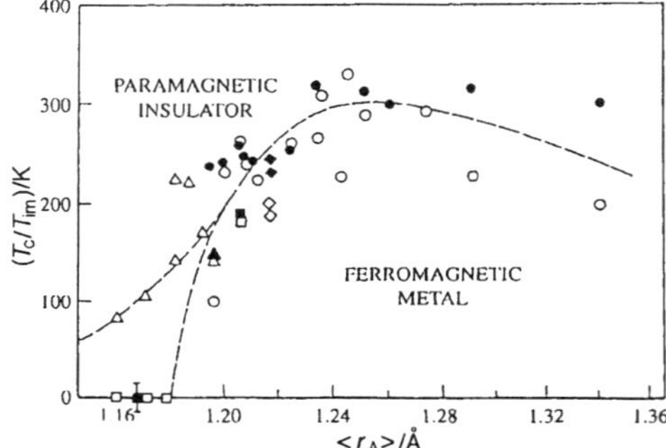

Fig. 8 Variation of the ferromagnetic T_c or the metal–insulator transition temperature, T_{im}, in $Ln_{1-x}A_xMnO_3$ with weighted average radius of the A site cations, $<r_A>$ [from Rao (reproduced with permission from ref. 8)].

wide temperature range. The low-temperature MR is significantly dependent on the grain size. For larger grain sizes, the MR at the lowest temperature decreases eventually becoming zero as in good single crystals. Samples of manganates with different particle sizes prepared with different heat treatments often show large differences between the ferromagnetic T_c and the temperature of the I–M transition. Grain boundaries appear to contribute substantially to magnetoresistance in polycrystalline materials, in particular at $T \ll T_c$.

The effects of dimensionality on CMR and related properties have been studied by examining the $(SrO)(La_{1-x}Sr_xMnO_3)_n$ family. The $n=1$ member which is two dimensional with the K_2NiF_4 structure, is an insulator, while the $n=\infty$ member is the three-dimensional perovskite showing CMR and other properties. The $n=2$ member exhibits the sharpest I–M transition and high MR. In $La_{1.4}Sr_{1.6}Mn_2O_7$, the current perpendicular to the MnO_2 planes is large, particularly close to T_c, and the barrier to transport provided by the intervening insulating layers is removed by the magnetic field as shown by Tokura et $al.$ Interlayer transport seems to involve tunneling. We see some comparable features between the layered cuprate superconductors and the layered manganates. The big difference is that electrons are spin polarized in the latter.

Studies of the electrical, magnetic and other properties of $Ln_{1-x}A_xMnO_3$ systems have revealed certain unusual features with respect to the charge carriers in these oxides. The manganates exhibit high resistivities even in the metallic state, particularly at low temperatures. The values of resistivities are considerably higher than Mott's value of maximum metallic resistivity in many of these materials, the resistivity reaching constant values as high as 10^3–10^4 ohm cm at low temperatures. A clear picture of the nature of carriers responsible for the high resistivity is yet to emerge. The cause of the observed temperature dependence of resistivity at low temperatures is also not entirely clear. Another unusual feature of the manganates is related to the occurrence of large MR in a small field at relatively high temperatures. The scale of Zeeman energy associated with a field of 1 T in these spin systems is ≈ 5 K. This is much too small compared to the scale of thermal energy (200–300 K) where one sees such a large MR.

That double exchange is necessary for CMR in the manganates is shown by a study of the $LnMn_{1-x}Cr_xO_3$ system where Cr^{3+} replaces Mn^{4+}. In fact, double exchange probably suffices to explain most properties of $La_{1-x}Sr_xMnO_3$. The basic question that remains is why these manganates are insulating when $T > T_c$. The anomalous resistance has been explained by considering electron–phonon interactions arising from the Jahn–Teller splitting of the Mn d-levels. Jahn–Teller distortion localizes conduction electrons as polarons, especially in narrow band materials like $La_{1-x}Ca_xMnO_3$. As the temperature decreases, the transfer integral t_{ij} increases and the ratio of the Jahn–Teller energy to t_{ij} decreases. The coupling varies with temperature and composition, and increases across the metal–insulator transition. The large effects of local lattice distortion and crystal structure symmetry on electron transport and magnetism show that one cannot explain the unusual behavior without taking the lattice into account, especially in systems such as $La_{1-x}Ca_xMnO_3$. The observed T_c ($ca.$ 250–400 K) in most manganates is considerably lower than that estimated from the measured spin-wave stiffness constant implying that the ferromagnetic exchange arising from the double exchange mechanism is not the only factor that affects T_c. The electrical resistivity of the rare earth manganates across the I–M transition is indeed difficult to understand especially at low temperatures based on double exchange or electron–phonon interaction alone. We also do not understand why CMR actually occurs in these oxides.[20] There is however every expectation that CMR will be exploited in the recording and sensor industries. We need to explore pure as well as composite materials showing CMR at ambient temperatures

at low magnetic fields. The exploration should not be limited to rare earth manganates since many other inorganic materials, including $Fe_{1-x}Cu_xCr_2S_4$ and EuO_x, exhibit large MR effects.

Marginal metallicity in oxide materials

We had earlier looked briefly at the metal–insulator transitions in oxides of the type $La_{1-x}Sr_xCoO_3$ or $LaNi_{1-x}Mn_xO_3$ where the temperature coefficient of resistance changes sign across a universal value of conductivity, $viz.$ Mott's minimum metallic conductivity. These oxides not only obey the minimum metallic conductivity criterion arising from consideration of disorder, but also the $n_c^{1/3}a \approx 0.25$ criterion ($n_c =$ critical carrier concentration at the transition) arising from electron interactions. The same situation holds for the cuprate superconductors which follow both these criteria. The rare earth manganates, $Ln_{1-x}A_xMnO_3$, exhibiting colossal magnetoresistance, on the other hand, are quite different in that many of them exhibit high residual resistivities far above Mott's value. The nature of carriers in the manganates would be much more difficult to describe than those in cuprate superconductors, which are themselves far from being understood. What is to be noted is that three of the most fascinating phenomena in materials, namely insulator–metal transitions, high temperature superconductivity and CMR are exhibited by oxide materials which are marginally metallic. Marginal metallicity appears to be a universal feature,[21] which requires further probing. It is interesting to speculate whether a universal minimum electron diffusivity due to localized states is present in these materials, besides the extended states.

Charge ordering in the manganates

Charge ordering is not new to transition metal oxides. Fe_3O_4 undergoes the Verwey transition with a resistivity anomaly due to charge ordering at 120 K and ferrimagnetic ordering at 860 K. Charge ordering of the Mn^{3+} and Mn^{4+} ions occurs in the perovskite manganates. $Ln_{1-x}A_xMnO_3$ is more interesting and is associated with dramatic changes in properties.[8,11,20] Charge ordering in the manganates is strongly dependent on the average radius of the A-site cations, $<r_A>$. The ferromagnetic T_c increases with increase in $<r_A>$ while charge ordering is favoured by small $<r_A>$. Charge ordering and double exchange are thus competing interactions. We shall examine charge ordering in $Ln_{0.5}A_{0.5}MnO_3$ with equal proportions of Mn^{3+} and Mn^{4+}.

When $<r_A>$ is sufficiently large, as in $La_{0.5}Sr_{0.5}MnO_3$ ($<r_A> = 1.26$ Å), the manganates become ferromagnetic and undergo an insulator–metal transition around T_c as described earlier. $Nd_{0.5}Sr_{0.5}MnO_3$ with a slightly smaller $<r_A>$ (1.236 Å), is ferromagnetic and metallic at 250 K (T_c), but transforms to a charge-ordered (CO), antiferromagnetic state at 150 K as shown by Tokura.[20] The charge-ordering transition temperature, T_{CO}, is the same as T_N. There is a significant specific heat anomaly at T_{CO} (T_N) and the resistivity increases markedly due to the metal–insulator (FM–CO) transition. This situation is to be contrasted with that of $Nd_{0.5}Ca_{0.5}MnO_3$ ($<r_A> = 1.17$ Å) which exhibits a T_{CO} of 250 K in the paramagnetic state exhibiting a small heat capacity anomaly. $Nd_{0.5}Ca_{0.5}MnO_3$ is an insulator at all temperatures. $Y_{0.5}Ca_{0.5}MnO_3$ with a $<r_A>$ of 1.13 Å also gets charge ordered in the paramagnetic state (T_{CO}, 240 K) and becomes antiferromagnetic at 140 K (T_N). In Fig. 9 we show a simple diagram where we delineate the different types of behavior of the manganates depending on the $<r_A>$. The charge ordered state of $Nd_{0.5}Sr_{0.5}MnO_3$ is melted by the application of a magnetic field, rendering the material metallic. On the other hand, a magnetic field of 6 T has no effect on the charge-ordered insulating state of $Y_{0.5}Ca_{0.5}MnO_3$. The charge ordered states in $Nd_{0.5}Sr_{0.5}MnO_3$ and $Nd_{0.5}(Y_{0.5})Ca_{0.5}MnO_3$ are

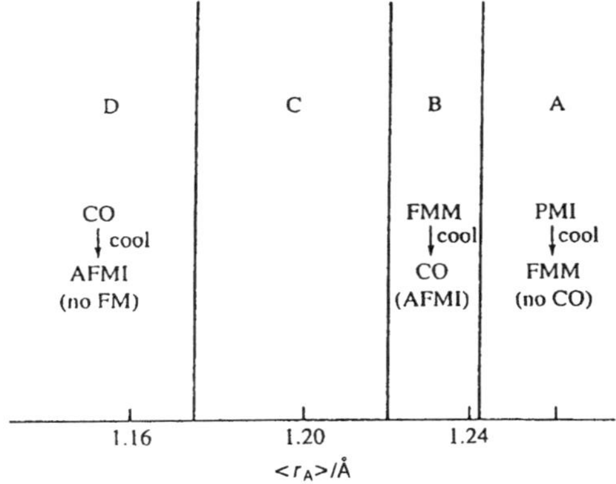

Fig. 9 Schematic diagram showing the different types of behavior in $La_{0.5}A_{0.5}MnO_3$. Region A corresponds to manganates showing CMR while region B corresponds to manganates of the type $Nd_{0.5}Sr_{0.5}MnO_3$ transforming from the ferromagnetic metallic (FMM) state to a charge-ordered (CO) state on cooling. Region D corresponds to manganates where the CO state is generally found in the paramagnetic insulating (PMI) state (*e.g.* $Nd_{0.5}Ca_{0.5}MnO_3$ and $Y_{0.5}Ca_{0.5}MnO_3$). Region C is expected to show complex behavior. AFMI stands for the antiferromagnetic insulating state [from Rao (reproduced with permission from ref. 8)].

clearly of different kinds, the difference being caused by the difference in $<r_A>$.

The occurrence of two types of CO states can be understood qualitatively in terms of the variation of the exchange couplings, J_{FM} and J_{AFM}, and the single-ion Jahn–Teller energy (E_{JT}) with $<r_A>$. While J_{FM} and J_{AFM} are expected to decrease with a decrease in $<r_A>$, albeit with different slopes, E_{JT} would be insensitive to $<r_A>$. In the small $<r_A>$ regime, a cooperative Jahn–Teller effect involving long-range elastic strain would dominate charge ordering, while at moderate values of $<r_A>$ (when $J_{AFM}>E_{JT}$, J_{FM}), the e_g electrons which are localized magnetically lower the configuration energy by charge ordering. Such a CO state, as exemplified by $Nd_{0.5}Sr_{0.5}MnO_3$, would be sensitive to magnetic fields unlike the manganates such as $Y_{0.5}Ca_{0.5}MnO_3$ in the small $<r_A>$ regime. In $Nd_{0.5}Sr_{0.5}MnO_3$, the gain in Zeeman energy resulting from the application of a magnetic field stabilizes the ferromagnetic metallic state over the antiferromagnetic insulating state. The general phase diagram in the $T–<r_A>$ plane has been constructed by Kumar and Rao based on these considerations.[8] Although we understand the gross features of charge-ordering and associated properties in manganates, we do not have a proper theory or model to explain all the varied aspects of these fascinating materials.

The high sensitivity of charge ordering in manganates to $<r_A>$ cannot be understood on the basis of the variation of the Mn–O–Mn angle and the Mn–O distance. This is because these geometrical changes do not give rise to sufficiently large changes in bandwidth as are necessary to explain the significant changes in the charge ordering temperature and the Curie temperature with $<r_A>$. The study of the complex regime (region C in Fig. 9), has yielded fascinating results. Thus, $Nd_{0.25}La_{0.25}Ca_{0.5}MnO_3$ ($<r_A> = 1.19$ Å) undergoes a reentrant transition from an incipient charge-ordered state to a FMM state on cooling, driven by a first order phase transition.[22] The charge-ordering gap collapses at the transition. Charge ordering in the manganates needs to be pursued further, especially regarding the marked effects of electric fields and chemical substitution. The role of the A-site ion mismatch on CO and FM states also needs attention.

Porous solids

Porous inorganic materials have many applications in catalytic and separation technologies. The subject has grown explosively with 8000 or more literature citations. The synthesis of microporous solids with connectivities and pore chemistry different from zeolitic materials has attracted considerable attention. A variety of such open framework structures, in particular Al and Ga phosphates as well as many other metal phosphates prepared in the presence of structure directing agents, have been characterized. The work of Cheetham, Ferey, Flanigen, Stucky and Thomas in this direction is noteworthy.[23a] Zeolites and zeolite-like materials[23,24] with micropores in the 5–20 Å range have contributed much to the advances in catalysis. There have been many break-throughs in the design and synthesis of these molecular seives with well defined crystalline architecture and there is a continuing quest for extra-large pore molecular sieves.[24] Several types of new materials including tunnel and layer structures of porous manganese oxides have been reported.[25] The progress has resulted mainly from our understanding of the role of structure-directing agents. The discovery of mesoporous silica (pore diameter 20–100 Å) by Mobil chemists added a new dimension to the study of porous solids. The synthesis of mesoporous materials also makes use of structure-directing surfactants[24,26] (cationic, anionic, and neutral) and a variety of mesoporous oxidic materials (*e.g.* ZrO_2, TiO_2, $AlPO_4$, aluminoborates), have been prepared and characterized.[26,27] In Fig. 10, the X-ray diffraction patterns of the three forms of $CaAlBO_4$ are shown.

An appropriate integration of hydrogen bond interactions at the inorganic–organic interface and the use of sol–gel and emulsion chemistry has enabled the synthesis of a large class of porous materials.[28a] Today, we have a better understanding of the structure, topology and phase transitions of mesoporous solids. Thus, block copolymers have been used to prepare mesoporous materials with large pore sizes[28a] (>30 nm). There is also some understanding of the lamellar–hexagonal-cubic phase transitions in mesoporous oxides[28b] (Fig. 11). Derivatized mesoporous materials have been explored for potential applications in catalysis and other areas.[29a] It has been found recently that transition metal complexes encapsu-

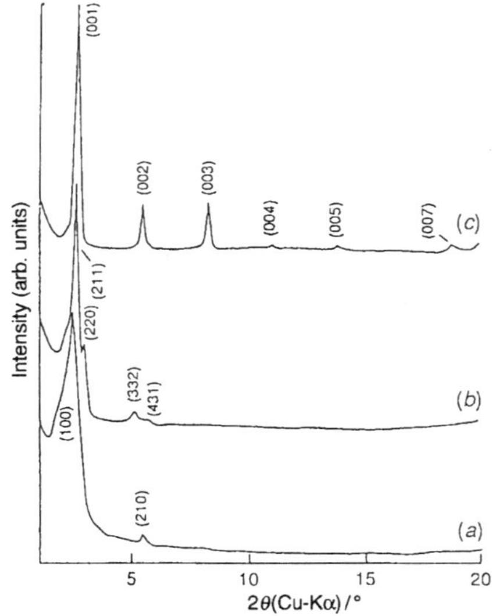

Fig. 10 X-Ray diffraction patterns of (*a*) hexagonal, (*b*) cubic and (*c*) lamellar forms of mesoporous calcium aluminoborate (from Ayyappan and Rao[27]).

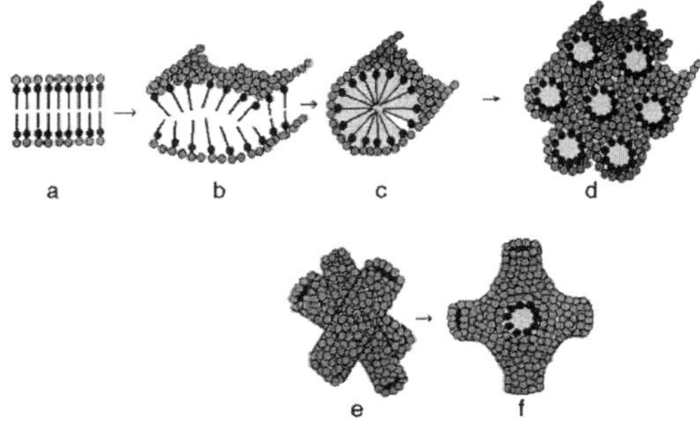

Fig. 11 Phase transitions in mesoporous solids: a–d, lamellar–hexagonal; e–f, hexagonal–cubic. The circular objects around the surfactant assemblies are the metal-oxo species (from Neeraj and Rao[28b]).

lated in cubic mesoporous phases show high catalytic activity in oxidation reactions.[29b]

Organic inclusion compounds and clathrates have been known for a long time. While these compounds are still being investigated, there have been efforts to synthesise novel organic structures by supramolecular means.[30] A recent example in this direction is the noncovalent synthesis of a novel channel structure formed by trithiocyanuric acid and bipyridine.[30b] The channels can accommodate benzene, xylenes and other molecules (Fig. 12) and the process is reversible.

Synthesis of inorganic materials

Although rational design and synthesis have remained primary objectives of materials chemistry, we are far from achieving this goal universally. There are many recent instances where rational synthesis has worked (*e.g.*, porous solids), but by and large the success has been limited. Thus, one may obtain the right structure and composition, but not the properties. There are very few instances where both the structure and properties of the material obtained are as desired. The synthesis of $ZrP_{2-x}V_xO_7$ solid solutions showing zero or negative thermal expansion is one such example.[31] The synthesis of $HgBa_2Ca_2Cu_3O_8$ with the highest superconducting T_c (*ca.* 160 K under pressure) is another example.[15,32] The preparation of modulation free bismuth cuprate superconductors by the substitution of Bi by Pb is also a good example.[33] One may

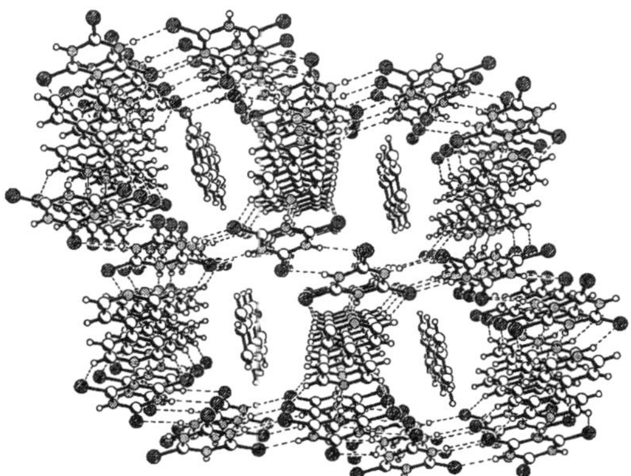

Fig. 12 Noncovalent synthesis of a layered network with large cavity based on hydrogen bonding between trithiocyanuric acid and bipyridine. Three-dimensional channels with benzene molecules can be seen [from Pedireddi *et al.* (reproduced with permission from ref. 30(*b*))].

also cite examples of the synthesis of ionic conductors and other materials. Most inorganic materials are still discovered accidentally, the cases of high T_c cuprates and CMR manganates being well known. What is noteworthy is the development of new strategies for synthesis, particularly those involving soft chemical routes. The subject has been reviewed adequately in the recent literature.[16,34,35]

The methods of soft chemistry include sol–gel, electrochemical, hydrothermal, intercalation and ion-exchange processes. Many of these methods are employed routinely for the synthesis of ceramic materials.[36,37] There have been recent reviews of the electrochemical methods,[38] intercalation reactions,[39] and the sol–gel technique.[40] The sol–gel method has been particularly effective with wide-ranging applications in ceramics, catalysts, porous solids and composites[41] and has given rise to fine precursor chemistry. Hydrothermal synthesis has been employed for the synthesis of oxidic materials under mild conditions[42] and most of the porous solids and open-framework structures using organic templates are prepared hydrothermally.[43] The advent of supramolecular chemistry has started to make an impact on synthesis,[16] mesoporous solids being well known examples.[43]

Many of the traditional methods continue to be exploited to synthesise novel materials. In the area of cuprates, the synthesis of a superconducting ladder cuprate and of carbonato- and halocuprates is noteworthy. High pressure methods have been particularly useful in the synthesis of certain cuprates and other materials.[36,44] The combustion method has come of age for the synthesis of oxidic and other materials.[45] Microwave synthesis is becoming popular[46] while sonochemistry has begun to be exploited.[47]

In the oxide literature, there are several reviews specific to the different families, such as layered transition metal oxides[48] and metal phosphonates.[49] Synthesis of metal nitrides has been discussed by Di Salvo.[50] Precursor synthesis of oxides, chalcogenides and other materials is being pursued with vigour.[16,36] Thus, single-molecule precursors of chalcogenide containing compound semiconductors have been developed.[51] Kanatzidis[52] has reviewed the application of molten poly-chalcophosphate fluxes for the synthesis of complex metal thiophosphates and selenophosphates.

In our own work over the years,[53] we have found the use of precursor carbonates to be effective in the synthesis of many oxide materials. Electrochemical techniques have been useful in oxidation reactions (*e.g.*, preparation of $SrCoO_3$, ferromagnetic $LaMnO_3$ containing *ca.* 30% Mn^{4+}, $La_2NiO_{4.25}$, $La_2CuO_{4+\delta}$). Intercalation and deintercalation of amines can be used to produce novel phases of oxides such as WO_3 and MoO_3. Topochemical oxidation, reduction and dehydration reactions also give rise to novel oxide phases, which cannot be prepared otherwise. Thus, $La_2Ni_2O_5$ can only be made by the reduction of $LaNiO_3$. Special mention must be made of the simple chemical technique of nebulized spray pyrolysis of solutions of organometallic precursors, to obtain epitaxial films of complex oxide materials; this technique can also be used to prepare nanoscale oxide powders.[54] Epitaxial films of PZT, $LaMnO_3$, $LaNiO_3$ and other oxide materials have been prepared by this method. Good films of copper and other metallic systems have also been obtained.

Molecular materials

Molecular materials, especially organic ones, have gained prominence in the last few years, with practical applications as optical and electronic materials already becoming possible. This rich area of research has benefitted from attempts to carry out rational design based on crystal engineering, including supramolecular chemistry[55,56] and the interest to investigate the structure and chemistry of the organic solid state.[57] The conventional areas of polymer research have become more

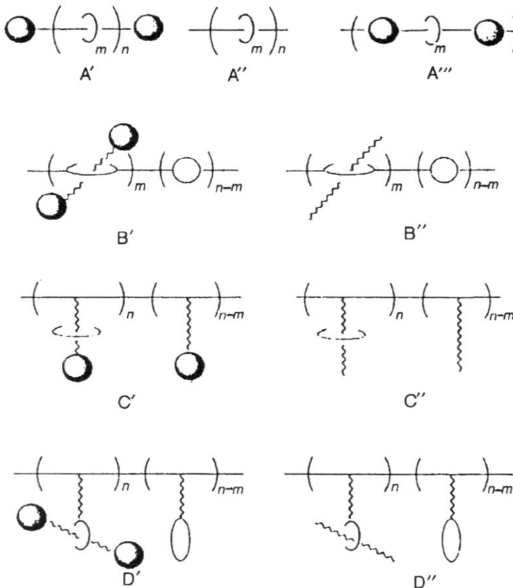

Fig. 13 Various types of main chain and side chain polyrotaxanes: blocking group (shaded circle); hollow circle (cyclic component); ellipses (cyclics threaded by linear species) [from Gong and Gibson (reproduced with permission from ref. 58)].

focussed on developing new synthetic methods and controlling polymerization. Since architecture and functionality at a molecular level control many of the properties of polymers, there are efforts to synthesise polymers with well defined topologies. Polymers involving rotaxanes, catenanes, rods, dendrimers and hyperbranched systems[58,59] are some recent examples of such efforts. Recognizing that hydrogen bonding is the driving force for the formation of crown ether based polyrotaxanes, mechanically linked polymers have been prepared, with the possibility of molecular shuttles at polymeric level (Fig. 13). Standard synthetic routes for dendrimers (Fig. 14) and hyperbranched polymers suggest many new possibilities.

Molecular magnetic materials constitute an interesting area of research, although practical applications may not be feasible in the short term. Synthesis and design of molecular ferromagnets, ferrimagnets and weak ferromagnets, providing permanent magnetization with fairly high T_c values are the main objectives of this effort. While purely organic ferromagnets have only reached modest T_cs below 2 K, materials incorporating transition metals have shown promise [e.g. V(TCNE)$_x$]. Molecular magnetic materials generally make use of free

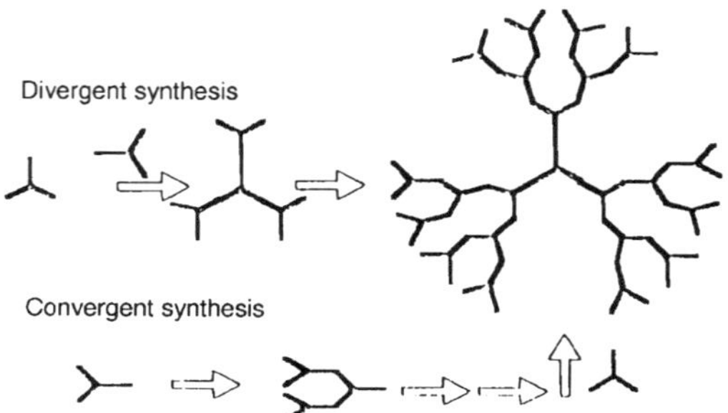

Fig. 14 Synthetic routes to dendrimers [from Hobson and Harrison (reproduced with permission from ref. 59)].

radicals such as nitroxides, high-spin metal complex magnetic clusters or a combination of radicals and metal complexes.[60,61] There has been a renaissance in spin-crossover materials with emphasis on active elements for memory devices and light-induced spin changes.[62] The study of diradicals and radical pairs has been strengthened by our understanding of how to stabilize the triplet state.[61] C$_{60}$–TDAE ($T_c \approx 16$ K) and related compounds have attracted some attention (TDEA = tetrakisdimethylaminoethylene). BEDT-TTF salts with magnetic anions have been discovered recently.[63] Organometallic materials based on neutral nitroxide radicals and charge-transfer (CT) complexes derived from the radicals have been examined in some detail.[64]

Semiconducting, metallic and superconducting molecular materials have been investigated by several workers in the last two decades. New types of TTF type molecules, transition metal complexes with elongated π ligands, neutral radicals and Te-containing π donors have been synthesized and an organic superconductor with an organic anion, β''-(ET)$_2$SF$_5$CH$_2$CF$_2$SO$_3$, has been discovered.[65] Cation radical salts of BEDT-TTF and similar donors as well as BEDT-TTF salts are still a fertile ground for the production of new materials.[63,65] Synthesis of linearly extended TTF analogues[66] as well as of new TCNQ and DCNQ acceptors[67] have been reported. The use of organic semiconductors as active layers in thin films holds promise, α-sexithiophene being an example.[68,69]

High performance photonic devices have been fabricated from conjugated polymers such as poly(p-phenylenevinylene), polyaniline and polythiophene.[70] The devices include diodes, light-emitting diodes, photodiodes, field effect transistors, polymer grid triodes, light emitting electrochemical cells, optocouplers and lasers. The performance levels of many of these organic devices have reached those of the inorganic counterparts. The high photoluminescence efficiency and large cross section for stimulated emission of semiconducting polymers persist upto high concentrations (unlike dyes). By combination with InGaN, hybrid lasers can now be fabricated. It is gratifying that plastic electronics is moving rapidly from fundamental research to industrial applications.

The area of molecular nonlinear optics has been rejuvenated by efforts to investigate three-dimensional multipolar systems, functionalized polymers as optoelectronic materials, near infrared optical parametric oscillators and related aspects.[71] There have been some advances in chromophore design for second-order nonlinear optical materials;[72] these include one-dimensional CT molecules, octopolar compounds and organometallics. Some of the polydiacetylenes and poly(p-phenylenevinylene)s appear to possess the required properties for use as third-order nonlinear optical materials for photonic switching.[73]

Increasing interest in molecular materials has made an impact on the direction of organic chemistry as well, giving rise to large-scale activity in this area. The discovery of fullerenes and fullerene-based materials has contributed to some interesting developments.[74–76] Organic–inorganic hybrids offer many interesting possibilities, with the composites exhibiting properties and phenomena not found in either component.[77a] Two types of materials have been classified, organic–inorganic units held together by weak bonds and organometallic polymers.[77b] Silicon based hybrids and related materials have been reviewed by Livage.[40] In Fig. 15, we show some examples of polyhedral oligomeric silsesquioxanes used as building blocks for the sol–gel synthesis of hybrid materials. It is indeed gratifying that organic molecular devices have become a reality. The concepts and directions suggested by Whitesides[78a] have been useful in designing devices. Special mention must be made of the devices developed by Willner[78b] making use of self-organized thiols and other molecular systems. In this connection, the device-related studies of functional polymers by Wegner and others are also relevant.

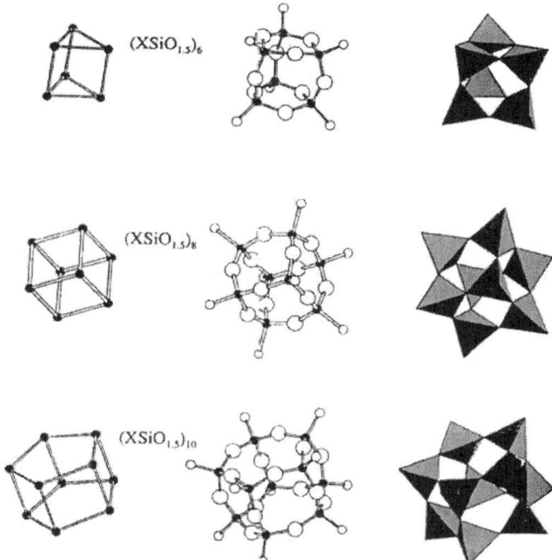

Fig. 15 Examples of silsesquioxanes used as building blocks for the sol–gel synthesis of hybrid materials [from Livage (reproduced with permission from ref. 60)].

Biomolecular materials

Biomolecular materials constitute a class of materials at the interface between biology and materials science. They not only provide an insight to Nature's ways, but also suggest new methods of synthesis and design of materials. Thus we can learn much from a study of the principles of structure–function relations in mineralized biological materials, the various forms of calcium carbonate forming the basis for many things that happen.[79,80] Some of the materials such as shells can, in principle, be considered to be part of traditional ceramic materials. The synthesis of mesoporous silica and related materials also owes its origins to biomineralization. Biomineralization has inspired the synthesis of nanocomposites, nanoparticles and nanoparticulate films.[81,82] Collagen fibrillar structure in mineralized and nonmineralized tissue, in particular the ultrastructure orientation of mineral crystals in bone relative to that of the collagen fibril, is an important problem receiving attention.[83]

Some of the recent developments in biomolecular materials include designed drug delivery systems consisting of self-assemblies of lipid and biocompatible lipids,[84] protein-based biological motors,[85] formation of new types of liposomes (*e.g.* spherical liposomes) and emulsions, tissue engineering, biogels, higher-order self-assemblies in biomaterials, and new methods of biopolymer synthesis.[86] Assembly of soft biofunctional and biocompatible interfaces of solids by the deposition of ultrathin soft polymer films, supported lipid–protein bilayers or membranes separated from solids by soft polymer cushions are other areas of vital interest.[87] The ability of polymer molecules attached at one end to a surface to prevent or enhance protein adsorption is being examined.[88]

Glasses and ceramics can bond to living tissues if there is bioactive layer. The development of a bioactive hydroxyapatite layer *in vivo* at body temperature is therefore an important problem.[89] Materials with the highest level of bioactivity develop a silica layer that enhances the formation of such a layer. Such sol–gel processes are used to produce bioactive coatings, powders and substrates which allow molecular control over the incorporation and behavior of proteins and cells with applications as sensors and implants. Sol–gel encapsulation of biomolecules within silica matrices has encompassed enzymes, metalloproteins, photoactive biomolecules and even whole cells.[78] Recently, interfacing electronic materials with lipids and proteins has shown promise in several fields, including biosensors.[90]

Nanomaterials

The synthesis, characterization and properties of nanomaterials have become very active areas of research in the last few years. In particular, nanostructured materials assembled by means of supramolecular organization offer many exciting possibilities. These include self-assembled monolayers and multilayers with different functionalities, intercalation in preassembled layered hosts and inorganic three-dimensional networks. The reader is referred to the special issue of *Chemistry of Materials*[91] for an overview of present day interests. There are many recent reviews on the varied aspects of nanomaterials. The work of Alivisatos[92] on the structural transitions, elec-

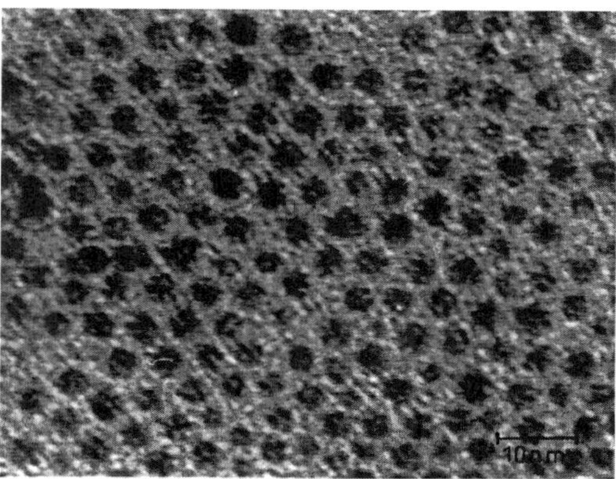

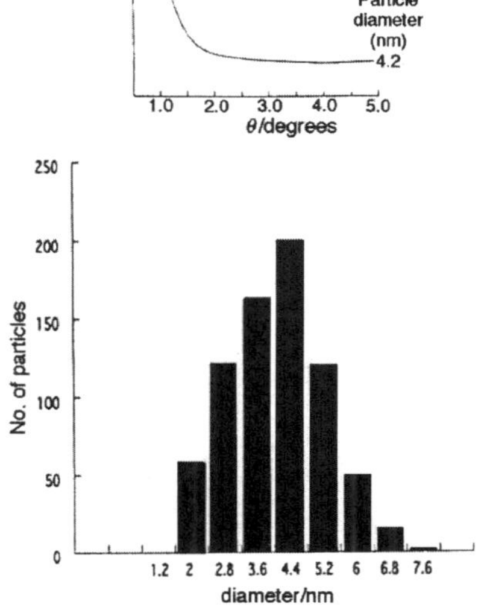

Fig. 16 TEM image of a nanocrystalline array of thiol-derivatized Au particles [from Sarathy *et al.* (reproduced with permission from ref. 96(*b*))].

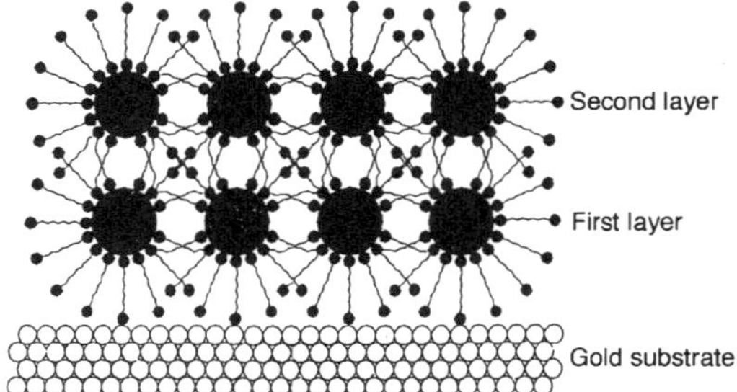

Fig. 17 Schematic representation of successive deposition of layers of metal (Pt or Au) nanoparticles and dithiols.

tronic properties and related aspects of nanoparticles of CdS, InAs and such materials is particularly noteworthy.

We have been interested in investigating the size-dependent electronic structure and reactivity of metal clusters deposited on solid substrates. Thus, we have shown that when the cluster size is small ($\lesssim 1$ nm), an energy gap opens up.[93a] Bimetallic clusters show additive effects due to alloying and cluster size[93b] in their electronic properties. Small metal clusters of Cu, Ni and Pd show enhanced chemical reactivity with respect to CO and other molecules.[94] Metal clusters and colloids, especially those with protective ligands, have been reviewed in relation to nanomaterials.[95] We have recently developed methods of preparing nanoparticles of various metals as well as nanocrystalline arrays of thiolized nanoparticles of Au, Ag and Pt.[96] In Fig. 16, we show the TEM image of thiol-derivatized Au

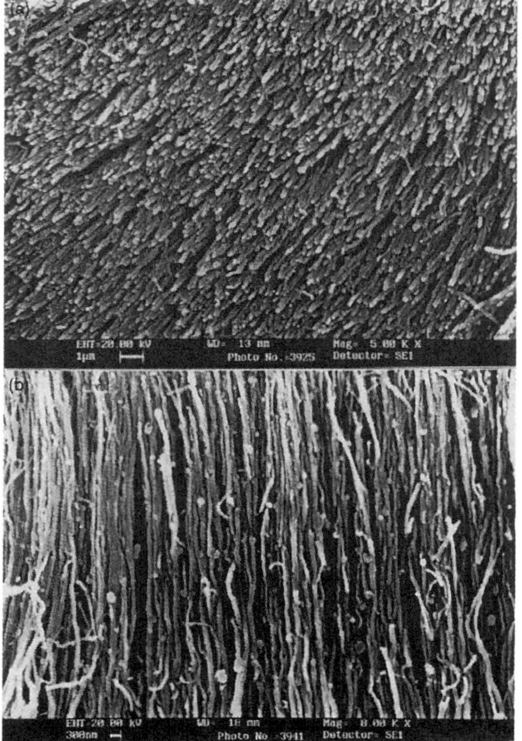

Fig. 18 SEM images of aligned-nanotube bundles obtained by the pyrolysis of ferrocene (from Rao *et al.*[99b]).

nanoparticles forming a nanocrystalline array. More interestingly, by using dithiols, it has been possible to accomplish layer-by-layer deposition of dithiol–metal nanoparticle films (Fig. 17). This is somewhat similar to the layer-by-layer self assembly of polyelectrolyte–inorganic nanoparticle sandwich films.[97] Such superlattices involving vertical organization of arrays of metal quantum dots may have novel properties.

Unprecedented interest has developed in carbon science ever since the discovery of fullerenes and nanotubes. Solid state properties of C_{60} and C_{70} have been of interest to the author since 1991. Some of the aspects investigated in this laboratory include orientational phase transitions, amorphization and polymerization under pressure, molecular magnetism of the TDAE derivative *etc.*[74] We have been working on the synthesis and characterization of carbon nanotubes for some time.[74] Opening, filling, closing and functionalizing carbon nanotubes have been accomplished.[98] Since metal particles are essential as catalysts to prepare nanotubes by the pyrolysis of hydrocarbons, we have employed organometallic precursors to generate nanotubes.[99a] Single-wall nanotubes have been obtained by the pyrolysis of metallocene or $Fe(CO)_5$–hydrocarbon mixtures under controlled conditions. It has also been possible to obtain copious quantities of aligned-nanotube bundles by the pyrolysis of ferrocene (Fig. 18) precursors. C–N and B–C–N nanotubes have been prepared by precursor pyrolysis as well.[99a] By using acid-treated carbon nanotubes as templates, ceramic oxide nanotubes have been prepared.[100] The procedure involves coating the acid-treated nanotubes by metal alkoxides and such precursors, and burning off the carbon.

Concluding remarks

In a brief overview of this type, it has not been possible to do justice to the work of many leading practitioners of materials chemistry or cover all the known classes of materials and phenomena. Thus I have not covered chalcogenides, nitrides and the like. An area that has not been discussed is noncrystalline materials, especially glasses, which have been of vital interest to materials chemists for many years. The subject is growing rapidly because of the increasing number of applications. Amorphous carbon, amorphous silicon, fluoride glasses, superionic glasses and metallic glasses are some of the amorphous materials of great interest.[101] The glass transition continues to arouse interest.[102] It is an unusual phase transition indeed. The subject of phase transitions itself is of great importance in materials chemistry and a proper understanding of the subject is essential to appreciate many of the materials properties.[103] Ionic conductors, energy storage materials and dielectrics constitute important areas of vital technological relevance. Recent developments in solid state electrochemistry are nicely covered in the book edited by Bruce.[104] An aspect of materials chemistry that is being increasingly exploited is computer simulation and modelling of structures, surfaces, processes and mechanisms.[105] Simulation and modelling techniques are particularly useful in understanding phenomena and structures in situations where experimentation is difficult (*e.g.* molecular processes in zeolites). Another development that deserves notice is the atomic layer-by-layer synthesis of inorganic materials,[106] which suggests the possibility of making use of oxides and other materials in integrated circuits. In spite of the brevity, I trust that the article has succeeded in communicating the nature of present-day materials chemistry. I believe that the references to many of the recent reviews will be useful to students, teachers and practitioners of the subject.

References

1 (*a*) *Modern Aspects of Solid State Chemistry*, ed. C. N. R. Rao, Plenum Press, New York, 1970; (*b*) C. N. R. Rao and

J. Gopalakrishnan, *New Directions in Solid State Chemistry*, first edition, Cambridge University Press, 1986 (second edition 1997).

2 (*a*) A. R. West, *Solid State Chemistry and Applications*, Wiley, Chichester, 1984; (*b*) *Solid State Chemistry*, ed. A. K. Cheetham and P. Day, Clarendon Press, Oxford 1987, 1992; (*c*) *Chemistry of Advanced Materials*, ec. C. N. R. Rao, Blackwell, Oxford, 1994; (*d*) L. V. Interrante and M. J. Hampden-Smith, *Chemistry of Advanced Materials*, Wiley-VCH, 1998.

3 (*a*) J. B. Goodenough, *Prog. Solid State Chem.*, 1971, **5**, 149; (*b*) C. N. R. Rao and B. Raveau, *Transition Metal Oxides*, second edition, Wiley-VCH, New York, 1998.

4 M. Greenblatt, *Chem. Rev.*, 1988, **88**, 31.

5 J. G. Bednorz and K. A. Müller, *Z. Phys. B*, 1986, **64**, 189.

6 (*a*) *Metal-insulator Transitions*, ed. P. P. Edwards and C. N. R. Rao, Taylor and Francis, London, 1985; (*b*) *Metal-insulator Transition Revisited*, ed. P. P. Edwards and C. N. R. Rao, Taylor and Francis, London, 1995.

7 (*a*) J. M. Honig and L. L. Van Zandt, *Annu. Rev. Mater. Sci.*, 1975, **5**, 225; (*b*) J. M. Honig and J. Spalek, *Proc. Ind. Nat. Sci. Acad. A*, 1986, **52**, 232.

8 C. N. R. Rao, *Philos. Trans. R. Soc. London*, 1998, **356**, 23.

9 P. Ganguly and C. N. R. Rao, *J. Solid State Chem.*, 1984, **53**, 193.

10 C. N. R. Rao, P. Ganguly, K. K. Singh and R. A. Mohanram, *J. Solid State Chem.*, 1988, **72**, 14.

11 C. N. R. Rao and A. K. Cheetham, *Adv. Mater.*, 1997, **9**, 1009 and references therein.

12 C. N. R. Rao and J. M. Thomas, *Acc. Chem. Res.*, 1985, **18**, 113.

13 M. K. Wu *et al.*, *Phys. Rev. Lett.*, 1987, **58**, 908.

14 C. N. R. Rao *et al.*, *Nature*, 1987, **326**, 856.

15 M. Hervieu, *Curr. Opin. Solid State Mater. Sci.*, 1996, **1**, 29.

16 J. Gopalakrishnan, N. S. P. Bhuvanesh and K. K. Rangan, *Curr. Opin. Solid State Mater. Sci.*, 1996, **1**, 285.

17 B. Raveau, C. Michel, M. Hervieu and D. Groult, *Crystal Chemistry of High T_c Superconducting Copper Oxides*, Springer, Berlin, 1991.

18 (*a*) C. N. R. Rao and A. K. Ganguli, *Chem. Soc. Rev.*, 1995, **24**, 1; (*b*) C. N. R. Rao, *Philos. Trans. R. Soc. London*, 1991, **336**, 595; (*c*) T. V. Ramakrishnan, in *Critical Problems in Physics*, Princeton University Press, 1996.

19 J. Rouxel, *Curr. Sci. (India)*, 1997, **73**, 31.

20 (*a*) *Colossal Magnetoresistance, Charge-ordering and Related Aspects of Manganese Oxides*, ed. C. N. R. Rao and B. Raveau, World Scientific, Singapore, 1998; (*b*) Y. Tokura in ref. 20(a).

21 C. N. R. Rao, *Chem. Commun.*, 1996, 2217.

22 A. Arulraj, A. Biswas, A. K. Raychaudhuri, C. N. R. Rao, P. M. Woodward, T. Vogt, D. E. Cox and A. K. Cheetham, *Phys. Rev. B*, 1998, **57**, R8115.

23 (*a*) S. Natarajan, M. Eswaramoorthy, A. K. Cheetham and C. N. R. Rao, *Chem. Commun.*, 1998, 1561 and references therein; (*b*) S. I. Zones and M. E. Davis, *Curr. Opin. Solid State Mater. Sci.*, 1996, **1**, 107; (*c*) J. M. Thomas, *Angew. Chem., Int. Ed. Engl.*, 1994, **33**, 913; (*d*) G. Ozin, *Adv. Mater.*, 1992, **4**, 612.

24 (*a*) C. J. Brinker, *Curr. Opin. Solid State Mater. Sci.*, 1996, **1**, 795; (*b*) M. E. Davis, *Chem. Eur. J.*, 1997, **3**, 1745.

25 S. L. Suib, *Curr. Opin. Solid State Mater Sci.*, 1998, **3**, 63.

26 J. S. Beck and J. C. Vartuli, *Curr. Opin. Solid State Mater. Sci.*, 1996, **1**, 76.

27 S. Ayyappan and C. N. R. Rao, *Chem. Commun.*, 1997, 575 and references therein.

28 (*a*) D. Zhao, P. Yang, Q. Huo, B. F. Chmelka and G. D. Stucky, *Curr. Opin. Solid State Mater. Sci.*, 1998, **3**, 111; (*b*) Neeraj and C. N. R. Rao, *J. Mater. Chem.*, 1998, **8**, 1631.

29 (*a*) T. Maschmeyer, *Curr. Opin. Solid State Mater. Sci.*, 1998, **3**, 71; (*b*) M. Eswaramoorthy, Neeraj and C. N. R. Rao, *Chem. Commun.*, 1998, 615.

30 (*a*) O. M. Yagi, G. Li and H. Li, *Nature*, 1995, **378**, 703; (*b*) V. R. Pedireddi, S. Chatterji, A. Ranganathan and C. N. R. Rao, *J. Am. Chem. Soc.*, 1997, **119**, 10867.

31 A. W. Sleight, *Endeavour*, 1995, **19**, 64.

32 E. V. Antipov, S. M. Loureno, C. Chaillout, J. J. Capponi, P. Bordet, J. L. Tholence, S. N. Putlin and M. Marezio, *Physica C*, 1993, **215**, 1.

33 V. Manivannan, J. Gopalakrishnan and C. N. R. Rao, *Phys. Rev. B*, 1991, **43**, 8686.

34 J. Gopalakrishnan, *Chem. Mater.*, 1995, **7**, 1265.

35 *Soft Chemistry Routes to New Materials—Chemie Douce*, ed. J. Rouxel, M. Tournoux and R. Brec, Trans. Tech. Publications, Aedermannsdorf, Switzerland, 1994.

36 (*a*) *Preparative Solid State Chemistry*, ed. P. Hagenmuller, Academic Press, New York, 1997, p. 2; (*b*) C. N. R. Rao, *Chemical Approaches to the Synthesis of Inorganic Materials*, John Wiley, Chichester, 1994; (*c*) J. D. Corbett, in *Solid State Chemistry—Techniques*, ed. A. K. Cheetham and P. Day, Clarendon Press, Oxford, 1987.

37 D. Segal, *J. Mater. Chem.*, 1997, **7**, 1297.

38 J. C. Grenier, M. Pouchard and A. Wattiaux, *Curr. Opin. Solid State Mater. Sci.*, 1996, **1**, 233.

39 G. Ouvrad and D. Guyomard, *Curr. Opin. Solid State Mater. Sci.*, 1996, **1**, 260.

40 J. Livage, *Curr. Opin. Solid State Mater. Sci.*, 1997, **2**, 132.

41 See special issue of *Chem. Mater.*, 1997, **9**, 2247–2670.

42 M. S. Whittingham, *Curr. Opin. Solid State Mater. Sci.*, 1996, **1**, 227.

43 M. E. Davis and I. E. Maxwell, *Curr. Opin. Solid State Mater. Sci.*, 1996, **1**, 55.

44 M. Takano and A. Onodera, *Curr. Opin. Solid State Mater. Sci.*, 1997, **2**, 166.

45 K. C. Patil, S. T. Aruna and S. Ekambaram, *Curr. Opin. Solid State Mater. Sci.*, 1997, **2**, 158.

46 (*a*) K. J. Rao and P. D. Ramesh, *Bull. Mater. Sci.*, 1995, **18**, 447; (*b*) D. M. P. Mingos, *Chem. Soc. Rev.*, 1998, **27**, 213.

47 D. Peters, *J. Mater. Chem.*, 1996, **6**, 1605.

48 P. A. Salvador, T. O. Mason, M. E. Hagerman and K. R. Poeppelmeier, in *Chemistry of Advanced Materials: An Overview*, ed L. V. Interrante and M. J. Hampden-Smith, Wiley—VCH, New York, 1998.

49 A. Clearfield, *Curr. Opin. Solid State Mater. Sci.*, 1996, **1**, 268.

50 F. J. Di Salvo, *Curr. Opin. Solid State Mater. Sci.*, 1996, **1**, 241.

51 P. O'Brien and R. Nomura, *J. Mater. Chem.*, 1995, **5**, 1761.

52 M. G. Kanatzidis, *Curr. Opin. Solid State Mater. Sci.*, 1997, **2**, 139.

53 C. N. R. Rao, *Pure Appl. Chem.*, 1994, **66**, 1765; 1997, **69**, 199.

54 (*a*) H. N. Aiyer, A. R. Raju, G. N. Subbanna and C. N. R. Rao, *Chem. Mater.*, 1997, **9**, 755; (*b*) P. Murugavel, M. Kalaiselvam, A. R. Raju and C. N. R. Rao, *J. Mater. Chem.*, 1997, **7**, 1433.

55 A. Gavezzotti, *Curr. Opin. Solid State Mater. Sci.*, 1996, **1**, 501.

56 G. R. Desiraju, *Curr. Opin. Solid State Mater. Sci.*, 1997, **2**, 451.

57 See special issue of *Chem. Mater.* (dedicated to M.C. Etter), 1994, **6**, 1087–1461.

58 C. Gong and H. W. Gibson, *Curr. Opin. Solid State Mater. Sci.*, 1997, **2**, 647.

59 L. C. Hobson and R. M. Harrison, *Curr. Opin. Solid State Mater. Sci.*, 1997, **2**, 683.

60 D. Gatteschi, *Curr. Opin. Solid State Mater. Sci.*, 1996, **1**, 192.

61 K. Matsuda and H. Iwamura, *Curr. Opin. Solid State Mater. Sci.*, 1997, **2**, 446.

62 O. Kahn, *Curr. Opin. Solid State Mater. Sci.*, 1996, **1**, 547.

63 P. Day and M. Kurmod, *J. Mater. Chem.*, 1997, **7**, 1291.

64 S. Nakatsuji and H. Auzai, *J. Mater. Chem.*, 1997, **7**, 2161.

65 H. Kobayasbi, *Curr. Opin. Solid State Mater. Sci.*, 1997, **2**, 440.

66 J. Roncali, *J. Mater. Chem.*, 1997, **7**, 2307.

67 N. Martin, J. Segura and C. Sevane, *J. Mater. Chem.*, 1997, **7**, 1661.

68 F. Garnier, *Curr. Opin. Solid State Mater. Sci.*, 1997, **2**, 455.

69 H. E. Katz, *J. Mater. Chem.*, 1997, **7**, 369.

70 A. J. Heeger and M. A. Diaz-Garcia, *Curr. Opin. Solid State Mater. Sci.*, 1998, **3**, 16.

71 J. Zyss and J-F. Nicoud, *Curr. Opin. Solid State Mater. Sci.*, 1996, **1**, 533.

72 T. Verbiest, S. Houbrechts, M. Kauranen, K. Clays and A. Persons, *J. Mater. Chem.*, 1997, **7**, 2175.

73 B. Luther-Davies and M. Samic, *Curr. Opin. Solid State Mater. Sci.*, 1997, **2**, 213.

74 C. N. R. Rao, R. Seshadri, R. Sen and A. Govindaraj, *Mater. Sci. Eng. Rep.*, 1995, **R15**, 209; also see *Curr. Opin. Solid State Mater. Sci.*, 1996, **1**, 279.

75 K. Prassides, *Curr. Opin. Solid State Mater. Sci.*, 1997, **2**, 433.

76 M. Prato, *J. Mater. Chem.*, 1997, **7**, 1097.

77 (*a*) Y. Chujo, *Curr. Opin. Solid State Mater. Sci.*, 1996, **1**, 806; (*b*) P. Judeinstein and C. Sanchez, *J. Mater. Chem.*, 1996, **6**, 511.

78 (*a*) See for example G. M. Whitesides, *Acc. Chem. Res.*, 1995, **28**, 37, 219; (*b*) I. Willner, *Acc. Chem. Res.*, 1997, **30**, 347.

79 *Biomimetic Materials Chemistry*, ed. S. Mann, VCH Publishers, Weinheim, 1996.

80 S. Weiner and L. Addadi, *J. Mater. Chem*, 1997, **7**, 689.

81 S. Manne and L. A. Aksay, *Curr. Opin. Solid State Mater. Sci.*, 1997, **2**, 358.

82 J.H. Fendler, *Curr. Opin. Solid State Mater. Sci.*, 1997, **2**, 365.

83 A. Veis, *Curr. Opin. Solid State Mater. Sci.*, 1997, **2**, 370.

84 D. D. Lasic and D. Papahadzopouls, *Curr. Opin. Solid State Mater. Sci.*, 1996, **1**, 392.

85 F. Gittes and C. F. Schmidt, *Curr. Opin. Solid State Mater. Sci.*, 1996, **1**, 412.

86 C. R. Safinya and L. Addadi, *Curr. Opin. Solid State Mater. Sci.*, 1996, **1**, 387.
87 J. Rädler and E. Sackmann, *Curr. Opin. Solid State Mater. Sci.*, 1997, **2**, 330.
88 I. Szleifer, *Curr. Opin. Solid State Mater. Sci.*, 1997, **2**, 337.
89 L. L. Hench, *Curr. Opin. Solid State Mater. Sci.*, 1997, **2**, 604.
90 J. A. Zasadzinski, *Curr. Opin. Solid State Mater. Sci.*, 1997, **2**, 345.
91 See special issue of *Chem. Mater.*, 1996, **8**, 1569–2193.
92 P. Alivisatos, *J. Phys. Chem.*, 1996, **100**, 13226.
93 (*a*) C. P. Vinod, G. U. Kulkarni and C. N. R. Rao, *Chem. Phys. Lett.*, 1998, **289**, 329. (*b*) K. R. Harikumar, S. Ghosh and C. N. R. Rao, *J. Phys. Chem.*, 1997, **101**, 536.
94 A. K. Santra, S. Ghosh and C. N. R. Rao, *Langmuir*, 1994, **10**, 3937.
95 G. Schmid and G. L. Hornyak, *Curr. Opin. Solid State Mater. Sci.*, 1997, **2**, 204.
96 (*a*) S. Ayyappan, R. S. Gopalan, G. N. Subbanna and C. N. R. Rao, *J. Mater. Res.*, 1997, **12**, 398; (*b*) K. V. Sarathy, G. Raina, R. T. Yadav, G. U. Kulkarni and C. N. R. Rao, *J. Phys. Chem.*, 1997, **101**, 9876.
97 J. H. Fendler, *Chem. Mater.*, 1996, **8**, 1616.
98 B. C. Satishkumar, A. Govindaraj and C. N. R. Rao, *J. Phys. B*, 1996, **29**, 4925.
99 (*a*) R. Sen, A. Govindaraj and C. N. R. Rao, *Chem. Mater.*, 1997, **9**, 2078; *Chem. Phys. Lett.*, 1998, **287**, 671; (*b*) C. N. R. Rao, R. Sen, B. C. Satishkumar and A. Govindaraj, *Chem. Commun.*, 1998, 1525.
100 C. N. R. Rao, B. C. Satishkumar and A. Govindaraj, *Chem. Commun.*, 1997, 1581.
101 S. R. Elliott and R. Street, *Curr. Opin. Solid State Mater. Sci.*, 1996, **1**, 555; 1997, **2**, 397.
102 C. A. Angell, *Curr. Opin. Solid State Mater. Sci.*, 1996, **1**, 578.
103 C. N. R. Rao, *Acc. Chem. Res.*, 1984, **17**, 83; see also *J. Mol. Struct.*, 1993, **292**, 229.
104 *Solid State Electrochemistry*, ed. P. G. Bruce, Cambridge University Press, 1995.
105 M. Stoneham and M. L. Klein, *Curr. Opin. Solid State Mater. Sci.*, 1996, **1**, 817; M. Stoneham and S. Panteldes, *Curr. Opin. Solid State Mater. Sci.*, 1997, **2**, 6.
106 A. Gupta, *Curr. Opin. Solid State Mater. Sci.*, 1997, **2**, 23.

Paper 8/04467H

5228

Reprinted from **The Journal of Physical Chemistry, 1995, 99.**
Copyright © 1995 by the American Chemical Society and reprinted by permission of the copyright owner.

The Metal−Nonmetal Transition: A Global Perspective

P. P. Edwards

The School of Chemistry, University of Birmingham, Birmingham, B 15 2TT, U.K.

T. V. Ramakrishnan

Department of Physics, Indian Institute of Science, Bangalore 560 012, India

C. N. R. Rao*

*CSIR Centre of Excellence in Chemistry, Indian Institute of Science, Bangalore 560 012, India, and
Department of Chemistry, University of Wales, Cardiff CF1 3TB, U.K.*

Received: October 24, 1994; In Final Form: January 26, 1995[⊗]

A wide range of condensed matter systems traverse the metal−nonmetal transition. These include doped semiconductors, metal−ammonia solutions, metal clusters, metal alloys, transition metal oxides, and superconducting cuprates. Certain simple criteria, such as those due to Herzfeld and Mott, have been highly successful in explaining the metallicity of materials. In this article, we demonstrate the amazing effectiveness of these criteria and examine them in the light of recent experimental findings. We then discuss the limitations in our understanding of the phenomenon of the metal−nonmetal transition.

1. Introduction

Of all the physical properties of condensed materials, the electrical conductivity exhibits the widest range, anywhere from 10^{-22} ohm^{-1} cm^{-1} in the best nonmetals to around 10^{10} ohm^{-1} cm^{-1} in pure metals (not in the superconducting state). There are several situations where condensed phases transform from the metallic to the nonmetallic state on changing thermodynamic parameters such as temperature, pressure, and composition, with the electrical conductivity changing by factors of 10^3-10^{14} over a small range of the thermodynamic parameter.[1-3] In Figure 1, we show the temperature−composition plane with the electron density varying between 10^{12} and 10^{30}, with the temperature going up to 10^{10} K. The normal experimental conditions where we find metals and semiconductors are indicated in the figure, as are also the conditions appropriate to stars. The elements H, Xe, Cs, and Hg shown by asterisks are at conditions close to the critical points attained by one of many routes such as shock waves, wire explosions, or MHD (magnetohydrodynamic implosion). The figure also shows the regime for a degenerate strongly coupled plasma, besides the values of the Wigner− Seitz radius, r_s, given by

$$r_s = \left(\frac{3}{4\pi na_0{}^3}\right)^{1/3} \qquad (1)$$

where n is the conduction electron density and a_0 is the Bohr radius. The largest value of r_s in metals[5] is found for Cs. In all the systems in Figure 1, the possibility of a thermodynamically induced transition from the metallic to the nonmetallic regime exists. Clearly, the metal−nonmetal transition is a

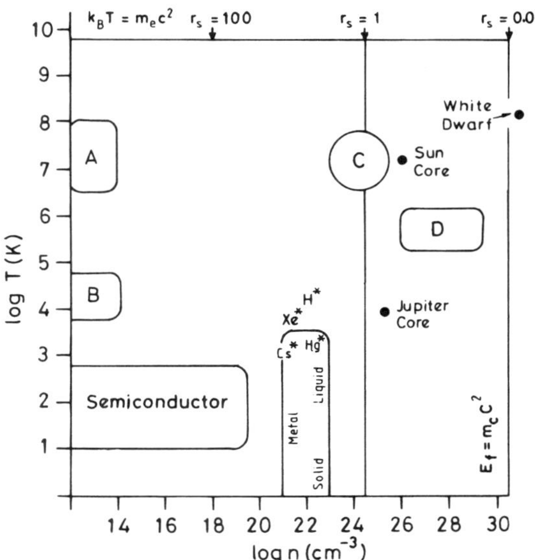

Figure 1. A nonrelativistic window of the temperature−composition plane, showing electron density (n) and temperature (T). Normal conditions (on earth) for semiconductors and elemental metals and conditions on the Sun, Jupiter, and the White Dwarf are shown. Experimental methods in A, B, C, and D are Tokamak, glow-discharge, laser fusion, and degenerate strongly coupled plasma, respectively. Wigner−Seitz radii, r_s, are also shown (adapted from Redmer[4]).

problem of vital interest, being concerned with a wide range of issues from the metallization of stars to the size-induced transition in small clusters of metals.[3,6,7] Between these two extrema is a myriad of examples in a variety of condensed matter systems.[1,2] The range of systems traversing the metal−nonmetal

* To whom correspondence should be addressed at CSIR Centre of Excellence in Chemistry, Indian Institute of Science, Bangalore 560 012, India.
⊗ Abstract published in *Advance ACS Abstracts*, March 15, 1995.

transition is continually increasing. Thus, oxides exhibiting high-temperature superconductivity are close to or at the metal−nonmetal boundary,[8,9] and some of them actually exhibit metal to nonmetal transitions with a change in composition.

In spite of the plethora of experimental findings, the status of our theoretical understanding of metal−nonmetal transitions is far from satisfactory. The difficulty is largely intrinsic to the phenomenon. Electronic states involved in charge transport(i.e., those near the Fermi energy) are spatially extended in the metal and are localized in the insulator. The localization may be due to static disorder (Anderson localization), to strong local electron−electron correlations which "freeze" the local electron number (Mott transition), or to strong electron−lattice coupling which traps the electron locally.[1,2,10,11] In all these cases, the natural modes of description of the electronic states in the different phases are diametrically opposite; it is difficult to find an approach which does both. Secondly, in many cases, more than one mechanism is operative, and one may reinforce the other. For example, in a disordered, strongly correlated oxide, Anderson localization due to disorder tends to increase the local correlation effect. In all electronic systems, the Coulomb interaction which is relatively weak and short ranged in a metal but strong and long ranged in an insulator (giving rise to bound electron−hole states) is present and can be important in promoting the insulating state. There is one mechanism for the metal−nonmetal (M−NM) transition, in a crystalline solid, that does not involve localized states. This is the transition of electrons from a fully filled band (insulator) to a partially filled band (metal) under pressure or structural change. This transition, however, appears to be uncommon. We discuss these mechanisms in some detail in section 6 (see also refs 1, 2, 10, and 11).

There are certain simple criteria for the occurrence of the metal−nonmetal transition, based on powerful physical concepts which turn out to be surprisingly successful. One such criterion is the idea due to Mott that in a low carrier density metal, the screened Coulomb attraction may be strong enough to bind an electron hole pair, thus destabilizing the metal.[2,12] Another useful criterion due to Mott[13] is the idea of a minimum conductivity, σ_{min}, that a metal can support, corresponding to the mean free path being equal to the de Broglie wavelength of the electron at the Fermi energy. Then, there is the Herzfeld metallization criterion of the dielectric catastrophe[14] which could occur for a dense collection of polarizable atoms. We discuss these criteria further in sections 3−5. We shall first present a brief overview of phenomena and systems associated with the M−NM transition.

2. Diverse Systems Exhibiting Metal−Nonmetal Transitions

As mentioned earlier, a large variety of systems exhibit M−NM transitions. The systems include[1−3] metal−ammonia (or amine) solutions, expanded metals, doped semiconductors, metal−noble gas films, metal−metal halide melts, alloys of gold with metals such as cesium, transition metal oxides and sulfides, and other inorganic and organic solids. These systems have been adequately reviewed, and we shall briefly examine only those findings that are new or directly relevant to the later discussion. Transition metal oxides[15] are especially noteworthy in that the M−NM transition in them can arise from one of many causes. Typical of the transitions found in oxide systems are the following: (i) pressure-induced transitions as in NiO, (ii) transitions as in Fe_3O_4 involving charge ordering, (iii) transitions as in $LaCoO_3$ that are initially induced because of the different spin configurations of the transition metal ion, (iv)

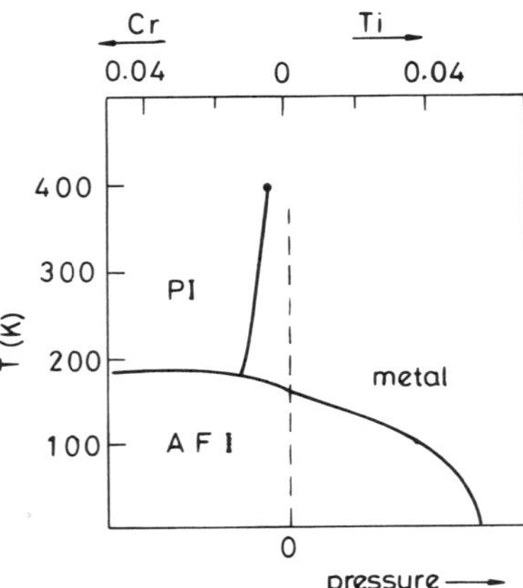

Figure 2. Phase diagram of the metal insulator transition in V_2O_3 as a function of doping with Cr or Ti (regarded as chemical pressure) as well as actual pressure, showing a critical point (after McWhan, D. B.; et al. *Phys. Rev. Lett.* **1971,** *27,* 941). The three phases are paramagnetic insulator (PI), metal, and antiferromagnetic insulator (AFI). The different kinds of resistivity behavior in Figure 3 correspond to different constant pressure (or Cr, Ti content) cuts in this phase diagram.

transitions as in EuO arising from the disappearance of spin polarization band-splitting effects when the ferromagnetic Curie temperature is reached, (v) compositionally induced transitions, as in $La_{1-x}Sr_xCoO_3$ and $LaNi_{1-x}Mn_xO_3$, in which changes of band structure in the vicinity of the Fermi level are brought about by a change in composition or are due to disorder-induced localization, (vi) transitions as in $K_{0.3}MoO_3$ due to charge-density waves, and (vii) temperature-induced transitions in a large class of oxides such as Ti_2O_3, VO_2, and V_2O_3 involving more than one mechanism.

The last category of M−NM transitions has attracted considerable attention. In Ti_2O_3, a second-order transition occurs around 410 K, accompanied by a gradual change in the rhombohedral *c/a* ratio and a 100-fold jump in conductivity; the oxide remains paramagnetic throughout. A simple band-crossing mechanism occurring with the change in the *c/a* ratio can explain this transition. Accordingly, substitution of Ti by V up to 10% in Ti_2O_3 makes the system metallic; the *c/a* ratio of this metallic solid solution and the high-temperature phase of Ti_2O_3 are similar. In VO_2, a first-order transition occurs around 340 K, accompanied by a change in structure (monoclinic to tetragonal) and a 10^4-fold jump in conductivity; the material remains paramagnetic throughout. A crystal distortion model wherein a gap opens up in the low-temperature low-symmetry structure adequately explains the transition. Substitution of trivalent ions such as Cr^{3+} and Al^{3+} for vanadium in VO_2 leads to a complex phase diagram with at least two insulating phases whose properties are significantly different from those of the insulating phase of pure VO_2. These phases are now fairly well understood.

The M−NM transition in V_2O_3 and its alloys has been the subject of a large number of publications.[1,2,15−19] Pure V_2O_3 undergoes a first-order transition (monoclinic−rhombohedral) at 150 K accompanied by a 10^7-fold jump in conductivity and an antiferromagnetical−paramagnetic transition. The carrier effective mass and other properties also show large changes at this transition. Application of pressure makes V_2O_3 increasingly metallic, thus suggesting that it is near a critical point. Accordingly, doping with Ti or Cr has a marked effect on the transition; the former has a positive pressure effect and the latter a negative pressure effect (Figure 2). V_2O_3 also shows a second-

5230 *J. Phys. Chem., Vol. 99, No. 15, 1995*

Edwards et al.

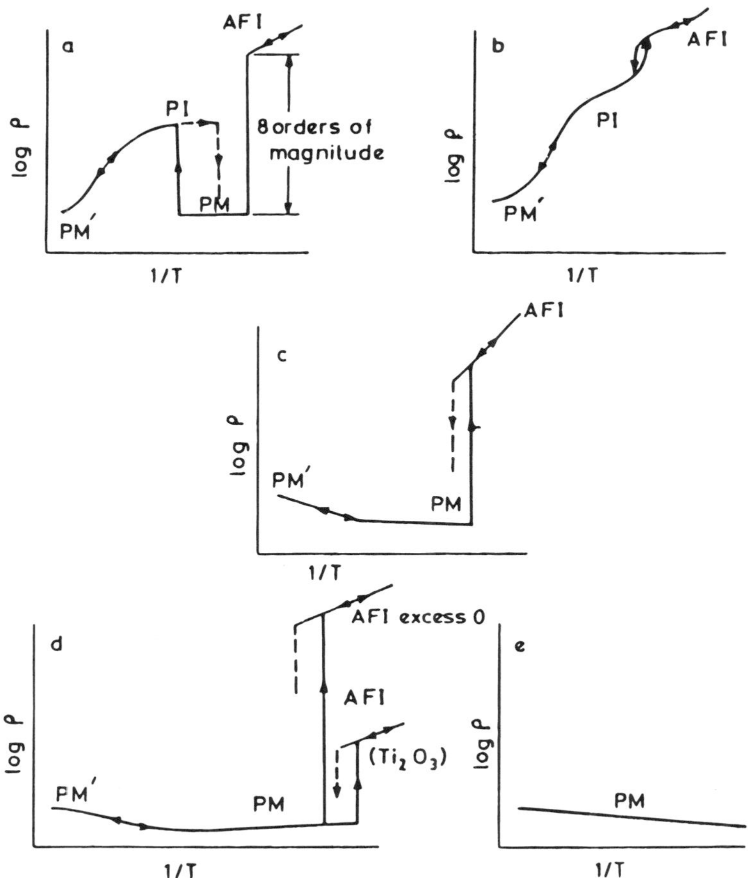

Figure 3. Schematic diagram depicting the changes of resistivity, ϱ, with temperature, T, in the V_2O_3 system: (a) V_2O_3 doped with 1.5% Cr_2O_3, (b) V_2O_3 doped with 3% Cr_2O_3, (c) pure V_2O_3, (d) V_2O_3 doped with 1% Ti_2O_3 and nonstoichiometric V_2O_3, and (e) 5.5% Ti_2O_3 doped V_2O_3. AFI, antiferromagnetic insulator; PM (PM'), paramagnetic metal; PI, paramagnetic insulator (after Honig and Spalek[16]).

order transition around 400 K with a small conductivity anomaly. Mere crystal distortion or magnetic ordering cannot explain the large connectivity jump at 150 K. The current status of the V_2O_3 transition can be represented in terms of Figure 3. This figure also serves to indicate the complexity of the metal−nonmetal transition in a relatively simple oxide system. There are many recent findings on V_2O_3 which are interesting. Thus, Carter et al.[17] have carefully measured the electrical and magnetic properties of single crystals of pure and doped V_2O_3 near the M−NM transition. Bao et al.[18] have examined the phase diagram of $V_{2-x}O_3$ in the x, P, T space and identified an incommensurate spin density wave (SDW) in metallic V_2O_3 close to the transition. The optical conductivity of V_2O_3, α (ω), has been investigated by Thomas et al.[19] In spite of extensive experimental and theoretical effort, a complete understanding of the transition in the V_2O_3 system is yet to emerge. We shall examine some of the factors responsible for this situation in section 6.

Compositionally controlled M−NM transitions in oxides are worthy of special mention. We shall examine two types of compositionally controlled transitions[20] as typified by $La_{1-x}A_xMO_3$ (A = Ca or Sr and M = V, Mn, or Co) and $LaNi_{1-x}M_xO_3$ (M = Mn or Fe). In $La_{1-x}A_xMO_3$, progressive substitution of trivalent La by divalent A brings about itinerant behavior of the d electrons, because every A ion creates an M^{4+} ion and promotes electron hopping from M^{3+} to M^{4+} ions (impurity-

band formation). This is to be contrasted with $LaNi_{1-x}M_xO_3$ where $LaNiO_3$ ($x = 0$), which is a correlated metal, transforms to an insulator on progressive substitution of Ni by M (somewhat like a deep-impurity situation). In Figure 4 we show typical electrical resistivity data in the two types of transitions. Unlike the above two systems, $A_{0.3}MoO_3$ (M = K or Rb) shows a metal−nonmetal transition associated with charge-density waves.[21]

Many of the high-temperature superconducting cuprates show compositionally controlled M−NM transitions.[8] Thus, in $TlCa_{1-x}Ln_xSr_2Cu_2O_y$ (Ln = Y or rare earth), the superconducting T_c shows a maximum at an optimal value of x (corresponding to the optimal value of the hole concentration). The system also shows a M−NM transition in the normal state as x is varied (Figure 5). Thus, the cuprate is metallic when $x = 0.25$ and insulating when $x = 1.0$. Rather interesting behavior occurs at $x = 0.75$ when the superconducting transition occurs from a seemingly semiconducting state. A compositionally controlled M−NM transition is also exhibited by $Bi_2Ca_{1-x}Ln_xSr_2Cu_2O_{8+\delta}$ (Figure 5) which has a maximum T_c at an optimal x value of 0.25. $La_{2-x}Sr_xCuO_4$ shows a similar metal−nonmetal transition ($x = 0$ is an insulator and $x = 0.3$ is a metal) with change in x and the maximum T_c is at $x = 0.20$. $La_{2-x}Sr_xCuO_4$ and a few other systems traverse the insulator−superconductor−metallic regimes with change in composition (increase in x from 0.0 to 0.3 in $La_{2-x}Sr_xCuO_4$). This suggests that the high-temperature

Feature Article

J. Phys. Chem., Vol. 99, No. 15, 1995 **5231**

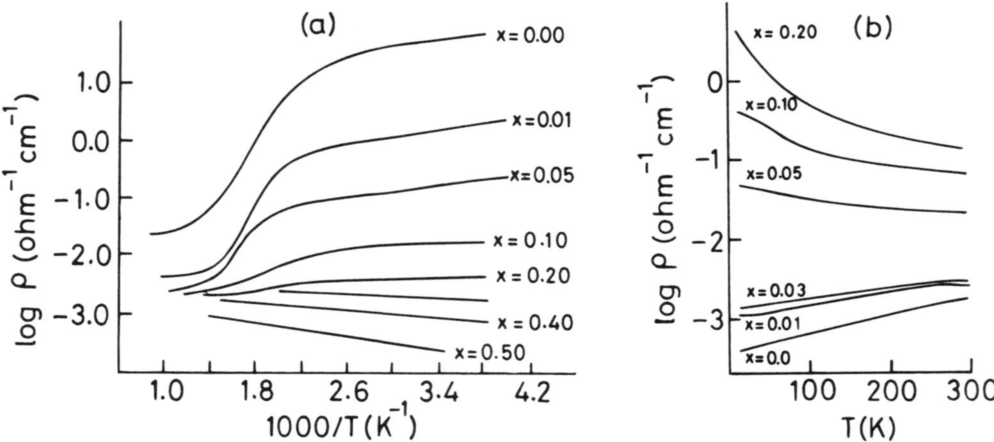

Figure 4. Compositionally controlled metal−nonmetal transitions in (a) $La_{1-x}Sr_xCoO_3$ and (b) $LaNi_{1-x}Mn_xO_3$ (from Rao and Ganguly[20]).

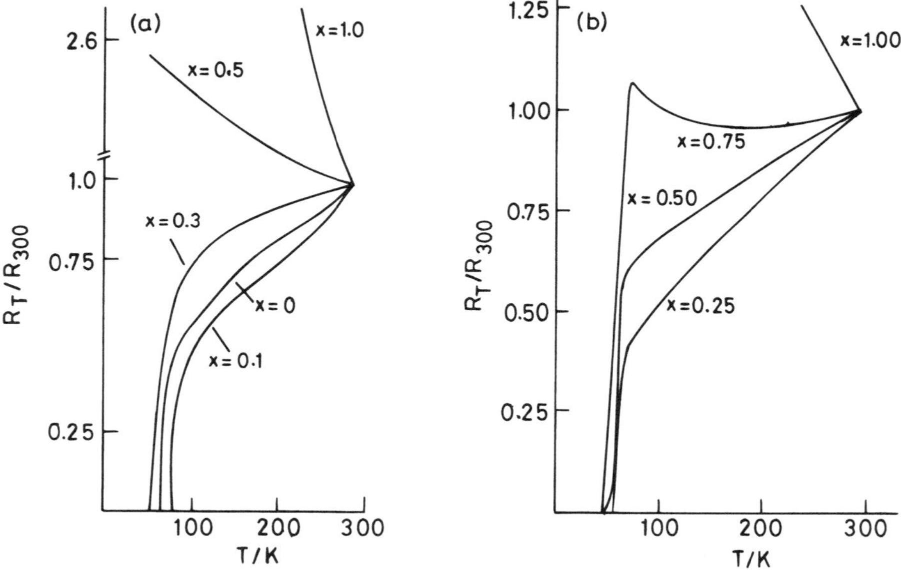

Figure 5. Compositionally controlled metal−nonmetal transition in superconducting cuprates: (a) $Bi_2Sr_2Ca_{1-x}Nd_xCu_2O_8$ and (b) $TlCa_{1-x}Nd_xSr_2CuO_4$ (after Rao[8]).

superconducting cuprates are at the boundary between metals and insulators.

One of the fundamental questions that has intrigued scientists is, "how many atoms maketh a metal?". The question is concerned with the possibility of a transition from the metallic to the nonmetallic state as the bulk metal is divided into finer particles. Recent studies of metal clusters[22,23] have attempted to answer the question. Careful investigations of gold clusters have shown that the binding energy of the core levels increases markedly (relative to the bulk metal value) when the cluster size decreases below 1 nm. That the effect is not merely due to final state effects but due to the occurrence of metal to nonmetal transitions as the cluster size decreases is reinforced by valence band and Bremstraahlung isochromat spectroscopic studies. Furthermore, tunneling conductance measurements show the existence of a gap in clusters smaller than 1 nm containing around 50 atoms or less; in Hg atom clusters, the 6s−6p atomic transition gives way to a collective metal-like plasmon absorption for a cluster size between 7 and 20

atoms.[23] The size-dependent transition from the metallic to the nonmetallic state does not occur abruptly. The possibility of a matrix-bound insulator−metal transition in alkali metal-doped zeolites at a critical stage of loading has been suggested by Edwards et al.[24] The dependence of the energetics and charge distribution of electron states on the cluster size and the dielectric constant has been examined by Rosenblit and Jortner,[7] to shed light on the cluster size-induced metal−nonmetal transition.

3. The Herzfeld Criterion

The earlier theoretical prediction of a M−NM transition is that derived from the work of Goldhammer[25] and Herzfeld.[14] These authors considered the effect of increasing density on the atomic polarizability and suggested that there would be a divergence in the polarizability or the dielectric constant causing the release of bound electrons. The Herzfeld criterion for dielectric catastrophe is given by,

Edwards et al.

$$\frac{4\pi N\alpha_0}{3V} = \frac{R}{V} = 1 \tag{2}$$

where α_0 is the low-density static polarizability of the atom, N Avogadro's number, V the molar volume, and R the molar polarizability. As a result of cooperative polarization effects, valence electrons get delocalized from the lattice sites at very high (metallic) densities[26] and the Drude free electron model becomes applicable. In the metallic regime, $(R/V) > 1$. In Figure 6 we show how this criterion excellently delineates metals from nonmetals in the periodic table.[27] Rao and Ganguly[28] have pointed out that the latent heat of vaporization, ΔH_v, of elements with metallic cohesion is larger than that of elements which are insulating because of relatively weak bonding (Figure 7). There are, however, exceptions such as strongly covalently bonded solids e.g., carbon.

A recent development is the realization of a link between the Herzfeld metallization view and the stress-induced transformations in solids, notably in semiconductors.[29] Under a diamond pressure indentor, ordinary (semiconducting) silicon transforms to the much denser β-tin (metallic) structure, the critical pressure being in the range $11-12$ GPa. This is consistent with the experiments which reveal a large, reversible drop in the electrical resistivity in a thin layer surrounding a microindentation in Si, as would be anticipated because of the metallic characteristics of the β-tin structure. Good correlations are found between the experimental metallization pressure and the values calculated from the Herzfeld polarization catastrophe criterion. Experimental transition pressures also correlate with Vickers hardness numbers and activation energies for dislocation motion. We show in Figure 8, a comparison of the calculated critical pressure (Herzfeld model) with the measured Vickers hardness (expressed in kilobars) for the group IV elements and SiC. Equally good correlations are obtained for all tetrahedrally bounded semiconductors and alkali and alkaline earth oxides.

Fujii et al.[30] have reported experimental evidence for the molecular dissociation process in Br_2 near 80 GPa. This transition, which is coincident with the onset of pressure-induced metallization, was first discovered in molecular/metallic iodine.[31] A diatomic molecular crystal loses its molecular character in the limit when the intermolecular distance becomes equal to the intramolecular bond length. Fujii et al.[30] applied the Herzfeld criterion to I_2 and Br_2 and estimated that the molar refractivity reaches the atomic limit around 20 GPa in I_2 and 80 GPa in Br_2. In both cases, the computed pressure coincides with that for molecular dissociation accompanied by metallization.

4. The Mott Criterion

The Herzfeld criterion considers the M$-$NM transition as viewed from the nonmetallic side. Over 30 years ago, Mott[12] proposed a simple model of the M$-$NM transition which considers how electron localization occurs as the transition is approached from the metallic side. In Figure 9, we show a schematic representation of a lattice of one-electron hydrogenic centers (P donors in Si) in the two limiting electronic regimes of high and low donor densities. At a sufficiently high density (small interparticle distance), the system would be a metal;[32] in this state, the system has a finite electrical conductivity at the absolute zero of temperature, i.e., $\sigma(T = 0) \neq 0$. At large interparticle distances, the system must surely become nonmetallic, or insulating, having a conductivity of 0 at $T = 0$ K, viz., $\sigma(T = 0) \rightarrow 0$. Mott argued that at a critical interdonor distance (d_c) a first-order (discontinuous) transition from metal to nonmetal would occur (Mott transition). A discontinuous

transition at the absolute zero of temperature will always remain a tantalizing theoretical prediction. Experimentally, however, even very close to $T = 0$ K (down to 0.03 K), the experimental situation is equivocal.[32-35] The physics of the problem is that, on the metallic side of the transition, the effective valence electron$-$cation potential in an atom is completely screened by the conduction electron gas and bound states are therefore nonexistent. However, if the conduction electron density, or equivalently the average separation between donor centers is changed, there comes a point at which bound levels (i.e., nonmetallic states) appear at some critical concentration of centers such that the following condition is satisfied.

$$n_c^{1/3} a_H^* \leq 0.25 \tag{3}$$

Here, a_H^* is the Bohr orbit radius of the isolated center and n_c is the critical carrier density at the M$-$NM transition. Another way of viewing the transition is that of an electronic instability which ensues when the trapping of an electron into a localized level also removes one electron from the Fermi gas of electrons. This must clearly lead to a further reduction in the screening properties (which are themselves directly related to the conduction electron density) and a catastrophic situation then ensures the localization of electrons from the previously metallic electron gas.

There appears to be little doubt that the Mott criterion given by eq 3 is an effective indicator of the critical condition at the M$-$NM transition itself. At the least, this simple criterion provides a numerical prediction for the metal$-$nonmetal transition in many situations. Figure 10 summarizes some of the experimental data.[34,36] Interestingly, besides doped semiconductors, metal$-$ammonia and metal$-$noble gas systems and superconducting cuprates all follow the linear relation given by eq 3. This is truly remarkable.

5. Minimum Metallic Conductivity at the Metal$-$Nonmetal Transition

Mott[13] has argued that the M$-$NM transition in a perfect crystalline material at $T = 0$ K is discontinuous (Figure 8) and proposed that, at the transition, there exists a minimum conductivity, σ_{min}, for which the material could still be viewed as metallic, prior to the localization of electrons.[2] Mott's ideas were based on arguments developed earlier by Ioffe and Regel[37] for the breakdown of the theory of electronic conduction in semiconductors. The conventional Boltzmann transport theory becomes meaningless when the mean-free path, l, of the itinerant conduction electrons becomes comparable to, or less than, the interatomic spacing, d. The Ioffe$-$Regel mean free path, l_{IR}, at the minimum metallic conductivity is equal to d. Abrahams et al.[38] have, however, predicted a continuous M$-$NM transition on the basis of a scaling theory of noninteracting electrons in a disordered system,[39] and their results question the existence of σ_{min} in both two and three dimensions.[32,39,40] The two possible scenarios of the transition are compared in Figure 11.

5.1. The Situation in Doped Semiconductors. There is an increasing belief amongst workers in the field that the M$-$NM transition is continuous, based on experimental measurements carried out at low temperatures down to 3 mK. In Figure 12, we show the experimental evidence in P-doped Si. Note that at a fixed (very low) temperature, the conductivity changes continuously with, for example, donor concentration. In addition, the extrapolated zero-temperature value of the conductivity ($\sigma(0)$) varies continuously with impurity concentration. An example showing the variation of the extrapolated "zero-temperature" conductivity[41] in the case of B-doped Si is

Feature Article

J. Phys. Chem., Vol. 99, No. 15, 1995 **5233**

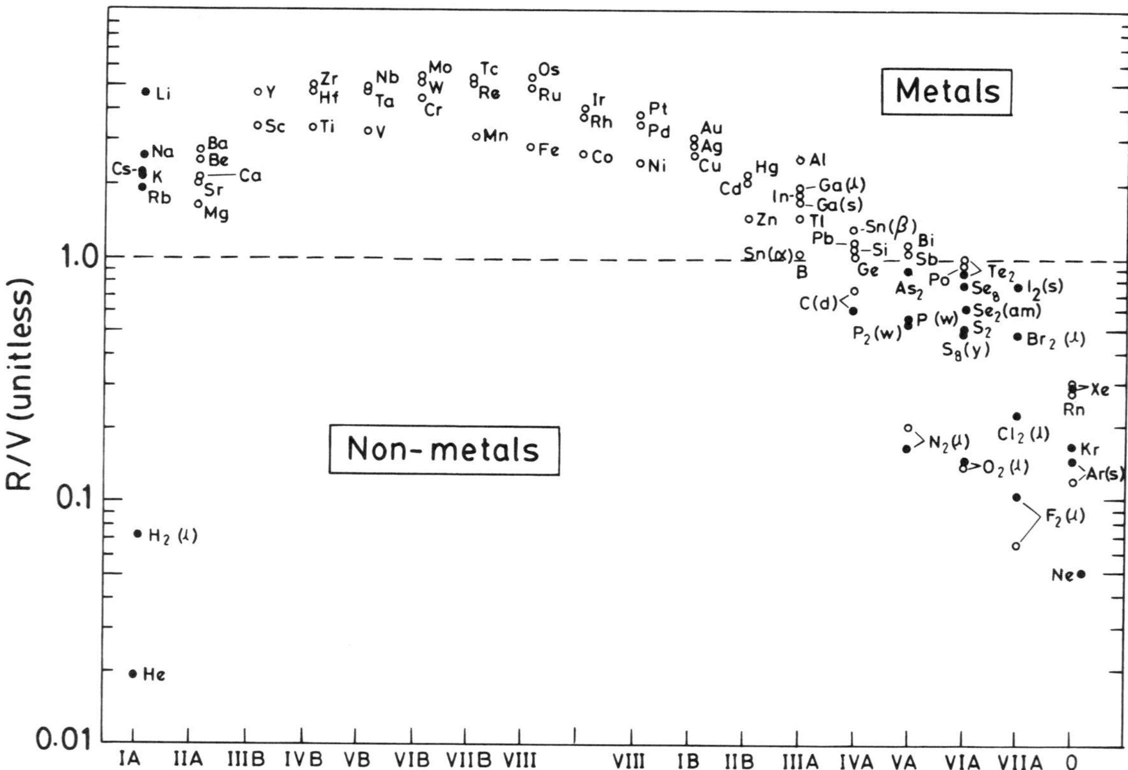

Figure 6. The occurrence of metallic vs nonmetallic character in the periodic table of the elements: the variation of R/V for naturally occurring elements of the s, p, and d blocks (adapted from Edwards and Sienko[27]).

shown in Figure 13. This could be taken as strong experimental evidence for a continuous M−NM transition in doped semiconductors at $T = 0$ K. Möbius,[35,42] however, questions the reliability of such 0 K extrapolations and suggests that these findings do not disprove the existence of finite σ_{min} at the transition. He argues that the data can be explained by a combination of σ_{min} on the metallic side and a Coulomb interaction dominated $\sigma(T) \simeq \sigma_{min} \exp\{-(T_0/T)^{1/2}\}$ on the insulating side, with $T_0 \rightarrow 0$ as the disorder decreases to the critical value. As noticed by him,[35,42] not all the measurements are consistent with this suggestion; furthermore, the clear observation of characteristic weak localization or precursor effects in transport and magnetotransport weakens the argument for a σ_{min}. However, this analysis points out the fact that while the continuous conductivity transition prediction is for *noninteracting* disordered electrons, just on the insulating side the Coulomb interaction is by definition of long range and hence qualitatively important. It is possible that 3 mK may not be a low enough temperature since $k_B T$ may exceed the activation energy for conduction even at this temperature.

While the presence of a high degree of disorder may wipe out the discontinuous nature of the M−NM transition,[43,44] inclusion of strong correlation in scaling models could conversely change a continuous transition to a discontinuous one. In spite of such difficulties, however, σ_{min} continues to be a useful experimental criterion[17,20,45] at least at the "high-temperature" limit. Earlier results providing experimental evidence for σ_{min} have been reviewed by several authors. In Figure 14, we show some of the results. We note that σ_{min} scales with n_c. As pointed out by Fritzsche,[46] σ_{min} appears to satisfactorily represent the value of conductivity where the

activation energy for conduction disappears. This aspect is specially borne out by investigations of transition metal oxide systems.

5.2. The Situation in Transition Metal Oxides. In many of the oxide systems,[15,20] especially those exhibiting compositionally controlled M−NM transitions (Figures 4 and 5), the temperature coefficient of the conductivity changes sign around σ_{min} ($\sim 10^3$ ohm^{-1} cm^{-1}). Most of these oxides, including the superconducting cuprates, follow the relation shown in Figure 14. The points corresponding to these oxides fall somewhere between those of fluid alkali metals and La$_{1-x}$Sr$_x$VO$_3$. The critical carrier concentrations in these materials from Figure 10 also give σ_{min} values close to the observed values. Accordingly, σ_{min} is often taken to represent the separation of localized and itinerant electron regimes.

Recent work of Raychaudhuri and co-workers[47,48] suggests the need to reevaluate the status of transition metal oxides with regard to σ_{min}. It appears that in these disordered oxide systems, genuine metallic states (with $\sigma(T = 0) \neq 0$) exist even when the conductivity is activated ($\sigma < \sigma_{min}$). This implies that the earlier values of the critical electron density at the transition in such oxides may be overestimates. In Figure 15, we show the behavior of Na$_x$W$_{1-y}$Ta$_y$O$_3$ where the temperature coefficient of the conductivity changes sign when $(x - y) \approx 0.20$, but the transition actually occurs at $(x - y) = 0.19$. The curves for $x = 0.35$ and 0.34 compositions both show "activated conductivity" at $T > 10$ K (i.e., a negative temperature coefficient of the resistivity), but the $\sigma(T)$ *saturates* at a fairly high residual value in the case of the $x = 0.35$ sample as expected of a metal ($\sigma(T = 0) \neq 0$). The $x = 0.34$ sample, however, shows a very much lower σ value at low T, tending to infinity, and fits an activated

Edwards et al.

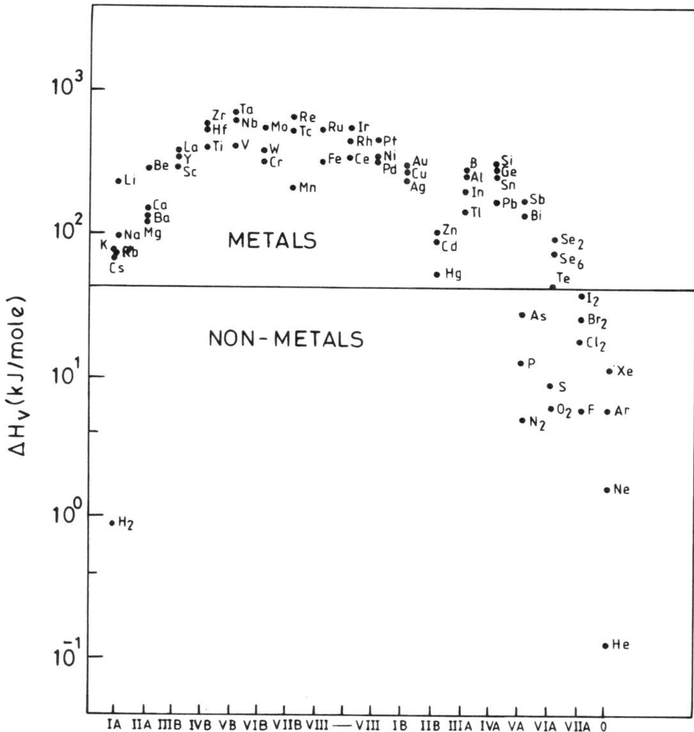

Figure 7. Plot of the latent heat of vaporization, ΔH_v, for metals and nonmetals of the periodic table (from Rao and Ganguly[28]).

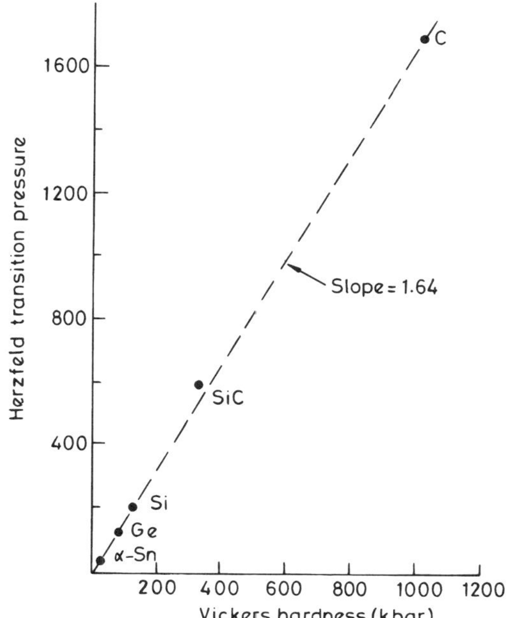

Figure 8. Correlation of the metallization pressure as calculated from the Herzfeld model and Vicker's indentation hardness number (from Gilman[29]).

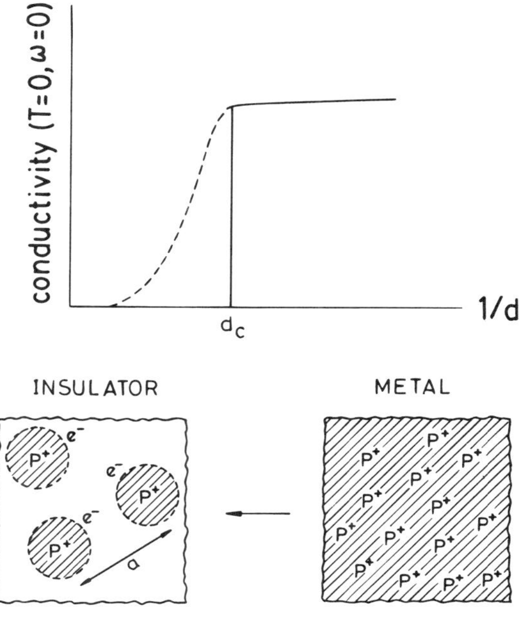

Figure 9. A schematic representation of the situation ($T = 0$ K) for P doped Si at both high and low donor densities. Also shown are two scenarios for the composition dependence of the electrical conductivity, showing the metal−nonmetal transition.

temperature dependence, with the behavior of a nonmetal ($\sigma(T = 0) = 0$). These results show that it is difficult to differentiate metallic *vs* nonmetallic behavior in oxides based on high-temperature measurements. What emerges, however, is that the

"high-temperature" σ_{min} value differentiates the (metallic *vs* nonmetallic) conductivity behavior in that temperature regime.

J. Phys. Chem., Vol. 99, No. 15, 1995 **5235**

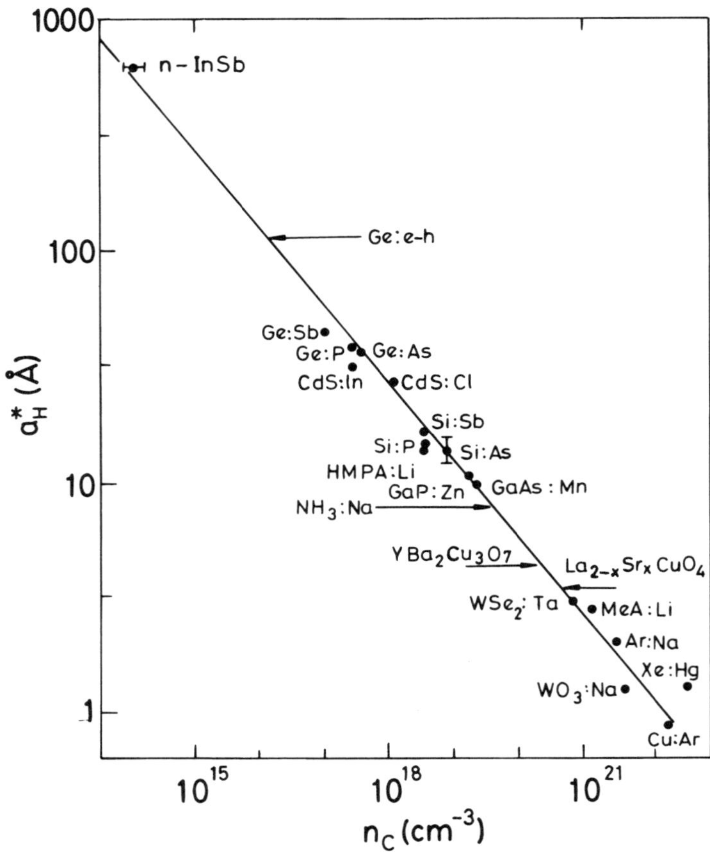

Figure 10. An experimental correlation of the critical density, n_c, with the Bohr radius, a_H^*. The solid line represents $n_c^{1/3} a_H^* = 0.26$ (from Edwards and Sienko[36] and Thomas[34]). Here, InSb refers to lightly doped n-InSb, Ge:e-h means germanium with photoexcited electrons and holes; Cu:Ar is a codeposited Cu and Ar mixture on a cold substrate.

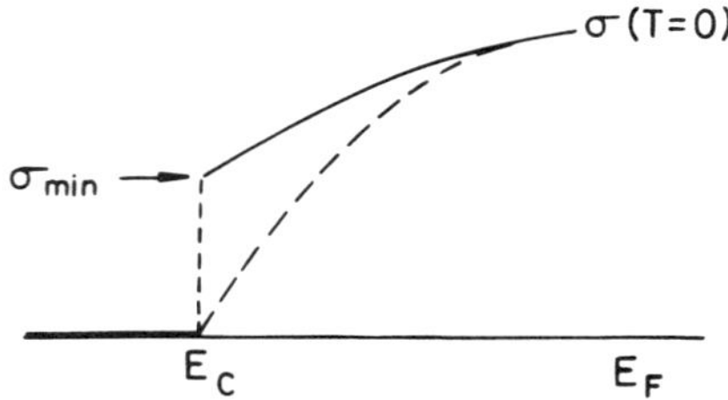

Figure 11. Schematic illustration of the two possibilities of a continuous or discontinuous metal–nonmetal transition at $T = 0$ K. The minimum metallic conductivity, σ_{min}, at the transition is also shown (from Lee and Ramakrishnan[39]). The conductivity at zero temperature (ordinate) and Fermi energy (abscissa) are shown. The discontinuous conductivity transition suggested by Mott is the full curve, with σ_{min} occurring as E_F crosses the mobility edge energy E_c. The dotted curve is the continuous conductivity transition predicted by the scaling theory.

6. Our Theoretical Understanding

The metal to nonmetal transition, a basic electronic change, has proven surprisingly difficult to understand in detail. The underlying reason is the diametrically opposite modes of description natural for the metal (extended electronic states) and for the insulator (localized states, often with local constraints on electron number). The observed diversity of systems and phenomena indicates that a number of causes may be at work, e.g., disorder, short-range electron correlation, long-range Coulomb interaction, and electron lattice coupling. The effects are often present simultaneously and interactively. Many reasonable scenarios as well as detailed theories of model systems do exist, but the distance between them and experiments is still large. In this section, we first look at the implications of the successful criteria for the M–NM transition. We then describe some theoretical models and approaches.

6.1. Criteria for the Transition. The three criteria discussed earlier are the dielectric catastrophe of Herzfeld, the Mott density criterion, and the idea (again due to Mott) that a metal is characterized by a minimum value of the electrical conductivity at the transition. The first two are related and consider how the Coulomb interaction and electron or atom density together can lead to the metal–insulator transition. The polarizability of an isolated atom depends on its size, being roughly proportional to its volume (a_H^{*3}). When such atoms are put together, the dielectric constant is enhanced because of local field effects, i.e., the large electric field at each atom due the collection of polarized atoms (or dipoles). At a density such that the polarizability proportional to a_H^{*3} is comparable to the volume per atom, i.e., when $a_H^* n^{1/3} \simeq 1$, the dielectric constant diverges. It is difficult to develop a detailed theory along these lines, because at such high densities electron overlap effects are large. The electron energy level structure (band structure) and wave functions are very different from those at the atomic limit, so that the dielectric constant is not easily related to properties of isolated atoms. Further, disorder and lattice type can also have a large quantitative effect.

5236 *J. Phys. Chem., Vol. 99, No. 15, 1995*   Edwards et al.

The Mott criterion, namely $a_H^* n^{1/3} \simeq 0.25$, is dimensionally identical to that of Herzfeld and is physically related. It considers the transition from the metallic side. If an electron—hole excitation is created in a metal, the attraction between them is screened, and they may not form a bound state unless the screened interaction is strong enough. This will happen at a low enough carrier density estimated by Mott to be such that $a_H^* n^{1/3} \lesssim 0.25$. The metal is then unstable against the formation of a bound electron—hole pair; a collection of these is an insulator or a nonmetal. Both the criteria thus point to the Coulomb interaction which is screened and effectively weak in the metal and is strong and long ranged in a nonmetal. A theory describing this crossover and its consequences does not exist. These approaches neglect electron spin and associated magnetic effects, *local* electron correlation, as well as disorder.

The minimum metallic conductivity criterion of Mott emphasizes the role of disorder. It was first realized by Anderson[49] that electronic states originally extended can become spatially localized if the disorder is strong enough. Mott argued that in a disordered metal, the minimum possible value of the mean free path is the interatomic spacing. For stronger disorder, the picture of electron waves scattered randomly is inappropriate; the electron state is localized, and one has a nonmetal. Thus, a minimum electrical conductivity, σ_{min}, characterizes the transition. The conductivity discontinuously jumps from this value to zero on the nonmetallic side. As mentioned earlier, this is a very good first approximation; very close to the critical disorder *and* at very low temperatures, $\sigma \to 0$ and values much less than σ_{min} (but tending to a nonzero limit at $T = 0$) have been observed. A detailed theory for the disorder caused continuous M—NM transition exists.[38,39] The minimum metallic conductivity criterion emphasizes disorder and is silent about the effect of interactions. The Mott density criterion focuses on the Coulomb interaction and completely neglects disorder effects. The success of both these criteria, often in the same system, implies that both interactions and disorder are significant in real systems.

6.2. Detailed Models. *6.2.1. Anderson Localization.* It was pointed out by Anderson[49] that if the randomness in electronic state energies at different sites is large enough, electrons do not diffuse from one end of the system to another but are localized at potential fluctuations. Since for weak disorder electrons do diffuse, there must be a critical disorder at which an electron of a particular energy (say, the Fermi energy) does not diffuse, i.e., the dc electrical conductivity is zero. The energy separating the band of extended states from localized states is called the mobility edge. As disorder or electron density changes, the Fermi energy and the mobility edge coincide, and the system makes a transition from a metal to a nonmetal; this is the disorder-driven Anderson transition.

Predictions for physical properties near the transition were first made by Mott. He argued for a nonzero minimum conductivity of the metal at the transition point and for a divergence of localization length in the nonmetal with a consequent divergence in the dielectric constant.

A detailed theory of the transition, which identified singular backscattering as the mechanism of localization, was developed by Abrahams et al.[38] Quantum mechanically, an electron in a random medium can go from one point to another via infinitely many paths with appropriate amplitudes. Of these, self intersecting paths in which the loop is traversed in opposite directions have the same amplitude. Thus the probability (the square of the amplitude) for return to the origin is enhanced because of constructive interference. Nearby paths also contribute, depending on dimensionality. This process thus tends to localize

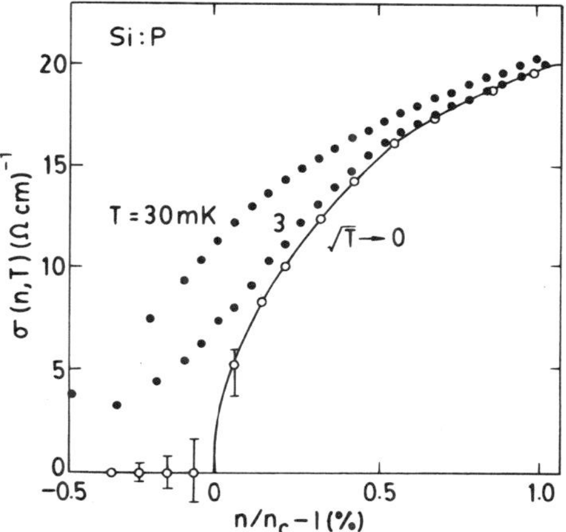

Figure 12. Electrical conductivity as a function of relative electron (donor) concentration in P doped Si to illustrate the metal—nonmetal transition at 0 K. The relative P density is varied by changing n_c with uniaxial stress on a single sample. The open circles are extrapolated to $T = 0$ K (from Thomas[33]).

electrons, depending on disorder and dimensionality. Detailed calculations show for example that the conductivity decreases from its value σ at atomic length scales (of order the mean free path l) to smaller values at a larger length scale, L, on account of interference effects. In two dimensions, the decrease is logarithmic in L.

$$\sigma(L) = \sigma_0 - \left(\frac{e^2}{\hbar\pi^2}\right)\ln(L/l) \tag{4}$$

This means, from a scaling analysis, that for large enough systems, the conductivity (or conductance) is zero. There are no disordered *metal* films at absolute zero. In three dimensions, $\sigma(L)$ has the form

$$\sigma_{3d}(L) = \sigma_0 - \frac{e^2}{\hbar\pi^3}\left(\frac{1}{l} - \frac{1}{L}\right) \tag{5}$$

This implies for example that when $\sigma(l)$ has the Mott minimum value ($e^2/\hbar\pi^3 l$), the dc or large length scale ($L \to \infty$) conductivity is zero. Thus, the conductivity goes continuously to zero at the Anderson transition, due to interference effects not envisaged in the Mott approximation. The exponent with which σ vanishes is the same as that with which the localization length diverges.

The scaling picture, starting from weak localization due to interference, puts in perspective both the minimum metallic conductivity σ_{min} and the possibility of a continuous conductivity transition. At nonzero temperatures, inelastic scattering limits the length scale L to $L_{in} \sim T^{-\alpha}$ up to which interference effects take place. If L_{in} is small, the dc conductivity even near critical disorder will be close to σ_{min}. As T decreases, L_{in} increases, the interference effects are stronger, and the conductivity decreases. This is considered "nonmetallic" behavior. However $\sigma(T = 0)$ is nonzero, so that one can naturally have strongly disordered metals with a negative temperature coefficient of the resistivity. The purely disorder driven transition is a continuous quantum ($T = 0$) transition. A number of predictions for quantities such as magnetoresistance, the Hall effect, and resistivity in the presence of magnetic and other scatterers have been made. These agree with experiment.[39,40] It is also realized that many of these phenomena are common to *all* waves in random media, e.g., light and sound waves.

J. Phys. Chem., Vol. 99, No. 15, 1995 **5237**

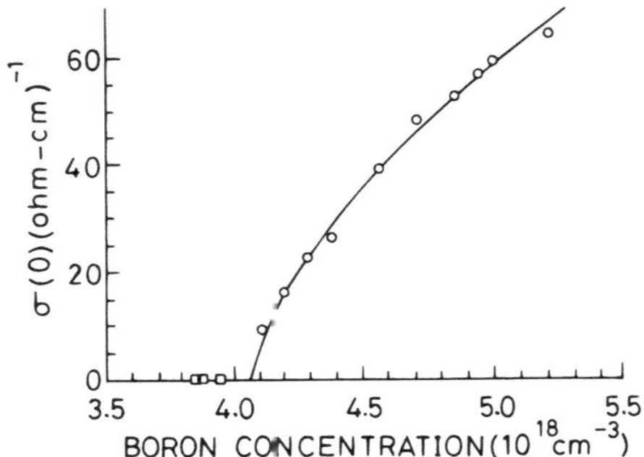

Figure 13. Zero-temperature conductivity $\sigma(T = 0)$ vs boron concentration in B doped Si. The circles represent metallic samples, while the three squares indicate samples which are nonmetallic (from Dai et al.[41]).

Real disordered electronic systems cannot be thought of as a collection of noninteracting particles moving independently in a random potential. With increasing disorder, electrons diffuse very slowly and so in effect interact strongly. The range of interaction increases until in the nonmetal it decreases only inversely with distance. It has proven very difficult to describe the passage of a disordered interacting electron gas to the insulating state. In the metallic regime, for effectively weak interaction, there are predictions for characteristic effects such as a dip in the density of states proportional to $(|E - E_F|)^{1/2}$ where E is the energy and E_F the Fermi energy. Scaling approaches to the transition starting from the perturbative regime show that there are several kinds of M−NM transitions, characterized by different critical behavior for density of states, conductivity, magnetic susceptibility, low-temperature specific heat, etc.[39,50] Unfortunately, the actual transition is often in the regime of strong coupling and so the perturbative scaling analysis is not reliable. Also, there is no theory which describes the Coulomb interaction realistically in the entire regime (weakly disordered metal−critical region−Coulomb glass). Experimentally, this transition has been recently explored in several systems.[47-48] The square root dip in the density of electronic states near the Fermi energy steepens, changes to a linear form, and becomes a soft quadratic gap just on the insulating side (i.e., the density of states goes as $(E - E_F)^2$). There are characteristic temperature and field dependences for the resistivity. These are not yet comprehensively understood.

6.2.2. Correlation Effects. The effects of electron correlation are strong in narrow band systems and can cause them to be insulating when they should be metallic according to band theory. Well-known examples are NiO and La_2CuO_4, which, with an odd number of electrons per site in the highest unfilled band, ought to be metallic. In the latter, for example, only the Cu−O configurations d^9p^6 and $d^{10}p^5$ are close in energy (the configuration d^8p^6 has very high energy). The former, containing one d hole per site, is more stable so that the system has a gap against transfer of one hole from Cu to O. This is a charge transfer insulator. An even simpler correlation driven one-band (d band) insulator requires the energy cost U of local configurational changes involving sites l, $l + 1$, i.e., the change ($d_l^n - d_{l+1}^n$ to $d_l^{n-1} - d_{l+1}^{n+1}$) to be greater than the kinetic energy gain, KE, due to electron hopping. Thus, as U decreases or increases (e.g., due to changes in physical conditions such as pressure) in relation to KE, the system can make a transition from insulator to metal. These ideas were extensively discussed by Mott. The one-band model of electrons on a lattice, with local correlations, is the well-known Hubbard model, and the correlation driven transition for one (or commensurate) electron per site, from a paramagnetic metal to a paramagnetic insulator is the Mott transition, mentioned earlier in section 4.

The Mott transition has proven difficult to describe in detail, though there are several plausible conclusions and results for spatial dimensionality $d = 1$ and ∞. On the metallic side close to the transition, for the case of one electron per site, the probability of two electrons being on the same site is small because of correlation. Using this as a variational parameter in a ground state wave function as was first proposed by Gutzwiller (see ref 40 for details) a number of conclusions have been drawn. As $U \to U_c$ (the critical correlation), electron hopping becomes infrequent so that the electron effective mass is large. Since each site has mostly one electron, local magnetic moments are very long lived, and the Pauli spin susceptibility is large, diverging as $U \to U_c$. On the insulating side, a gap (the Mott−Hubbard gap) opens up for charge excitations. This gap is different from that in a semiconductor. The former is

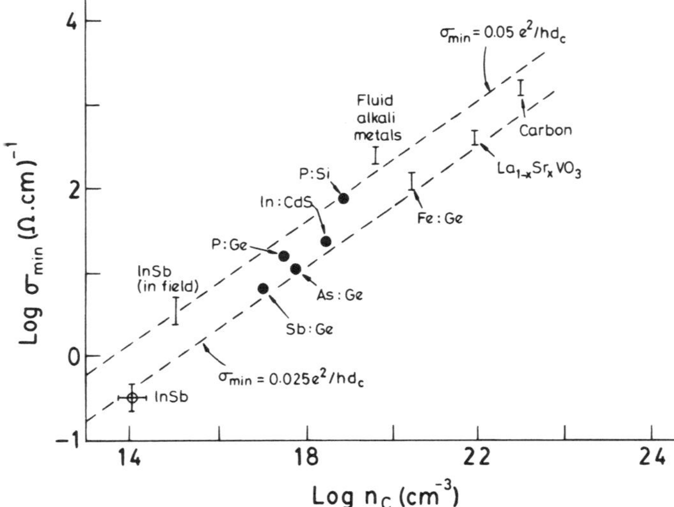

Figure 14. A plot of the minimum metallic conductivity, σ_{min}, against log n_c, the critical electron density at the metal−nonmetal transition. The two straight lines correspond to the values of the constant 0.05 and 0.026 in the equation relating σ_{min} and $n_c^{1/3}$ (adapted from Mott[44] and Fritzche[46]).

5238 *J. Phys. Chem., Vol. 99, No. 15, 1995*

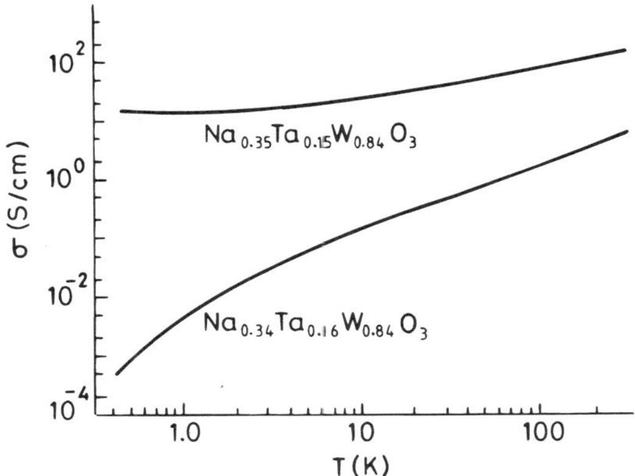

Figure 15. Electrical conductivity versus temperature for two samples of $Na_xTa_yW_{1-y}O_3$, one having metallic properties for $T \rightarrow 0$ (top curve) and one having nonmetallic status for $T \rightarrow 0$ (bottom curve). Notice the extremely small differences in concentration for the metallic and nonmetallic samples (from Raychaudhuri[47]).

due entirely to local correlations and the latter to periodicity. On the insulating side, local moments are well defined and interact antiferromagnetically. There is experimental evidence in metallic V_2O_3, near the M—NM transition, for mass enhancement, increased Pauli paramagnetism, a large T^2 term in the resistivity etc.[17-19,40]

Recently, there has been considerable progress in developing a theory of the Hubbard model in high spatial dimensions.[51] There are reasons to believe that the results can be usefully extrapolated to $d = 3$. The development of a Mott—Hubbard gap is due to the size and time dependence of the field on an electron at a given site. This field is calculated self-consistently. Some results are as follows: In the (U,T) plane, there are three phases, metallic (PM), insulating paramagnetic (PI), and insulating antiferromagnetic (AFI). By suppressing the intermoment exchange coupling, the Mott transition (PM—PI) can be located and followed. It is discontinuous at $T \neq 0$ and is believed to become continuous as $T \rightarrow 0$. The physical properties of the heavy fermion metal for $U \lesssim U_c$ can also be calculated and agree with expectations. Electron—lattice coupling can drive the Mott transition discontinuous, with a $T \neq 0$ critical point.[52] Many of these results are similar to those found in V_2O_3.

The large d approach developed so far concentrates exclusively on the single-site time-dependent electron potential. The other low-lying fluctuations (charge and spin) are integrated out so that we do not have a detailed picture of the quantum dynamics. There is little understanding of the effects of deviations from integer site occupation, i.e., of doping. This lacuna is serious because cuprate superconductors can be thought of as doped Mott insulators. The effect of disorder is also not known.

7. Concluding Remarks

The transition of a metal to an insulator is caused and accompanied by changes in the nature of chemical bonding as well as by changes in physical properties. The transition occurs in a wide variety of systems, as we have tried to indicate. In spite of this variety, simple criteria reflecting a basic competition between potential and kinetic energies have been remarkably successful in locating broad conditions under which such transitions may occur. A detailed understanding of transport, magnetic, optical, and other properties in the transition regime has proven difficult, since several causes seem to be at work

simultaneously. However, a number of theoretical models valid for various limiting situations have given us valuable qualitative and quantitative insights.

The entire area has experienced a rebirth in the last few years, for several reasons. The most important perhaps is the realization that cuprate superconductors are not far from a largely correlation driven metal—insulator transition. Another is the appreciation that the transition is of wide occurrence, some recent examples being the superconductor—insulator, the quantum Hall fluid—insulator, and the stable quasicrystal metal—insulator transitions. Hopefully, the next few years should see considerable progress in the experimental and theoretical description of this basic electronic transition.

References and Notes

(1) Edwards, P. P., Rao, C. N. R., Eds. *The Metallic and the Nonmetallic States of Matter*; Taylor and Francis: London, 1985.

(2) Mott, N. F. *Metal-Insulator Transitions*, 2nd ed.; Taylor and Francis: London, 1990.

(3) Rao, C. N. R.; Edwards, P. P. *Proc. Indian Acad. Sci. Chem. Sci.* **1986**, *96*, 473.

(4) Redmer, R. Thermodynamische und Transporteigenschafter dicter Alkali plasmen. Habilitationsschrift, Universitat Rostock, 1991.

(5) Ashcroft, N. W.; Mermin, N. D. *Solid State Physics*; Saunders College: Philadelphia, 1976.

(6) See articles of K. Kimura and P. P. Edwards in *Chemical Processes in Inorganic Materials: Metal and Semi-conductor Clusters and Colloids*; Persans, P. D., Bradley, J. S., Chianelli, R. R., Schmid, S., Eds.; *Mat. Res. Soc. Symp. Proc.* **1992**, *272*, 193, 311.

(7) Rosenblit, M.; Jortner, J. *J. Phys. Chem.* **1994**, *98*, 9365.

(8) Rao, C. N. R. *Philos. Trans. R. Soc. London* **1991**, *A336*, 595.

(9) Iye, Y. In *Physical Properties of High Temperature Superconductors*; Ginsberg, D. M., Ed.; World Scientific: Singapore, 1992; Vol. II.

(10) Logan, D. E.; Tusch, M. A. *J. Non-Cryst. Solids* **1993**, *156—158*, 639.

(11) Ramakrishnan, T. V. *J. Solid State Chem.* **1994**, *111*, 4.

(12) Mott, N. F. *Philos. Mag.* **1961**, *6*, 287.

(13) Mott, N. F. *Philos. Mag.* **1972**, *26*, 1015.

(14) Herzfeld, K. F. *Phys. Rev.* **1927**, *29*, 701.

(15) Rao, C. N. R. *Annu. Rev. Phys. Chem.* **1989**, *40*, 291.

(16) Honig, J. M.; Van Zandt, L. L. *Annu. Rev. Mater. Sci.* **1975**, *5*, 225. Honig, J. M.; Spalek, J. *Proc. Indian Natl. Sci. Acad.* **1986**, *A52*, 232.

(17) Carter, S. A.; Rosenbaum, T. F.; Honig, J. M.; Spalek, J. *Phys. Rev. Lett.* **1991**, *67*, 3440. Carter, S. A.; Rosenbaum, T. F.; Metcalf, P.; Honig, J. M.; Spalek, J. *Phys. Rev.* **1993**, *B48*, 16841.

(18) Bao, W.; Broholm, C.; Carter, S. A.; Rosenbaum, T. F.; Aeppli, G.; Trevino, S. F.; Metcalf, P.; Honig, J. M.; Spalek, J. *Phys. Rev. Lett.* **1993**, *71*, 766.

(19) Thomas, G. A.; et al. *Phys. Rev. Lett.* **1994**, *73*, 1529.

(20) Rao, C. N. R.; Ganguly, P. In *The Metallic and the Non-metallic States of Matter*; Edwards, P. P., Rao, C. N. R., Eds.; Taylor and Francis: London, 1985.

(21) Schlenker, C.; Dumas, J.; Escribe-Filippini, G. H.; Marcus, J.; Fourcandot, G. *Philos. Mag.* **1985**, *B52*, 643.

(22) Vijayakrishnan, V.; Chainani, A.; Sarma, D. D.; Rao, C. N. R. *J. Phys. Chem.* **1992**, *96*, 8679.

(23) Rao, C. N. R.; Vijayakrishnan, V.; Aiyer, H. N.; Kulkarni, G. U.; Subbanna, G. N. *J. Phys. Chem.* **1993**, *97*, 11157. Haberland, H.; von Issendorff, B.; Yufeng, Ji; Kolar, T. *Phys. Rev. Lett.* **1992**, *69*, 3212.

(24) Edwards, P. P.; Anderson, P. A.; Armstrong, A. R.; Slaski, M.; Woodall, L. J. *Chem. Soc. Rev.* **1993**, *22*, 305.

(25) Goldhammer, D. A. *Dispersion via Absorption des Lichtes*; Teubner: Leipzig, 1911.

(26) Sommerfeld, A. In *Gesammelten Schriften*; Vieweg: Braunschweig, 1968; Vol. 4, as cited by Ehrenreich, H. E. *Science* **1987**, *235*, 1029.

(27) Edwards, P. P.; Sienko, M. J. *J. Chem. Educ.* **1983**, *60*, 691.

(28) Rao, C. N. R.; Ganguly, P. *Solid State Commun.* **1986**, *57*, 5.

(29) Gilman, J. J. *J. Mater. Res.* **1992**, *7*, 535; see also *Philos. Mag.* **1993**, *67*, 207.

(30) Fujii, Y.; Hase, K.; Ohishi, Y.; Fujihisa, H.; Hamaya, N. H.; Takemura, K.; Shimomura, O.; Kikegawa, T.; Amemiya, Y.; Matsushita, T. *Phys. Rev. Lett.* **1989**, *63*, 536.

(31) Balchan, A. S.; Drickamer, H. G. *J. Chem. Phys.* **1961**, *34*, 1948.

(32) For a review see: Milligan, R. F.; Thomas, G. A. *Annu. Rev. Phys. Chem.* **1985**, *36*, 139.

(33) Thomas, G. A. *J. Phys. Chem.* **1983**, *88*, 3749.

(34) Thomas, G. A. In *Proceedings of the 39th Scottish Universities Summer School in Physics*; Tunstall, D. P., Barford, W., Eds.; Adam Hilger: Edinburgh, 1991.

J. Phys. Chem., Vol. 99, No. 15, 1995 **5239**

(35) Möbius, A. *J. Phys. C: Solid State Phys.* **1985**, *18*, 4639.

(36) Edwards, P. P.; Sienko, M. J. *Phys. Rev.* **1978**, *B17*, 2575.

(37) Ioffe, A. F.; Regel, A. R. *Prog. Semicond.* **1960**, *4*, 237.

(38) Abrahams, E.; Anderson, P. W.; Liccardello, D. C.; Ramakrishnan, T. V. *Phys. Rev. Lett.* **1979**, *42*, 693.

(39) Lee, P. A.; Ramakrishnan, T. V. *Rev. Mod. Phys.* **1985**, *57*, 287.

(40) Ramakrishnan, T. V. In *The Metallic and Non-metallic States of Matter*; Edwards, P. P., Rao, C. N. R., Eds.; Taylor and Francis: London, 1985.

(41) Dai, P.; Zhang, Y.; Sarachik, M. P. *Phys. Rev. Lett.* **1991**, *66*, 1914.

(42) Möbius, A. *Phys. Rev.* **1989**, *B40*, 4194.

(43) Mott, N. F. *Philos. Mag.* **1978**, *B37*, 377.

(44) Mott, N. F. *Proc. R. Soc.* **1982**, *A382*, 1.

(45) Edwards, P. P.; Sienko, M. J. *Int. Rev. Phys. Chem.* **1983**, *3*, 83.

(46) Fritzche, H. In *The Metal-Nonmetal Transitions in Disordered Systems*; Friedman, L. R., Tunstall, D. P., Eds.; Scottish Universities Summer School in Physics, 1978.

(47) Raychaudhuri, A. K. *Phys. Rev.* **1991**, *B44*, 8572.

(48) Raychaudhuri, A. K.; Rajeev, K. P.; Srikanth, H.; Mahendiran, R. *Physica B* **1994**, *197*, 124.

(49) Anderson, P. W. *Phys. Rev.* **1958**, *109*, 1492.

(50) Belitz, D.; Kirkpactrick, T. R. *Rev. Mod. Phys.* **1994**, *66*, 261.

(51) Georges, A.; Krauth, W. *Phys. Rev. B* **1993**, *48*, 7167. Rozenberg, M. J.; Zhang, X. Y.; Kotliar, G. *Phys. Rev. B* **1994**, *49*, 10181.

(52) Majumdar, P.; Krishnamurthy, H. R. *Phys. Rev. Lett.* **1994**, *73*, 1525.

JP9428496

Virtues of marginally metallic oxides

C. N. R. Rao

CSIR Centre of Excellence in Chemistry & Solid State and Structural Chemistry Unit, Indian Institute of Science, Bangalore 560 012, India and Materials Research Laboratory, University of California, Santa Barbara, CA 93106, USA

Transition-metal oxides at the metal–insulator boundary, especially those belonging to the perovskite family, exhibit fascinating phenomena such as insulator–metal transitions controlled by composition, high-temperature superconductivity and giant magnetoresistance (GMR). Interestingly, many of these marginally metallic oxides obey the established criteria for metallicity and have a finite density of states at the Fermi level. The perovskite manganates exhibiting GMR, on the other hand, are unusual in that they possess very high resistivities in the 'metallic' state and show no significant density of states at the Fermi level. Marginal metallicity in oxide systems is a problem of great complexity and contemporary interest and its understanding is of crucial significance to the diverse phenomena exhibited by these materials.

Properties of conventional metals such as gold and copper are familiar to most of us. One of these properties is the low electrical resistivity of metals. Some transition-metal compounds, in particular the oxides, exhibit low electrical resistivities just like conventional metals. The resistivity behaviour of the oxides such as ReO_3 and RuO_2 is indeed comparable to that of copper (Fig. 1). Metal oxides, however, exhibit a very wide range of electrical resistivity, anywhere between 10^{20} to 10^{-10} ohm cm, probably representing the widest range in any physical property known to date.[1] Besides the oxides with metallic resistivities and those with insulating properties, there are oxides which exhibit transitions from the insulating state to the metallic state. The insulator-to-metal (I–M) transition in V_2O_3 is well known for the unparalleled 10 or 100 million-fold jump in resistivity at 150 K; it is as if wood becomes copper at the transition temperature. The I–M transition in oxides can be brought about by changing the temperature, pressure or composition.[1,2] The perovskite oxides which exhibit compositionally controlled I–M transitions are especially noteworthy.[3] $La_{1-x}Sr_xCoO_3$ is an example of an oxide system which becomes metallic at a particular Sr concentration, the parent $LaCoO_3$ ($x = 0.0$) being an insulator at ordinary temperatures.

$LaNi_{1-x}Mn_xO_3$ is an example of an oxide system which becomes insulating as x is increased, $LaNiO_3$ ($x = 0.0$) being metallic. I shall discuss briefly the nature of such metal oxides to point out how the I–M transition occurs across a state which can be considered as barely metallic and yet obey the known criteria for metallicity.†

The cuprates which exhibit high-temperature superconductivity[4,5] are marginally metallic in the normal (non-superconducting) state and exhibit certain anomalous properties. What is even more interesting is that perovskite manganates of the general formula $La_{1-x}A_xMnO_3$ (Ln = rare earth, A = divalent ion such as Ca, Sr, Ba or Pb) exhibit high resistivities in what is considered to be the metallic state and do not seem to obey the criteria for metallicity. Marginal metallicity seems to be generally associated with all the oxides exhibiting giant magnetoresistance.[6,7] Thus, marginal metallicity of metal oxides is closely linked to three of the most important aspects of present-day solid-state science, namely insulator–metal transitions, superconductivity and giant magnetoresistance. In Fig. 2 we compare the temperature variation

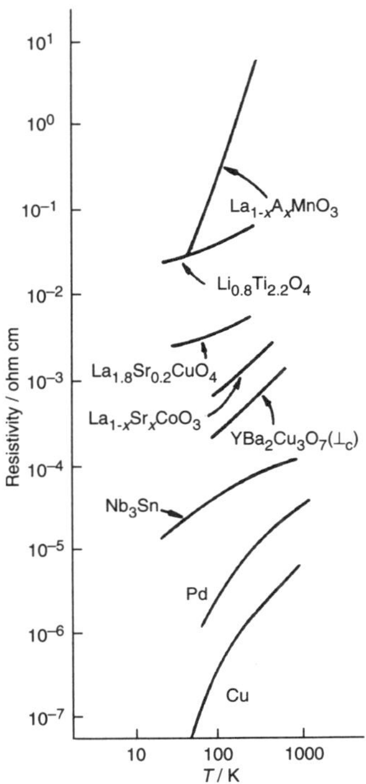

Fig 2 Resistivities of some of the oxides exhibiting marginal metallicity. Data on metallic Nb_3Sn, Pd and Cu are shown for comparison.

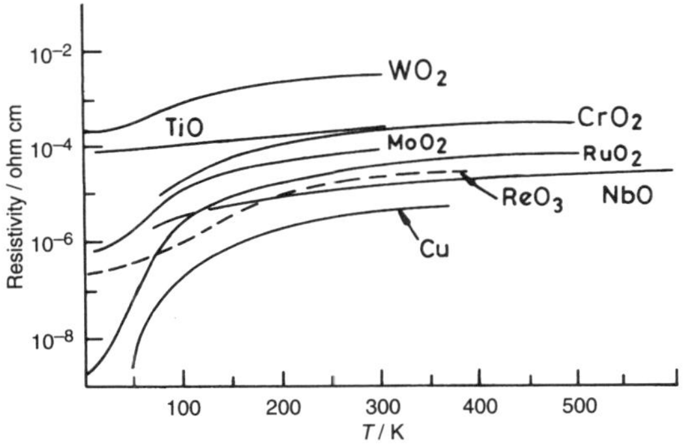

Fig 1 Resistivities of highly conducting (metallic) oxides

of resistivity of cuprates and other oxides to give an indication of the resistivity ranges encountered in marginally metallic oxides. In this article, I examine the nature of marginal metallicity encountered in transition-metal oxide systems after a brief discussion of the criteria for metallicity.

Operational criteria for metallicity

An important criterion for metallicity is the existence of the Fermi surface or the presence of a finite density of states at the Fermi level.† Two of the successful criteria for metallicity derived from theories of electron transport in solids are due to Mott.[8] Several years ago, Mott[9] pointed out how a material would be metallic at high charge carrier (electron) densities or small interparticle distances; in this state, the material would have a finite electrical conductivity at $T = 0$ K. On the other hand, at low carrier densities or large interparticle distances, the material would be an insulator with zero conductivity at 0 K. A discontinuous transition from the metallic state to the insulating state (Mott transition) can occur at a critical distance or a critical carrier concentration. At the transition, $n_c^{1/3}a_H \approx 0.25$, where n_c is the critical carrier concentration and a_H is the Bohr orbit radius of an isolated centre. A large number of systems, including doped semiconductors and metal–ammonia solutions, obey this relation.[2] The metallicity criterion as defined by this expression is essentially a manifestation of electron interactions or Coulomb correlation resulting in an energy gap at low densities.

Another criterion for metallicity arises from electron localization induced by disorder proposed by Anderson.[10] Electrons diffuse when the disorder is small, but at a critical disorder they do not diffuse (giving rise to zero conductivity). A transition from the metallic to the insulating state occurs as the disorder increases. Mott[11] proposed that there exists a minimum metallic conductivity, σ_{min}, or maximum metallic resistivity, ρ_{max}, for which the material may still be viewed as being metallic, prior to the localization of electrons due to disorder. The σ_{min} is given by $\rho^2/h\pi^3l$, where l is the mean free path of the electrons. Mott's σ_{min} criterion assumes that the disorder-driven metal–insulator transition is discontinuous, but the scaling theory[12] predicts it to be continuous. It is, however, found that σ_{min} or ρ_{max} is a useful experimental criterion and represents the value of conductivity where the temperature coefficient of resistance changes sign from metal-like (+) to insulator-like (−) behaviour.[2,3] The value of Mott's ρ_{max} is around 1–10 mohm cm in oxide systems.

A useful way of describing and classifying transition-metal oxides in terms of their electronic properties is that due to Zaanen, Sawatzky and Allen (ZSA)[13] who consider the intra-atomic Coulomb strength, U, the ligand–metal charge-transfer energy, Δ, and the metal (d)–oxygen(p) hybridization strength, t_{pd}, to provide a electronic phase diagram (with U/t_{pd} plotted against Δ/t_{pd}). This diagram has been modified by Sarma to account for magnetic properties and doping effects. Such a modified phase diagram[14] is shown in Fig. 3. In this phase diagram, region D corresponds to d-band metals with $U < \Delta$, and $U < W$ where W is the d-band width and region C represents mixed-valent or p-type metals with $U > \Delta$. Mott insulators with $\Delta > U > W$ wherein U essentially determines the band gap are in region B. Charge-transfer insulators with $U > \Delta$ wherein the band gap is mainly controlled by Δ are in region A. Covalent insulators in region E are those where t_{pd} plays a crucial role; this region had not been identified in the ZSA phase diagram. Amongst the binary oxides, NiO is a charge-transfer insulator (region A) while ReO$_3$ is a d-band metal (D). Amongst the ternary oxides, LaTiO$_3$ is best described as a Mott insulator (B) while LaNiO$_3$ is a metal in the C region. Both LaCoO$_3$ and LaMnO$_3$ have mixed-valent ground states, but LaCoO$_3$ is closer to A than B, probably on the borderline, while LaMnO$_3$ is closer to B than A. La$_2$CuO$_4$, which is the parent cuprate for the numerous superconductor families, is described as a charge-transfer insulator with $\Delta \ll t_{pd}$.

Compositionally controlled metal–insulator transitions in perovskite oxides

In Fig. 4 we show typical compositionally controlled I–M transitions in two perovskite oxide systems. In La$_{1-x}$Sr$_x$CoO$_3$, the material becomes metallic as x is increased beyond a critical composition. A similar behaviour is seen in La$_{1-x}$Sr$_x$VO$_3$. Metallicity arises in these oxides because of fast hopping of

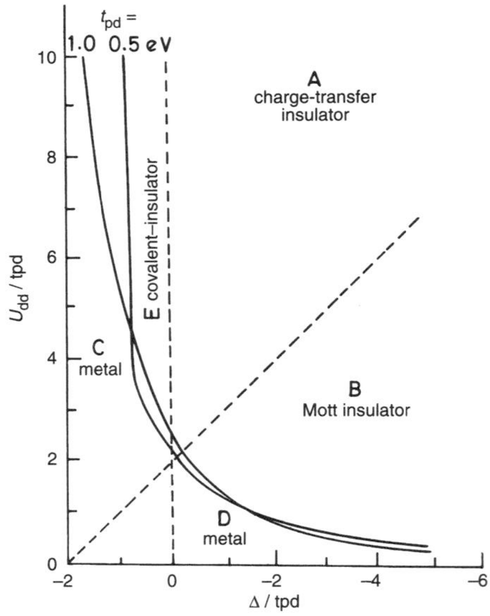

Fig. 3 ZSA phase diagram as modified by Sarma.[14] The solid lines separate the metallic and insulating regimes.

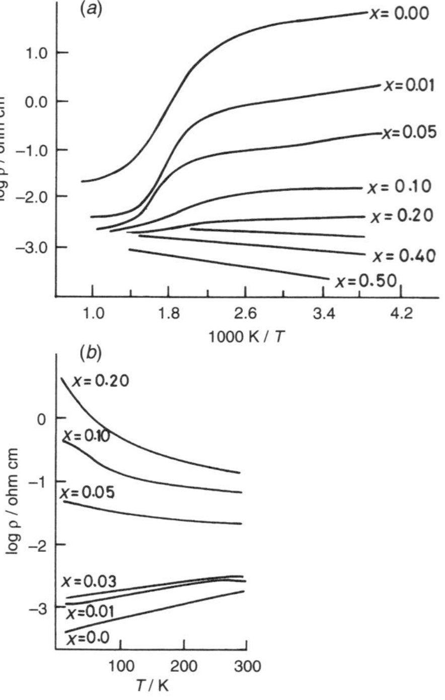

Fig. 4 Compositionally controlled insulator metal transitions in (a) La$_{1-x}$Sr$_x$CoO$_3$ and (b) LaNi$_{1-x}$Mn$_x$O$_3$ (from the author's laboratory)

electrons between the 3+ and 4+ states of the transition metal. It is noteworthy that the temperature coefficient of resistance changes sign around a resistivity value close to Mott's maximum resistivity, ρ_{max}. In $LaNi_{1-x}Mn_xO_3$ where the material becomes insulating with increasing x, the change from the metallic to the insulating state occurs again around Mott's ρ_{max}. The value of ρ_{max} or σ_{min} is known to scale with the critical carrier concentration,[15] n_c, as shown in Fig. 5 where log ρ_{max} is plotted against log n_c for a variety of systems including doped semiconductors. The points corresponding to the metal oxides exhibiting compositionally controlled metal–insulator transitions fall in line with the other electronic materials at the threshold of metallicity. It is noteworthy that these oxides obey the ρ_{max} or the σ_{min} criterion as well as the $n_c^{1/3}a_H \approx 0.25$ criterion. Although the oxides obey the σ_{min} criterion, it is possible that some of them may show a departure from this behaviour and become insulating at very low temperatures (say < 50 K), with the resistivity increasing with decreasing temperature[16] (*i.e.* conductivity is zero at 0 K). If so, σ_{min} serves as the high-temperature limit for the conductivity of these marginal metals.

Photoelectron spectroscopic studies[14] of $La_{1-x}Sr_xCoO_3$ in the valence band region show the presence of a finite density of states at the Fermi level in these marginally metallic oxides (Fig. 6). The oxygen K-absorption spectra of $La_{1-x}Sr_xCoO_3$ show a feature increasing in intensity and width with increasing x, suggesting a progressive creation of doped hole states with substantial oxygen p-character as expected of a charge-transfer insulator. These states overlap the top of the valence band above a critical concentration. A finite density of states at the Fermi level is seen in $LaNi_{1-x}Mn_xO_3$ as well, even slightly beyond the critical x value above which the material is insulating.

Cuprate superconductors

Most of the cuprates which exhibit high-temperature superconductivity exhibit metallic resistivity (with the resistivity decreasing with temperature) before they become superconducting (Fig. 7). The values of resistivities in the metallic regime before they become superconducting is in the range 2–5 mohm cm, not far from Mott's ρ_{max}. Interestingly, the cuprates conform to the $n_c^{1/3}a_H \approx 0.25$ relation (Fig. 8) and also follow the linear log ρ_{max}–log n_c plot (Fig. 5). Although these observations would suggest that the cuprates are similar to $La_{1-x}Sr_xVO_3$ and such oxides showing I–M transitions, they exhibit certain anomalous properties[18] in the metallic state, one of them being the linear variation of resistivity with temperature in the metallic state (see Fig. 7). High-energy spectroscopic studies[19] of the cuprates show that the oxygen 2p contribution in the hole doped states near the Fermi level is significant. The

cuprates have strong correlation effects and there is as yet no unanimous view as to the best way of describing the marginally metallic state.

A novel feature of the cuprates is that they exhibit compositionally controlled metal–insulator transitions in the normal state ($T > T_c$) as can be seen in Fig. 9(*a*), (*b*). It is indeed curious that the insulating state transforms directly to the superconducting state in some compositions [Fig. 9(*b*)]. Such an insulator–metal transition is found in $La_{2-x}Sr_xCuO_4$ as well. The question that arises is whether there can be insulating and superconducting ground states without a metallic ground state in between. A careful comparison[20] shows that the Hall coefficients and other properties of such cuprate compositions are different from those of oxide systems such as $La_{1-x}Sr_xTiO_3$ which exhibit insulator–metal transitions. Clearly, the superconducting cuprates are associated with a unique kind of a marginal metallic state.

Perovskite manganates exhibiting GMR

Of all the oxide systems considered here, the manganates of the general formula $Ln_{1-x}A_xMnO_3$ (Ln = La, Pr, Nd; A = Ca, Sr, Ba, Pb) possessing the perovskite structure are most unusual in

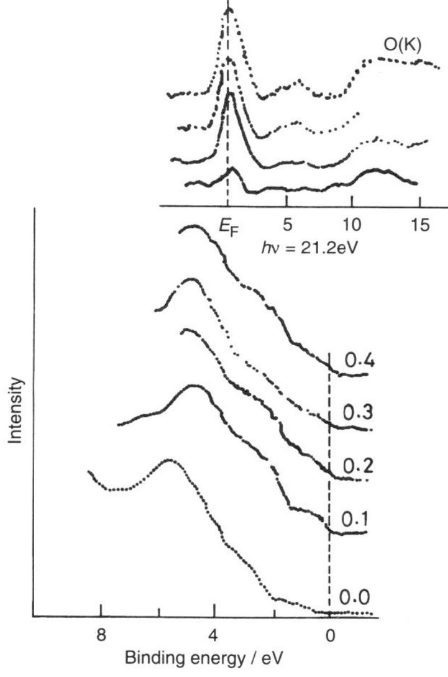

Fig. 6 Photoelectron spectra of $La_{1-x}Sr_xCoO_3$ in the valence region. The inset shows the oxygen K-absorption spectra (from ref. 14).

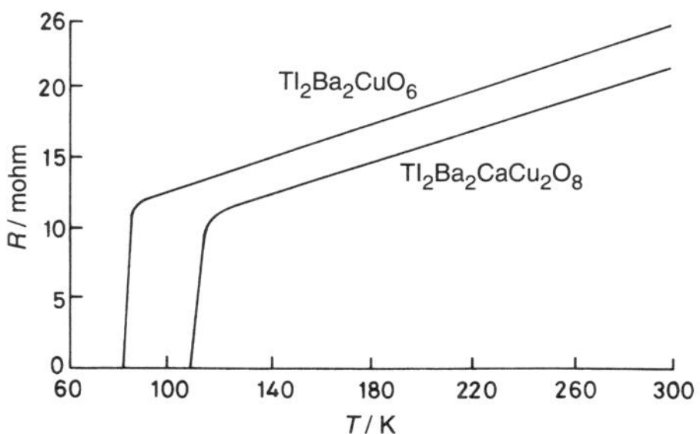

Fig. 7 Resistivity behaviour of cuprates showing linear temperature variation of resistivity in the normal (metallic) state. The linearity can be extended down to 0 K.

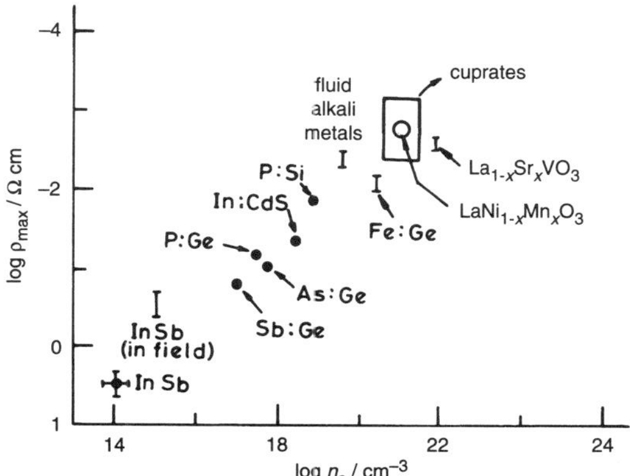

Fig. 5 A log–log plot of the maximum metallic resistivity, ρ_{max}, against the critical carrier density, n_c, at the insulator–metal transition

that they possess high resistivities in the metallic state which cannot be accounted for or explained on the basis of the criteria discussed earlier. Let us briefly examine the nature of the metallic state in these materials. In $La_{1-x}A_xMnO_3$, the Mn^{4+} content increases with x (when $x = 0.0$, Mn will be nominally in the 3+ state).‡ The Mn^{3+}–O–Mn^{4+} interaction is ferromagnetic, unlike the Mn^{3+}–O–Mn^{3+} and Mn^{4+}–O–Mn^{4+} interactions which are antiferromagnetic. The manganates become ferromagnetic above a critical concentration of Mn^{4+} (or value of x) and simultaneously exhibit metallic conductivity.§ The primary process involved is the hopping of a charge carrier as described by,

$$Mn^{3+}(i)-O-Mn^{4-}(j) \rightarrow Mn^{4+}(i)-O-Mn^{3+}(j)$$

where i and j are nearest neighbours. Zener[22] suggested that a paramagnetic to ferromagnetic transition should occur with a transition temperature T_c according to the relation $k_BT_c \approx x_htZ$, where x_h is the Mn^{4+}(hole) concentration, t is the hopping amplitude and Z is the number of nearest neighbours. These materials exhibit a transition from an insulating to a metallic state as the temperature is lowered, the transition occurring close to the ferromagnetic Curie temperature, T_c. The material is insulating at $T > T_c$ because thermal fluctuations of the magnetic moments impede the motion of holes. The I–M transition in an $La_{1-x}Ca_xMnO_3$ composition is shown in Fig. 10. The resistivity peak in this figure arises from the I–M transition, with the temperature variation of resistivity being metal-like at $T < T_c$ (T_{IM}) and insulator-like when $T > T_c$ (T_{IM}).

When a magnetic field is applied to such a material the resistivity decreases enormously[23] (Fig. 10). This phenomenon, called giant magnetoresistance (GMR), is of great importance in magnetic recording and other technological applications. GMR-related aspects of the manganates are not within the scope of this article¶ and I shall restrict myself to the metallic state of these materials.

The resistivity of the manganates at the I–M transition is generally high, anywhere from 10 mohm cm to several ohm cm (see Fig. 2). No metal can have such resistivities, the observed values being much higher than Mott's ρ_{max}. Furthermore, the resistivity in the metallic regime ($T < T_c$) in some of the manganates is comparable to or even higher than that in the insulating state ($T > T_c$).‖ Residual resistivities at low temperatures (ca. 4 K) are also very large (several mohm cm), larger than in disordered metals. While the effects of the lattice and of magnetic polarons may contribute to such high resistivities, we note that electron correlation would play a major role in a Mott insulator such as $LaMnO_3$. What is curious is that photoelectron spectra in the valence-band region[14] of $La_{1-x}Sr_xMnO_3$ show the absence of any significant density of states in the 'metallic' compositions at ordinary temperatures as can be seen from Fig. 11. The oxygen K-absorption spectra show a progressive formation of hole states with the increase in x ca. 1 eV above the Fermi level (Fig. 11).

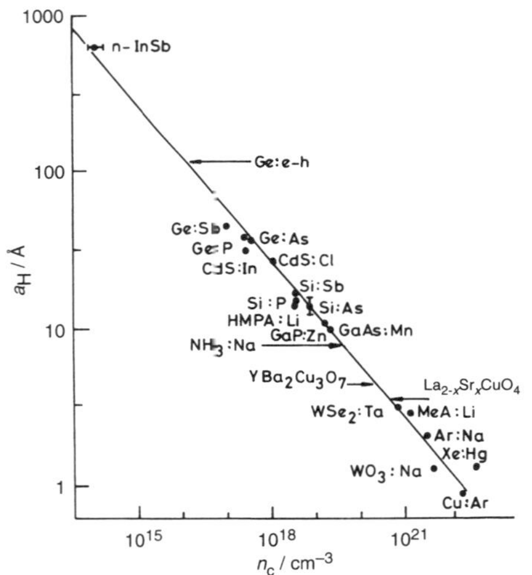

Fig. 8 Relation between the critical carrier concentration, n_c, and the Bohr radius, a_H. The solid line represents $n_c^{1/3}a_H = 0.26$ (from ref. 17).

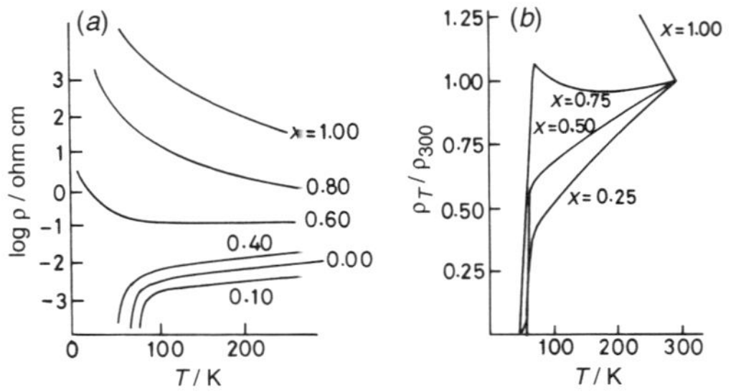

Fig. 9 Temperature variation of the resistivity of (a) superconducting $Bi_2Ca_{1-x}Y_xSr_2Cu_2O_{8+\delta}$ and (b) $TlCa_{1-x}Nd_xSr_2CuO_6$ (from the author's laboratory)

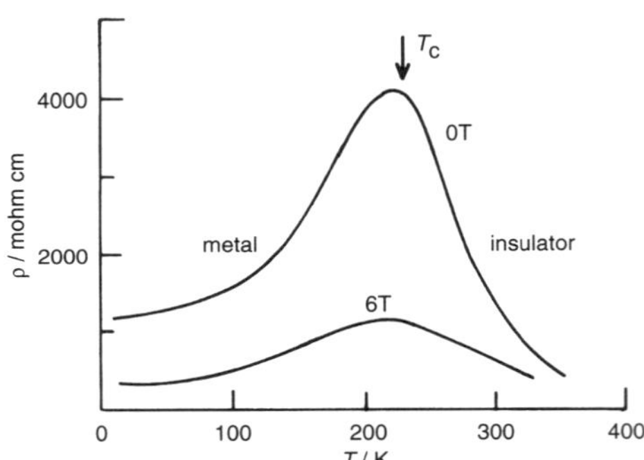

Fig. 10 Resistivity behaviour of $La_{0.9}Ca_{0.1}MnO_3$ at zero field (0 T) and in a magnetic field (6 T) (from ref. 23)

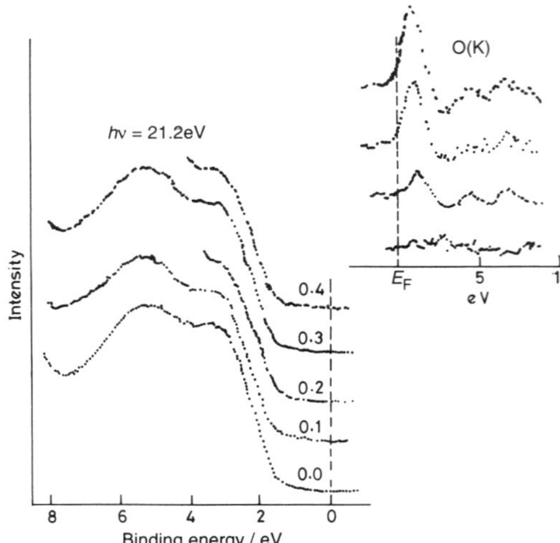

Fig. 11 Photoelectron spectra of $La_{1-x}Sr_xMnO_3$ in the valence region. Inset shows the oxygen K-absorption spectra (from ref. 14).

Concluding remarks

I have discussed three types of complex metal oxides which are marginally metallic. The first category of oxides exhibiting compositionally controlled insulator–metal transitions have a finite density of states at the Fermi level in the metallic state and obey the known criteria for metallicity. The second category involves the superconducting cuprates. They are marginally metallic in the normal state, obey the criteria for metallicity and undergo unusual insulator–metal and insulator–superconductor transitions, besides exhibiting certain anomalous properties in the metallic state. The first two categories of oxides seem to obey both the criteria of Mott arising from considerations of electron interactions (correlation) and of disorder. It is obvious that both these factors should be considered in describing these oxides. The importance of disorder in these oxides is also indicated by the fact that the phenomena are dependent on a critical concentration of one of the component species. Correlation is certainly important in all the systems considering that the parent oxides, $LaCoO_3$, $LaMnO_3$ and La_2CuO_4, all have mixed-valent ground states and fall in the category of Mott or charge-transfer insulators (Fig. 3). Yet, those who work on electron correlation models of oxides ignore disorder and *vice versa*. Unlike the two categories of oxides mentioned above, the manganates exhibiting giant magnetoresistance have high resistivities beyond the range of all known metallic oxides and show no evidence for any significant density of states at the Fermi level in the so-called 'metallic' state. The metallicity of the manganates is clearly of an entirely different category, not encountered hitherto in any other oxide system.

Clearly, there is need for models which include both electron interactions and disorder to describe marginally metallic oxides. This is not an easy task and requires new ideas. The problem does not end here. Besides electron interactions and disorder, other factors such as electron–lattice interactions, magnetic effects and the effect of finite temperature would have to be taken into account depending on the situation as shown schematically in Fig. 12. The theoretical complexity of the marginally metallic state is truly formidable, a proper description of either one of the factors alone (*e.g.* pure correlation or pure disorder) being fraught with many difficulties. Some workers have considered simple models which essentially assume the coexistence of localized and itinerant electrons to describe insulator–metal transitions[25,26] and superconductivity[26] in oxides. For example, Burdett[26] makes use of the interaction or crossing of two diabatic potential-energy curves for the insulating and metallic states to describe the two phenomena. Although educative, such models do not throw light on the nature of marginal metallicity. An effort to combine both electron interactions and disorder has been made recently by Logan *et al.*[27] Whether the metallic state has a universal minimum (electron) diffusity and hence minimum conductivity[28] due to localized states (in addition to extended electron states) is another question that needs to be explored.

Acknowledgements

The author thanks the Science Office of the European Union and the Department of Science and Technology, Government of India, for support of this research.

C. N. R. Rao is Albert Einstein Research Professor at the Indian Institute of Science and President of the Jawaharlal Nehru Centre for Advanced Scientific Research, Bangalore. He was born in Bangalore and educated in Mysore, Banaras and Purdue Universities. He was Visiting Professor at the Oxford and Cambridge Universities and is Honorary Professor at the University of Wales, Cardiff. He is a Fellow of the Royal Society, Foreign Associate of the US National Academy of Sciences and Member of the Pontifical Academy of Sciences. He is the recipient of the Marlow Medal and the Solid State Chemistry Medal of the Royal Society of Chemistry, of which he is an honorary fellow.

Footnotes

† An important experimental criterion for metallicity is the presence of a finite density of states at the Fermi level, which is readily established by photoelectron spectroscopy in the valence region and other high-energy spectroscopic techniques. Two of the most useful operational criteria for metallicity are those due to Mott discussed in the next section.

‡ $LaMnO_3$ as prepared in the laboratory by the solid-state reaction of the oxides and carbonates of the component metals at high temperatures, generally contains around 10% Mn^{4+}. The origin of Mn^{4+} is the presence of cation vacancies in both the A(La) and B(Mn) sites in roughly equal proportions. It cannot be due to oxygen excess since the perovskite structure cannot accommodate excess oxygen.[21]

§For small x, the material is an antiferromagnetic insulator; the same is true for large x ($x \geqslant 0.5$).

¶ The ferromagnetic T_c in $Ln_{1-x}A_xMnO_3$ increases markedly with the increase in the average radius of the A-site cations, $<r_A>$; accordingly, T_{IM} also increases with $<r_A>$. The magnitude of GMR as well as the peak resistivity at the I–M transition decrease with the increase in $<r_A>$. Increasing $<r_A>$ in these perovskites is analogous to increasing pressure.[24] It should also be noted that GMR is not confined to perovskites or manganates.[7]

‖ The magnitude of GMR itself increases with an increase in the peak resistivity (at T_{IM}). In other words, one needs a bad metal to observe good GMR. Furthermore, even formally antiferromagnetic compositions ($x > 0.5$) show GMR because of the presence of ferromagnetic clusters.

References

1 C. N. R. Rao and B. Raveau, *Transition Metal Oxides*, VCH, Cambridge, 1995; C. N. R. Rao, *Annu. Rev. Phys. Chem.*, 1989, **40**, 291.

2 P. P. Edwards, T. V. Ramakrishnan and C. N. R. Rao, *J. Phys. Chem.*, 1995, **99**, 5228.

3 C. N. R. Rao and P. Ganguly, in *The metallic and the nonmetallic states of matter*, ed. P. P. Edwards and C. N. R. Rao, Taylor and Francis, London, 1985.

4 C. N. R. Rao, *Philos. Trans. R. Soc. London, A*, 1991, **336**, 595.

5 C. N. R. Rao and A. K. Ganguli, *Chem. Soc. Rev.*, 1995, **24**, 1; *Acta Crystallogr., Sect. B*, 1995, **51**, 604.

6 C. N. R. Rao and A. K. Cheetham, *Science.*, 1996, **272**, 369.

7 G. Briceno, H. Chang, X. Sun, P. G. Schultz and X. D. Xiang, *Science*, 1995, **270**, 273; T. Shimakawa, Y. Kubo and T. Manako, *Nature (London)*, 1996, **379**, 53.

8 N. F. Mott, *Metal–insulator transitions*, 2nd edn., Taylor and Francis, London, 1990.

9 N. F. Mott, *Philos. Mag.*, 1961, **6**, 287.

10 P. W. Anderson, *Phys. Rev.*, 1958, **109**, 1492.

11 N. F. Mott, *Philos. Mag.*, 1972, **26**, 1015.

12 E. Abrahams, P. W. Anderson, D. C. Liccardello and T. V. Ramakrishnan, *Phys. Rev. Lett.*, 1979, **42**, 693.

13 J. Zaanen, G. A. Sawatzky and J. W. Allen, *Phys. Rev. Lett.*, 1985, **55**, 418.

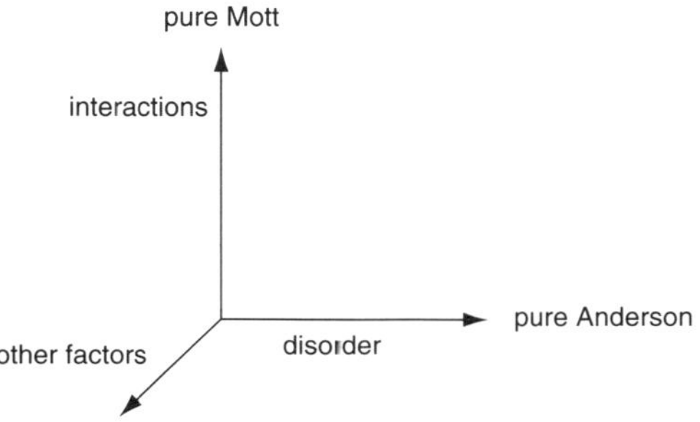

Fig. 12 Complexity of the problem of marginal metallicity (adapted from ref. 27). The oxides discussed in this article fall somewhere in the three-dimensional space indicated here. The 'other factors' include electron–lattice interaction, magnetic polaron and finite temperature effects.

14 D. D. Sarma, in *Metal–insulator transitions revisited*, ed. P. P. Edwards and C. N. R. Rao, Taylor and Francis, London, 1995.
15 N. F. Mott, *Proc. R. Soc. London, A*, 1982, **382**, 1.
16 A. K. Raychaudhuri, *Phys. Rev. B*, 1991, **44**, 8572.
17 G. A. Thomas, *J. Phys. Chem.*, 1983, **88**, 3749; P. P. Edwards and M. J. Sienko, *Phys. Rev. B*, 1978, **17**, 2575.
18 T. V. Ramakrishnan and C. N. R. Rao, *J. Phys. Chem.*, 1989, **93**, 4414.
19 D. D. Sarma, in *Chemistry of high temperature superconductors*, ed. C. N. R. Rao, World Scientific, Singapore, 1991.
20 Y. Iye, in *Metal–insulator transitions revisited*, ed. P. P. Edwards and C. N. R. Rao, Taylor and Francis, London, 1995.
21 M. Hervieu, R. Mahesh, N. Rangavittal and C. N. R. Rao, *Eur. J. Solid State Inorg. Chem.*, 1995, **32**, 79.
22 C. Zener, *Phys. Rev.*, 1951, **82**, 403.
23 It is possible to increase the Mn^{4+} content in the parent $LaMnO_3$ and render it ferromagnetic. $LaMnO_3$ with $\geqslant 20\%$ Mn^{4+} shows the I–M transition and GMR: R. Mahesh, K. R. Kannan and C. N. R. Rao, *J. Solid State Chem.*, 1995, **114**, 294; R. Mahendiran, R. Mahesh, S. K. Tewari, N. Rangavittal, A. K. Raychaudhuri, T. V. Ramakrishnan and C. N. R. Rao, *Phys. Rev. B*, 1995, **53**, 3348.
24 R. Mahesh, R. Mahendiran, A. K. Raychaudhuri and C. N. R. Rao, *J. Solid State Chem.*, 1995, **120**, 204.
25 J. M. Honig and J. Spalek, in *Advances in Solid State Chemistry*, ed. C. N. R. Rao, Indian National Science Academy, New Delhi, 1986.
26 J. K. Burdett, *Acc. Chem. Res.*, 1995, **28**, 227.
27 D. E. Logan, Y. H. Szczechand and M. A. Tusch, in *Metal–insulator transitions revisited*, ed. P. P. Edwards and C. N. R. Rao, Taylor and Francis, London, 1995.
28 N. Kumar and A. M. Jayannavar, in *Metal–insulator transitions revisited*, ed. P. P. Edwards and C. N. R. Rao, Taylor and Francis, London, 1995.

Received, 21st March 1996; 6/01957I

Phase Separation in Metal Oxides

C. N. R. Rao,*[a] P. V. Vanitha,[a] and Anthony K. Cheetham[b]

Abstract: A fascinating phenomenon, recently found to occur in certain transition-metal oxides, is phase separation wherein pure, nominally monophasic oxides of transition metals with well-defined compositions separate into two or more phases over a specific temperature range. Such phase separation is entirely reversible, and is generally the result of a competition between charge-localization and -delocalization, the two situations being associated with contrasting electronic and magnetic properties. Coexistence of more than one phase, therefore, gives rise to electronic inhomogeneity and a diverse variety of magnetic, transport, and other properties, not normally expected of the nominal monophasic composition. An interesting feature of phase separation is that it covers a wide range of length scales anywhere between 1–200 nm. While cuprates and manganates, especially the latter, provide excellent examples of phase separation, it is possible that many other transition-metal compounds with extended structures will be found to exhibit phase separation.

Keywords: magnetic properties · metal oxides · phase separation · rare earth manganates

Introduction

It is generally assumed that a single crystal of a compound is pure and monophasic, with an exact composition and a well-defined set of properties. While this is indeed true of most molecular and extended solids, it has been found in recent years that a few of the transition-metal oxides negate this

presumption, and exhibit compositional and electronic inhomogeneities arising from the existence of more than one phase in crystals of nominally monophasic composition. The different phases in such materials have comparable compositions, and the phenomenon is commonly referred to as phase separation. One may be tempted to think that there is little new in this phenomenon. Afterall, there are many oxide systems in which there is intergrowth of related phases, as exemplified by the Aurivillius family of oxides of the general composition, $Bi_4A_{m+n-2}B_{m+n}O_{3(m+n)+6}$ (A = Ba or Bi, B = Ti, Nb etc).[1, 2] In the oxides of this family, it is common to find the presence of unit cells of the wrong $(m+n)$ values to intergrow along with the major phase (e.g., $(m+n)$ of 1 or 2 in a major phase corresponding to $(m+n)$ of 3). Similar intergrowths occur in cuprates of the type $Bi_2(CaSr)_{n+1}Cu_nO_{2n+4}$ and $Tl_2A_{n+1}Cu_nO_{2n+4}$ (A = Ca, Sr, Ba), as well as in the Ruddlesden–Popper series of oxides of the type $Sr_{n+1}Ti_nO_{3n+1}$.[1, 2] While the presence of intergrowths in the metal oxides such as the above affects the properties of the major phase, no new property or phenomenon seems to result from it. Our interest in this article is in phase-separated solids whose behavior is significantly different from what is expected of the nominal composition, and reflects a combination of disparate electronic, magnetic, and other properties associated with the different phases coexisting together. In certain cases, phase separation does indeed give rise to entirely new properties.

A good example of phase separation is that in $La_2CuO_{4+\delta}$, in which two phases with different oxygen stoichiometries coexist over a specific temperature range.[3] Thus, the cuprate with a nominal $\delta \approx 0.03$ separates into two phases below a certain temperature, one with a small oxygen excess ($\delta \sim 0.01$) and another with a higher oxygen excess ($\delta \sim 0.06$).[4] The two phases possess entirely different magnetic and electrical properties. Rare-earth manganates of the general composition, $Ln_{1-x}A_xMnO_3$ (Ln = rare earth, A = alkaline earth), display a variety of effects due to phase separation, giving rise to novel electronic and magnetic properties. The rare-earth manganates became popular because of the colossal magnetoresistance (CMR) exhibited by them.[5] CMR and related properties are generally explained on the basis of the double-exchange mechanism of electron hopping between the Mn^{3+} ($t_{2g}^3e_g^1$) and Mn^{4+} ($t_{2g}^3e_g^0$) ions. In this mechanism, the ferromagnetic alignment of the spins of the incomplete e_g orbitals of adjacent Mn ions is directly related to the rate of

[a] Prof. Dr. C. N. R. Rao, P. V. Vanitha
CSIR Centre of Excellence in Chemistry and
Chemistry and Physics of Materials Unit
Jawaharlal Nehru Center for Advanced Scientific Research
Jakkur P.O., Bangalore, 560 064 (India)
Fax: (+91) 80-846-270
E-mail: cnrrao@jncasr.ac.in

[b] Prof. Dr. A. K. Cheetham
Materials Research Laboratory
University of California
Santa Barbara, CA-93106 (USA)

hopping of the electrons, giving rise to an insulator–metal transition at the ferromagnetic Curie temperature (T_C). In the ferromagnetic state ($T < T_C$), the material is metallic, but is an insulator in the paramagnetic state ($T > T_C$). Jahn–Teller distortion associated with the Mn^{3+} ions and charge-ordering of the Mn^{3+} and Mn^{4+} ions compete with double exchange and promote the insulating behavior and antiferromagnetism.[5] Charge-ordering in these materials is also closely linked to the ordering of the e_g orbitals. The nature of phase separation in the manganates depends on the average size of the A-site cations, carrier concentration or the composition (value of x), temperature, and other external factors such as magnetic and electric fields. Phases with different charge densities (carrier concentrations) as well as magnetic and transport properties coexist as carrier-rich ferromagnetic (FM) clusters or domains along with a carrier-poor antiferromagnetic (AFM) phase. Such an electronic phase separation giving rise to microscopic or mesoscopic inhomogeneous distribution of electrons results in rich phase diagrams that involve various types of magnetic structures.[6] What is noteworthy is that electronic phase separation is likely to be a general property of solids with correlated electrons such as the large family of transition-metal oxides. There are indications that many of the unusual magnetic and transport properties of oxide materials arise from phase separation.[3, 6]

The term phase separation or segregation implies the presence of at least two distinct phases in the sample, but the relative fractions may vary anywhere from a dilute regime, involving small domains of the minor phase (or clusters) in the matrix of the major phase, to a situation in which the fractions of the two phases is comparable. Thus, FM clusters present randomly in an AFM host matrix often give rise to a glassy behavior. As the FM clusters in an AFM matrix grow in size to become reasonably sized domains, due to effect of temperature, composition, or an applied magnetic field, the system acquires the characteristics of a genuine phase-separated system. We shall discuss the various scenarios in phase-separated rare-earth manganates in this article by presenting some of the recent results in the form of phase diagrams and schematic illustrations of the spatial distribution of the coexisting phases. We shall also briefly examine the features of phase separation in oxygen excess La_2CuO_4. Before discussing the results on these oxide systems, we shall examine the general features of phase separation in solids.[3, 6]

Phase Separation: The Phenomenon

Thermodynamic, equilibrium phase separation is distinct from phase separation caused by inhomogeneties in chemical composition, such as those due to non-uniformity in impurity distribution, as it can have an electronic origin or could also arise from the presence of magnetic impurities. Such phase separation can be controlled or changed by temperature, magnetic fields, and other external factors. In both these types of phase separation, a high carrier density favors ferromagnetic ordering and/or metallicity. If the carrier concentration is not sufficient, ferromagnetic ordering can occur in one portion of the crystal, the rest of the crystal remaining insulating and antiferromagnetic. In electronic phase separation, the concentration of the charge carriers giving rise to ferromagnetism and/or metallicity in a part of the crystal causes mutual charging of the two phases. This gives rise to strong Coulomb fields, which may mix the conducting ferromagnetic and insulating antiferromagnetic phases in order to lower the Coulomb energy. When the carrier concentration is small, the conducting ferromagnetic regions are separated forming droplets as in Figure 1a. At higher carrier concentrations, the volume of the ferromagnetic phase

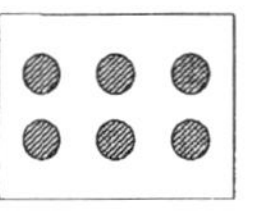 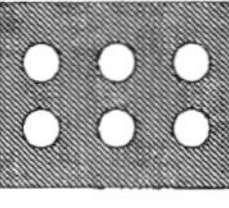 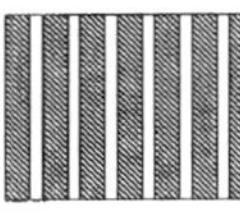 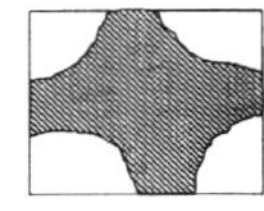

a) b) c) d)

Figure 1. Microscopic phase-separated state giving rise to a) an insulating state with ferromagnetic (conducting/metallic) droplets, b) a conducting state where a ferromagnetic (conducting) part of the crystal has separated (insulating) droplets, and c) charge-stripes. A macroscopic phase separation state is shown in d). Hatched portions represent the ferromagnetic regions with high carrier concentration (and are therefore conducting)

increases rendering the droplets to coalesce, giving rise to a situation shown in Figure 1b. Electronic phase separation has been observed in magnetic semiconductors such as heavily doped EuSe and EuTe. Here, the main portion of the crystal is antiferromagnetic at low temperatures and the conducting electrons occur in the ferromagnetic droplets. A magnetic-field-induced insulator to metal transition, giving rise to high magnetoresistance, is accompanied by an increase in the size of the ferromagnetic droplets.

Impurity phase separation is different from electronic phase separation in that there is no mutual charging of the phases in the former. Here also, magnetic fields increase the size of the magnetic domains or regions. Impurity atom diffusion has to be sufficiently large to give rise to phase separation in such systems. The case of oxygen-excess La_2CuO_4 is one such example.

The case of rare-earth manganates, $Ln_{1-x}A_xMnO_3$ (Ln = rare earth, A = alkaline earth), is one in which the ferromagnetic phase is conducting and the antiferromagnetic phase is insulating. Depending on x or the carrier concentration, we can have a situation such as that shown in Figure 1a or b. The phase separation scenario here is somewhat complex because the transition from the metallic to the insulating state is not sharp, and the domains of the two phases are often sufficiently large to give rise to well-defined signatures in neutron scattering or diffraction experiments. However, one may consider the large magnetoresistance in these systems to be a consequence of the electronic phase separation. In the presence of Coulomb interaction, the microscopically charged inhomogeneous state is stabilized, giving rise to clusters of one phase embedded in another. The size of the clusters depends on the competition between double exchange and Coulomb forces. Electronic phase separation with phases of different charge densities is generally expected to give rise to nanometer scale clusters. This is because large phase-separated domains would break up into small pieces because of Coulomb interactions. The shapes of these pieces could be droplets or stripes (Figure 1a–c).

One can visualize phase separation arising from disorder as well. The disorder can arise from the size mismatch of the A-site cations in the perovskite structure.[7] Such phase separation is seen in the $(La_{1-y}Pr_y)_{1-x}Ca_xMnO_3$ (LPCM) system in terms of a metal–insulator transition induced by disorder. The size of the clusters depends on the magnitude of disorder. The smaller the disorder, the larger would be the size of the clusters. This could be the reason why high magneto-resistance occurs in systems with small disorder.

Microscopically homogeneous clusters are usually of the size of 1–2 nm in diameter dispersed in an insulating or charge-localized matrix as seen in Figure 1a. Such a phase-separation scenario bridges the gap between the double-exchange model and the lattice distortion models. A number of recent papers on manganates show that in addition to microscopic phase separation there can also be mesoscopic phase separation whereby the length scale is between 30–200 nm, arising from the comparable energies of the ferro-magnetic metallic and antiferromagnetic insulating states. In certain manganate compositions, mesoscopic as well as microscopic phase separation has been observed. In LPCM and other manganates, the occurrence of multiple phases has also been noticed.

The techniques required to identify phase separation in different length scales vary. For example, diffraction techniques can be used to examine macroscopic phase separation (Figure 1d), for which distinct features occur in the diffraction patterns due to the different phases in the system. Techniques such as NMR spectroscopy, on the other hand, give information on the local environment at a microscopic level. It is often difficult to identify electronic phase separation based on magnetic measurements because of the sensitivity of phase separation to magnetic fields. Thus, magnetic fields transform the antiferromagnetic insulating state to the ferromagnetic metallic state. However, transport measurements, under favorable conditions, can provide valuable information on phase separation.

Lanthanum Cuprate

La_2CuO_4 is an insulating layered oxide that contains CuO_2 sheets.[2] It is antiferromagnetic, but the AFM order is destroyed on doping it with holes. Hole doping is accomplished by introducing excess oxygen between the LaO layers or by substituting La partly by a divalent cation such as Sr^{2+}. A neutron diffraction study[4] of $La_2CuO_{4.03}$ synthesized under high oxygen pressure showed that it undergoes a reversible macroscopic phase separation below a certain temperature, T_{ps}, into two nearly identical orthorhombic phases with δ values of ~ 0.01 and ~ 0.08. Phase separation in this oxygen-excess cuprate is evidenced from resistivity, magnetic susceptibility, specific heat, NMR, and NQR measurements, which show anomalies around T_{ps}.[8] In Figure 2, we show the phase diagram of $La_2CuO_{4+\delta}$ (upto $\delta = 0.07$) constructed by Chou and Johnston.[8] The $\delta = 0.01$ and 0.08 phases possess entirely

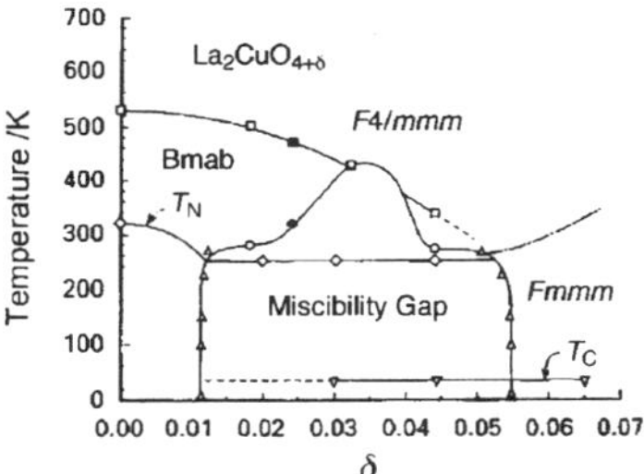

Figure 2. Phase diagram of $La_2CuO_{4+\delta}$ (reproduced with permission from ref. [8]).

different properties, the former being an AFM insulator ($T_N = 250$ K) and the latter a superconductor ($T_C \approx 35$ K). Accordingly, magnetic susceptibility measurements on a sample of $La_2CuO_{4+\delta}$ show both the AFM and superconducting transitions[8] as shown in Figure 3. The transitions exhibit

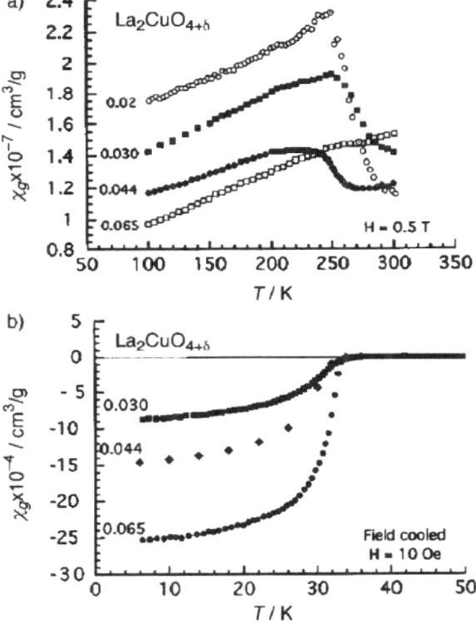

Figure 3. Temperature dependence of magnetic susceptibility of $La_2CuO_{4+\delta}$ showing antiferromagnetic and superconducting transitions (reproduced with permission from ref. [8]).

considerable thermal hysteresis. TEM images reveal the domain structure of the oxygen-rich and -poor phases with a minimum dimension of 30–150 nm.[9] Phase separation in $La_2CuO_{4+\delta}$ can be visualized from the spatial distribution of the AFM and superconducting regions presented in Figure 4.[10]

Inhomogeneous doping of the CuO_2 planes gives rise to hole-rich (metallic) one-dimensional features (charge stripes) due to the concentration of holes in periodic walls superposed on AFM stripes (Figure 1c). The stripes are a consequence of phase separation, arising from preferred hole dopings with independent dispersions. The stripes are charge-driven rather than spin-driven.[11, 12] Such stripes due to the presence of hole-rich and hole-poor regions are also seen in nickelates of

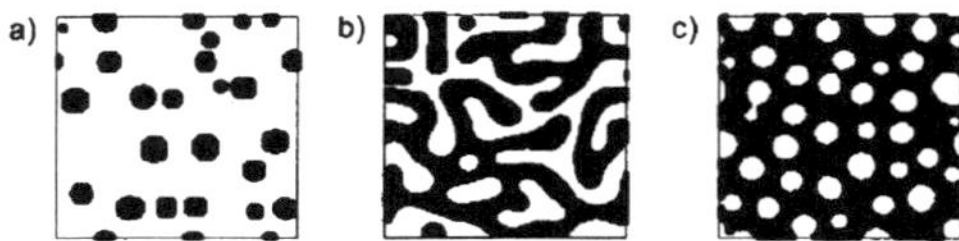

Figure 4. Schematic representation of the growth of the antiferromagnetic (white) and superconducting (black) phases with increasing doping or δ value (reproduced with permission from ref. [10]).

the type $La_2NiO_{4+\delta}$.[11] Phase separation and stripes are found in $La_{2-x}Sr_xCuO_4$ ($0.0 < x < 0.15$) as well.[13, 14] In adiition, this system shows another kind of phase separation. When the hole concentration is greater than 0.085, the cuprate exhibits two superconducting transitions at 15 K and 30 K.[14] When it is less than 0.02, AFM order coexists with a spin-glass phase at low temperatures.[15]

Rare-Earth Manganates

A wide range of compositions of the rare-earth manganates, $Ln_{1-x}A_xMnO_3$ (Ln = rare earth, A = alkaline earth), exhibit charge-ordering.[5] Charge- and AFM-ordering may occur at the same temperature or at different temperatures. The Mn^{3+} (d_{z^2}) orbitals and the associated lattice distortions develop long-range order, and such orbital-ordering occurs with charge ordering in many of the manganates, but it is always accompanied by AFM ordering. In $Ln_{1-x}A_xMnO_3$, small Ln and A ions stabilize the charge-ordered state. Thus, $Pr_{0.7}Ca_{0.3}MnO_3$, with an average A-site cation radius, $\langle r_A \rangle$, of 1.18 Å, charge-orders around 230 K (T_{CO}) in the paramagnetic state, becoming antiferromagnetic at 170 K; it is an insulator and does not show ferromagnetism in the absence of a strong magnetic field. $La_{0.7}Ca_{0.3}MnO_3$ ($\langle r_A \rangle = 1.27$ Å), on the other hand, is an FM metal below the T_C ($T_C \approx 250$ K) and a paramagnetic insulator above the T_C. $La_{0.5}Sr_{0.5}MnO_3$ ($\langle r_A \rangle = 1.26$ Å) is metallic both in the FM and paramagnetic states, whereas $Nd_{0.5}Sr_{0.5}MnO_3$ ($\langle r_A \rangle = 1.24$ Å) shows a transition from a FM metallic state to an AFM charge-ordered state around 150 K. The charge-ordered states in these manganates are associated with CE-type AFM-ordering, whereby the Mn^{3+} and Mn^{4+} ions are arranged as in a checker board. The CE-type AFM charge-ordered state in $Ln_{1-x}A_xMnO_3$ is associated with the ordering of $3x^2 - r^2$ or $3y^2 - r^2$ type orbitals at the Mn^{3+} site. The Jahn–Teller distortion that accompanies orbital ordering stabilizes the CE-type AFM state relative to the FMM state. Orbital and spin ordering occur without charge ordering in the manganates that show A-type antiferromagnetism.

Evidence for charge ordering in the rare-earth manganates is found in the crystal structures at low temperatures. A charge-ordering transition is marked by a resistivity anomaly, specially if the transition is first order.[5] Magnetization shows an abrupt change or a peak depending on the nature of the transition. Accordingly, the transition in $Nd_{0.5}Sr_{0.5}MnO_3$ at 150 K from the FM state to the AFM charge-ordered state ($T_{CO} = T_N$) is accompanied by a sharp increase in resistivity and a decrease in the magnetization (Figure 5). In $Pr_{0.6}Ca_{0.4}$-

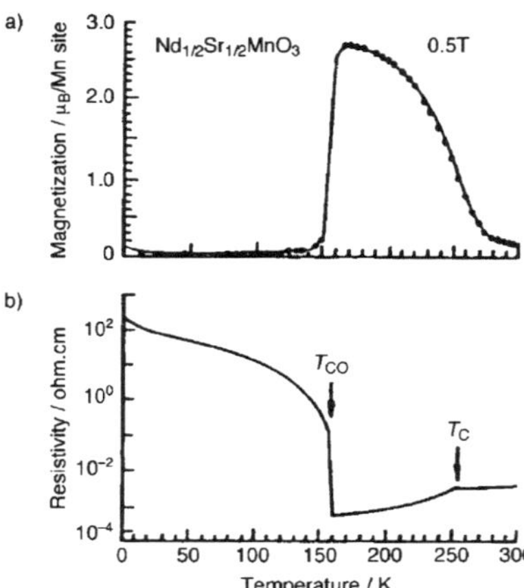

Figure 5. Temperature variation of a) magnetization and b) resistivity of $Nd_{0.5}Sr_{0.5}MnO_3$ (reproduced with permission from ref. [5b]).

MnO_3 ($\langle r_A \rangle = 1.18$ Å) the paramagnetic ground state is charge-ordered and becomes antiferromagnetic on cooling (Figure 6). This manganate shows a small peak in magnetization at T_{CO}. Furthermore, well below T_N the magnetic susceptibility shows clear indications for the presence of

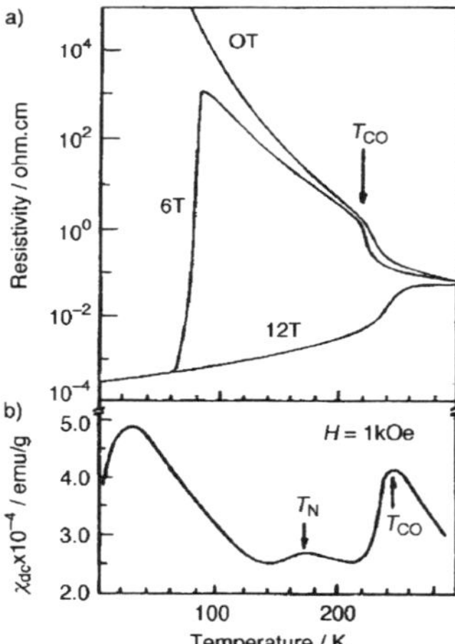

Figure 6. Temperature variation of a) resistivity and b) magnetic susceptibility of $Pr_{0.6}Ca_{0.4}MnO_3$ (reproduced with permission from ref. [5b]).

ferromagnetic interactions (see Figure 6). The charge-ordered states in the manganates can be transformed to a FM metallic state by the application of magnetic fields, the strength of the field depending on the robustness of the charge-ordered state. The charge-ordered state in the manganates with small A-site cations is often unaffected by strong magnetic fields or by doping the Mn site by cations such as Ru^{4+} and Cr^{3+}. These dopants transform the charge-ordered state in both $Nd_{0.5}Sr_{0.5}MnO_3$ and $Pr_{0.6}Ca_{0.4}MnO_3$ into a FM metallic state.

© 2003 WILEY-VCH Verlag GmbH & Co. KGaA, Weinheim   0947-6539/03/0904-0832 $ 20.00+.50/0   *Chem. Eur. J.* **2003**, *9*, No. 4

In Figure 7, we show the phase diagrams of $La_{1-x}Ca_xMnO_3$ and $Pr_{1-x}Ca_xMnO_3$. In the former, charge ordering occurs in the $x \approx 0.5-0.8$ range, while in the latter charge ordering

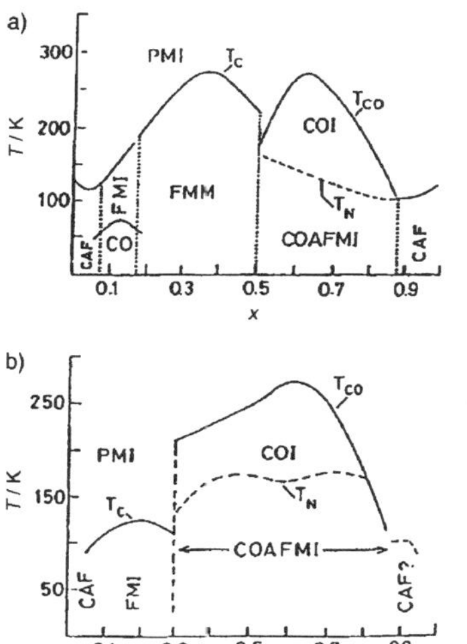

Figure 7. Phase diagrams of a) $La_{1-x}Ca_xMnO_3$ and b) $Pr_{1-x}Ca_xMnO_3$. CAF, canted antiferromagnet; CO, charge-ordered state; FMI, ferromagnetic insulator; FMM, ferromagnetic metal; PMI, paramagnetic insulator; COI, charge-ordered insulator (paramagnetic); COAFMI, charge-ordered antiferromagnetic insulator (reproduced with permission from ref. [16]).

occurs over the $x \approx 0.3-0.8$ range. These differences are essentially due to the effect of the size of the A-site cations, the smaller size favoring the charge-ordered insulating state. From Figure 7, we see that the charge-ordered regime is prominent at large values of x. Thus, the $x > 0.5$ compositions in $Ln_{1-x}Ca_xMnO_3$ are almost entirely in the charge-ordered regime when $Ln = La$ or Pr. This regime is referred to as the electron-doped regime and the $x < 0.5$ compositions as the hole-doped regime. There is marked electron-hole asymmetry in these manganates, ferromagnetism and metallicity not being encountered in the electron-doped regime.[16]

At low dopant levels (low x values, say $x < 0.1$), FM clusters are found to be present in an antiferromagnetic host matrix in $Ln_{1-x}Ca_xMnO_3$, often giving rise to a spin-glass behavior. A similar situation obtains when $x > 0.9$. The coexistence of charge-ordered (AFM) and FM domains, the sizes of which are affected by the composition (x value), relative sizes of A-site cations, temperature, magnetic field, and cation doping, arises from electronic phase separation,[6, 17, 18] giving rise to anomalous magnetic and transport properties. In $La_{1-x-y}Pr_yCa_xMnO_3$ ($x = 0.37$), mesoscopic phase separation into submicrometer-scale charge-ordered regions (3–20 nm) and FM metallic domains have been observed in electron microscopic images.[19] In $Nd_{0.5}Sr_{0.5}MnO_3$, phase segregation of the FM metallic and CE-AFM charge-ordered phases along with a A-type AFM phase has been observed.[20] The point to note is that phase separation in all these materials is a consequence of the competition between charge localization

and charge-delocalization, the two being associated with contrasting magnetic and transport properties.[6, 17]

As mentioned earlier, coexistence of the FM metallic and the AFM insulator phases due to electronic phase separation (inhomogeneous distribution of charge densities) cannot be a long-range phenomenon owing to the high coulomb energy cost. Therefore, we observe a microscopically inhomogeneous state with FM clusters of 1–2 nm in diameter dispersed in an insulating (charge-localized) matrix as in Figure 1a. Evidence for such microscopic phase separation in the manganates is found from various physical measurements[6, 17, 21] (e.g., spin-glass behavior or canted-spin ordering as seen at low temperatures in $Pr_{0.6}Ca_{0.4}MnO_3$, see Figure 6). The cause of mesoscopic phase separation found in some manganate compositions lies in the comparable energies of the FM metallic and insulating phases, and the large strain mismatch of the domains of the two phases.[22]

Some Recent Results

We shall now briefly examine some of the recent findings on rare earth manganates. Based on neutron scattering and diffraction studies, Radaelli et al.[23, 24] have shown tunable mesoscopic phase separation in $Pr_{0.7}Ca_{0.3}MnO_3$. Intragranular strain-driven mesoscopic phase segregation (5–20 nm) between two insulating phases (one charge-ordered and another spin-glass) occurs below T_{CO}. The charge-ordered phase orders antiferromagnetically and the other remains a spin-glass. On the application of a magnetic field, most of the material goes to a FM state. Microscopic phase separation (0.5–2 nm) is present at all temperatures, especially in the spin-glass phase at low temperatures. These results are shown in the form of a phase diagram in Figure 8.

As mentioned earlier, submicrometer-sized phase separation involving FM and charge-ordered AFM domains has been found in $La_{5/8-y}Pr_yCa_{3/8}MnO_3$. By varying y, the volume fraction and the domain size of the FM and charge-ordered phases can be varied.[19] In Figure 9, a schematic diagram to describe the coexistence of the two phases is shown. Electrical conduction in this manganate occurs through a percolative mechanism because of phase separation. The phase diagram of $(La_{1-y}Pr_y)_{0.7}Ca_{0.3}MnO_3$ showing the dependence of phase separation on the composition and temperature[25] is shown in Figure 10. Clearly, phase separation is sensitive to the cation size and size disorder.

$La_{0.5}Ca_{0.5}MnO_3$ changes to a FM phase on cooling to 220 K (T_C) and then to a charge-ordered AFM phase around 150 K (T_{CO}). This manganate is best described as magnetically phase separated over a wide range of temperatures.[26, 27] At low temperatures ($T < T_{CO}$), FM metallic domains are trapped in the charge-ordered AFM matrix, giving rise to percolative metallic conduction. The fraction of the FM phase at low temperatures is highly dependent on the thermal treatment. Even within the FM phase, in the $T_{CO} < T < T_C$ region, there is phase separation. A second crystallographic phase, probably without magnetic order, has been identified.[27] The phase diagram of $La_{1-x}Ca_xMnO_3$ in the $0.47 \leq x \leq 0.5$ range (Figure 11) reveals the nature of phase separation. Magnetization

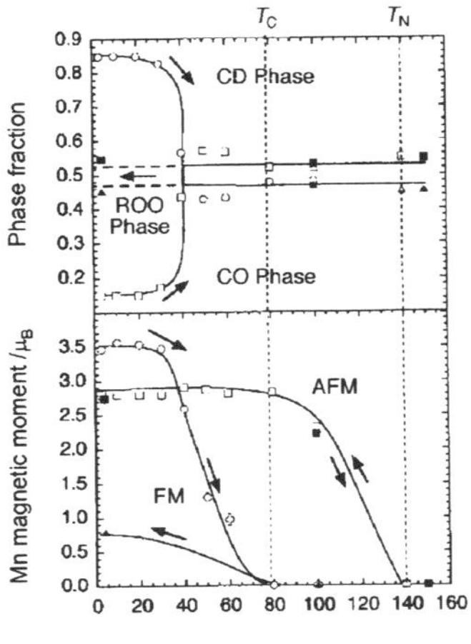

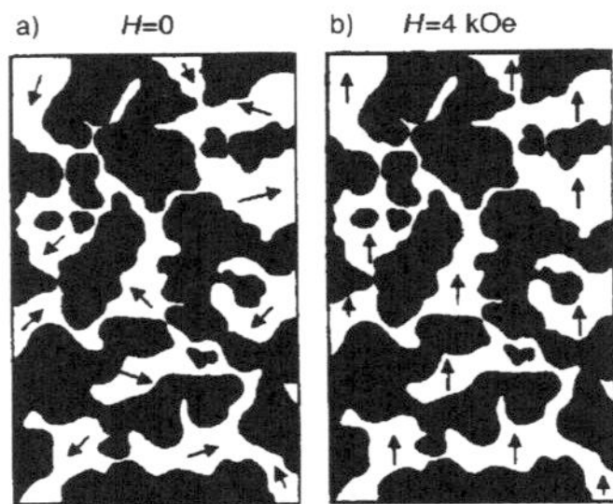

Figure 8. Fractions and magnetism of the coexisting phases $Pr_{0.7}Ca_{0.3}MnO_3$ from multiphase Rietveld refinements of neutron data. Top: Phase fractions of charge-ordered (AFM) phase (squares), ferromagnetic charge-delocalized (CD), and reverse orbital-ordered (ROO) weakly ferromagnetic spin-glass phase (triangles) on zero-field cooling (filled symbols) and warming after 7 T field processing at 3 K (open symbols). Arrows indicate the direction of cooling/heating. Bottom: magnetic moments per manganese atom for the individual phases (symbols as in top panel) (reproduced with permission from ref. [24]).

Figure 9. Schematic diagram showing the coexistence of the charge-ordered insulating (dark area) and FM metallic (white area) domains in $La_{1-x-y}Pr_yCa_xMnO_3$. Arrows show alignment in a magnetic field (reproduced with permission from ref. [19]).

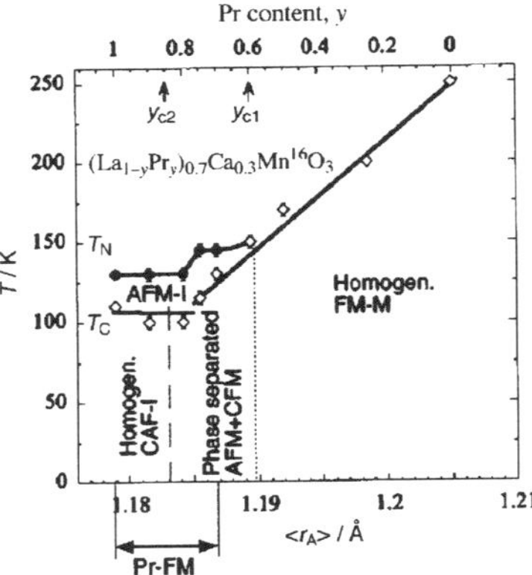

Figure 10. Phase diagram of the $(La_{1-y}Pr_y)_{0.7}Ca_{0.3}MnO_3$. The diamonds and circles show the T_C and T_N values. The bottom x axis shows the average A cation radius $\langle r_A \rangle$. The low-temperature state is homogeneous for $y > 0.8$ (canted AFM insulator, CAF-I) and for $y < 0.6$ (FM metal). In the range $0.6 \leq x \leq 0.8$ the magnetic state is an inhomogeneous mixture of slightly canted ferromagnetic (CFM) and AFM regions. The T_N and T_C transition temperatures coincide for $x = 0.6$. A ferromagnetic contribution of Pr moments is found in the interval of the Pr concentration marked as Pr-FM (reproduced with permission from ref. [25]).

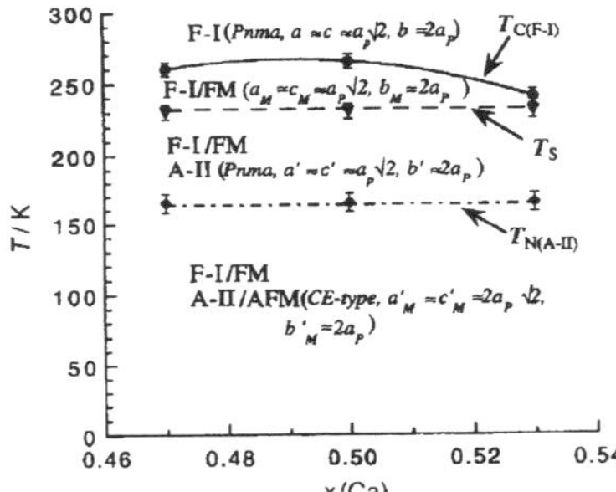

Figure 11. Phase diagram of $La_{1-x}Ca_xMnO_3$ in the range of compositions $0.47 \leq x \leq 0.53$. The horizontal curves separate, going from top to bottom: i) the FM transition of the F-I crystallographic phase at ~ 260 K; ii) the formation of the low-temperature A-II phase, which appears at ~ 230 K; and iii) the AFM transition that occurs in the A-II structure at ~ 160 K (T_N). As shown in the diagram, the ferromagnetically ordered F-I phase and the antiferromagnetically ordered A-II phase coexist at low temperatures (reproduced with permission from ref. [27]).

studies show three phase separated regimes: $T_C > T > T_O$, $T_O > T > T_{CO}$ and $T < T_{CO}$, in which T_O is the onset temperature below which the cooling field plays an unbalancing role in favor of the FM state.[26] It is in the last regime that minority FM domains are embedded in the AFM matrix. In the first regime near the T_C, the FM phase grows freely with the application of a magnetic field.

The occurrence of a phase-separated state below T_{CO} (T_N) in some of the manganate compositions was pointed out earlier. In Figure 12, we illustrate the phenomenon schematically in the case of $Nd_{0.55}(Sr_{0.17}Ca_{0.83})_{0.45}MnO_3$.[28] The situation is even more complex in $Nd_{0.5}Sr_{0.5}MnO_3$. High resolution X-ray and neutron diffraction investigations show that $Nd_{0.5}Sr_{0.5}MnO_3$ separates into three macroscopic phases at low temperatures.[20] The phases involved are the high-

temperature FMM phase, the orbitally ordered A-type AFM phase, and the charge-ordered CE-type AFM phase. On cooling this manganate, the A-type AFM phase starts manifesting itself around 220 K, with the charge-ordered AFM phase appearing at 150 K (as expected from Figure 5). At the so-called FM metallic-charge-ordered AFM transition, all the three phases coexist, and this situation continues down to very low temperatures as shown in Figure 13. In Figure 14, we show the percentage volume fraction of the different phases in the presence and absence of a magnetic field.[29] Phase segregation in this system seems to depend crucially on the Mn^{4+}/Mn^{3+} ratio, a ratio slightly greater than unity stabilizes the A-type AFM phase. Thus, $Nd_{0.45}Sr_{0.55}MnO_3$ has the A-type AFM structure.

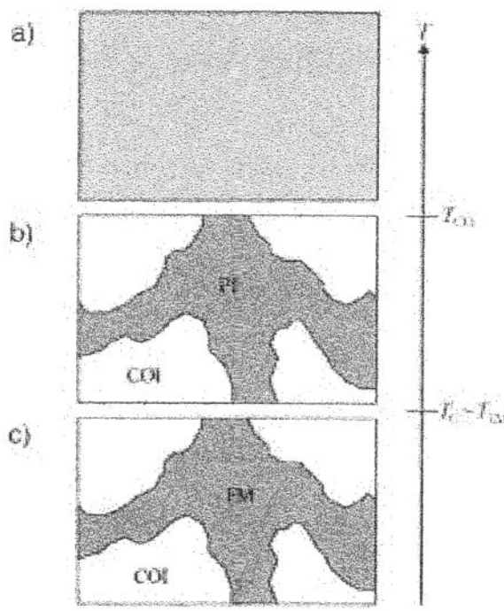

Figure 12. Schematic illustrations for the temperature variation of phase separated state in $Nd_{0.55}(Sr_{0.17}Ca_{0.83})_{0.45}MnO_3$: a) $T > T_{CO}$, b) $T_C < T < T_{CO}$, c) $T < T_C$. COI, PI, and FM, stand for the charge-ordered insulating, paramagnetic insulating, and ferromagnetic metallic phases, respectively (reproduced with permission from ref. [28]).

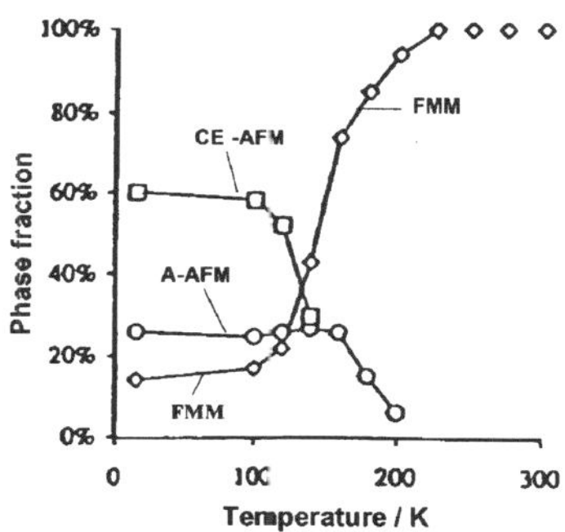

Figure 13. Variation in the percentage of the different phases of $Nd_{0.5}Sr_{0.5}MnO_3$ with temperature (reproduced with permission from ref. [20]).

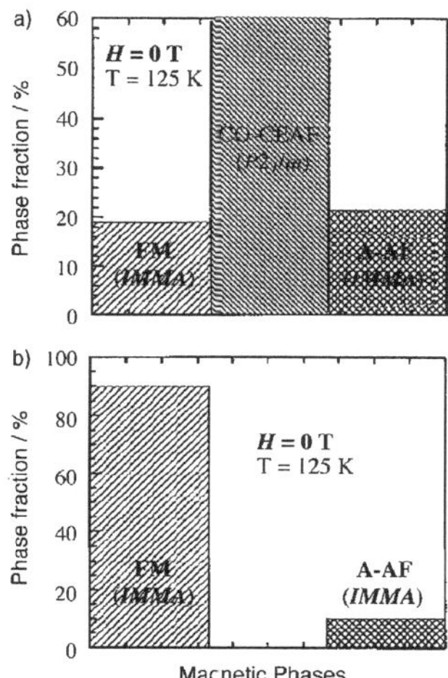

Figure 14. Schematic diagram of the percentage volume fractions of different phases of $Nd_{0.5}Sr_{0.5}MnO_3$ under a) $H = 0$ T and b) 6 T (reproduced with permission from ref. [29]).

The few case studies of the rare-earth manganates described above should suffice to illustrate how phase separation is of common occurrence in this family of oxides. Phase separation occurs in layered manganates,[30] electron-doped manganates,[31] $Ln_{1-x}A_xMnO_3$ $(x > 0.5)$, and in $Ca_{1-x}Bi_x$-MnO_3.[32]

Conclusion

Phase separation in metal oxide systems has emerged to become a phenomenon of importance, because of the diversity of properties found in the rare earth manganates. In Figure 1, we showed a few possible scenarios of phase separation schematically. These scenarios roughly represent the experimental observations in the rare-earth manganates. Phase separation has been observed recently in real space with atomic-scale resolution.[33] It seems likely that phase separation will be found in materials in which the electronic or magnetic properties vary strikingly over extremely narrow composition ranges. For most classes of materials this is not the case, but in highly correlated systems, instability towards phase separation and formation of inhomogeneous states may be an intrinsic property.[34] For example, phase separation is suspected to be responsible for the occurrence of two magnetic transitions in $Sr_3CuIrO_{6+\delta}$.[35] The compositional FM–AFM transition accompanying changes in electron bandwidth found in $La_{1-x}Y_xTiO_3$, coexistence of FM and paramagnetic phases in $La_{1-x}Sr_xCoO_3$, and the compositional AFM–FM transition in $Ca_{2-x}La_xRuO_4$ are all likely to be associated with phase separation.[6] What is important to note is that phase separation occurs over a large length scale from a few angstroms to a few hundred nanometers. The domain sizes of the component phases clearly determine the properties of the material.

Acknowledgement

This work was supported by BRNS (DAE, India) and the MRSEC program of the National Science Foundation under Award No. DMROO-80034.

[1] C. N. R. Rao, J. M. Thomas, *Acc. Chem. Res.* **1985**, *18*, 113.

[2] a) C. N. R. Rao, B. Raveau, *Transition Metal Oxides*, 2nd ed., Wiley-VCH, Weinehim, **1999**; b) C. N. R. Rao, J. Gopalakrishnan, *New Directions in Soild State Chemistry*, 2nd ed., Cambridge University Press, **1997**.

[3] a) *Phase Separation in Cuprate Superconductors* (Eds.: K. A. Müller, G. Benedek), World Scientific, Singapore, **1993**; b) *Phase Separation in Cuprate Superconductors* (Eds.: E. Sigmund, K. A. Müller), Springer, Heidelberg, **1994**; c) *Stripes and Related Phenomena* (Eds.: A. Bianconi, N. L. Saini), Kluwer Academic, Dordrecht, **2000**.

[4] a) J. D. Jorgensen, B. Dabroski, S. Pei, D. G. Hinks, L. Soderholm, B. Morosin, J. E. Schirber, E. L. Venturini, D. S. Ginley, *Phys. Rev. B* **1988**, *38*, 11337; b) B. Dabrowski, J. D. Jorgensen, D. G. Hinks, S. Pei, D. R. Richards, H. B. Vanfleet, D. L. Decker, *Physica C* **1989**, *162–164*, 99.

[5] a) *Colossal Magnetoresistance, Charge-Ordering and Related Properties of Manganese Oxides* (Eds.: C. N. R. Rao, B. Raveau), World Scientific, Singapore, **1999**; b) C. N. R. Rao, A. Arulraj, A. K. Cheetham, B. Raveau, *J. Phys.: Condens. Matter* **2000**, *12*, R83; c) Y.

Tokura, *Colossal Magnetoresistive Oxides*, Gordon and Breach, New York, **2000**; d) A. P. Ramirez, *J. Phys.: Condens. Matter* **1997**, *9*, 8171; e) C. N. R. Rao, A. K. Cheetham, R. Mahesh, *Chem. Mater.* **1996**, *8*, 2421; f) C. N. R. Rao, *Chem. Eur. J.* **1996**, *2*, 1499.

[6] a) E. Dagotto, T. Hotta, A. Moreo, *Phys. Rep.* **2001**, *344*, 1; b) A. Moreo, S. Yunoki, E. Dagotto, *Science* **1999**, *283*, 2034; c) E. L. Nagaev, *Phys.-Usp.* **1996**, *39*, 781.

[7] a) L. M. Rodriguez-Martinez, J. P. Attfield, *Phys. Rev. B* **1996**, *54*, R15622; b) J. P. Attfield, *Chem. Mater.* **1998**, *10*, 3239.

[8] F. C. Chou, D. C. Johnston, *Phys. Rev. B* **1996**, *54*, 572, and references therein.

[9] J. Ryder, P. A. Midgley, R. Exley, R. J. Beynon, D. L. Yalis, L. Afalfiz, J. A. Wilson, *Physica C* **1991**, *173*, 9.

[10] J. Wang, D. Y. Xing, J. Dong, P. H. Hor, *Phys. Rev. B* **2000**, *62*, 9827.

[11] O. Zachar, S. A. Kivelson, V. J. Emery, *Phys. Rev. B* **1998**, *57*, 1422.

[12] a) C. N. A. van Duin, J. Zaanen, *Phys. Rev. Lett.* **1998**, *80*, 1513; b) R. S. Markiewicz, *Phys. Rev B* **2000**, *62*, 1252, and references therein; c) V. V. Moshchalkov, J. Vancken, L. Trappeniers, *Phys. Rev. B* **2001**, *64*, 214504 and references therein.

[13] N. A. Nemov, V. R. Belosludov, *Phys. C* **1998**, *308*, 55.

[14] B. Lorenz, Z. G. Li., T. Honma, P.-H. Hor, *Phys. Rev. B* **2002**, *65*, 144522.

[15] M. Matsuda, M. Fujita, K. Yamada, R. J. Birgeneau, Y. Endoh, G. Shirane, *Phys. Rev. B* **2002**, *65*, 134515.

[16] K. V. Sarathy, P. V. Vanitha, R. Seshadri, A. K. Cheetham, C. N. R. Rao, *Chem. Mater.* **2001**, *13*, 787.

[17] C. N. R. Rao, P. V. Vanitha, *Curr. Opin. Solid State Mater. Sci.* **2002**, *6*, 97.

[18] B. Raveau, M. Hervieu, A. Maignan, C. Martin, *J. Mater. Chem.* **2001**, *11*, 29.

[19] M. Uehara, S. Mori, C. H. Chen, S.-W. Cheong, *Nature* **1999**, *399*, 560.

[20] P. M. Woodward, D. E. Cox, T. Vogt, C. N. R. Rao, A. K. Cheetham, *Chem. Mater.* **1999**, *11*, 3528.

[21] J. M. de Teresa, M. R. Ibarra, P. A. Algarabel, C. Ritter, C. Marquina, J. Blasco, J. Garcia, A. del Moral, Z. Arnold, *Nature* **1997**, *386*, 256.

[22] P. B. Littlewood, *Nature* **1999**, *399*, 529.

[23] D. E. Cox, P. G. Radaelli, M. Marezio, S.-W. Cheong, *Phys. Rev. B* **1998**, *57*, 3305.

[24] P. G. Radaelli, R. M. Ibberson, D. N. Argyriou, H. Casalta, K. H. Andersen, S.-W. Cheong, J. F. Mitchell, *Phys. Rev. B* **2001**, *63*, 172419.

[25] A. M. Balagurov, V. Yu. Pomjakushin, D. V. Sheptyakov, V. L. Aksenov, P. Fischer, L. Keller, O. Yu. Gorbenko, A. R. Kaul, N. A. Babushkina, *Phys. Rev. B* **2001**, *64*, 024420.

[26] a) R. S. Freitas, L. Ghivelder, P. Levy, F. Parisi, *Phys. Rev. B* **2002**, *65*, 104403 and references therein; b) F. Parisi, P. Levy, L. Givelder, G. Polla, D. Vega, *Phys. Rev. B* **2001**, *63*, 144419.

[27] Q. Huang, J. W. Lynn, R. W. Erwin, A. Santoro, D. C. Dender, V. N. Smolyaninova, K. Ghosh, R. L. Greene, *Phys. Rev. B* **2000**, *61*, 8895.

[28] A. Machida, Y. Moritomo, E. Nishibori, M. Takata, M. Sakata, K. Ohoyama, S. Mori, N. Yamamoto, A. Nakamura, *Phys. Rev. B* **2000**, *62*, 3883.

[29] C. Ritter, R. Mahendiran, M. R. Ibarra, L. Morellon, A. Maignan, B. Raveau, C. N. R. Rao, *Phys. Rev. B* **2000**, *61*, R9229.

[30] S. H. Chun, Y. Lyanda-Geller, M. B. Salamon, R. Suryanarayanan, G. Dhalene, A. Revcolevschi, *J. Appl. Phys.* **2001**, *90*, 6307.

[31] a) M. Respaud, J. M. Broto, H. Rakoto, J. Vanacken, P. Wagner, C. Martin, A. Maignan, B. Raveau, *Phys. Rev. B* **2001**, *63*, 144426; b) P. A. Algarabel, J. M. de Teresa, B. Garcia-Landa, L. Morellon, M. R. Ibarra, C. Ritter, R. Mahendiran, A. Maignan, M. Hervieu, C. Martin, B. Raveau, A. Kurbakov, V. Tournov, *Phys. Rev. B* **2002**, *65*, 104437.

[32] P. N. Santhosh, J. Goldberger, P. M. Woodward, T. Vogt, W. P. Lee, A. J. Epstein, *Phys. Rev. B* **2000**, *62*, 14928.

[33] Ch. Renner, G. Aeppli, B.-G. Kim, Y.-A. Soh, S.-W. Cheong, *Nature* **2002**, *416*, 518.

[34] D. Khomskii, *Physica B* **2000**, *280*, 325.

[35] A. Niazi, P. L. Paulose and E. V. Sampathkumaran, *Phys. Rev. Lett.* **2002**, *88*, 107202.

Science and technology of nanomaterials: current status and future prospects

C. N. R. Rao*[a,b] and A. K. Cheetham[a]

[a] Materials Research Laboratory, University of California, Santa Barbara, CA 93106, USA.
E-mail: cnrrao@jncasr.ac.in
[b] Chemistry and Physics of Materials Unit, Jawaharlal Nehru Centre for Advanced Scientific Research, Jakkur, Bangalore, 560 064, India

Received 8th June 2001, Accepted 7th August 2001
First published as an Advance Article on the web 10th October 2001

The science and technology of nanomaterials has created great excitement and expectations in the last few years. By its very nature, the subject is of immense academic interest, having to do with very tiny objects in the nanometer regime. There has already been much progress in the synthesis, assembly and fabrication of nanomaterials, and, equally importantly, in the potential applications of these materials in a wide variety of technologies. The next decade is likely to witness major strides in the preparation, characterization and exploitation of nanoparticles, nanotubes and other nanounits, and their assemblies. In addition, there will be progress in the discovery and commercialization of nanotechnologies and devices. These new technologies are bound to have an impact on the chemical, energy, electronics and space industries. They will also have applications in medicine and health care, drug and gene delivery being important areas. This article examines the important facets of nanomaterials research, highlighting the current trends and future directions. Since synthesis, structure, properties and simulation are important ingredients of nanoscience, materials chemists have a major role to play.

Introduction

Nanotechnology is the term used to describe the creation and exploitation of materials with structural features in between those of atoms and bulk materials, with at least one dimension in the nanometer range ($1 \text{ nm} = 10^{-9}$ m). Table 1 lists typical nanomaterials of different dimensionalities. The properties of materials with nanometric dimensions are significantly different from those of atoms or bulk materials. Suitable control of the properties of nanometer-scale structures can lead to new science as well as new products, devices and technologies. The underlying theme of nanotechnology is miniaturization, the importance of which was pointed out by Feynman[1] as early as 1959 in his often-cited lecture entitled "There is plenty of room at the bottom". The challenge is to beat Moore's law[2] and accommodate 1000 CDs in a wristwatch.[3]

There has been explosive growth of nanoscience and technology in the last decade, primarily because of the availability of new methods of synthesizing nanomaterials, as well as tools for characterization and manipulation (Table 2). Several innovative methods for the synthesis of nanoparticles and nanotubes, and their assemblies, are now available. There is a better understanding of the size-dependent electrical, optical and magnetic properties of individual nanostructures of semiconductors, metals and other materials. Besides the established techniques of electron microscopy, crystallography and spectroscopy, scanning probe microscopies have provided powerful tools for the study of nanostructures. Novel methods of fabricating patterned nanostructures, as well as new device and fabrication concepts, are constantly being discovered. Nanostructures have also been ideal for computer simulation and modelling, their size being sufficiently small to accommodate considerable rigour in treatment. In computations on nanomaterials,[4] one deals with a spatial scaling from 1 Å to

Table 1 Typical nanomaterials

	Size (approx.)	Materials
(a) Nanocrystals and clusters (quantum dots)	diameter 1–10 nm	Metals, semiconductors, magnetic materials
Other nanoparticles	diameter 1–100 nm	Ceramic oxides
(b) Nanowires	diameter 1–100 nm	Metals, semiconductors, oxides, sulfides, nitrides
Nanotubes	diameter 1–100 nm	Carbon, layered metal chalcogenides
(c) 2-Dimensional arrays (of nano particles)	several nm^2–μm^2	Metals, semiconductors, magnetic materials
Surfaces and thin films	thickness 1–1000 nm	Various materials
(d) 3-Dimensional structures (superlattices)	Several nm in all three dimensions	Metals, semiconductors, magnetic materials

Table 2 Different methods of synthesis and investigation of nanomaterials

Scale (approx.)	Synthetic methods	Structural tools	Theory and simulation
0.1–10 nm	Covalent synthesis	Vibrational spectroscopy NMR Diffraction methods Scanning probe microscopies (SPM)	Electronic structure
<1–100 nm	Self-assembly techniques	SEM, TEM, SPM	Molecular dynamics and mechanics
100 nm–1 μm	Processing, modifications	SEM, TEM	Coarse-grained models, hopping *etc.*

DOI: 10.1039/b105058n

1 μm and temporal scaling from 1 fs to 1 s, the limit of accuracy going beyond 1 kcal mol^{-1}. There are many examples which demonstrate the current achievements and paradigm shifts in this area; STM images of quantum dots (*e.g.*, a germanium pyramid on a silicon surface) and of the quantum corral of 48 Fe atoms placed in a circle of 7.3 nm radius being familiar ones. Ordered arrays or superlattices of nanocrystals of metals and semiconductors have been prepared by several workers. Nanostructured polymers formed by the ordered self-assembly of triblock copolymers and nanostructured high-strength materials (*e.g.*, Cu/Cr nanolayers) are other examples. Prototype circuits involving nanoparticles and nanotubes for nanoelectronic devices have been fabricated. Quantum computing[6] has made a good start and appropriate quantum algorithms have been developed.

Not everything in nanoscience is new; many existing technologies employ nanoscale processes, catalysis and photography being well-known examples. Our capability to synthesize, organize and tailor-make materials at the nanoscale is, however, of very recent origin. The immediate goals of the science and technology of nanomaterials must be to fully master the synthesis of isolated nanostructures (building blocks) and their ensembles and assemblies with the desired properties, to explore and establish nanodevice concepts and systems architecture, to generate new classes of high-performance materials, including biologically-inspired systems, to connect nanoscience to molecular electronics and biology, and to improve known investigative methods while discovering better tools for the characterization of nanostructures.[4,5] Some of the potential applications of nanotechnology which will have the greatest societal and economic impact are in the production of novel materials and devices, in nanoelectronics and computer technology, as well as in medicine and health care. These wide-ranging applications may indeed usher in a new direction in manufacturing and in the electronics, space, chemical and energy industries.

In this article, we will present the important highlights in the field of nanomaterials related to their synthesis and assembly, experimental tools for their investigation, modelling and simulation, and their potential applications. We have endeavoured to make the article concise and illustrative rather than encyclopedic, and have cited key, representative references to help describe the present status of the subject.

Synthesis and assembly

The synthesis of nanomaterials spans inorganic, organic and biological systems, all with control of size, shape and structure. Assembling the nanostructures into ordered arrays is often necessary to render them functional and operational. It is by a combination of novel nanobuilding units and strategies for assembling them that nanostructured materials and devices with new capabilities can be generated. In the last decade, nanoparticles (powders) of ceramic materials have been produced on a large scale by employing both physical and chemical methods. In addition, there has also been considerable progress in the preparation of nanocrystals of metals, semiconductors and magnetic materials by employing colloid chemical methods.[7–10] Nearly monodisperse nanocrystals of materials with narrow size distributions have been prepared by several workers, with control of the shape in some instances. To illustrate this aspect, transmission electron microscopy (TEM) images of CdSe nanorods are shown in Fig. 1. The availability of such nanocrystals has enabled better studies of size-dependent properties such as size-controlled light emission properties of semiconductor nanocrystals, the nonmetallic gap in metal nanocrystals, size-dependent Coulomb blockade and the associated scaling laws. Fig. 2 and 3 illustrate some typical size-dependent properties of materials. Size-dependent

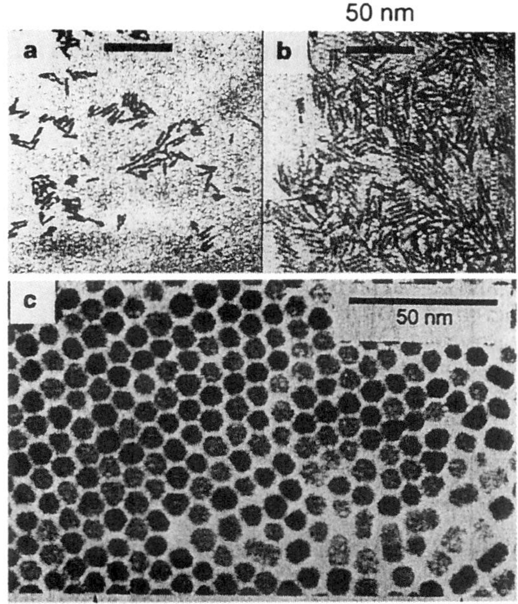

Fig. 1 TEM images of CdSe quantum rods: (a), (b) low-resolution images of quantum rods of different aspect ratios; (c) 3-dimensional orientation. (Reproduced by permission from ref. 10, X. Peng, L. Manna, W. Yang, J. Wickham, E. Scher, A. Kadavanich and A. P. Alivisatos, *Nature*, 2000, **404**, 59. Copyright Nature 2000.)

properties have been exploited for biological tagging, for example using semiconductor nanocrystals as fluorescent biological labels.[11] There is a qualitative change in some facets of synthesis, the preparation of thin films being one such. For example, it is now possible to carry out atomic layer-by-layer deposition of films, thereby affording new heterostructures and metastable phases.[12]

Since the discovery of carbon nanotubes,[13] much has been achieved in the synthesis of multiwalled and single-walled nanotubes and bundles of aligned nanotubes.[14] Fig. 4 shows an electron microscopy image of aligned multiwalled nanotubes. Carbon nanotubes have been doped with nitrogen and boron. Especially noteworthy is the recent synthesis of Y-junction carbon nanotubes (Fig. 5), which can act as vital components in nanoelectronics[15] since, unlike ordinary nanotubes which can be metallic or insulating, kinks or bends will act as metal–insulator junctions. Nanotubes of other inorganic materials, in particular those of layered metal chalcogenides (*e.g.* MoS$_2$,

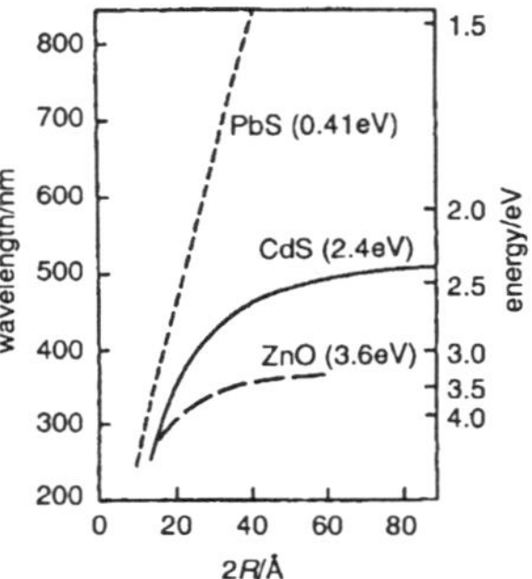

Fig. 2 Size-dependence of the wavelength of the absorption threshold in semiconductors. (Reproduced by permission from A. Henglein, *Ber. Bunsen-Ges. Phys. Chem.*, 1995, **99**, 903. Copyright VCH 1995.)

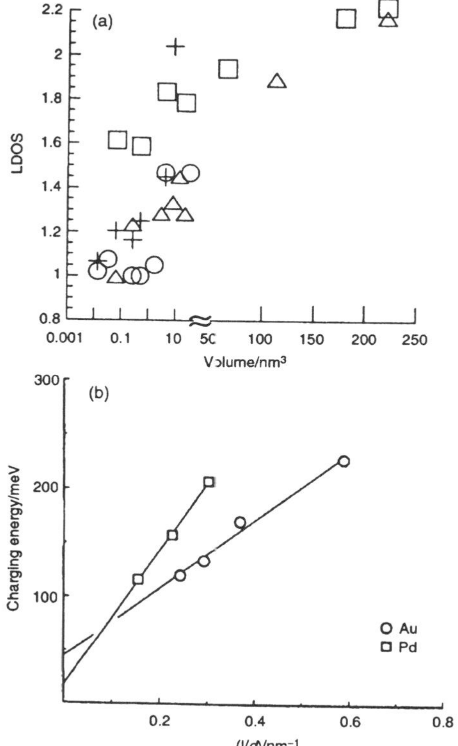

Fig. 3 (a) Size-dependence of the nonmetallic gap (local density of states) with the volume of metal nanocrystals. (Reproduced by permission from C. P. Vinod, G. U. Kulkarni and C. N. R. Rao, *Chem. Phys. Lett.*, 1998, **289**, 329. Copyright Elsevier 1998.) (b) Size-dependence of the charging capacity (Coulomb blockade) in metal nano crystals. (Reproduced by permission from P. J. Thomas, G. U. Kulkarni and C. N. R. Rao, *Chem. Phys. Lett.*, 2000, **321**, 163. Copyright Elsevier 2000.)

WS_2, $MoSe_2$, NbS_2), have been synthesized.[14,16,17] These materials are likely to possess interesting and useful properties. Nanowires of metals, semiconductors, oxides, nitrides, sulfides and other materials have been prepared by various methods.[14,18] In the area of polymers, the engineering of new polymeric structures by using dendrimers and block copolymers deserves special mention[19] (see Fig. 6, for example). Nanosized polymers of different shapes could have many applications.

The construction of ordered arrays of nanostructures by employing organic self-assembly techniques provides alternative strategies for the production of nanodevices. Two- and three-dimensional arrays of nanocrystals of semiconductors, metals and magnetic materials have been assembled by using suitable organic reagents.[20] Strain-directed assembly of nanoparticle arrays (*e.g.*, of semiconductors) provides the means to introduce functionality into the substrate that is coupled to that on the surface.[21] Fig. 7 shows a TEM image of Pd nanocrystals containing 1415 atoms. By varying the spacer length (or crystal size), arrays of metal nanocrystals can be made to undergo a metal–non-metal transition of the Mott–Hubbard type.[8] Monodisperse FePt nanoparticles and ferromagnetic FePt nanocrystal superlattices have been described recently[22] (Fig. 8). Spin-dependent tunnelling has been observed in self-assembled cobalt nanocrystal superlattices.[22] Nanocrystals of uniform size suitably coated with a polymer are found to assemble to form giant nanocrystals with magic nuclearity.[23] It

is noteworthy that self-assembly of objects of varying sizes (nm to mm) occurs through weak interactions.[24] Single crystals of single-walled carbon nanotubes (Fig. 9) have been formed by self-assembly.[25] Lithography-induced self-assembly (LISA) is likely to open up new possibilities in the near future.[26]

The area of nanoporous solids has witnessed many major advances. There has been a constant quest for crystalline solids with giant pores and several have been synthesized recently.[27] It has become possible to control pore size in zeolites and other nanoporous materials, and the shape-selective catalysis afforded by nanoporous solids continues to motivate much of the work in this area. Since the discovery of mesoporous MCM 41 by Mobil chemists, a variety of mesoporous inorganic solids with pore diameters in the 2–20 nm range have been prepared and characterized.[28] Mesoporous fibers and spheres of silica and other materials have also been prepared. A variety of inorganic, organic and organic–inorganic hybrid open-framework materials with different pore architectures have been synthesized in the last few years, particularly noteworthy being the large variety of open-framework metal phosphates.[29] Equally interesting are the macroporous solids (pore diameters 100–1000 nm) prepared by templating crystalline arrays of silica or polymer spheres.[30]

Many new nanostructured systems will undoubtedly be built in the next few years using a variety of nanobuilding units, and there is considerable scope for the discovery of newer and better methods of synthesis, assembly and processing. Robotic self-assembly of nanostructures may be employed more commonly. It is also likely that in the next few years, more effective interfaces with biology will emerge. There is already evidence to show that DNA-directed assembly has the potential for nanofabrication.[31] DNA chips and microarrays have imminent applications in diagnostics and genetic research. Preparation of nanoparticles of therapeutic drugs would directly help in drug targeting. Similarly, gene delivery may become possible through the use of nanoparticles of lipid/polymer–DNA complexes.

Experimental tools

As mentioned earlier, the rapid pace of progress in the area of nanomaterials has been made possible in part because of improved methods of measurement, characterization and manipulation. While standard methods of measurement and characterization are routinely employed for investigation, the increasing use of scanning probe microscopies (spatial resolution 1 nm), combined with high-resolution electron microscopy, has enabled direct images of structures to be obtained and aided the study of properties. Thus, scanning tunnelling spectroscopy provides important information on electronic structure and properties. Scanning probe microscopies are now employed at low temperatures, under vaccum or in a magnetic field. Computer-controlled scanning probe microscopy is useful for nanostructure manipulation in real time, and nanomanipulators are being used with scanning and transmission electron microscopes.[32] Newer versions of nanomanipulators will have to be developed by using technologies such as nanoelectromechanical systems (NEMS).[32]

Techniques such as magnetic force microscopy and surface force microscopy are of great value in specific situations. While magnetic force microscopy directly images magnetic domains, magnetic resonance microscopes can detect nuclear or electron spin resonance with submicron spatial resolution. Near field scanning optical microscopy allows optical access to sub-wavelength scales (50–100 nm) by breaking the diffraction limit.[33]

Nanomechanics performed in the atomic force microscope enables the study of single molecules, and is valuable in understanding folding and related problems in biological

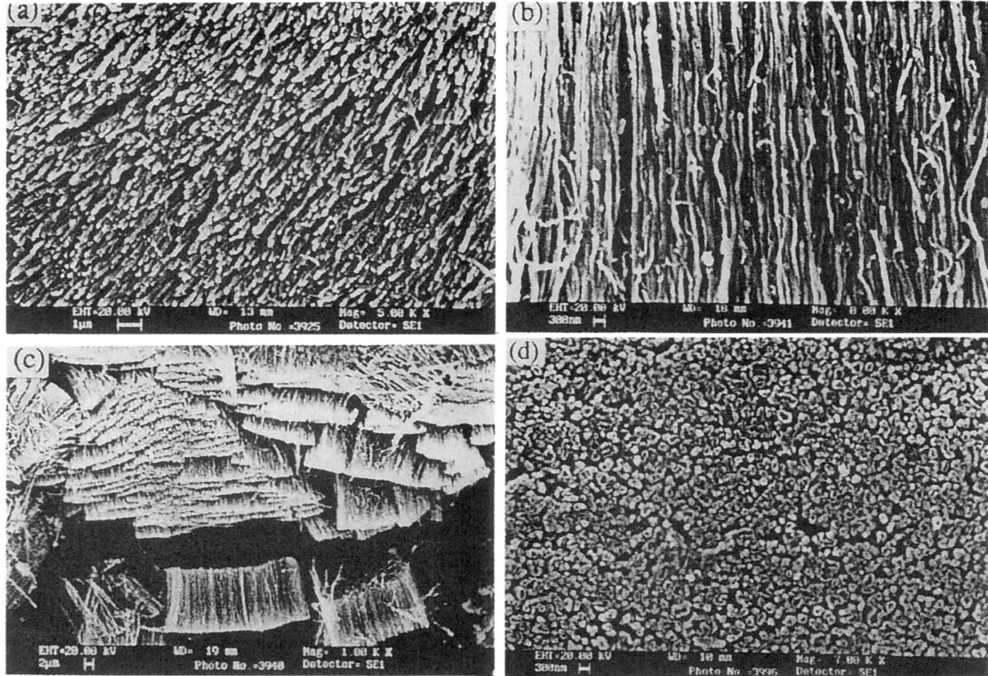

Fig. 4 TEM images of aligned multi-walled carbon nanotubes; (a), (b) and (d) show nanotubes in different orientations, (c) shows several bunbles. (Reproduced by permission from C. N. R. Rao, A. Govindaraj, R. Sen and B. C. Satishkumar, *Mater. Res. Innovations*, 1998, **2**, 128. Copyright Springer 1998.)

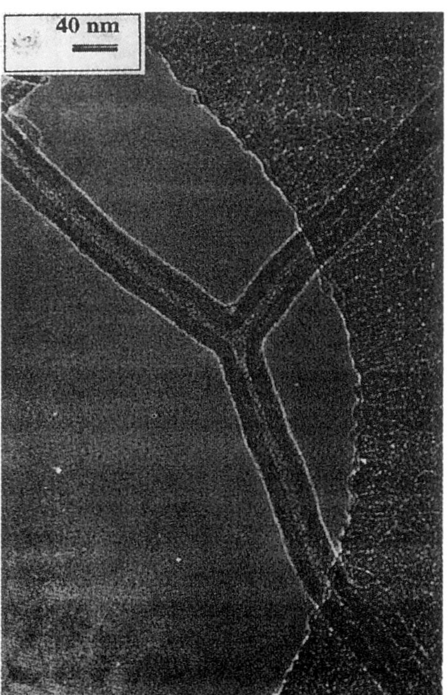

Fig. 5 TEM image of a Y-junction nanotube. (Reproduced by permission from ref. 15, B. C. Satishkumar, P. J. Thomas, A. Govindaraj and C. N. R. Rao, *Appl. Phys. Lett.*, 2000, **77**, 2530. Copyright AIP 2000.)

molecules. In order to circumvent the limitations of scanning probe microscopy, cantilever probes have been developed to enable high-speed nanometer scale imaging.[34] Clearly, there is much scope for the development of miniaturized instruments for the study of individual molecules and nanounits.

Optical tweezers provide an elegant means to investigate the dynamics of particles and molecules.[35] Thus, force measurement of complementary DNA binding provides a selective and sensitive sensor.[35]

Microfabricated chips for DNA analysis and polymerase

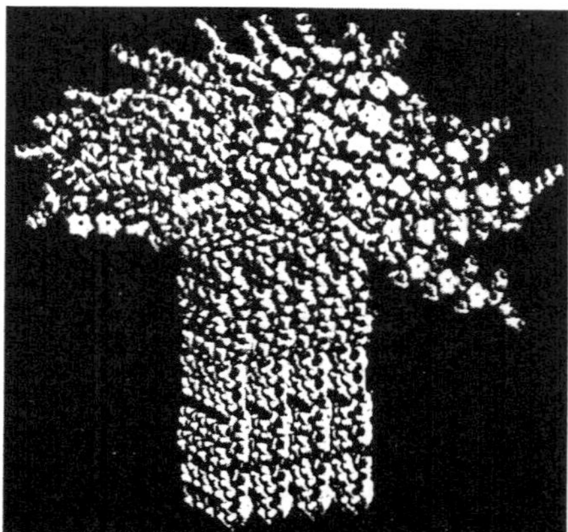

Fig. 6 Supramolecular nanostructure formed by the self-assembly of triblock copolymers. (Reproduced by permission from ref. 19, S. I. Stupp, V. LeBouheur, K. Walker, L. S. Li, K. E. Huggins, M. Keser and A. Amstuz, *Science*, 1997, **276**, 384. Copyright AAAS 1997.)

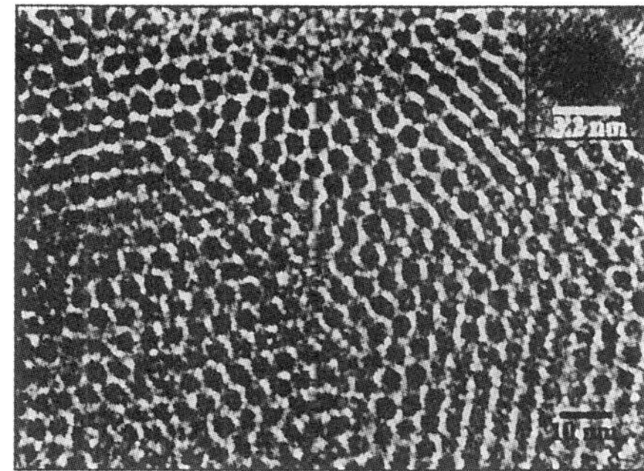

Fig. 7 TEM image of a two-dimensional array of Pd_{1415} nanocrystals (diameter 3.2 nm). (Reproduced from ref. 8, C. N. R. Rao, G. U. Kulkarni, P. J. Thomas and P. P. Edwards, *Chem. Soc. Rev.*, 2000, **29**, 27.)

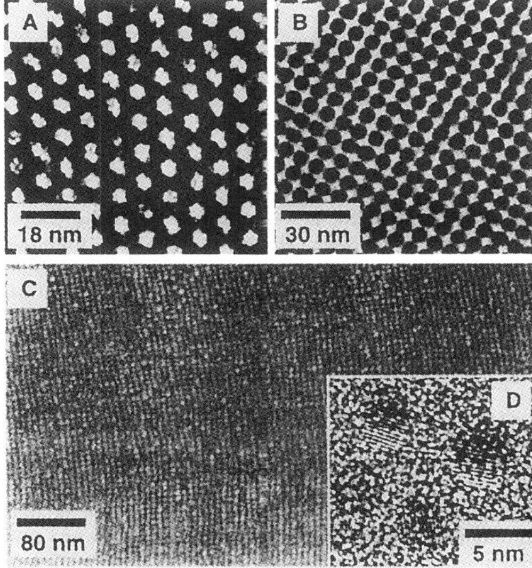

Fig. 8 (A) TEM image of a 3-D assembly of $Fe_{50}Pt_{50}$ nanocrystals (diameter 6 nm). (B) TEM image after replacing the capping oleic acid with hexanoic acid. (C) High-resolution scanning electron microscopy (HRSEM) image of 180 nm thick $Fe_{52}Pt_{48}$ nanocrystals (diameter 4 nm). (D) HRTEM image of $Fe_{52}Pt_{48}$ nanocrystals. (Reproduced by permission from ref. 22a, S. Sun, C. B. Murray, D. Weller, L. Folks and A. Moser, *Science*, 2000, **287**, 1989. Copyright AAAS 2000.)

chain reactions have been developed.[36] It would be of great benefit if appropriate tools for 3-D imaging and microscopy as well as for chemical analysis of materials in nanometric dimensions become available.

Simulation and modelling

A variety of computational techniques have been employed to simulate and model nanomaterials. Since the relaxation times can vary anywhere between picoseconds to hours, it becomes necessary to employ Langevin dynamics besides molecular dynamics in the calculations. Simulation of nanodevices as a whole, through the optimization of the various components and functions, is challenging, as in the simulation of large systems and coupled phenomena. There are many instances where simulation and modelling have yielded impressive

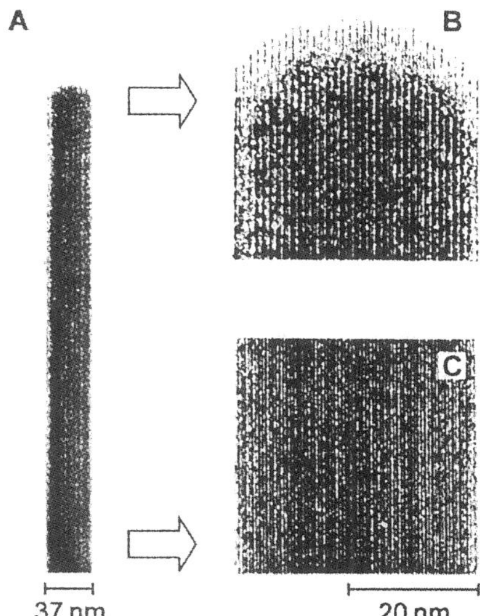

Fig. 9 (A) HRTEM image of a single-walled nanotube crystal (diameter 37 nm). (B) and (C) Lattice resolution of the crystal. (Reproduced by permission from ref. 25, R. R. Schlittler, J. W. Seo, J. K. Gimzewski, C. Durkan, M. S. M. Saifullah and M. E. Welland, *Science*, 2001, **292**, 1136. Copyright AAAS 2001.)

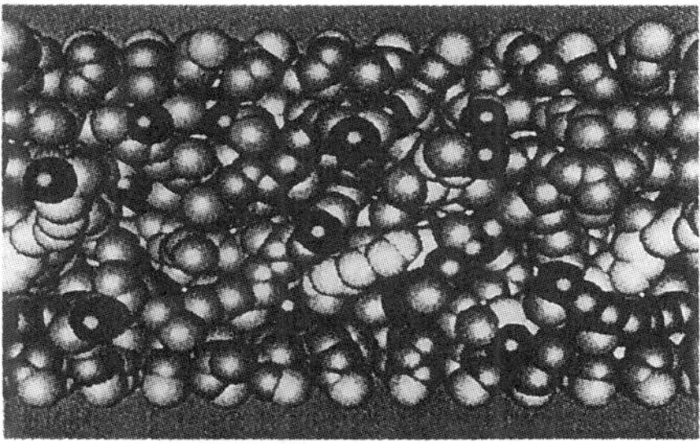

Fig. 10 Snapshot of squalane under shear flow confined between two walls with tethered butane chains. (Reproduced by permission from A. Gupta , H. D. Cochran and P. T. Cummins, *J. Chem. Phys.*, 1997, **107**, 10327. Copyright AIP 1997.)

results, nanoscale lubrication being one such example (Fig. 10). Simulation of the molecular dynamics of DNA has met with some success[37] and quantum dots and nanotubes have been modelled satisfactorily.[38,39] Nevertheless, first principles calculations of nanomaterials can be problematic if the clusters are too large to treat by Hartree–Fock methods and too small for density functional theory.

Applications

By employing sol–gels and aerogels, high-surface area inorganic oxide materials with improved absorption and other properties are being produced. The preparation of superadsorbents for toxic chemicals is becoming possible by these methods.[40] Consolidated nanocomposites and nanostructures will yield ultrahigh-strength, tough structural materials, strong ductile cements and novel magnets. Significant developments will occur in the sintering of nanophase ceramic materials and in textiles and plastics containing dispersed nanoparticles, the latter being useful in the design of miniature lithium-ion batteries; nanostructured electrode materials would improve the capacity and performance of Li-ion batteries. Nanostructured tin oxide also appears to attain higher reversible capacities.[41] Shipway *et al.*[42] have recently reviewed nanoparticle-based applications.

Both known and new types of nano-, meso- and macroporous materials will be put to use for inorganic synthesis and industrial catalysis. Porous films (*e.g.* mesoporous silica) can be used as molecular sieving membranes and in catalysis. The chemical industry may indeed get involved to a greater extent in the design of catalysts containing different types of nanometric particles, since nanoscale catalysis could provide greater selectivity. Nanocatalysis by gold illustrates how nanoparticles can have entirely different catalytic properties. Electron transfer in nanostructures can also be exploited for applications. There is a good possibility that nanomaterials such as carbon nanotubes will be used for storage of hydrogen and other gases.[14,43] It has been reported that carbon nanotubes can store hydrogen up to *ca.* 4 wt% or more and that the stored hydrogen is released easily without fatigue, but some doubts have been raised about these findings.[43] Nanoporous polymers are useful for water purification, while MoS_2 nanotubes and fullerenes are good solid lubricants.

The techniques of nanoimprint-lithography and soft lithography are sufficiently developed[44] and a combination of self-assembly with patterning tools could enable new nanolithographic patterns to be obtained.[45] Examples of nanoimprint lithography and the LISA process are shown in Fig. 11. Thin-film electrets patterned with trapped charge provides another method of patterning which may be useful in

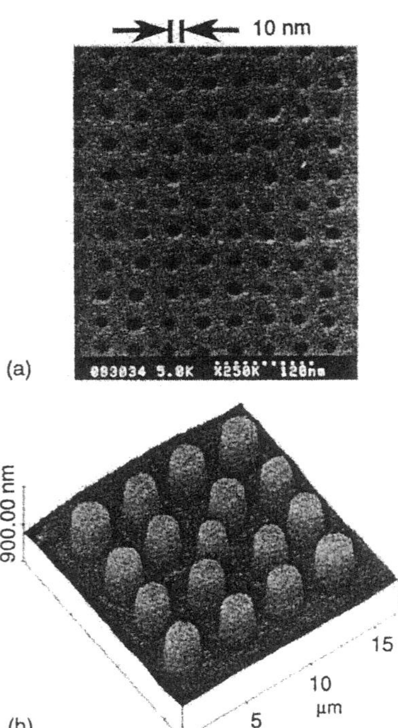

Fig. 11 (a) SEM micrograph of a top view of holes (10 nm diameter, 40 nm period) imprinted into poly(methyl methacrylate) (PMMA) (60 nm deep). (Reproduced by permission from S.-Y. Chou, P. R. Krauss, W. Zhang, L Guo and L. Zhuong, *J. Vac. Sci. Technol., B*, 1997, **15**, 2897. Copyright Slack 1997.) (b) AFM image of a PMMA/LISA pillar array formed under a square pattern. (Reproduced by permission from ref. 26, S.-Y. Chou and L. Zhang, *J. Vac. Sci. Technol., B*, 1999, **17**, 3197. Copyright Slack 1999.)

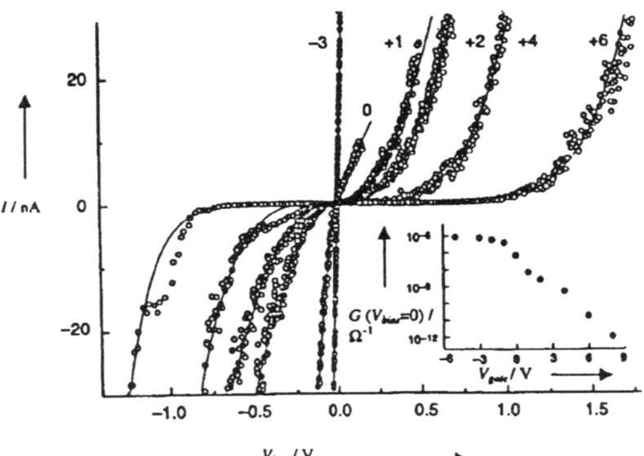

Fig. 12 Current–potential characteristics at different gate voltages of a field-effect transistor based on a single-walled nanotube. (Reproduced by permission from S. J. Tans, A. R. M. Vercsheuren and C. Dekker, *Nature*, 1998, **393**, 49. Copyright Nature 1998.)

high-density charge-based data storage and high-resolution printing.[46]

Carbon nanotubes are being used as tips in scanning microscopes and also as efficient field emitters for possible use in display devices. Nanotubes have been shown to act as field-effect transistors, as illustrated in Fig. 12. Newer properties of nanotubes continue to be discovered. For example, nanotubes have been found to exhibit optical limiting properties.[14] Modulated chemical doping of individual nanotubes yields an intramolecular wire electronic device,[47] and magnetic clusters deposited on single-walled nanotubes exhibit the Kondo effect.[48]

Colloidal gold particles attached to DNA strands can be employed to assay specific complementary DNAs.[49] The technology of DNA microchip arrays, involving lithographic patterning, which has been used effectively for mapping genetic information in DNA and RNA, is bound to see further improvement. Similarly, drug and gene delivery will be rendered more effective with the use of nanoparticles and nanocapsules. New miniaturized devices could become useful not only in drug delivery,[50] but also in diagnostics and other applications. Molecular motors, such as the protein F_1-AT phase, are already known, but it may become practical to power an inorganic nanodevice with such a biological motor.[51] Other areas in biology where nanomaterials can have an impact are in the monitoring of the environment and living systems by the use of nanosensors and in the improvement of prosthetics used to repair or replace parts of the human body. Water-based ferrofluids, prepared by using organic templates, are useful as coloured magnetic inks. Soft materials can also be used for nanofabrication.[52]

The most important and far-reaching applications of nanomaterials will be in nanodevices and nanoelectronics. There are already significant advances in these areas to justify greater expectations. Typical of the advances to date are the demonstration of single electron memory, Coulomb blockade and quantum effects, scanning probe tips in arrays, logic elements and sensors. The number of applications for semiconductor nanostructures, in particular those of the group III–V nitrides (*e.g.* InGaN) as LEDs and laser diodes,[53] are impressive, and quantum dots and wires of these materials will have a great number of uses. In Fig. 13, the essential features of nanodevice fabrication are shown schematically. A noteworthy development is the possible design of defect-free molecular electronics involving chemically fabricated systems, wherein an electronically addressable molecular switch operates.[54] Electron transport studies have been carried out on

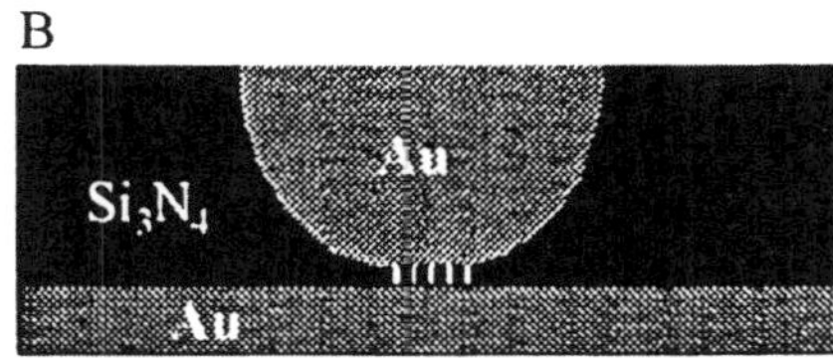

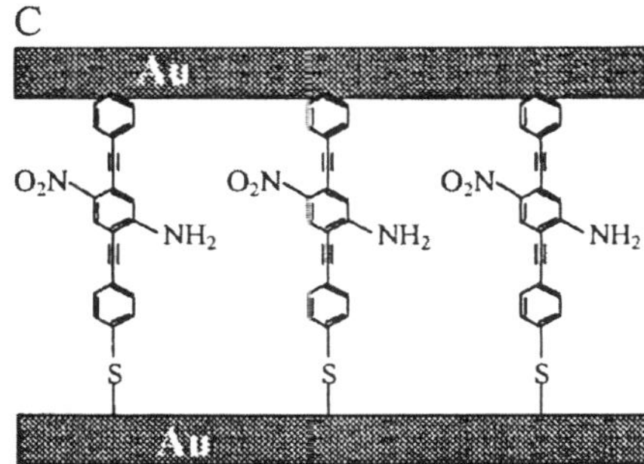

Fig. 13 Schematics of nanodevice fabrication: (A) cross-section of a silicon wafer with a nanopore etched through a silicon nitride membrane; (B) gold–self-assembled monolayer–gold junction in the pore area; (C) enlargement of the active self-assembled monolayer region with the molecules sandwiched in the junction. (Reproduced by permission from J. Chen, M. A. Reed, M. A. Rawlett and J. M. Tour, *Science*, 1999, **286**, 1550. Copyright AAAS 1999.)

molecular-scale systems employing advanced microfabrication and self-assembly techniques.[55]

Nanocircuits and nanocomputers making use of carbon nanotubes have already been described.[14] Recently, metallic and semiconducting properties of multi-walled nanotubes have been engineered by stepwise burning of layers or by tailoring the chirality.[56] Such strategies directly help in the use of nanotubes in nanocircuitry. Use of Y-junction nanotubes may help to make further advances in this direction.

Resonant tunnelling devices in nanoelectronics deserve special mention, since they have already demonstrated success in multivalued logic and memory circuits. Functional devices based on quantum confinement would be of use in photonic switching and optical communications.

It is altogether likely that chips for chemical analysis, nanorobotic systems, spintronic memory based on magnetic semiconductors and a variety of other technologies will become realities in the next few years. Our ability to carry out studies of nanometer-scale objects will be effectively improved by integrating nanometer-scale control electronics into micro/nanomachines.

Concluding remarks

The preceding discussion of the current status and future prospects of the science and technology of nanomaterials should suffice to demonstrate the great vitality of the subject and the immense opportunities it offers. It is a truly interdisciplinary area, encompassing physics, chemistry, biology, materials science and engineering. Interaction amongst scientists with different backgrounds will undoubtedly create new science, and in particular new materials, with unforeseen technological possibilities. While there is already considerable effort being devoted to nanotechnology in various academic and industrial laboratories, there is certainly a need to establish dedicated centres with the required infrastructure and experimental facilities. The subject has caused excitement in the advanced as well as the developing nations, and is one where international cooperation would be fruitful. What is also noteworthy is that nanotechnology will benefit not only the electronics industry, but also the chemical and space industries, as well as medicine and health care. Materials chemists have a major role to play in developing the various facets of nanoscience, such as design, synthesis, assembly and catalysis. They can also make valuable contributions to theory, modelling and computer simulation.

References

1 R. P. Feynman, *Miniaturization*, Reinhold, New York, 1961.
2 One form of Moore's law states that the size of microelectronic devices shrinks by half every four years. This would mean by 2020, the size will be in the nm scale. See: P. L. Packman, *Science*, 1999, **285**, 2079.
3 G. M. Whitesides, *Nanotechnology: Art of the Possible, Technology Review*, Technology Review Inc., Cambridge, MA, Nov/Dec 1998.
4 *Nanostructure Science and Technology*, National Science & Technology Council Report, ed. R. W. Seigel, E. Hu and M. C. Roco, Kluwer Academic Publishers, Boston, 1999; M. C. Roco, R. S. Williams and A. P. Alivisatos, *Nanotechnology Research Directions*, National Science & Technology Council Report, Kluwer Academic Publishers, Boston, 2000.
5 "Issues in Nanotechnology", *Science*, 2000, **290**, 1523–1555.
6 N. Gershenfeld and I. L. Chuang, *Science*, 1997, **275**, 350; A. O. Orlav, *Science*, 1997, **277**, 928.
7 L. Brus, *Curr. Opin. Colloid Interface Sci.*, 1996, **1**, 197.
8 C. N. R. Rao, G. U. Kulkarni, P. J. Thomas and P. P. Edwards, *Chem. Soc. Rev.*, 2000, **29**, 27.
9 C. R. Martin, *Science*, 1994, **266**, 2961.
10 X. Peng, L. Manna, W. Yang, J. Wickham, E. Scher, A. Kadavanich and A. P. Alivisatos, *Nature*, 2000, **404**, 59.
11 M. M. Bruchez, P. Moronne, P. Gin, S. Weiss and A. P. Alivisatos, *Science*, 1998, **281**, 2013; W. C. W. Chan and S. M. Nie, *Science*, 1998, **281**, 2016.
12 A. Gupta, *Curr. Opin. Solid State Sci.*, 1997, **2**, 23.
13 S. Iijima, *Nature*, 1993, **363**, 603.
14 C. N. R. Rao, B. C. Satishkumar, A. Govindaraj and M. Nath, *Chem. Phys. Chem.*, 2001, **2**, 78.
15 B. C. Satishkumar, P. J. Thomas, A. Govindaraj and C. N. R. Rao, *Appl. Phys. Lett.*, 2000, **77**, 2530.
16 Y. Feldman, E. Wasserman, D. J. Srolovitch and R. Tenne, *Science*, 1995, **267**, 222.
17 M. Nath, A. Govindaraj and C. N. R. Rao, *Adv. Mater.*, 2001, **13**, 283; M. Nath and C. N. R. Rao, *J. Am. Chem. Soc.*, 2001, **123**, 4841.
18 X. Duan and C. M. Lieber, *J. Am. Chem. Soc.*, 2000, **122**, 188; Y. Cui and C. M. Lieber, *Science*, 2001, **291**, 851; C. N. R. Rao, A. Govindaraj, F. L. Deepak, N. A. Gunari and M. Nath, *Appl. Phys. Lett.*, 2001, **78**, 1853; M. P. Zach, K. H. Ng and R. M. Penner, *Science*, 2000, **280**, 2120.
19 S. I. Stupp, V. LeBouheur, K. Walker, L. S. Li, K. E. Huggins, M. Keser and A. Amstuz, *Science*, 1997, **276**, 384; O. A. Mathews, A. N. Shipway and J. F. Stoddart, *Prog. Polym. Sci.*, 1998, **23**, 56.
20 C. P. Collier, T. Vossmeyer and J. R. Health, *Annu. Rev. Phys. Chem.*, 1998, **49**, 371.
21 R. A. Kiehl, M. Yamaguchi, O. Ueda, N. Horiguchi and N. Yokoyama, *Appl. Phys. Lett.*, 1996, **68**, 478.
22 (*a*) S. Sun, C. B. Murray, D. Weller, L. Folks and A. Moser, *Science*, 2000, **287**, 1989; (*b*) C. T. Black, C. B. Murray, R. L. Sandstrom and S. Sun, *Science*, 2000, **290**, 1131.
23 P. J. Thomas, G. U. Kulkarni and C. N. R. Rao, *J. Phys. Chem.*, 2001, **105**, 2515.
24 A. Terfort, N. Bowden and G. M. Whitesides, *Nature*, 1997, **386**, 162; J. S. Choi, N. Bowden and G. M. Whitesides, *J. Am. Chem. Soc.*, 1999, **121**, 1754.
25 R. R. Schlittler, J. W. Seo, J. K. Gimzewski, C. Durkan, M. S. M. Saifullah and M. E. Welland, *Science*, 2001, **292**, 1136.
26 S. Y.-Chou and L. Zhang, *J. Vac. Sci. Technol. B*, 1999, **17**, 3197.

27 G. Ferey and A. K. Cheetham, *Science*, 1999, **283**, 1125; G. Ferey, *Science*, 2001, **291**, 994.
28 D. M. Antonelli and J. Y. Ying, *Curr. Opin. Colloid Interface Sci.*, 1996, **1**, 523.
29 A. K. Cheetham, G. Ferey and T. Loiseau, *Angew. Chem., Int. Ed.*, 1999, **38**, 3268; H. Li, A. Laine, M. O' Keepa and O. M. Yaghi, *Science*, 1999, **283**, 1145; S. S. Y. Chui, S. M. F. Lo, J. P. H. Charmant, A. G. Orpen and I. D. Williams, *Science*, 1999, **283**, 1148.
30 B. T. Holland, C. F. Blanford and A. Stein, *Science*, 1998, **281**, 538; G. Gundiah and C. N. R. Rao, *Solid State Sci.*, 2000, **2**, 877.
31 R. C. Mucic, J. J. Storhoff, C. A. Mirkin and R. L. Letsinger, *J. Am. Chem. Soc.*, 1998, **120**, 12674; E. Braun, Y. Eichen, U. Sivan and G. Ben Yoseph, *Nature*, 1998, **391**, 775; A. P. Alivisatos, K. P. Johnson, X. G. Peng, T. E. Wilson, C. J. Loweth, M. P. Bruchez and P. G. Schultz, *Nature*, 1996, **382**, 609.
32 M. F. Yu, M. J. Dyer, G. D. Skidmore, H. W. Rohrs, X. K. Lu and K. D. Ausman, *Nanotechnology*, 1999, **10**, 244; H. G. Craighead, *Science*, 2000, **290**, 1532.
33 E. Betzig, J. K. Trautman, T. D. Harris, J. S. Weiner and R. S. Kostelak, *Science*, 1991, **251**, 1468; D. W. van der Weide, *Appl. Phys. Lett.*, 1997, **70**, 667.
34 S. C. Minne, G. Yaralioglu, S. R. Manalis, J. D. Adams, J. Zesch, A. Atalar and C. F. Quate, *Appl. Phys. Lett.*, 1998, **72**, 2340.
35 A. D. Mehta, M. Rief, D. A. Spudnich and R. M. Simmons, *Science*, 1999, **283**, 1689; D. R. Baselt, G. U. Lee and R. J. Colton, *J. Vac. Sci. Technol. B*, 1996, **14**, 789.
36 K. M. Kurian, C. J. Watson and A. H. Wyllie, *J. Pathology*, 1999, **187**, 267; M. U. Kopp, A. J. de Mello and A. Manz, *Science*, 1998, **280**, 1046.
37 H. Jian, T. Schlick and A. Vologodskii, *J. Mol. Biol.*, 1998, **284**, 287.
38 M. Menon and P. Srivastava, *Phys. Rev. Lett.*, 1997, **79**, 4453.
39 S. Ogut, J. R. Chelikowsky and S. G. Louie, *Phys. Rev. Lett.*, 1997, **79**, 1770.
40 O. Kofer, I. Lagadic, A. Volodin and K. J. Klabunde, *Chem. Mater.*, 1997, **9**, 2468.
41 J. M. McGraw, C. S. Bahn, P. A. Parilla, J. D. Perkins, D. W. Readey and D. S. Ginley, *Electrochim. Acta*, 1999, **45**, 187; A. H. Whitehead, J. M. Eliott and J. R. Owen, *J. Power Sources*, 1999, **81–82**, 33; T. Brousse, R. Retoux, U. Herterich and D. M. Schleich, *J. Electrochem. Soc.*, 1998, **145**, 1.
42 A. N. Shipway, E. Katz and I. Willner, *Chem. Phys. Chem.*, 2001, **1**, 18.
43 M. S. Dresselhaus and P. C. Eklund, *MRS Bull.*, 1999, **24**, 45; C. Zandonella, *Nature*, 2001, **410**, 734.
44 S. Y. Chou, P. R. Krauss and P. J. Renstrom, *Science*, 1996, **272**, 85.
45 Y. Xia, J. A. Rogers, K. E. Paul and G. M. Whitesides, *Chem. Rev.*, 1999, **99**, 1823.
46 H. O. Jacobs and G. M. Whitesides, *Science*, 2001, **291**, 1763.
47 C. Zhou, J. Kong, E. Yenilmez and H. Dai, *Science*, 2000, **290**, 1552.
48 T. W. Odom, J. L. Huang, C. L. Cheung and C. M. Lieber, *Science*, 2000, **290**, 1549.
49 C. A. Mirkin, R. L. Letsinger, R. C. Mucic and J. L. Storhoff, *Nature*, 1996, **382**, 607.
50 J. T. Santini Jr., M. J. Cima and R. Langer, *Nature*, 1999, **397**, 335.
51 H. Noji, *Science*, 1998, **282**, 1844; R. K. Soong, G. D. Bachand, H. P. Never, A. G. Olkhovets, H. G. Craighead and C. D. Montemagno, *Science*, 2000, **290**, 1555.
52 S. R. Quake and A. Scherer, *Science*, 2000, **290**, 1536.
53 S. Nakamura and G. Fasol, *The Blue Laser Diode*, Springer Verlag, Heidelberg, 1997.
54 J. R. Heath, P. J. Kuekes, G. S. Snider and R. S. Williams, *Science*, 1998, **280**, 1716; C. P. Collier, E. W. Wong, M. Belohradsky, F. M. Raymo, J. F. Stoddart, P. J. Kuekes, R. S. Williams and J. R. Heath, *Science*, 1999, **285**, 391.
55 *Molecular Electronics: Science and Technology*, ed. A. Aviram and M. Ratner, New York Academy of Sciences, New York, 1998; M. A. Reed, A. M. Rawlett and J. M. Tour, *Science*, 1999, **286**, 1550.
56 P. Avouris and P. G. Collins, *Science*, 2001, **292**, 706; M. Ouyang, J. L. Huang, C. L. Cheung and C. M. Lieber, *Science*, 2001, **292**, 702.

II. Transition Metal Oxides
(including Cuprate Superconductors)

C.N.R. Rao

CSIR Centre of Excellence in Chemistry,
Chemistry & Physics of Materials Unit
Jawaharlal Nehru Centre for Advanced Scientific Research
Jakkur P.O., Bangalore-560 064, INDIA
cnrrao@jncasr.ac.in

Transition metal oxides constitute probably one of the most interesting classes of solids, exhibiting a variety of structures and properties.[1,2] The nature of metal-oxygen bonding can vary between nearly ionic to highly covalent or metallic. The unusual properties of transition metal oxides are clearly due to the unique nature of the outer d-electrons. The phenomenal range of electronic and magnetic properties exhibited by transition metal oxides is especially noteworthy. Thus, we find oxides with metallic properties (e.g., RuO_2, ReO_3, $LaNiO_3$) at one end and insulating oxides (e.g., $BaTiO_3$) at the other. There are also oxides that traverse both these regimes with change of temperature, pressure, or composition (e.g., V_2O_3, $La_{1-x}Sr_xCoO_3$). Interesting electronic properties also arise from charge density waves (e.g., $K_{0.3}MoO_3$), charge ordering (e.g., Fe_3O_4), and defect ordering (e.g., $Ca_2Fe_2O_5$). Oxides with diverse magnetic properties anywhere from ferromagnetism (e.g., CrO_2, $La_{0.5}Sr_{0.5}MnO_3$) to antiferromagnetism (e.g., NiO, $LaCrO_3$) are known. Many oxides possess switchable orientation states, as in ferroelectric (e.g., $BaTiO_3$, $KNbO_3$) and ferroelastic (e.g., $Gd_2(MoO_4)_3$) materials. No discovery in solid state science has created as much sensation, however, as that of high-temperature superconductivity in cuprates. Although superconductivity in transition metal oxides has been known for some time, the highest T_c reached was around 13K; we now have oxides with T_c's approaching 160K. The discovery of high T_c oxides has focused worldwide attention on the chemistry of metal oxides and at the same time revealed how inadequate our understanding is of these fascinating materials. This section has a few representative articles on metal oxides, covering not only synthesis, structure and properties, but also high-temperature superconductivity.[3,4] Specially interesting is the article on biferroic $BiMnO_3$ which is ferromagnetic due to orbital ordering.

References

1. C.N.R. Rao, *Ann. Rev. Phys. Chem.* **40**, 291 (1989).

2. C.N.R. Rao and B. Raveau, *Transition Metal Oxides*, Wiley-VCH, Second Edition, 1999.

3. T.V. Ramakrishnan and C.N.R. Rao, *Superconductivity Today*, Second Edition, University Press, Hyderabad, India, 1999.

4. C.N.R. Rao (Ed.), *Chemistry of High Temperature Superconductors*, World Scientific, Singapore, 1991.

10 Chemical Strategies for the Synthesis of Metal Oxides

C.N.R. RAO

Solid State and Structural Chemistry Unit and CSIR Centre of Excellence in Chemistry, Indian Institute of Science, Bangalore 560012, India

1 Introduction

Tailor-making materials of the desired structure and properties is one of the main goals of materials chemistry, but it is not always possible to do so by conventional methods. A variety of chemical strategies have been employed for the synthesis of solid materials in recent years by making use of subtle relationships between structure and reactivity. Whilst rational synthesis has provided a variety of materials such as SIALON, NA-SICON and a large number of microporous materials, a rational approach to materials synthesis almost always yields thermodynamically stable materials, but may miss new and novel metastable ones. In this chapter, we shall be mainly concerned with chemical aproaches to the synthesis of oxide materials (Rao 1994) and shall discuss the various methods with examples and specially examine soft chemistry routes to illustrate the control of structure of solids through chemical means. We shall first briefly examine the traditional ceramic method.

2 Ceramic method

The most common method of preparing inorganic solid materials is by the reaction of the component materials in the solid state at high temperatures (Rao and Gopalakrishnan 1986). In some instances, as in the preparation of chalcogenides or of materials where the products or reactants are volatile, the reaction is carried out in sealed tubes. In preparing oxides one generally palletizes the reacting materials and repeats the grinding, palletizing and heating operations several times. Yet, the completion of the reaction or the phasic purity of the product is not assured. The ceramic method suffers from several disadvantages. Various modifications of the ceramic technique have been employed to overcome some of the limitations. An important effort has been to decrease diffusion path lengths. In a polycrystalline mixture of reactants, individual particles are approximately $10\,\mu m$ in size, representing diffusion distances of roughly 10 000 unit cells. By using freeze–drying, spray–drying, coprecipitation, sol-gel and other techniques, it is possible to reduce the particle size to a few hundred ångströms and thus effect a more intimate mixing of the reactants. In coprecipitation, the required metal cations, taken as soluble salts (for example, nitrates), are coprecipitated from a common medium, usually as hydroxides, carbonates, oxalates or citrates. In actual practice, one takes oxides or carbonates of the relevant metals, digests them with an acid and then the precipitating reagents are added to the solution. The precipitate obtained after drying is heated to the required temperature in a desired atmosphere to produce the final product. The decomposition temperatures of the precipitates are generally lower than the temperatures employed in the ceramic method.

3 Soft chemistry routes

Soft chemistry routes essentially employ simple reactions that can be carried out under mild conditions at relatively low temperatures. Generally, at least one of the steps of synthesis involves a reaction in solution. A few illustrative examples would help to understand the spirit of soft chemistry. Tournoux and co-workers (Marchand *et al.* 1980) prepared a new form (B) of TiO_2 by the dehydration of $H_2Ti_4O_9.xH_2O$, obtained from $K_2Ti_4O_9$ by exchange of K^+ by H^+. We illustrate the kind of transformations involved in Fig. 10.1. Raveau and co-workers (Rebbah *et al.* 1979) prepared $Ti_2Nb_2O_9$ by the dehydration of $HTiNbO_5$, which in turn was prepared by cation exchange with $ATiNbO_5$ (A = K, Rb). $Ni(OH)_2.xH_2O$, with a large intersheet distance of 0.78 nm, has been prepared by the hydrolysis of $NaNiO_2$, followed by the subsequent reduction of NiOOH. Similar transformations of Fe- and Co-doped $Ni(OH)_2.xH_2O$ have been examined (Delmas and Borthomieu 1993). In Fig. 10.2, we show the nature of this kind of soft chemical synthesis. Intercalation and deintercalation reactions are soft chemical routes for the synthesis of many solids. Thus, deintercalation of $LiVS_2$ gives VS_2 which cannot be prepared otherwise; deintercalation of $LiVO_2$ similarly gives metastable VO_2. Recently, we have used deintercalation of amine intercalates of WO_3 to prepare WO_3 in perfect cubic form (ReO_3 type) and also in other metastable forms. In Fig. 10.3 we show the X-ray diffraction patterns of stable WO_3, $WO_3.0.5NEt_3$ and WO_3 (cubic). Heating WO_3 (cubic) at 770 K transforms it to the stable monoclinic form. Acid leaching the NEt_3 intercalate gives another metastable form of WO_3 as shown in Fig 10.4.

Intercalation reactions involve the insertion of a guest species into a solid host lattice without any major rearrangement of the solid structure (Rao and Gopalakrishnan 1986; Jacobson 1992). Many intercalation reactions are known, with a variety of host

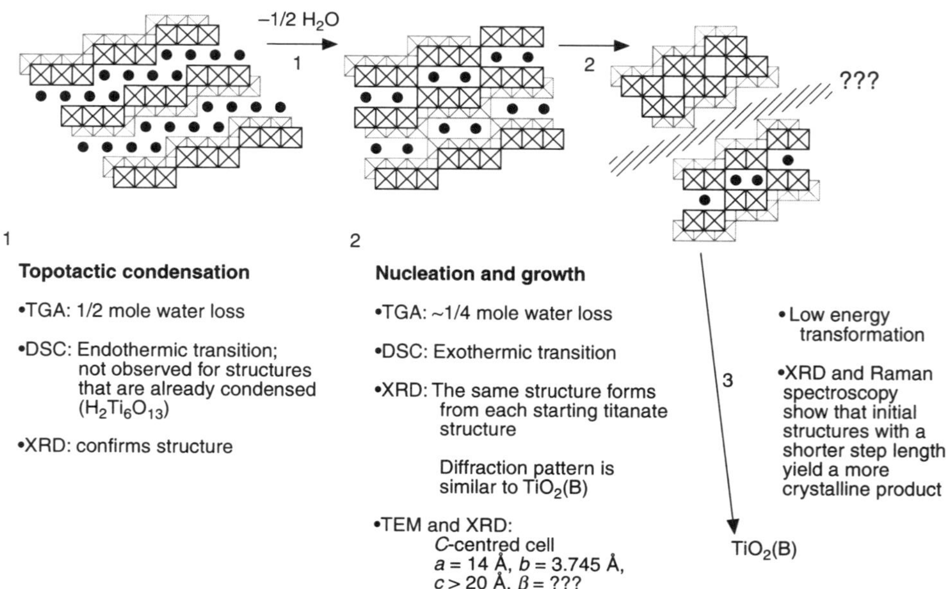

Figure 10.1 Formation of metastable $TiO_2(B)$ from $K_2Ti_4O_9$. (From Feist and Davis 1992.)

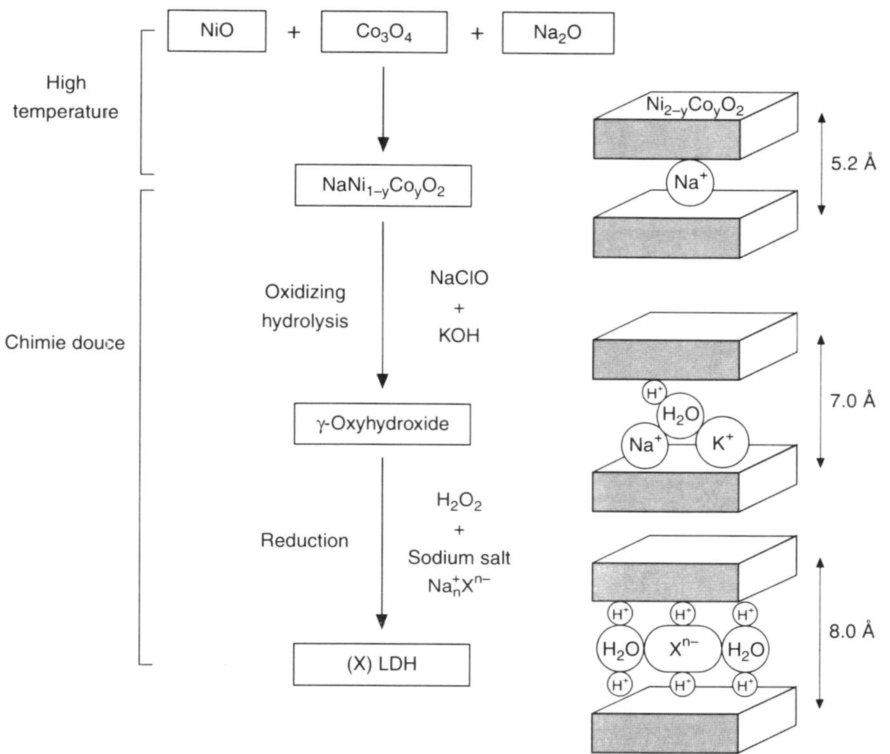

Figure 10.2 Preparation of layered double hydroxides (LDH). (From Delmas and Borthomieu 1993.)

materials which include graphite, oxides and chalcogenides. Thus, Li^+ has been intercalated in a large number of hosts (for example, VS_2, TiS_2, VO_2, Fe_2O_3, Fe_3O_4, MnO_2, etc.). W bronzes may be considered to be intercalation compounds. Chevrel phases are also intercalation compounds since Cu and such cations can be leached out to produce the chalcogenides. Novel materials can be obtained by restacking exfoliated single-layer MoS_2 and other layered materials with molecular units between the layers (Fig. 10.5). A variety of molecules have been incorporated between such single layers of MoS_2, MoO_3 and other layered materials (Divigalpitiya *et al.* 1989). Many of the topochemical reactions are gentle, although they may occur entirely in the solid state.

4   Precursor methods

The best way of reducing diffusion distances in solid state synthesis, in particular by ceramic procedures is to have precursors wherein the relevant metal ions are a short distance apart (Fig. 10.6). Precursor compounds to prepare a variety of perovskite oxides have been known for some time (for example, $Ba[TiO(C_2O_4)_2]$ for $BaTiO_3$, $Li[Cr(C_2O_4)_2.(H_2O)_2]$ for $LiCrO_2$, MCr_2O_4 from $(NH_4)_2M(CrO_4)_2.6H_2O$ and $LaCoO_3$ from $LaCo(CN)_6.5H_2O$). Hydrazinate precursors have been employed to prepare several complex oxides.

Especially noteworthy are the carbonate solid solutions of calcite structure which on

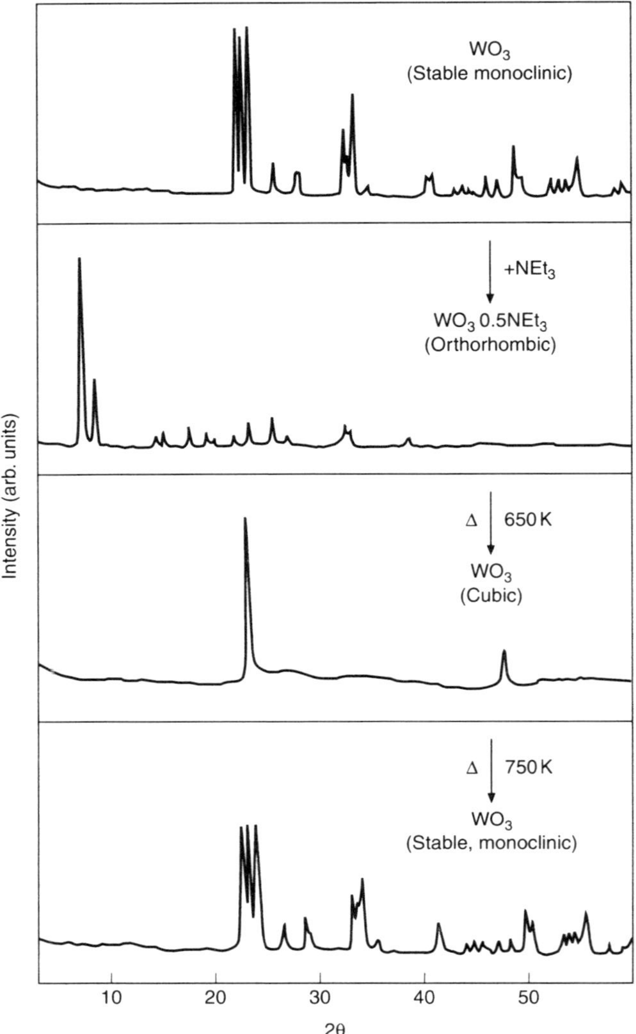

Figure 10.3 X-ray diffraction patterns of stable (monoclinic) WO_3, $WO_3.NEt_3$ and cubic WO_3 (ReO$_3$ type), obtained by thermal deintercalation of $WO_3.0.5NEt_3$ and WO_3 (monoclinic), obtained by heating the metastable cubic WO_3 phase. (From Ayyappan *et al.* 1995.)

decomposition give the oxide with the desired cation ratios (Rao *et al.* 1986). A variety of complex oxides have been prepared by this method and many of these cannot be prepared otherwise. A large number of carbonate solid solutions of calcite structure containing two or more cations in different proportions and these solid solutions are excellent precursors for the synthesis of oxides since the diffusion distances are considerably lower than in the ceramic procedure. The rhombohedral unit cell parameter, a_R, of the carbonate solid solutions varies systematically with the weighted mean cation radius. (Fig.10.7.) Carbonate solid solutions are ideal precursors for the synthesis of monoxide solid solutions of rocksalt structure. For example, the carbonates are decomposed in vacuum or in flowing dry N_2, to obtain monoxides of the type $Mn_{1-x}M_xO$ (M = Mg, Ca, Co or Cd), of rocksalt structure. Oxide solid solutions for

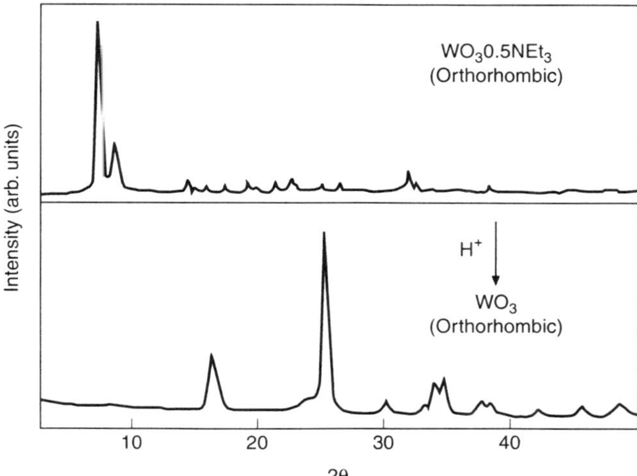

Figure 10.4 X-ray diffraction patterns showing the formation of a metastable orthorhombic phase of WO_3 by the acid leaching of the intercalated amine in $WO_3.0.5NEt_3$ at room temperature. (From Ayyappan *et al.* 1995.)

$M = Mg$, Ca and Co would require 770–970 K for their formation, whilst those containing Cd are formed at still lower temperatures. The facile formation of oxides of rocksalt structure by the decomposition of carbonates of calcite structure is due to the close (topotactic) relationship between the structures of calcite and rocksalt. The monoxide solid solutions can be used as precursors for preparing spinels and other complex oxides.

A number of ternary and quarternary oxides of novel structures have been prepared by decomposing carbonate precursors containing the different cations in the required proportions. Thus, one can prepare $Ca_2Fe_2O_5$ and $CaFe_2O_4$ by heating the corresponding carbonate solid solutions in air at 1070 (797) and 1270 K respectively, for about 1 hour. $Ca_2Fe_2O_5$ is a defect perovskite with ordered oxide ion vacancies and has the well-known brownmillerite structure with the Fe^{3+} ions in alternate octahedral (O) and tetrahedral (T) sites. Co oxides of similar compositions, $Ca_2Co_2O_5$ and $Ca_2Co_2O_4$, have been prepared by decomposing the appropriate carbonate precursors at around 940 K. Unlike in $Ca_2Fe_2O_5$, anion-vacancy ordering in $Ca_2Mn_2O_5$ gives rise to a square-pyramidal (SP) coordination around the transition metal ion. One can also synthesize quarternary oxides, Ca_2FeCoO_5, $Ca_2Fe_{1.5}Mn_{1.5}O_{8.25}$, $Ca_3Fe_2MnO_8$, etc., belonging to the $A_nB_nO_{3n-1}$ family, by the carbonate precursor route (Fig. 10.8). In the Ca–Fe–O system, there are several other oxides such as $CaFe_4O_7$, $CaFe_{12}O_{19}$ and $CaFe_2O_4(FeO)_n$ ($n = 1,2$ or 3) which can, in principle, be synthesized starting from the appropriate carbonate solid solutions and decomposing them in a proper atmosphere.

Hydroxide, nitrate and cyanide solid solutions have also been employed as precursors for oxides. Recently, a precursor has been found to prepare Chevrel phases (Nanjundaswamy *et al.* 1987) of the type $A_xMo_6S_8$ by the reaction:

$$2A_x(NH_4)_yMo_3S_9 + 10H_2 \Rightarrow A_{2x}Mo_6S_8 + 10H_2S + 2yNH_3 + yH_2 \tag{10.1}$$

Precursors for the synthesis of $YBa_2Cu_3O_7$ have been described (Rao *et al.* 1993a).

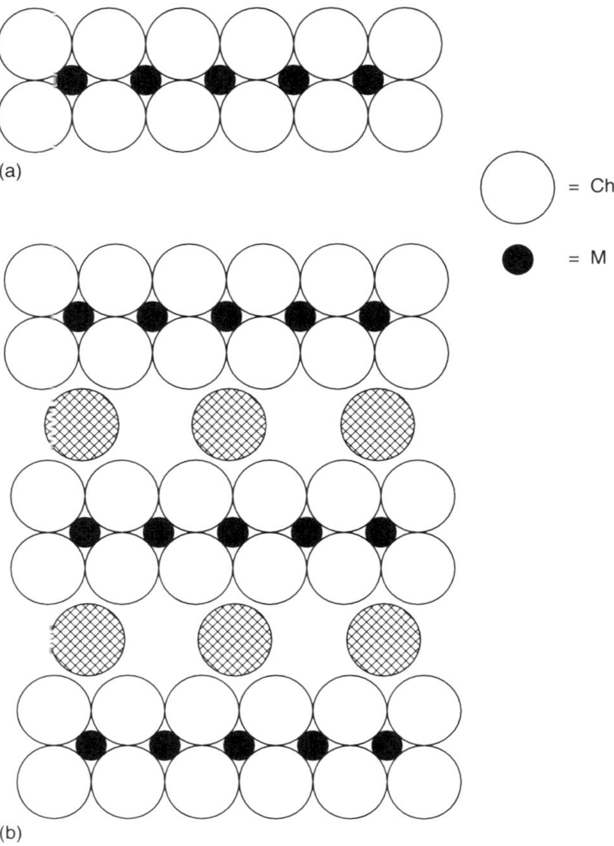

Figure 10.5 Single layers of $MoCl_2$ such as layered materials obtained by exfoliation (a) are employed to obtain new intercalated materials (b). (From Divigalpitiya *et al.* 1989.)

5 Topochemical reactions

Intercalation reactions are generally topochemical in nature. In topochemical reactions the reactivity is controlled by the crystal structure and there is an orientational relationship between the product and the parent just as in Martensitic transformations (Rao and Gopalakrishnan 1986). Dehydration of β-$Ni(OH)_2$ to NiO, as well as the oxidation of $Ni(OH)_2$ to NiOOH, are both topochemical (Figlarz *et al.* 1996). γ-FeOOH topochemically transforms to γ-Fe_2O_3 on treatment with an organic base. Dehydration of many hydrates such as $VOPP_4.2H_2O$ and $MoO_3.H_2O$ is topochemical. The topochemical nature of dehydration has been exploited to prepare MoO_3 in the metal stable ReC_3-type structure (Rao *et al.* 1986). $WO_3.1/3H_2O$ seems to yield WO_3 in different structures depending on the temperature of dehydration (Figlarz *et al.* 1996), as shown in Fig. 10.9. Topochemical reduction and oxidation reactions are known. Thus, the reduction of $YBa_2Cu_3O_7$ to $YBa_2Cu_3O_6$ is topochemical. $La_2Ni_2O_5$ can only be made by the topochemical reduction of $LaNiO_3$.$La_2Co_2O_5$ can similarly be prepared only from $LaCoO_3$ (Vidyasagar *et al.* 1985); in Fig. 10.10 we show the nature of structural transformation involved in the reduction process.

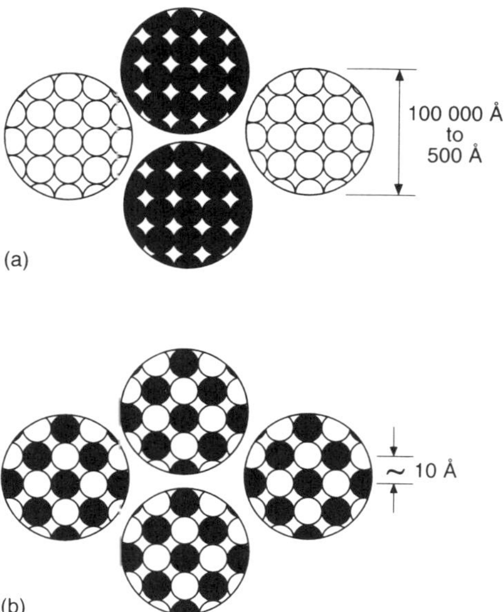

Figure 10.6 Large diffusion distances involved in ceramic preparations involving two types of cations (closed circles and open circles represent two cations) compared to a precursor (the two cations are present in proximity).

6 Ion exchange and alkali flux methods

Ion exchange is an important property of fast-ion conductors such an β-alumina. Ion exchange also provides a means of preparing other materials. Typical examples are the preparation of $LiCrO_2$ from $NaCrO_2$ and $LiNO_3$, $AgAlO_2$ from $KAlO_2$ and $AgNO_3$, and $CuFeO_2$ from $LiFeO_2$ and $CuCl$. These reactions are carried out in molten state or in aqueous solution. It is through proton exchange with oxide materials containing

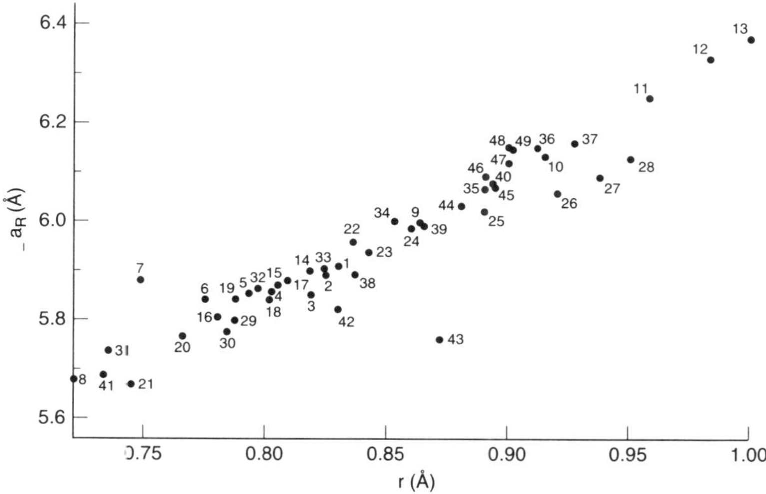

Figure 10.7 Plot of the rhombohedral lattice parameters of a binary and ternary carbonate of calcite structure against the mean cation radius. (From Rao *et al.* 1986.)

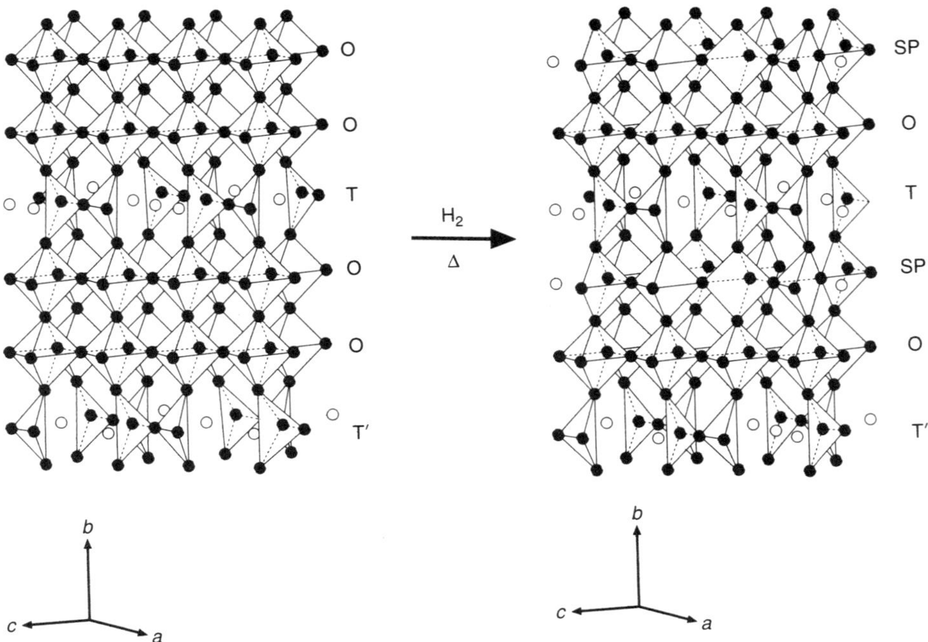

Figure 10.8 $Ca_3Fe_2MnO_{7.5}$ with three types of transition metal coordinations obtained by the topochemical reduction of $Ca_3Fe_2MnO_8$ with only octahedral and tetrahedral coordinations. The latter oxide is obtained by the thermal decomposition of the precursor carbonate, $Ca_2Fe_{4/3}Mn_{2/3}(CO_3)_4$. (From Rao *et al.* 1986.)

alkali metal ions (especially Li^+) that one obtains several materials used as precursors to prepare some of the oxides. We examined some examples whilst discussing soft chemistry routes. Another example is the conversion of $LiNbO_3$ and $LiTaO_3$ to $HNbO_3$ and $HTaO_3$ (Rice and Jackal 1982).

Use of strong alkaline media in the form of solid fluxes or molten solutions is helpful in preparing oxides, especially if one of the metal ions is required in a high-oxidation state. Thus, $Pb_2(Ru_{2-x}Pb_x)O_7-y$ with Pb in the 4^+ state is prepared in a highly alkaline media. Molten alkali has been used to prepare superconducting $La_2CuO_{4+\delta}$ (Rao *et al.* 1993a).

7   Electrochemical methods

Electrochemistry in aqueous or molten media has yielded a large number of inorganic materials including carbides, borides, silicides, oxides and sulphides. Many of the oxide bronzes are prepared electrochemically, just as other alkali metal intercalation compounds (for example, Li_xMS_2). The electrochemical method provides the best means of preparing oxides where a transition metal ion is required to be present in a high-oxidation state. In Fig. 10.11 we show the simple electrode systems employed for the purpose. Thus, superconducting $La_2CuO_{4+\delta}$ as well as $SrFeO_3$-type oxides have been prepared electrochemically (Grenier *et al.* 1991). Recently, oxygen-excess $La_8Ni_4O_{17}$ with unusual oxide has been prepared species electrochemically (Demourgues *et al.*

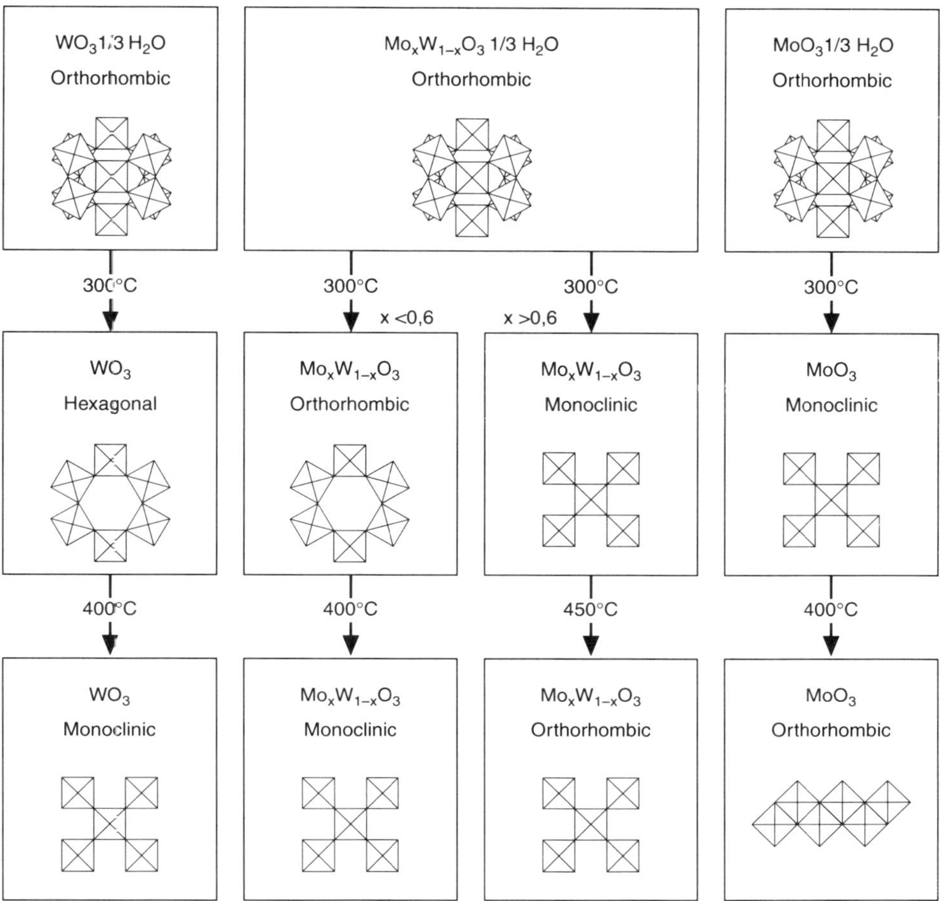

Figure 10.9 Different forms of WO$_3$ obtained from topochemical dehydration of WO$_3$. 1/3H$_2$O and related hydrates. (From Figlarz *et al.* 1990.)

1993), whilst we have prepared ferromagnetic cubic LaMnO$_{3+\delta}$ with more than 40% Mn^{4+} content. In Fig. 10.12 we show the progressive oxidation of LaMnO$_3$ from the orthorhombic phase to the cubic phase via the rhombohedral phase.

8　Nebulized spray pyrolysis

Pyrolysis of sprays is a well-known method for depositing films. Thus, one can obtain films of oxidic materials such as CoO, ZnO and YBa$_2$Cu$_3$O$_7$ by the spray pyrolysis of solutions containing salts (for example, nitrates) of the cations. A novel improvement in this technique is the so-called pyrosol process or nebulized spray pyrolysis, involving the transport and subsequent pyrolysis of a spray generated by an ultrasonic atomizer as demonstrated by Joubert and co-workers (Langlet and Joubert 1992). Xu *et al.* (1990) as well as Raju *et al.* (1995), have employed this method to prepare films of a variety of oxides. When high-frequency (100 kHz–10 MHz range) ultrasonic beam is directed at a gas–liquid interface, a geyser is formed and the height of the geyser is

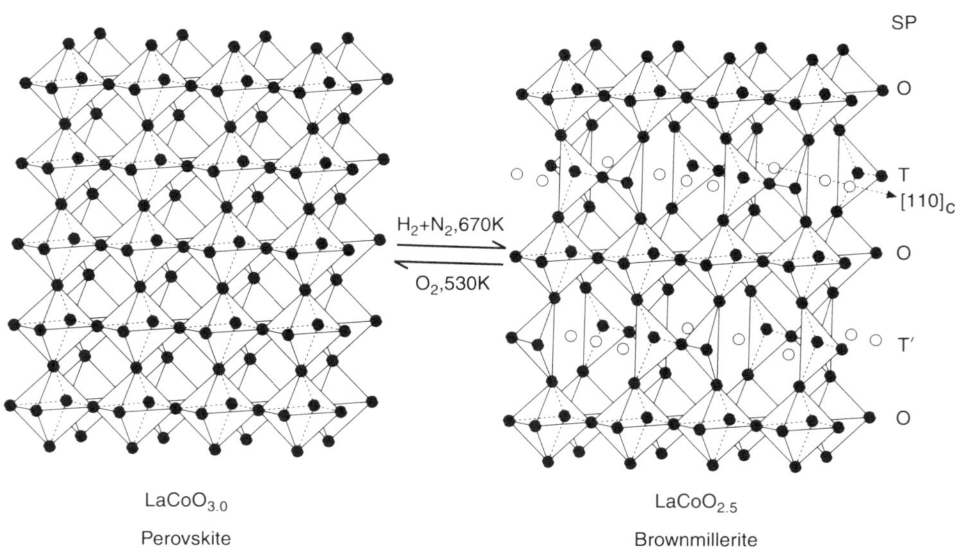

Figure 10.10 Reduction of $LaCoO_3$ to $La_2Co_2O_5$. (From Vidyasagar *et al.* 1985.)

proportional to the acoustic intensity. Its formation is accompanied by the generation of a spray, resulting from the vibration at the liquid surface and cavitation at the liquid–gas interface. The quantity of spray is a function of the intensity. Ultrasonic atomization is accomplished by using an appropriate transducer made of PZT located at the bottom of the liquid container. A 500–1000-kHz transducer is generally adequate. The atomized spray, which goes up in a column fixed to the liquid container, is deposited on a suitable solid substrate and then heat treated to obtain the film of the concerned material. The flow rate of the spray is controlled by the flow rate of air or any other gas. The liquid is heated to some extent, but its vaporization should be avoided. In Fig. 10.13 we show the apparatus employed in the pyrosol method.

The source liquid would contain the relevant cations in the form of salts dissolved in an organic solvent. Organometallic compounds (for example, acetates, alkoxides, β-diketonates, etc.) are generally used for the purpose. Proper gas flow is crucial to obtain satisfactory conditions for obtaining a good liquid spray. Nebulized spray pyrolysis is somewhere between MOCVD and spray pyrolysis, but the choice of source compounds for the pyrosol process is much larger than available for MOCVD. Furthermore, the use of a solvent minimizes or eliminates the difficulties faced in MOCVD. Films of a variety of oxide materials, such as $(Pb,Zr)TiO_3$, $YBa_2Cu_3O_7$ and $LaNiO_3$, have been obtained by the pyrosol method. Nebulized spray pyrolysis is truly inexpensive compared to CVD/MOCVD. The thickness of the films obtained by this method can be anywhere between a few hundred ångströms to a few microns. In Table 10.1, we show typical materials prepared by nebulized spray pyrolysis.

9 Cuprate superconductors

The discovery of a superconducting cuprate with a transition temperature that enabled conduction (T_c) above 77 K created a sensation in early 1987. Wu *et al.* (1987) who

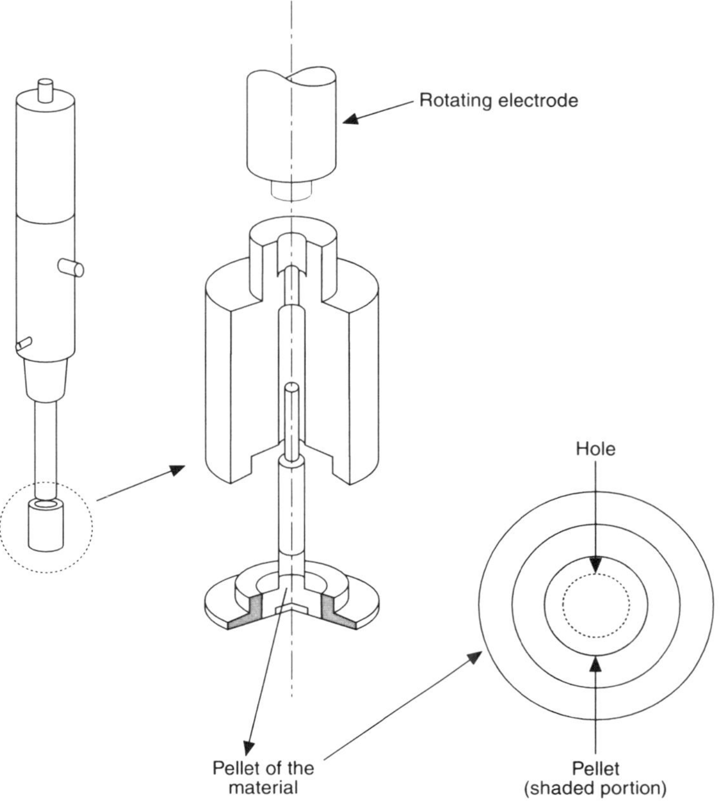

Figure 10.11 Simple electrode system for oxidation studies. (From Mahesh *et al.* 1995.)

announced this discovery first, made measurements on a mixture of oxides containing Y, Ba and Cu, obtained in their efforts to obtain the Y analogue of $La_{2-x}Ba_xCuO_4$. In our laboratory, we worked independently on the Y–Ba–Cu–O system on the basis of solid state chemistry. We knew that Y_2CuO_4 could not be made and that substituting Y by Ba in this cuprate was not the way to proceed. We therefore tried to make $Y_3Ba_3Cu_6O_{14}$ by analogy with the known $La_3Ba_3Cu_6O_{14}$, and varied the Y/Ba ratio as in $Y_{3-x}Ba_{3+x}Cu_6O_{14}$ (Rao *et al.* 1987). By making $x = 1$, we obtained $YBa_2Cu_3O_7$ ($T_c \approx 90$ K). We knew the structure had to be that of a defect perovskite from the beginning because of the route we adopted for the synthesis.

Preparative aspects of the various types of cuprate superconductors have been recently reviewed (Rao *et al.* 1993a). The cuprates are ordinarily made by the traditional ceramic method (mix, grind and heat), which involves thoroughly mixing the various oxides and/or carbonates in the desired proportions and heating the mixture at a high temperature. The mixture is ground again after some time and reheated until the desired product is formed, as indicated by X-ray diffraction. This method may not always yield the product with the desired structure, purity or O_2 stoichiometry. Variants of this method are often employed. For example, decomposing a mixture of nitrates has been found, by some, to yield a better product in the case of the 123 compounds; others prefer to use BaO_2 in place of $BaCO_3$ for the synthesis. One of the

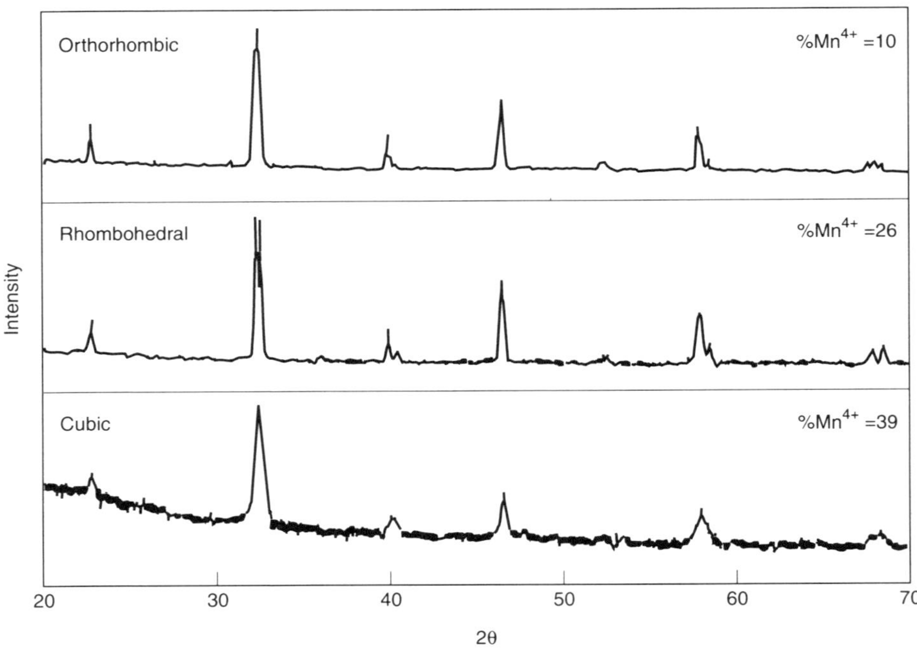

Figure 10.12 Rhombohedral and cubic $LaMn_{1-x}^{3+}Mn_x^{4+}O_{3+\delta}$ from orthorhombic $LaMnO_3$. (From Mahesh *et al.* 1995.)

Table 10.1 Typical films prepared by nebulized spray pyrolysis

Material	Compound used	Solvent	Gas	Substrate*
Pt	Pt acetylacetonate	Acetylacetone	Air	Glass, Al_2O_3,Si (670 K)
ZnO	Zn acetate	Methanol	Air	Glass, Al_2O_3,Si (770 K)
In_2O_3	In acetylacetonate	Acetylacetone	Air	Glass, Al_2O_3,Si (770 K)
SnO_2	$SnCl_4$	Methanol	Air, N_2	Glass, Al_2O_3 (670 K)
$La_4Ni_3O_{10}$,$LaNiO_3$	La + Ni Acetylacetonates	Ethanol	Air/O_2	Si, Al_2O_3 (770 K)
$CdIn_2O_4$	In acetylacetonate, + Cd acetate	Acetylacetone, methanol	Air	Glass, Al_2O_3 (710 K)
TiO_2	Butylorthotitanate	Butanol, acetylacetone	Air, N_2	Glass, steel (670 K)
γ-Fe_2O_3	Fe acetylacetonate	Butanol	Air/argon	Glass, Al_2O_3 (760 K)
(Ni–Zn) Fe_2O_4	Ni, Zn, Fe acetylacetonates	Butanol	Air	Glass (770 K)
Al_2O_3	Al isopropoxide	Butanol	Air	Glass (920 K)
$YBa_2Cu_3O_7$	Dipivaloylmethane derivatives	Butanol	Air/O_2	MgO, $SrTiO_3$ (870 K)

*Substrate temperature is shown.

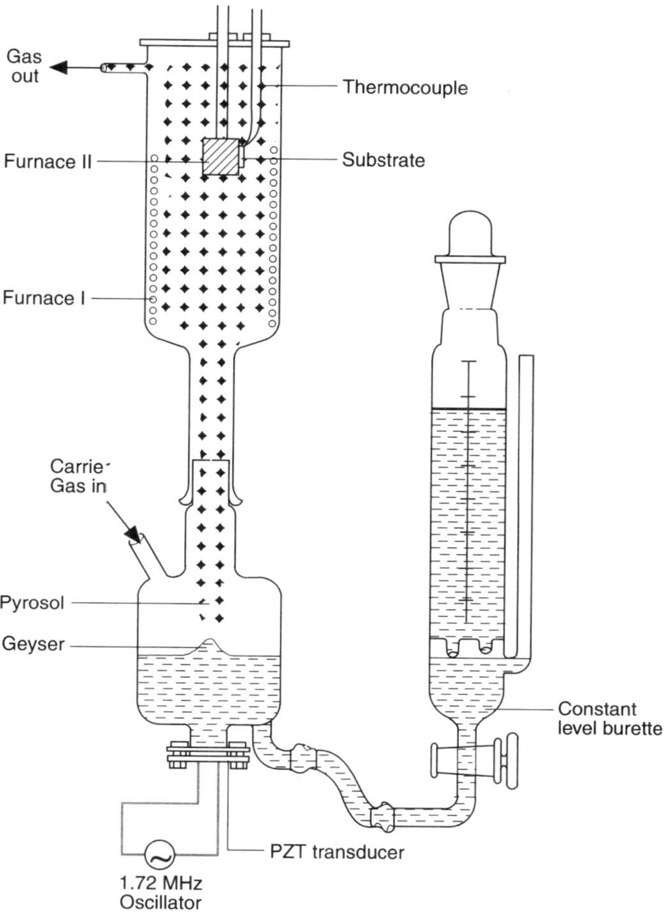

Figure 10.13 Apparatus for nebulized spray pyrolysis. (From Raju *et al.* 1995.)

problems with the Bi cuprates is the difficulty in obtaining phasic purity (minimizing intergrowth of the different layered phases). The glass or the melt route has been employed to obtain better samples. Sealed tube reactions are essential in the case of Tl and Hg cuprates.

Heating oxidic materials under high O_2 pressures or in flowing O_2 often becomes necessary to attain the desired O_2 stoichiometry. Thus, La_2CuO_4 and $La_2Ca_{1-x}Sr_xCu_2O_6$ heated under high O_2 pressures become superconducting, with T_c values of 40 and 60 K, respectively. We have obtained an analogous cuprate with two CuO_2 sheets ($T_c \sim 60$ K) by using $KClO_3$ in the preparation. In the case of the 123 compounds, one of the problems is that they lose O_2 easily. It is necessary to heat these materials in an O_2 atmosphere below the orthorhombic–tetragonal transition temperature. The 124 superconductors were first prepared under high O_2 pressures, but it was later found that heating the oxide or nitrate mixture in the presence of Na_2O_2 in flowing O_2 is sufficient to obtain 124 compounds. Superconducting Pb cuprates, however, can only be prepared in the presence of very little O_2. In the case of the electron superconductor

$Nd_{2-x}Ce_xCuO_4$, it is necessary to heat the material in an O_2-deficient atmosphere. Many of the Tl cuprates have to be heated in vacuum, N_2 or H_2 atmospheres to make them superconducting.

The sol-gel method has been conveniently employed for the synthesis of 123 compounds and Bi cuprates. Materials prepared by such low-temperature methods have to be heated under suitable conditions to obtain the desired O_2 stoichiometry as well as the characteristic high T_c value. 124-cuprates, Pb cuprates and even Tl cuprates have been made by the sol-gel method; the first two are particularly difficult to make by the ceramic method. Coprecipitation of all the cations in the form of a sparingly soluble salt such as carbonate in a proper medium (for example, using tetraethylammonium oxalate), followed by thermal decomposition of the dried precipitate has been employed by many workers to prepare cuprates. Several other novel strategies employed for the synthesis of superconducting cuprates were mentioned whilst discussing the various methods. Strategies where structure and bonding considerations are involved in the synthesis are generally more interesting. One such example is the synthesis of modulation-free superconducting Bi cuprates. Special mention should be made of oxyanion derivatives of cuprates, some of which are superconducting. Anions such as CO_3^{2-}, SO_4^{2-}, PO_4^{3-} and BO_3^{3-} replace the CuO_4 units in the 123 compounds as well as other cuprates (Konishita and Yamada 1993; Ayyappan *et al.* 1993; Huve *et al.* 1993; Maignan *et al.* 1993; Rao *et al.* 1993b; Nagarajan *et al.* 1994).

10   Intergrowth structures

There are several metal oxides exhibiting well-defined recurrent intergrowth structures with large periodicities, rather than forming random solid solutions with variable composition. Such ordered intergrowth structures themselves, however, frequently show the presence of wrong sequences. High-resolution electron microscopy (HREM) enables a direct examination of the extent to which a particular ordered arrangement repeats itself and the presence of different sequences of intergrowth, often of unit cell dimensions. Many systems forming ordered intergrowth structures have come to be known in recent years (Rao and Thomas 1985). These systems generally exhibit homology. The Aurivillius family of oxides of the general formula $Bi_2A_{n-1}B_nO_{3n+3}$ form intergrowth structures of the general formula $Bi_4A_{m+n-2}B_{m+n}O_{3(m+n)+6}$, involving alternate stacking of two Aurivillius oxides with different n values. The method of preparation involves simply heating a mixture of the component metal oxides at about 1000 K. Ordered intergrowth structures with (m,n) values of (1,2), (2,3) and (3,4) have been fully characterized. Intergrowth bronzes and hexagonal Ba ferrites are also examples of ordered intergrowths. What is surprising is that such periodicity occurs even when neither unit involved in the intergrowth is stable by itself.

11   Acknowledegements

The author thanks the EU Science Office and the US National Science Foundation for support.

12 References

Ayyappan, S., Manivannan, V. & Rao, C.N.R. (1993) *Solid State Commun* **87**, 551.

Ayyappan, S., Subbanna, G.N. & Rao, C.N.R. (1995) *Chem Euro J* **1**, 165.

Delmas, C. & Borthomieu, Y. (1993) *J Solid State Chem* **104**, 345–355.

Demorgues, A., Weill, F., Darriet, B. *et al.* (1993) *J Solid State Chem* **106**, 317–338.

Divigalpitiya, W.M.R., Frindt, R.F. & Morrison, S.R. (1989) *Science* **246**, 369–372.

Feist, T.P. & Davis, P.K. (1992) *J Solid State Chem* **101**, 275–287.

Figlarz, M., Gerard, B., Delahaye-Vidal, A. *et al.* (1990) *Solid State Ionics* **43**, 143–155.

Grenier, J.C., Wattiaux, A., Lagueyte, N. *et al.* (1991) *Physica C* **173**, 139–146.

Huve, M., Michel, C., Maignan, A. *et al.* (1993) *Physica C* **205**, 219.

Jacobson, A.J. (1992) In: *Solid State Chemistry* (eds A.K. Cheetham & P. Day). Clarendon Press, Oxford.

Konishita, K. & Yamada, T. (1993) *Nature* **357**, 313–315.

Langlet, M. & Joubert, J.C. (1992) In: *Chemistry of Advanced Materials* (ed C.N.R. Rao), pp. 55–71. Blackwell Scientific Publications, Oxford.

Mahesh, R., Kannan, K.R. & Rao, C.N.R. (1995) *J Solid State Chem* **114**, 294–297.

Maignan, A., Hervieu, M., Michel, C. & Raveau, B. *et al.* (1993) *Physica C* **208**, 116.

Marchand, R., Brohan, L. & Tournoux, M. (1980) *Mat Res Bull* **15**, 1229–1235.

Nagarajan R., Ayyappan, S. & Rao, C.N.R. (1994) *Physica C* **220**, 373–377.

Nanjundaswamy, K.S., Gopalakrishnan, J. & Rao, C.N.R. (1987) *Inorg Chem* **26**, 4286–4288.

Raju, A.R., Aiyer, H.N. & Rao, C.N.R. (1995) *Chem Mater* **7**, 225–231.

Rao, C.N.R. (1994) *Chemical Approaches to the Synthesis of Inorganic Materials*, John Wiley & Sons, Chichester.

Rao, C.N.R. & Gopalakrishnan, J. (1986) *New Directions in Solid State Chemistry*. Cambridge University Press, Cambridge.

Rao, C.N.R. & Thomas, J.M. (1985) *Acc Chem Res* **18**, 113–121.

Rao, C.N.R., Nagarajan, R. & Vijayaraghavan, R. (1993a) *Supercond Sci Tech* **6**, 1–12.

Rao, C.N.R., Nagarajan, R., Ayyappan, S. & Mahesh, R. (1993b) *Solid State Commun* **88**, 757–761.

Rao, C.N.R., Ganguly, P., Raychaudhri, A.K., Mohan Ram, R.A. & Sreedhar, K. (1987) *Nature* **326**, 856.

Rao, C.N.R., Gopalakrishnan, J., Vidyasagar, K., Ganguli, A.K., Ramanan, A. & Ganapathi, L. (1986) *J Mat Res* **1**, 280–292.

Rebbah, H., Desgardin, G. & Raveau, B. (1979) *Mat Res Bull* **14**, 1131–1136.

Rice, C.E. & Jackal, J.L. (1982) *J Solid State Chem* **41**, 308–314.

Vidyasagar, K., Reller, A., Gopalakrishnan, J. & Rao, C.N.R. (1985) *J Chem Soc Chem Commun* 7–8.

Wu, M.K., Ashburn, J.R., Torng, C.J. *et al.* (1987) *Phys Rev Lett* **58**, 908–911.

Xu, W.W., Kershaw, R., Dwight, K. & Wold, A. (1990) *Mat Res Bull* **25**, 1385–1389.

1 Solid–Solid Interfaces in the Epitaxial Films of Complex Oxides Deposited by Chemical Methods

A.R. RAJU, H.N. AIYER* and C.N.R. RAO

Materials Research Centre and CSIR Centre of Excellence in Chemistry, Indian Institute of Science, Bangalore 560 012, India and Jawaharlal Nehru Centre for Advanced Scientific Research, Jakkur, Bangalore 560 064, India

** Materials Research Centre and CSIR Centre of Excellence in Chemistry, Indian Institute of Science, Bangalore 560 012, India*

1 Introduction

The term *epitaxy* literally means 'arrangement upon'; it describes the oriented growth of one crystalline material (guest) on the surface of a single crystal of a different material (substrate). The adjective *epitaxial* implies not only that one particular crystal plane of the deposited guest crystal (deposited film) comes into contact with the surface of the host crystal, but also that one particular crystallographic direction in the contact plane of the guest crystal is parallel to a specific crystallographic direction in the contact plane of the host crystal. This is usually described in terms of Miller indices of the crystal planes and directions. For instance, (001)F [100]F//(001)S [100]S means that the (001) plane of the deposited film is in contact or aligned with the (001) plane of the substrate surface and that the [100] film growth direction coincides with the [100] direction of the substrate surface. For epitaxial growth to occur, it is desirable to have good lattice matching between the material of the film and the substrate, which would alleviate the problems of stress and relaxation through the formation of misfit dislocations. Another requirement for epitaxial nucleation is that the surface energy of the nucleus–substrate interface should be lower for the epitaxial orientation than for other orientations, so that the nucleation rate for the epitaxial orientation is greater. However, epitaxy can also occur due to different growth processes.

Preparation of single-crystalline films of complex oxides by means of epitaxial growth on suitable single-crystal substrates is an important area of research today because of the extensive, sophisticated technological applications of the epitaxial films in various types of devices. The single-crystalline or epitaxial nature of a film is established by X-ray diffraction techniques and more particularly by direct observation of the interface between the substrate and the deposited film by means of high-resolution transmission electron microscopy (HRTEM). By studying HRTEM images, one can determine how good the lattice match is between the lattice of the deposited film and that of the substrate. One can estimate the strain at the interface based on the mismatch, besides observing structural defects at the interface. In recent years, excellent single-crystalline (epitaxial) films of ferroelectric oxides such as $PbTiO_3$ and $PbZr_xTi_{1-x}O_3$ (PZT), superconducting $YBa_2Cu_3O_{7-x}$ and other complex oxide systems have been prepared on different single-crystal substrates by several techniques [1–8]. A few important deposition techniques are: (i) pulsed laser deposition (PLD); (ii) molecular beam epitaxy (MBE); and (iii) metal–organic chemical vapour deposition (MOCVD). Of these three methods, MOCVD is truly a chemical method involving reactions of precursors, while PLD involves the physical process of transferring the material from the target to the substrate. MBE involves vaporisation and monolayer deposition, followed by chemical reaction of the layers at the atomic scale. All three methods have yielded high-quality films, but they are all expensive and need sophisticated instrumentation. This is especially true of MBE.

In our laboratory, we have established inexpensive but powerful techniques for preparing oriented films of complex oxides by nebulised spray pyrolysis. In this paper, we shall briefly describe the PLD, MBE and MOCVD methods and illustrate results obtained by them with respect to epitaxial films of complex oxides on single-crystal substrates. The single-crystal substrates are generally oxides such as MgO, $SrTiO_3$ (STO) and $LaAlO_3$ (LAO), giving rise to oxide–oxide (solid–solid) interfaces. We provide a detailed description of the results on thin films of complex oxides obtained by nebulised spray pyrolysis and demonstrate how this inexpensive technique gives excellent-quality epitaxial films of complex oxides of technological importance, which include ferroelectric $PbZr_xTi_{1-x}O_3$ ($x = 0.5$), metallic $LaNiO_3$, and $LaMnO_3$ which exhibits giant magnetoresistance (GMR). In so doing, we describe the method in some detail and provide the HRTEM evidence for the epitaxial nature of the films obtained by this technique, showing the presence of an excellent lattice match between the deposited film and the substrate.

2   Pulsed laser deposition

In PLD, the material is evaporated from a target composed of the stoichiometric solid by means of an intense pulsed laser beam [9–12]. An excimer laser (KrF type, with a wavelength of 248 nm and a pulsewidth of c. 30 ns, operated at 100–1000 mJ pulse energy and a typical repetition rate of 1–10 Hz) is generally employed. A lower-power HeNe laser is employed to align and focus the excimer laser on to the target. The PLD process can be classified into three regions.

1 Absorption of the laser radiation by the target material resulting in the ablation (evaporation) of the surface layer.

2 Interaction of the evaporated material with the laser beam during which the evaporated ions and atoms absorb the photon energy.

3 Anisotropic adiabatic expansion of the plasma resulting in a forward-directed plume.

The epitaxial growth of the oxide superconductors on single-crystal substrates is required for microwave and other applications. LAO is a strong candidate as a substrate for superconducting thin films for microwave applications due to its low dielectric constant and loss tangent [13]. Although the actual structure of $LaAlO_3$ at room temperature is rhombohedral, its unit cell can be described as a slightly distorted pseudocubic perovskite providing a close lattice match with the unit cell of YBCO. Highly epitaxial thin films of YBCO have been obtained on (001) single-crystal LAO by PLD [14]. An HRTEM image of the interface between the YBCO film and the LAO substrate is presented in Fig. 1.1. The image shows near-perfect epitaxial c perpendicular growth of YBCO on (001) LAO. This interface microstructure also reveals the presence of regions of strain which extend into both the film and the substrate. These strained regions correspond to interfacial dislocations. The presence of interfacial dislocations is due to the slight mismatch of the lattice parameters of YBCO and LAO.

Figure 1.1. HRTEM image of the interface between the YBCO film deposited by PLD and the LaAlO$_3$(001) substrate. The inset SAED pattern taken from an area including both the film and the substrate confirms the epitaxial growth. The presence of an interfacial dislocation is seen in form of an extra half-plane in the substrate within the marked regions. From [14].

The epitaxial c perpendicular growth of the YBCO film is further confirmed by the selected area electron diffraction (SAED) pattern which is taken from an area at the interface including both the film and the substrate (see inset Fig. 1.1). The coincidence of the diffraction spots of the film and the substrate shows the epitaxial nature of the film. The epitaxial relationship is (001)YBa$_2$Cu$_3$O$_{7-x}$//(001)LaAlO$_3$. Another example of an epitaxial film obtained by PLD is illustrated in Fig. 1.2 [15]. The figure shows an HRTEM image of a PbTiO$_3$ film on an (001)STO substrate. This image reveals a perfect in-plane matching and a complete absence of a-axis-oriented domains establishing the c-axis orientation of the film. The SAED pattern shown in the inset also confirms the c-axis orientation of the film. The orientation relationship of the film and the substrate is (001)PbTiO$_3$//(001)SrTiO$_3$.

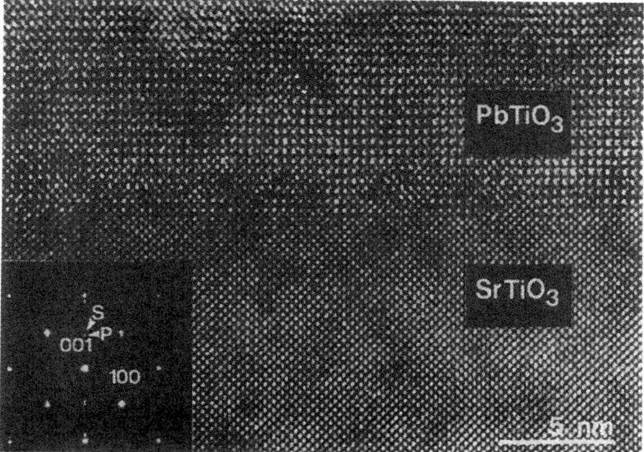

Figure 1.2. HRTEM image of the interface between a PbTiO$_3$(P) film deposited on the SrTiO$_3$(001) substrate by PLD showing perfect in-plane matching. The SAED pattern shown in the inset also evidences the c orientation of the PbTiO$_3$ film. From [15].

3   Molecular beam epitaxy

MBE is a versatile technique for growing epitaxial films of metals, semiconductors, insulators and oxide materials [16,17]. In this technique, molecular beams of the constituent atoms formed from Knudsen effusion cells are allowed to impinge on a heated substrate under ultra-high vacuum ($c.$ 10^{-10} torr). The heated substrate provides the impinging molecules with sufficient energy to be mobile on the substrate so that these atoms migrate on the surface until they encounter an appropriate site on which to condense. Under satisfactory conditions, one obtains the growth of a single-crystal epitaxial film on a single-crystal substrate. Typical rates used for deposition in MBE are about one atomic layer per second. The rate of deposition is controlled by the intensity of the molecular beam. Therefore, the uniformity of the film depends on the uniformity of the beams. Thus, the MBE process offers a high degree of uniformity, controllability and reproducibility. In the MBE system, several beams of different substances can be operated in the same chamber and directed on to the substrate surface. Given the independent control of the intensities of the beams and the control of the substrate temperature, structures of the prescribed composition with a variety of layers can be deposited in the form of a film. MBE has been used to prepare high-quality epitaxial films of a variety of oxides as well as superlattices on different single-crystal substrates. One of the important applications of the ferroelectric perovskite $BaTiO_3$ is in electro-optic devices. The large electro-optic coefficient of $BaTiO_3$, coupled with the small microwave dielectric constant of a substrate such as MgO, could substantially improve the device speed and efficiency. Devices based on these materials in thin-film form require long-range structural coherence at the interface, which can be realised in epitaxial thin films. Epitaxial $BaTiO_3$ films have been prepared on a MgO substrate by using MBE [18]. Figure 1.3 shows a scanning electron micrograph of $BaTiO_3$ on MgO(001) in cross-sectional view. We see that the film is 0.6 µm thick, crack-free, dense, single-phase and optically transparent. Figure 1.4 shows an HRTEM image of the $BaTiO_3$–MgO interface. The cube-on-cube epitaxy can be readily observed in this image. At a larger scale, the HRTEM images showed dislocations nucleated to relieve the interfacial strain.

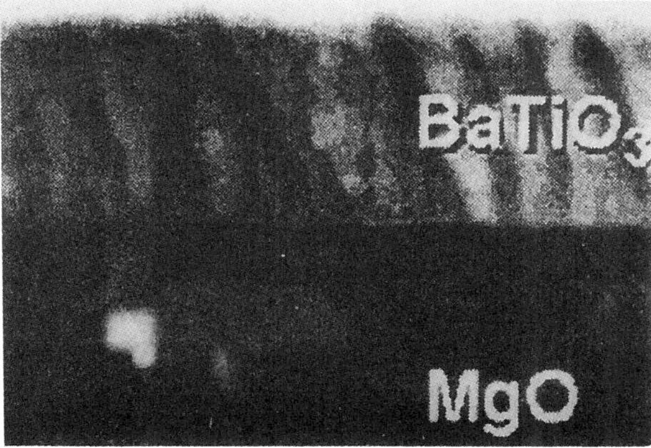

Figure 1.3. Cross-sectional scanning electron micrograph of $BaTiO_3$ film deposited on MgO(001) by MBE. Film thickness is 0.6 µm. From [18].

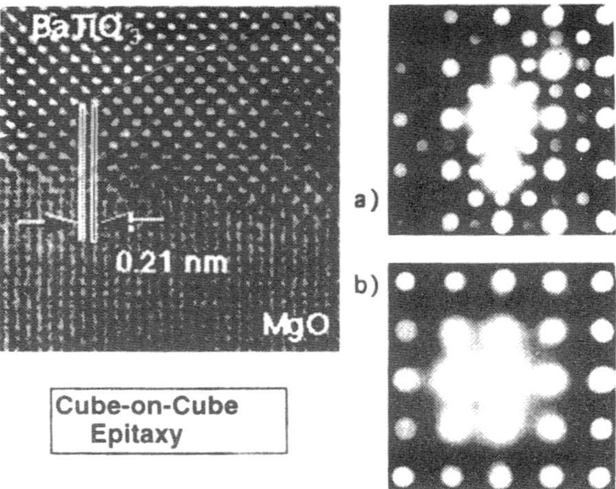

Figure 1.4. HRTEM image of the $BaTiO_3$–MgO(001) interface (in a film obtained by MBE) with the SAED pattern inserts from: (a) $BaTiO_3$; (b) MgO. From [18].

4 Metal–organic chemical vapour deposition

In MOCVD, a metal–organic precursor compound or combination of precursor compounds from the gas phase condenses on a substrate surface where a chemical reaction occurs, leading to the formation of a solid film [19]. If the (source) material to be deposited is not in the vapour state, it can be converted into the vapour state by volatilising from either the solid or the liquid source. In its most common form, the CVD process relies upon elevated substrate temperatures to pyrolyse the gaseous precursor for film growth. However, when high growth temperatures are incompatible with the substrate material, or when metastable materials are to be deposited, it is necessary to activate the chemical reaction near or on the surface to facilitate film growth at reduced temperatures. This is accomplished by the application of a radio-frequency (RF) field (plasma excitation), light (photoexcitation) or by direct heating of the gaseous source (thermal excitation). MOCVD has been widely employed for the deposition of epitaxial ferroelectric films. An epitaxial ferroelectric $BaTiO_3$ film is obtained on a MgO(100) substrate by MOCVD at 875 K by using titanium isopropoxide and bis(2,2,6,6,-tetramethyl 3,5 heptanedionato)barium, i.e. $Ba(thd)_2$, as precursors [20]. An HRTEM image of this $BaTiO_3$ film, showing the $BaTiO_3$–MgO interface, is shown in Fig. 1.5(a). We see that the (200) lattice planes are continuous across the interface, demonstrating the epitaxial nature of the film. The image also shows that the film–substrate interface is sharp and contains no second phases. Misfit dislocations in the $BaTiO_3$ film were observed at the interface. One such dislocation is indicated by an arrow in the micrograph. Formation of the dislocations probably occurred during the deposition process in order to accommodate the 5.4% lattice mismatch between the substrate and the growing film at the growth temperature. The SAED pattern of the $BaTiO_3$–MgO interface shown in Fig. 1.5(b) shows a cube-on-cube orientation between $BaTiO_3$ and MgO (i.e. alignment of both normal and in-plane lattice vectors of the film with that of the substrate) confirming that the film is epitaxial. Higher-order diffraction spots showed some arcing, indicating that the different regions of the film were slightly

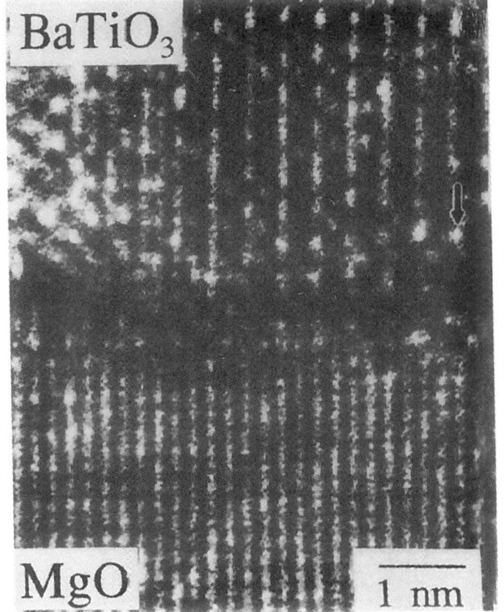

Figure 1.5. (a) HRTEM image of the interface between the BaTiO$_3$ film deposited on MgO(100) by MOCVD. (b) SAED pattern from the BaTiO$_3$–MgO(100) interface showing the 'cube-on-cube' orientation relationship between the film and the substrate. The arrow indicates the direction of the normal (n) to the film–substrate interface. From [20].

misoriented relative to the $\langle 100 \rangle$ direction in the MgO substrate. An upper limit of 1% for the misorientation angle was estimated from the extent of arcing. Lattice parameters calculated from the pattern confirmed that the film was epitaxial and oriented with the a-axis normal to the substrate surface.

Epitaxial thin films of PbTiO$_3$ have recently been deposited on SrRuO$_3$-buffered SrTiO$_3$(001) substrates by MOCVD [21]. Tetraethyl-lead and titanium isopropoxide were used as precursors in this study. The SrRuO$_3$ buffer layer was prepared by RF sputtering. In Fig. 1.6, the cross-sectional TEM image of the PbTiO$_3$–SrRuO$_3$–SrTiO$_3$

Figure 1.6. Cross-sectional TEM image of a PbTiO$_3$ film prepared by MOCVD deposited on a SrTiO$_3$ substrate with a SrRuO$_3$(001) epitaxial buffer layer. The image shows that the dominant defects in the film are 90° domains and threading dislocations. From [21].

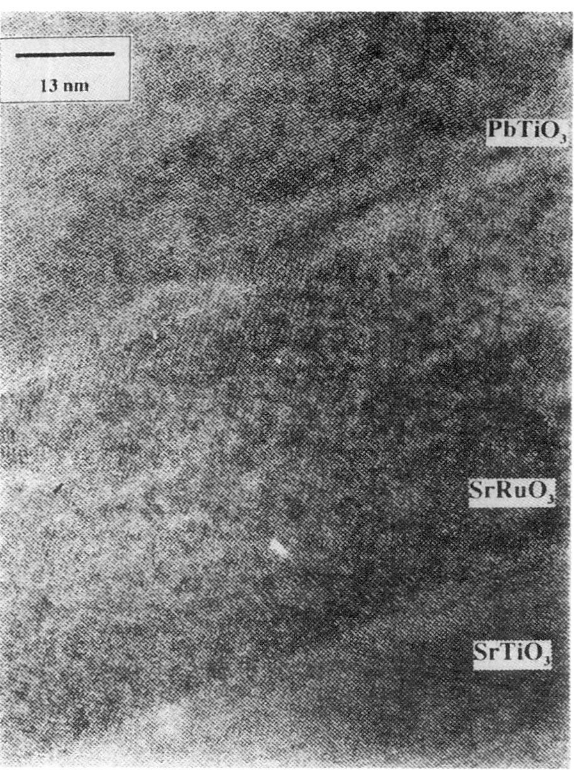

Figure 1.7. HRTEM image of $PbTiO_3(001)$–$SrRuO_3(001)$–$SrTiO_3(001)$ interfaces showing atomically sharp interfaces between individual layers. From [21].

film is shown. The image shows that the film is epitaxial and c-axis oriented, with $90°$ domains, with threading dislocations as the primary structural defects. The thickness of the $SrRuO_3$ buffer layer is about 33 nm. The $90°$ domains visible in the image nucleate at the structural defects in the substrate. The strain contrast associated with the substrate defect sites appears to propagate directly through the $SrRuO_3$ layer into the $PbTiO_3$ layer. The threading dislocations appear normal to the substrate–film interface and propagate through the $90°$ domains. This is indicative of the formation of dislocations prior to the ferroelectric phase transition, while the film is in the cubic state. A cross-sectional HRTEM image of the $PbTiO_3(001)$–$SrRuO_3(001)$–$SrTiO_3(001)$ interfaces is shown in Fig. 1.7. The image shows atomically sharp interfaces. The $PbTiO_3(001)$–$SrRuO_3(001)$ interface appears to be cleaner than the $SrRuO_3(001)$–$SrTiO_3(001)$ interface, indicating that the deposition of the buffer layer appears to improve the quality of the substrate surface, resulting in an improved ferroelectric film.

5 Nebulised spray pyrolysis

The technique of nebulised spray pyrolysis developed in our laboratory does not employ a vacuum. It is a simpler, low-cost alternative for the deposition of high-quality thin films of novel complex oxides [22–24]. Pyrolysis of sprays is a well-known method for depositing films. A novel improvement in this technique is the so-called pyrosol process or nebulised spray pyrolysis of a spray generated by an ultrasonic atomiser. In nebulised spray pyrolysis, a solution containing the organometallic derivatives of the relevant metals in a suitable solvent (source liquid) is nebulised by making use of a PZT transducer operated at a frequency of 1.72 MHz. The nebulised spray is slowly deposited on a solid substrate at a relatively low temperature with sufficient control of the rate of deposition to yield the oxide film of the desired stoichiometry. The rate of deposition is controlled by means of the flow rate of the carrier gas used, the temperature of the substrate and the concentration of the organometallic precursors. The film thickness can vary from few hundred nanometres (c. 200 nm) to a few micrometres, depending upon the solution concentration, deposition time, etc.

Figure 1.8 is a schematic diagram showing the different parts of our apparatus for ultrasonically nebulised spray pyrolysis. It consists of two independent zones: the atomisation chamber and the pyrolysis reactor. The source liquid is kept in the atomisation chamber. The atomisation chamber is designed so that the bottom of the chamber has a cylindrical opening to fit the PZT transducer of 20 mm diameter (1 mm thick). The special feature of this design is that the upper portion of the PZT transducer is in direct contact with the source liquid. This gives the highest energy transfer to the liquid when compared to the non-contact method. Its disadvantage is that the pH of the liquid has to be maintained at around 7. The liquid level in the atomisation chamber is maintained by using a constant level burette which allows the measurement of the

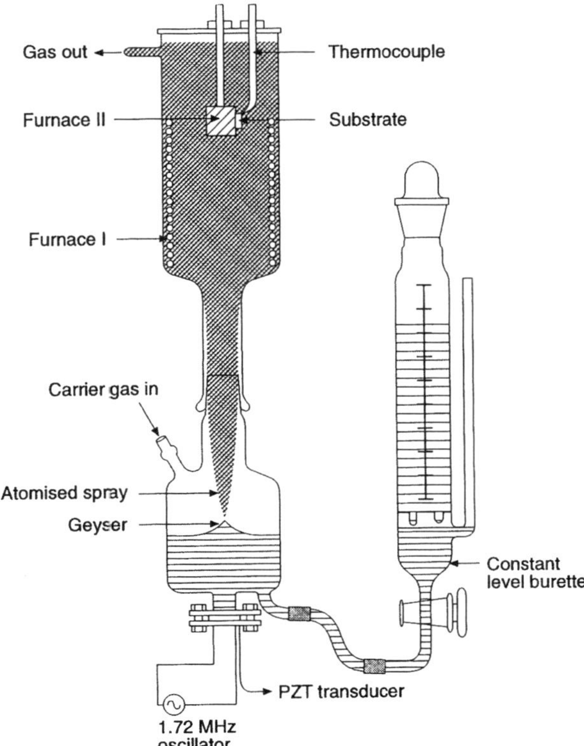

Figure 1.8. Schematic drawing of the apparatus used for nebulised spray pyrolysis.

volume of the liquid nebulised. When a high-frequency ultrasonic beam is directed at the gas–source-liquid interface through the PZT transducer, a geyser forms at the surface, the height of which is proportional to the acoustic intensity and the physical properties of the liquid (vapour pressure, viscosity and surface tension). The geyser formation is accompanied by cavitation at the gas–source-liquid interface. When the amplitude of the acoustic vibrations exceeds a certain threshold value, liquid atomisation occurs. Above this threshold, a continuous and regular mist (nebulised spray) is generated. The nebulised spray produced in the first chamber is transported by a carrier gas introduced through a side port. The second zone is a long glass tube which consists of a heater (nichrome wire wound on a ceramic tube), a substrate holder and a chromel–alumel thermocouple fixed to the substrate holder. The nebulised spray is decomposed in the pyrolysis reactor zone on a hot substrate at an appropriate temperature and the pyrolysis reaction product consists of a thin film whose composition, adherence and morphology depend on the experimental conditions.

5.1 *Ferroelectric oxide films*

Epitaxial thin films of ferroelectric materials such as PZT have received much attention because of their applications in non-volatile memories, sensor and actuator devices, etc. [25]. Using the nebulised spray pyrolysis technique, we have deposited films of $PbZr_xTi_{1-x}O_3$ with $x = 0.5$ on STO(100) substrates at 625 K for 3 hours, which were further annealed for 12 hours in air at the same temperature. The precursors employed were lead acetate, zirconium isopropoxide and titanium isopropoxide. The solution containing the stoichiometric quantities of the precursors in methanol solvent (0.1 mol) was nebulised to obtain the PZT film. The films were first subjected to energy dispersive X-ray analysis (EDAX), which showed a Pb : Zr : Ti ratio close to 1 : 0.5 : 0.5, confirming the film stoichiometry. An X-ray Θ–Θ scan of the PZT film deposited on the STO(100) substrate is shown in Fig. 1.9. The $PbZr_{0.5}Ti_{0.5}O_3$ corresponds to a composition on the tetragonal side of the $PbTiO_3$–$PbZrO_3$ phase diagram. The only phase

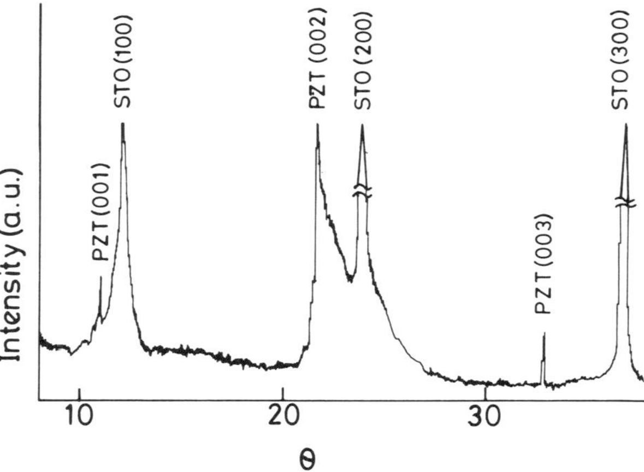

Figure 1.9. X-ray diffraction pattern (Θ–Θ scan) of a $PbZr_{0.5}Ti_{0.5}O_3$ film deposited on $SrTiO_3$(100) substrate by nebulised spray pyrolysis.

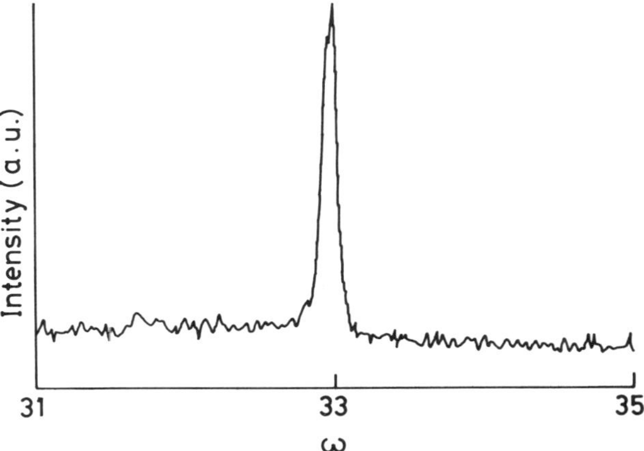

Figure 1.10. X-ray rocking curve (ω scan) of the (003) reflection of a PZT film deposited on STO(100) by nebulised spray pyrolysis.

observed is that of the perovskite type $PbZr_xTi_{1-x}O_3$. The X-ray diffraction (XRD) scan showed that the film on the STO(100) is c-axis oriented since only the reflections appearing in the pattern are those from the (001) plane of PZT along with the (h00) reflections of the STO substrate.

X-ray rocking curve measurements were performed to obtain information on the crystalline perfection of these films and to quantify their planar orientation of these films. Both Θ and ω scans were carried out on the preselected film peaks known from the Θ–Θ XRD pattern. The full width at half maximum (FWHM) of the X-ray rocking curves of the (003) reflection of PZT (Fig. 1.10, ω scan) was c. 0.2°, indicating an excellent alignment in the growth direction among the c-axis oriented grains and implying that the film deposited was indeed epitaxial. This FWHM value compares well with that of epitaxial ferroelectric (PZT, PT) films obtained by processes such as RF magnetron sputtering, MOCVD, etc. This epitaxial deposition of a PZT film on an

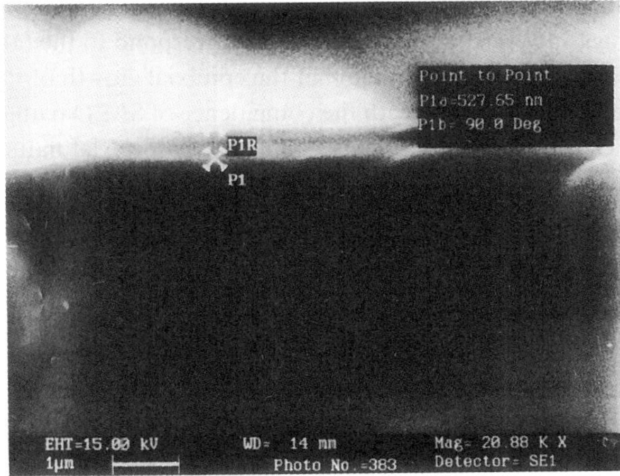

Figure 1.11. Cross-sectional scanning electron micrograph of the PZT film deposited on the STO(100) substrate by nebulised spray pyrolysis. Film thickness is about 530 nm.

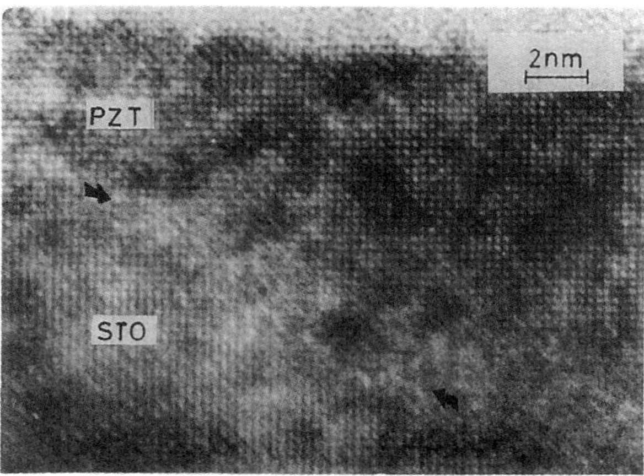

Figure 1.12. HRTEM image of the PZT–STO interface revealing the epitaxial (c-oriented) growth of PZT film. The interface is indicated by the arrows.

STO substrate is realised because of reasonably good lattice matching between STO ($a = 3.905$ Å) and PZT ($c = 4.1$ Å). A cross-sectional scanning electron micrograph of the PZT film deposited on the STO substrate is shown in Fig. 1.11. The micrograph shows the film to be smooth, uniform and densely packed, having a thickness of 530 nm.

Cross-sectional transmission electron microscopy (TEM) of PZT–STO revealed the film to be a single domain. No 90° domains were observed in the bright field images. This may be due to the fact that PZT films in the present case have been deposited below the Curie temperature T_c. Figure 1.12 shows a high-resolution lattice image of the PZT–STO interface. A direct correspondence between the crystallography in the film and that of the substrate is evident. The (001) planes of both PZT and $SrTiO_3$ are parallel to the interface and the a- and b-axes, i.e. the $\langle 100 \rangle$ and $\langle 010 \rangle$ directions of the PZT are aligned with the corresponding directions of STO in the interfacial plane. This clearly demonstrates the epitaxial growth of the PZT on STO. The SAED pattern from the film–substrate (PZT–STO) interface is shown in Fig. 1.13. The interplanar spacings derived from the Bragg spot separations (about 4 Å) correspond to the (100) and (001) reflections. From this analysis, we conclude that the epitaxial growth of $PbZr_{0.5}Ti_{0.5}O_3$ on STO is c-axis oriented. This, along with the coincidence of SAED pattern of the PZT film with that of STO(100) substrate, further confirms the epitaxial nature of the PZT film. The epitaxial relationship is (001)PZT//(100)STO.

The epitaxial PZT films deposited by nebulised spray pyrolysis show a hysteresis loop characteristic of ferroelectric films with reasonably good values of coercive field, E_c (c. 80 kV) and remnant polarisation, P_r (c. 22 C cm^{-2}).

5.2   Metallic LaNiO₃ films

One of the problems in integrating ferroelectric thin films into devices is the difficulty of growing a single-crystalline thin film with an appropriate electrode material. Perovskite type metallic oxides such as $LaNiO_3$ (LNO) lattice-matched to ferroelectric oxides would be good candidates for this purpose [26]. It is therefore important to study

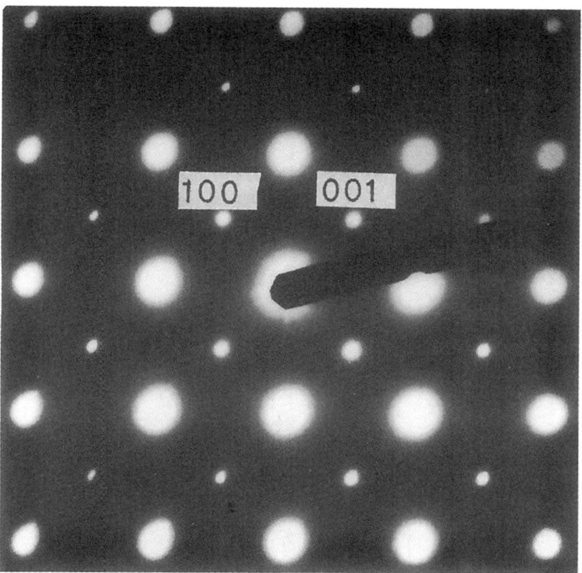

Figure 1.13. SAED pattern from the PZT–STO(100) interface region. The interplanar spacing derived from the spot separations is *c.* 4 Å.

epitaxial thin films of LNO. We have deposited LNO films on STO(100) substrates at 675 K (8 hours) by the nebulised spray pyrolysis of acetylacetonate precursors of lanthanum and nickel.

The deposited films of LNO were stoichiometric, as confirmed by EDAX analysis, which gave the La : Ni ratio as 1 : 1. The X-ray diffraction pattern of LNO deposited on STO (Fig 1.14) distinctly shows the (h00) reflections of LNO besides the substrate (h00) reflections, indicating that the film deposited is clearly highly oriented or textured. The crystallinity of the LNO film in the growth direction was ascertained by means of the X-ray rocking curve. The ω scan (Fig. 1.15) of the LNO(100) reflection has an FWHM of 0.4°, indicating very good crystallinity and the epitaxial nature of the

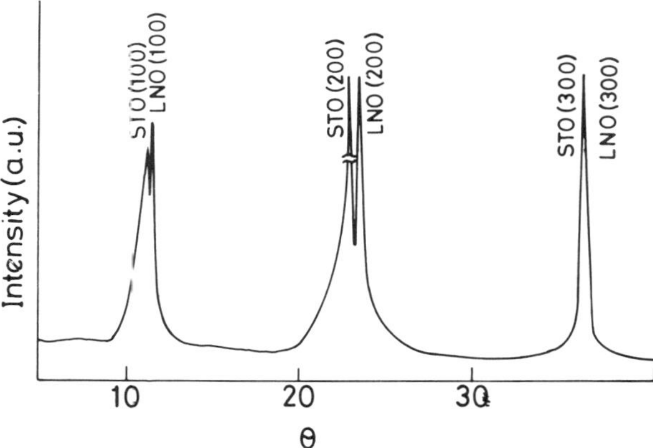

Figure 1.14. X-ray diffraction pattern (Θ–Θ scan) of an LNO film deposited on STO(100) by nebulised spray pyrolysis.

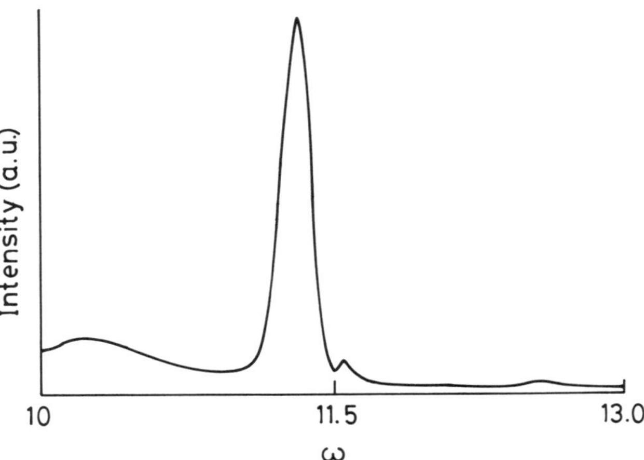

Figure 1.15. X-ray rocking curve (ω scan) of the (100) reflection of the LNO film deposited on STO(100).

film. The STO(100) substrate promotes the epitaxial growth of the LNO film due to closeness of the lattice constants between STO ($a = 3.905$ Å) and that of the LNO ($a = 3.846$ Å, pseudocubic), i.e. a lattice mismatch of only 1.66%.

A cross-sectional scanning electron micrograph of an LNO–STO film revealing the film and the substrate regions is shown in Fig. 1.16. The LNO film is highly dense, without any porosity. The thickness of the LNO film obtained from the micrograph was 19.8 μm.

In Fig. 1.17, we show a cross-sectional low-magnification TEM image (bright field image) of LNO–STO. The interface is clearly visible and is well marked by the presence of the extinction contours arising out of the differential thinning process of the film and the substrate. The film is continuous and is devoid of any grain boundaries or defects. The high-resolution cross-sectional image of the LNO–STO interface is shown in

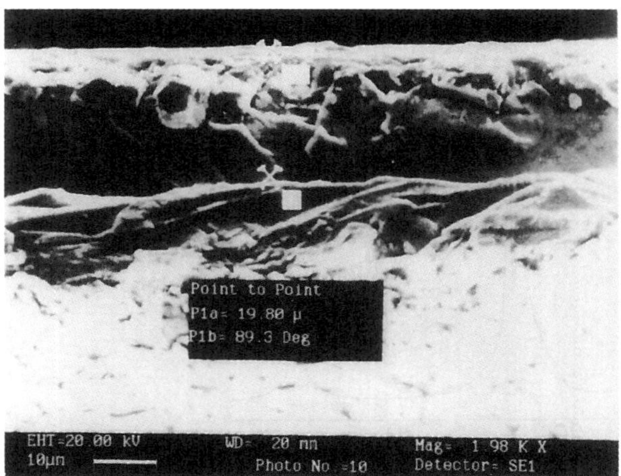

Figure 1.16. Cross-sectional scanning electron micrograph of the LNO film deposited on STO(100) substrate. Film thickness is *c.* 19 μm.

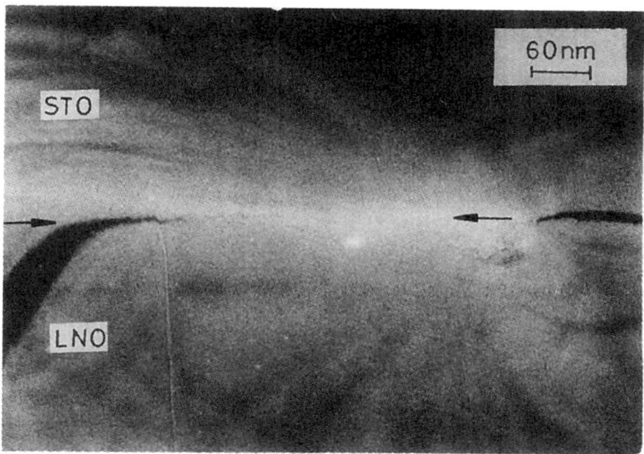

Figure 1.17. Cross-sectional low-magnification TEM bright field image of the LNO film deposited on STO(100) showing the interface. The interface is indicated by the arrows.

Fig. 1.18. The bright spots on either side of the interface are c. 2.7 Å apart, showing good lattice matching. The SAED pattern (Fig. 1.19), including diffraction spots of both the LNO film and the STO substrate, shows intense Bragg spots corresponding to the $\sqrt{2}$ type of the perovskite cell. The weak spots corresponding to c. 3.9 Å are actually split, thereby showing the slight mismatch of the a parameters of the film and the underlying substrate. Thus, the cross-sectional electron microscopic studies clearly show that the films deposited are truly epitaxial and that the epitaxial relationship is (100)LNO//(100)STO.

The epitaxial LNO film showed the expected metallic nature of the film in the four-probe electrical resistivity measurements.

5.3   *Films of LaMnO$_3$ exhibiting giant magnetoresistance*

Recently, large negative magnetoresistance effects have been observed in doped perovskite La$_{1-x}$A$_x$MnO$_3$ (A = Ca, Sr, Ba) epitaxial thin films, which are significantly larger

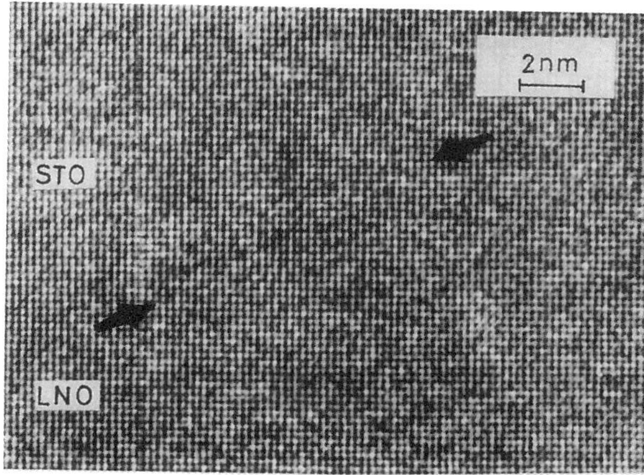

Figure 1.18. HRTEM image of LNO–STO(100) interface showing perfect lattice matching.

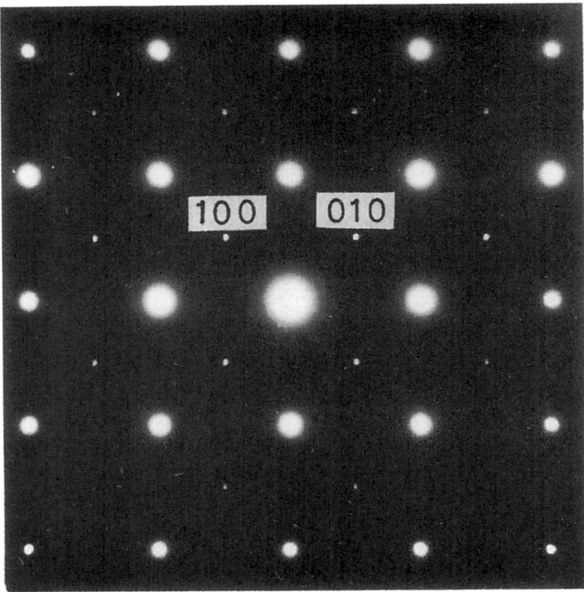

Figure 1.19. SAED pattern from the LNO–STO(100) interface region.

than their bulk counterparts [27–30]. This can be useful for various device applications such as magnetic recording, magnetoresistive sensors and magnetoresistive microphones. In this context, we sought to obtain epitaxial films of lanthanum manganates by nebulised spray pyrolysis. We deposited $LaMnO_3$ (LMO) films on STO(100) substrates at 675 K (5 h) using lanthanum acetylacetonate and manganese acetylacetonate as precursors. The film stoichiometry was first confirmed by EDAX analysis. Figure 1.20 gives the XRD Θ–Θ scan of the LMO–STO film. This pattern reveals the film to be of

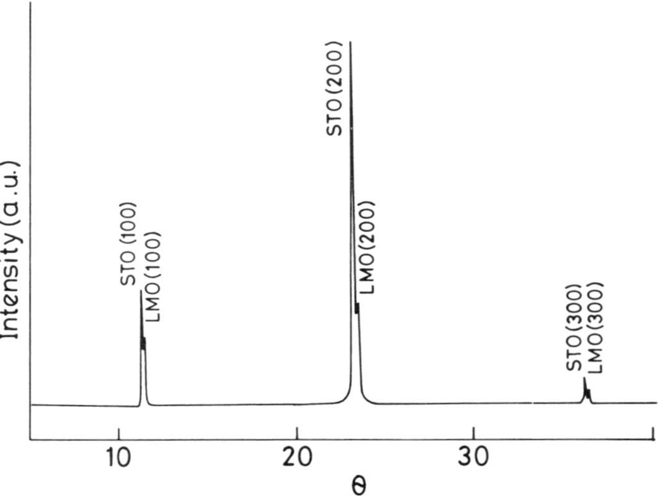

Figure 1.20. X-ray diffraction pattern (Θ–Θ scan) of an LMO film on STO(100) substrate obtained by nebulised spray pyrolysis.

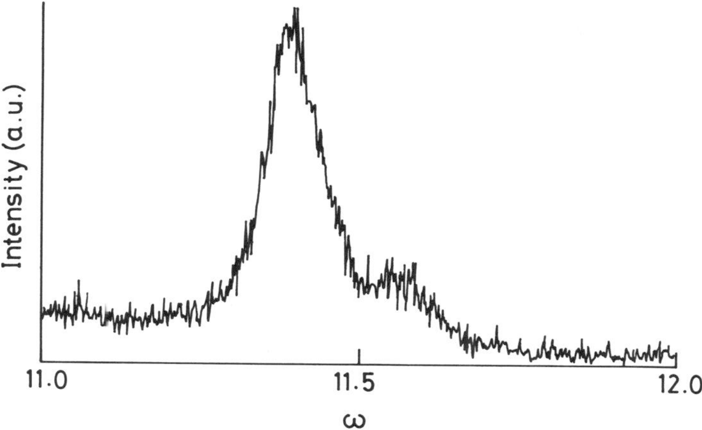

Figure 1.21. X-ray rocking curve (ω scan) of the (100) reflection of the LMO film deposited on STO(100) substrate.

a cubic perovskite phase with $a = 3.88$ Å and to be highly oriented along the (100) direction on the STO(100) substrate.

LMO containing around 30% Mn^{4+} has been shown to have a cubic structure [31,32]. The Mn^{4+} ions arise not because of the oxygen excess but due to the presence of a roughly equal number of vacancies on both the La and the Mn sites. Cation vacancies are present randomly but do not give rise to extended defects. The cubic LMO phase obtained by us can be approximately described as $La_{0.95}Mn_{0.95}O_3$ [33]. We have carried out electrical resistivity measurements on some of these films. The cubic LMO phase prepared by us showed a metal–insulator transition at around 200 K, in agreement with the literature [34].

The X-ray rocking curve (ω scan) of the (100)LMO reflection is shown in Fig. 1.21. The FWHM value of the curve is about 0.15°, indicating a very high degree of alignment/perfection of the film growth plane with respect to the STO(100) substrate, confirming the epitaxial nature of the film. Fig. 1.22 shows an HRTEM image of the

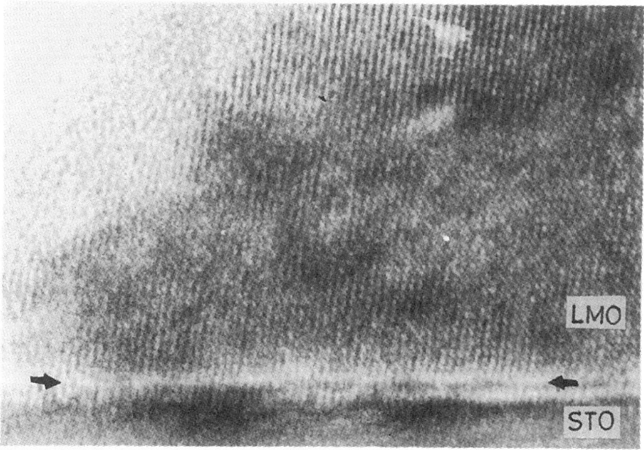

Figure 1.22. HRTEM lattice image of the LMO film deposited on STO(100) substrate showing the epitaxial nature of the film. The interface is indicated by the arrows.

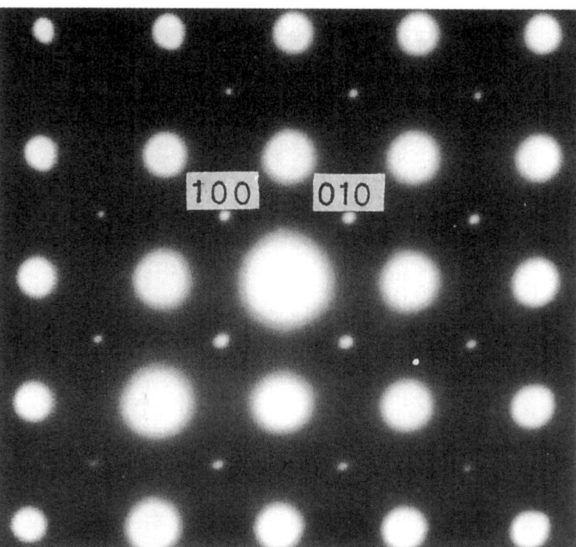

Figure 1.23. SAED pattern from the LMO–STO(100) interface region.

LMO–STO interface. The interfacial boundary region does not show any extra phases. The LMO lattice lines are well resolved and are seen to extend all the way down to the STO surface, closely matching with the STO lattice. The SAED pattern of the LMO–STO interface is given in Fig. 1.23. In this pattern also, the Bragg spots corresponding to *c.* 3.9 Å are split, showing the slight mismatch in the *a* parameters. These HRTEM observations establish the epitaxial growth of the LMO film on STO(100) and the epitaxial relationship is (100)LMO//(100)STO.

We have also successfully deposited LMO films on an LAO(100) substrate. The XRD pattern (Θ–Θ scan) of the LMO–LAO film is shown in Fig. 1.24. The epitaxial nature of

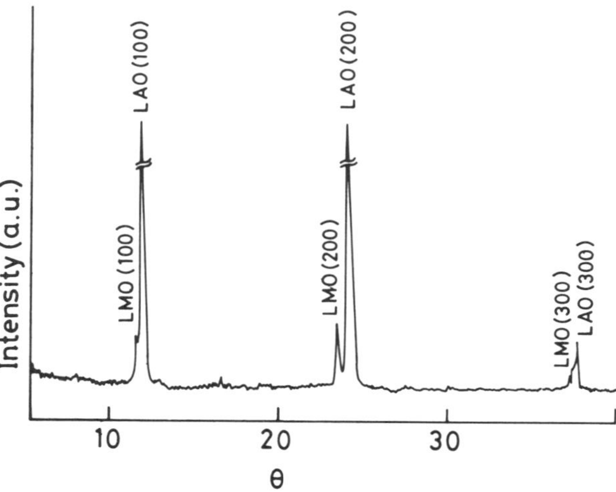

Figure 1.24. X-ray diffraction pattern (Θ–Θ scan) of an LMO film deposited on LAO(100) substrate by nebulised spray pyrolysis.

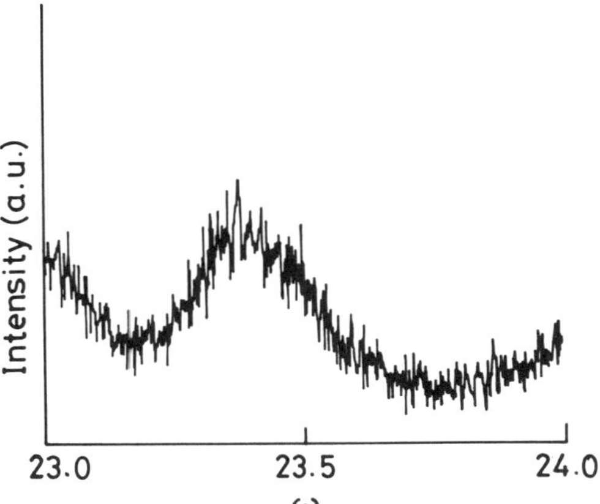

Figure 1.25. X-ray rocking curve (ω scan) of (200) reflection of an LMO film deposited on LAO(100).

the LMO film is clearly evidenced by the presence of only the (h00) pseudocubic reflections of LMO in the Θ–Θ scan and also from the rocking curve of the (200) reflection of LMO (Fig. 1.25), which shows an FWHM of 0.3°.

6   Conclusions

The above discussion of the results on films of various oxide systems deposited by different methods should suffice to demonstrate how one is able to get excellent single-crystalline (epitaxial) films with good interface properties as determined by lattice matching. What is particularly noteworthy is that an inexpensive, chemical method, namely nebulised spray pyrolysis, gives epitaxial films of technologically important oxides. The interface features, as well as the quality of the films obtained by nebulised spray pyrolysis, are as good as those obtained by the more difficult and expensive methods such as PLD and MBE. One can readily verify this observation by comparing the results on ferroelectric films obtained by the different methods presented earlier. Clearly, nebulised spray pyrolysis is likely to emerge as a viable technique for the preparation of epitaxial films of complex metal oxides.

7   References

1 Bai GR, Chang HLM, Foster CM, Shen Z, Lam DJ. *J. Mater. Res.* 1994; **9**: 156.

2 De Veirman AEM, Timmers J, Hakkens FJG, Cillesen JFM, Wolf RM. *Philips J. Res.* 1993; **47**: 185.

3 Greek J, Linker G, Mayer O. *Mater. Sci. Rep.* 1989; **4**: 193.

4 Kentgens APM, Carim AH, Dam B. *J. Cryst. Growth* 1988; **91**: 355.

5 Hseigh YF, Siegal MP, Hull R, Phillips JM. *Appl. Phys. Lett.* 1990; **57**: 2268.

6 Hwang DM, Nazar L, Venkatesan T, Wu XD. *Appl. Phys. Lett.* 1988; **57**: 1834.

7 Lee ST, Chen S, Hung LS, Braunstein G. *Appl. Phys. Lett.* 1989; **55**: 286.

8 Gupta A, Hussey BW, Guloy AM *et al. J. Solid State Chem.* 1994; **108**: 202.

9 Chrisey DB, Hubler GK. *Pulsed Laser Deposition of Thin Films.* New York: Wiley, 1994.

10 Blank DHA, Ijsselteijn RPJ, Out PG, Kuiper HJH, Flokstra J, Rogalla H. *Mater. Sci. Engg.* 1992; **B13**: 67.

11 Singh RK, Narayan J. *Phys. Rev.* 1990; **B41**: 8843.

12 Singh RK, Narayan J. *Mater. Sci. Engg.* 1989; **B3**: 217.

13 Lee AE, Platt CE, Burch JF, Simon RW, Goral JP, Al-Jassim MM. *Appl. Phys. Lett.* 1990; **57**: 2019.

14 Basu SN, Carim AH, Mitchell TE. *J. Mater. Res.* 1991; **6**: 1823.

15 De Veirman AEM, Cillesen JFM, Dekeijser M *et al. Mater. Res. Soc. Symp. Proc.* 1994; **341**: 329.

16 Herman MA, Sitter H (eds). *Molecular Beam Epitaxy: Fundamentals and Current Status.* Berlin: Springer-Verlag, 1989.

17 Joyce BA. *Contemp. Phys.* 1991; **32**: 21.

18 McKee RA, Walker FJ, Specht ED, Alexander KB. *Mater. Res. Soc. Symp. Proc.* 1994; **341**: 309.

19 Wessels BW. *Ann. Rev. Mater. Sci.* 1995; **25**: 525.

20 Kaiser DL, Vaudin MD, Rotter LD *et al. Mater. Res. Soc. Symp. Proc.* 1991; **361**: 355.

21 Foster CM, Csencsits R, Baldo PM *et al. Mater. Res. Soc. Symp. Proc.* 1995; **361**: 307.

22 Langlet M, Joubert JL. In Rao CNR (ed.) *Chemistry of Advanced Materials.* Oxford: Blackwell Scientific, 1993.

23 Raju AR, Aiyer HN, Rao CNR. *Chem. Mater.* 1995; **7**: 225.

24 Raju AR, Rao CNR. *Appl. Phys. Lett.* 1995; **66**: 896.

25 Lines ME, Glass AM (eds). *Principles and Applications of Ferroelectrics and Related Materials.* Oxford: Clarendon Press, 1977.

26 Klein JD, Yen A, Clauson SL. *Mater. Res. Soc. Symp. Proc.* 1994; **341**: 393.

27 Chahara K, Ohno T, Kasai M, Kozono Y. *Appl. Phys. Lett.* 1993; **63**: 1990.

28 von Helmolt R, Wecker J, Samwer K, Haupt L, Barner K. *J. Appl. Phys.* 1994; **76**: 6925.

29 Ju HL, Kwon C, Li Q, Greene RL, Venkatesan T. *Appl. Phys. Lett.* 1994; **65**: 2108.

30 Cheetham AK, Rao CNR. *Science* 1996; **272**: 369.

31 Verelst M, Rangavittal N, Rao CNR, Rousset A. *J. Solid State Chem.* 1993; **104**: 74.

32 Mahesh R, Kannan KR, Rao CNR. *J. Solid State Chem.* 1995; **114**: 294.

33 Hervieu M, Mahesh R, Rangavittal N, Rao CNR. *Euro. J. Solid State Inorg. Chem.* 1995; **32**: 79.

34 Mahendiran R, Tiwary SK, Raychaudhuri AK *et al. Phys. Rev.* 1996; **B53**: 3348.

Sub-micrometre spherical particles of TiO$_2$, ZrO$_2$ and PZT by nebulized spray pyrolysis of metal–organic precursors

P. Murugavel, M. Kalaiselvam, A. R. Raju and C. N. R. Rao*

Jawaharlal Nehru Centre for Advanced Scientific Research, Jakkur, Bangalore—560 064, India and Materials Research Centre, Indian Institute of Science, Bangalore—560 012, India

Nebulized spray pyrolysis of metal–organic precursors in methanol solution has been employed to prepare powders of TiO$_2$, ZrO$_2$ and PbZr$_{0.5}$Ti$_{0.5}$O$_3$(PZT). This process ensures complete decomposition of the precursors at relatively low temperatures. The particles have been examined by scanning and transmission electron microscopy as well as X-ray diffraction. As prepared, the particles are hollow agglomerates of diameter 0.1–1.6 μm, but after heating to higher temperatures the ultimate size of the particles comprising the agglomerates are considerably smaller (0.1 μm or less in diameter) and crystalline.

Spray pyrolysis has been widely used to prepare a variety of materials.[1-3] Thus, micrometre-sized particles of Al$_2$O$_3$ and CaAl$_2$O$_4$ were prepared by Rcy *et al.*[4] Several other oxides have also been prepared by spray pyrolysis by employing several variants of the techniques for atomization of the precursor solution which include pneumatic, ultrasonic and electrostatic methods. The droplet size, size distribution and the rate of atomization differ with the method employed for atomization. Spray pyrolysis consists of various steps which are atomization, precipitation, drying, thermolysis and sintering in the given order. Ultrasonic nebulization is a spraying technique in which ultrasound is focused at the surface of the liquid to produce micrometre sized droplets. This is a simple low-cost alternative for the deposition of high quality thin films of novel complex oxides.[5-7] In this laboratory, we have used nebulized spray pyrolysis to prepare oriented epitaxial films of several complex metal oxides.[8,9] This method essentially uses metal–organic precursors just as in MOCVD. Ultrasonic nebulizers have been recently used to prepare oxide materials such as ZnO,[10] stabilized zirconia,[11] BaTiO$_3$[12] and mullite.[13] In all these preparations, aqueous solutions of the relevant metal compounds have been employed and the particle size of the final oxide powders is of the order of 1 μm or higher. The use of aqueous solutions necessarily requires that the transducer employed in ultrasonic nebulization is outside the container so that it does not come in direct contact with the solution. However, if one uses metal salts in non-aqueous solvents, it would be possible to keep the solution in contact with the transducer which would lower the power required for nebulization. Besides, the use of organometallic precursors in organic solvents may be better for preparing powders of complex oxides with greater compositional homogeneity. We have employed nebulized spray pyrolysis of organometallic precursors and prepared submicrometre, spherical particles of TiO$_2$, ZrO$_2$ and PbZr$_{0.5}$Ti$_{0.5}$O$_3$.

Experimental

We have employed titanium isopropoxide and zirconium isopropoxide precursors in methanol solution (0.1 M) for preparing TiO$_2$ and ZrO$_2$ powders respectively. In order to prepare

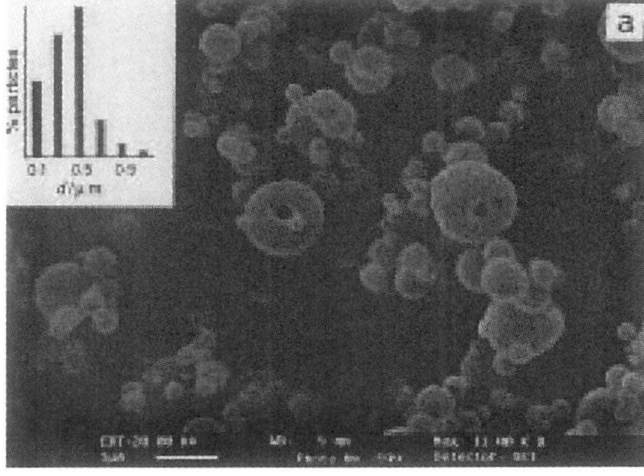
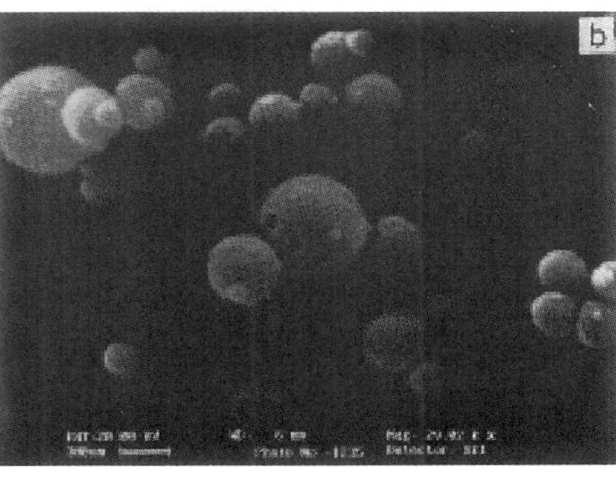

Fig. 2 Scanning electron micrographs of TiO$_2$ particles: (a) as obtained by nebulized spray pyrolysis at 723 K (the inset shows the particle size distribution of the powder) and (b) after heating the powder at 773 K for 12 h

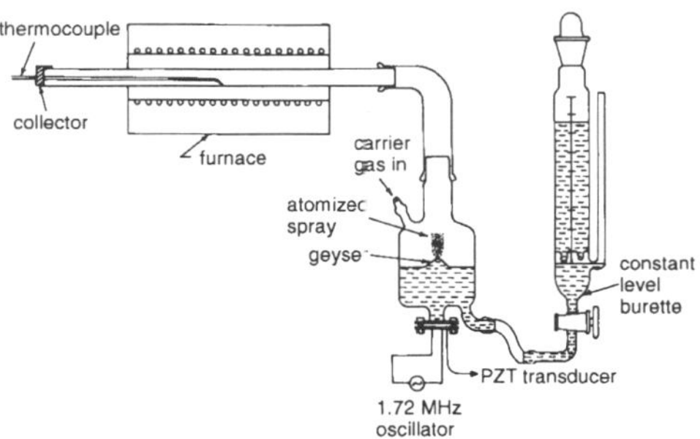

Fig. 1 Schematic diagram of the nebulized spray pyrolysis apparatus used for the preparation of spherical oxide particles

powders of $PbZr_{0.5}Ti_{0.5}O_3$ (PZT), a methanol solution containing stoichiometric quantities of lead acetate, zirconium isopropoxide and titanium isopropoxide (0.05 M each) was employed. Dry air was used (5 l min^{-1}) as a carrier gas.

A schematic diagram showing different parts of the locally fabricated apparatus for ultrasonically nebulized spray pyrolysis is shown in Fig. 1. It consists of two zones: (a) the atomization chamber and (b) the pyrolysis reactor. The source liquid is kept in the atomization chamber and it is designed such that the bottom of the chamber has a cylindrical opening to fit the PZT transducer which oscillates at a frequency of 1.72 MHz. The special feature of this design is that the upper electrode of the PZT transducer is in direct contact with the source liquid. This gives the highest energy transfer to the liquid when compared to the non-contact method, which is used when there is a possibility of the electrode material

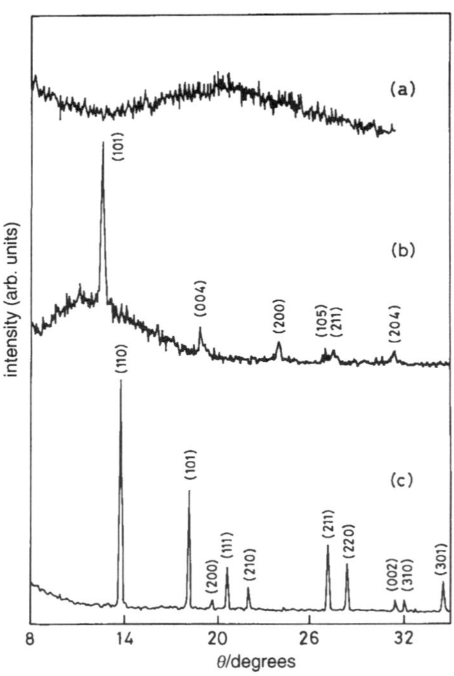

Fig. 4 X-Ray diffraction patterns of TiO_2 powders: (a) as obtained at 723 K, (b) after heating at 773 K for 12 h and (c) after heating at 1073 K for 12 h

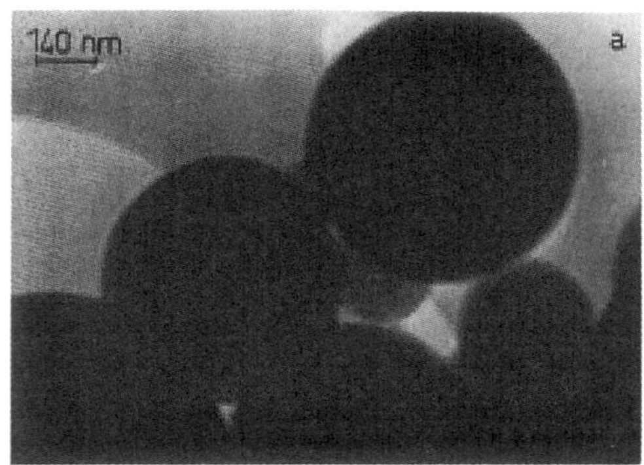

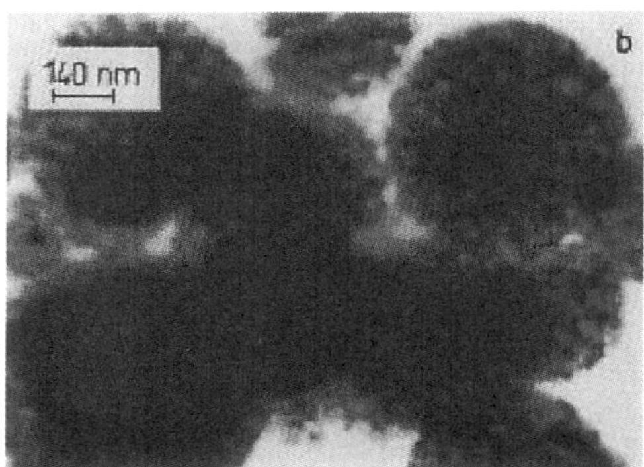

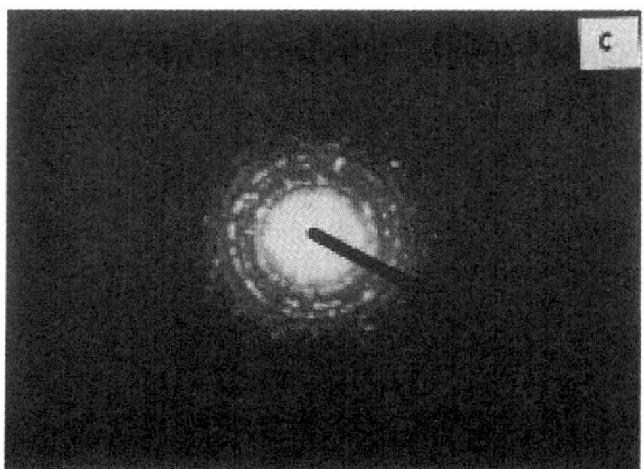

Fig. 3 Transmission electron micrographs of TiO_2 particles: (a) as obtained by nebulized spray pyrolysis at 723 K, (b) after heating the powder at 773 K for 12 h and (c) electron diffraction pattern of the powder heated at 773 K for 12 h

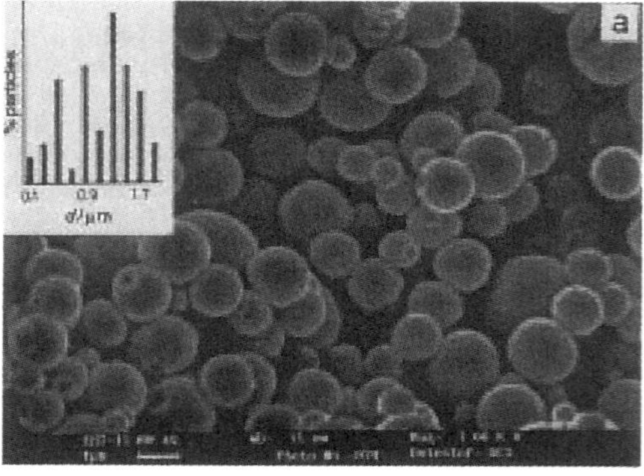

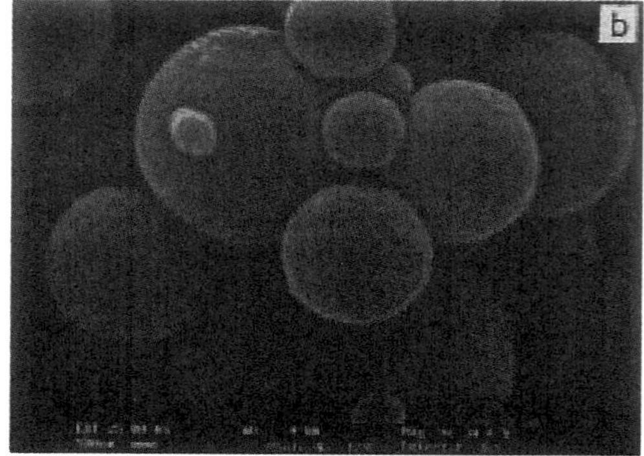

Fig. 5 Scanning electron micrographs of ZrO_2 particles: (a) as obtained by nebulized spray pyrolysis at 723 K (the inset shows the particle size distribution of the powder) and (b) after heating the powder at 773 K for 12 h

reacting with the solution. A minor disadvantage is that the liquid pH has to be maintained around 7. The liquid level in the atomization chamber is maintained by using a constant level burette which also allows the measurement of the volume of nebulized liquid.

When a high-frequency ultrasonic beam is directed at the liquid/gas interface through the PZT transducer a geyser forms at the surface of the liquid and the height of the geyser is proportional to the acoustic intensity and the physical properties of the liquid (vapour pressure, viscosity, density and surface tension). The geyser formation is accompanied by the cavitation at the liquid/gas interface. When the amplitude of the acoustic vibrations exceeds a certain threshold value, liquid atomization occurs. Above this threshold a continuous and regular mist (nebulized spray) is generated. The nebulized spray produced in the first chamber is transported by a carrier gas, which was dry air in all the experiments, introduced

through a side-port. The second zone is a long quartz tube which is kept in a furnace at a preset temperature. The processes of drying and thermolysis take place in this zone and sintering is done in a separate furnace. The reason for having a single zone for drying and thermolysis is the following. If precipitation and drying occur slowly, there is a tendency for different components in the droplet to precipitate at different times which would lead to inhomogenity in the composition. The product consisting of fine powder is collected in a collector fixed at the other end of the quartz tube.

The morphology and particle size distribution were analyzed by employing a Leica S440i scanning electron microscope and a Quantimet Q500MC image analyser respectively and energy dispersive X-ray analysis was carried out by employing an Oxford instruments ISIS Si detector. Transmission electron microscope images were recorded with a JEOL 3010 microscope. The phase compositions of the powders obtained by the nebulized spray pyrolysis and subsequent heating at different temperature were determined by employing a Seifert 3000 X-ray powder diffractometer with Cu-Kα radiation (θ–θ geometry).

Results and Discussion

TiO$_2$

Fig. 2(a) shows an SEM micrograph of the TiO$_2$ powder obtained after nebulized spray pyrolysis of a 0.1 M solution of

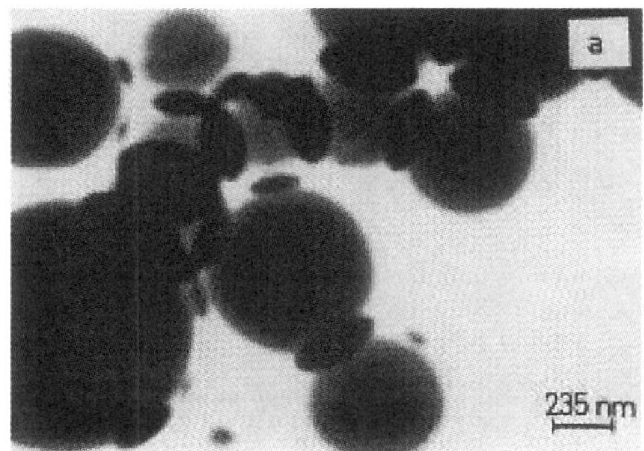

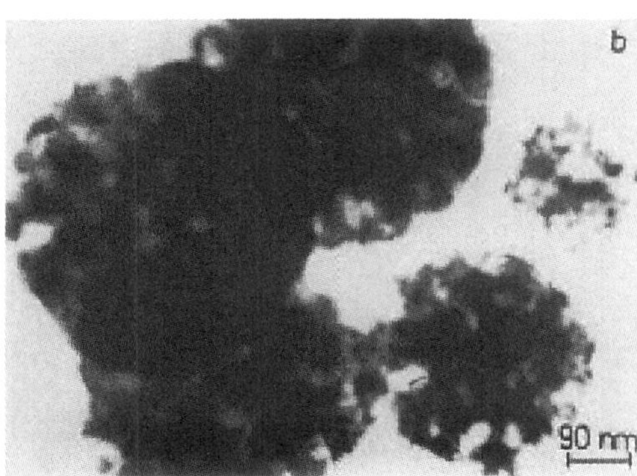

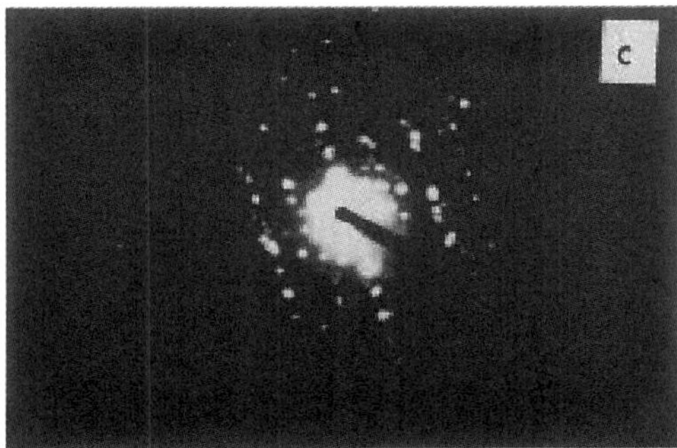

Fig. 6 Transmission electron micrographs of ZrO$_2$ particles: (a) as obtained by nebulized spray pyrolysis at 723 K, (b) after heating the powder at 773 K and (c) electron diffraction pattern of the powder heated at 1023 K for 12 h

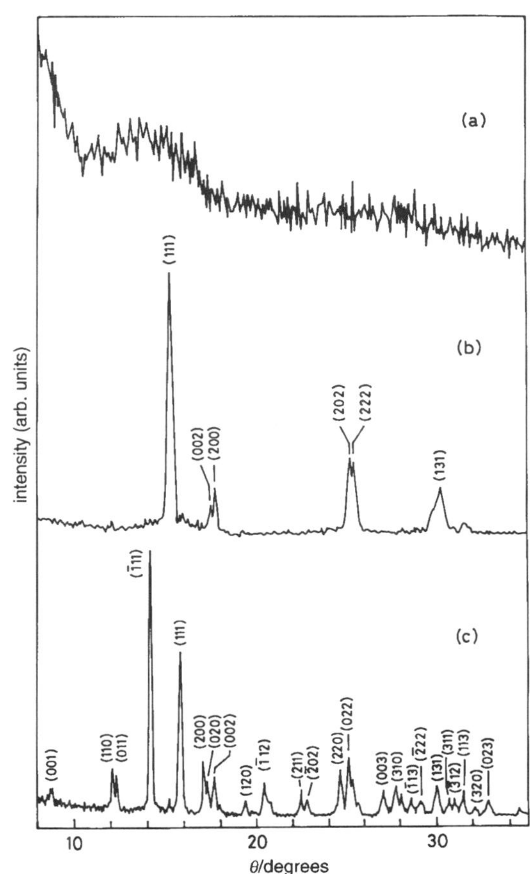

Fig. 7 X-Ray diffraction patterns of ZrO$_2$ powders: (a) as obtained at 723 K, (b) after heating at 773 K for 12 h and (c) after heating at 1073 K for 12 h

titanium isopropoxide in methanol at 723 K. The particles obtained are spherical or doughnut shaped. The particle size distribution shown in Fig. 2(a) suggests that most of the particles have a diameter in the range 0.3–0.5 µm. Some of the particles have cavities in them which are likely to result from predominant surface precipitation and fast evaporation of solvent. The particles are actually hollow spherical shells and shell fragments and the particle diameter of 0.3–0.5 µm truly corresponds to agglomerates of small individual particles of <0.1 µm diameter. A TEM image of the TiO_2 particles as obtained from nebulized spray pyrolysis is shown in Fig. 3(a). This image also confirms the spherical nature of the particles. Electron diffraction patterns show them to be amorphous and this is also confirmed by the X-ray diffraction pattern [Fig. 4(a)]. However, after heating the as-obtained particles at 773 K for 12 h they become crystalline. The X-ray diffraction pattern [Fig. 4(b)] shows the structure to correspond to that of anatase phase (JCPDS No. 21-1272) with $a = 3.7986$ A and $c = 9.5544$ A. The electron diffraction pattern shows spotty patterns [Fig. 3(c)] confirming the crystallization of TiO_2 particles on heating to 773 K. The SEM image of the anatase phase particles is shown in Fig. 2(b) and the corresponding TEM image in Fig. 3(b). These images establish that the particles remain spherical after crystallization. The TEM image also reveals how the large spherical particles are agglomerates and are actually composed of smaller particles (*ca.* 20 nm). We obtained the rutile phase of TiO_2 (JCPDS No. 22-1276) with $a = 4.5949$ A and $c = 2.9540$ A by heating the anatase powder at 1073 K for 12 h as indicated by the X-ray diffraction pattern in Fig. 4(c).

ZrO_2

Fig. 5(a) shows the SEM micrograph of the ZrO_2 powder obtained by the nebulized spray pyrolysis of the precursor solution at 723 K. The particles are spherical [Fig. 5(a)] and the size distribution of the particles indicates the diameter to be in the range 0.8–1.6 µm. The TEM micrograph in Fig. 6(a) also shows the spherical nature of these particles. The powder is amorphous as revealed by the electron and X-ray diffraction patterns and the latter is shown in Fig. 7(a). Heating the powder at 823 K for 12 h gives a tetragonal phase (JCPDS No. 17-923) with $a = 5.0922$ A and $b = 5.1958$ A [Fig. 7(b)]. SEM and TEM images of the powder heated to 823 K are shown in Fig. 5(b) and 6(b), respectively. The images show that the particles remain spherical after the heat treatment and that they are in turn composed of smaller particles (*ca.* 20 nm). Heating the tetragonal powder phase at 1023 K for 12 h resulted in the monoclinic phase (JCPDS No. 36-420) $a = 5.3165$ A, $b = 5.1867$ A, $c = 5.1445$ A and $\beta = 99.151°$, as confirmed by the X-ray diffraction pattern shown in Fig. 7(c).

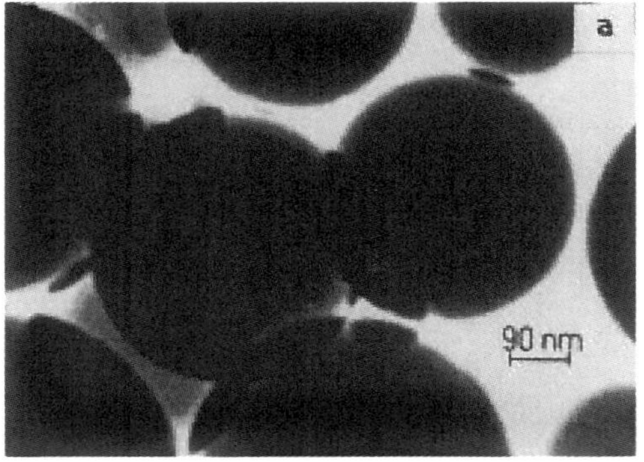

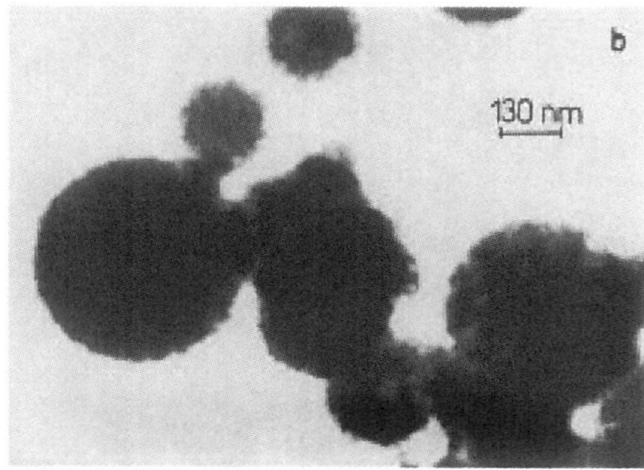

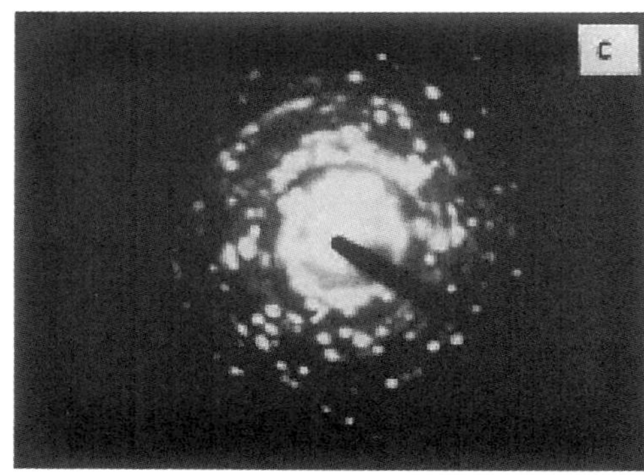

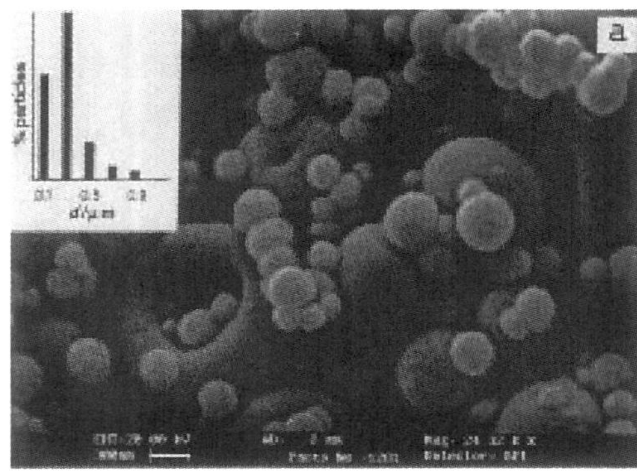

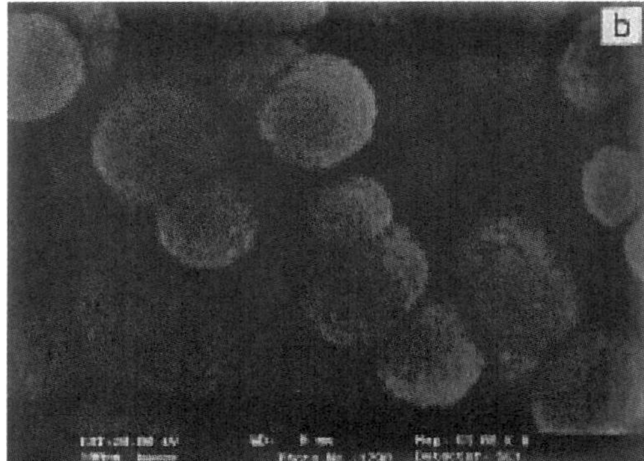

Fig. 8 Scanning electron micrographs of PZT particles: (a) as obtained by nebulized spray pyrolysis at 673 K (the inset shows the particle size distribution of the powder) and (b) after heating the powder at 823 K for 12 h

Fig. 9 Transmission electron micrographs of PZT particles: (a) as obtained by nebulized spray pyrolysis at 673 K, (b) after heating the powder at 823 K and (c) electron diffraction pattern of the powder after heating at 823 K for 12 h

PZT

The SEM and TEM images of the powders of PZT as obtained by nebulized spray pyrolysis of the precursor solution at 673 K are shown in Fig. 8(a) and 9(a), respectively. The particle size distribution shown as an inset in Fig. 8(a) suggest the diameter to be in the range 0.1–0.5 µm. This powder is amorphous as revealed by the electron and X-ray diffraction patterns and the latter is shown in Fig. 10(a). After heating the amorphous powder at 823 K for 12 h, it transformed into the stable tetragonal phase (JCPDS No. 33-784) with $a = 4.0655$ A and $c = 4.0646$ A. The X-ray diffraction pattern of the powder heated to 823 K is shown in Fig. 10(b). The SEM and TEM micrographs of these particles in Fig. 8(b) and 9(b) show them to be spherical even after crystallization. The large spherical particles are composed of smaller particles (20–30 nm) just as in the case of TiO_2 and ZrO_2. Energy dispersive X-ray analysis (Fig. 11) of the PZT powder was analyzed for quantitative elemental composition by employing the SEMQuant software with ZAF correction procedure. This gave a Pb:Zr:Ti ratio of $1:0.5:0.5$, confirming the composition to be $PbZr_{0.5}Ti_{0.5}O_3$.

Fig. 12(a) shows a TEM image to illustrate how the particles

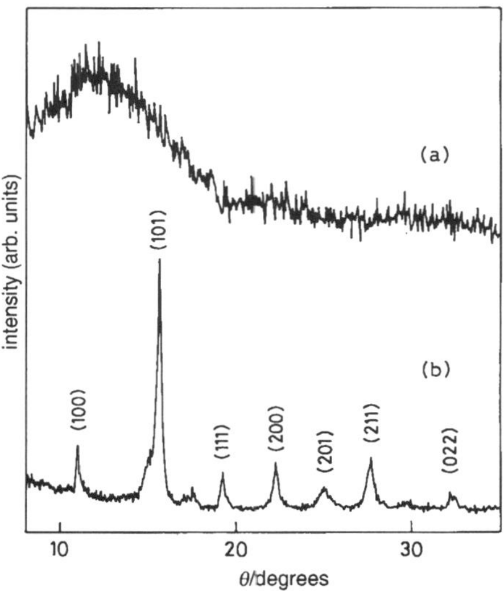

Fig. 10 X-Ray diffraction patterns of PZT powder: (a) as obtained at 723 K, (b) after heating at 823 K for 12 h

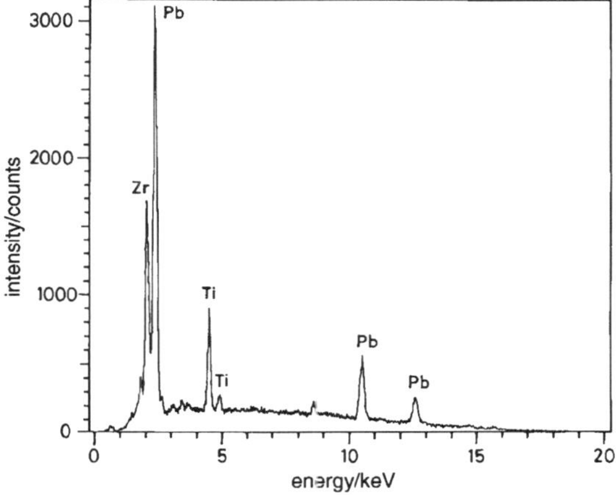

Fig. 11 EDAX spectrum of the $PbZr_{0.5}Ti_{0.5}O_3$ powder

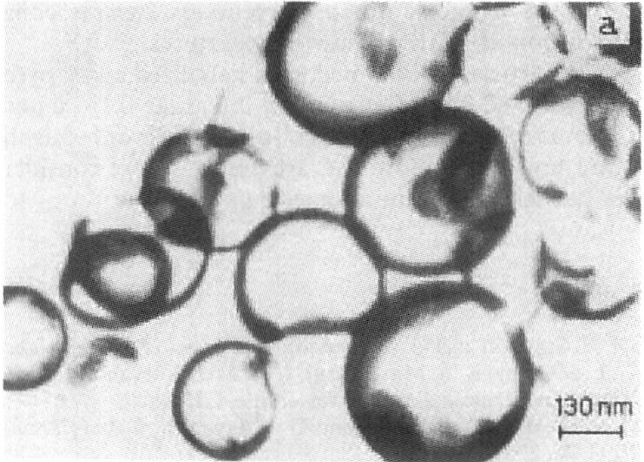

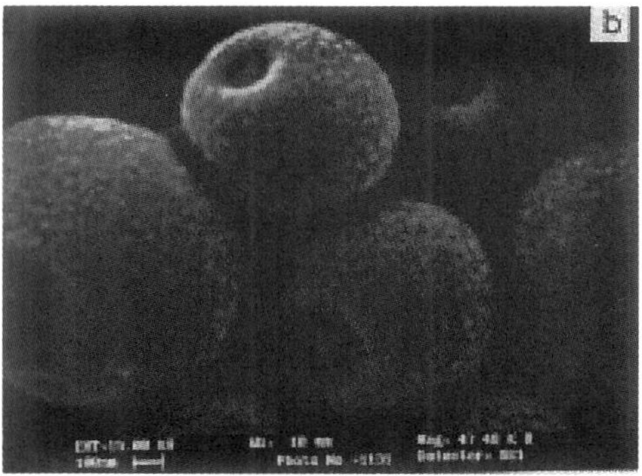

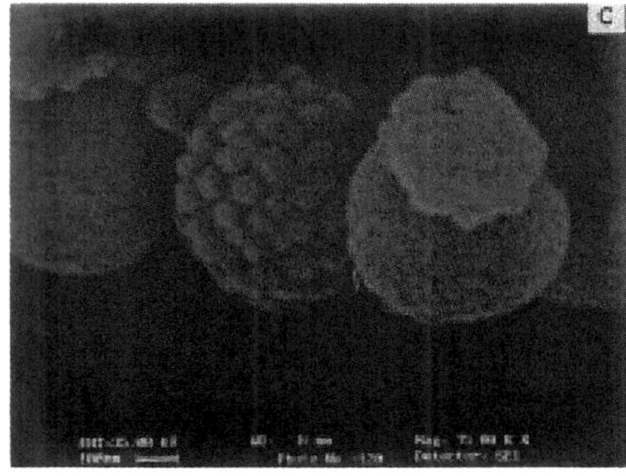

Fig. 12 (a) TEM of PZT spheres and shells as obtained from nebulized spray pyrolysis, (b) high magnification SEM image of ZrO_2 heated to 773 K, (c) high magnification SEM image of PZT heated to 823 K

are hollow. These hollow spherical particles obtained by pyrolysis are generally composed of smaller particles as mentioned earlier. This is clearly seen from Fig. 12(b) and (c) where high magnification SEM images of the crystalline particles of ZrO_2 and PZT are shown. It thus appears that although the particle size distributions in Fig. 2(a), 5(a) and 8(a) give diameters up to 1 µm or more, the ultimate size of the particles is much smaller. Thus, the size of the larger agglomerated spherical particles of ZrO_2 in Fig. 12(b) and of PZT in Fig. 12(c) are in the range 0.5–1.5 µm, but the small particles comprising these are fairly uniform with mean diameters of <0.1 µm.

Conclusion

The present study of the preparation of metal oxide powders by nebulized spray pyrolysis of metal–organic precursors establishes the efficacy of the method. It has certain noteworthy features as detailed overleaf.

(*i*) The use of metal–organic precursors ensures complete decomposition at relatively low temperatures.

(*ii*) The particles as-obtained from nebulized spray pyrolysis are spherical and have diameters in the range 0.1–1.6 μm.

(*iii*) Some of the particles are collapsed shells or hemispheres. Most are however hollow and are composed of considerably smaller particles of 0.1 μm diameter or less.

References

1 D. W. Sproson and G. L. Messing, in *Ceramic Powder Science*, ed. G. L. Messing, K. S. Mazdiyasni, J. W. McCauley and R. A. Haber, American Ceramic Society, Westerville, OH, 1987.
2 G. L. Messing, S. C. Zhang and G. V. Jayanthi, *J. Am. Ceram. Soc.*, 1993, **76**, 2707.
3 R. Stevens, *Trans. J. Br. Ceram. Soc.*, 1981, **80**, 81.
4 D. M. Roy, R. R. Neurgaonkar, T. P. O'Holleran and R. Roy, *Am. Ceram. Soc. Bull.*, 1981, **56**, 1023.
5 W. W. Xu, R. Kershaw, K. Dwight and A. Wold, *Mater. Res. Bull.*, 1990, **25**, 1385.
6 M. Langlet and J. C. Joubert, in *Chemistry of Advanced Materials*, ed. C. N. R. Rao, IUPAC, 21st Century Monogr. Ser., Blackwell, Oxford, 1993.
7 A. R. Raju, H. N. Aiyer and C. N. R. Rao, *Chem. Mater.*, 1995, **7**, 225.
8 A. R. Raju, H. N. Aiyer and C. N. R. Rao, in *Interfacial Science*, ed. M. W. Roberts, IUPAC, 21st Century Monogr. Ser., Blackwell, Oxford, 1997.
9 H. N. Aiyer, A. R. Raju and C. N. R. Rao, *Chem. Mater.*, 1997, **9**, 755.
10 T. Q. Liu, O. Sakurai, N. Mizutani and M. Kato, *J. Mater. Sci.*, 1986, **21**, 3698.
11 B. Dubois, D. Ruffier and P. Odier, *J. Am. Ceram. Soc.*, 1989, **72**, 713.
12 O. Milosevic and D. Uskokovic, *Mater. Sci. Eng. A*, 1993, **168**, 249.
13 Dj. Janackovic, V. Jokanovic, Lj. Kostic-Gvozdenovic, Lj. Zivkovic and D. Uskokovic, *J. Mater. Res.*, 1996, **11**, 1706.

Paper 7/00301C; Received 13th January, 1997

Eur. J. Solid State Inorg. Chem.
t. 31, 1994, p. 409-422

A study of cubic bismuth oxides of the type $Bi_{26-x}M_xO_{40-\delta}$ (M = Ti, Mn, Fe, Co, Ni or Pb) related to γ-Bi_2O_3

N. RANGAVITTAL, T. N. GURU ROW and C. N. R. RAO*

Solid State and Structural Chemistry Unit
Indian Institute of Science, Bangalore-560 012, India

(E. K., received May 25, 1994; accepted June 18, 1994)

ABSTRACT. – A structural investigation of cubic oxides (space group I23) of the formula $Bi_{26-x}M_xO_{40-\delta}$ (M = Ti, Mn, Fe, Co, Ni and Pb) related to the γ-Bi_2O_3 phase has been carried out by the Rietveld profile analysis of high-resolution X-ray powder diffraction data in order to establish the cation distributions. Compositional dependence of the cation distribution has been examined in the case of $Bi_{26-x}Co_xO_{40-\delta}$ (1 < x < 16). The study reveals that in $Bi_{26-x}M_xO_{40-\delta}$ with M = Ti, Mn, Fe, Co or Pb, the M cations tend to occupy tetrahedral (2a) sites when x < 2 while the octahedral (24f) sites are shared by the excess Co or Ni cations with Bi atoms when x > 2. Also experimental magnetic moments of Mn, Co and Ni derivatives have been used to establish the valence state and distribution of these cations.

* To whom all correspondence should be addressed.

ISSN 0992-4361/94/05/$4.00/© Gauthier-Villars

INTRODUCTION

Aurivillius and Sillen [1] long ago reported the formation of body-centered cubic oxides (space group I23) of the general formula $Bi_{12}MO_{20}$ or $Bi_{24}M_2O_{40}$ (M = Al, Si, Fe, Zr and Ce) and this was further extended by Levin and Roth [2]. The first X-ray crystallographic study of this family of bismuth oxides was carried out by Abrahams et al [3] in $Bi_{12}GeO_{20}$ and this was followed by an investigation of $Bi_{25}FeO_{40}$ and $Bi_{38}ZnO_{60}$ by Craig and Stephenson [4]. Ramanan et al [5] reported the reaction of Bi_2O_3 with MgO, NiO, Co_3O_4 and Al_2O_3 giving rise to the corresponding ternary bismuth oxides, $Bi_{18}Mg_8O_{38}$, $Bi_{18}Ni_8O_{36}$, $Bi_{20}Co_6O_{39}$ and $Bi_{24}Al_2O_{39}$ crystallizing in the BCC structure related to $\gamma-Bi_2O_3$ [6]. The structure of $\gamma-Bi_2O_3$ is best represented as $Bi_{26}O_{40}$ with 24 Bi^{3+} ions in octahedral coordination with one Bi^{3+} and one Bi^{5+} ion in each of the two tetrahedral sites. A neutron powder diffraction investigation of $Bi_{12}[Bi^{+V}_{0.75}V^{+V}_{0.05}\ [\ ]_{0.20}]O_{20}$ was carried out by Soubeyroux et al [7]. This compound crystallizes in the I23 space group with a = 10.265 Å. Devalette et al [8] reported the study of AO_4 and BO_4 tetrahedral groups in the $Bi_{12}[A^{+III}_{1/2}\ B^{+V}_{1/2}]O_{20}$ (A = Fe, Co, Ga and B = P, As, V, Bi) by vibrational spectroscopy and magnetic measurements. Bismuth in these oxides is in an unusual (5 + 2) oxygen coordination (5 Bi-O, 2.07 - 2.60 Å and 2Bi - O, 3.08 - 3.17 Å) arising from the presence of the $6s^2$ lone pair on the Bi^{3+} ions.

We considered it interesting to investigate the cation distribution in the octahedral (24f) and the tetrahedral (2a) sites in several of the $\gamma-Bi_2O_3$-like phases of the general formula $Bi_{26-x}M_xO_{40-\delta}$ and also the compositional dependence of the cation distribution. For this purpose, we have investigated the structures of $Bi_{24}Ti_2O_{40-\delta}$, $Bi_{24}Mn_2O_{40-\delta}$, $Bi_{25}FeO_{40}$, $Bi_{10}Co_{16}O_{40-\delta}$, $Bi_{20}Ni_6O_{40-\delta}$ and $Bi_{24}Pb_2O_{40-\delta}$ by Rietveld profile refinement of high-resolution X-ray powder diffraction data. In addition, we have examined the cation distribution in $Bi_{26-x}Co_xO_{40-\delta}$ with x varying between 1 and 16. We have carried out magnetic susceptibility studies of the Mn, Co and Ni derivatives to establish the oxidation state and distribution of the cation in the octahedral and tetrahedral sites.

EXPERIMENTAL

Oxides of the general formula $Bi_{26-x}M_xO_{40-\delta}$ were prepared as follows. $Bi_{24}Ti_2O_{40-\delta}$, $Bi_{24}Mn_2O_{40-\delta}$, $Bi_{25}FeO_{40}$ and $Bi_{24}Pb_2O_{40-\delta}$ were prepared by the reaction of Bi_2O_3 with the other component oxides. The oxide mixture was heated at 1050 K for 72 hr with two grindings inbetween. The oxides of the series $Bi_{26-x}Co_xO_{40-\delta}$ were prepared starting with oxide mixtures with Bi:Co ratios of $\simeq$ 25:1, 24:2, 20:6, 17:9, 13:13 were also prepared by this procedure. the products with Bi:M ratio of 20:6,17:9 and 13:13 had minor quantities of Co_3O_4. $Bi_{10}Co_{16}O_{40-\delta}$ and $Bi_{20}Ni_6O_{40-\delta}$ were prepared by a low temperature method employing a strong alkaline medium. The alkaline solution served both as a precipitating agent and as a reaction medium for the formation of the oxides with higher content of the transition metal. An aqueous solution of bismuth and transitional metal nitrates was added dropwise to a strong KOH solution under constant stirring. The pH of the reaction mixture was adjusted to be between 9.0 and 12.0. Stabilization of the γ-Bi_2O_3 like phase was achieved in a period of 24 hours by maintaining the stirred oxygen spurged reaction mixture at a temperature of 350 K.

X-ray diffraction patterns of the $Bi_{26-x}M_xO_{40-\delta}$ series were recorded with a STOE/STADI high-resolution X-ray powder diffraction. The diffraction data were recorded at intervals of 0.02° from 8 - 80° in 2θ at the rate of 4 sec/per step with a linear position sensitive detector. The indexing and the profile refinements on these samples were carried out using the STOE STADIP SOFTWARE [9]. The Bi:M ratios in the samples prepared by solution method were determined using energy dispersive X-ray emission (EDX) analysis carried out using a scanning electron microscope fitted with a link A - 1- EDX analyzer. The EDX analysis on the samples $Bi_{26-x}M_xO_{40-\delta}$ (M = Co and Ni) showed the Bi:M ratio to be 37:63 and 77:23 for M = Co and Ni respectively. Magnetic susceptibility measurements of $Bi_{26-x}M_xO_{40-\delta}$ (M = Mn, Co, Ni) were carried out with a model - 300 Lewis-Coil magnetometer at an applied field of 5000 Oe from 30 to 300 K.

412   N. RANGAVITTAL, T. N. GURU ROW AND C.N.R. RAO

RESULTS

X-ray powder diffraction patterns of the compounds with the general formula $Bi_{26-x}M_xO_{40-\delta}$ could be indexed using 30 well resolved peaks on the basis of a BCC structure in the space group I23. High-resolution X-ray powder diffraction data of $Bi_{26-x}M_xO_{40-\delta}$ were subjected to deconvolution and the intensities thus obtained were used to refine the structural parameters. In Figures 1-4 we show the experimental X-ray powder diffraction patterns for the compounds $Bi_{24}Mn_2O_{40-\delta}$, $Bi_{10}Co_{16}O_{40-\delta}$, $Bi_{20}Ni_6O_{40-\delta}$ and $Bi_{20}Co_6O_{40-\delta}$. These oxides represent the different types of γ-Bi_2O_3-like phases investigated by us. For the compositions, $Bi_{24}M_2O_{40-\delta}$ (M = Ti, Mn, Pb), structure factor calculations were carried out with two M atoms present in the tetrahedral (2a) sites and 24 Bi atoms present in the octahedral (24f) sites resulting in the γ-Bi_2O_3 - like structure. The oxygen atoms were distributed in 24f and two sets of 8c sites in the I23 space group. The agreement of the calculated X-ray profile with the experimental one was excellent (Figure 1).

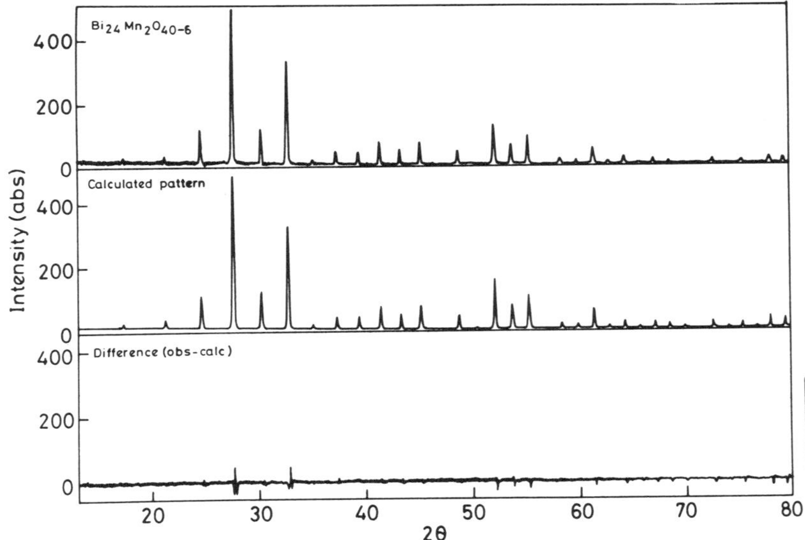

Fig. 1 - Observed , calculated and difference X-ray diffraction profiles for $Bi_{24}Mn_2O_{40-\delta}$

This starting model for structural refinement is also consistent with the structure of $Bi_{24}Ge_2O_{40}$ [3]. In the case of $Bi_{25}FeO_{40}$, it was assumed that 24 Bi atoms occupy octahedral (24f) sites with one Bi sharing the tetrahedral (2a) site with one Fe atom. This model gave a satisfactory agreement between the calculated and experimental diffraction profiles.

Next, the structural refinements on the compositions $Bi_{10}Co_{16}O_{40-\delta}$ and $Bi_{20}Ni_6O_{40-\delta}$ showed partial occupancy of Co and Ni in the octahedral (24f) sites, the tetrahedral (2a) sites being fully occupied by these transition metal cations. In these two oxides, the cation occupancies in octahedral (24f) sites were refined in addition to the atomic coordinates. The structure factor calculations based on this resulted in good agreement of the calculated X-ray diffraction patterns with the experimental ones (Figure 2 & 3).

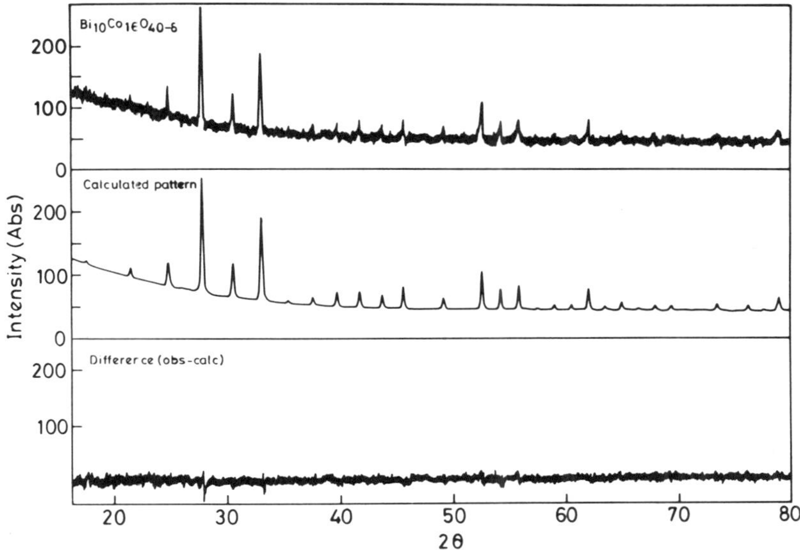

Fig. 2 - Observed , calculated and difference X-ray diffraction profiles for $Bi_{10}Co_{16}O_{40-\delta}$

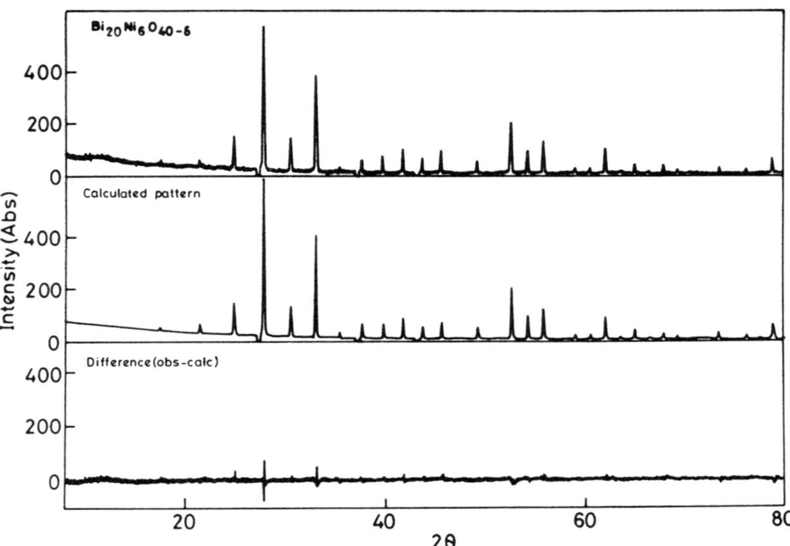

Fig. 3 - Observed , calculated and difference X- ray diffraction profiles for $Bi_{20}Ni_6O_{40-\delta}$

The refined atomic parameters with the derived compositions are listed in Table I and the details of the distribution of the M cations are shown in Table II .

In the case of $Bi_{26-x}Co_xO_{40-\delta}$, refinements were carried out assuming that the 24f Bi-sites in the γ-Bi_2O_3 structure are occupied randomly by Bi and Co atoms. However in $Bi_{25}CoO_{40-\delta}$ the calculations were based on the extra bismuth present in the tetrahedral (2a) site. The cation occupancies in octahedral and tetrahedral sites were refined to convergence. The comparison of the calculated X-ray diffraction pattern at this stage of refinement showed good agreement with the experimental X-ray diffraction pattern (Figure 4).

The derived compositions on the basis of cation occupancy refinement is shown in Table III. The details of Co distribution in the octahedral (24f) and tetrahedral (2a) sites are given in Table IV.

TABLE I – Atomic parameters for $Bi_{26-x}M_xO_{40-\delta}$ (M = Ti, Mn, Fe, Co, Ni, Pb) in the space group I23

	M = Ti, x = 2	M = Mn, x = 2	M = Fe, x = 1	M = Pb, x = 2	M = Co, x = 15.9	M = Ni, x = 6.4
a,Å	10.202(5)	10.230(7)	10.210(5)	10.292(8)	10.138(2)	10.141(7)
Bi1/M1 (24f)						
x	0.1760(7)	0.1777(9)	0.1754(8)	0.1768(9)	0.169(8)	0.1744(9)
y	0.318(6)	0.3186(9)	0.3170(7)	0.3209(10)	0.3150(7)	0.3175(9)
z	0.0151(6)	0.0174(8)	0.0141(7)	0.0171(9)	0.020(7)	0.0167(8)
n Bi1/M1	1.0/0.0	1.0/0.0	1.0/0.0	1.0/0.0	0.42(3)/0.58	0.82(4)/0.18
U	0.0181(9)	0.0439(12)	0.0258(20)	0.0326(20)	0.0175(6)	0.0178(20)
M2/Bi2(2a)						
x,y,z	0.0	0.0	0.0	0.0	0.0	0.0
M2/Bi2 U	0.0162(10)	0.0175(30)	0.0240(14)	0.0383(15)	0.04	0.04
n	1.0/0.0	1.0/0.0	0.42(4)/0.52	1.0/0.0	1.0/0.0	1.0/0.0
^{a}O1 (8c)						
(x,x,x)	0.664(8)	0.684(9)	0.689(9)	0.674(14)	0.688(9)	0.698(9)
^{a}O2 (24f)						
x	0.639(6)	0.627(9)	0.644(10)	0.644(10)	0.634(16)	0.629(8)
y	0.753(7)	0.756(8)	0.766(8)	0.744(12)	0.761(16)	0.757(9)
z	0.989(12)	0.988(10)	0.978(12)	0.979(12)	0.987(10)	0.994(19)
^{a}O3 (8c)						
(x,x,x)	0.927(7)	0.892(9)	0.896(2)	0.917(10)	0.822(20)	0.896(14)
$R_{(p)}$,$R_{(wp)}$	0.1011,0.1380	0.1035,0.1420	0.096,0.1265	0.092,0.1392	0.0759,0.1016	0.1121,0.1326
$R_{(I,hkl)}$	0.0781	0.1090	0.0816	0.1035	0.1866	0.0957

^{a} Temperature factors of oxygens fixed at 0.04 $Å^2$

TABLE II - Distribution of the M cations in $BI_{26-x}M_xO_{40-\delta}$ (M = Ti, Mn,Fe,Co,Ni,Pb) in octahedral (24f) **and** tetrahedral (2a) sites.

	composition	occupancy	
		24f	2a
(a)	M = Ti, x = 2	–	2
	M = Mn, x = 2	–	2
	M = Fe, x = 1	–	1
	M = Co, x = 15.9	13.9(3)	2
	M = Ni, x = 6.4	4.4 (4)	2
	M = Pb, x = 2	–	2

(a)   In $Bi_{25}FeO_{40-\delta}$ 24 Bi occupy octahedral site and one extra Bi goes to tetrahedral site.

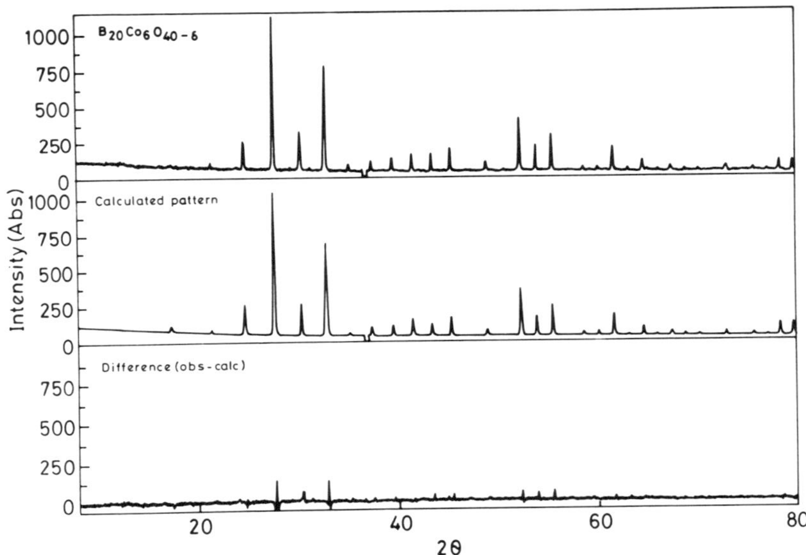

Fig. 4 – Observed , calculated and difference X-ray diffraction profiles for $Bi_{20}Co_6O_{40-\delta}$

TABLE III : Atomic parameters for $Bi_{26-x}Co_xO_{40-\delta}$ in the space group I23

	$Bi_{25}CoO_{40-\delta}$	$Bi_{23.6}Co_{1.8}O_{40-\delta}$	$Bi_{19.7}Co_{6.2}O_{40-\delta}$	$Bi_{16.3}Co_{8.9}O_{40-\delta}$	$Bi_{12.2}Co_{12.8}O_{40-\delta}$	$Bi_{10.1}Co_{15.9}O_{40-\delta}$
a, Å	10.215(1)	10.217(1)	10.210(1)	10.199(1)	10.190(1)	10.138(2)
Bi1/Co1(24f)						
x	0.1756(8)	0.1758(8)	0.1755(9)	0.1765(9)	0.1757(11)	0.1690(8)
y	0.3176(7)	0.3174(7)	0.3177(8)	0.3187(10)	0.3157(13)	0.315(7)
z	0.0148(6)	0.0122(9)	0.0138(9)	0.0111(14)	0.0137(14)	0.020(7)
n Bi1/Co1	1.0/0.0	0.98(5)/0.0	0.82(9)/0.18	0.68(6)/0.29	0.51(6)/0.45	0.42(3)/0.58
U	0.273(18)	0.0236(18)	0.0196(20)	0.0349(24)	0.0286(32)	0.0175(6)
Co2/Bi2(2a)						
x,y,z	0.0	0.0	0.0	0.0	0.0	0.0
n Co2/Bi2	0.5/0.48(4)	0.90(5)/0.0	0.94(5)/0.0	0.97(6)/0.0	1.0/0.0	1.0/0.0
U	0.0493(14)	0.0056(24)	0.0051(24)	0.04	0.04	0.04
^{a}01 (8c)						
(x,x,x)	0.697(8)	0.643(9)	0.634(10)	0.656(22)	0.683(10)	0.688(2)
^{a}02 (24f)						
x	0.635(7)	0.631(8)	0.641(8)	0.634(8)	0.645(8)	0.634(10)
y	0.775(7)	0.755(8)	0.740(8)	0.763(8)	0.757(8)	0.761(17)
z	0.988(2)	0.967(9)	0.952(9)	0.974(10)	0.968(10)	0.987(20)
^{a}03 (8c)						
(x,x,x)	0.886(12)	0.838(10)	0.816(9)	0.802(8)	0.839(9)	0.822(20)
$R_{(p)},R_{(wp)}$	0.1097,0.1173	0.0737,0.0949	0.1035,0.1355	0.0666,0.0873	0.0591,0.0769	0.0759,0.1016
$R_{(I,hkl)}$	0.0503	0.0826	0.1001	0.1125	0.1351	0.1866

^{a} Temperature factors of oxygens fixed at 0.04 $Å^2$

TABLE IV - Distribution of Co in $Bi_{26-x}Co_xO_{40-\delta}$ in the octahedral (24f) and tetrahedral (2a) sites.

x	occupancy	
	24f	2a
1.0	-	1.0(4)
1.8	-	1.8(5)
6.2	4.3(9)	1.9(5)
8.9	7.0(6)	1.9(6)
12.8	10.8(6)	2.0
15.9	13.9(3)	2.0

In Figure 5 we have plotted the site occupancies in the octahedral (24f) and tetrahedral (2a) sites of M cations in the $Bi_{26-x}M_xO_{40-\delta}$ phases as a function of the M:Bi ratio. We see from the plot that the tetrahedral occupancy becomes 2 when M/Bi rat o is 0.5 or above. The metal occupancy of the octahedral site is clearly evidenced when the M/Bi ratio is 0.3 and increases with increase in this ratio.

In Table V we compare the Bi-O bond distances in $Bi_{26-x}M_xO_{40-\delta}$ with those in γ-Bi_2O_3 and α- Bi_2O_3. The M-O distances within the tetrahedral units are also listed in the table. The similarity in the distances of the BCC ternary oxides and γ-Bi_2O_3 is obvious.

Magnetic susceptibility measurements on the Mn, Co and Ni derivatives of these BCC phases provide further information on the oxidation state and site distribution of the transition metal ions in $Bi_{26-x}M_xO_{40-\delta}$. The χ_M^{-1} - T plots showed Curie-Weiss behaviour. The experimental effective magnetic moment per M ion and the derived compositions based on these values are listed in Table VI.

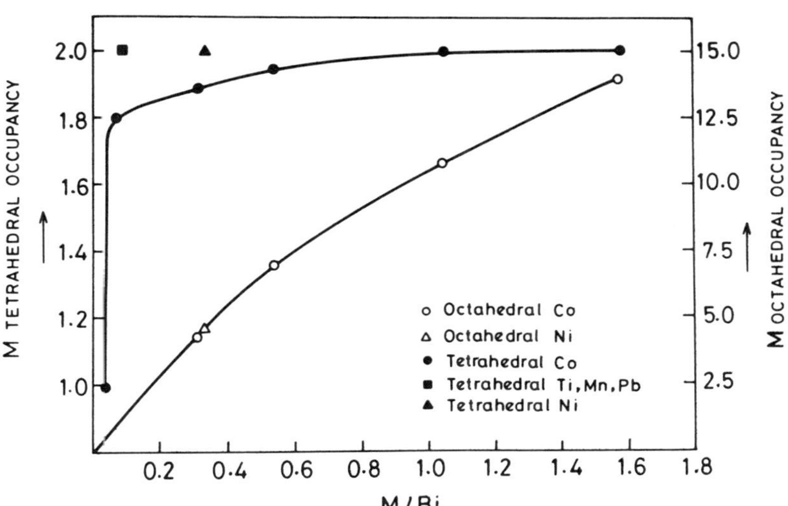

Fig. 5 - Plot showing the variation of the tetrahedral and octahedral occupancies of the M cation against the M/Bi ratio.

TABLE V - Bond distances (Å) in $Bi_{26-x}M_xO_{40-\delta}$, γ-Bi_2O_3,α- Bi_2O_3

	$Bi_{20}Co_6O_{40-\delta}$	$Bi_{10}Co_{16}O_{40-\delta}$	$Bi_{20}Ni_6O_{40-\delta}$	γ-Bi_2O_3 [a]	α- Bi_2O_3 [b]	
octahedral						
					(Bi1-0)	(Bi2-0)
Bi --- 01	2.37(08)	2.16(07)	2.25(08)	2.23(02)	2.08	2.14
--- 02	2.07(09)	2.13(08)	2.13(09)	2.13(04)	2.17	2.22
--- 02	2.33	2.28	2.26	2.54	2.21	2.29
--- 02	2.83	2.70	2.60	2.65	2.54	2.48
--- 02	3.05	2.95	3.01	3.12	2.63	2.54
--- 02	3.19	3.11	3.16	3.20	3.25	2.80
--- 02	2.56	2.54	2.56	2.66		
tetrahedral						
M --- 03	1.95(09)	1.87(08)	1.88(07)	1.74(04)		
average octahedral						
Bi --- 0	2.62	2.55	2.57	2.64		

(a) From Harwig [6] (b) From Malmros' [10]

TABLE VI - Magnetic susceptibility data for $Bi_{26-x}M_xO_{40-\delta}$
(M =Mn,Co,Ni)

t: tetrahedral (2a) site   o: octahedral (24f) site

ccmpound	μ_{eff}(exp)	derived composition
$Bi_{24}Mn_2O_{40-\delta}$	4.4	$(Mn^{3+}_2)_t[Bi^{3+}_{24}]_oO_{39}$ (a)
$Bi_{25}CoO_{40-\delta}$	4.6	$(Bi^{5+}Co^{3+})_t[Bi^{3+}_{24}]_oO_{40}$
$Bi_{10}Co_{16}O_{40-\delta}$	4.5	$(Co^{2+}_2)_t[Bi^{3+}_{10}\ Co^{3+}_{14}]_oO_{38}$
$Bi_{20}Ni_6O_{40-\delta}$	3.4	$(Ni^{2+}_2)_t[Bi^{3+}_{20}\ Ni^{3+}_4]_oO_{38}$

(a) small proportion of the Mn could be in the 4+ state.

In the case of $Bi_{24}Mn_2O_{40-\delta}$, the observed magnetic moment is
consistent with a majority of the Mn ions being present in the
tetrahedral (2a) sites in the 3+ state. In case of $Bi_{25}CoO_{40-\delta}$
the experimental μ_{eff} would be close to the calculated μ_{eff} if
one Co^{3+} is present in the tetrahedral (2a) site. It is rather
unusual to have Co^{3+} in a tetrahedral environment. In the case
of $Bi_{10}Co_{16}O_{40-\delta}$, the magnetic moment per cobalt of 4.5 B.M.
indicates that the tetrahedral sites are occupied by two Co^{2+}
ions while the Co^{3+} ions are present in the octahedral sites
in the high-spin state , possibly due to the unsymmetrical oxygen
coordination in the octahedral (24f) sites of the γ-Bi_2O_3
structure. In the case of $Bi_{20}Ni_6O_{40-\delta}$ the experimental μ_{eff}
value of around 3.4 B.M. suggests that the tetrahedral ions are
occupied by Ni^{2+}, while the Ni^{3+} ions are in the octahedral
(24f) sites.

A STUDY OF CUBIC BISMUTH OXIDES   421

DISCUSSION

The structure of the $Bi_{26-x}M_xO_{40-\delta}$ phases consist of a 7-coordinated Bi-O arrangement with bismuth in the bipyramidal coordination. The base of the bipyramid has two short bonds, 2.30 Å and 2.20 Å and two relatively larger bonds, 2.70 Å and 2.60 Å. The apex of the bipyramid is an exceptionally short bond (2.13 Å). Below the base are two additional oxygen ions at distances of 3.00 and 3.16 Å. This unusual arrangement is obtained because of non bonding lone pair of $6s^2$ electrons on the Bi^{3+} ion which are oriented between the extremely long bonds. The $Bi^{3+} - Bi^{3+}$ distance is about 3.73 Å so that the electron pair is about 1.87 Å from Bi^{3+} ions. The 24 octahedra share corners to form a cage of $Bi_{24}O_{40}$ atoms and within this cage there are two tetrahedral sites. In Figure-6 we show the coordination in the compound $Bi_{24}Mn_2O_{40-\delta}$.

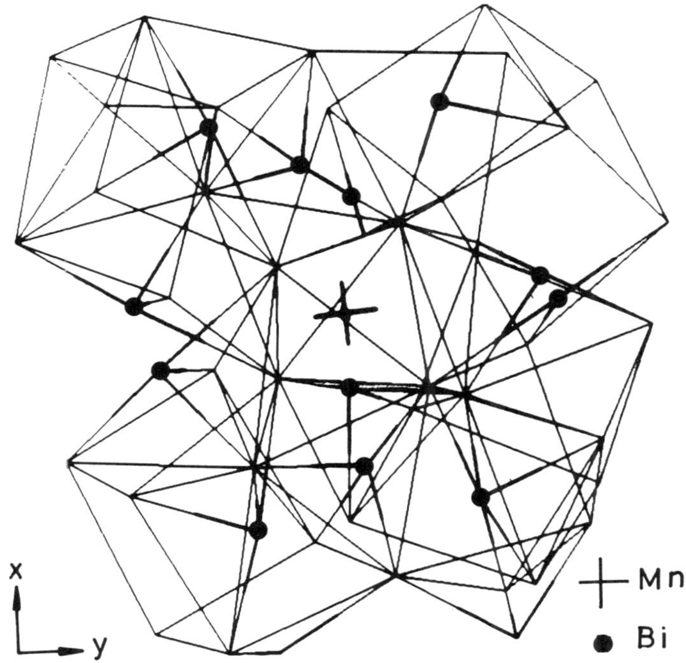

Fig.6 - A schematic diagram of the coordination geometry in $Bi_{24}Mn_2O_{40-\delta}$.

The tetrahedral unit, with Mn atom in the center, is surrounded by 12 bipyramidal units with Bi atoms connected to 4 inplane oxygens and one apical oxygen. The two oxygens below the plane of the bipyramid are not shown. The refinement of the cation occupancy using high-resolution X-ray diffraction profiles and the effective magnetic moment calculated for $Bi_{26-x}M_xO_{40-\delta}$ (M = Mn, Co, Ni) using magnetic susceptibility data show that the M cations occupy the tetrahedral sites fully. The tetrahedral arrangement is regular with the metal to oxygen distance of around 1.87 Å. Thus, the body centered cubic structure can accommodate a variety of different M atoms in the formula $Bi_{24}M_2O_{40-\delta}$ because the tetrahedron of oxygen atoms surrounding the M atom is able to expand or contract without a major effect on the remaining atomic arrangement. Essentially, the Bi-O network in the compounds of the formula $Bi_{26-x}M_xO_{40-\delta}$ remains nearly unchanged as the amount of M atom is varied.

REFERENCES

[1] B. AURIVILLIUS and L.G. SILLEN, *Nature.*, 1945, **155**, 305.

[2] E.M. LEVIN and R.S. ROTH, *J. Res. Natl. Bur. Std.*, 1964, **68A**, 197.

[3] S.C. ABRAHAMS, P.B. JAMIESON and J.L. BERNSTEIN, *J.Chem. Phys.*, 1967, **47**, 4034.

[4] D.C. CRAIG and N.C. STEPHENSON, *J. Solid.State.Chem.*, 1975, **15**, 1.

[5] A. RAMANAN, J. GOPALAKRISHNAN and C.N.R. RAO, *Mater. Res. Bull.*, 1981, **16**, 169.

[6] H.A. HARWIG, *Z. Anorg. Allg. Chem.*, 1978, **444**, 51.

[7] J.L. SOUBEYROUX, M. DEVALETTE, N. KHACHANI and P. HAGENMULLER, *J. Solid. State. Chem.*, 1990, **86**, 59.

[8] M. DEVALLETTE, J. DARRIET, M. COUZI, C. MAZEAU and P. HAGENMULLER, *J. Solid. State. Chem.*, 1982, **43**, 45.

[9] H.M. RIETVELD, *J. Appl. Cryst.*, 1969, **2**, 65.

[10] G. MALMROS, *Acta. Chem. Scand.*, 1970, **24**, 384.

1666

Chem. Mater. **2000**, *12*, 1666–1670

Effect of Cation Size and Disorder on the Structure and Properties of the Rare Earth Cobaltates, $Ln_{0.5}A_{0.5}CoO_3$ (Ln = Rare Earth, A = Sr, Ba)

P. V. Vanitha, Anthony Arulraj, P. N. Santhosh, and C. N. R. Rao*

*Chemistry and Physics of Materials Unit and CSIR Center of Excellence in Chemistry,
Jawaharlal Nehru Center for Advanced Scientific Research, Jakkur P.O.,
Bangalore 560 064, India*

Received May 5, 1999. Revised Manuscript Received March 28, 2000

The structure of $Ln_{0.5}Sr_{0.5}CoO_3$ is rhombohedral ($R\bar{3}c$) when Ln = La, Pr, or Nd, but orthorhombic (*Pnma*) when Ln = Gd. The $Ln_{0.5}Ba_{0.5}CoO_3$ compounds, except for Ln = La, are orthorhombic (*Pmmm*). The ferromagnetic Curie temperature, T_C, of $Ln_{0.5}A_{0.5}CoO_3$ increases with the average size of the A-site cation up to an $\langle r_A \rangle$ of 1.40 Å, and decreases thereafter due to size mismatch. Disorder due to cation-size mismatch has been investigated by studying the properties of two series of cobaltates with fixed $\langle r_A \rangle$ and differing size variance, σ^2. It is found that T_C decreases linearly with σ^2, according to the relation, $T_C = T°_C - p\sigma^2$. When σ^2 is large (>0.012 Å^2), the material becomes insulating, providing evidence for a metal–insulator transition caused by cation-size disorder. Thus, $Gd_{0.5}Ba_{0.5}CoO_3$ with a large σ^2 is a charge-ordered insulator below 340 K. The study demonstrates that the average A-cation radius, as well as the cation-size disorder, affects the magnetic and transport properties of the rare earth cobaltates significantly.

Introduction

Rare earth cobaltates of the formula $Ln_{0.5}Sr_{0.5}CoO_3$ (Ln = rare earth) are known as metallic ferromagnets and their electrical and magnetic properties have been described adequately in the literature.[1] The ferromagnetic Curie temperature, T_C, in these materials decreases significantly with the decrease in the size of the rare earth ion. The magnetic and electrical properties of the barium-substituted cobaltates of the formula $Ln_{0.5}Ba_{0.5}CoO_3$ are different from those of the strontium analogues. Recent studies show that the barium-substituted cobaltates exhibit a insulator–metal transition when the radius of the rare earth ion is small.[2,3] Thus, $Gd_{0.5}Ba_{0.5}CoO_3$ is a charge-ordered insulator, although there are some differences in the published reports regarding its electrical properties.[2–4] Structures of the $Ln_{0.5}A_{0.5}CoO_3$ (A = Sr, Ba) compounds also vary with Ln and A, although there are conflicting structural assignments in the literature. Thus, $Ln_{0.5}Sr_{0.5}CoO_3$ have generally been considered to be cubic, although the possible orthorhombicity has been indicated recently.[5]

In the corresponding Ba compounds, cubic, tetragonal, and orthorhombic structures have been assigned depending on the Ln.[2–4] An examination of the electrical and magnetic properties of the various rare earth cobaltates, $Ln_{0.5}A_{0.5}CoO_3$ (A = alkaline earth), suggests that the properties are likely to be affected by both the average radius of the A-site cation and the cation disorder arising from size mismatch. Both the A-site cation radius and the size variance, σ^2, of the A-site cations are known to markedly affect the magnetic properties of the rare earth manganates of the general formula $Ln_{0.7}A_{0.3}MnO_3$.[6–8] Here, the variance, σ^2, is defined by

$$\sigma^2 = \sum x_i r_i^2 - \langle r_A \rangle^2 \qquad (1)$$

where x_i is the fractional occupancy of A-site ions, r_i is the corresponding ionic radii and $\langle r_A \rangle$ is the weighted average radius calculated from the r_i values. In the manganates, the size variance has a marked effect on the ferromagnetic properties of $Ln_{0.7}A_{0.3}MnO_3$, but appears to affect charge-ordering in $Ln_{0.5}A_{0.5}MnO_3$ only marginally.[9] In this article, we report the structures of several members of the $Ln_{0.5}A_{0.5}CoO_3$ family with A = Sr and Ba, based on Rietveld analysis of powder X-ray diffraction patterns. To understand the dependence of the ferromagnetic Curie temperature of $Ln_{0.5}A_{0.5}CoO_3$

* To whom correspondence should be addressed. E-mail: cnrrao@jncasr.ac.in.

(1) Jonker, G. H.; van Santen, J. H. *Physica* **1953**, *19*, 120. (b) Raccah, P. M.; Goodenough, J. B. *J. Appl. Phys.* **1968**, *39*, 1209. (c) Rao, C. N. R.; Prakash, O.; Bahadur, D.; Ganguly, P.; Nagabhushana, S. *J. Solid State Chem.* **1977**, *22*, 353. (d) Señaris-Rodríguez, M. A.; Goodenough, J. B. *J. Solid State Chem.* **1995**, *118*, 323.

(2) Moritomo, Y.; Takeo, M.; Liu, X. J.; Akimoto, T.; Nakamura, A. *Phys. Rev.* **1998**, *B58*, R13334.

(3) Troyanchuk, I. O.; Kasper, N. V.; Khalyavin, D. D.; Szymczak, H.; Szymczak. R. Baran, M. *Phys. Rev.* **1998**, *B58*, 2418.

(4) Troyanchuk, I. O.; Kasper, N. V.; Khalyavin, D. D.; Szymczak, H.; Szymczak. R. Baran, M. *Phys. Rev. Lett.* **1998**, *80*, 3380.

(5) Brinks, H. W.; Fjellvåg, H.; Kjekshus, A.; Hauback, B. C. *J. Solid State Chem.* **1999**, *147*, 464.

(6) Rodriguez-Martinez, L. M.; Attfield, J. P. *Phys. Rev.* **1996**, *B54*, R15622.

(7) Attfield, J. P. *Chem. Mater.* **1998**, *10*, 3239.

(8) Damay, F.; Martin, C.; Maignan, A.; Raveau, B. *J. Appl. Phys.* **1997**, *82*, 6181.

(9) Vanitha, P. V.; Santhosh, P. N.; Singh, R. S.; Rao, C. N. R.; Attfield, J. P. *Phys. Rev.* **1999**, *B59*, 13539.

10.1021/cm990268t CCC: $19.00   © 2000 American Chemical Society
Published on Web 05/20/2000

Table 1. Structure and Properties of $Ln_{0.5}A_{0.5}CoO_3$

Ln	A	space group	$\langle r_A \rangle$ (Å)	σ^2 (Å²)	lattice parameter (Å) a	b	c	T_C (K)
La	Sr	$R\bar{3}c^a$	1.400	0.0016	5.4152			258
Nd	Sr	$R\bar{3}c^b$	1.355	0.0072	5.3770			226
Gd	Sr	*Pnma*	1.329	0.0123	5.3746	7.5601	5.3723	162
La	Ba	$R\bar{3}c^c$	1.485	0.0156	5.4997			219
Nd	Ba	*Pmmm*	1.317	0.0240	11.7100	11.6778	7.6130	178
Gd	Ba	*Pmmm*	1.288	0.0330	11.7196	11.6323	7.5392	

[a] $\alpha = 60.17°$. [b] $\alpha = 60.28°$. [c] $\alpha = 60.04°$.

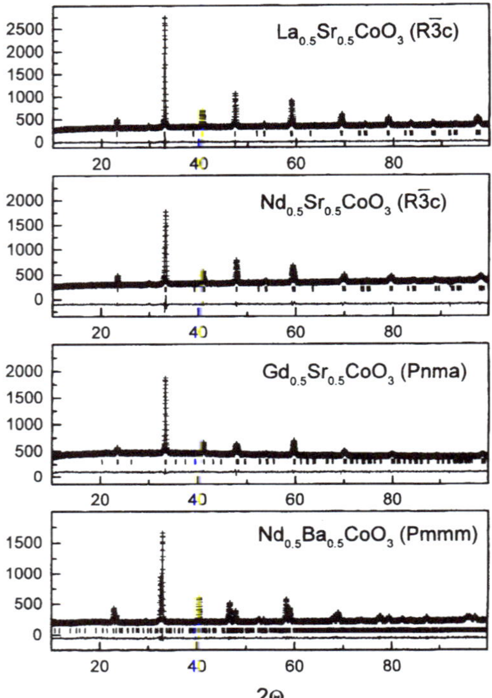

Figure 1. XRD patterns for the representative members of $Ln_{0.5}A_{0.5}CoO_3$. The calculated profile and the difference curve obtained from Rietveld analysis are also shown.

on $\langle r_A \rangle$ as well as cation disorder, we have examined the magnetic and electrical properties of several of these materials. In particular, we have investigated the variation of the properties of the cobaltates with constant $\langle r_A \rangle$, but variable σ^2.

Experimental Section

Cobaltates of the general formula $Ln_{0.5}A_{0.5}CoO_3$ (Ln = rare earth, A = Sr or Ba) were prepared by the ceramic route. Stoichiometric quantities of the respective rare earth oxides, the carbonates of the alkaline earth elements, and cobalt oxide (Co_3O_4) were ground and prefired at 900 °C for 12 h in air. The powder so obtained was ground thoroughly and heated at 1000 °C for 12 h, and the pellets were finally sintered at 1200 °C. The samples were heated in oxygen atmosphere at a lower temperature to improve the oxygen stoichiometry. The phase purity of the samples was established by recording the X-ray diffraction patterns with a SEIFERT 3000 TT diffractometer. X-ray diffraction patterns of all the $Ln_{0.5}Sr_{0.5}CoO_3$ and $Ln_{0.5}Ba_{0.5}CoO_3$ compositions gave sharp reflections (fwhm $\approx$ 0.07°). Rietveld analysis of the powder X-ray diffraction data

Table 2. Atomic Coordinates of a Few Members of the $Ln_{0.5}A_{0.5}CoO_3$ Family

Ln	A	atom	site	x	y	z	U_{iso}
La	Sr	La/Sr	2a	0.25000	0.25000	0.25000	0.00350
		Co	2b	0.00000	0.00000	0.00000	0.00376
		O	6e	0.28006	0.21994	0.75000	0.01221
Nd	Sr	Nd/Sr	2a	0.25000	0.25000	0.25000	0.00611
		Co	2b	0.00000	0.00000	0.00000	0.00606
		O	6e	0.29662	0.20338	0.75000	0.00118
Gd	Sr	Gd/Sr	4c	0.00407	0.25000	0.00490	0.02818
		Co	4b	0.00000	0.00000	0.50000	0.02018
		O	8d	0.30840	−0.00712	0.25408	0.05097
		O	4c	0.08477	0.2500	0.50490	−0.03259
Nd	Ba	Nd/Ba	1c	0.00000	0.00000	0.50000	0.00959
		Nd/Ba	2n	0.00000	0.67102	0.50000	0.01407
		Nd/Ba	2j	0.31649	0.00000	0.50000	0.04112
		Nd/Ba	4z	0.33870	0.67146	0.50000	0.00225
		Nd/Ba	1a	0.00000	0.00000	0.00000	−0.00658
		Nd/Ba	2m	0.00000	0.66496	0.00000	−0.01020
		Nd/Ba	2I	0.33424	0.00000	0.00000	−0.01217
		Nd/Ba	4y	0.32853	0.66496	0.00000	0.00080
		Co	2t	0.50000	0.50000	0.74589	−0.04881
		Co	4v	0.50000	0.82006	0.73592	0.04939
		Co	4x	0.84936	0.50000	0.74713	−0.00517
		Co	8μ	0.82403	0.16557	0.25343	−0.00720
		O	1h	0.50000	0.50000	0.50000	0.13774
		O	2p	0.50000	0.84084	0.50000	−0.03272
		O	2l	0.15382	0.50000	0.50000	0.06328
		O	4z	0.18089	0.84139	0.50000	0.09803
		O	2s	0.50000	0.00000	0.70000	−0.00021
		O	4v	0.50000	0.66554	0.69596	−0.07491
		O	4w	0.86807	0.00000	0.76102	−0.02076
		O	8μ	0.83519	0.33170	0.27143	−0.01216
		O	2r	0.00000	0.50000	0.71388	−0.00142
		O	4u	0.00000	0.83147	0.69421	0.02978
		O	4x	0.66299	0.50000	0.73289	0.04036
		O	8μ	0.67560	0.16653	0.28503	0.03245
		O	1f	0.50000	0.50000	0.00000	−0.04651
		O	2o	0.50000	0.81642	0.00000	−0.07971
		O	2k	0.18571	0.50000	0.00000	−0.04482
		O	4y	0.16576	0.85300	0.00000	0.07722

was carried out using the GSAS software suite.[10] The unit cell parameters of representative cobaltates are listed in Table 1. In Table 1, we have also listed the weighted average radius, $\langle r_A \rangle$, and the σ^2 values of these materials. The $\langle r_A \rangle$ values were calculated using the Shannon-radii, r_i, for 12-coordination in the case of the rhombohedral cobaltates and for 9-coordination in the case of orthorhombic ones. Two series of rhombohedral cobaltates of the general formula $Ln_{0.5-x}Ln'_xA_{0.5-y}A'_yCoO_3$ with constant $\langle r_A \rangle$ values of ~1.357 Å and 1.369 Å were prepared to study the effect of the A-site cation-size mismatch on the properties. These materials were prepared in the manner described earlier.

Electrical resistivity measurements were carried out from 300 to 20 K by the four-probe method. Magnetization measurements were carried out in the temperature range 300–80 K by means of a vibrating sample magnetometer (Lakeshore VSM 7300). The oxygen stoichiometry was determined by iodometric titrations. The error in oxygen content was ±0.02. The oxygen stoichiometry in the cobaltates studied by us was generally within this experimental error. Magnetoresistance measurements were carried out using a cryocooled closed cycle superconducting magnet.

Results and Discussion

Rietveld analysis of the powder X-ray diffraction data of the cobaltates gave good fits as shown for four of the materials in Figure 1. The atomic coordinates for some of the cobaltates are given in Table 2. The structural

(10) Larson, A. C.; Von Dreele, R. B. *GSAS: General Structural Analysis System*; LANSCE, Los Alamos National Laboratory: Los Alamos, NM, 1994.

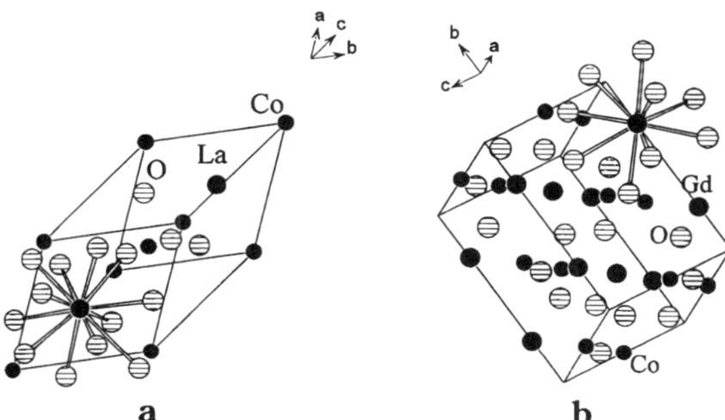

Figure 2. Rhombohedral and orthorhombic structures of representative members of $Ln_{0.5}A_{0.5}CoO_3$, showing different coordinations of the A-site cations: (a) La, Sr and (b) Gd, Sr.

Table 3. Important Structural Parameters of a Few Members of $Ln_{0.5}A_{0.5}CoO_3$

$Ln_{0.5}A_{0.5}CoO_3$		bond distances (Å) Co–O	bond angles (deg) Co–O–Co
Ln	A		
La	Sr	6 × 1.9336	6 × 164.96
Nd	Sr	6 × 1.9217	6 × 164.94
Gd	Sr	2 × 1.7107	4 × 165.23
		2 × 1.9444	2 × 152.85
		2 × 2.1203	
Nd	Ba[a]	1 × 2.0989	2 × 180.00
		1 × 2.0039	2 × 174.19
		1 × 1.9488	2 × 159.31
		1 × 1.9448	
		1 × 1.8794	
		1 × 1.7546	
Gd	Ba[a]	1 × 1.9013	1 × 174.72
		1 × 1.8984	1 × 172.08
		2 × 1.8932	1 × 169.15
		1 × 1.8749	2 × 165.93
		1 × 1.8724	1 × 151.58

[a] There are other unique sets of six Co–O distances in this compound, and we have only listed one representative set.

data reveal that when A = Sr, the structure is rhombohedral (space group: $R\bar{3}c$) up to Ln = Nd. In the case of $Nd_{0.5}Sr_{0.5}CoO_3$, we could get an equally good fit for the *Pnma* and $R\bar{3}c$ space groups, and we have, therefore, preferred the space group with higher symmetry as per the normal practice. When Ln = Gd, the structure is orthorhombic (space group: *Pnma*). The corresponding Ba compounds, except for La, are orthorhombic with the space group *Pmmm*. The A-site coordination number is 12 when the structure is rhombhohedral, and 9 when it is orthorhombic. In Figure 2, we show rhombohedral and orthorhombic structures for purpose of illustration. In Table 3 we list some of the structural parameters. The Co–O bond distances listed in the table show how the CoO_6 octahedra are distorted in the orthorhombic structure, especially in the *Pmmm* space group of the Ba compounds.

In Figure 3a, we show the temperature variation of the magnetization of a few compositions of $Ln_{0.5}A_{0.5}CoO_3$. The T_C values obtained from the magnetization data are listed in Table 1. In Figure 4, T_C values obtained by us are plotted against $\langle r_A \rangle$, along with some of the data from the literature. The T_C increases up to a $\langle r_A \rangle$ value of 1.40 Å and decreases thereafter. The shape of the curve in Figure 4 is similar to the T_C – $\langle r_A \rangle$ plot for $Ln_{0.7}A_{0.3}MnO_3$.[11,12] The decrease in T_C for $\langle r_A \rangle$ > 1.40 Å is likely to arise from A-site cation disorder, just as in the case of the manganates.[6–8]

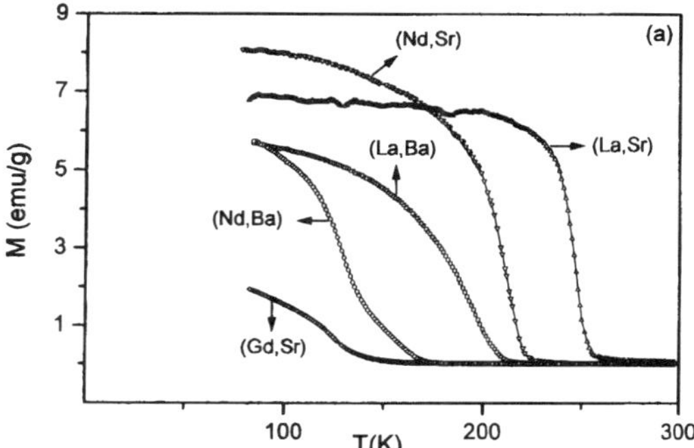

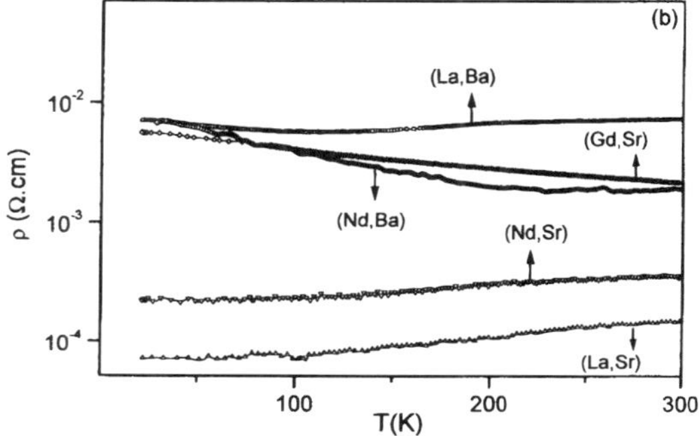

Figure 3. Temperature variation of (a) the magnetization, M, and (b) the electrical resistivity, ρ, of the cobaltates $Ln_{0.5}A_{0.5}CoO_3$. Ln and A are indicated in the figure.

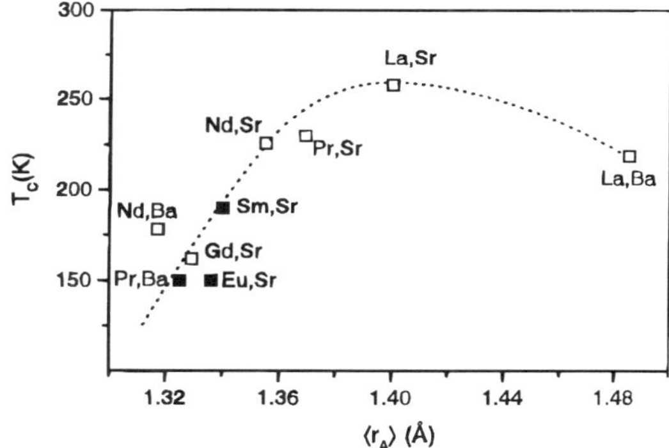

Figure 4. Variation of the ferromagnetic Curie temperature, T_C, with $\langle r_A \rangle$ in $Ln_{0.5}A_{0.5}CoO_3$. Ln and A are indicated in the figure. The filled symbols are the data from the literature. The broken curve is drawn as a guide to the eye.

Electrical resistivity data of the cobaltates also reflect the effect of cation size and disorder. In Figure 3b, we show the temperature variation of the electrical resistivity of the $Ln_{0.5}A_{0.5}CoO_3$ compounds. Both $La_{0.5}Sr_{0.5}CoO_3$ and $Nd_{0.5}Sr_{0.5}CoO_3$ show the expected metallic behavior, but orthorhombic $Gd_{0.5}Sr_{0.5}CoO_3$ shows a slight departure from metallic behavior. While $La_{0.5}Ba_{0.5}CoO_3$ is metallic, $Nd_{0.5}Ba_{0.5}CoO_3$ is an insulator. $Pr_{0.5}Ba_{0.5}CoO_3$ is reported to be an insulator.[3] It is

(11) Mahesh, R.; Mahendiran, R.; Raychaudhuri, A. K.; Rao, C. N. R. *J. Solid State Chem.* **1995**, *120*, 204.
(12) Hwang, H. Y.; Cheong, S. W.; Radaelli, P. G.; Marezio, M.; Batlogg, B. *Phys. Rev. Lett.* **1995**, *75*, 914.

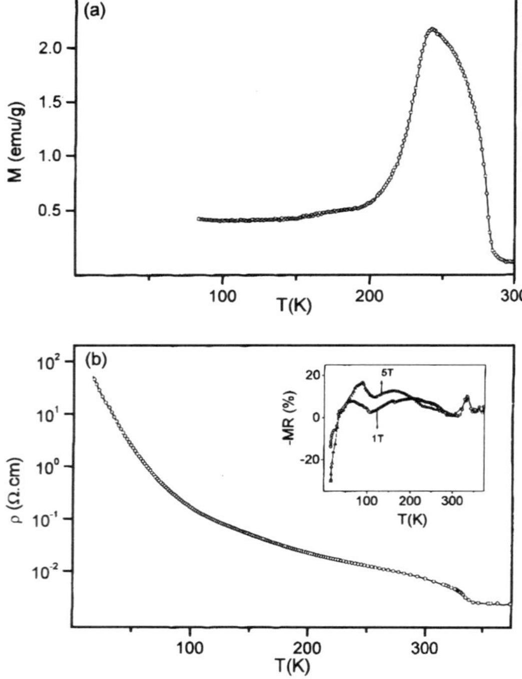

Figure 5. Temperature variation of (a) magnetization, *M*, and (b) resistivity, ρ, of the cobaltate, $Gd_{0.5}Ba_{0.5}CoO_3$. The inset in b shows the $-MR$ (%) versus temperature plot in a field of 1 and 5 T.

Table 4. Structure and Properties of $Ln_{0.5-x}Ln'_xA_{0.5-y}A'_yCoO_3$ with a Fixed $\langle r_A \rangle$ of 1.357 Å

		lattice parameter		
composition	σ^2 (Å^2)	a (Å)	α (deg)	T_C (K)
$Pr_{0.15}La_{0.35}Ca_{0.16}Sr_{0.34}CoO_3$	0.0026	5.3929	60.15	236
$Nd_{0.15}La_{0.35}Sr_{0.355}Ca_{0.145}CoO_3$	0.0034	5.3913	60.23	233
$Nd_{0.1}Pr_{0.4}Sr_{0.45}Ca_{0.05}CoO_3$	0.0053	5.3858	60.37	227
$Gd_{0.11}Pr_{0.39}Sr_{0.5}CoO_3$	0.0070	5.3887	60.32	222
$Nd_{0.5}Sr_{0.5}CoO_3$	0.0072	5.3770	60.28	226
$Sm_{0.315}La_{0.185}Sr_{0.5}CoO_3$	0.0077	5.3883	60.24	223
$Gd_{0.245}La_{0.255}Sr_{0.5}CoO_3$	0.0081	5.3948	60.16	217
$Nd_{0.35}Gd_{0.15}Sr_{0.45}Ba_{0.05}CoO_3$	0.0118	5.3882	60.24	207
$Sm_{0.5}Sr_{0.405}Ba_{0.095}CoO_3$	0.0157	5.3896	60.10	191
$Nd_{0.2}Gd_{0.3}Sr_{0.395}Ba_{0.105}CoO_3$	0.0168	5.3943	60.01	181

Table 5. Structure and Properties of $Ln_{0.5-x}Ln'_xA_{0.5-y}A'_yCoO_3$ with a Fixed $\langle r_A \rangle$ of 1.369 Å

		lattice parameter		
composition	σ^2 (Å^2)	a (Å)	α (deg)	T_C (K)
$Pr_{0.5}Sr_{0.5}CoO_3$	0.0051	5.3839	60.41	230
$La_{0.235}Sm_{0.265}Sr_{0.5}CoO_3$	0.0070	5.3942	60.13	214
$La_{0.315}Dy_{0.185}Sr_{0.5}CoO_3$	0.0085	5.3967	60.10	200
$Pr_{0.3}Gd_{0.2}Sr_{0.41}Ba_{0.09}CoO_3$	0.0134	5.3982	60.08	176
$Pr_{0.2}Gd_{0.3}Sr_{0.36}Ba_{0.14}CoO_3$	0.0178	5.4021	60.08	143

interesting that $Pr_{0.5}Ba_{0.5}CoO_3$ and $Nd_{0.5}Ba_{0.5}CoO_3$ are insulating, despite the large $\langle r_A \rangle$, unlike $La_{0.5}Ba_{0.5}CoO_3$ and the strontium analogues. This is likely due to the large cation-size disorder, with $\sigma^2 > 0.01$ Å^2. $Gd_{0.5}Sr_{0.5}CoO_3$ with $\sigma^2 \approx 0.012$ Å^2 tends to be insulating because it also has a relatively small $\langle r_A \rangle$. These results suggest that cation-size disorder can render a material insulating even though the $\langle r_A \rangle$ is considerably large, as in the

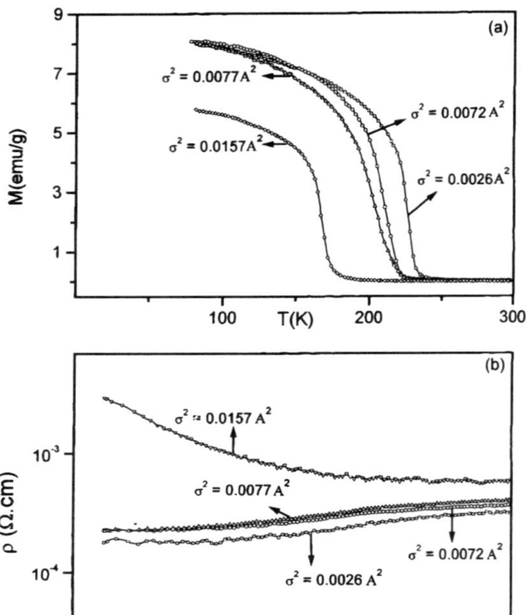

Figure 6. Temperature variation of (a) magnetization, *M*, and (b) resistivity, ρ, of rhombohedral $Ln_{0.5}A_{0.5}CoO_3$ with fixed $\langle r_A \rangle$ of 1.357 Å and variable σ^2. The σ^2 values are shown in the figure.

case of $Ln_{0.5}Ba_{0.5}CoO_3$ (Ln = Pr, Nd). These results show the important role of cation size and disorder on both the magnetic and the electrical properties.

$Gd_{0.5}Ba_{0.5}CoO_3$ ($\langle r_A \rangle = 1.288$ Å), with a large value of σ^2 (0.033 Å^2) as well as orthorhombic distortion, exhibits unique magnetic and transport properties. It is not ferromagnetic as the La, Pr, and Nd derivatives of the same series. Instead, it shows a metamagnetic type of transition around 240 K as illustrated in Figure 5a. A similar transition has been reported by Troyanchuk et al.[4] Across the magnetic transition, the material remains an insulator. $Gd_{0.5}Ba_{0.5}CoO_3$ also exhibits a small resistivity transition at 340 K as revealed in Figure 5b, in agreement with the reports in the literature.[2,4] This transition has been attributed to charge ordering of the Co^{3+} and Co^{4+} ions.[2] We do not however see the resistivity anomaly around 250 K reported by Troyanchuk et al.[4] Magnetoresistance measurements show about 17% negative magnetoresistance (MR) at 75 K in a field of 5 T, but no MR maximum at 250 K as reported by Troyanchuk et al.[4] There is a small increase in MR at 340 K where the resistivity transition occurs (Figure 5b).

To investigate the effect of the A-site cation-size mismatch on the magnetic and electrical properties of the cobaltates, we have systematically varied σ^2 in two series of compounds with fixed $\langle r_A \rangle$ values of 1.357 and 1.369 Å. We list the lattice parameters and σ^2 values in Tables 4 and 5. We show typical magnetization data of the series with $\langle r_A \rangle = 1.357$ Å in Figure 6a to illustrate how the ferromagnetic T_C decreases with the increase in σ^2. We have listed the T_C values of the two series of cobaltates in Tables 4 and 5 and plotted the T_C values

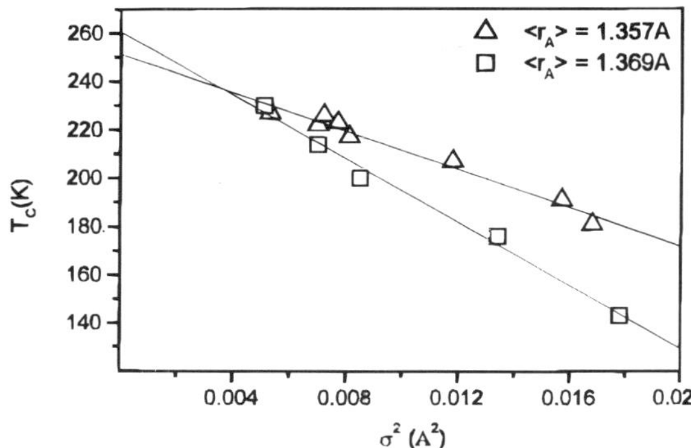

Figure 7. Variation of the ferromagnetic Curie temperature, T_C with σ^2 in rhombohedral $Ln_{0.5}A_{0.5}CoO_3$ for fixed $\langle r_A \rangle$ values of 1.357 Å and 1.369 Å.

against σ^2 in Figure 7. The plots are fairly linear. We can write the relation as

$$T_C = T^{\circ}_C - p\sigma^2 \qquad (2)$$

where the value of the intercept, T°_C is an estimate of the ideal ferromagnetic Curie temperature that would have been observed in the absence of A-site cation-size disorder ($\sigma^2 = 0$). We find T°_C to be 251 ± 3 K and 261 ± 4 K respectively for $\langle r_A \rangle$ values of 1.357 and 1.369 Å respectively, the corresponding slopes being 3961 ± 270 K Å^{-2} and 6558 ± 369 K Å^{-2}. There is some variation in T°_C in the cobaltates even for a change of 0.012 Å in $\langle r_A \rangle$, compared to the manganates where marked changes in T°_C occur for larger changes in $\langle r_A \rangle$.[7,8] Furthermore, the slope of the $T_C - \sigma^2$ plot increases with $\langle r_A \rangle$ in the cobaltates while it decreases in the manganates.

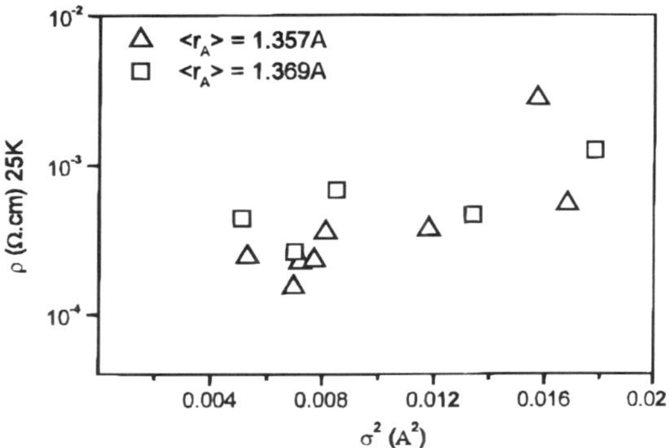

Figure 8. Variation of the resistivity of rhombohedral $Ln_{0.5}A_{0.5}CoO_3$ at 25 K with σ^2.

In Figure 6b, we have plotted the electrical resistivity of the cobaltates in Table 4 with a fixed $\langle r_A \rangle$ of 1.357 Å. The materials are metallic for $\sigma^2 < 0.012$ Å^2 and become insulating at higher σ^2, as also corroborated by the data in Figure 3b. This metal−insulator transition is brought about entirely by size disorder. The rare earth cobaltates are distinct from the rare earth manganates in that the latter show an insulator−metal transition around the ferromagnetic T_C and the transition temperature decreases with increase in σ^2. We have plotted the resistivity of $Ln_{0.5}A_{0.5}CoO_3$ at 25 K against σ^2 in Figure 8. We see a general increase in resistivity with increase in σ^2 in the cobaltates studied.

Acknowledgment. The authors thank Dr. A. R. Raju for assistance with the magnetoresistance measurements.

CM990268T

PHYSICAL REVIEW B **66**, 064425 (2002)

Orbital ordering as the determinant for ferromagnetism in biferroic $BiMnO_3$

A. Moreira dos Santos and A. K. Cheetham

Materials Department, University of California, Santa Barbara, California 93106-5050

T. Atou, Y. Syono, Y. Yamaguchi, K. Ohoyama, and H. Chiba

IMR, Tohoku University, Katahira 2-1-1, Aoba-ku, Sendai 980-8577, Japan

C. N. R. Rao

Chemistry and Physics of Materials Unit, Jawaharlal Nehru Center for Advanced Scientific Research, Bangalore 560 064, India

(Received 28 March 2002; revised manuscript received 19 June 2002; published 20 August 2002)

The ferromagnetic structure of $BiMnO_3$, $T_c = 105$ K, has been determined from powder neutron-diffraction data collected at 20 K on a sample synthesized at high pressures using a cubic anvil press. $BiMnO_3$ is a distorted perovskite that crystallizes in the monoclinic space group $C2$ with unit-cell parameters $a = 9.5317(7)$ Å, $b = 5.6047(4)$ Å, $c = 9.8492(7)$ Å, and $\beta = 110.60(1)°$ ($Rp = 6.78\%$, $wRp = 8.53\%$, reduced $\chi^2 = 1.107$). Data analysis reveals a collinear ferromagnetic structure with the spin direction along [010] and a magnetic moment of $3.2\mu B$. There is no crystallographic phase transition on cooling the polar room-temperature structure to 20 K, lending support to the belief that ferromagnetism and ferroelectricity coexist in $BiMnO_3$. Careful examination of the six unique Mn-O-Mn superexchange pathways between the three crystallographically independent Mn^{3+} sites shows that four are ferromagnetic and two are antiferromagnetic, thereby confirming that the ferromagnetism of $BiMnO_3$ stems directly from orbital ordering.

DOI: 10.1103/PhysRevB.66.064425   PACS number(s): 75.50.Dd, 75.25.+z, 77.84.Dy

I. INTRODUCTION

The remarkable magnetoelectric properties of $BiMnO_3$ have attracted considerable attention during the last three years. It is well established that $BiMnO_3$ becomes ferromagnetically ordered on cooling below 110 K,[1–3] and there is good reason to believe that the ferromagnetism coexists with ferroelectricity. This is to be contrasted with the behavior of the analogous $LaMnO_3$, which orders antiferromagneticaly at 150 K.[4] Such biferroic behavior is very rare in a single phase material.[5–7] The evidence for ferroelectricity, however, has largely been based upon informed speculation,[8] although a structure determination by powder diffraction has indicated that the space group at room temperature is $C2$,[9] which would be consistent with ferroelectricity. The reason for the uncertainty in this area is that the preparation of bulk $BiMnO_3$ requires high pressures and temperature, so it is difficult to make large samples of high quality material. However, evidence for magnetoelectric behavior has recently been garnered from measurements on thin films and impure bulk samples, both of which show ferroelectric hysterisis loops below the Curie temperature.[10] The low-temperature crystal and magnetic structures of $BiMnO_3$ have not been reported, but would shed further light on the coexistence of ferroelectricity and ferromagnetism. In the present work, we describe the determination of the crystal and magnetic structures of $BiMnO_3$ from a low-temperature, powder neutron-diffraction study.

II. EXPERIMENT

The preparation of a polycrystalline sample of $BiMnO_3$ was carried out using a cubic anvil press at 973 K and 6 GPa,

as reported previously.[9] Powder neutron-diffraction data were collected on the Hermes diffractometer on the JRR-3M facility at the Japan Atomic Energy Research Institute with a mean neutron wavelength of 1.8196 Å. Data were analyzed by Rietveld refinement using the GSAS suite of programs.[11] Neutron-scattering lengths of 8.531, -3.730, and $5.803(10^{-15}$ m) were used for Bi, Mn, and O, respectively. The magnetic form factor of Mn^{3+} was based upon the free-ion calculations of Watson and Freeman.[12]

III. RESULTS AND DISCUSSION

A careful comparison between the high angle portions of the neutron-diffraction data collected at 20 K and room temperature[9] revealed no evidence for a structural phase

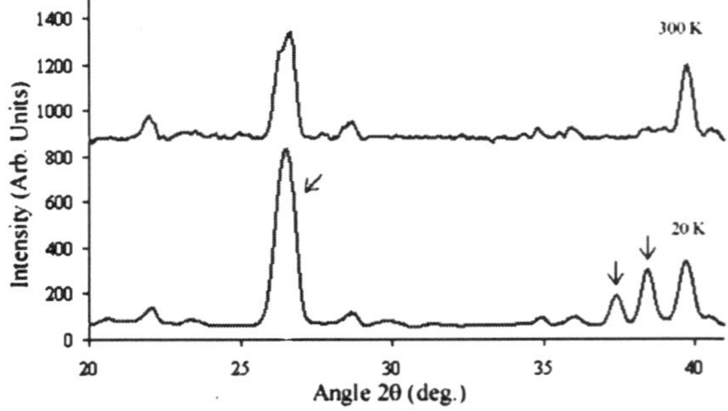

FIG. 1. A comparison of the low angle regions of the neutron powder-diffraction pattern of $BiMnO_3$ at 300 and 20 K. Note the appearance of additional reflections at 20 K due to the onset of ferromagnetic ordering.

A. MOREIRA DOS SANTOS *et al.*   PHYSICAL REVIEW B **66**, 064425 (2002)

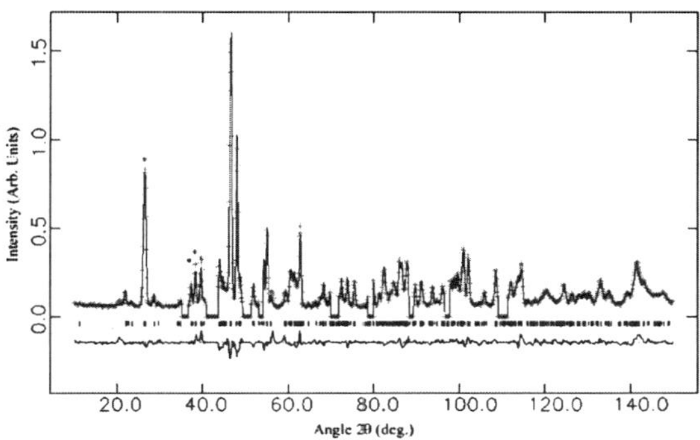

FIG. 2. Observed (dots) and calculated (line) profile of BiMnO$_3$ at 20 K. A difference curve is also shown. The magnetic peaks are indicated by an asterisk.

transition on cooling to low temperatures. The crystal structure of BiMnO$_3$ at 20 K was therefore refined using the room-temperature structure in space group $C2$ as a starting model. The low angle portion of the pattern ($<40°\,2\theta$), however, contained a small number of magnetic Bragg reflections that could be indexed on the basis of the crystallographic unit cell; the most intense features are the (-202), (-113), (311), and (202) reflections (see Fig. 1). These observations are consistent with the onset of ferromagnetic ordering, and we attempted to model the intensities of these magnetic peaks on the basis of a variety of plausible collinear spin models. An excellent fit was obtained with a model in which the spin vector was aligned along the unique [010] axis of the monoclinic crystal structure, and a full Rietveld refinement of the crystal and magnetic structure was carried out on this basis (see Fig. 2 and Table I). Lines from small amounts of impurity phases were excluded from the analysis. One of the impurity phases was identified as Bi$_2$(CO)$_3$O$_2$; the identity of the second phase could not be established. Isotropic temperature factors for each atom type were constrained to be equal. The refined magnetic moment was

found to be 3.2(3) μB, which is in the expected range for a high spin d^4 Mn^{3+} ion and consistent with the previously reported magnetization curve of BiMnO$_3$ measured at 5 K.[13]

Selected bond lengths and bond angles for BiMnO$_3$ are given in Tables II and III, and the coordination environments for the Bi and Mn atoms are shown in Figs. 3 and 4, respectively. The BiO$_n$ polyhedra are unsymmetrical as a consequence of the stereochemical activity of the lone pairs of electrons on the Bi^{3+} ions. This asymmetry is believed to be the major driving force behind the ferroelectric properties of BiMnO$_3$, as it is in the behavior of Pb(Zr$_x$Ti$_{1-x}$)O$_3$ (PZT) and related phases. In both BiMnO$_3$ and PZT for example, the A cations are displaced approximately along the body diagonal of the cubic perovskite subcell. The fact that there is no structural phase transition and that the distortions remain substantially the same on cooling from room temperature to 20 K lends strong support to the belief that BiMnO$_3$ remains ferroelectric in the ferromagnetic phase.

TABLE I. Refined structural parameters for BiMnO$_3$ at 20 K. Space group $C2$, $a=9.5317(7)$ Å, $b=5.6047(4)$ Å, $c=9.8492(7)$ Å, $\beta=110.60(1)°$, $V=492.54(7)$ Å^3. There were 1249 observations. $Rp=6.78\%$, $wRp=8.53\%$. Reduced χ^2 = 1.107 for 20 variables.

Atom	x	y	z	$Ui/Ue*100$
Bi(1)	0.136(1)	0.004(7)	0.373(1)	0.4
Bi(2)	0.363(1)	0.068(8)	0.116(1)	0.4
Mn(1)	0	0	0	0.25
Mn(2)	0.250(4)	0.036(9)	0.754(3)	0.25
Mn(3)	0.5	0.078(4)	0.5	0.25
O(1)	0.092(2)	$-0.025(9)$	0.834(2)	0.9
O(2)	0.397(2)	0.128(8)	0.673(2)	0.9
O(3)	0.139(2)	0.367(8)	0.622(2)	0.9
O(4)	0.355(2)	0.316(8)	0.420(2)	0.9
O(5)	0.300(2)	0.225(8)	0.907(2)	0.9
O(6)	0.154(2)	0.225(8)	0.115(2)	0.9

TABLE II. Selected bond lengths for BiMnO$_3$ at 20 K.

Mn1–O1	2.11(1)×2	Mn2–O1	1.97(4)
Mn1–O5	2.08(4)×2	Mn2–O2	1.90(4)
Mn1–O6	1.97(3)×2	Mn2–O3	2.30(4)
Mn3–O2	2.27(1)×2	Mn2–O4	2.06(3)
Mn3–O3	1.86(4)×2	Mn2–O5	1.82(3)
Mn3–O4	1.88(4)×2	Mn2–O6	2.17(4)
Bi1–O1	2.43(1)	Bi2–O1	2.36(2)
Bi1–O2	2.16(2)	Bi2–O1'	3.02(2)
Bi1–O3	2.25(1)	Bi2–O2	2.53(2)
Bi1–O3'	3.20(2)	Bi2–O3	2.83(2)
Bi1–O4	2.29(2)	Bi2–O5	2.17(1)
Bi1–O4'	2.64(2)	Bi2–O5'	2.74(2)
Bi1–O4"	3.06(1)	Bi2–O5"	2.95(1)
Bi1–O5	3.21(1)	Bi2–O6	2.18(1)
Bi1–O6	3.20(1)	Bi2–O6'	2.87(2)

TABLE III. Selected bond angles for BiMnO$_3$ at 20 K.

O1–Mn1–O1	172.5(3)	O1–Mn2–O2	174.1(3)
O1–Mn1–O5	91.2(1)	O1–Mn2–O4	90.7(2)
O1–Mn1–O5	83.2(1)	O1–Mn2–O5	90.9(2)
O1–Mn1–O6	93.4(1)	O1–Mn2–O6	80.9(1)
O1–Mn1–O6	91.5(1)	O2–Mn2–O4	91.8(2)
O1–Mn1–O5	83.2(1)	O2–Mn2–O5	86.8(2)
O1–Mn1–O5	91.2(1)	O2–Mn2–O6	104.4(2)
O1–Mn1–O6	91.5(1)	O4–Mn2–O5	177.8(3)
O1–Mn1–O6	93.4(1)	O4–Mn2–O6	89.6(2)
O5–Mn1–O5	84.6(2)	O5–Mn2–O6	89.1(1)
O5–Mn1–O6	170.7(2)	O3–Mn3–O3	101.5(3)
O5–Mn1–O6	87.9(1)	O3–Mn3–O4	165.8(1)
O5–Mn1–O6	87.9(1)	O3–Mn3–O4	85.8(7)
O5–Mn1–O6	170.7(2)	O3–Mn3–O4	85.8(7)
O6–Mn1–O6	100.0(2)	O3–Mn3–O4	165.8(1)
		O4–Mn3–O4	89.7(2)

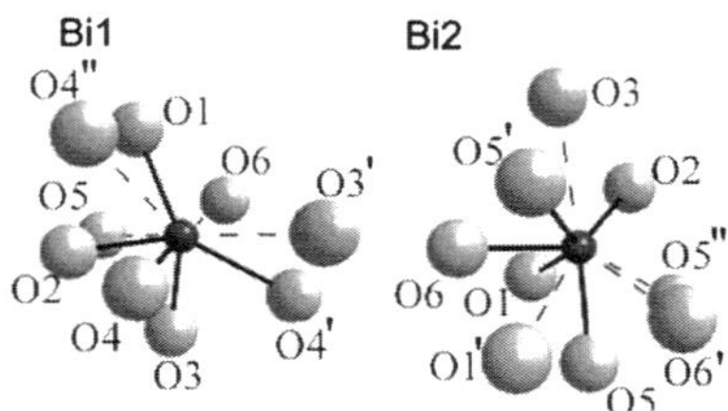

FIG. 3. Coordination environments of Bi1 and Bi2; showing the shortest distances in bold and longer ones in dashed lines.

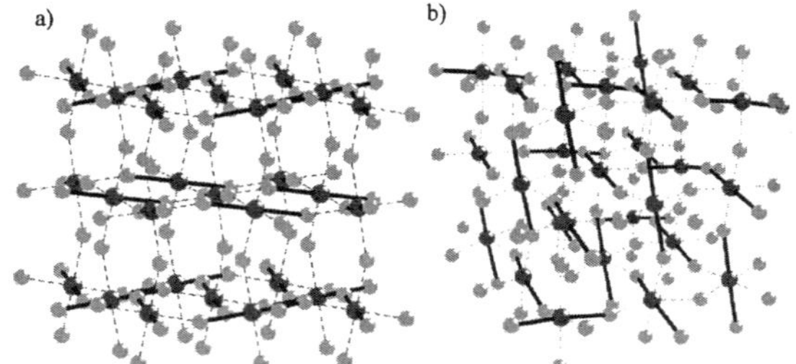

FIG. 6. The two-dimensional orbital ordering in (a) $LaMnO_3$ is compared with the three-dimensional orbital ordering seen in (b) $BiMnO_3$. Bold lines represent the orientation of the d_{z^2} orbitals, as revealed by the elongations of the MnO_6 octahedra.

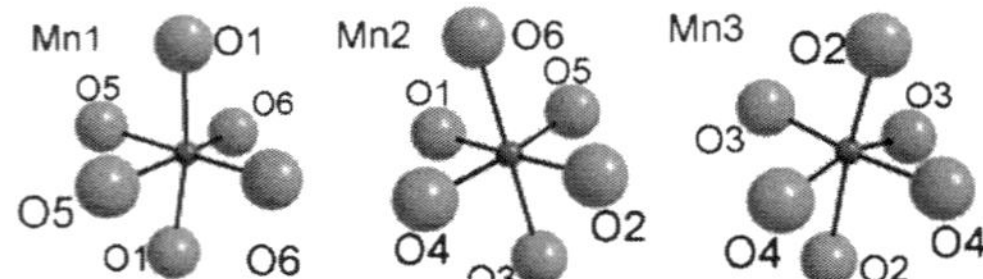

FIG. 4. Coordination environment of the Jahn-Teller distorted Mn cations; see Tables II and III.

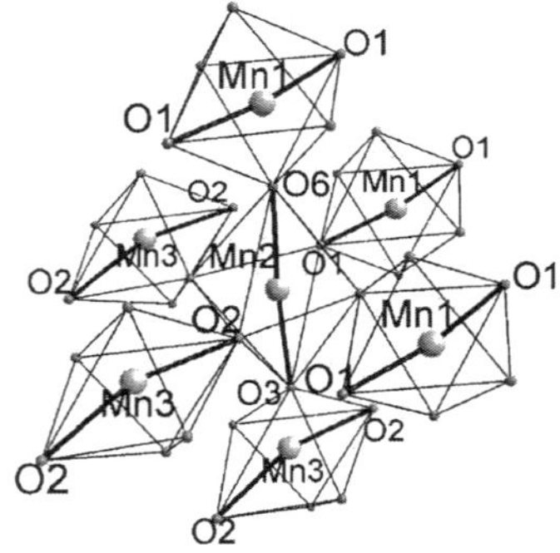

FIG. 5. Three-dimentional magnetic exchange between the Mn atoms; the thick lines correspond to the occupied d_{z^2} orbitals.

TABLE IV. Primary superexchange interactions in $BiMnO_3$ at 20 K. The bond angle estimated standard deviations are $\sim 0.3°$.

Pathway	Angle°	Interaction
Mn(1)–O(1)–Mn(2)	154.8	FM
Mn(2)–O(2)–Mn(3)	147.0	FM
Mn(2)–O(3)–Mn(3)	160.4	FM
Mn(2)–O(4)–Mn(3)	148.8	AFM
Mn(1)–O(5)–Mn(2)	149.5	AFM
Mn(1)–O(6)–Mn(2)	158.7	FM

Each of the three MnO_6 polyhedra shows the axial elongation that is typical of Jahn-Teller distorted d^4 cations in perovskite systems. The orbital ordering that is associated with these distortions in $BiMnO_3$ is the same as that observed at room temperature (Fig. 5), resulting in superexchange interactions that are largely ferromagnetic (Table IV). In four of the six pathways, the orbital ordering ensures that half filled d_{z^2} orbitals point towards the empty $d_{x^2-y^2}$ orbitals on the next manganese; such interactions are predicted to be ferromagnetic according to the rules proposed by Goodenough[14,15] and Kanamori;[16] they are strongest when the M-O-M bond angle is close to $180°$. We note that three of the four ferromagnetic Mn-O-Mn angles are significantly larger than the antiferromagnetic ones (Table IV). There is no instance in which a half filled d_{z^2} orbital points towards another half filled d_{z^2} orbital (this would be strongly antiferromagnetic), but two cases in which empty $d_{x^2-y^2}$ orbitals point towards each other [through O(4) and O(5)]. Optimally, these interactions would be weakly antiferromagnetic, but this cannot be accommodated in combination with the constraints of the strong ferromagnetic interactions, so the system must be slightly frustrated.

It is interesting to compare the ferromagnetic structure of $BiMnO_3$ with the A-type antiferromagntic structure found in the related perovskite, $LaMnO_3$. This comparison has recently been discussed by Hill and Rabe[17] and Woo *et al.*[18] $LaMnO_3$ is also orbitally ordered,[19] but in a simple manner which leads to ferromagnetic sheets that are antiferromagnetically aligned with respect to each other. Half filled d_{z^2} orbitals point towards empty $d_{x^2-y^2}$ orbitals within the ferromagnetic sheets, but this results in empty $d_{x^2-y^2}$ orbitals facing each other via the oxygens between the sheets, leading to the overall antiferromagnetic structure (see Fig. 6). It is the substitution of La by Bi that is responsible for stabilizing a different structural distortion with different orbital ordering that leads to ferromagnetism. There are several factors that may be responsible for this crucial difference. The greater covalence of the Bi-O bonds compared with La-O bonds will have an impact on the Mn-O-Mn interactions. Second, the bismuth lone pairs lead to A cations displacements along the $\langle 111 \rangle$ directions of the cubic perovskite sub-

A. MOREIRA DOS SANTOS *et al.*

PHYSICAL REVIEW B **66**, 064425 (2002)

cell that are imcompatible with the two-dimensional orbital ordering found in $LaMnO_3$. The more complex orbital ordering pattern found in $BiMnO_3$, which must stem from the distortions imposed by the Bi lone pairs, leads to three-dimensional ferromagnetic interactions that result in the observed spin structure.

ACKNOWLEDGMENTS

The work was supported by the MRSEC Program of the National Science Foundation under Award No. DMR00-80034. A. Moreira dos Santos wishes to thank Dr. Angus Lawson for help with the magnetic refinement.

[1] I. Troyachuk, N.V. Samsonenko, E.F. Shapovalova, I.M. Kolesova, and H. Shymczak, J. Phys.: Condens. Matter **8**, 11 205 (1996).

[2] F. Sugawara, S. Iida, Y. Syono, and S.-i. Akimoto, J. Phys. Soc. Jpn. **20**, 1529 (1965).

[3] V. Bokov, I.E. Myl'nikova, S.A. Kizaev, M.F. Bryzhina, and N.A. Grigoryan, Fiz. Tverd. Tela (Leningrad) **7**, 2993 (1966) [Sov. Phys. Solid State **7**, 2993 (1966)].

[4] C.N.R. Rao, A.K. Cheetham, and R. Mahesh, Chem. Mater. **8**, 2421 (1996).

[5] N. Hill, J. Phys. Chem. B **104**, 6694 (2000).

[6] H. Schmidt, Ferroelectrics **162**, 317 (1994).

[7] S. Skinner, IEEE Trans. Parts Mater. Packag. **6**, 68 (1970).

[8] V. Bokov, N. Grigoryan, M. Bryzhina, and V. Kazary, Izv. Akad. Nauk Arm. SSR, Fiz. **33**, 1164 (1969).

[9] T. Atou, H. Chiba, K. Ohoyama, Y. Yamaguchi, and Y. Syono, J. Solid State Chem. **145**, 639 (1999).

[10] A. Moreira dos Santos, S. Parashar, A.R. Raju, A.K. Cheetham, and C.N.R. Rao, Solid State Commun. **122**, 49 (2002).

[11] A. Larson and R.B. von Dreele, GSAS Manual, 1986.

[12] R. Watson and A. Freeman, Acta Crystallogr. **14**, 27 (1961).

[13] H. Chiba, T. Atou, and Y. Syono, J. Solid State Chem. **132**, 139 (1997).

[14] J.B. Goodenough, Phys. Rev. **100**, 564 (1955).

[15] J.B. Goodenough, J. Phys. Chem. Solids **6**, 287 (1958).

[16] J. Kanamori, J. Phys. Chem. Solids **10**, 87 (1959).

[17] N.A. Hill and K. Rabe, Phys. Rev. B **59**, 8759 (1999).

[18] H. Woo, T.A. Tyson, M. Croft, S.-W. Cheong, and J.C. Woicik, Phys. Rev. B **63**, 134412 (2001).

[19] L.E. Gonchar and A.E. Nikiforov, Fiz. Tverd. Tela (St. Petersburg) **42**, 1070 (2000) [Phys. Solid State **42**, 1070 (2000)].

Structure–Property Relationships in Superconducting Cuprates

C. N. R. Rao*

Department of Chemistry, University of Wales, Cardiff CF1 3TB, U.K., and Solid State and Structural Chemistry Unit, Indian Institute of Science, Bangalore 560 012, India

A. K. Ganguli

CSIR Centre of Excellence in Chemistry, Indian Institute of Science, Bangalore 560 012, India

1 Introduction

The highest superconducting transition temperature known till 1986 was 23K and it seemed as though this barrier would not be broken.[1] The discovery of 30K superconductivity in an oxide of the La–Ba–Cu–O system by Bednorz and Müller[2] changed the picture. A variety of superconducting oxides, especially cuprates, have since been synthesized and characterized,[3—6] the highest transition temperature as of today being 155K in $HgBa_2Ca_2Cu_3O_{8+\delta}$. Studies of the various families of cuprates have shown many commonalities and unifying features.[6,7] Properties of the cuprates have been related to certain structural and electronic parameters. Although there is no simple relationship between the superconducting transition temperature and any specific structural feature of the cuprates, the various correlations help us to understand these materials better and to design newer ones. In this article, we shall briefly examine some of the significant structure–property relationships in superconducting cuprates along with the structural commonalities.

2 Common Structural Features

Many cuprate families have been discovered in the past seven years. The general features of the cuprates are shown schematically in Figure 1. The major families of cuprates are:
(a) $La_{2-x}A_xCuO_4$ (A = alkaline earth) possessing the K_2NiF_4(T) structure; (b) $LnBa_2Cu_3O_{7-\delta}$ (Ln = Y or rare-earth other than Ce, Pr, and Tb) referred to as the 123 type (Figure 2) and the related $LnBa_2Cu_4O_8$ (124) and $Ln_2Ba_4Cu_7O_{15}$ (247) cuprates containing perovskite layers with CuO_2 sheets as well as Cu–O chains; (c) $Bi_2(Ca,Sr)_{n+1}Cu_nO_{2n+4}$ containing two BiO layers and perovskite layers with CuO_2 sheets (Figure 2); (d) $Tl_2A_{n+1}Cu_nO_{2n+4}$ and $TlA_{n+1}Cu_nO_{2n+3}$ (A = Ca, Ba, Sr *etc.*) containing Tl–O layers and perovskite layers with CuO_2 sheets (Figure 2); (e) lead-based superconducting cuprates such as $Pb_2Sr_2LnCu_3O_8$ containing CuO_2 sheets and CuI–O sticks; (f) Tl, Bi, and Pb cuprates containing fluorite

* For correspondence

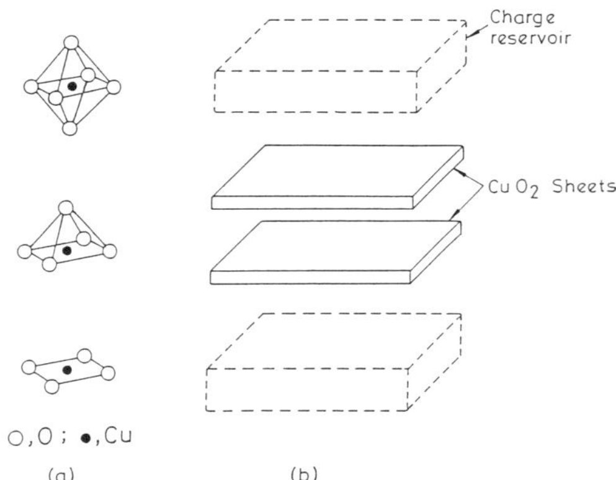

Figure 1 (a) The three types of Cu–O polyhedra found in the superconducting cuprates (b) Schematic representation of cuprates.

layers and CuO_2 sheets; (g) mercury cuprates of the type $HgCa_{n-1}Ba_2Cu_nO_{2n+2+\delta}$; and (h) infinite layer cuprates such as $ACuO_2$ (A = Ca, Sr, Ba) and $Ln_{1-x}A_xCuO_2$ (Ln = Nd, Pr; A = Sr, Ba). All these cuprates have holes as charge carriers. The only well established superconducting cuprates with electrons as charge carriers are $Nd_{2-x}M_xCuO_4$ (M = Ce, Th) and related compounds with the T' structure possessing square-planar CuO_4 units instead of the octahedra in the T structure (Figure 1).

A. K. Ganguli is a scientific officer at the Centre of Excellence in Chemistry at the Indian Institute of Science. He has an M.Sc. degree from the University of Delhi and a Ph.D. degree from the Indian Institute of Science and has carried out post-doctoral work at du Pont and Iowa State University.

C. N. R. Rao is a Professor of Chemical Science at the Indian Institute of Science, and President of the Jawaharalal Nehru Centre for Advanced Scientific Research, Bangalore, India. He is an Honorary Professor of Chemistry at the University of Wales, Cardiff. He was Commonwealth Visiting Professor at the University of Oxford and Nehru Professor at the University of Cambridge. He is a Fellow of the Royal Society, London, Foreign Associate of the U.S. National Academy of Sciences, and Foreign member of several other academies. He is a member of the Pontifical Academy of Sciences and an Honorary Fellow of the Royal Society of Chemistry. He is a recipient of the Marlow Medal of the Faraday Society and the RSC Medal for solid-state chemistry. His main research interests are in solid-state chemistry, spectroscopy, and molecular structure and surface science. He is the author of over 500 research papers and several books in solid state chemistry.

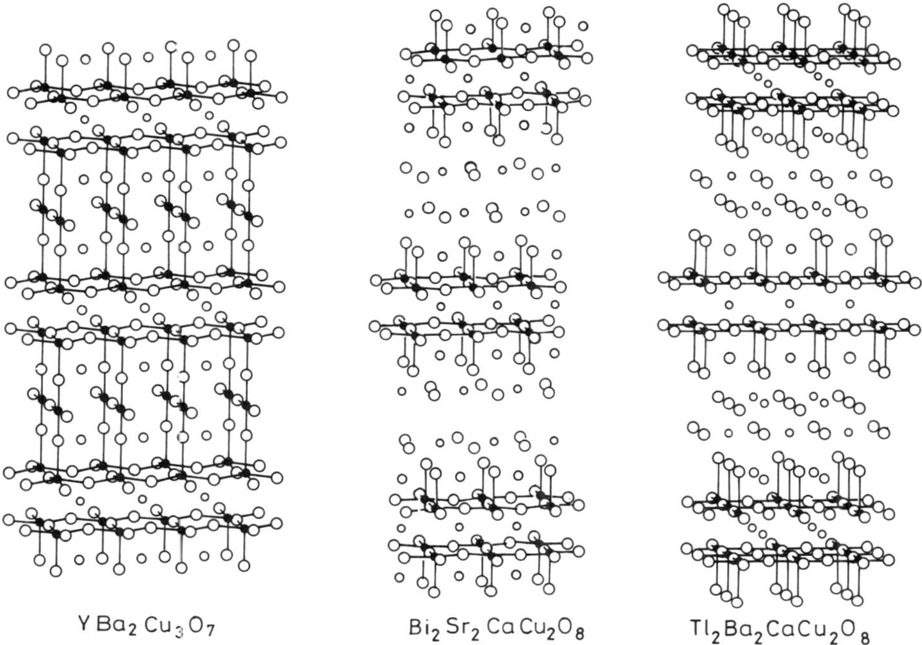

Figure 2 Structures of $YBa_2Cu_3O_7$, $Bi_2CaSr_2Cu_2O_8$, and $Tl_2CaBa_2Cu_2O_8$. Notice the presence of CuO_2 sheets containing CuO_5 square-pyramids. Additionally, there are Cu–O chains in $YBa_2Cu_3O_7$.

Many of the cuprate families have an antiferromagnetic insulator member at one end of the composition (*e.g.* La_2CuO_4 in $La_{2-x}A_xCuO_4$, $YBa_2Cu_3O_6$ in $YBa_2Cu_3O_{7-\delta}$, and $TlYSr_2Cu_2O_7$ in $TlY_{1-x}Ca_xSr_2Cu_2O_{7-\delta}$. What is more interesting is that all the cuprates are at a metal–insulator boundary. Some of them undergo a metal–non-metal transition as a function of composition (*e.g.* $Bi_2Ca_{1-x}Y_xSr_2Cu_2O_8$ and $TlY_{1-x}Ca_xSr_2Cu_2O_7$ with change in *x*).

The cuprates can be described on the basis of certain structural features common to many of them. For example, the structures of $La_{2-x}A_xCaO_4$, $Bi_2(Ca, Sr)_{n+1}Cu_nO_{2n+4}$ and the thallium cuprates can be considered to be intergrowths of oxygen-deficient perovskite layers, $ACuO_{3-x}$, with AO-type rock-salt layers.[7] The cuprates contain different types of Cu–O polyhedra with the hole superconductors necessarily having CuO_5 or CuO_6 units and the electron-superconducting $Nd_{2-x}M_xCuO_4$ containing only CuO_4 square-planar units (Figures 1 and 2). Thus, the essential feature of the cuprates is the presence of CuO_2 sheets with or without apical oxygens. The mobile charge carriers in the cuprates are in the CuO_2 sheets. All the cuprates have charge reservoirs as exemplified by the Cu–O chains in the 123 and 124 cuprates and the TlO, BiO, and HgO layers in the other cuprates. That the CuO_2 sheets are the seat of high-temperature superconductivity is demonstrated by the fact that intercalation of iodine between BiO layers in the bismuth cuprates does not affect the superconducting transition temperature while introduction of fluorite layers between the CuO_2 sheets adversely affects superconductivity. In the different series of cuprates with varying number of CuO_2 sheets studied hitherto, the T_c reaches a maximum when $n = 3$ except in single thallium layer cuprates where the maximum is at $n = 4$. The infinite layered cuprates, where the CuO_2 sheets are separated by alkaline earth and other cations, show T_c's in the 40—110K range.[8] Superconductivity in these materials appears to be due to the presence of Sr–O defect layers corresponding to the insertion of $Sr_3O_{2\pm x}$ blocks.[8]

Based on the interplanar Cu–Cu distances, one can classify cuprates into two categories.[9] In one category, r(Cu–Cu) lies between 3.0 and 3.6Å with T_c's varying between 50 and 133K and in another it is between ~ 6 and 12.5Å encompassing superconductors with lower T_c($< 50K$), except $Tl_2Ba_2CuO_6$ and $HgBa_2CuO_{4.065}$ with T_c's of $\sim 90K$. In the first category with r(Cu–Cu) < 3.6Å, the T_c increases as the Cu–Cu distance decreases. In the 2222-type fluorite-based superconductors, there are three copper oxygen sheets, $[CuO_2–CuO_\delta–CuO_2–$fluorite–$CuO_2–CuO_\delta–CuO_2]$, each block of three sheets separated by a Ln_2O_2 fluorite layer. The r(Cu–Cu) relevant to these compounds would be the distance between the CuO_2 sheets across the fluorite layer (~ 6.2Å) and not the distance between two neighbouring sheets. Accordingly, these cuprates exhibit a low T_c(45—50K). It therefore appears that the distance between the CuO_2 sheets is a factor in determining the value of T_c, indicating that there is some interaction between the closely spaced CuO_2 sheets although the cuprates have quasi two-dimensional character.

Oxygen stoichiometry and ordering play a crucial role in determining the structure and properties of cuprates. The dependence of the structure and properties of $YBa_2Cu_3O_{7-\delta}$ on oxygen content has been studied in detail. Thus, $YBa_2Cu_3O_{7-\delta}$ which is orthorhombic ($c \simeq 3$ *b*) with a T_c of 90K when $0.0 \lesssim \delta \lesssim 0.25$, assumes another orthorhombic structure ($c \neq 3b$) when $0.3 \lesssim \delta \lesssim 0.4$ with a T_c of 60K. When $\delta = 1.0$, all the oxygens in the CuO chains are depleted and the structure becomes tetragonal and the material is non-superconducting. When $\delta = 0.5$, there is an ordered arrangement of oxygen vacancies with the presence of fully oxidized (O_7) and fully reduced (O_6) chains alternately. The compositions showing 60K superconductivity are metastable and transform to a 124-type phase on heating at low temperatures.[10]

Oxygen-excess La_2CuO_4 is biphasic, consisting of the stoichiometric antiferromagnetic phase and an oxygen-excess superconducting phase.[11] In bismuth cuprates, excess oxygen in the BiO layers gives rise to incommensurate modulation. Modulation-free superconducting bismuth cuprates have been made[12] by replacing one Bi^{3+} by Pb^{2+}. In $HgBa_2CuO_{4+\delta}$ oxygen excess in the Hg plane is necessary to render it superconducting.

3 The Relationship between T_c and the Hole Concentration

As mentioned earlier, a majority of the cuprates have holes as charge carriers. These holes are created by the extra positive charge on copper (*e.g.* Cu^{3+}) or on oxygen (*e.g.* O^{1-}). The excess positive charge can be represented in terms of the formal valence of copper, which in the absence of holes will be $+2$ in the CuO_2 sheets. In hole superconducting cuprates, it is generally around $+2.2$. In electron superconductors, it would be less than $+2$ as expected. The actual concentration of holes, n_h, in the CuO_2 sheets in $La_{2-x}A_xCuO_4$, $YBa_2Cu_3O_7$, and Bi cuprates is readily determined by redox titrations. In the 123 cuprates, the concentration of mobile holes in the CuO_2 sheets can be delineated from that in the Cu—O chains.[13] Determination of n_h in thallium cuprates poses some problems, but in single Tl—O layer cuprates, chemical methods have been developed to obtain reasonable estimates.[14] Generally, T_c in a given family of cuprates reaches a maximum value at an optimal value of n_h as shown in Figure 3; the maximum is around $n_h \sim 0.2$ in most cuprates.[15] Notice that the points in the underdoped region in Figure 3 fall close to a straight line. Deviations occur in the overdoped region. Single layer thallium cuprates also show this behaviour. In $Tl_{1-y}Pb_yY_{1-x}Ca_xSr_2Cu_2O_7$ where the substitution of Tl^{3+} by Pb^{4+} has an effect opposite to that due to the substitution of Y^{3+} by Ca^{2+}, the T_c becomes a maximum at an optimal value of $(x-y)$, which is a measure of the hole concentration[16] (Figure 4). By suitably manipulating x and y, the T_c of this system can be increased from 85—90K up to 110K.

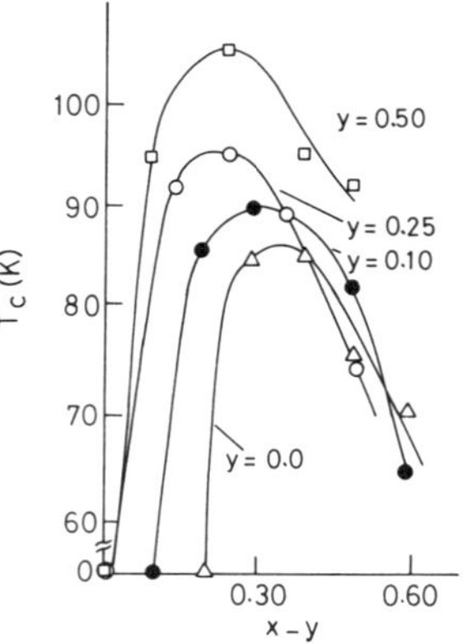

Figure 4 Plots of T_c against the effective hole concentration $(x-y)$ in $Tl_{1-y}Pb_yY_{1-x}Ca_xSr_2Cu_2O_7$ (from reference 16).

Another way of representing the variation[17] of T_c with n_h is to plot the reduced T_c (T_c observed/maximum value of T_c) against n_h as shown in the inset of Figure 3. The ends of the plateau region in this curve correspond to insulating (possibly antiferromagnetic) and metallic regimes of the materials.

4 Relation between T_c and the In-plane Cu—O Distance

The Cu—O bonds in the CuO_2 sheets involve an antibonding π interaction and doping with holes reduces the bond distance. The in-plane Cu—O bond distance $r(Cu-O)$, therefore reflects the hole concentration and a variation of T_c with $r(Cu-O)$ represents an alternative way of examining the T_c–n_h relationship. In cuprates where n_h cannot be determined, as for example in Tl cuprates, the T_c *vs.* $r(Cu-O)$ plots show maxima at an optimal distance. The value $r(Cu-O)$ is around half that of the a-parameter in most cuprates. Whangbo *et al.*[18] find three distinct T_c–$r(Cu-O)$ relationships depending on the cation located above and below the CuO_2 sheets, with each exhibiting a T_c maximum at an optimal value of the distance (Figure 5). If we plot the reduced T_c against $r(Cu-O)$, we get the curves shown in Figure 6 where the highest T_c values occur in the 1.89—1.94 Å

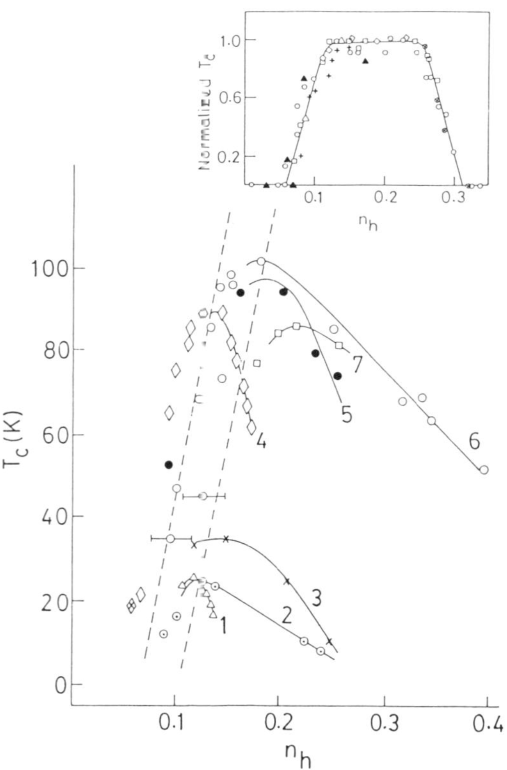

Figure 3 Variation of T_c with hole concentration n_h in superconducting cuprate families (adapted from reference 15). Deviations in overdoped region are similar to those in the plots of muon spin relaxation rate (n_h/m^*) against T_c of Uemura *et al.* (*Phys. Rev. Lett.*, 1991, **66**, 2665). 1, 2, 6, and 7, Bi cuprates; 5, 123 cuprates; 3, $La_{2-x}Sr_xCuO_4$, 4, T_c cuprates. The variation of normalized T_c with n_h is shown in the inset (from reference 17).

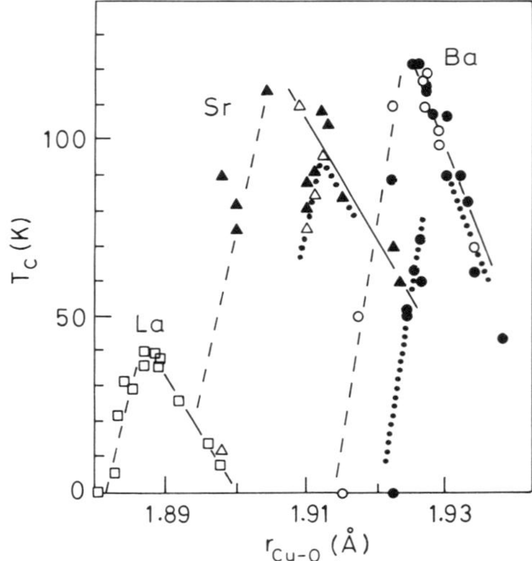

Figure 5 Variation of T_c with in-plane Cu—O distances in various families of cuprates (from reference 18).

4

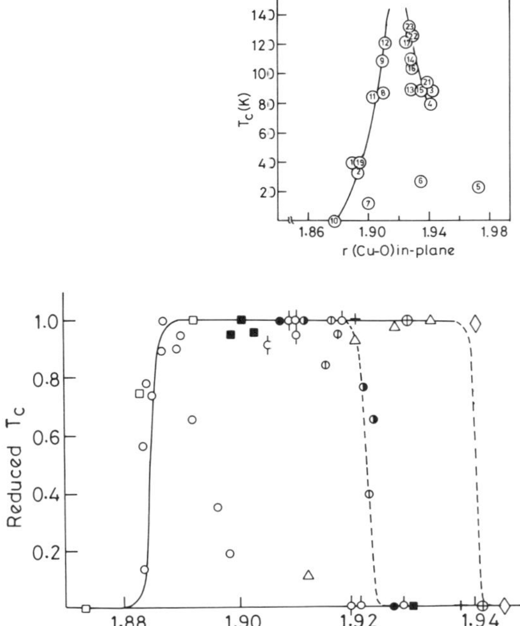

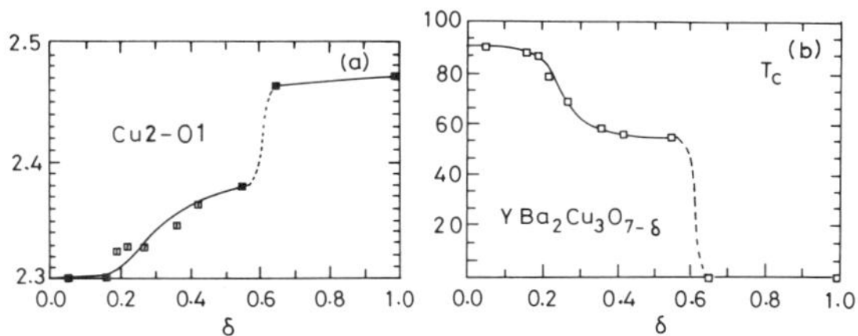

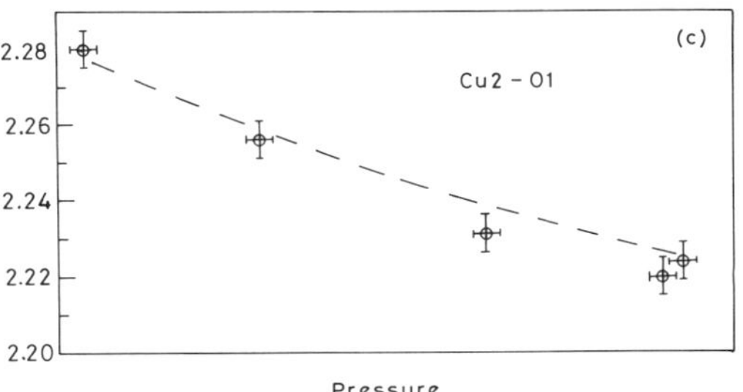

Figure 6 Plot of reduced T_c with in-plane $r(Cu-O)$ distance: $La_{2-x}Sr_x$-CuO_4 (open circles), $YBa_2Cu_3O_{7-\delta}$ (circles with cross), Bi_2 (Ca, Y, Sr)$_3Cu_2O_8$ (half-shaded circles), $Bi_2Ca_{1-x}Y_xSr_2Cu_2O_8$ (circles with a line in the centre), modulation-free Bi-cuprates (filled squares), $TlSr_{2-x}La_xCuO_5$ (open squares). $Tl_2Ba_{2-x}Sr_xCuO_6$ (open triangles), $TlCa_{1-x}Y_xBa_2Cu_2O_7$ (crosses), $TlCa_{1-x}Ln_xSr_2Cu_2O_7$ (circles with one line on top), and $Tl_{0.5}Pb_{c.5}Ca_{1-x}Y_xSr_2Cu_2O_7$ (circles with lines above and below), $Pb_2Sr_2Y_{1-x}Ca_xCu_3O_{8+\delta}$ (filled circles), $HgBa_2CuO_{4+\delta}$ (open diamonds). Inset shows the variation of T_c with the in-plane $Cu-O$ distance where only the cuprate compositions showing the maximum T_c in each family are taken into account. (1) $La_{1.85}Sr_{0.15}CuO_4$. (2) $La_{1.85}Ba_{0.15}CuO_4$. (3) $YBa_2Cu_3O_{6.91}$. (4) $YBa_2Cu_4O_8$. (5) $Nd_{1.85}Ce_{0.15}CuO_4$. (6) $Nd_{1.4}Ce_{0.2}Sr_{0.4}CuO_4$. (7) $Bi_2Sr_2CuO_6$. (8) $Bi_2CaSr_2Cu_2O_8$. (9) $Bi_2Ca_2Sr_2Cu_3O_{10}$. (10) $Tl_{0.5}Pb_{0.5}Sr_2CuO_5$. (11) $Tl_{0.5}Pb_{0.5}CaSr_2Cu_2O_7$. (12) $Tl_{0.5}Pb_{0.5}Ca_2Sr_2Cu_3O_9$. (13) $TlCaBa_2Cu_2O_7$. (14) $TlCa_2Ba_2Cu_3O_9$. (15) $Tl_2Ba_2CuO_6$. (16) $Tl_2CaBa_2Cu_2O_8$. (17) $Tl_2Ca_2Ba_2Cu_3O_{10}$. (18) $Tl_2Ca_3Ba_2Cu_4O_{12}$. (19) $TlSrLaCuO_5$. (2) $TlCa_{0.5}La_{0.5}Sr_2Cu_2O_7$. (21) $HgBa_2CuO_{4.065}$. (22) $HgCaBa_2Cu_2O_{6.22}$. (23) $HgCa_2Ba_2Cu_3O_{8.41}$.

range. When $r(Cu-O) < 1.88$ Å, the material is metallic; those with $r(Cu-O) > 1.94$ Å are certainly insulating, but there are different insulating boundaries for the different cation families, somewhat like in Figure 5. However, if we consider only the maximum T_c value in each cuprate family and the corresponding $r(Cu-O)$, we get the curve shown in the inset of Figure 6 which peaks at $r \approx 1.92$ Å. $Bi_2Sr_2CuO_6$ with a T_c of 12K would not fall on the curve, but $Bi_2Sr_{2-x}La_xCuO_{6+\delta}$ with a T_c of 30K would.

5 The Relationship between T_c and the Apical Cu–O Distance

All the cuprates which are hole superconductors have apical oxygens which act as the link between the charge reservoirs and the CuO_2 sheets. (Note that electron superconductors such as $Nd_{2-x}Ce_xCuO_4$ contain only CuO_4 units without apical oxygens.) In $YBa_2Cu_3O_{7-\delta}$, the T_c–$r(Cu2-O1)$ relationship (Figures 7a and b) mirrors the T_c–δ relationship.[19] In $YBa_2Cu_4O_8$, the T_c increases with pressure from 80K to 90K, as the apical $Cu-O$ distance decreases[20] (Figure 7c). We have sought to find relationships between T_c and apical $Cu-O$ distance in

Figure 7 Variation of (a) the apical $Cu-O$ distance with δ in YBa_2-$Cu_3O_{7-\delta}$ (b) T_c with δ and (c) variation of the $Cu-O$ apical distance with pressure in $YBa_2Cu_4O_8$ as shown by the points (the dashed line is obtained by the oxidation of $YBa_2Cu_3O_{7-\delta}$) (from references 19 and 20).

different cuprate families (Figure 8). Within a series of cuprates with varying number of CuO_2 sheets, (*e.g.*, $TlCa_{n-1}Ba_2Cu_nO_{2n+3}$), the apical $(Cu-O)$ distance decreases with the increase in n while T_c increases linearly with the decrease in the apical distance. The slope of the T_c *vs.* the apical $Cu-O$ distance plot is nearly the same in $TlCa_{n-1}Ba_2Cu_nO_{2n+3}$, $Tl_{0.5}Pb_{0.5}Ca_{n-1}Sr_2Cu_nO_{2n+3}$, and $HgCa_{n-1}Ba_2Cu_nO_{2n+3}$, all of them having a single rock-salt layer (TlO, $Tl_{0.5}Pb_{0.5}O$, or HgO_δ). The $Tl_2Ca_{n-1}Ba_2Cu_nO_{2n+4}$ family also shows increasing T_c with the decrease in the apical $Cu-O$ distance, but with a different slope. The mercury-based superconductors have larger apical distances compared to the other cuprates and they also show a large pressure dependence of T_c. Interestingly, if we consider the maximum T_c points at the top of the plots for the different

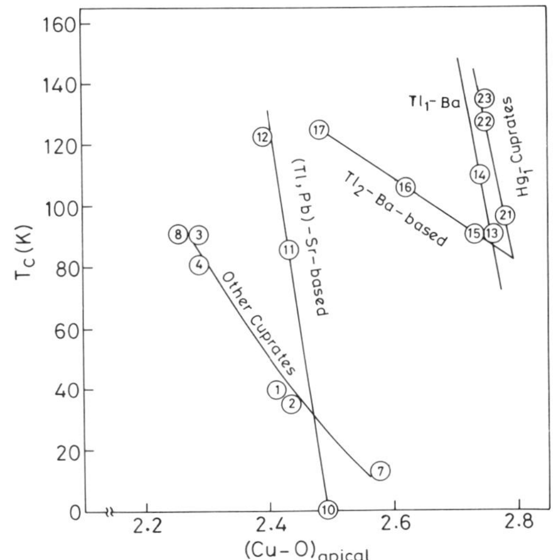

Figure 8 Variation of T_c with the apical $Cu-O$ distance in cuprate superconductors.

groups of cuprates in Figure 8, we see that the T_c increases with the increase in the apical Cu–O distance.

6 Covalency of the Charge Reservoir

All the superconducting cuprates have charge reservoirs and any damage to these reservoirs adversely affects superconductivity. The nature of the charge reservoir determines the carrier concentration and the ease of charge-transfer to the CuO$_2$ sheets. Covalent charge reservoirs can redistribute charge effectively through the apical oxygen of the CuO$_5$ square pyramids giving rise to high T_c's. Ionic charge reservoirs, on the other hand, would be less flexible with regard to the charge states and do not favour high T_c's. Structural mismatch as well as disorder in the reservoirs also adversely affect the superconducting properties. The covalency of the Hg–O bond could be related to the high T_c of Hg cuprates.

The effect of covalency of the charge reservoir is clearly seen in $Tl_{0.5}Pb_{0.5}Sr_2Y_{1-x}Ca_xCu_2O_7$ which shows a T_c of 110K at an optimal x value, the material being an insulator when $x = 0.0$. The a parameter decreases with the increase in x, because the population of the antibonding Cu $3d_{x^2-y^2}$ orbitals decreases with the increase in x, causing a strengthening of the Cu–O bond. The puckering of the CuO$_2$ sheets decreases with increasing hole concentration. The displacement of the apical oxygen is around 0.06 Å when $x = 1.0$ and 0.20 Å when $x = 0.0$. An increase in Y content (increased electron population), however, increases the puckering and pulls the apical oxygen away from the base of the pyramid.

7 The Relationship between T_c and Madelung Potentials

The role of the Madelung site potential in the hole conductivity of the cuprate superconductors was first pointed out by Torrance and Metzger.[21] Two classes of cuprates can be delineated depending on the value of ΔV_M which is the difference in Madelung site potential for a hole on a Cu site and that on an oxygen site. Those with high ΔV_M ($\geq$ 47eV) are metallic and superconducting; those with lower ΔV_M are semiconducting with localized holes. It is possible to define a term ΔV_A which is the difference in the Madelung site potentials for a hole between the apex and the in-plane oxygen atoms and provides a measure of the position of the energy level of the p_z-orbital on the apical oxygens.[22] When the maximum values of T_c of hole-doped superconductors are plotted against ΔV_A (Figure 9), one finds that nearly all the cuprates are located on a curve (with some width), the cuprates with large ΔV_A exhibiting high T_c's. It appears that the energy level of the apical oxygen plays a significant role in the electronic states of the doped holes,

thereby affecting the T_c. The correlation probably owes its origin to the stability of local singlet states made up of two holes in the Cu $3d_{x^2-y^2}$ and O $2p$ orbitals in the CuO$_2$ sheet. The local singlet is well defined and stable when the energy level of the apical oxygen atom is sufficiently high. A comparison of the correlations of T_c with ΔV_A and ΔV_M indicates that ΔV_A scales better with T_c.

It is instructive to correlate T_c simultaneously with ΔV_M and the in-plane Cu–O bond length, d_p, in the CuO$_2$ planes. The maximum T_c for each cuprate is shown in the d_p *vs.* ΔV_M plot[23] in Figure 10. The data points are confined to a narrow strip running from the top left (high d_p and low ΔV_M) to the bottom right corner (low d_p and high ΔV_M). The T_c value increases as we go from right to left (larger to smaller value of ΔV_M) or from top to bottom (from higher d_p to lower d_p). An important observation is that T_c changes to a small extent as one goes from the top left to bottom right (d_p decreasing and ΔV_M increasing) while T_c increases drastically when traversing from the top right to bottom left of the map (both d_p and ΔV_M decreasing). Clearly, the T_c is governed by both ΔV_M and d_p. It appears that the difference in T_c is a result of the difference in internal stress of the crystal (high T_c when the CuO$_2$ planes are under compression and low T_c when they are strained).

8 The Relationship between Bond Valence Sums and T_c

The bond-valence sum is a measure of the total charge on an atom in a structure. Its value changes with oxygen doping, cation substitution, or applied pressure indicating the occurrence of charge transfer within the structure. One defines, $V_- = 2 + V_{Cu2} - V_{O2} - V_{O3}$ where V_- is the total excesss charge in the planes (O2 and O3 are the oxygens in the plane) and $V_+ = 6 - V_{Cu2} - V_{O2} - V_{O3}$. The T_c *vs.* V_- plot for YBa$_2$Cu$_3$O$_{7-\delta}$ is similar to the T_c vs. δ plot; the plot of T_c against V_+ is linear[24] (Figure 11).

9 The Importance of the Cu–O Charge-Transfer Energy

One of the unique features of the cuprates is the relatively small Cu–O charge-transfer energy. It is therefore of significance to relate this property with superconductivity. X-Ray photoemission spectra of cuprates in the Cu $2p_{3/2}$ region show a characteristic two-peak feature. The peak around 933 eV (main peak) is due to a final state with primarily $3d^{10}$ character, while the peak of weaker intensity at about 941 eV binding energy is mainly due to a $3d^9$ final state (satellite). Model calculations of the Cu $2p$ core-level photoemission within a CuO$_4$ square-planar cluster show that the I_s/I_m ratio (relative intensity of the satellite to the main peak) is related to Δ/t_{pd} where Δ is the charge-transfer excitation energy and t_{pd} is the hybridization strength between the Cu $3d$ and O $2p$ orbitals. Thus, any relationship between the experimentally observed I_s/I_m ratio with n_h would suggest a link between n_h (T_c) and the Δ/t_{pd} ratio. It is indeed found that the I_s/I_m ratio is related to the hole concentration n_h in many of the superconducting cuprates.[25]

In Figure 12(a), we show the relative intensity of the satellite (I_s/I_m) as a function of x in La$_{2-x}$Sr$_x$CuO$_4$. In the same plot, we also show the variation of the experimentally determined hole concentration with x in this series. The inset of Figure 12(a) shows the dependence of T_c on n_h exhibiting a maximum around $n_h = 0.15$. We see that the value of I_s/I_m decreases markedly with increasing x until about 0.3, after which it changes slope. What is significant is that I_s/I_m decreases as the hole concentration increases. In Figure 12(b) we show the variation of n_h and I_s/I_m with x in BiPbSr$_2$Y$_{1-x}$Cu$_2$O$_8$. In the inset we show the variation of T_c with n_h. Here also we see that the I_s/I_m ratio decreases as the hole concentration increases. Figure 13 shows how in three series of high T_c cuprates, the I_s/I_m ratio decreases monotonically with the increase in magnitude of hole doping. The

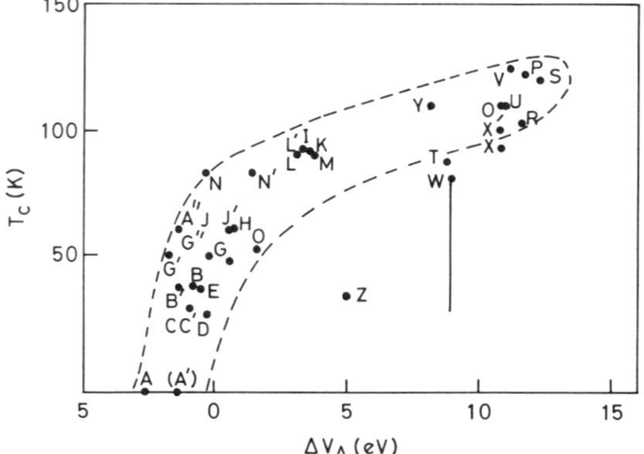

Figure 9 Variation of T_c with ΔV_A. See reference 22 for the explanation of the letter symbols.

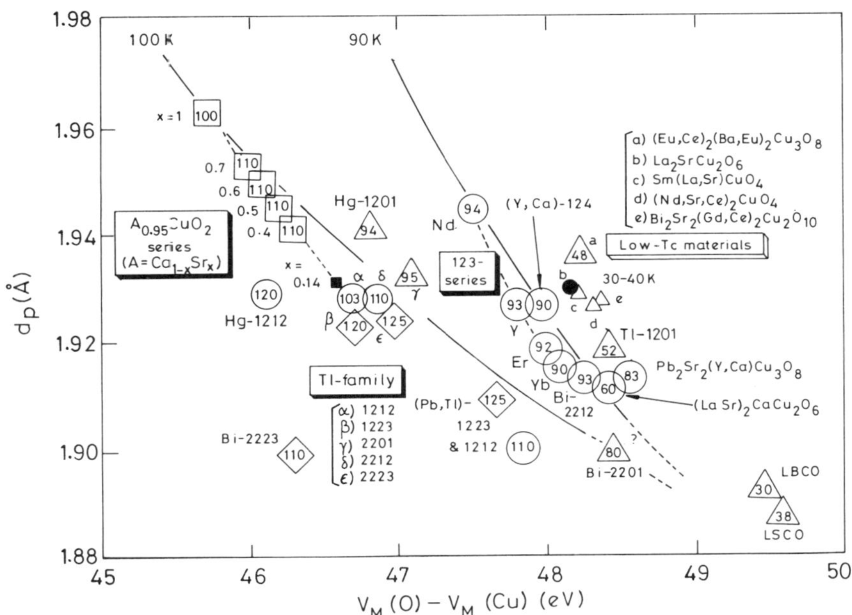

Figure 10 Plot of d_p *vs.* ΔV_M showing region of high T_c and low T_c superconductors (from reference 23).

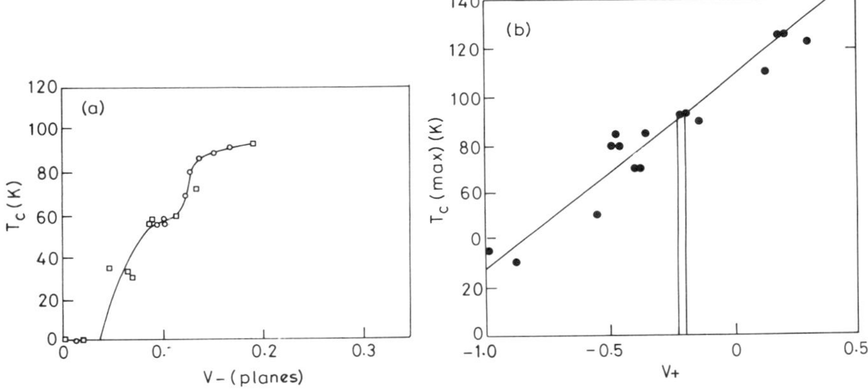

Figure 11 Variation of (a) T_c with V_- (see text) and (b) T_c with V_+ for YBa$_2$Cu$_3$O$_{7-\delta}$ (from reference 24).

variation is smooth across the insulator–superconductor–metal boundaries. Calculations show that the I_s/I_m ratio increases with the Δ/t_{pd} ratio in the cuprates. Accordingly, at a given n_h value where the T_c is maximum, increasing I_s/I_m is accompanied by an increase in T_c. Thus, Bi$_2$Ca$_{1-x}$Ln$_x$Sr$_2$Cu$_2$O$_{8+\delta}$ exhibits the highest T_c of around 100K while La$_{2-x}$Sr$_x$CuO$_4$ has the lowest T_c value. The observation of a maximum T_c around an optimal value of n_h or Cu–O distance can therefore be traced to the optimal Δ/t_{pd} value or of the Cu–O charge-transfer energy.

10 Concluding Remarks

We have discussed several interesting correlations between the superconducting transition temperature and some of the crucial structural and electronic parameters. These correlations along with the structural commonalities in the cuprates show how superconductivity is intimately connected with the structural

chemistry of these materials. They may also suggest many other relationships, some of which may be even more significant.

Acknowledgement. The authors thank the Science Office of the European Union, the Department of Science and Technology, and the University Grants Commission for support of this research.

7 References

1 C. N. R. Rao, K. J. Rao, and J. Gopalakrishnan, *Ann. Rep. Prog. Chem., Sect C*, 1985, **82**, 193.

2 J. Bednorz and K. A. Müller, *Z. Phys.*, 1986, **B64**, 189.

3 B. Raveau, C. Michel, M. Hervieu, and D. Groult, 'Crystal Chemistry of High T_c Superconducting Copper Oxides', Springer-Verlag, Berlin, 1991.

4 'Chemistry of High-Temperature Superconductors', ed. C. N. R. Rao, World Scientific, Singapore, 1991.

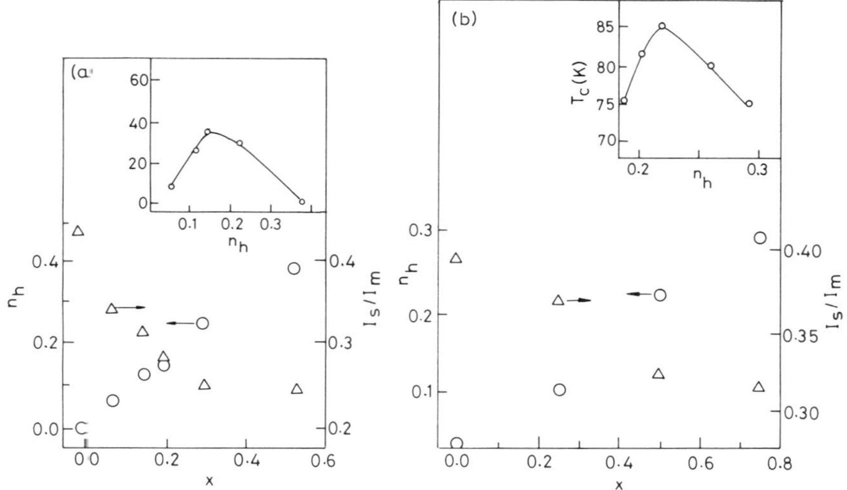

Figure 12 (a) Variation of I_S/I_M ratio with x in $La_{2-x}Sr_xCuO_4$. (b) Also shown is the variation of n_h with x and I_S/I_M and n_h with x in $BiPbSr_2Y_{1-x}Ca_xCu_2O_8$ (from reference 25).

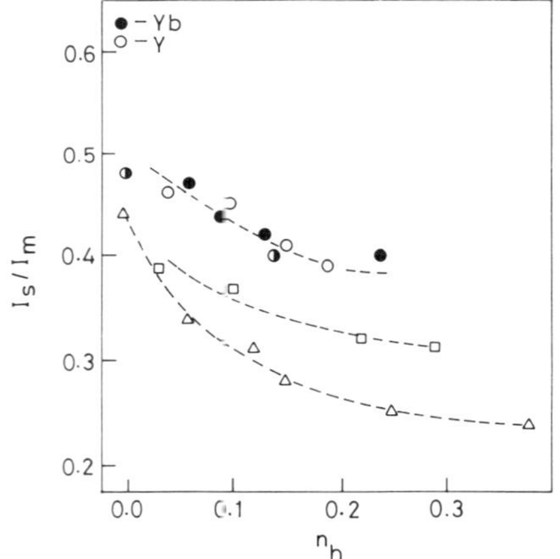

Figure 13 Variation of the I_S/I_M ratio with hole concentration (n_h): $La_{2-x}Sr_xCuO_4$ (triangles), $BiPbSr_2Y_{1-x}Ca_xCu_2O_8$ (squares), and $Bi_2Ca_{1-x}L_xSr_2CuO_2O_8$ (circles) (from reference 25).

5 M. Nunez-Regueiro, J. L. Tholence, E. V. Antipov, J. J. Capponi, and M. Marezio, *Science*, 1993, **262**, 976.
6 A. R. Armstrong and P. P. Edwards, *Ann. Rep. Prog. Chem., Sect. C*, 1991, **88**, 259.
7 C. N. R. Rao, *Philos. Trans. R. Soc. London*, 1991, **A336**, 595, 106.
8 H. Zhang, Y. Y. Wang, V. P. Dravid, L. D. Marks, P. D. Han, D. A. Payne, P. G. Radaelli, and J. D. Jorgensen, *Nature*, 1994, **370**, 352, and references therein.
9 H. Nobumasa, K. Shimizu, and T. Kawai, *Physica C*, 1990, **167**, 515.
10 R. Nagarajan and C. N. R. Rao, *J. Solid State Chem.*, 1993, **103**, 533.
11 J. D. Jorgensen, *Physics Today*, 1991, **44**, 34
12 V. Manivannan, J. Gopalakrishnan, and C. N. R. Rao, *Phys. Rev. B*. 1991, **43**, 8686.
13 R. Nagarajan and C. N. R. Rao, *J. Mater. Chem.*, 1993, **3**, 969.
14 C. N. R. Rao in 'Thallium-based High Temperature Superconductors', ed. A. M. Hermann and J. V. Yakhmi, Marcel Dekker, New York, 1994.
15 C. N. R. Rao, J. Gopalakrishnan, A. K. Santra, and V. Manivannan, *Physica C*, 1991, **174**, 11.
16 R. Vijayaraghavan, N. Rangavittal, G. U. Kulkarni, E. Grantscharova, T. N. Guru Row, and C. N. R. Rao, *Physica C*, 1991, **179**, 183.
17 H. Zhang and H. Sato, *Phys. Rev. Lett.*, 1993, **70**, 1697.
18 M. H. Whangbo, D. B. Kang, and C. C. Torardi, *Physica C*, 1989, **158**, 371.
19 R. J. Cava, A. W. Hewat, E. A. Hewat, B. Batlogg, M. Marezio, K. B. Rabe, J. J. Krajewski, W. F. Peck, and L. W. Rupp, *Physica C*, 1990, **165**, 419.
20 R. J. Nelmses, E. Loveday, E. Kaldis, and J. Karpinski, *Physica C*, 1992, **172**, 311.
21 J. B. Torrance and R. M. Metzger, *Phys. Rev. Lett.*, 1989, **63**, 1515.
22 Y. Ohta, T. Tohyama, and S. Maekawa, *Physica C*, 1990, **167**, 515.
23 M. Muroi, *Physica C*, 1994, **219**, 129.
24 J. L. Tallon and G. V. M. Williams, *J. Less Common Metals*, 1990, **164—165**, 60.
25 A. K. Santra, D. D. Sarma, and C. N. R. Rao, *Phys. Rev. B*, 1991, **43**, 5612.

Stripes and Superconductivity in Cuprates – Is there a Connection?

N. Kumar[a] and C. N. R. Rao*[b]

KEYWORDS:

cuprates · scanning probe microscopy · soft X-ray scattering · stripes · superconductors

Charge stripes can be traced back to the 1990s or even earlier.[1] Stripes are a special case of the general phenomenon of electronic phase separation in strongly correlated transition metal oxide systems wherein regions with a high concentration of mobile carriers are separated from those with a low concentration due to Coulomb interactions. The carrier-rich regions are generally metallic and/or ferromagnetic, while the carrier-poor regions are antiferromagnetic and insulating. Such phase separation manifests itself in the form of metallic droplets in an antiferromagnetic insulating matrix or self-organization, for example, in one dimension as stripes.[1, 2] The phase-separated regions tend to be of nanometric dimensions.

Over the years, experimental evidence[1–12] has been steadily mounting in support of the theoretical[13–19] proposition that the cuprate superconductors may be intrinsically inhomogeneous in the distribution of their spins and charges in the CuO_2 sheets, and that this inhomogeneity may hold a clue to the mechanism of high-T_c (high critical temperature) superconductivity in these materials. The growing realization that stripes matter has enlivened the debate on high temperature superconductivity in these ceramic materials, and emphasizes the incompleteness in our understanding of the phenomenon discovered in 1986 by Bednorz and Müller.[20]

Stripes have been reported in at least three families of hole-doped high-T_c cuprates, all of which contain the CuO_2 sheets. These are the single-layered cation-doped 214 compound LSCO ($La_{2-x}Sr_xCuO_4$) or the oxygenated (anion-doped) LCO214 (La_2CuO_{4+x}), the 123 compound YBCO ($YBa_2Cu_3O_{7-x}$), which contains CuO chains in addition to the CuO_2 sheets, and the two-layered BSCCO-2212 ($Bi_2Sr_2CaCu_2O_{8+x}$) with two CuO_2 layers per unit cell. The strongest case for the stripes so far has been made for the model system NdLSCO ($La_{1.6-x}Nd_{0.4}Sr_xCuO_4$),[2] in which the stripes are structurally pinned, nearly static. The commonality of the all-important CuO_2 sheets, and the associated low-dimensional physics,[21–23] which is known to favor solitons and the solitonic doping, has been the compelling reason to believe that stripes may be generic to the layered cuprates—at least in the underdoped region of the phase diagram, which is shown schematically in Figure 1.

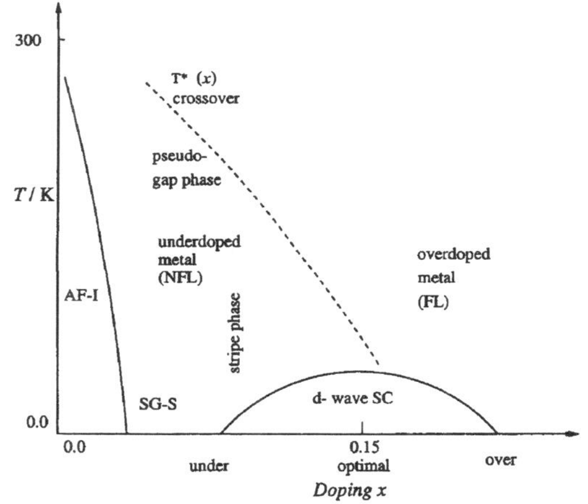

Figure 1. Schematic temperature (T) – doping (x) phase diagram typical of hole-doped high-T_c layered cuprate superconductors. AF-I denotes antiferromagnetically ordered insulator; SG-S, the spin-glass semiconductor; d-wave SC, the $d_{x^2-y^2}$ ordered superconductor; and T*(x), the crossover temperature to the pseudo-gapped, abnormal metallic underdoped phase. Stripe phase occupies an underdoped pseudogapped region, possibly extending into the superconducting phase. The overdoped region is a normal metal.

Charge stripes have also been found in a few other transition metal oxides and were first reported in the isostructural hole-doped nickelates, $La_{2-x}Sr_xNiO_4$ and La_2NiO_{4+x} which contain modulated NiO_2 sheets.[24] But there is a striking difference between the nickelates and the cuprates in that the stripe phase in the former is insulating, much less superconducting, unlike in

[a]   Prof. Dr. N. Kumar
      Raman Research Institute
      Bangalore 560 080 (India)

[b]   Prof. Dr. C. N. R. Rao
      CSIR Center of Excellence in Chemistry and Chemistry & Physics of Materials Unit
      Jawaharlal Nehru Center for Advanced Scientific Research
      Bangalore 560 064 (India)
      Fax: (+ 91) 80846 2760
      E-mail: cnrrao@jncasr.ac.in

the cuprates where it is conducting or superconducting, thus providing a case for a study in contrast between two related layered oxides systems with the spin-$1/2$ containing Cu^{2+}:$3d^9$ and the spin-1 containing Ni^{2+}:$3d^8$. Although cuprates are superconducting, while the nickelates are insulators, the characteristics of the stripes in the two materials are comparable in that for example, the stripes in both the systems are fluctuating, strongly correlated fluids. Interestingly, stripes have been reported in the rare earth manganates of the general formula, $Ln_{1-x}A_xMnO_3$ (Ln = rare earth metal, A = alkaline earth metal) as well.[25] Here, ferromagnetic metallic stripes are bounded by antiferromagnetic insulating stripes.

While the existence of charged stripes is not much in doubt, their relevance to high-temperature superconductivity (HTSC) is by no means established, particularly in the light of the recent anomalous X-ray scattering measurements on $La_2CuO_{4+\delta}$[26] (see, however, refs.[27, 28]). The presence of stripes raises many questions. Thus, we may ask whether they constitute a truly thermodynamic phase. Does the stripe order compete with superconductivity in the layered cuprates? Does it promote HTSC? Does it coexist with superconductivity—trivially, spatially separated from it, or nontrivially, overlapping with it? Or, as is likely, do the stripe order and the HTSC have a common underlying physics? How can the charged stripes, with 1D topology, subtend a 2D electrical conductivity, let alone superconductivity? Or, is it possible that the 1D stripes merely provide a solitonic route to hole doping into the 2D CuO_2 sheets at low levels of doping, as in its low-dimensional analogs in which point defects (zero-dimensional domain walls) provide solitonic doping, for example, polyacetylene.[21–23] Herein, we address some of these and related issues, including the provenance of stripes.

Figure 2 depicts schematically the geometry and the spin-charge order for an in-plane stripe as derived from low-energy neutron diffraction/scattering in a typical hole-doped cuprate

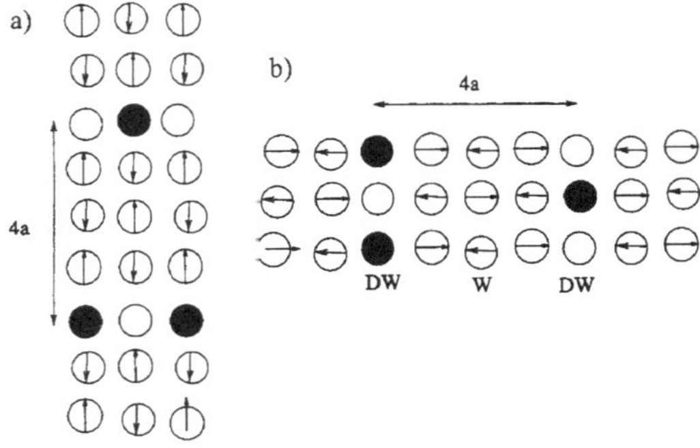

Figure 2. A schematic diagram of the stripe showing spin-charge order in CuO_2 plane for hole doping $x = 1/8$. AFM ordered planar domains separated by charged linear domain walls (DW)—the charge stripes. The domain walls are antiphase: with AFM order phase shifted by π across it, making the spacing of magnetic stripe (8a) twice that of the charged stripes (4a). The open circle with arrow denotes a spin $1/2$ on copper, and the filled circle denotes a hole nominally on copper (actually on oxygen, not shown) with one hole per two copper atoms in the linear DW. The stripes in the adjacent planes (a) and (b) are orthogonal to each other.

superconductor, for example, $La_{2-x}Sr_xCuO_4$ (LSCO), or NdLSCO ($La_{2-x}Nd_ySr_xCuO_4$) for $x = 1/8$ per CuO_2 unit.[2] The charge carriers (holes), which are introduced by doping, are segregated into linear (1D) domain walls—the charged stripes—that are barely a few atoms wide and separate the antiferromagnetically (AFM) ordered planar (2D) domains in all the CuO_2 sheets. Furthermore, the domain walls are antiphase, so that the AFM domain order is phase shifted by π across the charged domain wall. This shift has the consequence of making the spatial period of the magnetic stripes twice that of the charge stripes. This factor of two follows from a general consideration of the ground state of the interacting spin–hole system. Such a real-space static stripe modulation gives the incommensurate elastic magnetic peaks, which are positioned in the reciprocal space at $(2\pi/a)$. The main Bragg diffraction peak for the parent AFM order is (0.5, 0.5). The incommensuration δ is a function of doping x (at low doping, $x \approx \delta$) and is related to the real interstripe spacing $(= a/2\delta)$. The tetragonal zone indexing (h,k) is appropriate to the planar geometry of the CuO_2 sheets.

In simplistic terms, the doped holes in the cuprates are mostly on the oxygen in the CuO_2 sheets creating oxygen 2p-holes, while the spins reside on the copper (the Cu^{2+}:$3d^9$-spins). In Figure 2 we have only shown the copper atoms. The charge stripes are aligned parallel to the Cu—O bonds, but may be centered on the sites (on the Cu—O—Cu legs), on the bonds (between two Cu—O—Cu legs), or in between. The stripe phase is singly modulated, but the stripes in the adjacent CuO_2 planes are at 90° because of the symmetry-lowering lattice distortions. Hence, no incommensurate peaks occur at $(2\pi/a)$ $(1/2 + \delta, 1/2 + \delta)$, or $(2\pi/a)$ $(1/2 - \delta, 1/2 - \delta)$. Figure 2 assumes the stripes to be static; this is the case for low levels of doping (for the underdoped samples), where one observes narrow diffraction (elastic) peaks. In the optimally doped superconducting samples, on the other hand, the stripes fluctuate, giving relatively broad inelastic peaks, still positioned incommensurately about the AFM diffraction peaks. In the overdoped samples the stripes simply fade away. We may recall that the strongest evidence for the stripes comes from NdLSCO ($La_{2-x-y}Nd_ySr_xCuO_4$, with $x = 0.12$, $y = 0.4$),[2] in which the stripes are immobilized by commensuration pinning to the lattice distortion of the low-temperature tetragonal (LTT) phase. It is the modulation of the lattice distortion induced by the stripe charge order that is picked up in neutron scattering as an indirect signature of the charged stripes.

Let us briefly examine the nature of the stripes in cuprates. Several of the reciprocal-space bulk scattering probes and real-space surface-imaging tools have been used to study the stripes. The two methods are somewhat complementary. The scattering probes can detect fluctuating stripes as well, while surface imaging can only probe the static stripes at atomic resolution. The geometry of the stripe order in the bulk is best seen by neutron diffraction. If the stripes can be immobilized through pinning due to impurity atoms and structural commensuration as in NdLSCO, the applicability of elastic scattering allows a more thorough investigation. In order to probe spatially modulated dynamical correlations, low-energy ($\approx$ few meV) inelastic neutron scattering is employed. The problem, however, is that neutrons couple directly to the electron spin and to the positions

MINIREVIEWS

of the nuclei, whose forced displacement in response to any charge-ordering gives only an indirect evidence for charge order. Thus, the incommensurate, split superlattice peaks at $(2\pi/a)$ $(2 \pm 2\delta)$ and their near absence at $(2\pi/a)$ $(2 \pm 2\delta)$ constitute indirect evidence for the charged stripes (with a spacing half of that for the magnetic stripes). For direct evidence, one must turn to X-rays that directly couple to the charge. X-rays, however, couple nonspecifically to all the charges, including those of the core electrons that outnumber the mobile charge carriers (doped holes) by a factor of ≈ 500 to 1, for example, in optimally doped LSCO. The normal X-ray diffraction cannot resolve this small proportion. This problem has been overcome recently by Abbamonte et al.[26] by the use of multiwavelength anomalous soft X-ray scattering, wherein the elastic (or the low-energy inelastic) scattering is resonantly and selectively enhanced many-fold (by over a factor of $\approx 10^3$) by operating close to an absorption pre-edge (pre-edge of the O K shell) because of the near-vanishing of the energy denominator that is associated with the virtual excitation of the O (1s) electron to the Fermi level vacated due to hole doping. Even this enhanced sensitivity and specificity have failed to detect incommensurate X-ray diffraction peaks corresponding to a charged-stripe modulation in La$_2$CuO$_{4+x}$ films of atomic perfection deposited on SrTiO$_3$ substrates. But, this new technique clearly needs to be tested out on samples proven to give the incommensurate magnetic peaks in neutron scattering. Also, Abbamonte et al.[26] studied the sample at a high temperature ($> T_c$), and the stripes may well appear below T_c. The issue is, therefore, not fully resolved.

Complementary to the soft X-ray scattering probe, which is a bulk probe of mesoscopic resolution, is the real-space surface imaging by scanning tunneling microscopy (STM) at atomic resolution. STM measures the differential tunneling conductance, which is in turn proportional to the local density of states (LDOS), $\alpha \sum_k | \varphi_k(r) |^2 \delta[E - e(k)]$, at the position (r) of the STM tip and at the energy E corresponding to the bias voltage for the charge-carrier injection. An earlier study of a single crystal of BSCCO 2212, cleaved along the BiO planes, reported charge-stripe modulations for optimal doping in the superconducting phase.[27] This result is subject to some doubt since STM studies find that the spatial modulation wavelength varies with the STM bias voltage, which suggests an extrinsic interpretation involving interference of the electron waves, which are scattered from defects/impurities—akin to Friedel oscillations or more accurately, von Laue-theoretic oscillations. Stripes should have given intrinsic voltage-independent spatial oscillations. It is, however, possible that the latter are masked by the extrinsic effect. It is important to note that, over and above the modulations due to any possible stripe order, there are microscopic inhomo-

geneities arising from local doping concentration (LDC) effects, because of the poor screening of the dopant potentials in these doped Mott insulators. The LDC effects have been isolated[29] by Fourier-filtering out of the stripe modulations, and remarkable correlations of, for example, the LDOS, the superconducting gap, and the doping, have been found; this is indeed intriguing.

The topological issue of one-dimensionality of the charge stripes, namely, that the mobile charge carriers must move along the linear tracks, defined laterally by the Mott-insulating AFM domains on the sides, is best probed through studies of charge motion in real space (for example, Hall effect measurements) and in the reciprocal (momentum) space (for example, through angle resolved photo electron spectroscopy; ARPES).

The 1D topology should block any off-diagonal Hall response making the Hall coefficient vanish, even for an arbitrary collection of nonpercolating stripes containing mobile holes. This is found to be the case in NdLSCO (La$_{1.4-x}$Nd$_{0.6}$Sr$_x$CuO$_4$).[6] At low doping ($x < 1/8$, the magic number at which the stripes are known to be structurally pinned in the LTT phase) the Hall coefficient decreases rapidly with decreasing temperature. For $x > 1/8$, however, where the stripes are known to be fluctuating, the Hall coefficient is large and featureless despite the charge-stripe order. The magic number $x = 1/8$ seems to mark a cross over from 1D to 2D-hole motion. The 1D electronic nature of the stripes with mobile carriers is strongly supported by the ARPES studies.[4]

ARPES measures the single-particle spectral function $A(\mathbf{k},w)$, within the photoionization weight factor (w), and gives, on frequency integration the in-plane momentum $(\mathbf{k})$ distribution, function

$$n(\mathbf{k}) = \int_{-\infty}^{+\infty} A(\mathbf{k},w)f(w)dw/2\pi,$$

from which one constructs the Fermi surface that contains the high spectral weight region. Figure 3a shows such a Fermi surface, which is redrawn schematically from ARPES measurements on NdLSCO.[4] Figure 3b shows the Fermi surface expected of the orthogonal 1D charge stripes of the type shown in Figure 2, where the vertical dotted lines are along [100] and the horizontal lines (solid lines) are along [010]. (Recall that the

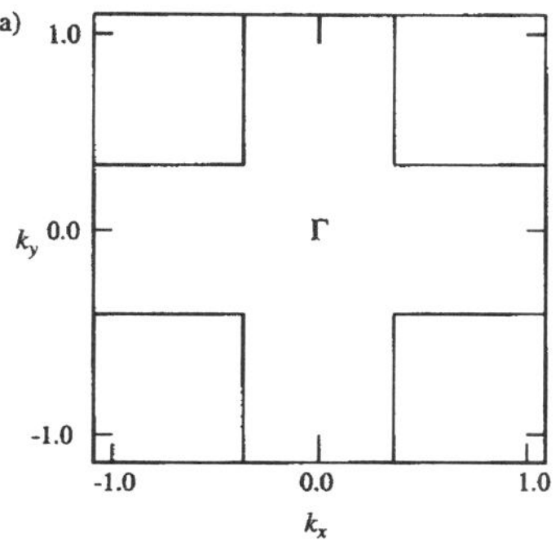
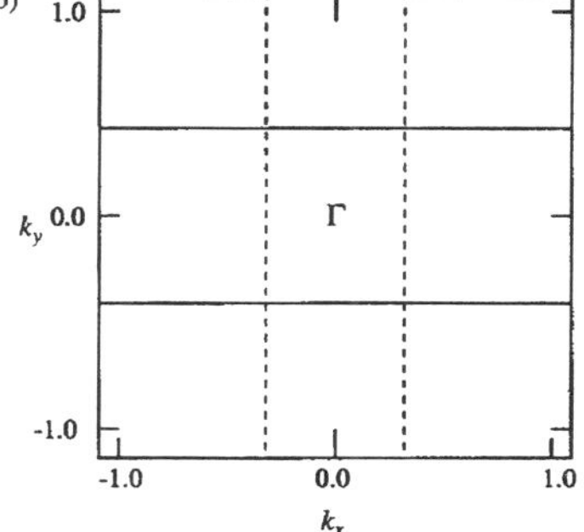

Figure 3. *a) A schematic of the ARPES results for the Fermi surface of NdLSCO for* x = 1/8, *taken and redrawn from ref.[4]. b) Fermi surfaces expected for 1/4-filled 1D bands, for comparison. Dashed lines (----) are for the horizontal stripes, and solid lines (——) for the vertical stripes.*

stripes in the adjacent singly modulated CuO_2 planes are orthogonal). The striking similarity suggests a $1/4$-filled 1D band for a $1/2$-filled charge stripe corresponding to one hole per two copper atoms in the linear stripe and satisfying the Luttinger theorem. The Fermi surfaces, or rather lines, are $(\pm \pi/4a, k_y)$ and $(k_x, \pm \pi/4a)$. A 2D homogeneous electronic structure clearly appears to be discredited, which thus suggests the striped nature of high-T_c cuprates.

Whether there is any relation between stripes and cuprate superconductivity is the important question facing us today. Systematics of the dependence of the magnetic incommensuration δ and the superconductivity transition temperature T_c on the hole concentration (the value of x) in LSCO and NdLSCO show striking correlations.[2, 3] First, superconductivity is anomalously suppressed (T_c goes to almost zero) for $x = 1/8$, the magic number at which the stripes are structurally pinned, showing up in low-energy neutron scattering as incommensurate peaks of small momentum width. Second, away from this magic number, the dependence of δ and of T_c on x from underdoping to optimal doping closely track each other. Thus, the critical doping for the onset of superconductivity and for the appearance of the magnetic stripes are the same; signalled by the shifting of the neutron diffraction peaks away from the AFM position to the incommensurate split-peak positions. The incommensurate, low-energy inelastic magnetic peaks remain narrow in momentum in the superconducting phase, except in the weakly superconducting overdoped region, where they become broad and incoherent. Thus, fluctuating stripes are correlated with, and certainly can coexist with, superconductivity. Static stripes, on the other hand, appear to suppress superconductivity. ^{63}Cu NQR (Nuclear Quadrupole Resonance) measurements that detect the charge dynamics in the proximate charge stripes through the electric field gradient they produce, however, indicate that static stripes and superconductivity can coexist.[30]

An intriguing aspect of the doping dependence of incommensuration δ is that the linear variation of δ with x saturates at $x = 1/8$, while T_c continues to increase up to optimal doping. While the linearity of δ with x implies that the added holes are accommodated in the linear stripes, occurrence of saturation suggests an alternative mode of accommodation of the holes. This is reminiscent of the low-dimensional polyacetylene-like system,[22] where solitonic doping gives way to band-type doping beyond a critical doping. This would suggest that the stripes may well be a solitonic route to hole doping at low levels.

We now turn to microscopic considerations of the stripes and their implication for a theory of HTSC. Incommensurate modulations, structural or electronic (charge and spin), can result from competing length scales, for example, the length scale set by the nested Fermi surface and that of the lattice periodicity as in the Peierls or spin-Peierls instability that leading to spatial modulations (and gapping of the Fermi surface), well known in low-dimensional systems as the CDW (charge density wave) and the SDW (spin density wave) orders.[21, 23] In the case of cuprates, however, it is known that charge ordering occurs at a higher temperature than spin ordering, and so the stripe phase is charge driven[2]—as in $La_{2-x}Sr_xCuO_4$ as well as in the nickelates. This is not consistent with the usual Fermi surface instabilities,

where the spins would order first. Incommensuration can also arise from competing interactions such as short-range attraction and the long-range repulsion, or short-range FM and the long-range AFM interactions in a spin-1 lattice gas that frustrate any macroscopic phase separation. Charge stripes, and the magnetic stripes driven by the charge order, ultimately result from the fact that the cuprates are essentially hole-doped Mott (or rather charge-transfer) insulators and the doped holes are energetically expelled from the AFM ordered domains, which they would otherwise disrupt through their motion. This basic physics is well captured by a short-range two-band 2D Hubbard model with strong on-site repulsion (the Hubbard correlation energy) on the copper atoms, and the $3d_{x^2-y^2}$-$2p_{x,y}$ hybridization between copper and oxygen, and the energy of charge transfer, Cu^{2+}:$3d^9 \rightarrow Cu^{1+}$:$3d^{10}$ + oxygen 2p-hole, in these strongly correlated insulators.[13] A mean-field treatment indeed gives charge-stripes, but they turn out to be insulating, as in the nickelates (unlike the conducting cuprate stripes). Antiphase spin-charge stripes of the form shown in Figure 2 result from a short-range one-band $t-J$ model.[19] Thus, a long-range repulsion is not a necessary condition for the occurrence of the stripes.

Next, we shall look at the possible connection between stripes and the pairing mechanism. Almost all the nonphonon pairing mechanisms proposed so far invoke interaction between spin fluctuations (magnetism) and hole motion (charge) in some form.[31] For a pairing mechanism involving spin–hole interaction, the electronic specific heat near the superconducting transition must track the temperature derivative of the average exchange energy of the spins. This follows from the thermodynamics of the superconducting phase transition, derived first in the context of the conventional electron–phonon pairing mechanism,[32] and generalized later to the spin-fluctuation mechanism.[33] This mechanism is really much the same as the temperature derivative of the spin-disorder resistivity that shows the critical behavior of the electronic specific heat in bad itinerant metals.[34] Recent inelastic neutron scattering measurements[35] of the spin-fluctuation spectrum of YBCO123, which are enhanced by the so-called π-resonance at the wavevector $(\pi/a, \pi/b, \pi/c)$, verify the above relation between the temperature dependence of the AFM exchange energy and the electronic specific heat.[36] While this does implicate magnetic fluctuations as a hole-pairing mechanism, it is too nonspecific to support any stripe-based theory.

A minimal model for the layered cuprates is the 2D short-range single-band Hubbard model that effectively describes the strongly correlated system.[37, 38] The undoped parent material, for example, La_2CuO_4, is a Mott insulator, as can be seen from the electron count, for example, $(La^{3+})_2(Cu^{2+})(O^{2-})_4$, that would give a half-filled band and is therefore a metal. It becomes conducting when doped away from stoichiometry as in, $La_{2-x}Sr_xCuO_4$. When the doping is not too small, the holes get unbound from the charged dopants as their potentials begin to overlap. The hole-doped Mott insulator is then well-described in the space of nondouble occupancy by the so-called $t-J$ Hamiltonian [Equation (1) and (2)][38]

$$H = -t \sum_{\langle ij \rangle, \sigma} (C_{i\sigma}^+ C_{1\sigma} + Hc) + J \sum_{\langle ij \rangle} \left(S_i S_j - \frac{n_i n_j}{4} \right) \tag{1}$$

MINIREVIEWS

$$S_i = \tfrac{1}{2}\sum_{\alpha,\beta} C_{i\beta}\sigma_{\beta\alpha}C_{i\alpha} \tag{2}$$

that describes the motion of the holes in the active background of spins. This spin-hole Hamiltonian suggests a kinetic hole-pairing mechanism driven by the lowering of the kinetic energy due to delocalization, as described below.

A long-range or even a mesoscopic AFM order (Néel ordered) encumbers the hole motion. (As a hole moves to the nearest-neighbouring site, the spin at that site concomitantly moves to the site previously occupied by the hole). A succession of such moves shift-registers a line of spins, which become laterally misaligned with respect to the spins on the sides (Figure 4). The broken AFM bonds shown by the dashed lines create a linearly rising potential that impedes free hole motion. A compact

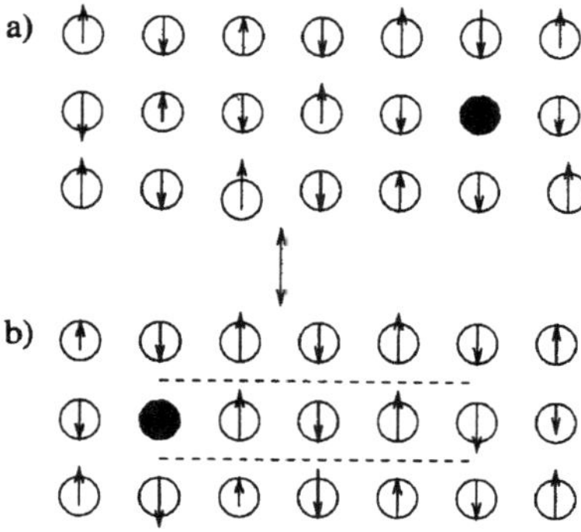

Figure 4. *a) A hole at the initial site in a Néel-ordered antiferromagnet; b) a hole at the final site after four hops. ---- show disrupted bonds between misaligned spins giving a linearly rising potential.*

pair of holes can, however, toboggan together, and thus delocalize quite freely. The trailing hole undoes what the leading hole does, thus healing the lateral misalignment of the spins. The lowering of the pair kinetic energy by the ease of pair delocalization then provides the kinetic energy mechanism for pairing.[39] (The quantum fluctuation of spins away from the Néel order must modify the above naive visualization, but one can still talk meaningfully about the hole-pair moving unencumbered by entanglement with the background spins). Clearly, for low doping the hole-hole repulsion will dominate over the kinetic mechanism that favors compact pairing. It is only for higher doping, when the mean hole separation matches the pair size that the kinetic pairing wins. For such a kinetic mechanism for pairing, stripe modulation in the zeroth approximation merely provides a modified spin background for the hole motion—not necessarily conducive to pairing. This basic idea of kinetic pairing seems to suggest a novel, highly elaborate quantum liquid-crystalline model for HTSC, in which the fluctuating stripes play a crucial role.[14–18]

The charge stripe is normally viewed as a 1D lane, along which the holes can freely move longitudinally, but are constrained in the transverse direction by the antiphase AFM-ordered magnetic domains. Indeed, the pinning at the magic number doping ($x =$

$^1/_8$) gives such a stripe which is static and gives a resolution-limited elastic neutron scattering. It is well known that an array of such static charge stripes will be insulating through a CDW instability. In the electronic quantum liquid-crystalline phase model of the doped Mott insulator the stripes show transverse quantum fluctuations that affect the motion of the charged particles which is guided by the fluctuating stripes. Thus, the charge stripes are viewed as a quantum mechanical deformable object with a topological integrity, and as having many degrees of freedom. The role of zero-point fluctuations was recognized quite early.[40] The mechanism for the mobile hole pairing here involves a magnetic analog of the superconducting proximity effect between the metallic stripe and the Mott insulators that bound it transversely. The mobile hole pairs virtually hop between the Mott-insulator domains and acquire a spin gap. This is in an underdoped system. At higher doping the hole pairs can tunnel between the stripes giving a global 3D phase stiffness to the local pairs. Several known features of HTSC such as the normal-state pseudogap in the underdoped samples, which have the same d-wave symmetry as the d-wave superconducting gap at lower temperatures, come out of this model. The electronic quantum liquid-crystalline model has many topological features of its underlying one-dimensionality that affect its charge and spin degrees of freedom. It certainly warrants a detailed examination—experimentally and theoretically, as something interesting in its own right.

Any discussion of the role of stripes in the high-T_c layered cuprates is necessarily inconclusive at present. It is possible, however, that stripes may well be a solitonic route to hole doping. In any case, as of now the stripes cannot confidently signify superconductivity.

[1] For a critical discussion of stripes and electronic phase separation see a) J. Zaanen, *J. Phys. Chem. Solids* **1998**, *59*, 1769; b) J. M. Tranquada, *Physica B* **1998**, *241–243*, 745; c) J. M. Tranquada, *J. Phys. Chem. Solids* **1998**, *59*, 2150; d) *Proceedings of the workshop on Phase Separation in Cuprate Superconductors* (Eds.: K. A. Müller, G. Benedek), World Scientific, Singapore, **1993**; e) *Phase Separation in Cuprate Superconductors* (Eds.: E. Sigmund, K. A. Müller), Springer-Verlag, Heidelberg, **1994**; f) E. Dagotto, T. Hotta, A. Moreo, *Phys. Rep.* **2001**, *344*, 1; g) C. N. R. Rao, P. V. Vanitha, A. K. Cheetham, *Chem. Eur. J.* **2003**, *9*, 828..

[2] a) J. M. Tranquada, S. J. Sternlieb, J. D. Axe, Y. Nakamura, S. Uchida, *Nature* **1995**, *375*, 561; b) J. M. Tranquada, J. D. Axe, N. Ichikawa, Y. Nakamura, S. Uchida, B. Nachumi, *Phys. Rev. B* **1996**, *54*, 7489; c) J. M. Tranquada, J. D. Axe, N. Ichikawa, A. R. Moodenbaugh, Y. Nakamura, S. Uchida, *Phys. Rev. Lett.* **1997**, *78*, 338; d) *Stripes and Related Phenomena* (Eds.: A. Bianconi, N. L. Saini), Kluwer Academic Publishers, Dordrecht, **2000**.

[3] K. Yamada, C. H. Lee, K. Kurahashi, J. Wada, S. Wakimoto, S. Ueki, H. Kimura, Y. Endoh, S. Hosoya, G. Shirane, R. J. Birgeneau, M. Greven, M. A. Kastner, Y. J. Kim, *Phys. Rev. B* **1998**, *57*, 6165.

[4] X. J. Zhou, P. Bogdanov, S. A. Kellar, T. Noda, H. Eisaki, S. Uchida, Z. Hussain, Z.-X. Shen, *Science* **1999**, *286*, 268.

[5] X. J. Zhou, T. Yoshida, S. A. Kellar, P. V. Bogdanov, E. D. Lu, A. Larzava, M. Nakamura, T. Noda, T. Takeshita, H. Eisaki, S. Uchida, A. Fujimori, Z. Hussain, Z.-X. Shen, *Phys. Rev. Lett.* **2001**, *86*, 5578.

[6] T. Noda, H. Eisaki, S.-I. Uchida, *Science* **1999**, *286*, 265.

[7] H. A. Mook, P. Dai, F. Dogan, *Phys. Rev. Lett.* **2002**, *88*, 097004.

[8] B. Lake, G. Aeppli, K. N. Clausen, D. F. McMorrow, K. Letmann, N. E. Hussey, N. Manykorntong, H. Nohava, H. Takagi, T. E. Mason, A. Schröder, *Science* **2001**, *291*, 1759.

CHEMPHYSCHEM *C. N. R. Rao and N. Kumar*

[9] B. Lake, H. M. Ronnow, N. B. Christensen, G. Aeppli, K. Lefmann, D. F. McMorrow, P. Verderwisch, P. Smeibidl, N. Mangkorntong, T. Susagawa, M. Nohara, H. Takagi, T. E. Mason, *Nature* **2002**, *415*, 299.

[10] F. Mitrovic, E. E. Sigmund, M. Eschrig, H. N. Bachman, W. P. Halperin, A. P. Reyas, P. Kuhns, W. G. Moulton, *Nature* **2001**, *413*, 501.

[11] Y. S. Lee, R. J. Birgeneau, M. A. Kastner, Y. Endoh, S. Wakimoto, K. Yamada, R. W. Erwin, S. H. Lee, G. Shirane, *Phys. Rev. B* **1999**, *60*, 3643.

[12] C. Howald, P. Fournier, A. Kapitulnik, *Phys. Rev. B* **2001**, *64*, R100504.

[13] J. Zaanen, O. Gunnarsson, *Phys. Rev. B* **1989**, *40*, 7391.

[14] V. J. Emery, A. Kivelson, *Strongly Correlated Electronic Materials: The Los Alamos Symposium 1993* (Eds.: K. S. Bedell, Z. Wang, D. E. Meltzer, A. V. Balatsky, E. Abrahams), Addision-Wesley, Redwood City, **1994**, p. 619.

[15] A. Kivelson, E. Fradkin, V. J. Emery, *Nature* **1998**, *393*, 550.

[16] V. J. Emery, S. A. Kivelson, O. Zachar, *Phys. Rev. B* **1997**, *56*, 6120.

[17] a) V. J. Emery, A. Kivelson, *Phys. C* **1993**, *209*, 597; b) V. Löw, V. J. Emery, K. Fabricius, S. A. Kivelson, *Phys. Rev. Lett.* **1994**, *72*, 1918.

[18] a) M. Salkola, V. J. Emery, S. A. Kivelson, *Phys. Rev. Lett.* **1996**, *77*, 155; b) V. J. Emery, S. A. Kivelson, J. M. Tranquada, *Proc. Natl. Acad. Sci. USA* **1999**, *96*, 8814.

[19] a) S. R. White, D. J. Scalapino, *Phys. Rev. Lett.* **1998**, *81*, 3227; b) S. R. White, D. J. Scalapino, *Phys. Rev. Lett.* **1998**, *80*, 1272.

[20] J. G. Bednorz, K. A. Müller, *Z. Phys. B* **1986**, *64*, 189.

[21] W. P. Su, J. R. Schrieffer, A. J. Heeger, *Phys. Rev. B* **1980**, *22*, 2099.

[22] E. M. Conwell, *Phys. Today*, **1985**, 48 (June).

[23] V. L. Pokhrovsky, A. L. Talapov, *Theory of Incommensurate Crystals, Soviet Scientific Reviews Suppl. Sec. A, Vol. 1*, (Ed.: I. M. Khalatnikov), Harwood Academics, Switzerland, **1984**.

[24] a) J. M. Tranquada, J. E. Lorenzo, D. J. Buttrey, V. Sachan, *Phys. Rev. B* **1995**, *52*, 3581; b) S. W. Cheong, H. Y. Hwang, C. H. Chen, B. Batlogg, L. W. Rupp, Jr., S. A. Carter, *Phys. Rev. B* **1994**, *49*, 7088; c) H. Kuwahara, Y. Tomioka, Y. Moritomo, A. Asamitsu, M. Kasai, R. Kumai, Y. Tokura, *Science* **1996**, *272*, 80; d) S. H. Lee, J. M. Tranquada, K. Yamada, D. J. Buttrey, Q. Li, S. W. Cheong, 24935 *Phys. Rev. Lett.* **2002**, *88*, 126401; e) M. Abu-Sheikah, O. Bakharev, H. B. Bram, J. Zaanen, *Phys. Rev. Lett.* **2001**, *87*, 237201.

[25] a) *Colossal magnetoresistance, Charge-ordering and Related Properties of Manganese Oxides* (Eds.: C. N. R. Rao, B. Raveau), World Scientific, Singapore, 1999; b) C. N. R. Rao, A. Arulraj, A. K. Cheetham, B. Raveau, *J. Phys. Condens. Matter* **2000**, *12*, R83; c) C. N. R. Rao, P. V. Vanitha, *Curr. Opin. Solid State Mater. Sci.* **2002**, *6*, 97.

[26] P. Abbamonte, L. Venema, A. Rusydi, G. A. Sawatzky, G. Logvenov, I. Bozovic, *Science* **2002**, *297*, 581.

[27] C. Howald, H. Eisaki, N. Kaneko, A. Kapitulnik, *cond-mat/0201546*.

[28] J. E. Hoffman, K. McElroy, D.-H. Lee, K. M. Lang, H. Eisaki, S. Uchida, J. C. Davis, *Science* **2002**, *297*, 1148.

[29] S. H. Pan, J. P. O'Neal, R. D. Badzey, C. Chamon, N. Ding, J. R. Engelbracht, Z. Wang, N. Eisaki, S. Uchida, A. K. Gupta, K.-W. Ng, E. W. Hudson, K. M. Lang, J. C. Davis, *Nature* **2001**, *413*, 282.

[30] P. M. Singer, A. W. Hunt, A. F. Cederstrom, T. Imai, *cond-mat/9906140*.

[31] J. R. Schrieffer, *J. Low Temp. Phys.* **1995**, *99*, 97.

[32] G. V. Chester, *Phys. Rev.* **1956**, *103*, 1693.

[33] D. J. Scalapino, S. R. White, *Phys. Rev. B* **1998**, *58*, 8222.

[34] L. Klein, J. S. Dodge, C. H. Ahn, G. J. Snyder, T. H. Geballe, M. R. Beasley, A. Kapitulnik, *Phys. Rev. Lett.* **1996**, *77*, 2774.

[35] a) P. Dai, H. A. Mook, S. M. Hayden, G. Aeppli, T. G. Perring, R. D. Hunt, F. Dogan, *Science* **1999**, *284*, 1344; b) E. Demler, S. C. Zhang, *Nature* **1998**, *396*, 733.

[36] J. W. Loram, K. A. Mirza, P. F. Freeman, *Phys. C* **1990**, *171*, 243.

[37] F. C. Zhang, T. M. Rice, *Phys. Rev. B* **1988**, *37*, 3759.

[38] E. Dagotto, *Rev. Mod. Phys.* **1994**, *66*, 763.

[39] a) J. E. Hirsch, *Phys. Rev. Lett.* **1987**, *59*, 228; b) M. M. Mohan, N. Kumar, *J. Phys. C* **1987**, *20*, L527; c) N. Kumar, *Phys. Rev. B* **1990**, *42*, 6138; d) Yu. A. Dimashko, *JETP Lett.* **1993**, *76*, 267.

[40] a) P. Prelovšek, X. Zotos, *Phys. Rev. B* **1993**, *47*, 5984; b) P. Prelovšek, I. Sega, *Phys. Rev. B* **1994**, *49*, 15241.

Received: November 28, 2002 [M601]

III. Colossal Magnetoresistance, Charge Ordering and Related Aspects of Rare Earth Manganates

C.N.R. Rao

CSIR Centre of Excellence in Chemistry,
Chemistry & Physics of Materials Unit
Jawaharlal Nehru Centre for Advanced Scientific Research
Jakkur P.O., Bangalore-560 064, INDIA
cnrrao@jncasr.ac.in

The impressive range of properties exhibited by transition metal oxides was pointed out in the previous section. This is specially true of oxides of perovskite structure. An important new addition to the fascinating phenomena exhibited by transition metal oxides is colossal magnetoresistance (CMR), discovered in 1993 in rare earth manganates of the formula $Ln_{1-x}A_xMnO_3$ (Ln = rare earth, A = Alkaline earth or other divalent ions). These manganates containing Mn in the +3 and +4 states become ferromagnatic and metallic due to the double exchange mechanism of electron hopping. At the ferromagnetic T_c, they show a large decrease in resistance on application of magnetic fields. The T_c and other properties of the manganates are sensitive to the average radius of the A-site cations, $\langle r_A \rangle$, as well as size disorder (mismatch). Besides CMR, the manganates undergo charge-ordering of the Mn^{3+} and Mn^{4+} ions depending on the $\langle r_A \rangle$, small values favoring charge-ordering.[1] Charge-ordering competes with double exchange and gives rise to anti-ferromagnatic insulating states. Besides the ordering of charges and spins, the ordering of the e_g orbitals of Mn^{3+} ions in the rare earth manganates plays a significant role. In this section, a small selection of articles on CMR, charge ordering and related aspects of rare earth manganates is presented. Other interesting aspects such as electron hole asymmetry and electronic phase separation are also examined in the papers. Electronic phase separation in solids is a phenomenon of importance being understood more clearly in the last few years.[2] It is likely that many transition metal oxides with correlated electrons will undergo phase separation.

References

1. C.N.R. Rao and B. Raveau (Eds.), *Colossal Magnetoresistance, Charge Ordering and Related Properties of Manganese Oxides*, World Scientific, Singapore, 1998.
2. C.N.R. Rao and P.V. Vanitha, *Current Opinion Mater. Solid State Sci.* **6**, 97 (2002).

COLOSSAL MAGNETORESISTANCE, CHARGE ORDERING AND OTHER NOVEL PROPERTIES OF MANGANATES AND RELATED MATERIALS

C.N.R. RAO* and A.K. RAYCHAUDHURI†

*Solid State & Structural Chemistry Unit and †Department of Physics
Indian Institute of Science, Bangalore 560012, India
and

Jawaharlal Nehru Centre for Advanced Scientific Research
Jakkur, Bangalore 560 064, India

Colossal magnetoresistance (CMR) and related properties of perovskite manganates of the general formula $Ln_{1-x}A_xMnO_3$ (Ln = rare-earth; A = divalent ion) are discussed in detail. The manganates are ferromagnetic at or above a certain value of x (or Mn^{4+} content) and become metallic at temperatures below the curie temperature, T_c. This behavior is commonly attributed to double-exchange. CMR is generally a maximum close to T_c or the insulator-metal (I-M) transition temperature, T_{im}. The T_c and %MR are markedly affected by the size of the A site cation, $<r_A>$, thereby affording a useful electronic phase diagram when T_c or T_{im} is plotted against $<r_A>$ or pressure. The commonalities and correlations found in the properties of manganates are examined along with certain unusual features in the electron-transport properties of these materials. Recent studies clearly indicate that double-exchange alone cannot explain the many fascinating features of the manganates. Some of the $Ln_{1-x}A_xMnO_3$ compositions exhibit charge-ordering and related effects. Charge ordering is crucially dependent on $<r_A>$ or the e_g bandwidth and the charge-ordered insulating state transforms to a metallic ferromagnetic state on the application of a magnetic field, charge-ordering and double-exchange being competing interactions. The importance of Jahn-Teller interaction in determining the various properties of manganates is highlighted. CMR found in a few inorganic solids other than the perovskite manganates is pointed out.

1. Introduction

Magnetoresistance is the relative change in the electrical resistance or resistivity of a material produced on the application of a magnetic field. It is generally defined by,

$$MR = [\ \Delta\rho/\rho\ (O)] = [\rho\ (H) - \rho\ (O)]/\rho\ (O) \tag{1}$$

2

where $\rho(H)$ and $\rho(O)$ are the resistances or resistivities at a given temperature in the presence and absence of a magnetic field, H, respectively. Magnetoresistance (MR) can be negative or positive. Most metals show a small MR (only a few percent). In non-magnetic pure metals and alloys MR is generally positive and MR shows a quadratic dependence on H. MR can be negative in magnetic materials because of the suppression of spin disorder by the magnetic field. Suppression of quantum mechanical interference can give rise to negative MR in highly resistive alloys. Large magnetoresistance, referred to as giant magnetoresistance (GMR), was first observed on the application of magnetic fields to atomically engineered magnetic superlattices (e.g. Fe/Cr)[1] and in magnetic semiconductors[2]. Besides some of the bimetallic or multimetallic layers, comprising ferromagnetic and antiferromagnetic or non-magnetic metals, GMR is also found in ferromagnetic granules dispersed in paramagnetic metal films (e.g. Co/Cu)[1]. The phenomenon of GMR is of considerable interest because of its potential technological applications in magnetic recording, actuators and sensors. GMR in layered and granular magnetic materials arises from the ability of magnetic fields to change and control the scattering of conduction electrons through the modification of the electron-orbit and spin-orbit interactions. GMR is a result of the reduction of the extra resistance due to the scattering of electrons by the non-aligned ferromagnetic components in zero magnetic field.

The discovery of negative GMR in rare-earth manganates,[3,4] $Ln_{1-x}A_xMnO_3$ (Ln = rare-earth, A = a divalent cation such as an alkaline earth) with the perovskite structure, has attracted wide attention. The magnitude of GMR in these materials can be very large, close to 100% as per eq. (1). For this reason, many workers prefer to call this colossal magnetoresistance (CMR), as distinct from GMR in layered or granular metallic materials. It is customary to give %MR by multiplying MR from eq. (1) by 100. Many workers give MR by using $\rho(H)$ in the denominator of eq. (1) instead of $\rho(O)$, giving rise to very large numbers. In metallic multilayers or granular alloys, the mechanism involves spin-polarized transport. In the manganates also, spin-polarized transport is responsible for the large negative MR, but it is distinctly different from what happens in metallic multilayers.

A variety of manganese oxides, in the form of polycrystalline powders, thin films and single crystals, have been investigated for CMR. These studies have provided valuable insight into the CMR phenomenon in the manganates and have also revealed several other fascinating features of these oxides. In this article, we present some of the highlights of the results pertaining to CMR and related properties of the manganates obtained hitherto and point out the significant general features. We examine certain novel aspects of the manganates, in particular, electron-lattice or Jahn-Teller interactions, charge-ordering and the associated properties arising from them. We also refer to other systems exhibiting CMR and indicate some future directions.

2. Introduction to Rare Earth Manganates

2.1. *Insulating parent manganates*

Without hole doping, parent compounds such as $LaMnO_3$, $PrMnO_3$ and on of $NdMnO_3$, are insulators at all temperatures. For example, in $LaMnO_3$ the optical gap is ~ 1.1eV. These oxides undergo an antiferromagnetic (AFM) transition with a Neél temperature, TN≈150K. The AFM ordering is of A-type where ferromagnetically aligned layers are coupled antiferromagnetically. The insulating nature of the parent compounds as well as the anisotropic magnetic interaction are related to their structure, in particular the Jahn-Teller (JT) distortion around Mn^{3+} ions. When these insulators are hole-doped, (i.e., Ln is partly substituted by a divalent ion), the Mn^{4+} ions decrease the cooperative JT distortion. The structure plays a crucial role in determining the electron transport and magnetic properties of these oxides.

There are two characteristic distortions which influence the perovskite structure of manganates. One consists of the cooperative tilting of MnO_6 octahedra which is essentially established below ~ 1000K. This distortion is a consequence of the mismatch of the ionic radii. For perovskites containing small central cations, the Goldschmidt tolerance factor, t, is typically < 1. The tilting of the octahedra is common in perovskites with t < 1. The other distortion, arising from the Jahn-Teller effect due to Mn^{3+}, distorts the MnO_6 octahedra in such a way that there are long and short Mn-O bonds. This occurs below a characteristic temperature (T_{JT}). The most effective distortion is the basal-plane distortion (called Q_2 mode) with one diagonally opposite O-pair displaced outwards and the other pair displaced inward. It has recently been shown that a JT distortion involving a displacement of oxygen ions $\gtrsim 0.1$ Å can split the e_g band of the manganate (which forms the conduction band) and opens a gap at the Fermi level.

4

In Fig. 1. We show a schematic of the band diagram of $LaMnO_3$ to elucidate how JT distortion splits the conduction band and makes the material insulating. We will see below that hole doping reduces JT distortion and makes the material metallic for hole filling ≥ 0.18. In Table 1 we show the lattice constants, distortion and the characteristic T_{JT} values. The perovskite structure has an orthorhombic distortion $(b > a > c/\sqrt{2})$ and the unit cell consisting of 4-formula units can be mapped into *Pbnm* (or *Pnma)* symmetry.

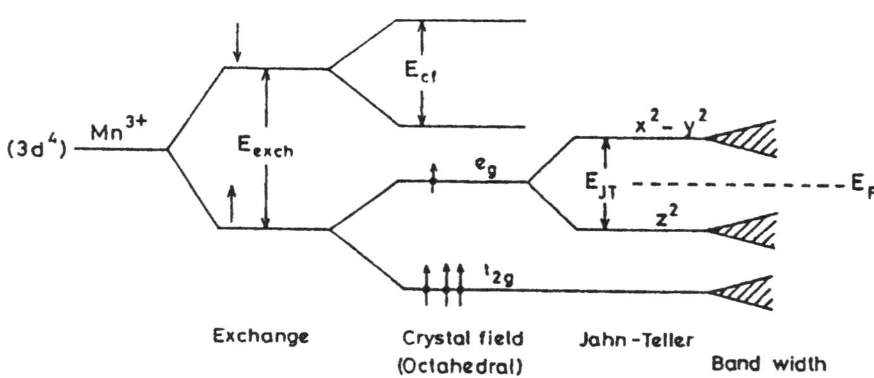

Fig. 1. Energy level scheme for $LaMnO_3$. The values of E_{ex}, E_{cf} and E_{JT} from density functional calculations are around 3eV, 2V and 1.5eV respectively. Typical bandwidth is 1.5 -2.0eV. Fermi level lies between JT split orbitals. (S. Satpathy et al, T. Appl. Phys. 79, 4555 (1996); I. Solovyev et al, Phys. Rev. Lett. 76, 4825 (1996).

Electronic structure calculations show that in the absence of JT distortion, if the crystal would have been cubic, the nearest neighbor exchange $J_{ab}^n = J_{ab}^n \cong 30$meV. With lattice distortion in $LaMnO_3$, the exchange is reduced and becomes anisotropic with $J_{ab}^n \approx 9$meV and $J_c^n \approx 3$meV. This is comparable to the next nearest neighbor exchange (J_c^{nn}) which is an AFM type giving $J_{total} = 2J_c^n + 8J_c^{nn} \cong 19$meV, which is close to $K_B T_N$ of the material. The ferromagnetic J_{ab}^n exchange and AFM J_{total} give rise to the

A-type AFM ordering in these materials. With doping, the structure changes along with the magnetic ordering.

Table 1: Properties of parent manganates

Material	a Å	b Å	$c/\sqrt{2}$ Å	$\tilde{D}^{(a)}$	V(Å)³	T_{JT} (K)
$LaMnO_3$ [b]	5.532	5.742	5.442	1.8%	244.5	875
$PrMnO_3$ [c]	5.454	5.809	5.365	3.2%	240.4	820
$NdMnO_3$ [d]	5.410	5.762	5.334	3.1%	235.2	1100

(a) $\tilde{D} = 1/3 \sum_{i=1}^{3} |(a_i - \tilde{a})| / \tilde{a}_i, \tilde{a} = (abc/\sqrt{2})^{1/3}$; $a_1 = a$, $a_2 = b$, $a_3 = c/\sqrt{2}$

(b) J.B.A. Elemao et al, *J. Solid State Chem.*, <u>3</u>, 288 (1971).

(c) K. Knizek et al, *J. Solid State Chem.*, <u>100</u>, 292 (1992).

(d) H. Taguchi et al, *J. Solid State Chem.*, <u>76</u>, 284 (1988).

2.2. *Hole-doped manganates*

$LaMnO_3$ with a small proportion of Mn^{4+} ($\leq 5\%$) becomes antiferromagnetically ordered at low temperatures ($\tilde{T}_N \sim 150K$). When the La^{3+} in $LaMnO_3$ is progressively substituted by a divalent cation as in $La_{1-x}A_xMnO_3$ (A = Ca, Sr or Ba), the proportion of Mn^{4+} increases and the orthorhombic distortion decreases. The material becomes ferromagnetic with a well-defined Curie temperature at a finite x, and metallic below T_c.[5] The saturation moment is typically ≈ 3.8 μ_B which is close to the theoretical estimate based on localized spin-only moments. This suggests that the conduction electrons are fully spin-polarized. An important quantity characterizing a spin wave is the stiffness constant D. From neutron scattering measurements on $La_{0.67}Ca_{0.33}MnO_3$, D has been found to be $\approx 150\text{-}170$ meV.$Å^{2}$,[6] The simple Heisenberg model predicts a D value of $\approx 18\text{-}20$ meV.$Å^2$ and a high T_c in the 900-1000K region. It has been argued that the lattice distortion brings down T_c to below 400K. Below T_c the manganate exhibits metal-like conductivity. Fig. 2 illustrates how the ferromagnetism and the insulator-metal (I-M) transition occur around T_c.

6

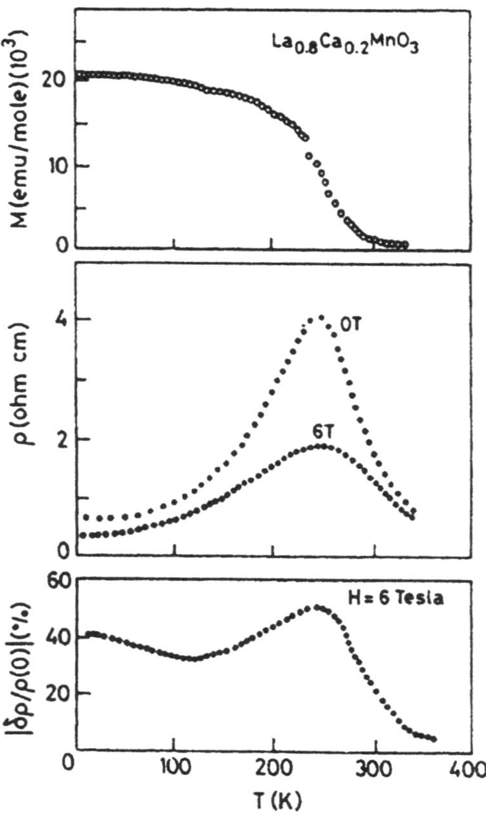

Fig. 2. Temperature variation of magnetization, resistivity (at H = 0T and 6T) and magnetoresistance of polycrystalline La$_{0.8}$Ca$_{0.2}$MnO$_3$ (adapted from Mahendiran et al[11]).

The simultaneous observation of itinerant electron behavior and ferromagnetism in the manganates is explained by Zener's double-exchange mechanism.[7] The basic process in this mechanism is the hopping of a d-hole from Mn^{4+} (d^3, t$_{2g}^3$, S = 3/2) to Mn^{3+} (d^4, t$_{2g}^3$e$_g^1$, S = 2) via the oxygen, so that the Mn^{4+} and Mn^{3+} ions change places:

$$Mn^{3+}O^{2-}Mn^{4+} \longrightarrow Mn^{4+}O^{2-}Mn^{3+} \tag{2}$$

This involves the transfer of an electron from the Mn^{3+} site to the central oxide ion and simultaneously the transfer of an electron from the oxide ion to the Mn^{4+} site. Such a transfer is referred to as *double-exchange (DE)*. The integral defining the exchange energy in such a system is nonvanishing only if the spins of the two $\underline{d}$-shells are parallel. That is, the lowest energy of the system is one with a parallel alignment of the spins on the Mn^{3+} and Mn^{4+} ions. This is indeed a novel situation where the lining up of the spins of the incomplete $\underline{d}$ orbitals of the adjacent Mn ions is accompanied by an increase in the rate of hopping of electrons and therefore by an increase in electrical conductivity. Thus, the mechanism which leads to enhanced electrical conductivity requires a ferromagnetic coupling. It is assumed here that the intra-atomic exchange, J, is large compared to the transfer integral, t_{ij}, between the two Mn sites.

The relation between the electrical conductivity and ferromagnetism by the DE mechanism as originally proposed by Zener is obtained as follows. The magnitude of the exchange energy, ϵ, is given by

$$\epsilon = h\nu/2 \qquad (3)$$

where ν is the frequency of oscillation of the electron between two Mn sites. The diffusion coefficient for an Mn^{4+} ion is defined by,

$$D = \underline{a}^2\epsilon/h \qquad (4)$$

where $\underline{a}$ is the lattice parameter. Making use of the Einstein equation relating conductivity, σ, and D, $\sigma = ne^2D/kT$, where n is the number of Mn^{4+} ions per unit volume, we obtain

$$\sigma = xe^2\epsilon/ahkT \qquad (5)$$

Here x is the fraction of Mn^{4+} ions in $La_{1-x}A_xMnO_3$. Since the ferromagnetic Curie temperature T_c is related to the exchange energy by the approximate relation, $\epsilon \approx kT_c$,

$$\sigma \approx (xe^2/ah)(T_c/T) \qquad (6)$$

Eq. (6) relates the electrical conductivity to ferromagnetic T_c and the fraction of Mn^{4+} ions. We would therefore expect the I-M transition in the manganates to occur at T_c. Double-exchange is strongly affected by structural parameters such as the Mn-O-Mn angle or the Mn-Mn transfer and integral.

The t_{2g}^3 electrons of the Mn^{3+} ion are localized on the Mn site giving rise to a local spin of 3/2, but the e_g state, which is hybridized with the oxygen 2p state, can be localized or itinerant and only those electrons which have their spins parallely aligned give rise to conductivity in the hopping process. There is strong Hund's rule interaction between the e_g and the t_{2g}^3 electrons. Goodenough[8] pointed out that ferromagnetism is

8

governed not only by double exchange, but also by the nature of the superexchange interactions. While delineating the nature of the double exchange and superexchange interactions, de Gennes[9] suggested that a non-collinear magnetic structure forms at intermediate concentrations between the antiferromagnetic and the ferromagnetic states. It may be recalled that the Mn^{3+}-O-Mn^{4+} superexchange interaction is ferromagnetic while the Mn^{3+}-O-Mn^{3+} and Mn^{4+}-O-Mn^{4+} interactions are both antiferromagnetic. The resulting magnetic exchange is strongly dependent on the structural distortion, as indeed shown by electronic structure calculations. While recent investigations bring out the essential role of DE, they also suggest the crucial role of the lattice and the electron-lattice interactions. In understanding the novel transport properties, we need to consider the overall lattice symmetry as well as local lattice distortions.

The $x = 0.0$ and the $x = 1.0$ members of the $La_{1-x}A_xMnO_3$ system (viz. $LaMnO_3$ and $AMnO_3$) are antiferromagnetic (AFM) insulators at low temperatures, with A- and G-type ordering, respectively.[10] In A type ordering, there are ferromagnetically aligned spins in the planes, but the planes are aligned antiferromagnetically. Addition of A ions reduces the JT distortion, in particular the Q_2 mode distortion which gives rise to long and short Mn-O distances, leading to anisotropic exchange. Removal of the distortion reduces the AFM interaction and this in turn destroys the AFM order. Ferromagnetic behavior starts to manifest itself when x is ~ 0.1, and the compositions up to x ~ 0.3 have both AFM and FM characteristics. The $x = 0.3$ composition is clearly ferromagnetic, while the $x > 0.5$ compositions are antiferromagnetic. The FM transition in the $La_{1-x}A_xMnO_3$ compositions is quite sharp[11] as can be seen from Fig. 2. Although manganates with x > 0.5 are essentially antiferromagnetic, ferromagnetic Mn^{3+}-O-Mn^{4+} clusters would be present in the AFM medium. Fig. 3 illustrates how the electrical and magnetic properties of $La_{1-x}Ca_xMnO_3$[12] and $La_{1-x}Sr_xMnO_3$[4] change with the composition. The phase diagrams are even richer than indicated in Fig. 3, giving rise to the fascinating properties and phenomena of these manganates. The material is generally a ferromagnetic metal (FMM) below T_c when ~ 0.2 < x < 0.5, becoming a paramagnetic insulator (PMI) when $T > T_c$. In the small x regime (x < ~ 0.2), a canted spin antiferromagnetic insulating state (CSI) or a ferromagnetic insulating state (FMI) is often encountered. Charge ordering can also occur in this composition range, but more commonly in the 0.3 < x < 0.7 range. When x > 0.5, the materials are generally antiferromagnetic insulators (AFMI) becoming paramagnetic metals (PMM) or insulators above T_N. It should be noted that $CaMnO_3$ and $SrMnO_3$ ($x = 1.0$) are antiferromagnetic insulators.

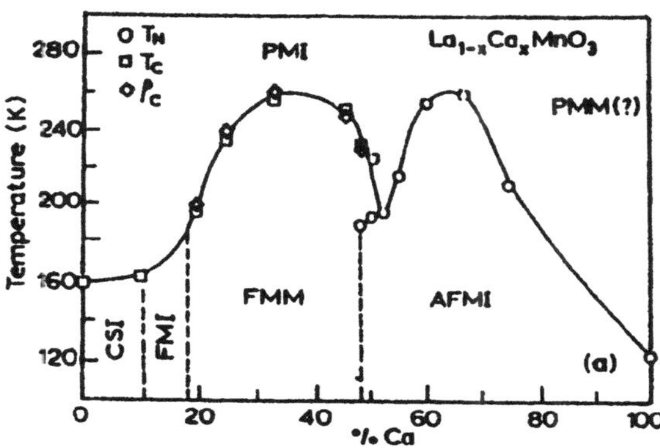

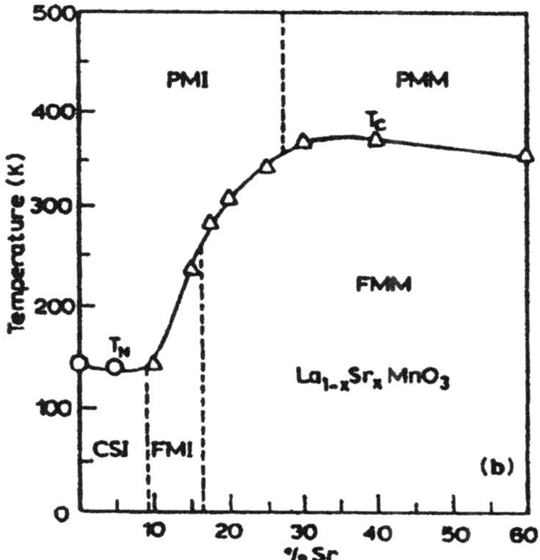

Fig. 3. Temperature-composition diagram of $La_{1-x}A_xMnO_3$ (a) = Ca (b) = Sr; CSI, canted-spin insulator; FMI, ferromagnetic insulator; FMM, ferromagnet metal; AFMI, antiferromagnetic insulator; PMI, paramagnetic insulator, PMM, paramagnetic metal (adapted from Schiffer et al[12] and Urushibara et al[4]).

10

The extent of the FMM regime depends on the A site cation, being greater when A = Sr than when A = Ca. Thus, many of the manganates exhibit charge ordering, its nature depending on the size of the A-site cations. We shall discuss this aspect later in this article.

As La in $LaMnO_3$ is progressively substituted by a divalent ion, the oxide becomes rhombohedral or pseudocubic at some value of x (typically x > 0.2). The structures of $La_{1-x}A_xMnO_3$ compositions are therefore described as orthorhombic (*Pbnm*), rhombohedral (*R 3c*), or pseudocubic. The same is broadly true of the other rare-earth manganates, $Ln_{1-x}A_xMnO_3$ (Ln = Pr, Nd etc.). The transfer interaction of e_g electrons is greater in the rhombohedral or the pseudocubic phase than in the orthorhombic phase because the Mn-O-Mn angle becomes closer to 180°. The size of the A-site cations plays an important role in manganates of the type $Ln_{1-x}A_xMnO_3$ (Ln = La, Pr, Y; A = Ca, Sr, Ba); the structure is always orthorhombic when the average size of the A-site cations, $<r_A>$, is small (< 1.22Å). With the increase in $<r_A>$, the structure becomes rhombohedral down to very low temperatures. However, in $La_{0.7}Ba_{0.3}MnO_3$, a new orthorhombic structure *(Imma)* has been observed below 150K, the $R\,3c \longrightarrow Imma$ transition being first order)[13]

It was pointed out that Jahn-Teller distortion due to the Mn^{3+} ions $(t_{2g}^3\,e_g^1)$ plays a key role in the manganates. $LaMnO_3$ has three Mn-O distances (1.91, 1.96 and 2.19Å) and a Mn-O-Mn angle of ~150°. The creation of Mn^{4+} ions removes the distortion leading to more cubic structures. X-ray crystallographic studies of $LaMnO_3$ suggest that the removal of the static Jahn-Teller distortion is a first order transition.[14] It has indeed been found that across the insulator-metal (I-M) transition occurring around T_c, the Jahn-Teller distortion decreases, the distortion being more prominent in the insulating phase.[15] Increasing the static coherent MnO_6 distortion favours the insulating behavior and decreases T_c. The structural parameters, in particular the oxygen thermal parameters, show significant changes across the I-M transition.[16]

As mentioned earlier, as-prepared $LaMnO_3$ generally contains some Mn^{4+}. The origin of Mn^{4+} has been examined in some detail. Since perovskites cannot accommodate excess oxygen, the defect chemistry of $LaMnO_3$ must involve the presence of cation vacancies in the La and Mn sites.[17] Accordingly, $LaMnO_3$ can be represented as $La_{1-\delta}\,Mn_{1-\delta}O_3$. If the Mn^{4+} content in $LaMnO_3$ is 33%, the formula will approximately be $La_{0.945}Mn_{0.945}O_3$. $LaMnO_3$ becomes rhombohedral and then pseudocubic as the Mn^{4+} content is increased by chemical or electrochemical means.

Creation of Mn^{4+} in $LaMnO_3$ leads to an increase of the tolerance factor, t, because of the smaller radius of the Mn^{4+} ions. For $\delta \approx 0.02$ (24 % Mn^{4+}), $t \geq 0.90$ and the transition occurs from the orthorhombic to the rhombohedral phase[18]. The $La_{1-\delta}MnO_3$ compositions exhibit ferromagnetism and the I-M transition up to $\delta = 0.3$, but $LaMn_{1-\delta'}O_3$ compositions do so only up to $\delta' = 0.05$.[19]

Since the presence of ~30% Mn^{4+} is essential for the ferromagnetism and the insulator-metal (I-M) transition, it is desirable to independently determine the Mn^{4+} content by means of redox titrations (Fe (II) sulphate + $KMnO_4$). It can be somewhat misleading to assume that the Mn^{4+} content is simply dictated by the formula $La_{1-x}A_xMnO_3$, since it can vary with the method and conditions of preparation. Clearly, Mn^{4+} plays a crucial role in this material. The role is essentially two-fold. It provides the double-exchange needed for ferromagnetism and metallic behavior and it also helps to remove the Jahn-Teller distortion of Mn^{3+} ions and makes the structure closer to cubic.

3. Transport Properties and Colossal Magnetoresistance (CMR) of Rare Earth Manganates

Most of the $La_{1-x}A_xMnO_3$ compositions are paramagnetic insulators at ordinary temperatures and exhibit an increase in electrical resistivity with a decrease in temperature. Compositions that are ferromagnetic show insulating behavior above T_c, but the resistivity decreases with decreasing temperature, as in metals, when they are cooled below T_c. This insulator-metal (I-M) transition is therefore associated with a peak in resistivity at a temperature T_{im}. A typical I-M transition in a polycrystalline $La_{1-x}Ca_xMnO_3$ sample was shown in Fig. 2. Such a transition is also found in $LaMnO_3$ with a sufficient proportion of Mn^{4+} (Fig. 4). Generally, T_{im} is somewhat lower than T_c.[11] The sharpness of the transition in polycrystalline samples and films is often dependent on the sample quality. For samples with grain size in excess of 5 μm, the transition is relatively sharp.

3.1. Transport in the paramagnetic phase (T>Tc)

The temperature dependence of the resistivity, ρ, of the manganates with $0.2 \leq x \leq 0.4$ in the paramagnetic phase generally shows two types of behavior. The manganates doped with larger cations such as Sr, Ba and Pb (i.e. large A-site cation radius, $\langle r_A \rangle$) have a larger bandwidth. As a

12

result the paramagnetic phase can be metallic (PMM) or insulating (PMI) with a small gap. This leads to a weak dependence of ρ on T. For

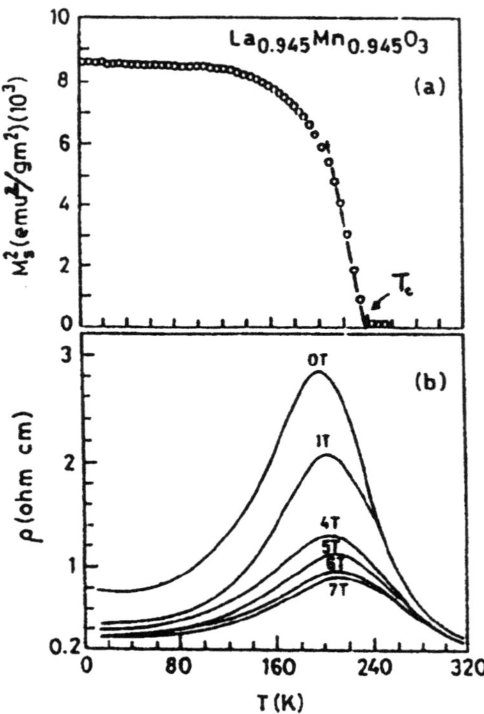

Fig.4. (a) Magnetization data of polycrystalline La$_{0.945}$Mn$_{0.945}$O$_3$ (b) resistivity-temperature plots for polycrystalline La$_{0.945}$Mn$_{0.945}$O$_3$ at different magnetic fields (after Mahendiran et al[11]).

manganates with relatively smaller cations like Ca (ie small $<r_A>$), the bandwidth would be small and the paramagnetic phase is generally insulating (PMI) with a well-defined activation energy E_a in the range $\approx$ 0.1- 0.25eV. There is an interesting dependence of E_a on the Mn-O distance[11] wherein it increases with the Mn-O distance. Thus the behavior above T_c (whether the high temperature phase is PMM or PMI) entirely depends on the bandwidth.

The important issue that remains is what makes the $T > T_c$ phase insulating. Is it lattice polaron formation from strong electron-lattice coupling due to JT effect or is it some other type of localization? This question has been investigated mainly by fitting ρ above T_c to different expressions. Some investigators find that a variable range hopping (VRH) type of expression $\rho \approx \rho_{oh} \exp (T_0/T)^{1/4}$ holds good above T_c. These investigations have been carried out in a limited temperature range above T_c. The problem, however, is that one obtains physically unrealistic values of T_0. Some authors argue in favor of lattice polaron formation and have fitted the resistivity to expressions of the type, $(\rho/T) \approx \rho_{o\rho} \exp (E_a/k_BT)$. Here, E_a is the small polaron hopping energy. From the resistivity data taken over a limited temperature range, it is often difficult to distinguish between VRH, small polaron hopping or the simple semiconductor form $\rho \sim \exp (E/k_BT)$. Analysis of the high temperature ρ data along with thermopower data tend to support the small polaron hopping picture. However, there is need for further investigations to settle the nature of electrical transport for $T > T_c$.

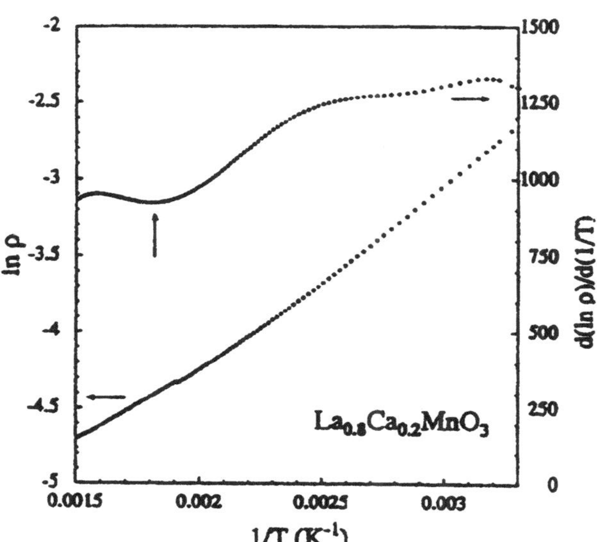

Fig. 5. Temperature dependence of the resistivity, ρ, (T > 300K) and $d\ln\rho \,/\, d(1/T)$ for $La_{0.8}Ca_{0.2}MnO_3$.

14

There are certain interesting observations when ρ is investigated at higher temperatures. In Fig. 5 we plot $d\ln\rho / d(1/T)$ vs $1/T$ in $La_{0.8}Ca_{0.2}MnO_3$. There are two regions with different activation energies. The two regions are demarkated by an enhancement of the activation energy ($d\ln\rho/d(1/T)$ α activation energy) at $T \simeq 500K$ - $600K$. A similar observation has been made in a single crystal of $La_{0.67}Ca_{0.33}MnO_3$ at $T \simeq 750K$. This change in activation energy can be due to the formation of polarons due to lattice distortion. This issue needs further investigation.

3.2. *The ferromagnetic metallic state $(T < T_c)$*

The electrical resistivity in the ferromagnetic region has been investigated by a number of authors. In the region just below T_c, ρ varies rapidly with T. For $T < 0.5\ T_c$, typically the variation is much less rapid, but it is

Table 2. Properties of $La_{1-x}A_xMnO_3$

Material	Physical State	Temp. Range	$\rho(T) = \rho_0 + \rho_2 T^n$	ρ_0 (mΩcm)	ρ_2/ρ_0 (10^{-6} K^{-n})	Reference
$La_{1-x}A_x$ $La_{0.67}Ca_{0.33}$	film/ crystal	$T/T_c < 0.4$	$n \cong 2$	0.1 - 0.3	60 - 100	J.G.Snyder et al Phys. Rev. B 54, 14434 (1996)
$La_{0.67}Ca_{0.33}$	Poly- crystal	$T/T_c < 0.5$	$n \simeq 2.5$	15	10	J.M. De Teresa et al, Phys. Rev. B54, 1187 (1996)
$La_{0.75}Ca_{0.25}$	Poly- crystal	$T/T_c < 0.5$	$n \simeq 2.5$	3	6	P. Schiffer et al Phys. Rev. Letts. 75, 3336 (1995)
$La_{0.8}Ca_{0.2}$	Poly- crystal	$T/T_c < 0.2$	3	24	12	A. Biswas (Private Comm.)
$La_{0.67}Sr_{0.33}$	film	$T/T_c < 0.4$	$n \simeq 2$	0.15	55	J.G. Snyder et al Phys. Rev. B 53, 14434 (1996)
$La_{0.7}Sr_{0.3}$	Single crystal	$T/T_c < 0.5$	$n \simeq 2$	0.1	150	A. Urushibara et al, Phys. Rev. B 51, 14103 (1995)
$La_{0.67}Ba_{0.33}$	film	$T/T_c < 0.4$	$n \simeq 2$	0.34	52	S.E. Lofland et al Phys. Rev. B 52, 15058 (1995)

different from what is generally seen in a metallic ferromagnet. In Table 2 we give a brief summary of the temperature dependence of ρ seen by different authors. The table is not exhaustive and gives typical data. In general, for $T < 0.2\ T_c$, a T^2 dependence seems to make the major contribution. At somewhat high temperatures, $T \simeq 0.5\ T_c$, contribution from an extra term faster than T^2 is needed. However, ρ_0 and ρ_2 of different authors show wide variations even for the same composition. Cearly, this is an indication that we are not dealing with the intrinsic temperature variation of ρ. For $0.5\ T_c < T < T_c$, the variation of ρ with T depends on the physical state of the samples. An interesting issue that needs to be settled in this region is how a temperature-dependent density of states at the Fermi level affects ρ (T).

3.3 Effect of magnetic field, CMR

Application of a magnetic field (up to, say, 6 Tesla) causes a significant decrease in the resistivity of the $La_{1-x}A_xMnO_3$ samples, particularly in the compositions with $0.1 < x < 0.5$; these materials are generally ferromagnetic with well-defined T_cs. The magnitude of the decrease in resistivity (i.e. the magnetoresistance, MR) is highest in the region of T_c or T_{im}. In Fig. 2 we compare the temperature variation of the resistivity of $La_{1-x}Ca_xMnO_3$ at 6T with that in zero field. Magnetoresistance in the manganates is generally negative and is highest around T_c or T_{im}. CMR close to 100% has been observed in many polycrystalline and single crystal $La_{1-x}A_xMnO_3$ compositions, but the applied field is quite high (5-6 Tesla). In $La_{1-x}Pb_xMnO_3$, high CMR is found around room temperature or above as shown in Fig. 6.[20] Self-doped $LaMnO_3$ samples with sufficient Mn^{4+} content (i.e. $La_{1-\delta}Mn_{1-\delta}O_3$) exhibit CMR similar to that observed in the $La_{1-x}A_xMnO_3$ compositions. In Fig. 4 we show the behavior of $La_{0.945}Mn_{0.945}O_3$, with 33% Mn^{4+}.[11] La-deficient, $La_{1-\delta}MnO_3$ compositions also exhibit CMR, but Mn-deficient $LaMn_{1-\delta'}O_3$ samples do not when $\delta' > 0.05$.[19] In Fig. 7 we show the resistivity behavior of La- and Mn- deficient compositions to illustrate how deficiency in La is tolerated readily without affecting the magnetotransport properties and CMR, unlike deficiency in Mn. This is understandable since the active electrical and magnetic network is made up of Mn-O-Mn bonds. As a result, any defect in the Mn lattice will affect the transport properties greatly. Substitution of other trivalent ions such as Fe^{3+} and Al^{3+} for Mn^{3+} adversely affects the system, lowering the T_c and increasing the resistivity.

16

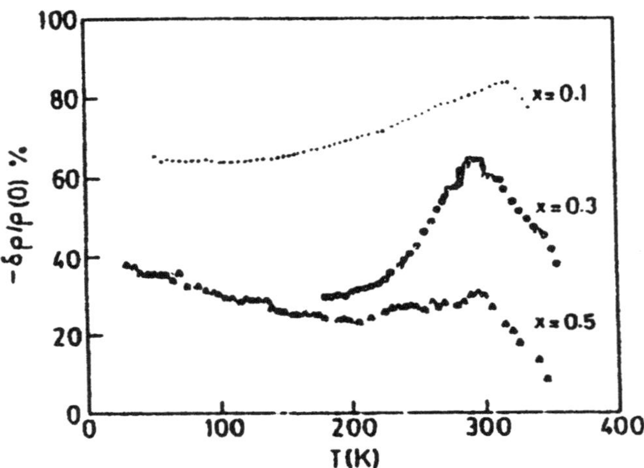

Fig. 6. Magnetoresistance of polycrystalline $La_{1-x}Pb_xMnO_3$ at 6T (from Mahendiran et al[20]).

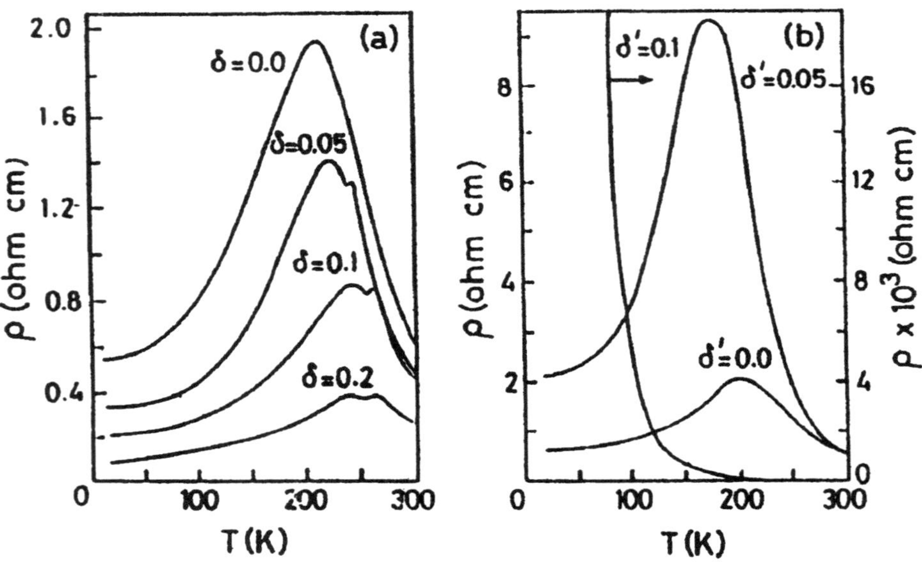

Fig. 7. Resistivity behavior of $La_{1-\delta}MnO_3$ and $LaMn_{1-\delta}O_3$ (from Arulraj et al[19]).

The variation of MR with the magnetic field at different temperatures is shown in Fig. 8.[11] The dependence of MR on H is a function of temperature of the measurement. The change is gradual close to T_c. For $(T/T_c) < 0.5$, the MR in polycrystalline manganates exhibits two distinct regimes as a function of magnetic field. For $H < 1T$, the MR changes rapidly with H followed by a gradual change as H is increased further. There are two aspects to the field dependence. One aspect has to do with ferromagnetic domain wall movement as in all ferromagnets and the other has to do with grain boundaries which contribute substantially to the MR of manganates at $T \ll T_c$. The magnitude of MR has a close relation to the magnetization (M) even in polycrystalline samples, showing a quadratic dependence[21]. It is interesting that the two distinct regimes seen in the MR -H curve for polycrystalline samples at $T \ll T_c$ is generally absent in single crystals. In single crystals, the magnitude of MR increases almost linearly with H.

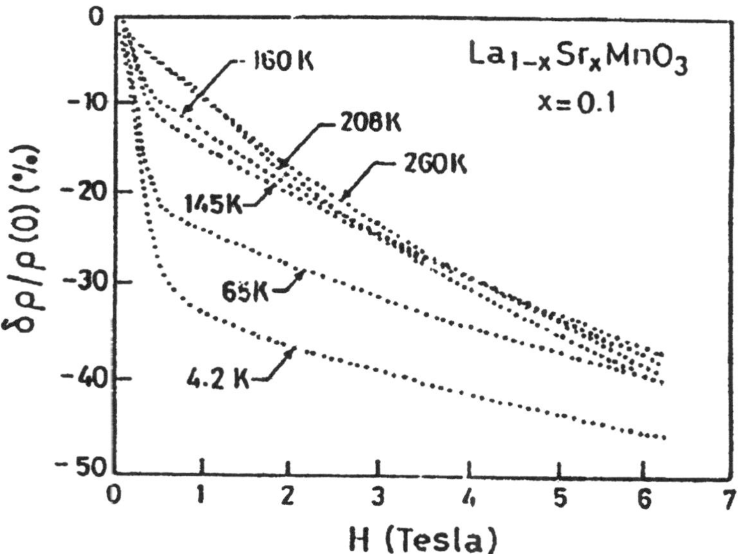

Fig. 8. Variation of magneto resistance of polycrystalline La$_{1-x}$Sr$_x$MnO$_3$ with magnetic field, H (from Mahendiran et al[11]).

CMR studies of single crystals of rare-earth manganates exhibit features in the transport properties which are not very different from those of the polycrystalline samples, although there are some subtle differences. For example, the I-M transition does not always show a distinct resistivity

18

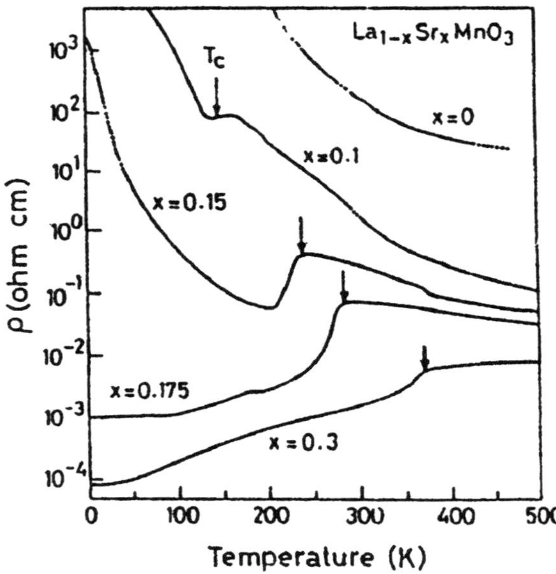

Fig. 9. Temperature variation of the resistivity of a single crystal of $La_{1-x}Sr_xMnO_3$ (after Urushibara et al[4]).

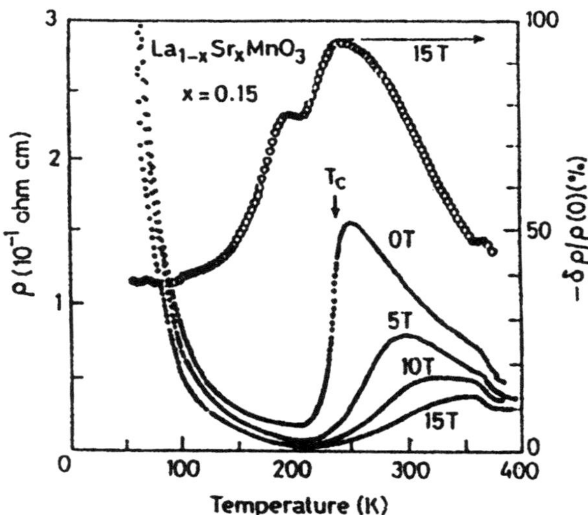

Fig. 10. Resistivity-temperature plots of a single crystal of $La_{0.85}Sr_{0.15}MnO_3$ at different magnetic fields (after Urushibara et al[4]).

peak as in the polycrystalline samples. In Figs. 9 and 10, we show the resistivity behavior of $La_{1-x}Sr_xMnO_3$ crystals at zero field and the effect of magnetic field on the resistivity.[5] In single crystals and good epitaxial films, CMR is substantial only near T_c and near zero as $T \rightarrow 0$. The reduction in the resistivity of $La_{1-x}Sr_xMnO_3$ crystal scales with the field-induced magnetization, M, as:

$$-\Delta\rho/\rho = C(M/M_s)^2 \qquad (7)$$

for $M/M_s < 0.3$, where M_s is the saturation magnetization. This dependence on magnetization is the same as that found in polycrystalline materials.

3.4. *Role of electron-lattice coupling*

Lattice polaron formation and local structural distortion have important implications on electron transport in these materials. In particular, they play a role in carrier localization and can be a source of the high resistivity seen in these materials. CMR is observed over a wide range of compositions of $La_{1-x}Ca_xMnO_3$ ($x < 0.5$), but the material shows lattice effects, particularly for low x. When $x = 0.2$, a local structural distortion occurs due to the formation of lattice polarons which arises from the strong electron-lattice coupling. EXAFS studies reveal significant changes in the local structure in the $x = 0.3$ composition in the 80-300K range which are attributed to the formation of small polarons due to the Jahn-Teller distortion when $T > T_c$.[22,23] These studies have shown that small polarons delocalize as the magnetization increases in the manganates which show the I-M transition. Channeling experiments close to T_c in $Nd_{0.7}Sr_{0.3}MnO_3$, $Pr_{0.7}Ba_{0.3}MnO_3$ and $La_{0.7}Ba_{0.3}MnO_3$ films have provided evidence for Jahn- Teller distortion.[24]

The spin and lattice dynamics of $La_{0.7}Sr_{0.3}MnO_3$ have been examined by neutron scattering.[25] The spin dynamics and the magnetic critical scattering seem to be comparable to those of typical metallic ferromagnets. In $La_{1-x}Sr_xMnO_3$ ($x = 0.17$), the local spin moments and charge carriers couple strongly to changes in the structure and the structure can be switched by the application of a magnetic field depending on the temperature. It is noteworthy that the magnetic field which causes this crossover from the orthorhombic to the rhombohedral structure is actually very low ($\sim 2T$). This observation is significant because it clearly shows that an extremely small energy difference exists between these two structures.[26a] Single crystal neutron diffraction experiments have confirmed that the structural transition in $La_{1-x}Sr_xMnO_3$ can be induced at constant temperature by the application of a magnetic

20

field.[26b] For $0.1 \leq x \leq 0.17$, ferromagnetic ordering suppresses the lattice distortion at and below T_c.[27a] Magnetostriction and linear thermal expansion experiments establish the strong correlation between magneto transport and magnetovolume properties.[27b] When $x = 0.1$ and 0.15, a polaron ordered phase is formed and the polaron lattice has a tendency to lock into a commensurate structure when $x = 0.125$.[28] Spontaneous formation of localized 12Å magnetic clusters at $T > T_c$ are found to grow in size but decrease in number on application of the magnetic field in $(La_{1-x}Y_x)_{0.67}Ca_{0.33}MnO_3$, showing thereby that magnetic polarons are present in the insulating phase.[29] While the role of lattice polarons seems clear, that of magnetic polarons requires further study.

3.5. *Effect of external and internal pressure*

Hydrostatic pressure stabilizes the ferromagnetic metallic state of $La_{1-x}A_xMnO_3$ (i.e. increases the T_c).[30] The effect of hydrostatic pressure is to enhance the Mn-O-Mn transfer integral through a change in the Mn-O-Mn angle. A similar effect can be brought about by increasing the radius of the A-site cation. There is a direct relationship between T_c and the average radius of the A-site cation, $<r_A>$, the T_c increasing with $<r_A>$.[31-33] The equivalence of increasing $<r_A>$ and increasing hydrostatic pressure can be seen in Fig. 11. It should be noted that small $<r_A>$ gives rise to a distortion of the MnO_6 octahedra by bending the Mn-O-Mn bond and in turn causing the narrowing of the e_g bandwidth to a greater extent. Accordingly, yttrium substitution in $La_{1-x}A_xMnO_3$ decreases T_c, and enhances the magnitudes of resistivity as well as magnetoresistance.[34,35] In $La_{1-x}A_xMnO_3$, magnetic frustration has been shown to increase with a decrease in $<r_A>$ or the bending of the Mn-O-Mn angle; the electronegativity of the A ion could also be a contributing factor.[36a] In Fig. 12 we show the resistivity behavior of $La_{0.5}(Ca_{0.5-z}Y_z)MnO_3$. In $La_{(2-x)/3}Nd_{x/3}La_{1/3}MnO_3$, the Mn-O-Mn angle bending increases with Nd content and the T_c decreases along with it.[36b] By plotting the T_c against $<r_A>$, one obtains a phase diagram separating the ferromagnetic metal and paramagnetic insulator regimes.[31,37] In Fig. 13 we show a typical phase diagram obtained in this manner. In the left hand bottom corner, we see the ferromagnetic insulator regime. CMR generally decreases with increase in $<r_A>$ just as the peak resistivity at T_c or at the insulator-metal transition as shown in Fig. 14. CMR is therefore favoured by low T_c and high peak resistivity. Enhanced CMR can be considered to arise from a reduced mobility of the doping holes and an increase in the coupling

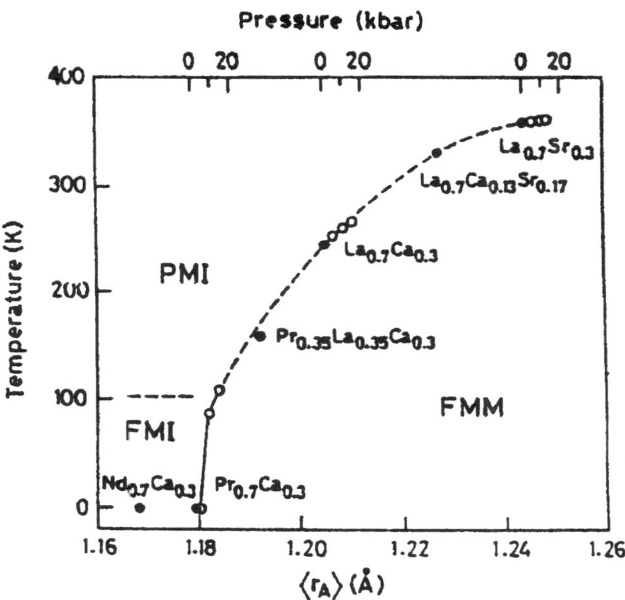

Fig. 11. Variation of $T_c(T_{im})$ of $Ln_{0.7}A_{0.3}MnO_3$ with hydrostatic pressure and the weighted average radius of the A-site cations (from Hwang et al[31]).

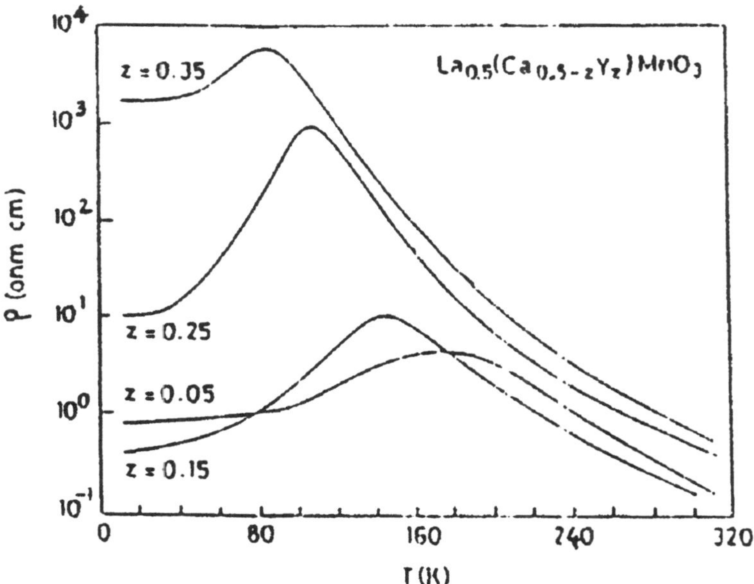

Fig. 12. Temperature variation of resistivity of $La_{0.5}(Ca_{0.5-z}Y_z)MnO_3$ (from Mahendiran et al[35]).

22

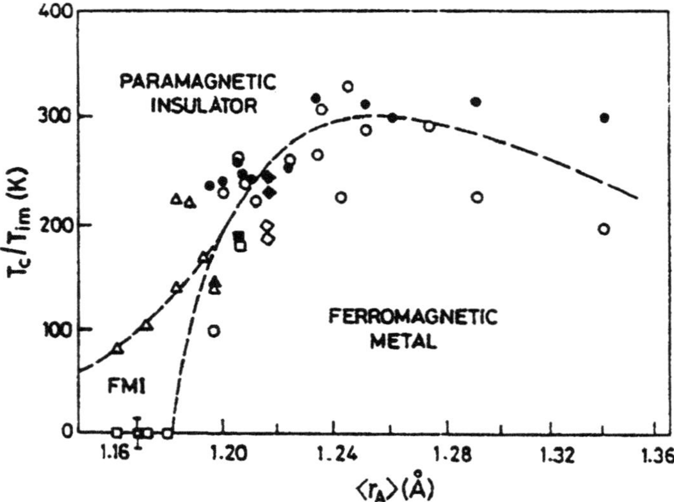

Fig. 13. Variation of $T_c(T_{im})$ of $Ln_{1-x}A_xMnO_3$ with the weighted average radius of the A-site cations, $\langle r_A \rangle$ (after Mahesh et al[37]).

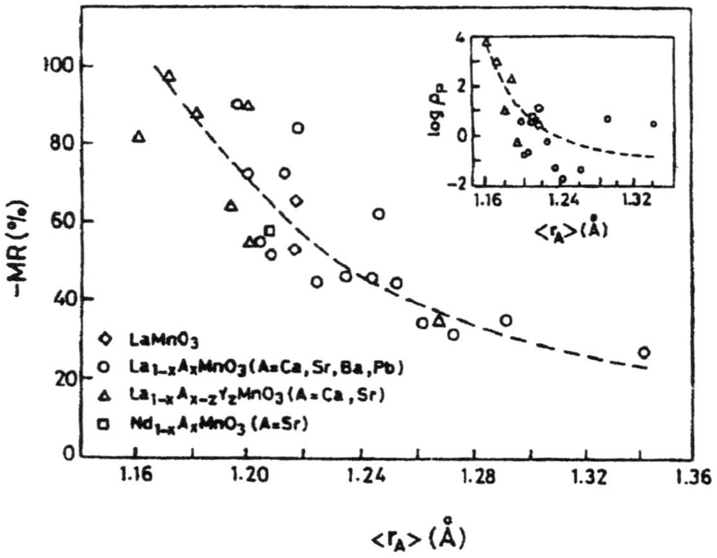

Fig. 14. Variation of the magnetoresistance of $Ln_{1-x}MnO_3$ with the weighted average radius of the A site cations, $\langle r_A \rangle$; the inset shows the variation of the logarithm of peak resistivity with $\langle r_A \rangle$ (after Mahesh et al[37]).

between the localized and itinerant electrons.[33] The enhanced CMR in materials with low T_c (and hence small bandwidth) can also arise from a stronger effect of the magnetic field on the electron-lattice interaction. The magnetic field in these materials can lead to significant carrier delocalization by lowering the lattice distortion or increasing the transfer integral.

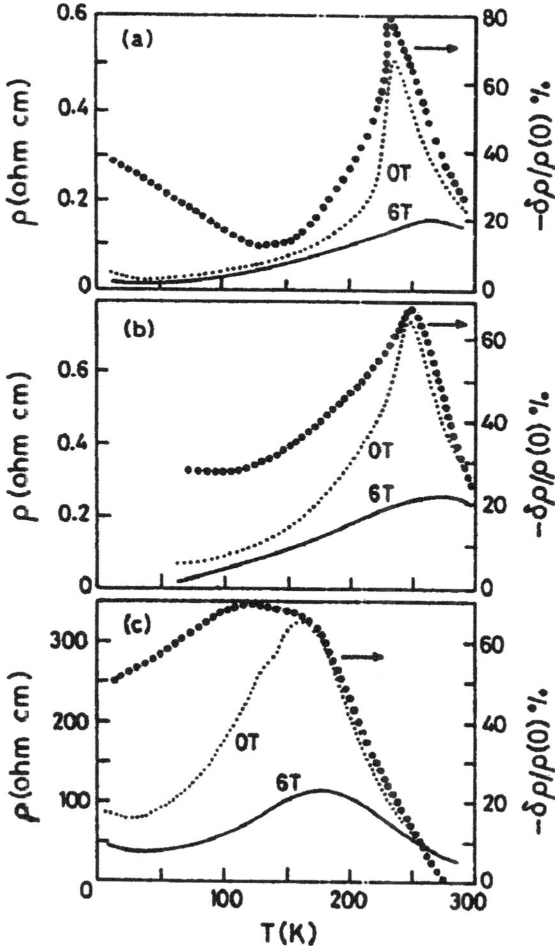

Fig. 15.   Temperature variation of resistivity (at 0T and 6T) and magnetoresistance of $La_{0.7}Ca_{0.3}MnO_{0.3}$ with particle diameters of (a) 3.51μm (b) 1.5μm and (c) 0.025μm (from Mahesh et al[38a]).

24

3.6. *Effect of grain boundaries*

The effect of particle size on the electron transport and magnetic properties of polycrystalline $La_{0.7}Ca_{0.3}MnO_3$ has been investigated.[38a] Although T_c decreases with decreasing particle size, the magnetoresistance is insensitive to particle size. We show this feature in Fig. 15 where the I-M transition becomes broader when the particle size is small, but the %MR is nearly constant over a wide temperature range. Fig. 15 also shows that the low temperature MR in these materials is significantly dependent on the grain size. For larger grain size, the MR at the lowest temperature decreases eventually becoming zero in good single crystals. Samples of manganates with different particle sizes prepared with different heat treatments often show large differences between the ferromagnetic transition temperatures and the temperatures of the I-M transition.[38b] A similar behavior appears to manifest itself when there is a large mismatch between the sizes of the A-site cations.[18b,37]

3.7. *Effect of Dimensionality*

The effect of dimensionality on CMR and related properties have been studied by examining the Ruddlesdon-Popper phases, (SrO) $(La_{1-x}Sr_xMnO_3)_n$ family.[39] The $n = 1$ member which is two dimensional with the K_2NiF_4 structure is an insulator, while the $n = \infty$ member is the three-dimensional perovskite showing CMR and other properties. The $n = 2$ member exhibits the sharpest I-M transition and high MR. In $La_{1.4}Sr_{1.6}Mn_2O_7$, the current perpendicular to the MnO_2 planes is large close to T_c (Fig. 16) and the barrier to transport provided by the intervening, insulating SrO rock salt layers can be removed by the application of a magnetic field; the interlayer transport appears to occur through tunneling.[40] The compressibility of the apical Mn-O bonds in $La_{1.2}Sr_{1.8}Mn_2O_7$, changes sign across the I-M transition. The Mn-O-Mn angle contracts with pressure in the paramagnetic insulating state, unlike in the ferromagnetic insulating state.[41a] These features are taken to support the presence of exchange striction due to a competition between superexchange and double-exchange. Neutron scattering studies of $La_{1.2}Sr_{1.8}Mn_2O_7$ show the existence of long-lived antiferromagnetic clusters along with ferromagnetic critical fluctuations.[41b] It appears that at least in two dimensions, Anderson localization has to be taken into account. Investigations of $Nd_{1+x}Sr_{2-x}Mn_2O_7$ (x = 0.0, 0.1) have revealed them to be biphasic, consisting of two closely similar Ruddlesdon-Popper phases, but they show CMR.[42]

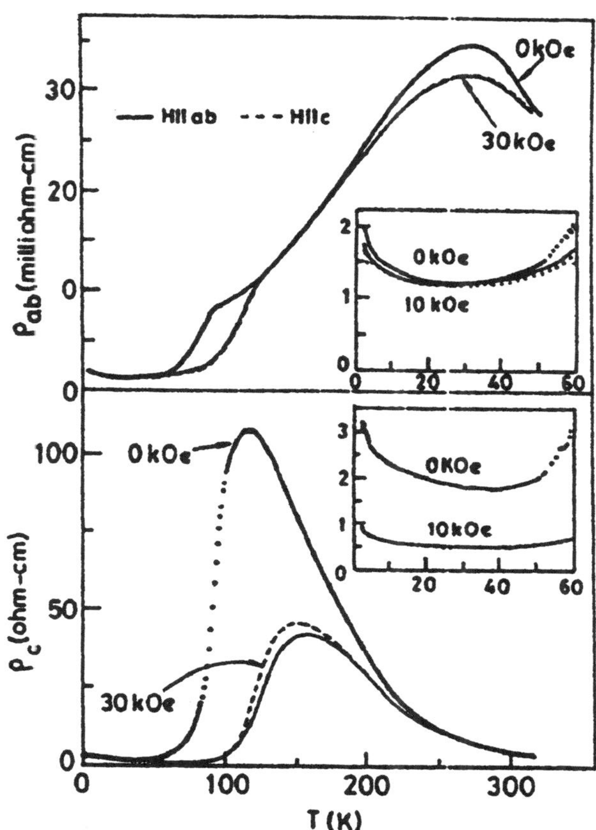

Fig. 16. Temperature variation of the resistivity of $La_{1.4}Sr_{1.6}Mn_2O_7$, in the <u>ab</u> plane and along the <u>c</u> axis under magnetic fields. Inset shows expanded low-temperature data (After Kimura et al[40]).

3.8. *Other observations*

Zero-field muon spin relaxation and resistivity experiments on $La_{0.7}Ca_{0.3}MnO_3$ show that the sublattice magnetization $v_\mu(T)$ is well described for $T \leq T_c$ by $(1-T/T_c)^\beta$, where $\beta = 0.345 \pm 0.015$.[43] Below T_c, v_μ and the zero-field resistivity ρ_o are correlated, with $v_\mu \propto -\ln \rho_o$. The rather unusual relaxational dynamics suggest spatially inhomogeneous Mn-ion correlation times.

26

For practical applications of the manganates, there is a need for enhancing the low field (< 0.01 T) response near room temperature. Two approaches have been tried so far as described below. An enhancement of MR in $La_{0.7}Ca_{0.3}MnO_3$ and related systems is accompanied by the substitution of La by a smaller rare earth as mentioned earlier. Thus, bulk $La_{0.33}Nd_{0.33}Ca_{0.33}MnO_3$ exhibits 96% MR at a relatively low field of ~ 0.7T.[44a] There is, however, one problem with such chemical substitution. The T_c is pushed to lower temperature, thereby shifting the peak in MR well below room temperature. Also, there is an increase in resistivity which increases the noise. An improvement in MR response has been obtained by sandwiching the manganates with soft ferromagnets.[44b] These observations suggest flexibility in material design for optimal CMR properties. A good strategy for material design for making sensors will be to couple the better magnetic properties of ferrites and the transport properties of manganates by making composites or multi layer films.

3.9. Concluding remarks

Some of the significant conclusions arising from the discussion in this section pertain to: (i) the key role of Mn^{4+} ion and the double-exchange interaction which provide a link between charge transport and magnetization; (ii) the importance of the packing of MnO_6 octahedra in determining the e_g electron bandwidth and the ferromagnetic exchange, which can be tuned by hydrostatic pressure or the radius of the A-site cations; (iii) local lattice distortion (which can be changed by temperature, chemical substitution as well as magnetization) which can give rise to polaron formation and charge localization; (iv) existence of structures with small energy differences so that a small pressure or a magnetic field can lead to a structural transition; (v) a substantial contribution of grain boundaries in MR of polycrystalline materials (in particular, at $T \ll T_c$).

4. Novel Features of Rare Earth Manganates

Electron transport, magnetic and other measurements on $La_{1-x}A_xMnO_3$ systems have revealed rather unusual features with respect to the charge carriers in these oxides. The manganates exhibit high resistivities even in the metallic state, particularly at low temperatures.[45] The values of resistivities are considerably higher than Mott's value of maximum metallic resistivity,[11] the resistivity reaching constant values as high as 10^3 - 10^4 ohm cm at low temperatures, in the so-called metallic state.

Coey et al[45] argue that $T < T_c$, the e_g electrons are delocalized on an atomic scale but the spatial fluctuations in the Coulomb and spin dependent potentials tend to localize the e_g electrons in wave packets larger than the Mn-Mn distance. A clear picture of the nature of the electrons responsible for such high resistivity is yet to emerge. It is also surprising that large MR occurs in manganates in a small magnetic field at a relatively high temperature. The scale of Zeeman energy associated with a field of 1T in these spin systems is $\approx$ 5K. This is much too small compared to the scale of thermal energy (200-300K) where one sees such a large MR.

Optical conductivity measurements on $La_{1-x}Sr_xMnO_3$ in the range 0-10eV show a band of 1.5eV at moderate doping and the spectral weight is transferred from this band with decreasing temperature. The 1.5eV band

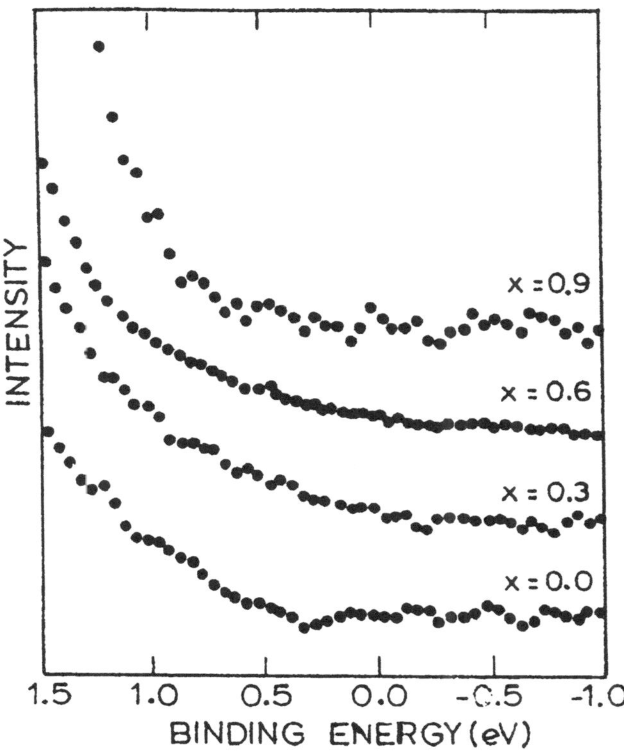

Fig. 17. Photoelectron spectra of $La_{1-x}Sr_xMnO_3$ in the valence region at room temperature (from Saitoh et al[49b])

28

is due to interband transitions between the exchange split spin polarized e_g bands. At $T < T_c$, intraband transitions in the e_g band dominate the spectrum.[46] Double-exchange alone cannot explain the optical conductivity data. A transfer of spectral weight from high energy to low energy occurs at $T < T_c$ in $Nd_{0.7}Sr_{0.3}MnO_3$.[47] A recent study of tunneling density of states by low temperature STM in $La_{0.8}Ca_{0.2}MnO_3$ has shown that at $T < T_c$, a large number of states become available near E_t.[48] Photoemission studies of $La_{1-x}Sr_xMnO_3$[49] have provided valuable information on the electronic structure of the manganates. An aspect that needs attention is concerned with the negligible density of states near the Fermi level in the metallic state of these systems. Very close to E_F, the density of states rises rapidly with energy. In Fig. 17 we show typical photoemission spectra to illustrate this feature. There is a transfer of spectral weight from the unoccupied states to occupied states with decrease in hole doping. (Note: This aspect is not shown by Fig. 17). The redistribution of spectral weight in photoemission spectra of $La_{0.6}Sr_{0.4}MnO_3$ is attributed to a change in the electron correlation strength induced by the changing degree of ferromagnetic order with temperature. The density of states at the Fermi level is much lower in the insulator regime than in the metallic regime of the manganates.[50] Mn 2p resonant photoemission and O 1s x-ray absorption studies have shown that the band gap collapses around T_c with an increase in the density of states near the Fermi level on cooling.[50] In the insulating state above T_c, small polaron effects become important.

The specific heat of a metal (the low temperature linear term, γ) provides information about the density of states at the Fermi level. The specific heat of the metallic samples $La_{0.7}Ba_{0.3}MnO_3$ and $La_{0.8}Ca_{0.2}MnO_3$ have been measured.[45,51a] The specific heat contains a linear term (γ) and a cubic term (β) as in an ordinary metal. The value of the γ term lies in the region 4-6 mJ / mole K^2. From the cubic term, one finds a Debye temperature $\theta_D \approx 400K$. The γ term is significantly enhanced from the free electron and the band values, the enhancement being most likely due to correlation effects and strong electron-lattice interaction. This γ value is comparable to that in a number of similar metallic ABO_3 perovskites. In short, the specific heat does not appear to be unusual and does not show any indication of depletion of the density of states at the Fermi level unlike that seen in photoemission studies. Low-temperature specific heat measurements on $La_{0.7}Sr_{0.3}MnO_3$ have shown that it is low-carrier density metal with conventional ferromagnetic spin waves.[51b] Raman scattering experiments on $La_{0.7}Sr_{0.3}MnO_3$ show a band at 2100 cm^{-1} whose intensity decreases with increasing temperature and this is taken to imply a reduction in the density of states at the Fermi energy with increasing temperature.[51c]

The Hall coefficient has been measured in epitaxial films of $La_{0.67}Ca_{0.33}MnO_3$ and $La_{0.67}Sr_{0.33}MnO_3$ grown by MOCVD and the carrier concentration estimated from the Hall effect is more than that calculated with the assumption that each Mn^{4+} contributes one hole and that all the holes are mobile.[52a] For the Sr-doped sample, it is 2.1 holes/Mn and for Ca-doped samples it is 0.9 holes/Mn. This observation is not compatible with the low density of states found from photoemission experiments.[49] It is to be noted that while specific heat and Hall measurements are bulk properties, photoemission measurements are sensitive to surface effects. The density of states at the Fermi level and the carrier concentration in the manganates are yet to be settled quantitatively. In particular, how these change as the samples are cooled through T_c need further investigation.

The Seebeck coefficient, S, in several rare earth manganates show a positive peak in the composition range and at a temperature where MR also peaks (around T_c), and becomes more negative with the increase in Mn^{4+} content.[11,52b] Seebeck coefficient measurements on $La_{1-x}Ca_xMnO_3$ along with self-doped $LaMnO_3$ with comparable Mn^{4+} contents carried out by us have shown the nature of hole doped states to be different in the two cases, being more localized in the latter.[52b] Measurements of S in $La_{1-x}Sr_xMnO_3$ show a change in sign from positive to negative at the I-M transition boundary or at $T < T_c$ depending on the composition.[53a] An anomalous change in electronic structure with spin-polarization may occur when $x = 0.2-0.3$. Based on the pressure dependence of S, it has been suggested that there is a crossover from a polaronic to extended state electronic conduction at T_c.[53b] While a positive S in the ferromagnetic state indicates transport involving polarized states, it is not clear why the values of S are so small in a regime where the resistivity changes by orders of magnitude.[54]

Furukawa[55] considers the origin of the large magnetoresistance observed near T_c in the manganates to essentially lie in the forced alignment of the local t_{2g} spins by the application of a magnetic field causing a reduction in the spin disorder scattering of the e_g electrons. This arises from the strong spin-charge coupling (Hund coupling) between the t_{2g} and e_g electrons on the same Mn atom. This intra-atomic ferromagnetic interaction is considerably large compared to the bandwidth and causes a splitting of the e_g band. The double-exchange model essentially explains the behavior of magnetoresistance with respect

30

to magnetization ($- \rho / \rho = C(M / M_s)^2$). Based on a study of the $LnMn_{1-x}Cr_xO_3$ system where isoelectronic Cr^{3+} replaces Mn^{4+}, it has indeed been shown recently that double-exchange is essential for the occurrence of large GMR in the manganates.[56] However, there is the basic question why these manganates are insulating when $T > T_c$, The anomalous resistance has been explained by Millis et al[57] by considering electron-phonon interaction arising from the Jahn-Teller splitting of the Mn d-levels. Jahn-Teller distortion localizes conduction electrons as polarons. As the temperature decreases, the transfer integral t_{ij} increases and the ratio of the Jahn-Teller energy to t_{ij} decreases. The coupling varies with temperature and composition and increases across the metal-insulator transition.[58] The large effect of local lattice distortion and crystal structure symmetry on electron transport and magnetism show that one cannot explain the unusual behavior without taking the lattice into account. The observed T_c ($\approx$ 250-400K) in most manganates is considerably lower than that estimated from the measured spin-wave stiffness constant[6] which implies that the ferromagnetic exchange arising from double exchange mechanism is not the only factor that affects T_c.[57] This has led to the proposal of Jahn-Teller distortion as another factor. The importance of electron-phonon interaction is also underscored by the observation of the large oxygen isotope shift of T_c in $La_{0.8}Ca_{0.2}MnO_3$.[59] Besides lattice effects, correlation effects also yield a large distribution of spectral weights.[60] Clearly, it is difficult to understand the electrical resistivity of the rare earth manganates across the I-M transition, especially at low temperatures based on double-exchange or electron-phonon interaction alone. We do not yet have a theory which explains most of the important features of the electronic transport in these materials. In particular, why CMR actually occurs has not yet been settled although we know some of the ingredients of the basic physics (like double-exchange, lattice distortion, magnetic localization etc.) which can lead to CMR. In terms of the chemistry of $Ln_{1-x}A_xMnO_3$ type manganates, there is always the problem of homogeneity. Thus, there would be ferromagnetic clusters at large or small x, even though the material may be considered to be antiferromagnetic. In $La_{(2-x)/3}Nd_{x/3}Ca_{1/3}MnO_3$, the presence of La-rich ferromagnetic domains along with semiconducting Nd-rich domains has been considered.[36b] The ordering of Ln and A ions *vis a vis* the ordering of Mn^{3+} and Mn^{4+} ions could also be a factor in determining some of the finer aspects of structure and properties.

5. Charge-ordering in Rare Earth Manganates

The study of CMR in $Ln_{1-x}A_xMnO_3$ has brought forth novel features related to the charge ordering in these oxides. Charge-ordering in the manganates is interesting because double-exchange gives rise to metallicity along with ferromagnetism while the charge-ordered state is associated with insulating and antiferromagnetic (or paramagnetic) behavior. What we have is a competition between ferromagnetism with metallic behavior and cooperative Jahn-Teller effect with charge-ordering. Lattice distortion plays a crucial role in charge-ordering, orthorhombic distortion stabilizing the charge-ordered state. The charge-ordered state can be melted into a metallic ferromagnetic state, by the application of a magnetic field. An examination of the observation on the rare earth manganates[61] reveals two types of scenarios, as exemplified by $Nd_{0.5}Sr_{0.5}MnO_3$ and $Pr_{1-x}Ca_xMnO_3$ ($0.3 \leq x \leq 0.5$). In $Nd_{0.5}Sr_{0.5}MnO_3$, a ferromagnetic metallic state ($T_c \sim 250K$) gives over to an antiferromagnetic charge-ordered state around 150K (T_{co}) accompanied by changes in lattice parameters[62] as shown in Fig. 18. The competition between the ferromagnetic double-exchange interaction and the antiferromagnetic charge-ordering instability has been examined in $(Nd,Sm)_{0.5}Sr_{0.5}MnO_3$.[63] This system exhibits large magnetoresistance which is strongly coupled to the lattice striction. The CMR phenomenon here is nothing but a first order I-M transition induced by a magnetic field, accompanied by a metamagnetic transition and a structural change. In $Pr_{1-x}Ca_xMnO_3$ ($x = 0.35$), an insulating charge-ordered state ($T_{co} \sim 200K$) becomes antiferromagnetic around 140K (T_N) and then undergoes a transformation to a canted-spin antiferromagnetic state at a still lower temperature ($\sim 110K$).[64] The application of a magnetic field melts the charge-ordered state in most of the charge-ordered manganates giving rise to metallic behavior (Fig. 19). The first-order I-M transition induced by magnetic fields is generally accompanied by considerable hysteresis as shown in Fig. 20.[62] The I-M transitions in the manganates where the one-electron bandwidth is controlled by the composition, show interesting electronic phase diagrams[61] in the temperature-magnetic field (T-H) plane as depicted in Fig. 21 for $Pr_{1-x}Ca_xMnO_3$ and $Nd_{0.5}Sr_{0.5}MnO_3$. The charge-ordering gap (0.3 eV) in $Nd_{0.5}Sr_{0.5}MnO_3$ has been measured by tunneling spectroscopy[65b] as shown in Fig. 22. Application of pressure increases the ferromagnetic T_c of $Nd_{0.5}Sr_{0.5}MnO_3$ and suppresses charge ordering; in the case of $Pr_{0.7}Ca_{0.3}MnO_3$, pressure increases the metallicity and suppresses the charge ordering transition.[65a]

32

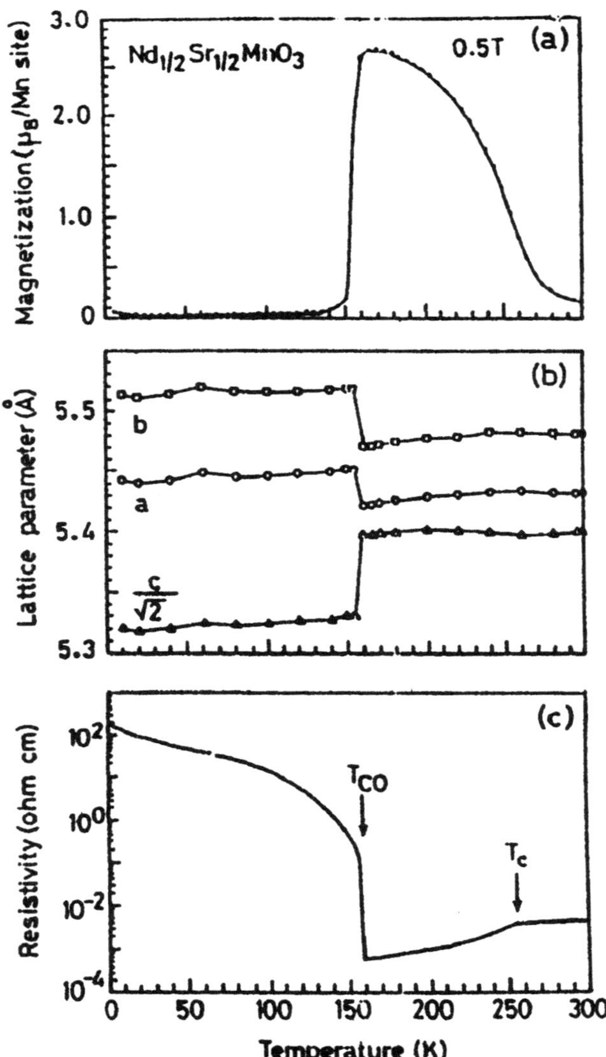

Fig. 18. The temperature variation of (a) magnetization (b) lattice parameters and (c) resistivity of Nd$_{0.5}$Sr$_{0.5}$MnO$_3$ single crystal (from Kuwahara et al[62]).

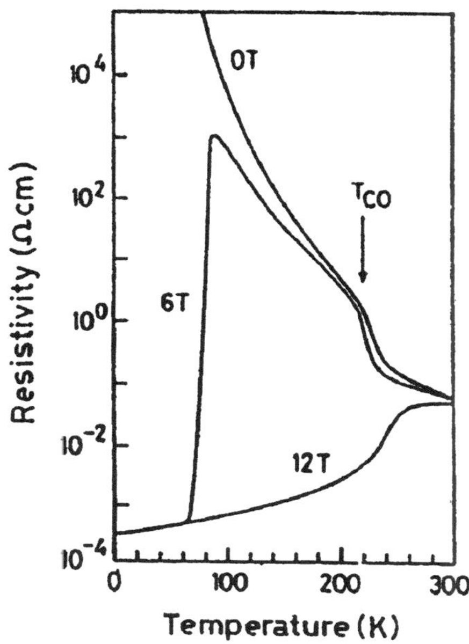

Fig. 19. The temperature variation of resistivity under H = 0T, 6T and 12T in $Pr_{1-x}Ca_xMnO_3$ (x = 0.35) single crystal (from Tomioka et al[64]).

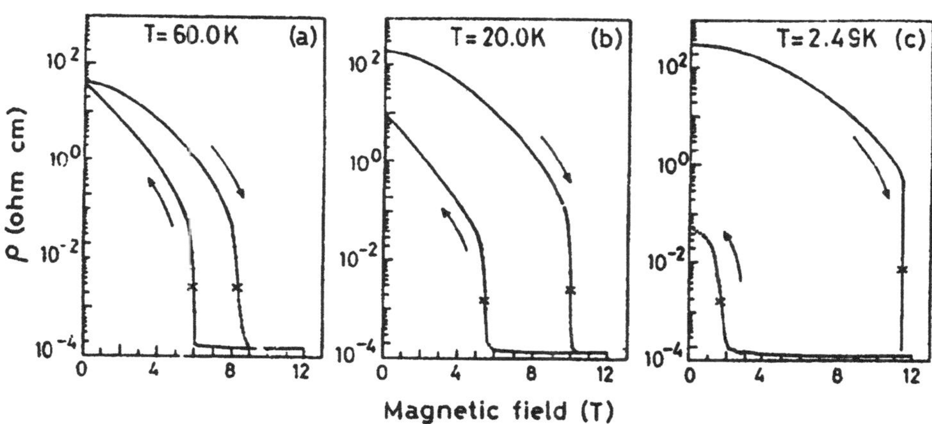

Fig. 20. Changes in the resistivity of $Nd_{0.5}Sr_{0.5}MnO_3$ with increasing and decreasing magnetic fields. Notice the sharp jump in resistivity at the lower and upper critical fields. (from Kuwahara et al[62]).

34

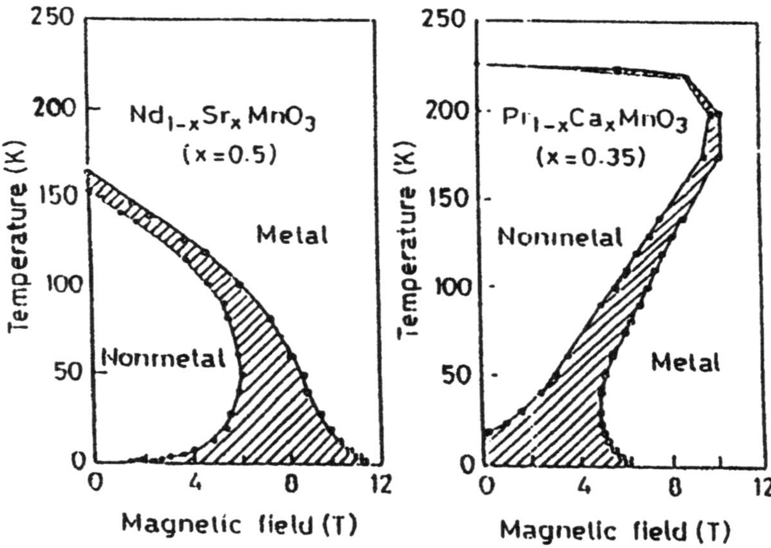

Fig. 21. The electronic phase diagrams in the T-H plane of (a) Nd$_{1-x}$Sr$_x$MnO$_3$ (x = 0.5) and (b) Pr$_{1-x}$Ca$_x$MnO$_3$ (x = 0.35) (from Tokura et al[61]).

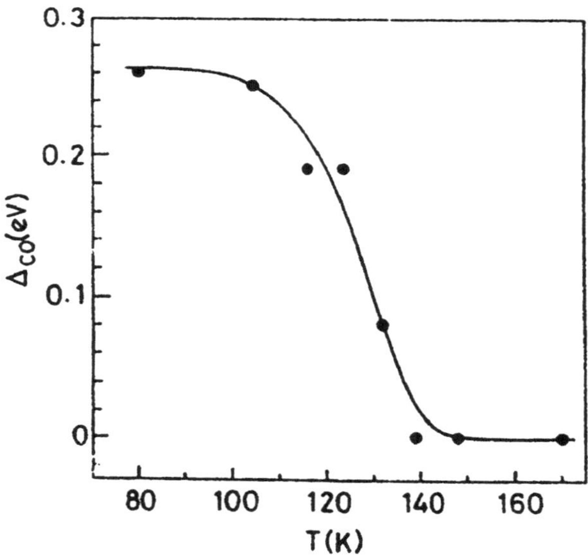

Fig. 22. Temperature dependence of the charge-ordering gap, Δ_{CO}, in Nd$_{0.5}$Sr$_{0.5}$MnO$_3$ (from Biswas et al[65b]).

Charge-ordering in the manganates is governed by the width of the e_g band which is directly determined by the weighted average radius of the A-site cations $\langle r_A \rangle$, or the tolerance factor. This is because a distortion of the Mn-O-Mn bond angle affects the transfer interaction of the e_g conduction electrons (holes). We can describe the spin and charge ordering phenomena in the rare-earth manganates in terms of the generalized phase diagram shown in Fig. 23a. The diagram shows that when $\langle r_A \rangle$ is large as in region A (e.g. $La_{1-x}Sr_xMnO_3$), only ferromagnetism and the associated I-M transition occur with no charge-ordering. With a slight decrease in $\langle r_A \rangle$ (region B), the ferromagnetic

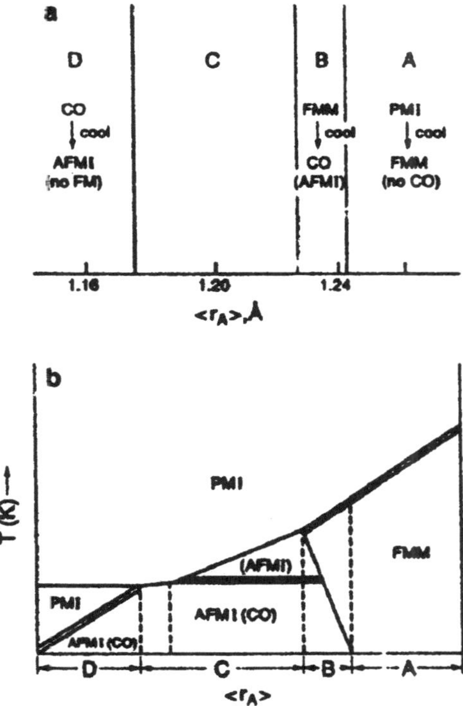

Fig. 23. (a) Schematic diagram showing the prevalence of charge-ordered (CO) and ferromagnetic states in $Ln_{1-x}A_xMnO_3$ depending on the weighted average radius of the A-site cation, $\langle r_A \rangle$ or the e_g bandwidth (b) Schematic temperature $\langle r_A \rangle$ diagram for $Ln_{0.5}A_{0.5}MnO_3$ where double line represents a second-order transition and a single line represents a first-order transition (from Kumar and Rao[66]).

36

(metallic) state transforms to the antiferromagnetic charge-ordered state on cooling as in $Nd_{0.5}Sr_{0.5}MnO_3$ ($\langle r_A \rangle = 1.236$Å). When $\langle r_A \rangle$ is very small as in region D as exemplified by $Pr_{0.7}Ca_{0.3}MnO_3$ and $Nd_{0.5}Ca_{0.5}MnO_3$ ($\langle r_A \rangle \sim 1.17$Å), no ferromagnetism is encountered and charge-ordering occurs in the paramagnetic state. The ferromagnetic metallic state can only be created by the application of the magnetic field in the charge-ordered state, depending on $\langle r_A \rangle$. In $Y_{0.5}Ca_{0.5}MnO_3$ with $\langle r_A \rangle = 1.13$Å, the charge ordered state is not perturbed by magnetic fields.

The dependence of charge ordering in $Ln_{1-x}A_xMnO_3$ on the radius of the A-site cations, $<r_A>$, shown in Fig. 23a is instructive and suggests at least two types of charge ordered states in regions B and D. It is remarkable that the range in $<r_A>$ giving such a vast variety of properties is only ~ 0.1Å. What actually happens, however, is that the Mn-O-Mn angle varies by $\sim 20^0$ over this region, the smallest angle of 148^0 being found in $Y_{0.5}Ca_{0.5}MnO_3$. The two types of charge ordering and the other. features of Fig. 23a can be understood in terms of the dependence of the exchange coupling constants J_{FM} and J_{AFM} on $<r_A>$, noting that the single ion Jahn-Teller energy is invariant.[66] We show a phase diagram worked out in this manner in Fig. 23b. The charge-ordered states in region D show some interesting features. Besides T_{CO} being in the paramagnetic state, the manganates in this small $<r_A>$ regime show negligible volume change at T_{CO}, and the unit cell volume decreases continuously from T_{CO} down to low temperatures. Furthermore, the bond valence sum shows charge localization only at low temperatures as shown in Fig. 24.[67]

Manganates in region C in the phase diagram (Fig. 23) are expected to show rather complex behavior. Thus, in $Ln_{1-x}Ca_xMnO_3$ ($0.63 \leq x \leq 0.67$) charge-ordering is accompanied by an increase in the sound velocity and anomalies in the heat capacity and derivative of the logarithm of resistivity.[68] Detailed studies[69] of $La_{0.5}Ca_{0.5}MnO_3$ have shown that Jahn-Teller distortion develops in the temperature region between 225K (T_c) and 155K (T_N), with the d_z^2 orbitals oriented perpendicular to the orthorhombic b-axis. Magnetic domain boundaries seem to break the coherence of the spin ordering on the Mn^{3+} sites while preserving the coherence of the spin ordering on the Mn^{4+} sublattice as well as the identity of the two sublattices. The similarity between these structures and the structural charge ordering and discommensuration domain boundaries suggests that in $La_{0.5}Ca_{0.5}MnO_3$, commensurate long-range charge ordering coexists with quasi-commensurate orbital ordering.

By subtle variations of $<r_A>$, one can bring manganates in the regions B and D of the phase diagram in Fig. 23 to region C. Thus, in

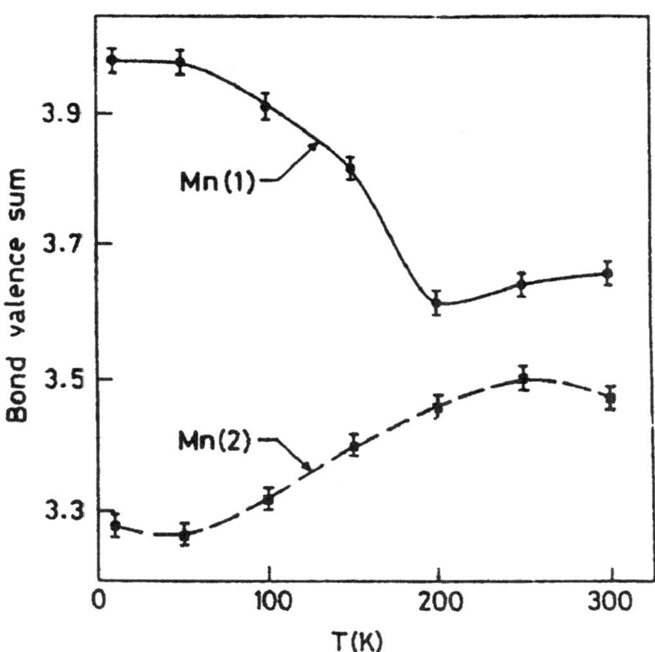

Fig. 24. Variation of bond valence sums in $Nd_{0.5}Ca_{0.5}MnO_3$ with temperature (from Vogt et al[67]).

$Nd_{0.5-x}Sm_xSr_{0.5}MnO_3$, a charge ordered state transforms to a ferromagnetic metallic state on cooling.[63] In $Nd_{0.25}La_{0.25}Ca_{0.5}MnO_3$ ($<r_A>$ = 1.19Å.), however, a novel reentrant ferromagnetic transition occurs from an incipient charge-ordered state on cooling; the transition is accompanied by a collapse of the charge order gap (Fig. 25), driven by a first order transition [70]. The incipient nature of the charge-ordered state is verified by the absence of superstructure peaks in the diffraction pattern.

Charge ordering has been found in quasi two-dimensional $La_{0.5}Sr_{1.5}MnO_4$ as well. Here, T_{CO} is 217K and T_N is 110K.[71] Below T_N, the magnetic moments get ordered with a $2\sqrt{2}$ a x $2\sqrt{2}$ a x $2c$ cell. Superlattice reflections corresponding to alternating Mn^{3+} and Mn^{4+} ions are seen below T_c. The size of the A-site cations would have negligible effect on the charge-ordering in these insulating two-dimensional manganates. It may be noted that in the perovskites, $Ln_{1-x}A_xMnO_3$ ($x \approx 0.5$), T_{CO} initially increases as $<r_A>$ decreases and then become nearly constant. Thus, T_{CO} = 0K in $La_{0.5}Sr_{0.5}MnO_3$ ($<r_A>$ = 1.23Å), 220K

38

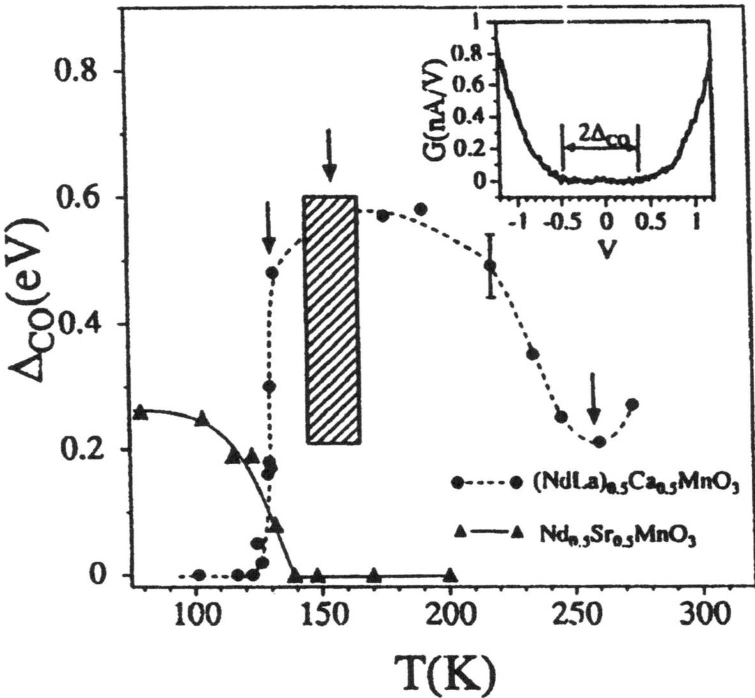

Fig. 25. Temperature variation of the charge-ordering gap, Δ_{CO}, measured by vacuum tunneling spectroscopy in $Nd_{0.5}Sr_{0.5}MnO_3$ (triangles) and $Nd_{0.25}La_{0.25}Ca_{0.5}MnO_3$ (circles). Inset shows a typical conductance curve (from Arulraj et al[70]).

in $La_{0.5}Ca_{0.5}MnO_3$ ($<r_A> = 120Å$) and 240K in $Nd_{0.5}Ca_{0.5}MnO_3$ ($<r_A> = 1.17Å$). The near constancy of T_{CO} is in the insulating regime.

Investigations on charge-ordering have established the intimate connection of the lattice distortion to CO. It is the lattice distortion associated with orbital ordering that appears to localize the charge and initiate charge-ordering. Eventually, the Coulomb interaction wins over the kinetic energy of the electrons to form long range CO state. The scale of the energy involved with CO as measured by the charge-ordering gap is around 0.5 - 1 eV. This is similar to the unscreened bare nearest neighbour Coulomb repulsion. This is also close to the approximate energy required to create ~1% orthorhombic distortion. It is likely that both the energy scales along with magnetic exchange decide the energy scale of the charge- ordering gap.

6. Other CMR Materials

CMR is found in $Tl_2Mn_2O_7$ which has only Mn^{4+} ions and hence without double-exchange. The material is an itinerant electron ferromagnet showing a FMM-PMM transition.[72] Recently, CMR has been found in a thiospinel, $Fe_{1-x}Cu_xCr_2S_4$, where there can be no double exchange.[73] The occurrence of CMR in nonstoichiometric EuO has been known for sometime[74]; in this oxide also, there is an insulator-metal transition around the ferromagnetic Curie temperature. Clearly, there is need to explore a variety of oxides, chalcogenides and other materials with or without double-exchange. Composites of manganates with other materials (see section 3.8) deserve further attention.

7. Future Prospects

Besides search for new and better CMR materials, there is scope of careful investigations of the various properties and phenomena exhibited by the manganates. These include a study of charge and orbital ordering[75] and the effect of electric fields and radiation on the charge ordered states. A transition from the insulating (charge ordered) state to the ferromagnetic metallic state has been induced in $Pr_{0.7}Ca_{0.3}MnO_3$ by x-ray illumination.[76] It has been possible to trigger the collapse of the charge-ordered state in $Pr_{0.7}Ca_{0.3}MnO_3$ by the application of an electric field as well.[77a] Laser irradiation is also found to induce an insulator-metal transition in this manganate.[77b] Structural studies of charge ordered manganates under magnetic fields would be interesting. Theoretical studies directed towards understanding the transport properties of manganates as well as charge-ordering[75] and related aspects are necessary.

Acknowledgement:

The authors thank the Science Office of the European Community, the Department of Science and Technology, Government of India and the CSIR (India) for support.

40

References

1. (a) P.M. Levy, *Solid StatePhys,* **47**, 367 (1994).
   (b) P.M. Levy and S. Zhang, J. *Mag. Mag. Mater.* **151**, 315 (1995).
2. N.B. Brandt andV.V. Moschalkov, *Adv. Phys.* 33, 193 (1984).
3. (a) R. von Helmolt et al, *Phys. Rev. Lett.* **71**, 2331 (1993).
   (b) K. Chahara et al, *Appl. Phys. Lett.* **63**, 1990 (1993).
   (c) M. McCormack et al, *Appl. Phys. Lett.* **64**, 3045 (1994).
   (d) R. Mahesh et al, *J. Solid State Chem.* **114**, 297 (1995).
   (e) C.N.R. Rao et al, *Chem. Mater.,* **8**, 2421 (1996) and the references therein.
4. A. Urushibara et al, *Phys. Rev.* **B 51**, 14103 (1995).
5. (a) G.H. Jonker and J.H. van Santen, *Physica* **16**, 377 (1950).
   (b) J.H. van Santen and G.H. Jonker, *Physica* **16**, 599 (1950).
6. J.W. Lynn et al, *Phys. Rev. Lett.* **76**, 4046 (1996).
7. C. Zener, *PhJs. Rev.* **82**, 403 (1951).
8. J.B. Goodenough, *Prog. Solid State Chem.* **5**, 149 (1971).
9. P.G. de Gennes, *Phys. Rev.* **118**, 141 (1960).
10. E.O. Wollan and W.C. Koehler, *Phys. Rev.* **100**, 545 (1955).
11. R. Mahendiran et al, *Phys. Rev.* **B 53**, 3348 (1996).
12. P. Schiffer et al, *Phys. Rev. Lett.* **75**, 3336 (1995).
13. P.G. Radaeli et al, *J. Solid State Chem.* **122**, 444 (1996).
14. I.O. Troyanchuk, *Sov. Phys. JETP* **75**, 132 (1992).
15. (a) J.L. Garcia-Munoz et al, J. *Phys. Rev.* **B 55**, 34 (1997).
    (b) L.M. Rodriguez-Martinez and J.P. Attfield, *Phys. Rev.* **B 54**, 15622 (1996).
16. P .G. Radaelli et al, *Phys. Rev.* **B 54**, 8992 (1996).
17. (a) J.A.M. van Roosmalen and E.H.P. Cordfunke, *J. Solid State Chem.* **110**, 109 (1994).
    (b) M. Hervieu et al, *Euro J. Solid State Chem.* **32**, 79 (1995).
18. R. Mahesh et al, *J. Solid State Chem.* **114**, 294 (1995).
19. A. Arulraj et al, *J. Solid State Chem.* **127**, 87 (1996).
20. R. Mahendiran et al, *J. Phys.* D. *Appl. Phys.* **28**, 1743 (1995).
21. R. Mahendiran et al, *Appl. Phys. Lett.* **66**, 233 (1995).
22. S.J.L. Brillinge et al, *Phys. Rev. Lett.* **77**, 715 (1996).
23. T.A. Tyson et al, *Phys. Rev.* **B 53**, 13985 (1996).
24. R.P. Sharma et al, *Phys. Rev.* **B 54**, 10014 (1996).
25. M.C. Martin et al, *Phys. Rev.* **B 53**, 14285 (1996).

26. (a) A. Asamitsu et al, *Phys. Rev.* **B 54**, 1716 (1996).
    (b) A.J. Campbell et al, *Phys. Rev.* **B 55**, 8622 (1997).
27. (a) H. Kaivano et al, *Phys. Rev.* **B 53**, 14709 (1996).
    (b) J.M. De Teresa et al, *J. Appl. Phys.* **B 79**, 5175 (1996).
28. Y. Yamada et al, *Phys. Rev. Lett.* **77**, 904 (1996).
29. J.M. De Teresa et al, *Nature* **386**, 256 (1997).
30. (a) Y. Moritomo et al, *Phys. Rev.* **B 51**, 16491 (1995).
    (b) J.J. Neumeier et al, *Phys. Rev.* **B 52**,7006 (1995).
31. H. Y. Hwang et al, *Phys. Rev. Lett.* **75**, 914 (1995).
32. B. Raveau et al, *J. Solid State Chem.* **117**, 424 (1995).
33. J. Fontcuberta et al, *Phys. Rev. Lett.* **76**, 1122 (1996).
34. (a) M.R.Ibarra et al, *Phys. Rev. Lett.* **75**, 3541 (1995).
    (b) A. Maignan et al, *J. Appl. Phys.* **79**, 7891 (1996).
35. R. Mahendiran et al, *Phys. Rev.* **B 53**, 12160 (1996).
36. (a) J.L. Garcia-Munoz et al, *Phys. Rev.* **B 55**, 668 (1997).
    (b) G.H. Rao et al, *Phys. Rev.* **B 55**, 3742 (1997).
37. R. Mahesh et al, *J. Solid State Chem.* **120**, 204 (1995).
38. (a) R. Mahesh et al, *Appl. Phys. Lett.* **68**, 2291 (1996).
    (b) R. Mahendiran et al, *Solid State Commun.* **99**, 149 (1996).
39. (a) Y. Moritomo et al, *Nature* **380**, 141 (1996).
    (b) R. Mahesh et al, *J. Solid State Chem.* **122**, 448 (1996).
40. T. Kimura et al, *Science* **274**, 1698 (1996).
41. (a) D.N. Argyriou et al, *Phys. Rev. Lett.* **78**, 1568 (1997).
    (b) T.G. Perring et al, *Phys. Rev. Lett.* **78**, 3197 (1997).
42. P.D. Battle et al, *Phys. Rev.* **B 54**, 15967 (1997).
43. R.H. Heffner et al, *Phys. Rev. Lett.* **77**, 1869 (1996).
44. (a) G.H. Rao et al, *Appl. Phys. Lett.* **69**, 424 (1996).
    (b) H.Y. Hwang et al, *Appl. Phys. Lett.* **68**, 3494 (1996).
45. J.M.D. Coey et al, *Phys. Rev. Lett.* **75**, 3910 (1995).
46. (a) Y. Okimoto et al, *Phys. Rev. Lett.* **75**, 109 (1995).
    (b) Y. Okimoto et al, *Phys. Rev.* **B 55**, 4206 (1997).
47. S.G. Kaplan et al, *Phys. Rev. Lett.* **77**, 2081 (1996).
48. A. Biswas and A.K. Raychaudhuri, *J. Phys. Condens. Matter.* **8**, L739 (1996).
49. (a) D.D. Sarma et al, *Phys. Rev.* **B 53**, 6873 (1996).
    (b) T. Saitoh et al, *Phys. Rev.* **B 51**, 13942 (1995).
50. (a) J.H. Park et al, *J. Appl. Phys.* **79**, 4558 (1996).
    (b) J.H. Park et al, *Phys. Rev. Lett.* **76**, 4215 (1996).

42

51. (a) J.J. Hamilton et al, *Phys. Rev.* **B 54**, 14926 (1996).
    (b) B.F. Woodfield et al, *Phys. Rev. Lett.* **78**, 3201 (1997).
    (c) R. Gupta et al, *Phys. Rev.* **B 54**, 14899 (1996).
52. (a) J.G. Snyder et al; *Phys. Rev.* **B53**,14434 (1996).
    (b) R. Mahendiran et al, *Solid State Commun.* **98**, 701 (1996); also see
       R. Mahendiran et al, *Phys. Rev.* **B 54**, 9604 (1996).
53. (a) A. Asamitsu et al, *Phys. Rev.* **B 53**, 2952 (1996).
    (b) W. Archibald et al, *Phys. Rev.* **B 53**, 14445 (1996).
54. B. Fisher et al, *Phys. Rev.* **B 54**, 9359 (1996).
55. N. Furukawa, *J. Phys. Soc. Jpn.* **63**, 3214 (1994).
56. R. Gundakaram et al, *J. Solid State Chem.* **127**, 354 (1996).
57. A.J. Millis et al, *Phys. Rev. Lett.* **74**, 5144 (1995).
58. H. Roder et al, *Phys. Rev. Lett.* **76**, 1356 (1996).
59. G. Zhao et al, *Nature* **381**, 676 (1996).
60. S. Satpathy et al, *Phys. Rev. Lett.* **76**, 960 (1996).
61. Y. Tokura et al, *J. Appl. Phys.* **79**, 5288 (1996).
62. H. Kuwahara et al, *Science* **270**, 961 (1995).
63. Y. Tokura et al, *Phys. Rev. Lett.* **76**, 3184 (1996).
64. Y. Tomioka et al, *Phys. Rev.* **B 53**, 1689 (1996).
65. (a) Y. Moritomo et al, *Phys. Rev.* **B 55**,7549 (1997).
    (b) A. Biswas et al, *J. Phys. Condens. Matter* **9**, L355 (1997).
66. N. Kumar and C.N.R. Rao, *J. Solid State Chern.* **129**, 363 (1997).
67. T. Vogt et al, *Phys. Rev.* **B 54**, 15303 (1996).
68. A.P. Ramirez et al, *Phys. Rev. Lett.* **76**, 3188 (1996).
69. P.G. Radaelli et al, *Phys. Rev.* **B 55**, 3015 (1997).
70. A. Arulraj et al, To be published.
71. B.J. Sternlieb et al, *Phys. Rev. Lett.* **76**, 2169 (1996).
72. (a) M.A. Subramanian et al, *Science* **273**, 81 (1996).
    (b) D.K. Seo et al, *Solid State Commun.* **101**, 417 (1997).
73. A.P. Ramirez et al, *Nature* **386**,186 (1997).
74. Y. Shapira et al, *Phys. Rev.* **B 8**, 2299, 2316 (1973).
75. C.N.R. Rao and A.K. Cheetham, *Science* **276**, 911 (1997).
76. V. Kiryukhin et al, *Nature* **386**, 813 (1997).
77. (a) A. Asamitsu et al, *Nature* **388**, 50 (1997).
    (b) K. Miyano et al, *Phys. Rev. Lett.* **78**, 4257 (1997).

J. Phys. Chem. B **2000**, *104*, 5877−5889

FEATURE ARTICLE

Charge, Spin, and Orbital Ordering in the Perovskite Manganates, $Ln_{1-x}A_xMnO_3$ (Ln = Rare Earth, A = Ca or Sr)

C. N. R. Rao*

Chemistry & Physics of Materials Unit and the CSIR Centre of Excellence in Chemistry, Jawaharlal Nehru Centre for Advanced Scientific Research, Jakkur P.O, Bangalore 560 064, India

Received: February 7, 2000; In Final Form: March 29, 2000

Charge ordering occurs in some mixed-valent transition metal oxides. The perovskite manganates of the formula $Ln_{1-x}A_xMnO_3$ (Ln = rare earth; A = Ca, Sr) are especially interesting because long-range ordering of the Mn^{3+} ($t_{2g}^3 e_g^1$) and Mn^{4+} ($t_{2g}^3 e_g^0$) ions in these materials is linked to antiferromagnetic spin ordering, and also to the long-range ordering of the Mn^{3+} (e_g) orbitals and the associated lattice distortions. Charge ordering occurs at a higher temperature than spin ordering in some of the manganates ($T_{CO} > T_N$), whereas in some others $T_{CO} = T_N$. Orbital ordering occurs without charge ordering in the A-type antiferromagnetic manganates, but in the manganates where charge ordering occurs, antiferromagnetism of CE-type is found along with orbital ordering. The subtle relations between charge, spin, and orbital ordering are discussed in the article, with special attention to the effects of cation size, chemical substitution, dimensionality, pressure, and magnetic and electric fields. Unusual features such as phase separation and electron−hole asymmetry are also examined.

Introduction

Charge ordering is a phenomenon generally observed in mixed-valent transition metal oxides. When differently charged cations (i.e., 2+ and 3+) in an oxide order on specific lattice sites, the hopping of electrons between the cations is no longer favored. One therefore observes an increase in the electrical resistivity at the charge-ordering transition, often accompanied by a change in crystal symmetry. Because transition metal ions also carry spins, it is interesting to examine the magnetic (spin) ordering in the solids in relation to charge ordering. A well-known example of charge ordering is found in Fe_3O_4 (magnetite) where it occurs at a temperature well below spin ordering.[1,2] Charge and spin ordering in real space in metal oxides received renewed attention because of their role in cuprate superconductors.[3] In recent years, charge and spin ordering have been discovered in a few other transition metal oxides,[4] typical examples being $La_{1-x}Sr_xFeO_3$, $La_{2-x}Sr_xNiO_4$, and $LiMn_2O_4$. Charge ordering in the rare earth manganates of the perovskite structure, with the general composition $La_{1-x}A_xMnO_3$ (Ln = rare earth; A = alkaline earth), is considerably more interesting, because it is closely associated with spin and orbital ordering, giving rise to fascinating properties.[5−7]

The perovskite manganates originally became popular because of the discovery of colossal magnetoresistance (CMR).[5,6] CMR and related properties essentially arise from the double-exchange mechanism of electron hopping between the Mn^{3+} ($t_{2g}^3 e_g^1$) and Mn^{4+} ($t_{2g}^3 e_g^0$) ions.[8] In this mechanism, lining up of the spins (ferromagnetic alignment) of the incomplete e_g orbitals of the adjacent Mn ions is directly related to the rate of hopping of the electrons, giving rise to an insulator−metal transition in the material at the ferromagnetic Curie temperature, T_c. In the

ferromagnetic phase ($T < T_c$), the material becomes metallic, but is an insulator in the paramagnetic phase ($T > T_c$). In the insulating phase, a Jahn−Teller distortion associated with the Mn^{3+} ions favors the localization of electrons. Charge ordering of the Mn^{3+} and Mn^{4+} ions competes with double-exchange and promotes insulating behavior and antiferromagnetism. It may be recalled that the $Mn^{3+}-O-Mn^{3+}$ and $Mn^{4+}-O-Mn^{4+}$ superexchange interaction, through the e_g orbitals, is antiferromagnetic. Although charge ordering in the rare earth manganates was investigated by Jirak et al.[9] as early as 1985, the subject has received renewed attention only in the past 5 years, for reasons described below.

Charge ordering occurs through a fairly wide range of compositions of $Ln_{1-x}A_xMnO_3$, provided the Ln and A ions are not too large. Large Ln and A ions (e.g., La, Sr) favor ferromagnetism and metallicity, whereas the smaller ones (e.g., La, Ca, or Pr, Ca) favor charge ordering. Charge ordering and spin (antiferromagnetic) ordering may or may not occur at the same temperature. Besides, the Mn^{3+} (d_{z^2}) orbitals and the associated lattice distortions develop long-range order (as illustrated later in this article). Such orbital ordering may or may not occur with charge ordering in the manganates, but it generally accompanies antiferromagnetic (spin) ordering. In this article, we discuss the interplay of charge, spin, and orbital ordering in the rare earth manganates in some detail, and highlight some of the important recent results. Before discussing the manganates, we shall briefly review the properties of a few of the other transition metal oxides that exhibit charge ordering.

Examples of Charge and Spin Ordering in Oxides. Fe_3O_4 has a spinel structure, represented as $Fe^{3+}[Fe^{2+}, Fe^{3+}]O_4$ in which one third of the cations are tetrahedrally coordinated (A-sites) and two thirds of the cations are octahedrally coordinated (B-sites). It is ferrimagnetic below 858 K (T_N), with the spins

---

* For correspondence: e-mail: cnrrao@jncasr.ac.in.

10.1021/jp0004866 CCC: $19.00   © 2000 American Chemical Society
Published on Web 06/02/2000

5878 *J. Phys. Chem. B, Vol. 104, No. 25, 2000*

Rao

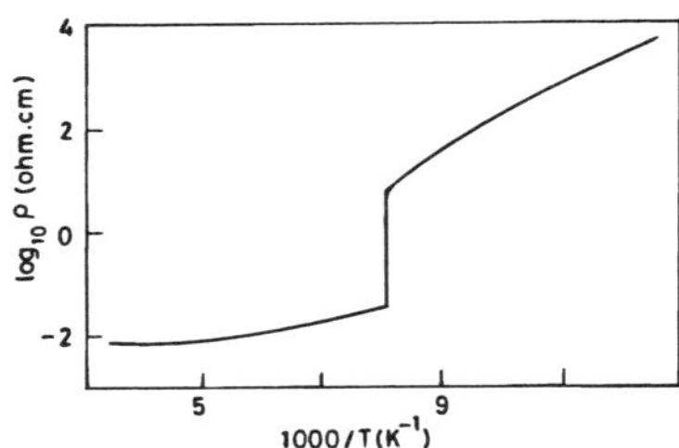

Figure 1. Verwey transition in Fe_3O_4 caused by charge ordering (from Honig[1b]).

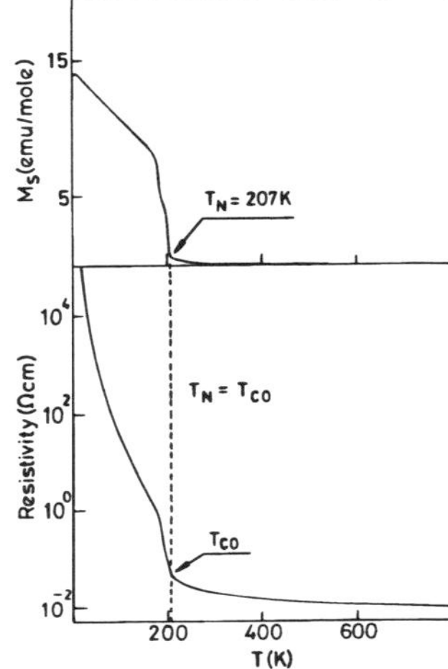

Figure 2. Temperature variation of electrical resistivity and magnetization of $La_{0.33}Sr_{0.67}FeO_3$ showing marked changes on charge ordering ($T_{CO} = T_N$) (from Park, Yamaguchi and Tokura, as quoted by Imada et al.[4]).

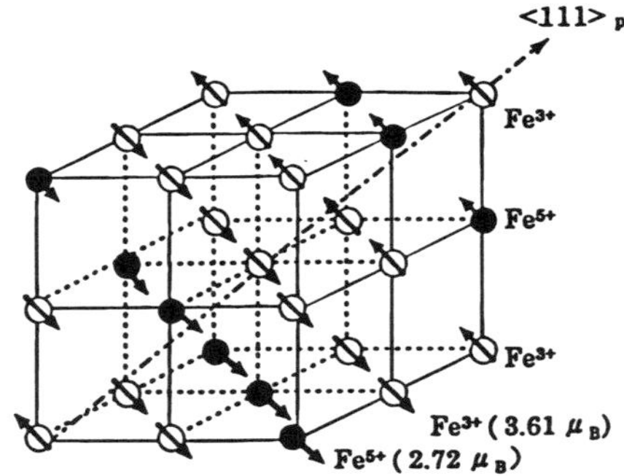

Figure 3. Structure showing charge and spin ordering in $La_{0.33}Sr_{0.67}$-FeO_3 (from Battle et al.[10b]).

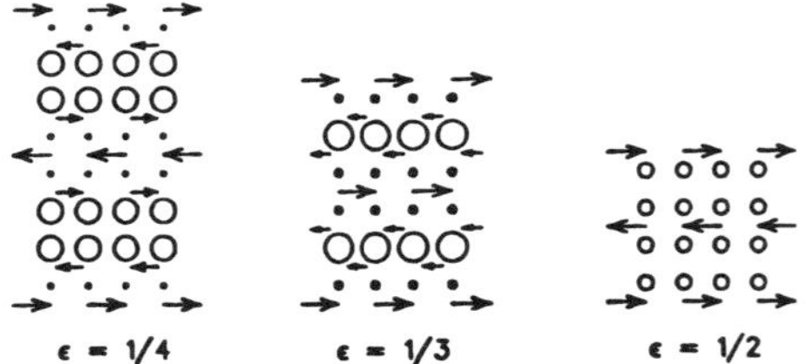

Figure 4. Spin and charge modulation in the NiO_2 planes of $La_{2-x}Sr_xNiO_{4+y}$ with increasing hole concentration. Here, the modulation wavenumber, ϵ, is equivalent to the hole concentration, $p = x + 2y$ (from Tranquada et al.[11a]). Spin density on nickel ions is shown by arrows, and charge density on oxygen ions is shown by circles.

of the A- and B-sublattices being antiparallel. Around 120 K (T_V), Fe_3O_4 shows a sharp increase in resistivity, commonly referred to as the Verwey transition (Figure 1). Below T_V, the Fe^{2+} and Fe^{3+} ions are considered to be ordered, thereby giving rise to high electrical resistivity. Above T_V, conduction occurs on the B site sublattice. It actually turns out, however, that charge ordering in Fe_3O_4 is much more complicated, with some short-range ordering present even above T_V. Even today, there is argument about the symmetry of the low-temperature ordered phase. The Verwey transition in this oxide has been reviewed excellently by Honig[1] and Tsuda et al.[2]

$La_{1-x}Sr_xFeO_3$ ($x = 0.67$) undergoes charge ordering at approximately 207 K (T_{CO}), at which temperature it also exhibits antiferromagnetic spin ordering. Thus, T_{CO} is also the Néel temperature (T_N) in this oxide. The resistivity shows a marked increase at $T_{CO} = T_N$ (Figure 2). Although the formula requires the presence of Fe^{3+} and Fe^{4+} ions, the Fe^{4+} ions disproportionate to give Fe^{3+} and Fe^{5+} ions.[10] The charge- and spin-ordered structure of $La_{0.33}Sr_{0.67}FeO_3$ is shown in Figure 3.

Quasi two-dimensional $La_{2-x}Sr_xNiO_{4+y}$ undergoes cooperative ordering of the Ni spins and of the charge carriers below a temperature (T_{CO}). Charge ordering occurs at nearly all doping levels ($p = x + 2y$), in the insulating regime of the material ($p < 0.7$). Spin and charge modulations are observed in the NiO_2 plane and these vary with the hole concentration (p value). This nickelate system is a typical instance of microscopic phase separation wherein the charge carriers (holes) localize in the domain walls in an antiferromagnetic system. Such phase

separation causes stripe modulations. Here, the ordering may be viewed as alternating stripes of antiferromagnetic and hole-rich regions.[11] In Figure 4, we show the spin and charge modulations in the NiO_2 planes at two hole concentrations. The $\epsilon = 1/4$ case corresponds to $La_2NiO_{4.125}$ ($p = 0.25$), where the interstitial oxygens form a superlattice with a unit cell of $3a \times 5b \times 5c$. The compositional dependence of resistivity of $La_{2-x}Sr_xNiO_4$ shows peaks at $x = 0.33$ and 0.25, owing to charge ordering, which in the $x = 0.33$ composition occurs at 240 K. At 240 K, various properties show anomalies, as depicted in Figure 5. Superlattice peaks show up in the electron diffraction pattern at this temperature. Antiferromagnetic spin ordering in the nickelate occurs at a lower temperature (180 K). It appears that ordering is driven by charge. Charge ordering of holes, accompanied by spin ordering or the segregation of holes and spins in the stripe form, also occurs in $La_{2-x}A_xCuO_4$ (A = Sr, Ba, $x = 0.125$), causing anomalous suppression of superconductivity.[3]

$LiMn_2O_4$ is a spinel with equal proportions of 3+ and 4+ Mn cations. It has been shown recently to undergo a Verwey-type transition with a resistivity anomaly at approximately 290 K where there is a structural (cubic-tetragonal) transition. Around 65 K, the material exhibits a complex long-range magnetic order.[13] Clearly, this charge-ordering transition requires further study.

A Brief Description of the Different Types of Ordering in the Manganates. $LaMnO_3$ has a layered antiferromagnetic structure, referred to as A-type antiferromagnetic ordering[14] (Figure 6a). Because there is a doubly degenerate e_g orbital in each Mn^{3+} ion ($t_{2g}^3 e_g^1$), $LaMnO_3$ and the other analogous rare earth compounds such as $NdMnO_3$ show a fine interplay between spin and orbital ordering. The orbital ordering is coupled to the Jahn–Teller (JT) distortion. Figure 6b describes the JT distortion in $LaMnO_3$. The distortion disappears above

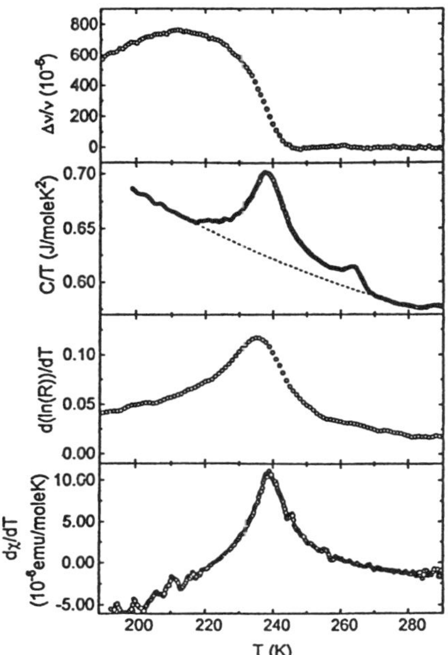

Figure 5. Normalized sound velocity ($\Delta v/v$), specific heat (c), as well as temperature derivatives of resistivity (ρ) and magnetic susceptibility (χ) showing anomalies at the charge ordering transition (240 K) in $La_{1.67}Sr_{0.33}NiO_4$ (from Ramirez et al.[12]).

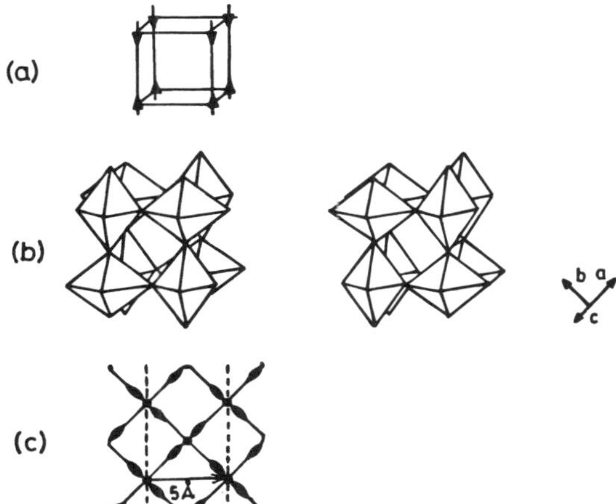

Figure 6. (a) A-type AFM ordering, (b) JT distortion, and (c) orbital ordering in $LaMnO_3$. The orientation of e_g orbitals shown in panel c is consistent with the 5 Å lattice spacing.

750 K. Orbital ordering of $3x^2 - r^2$ or $3y^2 - r^2$ type accompanied by the JT distortion leads to a superexchange coupling in $LaMnO_3$, which is ferromagnetic (FM) in the planes and antiferromagnetic (AFM) between the planes (Figure 6a).[15] Orbital ordering in $LaMnO_3$ is shown in Figure 6c. Without the JT distortion, $LaMnO_3$ would have been a FM insulator; it is an A-type AFM insulator instead.

In $Ln_{1-x}A_xMnO_3$, besides orbital and spin ordering, we can have charge ordering because of the presence of Mn^{3+} and Mn^{4+} ions. Small Ln and A ions stabilize the charge-ordered (CO) state. Thus, $Pr_{0.7}Ca_{0.3}MnO_3$, charge-orders around 230 K in the paramagnetic state, becoming AFM at 170 K; it is an insulator and does not show ferromagnetism in the absence of a strong

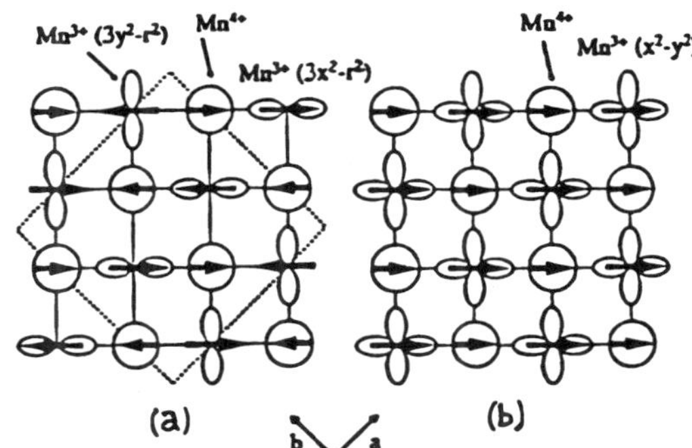

Figure 7. Charge, spin, and orbital ordering in (a) CE-type and (b) A-type AFM $Ln_{1-x}A_xMnO_3$. In part a, the broken line shows the unit cell for the CE-type AFM CO order. Mn^{4+} ions are shown by circles.

magnetic field. $La_{0.7}Ca_{0.3}MnO_3$, on the other hand, is a FM metal below the Curie temperature ($T_c \approx 230$ K) and a paramagnetic insulator above T_c. $La_{0.5}Sr_{0.5}MnO_3$ is metallic both in the FM and paramagnetic states, whereas $Nd_{0.5}Sr_{0.5}MnO_3$ shows a transition from a ferromagnetic metallic (FMM) state to an AFM charge-ordered state on cooling. The CO states in these manganates are associated with CE-type AFM ordering.[14] In the CE-type ordering, Mn^{3+} and Mn^{4+} ions are arranged as in a checker board and the Mn^{3+} sites are JT distorted.[15] Along the c-axis, the in-plane arrangement mentioned above gets stacked and the neighboring planes are coupled antiferromagnetically. The exchange coupling between the Mn^{3+} and Mn^{4+} ions depends on the type of e_g orbital at the Mn^{3+} site, and hence the nature of orbital ordering becomes important. The CE-type AFM CO state in $Ln_{1-x}A_xMnO_3$ is associated with the ordering of $3x^2 - r^2$ or $3y^2 - r^2$ type orbitals at the Mn^{3+} site. The JT distortion that follows such orbital ordering stabilizes the CE-type AFM state (relative to the FMM state). In Figure 7a, we show the spin, charge, and orbital ordering in the CE-type AFM state. The CO states in the manganates exhibit CE-type AFM ordering at the same temperature as the charge-ordering transition or at a lower temperature ($T_{CO} \geq T_N$). That is, spin ordering occurs concurrently or after charge ordering. Complete orbital ordering is achieved when there is both charge and spin ordering. Orbital and spin ordering occur without charge ordering in some of the manganates showing A-type antiferromagnetism.

The A-type AFM state described earlier in relation to $LaMnO_3$ (Figure 6a) is also encountered in the $Ln_{1-x}A_xMnO_3$ system. This state generally is not accompanied by charge ordering, because some electron transfer can occur between the Mn cations in the ab plane. Orbital ordering in A-type AFM ordering is depicted in Figure 7b. Here, the $x^2 - y^2$ type orbital is present at the Mn^{3+} site. $Pr_{0.5}Sr_{0.5}MnO_3$ transforms from a FMM state to a A-type AFM state on cooling. $Nd_{0.45}Sr_{0.55}MnO_3$ is a A-type antiferromagnet, unlike $Nd_{0.5}Sr_{0.5}MnO_3$, which is a CE-type antiferromagnet exhibiting charge ordering at low temperatures; the former shows conductivity in the ab plane and is not charge-ordered.

Evidence for charge ordering can be observed in crystal structures at low temperatures. Thus, in $La_{0.5}Ca_{0.5}MnO_3$, the Mn^{4+} environment is nearly isotropic, with all the Mn–O distances being nearly equal ($\sim$1.92 Å). In the $Mn^{3+}O_6$ octahedra, one of the Mn–O bonds is much longer ($\sim$2.07 Å) than the other bonds ($\sim$1.92 Å), consistent with orbital ordering. It must be noted that the Mn–O distances in the ab plane of the manganates are much longer than in the c-direction,

5880 *J. Phys. Chem. B, Vol. 104, No. 25, 2000*

Rao

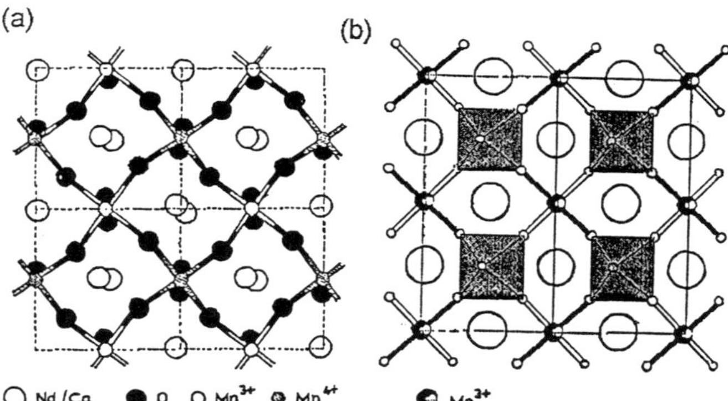

Figure 8. (a) Structure of charge-ordered $Nd_{0.5}Ca_{0.5}MnO_3$ in the *ab* plane at 10 K. Mn^{4+} (black) located at (1/2, 0, 0) and Mn^{3+} (grey) located (0, 1/2, 0) can be distinguished. The structure has zigzag chains with alternate long and short Mn–O distances. The distances are 1.921 and 2.021 Å $\langle 110 \rangle$ and 1.881 and 2.020 Å along $\langle -110 \rangle$ (from Vogt et al.[16]). (b) Charge-ordered structure of $Nd_{0.5}Sr_{0.5}MnO_3$. The Mn^{3+} ions are the filled circles, and Nd/Sr ions are large open circles. Oxygens are small open circles. The long Mn^{3+}–O bonds shown in black also represent the orientation of the e_g orbitals. Mn^{4+}–O octahedra are shown in polyhedral representation (from Woodward et al. [50]).

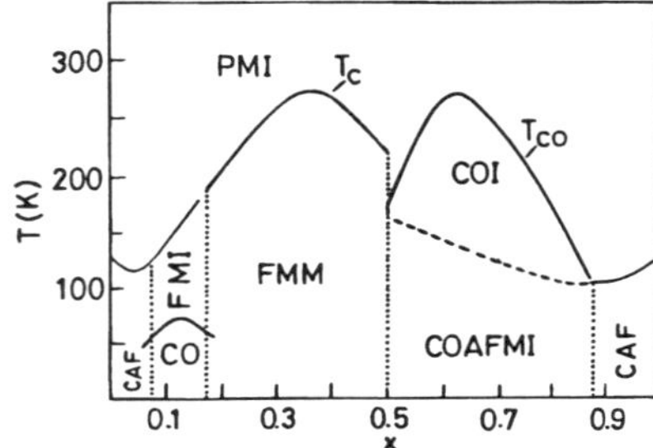

Figure 9. Phase diagram of $La_{1-x}Ca_xMnO_3$ (following Cheong). CAF, canted antiferromagnet; CO, charge-ordered phase; FMI, ferromagnetic insulator; PMI, paramagnetic insulator; FMM, ferromagnetic metallic state; CO AFMI, charge-ordered antiferromagnetic insulator; COI, charge-ordered insulator. Notice electron hole asymmetry by comparing the $x < 0.5$ and $x > 0.5$ regimes.

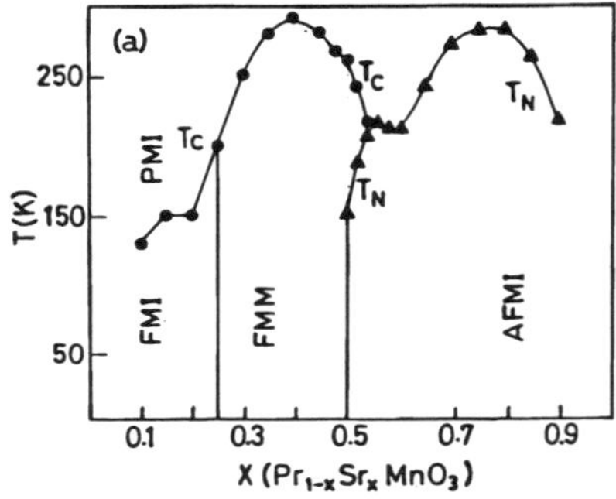

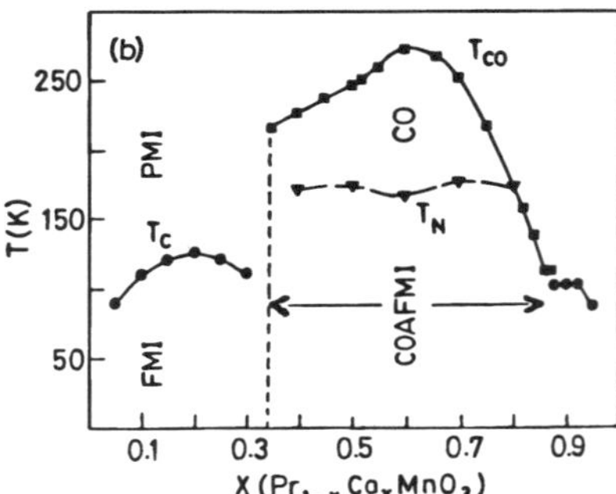

Figure 10. Phase diagrams of (a) $Pr_{1-x}Sr_xMnO_3$ and (b) $Pr_{1-x}Ca_xMnO_3$. Notice the wide charge ordering regime and electron−hole asymmetry in part b and the absence of charge ordering in part a. In part b, there is spin-glass or CAF behavior when $x \geq 0.8$.

particularly in the AFM state. In Figure 8a, we show the projection of the structure of charge-ordered $Nd_{0.5}Ca_{0.5}MnO_3$ down the *c*-axis to illustrate the definitive identification of the unique sites occupied by Mn^{3+} and Mn^{4+} ions in the CO state. The structure of charge-ordered $Nd_{0.5}Sr_{0.5}MnO_3$, where the Mn^{4+}-centered octahedra are represented in polyhedral notation, is shown in Figure 8b.

Representative Phase Diagrams of the Manganates. In Figure 9 we show the phase diagram of $La_{1-x}Ca_xMnO_3$. In this system, charge ordering occurs in the $x \approx 0.5-0.8$ range. In Figure 10 we show the phase diagrams of $Pr_{1-x}Ca_xMnO_3$ and $Pr_{1-x}Sr_xMnO_3$. The latter system shows no charge ordering, but $Pr_{1-x}Ca_xMnO_3$ exhibits charge ordering over the $x \approx 0.3-0.8$ range. Note that $Pr_{0.7}Ca_{0.3}MnO_3$ exhibits charge ordering but $La_{0.7}Ca_{0.3}MnO_3$ does not. All such variations are essentially due to the effect of the size of the A-site cations, the smaller size favoring charge ordering. From Figures 9 and 10, we also see that the CO regime is prominent at large *x*. In fact, the $x > 0.5$ compositions in $Ln_{1-x}Ca_xMnO_3$ are almost entirely in the CO regime both when Ln = La and Pr. This regime can be considered as the electron-doped regime (substitution of trivalent rare earth in $CaMnO_3$), whereas the $x < 0.5$ compositions may be considered as the hole-doped regime (substitution of divalent Ca in $LnMnO_3$). Clearly, there is electron−hole asymmetry in these manganates.

It is surprising that ferromagnetism is not encountered in the electron-doped regime ($x > 0.5$). Effects of magnetic and electric

fields on the hole- and electron-doped manganates (e.g., $Pr_{0.7}Ca_{0.3}MnO_3$ and $Pr_{0.3}Ca_{0.7}MnO_3$) are also different. There are some similarities between the hole- and electron-doped regimes. For example, in $Pr_{1-x}Ca_xMnO_3$, the charge-ordering transition temperature increases with hole concentration in the $0.3 < x \leq 0.5$ regime and with electron concentration in the $0.5 \leq x \leq 0.8$ regime.

Case Studies. To understand typical scenarios of charge-ordered manganates, it is instructive to examine the properties of two manganates with different sizes of the A-site cations. For this purpose, we choose $Nd_{0.5}Sr_{05}MnO_3$ with a weighted average radius of the A-site cations, $\langle r_A \rangle$, of 1.24 Å and $Pr_{0.6}Ca_{0.4}MnO_3$ with an $\langle r_A \rangle$ of 1.17 Å (Shannon radii are used here).

J. Phys. Chem. B, Vol. 104, No. 25, 2000 **5881**

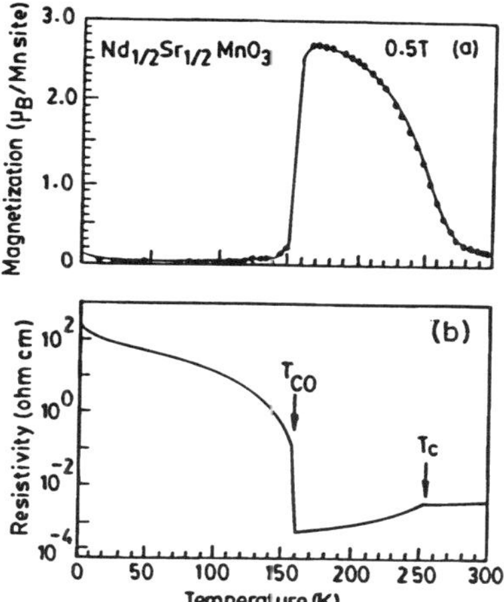

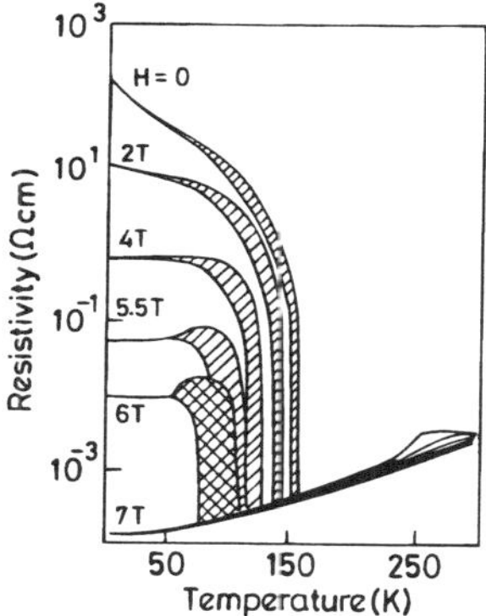

Figure 11. Temperature variation of (a) the magnetization and (b) the resistivity of $Nd_{0.5}Sr_{0.5}MnO_3$ (from Kuwahara et al.[17]).

Figure 12. Effect of magnetic fields on the charge ordering transition of $Nd_{0.5}Sr_{0.5}MnO_3$ (from Kuwahara et al.[17]).

$Nd_{0.5}Sr_{0.5}MnO_3$ is a ferromagnetic metal with a T_c of $\sim$250 K and transforms to an insulating CO state at about 150 K (Figure 11). The CO transition is accompanied by spin ordering, and the CO insulator is a CE-type antiferromagnet.[17] The orbital ordering therefore involves $3x^2 - r^2/3y^2 - r^2$ orbitals as in Figure 7a. Application of a magnetic field of 7T destroys the CO state, and the material becomes metallic (Figure 12), the sharpness of the transition decreasing with increasing strength of the field. The transition is first-order, showing hysteresis, and is associated with changes in unit cell parameters. The unit cell volume of the CO state is considerably smaller than that of the FMM state. The properties of $Nd_{0.5}Sr_{0.5}MnO_3$ can be described in terms of the phase diagram shown in Figure 13. The *Imma* space group of this manganate renders the Mn$-$O$-$Mn angle in the *ab* plane closer to 180°, promoting the overlap of the Mn(e_g) and O(2p) orbitals. Vacuum tunneling measurements[19] show that a gap of 250 meV opens up below T_{CO} (Figure 14). The gap collapses on applying a magnetic field, suggesting that a gap in the density of states at E_F is necessary for the stability of the CO state. Photoemission studies indicate a sudden change in electron states at the transition and give an estimate of 100 meV for the gap.[20] These estimates of the gap are considerably larger than the T_{co} (12 meV); it is not clear how a magnetic field of 6T (1.2 meV) can destroy the CO state. $Nd_{0.5}Sr_{0.5}MnO_3$ shows anomalous magnetostriction behavior, with a large positive magnetovolume effect (Figure 15), owing

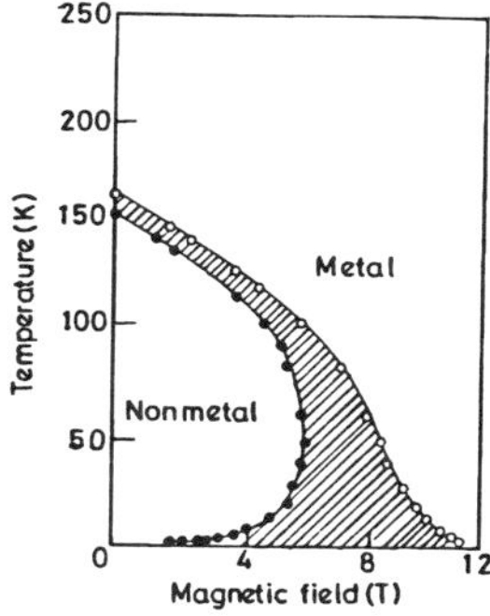

Figure 13. Temperature-magnetic field phase diagram for $Nd_{0.5}Sr_{0.5}$-MnO_3 (from Tokura et al.[18]).

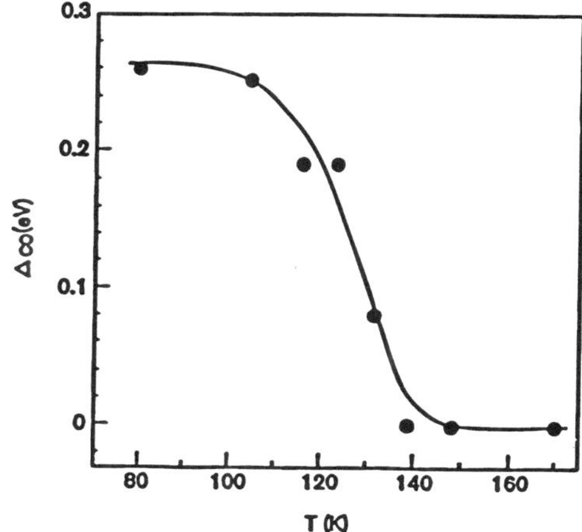

Figure 14. CO gap in $Nd_{0.5}Sr_{0.5}MnO_3$ as revealed by vacuum tunneling measurements (from Biswas et al.[19]).

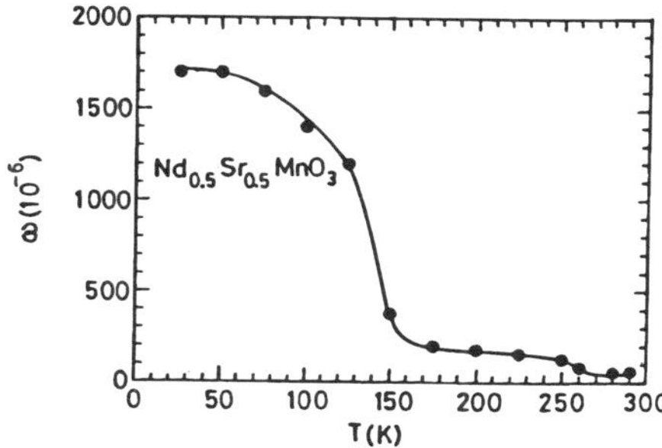

Figure 15. Temperature variation of the maximum volume magnetostriction in $Nd_{0.5}Sr_{0.5}MnO_3$ at 13.7T (from Mahendiran et al.[21]).

to the magnetic field-induced structural transition accompanying a change from the AFM CO state to the FMM state.[21]

$Pr_{0.6}Ca_{0.4}MnO_3$ is an insulator at all temperatures and becomes charge-ordered at about 230 K (T_{CO}). At this temperature, anomalies are found in the magnetic susceptibility, as well as in the resistivity, as shown in Figure 16. In the CO state, the Mn^{3+} and Mn^{4+} ions are regularly arranged in the *ab* plane with the associated ordering of the $3x^2 - r^2/3y^2 - r^2$ orbitals. On cooling, AFM ordering (CE-type) occurs at 170 K (T_N). At about 40 K, $Pr_{0.6}Ca_{0.4}MnO_3$ exhibits canted AFM ordering. Application of an external magnetic field transforms the CO state to a FMM state, as shown in Figure 16, but the field required is much larger than in $Nd_{0.5}Sr_{0.5}MnO_3$. The transition

5882 *J. Phys. Chem. B, Vol. 104, No. 25, 2000*   Rao

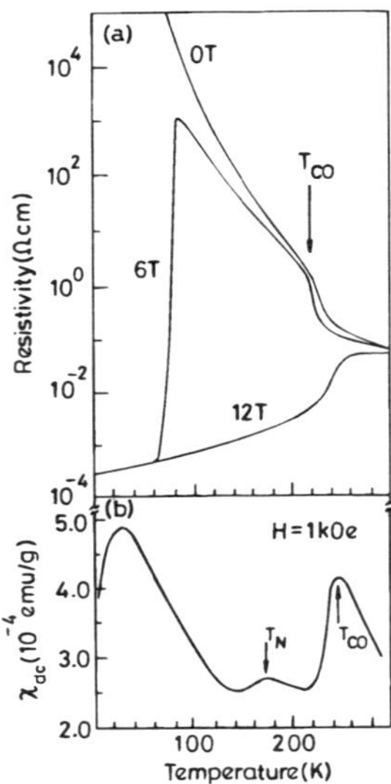

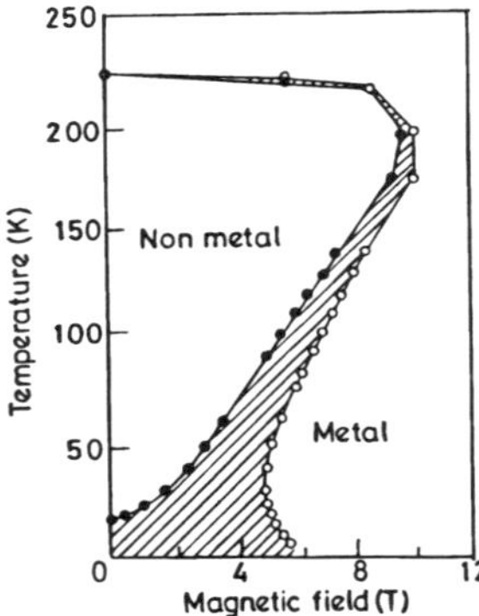

Figure 16. Temperature variation of (a) resistivity and (b) magnetic susceptibility of $Pr_{0.6}Ca_{0.4}MnO_3$ (from Tomioka et al.[22a] and Lees et al.[22b]).

Figure 17. Temperature-magnetic field phase diagram for $Pr_{0.6}Ca_{0.4}MnO_3$ (from Tokura et al.[18]).

is associated with hysteresis. The properties of $Pr_{0.6}Ca_{0.4}MnO_3$ can be represented by the phase diagram in Figure 17. The basic features of the CO state in $Pr_{0.6}Ca_{0.4}MnO_3$ are exhibited by several other rare earth manganates with relatively small A-size cations, in that the CO state is the ground state. Thus, $Nd_{0.5}Ca_{0.5}MnO_3$ ($\langle r_A \rangle = 1.17$ Å) is a paramagnetic insulator with a charge-ordering transition at about 240 K.

Charge ordering occurs in the paramagnetic state in $Pr_{1-x}Ca_xMnO_3$ ($0.35 \leq x \leq 0.5$) with the T_{CO} increasing with x. The paramagnetic state is characterized by FM spin fluctuations with a small energy scale.[23] At T_{CO}, these fluctuations decrease and disappear at T_N. Electron diffraction and dark-field transmission electron microscopy (TEM) images show the presence of incommensurate charge ordering in the paramagnetic insulating state (180−260 K) of the $x = 0.5$ composition.[24] At T_N, there is an incommensurate−commensurate CO transition. In the incommensurate CO structure, partial orbital ordering is likely to be present. Similar charge, orbital, and spin ordering has been found in the 0.3 composition as well.[25] Optical conductivity spectra of the $x = 0.4$ composition show evidence of spatial charge and orbital ordering at 10 K.[26] The CO state

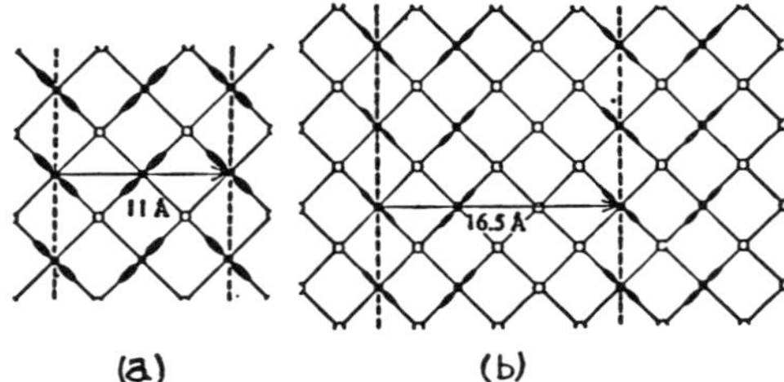

Figure 18. Charge and orbital ordering in $La_{1-x}Ca_xMnO_3$: (a) $x = 0.50$, (b) $x = 0.67$. Open circles represent Mn^{4+} and lobes show e_g orbitals of Mn^{3+}. Charge modulation wavelengths are $\sim$11 and 16.5 Å for $x = 0.50$ and 0.67, respectively (following Cheong).

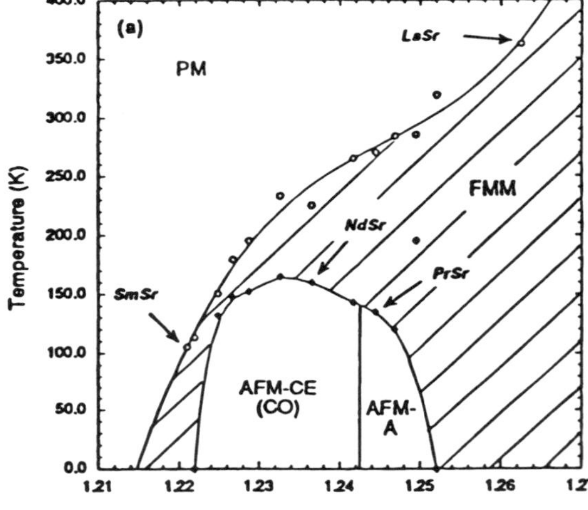

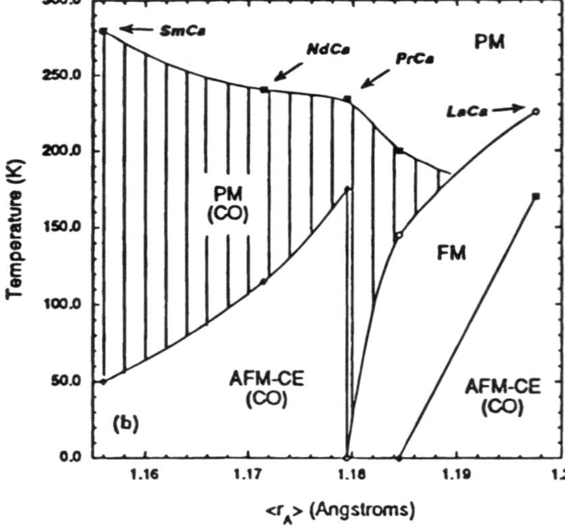

Figure 19. Temperature−$\langle r_A \rangle$ phase diagrams for (a) $Ln_{0.5}Sr_{0.5}MnO_3$ and (b) $Ln_{0.5}Ca_{0.5}MnO_3$ (from Woodward et al.[31]).

has a gap of $\sim$0.2 eV, and the gap remains up to 4.5T. The gap value is the order parameter of the CO state and couples with spin ordering. $Pr_{0.67}Ca_{0.33}MnO_3$ shows thermal relaxation effects from the metastable FMM state (produced by the application of 10T magnetic field) to the CO state.[27] A metal−insulator transition is observed as an abrupt jump in resistivity at a well-

J. Phys. Chem. B, Vol. 104, No. 25, 2000 **5883**

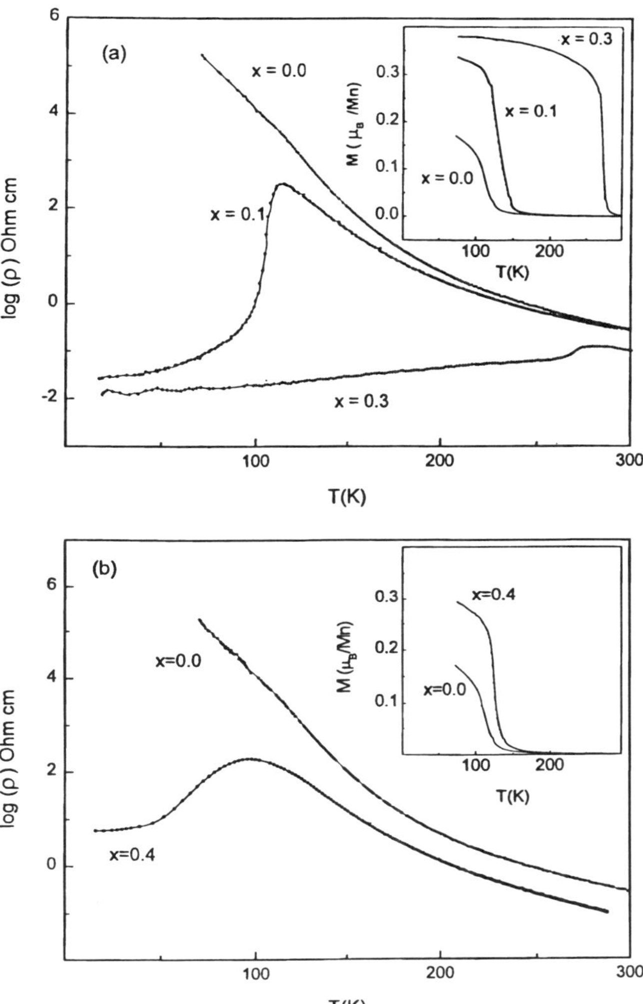

Figure 20. Effect of internal pressure on the properties of (a) $Pr_{0.7}Ca_{0.3-x}Sr_xMnO_3$ and (b) $Pr_{0.7-x}La_xCa_{0.3}MnO_3$. Note that substitution by a larger A-site cation renders the material FM (see insets). An insulator$-$metal transition is also observed (from Rao et al.[33]).

defined time, depending on the temperature. This observation seems to indicate a percolative nature of current transport.

Charge and orbital ordering in the manganates give rise to stripes.[28] In Figure 18 we show the modulation that can arise in $La_{1-x}Ca_xMnO_3$ with $x = 0.50$ and 0.67. At $x = 0.50$, the same number of Mn^{3+} and Mn^{4+} ions exist and diagonal stripes with a spacing of 11 Å are, therefore, to be expected (Figure 18a). When $x = 0.67$, there are twice as many Mn^{4+} ions as Mn^{3+} ions. Ordering of diagonal rows of Mn^{3+} and Mn^{4+} ions, besides the orientational ordering of the orbitals, would give rise to the striped pattern in Figure 18b, with a periodicity of 16.5 Å.

Paired JT distorted stripes or bistripes are believed to be present in $La_{0.33}Ca_{0.67}MnO_3$.[29] A supercell based on such ordering has been proposed, but a detailed structural investigation[30] has shown that the diffraction data can be explained satisfactorily without bistripes. Apparently, owing to orbital disordering, a mixture of paired and unpaired stripes seems to occur in $La_{0.5}Ca_{0.5}MnO_3$. Because such charge stripes generally are observed by electron microscopy, it is not clear whether these structures truly represent the bulk composition.

Cation Size Effects. In the manganates exhibiting CMR, the ferromagnetic T_c increases with the average radius of the A-site cations, $\langle r_A \rangle$. Increasing $\langle r_A \rangle$ is equivalent to increasing the hydrostatic pressure and is therefore accompanied by an increase in the Mn $-$ O $-$ Mn angle and the e_g bandwidth. If there is considerable mismatch in the radii of the different A-site cations, however, the T_c does not increase with $\langle r_A \rangle$. Charge ordering is also highly sensitive to $\langle r_A \rangle$ and the T_{CO} generally increases with decrease in $\langle r_A \rangle$. The sensitivity of T_{CO} to $\langle r_A \rangle$ has been examined[7,31,32] and is generally attributed to an increased tilting of the MnO_6 octahedra as the $\langle r_A \rangle$ decreases. In Figure 19 we show that the phase diagrams[31] of $Ln_{0.5}Sr_{0.5}MnO_3$ and $Ln_{0.5}Ca_{0.5}MnO_3$ illustrate some of the important features. (Here, the $\langle r_A \rangle$ is varied by changing the Ln ion.) The Mn$-$O(eq)$-$Mn and Mn$-$O(ax)$-$Mn bonds are identical in $Ln_{0.5}Ca_{0.5}MnO_3$ (excect when Ln = La). For $La_{0.5}Ca_{0.5}MnO_3$ and all the $Ln_{0.5}Sr_{0.5}MnO_3$ compounds, the Mn$-$O(eq)$-$Mn angle is significantly larger (by $2-6°$) than the Mn$-$O(ax)$-$Mn angle. Although the $Ln_{0.5}Ca_{0.5}MnO_3$ manganates crystallize in the *Pnma* symmetry, there is an evolution from *Pnma* to *I4/mcm* through *Imma* in the $Ln_{0.5}Sr_{0.5}MnO_3$ manganates, with increase in $\langle r_A \rangle$. The changes in the octahedral tilt system have consequences on the low-temperature magnetic structure. This is seen in $Nd_{0.5}Sr_{0.5}MnO_3$ where the charge ordering in the CE-type AFM state is associated with the *Imma* structure.

As pointed out earlier, the CO states of $Nd_{0.5}Sr_{0.5}MnO_3$ ($\langle r_A \rangle$ = 1.24 Å) and $Pr_{0.6}Ca_{0.4}MnO_4$ ($\langle r_A \rangle$ = 1.17 Å) are destroyed by magnetic fields. The field required to melt the CO state varies with $\langle r_A \rangle$ and the manganates with very small $\langle r_A \rangle$ remain charge-ordered even on the application of high magnetic fields.[32] Thus, $Y_{0.5}Ca_{0.5}MnO_3$ ($\langle r_A \rangle$ = 1.13 Å) has a robust CO state that is not affected by very high magnetic fields (>25T). We can thus distinguish three different categories of manganates with respect to their sensitivity to magnetic fields: (a) manganates that are FM and become charge-ordered at low temperatures (e.g., $Nd_{0.5}Sr_{0.5}MnO_3$ when $T_{CO} = T_N$), with the CO state transforming to a FMM state on the application of a moderate magnetic field; (b) manganates that are charge-ordered in the paramagnetic state ($T_N < T_{CO}$), and do not exhibit an FMM state, but transform to a FMM state under a magnetic field (e.g., $Pr_{1-x}Ca_xMnO_3$); and (c) those that are charge-ordered in the paramagnetic state ($T_N < T_{CO}$) as in b, but are not affected by magnetic fields up to 15T or greater (e.g., $Y_{0.5}Ca_{0.5}MnO_3$). Category c is encountered when $\langle r_A \rangle \leq 1.17$ Å. The apparent one-electron bandwidth estimated on the basis of the experimental Mn$-$O$-$Mn angle and the average Mn$-$O distance in $Ln_{0.5}A_{0.5}MnO_3$ does not vary significantly with $\langle r_A \rangle$, which suggests that other factors may be responsible for the sensitivity of the CO state to $\langle r_A \rangle$. One possibility is a competition between the A- and B-site cations for covalent mixing with the O(2p) orbitals.[32]

By increasing the size of the A-site cations or by the application of pressure, the CO state in the manganates can be transformed to the FMM state.[33,34] In Figure 20, we show the effect of internal pressure on the CO state of $Pr_{0.7}Ca_{0.3}MnO_3$ wherein Ca is substituted by the larger Sr or Pr is substituted by La. The T_c in the $Pr_{0.5}Sr_{0.5-x}Ca_xMnO_3$ system, decreases with an increase in x or a decrease in $\langle r_A \rangle$ up to $x = 0.25$; $T_{CO} = T_N$ from $x = 0.09$ to 0.25. When $\langle r_A \rangle$ is decreased further, T_{CO} increases from 180 K for $x = 0.25$ to 250 K for $x = 0.30$; for $0.30 \leq x \leq 0.50$, $T_{CO} > T_N$.[35]

The effect of cation size disorder on the ferromagnetic T_c of rare earth manganates exhibiting CMR has been investigated in detail. The disorder is quantified in terms of the variance in the distribution $\langle r_A \rangle$, as defined by Attfield.[36] The variance σ^2 is defined by, $\sigma^2 = \Sigma x_i r_i^2 - \langle r_A \rangle^2$, where x_i is the fractional occupancy of A-site ions and r_i is the ionic radius. The ferromagnetic T_c decreases significantly with the variance, σ^2, based on the studies of rare earth manganates with fixed $\langle r_A \rangle$. A similar study of the variation of T_{CO} with σ^2 in $Ln_{0.5}A_{0.5}MnO_3$, for fixed $\langle r_A \rangle$ values of 1.17 and 1.24 Å, has shown that T_{CO} is not very sensitive to the size mismatch.[37] It appears that JT distortion and Coulomb interactions play a prominent role in determining the nature of the CO state in these materials.

Considering that the rare earth manganates with large $\langle r_A \rangle$, as exemplified by $Nd_{0.5}Sr_{0.5}MnO_3$ ($\langle r_A \rangle$ = 1.17 Å), exhibit entirely different characteristics of the CO state, and that $\langle r_A \rangle$ = 1.17 Å categorizes the manganates with respect to their insensitivity to magnetic fields, we would expect interesting and unusual properties in the intermediate range of $\langle r_A \rangle$ of 1.20 $\pm$ 0.20 Å. In this regime T_{CO} approaches T_c, leading to a competition between charge ordering and ferromagnetism. Thus, $La_{0.5}Ca_{0.5}MnO_3$ ($\langle r_A \rangle$ = 1.20 Å) exhibits a region of coexistence of ferromagnetism and charge ordering ($T_c = 225$ K, $T_{CO} = 135$ K). At 135 K, the material also becomes AFM (CE-

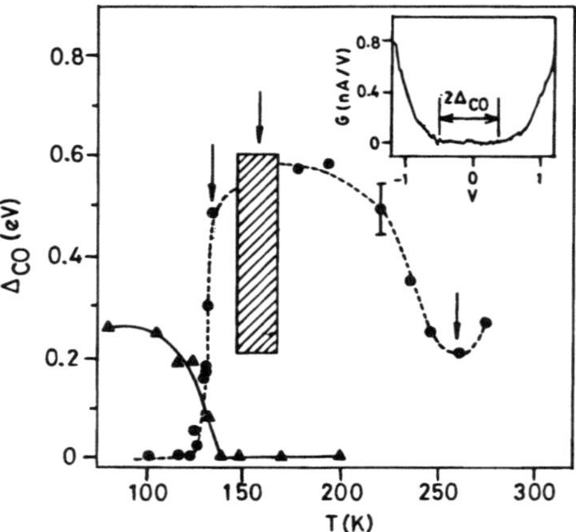

Figure 21. Temperature variation of the CO gap in $Nd_{0.25}La_{0.25}Ca_{0.5}$-$MnO_3$ (closed circles) and $Nd_{0.5}Sr_{0.5}MnO_3$ (closed triangles). Inset shows a typical tunneling conductance curve (from Arulraj et al.[40]). Shaded region represents the coexistence regime.

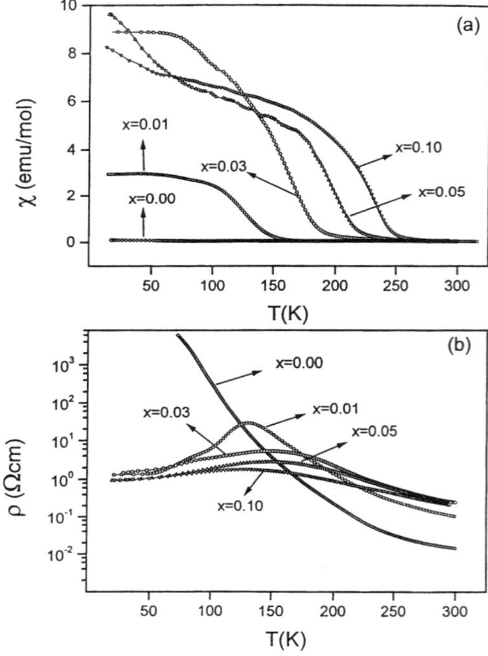

Figure 22. Temperature variation of the (a) magnetic susceptibility and (b) resistivity of $Nd_{0.5}Ca_{0.5}Mn_{1-x}Ru_xO_4$ (from Vanitha et al.[46b]).

type) and the orbital ordering becomes commensurate.[38] In $(Nd_{1-x}Sm_x)_{0.5}Sr_{0.5}MnO_3$, T_c decreases from 255 K to 115 K as x increases from 0.0 to 0.875, as expected of a decrease in $\langle r_A \rangle$.[39] The T_{CO} also decreases from 158 K to 0 K as x changes from 0.0 to 0.875. This unusual behavior wherein T_{CO} is suppressed as T_c approaches T_{CO} is interesting.

$Nd_{0.25}La_{0.25}Ca_{0.5}MnO_3$ ($\langle r_A \rangle$ = 1.19 Å) reveals an intriguing sequence of phase transitions.[40] On cooling, this manganate develops an incipient CO state below 220 K. The formation of this state is accompanied by an increase in electrical resistivity and the opening up of a gap in the density of states near E_F. The orthorhombic distortion also increases, as a consequence of cooperative JT distortion of the lattice and short-range

J. Phys. Chem. B, Vol. 104, No. 25, 2000 **5885**

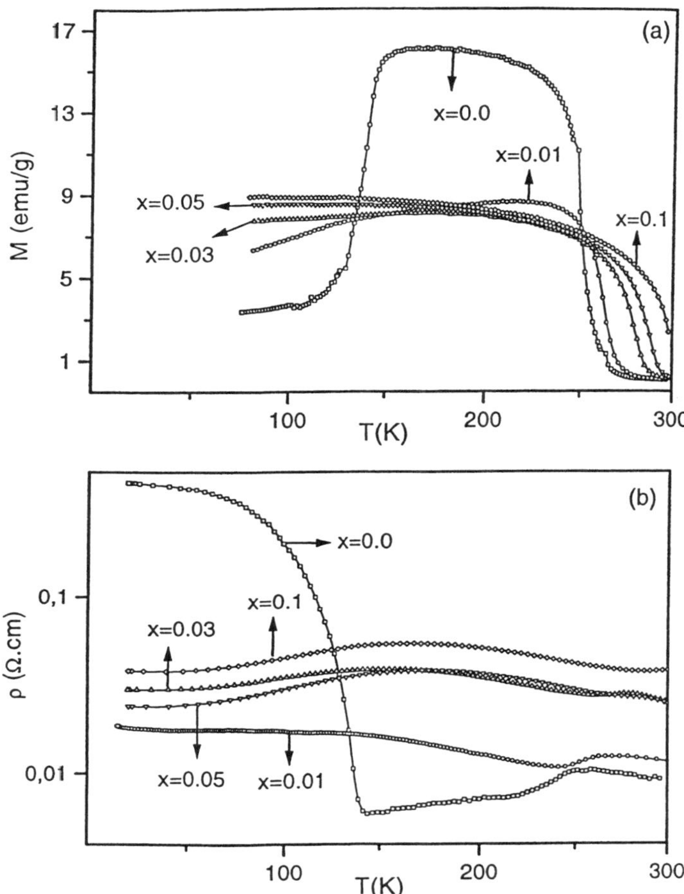

Figure 23. Temperature variation of the (a) magnetization and (b) resistivity of $Nd_{0.5}Sr_{0.5}Mn_{1-x}Ru_xO_4$ (from Vanitha et al.[46b]).

ordering of the Mn^{3+} and Mn^{4+} ions. At about 150 K, the incipient CO state becomes unstable and the material undergoes a reentrant transition to the FMM state. The transition is characterized by a sharp decrease in resistivity, collapse of the CO gap, development of FM ordering, and an abrupt decrease in the orthorhombic distortion. There is a two-phase coexistence region (150−220 K) around the CO−FMM transition. We show the fascinating reentrant transition in Figure 21 where the CO gap, obtained by vacuum tunneling measurements,[40] is plotted against temperature.

Effects of Substitution in the Mn Site and by ^{18}O. Substitution of ^{16}O by ^{18}O (generally up to 85−90%) has a marked effect on the magnetic properties and CMR of the manganates, indicating the important role of electron−phonon coupling. Substituting ^{18}O for ^{16}O in $La_{0.175}Pr_{0.525}Ca_{0.3}MnO_3$ destroys the insulator−metal transition and renders the material insulating down to 4 K.[41] The CO transition in $La_{0.5}Ca_{0.5}MnO_3$ increases by 9 K upon replacing ^{16}O by ^{18}O; the isotope shift increases with the magnetic field in $La_{0.5}Ca_{0.5}MnO_3$ and in $Nd_{0.5}$-$Sr_{0.5}MnO_3$.[42] In $Pr_{0.67}Ca_{0.33}MnO_3$, the magnetic field−induced insulator−metal transition occurs at a higher field on ^{18}O substitution; the heavier isotope favors the insulating state.[43] The isotope effect on T_{CO} is greater in $Nd_{0.5}Sr_{0.5}MnO_4$ than $Pr_{0.5}$-$Ca_{0.5}MnO_3$, indicating the role of $\langle r_A \rangle$ as well.[44]

Substitution of Mn by cations such as Al^{3+} and Fe^{3+} in charge-ordered manganates destroys charge ordering at moderate doping ($x \geq 0.03$), but the materials remain insulating. However, substitution by Cr^{3+} readily destroys charge ordering and renders the material FM and metallic.[45,46] In $Sm_{0.5}Ca_{0.5}Mn_{1-x}Cr_xO_3$, the T_{CO} (275 K when $x = 0.0$) decreases with increasing x and charge ordering disappears at $x = 0.05$. The $x = 0.05$ composition shows an insulator−metal transition when the material becomes FM.[47] The effectiveness of Cr^{3+} in destroying the CO state is considered to be due to its favorable electronic configuration (t_{2g}^3), which is the same as that of Mn^{4+}. However, Cr^{3+} in the Mn^{3+} site would be surrounded immediately by Mn^{4+} ions, which would not allow for near-neighbor electron hopping. Hopping would be more favored if Mn^{4+} were substituted by an appropriate quadrivalent cation such as Ru^{4+} ($t_{2g}^4 e_g^0$). Recent studies have shown that substitution of Mn by Ru in $Nd_{0.5}Ca_{0.5}MnO_3$ destroys charge ordering and renders the material FM, with the T_c increasing with Ru content; the material also shows an insulator−metal transition[46] (Figure 22). The marked effect of Ru substitution is also seen in $Nd_{0.5}Sr_{0.5}MnO_3$ where the T_c increases with the Ru content to well above 300 K, but charge ordering is destroyed (Figure 23).

Phase Separation. The formation of FM clusters in an AFM host matrix in the rare earth manganates has been noticed by many workers. Thus, spin glass behavior has been encountered in the $Ln_{1-x}A_xMnO_3$ system at either extreme, corresponding to large or small x. Electronic phase separation is also evidenced

5886 *J. Phys. Chem. B, Vol. 104, No. 25, 2000*

Rao

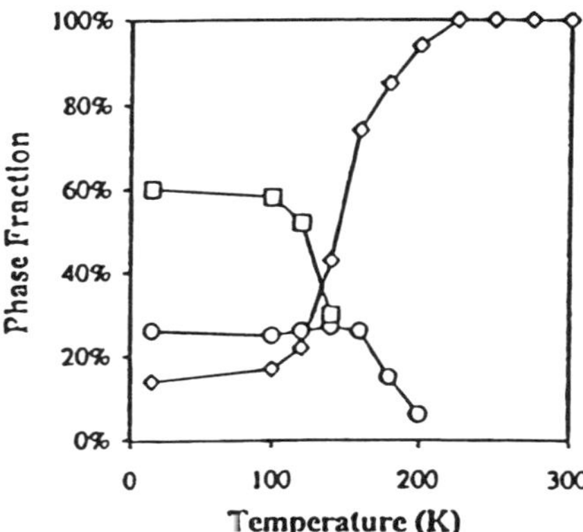

Figure 24. Variation in the percentage of the different phases of $Nd_{0.5}$-$Sr_{0.5}MnO_3$ with temperature: FMM phase (diamonds); orbitally ordered A-type AFM phase (circles); charge-ordered CE-type AFM phase (squares) (from Woodward et al.[50]).

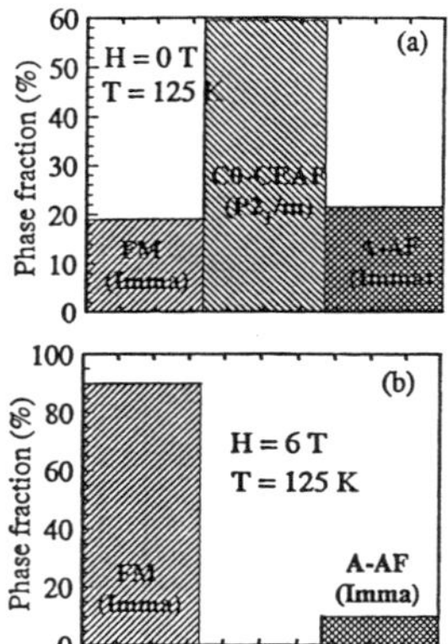

Figure 25. Effect of magnetic field (6T) on phase-separated $Nd_{0.5}$-$Sr_{0.5}MnO_3$ at 125 K (from Ritter et al.[52]).

in the manganates. Thus an electron microscopic study of $La_{1-x-y}Pr_yCa_xMnO_3$ ($x = 0.375$) has shown electronic phase separation into a submicrometer-scale mixture of CO insulator regions and FMM domains. Perculative transport could occur between the two states.[48] The coexistence of FM and CO states has been observed in $La_{0.5}Ca_{0.5}MnO_3$ and $Nd_{0.25}La_{0.25}Ca_{0.5}$-$MnO_3$.[38,40] Some phase separation is likely in many of the charge-ordered manganates. In Cr-doped $Nd_{0.5}Ca_{0.5}MnO_3$, sub-micrometer FMM domains are embedded in an AFM CO state, so the material shows a relaxor behavior.[49]

$Nd_{0.5}Sr_{0.5}MnO_3$, which shows evidence for separation into three macroscopic phases, is particularly interesting. These are the high-temperature FMM phase (*Imma*), the A-type AFM intermediate-temperature phase (*Imma*), and the CE-type AFM CO low-temperature phase ($P2_1/m$).[50] The A-type AFM phase starts manifesting itself around 220 K, whereas the CE-type CO phase first appears at 150 K. In Figure 24, we show the phase compositions at different temperatures. There are three phases at the so-called CO transition at 150 K. The presence of the high-temperature FMM phase, even at very low temperatures, is noteworthy. These results are of significance in interpreting many of the properties of this manganate. The fact that such phase separation is seen by X-ray/neutron diffraction implies that even the minority phases have large domains (≥ 100 nm). The FMM phase of $Nd_{0.5}Sr_{0.5}MnO_3$ has a larger volume than the average volume or the volume of the low-temperature CO phase. The phase-separation behavior of this system and the relative stabilities of the structures seem to depend crucially on the Mn^{4+}/Mn^{3+} ratio. Thus, the regime $Mn^{4+}/Mn^{3+} > 1$ seems to stabilize the orbitally ordered AFM (A-type) phase. Unlike $Nd_{0.5}Sr_{0.5}MnO_3$, $Nd_{0.45}Sr_{0.55}MnO_3$ has an A-type AFM state below T_N (220 K) and metallicity is confined to the ferromagnetic ab plane below T_N, whereas the material is insulating along the c-axis. The large anisotropy in this material implies that the carriers are confined to the FM layer by magnetic and orbital ordering.[51] An interesting experiment on phase-separated $Nd_{0.5}$-$Sr_{0.5}MnO_3$ at low temperatures was performed recently wherein the phase composition was determined by neutron diffraction in the absence and presence of a magnetic field.[52] The results show how on applying a magnetic field, the CO state at low temperatures, by and large, transforms to the FMM state (Figure 25). For some reason, a small proportion of the intermediate-temperature A-type AFM phase persists.

Effect of Electric Fields. Charge-ordered manganates such as $Pr_{1-x}Ca_xMnO_3$ ($x \approx 0.3-0.4$) undergo a CO−FMM transition

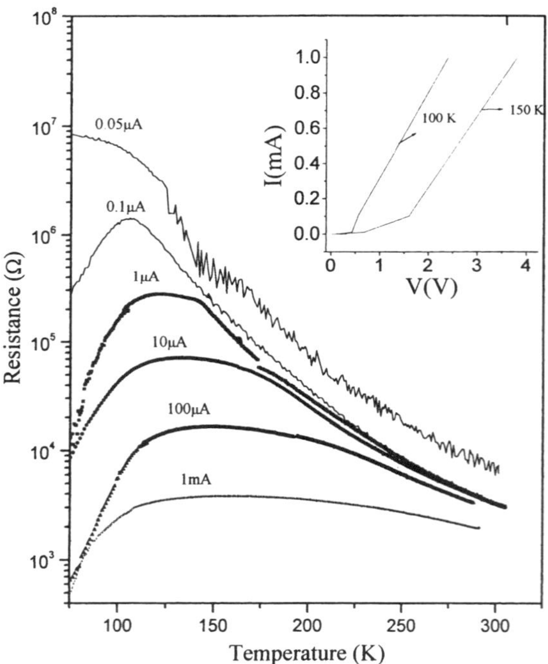

Figure 26. Electric current induced insulator−metal transition in $Nd_{0.5}$-$Ca_{0.5}MnO_3$ films deposited on Si(100) at different values of the current. Inset show I−V curves at different temperatures (from Rao et al.[54]).

on application of moderate magnetic fields. However, manganates with small $\langle r_A \rangle$ such as $Y_{0.5}Ca_{0.5}MnO_3$ are, for all practical purposes, unaffected by magnetic fields. Laser irradiation has been reported to cause an insulator−metal transition in $Pr_{0.7}$-$Ca_{0.3}MnO_3$, generating a localized conduction path, although the bulk of the sample is insulating.[53] It has been found recently that small d.c. currents induce insulator−metal transitions in thin films of several charge-ordered rare earth manganates, including $Y_{0.5}Ca_{0.5}MnO_3$, $Nd_{0.5}Ca_{0.5}MnO_3$, and $Pr_{0.7}Ca_{0.3}$-MnO_3.[54] The current−voltage characteristics are nonohmic and show hysteresis. The I−M transition temperature decreases with increasing current (Figure 26). The hysteretic I−M transition in Figure 27 is specially noteworthy in that there is a reproducible memory effect in the cooling and heating cycles. The current-induced I−M transition occurs even in $Y_{0.5}Ca_{0.5}MnO_3$,

J. Phys. Chem. B, Vol. 104, No. 25, 2000 **5887**

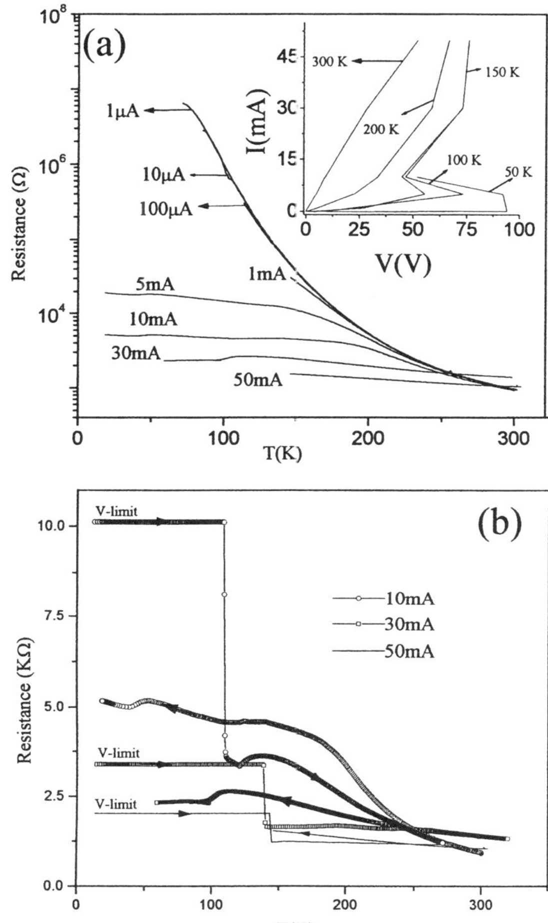

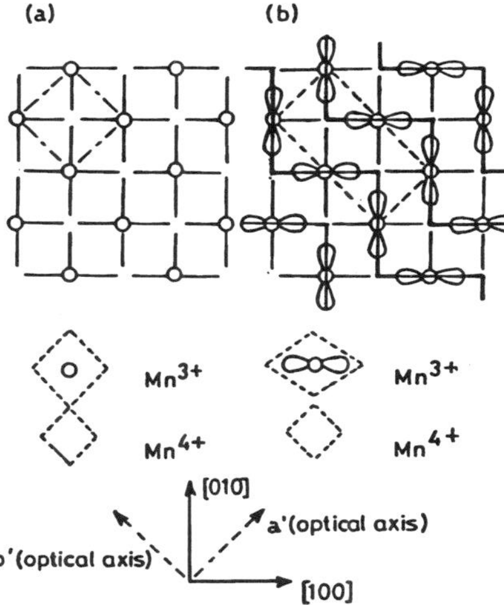

Figure 28. Charge ordering in $La_{0.5}Sr_{1.5}MnO_4$ showing oxide ion displacements. Mn^{3+} (e_g) orbitals are shown (from Ishikawa et al.[56a]).

Figure 27. (a) Temperature variation of resistance of an oriented $Nd_{0.5}$-$Ca_{0.5}MnO_3$ film deposited on $LaAlO_3(001)$ for different values of the current; (b) resistance−temperature plots for three current values recorded over cooling and heating cycles showing memory effect. Inset in part a shows I−V curves at different temperatures (from Rao et al.[54]).

which is not affected by large magnetic fields. Furthermore, there is no need for prior laser irradiation to observe the current-induced I−M transitions. It is proposed that electric fields cause depinning of the randomly pinned charge solid. There appears to be a threshold field in the CO regime beyond which nonlinear conduction sets in along with a large broad-band conductivity noise.[55] Threshold-dependent conduction disappears around T_{CO}, which suggests that the CO state gets depinned at the onset of nonlinear conduction. At small currents or low magnetic fields, resistance oscillations occur because of temporal fluctuations between resistive states.

Layered Manganates. The interplay of spin, orbital, and charge ordering in the rare earth manganates in determining their properties becomes even more prominent in the layered manganates. In the Ruddlesden−Popper series of manganates, $(Ln, A)_{n+1}Mn_nO_{3n+1}$, the $n = \infty$ phases are the three-dimensional perovskites. The $n = 2$ phases, such as $La_{2-2x}Sr_{1+2x}Mn_2O_7$, show a metallic ground state for $x \geq 0.17$ and high CMR; they also exhibit interesting properties arising from the layered nature of the materials. The $n = 1$ member, $Ln_{1-x}Sr_{1+x}MnO_4$, with the quasi two-dimensional K_2NiF_4 structure, becomes metallic only

when $x = 0.7$. $La_{0.5}Sr_{1.5}MnO_4$ shows ordering of the Mn^{3+} and Mn^{4+} ions at about 220 K, accompanied by the ordering of the e_g orbitals (Figure 28). The material becomes antiferromagnetic at 110 K and exhibits anisotropic properties arising from orbital ordering.[56] At the 220 K CO transition, a significant change occurs in the conductivity spectrum. The order parameter for orbital ordering increases at T_{CO}, grows further with decreasing temperature, and decreases on spin ordering at T_N. The order parameters for charge and orbital ordering seem to evolve together. Application of high magnetic fields gives rise to metamagnetic transitions below T_{CO}, and charge and orbital ordering are destroyed by the magnetic field. Charge ordering in $LaSr_2Mn_2O_7$ ($n = 2$) has been observed by electron diffraction.[57] There is a need to study charge ordering in this as well as in other rare earth analogues, including the calcium derivatives.

Concluding Remarks

The rare earth manganates exhibit a variety of properties and phenomena, with an extraordinary sensitivity to various factors such as cation size, pressure, and magnetic and electric fields. In particular, the mutual relations between orbital ordering, charge ordering, and spin ordering in the rare earth manganates is truly fascinating. It is because of this interplay that charge ordering in these materials has turned out to be such an attractive area of research. In manganates such as $Pr_{0.6}Ca_{0.4}MnO_3$ and $Nd_{0.5}Ca_{0.5}MnO_3$ ($\langle r_A \rangle \approx 1.17$ Å), charge ordering occurs in the paramagnetic state, and commensurate orbital ordering accompanies a transition to the CE-type AFM state. In the paramagnetic CO state, there could be some orbital ordering that would be incommensurate, and FM correlations or clusters are generally present. In $Nd_{0.5}Sr_{0.5}MnO_3$ ($\langle r_A \rangle = 1.24$ Å), where charge ordering occurs on cooling a FMM state, however, it is possible to delineate transitions associated with orbital and charge ordering. In $La_{0.5}Ca_{0.5}MnO_3$ ($\langle r_A \rangle = 1.20$ Å), charge ordering and orbital ordering occur when the FM state transforms to an AFM state (CE type) on cooling. In the regime where the FM and CO states coexist, there is no long-range orbital ordering. As expected, the $La_{0.5}Ca_{0.5}MnO_3$ system gives rise to complex lattice images. In $La_{0.33}Ca_{0.67}MnO_3$, however, there is a clear-cut charge ordering transition (260 K) accompanied by orbital ordering, and followed by a transition to a CE-type AFM state, and paired stripes of $Mn^{3+}O_6$ octahedra have been observed in the lattice images. In manganates such as $Nd_{0.5}Ca_{0.5}MnO_3$ and $Y_{0.5}Ca_{0.5}MnO_3$ with a small $\langle r_A \rangle$, it is not clear whether ordered single or bistripes would be present. It is important to ensure that bistripes and other features found by electron miscroscopy truly represent the bulk structure. Low-temperature lattice imaging of these materials with high-resolution electron microscopy would be of great value.

Although we are able to distinguish spin, charge, and orbital ordering transitions, we co not understand some aspects. It is not entirely clear whether commensurate orbital ordering occurs only in the AFM (CE) state. The electron−hole asymmetry in the manganates, with respect to the FM and CO states, needs to be explained properly. Is it certain that we can never encounter ferromagnetism in the electron-doped materials? Also, what is the type of AFM ordering in the electron-doped CO manganates?

Phase separation in the manganates is especially interesting. Phase separation generally involves the formation of clusters of one phase in another (e.g., FM clusters in a CO or AFM phase). Macroscopic phase separation involving the presence of distinct domains of fairly large sizes (≥ 100 nm) with Bragg peaks in the diffraction patterns is, however, different from the formation of clusters or stripes. Such inhomogeneous distributions of CO or FM (charge and spin) phases in a material requires understanding. Coulomb forces would prevent accumulation of charge in a phase-separated regime in the absence of the means to compensate the charge. Electronic phase separation is known in $La_2NiO_{4+\delta}$ and $La_2CuO_{4+\delta}$.[3,11,28] In the rare earth manganates, there is increasing evidence for the coexistence of FM and AFM (CO) phases. Although it is possible that small changes in the Mn^{3+}/Mn^{4+} ratio could affect phase separation, we co not understand the mechanism. Although one can consider the presence of large domains (e.g., CO and FM) of different phases of a given composition as a criterion for phase separation, the effects of phase separation, in contrast to those of chemical inhomogenieties or cluster formation, need to be investigated in greater detail.

The $\langle r_A \rangle$ regime of 1.20 ± 0.02 Å in $Ln_{0.5}A_{0.5}MnO_3$, which exhibits complex phenomena and properties, including reentrant transitions, deserves further study. It is useful to examine such systems with fixed $\langle r_A \rangle$ and variable cation size mismatch. Charge ordering in the layered manganates has to be investigated in detail. We do not yet have a full understanding of the extraordinary effect of electric fields on the CO state.

Acknowledgment. The author is thankful to the Department of Science and Technology and the Science Office of the European Union for support of this work.

References and Notes

(1) (a) Honig, J. M. In *The Metallic and the Nonmetallic States of Matter*; Edwards, P. P., Rao, C. N. R., Eds.; Taylor and Francis: London, 1985. (b) Honig, J. M. *Proc. Indian Acad. Sci., Chem. Sci.* **1986**, *96*, 391.

(2) Tsuda, N.; Nasu, K.; Yanase, A.; Siratori, K. *Electronic Conduction in Oxides*; Springer: Berlin, 991.

(3) (a) Kivelson, S. A.; Fradkin, E.; Emery, V. J. *Nature (London)* **1998**, *393*, 550. (b) Tranquada, J. M.; Sternleib, B. J.; Axe, J. D.; Nakamura, Y.; Uchida, S. *Nature (London)* **1995**, *375*, 561.

(4) Imada, M.; Fujimori, A.; Tokura, Y. *Rev. Mod. Phys.* **1998**, *70*, 1039.

(5) (a) von Helmolt, R.; Holzapfel, B.; Schultz, L.; Samwer, K. *Phys. Rev. Lett.* **1993**, *73*, 2331. (b) Chahara, K.; Ohno, T.; Kasai, M.; Kozono, Y. *Appl. Phys. Lett.* **1993**, *63*, 1990.

(6) (a) Rao, C. N. R.; Raveau, B., Eds. *Colossal Magnetoresistance, Charge Ordering and Related Properties of Manganese Oxides*; World Scientific: Singapore, 1998. (b) Ramirez, A. P. *J. Phys: Condens. Matter* **1997**, *9*, 8171.

(7) Rao, C. N. R.; Arulraj, A.; Santosh, P. N.; Cheetham, A. K. *Chem. Mater.* **1998**, *10*, 2714.

(8) Zener, C. *Phys. Rev.* **1951**, *82*, 403.

(9) Jirak, Z.; Krupicka, S.; Simsa, Z.; Dlouha, M.; Vratislav, S. *J. Magn. Magn. Mater.* **1985**, *53*, 153.

(10) (a) Takano, M.; Takeda, Y. *Bull. Inst. Chem. Res.* **1983**, *61*, 406. (b) Battle, P. D.; Gibb, T. C.; Lightfoot, P. *J. Solid State Chem.* **1990**, *84*, 271.

(11) (a) Tranquada J. M.; Loronzo, J. E.; Buttrey, D. J.; Sachan, V. *Phys. Rev.* **1995**, *B52*, 3581. (b) Tranquada, J. M.; Loronzo, J. E.; Buttrey, D. J.; Sachan, V. *Phys. Rev.* **1996**, *B54*, 12318.

(12) Ramirez, A. P.; Gammel, P. L.; Cheong, S. W.; Bishop, D. J.; Chandra, P. *Phys. Rev. Lett.* **1996**, *76*, 447.

(13) (a) Rodriguez-Carvajal, J.; Rousse, G.; Masquelier, C.; Hervieu, M. *Phys. Rev. Lett.* **1998**, *81*, 4660. (b) Willis, A. S.; Raju, N. P.; Greedan, J. E. *Chem. Mater.* **1999**, *11*, 1510.

(14) Wollan, E. O.; Koehler, W. C. *Phys. Rev.* **1955**, *100*, 545.

(15) (a) Goodenough, J. B. *Phys. Rev.* **1955**, *100*, 564. (b) Goodenough, J. B.; Wold, A.; Arnolt, R. J.; Menyuk, N. *Phys. Rev.* **1961**, *124*, 373.

(16) Vogt, T.; Cheetham, A. K.; Mahendiran, R.; Raychaudhuri, A. K.; Mahesh, R.; Rao, C. N. R. *Phys. Rev.* **1996**, *B54*, 15303.

(17) Kuwahara, H.; Tomioka, Y.; Asamitsu, A.; Moritomo, Y.; Tokura, Y. *Science* **1995**, *270*, 961.

(18) Tokura, Y.; Tomioka, Y.; Asamitsu, A.; Moritomo, Y.; Kasai, M. *J. Appl. Phys.* **1996**, *79*, 5288.

(19) Biswas, A.; Raychaudhuri, A. K.; Mahendiran, R.; Guha, A.; Mahesh, R.; Rao, C. N. R. *J. Phys. Condens. Matter* **1997**, *9*, L355.

(20) Sekiyama, A.; Suga, S.; Fujikawa; Imada, S.; Iwasaki, T.; Matsuda, K.; Matsushita, T.; Kaznacheyev, K. V.; Fujimori, A.; Kuwahara, H.; Tokura, Y. *Phys. Rev.* **1999**, *B59*, 15528.

(21) Mahendiran, R.; Ibarra M. R.; Maignan, A.; Millange, F.; Arulraj, A.; Mahesh, R.; Raveau, B.; Rao, C. N. R. *Phys. Rev. Lett.* **1999**, *82*, 2191.

(22) (a) Tomioka, Y.; Asamitsu, A.; Kuwahara, H.; Moritomo, Y.; Tokura, Y. *Phys. Rev.* **1996**, *B53*, 1689. (b) Lees, M. R.; Barrat, J.; Balakrishnan, G.; Paul D, M. C. K.; Yethiraj, M. *Phys. Rev.* **1995**, *B52*, 14303.

(23) Kajimoto R.; Kakeshita T.; Oohara Y.; Yoshizawa H.; Tomioka Y.; Tokura, Y. *Phys. Rev.* **1998**, *B58*, R11837.

(24) Mori, S.; Katsufuji, T.; Yamamoto, N.; Chen, C. H.; and Cheong, S. W. *Phys. Rev.* **1999**, *B59*, 13573.

(25) Cox, D. E.; Radaelli, P. G., Marezio, M.; Cheong, S. W. *Phys. Rev.* **1998**, *B57*, 3305.

(26) Okimoto Y.; Tomioka, Y.; Onose, Y.; Otsuka, Y.; Tokura, Y. *Phys. Rev.* **1999**, *B59*, 7401.

(27) Anane, A.; Renard, J. P.; Reversat, L.; Dupas, C.; Veillet P.; Viret M.; Pinsard, L.; Revcolevsci, A. *Phys. Rev.* **1999**, *B59*, 77.

(28) Cheong, S. W.; Chen, C. H. In *Colossal Magnetoresistance, Charge Ordering and Related Properties of Manganese Oxides*; Rao, C. N. R., Raveau, B, Eds. World Scientific: Singapore, 1998.

(29) Mori, S.; Chen, C. H.; Cheong, S. W. *Nature (London)* **1998**, *392*, 473.

(30) Radaelli, P. G.; Cox, D. E.; Capogna, L.; Cheong, S. W.; Marezio, M. *Phys. Rev.* **1999**, *B59*, 14440.

(31) Woodward, P. M.; Vogt, T.; Cox, D. E.; Arulraj, A.; Rao, C. N. R.; Karen, P.; Cheetham, A. K. *Chem. Mater.* **1998**, *10*, 3652.

(32) Arulraj, A.; Santosh, P. N.; Gopalan, R. S.; Guha, A.; Raychaudhuri, A. K.; Kumar, N.; Rao, C. N. R. *J. Phys: Condens. Matter* **1998**, *10*, 8497.

(33) Rao, C. N. R.; Santosh, P. N.; Singh, R. S.; Arulraj, A. *J. Solid State Chem.* **1998**, *135*, 169.

(34) Moritomo, Y.; Kuwahara, H.; Tomioka, Y.; Tokura, Y. *Phys. Rev.* **1997**, *B55*, 7549.

(35) Damay, F.; Martin, C.; Maignan, A.; Hervieu, M.; Raveau, B.; Jirak, Z.; Andre, G.; Bouree, F. *Chem. Mater.* **1999**, *11*, 536.

(36) Attfield, J. P. *Chem. Mater.* **1998**, *10*, 3239.

(37) Vanitha, P. V.; Santosh, P. N.; Singh, R. S.; Rao, C. N. R.; Attfield, J. P. *Phys. Rev.* **1999**, *B59*, 13539.

(38) Radaelli, P. G.; Cox, D. E.; Marezio, M.; Cheong, S. W. *Phys. Rev.* **1997**, *B55*, 3015.

(39) Tokura, Y.; Kuwahara, H.; Moritomo, Y.; Tomioka, Y.; Asamitsu, A. *Phys. Rev. Lett.* **1996**, *76*, 3184.

(40) Arulraj, A.; Biswas, A.; Raychaudhuri, A. K.; Rao, C. N. R.; Woodward, P. M.; Vogt, T.; Cox, D. E.; Cheetham, A. K. *Phys. Rev.* **1998**, *B57*, R8115.

(41) Babushkina, N. A.; Belova, L. M.; Gorbenko, O. Yu; Kaul, A. R.; Bosak, A. A.; Ozhogin, V. I.; Kugel, Kl. *Nature (London)* **1998**, *391*, 159.

(42) Zhao, G.; Ghosh, K.; Keller, H.; Greene, R. L. *Phys. Rev.* **1999**, *B59*, 81.

(43) Garcia-Landa, B.; Ibarra, M. R.; De Teresa, J. M.; Zhao, G.; Conder, K.; Keller, H. *Solid State Commun.* **1998**, *105*, 567.

(44) Mahesh, R.; Itoh, M. *J. Solid State Chem.* **1999**, *144*, 232.

(45) Raveau, B.; Maignan, A.; Martin, C.; Hervieu, M. *Chem. Mater.* **1998**, *10*, 2641.

(46) (a) Vanitha, P. V.; Singh, R. S.; Natarajan, S.; Rao, C. N. R. *Solid State Commun.* **1999**, *109*, 135. (b) Vanitha, P. V.; Arulraj, A.; Raju, A. R.; Rao, C. N. R. *C. R. Acad. Sci. Ser. II*, t.2 **1999**, 595.

(47) Barnabe, A.; Hervieu, M.; Martin, C.; Maignan, A.; Raveau, B. *J. Mater. Chem.* **1998**, *8*, 1405.

(48) Uehara M., Mori, S.; Chen, C. H.; Cheong, S. W. *Nature (London)* **1999**, *399*, 560.

(49) Kimura, T.; Tomioka, Y.; Kumai, R.; Okinoto, Y.; Tokura, Y. *Phys. Rev. Lett.* **1999**, *83*, 3940.

(50) Woodward, P. M.; Cox, D. E.; Vogt, T., Rao, C. N. R.; Cheetham, A. K. *Chem. Mater.* **1999**, *11*, 3528.

(51) Kuwahara, H.; Okuda, T.; Tomioka, Y.; Asamitsu, A.; Tokura, Y. *Phys. Rev. Lett.* **1999**, *82*, 4316.

(52) Ritter, C.; Mahendiran, R.; Ibarra, M. R.; Morellon, L.; Maignan, A.; Raveau, B.; Rao, C. N. R. *Phys. Rev.* **2000**, *B61*, R9229.

(53) Feibig, M.; Miyano, K.; Tomioka, Y.; Tokura, Y. *Science* **1998**, *280*, 1925.

(54) Rao, C. N. R.; Raju, A. R.; Ponnambalam. V.; Parashar, S.; Kumar, N. *Phys. Rev.* **2000**, *B61*, 591.

(55) Guha, A.; Ghosh, A.; Raychaudhuri, A. K.; Parashar, S.; Raju, A. R.; Rao, C. N. R. *Appl. Phys. Lett.* **1999**, *75*, 3381.

(56) (a) Ishikawa, T.; Ookura, K.; Tokura, Y. *Phys. Rev.* **1999**, *B59*, 8367. (b) Tokunaga, M.; Miura, N.; Moritomo, Y.; Tokura, Y. *Phys. Rev.* **1999**, *B59*, 11151.

(57) Li, J. Q.; Matsui, Y.; Kimura, T.; Tokura, Y. *Phys. Rev.* **1998**, *B57*, 3205.

Chem. Mater. **2001**, *13*, 787–795

Electron–Hole Asymmetry in the Rare-Earth Manganates: A Comparative Study of the Hole- and the Electron-Doped Materials

K. Vijaya Sarathy,[†] P. V. Vanitha,[†] Ram Seshadri,[‡] A. K. Cheetham,[§] and C. N. R. Rao*[†,‡,§]

Chemistry and Physics of Materials Unit, Jawaharlal Nehru Centre for Advanced Scientific Research, Jakkur P.O., Bangalore 560 064, India, and Solid State and Structural Chemistry Unit, Indian Institute of Science, Bangalore 560 012, India, and Materials Research Laboratory, University of California, Santa Barbara, California 93106

Received June 6, 2000. Revised Manuscript Received December 18, 2000

Properties of the hole-doped $Ln_{1-x}A_xMnO_3$ (Ln = rare earth, A = alkaline earth, $x < 0.5$) are compared with those of the electron-doped compositions ($x > 0.5$). Charge ordering is the dominant interaction in the latter class of manganates unlike ferromagnetism and metallicity in the hole-doped materials. Properties of charge-ordered (CO) compositions in the hole- and electron-doped regimes, $Pr_{0.64}Ca_{0.36}MnO_3$ and $Pr_{0.36}Ca_{0.64}MnO_3$, differ markedly. Thus, the CO state in the hole-doped $Pr_{0.64}Ca_{0.36}MnO_3$ is destroyed by magnetic fields and by substitution of Cr^{3+} or Ru^{4+} (3%) in the Mn site, while the CO state in the electron-doped $Pr_{0.36}Ca_{0.64}MnO_3$ is essentially unaffected. It is not possible to induce long-range ferromagnetism in the electron-doped manganates by increasing the Mn–O–Mn angles up to 165 and 180 ° as in $La_{0.33}Ca_{0.33}Sr_{0.34}MnO_3$; application of magnetic fields and Cr/Ru substitution (3%) do not result in long-range ferromagnetism and metallicity. Application of magnetic fields on the Cr/Ru-doped, electron-doped manganates also fails to induce metallicity. These unusual features of the electron-doped manganates suggest that the electronic structure of these materials is likely to be entirely different from that of the hole-doped ones, as verified by first-principles linearized muffin-tin orbital calculations.

Introduction

Rare-earth manganates of the formula $Ln_{1-x}A_xMnO_3$ (Ln = rare earth and A = alkaline earth) exhibit extraordinary properties such as colossal magnetoresistance (CMR) and charge ordering.[1] The manganates with $x < 0.5$, where the trivalent Ln ions are substituted by divalent A ions, have come to be designated as hole-doped. Accordingly, the manganates with $x > 0.5$, where the Ln ion substitutes a divalent A ion, are being referred to as electron-doped.[2] The manganates exhibiting CMR, by and large, have compositions in the range $0.1 < x < 0.5$, wherein the average radius of the A-site cations, $\langle r_A \rangle$, is fairly large. These manganates become ferromagnetic because of the double-exchange mechanism of electron hopping between Mn^{3+} ($t_{2g}^3e_g^1$) and Mn^{4+} ($t_{2g}^3e_g^0$) via the oxygen and undergo an insulator–metal transition around the ferromagnetic T_C. CMR is generally highest around T_C, and the T_C is sensitive to $\langle r_A \rangle$. When $\langle r_A \rangle$ is sufficiently small, the materials do not ordinarily exhibit ferromagnetism but instead become charge-ordered insulators. Thus, charge ordering and double exchange are competing interactions in the manganates. As x in $Ln_{1-x}A_xMnO_3$ increases, crossing over from the hole-doped regime to the electron-doped regime ($x > 0.5$), charge ordering becomes the dominant interaction and ferromagnetism does not appear to occur in any of the compositions. In this regime, CMR occurs over a narrow range of compositions, $0.80 < x < 1.0$, but there is no long-range ferromagnetism or metallicity associated with the materials.[2] The various features of the manganates are nicely borne out by the approximate phase diagrams shown in Figure 1, prepared on the basis of available data. What is noteworthy is the marked absence of symmetry in these phase diagrams. While the presence of electron–hole asymmetry in the manganates is not surprising, considering that the introduction of e_g electrons increases the lattice distortion and their removal would have the opposite effect, the asymmetry has some unusual features.

Electron–hole asymmetry is encountered in cuprate superconductors.[3] In the cuprates, superconductivity occurs in the electron-doped regime, although not as prominently as in the hole-doped regime. The electron–

* To whom correspondence should be addressed. E-mail: cnrrao@jncasr.ac.in.
† Jawaharlal Nehru Centre for Advanced Scientific Research.
‡ Indian Institute of Science.
§ University of California.

(1) (a) Ramirez, A. P. *J. Phys.: Condens. Matter* **1997**, *9*, 8171. (b) Rao, C. N. R.; Raveau, B., Eds. *Colossal Magnetoresistance, Charge ordering and Related Properties of Manganese Oxides*; World Scientific: Singapore, 1998. (c) Rao, C. N. R.; Arulraj, A.; Cheetham, A. K.; Raveau, B. *J. Phys.: Condens. Matter* **2000**, *12*, R83.
(2) (a) Maignan, A.; Martin, C.; Damay, F.; Raveau, B. *Chem. Mater.* **1998**, *10*, 950. (b) Santhosh, P. N.; Arulraj, A.; Vanitha, P. V.; Singh, R. S.; Sooryanarayana, K.; Rao, C. N. R. *J. Phys.: Condens. Matter* **1999**, *11*, L27.
(3) Ramakrishnan, T. V.; Rao, C. N. R. *Superconductivity Today*, 2nd ed.; University of India Press: India, 1999.

10.1021/cm000464w CCC: $20.00   © 2001 American Chemical Society
Published on Web 02/06/2001

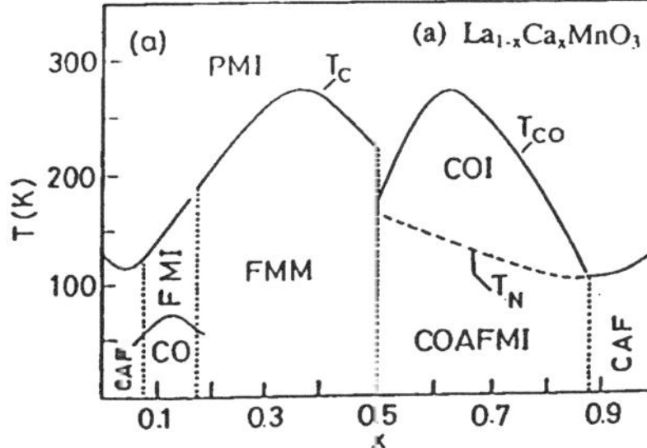

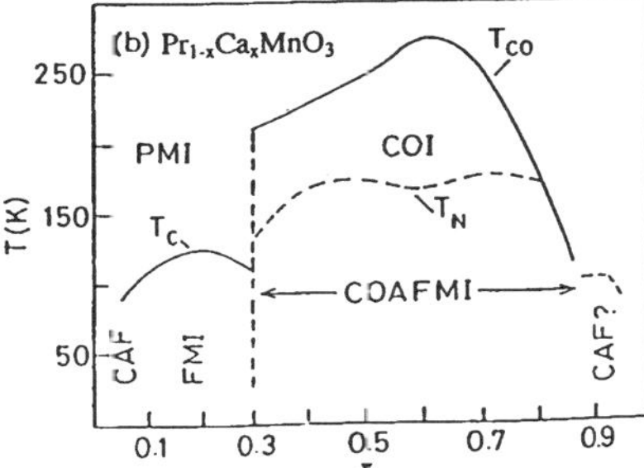

Figure 1. Phase diagrams of (a) $La_{1-x}Ca_xMnO_3$ and (b) $Pr_{1-x}Ca_xMnO_3$: CAF, canted antiferromagnet; CO, charge-ordered state; FMI, ferromagnetic insulator; FMM, ferromagnetic metal; PMI, paramagnetic insulator; COI, charge-ordered insulator (paramagnetic); COAFMI, charge-ordered antiferromagnetic insulator. These diagrams have been prepared based on the available data in the literature and reflect the properties of the systems fairly satisfactorily. The source material for diagram a can be found from ref 1c and from Cheong and Chen in ref 1b. The source material for diagram b can be found in ref 1b,c and in Maignan et al. *Phys. Rev. B* **1999**, *B60*, 12191.

hole asymmetry in the rare-earth manganates, involving the total absence of the ferromagnetic metallic (FMM) state in the electron-doped regime, is therefore worthy of investigation. We have studied this interesting problem by comparing the structural properties of the hole- and electron-doped manganates of similar compositions in the $Pr_{1-x}Ca_xMnO_3$ system ($x = 0.36$ and 0.64). Although there have been several studies on hole-doped compositions of this system,[1] a careful comparison of hole- and electron-doped materials seemed to be necessary. In particular, we have examined whether ferromagnetism can be induced in the electron-doped material by appropriate cation substitution in the B-site and/or application of magnetic fields. The cations chosen for this purpose are Cr^{3+} ($t_{2g}^3e_g^0$) and Ru^{4+} ($t_{2g}^4e_g^0$), which have been found to be effective in destroying charge ordering in materials such as $Nd_{0.5}Ca_{0.5}MnO_3$ and $Sm_{0.5}Ca_{0.5}MnO_3$.[4] To ensure that the Mn−O−Mn angle is not the limiting factor, we have prepared a manganate of the composition $La_{0.33}Ca_{0.33}Sr_{0.34}MnO_3$ with significantly large Mn−O−Mn angles and studied the effects of B-site substitution and magnetic fields on the properties of this material. We have also carried out first principles

electronic structure calculations to understand what makes the hole- and electron-doped manganates different.

Experimental Section

Polycrystalline samples of the manganates were prepared by the ceramic method. Stoichiometric quantities of the respective rare-earth oxides, alkaline-earth carbonates, MnO_2 or Mn_3O_4, and the dopant transition-metal oxide (Cr_2O_3 or RuO_2) were mixed and preheated at 1173 K for 12 h in air. They were subsequently ground and heated at 1473 K for another 12 h in air. The mixture so obtained was pelletized and heated at 1673 K. The X-ray diffraction pattern recorded (using a Seifert XRD 3000TT instrument) showed a single phase for all of the compositions prepared. Rietveld analysis was carried using the structure refinement program GSAS. Data were collected between $2\theta = 10°$ and $100°$ with a scan step of $0.02°$.

Electrical resistivity measurements were carried out on pressed pellets of polycrystalline materials by the four-probe method between 300 and 20 K. Magnetic measurements were carried out using a vibrating sample magnetometer (VSM-7300, Lakeshore Inc.) between 300 and 50 K employing a field of 0.01 T. Magnetoresistivity measurements were carried out using a cryocooled closed-cycle superconducting magnet designed by us along with Cryo Industries of America, Manchester, MA.

Results and Discussion

The phase diagrams of two rare-earth manganates, $Ln_{1-x}Ca_xMnO_3$ (Ln = La and Pr) in Figure 1, clearly demonstrate the electron−hole asymmetry present in these materials and also the preponderance of the charge-ordered (CO) state in the electron-doped regime ($x > 0.5$). The FMM state, generally found in the hole-doped regime ($x \leq 0.5$), is favored by the large size of the A-site cations.[5] Thus, the FMM state occurs up to $x = 0.5$ in $La_{1-x}Ca_xMnO_3$ (Figure 1a). In $Pr_{1-x}Ca_xMnO_3$, the FMM state is not found at any composition, and there is only a ferromagnetic insulating (FMI) state when $0.1 \leq x \leq 0.3$. The charge-ordering regime in $La_{1-x}Ca_xMnO_3$ is $0.5 \leq x \leq 0.85$ but is considerably wider ($0.3 \leq x \leq 0.85$) in $Pr_{1-x}Ca_xMnO_3$. Accordingly, charge ordering occurs in both the hole- and electron-doped compositions of the latter system. Charge ordering in these systems is ascertained by the appearance of superlattice reflections in the diffraction patterns and also by the occurrence of anomalies (observation of maxima) in the temperature variation of magnetic susceptibility and the activation energy for conduction.[1c]

Effects of Cation Size and Size Disorder. In Figure 2, we plot the charge-ordering transition temperatures, T_{CO}, in $Ln_{0.5}Ca_{0.5}MnO_3$ and $Ln_{0.36}Ca_{0.64}MnO_3$ against the average radius of the A-site cations $\langle r_A \rangle$. Here, $\langle r_A \rangle$ is varied by changing the Ln ion. While T_{CO} increases with a decrease in $\langle r_A \rangle$ in the case of $Ln_{0.5}Ca_{0.5}MnO_3$, it is not very sensitive to $\langle r_A \rangle$ in electron-doped $Ln_{0.36}Ca_{0.64}MnO_3$. In Table 1, we compare the T_{CO} values and other properties of $Ln_{0.64}Ca_{0.36}MnO_3$ and $Ln_{0.36}Ca_{0.64}MnO_3$. The T_{CO} value is generally higher in the latter system compared to that of the hole-doped materials, but there is little variation with $\langle r_A \rangle$ in both

(4) (a) Barnabe, A.; Maignan, A.; Hervieu, M.; Raveau, B. *Eur. Phys. J.* **1998**, *B1*, 145. (b) Barnabe, A.; Hervieu, M.; Martin, C.; Maignan, A.; Raveau, B. *J. Mater. Chem.* **1998**, *8*, 1405. (c) Martin, C.; Maignan, A.; Damay, F.; Hervieu, M.; Raveau B. *J. Solid State Chem.* **1997**, *134*, 198. (d) Vanitha, P. V.; Arulraj, A.; Raju, A. R.; Rao, C. N. R. *C. R. Acad. Sci. Paris* **1999**, *2*, 595.

(5) (a) Hwang, H. Y.; Cheong, S. W.; Radaelli, P. G.; Marezio, M.; Batlogg, B. *Phys. Rev. Lett.* **1995**, *75*, 914. (b) Mahesh, R.; Mahendiran, R.; Raychaudhuri, A. K.; Rao, C. N. R. *J. Solid State Chem.* **1995**, *120*, 204.

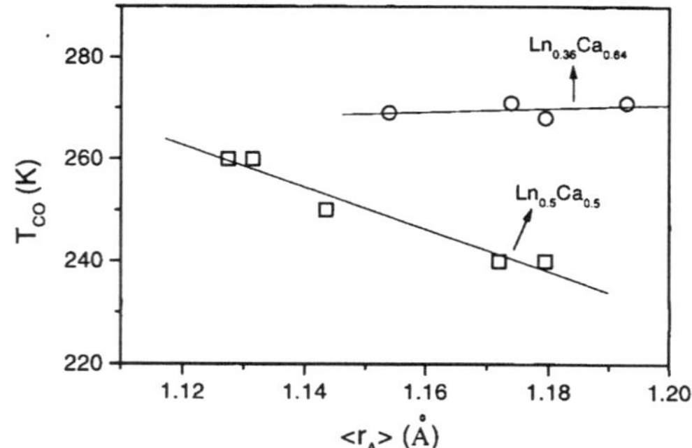

Figure 2. Variation of the charge-ordering transition temperature, T_{CO}, with the average size of the A-site cation, $\langle r_A \rangle$.

Table 1. Effect of $\langle r_A \rangle$ on Charge Ordering in $Ln_{1-x}Ca_xMnO_3$ (x = 0.36 and 0.64)

			lattice parameter (Å)			
Ln	<r_A> (Å)	σ^2 (Å^2)	a	b	c	$T_{CO}{}^a$ (K)
			$Ln_{0.64}Ca_{0.36}MnO_3$			
La	1.203	0.0003	5.454	5.468	7.704	b
Pr	1.179	0.0000	5.413	5.442	7.676	210
Nd	1.169	0.0001	5.407	5.458	7.646	212
			$Ln_{0.36}Ca_{0.64}MnO_3$			
La	1.193	0.0013	5.390	5.391	7.588	271
Pr	1.179	0.0000	5.374	5.369	7.576	268
Nd	1.174	0.0001	5.388	5.361	7.570	271

^{a} From magnetization measurements (H = 100 G). ^{b} This compound exhibits ferromagnetism and an insultor−metal transition around the T_C (~250 K).

series of compounds. In the series of manganates listed in Table 1, the cation size disorder,[6a] as measured by the variance, σ^2, is quite small. It may be noted that in the electron-doped manganates T_{CO} increases with electron concentration, x, but the ferromagnetic component[2a] (in the cluster regime 0.0 < x < 0.2 in $Ca_{1-x}Ln_xMnO_3$) is only slightly affected by $\langle r_A \rangle$ for a fixed value of x.

The effect of cation size disorder on the charge-ordering transition in $Ln_{0.5}Ca_{0.5}MnO_3$ has been investigated by keeping the average radius of the A-site cation fixed and varying σ^2. The value of σ^2 is varied by making different combinations of the Ln and alkaline-earth ions.[6b] The slope of the linear $T_{CO}-\sigma^2$ plot for $Ln_{0.5}Ca_{0.5}MnO_3$ is 10 975 K Å^{-2} and the intercept of the plot, $T_{CO}{}^0$, is 236 K. The value of $T_{CO}{}^0$ corresponds to that of the disorder-free manganate. We have not been able to obtain sufficient reliable data on the variation of T_{CO} with σ^2 (at fixed $\langle r_A \rangle$) in hole-doped compositions of the type $Ln_{0.64}Ca_{0.36}MnO_3$, but the limited data available show only small changes. In the case of the electron-doped $Ln_{0.36}Ca_{0.64}MnO_3$, however, we have obtained reliable data for two series of manganates with fixed $\langle r_A \rangle$ values of 1.180 and 1.174 Å, respectively, corresponding to $Pr_{0.36}Ca_{0.64}MnO_3$ and $Nd_{0.36}Ca_{0.64}MnO_3$. These compounds, along with their structural data and T_{CO} values, are listed in Table 2. We show the plots of T_{CO} against σ^2 for the two series of manganates in Figure 3. The plots are linear, giving slopes 6408 and 5813 K Å^{-2} for fixed $\langle r_A \rangle$ of 1.180 and 1.174 Å, respec-

tively. The intercept, $T_{CO}{}^0$, is around 266 K in both of the cases, a value considerably higher than that in $Ln_{0.5}A_{0.5}MnO_3$.[6b] An examination of the phase diagram in $La_{1-x}Ca_xMnO_3$ in Figure 1a shows that in $La_{1-x}Ca_xMnO_3$ T_{CO} reaches a maximum value around x = 0.65. We see a similar maximum in the phase diagram of $Pr_{1-x}Ca_xMnO_3$ as well (Figure 1b). In the latter system, however, T_{CO} increases with the hole concentration, x, in the hole-doped regime (0.3 < x ≤ 0.5) and with the electron concentration, $1 - x$, in the electron-doped regime (0.6 < x ≤ 0.85). Despite some of these apparent similarities, the hole-doped and electron-doped compositions exhibit significant differences in their electronic and magnetic properties, as detailed in the following sections.

Comparison of the Hole- and Electron-Doped $Pr_{1-x}Ca_xMnO_3$ (x = 0.36 and 0.64). To understand the nature of electron−hole asymmetry in the rare-earth manganates, it is useful to compare the electronic and magnetic properties of comparable compositions of the hole- and electron-doped materials. Thus, the hole-doped $La_{0.7}Ca_{0.3}MnO_3$ becomes ferromagnetic around 250 K, at which temperature it exhibits an insulator−metal transition. The electron-doped $La_{0.3}Ca_{0.7}MnO_3$, on the other hand, gets charge-ordered at 271 K and does not exhibit the FMM state at any temperature. A better appreciation of the differences in the properties of the hole- and electron-doped manganates is obtained by comparing the properties of $Pr_{1-x}Ca_xMnO_3$ at the same carrier concentration (equal values of $1 - x$ and x). We have carried out detailed studies on $Pr_{0.64}Ca_{0.36}MnO_3$ and $Pr_{0.36}Ca_{0.64}MnO_3$, both of which are charge-ordered.

$Pr_{0.64}Ca_{0.36}MnO_3$, I, and $Pr_{0.36}Ca_{0.64}MnO_3$, II, are both orthorhombic (*Pbnm*), but the unit cell is larger in the former as expected on the basis of the relative sizes of Mn^{3+} and Mn^{4+}. In Table 3, we list the atomic coordinates and the lattice parameters of the two manganates obtained from the Rietveld analysis. The Mn−O distances in I are longer than those in II, but the Mn−O−Mn angles in the two are comparable.

Both $Pr_{0.64}Ca_{0.36}MnO_3$ and $Pr_{0.36}Ca_{0.64}MnO_3$ get charge-ordered in the paramagnetic state, with transition temperatures (T_{CO}) of 210 and 268 K, respectively. They show maxima in the magnetization curves at the charge-ordering transition temperatures (Figure 4). $Pr_{0.64}Ca_{0.36}MnO_3$ also shows an antiferromagnetic transition around 140 K. The nature of the transitions has been ascertained independently by diffraction studies as well as EPR and other measurements.[1,7] Both $Pr_{0.64}$-$Ca_{0.36}MnO_3$ and $Pr_{0.36}Ca_{0.64}MnO_3$ are insulators down to low temperatures, as is expected of charge-ordered compositions, but the electron-doped composition shows a more marked change in resistivity at T_{CO} (Figure 5). The difference between the two lies in the effect of magnetic fields. Application of a magnetic field of 12 T melts the CO state to a metallic state in the case of $Pr_{0.64}Ca_{0.36}MnO_3$ (Figure 5a). A 12 T magnetic field has no effect whatsoever on the resistivity of $Pr_{0.36}Ca_{0.64}$-MnO_3 (Figure 5b).

Substitution of Cr^{3+} or Ru^{4+} in the Mn site of certain charge-ordered manganates is known to destroy the CO state, rendering them ferromagnetic and metallic.[1c,4,8] The effect of Cr^{3+} doping on $Pr_{1-x}Ca_xMnO_3$ has been

(6) (a) Rodriguez-Martinez, L. M.; Attfield, J. P. *Phys. Rev.* **1996**, *B54*, R15622; **1998**, *B58*, 2426. (b) Vanitha, P. V.; Santhosh, P. N.; Singh, R. S.; Rao, C. N. R.; Attfield, J. P. *Phys. Rev.* **1999**, *B59*, 13539.

(7) Gupta, R.; Joshi, J. P.; Bhat, S. V.; Sood, A. K.; Rao, C. N. R. *J. Phys.: Condens. Matter* **2000**, *12*, 6919.

Table 2. Structure and Properties of $Ln_{0.36-x}Ln'_xCa_{0.64-y}Sr_yMnO_3$ with Fixed $\langle r_A \rangle$ Values[a]

composition	σ^2 (Å²)	lattice parameter (Å)			% D (300 K)	T_{CO} (K)
		a	b	c		
		$\langle r_A \rangle = 1.174$ Å				
$Nd_{0.36}Ca_{0.64}MnO_3$	0.0001	5.388	5.361	7.570	0.33	271 (261)
$Pr_{0.28}Gd_{0.08}Ca_{0.64}MnO_3$	0.0003	5.353	5.366	7.548	0.18	279 (270)
$La_{0.185}Gd_{0.175}Ca_{0.64}MnO_3$	0.0011	5.354	5.369	7.537	0.27	257 (253)
$La_{0.225}Y_{0.135}Ca_{0.64}MnO_3$	0.0017	5.353	5.370	7.563	0.15	249 (248)
$Sm_{0.36}Ca_{0.554}Sr_{0.086}MnO_3$	0.0022	5.355	5.372	7.569	0.15	263 (258)
$Nd_{0.1}Gd_{0.26}Ca_{0.528}Sr_{0.112}MnO_3$	0.0033	5.355	5.364	7.580	0.06	245 (243)
$Gd_{0.15}Y_{0.21}Ca_{0.433}Sr_{0.207}MnO_3$	0.0066	5.346	5.374	7.567	0.21	~229 (223)
		$\langle r_A \rangle = 1.18$ Å				
$Pr_{0.36}Ca_{0.64}MnO_3$	0.0000	5.374	5.369	7.576	0.11	267 (267)
$Nd_{0.18}Sm_{0.18}Ca_{0.553}Sr_{0.087}MnO_3$	0.0019	5.369	5.370	7.580	0.07	~256 (251)
$Nd_{0.18}Gd_{0.18}Ca_{0.518}Sr_{0.122}MnO_3$	0.0031	5.363	5.376	7.573	0.14	250 (245)
$La_{0.16}Y_{0.2}Ca_{0.526}Sr_{0.114}MnO_3$	0.0043	5.364	5.370	7.578	0.07	~241 (~238)
$La_{0.1}Y_{0.26}Ca_{0.46}Sr_{0.18}MnO_3$	0.0060	5.386	5.369	7.550	0.32	234 (222)

[a] The values in parentheses are obtained from resistivity data; % D is the orthorhombic distortion.

Table 3. Atomic Coordinates and Structural Parameters of I, $Pr_{0.64}Ca_{0.36}MnO_3$,[a] and II, $Pr_{0.36}Ca_{0.64}MnO_3$[b]

atom	site	x	y	z	frac	U_{ISO}
Pr	4c	−0.0083 (0.5027)	0.0315 (0.4746)	0.2500 (0.2500)	0.6400 (0.3600)	0.0049 (−0.0033)
Ca	4c	−0.0083 (0.5027)	0.0315 (0.4746)	0.2500 (0.2500)	0.3600 (0.6400)	0.0049 (−0.0033)
Mn	4b	0.5000 (0.5000)	0.0000 (0.0000)	0.0000 (0.0000)	1.0000 (1.0000)	0.0040 (0.0063)
O	8d	0.0529 (0.0544)	0.4939 (0.5213)	0.2500 (0.2500)	1.0000 (1.0000)	0.0405 (0.0901)
O	4c	−0.7115 (0.2854)	0.2809 (0.2717)	0.0354 (−0.0460)	1.0000 (1.0000)	0.0122 (−0.0367)

bond	distance (Å)	bond	angle (deg)
Mn−O	2 × 1.935 (2 × 1.892)	Mn−O−Mn	4 × 157.6 (4 × 155.6)
	2 × 1.941 (2 × 1.993)		2 × 162.9 (2 × 161.2)
	2 × 1.989 (2 × 1.916)		

[a] $a = 5.4310$ Å, $b = 5.4573$ Å, $c = 7.6761$ Å; *Pbnm*; $R_{wp} = 3.45\%$. [b] $a = 5.3664$ Å, $b = 5.3746$ Å, $c = 7.5603$ Å; *Pbnm*; $R_{wp} = 3.28\%$. The values in brackets correspond to those of II.

investigated in the composition region $0.6 \leq x \leq 0.7$, with the Cr^{3+} content going up to 12%.[4a] These workers find a marked effect when Cr^{3+} is around 10%, at which composition one would expect clustering of the dopant ions leading to superexchange-induced ferromagnetism. We have substituted Mn by Cr^{3+} or Ru^{4+}, keeping the dopant concentration at 3% to avoid clustering. On doping with 3% Cr^{3+}, $Pr_{0.64}Ca_{0.36}MnO_3$ becomes ferromagnetic with a T_C of 130 K, but $Pr_{0.36}Ca_{0.64}MnO_3$ remains paramagnetic and charge-ordered, albeit with a slightly lower T_{CO} (215 K) as shown in Figure 6a. The 3% Ru doping shows similar differences between the two manganates (Figure 6b).

In Figure 7, we show the effect of 3% Cr^{3+} doping on the resistivity of $Pr_{0.64}Ca_{0.36}MnO_3$ and $Pr_{0.36}Ca_{0.64}MnO_3$. The former exhibits an insulator−metal (I−M) type transition around 80 K, but the latter remains an insulator. The I−M transition in the hole-doped system is shifted to higher temperatures on applying magnetic fields. Application of magnetic fields does not render the Cr^{3+}-doped $Pr_{0.36}Ca_{0.64}MnO_3$ metallic (Figure 7) at any temperature, indicating that it may not be possible to induce the FMM state in this electron-doped material even under favorable conditions. Results with 3% Ru^{4+} doping in the two manganates are similar, in that the hole-doped material becomes FMM while the CO state in the electron-doped material is essentially unaffected. In the insets of Figure 7, we show the results obtained with 3% Ru^{4+}-doped $Pr_{0.64}Ca_{0.36}MnO_3$ and $Pr_{0.36}Ca_{0.64}-MnO_3$. The failure to destroy the CO state of $Pr_{0.36}Ca_{0.64}-MnO_3$ with Cr/Ru doping as well as with a magnetic

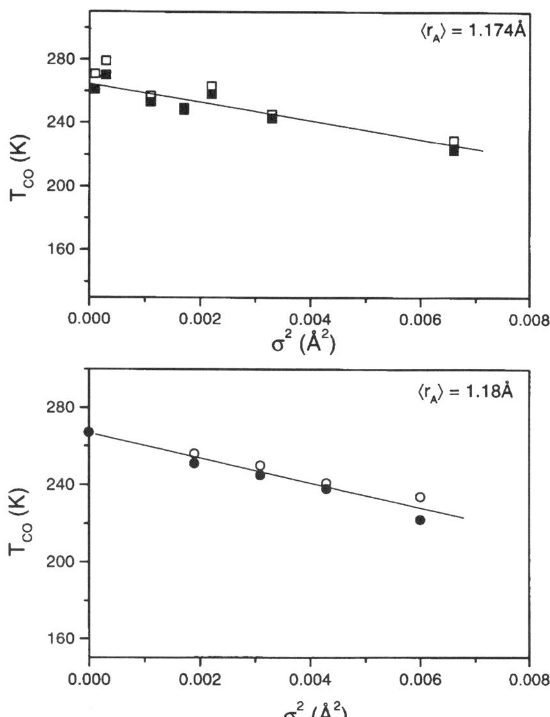

Figure 3. Variation of the charge-ordering temperature in $Ln_{0.36}Ca_{0.64}MnO_3$ with σ^2 for fixed $\langle r_A \rangle$ values. Open symbols represent data obtained from magnetic measurements, and the corresponding closed symbols are from resistivity measurements.

(8) (a) Raveau, B.; Maignan, A.; Martin, C.; Hervieu, M. *Chem. Mater.* **1998**, *10*, 2641. (b) Vanitha, P. V.; Singh, R. S.; Natarajan, S.; Rao, C. N. R. *J. Solid State Chem.* **1998**, *137*, 365.

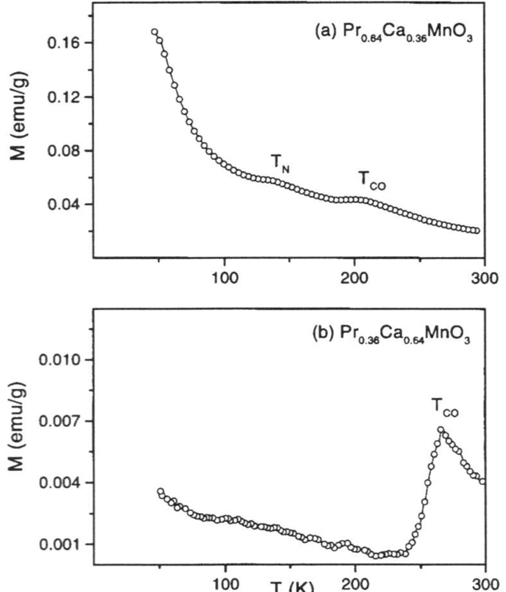

Figure 4. Temperature variation of the magnetization of (a) $Pr_{0.64}Ca_{0.36}MnO_3$ and (b) $Pr_{0.36}Ca_{0.64}MnO_3$.

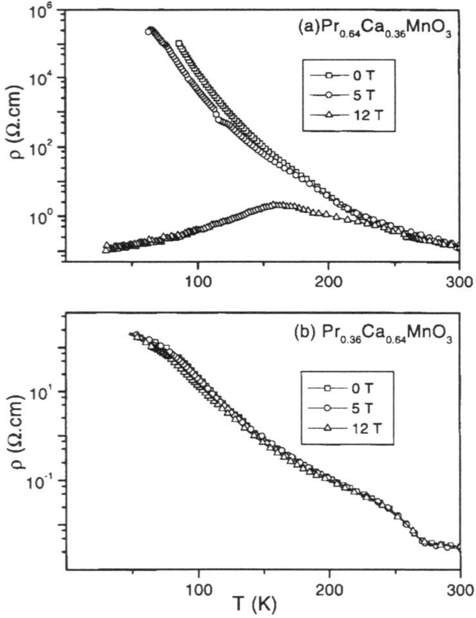

Figure 5. Temperature variation of the electrical reisistivity of (a) $Pr_{0.64}Ca_{0.36}MnO_3$ and (b) $Pr_{0.36}Ca_{0.64}MnO_3$. The effect of magnetic fields is shown.

field of 12 T is noteworthy. Raveau et al.[9] have recently observed that Ru substitution in $Ln_{0.4}Ca_{0.6}MnO_3$ gives rise to FMM clusters, but the effect is prominent at high Ru concentrations (>3%). The observed effect is probably due to clusters or domains containing Ru ions, as suspected by these authors. Furthermore, the observed

(9) Raveau, B.; Maignan, A.; Martin, C.; Mahendiran, R.; Hervieu, M. *J. Solid State Chem.* **2000**, *151*, 330.

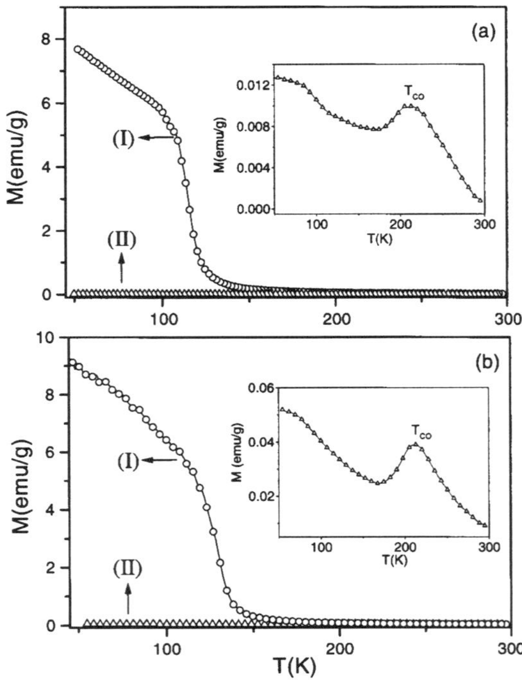

Figure 6. Effect of (a) 3% Cr doping and (b) 3% Ru doping in the Mn site of $Pr_{0.64}Ca_{0.36}MnO_3$, I, and $Pr_{0.36}Ca_{0.64}MnO_3$, II, on the magnetization. The inset gives the magnetization data of II on an enlarged scale to show T_{CO}.

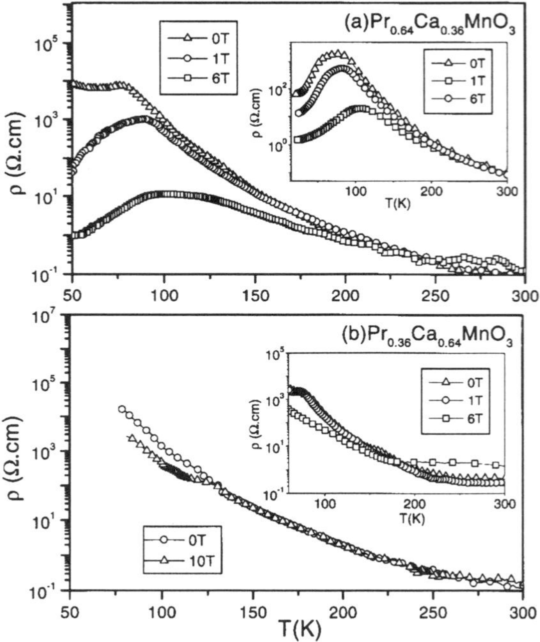

Figure 7. Effect of 3% Cr^{3+} doping on the electrical resistivity of (a) $Pr_{0.64}Ca_{0.36}MnO_3$ and (b) $Pr_{0.36}Ca_{0.64}MnO_3$. The effect of magnetic fields is also shown. Insets in a and b show the effect of 3% Ru doping.

magnetization in these samples is low. This raises the question as to whether long-range ferromagnetism can ever occur in the electron-doped manganates. We note

Table 4. Atomic Coordinates and Structural Parameters of La$_{0.33}$Ca$_{0.33}$Sr$_{0.34}$MnO$_3$[a]

atom	site	x	y	z	frac	U_{ISO}
La	4b	0.0000	0.5000	0.2500	0.3300	−0.0134
Ca	4b	0.0000	0.5000	0.2500	0.3300	0.0245
Sr	4b	0.0000	0.5000	0.2500	0.3400	0.0212
Mn	4c	0.0000	0.0000	0.0000	1.0000	0.0007
O	4a	0.0000	0.0000	0.2500	1.0000	0.0051
O	8h	0.2799	0.7799	0.0000	1.0000	0.0051

bond	distance (Å)	bond	angle (deg)
Mn−O	4 × 1.909	Mn−O−Mn	4 × 166.4
	2 × 1.949		2 × 180.0

[a] $a = b = 5.3608$ Å, $c = 7.7857$ Å; $I4/mcm$; $R_{wp} = 11.82\%$.

here that Maignan et al.[10] find a cluster glass state (but no long-range ferromagnetism) in Ca$_{1-x}$Sm$_x$MnO$_3$ ($0 \leq x \leq 0.12$). Neumeier and Cohn[11] observe only local ferromagnetic regions within an antiferromagnetic host in Ca$_{1-x}$La$_x$MnO$_3$ ($0 \leq x \leq 0.2$), as was indeed observed earlier by Mahendiran et al.[12]

Absence of Long-Range Ferromagnetism in Electron-Doped Manganates. To answer the above question, it is important to ensure that the Mn−O−Mn angle is not a limiting factor. This is because the Mn−O−Mn angle in Pr$_{1-x}$Ca$_x$MnO$_3$ is generally in the 156−162 ° range (Table 3), which may be considered to be somewhat small. For ferromagnetism to be favored in the electron-doped manganates, it is important to have a material with a much larger Mn−O−Mn angle. To decide on the composition of such a material, the following considerations are relevant. The Mn−O−Mn angles in CaMnO$_3$ and La$_{0.5}$Ca$_{0.5}$MnO$_3$ are around 158 ° and 160 °, respectively.[13,14] Therefore, substitution of Ca by any of the rare earths would not increase the Mn−O−Mn angle beyond 160 °. It is, however, possible to increase the angle by Sr substitution, as in Ca$_{1-x}$Sr$_x$MnO$_3$ and La$_{0.5}$Ca$_{0.5-x}$Sr$_x$MnO$_3$.[15,16] We therefore prepared a manganate of the composition La$_{0.33}$Ca$_{0.33}$Sr$_{0.34}$MnO$_3$. The structure of the manganate is tetragonal ($I4/mcm$). On the basis of a Rietveld analysis of the powder X-ray diffraction data, we have obtained the atomic coordinates and structural parameters listed in Table 4. The two Mn−O−Mn distances are around 1.91 and 1.95 Å, while the angles are 166.4 ° and 180 °. These values of the angles are comparable to those found in some of the ferromagnetic and metallic compositions of the hole-doped manganates.

In Figure 8 we show the magnetization and electrical resistivity data of La$_{0.33}$Ca$_{0.33}$Sr$_{0.34}$MnO$_3$. The material is a paramagnetic insulator down to 25 K, with a charge-ordering transition of around 220 K. Application of magnetic fields up to 12 T has no effect on the electrical

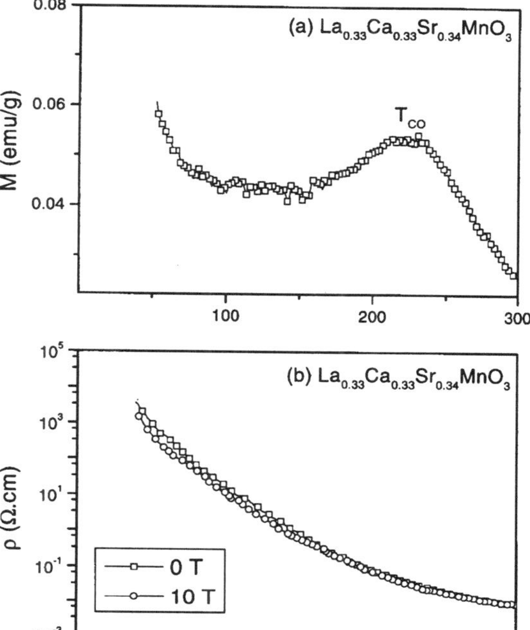

Figure 8. Temperature variation of (a) the magnetization and (b) the resistivity of La$_{0.33}$Ca$_{0.33}$Sr$_{0.34}$MnO$_3$. The effect of magnetic fields on the resistivity is shown.

resistivity (Figure 8b). Clearly, the large Mn−O−Mn angles do not help to make this manganate ferromagnetic. Furthermore, 3% Cr^{3+} doping of La$_{0.33}$Ca$_{0.33}$Sr$_{0.34}$MnO$_3$ does not transform it to a ferromagnetic metal (Figure 9). What is surprising is that the application of magnetic fields up to 10 T does not induce an I−M transition in the Cr-doped material. The same holds for the 3% Ru^{4+}-doped material, as shown in the insets of Figure 9a,b. It must be remembered that the analogous hole-doped composition La$_{0.67}$A$_{0.33}$MnO$_3$ (A = Ca/Sr) becomes a ferromagnetic metal at fairly high temperatures (230−300 K).[1] These results suggest that it may not be possible to make the electron-doped rare-earth manganates exhibit long-range ferromagnetism and metallicity. As observed earlier, other studies also show at best local ferromagnetic interactions or the presence of FM clusters.[9−12]

The absence of long-range ferromagnetism and metallicity in the electron-doped manganates is difficult to comprehend. One difference that is noteworthy is that the hole-doped manganates possess a higher proportion of e$_g$ electrons relative to the degenerate e$_g$ orbitals. There are also some intrinsic differences between the Mn^{3+} (d^4) and Mn^{4+} (d^3) ions. Although the structure of the parent LnMnO$_3$ compounds is influenced by the Jahn−Teller distorted Mn^{3+}, the stability of this state is not sufficient to inhibit the electron-transfer process when Mn^{4+} ions are introduced. The facile electron transfer required for double exchange, and hence ferromagnetism, can be sustained in such a hole-doped system. By contrast, the high ligand-field stabilization of the preponderant Mn^{4+} ions in the electron-doped materials can inhibit electron transfer. To unravel the

(10) Maignan, A.; Martin, C.; Damay, F.; Raveau, B.; Hejtmanek, J. *Phys. Rev.* **1998**, *B58*, 2758.

(11) Neumeier, J. J.; Cohn, J. L. *Phys. Rev.* **2000**, *B61*, 14319.

(12) Mahendiran, R.; Tiwary, S. K.; Raychaudhuri, A. K.; Ramakrishnan, T. V.; Mahesh, R.; Rangavittal, N.; Rao, C. N. R. *Phys. Rev.* **1996**, *B53*, 3348.

(13) Poeppelmeirer, K. R.; Leonowicz, M. E.; Scanlon, J. C.; Longo, J. M.; Yellon, W. B. *J. Solid State Chem.* **1982**, *45*, 71.

(14) Radaelli, P. G.; Cox, D. E.; Marezio, M.; Cheong, S. W.; Schiffer, P. E.; Ramirez, A. P. *Phys. Rev. Lett.* **1995**, *75*, 4488. (b) Radaelli, P. G.; Cox, D. E.; Capogna, L.; Cheong, S.-W.; Marezio, M. *Phys. Rev.* **1999**, *B59*, 14440.

(15) Taguchi, H.; Sonoda, M.; Nagao, M. *J. Solid State Chem.* **1998**, *137*, 82.

(16) Sundaresan, A.; Paulose, P. L.; Mallik, R.; Sampathkumaran, E. V. *Phys. Rev.* **1998**, *B57*, 2690.

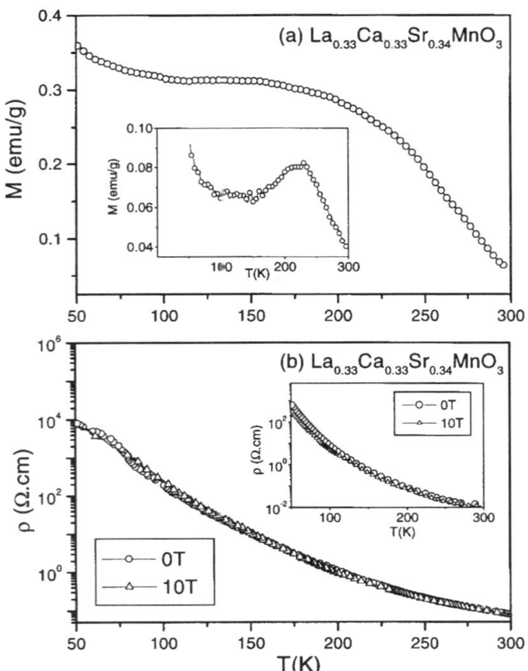

Figure 9. Effect of 3% Cr^{3+} doping on (a) the magnetization and (b) the resistivity of $La_{0.33}Ca_{0.33}Sr_{0.34}MnO_3$. The effect of magnetic fields on the resistivity is also shown. Insets in a and b show the effect of 3% Ru doping.

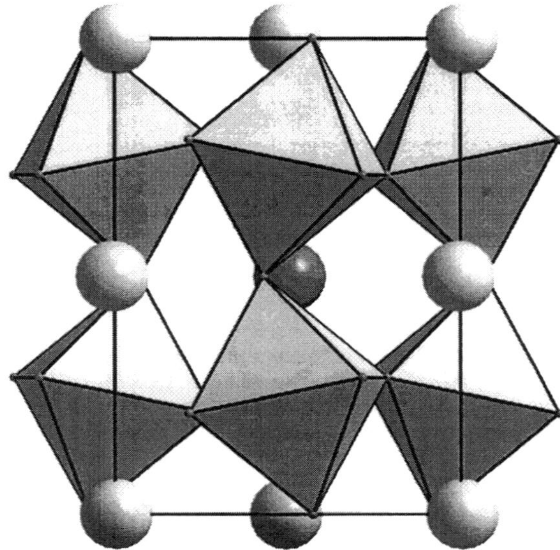

Figure 10. Monoclinic (space group *Pm*) structure of La_3-$CaMn_4O_{12}$. The dark spheres are La, and the light sphere is Ca. The view is looking down the short $a = a_P\sqrt{2}$ axis.

cause(s) for the absence of metallicity in the electron-doped managanates, we have carried out some electronic structure calculations.

Electronic Structure Calculations. We have used first-principles electronic structure calculations as manifest in the (spin) density functional linearized muffin-tin orbital method to examine whether the asymmetry in properties is reflected in a corresponding asymmetry in the one-electron band structure. While in a more complete analysis explicit electron correlation of the Hubbard U type would be intrinsic to the calculation,[17] we have taken the view that one-electron bandwidths point to the possible role that correlation might play and that correlation could be a consequence of the one-electron band structure rather than an integral part of the electronic structure. We have chosen the $La_{1-x}Ca_xMnO_3$ system for our calculations to avoid complications due to 4f electrons in the corresponding Pr system.

For our calculations on $La_3CaMn_4O_{12}$ ($La_{0.75}Ca_{0.25}$-MnO_3), we have used the structure reported by Radaelli et al.[14] refined from powder neutron diffraction data on 20 K. Ordering La and Ca in a supercell yields a structure with the same lattice parameters but in the space group *Pm* with 14 atoms (rather than 4) in the asymmetric unit. This structure is displayed in Figure 10. La and Ca have closely similar radii, and ordering them over crystallographically distinct sites is unphysical. However, the bond lengths and angles in the structures used in our calculations closely follow the experiments. For the electron-doped $La_{0.25}Ca_{0.75}MnO_3$

($LaCa_3Mn_4O_{12}$), we have used the cell and positional parameters of the refined 300 K neutron powder diffraction structure of $La_{0.33}Ca_{0.67}MnO_3$ from Radaelli et al.[14] Again, the supercell is in the space group *Pm* rather than in the orthorhombic *Pnma* space group. Calculations were performed using the Stuttgart TB-LMTO-ASA program.[18] The basis sets consisted of 6s, 5d, and 4f orbitals for La, 4s and 3d orbitals for Ca, 4s, 4p, and 3d orbitals for Mn, and 2p orbitals for O. The atomic sphere approximation (ASA) relies on the partitioning of space into atom-centered spheres as well as empty spheres, with the latter being critical in structures that are not closely packed. The basis for the empty spheres is 1s orbitals, with the 2p component treated using downfolding. The spheres are chosen so that the atom-centered spheres do not have a volume overlap of more than 16%. The calculations used 108K points in the primitive Brillouin zone for achieving convergence. Because of the *Pm* supercell employed, the nature of the magnetism becomes a little more complex and perhaps artificial for both of the systems studied. Indeed, there are now two types of somewhat indistinct Mn atoms in the unit cell. We have found a tendency for a ferrimagnetic ground state in both of the manganates, with the two Mn having opposite spins. For simplifying the comparison of the two electronic structures, the Mn atoms were provided similar polarization at the start of the calculations; self-consistency yielded a ferromagnet in both of the cases.

The calculations yielded a ferromagnetic ground state with a magnetic moment of 3.2 μ_B per Mn for La_3Ca-Mn_4O_{12} (spin-only value 3.75 μ_B). Such a reduction in the magnetic moment from the expected value can arise because of an infelicitous choice of sphere radii and does not merit interpretation. The refined neutron moment

(17) Anisimov, V. I.; Zaanen, J.; Andersen, O. K. *Phys. Rev.* **1991**, *B48*, 943.

(18) Andersen, O. K.; Jepsen, O.; et al. *The Stuttgart TB-LMTO-ASA-47 Program*; MPI für Feukörperforschung: Stuttgart, 1998. (b) Skriver, H. L. *The LMTO Method*; Springer: Berlin, 1984.

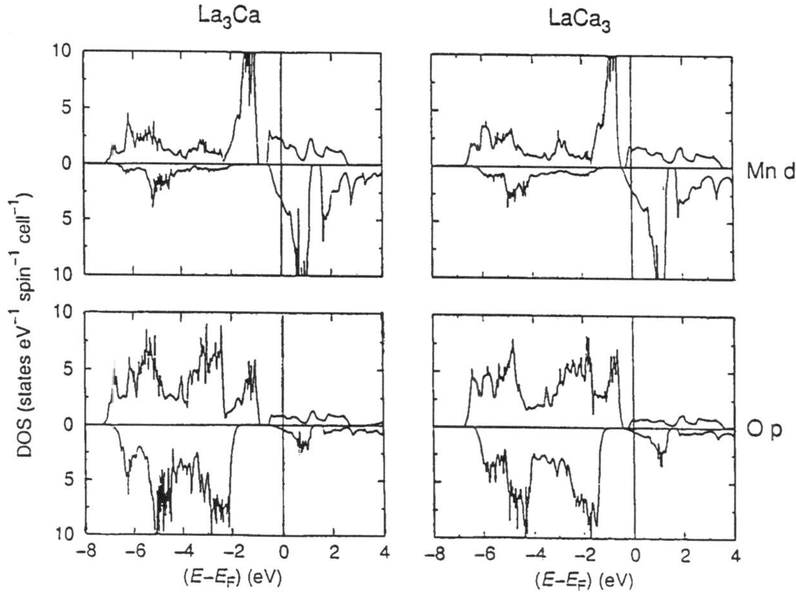

Figure 11. LMTO DOS for Mn and O in La$_3$CaMn$_4$O$_{12}$ and LaCa$_3$Mn$_4$O$_{12}$ near the Fermi energy. The upper halves of each panel display up-spin states, and the lower halves, down-spin states.

on Mn is 3.5 μ_B.[14] For ferromagnetic LaCa$_3$Mn$_4$O$_{12}$, the calculated magnetic moment was 3 μ_B (spin-only value 3.25). Figure 11 compares the spin-polarized densities of state (DOS) in the two spin directions for Mn and O in the two compositions. In each panel, the majority up-spin states are depicted in the upper half and the minority down-spin states in the lower half.

In both of the manganates, we find filled localized t_{2g}^3 (↑) states with a small bandwidth, of the order of 1.5 eV, polarizing a broader conduction band that is essentially e_g (↑) derived, although some t_{2g} (↓) states also occur at the Fermi energy. In La$_3$CaMn$_4$O$_{12}$, the Fermi energy E_F lies in a relatively broad conduction band. Both Mn and O states are present at E_F, indicating some covalency. Of note is the spin differentiation at the Fermi energy, with there being significantly more down-spin Mn states at the E_F than up-spin states. In the case of LaCa$_3$Mn$_4$O$_{12}$, E_F again lies in an e_g (↑) derived conduction band, but because of the smaller number of e_g electrons, E_F is at the band edge. Through examination of the so-called "fat-bands" (energy bands that have been decorated with the character of the corresponding orthonormal orbitals), we know that the e_g states in the perovskite systems[19] are derived from the combination of the narrow d$_{z^2}$ bands and the broader bands formed by the strong covalent overlap between O p$_x$ and p$_y$ and metal d$_{x^2-y^2}$ orbitals. A scheme depicting the nature of the overlap is displayed in Figure 12. In LaCa$_3$Mn$_4$O$_{12}$, where there is a much smaller filling of the e_g, it is the relatively narrow d$_{z^2}$-derived band that is mostly occupied. The finding that E_F lies on a band edge suggests that this oxide would be susceptible to the opening of a gap in the DOS at E_F, through correlations of the Hubbard type. At the same time, the propensity for the

(19) Felser, C.; Seshadri, R.; Leist, A.; Tremel, W. *J. Mater. Chem.* **1998**, *8*, 787.

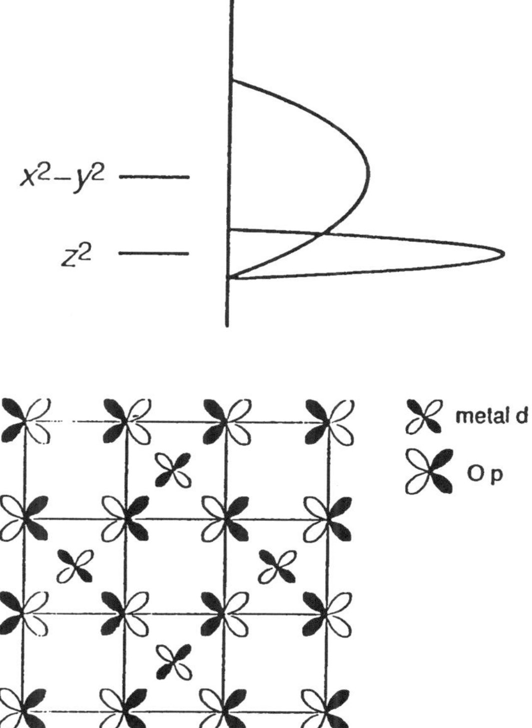

Figure 12. Scheme displaying the Jahn−Teller distorted e_g states becoming bands in solids such as perovskite manganese oxides. Because of overlap between metal d$_{x^2-y^2}$ and O p$_x$ and p$_y$, the d$_{x^2-y^2}$-derived bands are significantly broader. A scheme for such overlap is displayed along the *ab* plane.

localized and delocalized states to be separated through a mobility edge (the formation of an Anderson insulator) is also increased by E_F lying on a band edge. In systems such as the present ones, disorder due to disparate ions occupying the A site of the perovskite structure cannot be avoided. A point of interest is that in $LaCa_3Mn_4O_{12}$ the DOS at E_F shows a smaller spin differentiation than that in $La_3CaMn_4O_{12}$. Spin differentiation is believed to be the key to the unusual magnetic field dependence of the electrical transport properties such as colossal magnetoresistance.[20] This suggests that the electron-doped manganates may be less interesting with respect to the CMR properties.

Conclusions

The electron-doped regime ($x > 0.5$) of the rare_earth manganates of the general formula $Ln_{1-x}Ca_xMnO_3$ is dominated by charge-ordering effects. While the effects of cation size and size disorder on the charge-ordered states of the hole- and electron-doped compositions are similar, their properties vary markedly as revealed by the comparison of the properties of the hole-doped $Pr_{0.64}Ca_{0.36}MnO_3$ and the electron-doped $Pr_{0.36}Ca_{0.64}MnO_3$. The CO state of the former is transformed to the FMM state by magnetic fields as well as by 3% Cr^{3+} or Ru^{4+} substitution in the Mn site, but none of these affects the CO state of the electron-doped material, which remains a paramagnetic insulator under all conditions. Increasing the $Mn-O-Mn$ angle from 158 to 162 °, as in the above two manganates, to $165-180$ ° in $La_{0.33}Ca_{0.33}Sr_{0.34}MnO_3$ does not result in ferromagnetism and metallicity. The CO state in this material is also unaffected by magnetic fields and Cr^{3+}/Ru^{4+} substitution in the Mn site. We are, therefore, prompted to conclude that it is not possible to induce long-range ferromagnetism in the electron-doped manganates by any means. First-principles calculations suggest that this may be because the Fermi level lies on a band edge in these materials.

Acknowledgment. This work was partially supported by the MRSEC Program of the NSF under Grant DMR 96-32716.

(20) Pickett, W. E.; Singh, D. J. *Phys. Rev.* **1996**, *B53*, 1146.

CM000464W

PHYSICAL REVIEW B | VOLUME 61, NUMBER 1 | 1 JANUARY 2000-I

Electric-field-induced melting of the randomly pinned charge-ordered states of rare-earth manganates and associated effects

C. N. R. Rao,* A. R. Raju, V. Ponnambalam, and Sachin Parashar
*Chemistry and Physics of Materials Unit, Jawaharlal Nehru Centre for Advanced Scientific Research, Jakkur P.O.,
Bangalore 560 064, India*

N. Kumar
Raman Research Institute, Bangalore 560 080, India

(Received 30 November 1998; revised manuscript received 31 August 1999)

Films of charge-ordered $Nd_{0.5}Ca_{0.5}MnO_3$, $Gd_{0.5}Ca_{0.5}MnO_3$, $Y_{0.5}Ca_{0.5}MnO_3$, and $Nd_{0.5}Sr_{0.5}MnO_3$ show insulator-metal transitions on the passage of small electrical currents. That such an electric-field-induced transition occurs even in $Y_{0.5}Ca_{0.5}MnO_3$ where the charge-ordered state is not affected by magnetic fields is noteworthy. The transition is attributed to the depinning of the randomly pinned charge solid. These materials also exhibit an interesting memory effect probably due to the randomness of the strength as well as the position of the pinning centers.

I. INTRODUCTION

Charge ordering in some of the compositions of rare-earth manganates of the formula $L_{1-x}A_xMnO_3$ (L = rare earth, A = alkaline earth) has been well-documented.[1,2] Charge ordering in these materials is interesting since it competes with double exchange, giving rise to several interesting properties. The charge-ordered state is insulating unlike the double-exchange regime of the manganates. Two types of charge ordering can be distinguished in the manganates.[2,3] In $Nd_{0.5}Sr_{0.5}MnO_3$ with a relatively large average radius of the A-site cations, $\langle r_A \rangle$, a ferromagnetic metallic (FMM) state ($T_c = 250$ K) transforms to a charge-ordered (CO) state on cooling to ~150 K.[4] Manganates with a small $\langle r_A \rangle$ ($\lesssim 1.17$ Å) do not exhibit the FMM state at any temperature and instead occur in the CO state even at relatively high temperatures. The CO state in a manganate with a relatively large A-site ion radius ($\langle r_A \rangle \gtrsim 1.17$ Å) can be transformed to the FMM state by the application of magnetic fields. On the other hand, even large magnetic fields have negligible effect on the CO state of $Y_{0.5}Ca_{0.5}MnO_3$ with a $\langle r_A \rangle$ of 1.13 Å.[5] The CO state in single crystals of $Pr_{1-x}Ca_xMnO_3$ has been transformed to the FMM state by applying electric fields and/or laser irradiation.[6,7] The effect of electric fields on the CO state of the manganates clearly requires a thorough investigation, not only because of interesting features of the phenomenon but also due to possible technological implications.

There has been little effort to prepare and characterize thin films of the charge-ordered manganates, unlike the thin films of the manganates showing CMR.[8–11] We have prepared thin films of charge-ordered manganates of the general composition $L_{0.5}A_{0.5}MnO_3$ on single-crystal substrates by the nebulized spray pyrolysis of organometallic precursors. More significantly, we have investigated the electric current-induced transition of the insulating CO states to metal-like states in these films. For the purpose of this study, we have chosen three manganates at half doping of the composition

$L_{0.5}Ca_{0.5}MnO_3$ with L = Nd, Gd, and Y with $\langle r_A \rangle$ values of 1.17, 1.14, and 1.13 Å, respectively, as well as $Nd_{0.5}Sr_{0.5}MnO_3$ with a $\langle r_A \rangle$ of 1.24 Å. It is noteworthy that the insulating CO state in all these manganates, including $Y_{0.5}Ca_{0.5}MnO_3$, is melted by passing small currents. Furthermore, the films show nonohmic behavior and interesting memory effects. We propose that the electric-field-induced insulator-metal transitions and associated effects are brought about by the depinning of the randomly pinned charge solids.

II. EXPERIMENTAL

Thin films of the manganates were deposited on Si(100) as well as on lanthanum aluminate, LAO(100), single crystal substrates by employing nebulized spray pyrolysis.[12] This technique involves the pyrolysis of a nebulized spray of organic derivatives of the relevant metals. Since the nebulized spray is deposited on a solid substrate at relatively low temperatures, and with sufficient control of the rate of deposition, the oxide films obtained possess good stoichiometry. Employing acetylacetonates of Nd, Gd, Y, Ca, and Mn, and dipivaloylmethanato strontium as the organometallic precursors, films of ~1000 nm thickness were deposited at 650 K by using air as the carrier gas (1.5 liters/min). The films so obtained were heated at 1000 K in oxygen. The films were characterized by employing x-ray diffraction and scanning electron microscopy. The compositions of the films as determined by EDAX were close to the stated compositions. The films deposited on Si(100) showed a polycrystalline nature while those deposited on LAO were oriented along the (100) direction. The orthorhombic lattice parameters of the materials agree with the literature values. Temperature-dependent resistivity measurements were carried out by employing a close cycle refrigerator and sputtered gold electrodes.

III. RESULTS

In Figs. 1(a) and 1(b), we show the temperature variation of the resistance of $Nd_{0.5}Ca_{0.5}MnO_3$ (NCM) films deposited

0163-1829/2000/61(1)/594(5)/$15.00

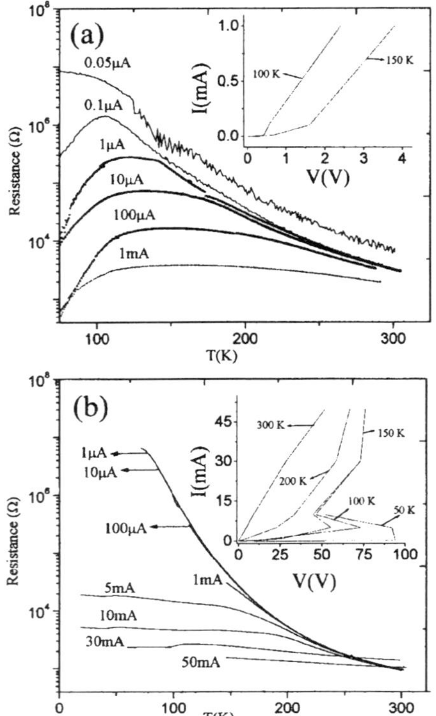

FIG. 1. Temperature variation of the resistance of $Nd_{0.5}Ca_{0.5}MnO_3$ (NCM) films deposited on (a) Si(100) and (b) LAO(100) for different values of current. The insets show I-V characteristics at different temperatures.

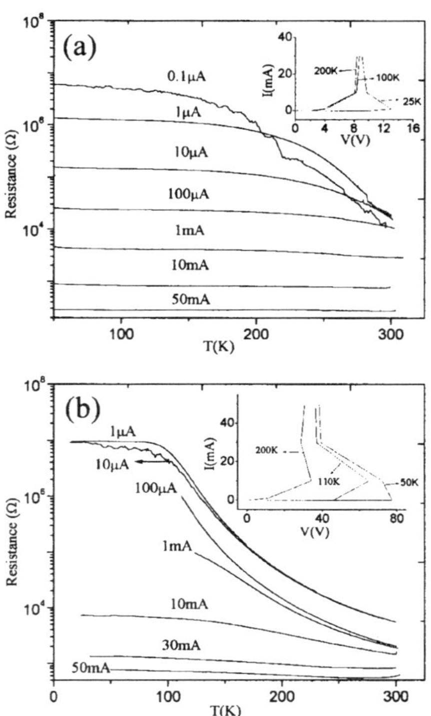

FIG. 2. Temperature variation of the resistance of $Gd_{0.5}Ca_{0.5}MnO_3$ (GCM) films deposited on (a) Si(100) and (b) LAO(100) for different values of current. The insets show I-V characteristics at different temperatures.

on Si(100) and LAO(100), respectively, for different values of the dc current passed. When the current is small (0.05 μA), the film on Si(100) shows insulating behavior. Upon increasing the current, we observe the occurrence of an insulator-metal (I-M) transition (Fig. 1(a)). It is noteworthy that even a current of 0.1 μA causes the I-M transition. (The values of the current density are 1.25 A cm^{-2} and 1.25×10^4 A cm^{-2} for currents of 1 μA and 10 mA, respectively.) The effects observed are not due to local Joule heating, which becomes appreciable only at high currents ($\gtrsim$ 50 mA). In fact, the irrelevance of the Joule heating can be clearly seen in the low-temperature metallic [temperature coefficient of resistance (TCR)>0] regime accessible in Fig. 1(a). Here the resistance for a given current increases with increasing temperature, while for a given temperature the resistance decreases with increasing current. Just the opposite would be true for a Joule heating. It is reasonable, therefore, to assume that the Joule heating is quite irrelevant to our transport results, qualitatively at least. The temperature of the I-M transition shifts from 100 to 150 K with increase in current. The I-V curves show nonohmic behavior as shown in the inset of Fig. 1(a). Measurements on the oriented NCM film deposited on LAO(100) also shows a marked decrease in resistance with increasing current (Fig. 1(b)). We do not clearly see a metal-like decrease in resistance at low temperatures, and the behavior is comparable to that of laser-irradiated $Pr_{1-x}Ca_xMnO_3$ crystals reported by

Ogawa *et al.*[7] The oriented NCM films also show nonohmic behavior [see inset of Fig. 1(b)]. On the LAO substrate, a higher current is required to reduce the resistance of the NCM film to the same extent as on the Si substrate.

In Figs. 2(a) and 2(b), we show the effect of increasing the electric current on the resistance of $Gd_{0.5}Ca_{0.5}MnO_3$ (GCM) films deposited on Si(100) and LAO(100) substrates, respectively. The behavior is comparable to that of NCM films, particularly on the LAO substrate. On the Si(100) substrate, the GCM films show essentially flat resistance curves with almost no change with temperature, reminiscent of degenerate materials. This is specially noticeable when the current is $\gtrsim$100 μA. On the LAO(100) substrate, such near constancy of resistance is seen when the current is greater than 10 mA. Nonohmic behavior is found in the GCM films as well, as can be seen from the insets in Figs. 2(a) and 2(b).

A particularly remarkable feature of the NCM and GCM films deposited on LAO(100) substrates is the occurrence of a hysteretic I-M transition driven by transport current. We show typical data on these films in Fig. 3. The resistance-temperature plots (at constant transport current) are as described earlier when the sample is cooled from room temperature. After attaining the lowest temperature of measurement ($\sim$20 K) the current is switched off, and switched on again to carry out measurements as the films are heated. Note that at the current switching, the large resistance increase overloads the available current source used.

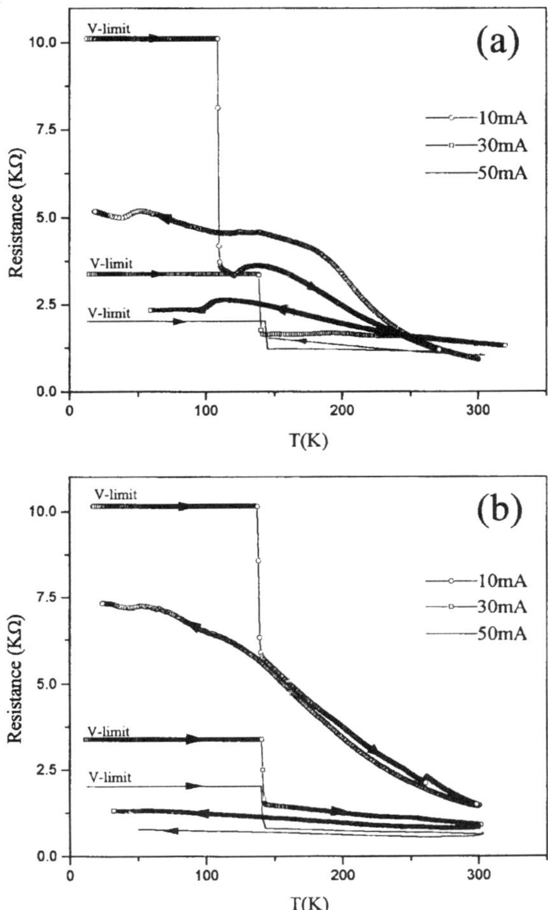

FIG. 3. Resistance versus temperature plots of (a) $Nd_{0.5}Ca_{0.5}MnO_3$ and (b) $Gd_{0.5}Ca_{0.5}MnO_3$ films deposited on LAO(100) for three different current values, recorded over cooling and heating cycles. The current was switched off after cooling curve was completed and turned on again to record the heating curve.

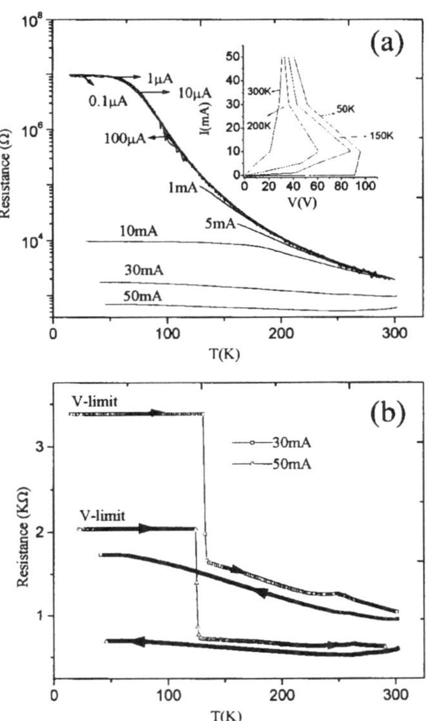

FIG. 4. (a) Temperature variation of the resistance of a $Y_{0.5}Ca_{0.5}MnO_3$ (YCM) film deposited on LAO(100) for different values of current. The inset shows I-V characteristics at different temperatures. (b) Cooling and heating curves obtained as described in the caption of Fig. 3.

Our current source overload was 105 V, thus causing overload when the resistance is 10.5 MΩ for a current of 10 mA. (We do not have a source of higher voltage.) The heating curves first show an abrupt increase in resistance followed by an equally abrupt drop around a temperature at which the charge solid apparently melts. The jump in resistance to the original value on the cooling curve in the charge liquid state is indeed remarkable. Such a memory of the current-specific resistance value registered in the charge liquid state is an interesting property.

In Fig. 4(a), we show the resistance vs temperature curves of a $Y_{0.5}Ca_{0.5}MnO_3$ (YCM) film on LAO(100) for different values of the current. We observe a substantial decrease in the resistance with increase in the current and the I-V behavior is nonohmic[(see inset of Fig. 4(a)]. The occurrence of a current-induced I-M-type transition in the YCM film is noteworthy as the CO state in this material is very robust, being

unaffected by magnetic fields or substitution of Mn^{3+} by Cr^{3+} and such ions.[3,5,13] The value of resistance at a given current varies as YCM>GCM>NCM, in the same order as the $\langle r_A \rangle$. The memory effect discussed earlier is also found in the YCM film, as shown in Fig. 4(b).

In Fig. 5, we show the temperature variation of resistance of a $Nd_{0.5}Sr_{0.5}MnO_3$ (NSM) film deposited on Si(100). The NSM films show metallic resistivity from $\sim$300 K down to $\sim$140 K and resistance increases below 140 K due to charge ordering. The increase in resistance in the CO state is not as sharp in the film as in a single crystal.[4] We, however, observe the high-temperature metallic behavior for all current values and the resistance decreases substantially in the charge-ordering regime, with increasing current. At 50 mA, the material remains metallic from 300 to 20 K, although there is a slight heating of the sample at this current value. The resistance values in NSM are considerably lower than in NCM and other films at similar currents. The NSM film also shows nonohmic behavior (see inset of Fig. 5). Mori[14] has recently observed effects in NSM crystals somewhat comparable to the memory effects [Figs. 3 and 4(b)] found by us.

IV. DISCUSSION

We discuss now the electric-field-induced insulator-metal transition and the nonohmic transport in the charge-ordered

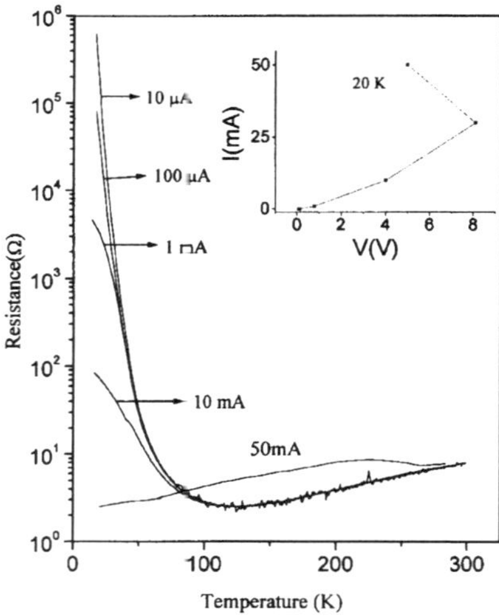

FIG. 5. Temperature variation of the resistance of a $Nd_{0.5}Sr_{0.5}MnO_3$ (NSM) film deposited on Si(100) for different values of current. The inset shows I-V characteristic at a low temperature.

rare-earth manganates. These materials are to be viewed as charge solids pinned randomly to the underlying lattice, and can be depinned by an externally applied electric field, and, of course, melted thermally. The random pinning is expected on general grounds, e.g , the L/A substitutional disorder in $L_{0.5}A_{0.5}MnO_3$ that acts as a quenched randomness, other lattice defects or possible inhomogeneities, including the occurrence of domains or clusters of CO and metallic phases. Charge ordering in the cubic manganates is, however, not a charge-density wave (CDW) arising from a nesting of the Fermi surface and the associated Peierls instability, as presumably is the case for the effectively low-dimensional layers system,[15] Sr_2IrO_4. Charge ordering in the manganates is driven by Coulomb interaction among the charge carriers, and is stabilized by the background lattice potential, with which it is ideally commensurate for the 1/1 ordered Mn^{3+}/Mn^{4+} case. The charge carriers themselves are expected to be polarons—specifically, the lattice polarons associated with the Jahn-Teller ions Mn^{3+} for the manganates with a small average A-cationic radius as, e.g., in NCM, YCM, or GCM. Charge transport in these manganates, again unlike a CDW with its sliding or unpinned condensate, is expected to proceed through the correlated motion of the polarons, depinned by the applied electric field. This picture is qualitatively consistent with the observed facts, namely, the observed threshold of electric field, or applied current, for electrical conduction, nonohmic transport, metal-insulator transition, negative differential resistance, and certain hysteretic and memory effects associated with the melting transition from a charge solid to a charge liquid as discussed below.

At the lowest temperature (77 K) and current ($I \lesssim 0.05\ \mu A$) employed, the NCM film [Fig. 1(a)] is highly insulating (TCR<0) with $R \sim 10^7\ \Omega$. The resistance then falls by an order of magnitude for a small increase of the current to $\sim 0.1\ \mu A$, but finally levels off to a gradual de-

crease around 1 mA. The corresponding variation in the applied voltage, and, therefore, of the electric field, is small, from ~ 0.5 to ~ 0.6 V. This nonlinear threshold conduction, qualitatively of the Zener type $R \propto \exp(E_0/E)$, is characteristic of a pinned charge solid (insulator), and its field-induced depinning gives a conductor. The field-induced depinning also accounts for the nonohmic decrease of resistance with increasing applied current at a given temperature as observed in these manganates. Let us next consider the temperature dependence of resistance at a given current. At low enough temperatures, the depinning energy far exceeds the thermal energy ($k_B T$) and the conduction is dominated by a coherent tunneling step. A positive TCR is to be expected because of the decohering thermal effects, as indeed observed [Fig. 1(a)]. At higher temperatures, the transport crosses over from this quantum coherent tunneling through pinning barrier to an incoherent thermal escape over the barrier giving a change in sign of the TCR (<0) as expected of a thermally activated process. This is again exactly what is observed [Fig. 1(a)]. Indeed, such a crossover from a low-temperature coherent conduction with a positive TCR to a higher temperature incoherent conduction with a negative TCR is well known for independent small polarons. In the present case, the single polaron is replaced by the polarons entrained in a correlation volume due to interaction. With increasing temperature, the coherence volume must decrease thereby lowering the depinning energy. It is very apt to point out here that the field-induced depinning of the randomly pinned charge solid has a close analogue in the well-known phenomenon of shear induced melting, which is not the result of heating.

The picture discussed above with reference to Fig. 1(a) for NCM/Si also covers the NCM/LAO, GCM/Si or LAO, and YCM/LAO films depicted in Figs. 1(b), 2, and 4(a), respectively. In these cases, the metallic-type regime with TCR>0 (found in NCM/Si at low temperatures) is absent. This suggests that the correlation volume, or the pinning barrier, is sufficiently small for the thermally activated depinning to dominate over tunneling even at low temperatures, thus making the regime with TCR>0 parametrically inaccessible. In the case of NSM (Fig. 5), charge ordering and its pinning involve antiferromagnetic (AFM) ordering, and hence the depinning occurs at relatively lower temperatures. The idea of pinning/depinning, and of the associated coherence volume entraining a number of charge carriers, is general and has a much wider applicability. Thus, it is applicable to the classic Wigner crystal pinned by random substrate imperfections, to which the CO state considered here approximates best—but with the proviso that the electrons have to be replaced by JT polarons whose higher effective mass favors charge solid formation in the parameter regime of interest.

The negative differential resistance in the case of the magnetic CO manganates (as in NSM) can be understood in terms of local ferromagnetic (FM) ordering forced by a sufficiently large transport current impressed as an external constraint. This is the spin-valve effect acting in the reverse, assuming that the Hund's coupling is far larger than the antiferromagnetic coupling. The transport-induced local FM order leads to a lower resistance because of the spin-valve effect, and hence the negative differential conductance. Such a transport driven (magnetic) structural change is a particular case of the general nonequilibrium phenomenon of ordering caused by transport. Similar transport driven instability is expected in the other systems showing negative differential conductance.

Finally, we turn to the remarkable hysteretic I-M transitions driven by the transport current in the CO systems shown in Figs. 3 and 4(b) and described in Sec. III. The R-T plot at constant transport current along the heating curve shows an abrupt drop at the transition temperature ($\sim$150 K) at which clearly the charge solid melts to a charge liquid. On the cooling curve, the charge liquid shows appreciable undercooling (stays liquid below $\sim$150 K) which is not surprising for a melting transition. What is surprising, however, is that on the heating curve (following the switching of the applied current at the lowest temperature) the resistance jumps back at melting to its original current-specific value on the cooling curve in the charge liquid melt. This memory of the current specific-resistance value registered in the charged

liquid state, and addressed uniquely by that specific value of the transport current, can be understood in terms of the quenched randomness of the strength as well as the position of the pinning referred to earlier.[16] Accordingly, at a given transport current, the charge solid is depinned only at a subset of pinning centers. We suggest that effectively only this depinned fraction melts cooperatively at the melting point, and then contributes to the conductance in the charge liquid state. This subset increases with the increasing applied current, giving therefore, lower resistance for higher currents in the charge liquid state, as is clearly seen in Figs. 2 and 4(b). (Here again, the Joule heating is irrelevant as is evident from the fact that the observed melting temperature is unaffected by the applied current magnitude.) Thus, the charge liquid just above the melting temperature also conducts nonohmically. It is the transport-current specificity of the charge liquid resistance that sets it apart from the hysteresis usually associated with the first-order melting transition.

---

*Electronic address: cnrrao@jncasr.ac.in

[1] Y. Tokura, in *Colossal Magnetoresistance, Charge-Ordering and Related Properties of Manganese Oxides*, edited by C. N. R. Rao and B. Raveau (World Scientific, Singapore, 1998); Y. Tokura, Curr. Opin. Solid State Mater. Sci. **4**, 175 (1998); Y. Tokura, Y. Tomioka, H. Kuwahara, A. Asamitsu, Y. Moritomo, and M. Kasai, J. Appl. Phys. **79**, 5288 (1996); A. P. Ramirez, J. Phys.: Condens. Matter **9**, 8171 (1997).

[2] C. N. R. Rao, A. Arulraj, P. N. Santosh, and A. K. Cheetham, Chem. Mater. **10**, 2714 (1998).

[3] A. Arulraj, P. N. Santosh, A. Guha, A. K. Raychaudhuri, N. Kumar, and C. N. R. Rao, J. Phys.: Condens. Matter **10**, 8497 (1998).

[4] H. Kuwahara, Y. Tomioka, A. Asamitsu, Y. Moritomo, and Y. Tokura, Science **270**, 961 (1995).

[5] A. Arulraj, R. Gundakaram, A. Biswas, N. Gayathri, A. K. Raychaudhuri, and C. N. R. Rao, J. Phys.: Condens. Matter **10**, 4447 (1998).

[6] A. Asamitsu, Y. Tomioka, H. Kuwahara, and Y. Tokura, Nature (London) **388**, 50 (1997).

[7] K. Ogawa, W. Wei, K. Miyano, Y. Tomioka, and Y. Tokura,

Phys. Rev. B **57**, R15 033 (1998); M. Fiebig, K. Miyano, Y. Tomiyoka, and Y. Tokura, Science **280**, 1925 (1998).

[8] J. Q. Wang, R. C. Barker, G. J. Cui, T. Tamagawa, and B. L. Halpern, Appl. Phys. Lett. **71**, 3418 (1997).

[9] A. R. Raju, H. N. Aiyer, B. V. Nagaraju, R. Mahendiren, A. K. Raychaudhuri, and C. N. R. Rao, J. Phys. D: Appl. Phys. **30**, L71 (1997).

[10] J. G. Snyder, R. Hiskes, S. DiCarolis, M. R. Beasley, and T. H. Geballe, Phys. Rev. B **53**, 14 434 (1996).

[11] G. C. Xiong, Q. Li, H. L. Ju, R. L. Green, and T. Venkatesan, Appl. Phys. Lett. **66**, 1689 (1995).

[12] H. N. Aiyer, A. R. Raju, G. N. Subbanna, and C. N. R. Rao, Chem. Mater. **9**, 755 (1997); A. R. Raju and C. N. R. Rao, Appl. Phys. Lett. **66**, 896 (1995).

[13] P. V. Vanitha, R. S. Singh, S. Natarajan, and C. N. R. Rao, J. Solid State Chem. **137**, 365 (1998).

[14] T. Mori, Phys. Rev. B **58**, 12 543 (1998).

[15] G. Grüner, Rev. Mod. Phys. **60**, 1129 (1988).

[16] G. Cao, J. Bolivar, S. McCall, J. E. Crow, and R. P. Guertin, Phys. Rev. B **57**, R11 039 (1998).

PERGAMON

Solid State Communications 125 (2003) 41–44

**solid
state
communications**

www.elsevier.com/locate/ssc

Occurrence of re-entrant ferromagnetic transitions in rare-earth manganates on cooling the charge-ordered states

Asish K. Kundu, P.V. Vanitha, C.N.R. Rao*

Chemistry and Physics Materials Unit, Jawaharlal Nehru Centre for Advanced Scientific Research, Jakkur P.O., Bangalore 560064, India

Received 22 August 2002; received in revised form 30 September 2002; accepted 3 October 2002 by A.K. Sood

Abstract

Some of the compositions of the half-doped rare-earth manganates, $La_{0.5-x}Ln_xCa_{0.5}MnO_3$ (Ln = Nd, Pr) and $Nd_{0.5}Ca_{0.5-x}Sr_xMnO_3$ with relatively small A-cation radii, $\langle r_A \rangle$, show an unusual behavior wherein they become ferromagnetic (FM) on cooling the charge ordered (CO) state ($T_{CO} > T_C$). With increase in $\langle r_A \rangle$, however, the T_C becomes greater than T_{CO}. Thus, plots of T_C and T_{CO} against $\langle r_A \rangle$ for $La_{0.5-x}Ln_xCa_{0.5}MnO_3$ (Ln = Nd, Pr) and $Nd_{0.5}Ca_{0.5-x}Sr_xMnO_3$ show cross-over from the $T_{CO} > T_C$ regime to the $T_C > T_{CO}$ regime around $\langle r_A \rangle$ values of 1.195 ± 0.003 and 1.200 ± 0.005 Å, respectively. Between T_C and T_{CO}, the CO and FM phases are likely to coexist. In $Nd_{0.5}Ca_{0.5}Mn_{1-x}M_xO_3$ (M = Cr, Ru), $T_{CO} > T_C$ when $x \leq 0.10$, suggesting the re-entrant nature of the FM transition.
© 2002 Elsevier Science Ltd. All rights reserved.

PACS: 71.30.+h; 75.30.Cr

Keywords: A. Magnetically ordered materials; D. Electronic transport

1. Introduction

There have been extensive investigations of the rare-earth manganates of the general formula $Ln_{1-x}A_xMnO_3$ (Ln = rare-earth, A = alkaline earth) over the last few years to understand the various phenomena and properties exhibited by them [1–4]. In particular, the competition between charge ordering and charge delocalization has received much attention. Charge ordering is specially favored in the manganates when the average radius of the A-site cations, $\langle r_A \rangle$ is relatively small [4]. Charge ordering is destroyed by the application of magnetic fields or by the substitution of certain cations in the B-site, provided $\langle r_A \rangle$ is sufficiently large ($\langle r_A \rangle \geq 1.17$ Å) [5–7]. If $\langle r_A \rangle$ is very small, such external factors have little effect on charge ordering. The $\langle r_A \rangle$ regime 1.17–1.20 Å, is specially interesting in that the manganates show charge ordering at ordinary temperatures and undergo a re-entrant transition to

a ferromagnetic (FM) state on cooling ($T_C < T_{CO}$). Thus, $Nd_{0.5}Ca_{0.5}MnO_3$ ($\langle r_A \rangle = 1.17$ Å) is a charge ordered (CO) insulator ($T_{CO} = 240$ K) and does not show ferromagnetism at any temperature [8]. $Nd_{0.25}La_{0.25}Ca_{0.5}MnO_3$ ($\langle r_A \rangle = 1.185$ Å), shows charge ordering at 239 K and undergoes a FM transition on cooling to ~ 140 K [9]. By contrast, $La_{0.5}Ca_{0.5}MnO_3$ with a $\langle r_A \rangle$ of 1.198 Å has a FM Curie temperature, T_C of 225 K and $T_{CO} = 135$ K [10,11]. In the intermediate temperature range between T_C and T_{CO}, these compositions show the coexistence of phases.

Investigations of the manganates of the type $La_{0.5-x}Ln_xCa_{0.5}MnO_3$ (Ln = Nd, Pr) in the $\langle r_A \rangle$ region of 1.17–1.20 Å showed that the T_C increases while T_{CO} decreases with increase in $\langle r_A \rangle$, the two curves intersecting around $\langle r_A \rangle = 1.195 \pm 0.003$ Å [12]. While this is an interesting observation, there are very few experimental points in the region around 1.195 Å. We have, therefore examined manganates of the type $La_{0.5-x}Ln_xCa_{0.5}MnO_3$ (Ln = Nd, Pr) in the $\langle r_A \rangle$ region of 1.194–1.197 Å. We also considered it important to explore whether a series of manganates obtained by substitution of Ca by Sr in $Ln_{0.5}Ca_{0.5}MnO_3$ would exhibit such re-entrant transition and if so how

* Corresponding author. Tel.: +91-80-856-3075; fax: +91-80-846-2760.

E-mail address: cnrrao@jncasr.ac.in (C.N.R. Rao).

Table 1
Structure and properties of $La_{0.5-x}Ln_xCa_{0.5}MnO_3$(Ln = Nd, Pr)

Composition	$\langle r_A \rangle$ (Å)	σ^2 (Å^2)	Lattice parameter (Å)			Mn^{4+} (%)	T_C (K)	T_p (K)
			a	b	c			
$La_{0.42}Nd_{0.08}Ca_{0.5}MnO_3$	1.194	0.00029	5.4105	7.6367	5.4258	45	218	~177
$La_{0.45}Nd_{0.05}Ca_{0.5}MnO_3$	1.195	0.00025	5.4149	7.6471	5.4241	46	229	169
$La_{0.45}Pr_{0.05}Ca_{0.5}MnO_3$	1.196	0.00021	5.4150	7.6372	5.4307	46	235	~172
$La_{0.48}Pr_{0.02}Ca_{0.5}MnO_3$	1.197	0.00022	5.4124	7.6447	5.4401	46	140	~138

the T_C and T_{CO} vary with $\langle r_A \rangle$. We have investigated the series of compositions of $Nd_{0.5}Ca_{0.5-x}Sr_xMnO_3$ for this purpose. It may be noted that earlier studies of this system of manganates have not paid attention to the re-entrant transition and relative values of T_C and T_{CO} [8,13]. The present study shows that there is indeed a re-entrant transition with $T_{CO} > T_C$, and that the curve $T_{CO}-\langle r_A \rangle$ and $T_C - \langle r_A \rangle$ curves intersect around a $\langle r_A \rangle$ of 1.20 ± 0.005 Å in $Nd_{0.5}Ca_{0.5-x}Sr_xMnO_3$. An examination of available data on $Nd_{0.5}Ca_{0.5}Mn_{1-x}M_xO_3$ (M = Cr, Ru), shows that $T_{CO} > T_C$ in these manganates as well, with a cross-over around same value of dopant concentration, although $\langle r_A \rangle$ remains constant [5].

2. Experimental procedure

Polycrystalline samples $La_{0.5-x}Nd_xCa_{0.5}MnO_3$, $La_{0.5-x}Pr_xCa_{0.5}MnO_3$ and $Nd_{0.5}Ca_{0.5-x}Sr_xMnO_3$ were prepared by the conventional ceramic method. Stoichiometric mixtures of the respective rare-earth oxides, alkaline earth carbonates, and MnO_2 were ground and heated at 900 °C in air followed by heating at 1000 and 1200 °C for 12 h each in air. The powders thus obtained were pelletized and the pellets sintered at 1400 °C for 12 h in air. To improve the oxygen stoichiometry, the samples were annealed in oxygen at 900 °C. The phase purity of the samples was established by recording the X-ray diffraction patterns in the 2θ range of 10–80° with a Seiferts 3000 TT diffractometer. Electrical resistivity (ρ) measurements were carried out from room temperature to 20 K by the four-probe method. DC magnetization (M) measurements were made with a vibrating sample magnetometer (Lakeshore 7300) in an applied magnetic field of 100 Oe.

In Tables 1 and 2, we have listed the values of $\langle r_A \rangle$, σ^2 and the lattice parameters along with the FM Curie temperature, T_C and the charge ordering transition temperature, T_{CO}. The T_C values were obtained from $M-T$ curves, taking the point at which there is an abrupt increase in M as the T_C. The T_{CO} values were also obtained from the magnetization measurements. In Tables 1 and 2, we have also listed the insulator–metal transition temperature, T_p, as obtained from the electrical resistivity measurements. The value of T_p corresponds to the maximum in the $\rho-T$ curve.

3. Results and discussion

We show the temperature variation of magnetization of $La_{0.5-x}Ln_xCa_{0.5}MnO_3$ (Ln = Nd, Pr) with composition in the $\langle r_A \rangle$ regime of 1.194–1.197 Å in Fig. 1(a). The data clearly show how the T_C increases with increase in $\langle r_A \rangle$ as expected. The plot of the inverse magnetization vs. temperature in the inset of Fig. 1(a) shows a clear minimum as evidence for charge ordering in the $\langle r_A \rangle = 1.194$ Å case, but we could not obtain reliable values of the charge ordering transition temperature, T_{CO}, in other samples. The temperature variation of resistivity of these materials (Fig. 1(b)) shows the occurrence of insulator–metal transitions in all the materials in this $\langle r_A \rangle$ range. Although

Table 2
Structure and properties of $Nd_{0.5}Ca_{0.5-x}Sr_xMnO_3$

Composition	$\langle r_A \rangle$ (Å)	σ^2 (Å^2)	Lattice parameter (Å)			Space group	Mn^{4+} (%)	T_C (K)	T_p (K)	T_{CO} (K)
			a	b	c					
$Nd_{0.5}Ca_{0.5}MnO_3$	1.172	0.0001	5.401	7.619	5.385	$Pnma$	47	–	–	240
$Nd_{0.5}Ca_{0.25}Sr_{0.25}MnO_3$	1.204	0.0038	5.4129	7.6331	5.4126	$Pnma$	46	155	(155)	214 ± 4
$Nd_{0.5}Ca_{0.2}Sr_{0.3}MnO_3$	1.211	0.0043	5.4189	7.6404	5.4143	$Pnma$	47	195	(178)	–
$Nd_{0.5}Ca_{0.15}Sr_{0.35}MnO_3$	1.217	0.0047	5.4167	7.6406	5.4345	$Imma$	49	218	202	110 ± 3
$Nd_{0.5}Ca_{0.1}Sr_{0.4}MnO_3$	1.224	0.0050	5.4205	7.6330	5.4542	$Imma$	49	235	220	124 ± 3
$Nd_{0.5}Ca_{0.05}Sr_{0.45}MnO_3$	1.230	0.0053	5.4257	7.6279	5.4655	$Imma$	48	250	236	144 ± 2
$Nd_{0.5}Sr_{0.5}MnO_3$	1.236	0.0054	5.426	7.634	5.475	$Imma$	49	267	249	150

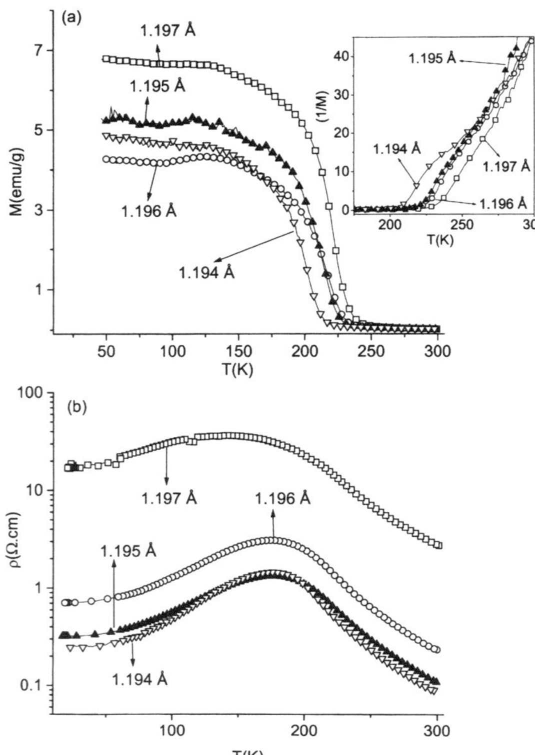

Fig. 1. Temperature variation of (a) the magnetization, M, and (b) the resistivity, ρ, of $La_{0.5-x}Ln_xCa_{0.5}MnO_3$ (Ln = Nd, Pr). The inset in (a) shows the variation of inverse magnetization with temperature.

the transition is broad, the data show that metallicity manifests itself when the material becomes FM. Clearly, the manganate compositions in this $\langle r_A \rangle$ regime are FM and metallic at low temperatures. The FM T_C values from Fig. 1(a) can be used along with the data from earlier studies

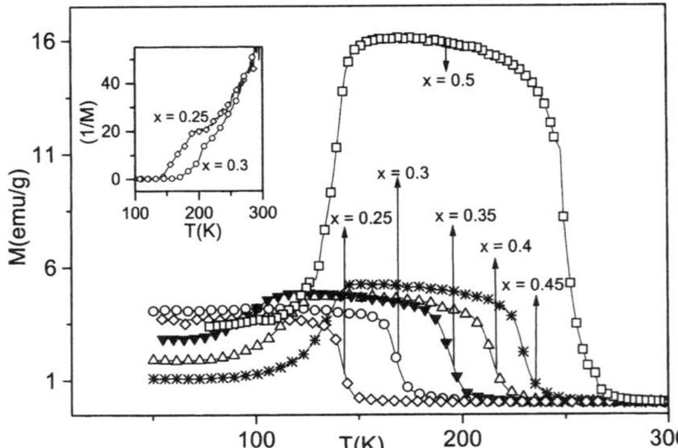

Fig. 3. Temperature variation of magnetization, M, of $Nd_{0.5}Ca_{0.5-x}Sr_xMnO_3$. The inset shows the variation of inverse magnetization with temperature.

to examine the relative variations of T_C and T_{CO} with $\langle r_A \rangle$ [12]. We compare the available T_C and T_{CO} data in Fig. 2. At $\langle r_A \rangle > 1.195$ Å, T_C values can be obtained reliably, but not the T_{CO}, the latter being associated with large uncertainties. Although the data suffer from the fact that we do not have sufficient T_{CO} data between $\langle r_A \rangle$ values 1.193 and 1.20 Å, the data do show that T_C and T_{CO} curves cross each other around $\langle r_A \rangle$ of 1.195 ± 0.003 Å.

We have carried out magnetization measurements on $Nd_{0.5}Ca_{0.5-x}Sr_xMnO_3$ system. When $x = 0.5$, the material shows the well known FM transition around 250 K and undergoes charge ordering transition on cooling at 150 K, the material becoming antiferromagnetic around the same temperature (Fig. 3). Both the FM and charge ordering transitions are sharp in this composition. As mentioned earlier, the $x = 0$ composition ($Nd_{0.5}Ca_{0.5}MnO_3$), shows only charge ordering ($T_{CO} = 240$ K), but no ferromagnetism. The $Nd_{0.5}Ca_{0.5-x}Sr_xMnO_3$, compositions with $x = 0.25–0.45$ show FM transitions in magnetization data (Fig. 3), with the T_C increasing with increase in x. Thus,

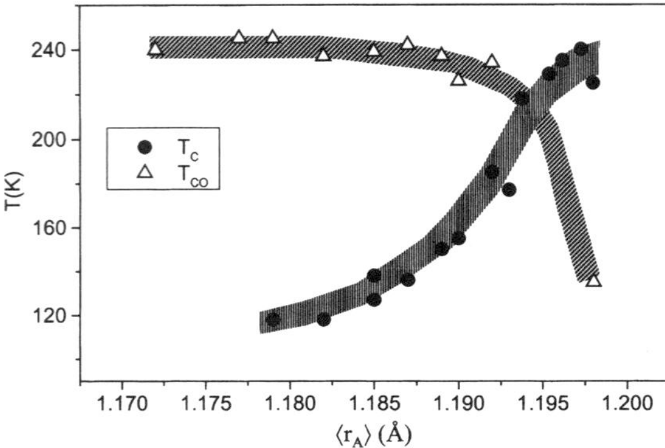

Fig. 2. Variation of the FM Curie temperature, T_C, and the charge ordering transition temperature, T_{CO}, with $\langle r_A \rangle$ in $La_{0.5-x}Ln_xCa_{0.5}MnO_3$ (Ln = Nd, Pr).

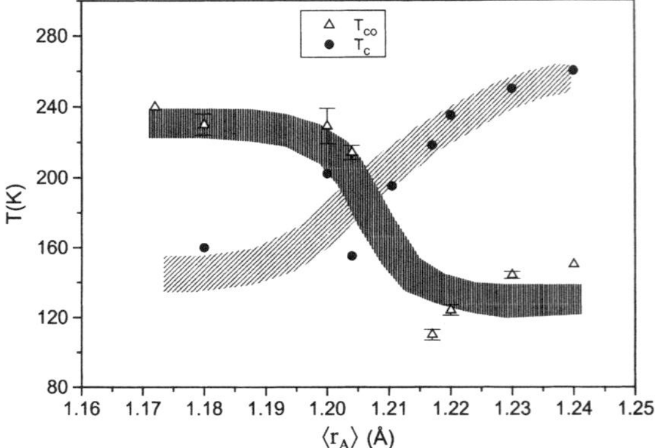

Fig. 4. Variation of FM Curie temperature, T_C, and the charge ordering transition temperature, T_{CO}, with $\langle r_A \rangle$ in $Nd_{0.5}Ca_{0.5-x}Sr_xMnO_3$.

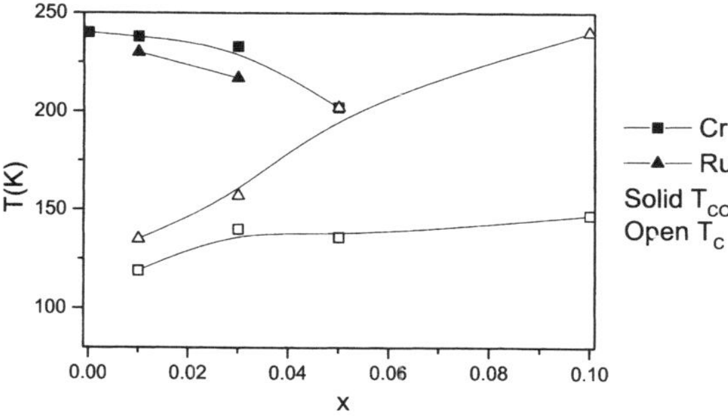

Fig. 5. Variation of FM Curie temperature, T_C, and the charge ordering transition temperature, T_{CO}, with x in $Nd_{0.5}Ca_{0.5}Mn_{1-x}M_xO_3$ (M = Cr, Ru).

the $x = 0.25$ composition has a T_C close to 150 K. However, it is noteworthy that when $x < 0.35$, there is no sharp drop in magnetization data corresponding to the charge ordering transition. We can however, get the T_{CO} values for $x = 0.35–0.45$ compositions, from the magnetization data. The compositions with $x = 0.25$ and 0.3 are more like $Nd_{0.5}Ca_{0.5}MnO_3$ and the temperature variation of the inverse magnetization show a dip corresponding to T_{CO} (see inset of Fig. 3). The T_{CO} values of $x = 0.25$ and $x = 0.1$ compositions are 214 and 230 K, respectively. In Fig. 4 we have plotted the data obtained by us on the T_C and T_{CO} of $Nd_{0.5}Ca_{0.5-x}Sr_xMnO_3$ against $\langle r_A \rangle$ along with data on two compositions $(0.0 < x < 0.25)$ from the literature [8]. Although there is some scatter in the points, the data clearly show that when $\langle r_A \rangle < 1.20$ Å, the $T_C < T_{CO}$, suggesting that we can consider the FM transition to be re-entrant in nature, just as the $La_{0.5-x}Ln_xCa_{0.5}MnO_3$ (Ln = Nd, Pr) (Fig. 2). Furthermore, the $T_C–\langle r_A \rangle$ and $T_{CO}–\langle r_A \rangle$ curves in Fig. 4, cross each other around $\langle r_A \rangle = 1.200 \pm 0.005$ Å, which is close to the cross-over $\langle r_A \rangle$ value found in $La_{0.5-x}Ln_xCa_{0.5}MnO_3$ (Fig. 2), within experimental error. It is possible that over the entire $\langle r_A \rangle$ range 1.17–1.24 Å, there is co-existence of the CO and FM phases. This would certainly be true below T_C as evidenced from the diffraction data [14].

In Fig. 5, we have plotted the available data for $Nd_{0.5}Ca_{0.5}Mn_{1-x}M_xO_3$ (M = Cr, Ru), where $\langle r_A \rangle$ is constant (1.17 Å) [5–7]. With the substitution of Cr and Ru in the Mn site, the material shows ferromagnetism as well as charge ordering. The T_{CO} generally decreases with increase in x, while T_C increases specially in the case of Ru substitution. It appears that this FM transition is re-entrant in nature in these manganate compositions. Fig. 5 also suggests that T_C and T_{CO} curves cross each other at a specific value of x.

Acknowledgements

The authors thank BRNS (DAE), India, for support of this research.

References

[1] A.P. Ramirez, J. Phys.: Condens. Matter 9 (1997) 8171.

[2] C.N.R. Rao, B. Raveau (Eds.), Colossal Magnetoresistance, Charge Ordering and Related Properties of Manganese Oxides, World Scientific, Singapore, 1998.

[3] C.N.R. Rao, A. Arulraj, A.K. Cheetham, B. Raveau, J. Phys.: Condens. Matter 12 (2000) R83.

[4] N. Kumar, C.N.R. Rao, J. Solid State Chem. 129 (1997) 363.

[5] H. Kuwahara, Y. Tomioka, A. Asamitgu, Y. Moritomo, Y. Tokura, Science 270 (1995) 961.

[6] P.V. Vanitha, R.S. Singh, S. Natarajan, C.N.R. Rao, J. Solid State Chem. 137 (1998) 365.

[7] P.V. Vanitha, A. Arulraj, A.R. Raju, C.N.R. Rao, C. R. Acad. Sci. Paris 2 (1999) 595.

[8] C.N.R. Rao, P.N. Santosh, R.S. Singh, A. Arulraj, J. Solid State Chem. 135 (1998) 169.

[9] A. Arulraj, A. Biswas, A.K. Roychaudhury, C.N.R. Rao, P.M. Woodward, T. Vogt, D.E. Cox, A.K. Cheetham, Phys. Rev. B 57 (1998) R8115.

[10] P.G. Radaelli, D.E. Cox, M. Marezio, S.W. Cheong, Phys. Rev. B 55 (1997) 3015.

[11] S. Mori, C.H. Chen, S.W. Cheong, Phys. Rev. Lett. 81 (1998) 3972.

[12] P.V. Vanitha, C.N.R. Rao, J. Phys.: Condens. Matter 13 (2001) 11707.

[13] C.W. Chang, A.K. Debnath, J.G. Lin, Phys. Rev. B 65 (2001) 24422.

[14] Y. Moritomo, Phys. Rev. B 60 (1999) 10347.

INSTITUTE OF PHYSICS PUBLISHING

JOURNAL OF PHYSICS: CONDENSED MATTER

J. Phys.: Condens. Matter **15** (2003) 3029–3040

PII: S0953-8984(03)59392-4

Electronic phase separation in the rare-earth manganates $(La_{1-x}Ln_x)_{0.7}Ca_{0.3}MnO_3$ (Ln = Nd, Gd and Y)

L Sudheendra and C N R Rao

Chemistry and Physics of Materials Unit, Jawaharlal Nehru Centre for Advanced Scientific Research, Jakkur PO, Bangalore-560064, India

E-mail: cnrrao@jncasr.ac.in

Received 10 February 2003
Published 6 May 2003
Online at stacks.iop.org/JPhysCM/15/3029

Abstract
Electron transport and magnetic properties of three series of manganates of the formula $(La_{1-x}Ln_x)_{0.7}Ca_{0.3}MnO_3$ with Ln = Nd, Gd and Y, wherein only the average A-site cation radius $\langle r_A \rangle$ and associated disorder vary, without affecting the Mn^{4+}/Mn^{3+} ratio, have been investigated in an effort to understand the nature of phase separation. All three series of manganates show saturation magnetization characteristic of ferromagnetism, with the ferromagnetic T_c decreasing with increasing x up to a critical value of x, x_c ($x_c = 0.6, 0.3, 0.2$ respectively for Nd, Gd, Y). For $x > x_c$, the magnetic moments are considerably smaller, showing a small increase around T_M, the value of T_M decreasing slightly with increase in x or decrease in $\langle r_A \rangle$. The ferromagnetic compositions ($x \leqslant x_c$) show insulator–metal transitions, while the compositions with $x > x_c$ are insulating. The magnetic and electrical resistivity behaviour of these manganates is consistent with the occurrence of phase separation in the compositions around x_c, corresponding to a critical average radius of the A-site cation, $\langle r_A^c \rangle$, of 1.18 Å. Both T_c and T_{IM} increase linearly when $\langle r_A \rangle > \langle r_A^c \rangle$ or $x \leqslant x_c$, as expected of a homogeneous ferromagnetic phase. Both T_c and T_M decrease linearly with the A-site cation size disorder as measured by the variance σ^2. Thus, an increase in σ^2 favours the insulating AFM state. Percolative conduction is observed in the compositions with $\langle r_A \rangle > \langle r_A^c \rangle$. Electron transport properties in the insulating regime for $x > x_c$ conform to the variable-range hopping mechanism. More interestingly, when $x > x_c$, the real part of dielectric constant (ε') reaches a high value (10^4–10^6) at ordinary temperatures dropping to a very small ($\sim$500) value below a certain temperature, the value of which decreases with decreasing frequency.

1. Introduction

Rare-earth manganates of the general formula $Ln_{1-x}A_xMnO_3$ (where Ln = rare earth, A = alkaline earth) exhibit many interesting properties such as colossal magnetoresistance, charge ordering and electronic phase separation [1–3]. Phase separation in these materials has attracted much interest in the last few years because of the fascinating features associated with the phenomenon [3]. Of the various manganite compositions, the $(La_{1-x}Pr_x)_{0.7}Ca_{0.3}MnO_3$ system has provided particularly valuable information on the electronic phase separation [4–6]. This is a convenient system for study of phase separation, since the Pr-rich compositions are antiferromagnetic (AFM) and charge ordered (CO), whereas the La-rich compositions are ferromagnetic metals (FMM). At a fixed Ca mole fraction of 0.3, a change in x leads to a transition from FMM behaviour to a charge-ordered insulator (COI) behaviour around an x-value of 0.7. That phase separation in $(La_{1-x}Pr_x)_{0.7}Ca_{0.3}MnO_3$ involves the FMM and COI domains has been demonstrated by various types of measurement. This system exhibits several unusual features. Thus, in spite of the drop in resistivity at the insulator–metal (IM) transition, the charge-ordered peaks in the x-ray diffraction pattern continue to grow at low temperatures [7]. Besides the COI phase, another phase is suggested to be present below the charge-ordering temperature [7]. On the basis of a neutron diffraction study, a phase diagram has been provided for this system wherein the $0.6 \leqslant x \leqslant 0.8$ region separates the homogeneous FMM and AFM phases [8].

Noting that the phase separation in $(La_{1-x}Pr_x)_{0.7}Ca_{0.3}MnO_3$ is triggered by A-site cation disorder, we considered it important to investigate a series of compounds of the type $(La_{1-x}Ln_x)_{0.7}Ca_{0.3}MnO_3$ over a range of compositions $0 \leqslant x < 1$ where the variation in Ln brings about a marked change in the average A-site cation radius, $\langle r_A \rangle$, as well as the size disorder. In this paper, we report the results of a systematic study of the electronic, magnetic and dielectric properties of three series of $(La_{1-x}Ln_x)_{0.7}Ca_{0.3}MnO_3$ with Ln = Nd, Gd and Y. It is to be noted that in all the compositions of the three series of manganates, the Mn^{4+}/Mn^{3+} ratio remains constant, the only variable being the average A-site cation radius and the associated disorder.

2. Experimental details

Compositions of the formula $(La_{1-x}Ln_x)_{0.7}Ca_{0.3}MnO_3$ (Ln = Nd, Gd or Y) were prepared by standard solid-state synthesis. A stoichiometric mixture of the rare-earth acetate, $CaCO_3$ and MnO_2 was mixed thoroughly in an agate mortar with the help of iso-propyl alcohol, and the mixture was first fired to 1273 K for 48 h with two intermediate grindings. This preheated sample was then further heated at 1473 K for 24 h, pressed into pellets and fired to 1648 K for 12 h to obtain the single-phase compounds. The samples were tested for phase purity by means of x-ray analysis using a Seifert 3000 diffractometer and the composition was checked by EDAX measurements. All the compositions were fitted to an orthorhombic unit-cell with the *Pnma* space group. The unit-cell dimensions of $(La_{1-x}Ln_x)_{0.7}Ca_{0.3}MnO_3$ decreased with increase in x in all the three series of compounds (figure 1), with a corresponding decrease in the unit-cell volume as expected when a large cation is replaced by a smaller cation. The dependence of the unit-cell dimensions and volume for three different compositions parallels the dependence of the average radius of the A-site cations, $\langle r_A \rangle$, on x. Thus a plot of the unit-cell volume for all the series against $\langle r_A \rangle$ is linear, as shown in the inset of figure 1.

Magnetization measurements were carried out in a Lakeshore vibrating sample magnetometer in the 300–50 K temperature range at an applied field of 4000 G. Electrical transport properties were measured by the standard four-probe method with silver epoxy as

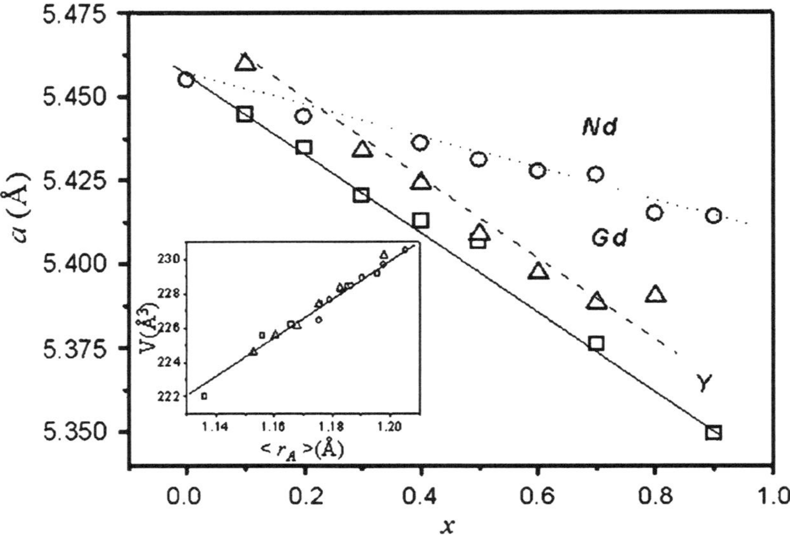

Figure 1. The variation of the unit-cell dimension a (Å) with x in $(La_{1-x}Ln_x)_{0.7}Ca_{0.3}MnO_3$. The inset shows the variation of the unit-cell volume with $\langle r_A \rangle$ (Å) in $(La_{1-x}Ln_x)_{0.7}Ca_{0.3}MnO_3$.

the electrodes in the 300–20 K temperature range. Dielectric measurements were carried out with the help of Agilent 4294A impedance analyser with gold coating on either side of the disc-shaped samples acting as electrodes. The data were collected from 85–300 K.

3. Results and discussion

In figure 2(a), we show the results of magnetic measurements on the $(La_{1-x}Nd_x)_{0.7}Ca_{0.3}MnO_3$ series, to show how the magnetic moment (μ_β) is sensitive to the substitution of the smaller Nd^{3+} cation in place of La^{3+}. We see a clear FM transition down to $x = 0.5$ with a saturation magnetic moment close to 3 μ_β. The ferromagnetic Curie temperature, T_c, shifts to lower temperatures with increase in x. We fail to see magnetic saturation in compositions with $x \geqslant 0.6$ and, instead, the maximum value of μ_β is far less than three at low temperatures. We designate the composition up to which ferromagnetism occurs as the critical composition x_c. The compositions with $x > x_c$ show a gradual, definitive increase in the magnetization or μ_β at a temperature T_M, the T_M-value decreasing with increasing x. It is possible that T_M represents the onset of canted antiferromagnetism (CAF). In $(La_{1-x}Gd_x)_{0.7}Ca_{0.3}MnO_3$, ferromagnetism is observed up to $x = 0.3$. T_c decreases with increase in x in the composition range $0.0 \leqslant x \leqslant 0.3$. For $x > 0.3$, we observe a gradual increase in the μ_β-value at low temperature around T_M, with the value of T_M decreasing with increase in x. Note that the value of x_c ($\sim$0.3) in the Gd series is considerably lower than that in the Nd series where it was 0.6, showing that x_c decreases with the decrease in the average radius of the A-site cation, $\langle r_A \rangle$. The results for the $(La_{1-x}Gd_x)_{0.7}Ca_{0.3}MnO_3$ series of manganates (figure 2(b)) obtained by us agree with those reported for the $x = 0.0$–0.25 compositions by Terashita and Neumeier [9]. In $(La_{1-x}Y_x)_{0.7}Ca_{0.3}MnO_3$, ferromagnetism is seen only for compositions with $x \leqslant 0.2$, T_c decreasing with increase in x. Compositions with $x > 0.2$ show a gradual increase in the μ_β-value below T_M and T_M decreases with increase in x (figure 2(c)). Thus, the x_c-value of

L Sudheendra and C N R Rao

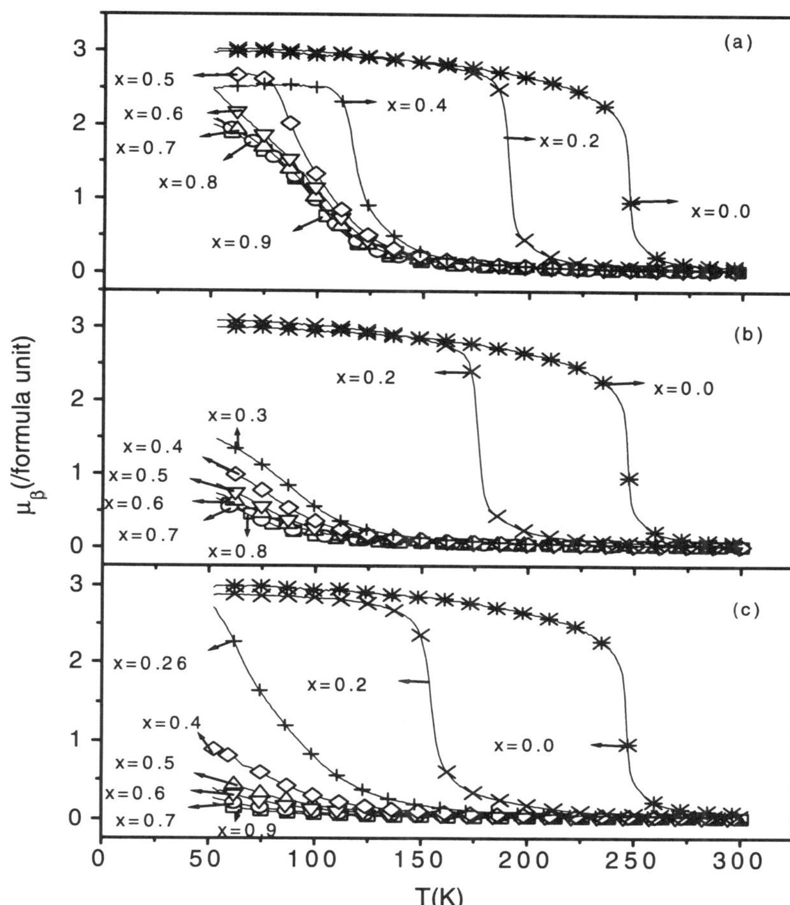

Figure 2. The temperature variation of μ_β in the manganites (a) $(La_{1-x}Nd_x)_{0.7}Ca_{0.3}MnO_3$, (b) $(La_{1-x}Gd_x)_{0.7}Ca_{0.3}MnO_3$ and (c) $(La_{1-x}Y_x)_{0.7}Ca_{0.3}MnO_3$.

$(La_{1-x}Ln_x)_{0.7}Ca_{0.3}MnO_3$ is 0.75, 0.6, 0.3 and 0.2 for Ln = Pr, Nd, Gd and Y respectively, showing a sensitive dependence of x_c on $\langle r_A \rangle$.

The magnetization data for $(La_{1-x}Ln_x)_{0.7}Ca_{0.3}MnO_3$ with Ln = Nd, Gd and Y in the composition range $x \geqslant x_c$ show certain features of significance. Thus, we find that for the compositions close to x_c ($x \sim 0.6$–0.7 when Ln = Nd; $x \sim 0.3$–0.4 in the case of Gd), the low-temperature μ_β-value is significant, reaching values anywhere between 1 and 2.5. It is only when x is very large ($x \gg x_c$) that the μ_β-value decreases to values less than unity. These relatively high values of μ_β, when Ln = Nd and Gd, are probably due to moments of the rare earths. When Ln = Y, the μ_β-values are all low, the highest value of ~ 1 μ_β being observed when $x \sim x_c$ (figure 2(c)). When $x > x_c$, the magnetic moment decreases below 0.5 μ_β. When $x \gg x_c$, the presence of small FM clusters cannot be ruled out, even though the CAF interactions may be dominant. The drastic change in the values of μ_β around x_c in the three series of manganates can be attributed to phase separation due to disorder caused by substitution of the smaller rare-earth cations in place of La. In the Pr system, phase separations

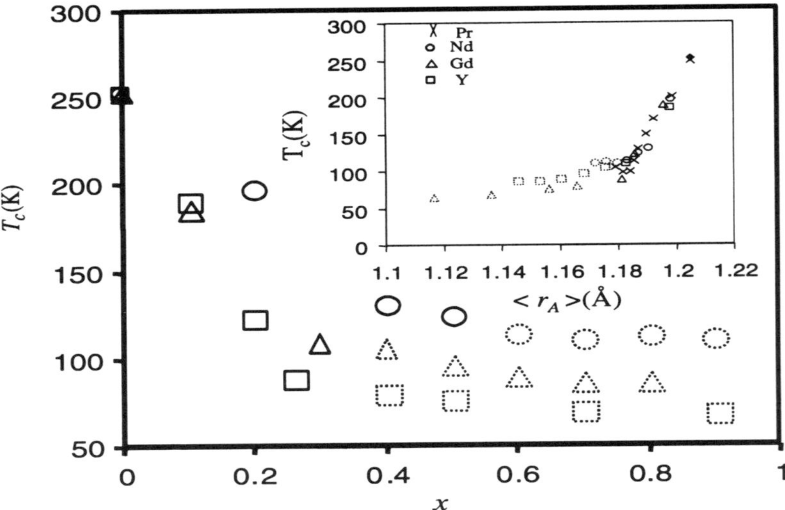

Figure 3. The variation of T_c with x for $(La_{1-x}Ln_x)_{0.7}Ca_{0.3}MnO_3$. Solid symbols represent real T_c and broken symbols represent T_M. The inset shows the variation of T_c with $\langle r_A \rangle$ (Å) for $(La_{1-x}Ln_x)_{0.7}Ca_{0.3}MnO_3$ (the $x = 0$ datum is shown as ◆). $(La_{1-x}Ln_x)_{0.7}Ca_{0.3}MnO_3$ data are taken from [8].

have been reported in the regime of $x \sim x_c$ ($x \sim 0.6$–0.8) [8]. To some extent, the results found here are somewhat comparable to those of de Teresa *et al* [10] who find FMM behaviour at low x and spin-glass behaviour at large x ($\geqslant 0.33$) in $(La_{1-x}Tb_x)_{0.67}Ca_{0.33}MnO_3$.

In figure 3, we have plotted the ferromagnetic T_c of the $(La_{1-x}Ln_x)_{0.7}Ca_{0.3}MnO_3$ composition against x ($x \leqslant x_c$). T_c decreases with increase in x, linearly. In figure 3, we have also plotted the T_M-values of the non-ferromagnetic compositions ($x > x_c$), to show how the T_M-value also decreases with increase in x, albeit with a considerably smaller slope. In the inset of figure 3, we have plotted the values of T_c and T_M against $\langle r_A \rangle$. The T_c-value increases with increase in $\langle r_A \rangle$ for $\langle r_A \rangle \geqslant 1.18$ Å (or $x < x_c$), while T_M increases with $\langle r_A \rangle$ with a smaller slope. The value of $\langle r_A \rangle$ of 1.18 Å marks the critical radius beyond which ferromagnetism manifests itself in $(La_{1-x}Ln_x)_{0.7}Ca_{0.3}MnO_3$. It must be noted that in a variety of manganates of the general composition $Ln_{1-x}A_xMnO_3$, an $\langle r_A \rangle$ of 1.18 Å marks the critical value below which charge ordering becomes robust, rendering it difficult to destroy it by applying magnetic fields or impurity substitutions [11]. An $\langle r_A \rangle$ of 1.18 Å also marks the critical value below which re-entrant FM transitions with $T_c < T_{co}$ are seen [12]. The compositions with $\langle r_A \rangle < 1.18$ Å are AFM insulators. In figure 4, we have plotted μ_β against $\langle r_A \rangle$ for the three series of $(La_{1-x}Ln_x)_{0.7}Ca_{0.3}MnO_3$. The μ_β-values when $\langle r_A \rangle < 1.18$ Å are rather small, well below 3 μ_β, generally less than 1 μ_β. Magnetization in this regime arises from FM interactions in a primarily AFM environment, probably due to the presence of magnetic clusters. Thus, the μ_β-values in the range of 1–2.5 μ_β in figure 4 correspond to the phase separation regime where the values of $\langle r_A \rangle$ are in the 1.18–1.165 Å range. This is consistent with results reported for $(La_{1-x}Pr_x)_{0.7}Ca_{0.3}MnO_3$ system [8]. When $\langle r_A \rangle \leqslant 1.165$ Å, the μ_β-values become considerably smaller, although ferromagnetic interactions due to clusters are likely to be present at low temperatures in this regime as well.

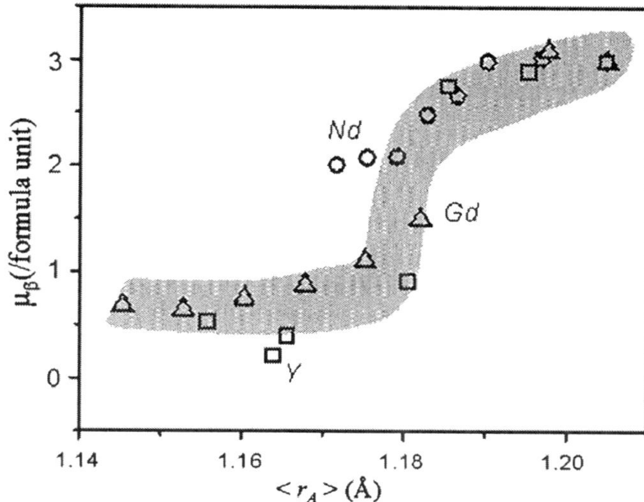

Figure 4. The variation of μ_β with $\langle r_A \rangle$ (Å) in $(La_{1-x}Ln_x)_{0.7}Ca_{0.3}MnO_3$ at 50 K.

The electrical resistivity data for $(La_{1-x}Ln_x)_{0.7}Ca_{0.3}MnO_3$ (Ln = Nd, Gd and Y) reflect the magnetization data, with the $x \leqslant x_c$ compositions showing IM transitions (figure 5). The compositions with $x > x_c$ are insulating and do not exhibit IM transitions. For $x \leqslant x_c$, the value of the resistivity at the IM transition increases with increase in x, and a change in the resistivity of 3–4 orders magnitude is observed at the transition. Thus, the value of resistivity at 20 K for $x \approx x_c$ is considerably higher than that for $La_{0.7}Ca_{0.3}MnO_3$. The temperature of the IM transition, T_{IM}, for $x \leqslant x_c$ compositions decreases linearly with increasing x (figure 6). The T_{IM} versus $\langle r_A \rangle$ plot is linear with a positive slope as expected (see the inset of figure 6). We have not observed any resistivity anomaly at T ($<T_{IM}$) for any of the compositions unlike Uehara *et al* [4] and Deac *et al* [13].

The small but finite magnetic moments and relatively large resistivities at low temperatures found in $(Ln_{1-x}Ln_x)_{0.7}Ca_{0.3}MnO_3$ for Ln = Pr, Nd, Gd and Y around x_c or $\langle r_A^c \rangle$ are a consequence of phase separation. Phase separation also causes thermal hysteresis in the resistivity behaviour around the IM transitions (see the insets in figure 5). The insets in figures 5(a)–(c) show that the resistivity in the warming cycle is lower than that in the cooling cycle up to a certain temperature beyond which the resistivities in the two cycles merge. Upon cooling the sample below the IM transition, the FMM phase grows at the expense of the AFM insulating phase, causing a decrease in the resistivity value. When the same sample is warmed, the insulating phase grows at the expense of the FMM phase, the latter providing the conductive path. The thermal hysteresis in resistivity is therefore due to the percolative conductivity in these manganates, the hysteresis decreasing with increase in $\langle r_A \rangle$ or decrease in x as expected. It must be noted that the compositions with $\langle r_A \rangle > 1.18$ Å which are in a homogeneous FM phase do not show any hysteresis around T_{IM}. Furthermore, the ratio of the peak resistivities in the cooling and heating cycles (ρ_c/ρ_w) is constant for samples with $\langle r_A \rangle > 1.18$ Å in the homogeneous FM regime, but increases significantly with decrease in $\langle r_A \rangle$.

Since phase separation in the $(La_{1-x}Ln_x)_{0.7}Ca_{0.3}MnO_3$ series of manganates is related to size disorder caused by the substitution of small ions in place of La, we have quantitatively examined the effect of size disorder in these manganates, in terms of the size variance of

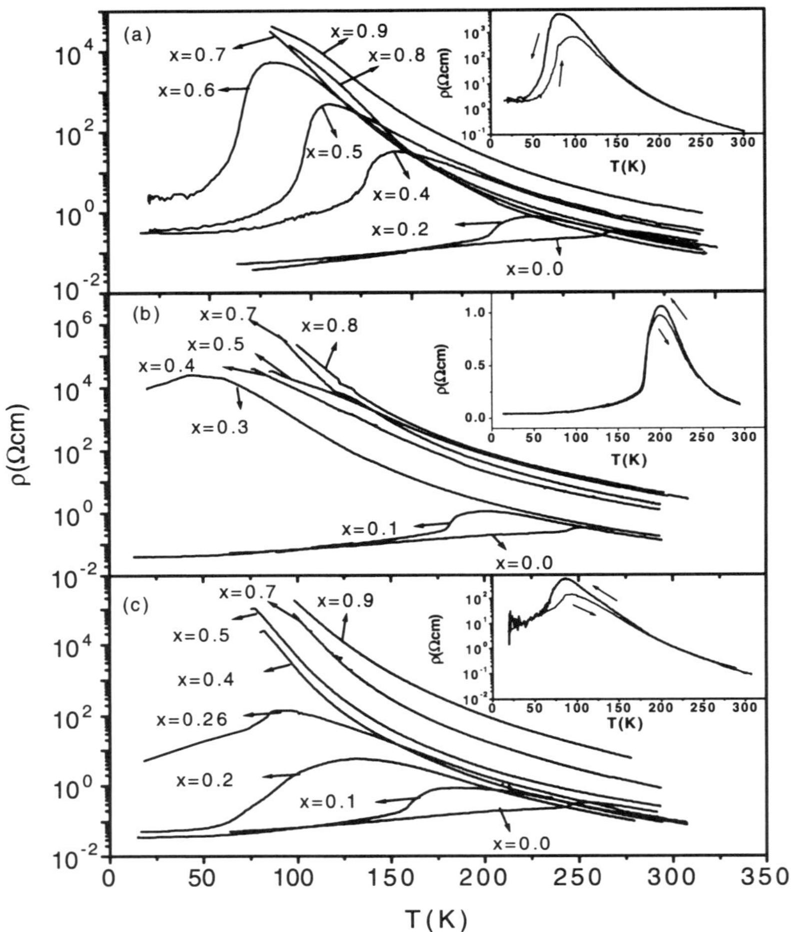

Figure 5. The temperature variation of the resistivity in (a) $(La_{1-x}Nd_x)_{0.7}Ca_{0.3}MnO_3$, (b) $(La_{1-x}Gd_x)_{0.7}Ca_{0.3}MnO_3$ and (c) $(La_{1-x}Y_x)_{0.7}Ca_{0.3}MnO_3$. The curves in the insets show warming cycle data.

the A-site cation radius distribution, σ^2 [14, 15]. For two or more A-site species with fractional occupancies x_i ($\sum x_i = 1$), the variance of the ionic radii r_i about the mean $\langle r_A \rangle$ is given by $\sigma^2 = (\sum x_i r_i^2 - \langle r_A \rangle^2)$. We have examined the electrical and magnetic properties for a series of $(La_{1-x}Ln_x)_{0.7}Ca_{0.3}MnO_3$ with fixed $\langle r_A \rangle$ and variable σ^2. In figure 7, we have shown typical plots of the temperature variation of μ_β for fixed values of $\langle r_A \rangle$ of 1.196 and 1.17 Å. We see a decrease in T_c with increase in σ^2. For $\langle r_A \rangle = 1.17$ Å the material is not ferromagnetic. The μ_β-values as well as T_M show a marked decrease with increase in σ^2. The resistivity decreases with increase in magnetization, and when $\langle r_A \rangle = 1.196$ Å, T_{IM} decreases with increase in σ^2. In figure 8, we have plotted T_c-values of the manganates against σ^2 for compositions with fixed values of $\langle r_A \rangle > 1.18$ Å, along with the T_M-value for fixed values of $\langle r_A \rangle \leqslant 1.18$ Å. Surprisingly, the slopes of the T_c- and T_M-plots are similar ($\sim$9000 K Å^{-2}). The slope of T_c versus σ^2 for $Ln_{0.5-x}La_xCa_{0.5}MnO_3$ compositions, where Ln = Nd and Pr, is reported to be $\sim$15 000 K Å^{-2} [12]. The similarity

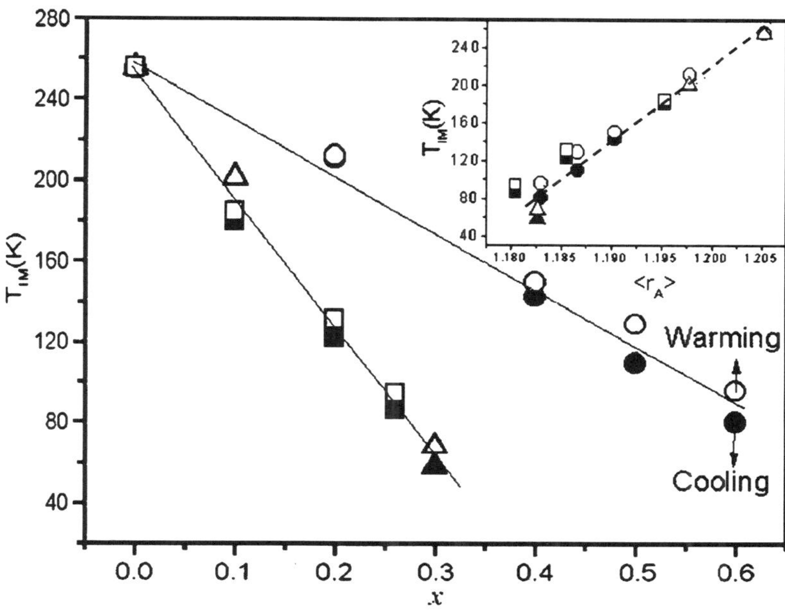

Figure 6. The variation of T_{IM} with x in $(La_{1-x}Ln_x)_{0.7}Ca_{0.3}MnO_3$ for $x \leqslant x_c$. The inset shows the variation of T_{IM} with $\langle r_A \rangle$ (Å) for the same composition range.

in behaviour of these different compositions may be due to presence of ferromagnetic correlations even when $\langle r_A \rangle \leqslant \langle r_A^c \rangle$ or $x > x_c$. By extrapolating the T_c- and T_M-values to $\sigma^2 = 0$, we obtain the intercepts T_c^0 and T_M^0, which represent the transition temperatures in the absence of any size disorder. We have plotted the values of T_c^0 and T_M^0 against $\langle r_A \rangle$ in the inset of figure 8. The values of T_c^0 and T_M^0 increase with increase in $\langle r_A \rangle$ as expected.

As can be seen from figure 5, the resistivities of the $(La_{1-x}Ln_x)_{0.7}Ca_{0.3}MnO_3$ compositions show an a IM transition up to x_c, and these compositions show a significant increase in resistivity with increase in x more prominently than the compositions with $x > x_c$. Electrical conductivity in the $x \approx x_c$ compositions is percolative at low temperatures and, accordingly, we are able to fit the low-temperature data for the compositions with $\langle r_A \rangle > 1.18$ Å to a percolative scaling law, $\log \rho \propto \log |\langle r_A \rangle - \langle r_A^c \rangle|$, as shown in figure 9. The slope of the plot is -2.63, a value close to the experimental and predicted values for percolative systems [3].

We have examined the resistivity data of the insulating compositions with $x > x_c$ in some detail to explore whether they conform to activated hopping defined by $\log \rho \propto 1/T^n$ where $n = 1, 2$ or 4. Here, $n = 1$ corresponds to a simple Arrhenius conductivity. When $n = 2$, the hopping is called nearest-neighbour hopping (NNH). The hopping is controlled by coulombic forces. When $n = 4$, the hopping is termed variable-range hopping (VRH) and the hopping dynamics is controlled by collective excitation of the charge carriers. Figure 10 shows typical fits of the resistivity data for $(La_{1-x}Y_x)_{0.7}Ca_{0.3}MnO_3$ for $x > x_c$. The resistivity data could be fitted to a $T^{-1/2}$-dependence with standard deviations varying from 0.008 to 0.04. The standard deviations for the $T^{-1/4}$-fits were 0.02–0.07. AC conductivity (σ_{AC}) measurements show a frequency dependence of the conductivity, the conductivity increasing with increasing frequency, ω, and decreasing x for a given Ln cation. A fit of σ_{AC} to ω^s gives a value of s in the

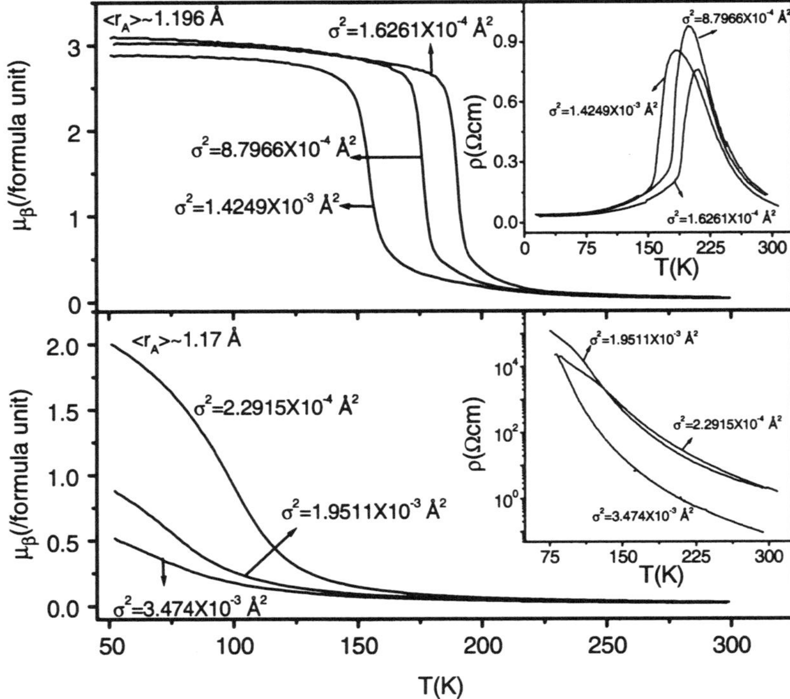

Figure 7. The temperature variation of μ_β in $(La_{1-x}Ln_x)_{0.7}Ca_{0.3}MnO_3$ with (a) $\langle r_A \rangle = 1.196$ Å, (b) $\langle r_A \rangle = 1.17$ Å. The insets show the corresponding variations in resistivities.

range 0.5–0.8, consistent with the VRH mechanism [16]. At high temperatures (>150 K), the conductivity exhibits a frequency independence indicating the dominance of DC conductivity.

In figure 11, we show the temperature response of the real part of the dielectric constant (ε') for $(La_{0.3}Y_{0.7})_{0.7}Ca_{0.3}MnO_3$ at different frequencies. ε' reaches high values at ordinary temperatures but decreases dramatically below ~ 120 K. The temperature at which the drop in ε' occurs increases with increasing frequency, exhibiting a relaxor-type behaviour. The dielectric relaxation obtained in the $(La_{1-x}Ln_x)_{0.7}Ca_{0.3}MnO_3$ compositions is similar to that found in copper titanate perovskite dielectrics [17]. The unusually high dielectric constant of $(La_{1-x}Ln_x)_{0.7}Ca_{0.3}MnO_3$ for $x > x_c$ can be understood by assuming the presence of conducting domains surrounded by insulating layers with an activated behaviour of the intra-domain (non-percolative) conductivity. Such a finite conductivity could be one of the reasons for the high dielectric constant beyond 120 K. The dielectric constant at any given frequency decreases with the decrease in conductivity for $x > x_c$ as expected [18]. Similar to the copper titanates, the $(La_{1-x}Ln_x)_{0.7}Ca_{0.3}MnO_3$ ($x > x_c$) compositions exhibit peaks in the dissipation factor which shifts to higher frequencies with increasing temperature. An Arrhenius plot of frequency against temperature gives an activation energy of 68 meV, close to the values for perovskite oxide dielectrics.

4. Conclusions

The present study of the electronic and magnetic properties of three series of rare-earth manganates of the type $(La_{1-x}Ln_x)_{0.7}Ca_{0.3}MnO_3$ where Ln = Nd, Gd and Y, has shown

                                           L Sudheendra and C N R Rao

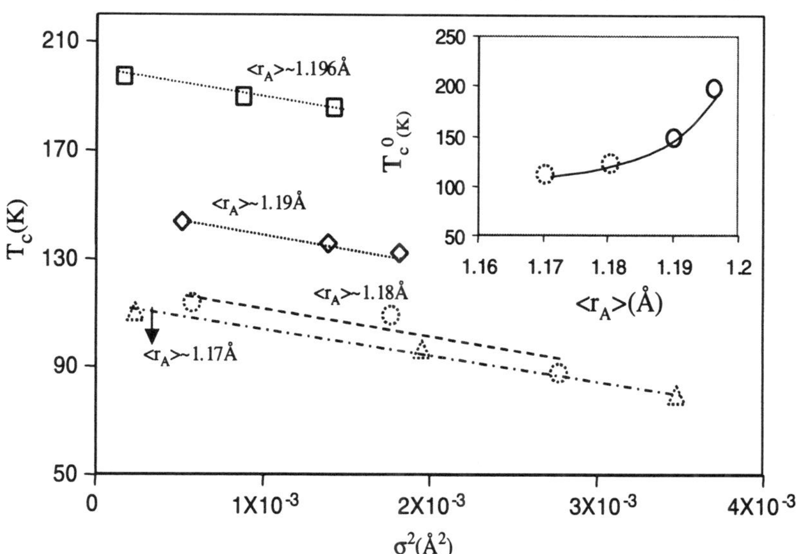

Figure 8. Variations of T_c (solid symbols) and T_M (broken symbols) against σ^2 (Å^2). The inset shows the variation of T_c^0 and T_M^0 against $\langle r_A \rangle$ (Å).

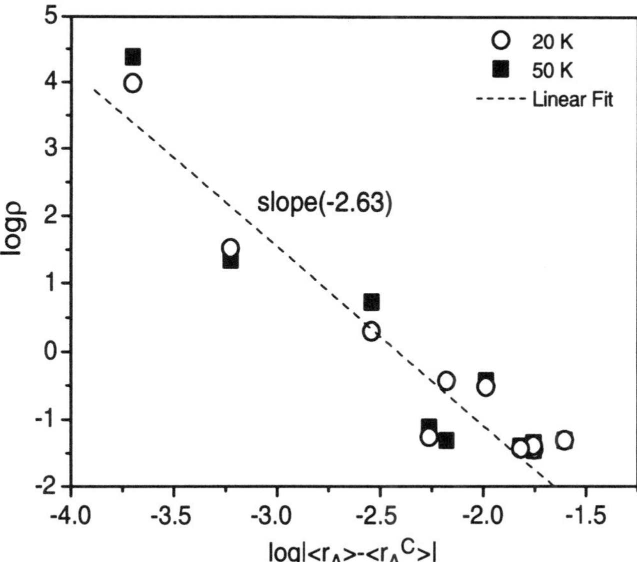

Figure 9. The linear scaling of $\log \rho$ with $\log |\langle r_A \rangle - \langle r_A^c \rangle|$ for $(La_{1-x}Ln_x)_{0.7}Ca_{0.3}MnO_3$ at 20 and 50 K.

that they become ferromagnetic when $x \leqslant x_c$, this change being accompanied by an IM transition. These manganates become AFM insulators when $x > x_c$, but show a small increase in magnetic moment at low temperatures ($T < T_M$). The values of T_c and T_M are both sensitive

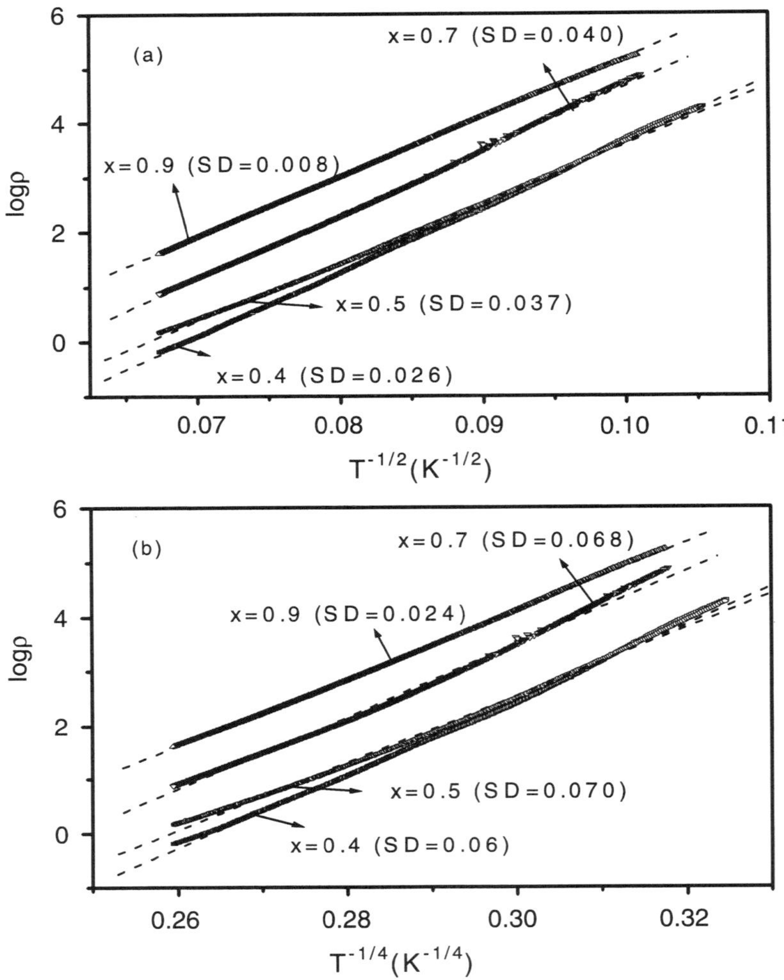

Figure 10. Fits of the resistivity data for $(La_{1-x}Y_x)_{0.7}Ca_{0.3}MnO_3$ for $x > x_c$ to (a) a $T^{-1/2}$-law and (b) a $T^{-1/4}$-law in the 90–220 K range. The open symbols are experimental data points and broken lines represent the corresponding linear fits.

to cation size disorder. The values of x_c are 0.6, 0.3 and 0.2 respectively for Nd, Gd and Y, corresponding to a unique value of the average size of the A-site cation, $\langle r_A \rangle$, of 1.18 Å ($\langle r_A^c \rangle$). It must be recalled that this value of $\langle r_A \rangle$ also marks the critical value below which the ground state is charge ordered. Phase separation is marked around x_c or $\langle r_A^c \rangle$ in all the series of compositions, although ferromagnetic clusters are likely to be present at low temperatures even when $1.0 > x > x_c$. The phase separation in these series of manganates is likely to be triggered by size disorder arising from the substitution of the smaller rare-earth cations in place of La. The results obtained in the present study are consistent with the accepted theoretical models for electronic phase separation in rare-earth manganates and other oxide materials [3, 19, 20].

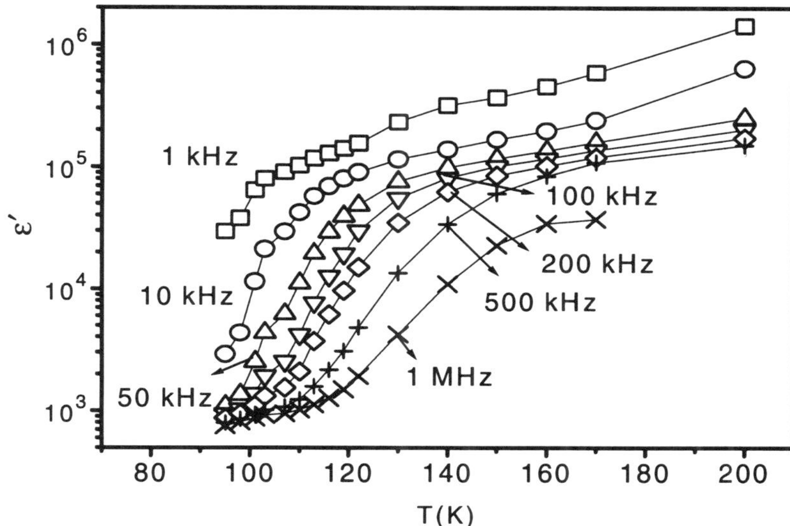

Figure 11. The temperature variation of the real part of the dielectric constant (ε') of $(La_{0.3}Y_{0.7})_{0.7}Ca_{0.3}MnO_3$ at different frequencies.

Acknowledgment

The authors thank BRNS (DAE) for support of this research.

References

[1] Rao C N R and Raveau B (ed) 1998 *Colossal Magnetoresistance, Charge Ordering and Related Properties of Manganese Oxides* (Singapore: World Scientific)

[2] Tokura Y (ed) 1999 *Colossal Magnetoresistance Oxides* (London: Gordon and Breach)

[3] Rao C N R and Vanitha P V 2002 *Curr. Opin. Solid State Mater. Sci.* **6** 97

[4] Uehara M, Mori S, Chen C H and Cheong S-W 1999 *Nature* **399** 560

[5] Babushkina N A, Taldenkov A N, Belova L M, Chistotina E A, Gorbenko O Yu, Kaul A R, Kugel K I and Khomskii D I 2000 *Phys. Rev.* B **62** R6081

[6] Lee H J, Kim K H, Kim M W, Noh T W, Kim B G, Koo T Y, Cheong S-W, Wang Y J and Wei X 2002 *Phys. Rev.* B **65** 115118

[7] Kiryukhin V, Kim B G, Podzorov V, Cheong S-W, Koo T Y, Hill J P, Moon I and Jeong Y H 2000 *Phys. Rev.* B **63** 024420

[8] Balagurov A M, Pomjakushin Y Yu, Sheptyakov D V, Aksenov V L, Fischer P, Keller L, Gorbenko O Yu, Kaul A R and Babushkina N A 2001 *Phys. Rev.* B **64** 024420

[9] Terashita H and Neumeier J J 2001 *Phys. Rev.* B **63** 174436

[10] de Teresa J M, Ibarra M R, García J, Blasco J, Ritter C, Algarabel P A, Marquina C and del Moral A 1996 *Phys. Rev. Lett.* **76** 3392

[11] Rao C N R, Arulraj A, Cheetham A K and Raveau B 2000 *J. Phys.: Condens. Matter* **12** R83

[12] Vanitha P V and Rao C N R 2001 *J. Phys.: Condens. Matter* **13** 11707

[13] Deac I G, Diaz S V, Kim B G, Cheong S-W and Schiffer P 2002 *Phys. Rev.* B **65** 174426

[14] Rodriguez-Martinez L M and Attfield J P 1996 *Phys. Rev.* B **54** R15622

[15] Vanitha P V, Santosh P N, Singh R S, Rao C N R and Attfield J P 1999 *Phys. Rev.* B **59** 13539

[16] Vijaya Sarathy K V, Parashar S, Raju A R and Rao C N R 2002 *Solid State Sci.* **4** 353

[17] Homes C C, Vogt T, Shapiro S M, Wakimoto S and Ramirez A R 2001 *Science* **293** 673

[18] He L, Neaton J B, Cohen M H, Vanderbilt D and Homes C C 2002 *Preprint* cond-mat/0110166

[19] Rao C N R, Vanitha P V and Cheetham A K 2003 *Chem. Eur. J.* **9** 828

[20] Dagotto E, Hotta T and Moreo A 2001 *Phys. Rep.* **344** 1

IV. Nanoparticles

C.N.R. Rao

CSIR Centre of Excellence in Chemistry,
Chemistry & Physics of Materials Unit
Jawaharlal Nehru Centre for Advanced Scientific Research
Jakkur P.O., Bangalore-560 064, INDIA
cnrrao@jncasr.ac.in

Nanoparticles constitute an important class of nanomaterials, and are not new to chemistry. One of the earliest experiments on nanoparticles was carried out by Michael Faraday who prepared sols of gold and other metals. Nanoparticles of a variety of metals, semiconductors and ceramic materials have been prepared and characterized in recent years.[1,2] An aspect of great interest concerns the size-dependent properties of nanoparticles. Quantum confinement in metal and semiconductor nanoparticles has been investigated widely. Thus, very small metal nanoparticles cease to be metallic and show modified reactivity compared to the bulk. Semiconductor nanoparticles exhibit size-dependent energy gaps and the resulting shifts in electronic absorption and emission spectra. Metal quantum dots exhibit Coulomb blockade. The articles in this section discuss the growth, physical properties, electronic structure, self-assembly and chemical reactivity of nanoparticles, specially those of metals.

References

1. C.N.R. Rao, G.U. Kulkarni, P.J. Thomas and P.P. Edwards, *Chem. Euro. J.* **8**, 28 (2002).
2. J. Jortner and C.N.R. Rao, *Pure Appl. Chem.* **74**, 1491 (2002).

J. Phys. Chem. **1995**, *99*, 5639-5644

Growth of Nanometric Gold Particles in Solution Phase

Ram Seshadri,[†] G. N. Subbanna,[‡] V. Vijayakrishnan,[†] G. U. Kulkarni,[§] G. Ananthakrishna,[‡] and C. N. R. Rao*[,†,§]

Solid State and Structural Chemistry Unit, CSIR Center of Excellence in Chemistry, and Materials Research Center, Indian Institute of Science, Bangalore 560 012, India

Received: January 26, 1995[⊗]

The time evolution of colloidal gold particles in the nanometric regime has been investigated by employing electron microscopy and electronic absorption spectroscopy. The particle size distributions are essentially Gaussian and show the same time dependence for both the mean and the standard deviation, enabling us to obtain a time-independent universal curve for the particle size. Temperature dependent studies show the growth to be an activated process with a barrier of about 18 kJ mol^{-1}. We present a phenomenological equation for the evolution of particle size and suggest that the growth process is stochastic.

Introduction

Interest in colloidal metal particles exemplified by aqueous gold sols dates back at least to the time of Michael Faraday, who coined the term "finely divided metals" and recognized that the different colors of gold sols could be indicative of different sizes or states of aggregation of the particles. In the state of matter comprising particles of a few tens to a few thousand atoms, the behavior is known to be distinct from the bulk as well as from individual atoms. In recent years, there has been a resurgence of interest in this intermediate state of matter from several points of view. There is considerable interest in nanoparticles as a distinct state of matter, with particular reference to their structure and reactivity.[1] A question that has attracted attention is whether there is a transformation from the metallic to the nonmetallic state with decrease in cluster size.[2,3] Secondly, aggregation of such particles into fractal-like structures has been the focus of some studies in the past decade or two.[4] An aspect of some importance in the context of metal particles is the phenomenon of nucleation and growth in the nanometer regime, which can be effectively investigated today by employing techniques such as high-resolution electron microscopy and scanning tunneling microscopy. The subject of nucleation and growth of particles of micrometer dimensions in condensed phases has been the focus of much attention.[5,6] To our knowledge, there have been very few such studies in the nanometer regime. We have chosen an aqueous gold sol for such a study in the belief that it is eminently preferred over traditional condensed phase systems in terms of control and manipulation. Subnanometer gold sols have been the focus of some recent investigation.[3,7] Another aspect that provoked our interest to examine the distribution of particle sizes at this length scale is whereas aggregates of metal particles exhibit a scaling regime over the length scale of 100–10 000 nm with a characteristic fractal dimension, the individual particles appear not to be in contact. There is thus a breakdown of the scaling regime at smaller length scales. The present study is significantly different from that of Turkevich et al.,[8] who studied bulk growth kinetics of gold sols in terms of volume fractions, rather than following the time evolution of the particle size distributions (PSDs) as we have done here.

In the present study, we have used chiefly transmission electron microscopy since *only* microscopy allows us to obtain the complete particle size histograms. Scanning tunneling microscopy (STM) of gold sols spotted on graphite substrates yield similar PSDs and this served as a check on the reliability of PSDs obtained from electron microscopy. The mean radii of the gold particles can be obtained from other techniques such as ultracentrifugation, light scattering, electrophoresis, and absorption spectroscopy, but none of these techniques yields the histogram of particles in the size range of interest to us. We have however employed electronic absorption spectroscopy to extend the results obtained by electron microscopy.

We have followed the evolution of the PSDs as a function of time and of temperature for nanometer-sized gold particles dispersed in an aqueous sol phase. Surprisingly, we find that the PSDs are well represented by Gaussians with the mean diameter $\langle d \rangle$ and the standard deviation σ showing the same time dependence, thus allowing us to represent the entire time evolution data by a universal curve. We have investigated this aspect in some detail, and we propose a phenomenological equation governing the evolution of the PSDs. Despite the fact that the temperature dependence of the growth could be studied over a small range due to experimental limitations intrinsic to the system, it has enabled us to confirm that the growth process is activated.

Experimental Section

Gold sols were prepared, following the method of Duff et al.,[7] using alkaline solutions of tetrakis(hydroxymethyl)phosphonium chloride (THPC from Fluka) to reduce dark-aged solutions of chloroauric acid (HAuCl$_4$, Loba-Chemie). The solutions were aged for at least a month. This ensures that hydrolysis is arrested and improves reproducibility between experiments. Typically, to 50 mL of aqueous 6.6 mM NaOH in a stoppered flask, 1 mL of THPC solution (1.2 mL of 80% THPC diluted to 100 mL) was added. After 2 min, 2 mL of 25 mM aged, aqueous chloroauric acid was added rapidly while shaking the flask. We avoided the use of Teflon stir bars as these scratch easily and are prone to contamination. The sols thus produced are typically very dilute, making the study of their growth a more tractable problem. The volume fraction of the particles was of the order of 10^{-4}–10^{-5}. The glassware (new and unscratched) was scrupulously washed with aqua regia followed by detergent–hydrofluoric acid mixtures before each experiment. Water used for the experiments and for making

---

† Solid State and Structural Chemistry Unit.
‡ Materials Research Center.
§ CSIR Center of Excellence in Chemistry.
* To whom correspondence should be addressed.
⊗ Abstract published in *Advance ACS Abstracts*, March 15, 1995.

5640 *J. Phys. Chem., Vol. 99, No. 15, 1995*

stock solutions was distilled once in metal or glass and then distilled twice in a quartz apparatus. For the electronic absorption spectroscopy experiments, a sol with larger particles was also prepared, using 3 times the quantity of chloroauric acid.

With passage of time, taking the addition of the chloroauric acid as zero time, the sol changes color from orange-brown eventually to dark red. After intervals of time, a drop of the sol was spotted on a carbon-coated copper grid for transmission electron microscopic investigations using a JEOL JEM 200CX. The uncertainty in time during the spotting was on the order of 2−3 min. The high dilution of the sol is necessary to ensure that the particles do not aggregate during drying. For recording UV/vis spectra, a 2−3 mL aliquot of the sol was periodically removed and its absorption spectra recorded in 1 cm cuvettes, following which the aliquot was discarded. The cuvettes were washed in aqua regia before and between measurements. The uncertainty in time was larger in these experiments, on the order of 5 min.

STM measurements (as a check on electron microscopy) were carried out by spotting the gold sols on a highly ordered pyrolytic graphite (HOPG) substrate (a gift of Dr. A. Moore, Union Carbide), using a NanoscopeII. The images were acquired in the constant current mode. The disadvantage of this technique over electron microscopy is that fewer particles can be imaged and therefore one obtains poorer statistics for the PSDs. Further details are given in ref 3.

The particle size distributions were obtained by considering the diameter of equivalent circles in the electron micrographs. In some cases, this was done using vernier calipers and magnifiers. Typically 100−180 well-separated particles from one or two micrographs were considered. The variable temperature experiments were carried out by preparing the sol in a double-walled vessel with alcohol circulation at constant temperature. A Julabo thermostat, with the alcohol temperature maintained at better than 0.1 K was used for this purpose. Only a limited temperature range could be probed, since at higher temperatures, there are problems of solvent evaporation and coagulation of the particles.

Results and Discussion

Structure and Morphology of the Gold Particles. A typical low-magnification transmission electron micrograph of the gold particles is shown in Figure 1a. At coarse magnification, the particles appear aggregated, and indeed, the structure resembles that studied by Weitz et al.[9] We have not attempted to obtain a fractal dimension for this aggregate, since our interest is more to do with distributions of particle size. Contrary to what one would expect from Figure 1a, the individual particles are seen to be distinctly separate from one another at higher magnification (Figure 1b), allowing us to determine their sizes. Such a crossover of the scaling regimes has been reported for silver particles in the literature.[10] At small times (diameters around 2−5 nm) the particles seem to be spherical. Reasons for this could include the possible fluid nature of the smaller particles,[11] i.e., the surface energies are not significant when the areas of the faces are very small. At larger times however (diameters ≈ 20 nm, corresponding to the tails of the PSDs) faceting is clearly observed. The projected morphologies include vertex-truncated triangles and squares similar to what is documented in the literature.[12]

Particle Size Distribution and Time Evolution of the Particle Size. Since most of the particles were spherical, we were able to obtain reliable distributions of the particle diameters from electron microscopy and other techniques. In Figure 2,

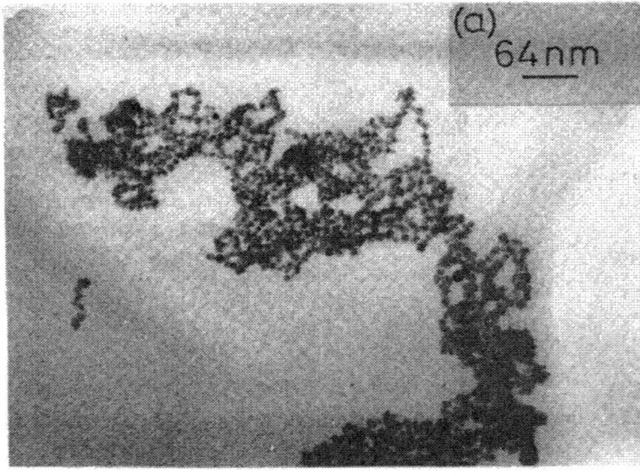

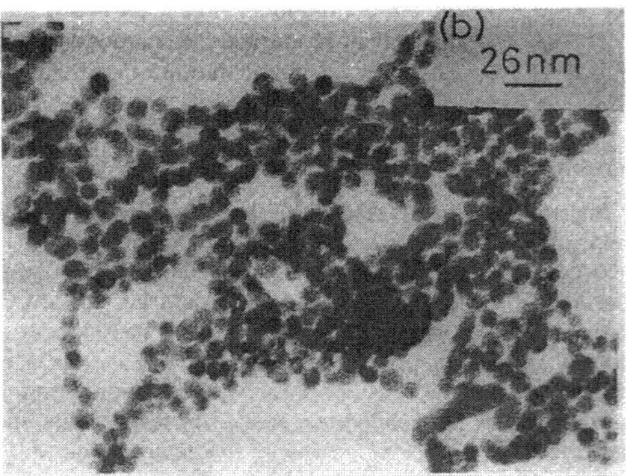

Figure 1. (a) Transmission electron micrograph of a gold sol grown for 60 min at 298 K shown at relatively low magnification. (b) The same particles at higher magnification showing well-separated individual particles.

we show the PSDs obtained from (a) electron microscopy and from (b) STM for a THPC reduced gold sol with particles in the subnanometer range. The similarity in the PSDs serves as a check on the reliability of the data. Electronic absorption spectroscopy is not suited for use in these small size ranges. In Figure 3, we show the electronic absorption spectrum in the visible region of a gold sol prepared by the THPC route at 0, 30, and 60 min the uncertainty in time being ≈5 min. For sols of very small particles, there are no distinct absorption maxima as can be seen in Figure 3 and the traditional Mie[13] theory cannot be used to interpret the spectra. Color has been a traditional characterization tool for gold sols, but usually in a slightly larger size range. The colors are largely due to surface plasmon excitations. In Figure 4, we show the absorption spectrum of a sol with a mean diameter of 11.4 nm along with the first term of the Mie[12] fit for small gold particles. The spectrum corresponds to particles shown in the electron micrograph in Figure 5, where the measured PSD with a Gaussian fit is shown as an inset. Although the Mie theory is not effective in this size regime as pointed out by Kreibig,[14] Doremus,[15] and others, we have obtained an average particle diameter from the ratio of the maximum absorbance to the absorbance at 440 nm, using the empirical correlation of Kreibig and Genzel,[14] and obtained a mean diameter of 10 nm. This size does indeed compare favorably with the value obtained from electron microscopy (11.4 nm).

Since the absorption spectrum is well defined for larger particles, we have followed the time evolution of the spectra of a gold sol prepared using a greater initial concentration of gold

Nanometric Gold Particles in Solution Phase

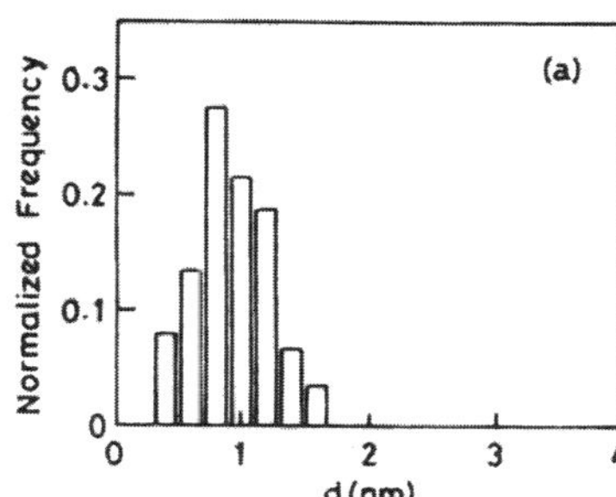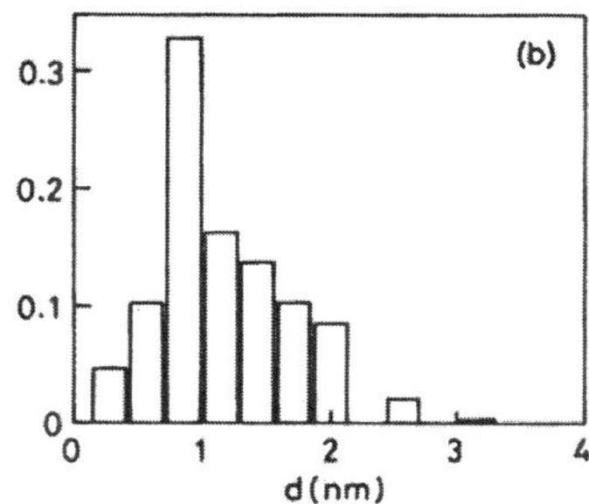

Figure 2. (a) Histogram of particle sizes (PSD) for a subnanometer (THPC reduced) gold sol obtained from analysis of transmission electron microscopy. (b) Histogram for the particles of the same sol obtained from analysis of STM images.

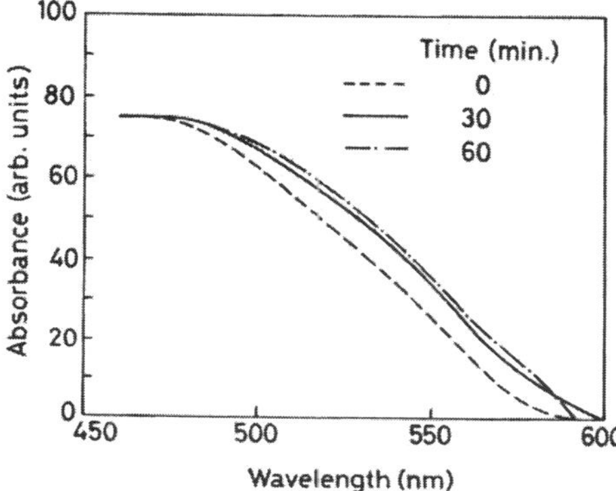

Figure 3. Absorption spectra of a gold sol with particles in the sub-10-nm range at different times, 0, 30, and 60 min.

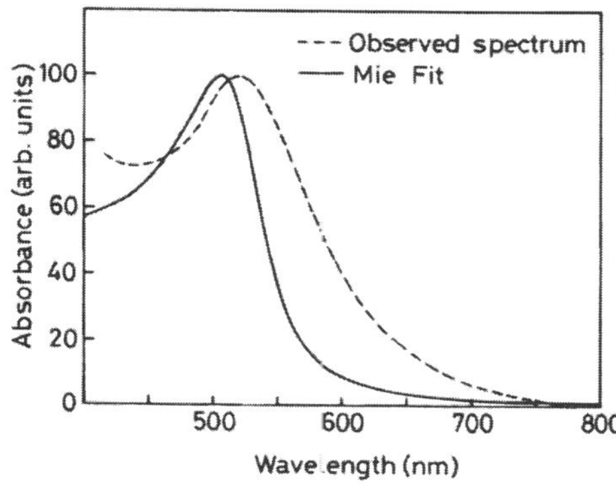

Figure 4. Absorption spectrum of a $\langle d \rangle = 11.4$ nm gold sol with the Mie fit. The y axis scale is arbitrary.

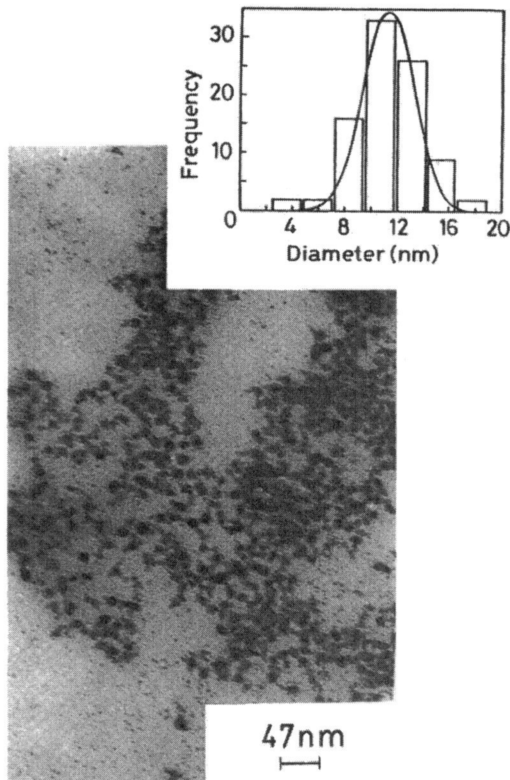

Figure 5. Transmission electron micrograph corresponding to the sol whose absorption spectrum is shown in Figure 4. The inset is a histogram of the particle size distribution. The smooth curve is a Gaussian fit.

(the initial chloroauric acid concentration was 75 mM, the other conditions being the same). Figure 6 shows representative spectra obtained at different times. The growth is initially very rapid and the uncertainty in time somewhat larger in these experiments. From the plot of the wavelength corresponding to the absorption maxima against the particle diameter as obtained from the Mie theory by Milligan and Morris[16] and Chow and Zukowski,[17] we have obtained the average particle diameters. We have plotted particle diameters against time plotted as an inset in Figure 6. We shall return to this inset later on in our discussion.

The bulk of our results is based on the PSD data obtained from electron microscopy. The particle size distributions

obtained by us at different times are shown in Figure 7a. At small times of about 30 min, the particles have a mean diameter of around 2 nm. The growth is initially quick, slowing down at longer times. After 240 min, the average particle diameter is about 12 nm. The distributions are nearly symmetric, suggesting that they can be fitted by Gaussians. Indeed, we find on doing so, that Gaussian distributions describe the PSDs very well. Attempts to use other distributions, for example, log-normal, were not as successful. We verified this by plotting the cumulative frequencies against the diameter, and against the logarithm of the diameter on a probability graph paper. This

5642 *J. Phys. Chem., Vol. 99, No. 15, 1995*

Seshadri et al.

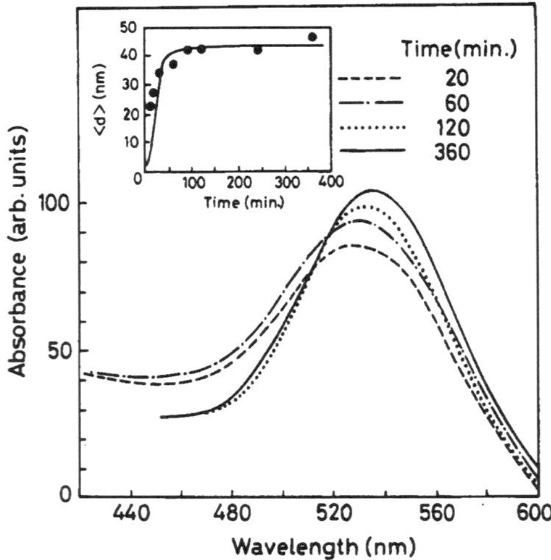

Figure 6. Time evolution of the absorption spectrum of a gold sol with higher initial gold concentrations and concomitant rapid growth. Representative spectra are shown. Inset: plot of mean diameters against time for diameters obtained from the peak positions in the absorption spectrum. The fit is from eq 4, for the parameters given in the text.

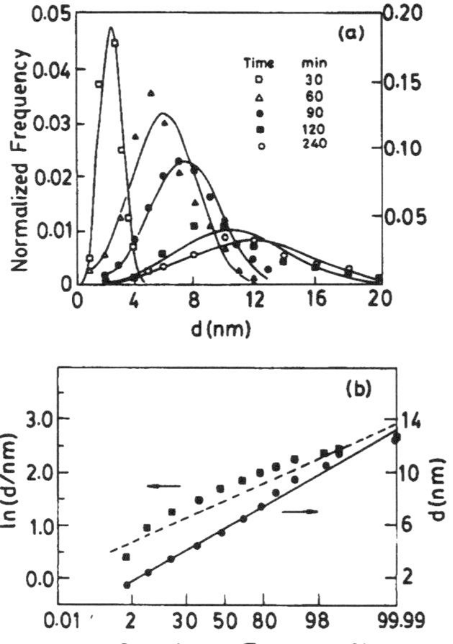

Figure 7. (a) Particle size distributions of the gold sol at different times obtained from electron microscopy. The curves are the fitted Gaussians distributions. The areas under the curve have been normalized to unity. (b) Plot of the diameter and the log diameter against cumulative frequency, on probability graph paper for a typical growth time of 60 min. The lines are guides to the eye.

is shown in Figure 7b for a typical time of 60 min. Small deviations from a straight line are seen in the plots of diameter against cumulative probability only at larger diameters. However, our large diameter statistics are less reliable because of the difficulty in resolving the larger particles due to overlapping images. The deviations in plots of the logarithm of the diameter against cumulative frequency (Figure 7b) may appear to be

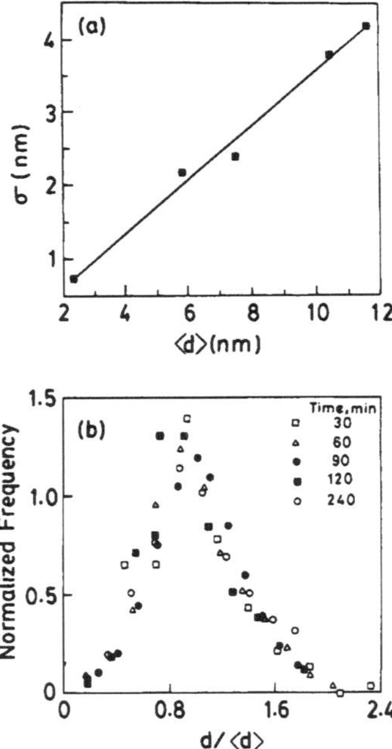

Figure 8. (a) Plot of the mean particle diameter, $\langle d \rangle$, against the standard deviation, σ, at different times showing linear behavior (R_{fit} = 0.99). (b) Universal curve of PSDs obtained by scaling out the mean diameter at different times.

small, but they are actually larger due to the logarithmic scale. A plot of the mean diameter, $\langle d \rangle$, as a function of $t^{1/3}$ gives a nearly straight line (R_{fit} = 0.91) which is the expected time dependence for Ostwald ripening. However, we prefer not to attempt such a power law fit since the data do not run over many decades. Indeed, that the process is *not Ostwald ripening* is clearly borne out by the fact that the PSDs do not display the sharp cutoff on the size-axis expected from the Lifshitz−Slyozov−Wagner (LSW) theory of Ostwald ripening.[18] A symmetric distribution from LSW theory is possible, but only at high volume fractions.[18] In our studies, the volume fraction is $\approx 10^{-4}$, so this does not apply. Since a Gaussian process essentially means that the underlying process is that of a sum of independent random events, the suggested growth process is stochastic.

Plotting $\langle d \rangle$ as a function of the standard deviation, σ, for different times, we find that the mean and the σ show the same time dependence (Figure 8a). This suggests that it should be possible to represent the PSDs by means of a variable scaled by the mean diameter. A plot of the normalized frequency (normalized to the area under the PSD curve) as a function of $d/\langle d \rangle$ for all times collapses onto a single curve, as shown in Figure 8b. To the best of our knowledge, this is the first growth kinetics which exhibits the same time dependence for both the mean and σ. Any theoretical model for the growth should therefore incorporate these unusual features.

Conventionally, growth kinetics are often described by partial differential equations for the distribution of particles. Such equations can be derived from a Master equation which in some way represents cluster−cluster aggregation where at most a few elementary clusters are mobile.[19,20] In the present situation, the growth kinetics is well described by a Fokker−Planck type equation, having the following form:

Nanometric Gold Particles in Solution Phase

J. Phys. Chem., Vol. 99, No. 15, 1995 **5643**

$$\frac{\partial N(d,t)}{\partial t} + \frac{\partial}{\partial d}\overline{(\alpha(d) - \beta(d))}N(d,t) = \frac{\partial^2}{\partial d^2} \overline{D(d,t)}\, N(d,t) \tag{1}$$

where $\overline{\alpha(d)}$ and $\overline{\beta(d)}$ are the probability of growth and shrinkage per unit time, and the overbars denote the averages over the distribution. Often $D = \overline{\alpha(d)} + \overline{\beta(d)}$ is used. This can be related to the mobility of the diffusing monomers. Since the growth process is new, identifying $\alpha(d)$ and $\beta(d)$ requires a clear understanding of the basic growth process which is lacking at present. We therefore proceed to guess the form of $\alpha(d)$, $\beta(d)$ and D to obtain the desired rate law. We identify $\alpha(d) - \beta(d) = d_0 f'(t)$ and $D(d) = 2Df(t)\, f'(t)$, where the prime denotes the derivative with respect to time. Normally, $\alpha(d)$, $\beta(d)$ and $D(d)$ are used as functions of d. However, for convenience of comparison, we have taken them to be functions of t. This gives

$$\frac{\partial N(d,t)}{\partial t} + d_0 f'(t)\frac{\partial N(d,t)}{\partial d} = 2Df(t)\, f'(t)\frac{\partial^2}{\partial d^2}N(d,t) \tag{2}$$

where d_0 is the initial diameter of the particle which grows with a functional form $f(t)$ and can be identified as the growth of the mean particle diameter. The PSD that one obtains from the solution of the above equation is

$$N(d,t) = (2\pi D f^2(t))^{-1/2}\exp\!\left(\frac{-[d-d_0 f(t)]^2}{2D f^2(t)}\right) \tag{3}$$

where d_0 is the initial value of the mean $\langle d\rangle$. The form of $f(t)$ has been obtained by plotting $\langle d\rangle$ vs t as shown in Figure 9. The continuous curve is given by a phenomenological equation:

$$<d> = \frac{d_0}{(1-d_0/d_\infty)\exp(-\gamma t) + d_0/d_\infty} \tag{4}$$

where d_∞ is the mean diameter at infinite time and γ is some rate. For the fit in Figure 9, we have used $d_0 = 1$ nm, $d_\infty = 12$ nm, and $\gamma = 0.035$ min^{-1}. We may regard d_0 as the upper bound for the critical size of a stable nucleus.

We now return to the inset in Figure 6. This inset shows the average particle diameter (from electronic absorption spectroscopy) as a function of time for experiments involving larger particles of gold, starting with a higher initial concentration of chloroauric acid. We obtain a reasonable fit of these data to the sigmoidal growth law given by equation 4. For this fit, we have used the parameters $\gamma = 0.125$ min.$^{-1}$, $d_0 = 1$ nm, and $d_\infty = 43$ nm. The larger value of γ is indicative of the more rapid growth, due to the higher initial gold concentration. (We have maintained the same value of $d_c = 1$ nm because this might be some critical nucleus size.) The absorption data do not fit the growth law as well as the PSDs from electron microscopy, probably more due to the intrinsic errors in time as well as in average diameters obtained from absorption spectroscopy. However the fit in the inset of Figure 6 provides additional support to the model proposed in the text.

It must be emphasized that the phenomenological growth law presented here is new although we are unable to present a better physical basis for it at the present time. We would expect that at small times, the particles are close to the minimum stable cluster size. For this reason, the particles with sizes beyond the critical size have fairly large inflow due to the large nucleation rates initially. In contrast, further growth of the particles (propagation step) is not likely to be equally fast since

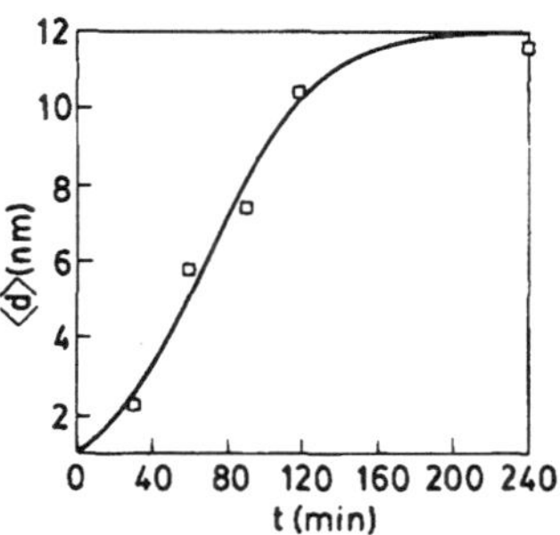

Figure 9. Plot of the mean particle diameter, $\langle d\rangle$, as function of time. The continuous curve is a fit to the model proposed in the text (eq 4).

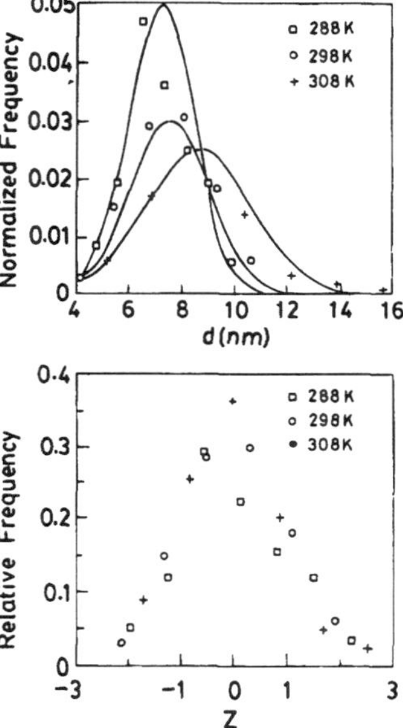

Figure 10. (a) PSDs of gold particles grown for 60 min at different temperatures. The curves are the fitted Gaussian distributions. (b) Data at different temperatures have been collapsed on to a universal curve, by using $z = (d - \langle d\rangle)/\sigma_0 \exp(-E/k_B T)$.

it requires diffusion of monomers. This would explain why the deviations from Gaussian behavior are seen for smaller times. At larger times, one would expect the nucleation rate to have slowed down sufficiently to give rise to more symmetric distributions.

Effect of Temperature on the PSD. We have studied the effect of temperature over the limited range 288–308 K, at a fixed time of 60 min, and obtained the respective PSDs (shown in Figure 10a) along with $\langle d\rangle$ and σ. We find that $\ln \sigma$ varies linearly with $1/k_B T$, implying the activated nature of this process within this temperature range. The activation energy works out to be of the order of 17.5 kJ mol^{-1}. Using the σ obtained from such a temperature dependence, we are the able to scale the independent Gaussian variable and collapse the PSDs at fixed time and different temperatures on to one curve as shown in Figure 10b.

Conclusions

The observations of a Gaussian distribution of particle sizes, having the same time dependence for both $\langle d\rangle$ and σ, suggest

5644 *J. Phys. Chem., Vol. 99, No. 15, 1995*

Figure 11. Scanning electron micrograph of a fractal structure formed by evaporating a gold sol on a glass slide. This structure had a fractal dimension of 1.7.

its time dependence is new and requires further investigation. The observed Gaussian distribution is not only consistant with the process being stochastic, but may also imply that the rates of nucleation and growth are well separated.

An aspect worthy of mention is the nature of the structures formed on allowing the gold sol particles to occlude at long times, for example, by evaporating the sol on a glass slide. Figure 11 shows a scanning electron micrograph of a typical branched fractal-like structure so formed. The measured fractal (capacity) dimension of this structure obtained by the box-counting method was 1.7. Such structures are to be contrasted with the micrographs in Figures 1 and 5, where the particles are well separated and do not seem to form ramified assemblies. Under different evaporation conditions, we are also able to obtain dendrites, usually with four arms and a smaller fractal dimension. The crossover from the fractal to the dendritic regime is of some interest and is presently being pursued by us.

that the basic processes contributing to the growth process at such small nanometric length scales are distinct from those at larger length scales, following traditional growth models such as Ostwald ripening. Earlier studies of evolutions of particle size distributions have chosen to assume Gaussian distributions and then follow their time dependence. Log-normal distributions are ubiquitous in situations such as growth from the vapor phase. As such, these unique features of growth of the nanometric gold particles could be characteristic of growth in solution phase, occurring over a relatively small temperature range. The temperature dependence implies that the process is activated. The small barrier is possibly associated with the typical length scales involved in these systems. The growth law provided to fit the observed particle size distribution and

References and Notes

(1) See for example: *Faraday Discuss. Chem. Soc.* **1991**, *92*. Henglein, A. *J. Phys. Chem.* **1993**, *97*, 5457.

(2) Halperin, W. P. *Rev. Mod. Phys.* **1986**, *58*, 533. Kubo, R.; Kawabata, A.; Kobayashi, S. *Annu. Rev. Mater. Sci.* **1984**, *14*, 49.

(3) (a) Rao, C. N. R.; Vijayakrishnan, V.; Aiyer, H. N.; Kulkarni, G. U.; Subbanna, G. N. *J. Phys. Chem.* **1993**, *97*, 11157. (b) Kulkarni, G. U.; Aiyer, H. N.; Vijayakrishnan, V.; Arunarkavalli, T.; Rao, C. N. R. *J. Chem. Soc., Chem. Commun.* **1993**, 1545.

(4) Vicsek, T. *Fractal Growth Phenomena*, 2nd ed.; World Scientific: Singapore, 1992.

(5) Gunton, J. D.; Miguel, M. S.; Sahni, P. S. In *Phase Transitions and Critical Phenomena*; Domb, C., Lebowitz, J. L., Eds.; Academic: New York, 1983; pp 267−482 and references therein.

(6) Nielson, A. E. *Kinetics of Precipitation*; Pergamon: Oxford, 1964.

(7) (a) Duff, D. G.; Baiker, A.; Edwards, P. P. *J. Chem. Soc., Chem. Commun.* 1993, 96. (b) Duff, D. G.; Baiker, A.; Edwards, P. P. *Langmuir* **1993**, *9*, 2301.

(8) Turkevich, J.; Stevenson, P. C.; Hillier, J. *Discuss. Faraday Soc.* **1951**, *11*, 55.

(9) (a) Weitz, D. A.; Oliviera, M. *Phys. Rev. Lett.* **1984**, *52*, 1433. (b) Weitz, D. A.; Huang, J. S.; Lin, M. Y.; Sung, J. *Phys. Rev. Lett.* **1985**, *54*, 1416.

(10) Mukherjee, M.; Saha, S. K.; Chakravorty, D. *Appl. Phys. Lett.* **1993**, *63*, 42.

(11) Iijima, S.; Ichihashi, T. *Phys. Rev. Lett.* **1986**, *56*, 616.

(12) Kirkland, A. I.; Edwards, P. P.; Jefferson, D. A.; Duff, D. G. *Annual Reports C*; The Royal Society of Chemistry: Cambridge, 1988; pp 247−304.

(13) Mie, G. *Ann. Phys.* **1908**, *25*, 377.

(14) Kreibig, U.; Genzel, L. *Surf. Sci.* **1985**, *156*, 679.

(15) Doremus, R. H. *J. Chem. Phys.* **1964**, *40*, 2389.

(16) Milligan, W. O.; Morris, R. H. *J. Am. Chem. Soc.* **1964**, *86*, 3461.

(17) Chow, M. K.; Zukowski, C. F. *J. Colloid Interface Sci.* **1994**, *165*, 97.

(18) Jayanth, C. S.; Nash, P. *J. Mater. Sci.* **1989**, *24*, 3041.

(19) van Kampen, N. G. *Stochastic Processes in Physics and Chemistry*; North-Holland: Amsterdam, 1981.

(20) Ananthakrishna, G. *Pramana J. Phys.* **1979**, *12*, 565.

JP950271+

Size-Dependent Chemistry: Properties of Nanocrystals

C. N. R. Rao,*[a] G. U. Kulkarni,[a] P. John Thomas,[a] and Peter P. Edwards*[b]

Abstract: Properties of materials determined by their size are indeed fascinating and form the basis of the emerging area of nanoscience. In this article, we examine the size dependent electronic structure and properties of nanocrystals of semiconductors and metals to illustrate this aspect. We then discuss the chemical reactivity of metal nanocrystals which is strongly dependent on the size not only because of the large surface area but also a result of the significantly different electronic structure of the small nanocrystals. Nanoscale catalysis of gold exemplifies this feature. Size also plays a role in the assembly of nanocrystals into crystalline arrays. While we owe the beginnings of size-dependent chemistry to the early studies of colloids, recent findings have added a new dimension to the subject.

Keywords: colloids · nanostructures · self-assembly · semiconductors

Introduction

Steric effects—arising from bulky groups—are well known as key factors in determining the reactivity of organic molecules. Otherwise, we do not ordinarily concern ourselves with the physical dimension of a system as a factor in determining its intrinsic properties except in intercalation chemistry or some such situation where the pore or cavity in a host lattice molecule can accommodate guest species of a particular size. Size, however, becomes *the* sole controlling factor while dealing with the science and application of the so-called nanoparticles, that cover a size range between 1–100 nm. Nanoparticles within this size domain are thus intermediate between the atomic and molecular size regimes on one hand, and the macroscopic, bulk on the other. In this regime, size-dependent properties manifest themselves when the size of an individual particle is sufficiently small. A key aspect of the rapidly developing area of nanoscience and nanotechnology concerns itself with the size dependence of intrinsic properties of materials. It is, therefore, instructive to look at typical properties of metals and semiconductors in the size-dependent regime of nanoparticles. We shall discuss three important aspect of the nanocrystals of these materials, electronic structure and properties, chemical reactivity and self-assembly. Size dependent structural and thermodynamic properties of nanoparticles such as bond lengths, melting point and specific heat have already been reviewed[1] in the literature and will not be discussed in any detail in the present article.

Two centuries ago the study of nanoscale solid particles, dispersed within a liquid host, played a pivotal role in establishing colloid science. During the final decades of the last century, colloid science was, perhaps, something of an intellectual backwater—with one or two notable exceptions. However, significant advances in both experimental and theoretical aspects of the subject, and, of course, the emergence of, and explosion of interest in, the broad area of nanoscience and nanotechnology, have now set the scene for a renaissance in colloid science. Interestingly, it does appear that many scientists in the modern field of nanoscience may not even recognize its colloidal source. This is unfortunate, since there is a wealth of information and expertise in that old and venerable science.

An important example of the parentage of the modern subject of metal nanoparticles derives from the work of Faraday in the 1850s. During that period, Faraday carried out groundbreaking studies of nanoscale gold particles in aqueous solution. He established the first quantitative basis for the area, noting that these colloidal metal sols ("pseudosolutions") are thermodynamically unstable, and that the individual gold nanoparticles must be stabilized kinetically against aggregation. Once the nanoparticles coagulate, the process cannot be reversed. Remarkably, Faraday also identified the very essence of the nature of colloidal, nanoscale particles of metals; specifically, for the case of gold, he concluded (in

[a] Prof. Dr. C. N. R. Rao, Prof. Dr. G. U. Kulkarni, P. J. Thomas
Chemistry and Physics of Materials Unit
Jawaharlal Nehru Center for Advanced Scientific Research
Jakkur P.O., Bangalore, 560 064 (India)
E-mail: cnrrao@jncasr.ac.in

[b] Prof. Dr. P. P. Edwards
School of Chemistry
University of Birmingham, Edgabaston,
Birmingham, B15 2TT (UK)
E-mail: p.p.edwards@bham.ac.uk

C. N. R. Rao, P. P. Edwards et al.

1857!) ... *"the gold is reduced in exceedingly fine particles which becoming diffused, produce a beautiful fluid ... the various preparations of gold whether ruby, green, violet or blue ... consist of that substance in a metallic divided state"*.

During that century, colloidal phenomena played a pivotal role in the genesis of physical chemistry by establishing a connection between descriptive chemistry and theoretical physics. For example, Einstein provided the relationship between Brownian motion and diffusion coefficient of colloidal particles.

Today, the overwhelming importance now associated with the nanoscale in both science and technology means that the scene is once again set for this key subject to impact upon the development of not only chemistry, but also physics and materials science. In the following sections we attempt to highlight a few of the key issues relating to nanparticles where size determines their properties.

Electronic structure and properties: The electronic structure of a nanocrystal critically depends on its very size. For small particles, the electronic energy levels are not continuous as in bulk materials, but discrete, due to the confinement of the electron wavefunction because of the physical dimensions of the particles (see Figure 1). The average electronic energy level spacing of successive quantum levels, δ, known as the so-called Kubo gap, is given by, $\delta = 4E_f/3n$, where E_f is the Fermi energy of the bulk material and n is total number of valence electrons in the nanocrystal. Thus, for an individual silver nanoparticle of 3 nm diameter containing approximately one

thousand silver atoms, the value of δ would be 5 – 10 meV. Since the thermal energy at room temperature, $kT \cong 25$ meV, a 3 nm particle would be metallic ($kT > \delta$). At low temperatures, however, the level spacings specially in small particles, may become comparable to kT, rendering them nonmetallic.[2] Because of the presence of the Kubo gap in individual nanoparticles, properties such as electrical conductivity and magnetic susceptibility exhibit quantum size effects.[3] The resultant discreteness of energy levels also brings about fundamental changes in the characteristic spectral features of the nanoparticles, especially those related to the valence band.

Extensive investigations of metal nanocrystals of various sizes obtained, for example by the deposition of metals on amorphised graphite and other substrates, by X-ray photoelectron spectroscopy and related techniques[4, 5] have yielded valuable information on their electronic structure. An important result from these experiments is that as the metal particle size decreases, the core-level binding energy of metals such as Au, Ag, Pd, Ni and Cu increases sharply. This is shown in the case of Pd in Figure 2, where the binding energy

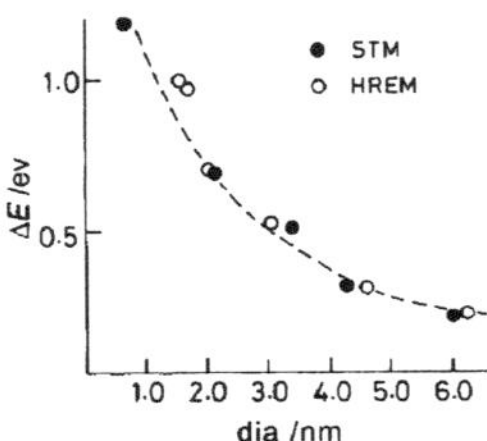

Figure 2. Variation of the shift, ΔE, in the core-level binding energy (relative to the bulk metal value) of Pd with the nanoparticle diameter. The diameters were obtained from HREM and STM images (reproduced with permission from ref. [3]).

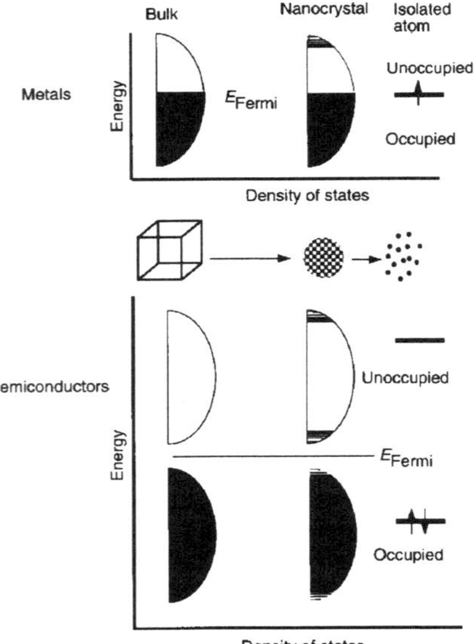

Figure 1. Density of states for metal (a) and semiconductor (b) nanocrystals. In each case, the density of states is discrete at the band edges. The Fermi level is in the center of a band in a metal, and so kT may exceed the electronic energy level spacing even at room temperatures and small sizes. In contrast, in semiconductors, the Fermi level lies between two bands, so that the relevant level spacing remains large even at small sizes. The HOMO – LUMO gap increases in semiconductor nanocrystals of smaller sizes.

increases by over 1 eV at small size. The variation in the binding energy is negligible at large coverages or particle size, since the binding energies are close to those of the bulk, macroscopic metals. The increase in the core-level binding energy in small particles occurs due to the poor screening of the core-hole and is a manifestation of the *size-induced metal-nonmetal transition* in nanocrystals. Further evidence for the occurrence of such a metal-nonmetal transition driven by the size of the individual particle is provided by other electron spectroscopic techniques such as UPS, BIS.[5–7] All these measurements indicate that an electronic gap manifests itself for a nanoparticle having diameters of 1 – 2 nm possessing 300 ± 100 atoms.

Photoelectron spectroscopic measurements[6] on mass-selected Hg_n nanoparticles ($n = 3$ to 250) in the gas phase reveal that the characteristic HOMO – LUMO (s – p) energy gap decreases gradually from ~3.5 eV for $n = 3$ to ~0.2 eV for $n = 250$, as shown in Figure 3. The band gap closure is predicted at $n \sim 400$. The metal – nonmetal transition in gaseous Hg nanoparticles was examined by Rademann and co-workers[7] by measuring the ionization energies (IE). For $n < 13$, the dependence of IE on n suggested a different type of bonding. A small Hg particle with atoms in the $6s^2 6p^0$ configuration

© WILEY-VCH Verlag GmbH, 69451 Weinheim, Germany, 2002   0947-6539/02/0801-0030 $ 17.50+.50/0   *Chem. Eur. J.* **2002**, 8, No. 1

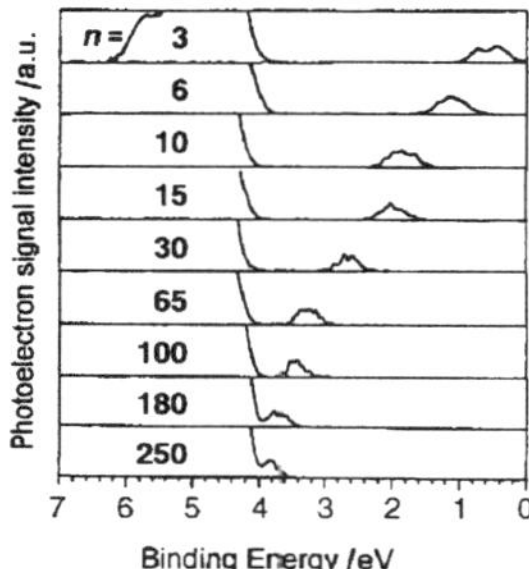

Figure 3. Photoelectron spectra of Hg clusters of varying nuclearity. The 6p feature moves gradually towards the Fermi level, emphasizing that the band gap shrinks with increase in cluster size (reproduced with permission from ref. [6]).

held together by relatively weak van der Waals forces, is essentially nonmetallic. As the nanoparticle grows in size, the atomic 6s and 6p levels broaden into bands and an insulator–metal transition appears to occur—driven by the physical dimensions of the individual particle. Note that this is the same element, Hg, behaving as either a metal or a nonmetal, depending upon its physical size!

The electronic absorption spectrum of metal nanocrystals in the visible region is dominated by the plasmon band. This absorption is due to the collective excitation of the itinerant electron gas on the particle surface and is characteristic of a metal nanocrystal of the given size. In colloids, surface plasmon excitations impart characteristic colours to the metal sols, the beautiful wine-red colour of gold sols being well-known. Gold nanocrystals of varying diameters between 2 and 4 nm exhibit distinct bands around ∼525 nm, the intensity of which increases with size.[9, 10] The intensity of this feature becomes rather small in the case of 1 nm diameter particles basically due to a reduced less number of "itinerant" electrons in the electron cloud. With a change in temperature, the intensity of the plasmon band decreases as seen in the case of Au nanocrystals.[10] In contrast to the situation for metals, exciton peaks dominate the absorption of semiconductor nanocrystals in the visible region.[11, 12a] Thus CdS, a yellow solid, exhibits an exciton peak around 600 nm, which gradually shifts into the UV region as the nanocrystal diameters are varied below 10 nm (see Figure 4).[12b] Similar effects have been observed in the case of PbS and ZnO.[12b]

Direct information on the gap states in nanocrystals of metals and semiconductors is obtained by scanning tunneling spectroscopy (STS). This technique provides the desired sensitivity and spatial resolution making it possible to carry out tunneling spectroscopic measurements on individual particles. A systematic STS study of Pd, Ag, Cd and Au nanoparticles of varying sizes deposited on a graphite substrate has been carried out under ultrahigh vacuum conditions, after having characterized the nanoparticles by XPS and STM.[13] The I-V spectra of bigger particles were featureless while those of the small particles (<1 nm) showed well-defined peaks on either side of zero-bias due to the presence of a gap. Ignoring gap values below 25 meV (∼kT), it is seen that small particles of ≤1 nm diameter are in fact

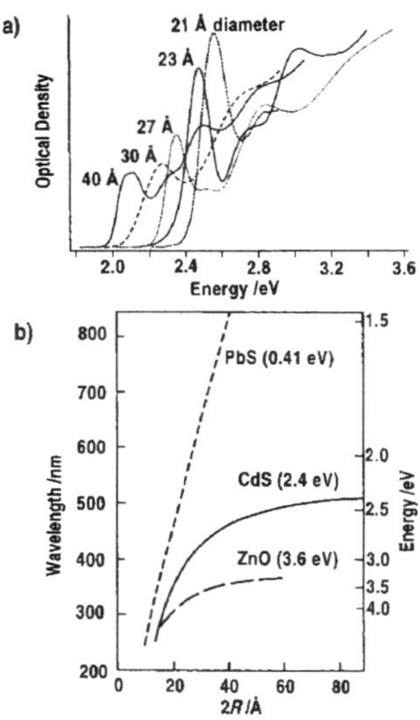

Figure 4. a) Absorption spectra of CdSe nanocrystals (at 10 K) of various diameters (reproduced with permission from ref. [12a]). b) Wavelength of the absorption threshold and band gap as a function of the particle diameter for various semiconductors. The corresponding energy gap in the bulk state is given in parenthesis (reproduced with permission from ref. [12b]).

nonmetallic! (Figure 5a) From the various studies discussed hitherto, it appears that the size-induced metal–insulator transition in metal nanocrystals (Figure 1) occurs in the range of 1–2 nm diameter or 300 ± 100 atoms. The band gap of CdS nanocrystals estimated by the above method yielded a value of 2.9 eV for a 3.1 nm diameter nanocrystal[14] and the gap increases with the decrease in size (Figure 5b).

Theoretical calculations of the electronic structure of metal and semiconductor nanocrystals throw light on the size-

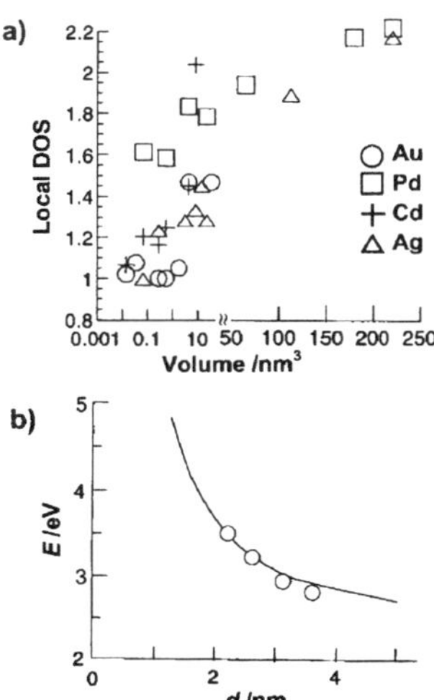

Figure 5. a) Variation of the nonmetallic band gap with nanocrystal size, b) in CdS nanocrystals (reproduced with permission from refs. [13, 14]).

induced changes in the electronic structure. Rosenblit and Jortner[15] calculated the electronic structure of a model metal cluster and predicted electron localization to occur in a cluster of diameter ~ 0.6 nm. A molecular orbital calculation on Au_{13} cluster[16] in icosahedral and cuboctahedral structures shows that the icosahedral structure undergoes Jahn–Teller distortion while the cuboctahedral structure does not distort. The onset of the metallic state is barely discernible in the Au_{13} cluster. Relativistic density functional calculations of gold clusters,[17] with $n = 6$ to 147 show that the average interatomic distance increases with the nuclearity of the cluster. The HOMO–LUMO electronic gap decreases with particle size from 1.8 eV for Au_6 (~ 0.5 nm diameter) to 0.3 eV for Au_{147} (~ 2 nm diameter). Ab initio molecular dynamics simulations of aluminum clusters,[18] with $n = 2-6$, 12, 13, 55 and 147 reveal that the minimum energy structures of Al_{13} and Al_{55} to be distorted icosahedra whereas Al_{147} is a near cuboctahedron. The HOMO–LUMO gap increases from ~ 0.5 eV for Al_2 to ~ 2 eV for Al_{13}; the gap is around 0.25 eV for Au_{55} and decreases to ~ 0.1 eV for Au_{147}. In the case of semiconductor nanocrystals, it is shown using tight binding approximation that the band gaps nearly reach the bulk values at sizes of around 5 nm.[19] The convergence of the cluster properties towards those of the corresponding bulk materials with increase in size is noteworthy.

In a bulk metal, the energy required to add or remove an electron is its work function. In a molecule, the corresponding energies, electron affinity and ionization potential, respectively, are, however, nonequivalent. Nanocrystals being intermediary, the two energies differ only to a small extent,[20] the difference being the charging energy, U. This is a Coulombic energy and is different from electronic energy gap. Further, Coulombic states can be similar for both semiconductor and metallic nanocrystals unlike the electronic states. A manifestation of single electron charging is the Coulomb staircase behaviour observed in the tunneling spectra,[21] when a nanocrystal, covered with an insulating ligand shell is held between two tunnel junctions. A typical staircase along with its theoretical fit is shown in Figure 6a. Such measurements have also been carried out on Pd and Au nanocrystals in the size range, $1.5-6.5$ nm.[22] The charging energies follow a scaling law[23] of the form, $U = A + B/d$, where A and B are constants, characteristic of the metal and d is the particle diameter(see Figure 6b).

Magnetic properties of nanoparticles of transition metals such as Co, Ni show marked variations with size. It is well known that in the nanometric domain, the coercivity of the particles tends to zero.[23] Thus, the nanocrystals behave as superparamagnets with no associated coercivity or retentivity. The blocking temperature which marks the onset of this superparamagnetism also increases with the nanocrystal size. Further, the magnetic moment per atom is seen to increase as the size of a particle decreases[25] (see Figure 7).

Chemical reactivity: The surface area of nanocrystals increases markedly with the decrease in size. Thus, a small metal nanocrystal of 1 nm diameter will have $\sim 100\%$ of its atoms on the surface. A nanocrystal of 10 nm diameter on the other hand, would have only 15% of its atoms on the surface. A

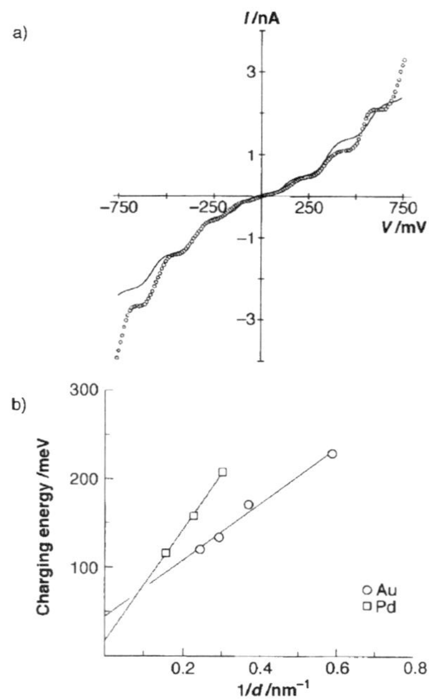

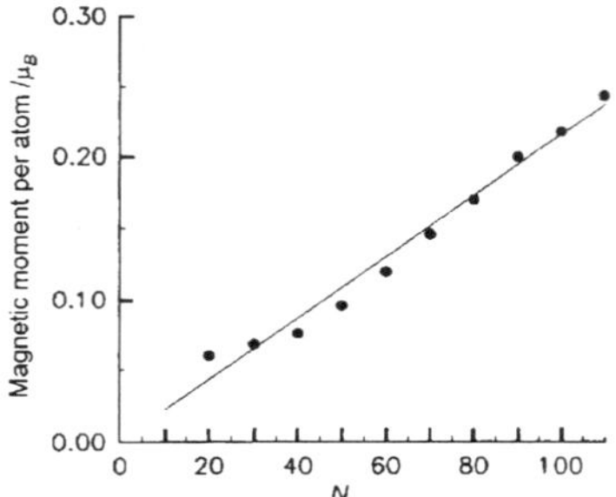

Figure 6. a) I-V characteristics of an isolated 3.3 nm Pd nanocrystal (dotted line) and the theoretical fit (solid line) obtained at 300 K using a semiclassical model according to which, the observed capacitance (C) may be resolved into two components $C1$ and $C2$ and the resistance (R) into $R1$ and $R2$, such that $C = C1 + C2$, $R = R1 + R2$. For $C1 \ll C2$ and $R1 \ll R2$, the model predicts steps in the measured current to occur at critical voltages, $V_c = n_c e/(C + (1/C)q_0 + e/2)$, where q_0 is the residual charge. b) Variation of the charging energies of Pd and Au nanocrystals with inverse diameters (d) (reproduced with permission from ref. [22]).

Figure 7. Variation of the magnetic moment per atom as a function of the number of atoms in Co clusters at $T = 170$ K and $H = 0.53$ T. The continuous curve is the theoretical value and the solid circles are the experimental results (reproduced with permission from ref. [25b]).

small nanocrystal with a higher surface area would be expected to be more reactive. Furthermore, the qualitative change in the electronic structure arising due to quantum confinement in small nanocrystals will also bestow unusual catalytic properties on these particles, totally different from those of the bulk metal. We illustrate these important aspects with a few examples from the recent literature.

A low temperature study[26] of the interaction of elemental O_2 with Ag nanocrystals of various sizes (Figure 8) has

 © WILEY-VCH Verlag GmbH, 69451 Weinheim, Germany, 2002   0947-6539/02/0801-0032 $ 17.50+.50/0   *Chem. Eur. J.* **2002**, *8*, No. 1

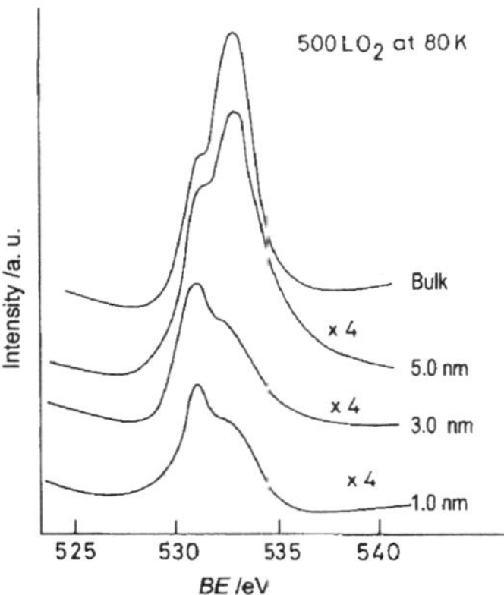

Figure 8. Change in the O(1s) spectra of Ag clusters exposed to 500 L O_2 at 80 K. The diameters of the clusters have been estimated from metal coverage. The lower binding energy peak at 531 eV corresponds to O^- while that at 533 eV arises due to molecular oxygen (reproduced with permission from ref. [26]).

revealed the capability of smaller nanocrystals to dissociate dioxygen to atomic oxygen species. On bulk Ag, the adsorbed oxygen species at 80 K is predominantly O_2^-. This interaction of O_2 with Ag—*dependent on its particle size*—is remarkable. Another important example is the reaction of H_2S with Ni nanocrystals giving rise to S^{2-} species, with nanocrystals of different sizes exhibiting different temperature profiles (see Figure 9). Unlike bulk nickel, small nanocrystals show less dependence in their catalytic activity on ambient temperature.

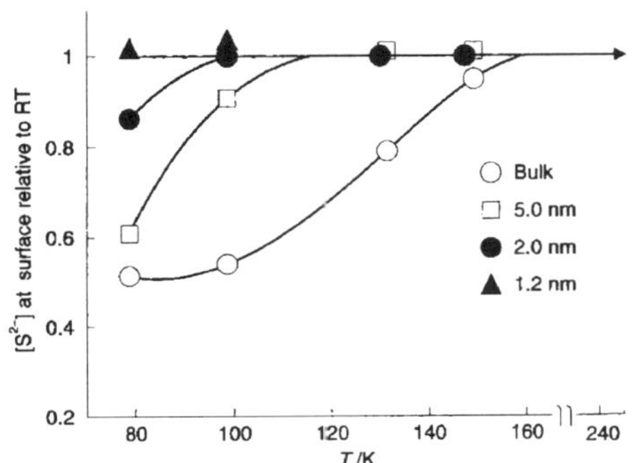

Figure 9. Variation of the normalized areas of the signal for the core-level transitions of S^{2-} with temperature for different sizes of Ni clusters deposited on graphite (reproduced with permission from ref. [26]).

The ability of Cu, Pd and Ni nanoparticles to absorb CO has been thoroughly investigated. Carbon monoxide from a bulk Cu surface desorbs above 250 K. Small Cu particles, however, retain CO up to much higher temperatures (Figure 10).[27] A similar observation has been made in case of Pd particles.[28] The results obtained with Ni particles are more even interesting. In addition to showing a trend similar to the above, small Ni particles are also capable of dissociating CO to form carbidic species on the particle surface (Figure 11).[29]

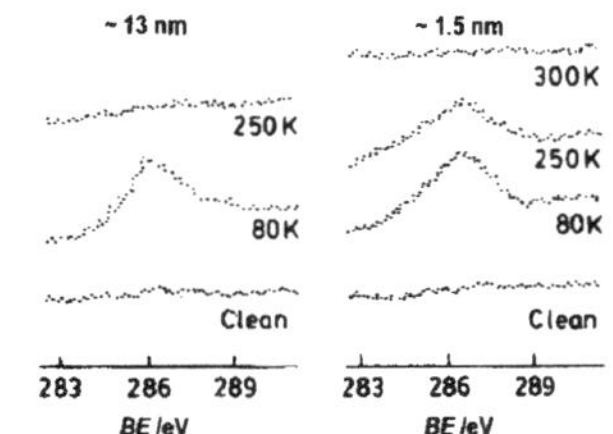

Figure 10. Change in the C(1s) spectra of CO adsorbed on Cu. The feature at 286 eV corresponds to molecularly adsorbed CO (reproduced with permission from ref. [27]).

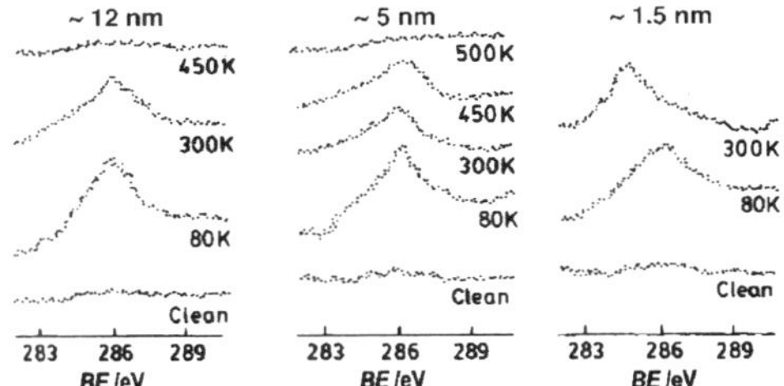

Figure 11. Change in the C(1s) spectra of CO adsorbed on Ni clusters with temperature. The feature at 286 eV corresponds to molecularly adsorbed CO while that at 284 eV arises due to the formation of carbidic species (reproduced with permission from ref. [27]).

This could be due to the Ni(3d) level in small clusters coming close to the anti-bonding energy level of $CO(2\pi^*)$.

Bulk Au is a noble metal. Goodman and co-workers,[30] however, found that Au nanocrystals supported on a titania surface show a marked size-effect in their catalytic ability for CO oxidation reaction, with Au nanoparticles in the range of 3.5 nm exhibiting the maximum chemical reactivity (Figure 12a). A metal to non-metal transition as observed in the I-V spectra (Figure 12b), as the cluster size is decreased below 3.5 nm^3 (consisting of ca. 300 atoms). This result is quite similar to that obtained with Pd particles supported on oxide substrate.[31] In another study of Au particles supported on a zinc oxide surface, smaller particles (<5 nm) exhibited a marked tendency to adsorb CO while those with diameters above 10 nm did not significantly adsorb CO (Figure 13).[32] The increased activity of these metal particles is attributed to the charge transfer between the oxide support and the particle surface.

Self-assembly of nanocrystals: Just as individual atoms aggregate to form crystals, nanocrystals themselves act as building units to form particle superlattices. Thus, monodispersed nanocrystals suitably covered by ligands such as alkane thiols, when transferred to a flat substrate, spontaneously assemble into two-dimensional lattices.[33–37] In Figure 14, we show typical arrays of 2.5 and 3.2 nm Pd nanocrystals coated with octane thiol. The diameter of the nanocrystal, d, and the length of the protecting ligand, l, play an important role in determining the very nature of the assembly.[38, 39] A study of the two-dimensional arrays formed by Pd nanocrystals of varying diameters covered with alkane thiols of different chain lengths has enabled to obtain an experimental stability

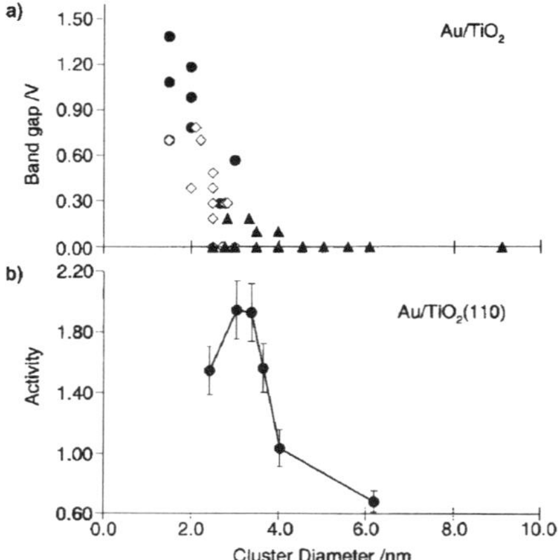

Figure 12. a) Cluster band gap measured by STS as a function of Au cluster size. The band gaps were obtained while the corresponding topographic scan was acquired on various Au coverages ranging from 0.2 to 4.0 ML. two-dimensional clusters ○; three-dimensional clusters, two atom layers in height □; three-dimensional clusters with three atom layers or greater in height △. b) The activity (*CO atoms/total Au atoms*) for CO oxidation at 350 K as a function of the Au cluster diameter supported on a TiO_2 surface. The CO/O_2 mixture was 1:5 at a total pressure of 40 Torr (reproduced with permission from ref. [30]).

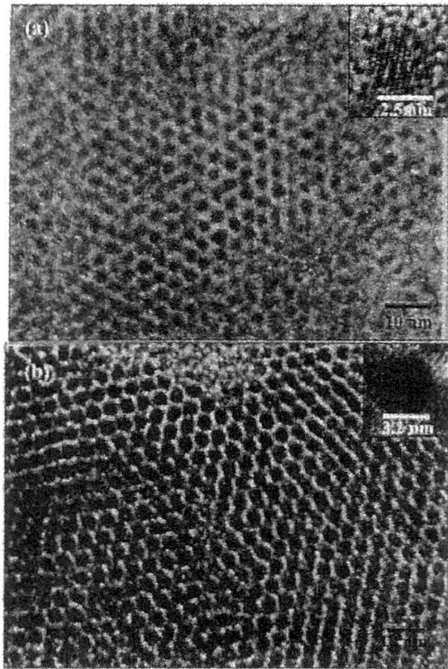

Figure 14. TEM micrograph showing hexagonal arrays of thiolized Pd nanocrystals: a) 2.5 nm, octane thiol, b) 3.2 nm, octane thiol (reproduced with permission from ref. [40]).

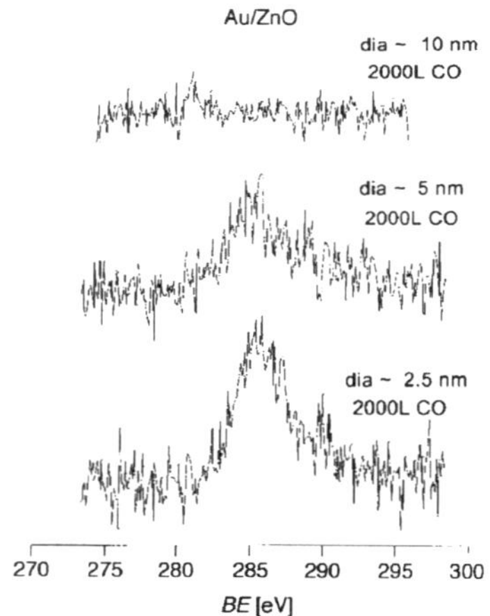

Figure 13. C(1s) core-level spectra of CO adsorbed on Au particles supported on a ZnO substrate. The feature at ∼285 eV corresponds to molecularly absorbed CO. The diameters have been obtained from the metal coverages.

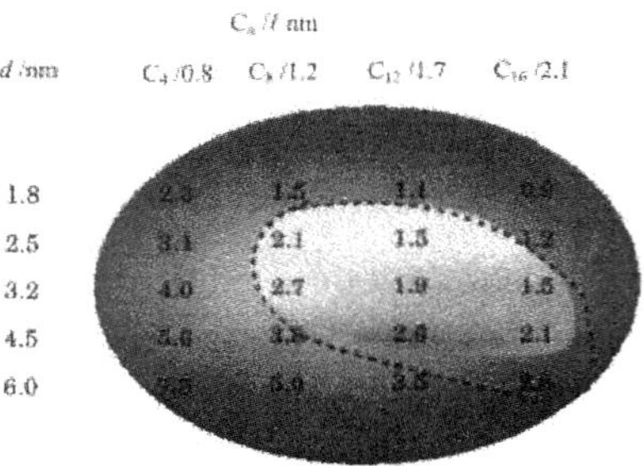

Figure 15. The $d-l$ phase diagram for Pd nanocrystals thiolized with different alkane thiols. The mean diameter, d, was obtained from the TEM measurements on as-prepared sols. The length of the thiol, l, is estimated by assuming an all-*trans* conformation of the alkane chain. The thiol is indicated by the number of carbon atoms, C_n. The bright area in the middle encompasses systems which form close-paced organizations of nanocrystals. The surrounding darker area includes disordered or low-order arrangements of nanocrystals. The area enclosed by the dashed line is derived from calculations from the soft sphere model (reproduced with permission from ref. [40]).

diagram (Figure 15) of the superlattices in terms of d and l.[40] In this Figure, the bright area in the middle is the most favourable d/l regime, corresponding to extended close packed organization, such as those illustrated in Figure 14. The d/l values for the most favorable regions are in the range 1.5–3.8. The area shaded dark in Figure 15 includes d/l regime giving rise to various short-range organizations; these are formed when the particles are small and the chain length is large or vice versa. The experimental results have been compared with empirical calculations based on a soft sphere

model,[41] involving an attractive van der Waals term and a repulsive steric term (see Figure 15).

The ability to synthesize lattices of nanocrystals have led to explorations of their collective physical properties. Thus, it is observed in the case of Co nanocrystals (5.8 nm) that, accompanying lattice formation, the blocking temperature increases.[42] FePt alloy nanocrystals yield ferromagnetic assemblies for which the coercivity is tunable by controlling the parameters such as Fe:Pt ratio and the particle size.[43] The evolution of collective electronic states in CdSe nanocrystals have been followed by optical spectroscopic methods. Compared with isolated nanocrystals, those in the lattice exhibited broader bands.[44] Various investigations have been carried out on the electrical transport properties of nanocrystalline lattices.[45, 46] Heath and co-workers have successfully demonstrated a reversible Mott–Hubbard metal–nonmetal transition with Ag nanocrystals (3 nm) capped with octane thiol.[47] Since then, this has been a subject of intense theoretical study.[48]

Conclusion

We have hitherto discussed in the earlier sections, electronic structure and properties, chemical reactivity and self-assembly of nanocrystals, particularly those of metals. The discussion should suffice to illustrate how size if a crucial factor in deciding the chemistry in the nano regime. These size dependent properties form the basis of nanoscience, where the properties are exploited for possible applications.

[1] a) J. S. Vermaak, L. W. Mays, D. Kuhlmann-Wilsdorf, *Surf. Sci.* **1968**, *12*, 128; b) G. Apai, J. F. Hamilton, J. Stohor, A. Thompson, *Phys. Rev. Lett.* **1979**, *40*, 165; c) A. A. Montano, G. K. Shenoy, E. E. Apl, W. Schulze, J. Urban, *Phys. Rev. Lett.* **1986**, *56*, 2076; d) H. J. Wasswemann, J. S. Vermaak, *Surf. Sci.* **1970**, *22*, 164; e) P. A. Buffat, J. P. Borel, *Phys. Rev. A* **1976**, *13*, 2287; f) W. P. Halperin, *Rev. Mod. Phys.* **1986**, *58*, 533.

[2] a) P. P.Edwards, R. L. Johnston, C. N. R. Rao in *Metal Clusters in Chemistry* (Eds.: P. Braunstein, G. Oro, P. R. Raithby), Wiley-VCH, **1998**; b) A. I. Kirkland, D. A. Jefferson, D. G. Duff, *Annual Reports C*, Royal Society of Chemistry, **1993** p. 247.

[3] H. N. Aiyer, V. Vijayakrishnan, G. N. Subanna, C. N. R. Rao, *Surf. Sci.* **1994**, *313*, 392.

[4] a) D. C. Johnson, R. E. Benfield, P. P. Edwards, W. J. H. Nelson, M. D. Vargas, *Nature* **1985**, *314*, 231; b) Y. Volokitin, J. Sinzig, L. J. de Jongh, G. Schmid, M. N. Vargaftik, I. I. Moiseev, *Nature* **1996**, *384*, 624.

[5] V. Vijayakrishnan, A. Chainani, D. D. Sarma, C. N. R. Rao, *J. Phys. Chem.* **1992**, *96*, 8679.

[6] R. Busani, M. Folker, O. Chesnovsky, *Phys. Rev. Lett.* **1998**, *81*, 3836.

[7] K. Rademann, O. D. Rademann, M. Schlauf, V. Even, F. Hensel, *Phys. Rev. Lett.* **1992**, *69*, 3208.

[8] H. Haberland, B. von Issendrof, Y. Yufeng, J. Kolar, G. Thanner, *Z. Phys. D: At. Mol. Clusters* **1993**, *26*, 8.

[9] K. V. Sarathy, G. Raina, R. T. Yadav, G. U. Kulkarni, C. N. R. Rao, *J. Phys. Chem. B* **1997**, *101*, 9876.

[10] S. Link, M. A. El-Sayed, *J. Phys. Chem. B* **1999**, *103*, 4212.

[11] A. P. Alivisatos, *J. Phys. Chem.* **1996**, *100*, 13226.

[12] a) D. M. Mittleman, R. W. Schoenlein, J. J. Shiang, V. L. Colvin, A. P. Alivisatos, C. V. Shank, *Phys. Rev. B* **1994**, *49*, 14435; b) A. Henglein, *Ber. Bunsenges. Phys. Chem.* **1995**, *99*, 903.

[13] C. P. Vinod, G. U. Kulkarni, C. N. R. Rao, *Chem. Phys. Lett.* **1998**, *289*, 329.

[14] M. Miyake, T. Torimoto, T. Sakata, H. Mori, S. Kuwabata, H. Yoneyama, *Langmuir* **1997**, *13*, 742.

[15] M. Rosenblit, J. Jortner, *J. Phys. Chem.* **1994**, *98*, 9365.

[16] R. A. Perez, A. F. Ramos, G. L. Malli, *Phys. Rev. B* **1989**, *39*, 3005.

[17] O. D. Haberlen, S. C. Chung, M. Stener, N. J. Rosch, *J. Chem. Phys.* **1997**, *106*, 5189.

[18] S. H. Yang, D. A. Drabold, J. B. Adams, A. Sachdev, *Phys. Rev. B* **1993**, *47*, 1567.

[19] P. E. Lippens, M. Lannoo, *Phys. Rev. B* **1989**, *39*, 10935.

[20] C. P. Collier, T. Vossmeyer, J. R. Heath, *Annu. Rev. Phys. Chem.* **1998**, *49*, 371.

[21] Single Charge Tunneling, Coulomb Blockade Phenomena in Nanostructures (Eds.: H. Grabert, M. H. Devoret), *NATO-ASI Ser. B* **1992**, *294*.

[22] P. J. Thomas, G. U. Kulkarni, C. N. R. Rao, *Chem. Phys. Lett.* **2000**, *321*, 163.

[23] J. Jortner, *Z. Phys. D: At. Mol. Clusters* **1992**, *24*, 247.

[24] C. P. Bean, J. D. Livingston, *J. Appl. Phys.* **1959**, *30*, 1208.

[25] a) Van de Heer, P. Milani, A. Chatelain, *Z. Phys. D: At. Mol. Clusters* **1991**, *19*, 241; b) S. N. Khanna, S. Linderoth, *Phys. Rev. Lett.* **1991**, *67*, 742.

[26] C. N. R. Rao, V. Vijayakrishnan, A. K. Santra, M. W. J. Prins, *Angew. Chem.* **1992**, *104*, 1110; *Angew. Chem. Int. Ed. Engl.* **1992**, *31*, 1062.

[27] A. K. Santra, S. Ghosh, C. N. R. Rao, *Langmuir* **1994**, *10*, 3937.

[28] E. Gillet, S. Channakhone, V. Matolin, M. Gillet, *Surf. Sci.* **1986**, *152/153*, 603.

[29] a) D. L. Doering, J. T. Dickinson, H. Poppa, *J. Catal.* **1982**, *73*, 91; b) D. L. Doering, H. Poppa, J. T. Dickinson, *J. Catal.* **1982**, *73*, 104.

[30] M. Valden, X. Lai, D. W. Goodman, *Science* **1998**, *281*, 1647.

[31] C. Xu, X. Lai, G. W. Zajac, D. W. Goodman, *Phys. Rev. B* **1997**, *56*, 13464.

[32] Unpublished results from our laboratory.

[33] C. N. R. Rao, G. U. Kulkarni, P. J. Thomas, P. P. Edwards, *Chem. Soc. Rev.* **2000**, *29*, 27.

[34] A. N. Shipway, E. Katz, I. Willner, *Chem. Phys. Chem.* **2000**, *1*, 18.

[35] M. P. Pileni, *New J. Chem.* **1998**, 693.

[36] G. Schmid, L. F. Chi, *Adv. Mater.* **1998**, *10*, 515.

[37] W. P. Wuelfing, F. P. Zamborini, A. C. Templeton, X. Wen, H. Yoon, R. W. Murray, *Chem. Mater.* **2001**, *13*, 87.

[38] R. L. Whetten, M. M. Shafigullin, J. T. Khoury, T. G. Schaaf, I. Vezmar, M. M. Alvarez, A. Wilkinson, *Acc. Chem. Res.* **1999**, *32*, 397.

[39] P. C. Ohara, D. V. Leff, J. R. Heath, W. M. Gelbart, *Phys. Rev. Lett.* **1995**, *75*, 3466.

[40] P. J. Thomas, G. U. Kulkarni, C. N. R. Rao, *J. Phys. Chem. B* **2000**, *104*, 8138.

[41] B. A. Korgel, S. Gullam, S. Conolly, D. Fitzmaurice, *J. Phys. Chem. B* **1998**, *102*, 8379.

[42] a) V. Russier, C. Petit, J. Legrand, M. P. Pileni, *Phys. Rev. B* **2000**, *62*, 3910; b) C. Petit, M. P. Pileni, *Appl. Surf. Sci.* **2000**, *162–163*, 519; c) C. Petit, A. Taleb, M. P. Pileni, *Adv. Mater.* **1998**, *10*, 259.

[43] S. Sun, C. B. Murray, D. Weller, L. Folks, A. Maser, *Science* **2000**, *287*, 1989.

[44] a) M. V. Artemyev, A. I. Bibik, L. I. Gurinovich, S. V. Gopenko, U. Woggon, *Phys. Rev. B* **1999**, *60*, 1504; b) C. R. Kagan, C. B. Murray, M. G. Bawendi, *Phys. Rev. B* **1996**, *54*, 8633; c) M. V. Artemyev, U. Woggon, H. Jaschinski, L. I. Gurinovich, S. V. Gopenko, *J. Phys. Chem. B* **2000**, *104*, 11617.

[45] M. Brust, D. Bethell, D. J. Schiffrin, C. J. Kiely, *Adv. Mater.* **1995**, *7*, 795.

[46] a) D. B. Janes, V. R. Kolagunta, R. G. Osifchin, J. D. Bielefeld, R. P. Andres, J. I. Henderson, C. P. Kubiak, *Superlattices Microstruct.* **1995**, *18*, 275; b) A. W. Snow, H. Wohltjen, *Chem. Mater.* **1998**, *10*, 947; c) Y. Liu, Y. Wang, R. O. Claus, *Chem. Phys. Lett.* **1998**, *298*, 315; d) H. M. Jaeger, personal communication, **2001**.

[47] a) G. Markovich, C. P. Collier, S. E. Hendricks, F. Remacle, R. D. Levine, J. R. Heath, *Acc. Chem. Res.* **1999**, *32*, 415; b) C. Mederios-Riberio, D. A. A. Phlberg, R. S. Williams, J. R. Heath, *Phys. Rev. B* **1999**, *59*, 1633.

[48] F. Remacle, R. D. Levine, *Chem. Phys. Chem.* **2001**, *2*, 20.

J. Phys. Chem. B **1999**, *103*, 399−401

Superlattices of Metal and Metal−Semiconductor Quantum Dots Obtained by Layer-by-Layer Deposition of Nanoparticle Arrays

K. Vijaya Sarathy, P. John Thomas, G. U. Kulkarni, and C. N. R. Rao*

Chemistry and Physics of Materials Unit, Jawaharlal Nehru Centre for Advanced Scientific Research, Jakkur, Bangalore-560 064, India

Received: September 23, 1998; In Final Form: November 12, 1998

Superlattices formed by arrays of Pt or Au nanoparticles have been obtained by layer-by-layer deposition by using dithiols as cross-linkers. The superlattices have been characterized by X-ray diffraction, photoelectron spectroscopy, and scanning tunneling microscopy. The core-level intensities of the metal and of the dithiol in the X-ray photoelectron spectra show the expected increase with successive depositions. The formation of such structures has been confirmed by depositing Pt and Au layers alternatively. Layers of metal and CdS nanoparticles have been deposited alternatively to obtain heterostructures.

The quest for nanoscale architectures has provided much impetus to investigate the self-assembly of metal and semiconductor particles as well as of other materials. A major motivation to construct such structures has been to exploit the novel properties that would manifest from size quantization which may allow band-gap engineering and to design nanoelectronic devices, sensors, and the like. Thus, Colvin et al.[1] have made use of a self-assembly of semiconductor nanocrystals to construct an optoelectronic device. Murray et al.[2] have demonstrated the self-organization of CdSe nanocrystallites into a three-dimensional superlattice. Multilayers of semiconductor CdS nanoparticles have been deposited on a gold substrate using the self-assembly of dithiol molecules.[3,4] There has been some effort to obtain regular arrangements of metal nanoparticles in different dimensions. A linear arrangement of Au clusters stabilized by phosphine ligands has been obtained by binding them to single-stranded DNA,[5] while ordered channels of porous alumina membranes have been filled with Au particles to obtain wires.[6] Two-dimensional arrays of metal nanoparticles are readily prepared by using alkanethiols.[7−9] Whetten et al.[9] have described gold nanocrystals stabilized by thiols. Although there are indications that multilayers of metal particles can be formed by use of dithiols,[4] there is no definitive evidence, based on spectroscopic, microscopic, and diffraction studies, for such superstructures. It was our interest to prepare superlattices of well-characterized metal quantum dots by use of spacer molecules. In this letter, we report the layer-by-layer deposition of nanocrystalline arrays of quantum dots of a single metal or of two metals by employing dithiol molecules to construct superlattices. Such lattices with alternate layers of metal and semiconductor nanoparticles can also be prepared.

The procedure employed to obtain metal quantum dot superlattices is simple. A clean polycrystalline metal substrate of gold or silver, prepared by the resistive evaporation of the metal onto a freshly cleaved mica substrate at 500 °C in a vacuum, was immersed in a 50 mM toluene solution of 1,10-decanedithiol. After 2 h, it was washed with toluene and dried in air. The dithiol-covered substrate was then immersed in a dilute dispersion of metal nanoparticles of the desired size (obtained by the controlled reduction of metal ions complexed

* Corresponding author.

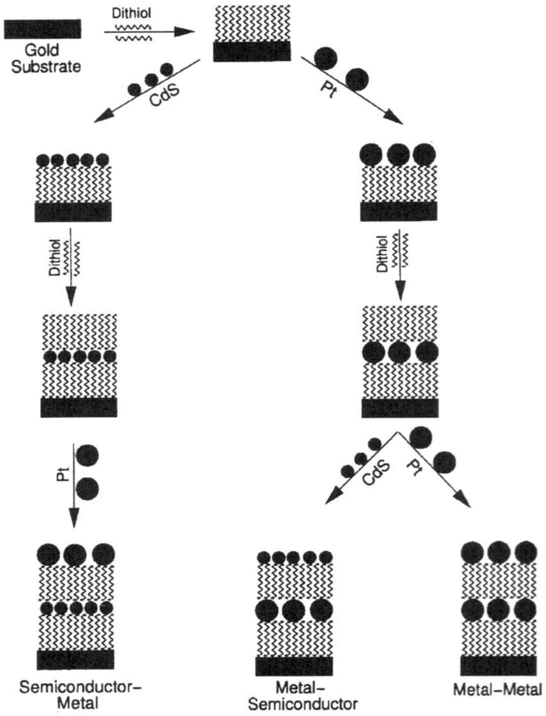

Figure 1. Schematic drawing depicting the layer-by-layer deposition of Pt nanoparticles onto a Au substrate, the layers being separated by dithiol molecules. Also shown is the formation of a heterostructure consisting of alternate layers of semiconductor and metal nanoparticles.

with tetra-*n*-octylammonium bromide in toluene using $NaBH_4$) in toluene for 12 h. After the formation of a particulate layer by this means, it was washed with toluene and dried. The above steps employed for depositing the first layer of nanoparticles were repeated to obtain multilayer superlattices. This is shown schematically in Figure 1. After each deposition, the nanostructures were characterized by X-ray and UV photoelectron spectroscopy (XPS and UPS), scanning tunneling microscopy

400 *J. Phys. Chem. B, Vol. 103, No. 3, 1999*

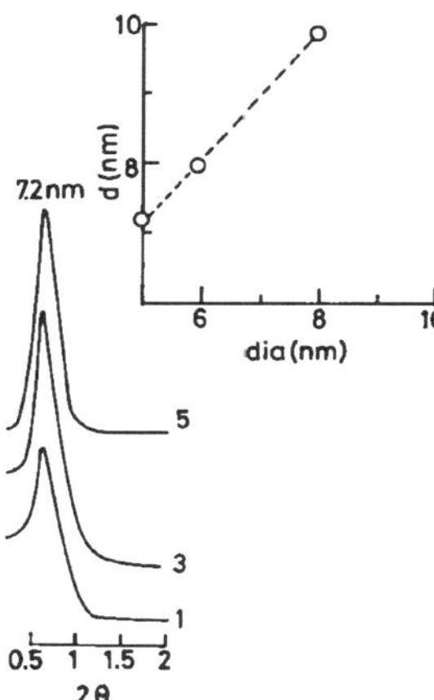

Figure 2. X-ray diffraction patterns of nanocrystalline arrays of the 5 nm Pt particles on a Au substrate after the first, third, and fifth depositions. The inset shows the variation in the *d*-spacing due to nanocrystallline arrays formed by the particles of different diameters. Layers of 1.5 nm Pt particles did not give a XRD pattern.

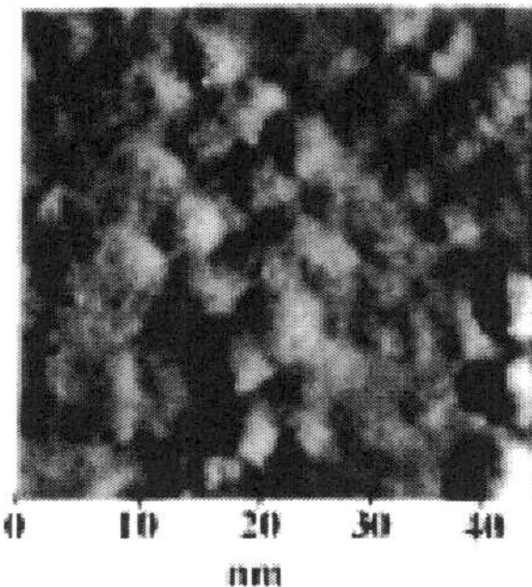

Figure 3. Scanning tunneling image of the first layer of the 5 nm Pt particles on a Au substrate.

(STM), and X-ray diffraction (XRD). The metal nanoparticles employed were characterized by transmission electron microscopy (TEM) to determine the size distribution.

In Figure 2, we show typical XRD patterns of lattices formed by Pt nanoparticles of a mean diameter of 5 nm (as determined by TEM) after the first, third, and fifth depositions on a gold substrate. We see a low-angle peak with a *d*-spacing of 7.2 nm corresponding to the center-to-center distance between two adjacent particles in the array. XRD patterns of such lattices obtained from the deposition of Pt or Au nanoparticles of different sizes reveal that the *d*-spacing of the low-angle feature increases with the mean diameter of the nanoparticle as shown in the inset of Figure 1. The intercept on the *y*-axis ($\sim$ 2 nm) approximately gives the spacing between two adjacent particles.

A scanning tunneling microscopic study of a layer of 5 nm Pt particles obtained after the first deposition shows regular arrays of nanoparticles extending over 300 nm which corresponds to the size of a typical flat terrace of the substrate. Imaging at lower scan size of $\sim$40 $\times$ 40 nm revealed the nearly regular spacing of 2 nm between the spherical particles of 5 nm diameter (Figure 3). This result corroborates the observations from XRD measurements. STM measurements after the deposition of the second layer also showed the presence of 5 nm Pt particles, although the particles were not as clearly resolvable.

Core-level X-ray photoelectron spectra (XPS) of the lattices formed by the 5 nm particles of Pt on a gold substrate are shown

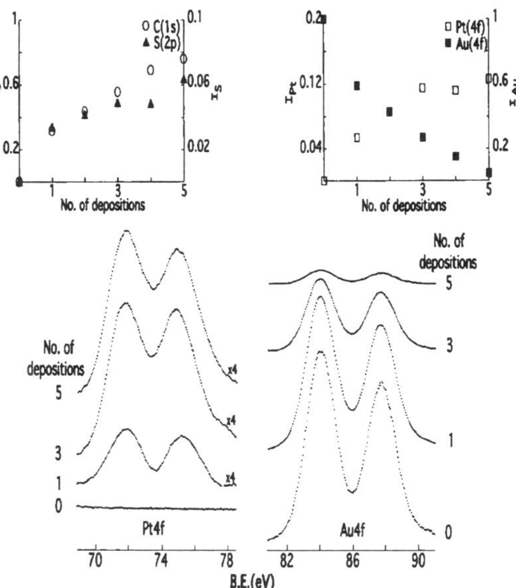

Figure 4. X-ray photoelectron spectra in the Pt(4f) and Au(4f) regions for the 5 nm Pt/Au system. Inset shows variations in the Pt(4f) and Au(4f) intensities as well as C(1s) and S(2p) intensities with the number of depositions.

in Figure 4 after the first, third, and fifth depositions. We observe that the intensity of the Pt(4f) feature increases with the number of depositions accompanied by a decrease in the Au(4f) intensity as the substrate gets increasingly shadowed due to the limited escape depth of the photoelectrons. The intensities of the C(1s) and S(2p) levels of the dithiol (at 285.0 and 163.6 eV, respectively) also increase with the increasing number of depositions. We have also carried out similar XPS measurements on lattices formed by 1.5 nm Pt nanoparticles and obtained similar results. XPS measurements on the superlattices formed by 6 nm Au particles on a Ag substrate showed an increase in the Au(4f) intensity accompanied by an increase in the C(1s) and S(2p) intensities due to the dithiol, with the increasing number of depositions. The Ag(3d) intensity decreases with increasing number of depositions as expected. The observed variation of the relevant core-level intensities with successive depositions is clearly indicative of superlattice formation as a result of layer-by-layer deposition. We have estimated the metal coverage after each deposition by employing the Seah–Dench formula[10] for metal overlayers. The plot of metal coverage versus the number of depositions gave a slope close to unity until the third deposition and increased thereafter, suggesting that we have indeed accomplished layer-by-layer deposition of the metal nanoparticles.

We have investigated the formation of a bimetallic superlattice by alternative depositions of 6 nm Au particles and 5 nm Pt particles on a Ag substrate, the particulate layers being separated by decanedithiol molecules. In Figure 5, we show the core-level spectra in the Ag(3d), Au(4f), and Pt(4f) regions. There is a decrease in the intensity of Ag(3d) with successive depositions, while the Au(4f) intensity appears after the first deposition and decreases after the deposition of Pt nanoparticles in the second layer. This experiment provides further proof for the formation of superlattices of arrays of metal nanoparticles by the technique employed by us.

He–I ultraviolet photoelectron spectra of the metal nanoparticles layers were recorded after each deposition. After the

Letters

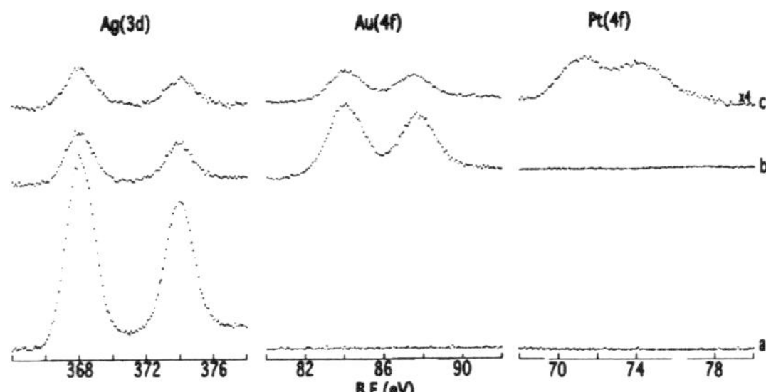

Figure 5. X-ray photoelectron spectra in the Ag(3d), Au(4f), and Pt(4f) regions for the Pt/Au/Ag system: (a) clean Ag substrate, (b) after the deposition of the 6 nm Au particles in the first layer, and (c) after the deposition of the 5 nm Pt particles in the second layer.

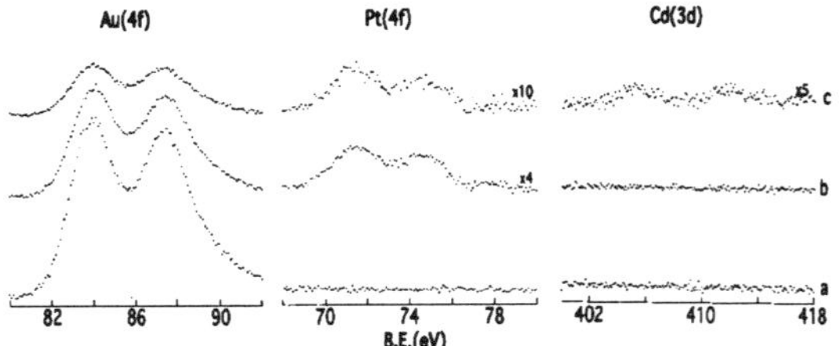

Figure 6. X-ray photoelectron spectra in the Cd(3d), Pt(4f), and Au(4f) regions for the CdS/Pt/Au system: (a) clean Au substrate, (b) after the deposition of the 5 nm Pt particles in the first layer, and (c) after the deposition of the 2.4 nm CdS particles in the second layer.

deposition of the first layer of 1.5 nm Pt nanoparticles, there were no features due to the dithiol-covered Au surface in the spectrum. Instead, there was a distinct feature at 3.5 eV but without the sharp increase in intensity at 0.5 eV found in the bulk Pt. There was a negligible density of states at the Fermi level. These features in the He—I spectra were found up to the fifth deposition. When the diameter of the Pt nanoparticles was larger (5 nm or more), features akin to those of bulk Pt could be seen in the He—I spectrum. The He—I spectra of layers of Au nanoparticles deposited on a Ag substrate showed general characteristics similar to those described for the Pt nanoparticles.

We have been able to prepare heterostructures consisting of alternate layers of semiconductor and metal nanoparticles. For this purpose, we first deposited a layer of 5 nm Pt nanoparticles on a dithiol-covered Au substrate, followed by immersion in dithiol solution, rinsing in toluene, and drying in air. It was then dipped into an inverse micellar solution of AOT (sodium bis(2-ethylhexyl)sulfosuccinate)/*n*-heptane/water containing CdS nanoparticles (2.4 nm) for 12 h in order to obtain a CdS—Pt heterostructure (Figure 1). The excess AOT present on the CdS layer was removed by repeatedly rinsing in *n*-heptane and drying in air. We show XPS evidence for the formation of the heterostructure in Figure 6. The intensity of the Au(4f) core level from the substrate decreases with each deposition. We also see that the core-level intensity in the Pt(4f) region decreases after the deposition of the CdS nanoparticles. The order of deposition of the Pt and CdS particle could be readily reversed to obtain a Pt—CdS—Pt-type heterostructure (Figure 1).

After we completed the present study, our attention was drawn to a recent paper of Brust et al.[11] who reported the formation of multilayers of Au nanoparticles using dithiols. These workers have confirmed layer-by-layer deposition of the Au nanoparticles by employing UV—vis spectroscopy and ellipsometry. They do

not, however, provide spectrochemical analysis in support of this result. By and large, the results of Brust et al. are in line with the findings of our study, although the techniques employed and the systems studied are different. Noteworthy features of the present study are the use of the dithiol C(1s) and S(2p) intensities along with the intensities of the metal core levels to establish the occurrence of layer-by-layer deposition of Au as well as Pt nanoparticles of different sizes. Deposition of bimetallic layers and metal—semiconductor layers is another aspect worthy of note.

Acknowledgment. The authors thank the Department of Science and Technology, Government of India, for support of this research.

References and Notes

(1) Colvin, V. L.; Schlamp, M. C.; Alivisatos, A. P. *Nature* **1994**, *370*, 354.

(2) Murray, C. B.; Kagan, C. R.; Bawendi, M. G. *Science* **1996**, *270*, 1335.

(3) Nakanishi, T.; Ohtani, B.; Uosaki, K. *J. Phys. Chem.* **1998**, *B102*, 1571.

(4) Brust, M.; Etchenique, R.; Calvo, E. J.; Gordillo, G. J. *Chem. Commun.* **1996**, 1949.

(5) Alvisatos, A. P.; Johnson, K. P.; Peng, X.; Wilson, T. E.; Loweth, C. J.; Burchez, M. P., Jr.; Schultz, P. G. *Nature* **1996**, *382*, 609.

(6) Hornyak, G. L.; Kröll, M.; Pugin, R.; Sawitowski, T.; Schmid, G.; Bovin, J.-O.; Karsson, G.; Hofmeister, H.; Hopfe, S. *Chem. Eur. J.* **1997**, *3*, 1951.

(7) Andres, R. P.; Beilefeld, J. D.; Henderson, J. I.; Janes, D. B.; Kolagunta, V. R.; Kubaik, C. P.; Mahoney, W. J.; Osifchin, R. G. *Science* **1996**, *273*, 1960.

(8) Vijaya Sarathy, K.; Raina, G.; Yadav, R. T.; Kulkarni, G. U.; Rao, C. N. R. *J. Phys. Chem.* **1997**, *B101*, 9876.

(9) Whetten, R. L.; Khoury, J. L.; Alvarez, M. M.; Murthy, S.; Vezmar, I.; Wang, Z.; Stephens, P. W.; Cleveland, C. L.; Luedtke, W. D.; Landmann, U. *Adv. Mater.* **1996**, *8*, 428.

(10) Seah, M. P.; Dench, W. A. *Surf. Interface Anal.* **1979**, *1*, 2.

(11) Brust, M.; Bethell, D.; Kiely, C. J.; Schiffrin, D. J. *Langmuir* **1998**, *14*, 5425.

536   *J. Phys. Chem. A* **1997**, *101*, 536−540

X-ray Photoelectron Spectroscopic Investigations of Cu−Ni, Au−Ag, Ni−Pd, and Cu−Pd Bimetallic Clusters[†]

K. R. Harikumar, S. Ghosh, and C. N. R. Rao*

Solid State and Structural Chemistry Unit and CSIR Center of Excellence in Chemistry, Indian Institute of Science, Bangalore 560 012, India

Received: June 27, 1996; In Final Form: September 30, 1996[⊗]

Core-level binding energies of the component metals in bimetallic clusters of various compositions in the Ni−Cu, Au−Ag, Ni−Pd, and Cu−Pd systems have been measured as functions of coverage or cluster size, after having characterized the clusters with respect to sizes and compositions. The core-level binding energy shifts, relative to the bulk metals, at large coverages or cluster size, ΔE_a, are found to be identical to those of bulk alloys. By substracting the ΔE_a values from the observed binding energy shifts, ΔE, we obtain the shifts, ΔE_c, due to cluster size. The ΔE_c values in all the alloy systems increase with the decrease in cluster size. These results establish the additivity of the binding energy shifts due to alloying and cluster size effects in bimetallic clusters.

Introduction

Clusters of metals such as Au, Ni, Cu, and Ag deposited on solid substrates have been studied extensively by employing X-ray photoelectron spectroscopy and cognate techniques.[1−6] These studies have shown that the increase in the metal core-level binding energy with the decrease in cluster size is not merely due to final-state effects but arises from the decrease in the core hole screening and the occurrence of a metal-to-nonmetal transition with decrease in cluster size.[7−9] We have been interested in investigating bimetallic clusters for some time. A preliminary study of certain specific compositions of three alloy systems, namely, Cu_7Ni_3, Ni_3Pd_2, and Cu_3Au, showed that the core-level binding energies of the component metals increase with the cluster size after accounting for the effect of alloying.[10] We considered it important to establish not only that alloying occurs in bimetallic clusters over a wide range of compositions but, more importantly, that the shifts in the core-level binding energies in the clusters due to alloying and cluster size are additive. For this purpose, we have carried out a detailed investigation of several compositions each of four alloy systems, including those compositions which are rich in either component. The alloy systems that we have investigated are Au−Ag, Cu−Ni, Ni−Pd, and Cu−Pd.

Bimetallic clusters of varying sizes were generated by resistive evaporation of the alloys under ultrahigh-vacuum conditions. The clusters were deposited on amorphized graphite surfaces and characterized by high-resolution electron micros-copy and EDAX analysis, before recording the core-level spectra. It may be noted that while the Au−Ag, Cu−Ni, and Ni−Pd systems form alloys over the entire range of composi-tions, the phase diagram of the Cu−Pd system is somewhat more complex, involving the formation of certain line phases.[11] In the present study, we find that the shifts in the core-level binding energies of the large bimetallic clusters of the various compositions of Cu−Ni (Ni_9Cu, Ni_3Cu_2, Ni_3Cu_7, $NiCu_9$), Au−

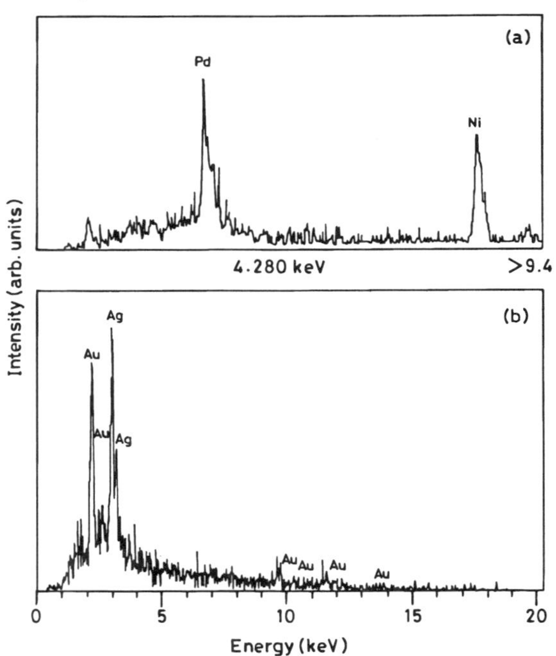

Figure 1. EDAX patterns of (a) Ni_3Pd_2 clusters corresponding to $I_{Ni} + I_{Pd}/I_C$ of 2.4 and (b) Ag_4Au clusters corresponding to $I_{Au} + I_{Ag}/I_C$ of 3.6.

Ag (Ag_9Au, Ag_4Au, $AgAu$, Ag_2Au_3), Ni−Pd (Ni_4Pd, $NiPd$, $NiPd_4$, Ni_3Pd_2), and Cu−Pd (Cu_4Pd, Cu_7Pd_3, Cu_3Pd_2, Cu_2Pd_3) systems truly correspond to those of the alloys. By substracting out the binding energy shifts due to alloying from the observed binding energy shifts, we obtain the shifts arising from the cluster size. Such additive effects of alloying and cluster size on the core-level binding energy shifts also support the view that the increase in core-level binding energy found at small cluster size is not due to final-state effects alone.[8]

* Corresponding author (Fax: +91 80 3346438).
[†] Dedicated to Professor Saburo Nagakura.
[⊗] Abstract published in *Advance ACS Abstracts*, December 1, 1996.

Spectroscopic Investigations of Bimetallic Clusters

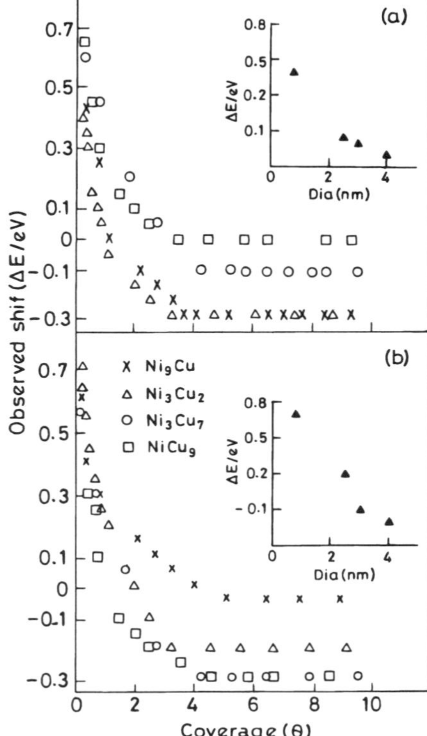

Figure 2. (a) Variation of the shift in the Cu(2p$_{3/2}$) binding energy (relative to the bulk metal), ΔE, of Cu$-$Ni bimetallic clusters of different compositions with the coverage as determined by the (I_{Cu} + I_{Ni})/I_C ratio. (b) Variation of the shift in the Ni(2p$_{3/2}$) binding energy, ΔE, with the coverage. Insets show variation of ΔE with actual cluster size in the case of Ni$_3$Cu$_2$.

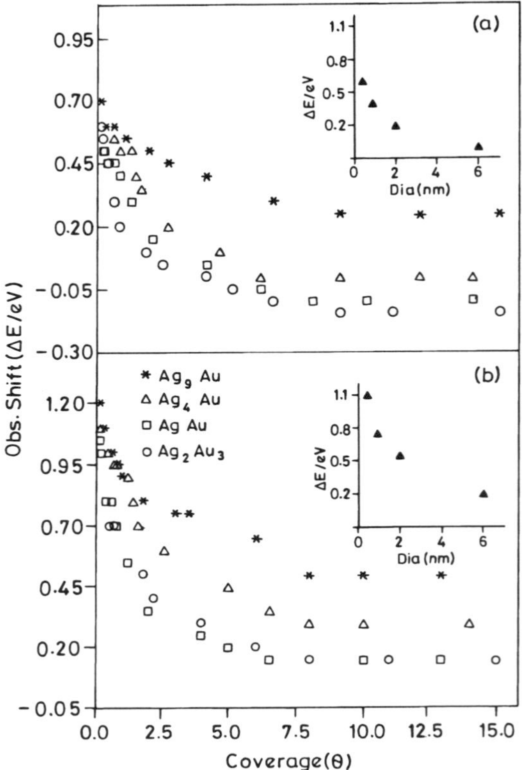

Figure 3. (a) Variation of the shift in the Ag(3d$_{5/2}$) binding enery (relative to the bulk metal), ΔE, of Au$-$Ag alloy clusters of different compositions with the metal coverage as determined by the (I_{Au} + I_{Ag})/I_C ratio. (b) Variation of the shift in the Au(4f$_{7/2}$) binding energy, ΔE, with the coverage. Insets show variation of ΔE with actual cluster size in the case of Ag$_4$Au.

Experimental Section

Electron spectroscopic measurements were carried out with a VG ESCALAB V spectrometer fitted with a sample preparation chamber at a base pressure of $\sim 2 \times 10^{-10}$ Torr. Al Kα (1486.6-eV) radiation was employed for the XPS measurements. Bimetallic clusters were deposited at room temperature under ultrahigh-vacuum conditions on amorphous graphite surfaces (obtained by Ar$^+$ ion bombardment of graphite surfaces) by means of resistive evaporation by taking an appropriate mixture of the two high-purity metals wound around a thoroughly degassed tungsten filament.[10] When the filament was heated in the preparation chamber of the spectrometer, the metals melted and formed an alloy. We take the metal coverage θ to be proportional to the intensity ratio (I_A + I_B)/I_C where I_A and I_B are the core-level intensities of the metals and I_C is the C(1s) intensity of the support. We have used these intensity ratios throughout the paper to describe the coverage, θ.

Analysis of the bimetallic clusters was carried out using quantitative energy dispersive X-ray (EDAX) analysis. A Leica S-440i microscope operating at 20 kV fitted with a link ISIS, Oxford, and the ZAF-4/FLS program was used for analysis. Typical EDAX patterns of two bimetallic clusters corresponding to the compositions Ni$_3$Pd$_2$ and Ag$_4$Au are shown in Figure 1. We have similarly obtained the compositions of the other alloys studied here. Cluster size distributions were obtained with the help of high-resolution electron microscope (HREM) images. A JEOL JEM 200 CX electron microscope operating at 200 kV was used for the study. Samples for electron microscopy

were prepared using amorphous carbon films as supports which were initially deposited on cleaned KBr crystals and subsequently transferred to a 200-mesh copper grid. The images were analyzed with respect to the size distribution by using an image analyzer (Leica Quantimet Q500MC). For example, in the case of the Au$-$Ag system, the cluster size distributions were fairly narrow. The cluster diameters are in the ranges 0.2$-$0.9, 0.5$-$1.8 and 5$-$10 nm for θ = 0.4, 0.9, and 13, respectively, with mean diameters of 0.4, 0.9, and 6 nm. In the case of the Cu$-$Ni system, the cluster diameter ranges are 1$-$5 and 2$-$8 nm for θ = 1.2 and 5, respectively, with mean diameters of 2.7 and 3.6 nm. These measurements confirmed that the θ values from surface spectroscopic methods adequately describe clusters of distinct size regimes.

Results and Discussion

We first investigated several compositions of the Cu$-$Ni and Au$-$Ag systems which readily form solid solutions over the entire composition range. In Figure 2, we show the plots of the observed shifts in the binding energies of the Cu(2p$_{3/2}$) and Ni(2p$_{3/2}$) levels, ΔE, relative to the bulk metals (933.1 and 852.9 eV, respectively) against the coverage for Ni$_9$Cu, Ni$_3$Cu$_2$, Ni$_3$Cu$_7$, and NiCu$_9$. The ΔE value should be close to zero at large coverages just as in monometallic clusters, but the behavior of the bimetallic clusters is different. In the Cu$-$Ni system, the ΔE of Cu(2p$_{3/2}$) becomes increasingly negative at large coverages as the Cu content in the alloy decreases. Thus, Ni$_9$Cu

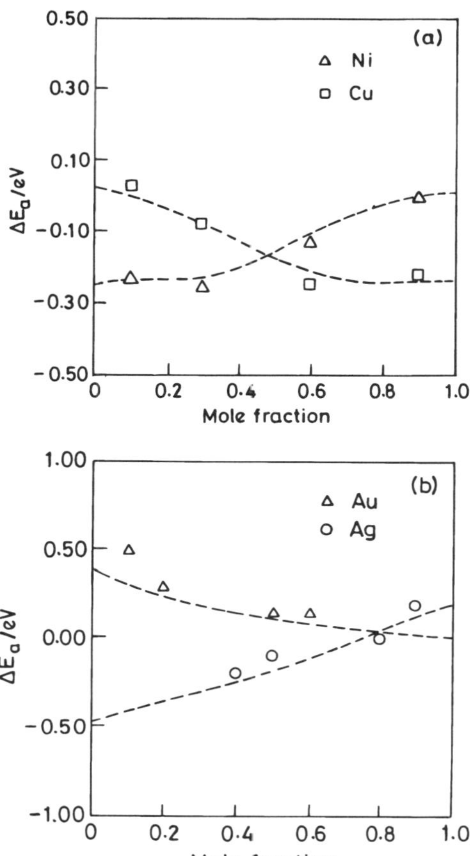

Figure 4. Plots of the core-level binding energy shifts (relative to the bulk metal value) at large coverages, ΔE_a, against the mole fraction for (a) Ni—Cu and (b) Au—Ag bimetallic clusters. Dashed lines represent the ΔE_a values for the bulk alloys, and the symbols denote the ΔE_a values for the bimetallic clusters at large coverages.

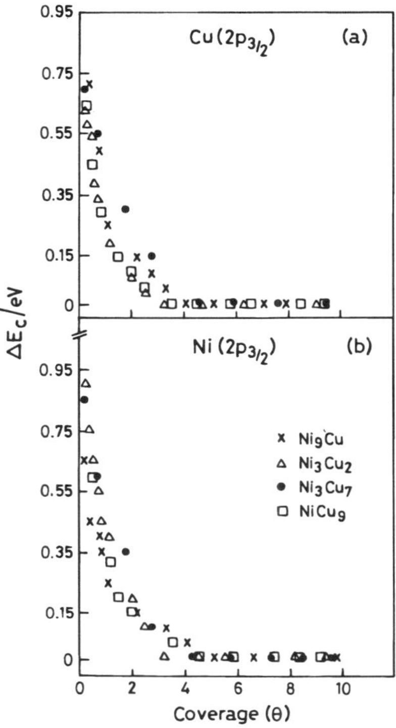

Figure 5. (a) Coverage or cluster size dependence of the shift in the $2p_{3/2}$ binding energy of Cu in Cu—Ni bimetallic clusters, ΔE_c. (b) Variation of the shift in Ni($2p_{3/2}$) binding energy, ΔE_c, due to cluster size effect as in (a).

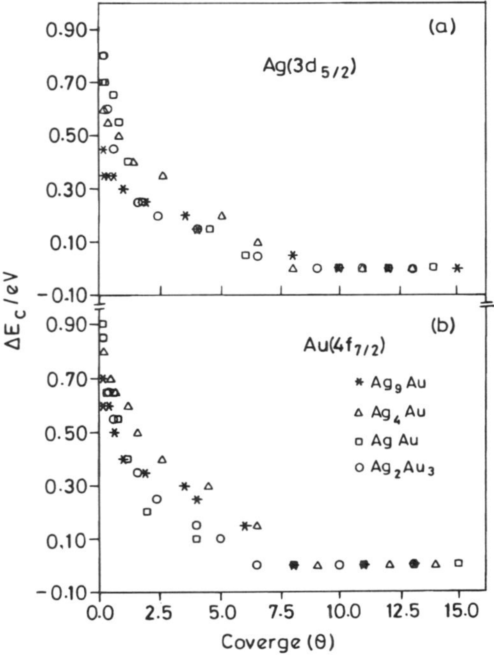

Figure 6. (a) Variation of the shift in the $3d_{5/2}$ binding energy of Ag in Au—Ag bimetallic clusters, ΔE_c, due to cluster size effect with the coverage. (b) Variation of the shift in Au($4f_{7/2}$) binding energy, ΔE_c, due to cluster size effect as in a.

shows a ΔE value of -0.3 eV at large coverages, while it is close to 0 in NiCu$_9$. The ΔE value of Ni($2p_{3/2}$) also becomes increasingly negative as the Ni content decreases, with the ΔE value being -0.3 eV in NiCu$_9$ and 0 in Ni$_9$Cu. The observed core-level binding energy shifts at large coverages are clearly due to alloying. From XPS studies of bulk alloys of Cu—Ni,[12–14] it is known that the magnitude of the shift in the core-level binding energy, of either component metal, increases as its content decreases, as is indeed observed in the bimetallic clusters in the present study.

In Figure 3, we have plotted the shifts in the binding energies, ΔE, of the Ag($3d_{5/2}$) and Au($4f_{7/2}$) levels of Ag$_9$Au, Ag$_4$Au, AgAu, and Ag$_2$Au$_3$ relative to the bulk metals (368.2 and 84 eV, respectively) against the coverage. The shift in the Ag($3d_{5/2}$) binding energy decreases as the Ag content in the alloy decreases. Thus, in Ag$_9$Au, the observed binding energy shift, ΔE, at large coverages is 0.3 eV, while in Ag$_2$Au$_3$ it is -0.2 eV. The ΔE of Au($4f_{7/2}$) increases markedly with the decrease in the Au content, with a value of 0.5 eV in Ag$_9$Au and 0.15 eV in Ag$_2$Au$_3$ at large coverages of these compositions.

It is instructive to compare the observed core-level binding energy shifts of bimetallic clusters at large coverages with those observed for similar compositions of bulk alloys. In Figure 4a, we have plotted the observed binding energy shifts of the Ni($2p_{3/2}$) and Cu($2p_{3/2}$) levels in the Ni—Cu bimetallic clusters

Spectroscopic Investigations of Bimetallic Clusters

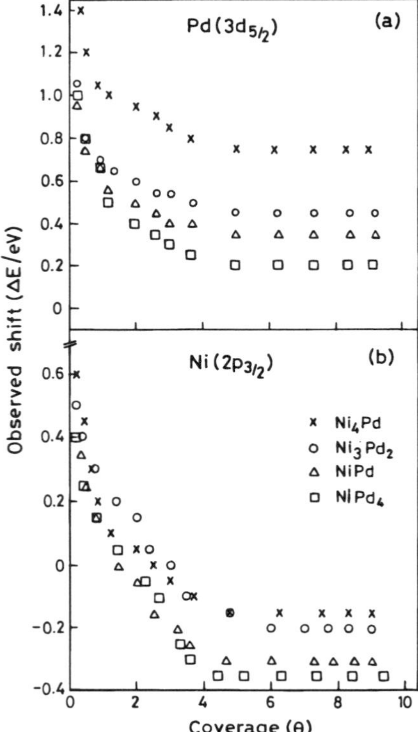

Figure 7. (a) Variation of the shift in the $Pd(3d_{5/2})$ binding energy (relative to the bulk metal), ΔE, of Ni−Pd bimetallic clusters of different compositions with the metal coverage as determined by the $(I_{Ni} + I_{Pd})/I_C$ ratio. (b) Variation of the shift in the $Ni(2p_{3/2})$ binding energy, ΔE, with the coverage.

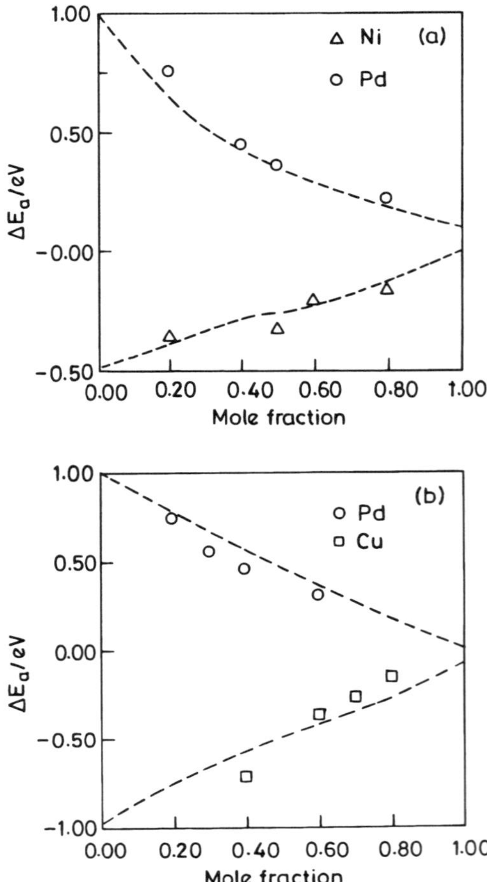

Figure 8. Plots of the core-level binding energy shifts (relative to the bulk metal values) at large coverages, ΔE_a, against the mole fraction for (a) Ni−Pd and (b) Cu−Pd bimetallic clusters. Dashed lines represent the ΔE_a values for bulk alloys, and the symbols denote the ΔE_a values for the clusters at large coverages.

at large coverages, ΔE_a, against the mole fraction. In the figure, we also show variations of ΔE_a of bulk alloys with the composition reported in the literature.[12] The observed shifts in the bimetallic clusters do indeed match closely with those of the bulk alloys. The plots in Figure 4b show the variation in the core-level binding energy shifts of $Au(4f_{7/2})$ and $Ag(3d_{5/2})$ in the Au−Ag bimetallic clusters at large coverages with mole fraction, along with the variation found in bulk alloys.[15] The agreement is good in this system as well. From these results, we establish that the bimetallic clusters of a wide range of compositions studied here are indeed those due to well-defined alloys without any significant surface segregation.

We obtain the binding energy shifts in the bimetallic clusters due to cluster size effect, ΔE_c, by subtracting the shifts due to alloying, ΔE_a, from the observed shifts, ΔE. The ΔE_c values so obtained are plotted against coverage (or cluster size) in Figure 5, for all the Cu−Ni compositions studied by us. We see that ΔE_c increases markedly with a decrease in coverage, reaching values of ∼0.75 and 0.95 eV at very small coverages for Cu and Ni, repectively. The ΔE_c values of Au and Ag also increase with a decrease in coverage, reaching values of ∼0.9 eV at very small coverages as shown in Figure 6. The ΔE_c values found at small coverages for the component metals in the Cu−Ni and Au−Ag systems are comparable to the values found in monometallic clusters of these metals.[6,8,11]

In Figure 7, we have shown the variation of ΔE values of $Pd(3d_{5/2})$ and $Ni(2p_{3/2})$ levels relative to the bulk metal values (335.1 and 852.9 eV, respectively) with coverage for the different compositions of the Ni−Pd system. The ΔE of $Pd(3d_{5/2})$ increases markedly from 0.2 eV in NiPd₄ to nearly 0.8 eV in Ni₄Pd at large coverages. The ΔE of $Ni(2p_{3/2})$ at

large coverage is around −0.4 eV for NiPd₄ and only −0.2 eV in Ni₄Pd. We have plotted the $Ni(2p_{3/2})$ and $Pd(3d_{5/2})$ binding energy shifts at large coverages, ΔE_a, against the mole fraction in Figure 8a and compared them to the variation of binding energy shifts in bulk alloys.[12,14] The ΔE values at large coverages correspond to those found in bulk alloys in the Ni−Pd system as well. The ΔE_c values of $Ni(2p_{3/2})$ and $Pd(3d_{5/2})$ obtained by subtracting the shifts due to alloying, ΔE_a, from the observed ΔE values show the expected variation with coverage, increasing markedly at small coverages, reaching up to 0.7−0.8 eV at the smallest coverages examined by us.

The bimetallic clusters in the Cu−Pd system show the variation in the $Cu(2p_{3/2})$ and $Pd(3d_{5/2})$ binding energy shifts with coverage (see Figure 9) similar to those in the Cu−Ni, Au−Ag, and Ni−Pd systems, in spite of the complexity of the phase diagram referred to earlier.[11] The ΔE values at large coverages are plotted against mole fraction and compared with the ΔE values of the bulk alloys in Figure 8b. Clearly, the Cu−Pd bimetallic clusters also represent real alloy compositions. The ΔE_c values of the Cu−Pd bimetallic clusters increase with decrease in coverage, reaching values of ∼0.6 and 0.7 eV, respectively, for Pd and Cu at the smallest coverages studied.

The above results on bimetallic clusters of varied compositions in the Cu−Ni, Ag−Au, Ni−Pd, and Cu−Pd systems

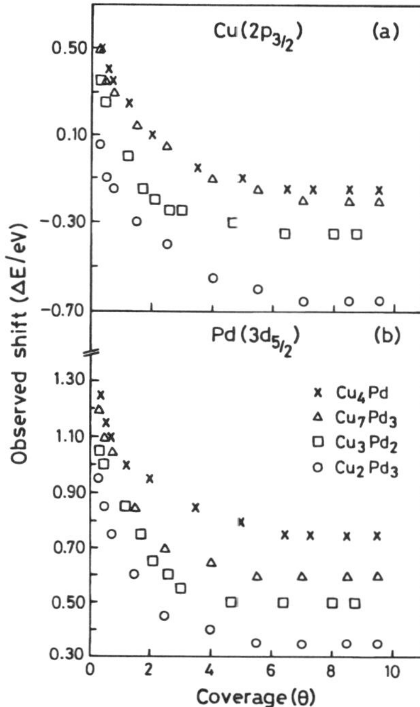

Figure 9. (a) Variation of the shift in Cu(2p$_{3/2}$) binding energy, ΔE, of Cu–Pd bimetallic clusters of different compositions with the metal coverage as determined by the ($I_{Cu} + I_{Pd}$)/I_C ratio. (b) Variation of the shift in the Pd(3d$_{5/2}$) binding energy, ΔE, with the coverage.

demonstrate that the observed core-level binding energy shifts of the component metals, ΔE (relative to the bulk metals), can be written as a sum of two effects, namely, those due to alloying and cluster size:

$$\Delta E = \Delta E_a + \Delta E_c$$

When the coverage is large, $\Delta E_c = 0$ and $\Delta E = \Delta E_a$. As the coverage is reduced, the contribution from ΔE_c due to cluster size increases.

References and Notes

(1) Wertheim, G. K.; DiCenzo, S. B.; Buchanan, D. N. E. *Phys. Rev. B* **1986**, *33*, 447.

(2) Mason, M. G. *Phys. Rev. B* **1983**, *27*, 748.

(3) Kohiki, S.; Ikeda, S. *Phys. Rev. B* **1986**, *34*, 3786.

(4) Eberhardt, W.; Fayet, P.; Cox, D. M.; Fu, Z.; Kaldor, A.; Sherwood, R.; Sondericker, D. *Phys. Rev. Lett.* **1990**, *64*, 780.

(5) Kuhrt, Ch.; Harsdorff, M. *Surf. Sci.* **1991**, *245*, 173.

(6) Vijayakrishnan, V.; Rao, C. N. R. *Surf. Sci.* **1991**, *255*, L516.

(7) Mason, M. G. In *Cluster Models for Surface and Bulk Phenomena*; Pacchioni, G., Eds.; Plenum: New York, 1992.

(8) Rao, C. N. R.; Vijayakrishnan, V.; Aiyer, H. N.; Kulkarni, G. U.; Subbanna, G. N. *J. Phys. Chem.* **1993**, *97*, 11157.

(9) Aiyer, H. N.; Vijayakrishnan, V.; Subbanna, G. N.; Rao, C. N. R. *Surf. Sci.* **1994**, *313*, 392.

(10) Santra, A. K.; Subbanna, G. N.; Rao, C. N. R. *Surf. Sci.* **1994**, *317*, 259.

(11) Hansen, H.; Anderko, K. *Constitution of Binary Alloys*: McGraw-Hill: New York, 1958.

(12) Steiner, P.; Hüfner, S. *Acta Metallurg.* **1981**, *29*, 1885.

(13) Steiner, P.; Hüfner, S.; Mårtensson, N.; Johansson, B. *Solid State Commun.* **1981**, *37*, 73.

(14) Hillebrecht, F. U.; Fuggle, J. C.; Bennett, P. A.; Zolnierek, Z.; Freiburg, Ch. *Phys. Rev. B* **1983**, *27*, 2179, 2194.

(15) Watson, R. E.; Hudis, J.; Periman, M. L. *Phys. Rev. B* **1971**, *4*, 4139.

J. Phys. Chem. B **2001**, *105*, 2515–2517

Magic Nuclearity Giant Clusters of Metal Nanocrystals Formed by Mesoscale Self-Assembly

P. John Thomas, G. U. Kulkarni, and C. N. R. Rao*

Chemistry and Physics of Materials Unit, Jawaharlal Nehru Center for Advanced Scientific Research, Jakkur, Bangalore-560 064, India

Received: June 16, 2000; In Final Form: December 5, 2000

Magic nuclearity giant clusters formed by the mesoscale self-assembly of Pd nanocrystals of 2.5 nm diameter (nuclearity, ∼561) have been identified by transmission electron microscopy. The clusters have discrete diameters, corresponding to those expected for magic nuclearity of 13, 55, 147, 309, 561, and 1415 and corresponding to closed shells of 1, 2, 3, 4, 5, and 7, respectively. Imaging at different tilt angles has provided confirmation of the spherical nature of these giant clusters. Giant clusters of magic nuclearity have also been found with Pd nanocrystals of ∼3.2 nm diameter (nuclearity, ∼1415).

Introduction

Mesoscale self-assembly of polyhedral objects is a topic of great current interest.[1] The occurrence of self-assembly in nanometric dimensions through weak forces has been well documented. Thus, cooperative assemblies of ligated metal and semiconductor nanocrystals as well as of colloidal polymer spheres appear to form through the mediation of electrostatic and capillary forces. Typical examples of mesoscale assemblies are provided by the ordering of nanocrystals of Au[2,3] and CdSe[4] and of polystyrene spheres.[5] Besides ordered two-dimensional arrays, such forces have also been exploited to obtain giant nanocrystal aggregates measuring tens of nanometers,[6] 2D rings of micrometer diameters,[7] and dendrimeric structures of Au nanocrystals.[8] The ability to engineer such assemblies may extend the reach of current lithographic techniques. In this context, the synthesis and programmed assembly of well-defined metal nanocrystals assumes significance.[9] Metal nanocrystals with magic number of atoms 13, 55, 309, 561, and 1415 corresponding to 1, 2, 4, 5, and 7 closed shells, respectively, have been prepared by chemical means.[10–13] We have obtained two-dimensional arrays of Pd_{561} and Pd_{1415} nanocrystals by employing alkanethiol spacers.[14] Schmid et al.[15,16] have arranged Au_{55} clusters on a polymer film and have isolated microcrystals of the same.

The assembly of magic nuclearity nanocrystals into giant clusters containing a magic number of initial nanocrystals has been a subject of fascination.[17] Initial electrophoresis experiments of Schmidt[17] with Au nanocrystals indicated the formation of $(Au_{13})_{13}$ types of superclusters. Further, a high mass secondary ion peak observed also seemed to support this contention.[18,19] Theoretical calculations based on the embedded atom method have indicated that such a growth of metal nanocrystals is indeed a possibility.[20] In Figure 1, is shown a schematic illustration of a $(M_{55})_{55}$ giant cluster.

In our experiments with monodisperse Pd nanocrystals in a ethanol−water mixture,[11] we observed some giant aggregates as a result of mesoscalar self-assembly. In this article, we report electron microscopic investigation of the aggregates of Pd_{561} and Pd_{1415} nanocrystals. The aggregates have diameters that could be assigned to values expected of giant clusters containing

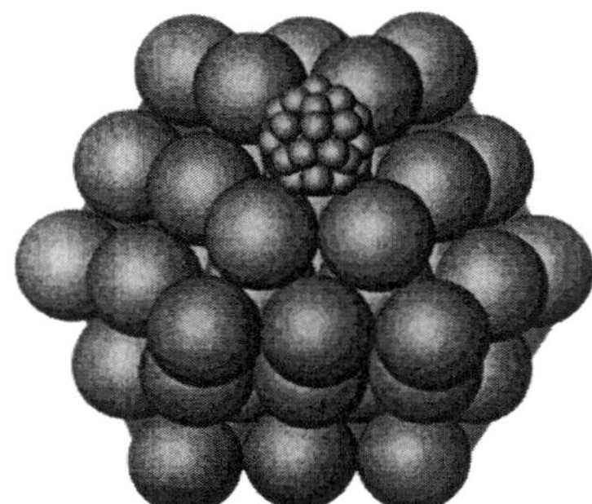

Figure 1. Schematic illustration of a $(M_{55})_{55}$ giant cluster. The structure of one of the nanocrystals is shown.

a magic number of the initial nanocrystals. Thus, it has been possible to obtain giant clusters containing approximately 561 nanocrystals, each of which comprises around 561 atoms.

Experimental Section

Monodisperse Pd nanocrystals were prepared in a ethanol−water mixture in the presence of poly(vinylpyrrolidone) (PVP) following the procedure of Teranishi et al.[12] Typically, a 15 mL of a 2.0 mM aqaueous solution of H_2PdCl_4 was reduced by refluxing it with a mixture of 15 mL of absolute ethanol and 25 mL of water containing 33.3 mg of PVP (M_w ∼ 40 000 g mol^{-1}) for ∼3 h. The obtained sols were examined with a JEOL-3010 transmission electron microscope (TEM), operating at 300 kV. Samples for TEM were prepared by depositing a drop of the sol on a holey carbon grid and allowing it to dry slowly in air and then in a desiccator overnight.

Results and Discussion

A scan of the grid revealed the presence of a large number of uniform spherical particles. Figure 2 shows a typical TEM image of the dispersion. In addition to isolated nanocrystals, we observed pairs of particles separated by a nearly uniform distance. A few were seen as sitting on top of each other. The histogram in Figure 2 shows that the metal cores are of a uniform diameter of 2.5 nm ($\sigma = 6\%$). Furthermore, high-

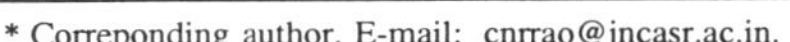

* Correponding author. E-mail: cnrrao@jncasr.ac.in.

10.1021/jp0021998 CCC: $20.00   © 2001 American Chemical Society
Published on Web 03/08/2001

2516 *J. Phys. Chem. B, Vol. 105, No. 13, 2001*

Thomas et al.

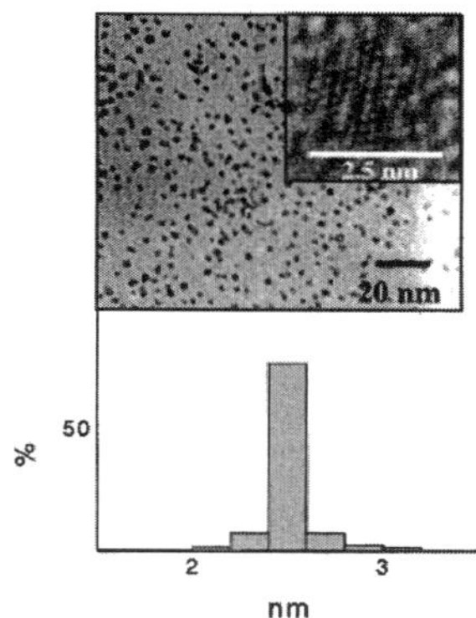

Figure 2. (a) TEM micrograph showing Pd$_{561}$ nanocrystals. The inset shows a high-resolution image of an individual nanocrystal. (b) Histogram showing the size distribution (in percentage).

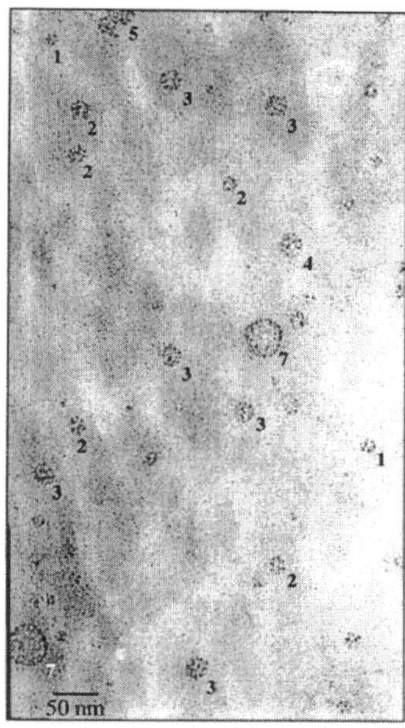

Figure 3. TEM image of Pd$_{561}$ nanocrystals forming giant clusters. The numbers correspond to the proposed number of nanocrystal shells, n.

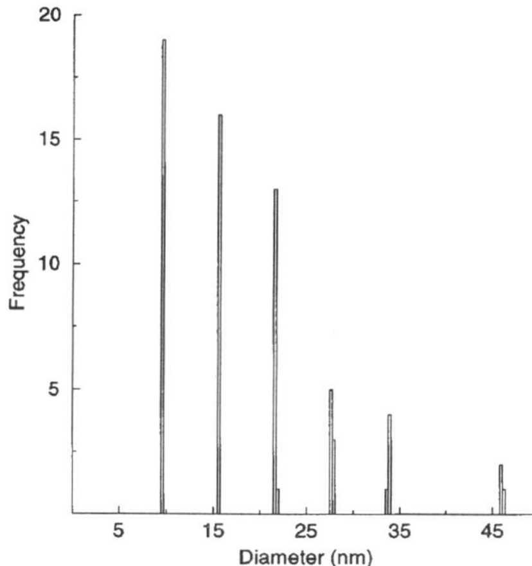

Figure 4. Histogram showing the distribution in the diameters of giant clusters.

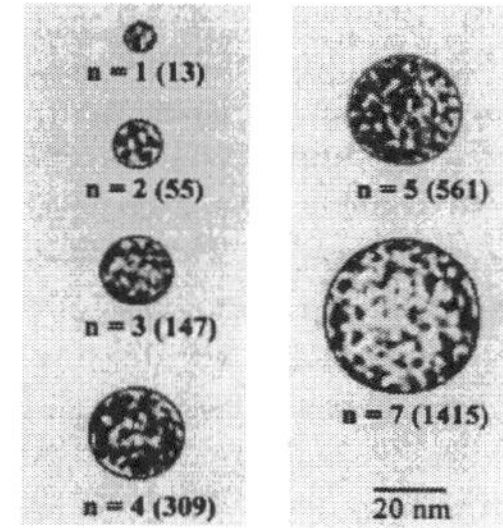

Figure 5. Giant clusters of different magic nuclearities, (Pd$_{561}$)$_n$, the circles corresponding to the diameters of the clusters calculated on the basis of the effective volume of an individual nanocrystal.

TABLE 1: Giant Clusters of Pd$_{561}$ Nanocrystals

diameters measd (nm)	no. of nanocrystals	diameters measd (nm)	no. of nanocrystals
9.6 ± 0.1	13 ± 1	27.7 ± 0.1	309 ± 34
15.6 ± 0.1	55 ± 6	33.8 ± 0.1	561 ± 61
21.6 ± 0.1	147 ± 16	46.0 ± 0.15	1415 ± 156

resolution TEM images (see inset of Figure 2) revealed 11 [111] lattice fringes with each nanocrystal,[13] indicating the atomic nuclearity to be close to 561.

Besides the 2.5 nm nanocrystals, the grid also contained bigger features resembling aggregates of nanocrystals. Such aggregates were spherical and seemed to exhibit an unusual preference for a few definite sizes. In Figure 3, a typical low magnification (×80 000) TEM image revealing the presence of giant aggregates of nanocrystals is shown. A histogram showing the frequency distribution of the observed diameters is given in Figure 4. Evidently, the giant clusters exhibit preference for specific diameters, 9.6, 15.6, 21.6, 27.7, 33.8, and 46.0 nm, with the smaller giant clusters being more abundant. To estimate the number of nanocrystals present in these aggregates, we calculated the effective volume of a nanocrystal on the basis of the distance between adjacent nanocrystals in TEM images (Figure 2). The pairs of particles were chosen carefully such that the

two metal cores were distinct. The mean value obtained from hundred such measurements was 4.1 ± 0.1 nm. This value is more suitable than the diameter of an isolated nanocrystal, the latter being 3.3 nm as estimated by STM measurements.[21] We assume that the effective volume thus estimated takes into account the free volume in the nanocrystal aggregate. In other words, a particle pair is taken to adequately describe interactions in the bigger aggregates. In Table 1, we list the observed diameters of the giant clusters along with the estimates of the nuclearities. To our surprise, the estimated values compare closely with the magic nuclearities of 13, 55, 147, 309, 561, and 1415 corresponding to closed shells 1, 2, 3, 4, 5, and 7, respectively. In Figure 5, we have projected the calculated perimeters enclosing the different giant clusters. The close agreement between the observed and calculated diameters is indeed gratifying.

The giant aggregates obtained are distinct from the two-dimensional rings obtained by other groups.[7,22] To establish the spherical nature of the giant clusters observed by us, they were imaged at different tilts (±17.5°) along two perpendicular axes in the focal plane. In Figure 6, we show such images obtained

Magic Nuclearity Giant Clusters

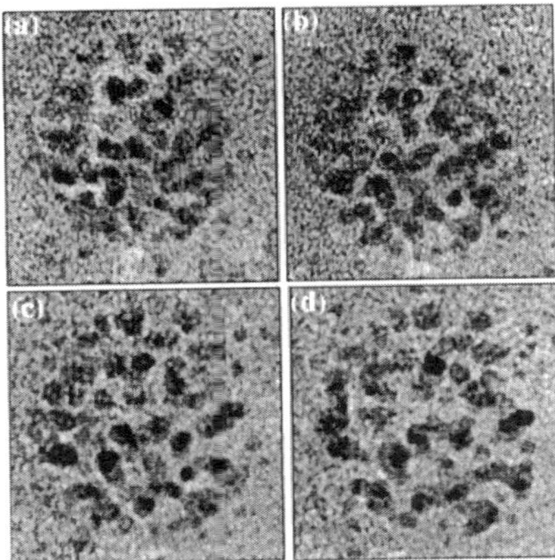

Figure 6. A $(Pd_{561})_{561}$ giant cluster imaged at various tilt angles along two perpendicular axes in the focal plane: (a) $-18.7°$, $1.4°$; (b) $-7.3°$, $-1.4°$; (c) $0°$, $1.4°$; (d) $12.4°$, $-1.4°$.

with a giant cluster of nuclearity 561. The projected image appears circular over the entire tilt range of $35°$, along both the directions. We observe a few individual nanocrystals to shift their positions in the image under different tilt conditions, though the exact trajectories could not be followed. A series of such tilt experiments confirmed the spherical nature of the aggregates. Furthermore, we were able to obtain lattice-resolved images of the individual Pd_{561} nanocrystals constituting the giant cluster. Such images revealed the characteristic 11 [111] lattice fringes. Attempts to image the internal structure of giant clusters were, however, not successful probably because of the interference from the polymeric ligand shell of the nanocrystals.

We have previously obtained two-dimensional arrays of metal nanocrystals by using monothiols[2,9,14,23] and three-dimensional layered assemblies with metal and semiconductor nanocrystals by employing dithiols.[24] Brushlike ligands in these cases direct the assembly process along definite directions. We believe that the giant clusters discussed in this paper are a result of self-assembly in three-dimensions facilitated by the relatively passive and isotropic PVP coating around the nanocrystals. Unlike in a two-dimensional assembly, where the solvent plays an important role,[24] the formation of giant clusters does not seem to be influenced by the choice of the solvent. We are able to obtain the giant clusters from a wide variety of solvents such as water, ethanol, ethanol water mixtures, or other types of solvents such as ethylene glycol. In all cases, there were no significant changes in the size distribution of the obtained giant nanocrystals. These giant clusters could be reproducibly obtained over several trials starting with sols of widely differing concentrations and employing PVP of molecular mass $\sim160,000$ g mol^{-1} as well.

We have made attempts to investigate the process of self-assembly of nanocrystals of a different magic nuclearity. Experiments with Pd nanocrystals of 3.2 nm diameter, corresponding to a nuclearity of 1415, have revealed the formation of giant clusters with possible nuclearities of 147 and 309, as shown in Figure 7. However, the slightly wider distribution in the pristine nanocrystal diameter (possibly due to a mixture of the 7 and 8 shell clusters[13]) seems to hinder the facile formation of uniform giant clusters. A majority of these clusters ($\sim60\%$) display elongation along one axis. A narrow size distribution in the initial nanocrystal sol appears to be essential for the formation of the magic nuclearity giant clusters.

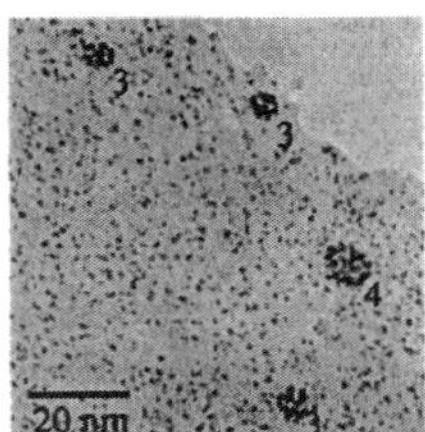

Figure 7. TEM image of giant clusters of Pd_{1415} nanocrystals. The numbers correspond to the proposed number of nanocrystal shells, n.

Conclusion

Giant clusters of Pd nanocrystals are obtained by the mesoscopic assembly of nanocrystals of uniform size. This observation provides an illustration of the principle of self-similarity. These clusters of metal nanocrystals are perhaps bound by the same laws that endow special stability to magic numbered nanocrystals.[15] It would therefore appear that these laws are invariant under scaling. The giant clusters of Pd_{561} nanocrystals obtained by us are different from other types of ligand directed nanocrystal aggregates which are formed mostly due to the directional nature of the ligand shell.[16,6]

References and Notes

(1) Terfort, A.; Bowden, N.; Whitesides, G. M. *Nature* **1997**, *386*, 162.

(2) Vijayasarathy, K.; Raina, G.; Yadav, R. T.; Kulkarni, G. U.; Rao, C. N. R. *J. Phys. Chem.* **1998**, *B101*, 9876.

(3) Whetten, R. L.; Khoury, J. T.; Alvarez, M. M.; Murthy, S.; Vezmar, I.; Wang, Z.; Stephens, P. W.; Clevend, L. Ch.; Luedtke, W. D.; Landman, U. *Adv. Mater.* **1996**, *8*, 428.

(4) Murray, C. B.; Kagan, C. R.; Bawendi, M. G. *Science* **1995**, *270*, 1335.

(5) Trau, M.; Saville, D. A.; Aksay, I. A. *Science* **1996**, *272*, 706.

(6) Boal, A. K.; Ilhan, F.; Derouchey, J. E.; Thurn-Albrecht, T.; Russull, T. P.; Rotello, V. M. *Nature* **2000**, *404*, 746.

(7) Ohara, P. C.; Heath, J. R.; Gelbart, W. M. *Angew. Chem., Int. Ed. Engl.* **1997**, *36*, 1078.

(8) Seshadri, R.; Subbanna, G. N.; Vijayakrishnan, V.; Kulkarni, G. U.; Ananthakrishna, G.; Rao, C. N. R. *J. Phys. Chem.* **1995**, *99*, 5639.

(9) Rao, C. N. R.; Kulkarni, G. U.; Thomas, P. J.; Edwards, P. P. *Chem. Soc. Rev.* **2000**, *29*, 27.

(10) Rao, C. N. R. *Chemical approaches to the Synthesis of Inorganic Materials*; Wiley Eastern: New Delhi, 1994.

(11) Vargaftik, M. N.; Moiseev, I. I.; Kochubey, D. I.; Zamareev, K. I. *Faraday Discuss.* **1991**, *92*, 13.

(12) Teranishi, T.; Hori, H.; Miyake, M. *J. Phys. Chem.* **1997**, *B101*, 5774.

(13) Schmid, G.; Harms, M.; Malm, J. O.; Bovin, J. O.; Ruitenbeck, J. V.; Zandbergen, H. W.; Fu, W. T. *J. Am. Chem. Soc.* **1993**, *113*, 2046.

(14) Thomas, P. J.; Kulkarni, G. U.; Rao, C. N. R. *J. Phys. Chem.* **2000**, *B104*, 8138.

(15) Schmid, G.; Bäumle, M.; Beyer, N. *Angew. Chem., Int. Ed. Engl.* **2000**, *39*, 181.

(16) Schmid, G.; Zaika, W. M.; Pugin, R.; Sawitowski, T.; Majoral, J.-P.; Caminade, A.-M.; Turrin, C.-O. *Chem. Eur. J.* **2000**, *6*, 1693.

(17) Schmid, G. *Polyhedron.* **1988**, *7*, 2321. (b) Schmid, G.; Klein, N. *Angew. Chem., Int. Ed. Engl.* **1986**, *25*, 922.

(18) Feld, H.; Leute, A.; Rading, D.; Benninghoven, A.; Schmid, G. *J. Am. Chem. Soc.* **1990**, *112*, 8166.

(19) McNeal, C. J.; Winpenny, R. E. P.; Hughes, J. M.; Macfarlane, R. D.; Pignolet, L. H.; Nelson, L. T. J.; Gardner, T. G.; Irgens, I. H.; Vigh, G.; Fackler, J. P. *Inorg. Chem.* **1993**, *32*, 5582.

(20) Fristche, H.-G.; Muller, H.; Fehrensen, B. *Z. Phys. Chem.* **1997**, *199*, 87.

(21) Thomas, P. J.; Kulkarni, G. U.; Rao, C. N. R. *Chem. Phys. Lett.* **2000**, *321*, 163.

(22) Shafi, K. V. P. M.; Felner, I.; Mastai, Y.; Gedanken, A. *J. Phys. Chem.* **1999**, *B103*, 3358.

(23) Vijayasarathy, K.; Kulkarni, G. U.; Rao, C. N. R. *Chem. Commun.* **1997**, 573.

(24) Vijayasarathy, K.; Thomas, P. J.; Kulkarni, G. U.; Rao. C. N. R. *J. Phys. Chem.* **1999**, *B103*, 399.

(25) Korgel, B. A.; Fitzmaurice, D. *Phys. Rev. Lett.* **1998**, *80*, 3531.

J. Phys. Chem. B **2002,** *106,* 4647−4651

Novel Effects of Metal Ion Chelation on the Properties of Lipoic Acid-Capped Ag and Au Nanoparticles

Sheela Berchmans, P. John Thomas, and C. N. R. Rao*

Chemistry and Physics of Materials Unit, Jawaharlal Nehru Centre for Advanced Scientific Research, Jakkur, Bangalore 560 064, India

Received: October 23, 2001; In Final Form: January 31, 2002

The effects of the interactions of metal ions with lipoic acid-capped Ag and Au nanoparticles have been studied by the combined use of electronic absorption spectroscopy and transmission electron microscopy. Three types of effects that are dependent on the metal ion concentration can be distinguished. First, in the dilute regime there is reversible chelation of the metal ions, causing a marked dampening of the plasmon resonance band of the nanoparticles, but there is no aggregation. The magnitude of plasmon dampening depends on the nature as well as the concentration of the metal ions. In the intermediate concentration regime, aggregation occurs, but in the high concentration regime, there is precipitation. These different regimes are clearly evidenced in the changes in the electronic spectra and in the electron micrographs.

Introduction

Nanoparticles of gold and silver have characteristic absorption bands in the electronic spectra because of plasmon resonance.[1,2] The plasmon absorption bands are found even when the nanoparticles are covered with various capping agents and have been utilized to probe the growth and agglomeration of the nanoparticles and chemisorption at their surfaces.[3–5] The dampening of the plasmon band of nanoparticles following the chemisorption of thiols has been examined by Linnert et al.[5] Flocculation rates of Au nanocrystals stabilized by thiols of the form $HS(CH_2)_nR$ (R = alkyl or carbonyl group) have been studied in terms of the flocculation parameter, which is the intergrated area between 600 and 800 nm in the absorption spectra.[6] The flocculation rates are pH-dependent.[7] When a mercaptocarboxylic acid is used as the capping agent, the thiol end binds to the metal nanoparticle with the carboxyl end sticking out radially, enabling chelation by metal ions.[8–11] Templeton et al.[9] have obtained a layer-by-layer assembly of Au nanocrystals on an Au surface, making use of the ability of the metal ion to coordinate between two layers of the nanocrystals. It has been suggested that the colorimetric response of such Au nanocrystals could be useful in sensing heavy metals that chelate with the carboxyl group.[10] Furthermore, pH titrations carried out by Simard et al.[11] reveal the reversible nature of the aggregates produced by chelation.

We have carried out a systematic investigation of the interaction of metal ions with α-lipoic acid (thioctic acid or 1,2-dithiolane-3-pentanoic acid) -capped Ag and Au nanoparticles by means of electronic absorption spectroscopy and transmission electron microscopy. In this paper, we report different types of changes that take place in metal colloids when the carboxyl functional group attached to the tail of the amphipilic capping agent interacts with metal ions. We show two possible ways metal ions can bind to the carboxyl end of lipoic acid in Figure 1. We observe three types of changes depending on the concentration of the metal ions used. The observed changes are specific to the metal ions and are useful in helping one to distinguish different metal ions present in a solution.

Experimental Section

Metal nanoparticles capped by α-lipoic acid were prepared by a slight modification of the procedure reported by Maya et al.[12] The procedure involves the addition of 15 mL of 2.5 mM $NaBH_4$ to a 15-mL aqueous solution containing 75 μmols each of $HAuCl_4$ and α-lipoic acid to obtain a wine-red sol containing nanoparticles of gold. The sol was thoroughly aerated with argon and repeatedly extracted with dicholromethane to remove unreacted products. Nanoparticles of Ag capped with α-lipoic acid could be obtained by starting with $AgNO_3$. The sols thus obtained were examined with a JEOL-3010 transmission electron microscope (TEM) operating at 300 kV and were further characterized by UV−visible spectroscopy carried out with a Perkin-Elmer lambda 900 spectrometer. Samples for TEM were prepared by depositing a drop of the sol on a holey carbon grid and allowing it to dry in air and then in a desiccator overnight. The mean diameter and the size distribution were determined from the TEM images and represent averages over a few hundred nanoparticles. The mean diameter of the nanocrystals could be varied by changing the starting metal/lipoic acid ratios. Thus, Au/lipoic acid molar ratios of 1:1 and 1:2.5 yielded particles of diameter 7.5 ± 0.7 nm and 3.0 ± 0.4 nm, respectively, and Ag/lipoic acid ratios of 1:2 and 1:5 yielded particles of diameter 5 ± 0.4 nm and 2.6 ± 0.4 nm, respectively.

Stock solutions of 1 mM $CuSO_4$, $FeSO_4$ (pH ≈ 3), $Ni(NO_3)_2$ and acetates of Mn, Cd, Zn, and Pb were used as sources of the metal ions, whose interactions with the lipoic acid-capped metal nanoparticles were studied. Stock solutions of higher concentrations were used when necessary. A few microliters of the stock solution were directly added to the sols in quartz cuvettes to follow the changes accompanying the increase in concentration of the metal ions. We have given metal-ion concentrations in units of μM. Aliquots of the sol containing metal ions were drawn at periodic intervals to carry out TEM measurements. The sols containing the metal ions remained stable and optically clear over periods of several weeks. The above experiments were

---

* Corresponding author. E-mail: cnrrao@jncasr.ac.in.

10.1021/jp013935q CCC: $22.00   © 2002 American Chemical Society
Published on Web 04/11/2002

4648   *J. Phys. Chem. B, Vol. 106, No. 18, 2002*   Berchmans et al.

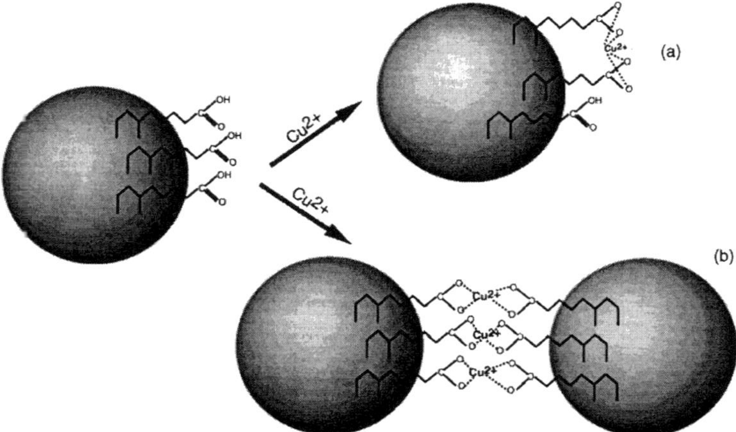

Figure 1. Schematic illustration showing the binding of metal ions to the carboxyl groups present at the tail of the amphiphile capping the metal nanoparticles. In (a), the metal ion is bound to the carboxyl end of one nanoparticle, but in (b), it binds to the carboxyls of two nanoparticles.

also carried out in the presence of poly(vinyl alcohol) (PVA, 5%). Quantitatively similar results were obtained with PVA except in the high concentration regime, as detailed in the Results and Discussion section.

Infrared spectra of the lipoic acid-covered nanoparticles before and after chelation (with excess Cu ions) showed the change from the carboxyl group to the carboxylate anion.

Results and Discussion

Figure 2 shows the changes in the absorption spectra of Ag nanoparticles of $\sim$5-nm diameter capped by α-lipoic acid following the addition of Cu^{2+} and Fe^{2+} ions. In the absence of metal ions, the plasmon band is observed at $\sim$430 nm. As Cu^{2+} or Fe^{2+} ions are added in the form of aqueous solutions, the plasmon band is dampened in proportion to the concentration of the metal ion (see Figure 2). After the addition of the metal ion, the plasmon-band intensity immediately reaches a new value, and the sol continues to exhibit the dampened band for extended periods of time (a few weeks) if left undisturbed. Several bivalent metal ions were examined. It was found that different metal ions cause the dampening of the plasmon band to different extents. For example, 80 μM Cu^{2+} solution brings about an $\sim$60% dampening of the plasmon intensity, but, on the other hand, 230 μM Fe^{2+} brings about a decrease of $\sim$50% (Figure 2). It is known that a change in the ionic strength causes aggregation of the sols, thus bringing about such changes in the plasmon band intensity. In the example above, the change in the ionic strength is around three times greater in case of Fe^{2+} ions than in the case of Cu^{2+} ions, but this change is not reflected in the dampening of the plasmon band. We therefore consider the observed dampening of the plasmon band to arise from a process that is selective to a particular metal ion, which involves the surface carboxyl group, rather than from the ionic strength of the medium. Accordingly, the sensitivity of the plasmon band varies from one metal ion to another.

To ascertain that this is indeed the case, we sought to remove the chelated metal ions by employing a ligand stronger than α-lipoic acid. Accordingly, by the addition of equimolar amounts of EDTA to Au and Ag sols containing chelated metal ions, it was possible to enhance the plasmon-band intensity to a value obtained in the absence of metal ions.

TEM studies shed further light on the above-mentioned change undergone by the colloid. In Figure 3, we show typical

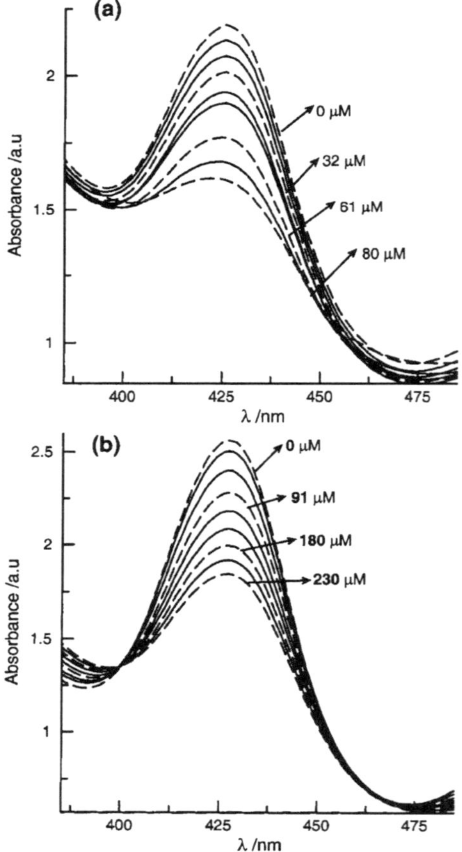

Figure 2. Electronic absorption spectra of $\sim$5-nm Ag nanoparticles showing changes accompanying the addition of (a) Cu^{2+} and (b) Fe^{2+}. The concentrations of the ions are indicated.

TEM micrographs of the $\sim$5-nm Ag nanoparticles before and after the addition of Cu^{2+} (80 μM) ions. We see that even in the presence of Cu^{2+} ions the nanoparticles remain isolated, with the aggregation[13] limited to very small areas (about 3% of

Effects of Chelation on Ag and Au Nanoparticles

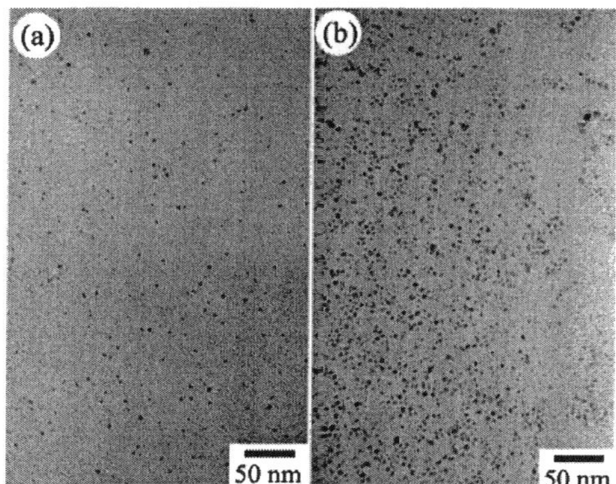

Figure 3. Electron micrographs of $\sim$5-nm Ag nanoparticles (a) before the addition of Cu^{2+} and (b) in the presence of 80 μM Cu^{2+}.

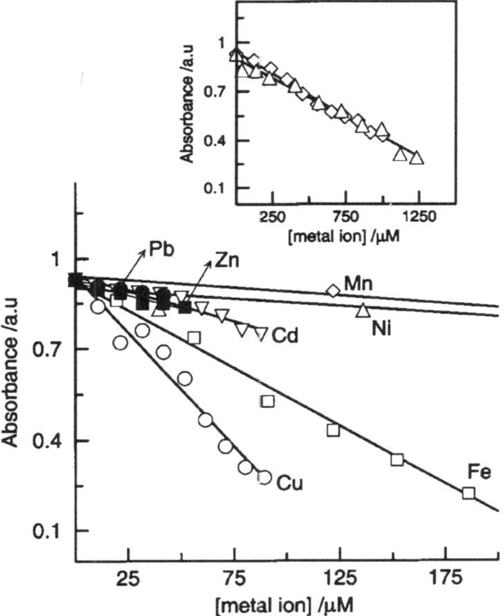

Figure 4. Variation in the absorbance of the plasmon band (at λ_{max}) of $\sim$5-nm Ag nanoparticles as a function of the concentration of metal ions. Linear fits are shown. The inset shows the data for Mn^{2+} and Ni^{2+} over an extended concentration range.

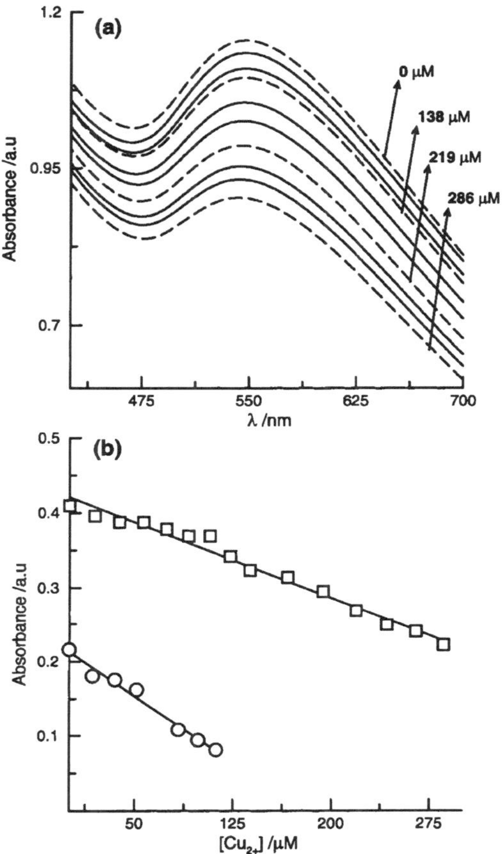

Figure 5. (a) Electronic absorption spectra of $\sim$7.5-nm Au nanoparticles showing changes accompanying the addition of Cu^{2+}. (b) Plots showing the variation of absorbance of the plasmon band (at λ_{max}) of Au nanoparticles accompanying the addition of Cu^{2+} ions. The circles correspond to nanoparticles of $\sim$7.5-nm diameter, and the squares correspond to $\sim$3.0 nm diameter nanoparticles.

the scanned area). This observation rules out aggregation as the cause of the observed dampening of the plasmon band. Theoretical investigations suggest that aggregates measuring a few hundred to a few thousand nanoparticles are required to obtain a substantive optical response.[14] The binding of metal ions appears to proceed more by path a than by path b in Figure 1.

Dampening of the plasmon band (along with accompanying changes in position and width) arising from the chemisorption of reactive molecules on the surfaces of nanoparticles[2,5,15] had earlier been attributed to charge transfer between the particle and the adsorbate. We have not found any change in the position or the width of the plasmon band accompanying the chelation of metal ions, but instead we observe only a decrease in the band intensity. The existence of an isobestic point at 399 nm in Figure 2b and the nearly constant absorbance at 460 nm corroborate this view. Even where there is no isobestic point

as in Figure 2a, there is little shift in the position of the width of the plasmon band. The trends were similar with different counteranions. Similar changes in the plasmon-band intensity were observed for Ag nanoparticles with a mean diameter of 2.6 nm following chelation. It can therefore be concluded that in the dilute metal-ion concentration regime the metal ions bind to the carboxyl functional group of α-lipoic acid, which caps the nanoparticles and thereby causes the observed dampening of the plasmon band.

The ability of metal ions to dampen the plasmon band has been quantified by making use of the slopes of the plots of the plasmon-band intensity against the metal-ion concentration. We present such plots of the absorbance of the plasmon band of Ag nanoparticles at the absorption maximum against the concentration of metal ions. We also show the corresponding linear fits in this Figure. The inset in Figure 4 shows that the intensity decrease can continue over an extended region of concentration in the case of weakly interacting metal ions such as Mn^{2+} and Ni^{2+}. The values of the slopes ($\times 10^3$ M^{-1}) for the interactions of Cu^{2+}, Fe^{2+}, Cd^{2+}, Zn^{2+}, Pb^{2+}, Mn^{2+}, and Ni^{2+} with $\sim$5-nm Ag nanoparticles are 7.5, 3.8, 2.0, 1.8, 1.2, 0.5, and 0.5 $\times$, respectively. The slopes ($\times 10^3$ M^{-1}) obtained for Ag nanoparticles with a mean diameter of 2.6 nm were 2.8,

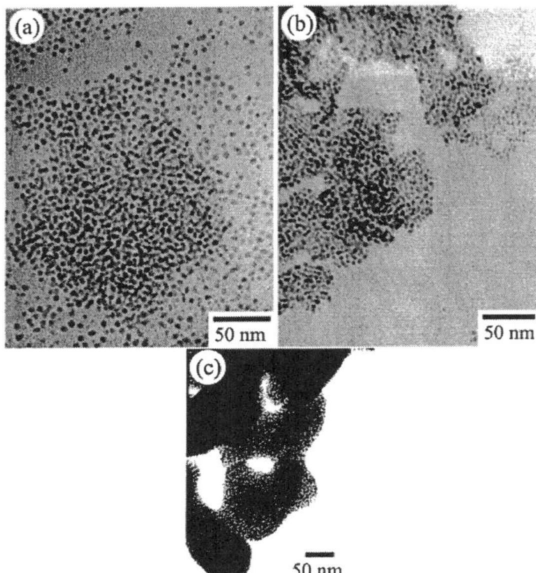

Figure 6. Electron micrographs of ∼5-nm Ag nanoparticles after interaction with aqueous Cu^{2+} solutions that are (a) 120 μM, (b) 200 μM, and (c) 500 μM.

0.6, 0.5, and 0.2 ×for Cu^{2+}, Fe^{2+}, Mn^{2+}, and Ni^{2+} ions, respectively. Similar results were obtained when the counterions of the Cu, Ni, and Mn ions were changed to chlorides. Although the slopes of the absorption−concentration plots are higher when the particle diameter is larger, the effectiveness of chelation varies in the order $Cu^{2+} > Fe^{2+} > Mn^{2+} > Ni^{2+}$ and is independent of the diameter of the Ag nanoparticles.

Figure 5 shows the results of our studies on the interaction of Cu^{2+} ions with Au nanoparticles. In Figure 5a, we present the changes in the absorption spectra of the ∼7.5-nm Au nanoparticles brought about by the addition of Cu^{2+} ions. A decrease in the plasmon-band intensity is observed just as in the case of Ag nanoparticles. In Figure 5b, we show the change in the absorbance with the Cu^{2+} concentration for 7.5- and 3.0-nm Au nanoparticles. The slope is slightly greater in the case of the larger Au nanoparticles.

Addition of Cu^{2+} ions to ∼5-nm Ag nanoparticles beyond the dilute regime (>90 μM) brings about different changes in the sol. The sols precipitate after about 200 μM Cu^{2+} has been added. The onset of precipitation is shifted to higher concentration upon addition of PVA(5%). TEM studies indicate that the nanoparticles form aggregates, with the extent of aggregation (measured as the fraction of particles belonging to an aggregate) increasing with the concentration of the metal ions, as shown by the micrographs in Figure 6. When the Cu^{2+} concentration is 120 μM (Figure 6a), ∼50% of the nanoparticles form aggregates, which is in contrast to the case of 80 μM Cu^{2+} where the aggregation is limited to ∼3% of the particles (see Figure 3b). Almost all the particles get aggregated when the concentration of Cu^{2+} is raised to 200 μM. We note that the sols showed little sign of opalescence and remained stable during all of the measurements discussed hitherto (in the Cu^{2+} concentration regime of 0−200 μM). Further addition of Cu^{2+} ions, however, leads to precipitation. TEM images of the precipitate showed multilayered, close-packed structures of nanoparticles (Figure 6c).

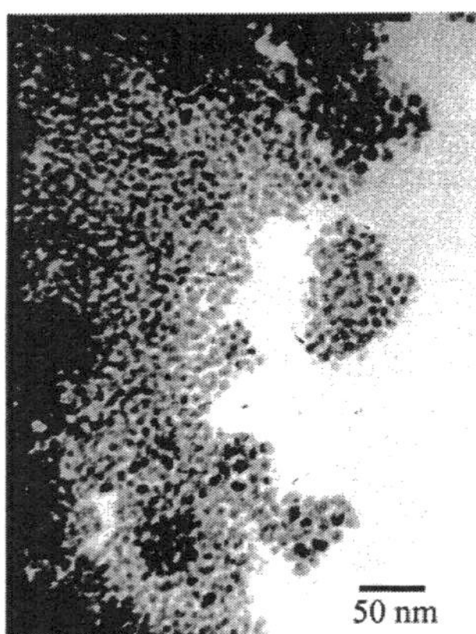

Figure 7. Electron micrographs of ∼7.5-nm Au nanoparticles after interaction with an aqueous 600 μM Cu^{2+} solution.

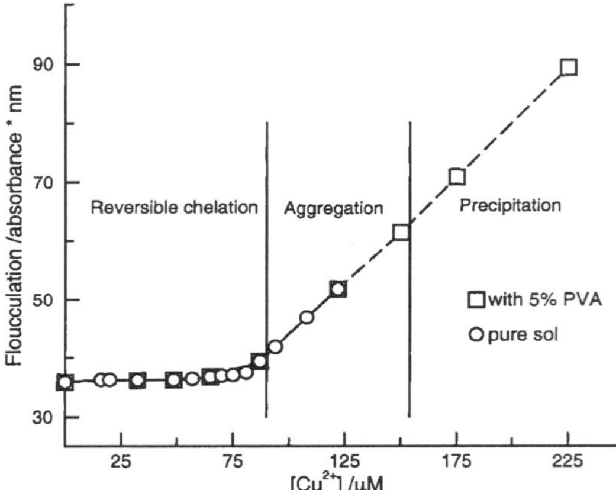

Figure 8. Variation of the floucculation parameter with the concentration of Cu^{2+} in the case of 5-nm Ag nanoparticles.

The above observations are supported by changes in the optical spectra of the nanoparticles. In the above concentration regime (90 μM < [Cu^{2+}] > 200 μM), the plasmon band is completely damped, and the scatter at longer wavelengths, which was essentially constant in the dilute regime, increases progressively. To quantify the changes, we employ a slightly modified version of the floucculation parameter defined previously by Weisbecker et al.[6] for Au colloids. The floucculation parameter, defined here for the case of Ag nanoparticles as the integrated area in the 550−750 nm region, shows an increase with the metal ion concentration that coincides with aggregation phenomena described previously. The change in the floucculation parameter could not be reversed by addition of EDTA, and thus the chelation is *irreversible*. Addition of PVA stabilizes the sol and permits further addition of Cu ions with higher values of the floucculation parameter than that of the native sol. We thus have a regime of chelation that is not reversible by addition of stronger chelating agents above 90 μM Cu^{2+}, followed by a regime where precipitation occurs (Figure 6c). In the case of Au nanoparticles, the region of irreversible chelation was also marked by aggregation, as can be seen from the TEM micrograph in Figure 7.

The different regimes of metal-ion interaction with α-lipoic acid-capped Ag nanoparticles, delineated on the basis of the floucculation parameter, are depicted in Figure 8. In the case

Effects of Chelation on Ag and Au Nanoparticles

J. Phys. Chem. B, Vol. 106, No. 18, 2002 **4651**

of other metal ions such as Fe^{2+} and Cd^{2+}, the different regimes occur at concentrations different from those found with Cu^{2+}, as expected on the basis of the slopes in the absorbance—concentration plots (Figure 4).

The observation that the aggregates and the precipitates are made of nanoparticles similar in dimensions to those of the colloid and the fact that the onset of aggregation depends on the metal ion used seem to suggest that the binding event (chelation) is responsible for aggregation and precipitation. However, given that the aggregation and precipitation are irreversible and that the stabilizing agent PVA is able to shift the onset of precipitation, we find that the ionic strength of the medium may indeed play a role. The actual factors leading to aggregation and precipitation, in the case of α-lipoic acid-capped Au and Ag nanoparticles, may well involve a combination of both of these factors.

Conclusions

The present study shows three distinct regimes of interactions of metal ions with α-lipoic acid-capped Ag nanoparticles, the regime of low metal-ion concentration being associated with reversible chelation and dampening of the plasmon band. Aggregation and irreversible chelation accompanied by an increase in the flocculation parameter characterize the regime of intermediate metal-ion concentration. Further increases in metal-ion concentration result in precipitation.

References and Notes

(1) Link, S.; El-Sayed, M. A. *J. Phys. Chem. B* **2001**, *105*, 1; Link, S.; El-Sayed, M. A. *Int. Rev. Phys. Chem.* **2001**, *19*, 409.

(2) Mulvaney, P. *Langmuir* **1996**, *788*, 12.

(3) Seshadri, R.; Subbanna, G. N.; Vijayakrishnan, V.; Kulkarni, G. U.; Ananthakrishna, G.; Rao, C. N. R. *J. Phys. Chem. B* **1995**, *99*, 5639.

(4) Mirkin, C. A.; Letsinger, R. L.; Mucic, R. C.; Storhoff, J. J. *Nature (London)* **1996**, *382*, 607.

(5) Linnert, T.; Mulvaney, P.; Henglein, A. *J. Phys. Chem. B* **1993**, *97*, 679.

(6) Weisbecker, C. S.; Merritt, M. V.; Whitesides, G. M. *Langmuir* **1996**, *12*, 3763.

(7) Mayya, K. S.; Patil, V.; Sastry, M. *Langmuir* **1996**, *13*, 3944.

(8) Schmitt, H.; Badia, A.; Dickson, L.; Reven, L.; Lennox, R. B. *Adv. Mater. (Weinheim, Ger.)* **1998**, *10*, 475.

(9) Templeton, C.; Zamborini, F. P.; Wuelfing, W. P.; Murray, R. W. *Langmuir* **2000**, *16*, 6682.

(10) Kim, Y.; Johnson, R. C.; Hupp, J. T. *Nano Lett.* **2001**, *1*, 165.

(11) Simard, J.; Briggs, C.; Boal, A. K.; Rotello, V. M. *Chem. Commun.* **2000**, 1943.

(12) Maya, L.; Muralidharan, G.; Thundat, T. G.; Kenik, E. A. *Langmuir* **2000**, *16*, 9151.

(13) Throughout this paper, we make no distinction between aggregation, which is the close, irreversible association of primary particles, and agglomeration, which is loose, reversible association. The floucculation parameter used here takes into account the contribution from both aggregates and agglomerates and follows the definition of Weisbecker et al.[6]

(14) Lazarides, A. A.; Schatz, G. C. *J. Phys. Chem. B* **2000**, *104*, 460; Lazarides, A. A.; Schatz, G. C. *J. Chem. Phys.* **2000**, *112*, 2987.

(15) Strelow, F.; Henglein, A. *J. Phys. Chem. B* **1995**, *99*, 11834; Strelow, F.; Fojtik, A.; Henglein, A. *J. Phys. Chem. B* **1994**, *98*, 3032; Henglein, A.; Meisel, D. *J. Phys. Chem. B* **1998**, *102*, 8364.

V. Nanotubes and Nanowires

C.N.R. Rao

CSIR Centre of Excellence in Chemistry,
Chemistry & Physics of Materials Unit
Jawaharlal Nehru Centre for Advanced Scientific Research
Jakkur P.O., Bangalore-560 064, INDIA
cnrrao@jncasr.ac.in

Soon after the laboratory synthesis of fullerenes in 1990, carbon nanotubes were discovered by Iijima. The discovery of the carbon nanotubes has given great impetus to investigate various one-dimensional materials.[1] The author initiated investigations of carbon nanotubes (CNTs) in the early 1990s. These include synthesis, opening, filling and functionalization of nanotubes. An important discovery was the precursor synthesis of carbon nanotubes by employing organometallics. This method also yielded aligned CNT bundles. Doping CNTs with nitrogen was accomplished for the first time in 1997, followed by boron doping. Several properties of CNT's, including those of single-walled nanotubes, were investigated, specially noteworthy being the hydrogen-uptake. In this section, representative articles covering some of the above investigations of CNTs are included.

In the last few years, nanotubes of several inorganic layered materials such as MoS_2, NbS_2 and BN have been prepared and characterized. The area of inorganic nanotubes has been reviewed in one of the articles presented here.[2]

Nanowires constitute a major family of one-dimensional materials. Besides confinement, they exhibit several interesting properties. Inorganic nanowires can be prepared by several methods, of which the use of CNTs is an important one.[1] A few articles dealing with the synthesis and characterization of nanowires of metals, oxides and nitrides are presented.[3]

References

1. C.N.R. Rao, B.C. Satishkumar, A. Govindaraj and M.Nath, *ChemPhysChem* **2**, 78 (2001).
2. C.N.R. Rao and M. Nath, *Dalton Transactions* 1 (2003).
3. C.N.R. Rao, F.L. Deepak, G. Gundiah and A. Govindaraj, *Progress in Solid State Chemistry* (2003).

Acc. Chem. Res. **2002**, *35*, 998–1007

ARTICLES

Carbon Nanotubes from Organometallic Precursors

C. N. R. RAO* AND A. GOVINDARAJ

Chemistry and Physics of Materials Unit and CSIR Centre of Excellence in Chemistry, Jawaharlal Nehru Centre for Advanced Scientific Research, Jakkur P. O., Bangalore 560 064, India

Received January 7, 2002

ABSTRACT

Multiwalled as well as single-walled carbon nanotubes are conveniently prepared by the pyrolysis of organometallic precursors such as metallocenes and phthalocyanines in a reducing atmosphere. More importantly, pyrolysis of organometallics alone or in mixture with hydrocarbons yields aligned nanotube bundles with useful field emission and hydrogen storage properties. By pyrolysis of organometallics in the presence of thiophene, Y-junction nanotubes are obtained in large quantities. The Y-junction tubes have a good potential in nanoelectronics. Carbon nanotubes prepared from organometallics are useful to prepare nanowires and nanotubes of other materials such as BN, GaN, SiC, and Si_3N_4.

Introduction

Carbon nanotubes were discovered as a microscopic miracle in the cathode deposits obtained in the arc evaporation of graphite.[1] The arc method has since been improved and modified to obtain good yields of both multiwalled and single-walled nanotubes. Carbon nanotubes are conveniently obtained by carrying out the pyrolysis of hydrocarbons such as ethylene and acetylene over nanoparticles of iron, cobalt, and other metals dispersed over a solid substrate.[2–4] The presence of nanoparticles is essential not only to form nanotubes but also to control the diameter of the nanotube to some extent.[5] Since a carbon source as well as metal nanoparticles is necessary for producing carbon nanotubes by the pyrolysis of hydrocarbons, the strategy of employing an appropriate organometallic precursor which can serve as a dual source of both the carbon and the metal nanoparticles was explored by us. The very first experiments carried out on the pyrolysis of organometallic precursors

such as metallocenes and iron pentacarbonyl were successful in producing multiwalled nanotubes.[6,7] We have employed this method not only to produce multiwalled and single-walled nanotubes but also to make aligned nanotube bundles and Y-junction nanotubes. In this Account, we present the salient features of the various types of nanotubes obtained by us in Bangalore by employing the organometallic route and also examine some of the properties of the nanotubes so produced. Aligned nanotube bundles are expected to have applications in electronic displays and hydrogen storage while Y-junction nanotubes could be useful as building blocks in nanoelectronics. The nanotubes produced from organometallics are also usefully employed as starting materials to prepare other types of nanostructures.

Multiwalled and Single-Walled Nanotubes

Sen et al.[6] prepared multiwalled carbon nanotubes (MWNTs) and metal-filled onionlike structures by the pyrolysis of metallocenes such as ferrocene, cobaltocene, and nickelocene under reducing conditions, wherein the precursor acts as the source of the metal as well as carbon. The pyrolysis setup consists of stainless steel gas flow lines and a two-stage furnace system fitted with a quartz tube (25 mm i.d.) as shown in Figure 1a, the flow rate of the gases being controlled by the use of mass flow controllers. In a typical preparation, a known quantity (100 mg) of the metallocene (presublimed 99.99% purity) is taken in a quartz boat and placed at the center of the first furnace, and a mixture of Ar and H_2 of the desired composition is passed through the quartz tube. The metallocene is sublimed by raising the temperature of the first furnace to 200 °C at a controlled heating rate (20 °C/min). The metallocene vapor so generated is carried by the Ar–H_2 gas stream into the second furnace, maintained at 900 °C, where the pyrolysis occurs. The main variables in the experiment are the heating rate of ferrocene, the flow rate of Ar gas, and the pyrolysis temperature. Ferrocene vapor carried by a 75% Ar + 25% H_2 mixture at a flow rate of 900 sccm (sccm = standard cubic centimeter per minute) into the furnace yields large quantities of carbon deposits, mainly containing carbon nanotubes. These deposits are examined by a scanning electron microscope (SEM) and transmission electron microscope (TEM). In Figure 2a, we show a TEM image of nanotubes produced by the pyrolysis of ferrocene. To increase the yield of the multiwalled carbon nanotubes with metallocene, vapors of an additional hydrocarbon source were mixed along with the metallocene vapor in the first furnace. Thus, pyrolysis of benzene in the presence of ferrocene or $Fe(CO)_5$ gives high yields of multiwalled nanotubes, the wall thickness of the nanotubes depending on the proportion of the carbon

C. N. R. Rao obtained his Ph.D. degree from Purdue University and D.Sc. degree from the University of Mysore. He is Linus Pauling Research Professor at the JNCASR and Honorary Professor at the Indian Institute of Science, Bangalore, India. He is a member of several academies including the Royal Society (London), U.S. National Academy of Sciences, French Academy of Sciences, and Japan Academy. His main research interests are in solid state and materials chemistry and nanomaterials. The most recent award received by him is the Hughes medal for physical sciences from the Royal Society.

A. Govindaraj obtained his Ph.D. degree from the University of Mysore and is a Scientific Officer at the Indian Institute of Science, Bangalore, India, and Honorary Senior Research Officer at the JNCASR. His main research interests are nanotubes and fullerenes. He is a recipient of the MRS (India) medal.

* To whom correspondence should be addressed. Fax: 91-80-8462760. E-mail: cnrrao@jncasr.ac.in.

10.1021/ar0101584 CCC: $22.00 © 2002 American Chemical Society
Published on Web 05/18/2002

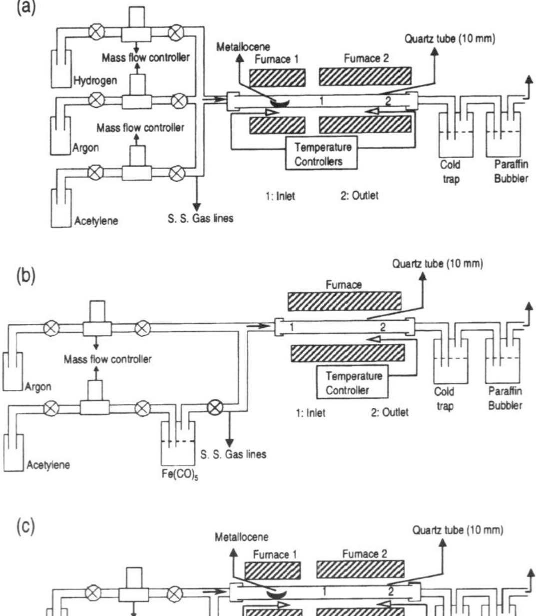

FIGURE 1. Pyrolysis apparatus employed for the synthesis of carbon nanotubes by pyrolysis of mixtures of (a) metallocene + C_2H_2, (b) $Fe(CO)_5$ + C_2H_2, and (c) metallocene + benzene or thiophene. The numbers 1 and 2 indicated in the figure represents inlet and outlet, respectively.[6]

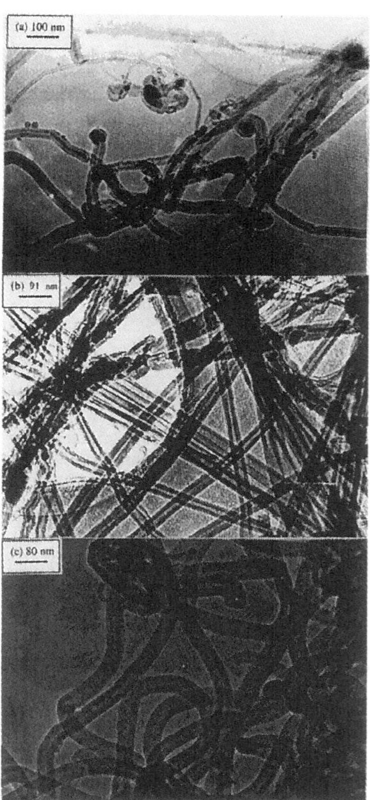

FIGURE 2. TEM image of a multiwalled carbon nanotube obtained by pyrolysis of (a) ferrocene at 900 °C in a mixture of 75% Ar/25% H_2 at a flow rate of 900 sccm, (b) a mixture of C_2H_2 (25 sccm) and ferrocene at 1100 °C at 1000 sccm Ar flow, and (c) a mixture of nickelocene and benzene at 900 °C in 85% Ar and 15% H_2 mixture at a flow rate of 1000 sccm.[6,7]

source and the metal precursor.[7] In Figure 2b, we show a TEM image of nanotubes obtained by the pyrolysis of a mixture of C_2H_2 (25 sccm) and ferrocene at 1100 °C in a Ar flow rate of 1000 sccm. The image clearly reveals that the addition of hydrocarbon not only increases the yield of hollow MWNTs but also reduces the amount of carbon-coated metal nanoparticles. In Figure 2c we show a TEM image of MWNTs obtained by the pyrolysis of a mixture of nickelocene and benzene at 900 °C in 85% Ar and 15% H_2 mixture at a flow rate of 1000 sccm. Besides metallocenes, one can also employ metal phthalocyanines as precursors to prepare MWNTs.[8,9]

To prepare single-walled nanotubes (SWNTs), alternate synthetic strategies have been explored. Under controlled conditions of pyrolysis, dilute hydrocarbon−organometallic mixtures yield SWNTs.[10,11] High-resolution TEM images of SWNTs, obtained by the pyrolysis of a nickelocene−acetylene mixture at 1100 °C,[11] are shown in the Figure 3a,b. The diameter of the SWNT in Figure 3a is 1.4 nm. It may be recalled that the pyrolysis of nickelocene in admixture with benzene under similar conditions primarily yields MWNTs. Acetylene appears to be a better carbon source for the preparation of SWNTs, since it contains a smaller number of carbon atoms. The bottom portion of the SWNT in Figure 3a shows an amorphous

carbon coating around the tube, commonly observed in many of the preparations. This can be avoided by reducing the proportion of the hydrocarbon and mixing hydrogen in the Ar stream. Pyrolysis of cobaltocene and acetylene under similar conditions gives rise to isolated SWNTs. In Figure 3c,d, we show the high-resolution electron microscope (HREM) images of the SWNTs obtained by the pyrolysis of ferrocene with methane at 1100 °C. Surprisingly, the pyrolysis of nickelocene or cobaltocene in admixture with methane under similar conditions did not give SWNTs in good yield. Pyrolysis of binary mixtures (1:1 by weight) of the metallocenes along with acetylene gives good yields of SWNTs, due to the beneficial effect of binary alloys.[12]

SWNTs are obtained in good yields by the pyrolysis of acetylene in mixture with $Fe(CO)_5$ at 1100 °C employing a setup of the type shown in Figure 1b. We show TEM images of the SWNTs so obtained in Figure 4. Pyrolysis of ferrocene−thiophene mixtures yield SWNTs, but the yield is somewhat low. Pyrolysis of a mixture of benzene and thiophene along with ferrocene (Figure. 1c), however, gives a high yield of SWNTs.[13]

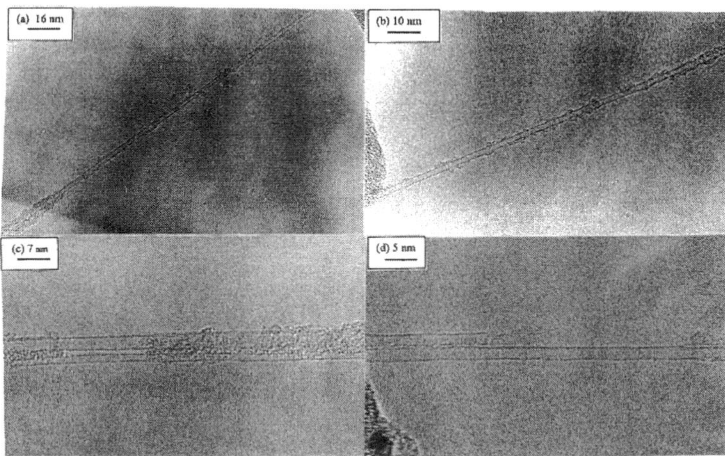

FIGURE 3. (a, b) HREM images of SWNTs obtained by the pyrolysis of nickelocene and C_2H_2 at 1100 °C in a flow of Ar (1000 sccm) with C_2H_2 flow rate of 50 sccm. (c, d) HREM images of SWNTs obtained by the pyrolysis of ferrocene and CH_4 at 1100 °C in a flow of Ar (990 sccm) with CH_4 flow rate of 10 sccm.[11]

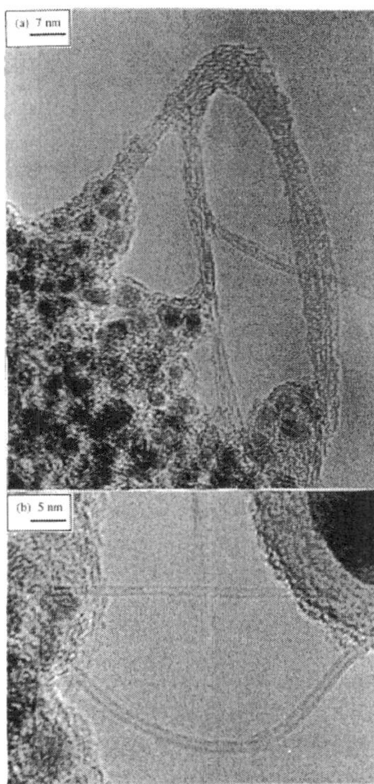

FIGURE 4. (a, b) TEM images of SWNTs obtained by the pyrolysis of $Fe(CO)_5$ and C_2H_2 (flow rate = 50 sccm) at 1100 °C in Ar (flow rate = 1000 sccm) flow.[10,11]

TEM examination of the various carbonaceous products obtained from the pyrolysis of hydrocarbons and organometallic precursors indicates that the size of the catalyst particle plays an important role with regard to the nature of the product. In the case of organometallic precursors, it seems that metal nanoclusters of ∼1 nm diameter are produced under controlled conditions. When the concentration of the organometallic precursor is high, MWNTs are formed around the metal particles of 5−20 nm diameter. This is true of carbon nanotubes obtained by the metallocene route.[6,7] In the higher size range of ≥50 nm, graphite-covered metal particles are formed predominantly.[15]

Aligned Carbon Nanotube Bundles

Since the pyrolysis of mixtures of organometallic precursors and hydrocarbons yields good quantities of multi-walled nanotubes,[6,7] we considered the possibility of obtaining aligned nanotube bundles under appropriate conditions. For this purpose, we carried out pyrolysis of metallocenes along with other hydrocarbon sources, in the apparatus shown in Figure 1a.[10,15,16] To obtain aligned nanotube bundles, a typical heating rate 50 °C/min of the first furnace and an Ar flow rate of 1000 sccm have been employed. Compact aligned nanotube bundles could be obtained by introducing C_2H_2 (50−100 sccm) during the sublimation of ferrocene. Scanning electron microscope images of aligned nanotubes obtained by the pyrolysis of ferrocene are shown in Figure 5. The image in Figure 5a shows the top view of the aligned nanotubes, wherein the nanotube tips are seen. The image in Figure 5b shows the side view. Aligned nanotubes obtained by the pyrolysis of ferrocene with different alkanes are shown in the SEM images in Figure 6. The average length of the nanotubes is generally around 60 μm with methane and acetylene. In the case of methane, the nanotubes are aligned, but the packing density is not high. The nanotube bundles obtained with ferrocene + acetylene mixtures appear to have a packing density greater than that obtained with Fe/silica catalysts.[17] SEM images of large bundles of the aligned nanotubes are shown in Figure 7. A small proportion of graphite-covered metal nanoparticles is often present along with the nanotubes. In Table 1, we sum-

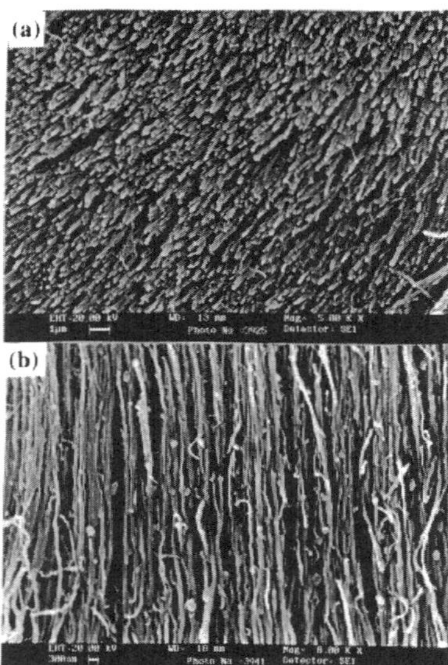

FIGURE 5. SEM images of aligned carbon nanotubes obtained by the pyrolysis of ferrocene. (a) and (b) show views of the aligned nanotubes along and perpendicular to the axis of the nanotubes, respectively.[16]

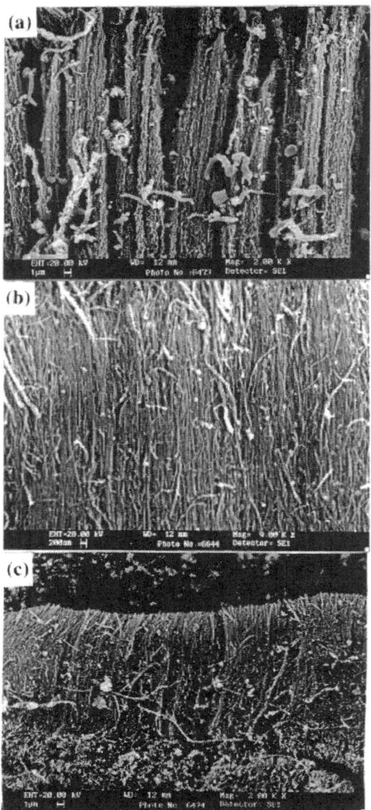

FIGURE 6. SEM images showing the bundles of aligned nanotubes obtained by the pyrolysis of ferrocene along with (a) methane, (b) acetylene, and (c) butane (hydrocarbon at 50 sccm) at 1100 °C in a Ar flow of 950 sccm.[15]

marize the products obtained by the pyrolysis of hydrocarbon + organometallic mixtures at 1100 °C in a stream of Ar + H_2.

Andrews et al.[18] have carried out the pyrolysis of ferrocene—xylene mixtures to obtain aligned carbon nanotubes. Pyrolysis of Fe(II) phthalocyanine also yields aligned nanotubes.[19] Chen et al.[20] have grown three-dimensional micropatterns of well-aligned carbon nanotubes on photolithographically prepatterned substrates, by the pyrolysis of iron(II) phthalocyanine in an Ar/H_2 atmosphere around 950 °C. They achieved the photopatterning by photolithographic cross-linking of a chemically amplified photoresist layer spin-cast on a quartz plate or a silicon wafer coupled with solution development. Owing to the appropriate surface characteristics, the patterned photoresist layer supports aligned nanotube growth. The difference in the chemical nature between the surfaces covered and uncovered by the photoresist film causes a region-specific growth of nanotubes with different tubular lengths and packing densities leading to the formation of three-dimensional aligned nanotube patterns.

Considering all aspects, we feel that the pyrolysis of organometallic precursors is the most convenient means of preparing aligned MWNTs. The other methods reported in the literature, such as hydrocarbon pyrolysis on patterned metal films,[21] are more difficult and are not amenable for large-scale synthesis. The advantage of the precursor method is that the aligned bundles are produced in one step, at a relatively low cost, without prior preparation of substrates. It is possible that the self-assembly of carbon nanotubes from precursor pyrolysis is influenced by the transition metal particles which take part in the nucleation and growth of the nanotubes. The metal often gets encapsulated to form nanorods or nanoparticles inside the carbon nanotubes. Ferromagnetism of these metal particles is also a property of significance. The iron nanorods encapsulated in the nanotubes exhibit a complex behavior with respect to magnetization reversal and could be useful as probes in magnetic force micros-

Table 1. Products Obtained by the Pyrolysis of Hydrocarbon—Organometallic Mixtures at 1100 °C

organometallic precursor	hydrocarbon	inlet	outlet
ferrocene	acetylene	aligned MWNTs, nanorods	metal nanoparticles*
nickelocene	acetylene	MWNTs	SWNTs
cobaltocene	acetylene	MWNTs	SWNTs
ferrocene + nickelocene	acetylene	aligned MWNTs	SWNTs
ferrocene + cobaltocene	acetylene	aligned MWNTs	SWNTs
nickelocene + cobaltocene	acetylene	MWNTs	SWNTs
Fe(CO)$_5$	acetylene	MWNTs	SWNTs (bundles)

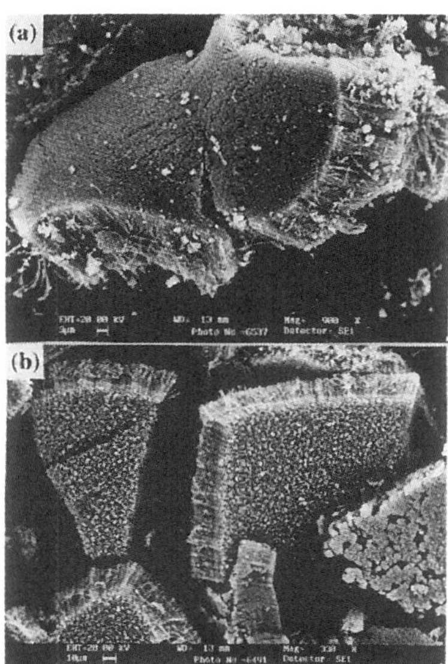

FIGURE 7. SEM images showing the bundles of aligned nanotubes obtained by pyrolyzing ferrocene along with (a) acetylene and (b) butane.[15]

copy. Aligned carbon nanotubes are potential candidates for use as field emitters,[14] and the easy synthesis from organometallic precursors is therefore of importance.

Y-Junction Carbon Nanotubes

Device miniaturization in semiconductor technology is expected to reach its limits due to the inherent quantum effects as one goes toward smaller size. In such a scenario, an alternative would be nanoelectronics based on molecules. The possible use of carbon nanotubes in nanoelectronics has aroused considerable interest. For such applications it is important to be able to connect the nanotubes of different diameters and chirality.[22–24] Complex three-point nanotube junctions have been proposed as the building blocks of nanoelectronics, and in this regard Y- and T-junctions have been considered as prototypes.[25,26] While we would expect an equal number of five- and seven-membered rings to create nanotube junctions, it appears that they can be created with an equal number of five- and eight-membered rings as well.[26] To date, there have been no practical devices made of real three-point nanotube junctions. However, junctions consisting of crossed nanotubes have been fabricated to study their transport characteristics.[27] Y-junction nanotubes have been produced by using Y-shaped nanochannel alumina as templates.[28] We have prepared Y-junction nanotubes in large quanties by carrying out the pyrolysis of a mixture of a metallocene with thiophene.[29,30]

The experimental setup employed by us for the synthesis of the Y-junction nanotubes employed a two-stage furnace system similar to that described earlier[29] (Figure 1c). A known quantity of metallocene was sublimed in the first furnace and carried along with a flow of argon (Ar)

gas to the pyrolysis zone in the second furnace. Simultaneously hydrogen was bubbled through thiophene and was mixed with the argon–metallocene vapors at the inlet of the furnace and carried to the pyrolysis zone. Pyrolyzing the mixed vapors at 1000 °C yielded Y-junction nanotubes in plenty. Pyrolysis of Ni/Fe phthalocyanine in mixture with thiophene was carried out in a similar manner taking phthalocyanine in place of the metallocenes to obtain good Y-junction nanotubes. Pyrolysis of various organometallics with sulfur-containing compounds has shown that pyrolysis of thiophene with nickelocene, ferrocene, and cobaltocene yields excellent Y-junction nanotubes. A TEM image of a Y-junction nanotube obtained by the pyrolysis of nickelocene/thiophene mixture is shown in Figure 8a. A TEM image revealing the presence of several Y-junction carbon nanotubes is shown in Figure 8b. Many of the nanotubes show multiple Y-junctions.

Pyrolysis of thiophene with Fe or Ni phthalocyanine or iron pentacarbonyl also yields Y-junction nanotubes. In Figure 9a, we show few Y-junctions, where as in Figure 9b we show the multiple junctions formed continuously by the pyrolysis of thiophene with Ni phthalocyanine. While the pyrolysis of nickelocene with CS_2 yields similar junctions, the yield and quality of the nanotubes is not satisfactory. At higher flow rates of CS_2, with ferrocene it pyrolyses and gives Y-junction carbon fibers. In Figure 9c we show the Y-junctions obtained by the pyrolysis of thiophene with Fe phthalocyanine. Pyrolysis of thiophene with $Fe(CO)_5$ carried out by bubbling H_2 (50–100 sccm) through the pentacarbonyl along with Ar (150–200 sccm) bubbled through thiophene showed the presence of interesting junction structures as revealed in Figure 9d. The pyrolysis of thiophene over a $Ni(Fe)/SiO_2$ catalyst provides an alternative procedure to the metallocene route, but the yield of Y-junction tubes is rather low. The availability of large quantities of Y-junctions should render them useful for exploitation in nanoelectronics. Compositional analysis at the Y-junction has shown absence of sulfur indicating that the junction is formed by the curvature caused by different carbon rings. The metal nanoparticles abstract the sulfur from thiophene forming a sulfide, the remaining carbon fragment probably being involved in ring formation.

HREM images show that the graphitic layers bend parallely with respect to the junction in many of these nanotubes. Our studies in collaboration with Torsteen Seeger and Manfred Ruhle have shown that the Y-juction tubes are entirely composed of carbon with no sulfur impurity. These observations suggest the presence of equal numbers of five- and seven-/eight-membered rings at the junctions. The metal particles formed in the pyrolysis contain both sulfur and carbon.

Scanning tunneling spectroscopic studies of Y-junction carbon nanotubes show interesting diodelike device characteristics at the junctions. A typical $I-V$ curve obtained from positioning the tip atop a Y-junction (the point of contact between the three arms) as well as on the individual arms of the Y-junction is shown in Figure 10. The $I-V$ plot at the junction is asymmetric (Figure 10b)

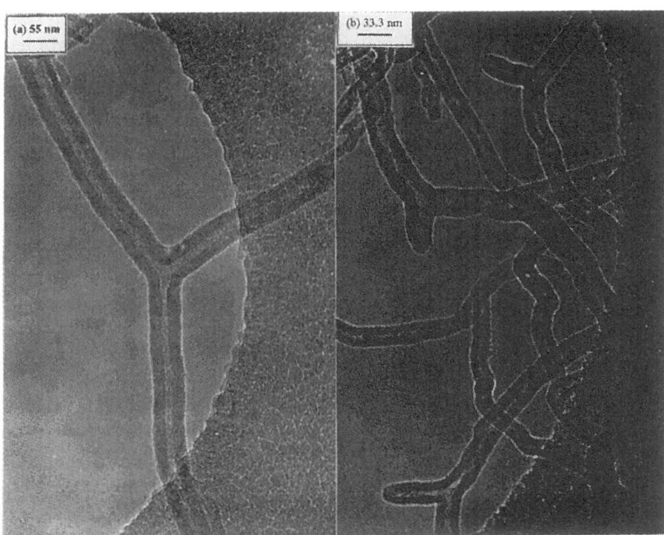

FIGURE 8. (a, b) TEM images of Y-junction carbon nanotubes obtained by the pyrolysis of nickelocene and thiophene at 1000 °C.[29]

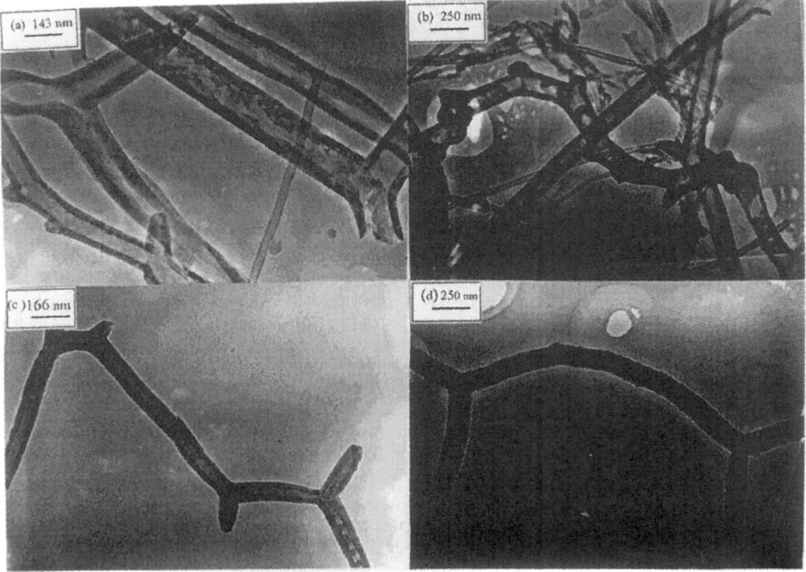

FIGURE 9. TEM images of Y-junction nanotubes: (a, b) obtained by the pyrolysis of thiophene along with Ni phthalocyanine at 1000 °C; (c) obtained by the pyrolysis of thiophene along with Fe phthalocyanine at 1000 °C; (d) obtained by the pyrolysis of an $Fe(CO)_5$–thiophene mixture at 1000 °C.[30]

with respect to bias polarity, unlike that along the arm (Figure 10a). The insets in Figure 10 gives a plot of differential conductance vs bias with respect to zero bias which is symmetric in Figure 10a and asymmetric in Figure 10b. Such asymmetry is characteristic of a junction diode, and this in turn indicates the existence of intramolecular junctions in the carbon nanotubes. The findings discussed above open up the possibility of assembling carbon nanotubes possessing novel devicelike properties[29,31-33] into multifunctional circuits and ultimately toward the realization of a carbon nanotube based

computer chip. Rueckes et al.[34] have described the concept of carbon-nanotube-based nonvolatile random access memory for molecular computing. The viability of the concept has been demonstrated.

Nanorods, Nanowires, and Nanotubes of Other Materials

Preparation of metal nanorods covered by carbon has been reported in the recent literature.[10,35-37] The organometallic precursor route provides a means of preparing

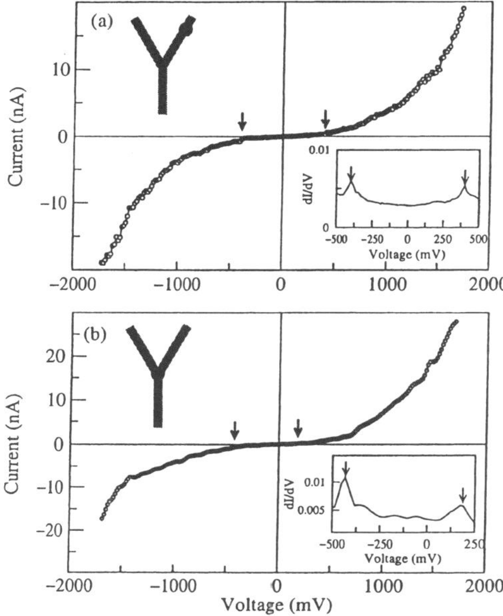

FIGURE 10. $I-V$ curves collected from different points in a Y-junction carbon nanotube (a) from one of the arms away from the junction and (b) from the junction of the three arms. Inset show plots of dI/dV vs bias voltage. The observed gaps are indicated by arrows.[29]

metal nanorods. The pyrolysis of ferrocene + hydrocarbon mixtures or of ferrocene alone yields iron nanorods encapsulated inside the carbon nanotubes as evidenced from TEM, the proportion of the nanorods depending on the proportion of ferrocene. Typical TEM images of such nanorods are shown in Figure 11a,b. The selected area electron diffraction (SAED) pattern of the nanorods (inset in Figure 11c) shows spots due to (010) and (011) planes of α-Fe. The HREM image of the iron nanorod in Figure 11c shows well-resolved (011) planes of α-Fe in single-crystalline form. X-ray diffraction patterns show the presence of α-Fe with a small portion of Fe_3C as the minor phase. In addition to the nanorods, iron nanoparticles (20−40 nm diameter) encapsulated inside the graphite layers are also obtained in many of the preparations. The iron nanorods and nanoparticles are well protected against oxidation by the graphitic layers. Ni/Fe phthalocyanines can be used in the place of metallocenes to prepare metal nanowires. However, the yields are not as good as in the case of metallocenes.

There are several reports on the preparation of SiC nanowires in the literature but fewer on the preparation of Si_3N_4 nanowires.[38,39] The methods employed for the synthesis of SiC nanowires have been varied. Since both SiC and Si_3N_4 are products of the carbothermal reduction of SiO_2, it should be possible to establish conditions wherein one set of specific conditions favor one over the other. We have been able to prepare Si_3N_4 nanowires,[40] by reacting multiwalled carbon nanotubes produced by ferrocene pyrolysis with ammonia and silica gel at 1360

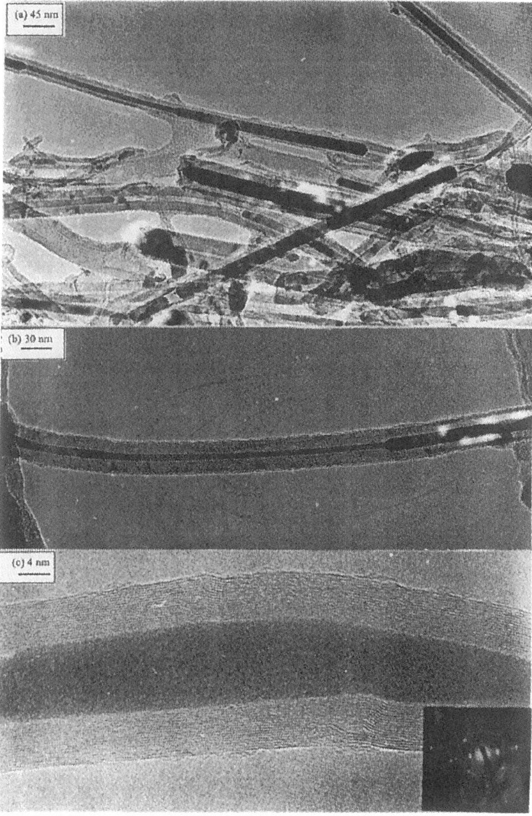

FIGURE 11. (a, b) TEM images of the iron nanorods encapsulated inside the carbon nanotubes from aligned nanotube bundles. (c) HREM image of a single-crystal iron nanorod encapsulated inside a carbon nanotube. The inset in (c) represents the selected area electron diffraction (SAED) pattern of a iron nanorod.[16,35]

°C. The MWNTs were used as a carbon source in the carbothermal reduction because they have higher thermal stability compared to activated carbon. The reaction of MWNTs with silica gel and NH_3 at 1360 °C yields a mixture of α- and β-Si_3N_4. SEM images of the product obtained by this reaction showed the nanowires to have large diameters (5−7 μm), with lengths of the order of hundreds of micrometers (Figure 12a). By addition of catalytic iron particles (0.5 at. %), we obtain β-SiC nanowires under the same conditions[40] (Figure 12b).

Several chemical methods of preparing boron nitride nanotubes and nanowires have been investigated.[41] The general methods involve reacting boric acid with ammonia in the presence of MWNTs. Good yields of clean BN nanotubes are obtained with MWNTs. Aligned BN nanotubes are obtained by heating aligned bundles of MWNTs with boric acid in the presence of ammonia at the 1000°− 1300 °C range. The SEM images of aligned BN nanotube bundles shown in Figure 13 a,b clearly reveal BN nanotubes aligned in two different orientations. The carbon MWNTs appear to not only take part in the reaction but also serve as templates. The average outer diameter of the

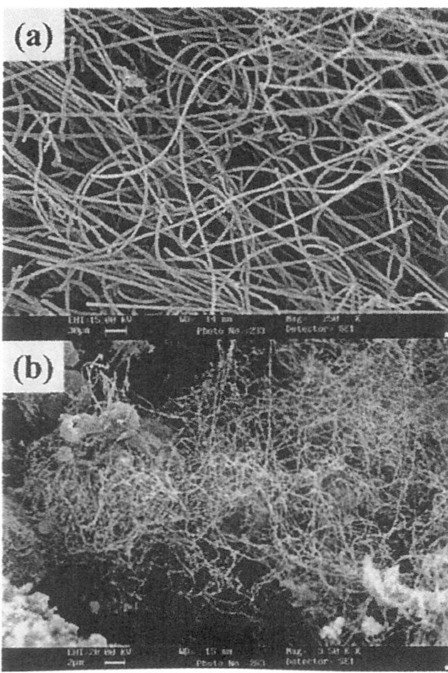

FIGURE 12. (a) SEM image of Si_3N_4 nanowires obtained by the reaction of aligned multiwalled nanotubes (produced by metallocene route) with silica gel at 1360 °C. (b) SiC nanowires.[40]

aligned BN nanotubes varies from 15 to 40 nm as revealed by the TEM image in Figure 13c. This suggests that, during the formation of BN nanotubes, the carbon MWNTs not only take part in the reaction but also serve as templates. It is noteworthy that the BN nanotubes can be produced at a temperature as low as 1000 °C by this procedure. On the basis of elemental analysis and X-ray diffraction, it is found that the carbon content of the BN nanotubes becomes marginal if the initial proportion of carbon nanotubes is kept low.

Gallium nitride nanowires have been prepared by us by employing several procedures involving the use of carbon nanotube templates.[42] We have employed gallium acetylacetonate as the precursor for the in-situ production of small particles of the oxide (GaO_x) which then react with NH_3 vapor around 900 °C in the presence of nanotubes. When multiwalled carbon nanotubes prepared by arc discharge are used as templates, the yield of the GaN nanowires is excellent and the diameter of the majority of the nanowires is in the 35–100 nm range. The length of the nanowires extends to a few micrometers. The linear nanowires are generally single crystalline, showing a layer spacing of 0.276 nm corresponding to the [100] planes. This variation in the diameter of the nanowires occurs because of the nonuniformity in the diameter of the nanotubes. The diameter of the nanowires could be reduced to 20 nm by using single-walled nanotubes in place of multiwalled nanotubes. The growth direction of the nanowires is nearly perpendicular to the [100] planes. The nanowires show satisfactory photoluminescence characteristics similar to those of bulk GaN. The non-carbon

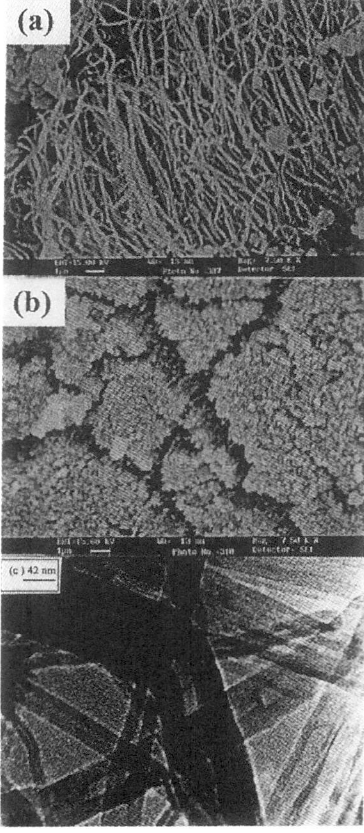

FIGURE 13. (a, b) SEM images of BN nanotubes obtained by heating aligned bundles of MWNTs with boric acid in the presence of ammonia at 1300 °C. (c) TEM image of the BN nanotubes obtained by the same procedure as in (a) and (b).[41]

nanowires and nanotubes have potential applications. For example, GaN nanowires, suitably doped, can have uses in nanoelectronics and in optical devices. BN nanotubes and Si_3N_4 nanowires are useful ceramic materials.

Properties

Field emission properties of carbon nanotubes have direct applications in vacuum microelectronic devices.[43–46] We have found that carbon nanotubes produced by the pyrolysis of ferrocene on a pointed tungsten tip exhibit high emission current densities with good performance characteristics.[47] In Figure 14a we show a typical $I–V$ plot for the carbon nanotube covered tungsten tip for currents ranging from 0.1 nA to 1 mA. The applied voltage was 4.3 kV for a total current of 1 μA and 16.5 kV for 1000 μA. The Fowler–Nordheim (F–N) plot shown in Figure 14b has two distinct regions. The behavior is metal-like in the low-field region, while it saturates at higher fields as the voltage is increased. We have obtained a field emission current density of 1.5 A cm^{-2} at a field of 290 V/mm, a value considerably higher than that found with planar cathodes.

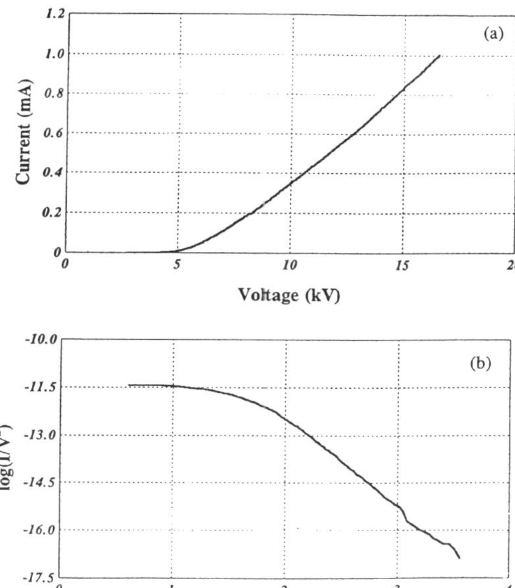

FIGURE 14. *I—V* characteristics showing field emission currents in the range of 0.1 nA to 1 nA. (b) Fowler—Nordheim plot corresponding to the data in (a).[47]

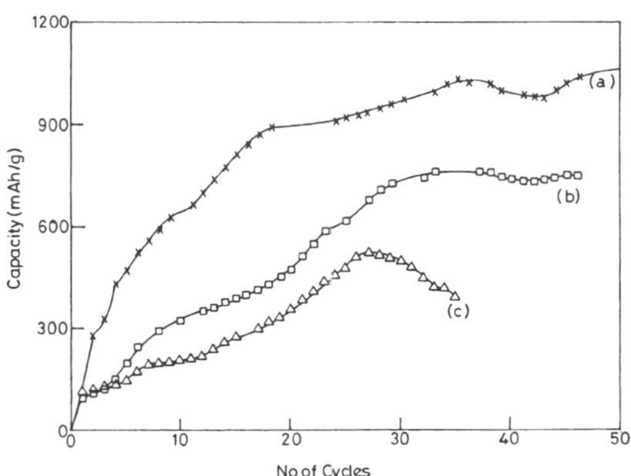

FIGURE 15. Comparison of electrochemical charging capacity of hydrogen of (a) aligned MWNT bundles, (b) SWNTs, and (c) MWNTs (arc-generated).[49]

Accordingly, the field enhancement factor calculated from the slope of the F—N plot in the low-field region is also large. The field emission micrographs reveal the lobe structure symmetries typical of carbon nanotube bundles. The emission current is remarkably stable over an operating period of more than 3 h for various current values in the 10—500 mA range. The relative fluctuations decrease with increasing current level, and the emitter can be operated continuously at the high current levels for at least 3 h without any degradation in the current. It appears that aligned carbon nanotubes from ferrocene pyrolysis hold promise as good field electron emission sources.

Carbon nanotubes are considered to be good hosts for hydrogen storage, although there is some controversy about the magnitude of the hydrogen uptake.[35,48] High-pressure adsorption experiments carried out by us with G. Gundiah show that the storage capacity of compact aligned nanotubes prepared by the pyrolysis of ferrocene— hydrocarbon mixtures is between 3 and 4 wt % (143 bar, 27 °C). Electrochemical hydrogen storage in these nanotubes is also substantial. In Figure 15 a—c we show (for comparison) the plots of electrochemical charging capacity of aligned nanotubes, SWNTs, and MWNTs (arc-generated), respectively.[49] Electrodes made out of aligned MWNTs clearly demonstrate higher electrochemical charging capacities up to 1100 mA h g^{-1} which correspond to a hydrogen storage capacity of 3.75 wt %. SWNTs and MWNTs (arc-generated), however, show capacity in the range of 2—3 wt %.

Concluding Remarks

Although arc evaporation of graphite has traditionally been found to yield both single-walled and multiwalled carbon nanotubes, the pyrolysis of organometallic precursors with or with out the presence of additional carbon sources seems to provide a direct and effective method of producing nanotubes of various kinds. A particularly important finding is the one-step synthesis of aligned carbon nanotubes and Y-junction nanotubes which cannot be made by arc evaporation or other methods easily. It is also noteworthy that nanotubes produced by organometallic precursors may also find applications in field emission and hydrogen storage. The successful synthesis of nanowires of gallium nitride and silicon nitride and of boron nitride nanotubes by using carbon nanotubes produced from organometallic precursors is also of interest.

The authors thank the Department of Science and Technology, Government of India, and the DRDO (India) for supporting this research. They acknowledge productive collaboration with Dr. B. C. Satishkumar, Dr. R. Sen, Mr. F. L. Deepak, and Mr. G. Gundiah.

References

(1) Iijima, S. Helical microtubules of graphitic carbon. *Nature* **1991**, *354*, 56.
(2) Jose-Yacaman, M.; Miki-Yoshida, M.; Rendon, L.; Santiesteban, T. G. Catalytic growth of carbon microtubles with fullerene structure. *Appl. Phys. Lett.* **1993** *62*, 202.
(3) Ivanov, V.; Nagy, J. B.; Lambin, Ph.; Lucas, A.; Zhang, X. B.; Zhang, X. F.; Bernaerts, D.; Van Tendeloo, G.; Amelinckx, S.; Van Landuyt, J. The study of carbon nanotubules produced by catalytic method. *Chem. Phys. Lett.* **1994**, *223*, 329.
(4) Hernadi, K.; Fonseca, A.; Nagy, J. B.; Bernaerts, D.; Riga, J.; Lucas, A. Catalytic synthesis and purification of carbon nanotubes. *Synth. Met.* **1996** *77*, 31.
(5) Rodriguez, N. M. A review of catalytically grown carbon nanofibers. *J. Mater. Res.* **1993** *8*, 3233.
(6) Sen, R.; Govindaraj, A.; Rao, C. N. R. Carbon nanotubes by metallocene route. *Chem. Phys. Lett.* **1997**, *267*, 276.
(7) Sen, R.; Govindaraj, A.; Rao, C. N. R. Metal-filled and hollow carbon nanotubes obtained by the decomposition of metal containing free precursor molecules. *Chem. Mater.* **1997**, *9*, 2078.
(8) Yudasaka, M.; Kikuchi, R.; Ohki, Y.; Yoshimura, S. Nitrogen-containing carbon nanotube growth from Ni-phthalocyanine by chemical vapor deposition *Carbon* **1997**, *35*, 195.
(9) Fan, S.; Chapline, M. C.; Franklin, N. R.; Tombler, T. W.; Cassel, A. M.; Dai, H. Self-oriented regular arrays of carbon nanotubes and their field emission devices. *Science* **1999** *283*, 512.

(10) Rao, C. N. R.; Govindaraj, A.; Sen, R.; Satishkumar, B. C. Synthesis of multiwalled and single-walled nanotubes, aligned bundles and nanorods by employing organometallic precursors. *Mater. Res. Innovations* **1998** *2*, 128.

(11) Satishkumar, B. C.; Govindaraj, A.; Sen, R.; Rao, C. N. R. Single-walled nanotubes by the pyrolysis of acetylene-organometallic precursors. *Chem. Phys. Lett.* **1998** *293*, 47.

(12) Seraphin, S.; Zhou, D. Single-walled carbon nanotubes produced at high yield by mixed catalysts. *Appl. Phys. Lett.* **1994** *64*, 2087.

(13) Cheng, H. M.; Li, F.; Su, G.; Pan, H. Y.; He, L. L.; Sun, X.; Dresselhaus, M. S. Large-scale and low-cost synthesis of single-walled carbon nanotubes by the catalytic pyrolysis of hydrocarbons. *Appl. Phys. Lett.* **1998** *72*, 3282.

(14) de Heer, W. A.; Bonard, J. M.; Fauth, K.; Chatelain, A.; Forro, L.; Ugarte, D. Electron field emitters based on carbon nanotube films. *Adv. Mater.* **1997**, *9*, 87.

(15) Satishkumar, B. C.; Govindaraj, A.; Rao, C. N. R. Bundles of aligned carbon nanotubes obtained by the pyrolysis of ferrocene-hydrocarbon mixtures: role of the metal nanoparticles produced in situ. *Chem. Phys. Lett.* **1999** *307*, 158.

(16) Rao, C. N. R.; Sen, R.; Satishkumar, B. C.; Govindaraj, A. Aligned nanotube bundles from ferrocene pyrolysis. *Chem. Commun.* **1998** 1525.

(17) Pan, Z. W.; Xie, S. S.; Chang, B. H.; Sun, L. F.; Zhou, W. Y.; Wang, G. Direct growth of aligned open carbon nanotubes by chemical vapor deposition. *Chem. Phys. Lett.* **1999** *299*, 97.

(18) Andrews, R.; Jacques, D.; Rao, A. M.; Derbyshire, F.; Qian, D.; Fan, X.; Dickey, E. C.; Chen, J. Continuous production of aligned carbon nanotubes: a step closer to commercial realization. *Chem. Phys. Lett.* **1999** *303*, 467.

(19) Huang, S.; Mau, A. W. H.; Turney, T. W.; White, P. A.; Dai, L. Patterned growth of well-aligned carbon nanotubes: A soft-lithographic approach *J. Phys. Chem. B* **2000** *104*, 2193.

(20) Chen. Q.; Dai, L. Three-dimensional micropatterns of well-aligned carbon nanotubes produced by photolithography. *J. Nanosci. Nanotechnol.* **2001** *1*, 43.

(21) Terrones, M.; Grobert, N.; Zhang, J. P.; Terrones, H.; Olivares, J.; Hsu, H. K.; Hare, J. P.; Cheetham, A. K.; Kroto, H. W.; Walton, D. R. M. Preparation of aligned carbon nanotubes catalysed by laser-etched cobalt thin films. *Chem. Phys. Lett.* **1998** *285*, 299.

(22) Chico, L.; Crespi, V. H.; Benedict, L. X.; Louie, S. G.; Cohen, M. L. Pure carbon nanoscale devices: Nanotube heterojunctions. *Phys. Rev. Lett.* **1996** *76*, 971.

(23) Kouwenhoven, L. Single-molecule transistors. *Science* **1997**, *275*, 1896.

(24) McEuen, P. L. Nanotechnology: Carbon-based electronics. *Nature (London)* **1998** *393*, 15.

(25) Menon, M.; Srivastava, D. Carbon nanotube "T-junctions": Nanoscale metal-semiconductor-metal contact devices. *Phys. Rev. Lett.* **1997**, *79*, 4453.

(26) Menon, M.; Srivastava, D. Carbon nanotube based molecular electronic devices. *J. Mater. Res.* **1998** *13*, 2357.

(27) Fuhrer, M. S.; Nygard, J.; Shih, L.; Forero, M.; Yoon, Y. G.; Mazzoni, M. S. C.; Choi, H. J.; Ihm, J.; Louie, S. G.; Zettl, A.; McEuen, P. L. Crossed nanotube junctions. *Science* **2000** *288*, 494.

(28) Li, J.; Papadopoulos, C.; Xu, J. Nanoelectronics: Growing Y-junction carbon nanotubes. *Nature (London)* **1999** *402*, 253.

(29) Satishkumar, B. C.; Thomas, P. J.; Govindaraj, A.; Rao, C. N. R. Y-junction carbon nanotubes. *Appl. Phys. Lett.* **2000** *77*, 2530.

(30) Deepak, F. L.; Govindaraj, A.; Rao, C. N. R. Synthetic strategies for Y-junction carbon nanotubes. *Chem. Phys. Lett.* **2001** *345*, 5.

(31) Tans, S. J.; Verschueren, A. R. M.; Dekker, C. Room-temperature transistor based on a single carbon nanotube *Nature* **1998** *393*, 49.

(32) Martel, R.; Schmidt, T.; Shea, H. R.; Hertel, T.; Avouris, Ph. Single- and multiwall carbon nanotube field-effect transistors. *Appl. Phys. Lett.* **1998** *73*, 2447.

(33) Yao, Z.; Postma, H. W. Ch.; Balents, L.; Dekker: C. Carbon nanotube intramolecular junctions. *Nature* **1999** *402*, 273.

(34) Rueckes, T.; Kim, K.; Joseluich, E.; Tsang, G. Y.; Cheung, C. L.; Leiber, C. M. Carbon nanotube-based nonvolatile random access memory for molecular computing. *Science* **2000** *289*, 94.

(35) Rao, C. N. R.; Satishkumar, B. C.; Govindaraj, A.; Nath, M. Nanotubes. *Chem. Phys. Chem.* **2001** *2*, 78.

(36) Morales, A. M.; Leiber, C. M. A laser ablation method for the synthesis of crystalline semiconductor nanowires. *Science* **1998** *279*, 208.

(37) Zhang, Y.; Suenaga, K.; Colliex, C.; Iijima, S. Coaxial nanocable: Silicon carbide and silicon oxide sheathed with boron nitride and carbon. *Science* **1998** *281*, 973.

(38) Han, W.; Fan, S.; Li, Q.; Gu, B.; Zhang, X.; Yu, D. Synthesis of silicon nitride nanorods using carbon nanotube as a template. *Appl. Phys. Lett.* **1997**, *71*, 2271.

(39) Wang, M. J.; Wada, H. Synthesis and characterization of silicon nitride whiskers. *J. Mater. Sci.* **1990** *25*, 1690. Wu, X. C.; Song, W. H.; Zhao, B.; Huang, W. D.; Pu, M. H.; Sun, Y. P.; Du, J. J. Synthesis of coaxial nanowires of silicon nitride sheathed with silicon and silicon oxide. *Solid State Commun.* **2000** *115*, 683.

(40) Gundiah, G.; Madhav, G. V.; Govindaraj, A.; Rao, C. N. R. Synthesis and characterization of silicon carbide, silicon oxynitride and silicon nitride nanowires. *J. Mater. Chem.* **2002** *12*, 1606−1611.

(41) Deepak, F. L.; Mukhopadhyay, K.; Vinod, C. P.; Govindaraj, A.; Rao, C. N. R. Boron nitride nanotubes and nanowires. *Chem. Phys. Lett.* **2002** *353*, 345.

(42) Deepak, F. L.; Govindaraj, A.; Rao, C. N. R. Single-crystal GaN nanowires. *J. Nanosci. Nanotechnol.* **2001** *1*, 303.

(43) de Heer, W. A.; Chatelain, A.; Ugarte, D. A carbon nanotube field-emission electron source. *Science* **1995** *270*, 1179.

(44) Wang, Q. H.; Corrigan, T. D.; Dai, J. Y.; Chang, R. P. H.; Krauss, A. R. Field emission from nanotube bundle emitters at low fields. *Appl. Phys. Lett.* **1997**, *70*, 3308.

(45) Bonard, J.-M.; Maier, F.; Stockli, T.; Chatelain, A.; de Heer, W. A.; Salvetat, J.-P.; Ferro, L. Field emission properties of multiwalled carbon nanotubes. *Ultramicroscopy* **1998** *73*, 7.

(46) Saito, Y.; Hamaguchi, K.; Hata, K.; Tohji, K.; Kasuya, A.; Nishina, Y.; Uchida, K.; Tasaka, Y.; Ikazaki, F.; Yumura, M. Field emission from carbon nanotubes; Purified single-walled and multiwalled tubes. *Ultramicroscopy* **1998** *73*, 1.

(47) Sharma, R. B.; Tondare, V. N.; Joag, D. S.; Govindaraj, A.; Rao, C. N. R. Field emission from carbon nanotubes grown on a tungsten tip. *Chem. Phys. Lett.* **2001** *344*, 283.

(48) Dresselhaus, M. S.; Williams, K. A.; Eklund, P. C. Hydrogen adsorption in carbon materials. *MRS Bull.* **1999** *24*, 45.

(49) Rajalakshmi, N.; Dhathathreyan, K. S.; Govindaraj. A.; Rao, C. N. R. To be submitted for publication.

AR0101584

26 May 2000

Chemical Physics Letters 322 (2000) 333–340

ELSEVIER

CHEMICAL PHYSICS LETTERS

www.elsevier.nl/locate/cplett

Production of bundles of aligned carbon and carbon–nitrogen nanotubes by the pyrolysis of precursors on silica-supported iron and cobalt catalysts

Manashi Nath [a,b], B.C. Satishkumar [a], A. Govindaraj [a,b], C.P. Vinod [a], C.N.R. Rao [a,b,*]

[a] *Chemistry and Physics of Materials Unit and CSIR Centre of excellence in Chemistry, Jawaharlal Nehru Centre For Advanced Scientific Research, Jakkur P.O., Bangalore-560 064, India*
[b] *Solid State and Structural Chemistry Unit, Indian Institute of Science, Bangalore-560012, India*

Received 22 February 2000; in final form 28 March 2000

Abstract

Pyrolysis of acetylene over iron or cobalt nanoparticles well dispersed on silica substrates gives copious yields of aligned carbon nanotube bundles. By carrying out the pyrolysis of pyridine over these catalyst surfaces, good quantities of aligned carbon–nitrogen nanotube bundles have been produced. The composition of the carbon–nitrogen nanotubes varies between $C_{10}N$ and $C_{33}N$, depending on the catalyst. © 2000 Elsevier Science B.V. All rights reserved.

1. Introduction

Carbon nanotubes and related materials have gained increasing importance in the last few years because of their possible technological applications [1]. Aligned carbon nanotube bundles, which are specially useful in some of the applications, have been synthesized using alumina membranes [2,3], mesoporous silica [4] and patterned silica substrates covered with Co [5]. Good quantities of aligned nanotube bundles have been obtained by the pyrolysis of ferrocene and other organometallic precursors [6,7]. In spite of these efforts, there is still a need for

newer methods for making aligned nanotube bundles. Pan et al. [8] reported recently that aligned carbon nanotubes can be obtained by carrying out the pyrolysis of acetylene over silica surfaces containing evenly positioned iron–silica nanocomposite particles while Mukhopadhyay et al. [9] have produced quasi-aligned carbon nanotubes by catalytic chemical vapour deposition. This led us to explore whether pyrolysis of simple organic precursor molecules over metal–silica catalyst surfaces can be usefully employed for this purpose. With this objective, we have carried out the pyrolysis of acetylene over the surfaces of silica-supported iron and cobalt catalysts prepared by the sol–gel route and obtained good results. More importantly, we have carried out the pyrolysis of pyridine over these catalyst surfaces to obtain carbon–nitrogen nanotube bundles. Car-

---

* Corresponding author. Fax: +91-80-846-2766; e-mail: cnrrao@jncasr.ac.in

 M. Nath et al. / Chemical Physics Letters 322 (2000) 333–340

bon–nitrogen nanotubes were first synthesized by Sen et al. [10] by the pyrolysis of pyridine over cobalt nanoparticles. Carbon–nitrogen nanotubes or nanofibres have since been prepared by means of a microwave-excited plasma of C_2H_2–N_2 mixtures [11], reactive/magnetron sputtering of graphite [12], pyrolysis of melamine on laser-patterned Fe/Ni substrates [13] and pyrolysis of ferrocene-melamine mixtures [14]. Some of these preparations yielded fully or partially aligned nanotubes. The present study reports good quantities of aligned carbon–nitrogen nanotubes obtained by pyridine pyrolysis over the catalysts. These n-type nanotube bundles are likely to be more useful because of their desirable electronic properties.

2. Experimental

The iron–silica catalyst surfaces were prepared by employing two procedures. In procedure **1**, 2 ml of tetraethylorthosilicate (TEOS) was mixed with 2 ml absolute ethanol and 9 ml of 1.5 M aqueous solution of $Fe(NO_3)_3$, and stirred for 20 min. This corresponds to a loading of 1.04 at.% of iron on silica. Few drops of concentrated HF solution were added to the mixture and stirred for another 20 min, to achieve slow gelation. The resulting gel was deposited on glass substrates and dried in an oven at 60°C for 12 h. The gel so deposited fragmented into smaller sections of 2–8 mm^2 area; it was then calcined at 450°C for 1 h under vacuum (10^{-3} Torr) and reduced in hydrogen at 500°C for 2 h. In procedure **2**, iron acetylacetonate was used as the metal precursor. A desired quantity of the acetylacetonate (1.5 mmol) was dissolved in 100 ml methanol and then mixed with TEOS (2 ml) and the gel obtained by a procedure similar to the one mentioned earlier. This procedure was also employed to prepare Co/silica catalysts starting with cobalt acetylacetonate. The catalyst samples were examined by scanning and transmission electron microscopy (SEM and TEM) using LEICA S440i and JEOL JEM-3010 microscopes, respectively.

In order to prepare bundles of aligned carbon nanotubes, acetylene was pyrolysed over the catalyst surfaces. The pyrolysis set-up consisted of stainless steel gas flow lines, fitted to a quartz tube (25 mm

inner diameter) placed in a furnace operating at a temperature in the 700–900°C range, depending on the requirement. The flow rates of the gases were controlled using UNIT mass flow controllers. Fragmented sections of the calcined catalyst were first

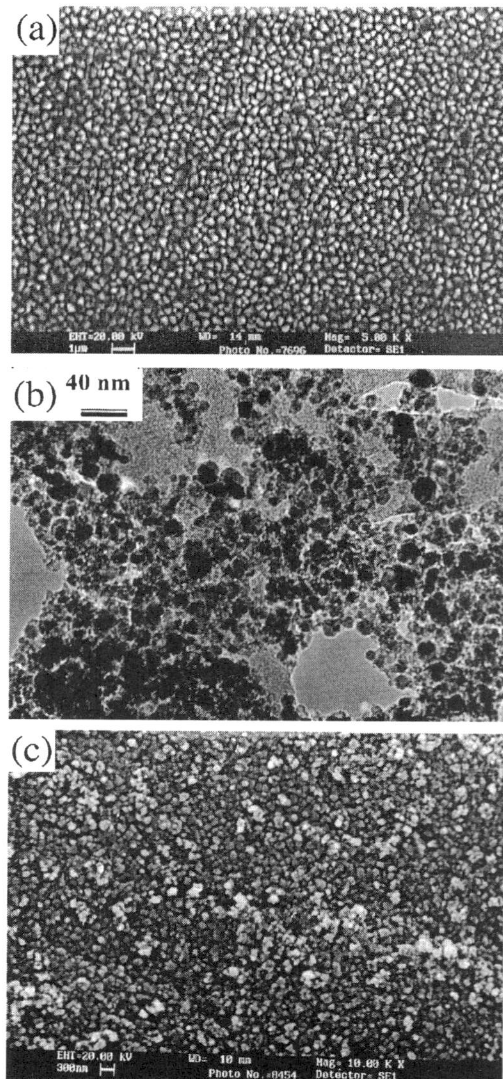

Fig. 1. SEM image of the Fe/silica catalyst (obtained from procedure **1** employing iron nitrate) showing the uniform distribution of Fe nanoparticles. (b) TEM image showing the uniform sized Fe nanoparticles obtained by the sol–gel process followed by calcination and reduction. (c) SEM image of the Fe/silica catalyst obtained by using iron acetylacetonate (procedure **2**).

Table 1
Conditions for the synthesis of aligned nanotubes

Catalyst preparation	Carbon nanotubes				Carbon–nitrogen nanotubes			
	Flow rates (sccm)		Pyrolysis temperature	Duration	Flow rates (sccm)		Pyrolysis temperature	Duration
	C_2H_2	Ar			Pyridine	Ar/H_2		
Procedure 1	15	85	700°C	1 h	30	120 (Ar)	900°C	1.5 h
Procedure 2 (Iron)	15	85	700°C	30 min	40	60 (H_2)	900°C	10 min
Procedure 2 (Cobalt)	15	85	700°C	30 min	45	55 (H_2)	900°C	30 min

placed in the quartz tube and reduced under hydrogen atmosphere at 500°C for 2 h. The temperature of the furnace was then raised to 700°C under a slow ramp rate and the pyrolysis of acetylene carried out with argon as the carrier gas. Aligned carbon–nitrogen nanotube bundles were prepared by a similar procedure except that pyridine was pyrolysed on the catalyst surface at 900°C. Typical pyrolysis conditions for the production of carbon and carbon–nitrogen nanotubes, are listed in Table 1.

The aligned nanotubes were observed by scanning and transmission electron microscopy. For TEM

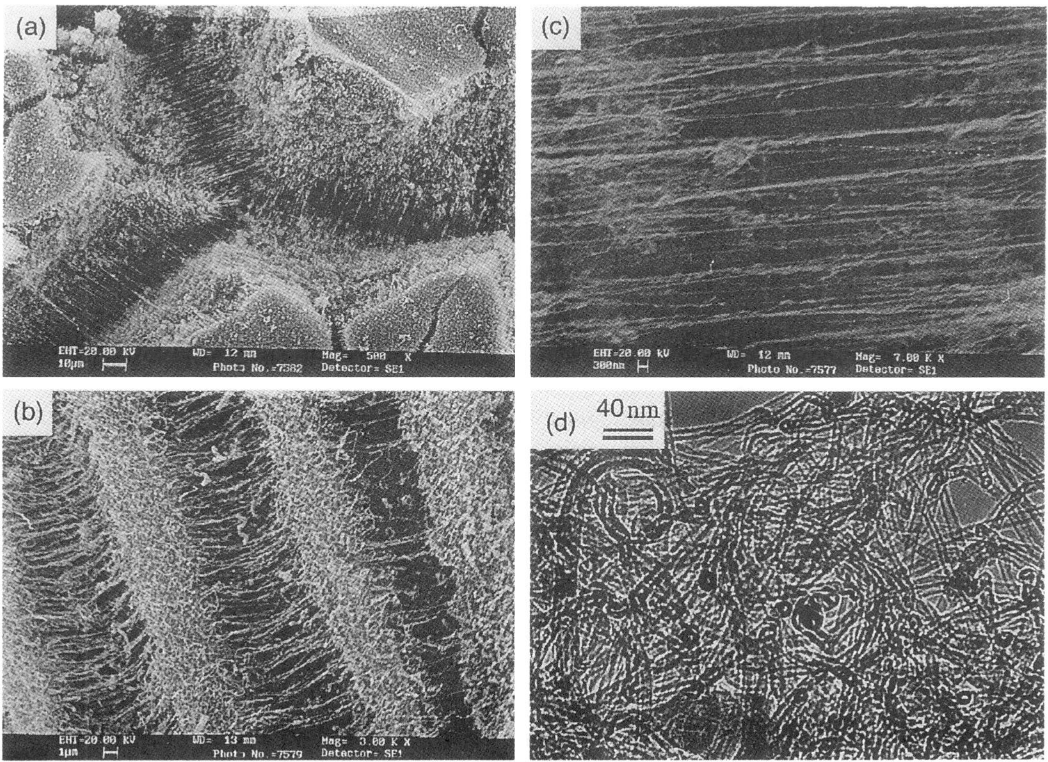

Fig. 2. SEM images of (a) aligned carbon nanotubes growing vertically from the iron–silica catalyst (prepared by procedure 1), obtained by the pyrolysis of acetylene, and (b) and (c) of aligned nanotubes in side view. (d) TEM image of the carbon nanotubes obtained with the Fe/silica catalyst.

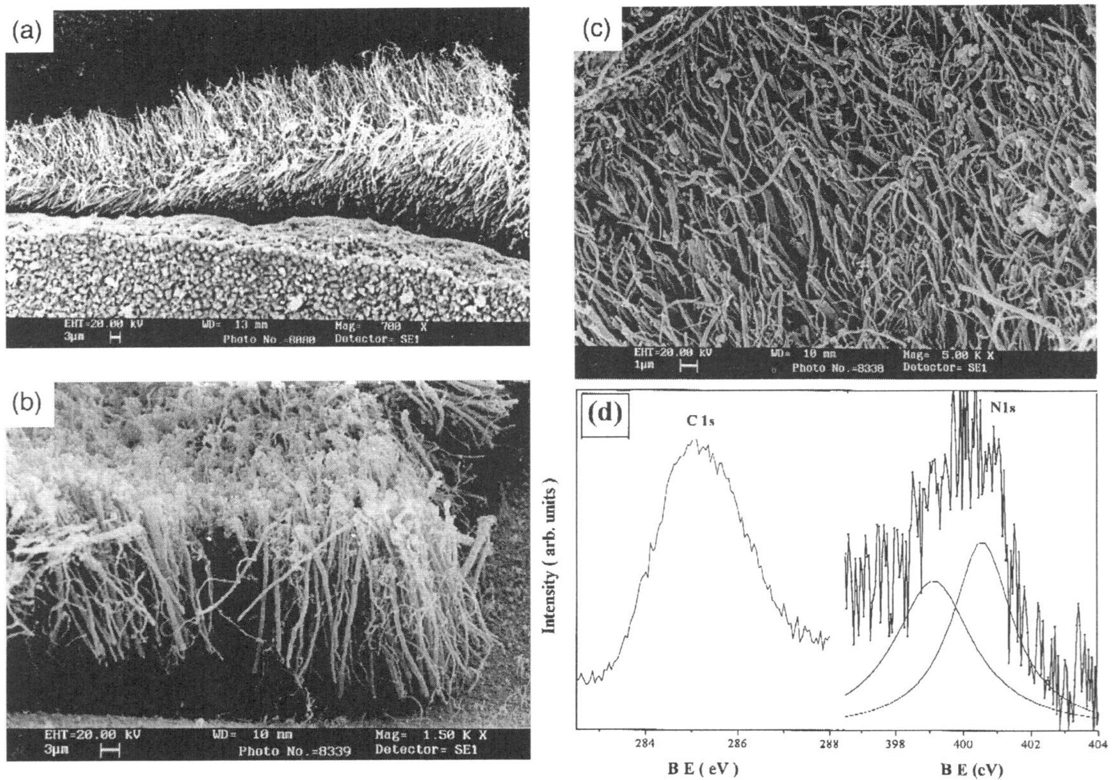

Fig. 3. SEM images (a), (b) and (c) of the aligned carbon–nitrogen nanotubes obtained by the pyrolysis of pyridine over Fe/silica substrates (prepared by procedure **1**). In (d) C 1s and N 1s XPS signals of the nanotubes are shown.

studies, the nanotube samples were ground, dispersed in carbon tetrachloride and deposited onto holey carbon copper grids. In the case of the aligned carbon–nitrogen nanotubes, the compositions were examined by X-ray photoelectron spectroscopy (XPS) using a VG scientific ESCA LAB V spectrometer. Electron energy loss spectroscopy (EELS) was also carried out concurrently with high-resolution electron microscopy.

3. Results and discussion

In Fig. 1a we show a SEM image of the iron–silica catalyst surface, obtained by procedure **1**, subjected

Fig. 4. SEM images of aligned nanotubes: Carbon nanotubes grown on (a) Fe/silica surfaces (prepared by procedure **2**), by pyrolysis of acetylene (flow rate, 15 sccm) at 700°C for 30 min under a flow of Ar (85 sccm), and (b) on Co/silica surfaces (by procedure **2**) by pyrolysis of acetylene (flow rate, 15 sccm) at 700°C for 30 min under Ar (70 sccm). Carbon–nitrogen nanotubes grown on (c) Fe/silica surface (obtained by procedure **2**) by pyrolysis of pyridine at 900°C for 10 min with Ar (40 sccm) as the carrier gas and under a flow of hydrogen (60 sccm), and (d) on Co/silica surfaces (by procedure **2**) by pyrolysis of pyridine at 900°C for 30 min with Ar (flow rate, 45 sccm) as the carrier gas under a flow of hydrogen (55 sccm). XPS N 1s signals of the samples from (c) and (d) are shown next to the SEM images.

to calcination and reduction. The image shows a uniform distribution of iron–silica nanoparticles on the surface. We show a TEM image of the catalyst surface in Fig. 1b. The image shows fairly uniform

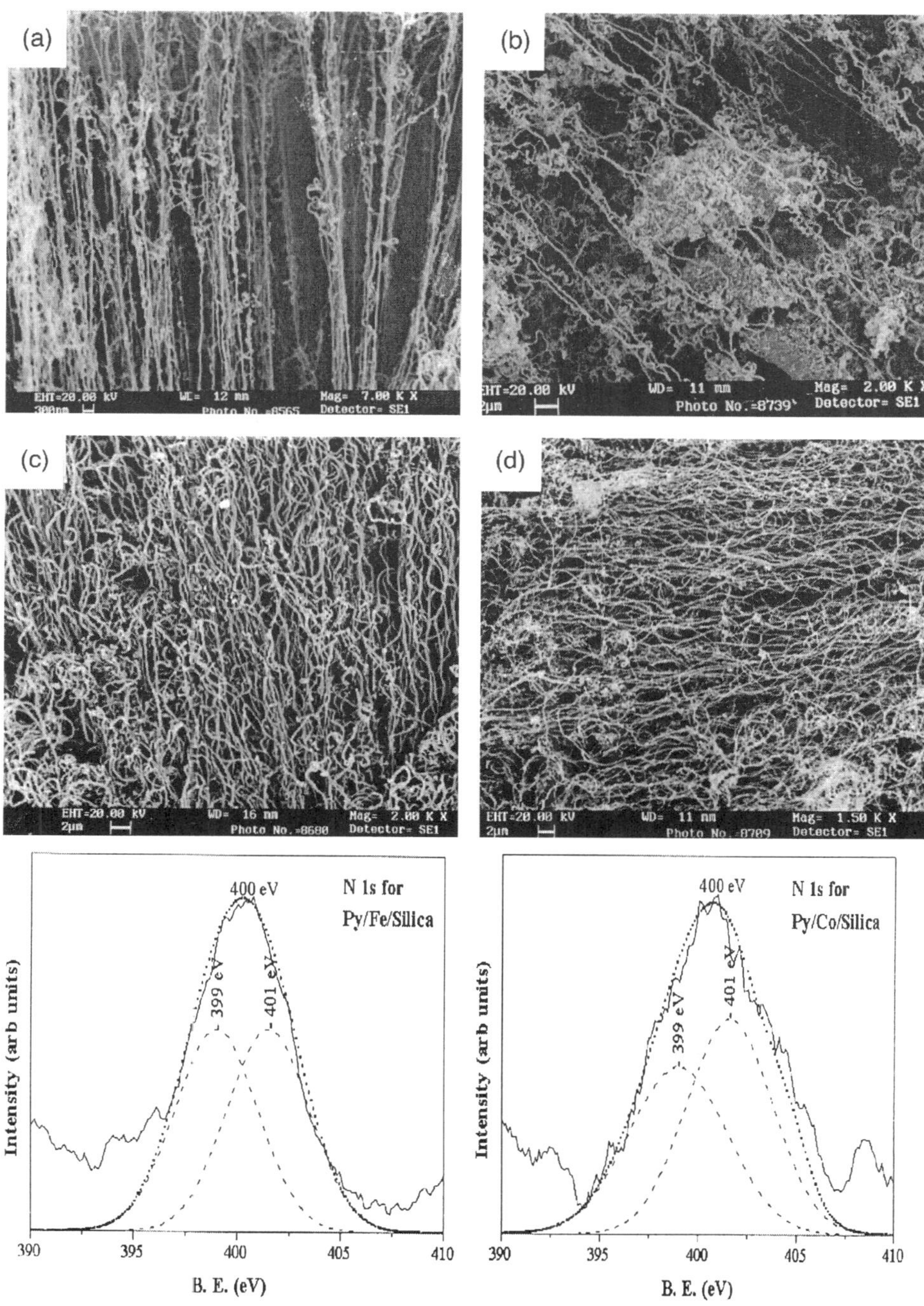

sized iron nanoparticles. An analysis of the TEM images showed that most of the particles had diameters in the 5–20 nm range. In Fig. 2, we show the SEM images of aligned carbon nanotube bundles obtained by the pyrolysis of acetylene (flow rate, 15 sccm; sccm, standard cubic centimeter per minute) over the iron–silica surface at 700°C for 1 h under argon flow of 85 sccm. The low magnification image in Fig. 2a shows the bundles of aligned carbon nanotubes growing out perpendicularly from the catalyst surface. Fig. 2b,c show the SEM images of compact aligned nanotube bundles.

In Fig. 3a we show a SEM image of the nanotubes obtained from the pyrolysis of pyridine (flow rate, 30 sccm) at 900°C for 1.5 h, under Ar (120 sccm) flow on the iron–silica catalyst surface. The image clearly reveals the alignment of the nanotubes, perpendicular to the catalyst surface. The SEM images in Fig. 3b,c show side and top views of the aligned nanotubes, respectively.

In order to characterize the aligned nanotubes obtained by pyridine pyrolysis, XPS analysis was carried out. In Fig. 3d we show the C 1s and N 1s signals from these nanotubes. This C 1s signal is at 285 eV and the N 1s signal around 400 eV. The carbon to nitrogen ratio was estimated by taking the ratio of the integrated peak areas under the C 1s and N 1s signals and dividing them by the respective photoionization cross-sections. The average composition for the aligned carbon–nitrogen nanotube turns out to be around C_6N. A Gaussian fit of the N 1s spectrum, however, shows the presence of two peaks with binding energies of 399 and 401 eV. The 399 eV feature is characteristic of pyridinic nitrogen (sp^2 hybridization), probably present at nanotube ends, while the peak centered at 401 eV is due to nitrogen present in the graphene sheets [10]. The latter corresponds to trivalent nitrogen replacing the carbon in the hexagonal structure. If we take the intensity of only the 401 eV feature, the composition turns out to be $C_{10}N$. EELS measurements confirmed the presence of nitrogen in the nanotube preparations. The compositions estimated from EELS are generally close to those from XPS [10,15].

It has been proposed that nanotubes prepared by precursor pyrolysis using catalyst particles grow by the tip-growth mechanism [16]. According to this mechanism, the metal particles stick at the tip of the

nanotube, rather than the growth starting at the surface of the particle. In order to examine this aspect, we carried out a TEM study of the aligned carbon nanotubes. In Fig. 2d, we show a typical TEM image of carbon nanotubes, with diameters in the range of 10–15 nm. The image shows the presence of few metal particles at the tips of the carbon nanotubes. Furthermore, the SEM image of the carbon nanotube bundles in Fig. 2a shows that the nanotubes lift the metal nanoparticles from the catalyst surface. The SEM image of carbon–nitrogen nanotube bundles in Fig. 3a also shows tip-growth. We see metal particles at the tips of some of the nanotubes in Fig. 3.

In Fig. 4a,b, we show the SEM images of bundles of carbon nanotubes obtained by the pyrolysis of acetylene over Fe and Co catalysts prepared from the acetylacetonates (procedure **2**). In Fig. 4c,d, we show the SEM images of carbon–nitrogen nanotube bundles obtained by pyridine pyrolysis over the Fe and Co catalysts prepared by procedure **2**. XPS analysis of the carbon–nitrogen nanotubes obtained with these catalysts shows broad signals due to N 1s around 400 eV. Based on the total N 1s intensities the compositions of the nanotubes work out to be C_9N and $C_{18}N$, respectively, with the iron and cobalt catalysts. However, if we take the intensity of the signal at 401 eV obtained after spectral decomposition, the compositions work out to be $C_{18}N$ and $C_{33}N$, respectively, with the iron and cobalt catalysts.

The TEM images of some of the carbon–nitrogen nanotubes exhibit interesting features. In Fig. 5 we present some of the unusual images of these nanotubes. The image in Fig. 5a shows a nanotube with bamboo-shaped morphology. Fig. 5b shows a nanotube having a nested cone-shaped cross-section, while that in Fig. 5c shows a somewhat unusual morphology involving a conical stacking sequence of the graphene sheets with intermittent hollow regions. The high-resolution TEM image in Fig. 5d shows that the graphene sheets are stacked in the form of nested cones. Clearly, doping with nitrogen modifies the morphology of the nanotubes drastically. It seems that doping favours the formation of quasi-continuous, cylindrical graphene fragments, which in turn facilitate the growth of the nanotube. It is to be noted that nitrogen can be present in pyridinic rings present at the edges of such graphene fragments. The conical

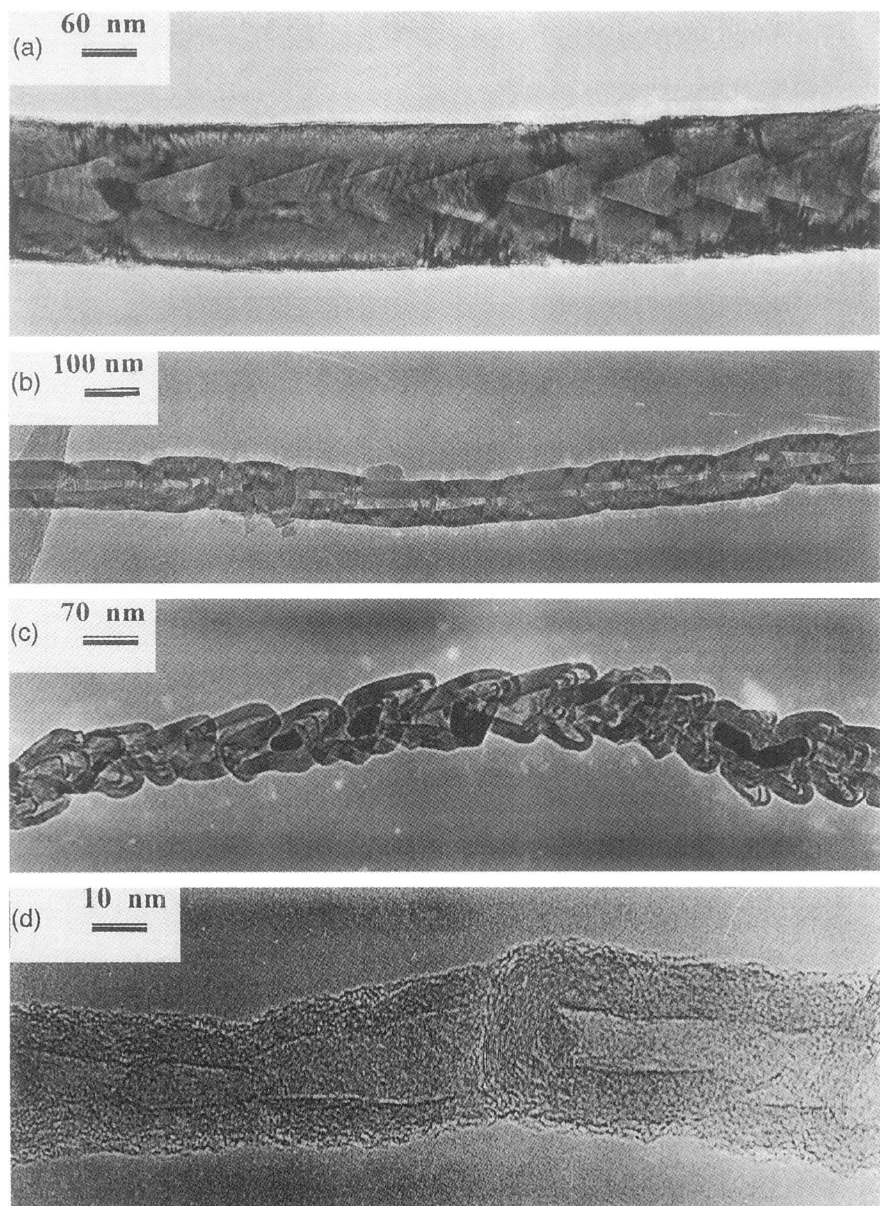

Fig. 5. (a)–(c) TEM images of carbon–nitrogen nanotubes obtained by the pyrolysis of pyridine (flow rate, 30 sccm) over Fe/silica substrates (prepared by procedure **1**) at 900°C for 1.5 h under Ar (120 sccm) flow, showing various morphologies. (d) HREM image of a nanotube showing the nested cone-type stacking of graphene sheets in the hollow region.

sections sometimes enclose the catalyst particle as shown in Fig. 5c, lending some support to the tip-growth mechanism. The conical curvature in the hollow regions could be due to the nitrogen doping, wherein nitrogen is present in the five-membered rings.

4. Conclusion

The present study shows that pyrolysis of appropriate organic precursors over iron–silica and cobalt/silica catalysts, prepared by the sol–gel method by using the metal nitrate or the acetylacetonate as the starting material, gives aligned bundles of carbon nanotubes as well as carbon–nitrogen nanotubes. The synthesis of copious quantities of aligned carbon–nitrogen nanotube bundles by this method is of significance since they are likely to possess more desirable electronic properties. The nanotube diameters obtained by us is in the range 15–150 nm, most of which have a stacked cone morphology. It would be therefore equally correct to consider them to be hollow carbon fibres, similar to those reported by Terrones et al. [13,14]. The carbon–nitrogen nanotube compositions obtained by us are in the range $C_{10}N$–$C_{33}N$. These are comparable to the compositions reported by Terrones et al. which are in the range $C_{13}N_x$ to $C_{49}N_x$ ($x \leqslant 1$).

References

[1] R. Dagani, Chem. Eng. News, Jan. 11 (1999) 31.

[2] G. Che, B.B. Laxmi, C.R. Martin, E.R. Fischer, R.S. Rouff, Chem. Mater. 10 (1998) 260.

[3] J. Li, C. Papadopoulos, J.M. Xu, M. Maskovits, Appl. Phys. Lett. 75 (1999) 367.

[4] W.Z. Li, S.S. Xie, L.X. Qian, B.H. Chang, B.S. Zou, W.Y. Zhou, R.A. Zhao, G. Wang, Science 274 (1996) 1701.

[5] M. Terrones, N. Grobert, J. Olivares, J.P. Zhang, H. Terrones, K. Kordatos, W.K. Hsu, J.P. Hare, P.D. Townsend, K. Prassides, A.K. Cheetham, H.W. Kroto, D.R.M. Walton, Nature 388 (1997) 52.

[6] C.N.R. Rao, A. Govindaraj, R. Sen, B.C. Satishkumar, Mater. Res. Innov. 2 (1998) 128.

[7] B.C. Satishkumar, A. Govindaraj, C.N.R. Rao, Chem. Phys. Lett. 307 (1999) 158.

[8] Z.W. Pan, S.S. Xie, B.H. Chang, L.F. Sun, W.Y. Zhou, G. Wang, Chem. Phys. Lett. 299 (1998) 97.

[9] K. Mukhopadhyay, A. Koshio, T. Sugai, N. Tanaka, H. Shinohara, Z. Konya, J.B. Nagy, Chem. Phys. Lett. 303 (1999) 117.

[10] R. Sen, B.C. Satishkumar, A. Govindaraj, K.R. Harikumar, G. Raina, J. P Zhang, A. K Cheetham, C.N.R. Rao, Chem. Phys. Lett. 287 (1998) 671.

[11] S.L. Sung, S.H. Tsai, C.H. Tseng, F.K. Chiang, X.W. Liu, H.C. Shih, J. Mater. Res. 15 (2000) 502.

[12] K. Suenaga, M.P. Johansson, N. Hellgren, E. Broitman, L.R. Wallenberg, C. Colliex, J.-E. Sundgren, L. Hultman, Chem. Phys. Lett. 300 (1999) 695.

[13] M. Terrones, Ph. Redlich, N. Grobert, S. Trasobars, W.K. Hsu, H. Terrones, Y.Q. Zhu, J.P. Hare, C.L. Reeves, A.K. Cheetham, M. Rühle, H.W. Kroto, D.R.M. Walton, Adv. Mater. 11 (1999) 655.

[14] M. Terrones, H. Terrones, N. Grobert, W.K. Hsu, Y.Q. Zhu, J.P. Hare, H.W. Kroto, D.R.M. Walton, Ph. Kohler-Redlich, M. Rühle, J.P. Zhang, A.K. Cheetham, Appl. Phys. Lett. 75 (1999) 3932.

[15] B.C. Satishkumar, A. Govindaraj, K.R. Harikumar, J.P. Zhang, A.K. Cheetham, C.N.R. Rao, Chem. Phys. Lett. 300 (1999) 473.

[16] S. Amelinckx, X.B. Zhang, D. Bernaerts, X.F. Zhang, V. Ivanov, J.B. Nagy, Science 265 (1995) 635.

8 May 1998

Chemical Physics Letters 287 (1998) 671–676

ELSEVIER

**CHEMICAL
PHYSICS
LETTERS**

B–C–N, C–N and B–N nanotubes produced by the pyrolysis of precursor molecules over Co catalysts

Rahul Sen [a], B.C. Satishkumar [a], A. Govindaraj [a], K.R. Harikumar [a], Gargi Raina [b], Jin-Ping Zhang [c], A.K. Cheetham [c], C.N.R. Rao [a,b,*]

[a] *CSIR Centre of Excellence in Chemistry, Solid State and Structural Chemistry Unit and Materials Research Centre, Indian Institute of Science, Bangalore 560012, India*
[b] *Chemistry and Physics of Materials Unit, Jawaharlal Nehru Centre for Advanced Scientific Research, Bangalore 560064, India*
[c] *Materials Research Laboratory, University of California, Santa Barbara, CA 93106, USA*

Received 27 January 1998; in final form 20 February 1998

Abstract

B–C–N, C–N and B–N nanotubes, prepared by the pyrolysis of the appropriate precursor compounds around 1000°C in an atmosphere of argon, have been examined by electron microscopy and other techniques. The compositions of the nanotubes have been analysed by electron energy loss spectroscopy and X-ray photoelectron spectroscopy. There is significant compositional variation in B–C–N and C–N nanotubes. Properties of the B–C–N and C–N nanotubes have been compared with those of carbon nanotubes, particularly with respect to the I–V characteristics as obtained from UHV–STM studies. The near absence of B–N nanotubes on pyrolysing the appropriate precursor compound, and other observations made in the present study, indicate the crucial role played by carbon in the initial nucleation and growth of nanotubes.
© 1998 Elsevier Science B.V. All rights reserved.

1. Introduction

Ever since the discovery of carbon nanotubes, there has been much interest in preparing nanotubes of other materials, in particular those of boron–nitrogen, boron–carbon–nitrogen and boron–carbon. Besides the well known technique involving arc evaporation of graphite [1], carbon nanotubes have been prepared by the pyrolysis of hydrocarbons over small particles of metals such as Co, Mo and Ni [2–4]. Since the presence of small metal particles is essential for the formation of carbon nanotubes by the pyrolysis of hydrocarbons, metallocenes and

$Fe(CO)_5$ have been used as precursors, along with hydrocarbons to produce the carbon nanotubes [5,6]. B–C–N nanotubes have been prepared by striking an arc between a graphite anode filled with BN and a pure graphite cathode in a helium atmosphere of 650 torr [7]. B–C–N nanotubes have also been made by the laser ablation of a composite target containing BN, carbon, Ni and Co at 1000°C under flowing nitrogen [8]. Terrones et al. [9], however, pyrolyzed the precursor addition compound, $CH_3CN:BCl_3$, over Co powder at 1000°C to obtain B–C–N nanotubes. We have been interested in carrying out a comparative study of nanotubes of B–C–N and B–N and indeed C–N, prepared by the pyrolysis of the appropriate precursor molecules. We have therefore inves-

---

* Corresponding author.

tigated the pyrolysis of pyridine over a Co catalyst [10], to establish definitively whether carbon nanotubes containing nitrogen are truly obtained by this means. If indeed C–N nanotubes are formed by this procedure, it was of interest to compare their properties with those of carbon nanotubes. We have carried out the pyrolysis of the 1:1 addition compound of BH_3 with $(CH_3)_3N$, as well as with NH_3, to examine the nanotubes produced, if any. The present study shows that the nanotubes obtained by pyridine pyrolysis are distinctly different from the carbon nanotubes, exhibiting unique Raman spectra and conducting properties. The B–C–N nanotubes obtained by us have compositions different from others reported in the literature. Furthermore, considerable variability exists in the composition in any given batch of B–C–N/C–N nanotubes obtained by the pyrolysis of precursors. Unlike the C–N and B–C–N nanotubes, B–N nanotubes are not readily formed by the pyrolysis of precursors. The study points to the seminal role of carbon in the formation of such nanotubes.

2. Experimental

In order to prepare C–N nanotubes, pyridine was pyrolyzed over Co powder as follows. Co powder containing particles of ∼ 50 nm diameter was prepared by the polyol process, wherein 5 g of polyvinly pyrrolidone (PVP) and 2.5 g of cobalt acetate were dissolved in 150 ml of ethylene glycol and refluxed in an argon atmosphere for 24h. The resulting Co powder was heated at 750°C to burn away the PVP and then treated with hydrogen at 700°C. Argon gas (flow rate of 50 sccm) was bubbled through pyridine at room temperature and the gas carrying the pyridine vapour was passed over the Co powder maintained at 1000°C in a tubular furnace. For preparing the B–C–N nanotubes, the 1:1 addition compound of BH_3 with $N(CH_3)_3$, as well as with pyridine, was pyrolysed over cobalt powder. $(CH_3)_3N:BH_3$ (Fluka) was first sublimed by heating it to 300°C and a mixture of the vapour with argon (flow rate, 50 sccm) was passed through a furnace containing the Co powder at 1000°C. Pyrolysis of $C_6H_5N:BH_3$ over cobalt powder at 1000°C gave carbon nanotubes and a yellow polymeric residue containing boron,

carbon and nitrogen [11]. The residue when heated with Co powder at 1000°C under Ar flow yielded some nanotubes, but they did not contain a sufficient proportion of B and N. In order to prepare the B–N nanotubes, pyrolysis of $H_3N:BH_3$ was carried out over cobalt powder at 1100°C under Ar flow. $H_3N:BH_3$ was prepared by the reaction of $LiBH_4$ and $(NH_4)_2SO_4$ in ether.

The nanotubes were characterized by X-ray photoelectron spectroscopy (XPS) with a VG scientific ESCA LAB V spectrometer, transmission electron microscopy (TEM) with a JEOL JEM-3010 microscope at 300 kV, scanning electron microscopy (SEM) with a LEICA S440i instrument, electron energy loss spectroscopy (EELS) and Raman spectroscopy. EELS measurements were carried out with a Gatan imaging filter system attached to a JEOL 2010 microscope and fitted with a CCD camera. The beam size was 10–150 nm, depending on the width of the nanotube, and the dispersion was 0.3–0.5 eV. Tunneling conductance of nanotubes was measured under ultra-high vacuum using a Omicron UHV–STM apparatus.

3. Results and discussion

The TEM images of the nanotubes obtained by the pyrolysis of pyridine over Co powder at 1000°C are shown in Fig. 1; no single-walled nanotubes were found in the TEM images. In Fig. 1(c), we show the image of a nanotube on which EELS analysis was carried out. The EEL spectra (Fig. 2) show characteristic edges at 284 eV and 400 eV corresponding to K-shell ionization of carbon and nitrogen, respectively. We see sharply defined fine structure due to π^* and σ^* pre-ionizations edges characteristic of the sp^2 hybridization in the case of carbon. Based on the EEL spectrum in Fig. 2(a), corresponding to the region marked in Fig. 1(c), the stoichiometry of the nanotube works out to be $C_{33}N$. EELS of another nanotube shown in Fig. 2(b) gave a composition of $C_{11}N$. In the inset of Fig. 2, we show the core-level XPS of the nanotubes obtained from pyridine pyrolysis, giving the C 1s signal at 285 eV and N 1s signal at 401 eV. The N 1s signal at 401 eV is characteristic of nitrogen atoms present in graphene sheets [12]. From the XPS band intensities we could obtain the

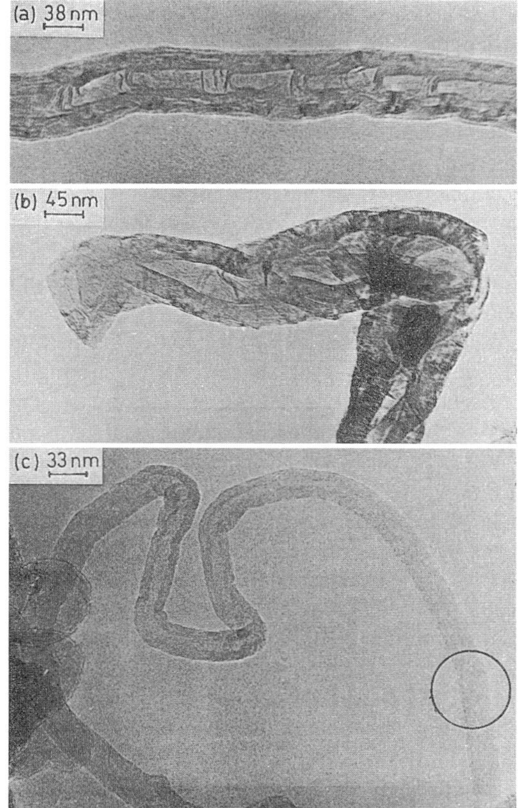

Fig. 1. TEM images of C–N nanotubes obtained by the pyrolysis pyridine over Co particles at 1000°C under Ar flow of 50 sccm: (a) shows a straight tube with cone-shaped features in the central hollow regions, (b) shows a thick bent tube with a metal particle inside, (c) shows a long curved tube. EELS analysis was carried out in the circled region.

C/N ratio, by making use of the ratio of the respective integrated peak areas and the photoionization cross section ratios for the 1s level. The stoichiometry of the C–N nanotubes by this method comes out to be $C_{38}N$, close to the composition from EELS in Fig. 2(a). While nitrogen appears to be a genuine constituent of these nanotubes, it is possible that their surface and bulk compositions are different, the variability in composition arising from the variability in the number of C–N layers. It is noteworthy that the nanotubes synthesised by the decomposition of a triazine derivative over cobalt contained some nitro-

gen [13]. We are unable to comment definitively on the nature of nitrogen in the C–N nanotubes, although the XPS results are consistent with sp^2 type nitrogens being present in graphene sheets. One can envisage nitrogen being accommodated at the open edges of nanotubes forming pyridine-like rings [14]. Any nitrogen that is accommodated as part of 5-membered rings would however have sp^3 hybridization.

The presence of nitrogen in the nanotubes obtained from pyridine pyrolysis manifests itself in the Raman spectrum, which is distinctly different from that of carbon nanotubes. Multi-walled carbon nanotubes give a single band at 1582 cm^{-1} with a full width at half-maximum (FWHM) of 12 cm^{-1}. The C–N nanotubes prepared by us give a relatively broad band around 1585 cm^{-1} with a shoulder at

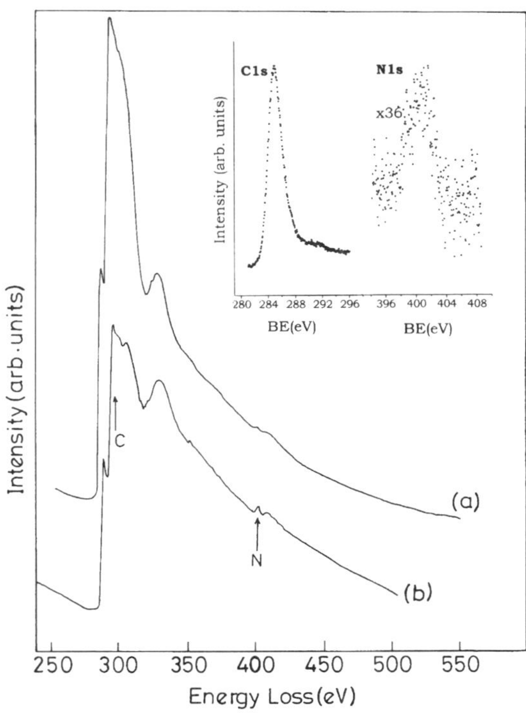

Fig. 2. EELS spectra of C–N nanotubes: The spectrum shown in (a) was obtained from the region marked by a circle in Fig. 1(c) and gives a stoichiometry of $C_{33}N$. The spectrum (b) gives a stoichiometry of $C_{11}N$. The inset shows the C 1s and N 1s signals obtained from XPS of C–N nanotubes.

R. Sen et al. / Chemical Physics Letters 287 (1998) 671–676

1623 cm^{-1}, with FWHMs of 16 cm^{-1} and 23 cm^{-1}, respectively. The feature at 1623 cm^{-1} could arise from the presence of C = N groups. The shift of the Raman band of graphite to a higher frequency on the incorporation of nitrogen has been mentioned in the literature [15]. X-ray diffraction patterns of C–N nanotubes show a (002) peak at 3.37 Å, close to the interlayer spacing in graphite. Thermal analysis of the C–N nanotubes shows that they start burning at a much lower temperature ($\sim$ 450°C) compared to multi-walled carbon nanotubes which burn at around 650°C.

Even more interesting about the C–N nanotubes are their electrical transport properties. We have measured the tunneling conductance of the C–N nanotubes in comparison with the carbon nanotubes under ultra-high vacuum. In Fig. 3 we compare the I–V curves of a C–N nanotube and a multi-walled carbon nanotube measured at different points along the length of the nanotubes. The I–V curves of the nanotubes vary from one position to another. However, the conductance of the C–N nanotubes is generally higher than that of the multi-walled carbon nanotube. Accordingly, the slope of the I–V curves in the case of C–N nanotube in the $\pm 0.1 - \pm 0.5$ volt range is around 1.2 nA/V and the corresponding slope in multi-walled carbon nanotubes is $\sim$ 0.5 nA/V. The slope in the case of single-walled carbon nanotubes is around 0.3 nA/V. The slope is a measure of the local density of states. Furthermore, the average gap as obtained from the normalized conductance ($d\ln I/d\ln V$ vs. V) curves, is $\sim$ 600 meV in the multi-walled nanotubes, while it is only $\sim$ 400 meV in the C–N nanotubes. The gap in the case of single-walled nanotubes is $\sim$ 650 meV.

TEM and SEM images of the nanotubes obtained by the pyrolysis of $(CH_3)_3N{:}BH_3$ over Co powder at 1000°C are presented in Fig. 4(a) and (b), respectively. The nanotubes are somewhat similar to the C–N nanotubes and there was no evidence for single-walled nanotubes. The EEL spectra gave edges at 188 eV, 284 ev and 400 eV due to K-shell ionization of B, C and N. Analysis based on the EEL spectrum in Fig. 5 showed that one of the nanotubes had a stoichiometry of $BC_{3.3}N_{0.1}$. The EEL spectra of some of the nanotubes gave an analysis close to $BC_{25}N_{0.5}$. These B/C and N/C ratios are smaller than the values reported by Terrones et al. [9]. Zhang et al. [8] have recently reported B–C–N nanotubes of the composition $BC_7N_{0.7}$. The variability in these compositions may be because the surface layers have different compositions than the bulk [8,16]. Based on the intensity of the B 1s, C 1s and N1s core level signals in the XPS, we have found the average composition of the nanotubes prepared by us to be $BC_{28}N$ (see inset of Fig. 5). It is to be noted that a small fraction of carbon nanotubes (without B and N) is generally found with the B–C–N nanotubes. The B–C–N nanotubes exhibit tunneling conductance which is lower or comparable to that of multi-walled carbon nanotubes. The Raman spectrum of the B–C–N nanotubes is comparable to that of multi-walled carbon nanotubes. The XRD pattern shows a peak around 3.35 Å, close to that in C–N nanotubes

Our experiments directed towards preparing B–N

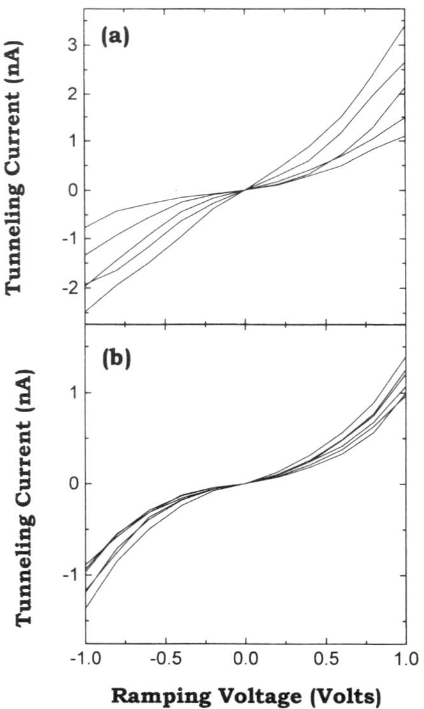

Fig. 3. I–V curves from UHV–STM of (a) a C–N nanotube and (b) a multiwalled carbon nanotube. The different curves were recorded at different points along the nanotube.

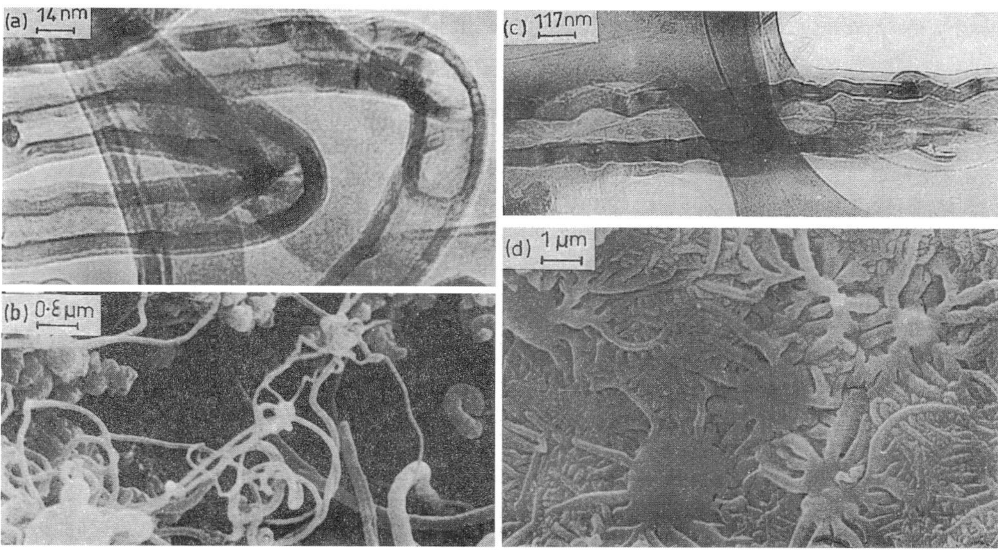

Fig. 4. (a) TEM and (b) SEM images of B–C–N nanotubes obtained by the pyrolysis of $(CH_3)_3N:BH_3$ over Co powder at 1000°C under Ar flow of 50 sccm. (c) TEM and (d) SEM images of B–N structures obtained by the pyrolysis of $H_3N:BH_3$ over Co powder at 1100°C under Ar flow of 50 sccm.

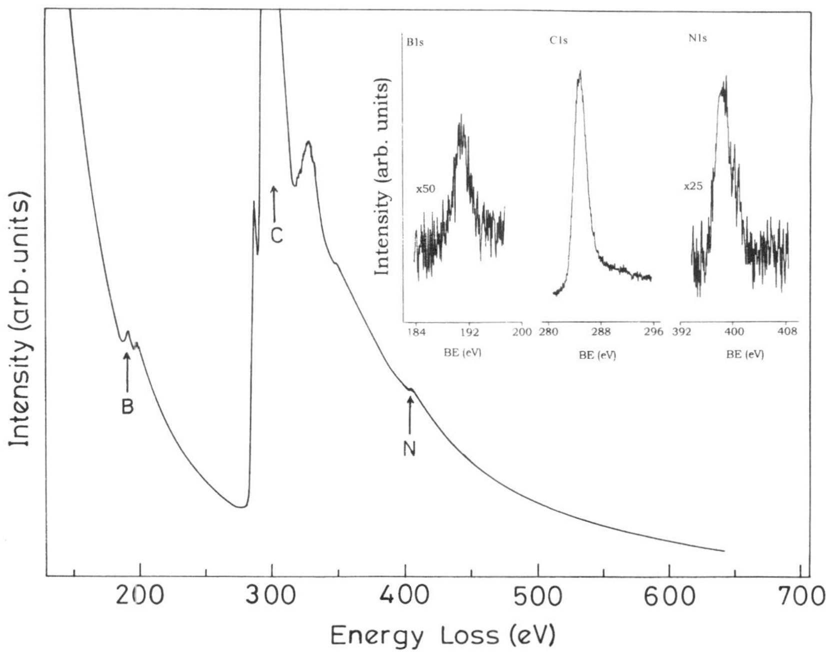

Fig. 5. EELS spectra of B–C–N nanotubes. The inset shows B 1s, C 1s and N 1s signals obtained from XPS of B–C–N nanotubes giving a composition of $BC_{28}N$.

nanotubes by the pyrolysis of $H_3N:BH_3$ over cobalt powder at 1100°C under Ar flow are revealing. The product obtained was essentially amorphous and contained very few fibres or nanotubes. In Fig. 4(c) we show the TEM image of a nanotube-type structure, occasionally found in the product. XPS analysis showed that the product of pyrolysis contained B and N. The SEM image in Fig. 4(d) shows that, unlike in the case of C–N or B–C–N nanotubes (see Fig. 4(b)), there are no nanotube-like structures, but only tube-like growth emanating from the outer surface of metal particles, but not incorporating them. It has been reported that the reaction of trichloroborazine with Cs as well as pyrolysis of borazine under high-pressure yield poorly crystalline BN fibers, some of which are hollow [17,18]. It appears that the graphitization of metal particles by carbon is a necessary initial step for the proper nucleation and growth of nanotubes In the pyrolysis experiments carried out by us with $H_3N:BH_3$, there was no obvious source of carbon in the gas mixture. This is probably the reason why there was negligible formation of nanotubes. Use of metals such as Hf and Ta, instead of cobalt, however appears to favour the formation of B–N nanotubes [19,20].

4. Conclusions

The present study demonstrates that B–C–N and C–N nanotubes can be prepared by the pyrolysis of the relevant precursor molecules over cobalt powder. The composition however shows variability, even in the same batch of nanotubes. XPS depth-profile analysis indeed showed that there was slight enrichment of carbon with successive etching in the case of C–N nanotubes. The same was true of B–C–N nanotubes, expect that the boron content essentially became zero after the third etching (~ 2–3 nm depth). There appears to be some commonality in the mechanism of formation of carbon nanotubes and the B–C–N/C–N nanotubes. The mechanism probably involves the initial graphitization of the metal particles followed by the growth of the nanotubes. It therefore becomes apparent why B–N nanotubes are not readily formed by the pyrolysis of precursor molecules.

Acknowledgements

The authors thank the Indo-French Centre for the Promotion of Advanced Scientific Research, New Delhi, for the support of this research.

References

[1] S. Iijima, Nature 354 (1991) 56.

[2] V. Ivanov, J.B. Nagy, Ph. Lambin, A. Lucas, X.B. Zhang, X.F. Zhang, D. Bernaerts, G. Van Tendeloo, S. Amelinckx, J. Van Landuyt, Chem. Phys. Lett. 223 (1994) 329.

[3] H. Dai, A.G. Rinzler, P. Nikolaev, A. Thess, D.T. Colbert, R.E. Smalley, Chem. Phys. Lett. 260 (1996) 471.

[4] A. Govindaraj, R. Sen, B.V. Nagaraju, C.N.R. Rao, Phil. Mag. Lett. 76 (1997) 363.

[5] R. Sen, A. Govindaraj, C.N.R. Rao, Chem. Phys. Lett. 267 (1997) 276.

[6] R. Sen, A. Govindaraj, C.N.R. Rao, Chem. Mater. 9 (1997) 2078.

[7] Z. Weng-Sieh, K. Cherrey, N.G. Chopra, X. Blase, Y. Miyamoto, A. Rubio, M.L. Cohen, S.G. Louie, A. Zettl, R. Gronsky, Phys. Rev. B 51 (1995) 11229.

[8] Y. Zhang, H. Gu, K. Suenaga, S. Iijima, Chem. Phys. Lett. 279 (1997) 264.

[9] M. Terrones, A.M. Benito, C. Mantega-Diego, W.K. Hsu, O.I. Osman, J.P. Hare, D.G. Reid, H. Terrones, A.K. Cheetham, K. Prassides, H.W. Kroto, D.R.M. Walton, Chem. Phys. Lett. 257 (1996) 576.

[10] R. Sen, B.C. Satishkumar, A. Govindaraj, K.R. Harikumar, M.K. Renganathan, C.N.R. Rao, J. Mater. Chem. 7 (1997) 2335.

[11] J. Bill, M. Friess, R. Riedel, Eur. J. Solid State Inorg. Chem. 29 (1992) 195.

[12] J. Casanovas, J.M. Ricart, J. Rubio, F. Illas, J.M. Jimenez-Mateos, J. Am. Chem. Soc. 118 (1996) 8071.

[13] M. Terrones, N. Grobert, J. Olivares, J.P. Zhang, H. Terrones, K. Kordatos, W.K. Hsu, J.P. Hare, P.D. Townsend, K. Prassides, A.K. Cheetham, H.W. Kroto, D.R.M. Walton, Nature 388 (1997) 52.

[14] A.T. Balaban, Match 33 (1996) 25.

[15] J.H. Kaufman, S. Metin, D.D. Saperstein, Phys. Rev. B 39 (1989) 13053.

[16] K. Suenaga, C. Colliex, N. Demoncy, A. Loiseau, H. Pascard, F. Willaime, Science 278 (1997) 653.

[17] E.J.M. Hamilton, S.E. Dolan, C.M. Mann, H.O. Colijn, C.A. McDonald, S.G. Shore, Science 260 (1993) 659.

[18] S.-I. Hirano, T. Yogo, S. Asada, S. Naka, J. Am. Ceram. Soc. 72 (1989) 66.

[19] A. Loiseau, H. Willaime, N. Demoncy, G. Hug, H. Pascard, Phys. Rev. Lett. 76 (1996) 4737.

[20] M. Terrones, W.K. Hsu, H. Terrones, J.P. Zhang, S. Ramos, J.P. Hare, R. Castillo, K. Prassides, A.K. Cheetham, H.W. Kroto, D.R.M. Walton, Chem. Phys. Lett. 259 (1996) 568.

5 February 1999

Chemical Physics Letters 300 (1999) 473–477

ELSEVIER

CHEMICAL PHYSICS LETTERS

Boron–carbon nanotubes from the pyrolysis of C_2H_2–B_2H_6 mixtures

B.C. Satishkumar [a], A. Govindaraj [a,b], K.R. Harikumar [a], Jin-Ping Zhang [c], A.K. Cheetham [c], C.N.R. Rao [a,b,*]

[a] Solid State and Structural Chemistry Unit, CSIR Centre of Excellence in Chemistry, Indian Institute of Science, Bangalore 560 012, India
[b] Jawaharlal Nehru Centre for Advanced Scientific Research, Jakkur P.O., Bangalore 560 064, India
[c] Materials Research Laboratory, University of California, Santa Barbara, CA 93106, USA

Received 16 November 1998; in final form 3 December 1998

Abstract

Boron-doped carbon nanotubes have been prepared by the pyrolysis of acetylene–diborane mixtures in a stream of helium and hydrogen. The nanotubes so obtained have compositions around $C_{35}B$. The presence of boron favours nanotube formation and the boron content does not vary with the depth. Considering that the nitrogen-doped nanotubes, reported recently (Chem. Phys. Lett. 287 (1997) 671) have the average composition $C_{35}N$, it may be possible to use a combination of these p-type and n-type nanotubes for certain applications. © 1999 Elsevier Science B.V. All rights reserved.

1. Introduction

One of the important advances in nanotube research has been the synthesis of carbon nanotubes containing other elements. Thus, nitrogen-doped carbon nanotubes have been recently prepared [1,2] and boron–carbon–nitrogen nanotubes have also been characterized [1,3]. Since, nitrogen-substitution in carbon nanotubes gives rise to n-type doping, it becomes important to prepare boron-doped carbon nanotubes, corresponding to p-type doping. In this context, it is noteworthy that boron carbides have been prepared in film form by chemical vapour

deposition [4]. We have attempted to prepare the boron-doped carbon nanotubes by carrying out the pyrolysis of appropriate precursors in an atmosphere of helium and hydrogen. The study has shown that boron–carbon nanotubes can be prepared starting with mixtures of acetylene and diborane.

2. Experimental

B_2H_6 was first prepared by the addition of boron trifloride diethyl etherate to sodium borohydride in tetraglyme [5]. We later changed to a better procedure, wherein B_2H_6 was prepared by the addition of iodine to $NaBH_4$ in tetraglyme [6]. We have employed this procedure for most of our studies. The experimental set-up for the synthesis of the B–C nanotubes by the pyrolysis of acetylene–B_2H_6 mix-

* Corresponding author. Fax: +91 80 331 1310; e-mail: cnrrao@sscu.iisc.emet.in

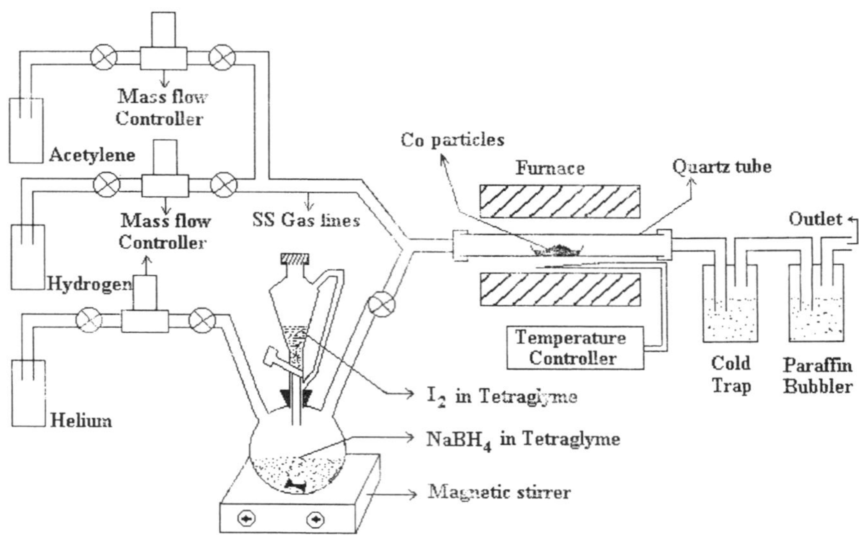

Fig. 1. Pyrolysis apparatus employed for the synthesis of boron–carbon nanotubes.

tures is shown in Fig. 1. Helium is used as the carrier gas and B_2H_6 passes through the quartz reaction tube along with C_2H_2 and H_2. The flow rates of the gases were controlled by UNIT mass flow controllers. Co nanoparticles (100 mg) of ~ 50 nm in diameter, produced by the polyol process, were placed in a quartz boat kept inside the reaction tube. The temperature of the furnace was ramped slowly up to 1000°C, under flowing hydrogen. Carbonaceous products resulting from the pyrolysis were examined with a Leica S440i scanning electron microscope (SEM). The samples were dispersed in carbon tetrachloride and deposited onto holey carbon copper grids for transmission electron microscope (TEM) observations with a JEOL JEM-3010 system.

The B–C nanotubes were characterized by X-ray photoelectron spectroscopy (XPS) using a VG scientific ESCALAB V spectrometer. Electron energy loss spectroscopy (EELS) measurements were carried out with a Gatan imaging filter system attached to a JEOL 2010 microscope and fitted with a CCD camera. The beam size was 10–150 nm, depending on the diameter of the nanotube and the dispersion was 0.3–0.5 eV. X-ray diffraction (XRD) patterns were recorded on the as obtained deposits using a SEIFERT XRD 3000TT diffractometer. Thermo-

gravimetric studies (TGA) were carried out using a Mettler Toledo TG-850 instrument.

3. Results and discussion

TEM images of the nanotubes obtained by the pyrolysis of acetylene and diborane over Co nanoparticles at 1000°C are shown in Fig. 2. The image in Fig. 2a shows the nanotubes obtained using B_2H_6 prepared by the addition of BF_3–etherate to $NaBH_4$ in tetraglyme, by maintaining the flow rate of acetylene at 20 sccm (standard cubic centimeter per minute) and a total flow rate of gases at 500 sccm. Fig. 2 are the TEM images of nanotubes obtained with B_2H_6 prepared by iodine addition to $NaBH_4$ in tetraglyme. An acetylene flow rate of 40 sccm and total flow rate of 500 sccm were employed in case of the nanotubes in seen in Fig. 2b. A flow rate of 20 sccm for acetylene with a total flow rate of 250 sccm were used for the nanotubes shown in Fig. 2c; in Fig. 2d, we show an SEM image of these nanotubes. The nanotubes have diameters in the range of 30–60 nm and were ~ 10 micron in length. They generally contain the nested cone-shaped graphene sheets in the tubular region and the metal

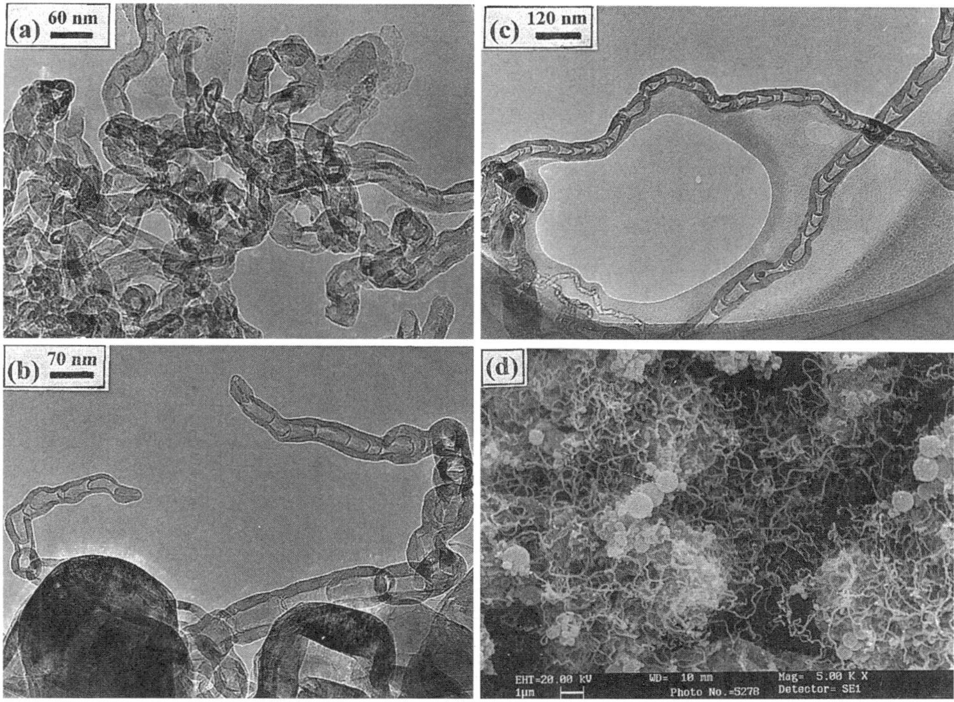

Fig. 2. TEM images of boron–carbon nanotubes prepared (a) from the pyrolysis of C_2H_2 and B_2H_6 (from BF_3–etherate addition to $NaBH_4$ in tetraglyme) at a C_2H_2 flow rate of 20 sccm and a total flow rate of 500 sccm, (b) from C_2H_2 and B_2H_6 (from iodine addition to $NaBH_4$ in tetraglyme) at a C_2H_2 flow rate of 40 sccm and a total flow rate of 500 sccm, (c) from C_2H_2 and B_2H_6 (from iodine addition to $NaBH_4$ in tetraglyme) at a C_2H_2 flow rate of 20 sccm and a total flow rate of 250 sccm and (d) an SEM image corresponding to the nanotubes in Fig. 2c.

Fig. 3. High resolution electron microscope image of nanotubes shown in Fig. 2c showing graphitic layers in the nested cone-shaped region.

 B.C. Satishkumar et al. / Chemical Physics Letters 300 (1999) 473–477

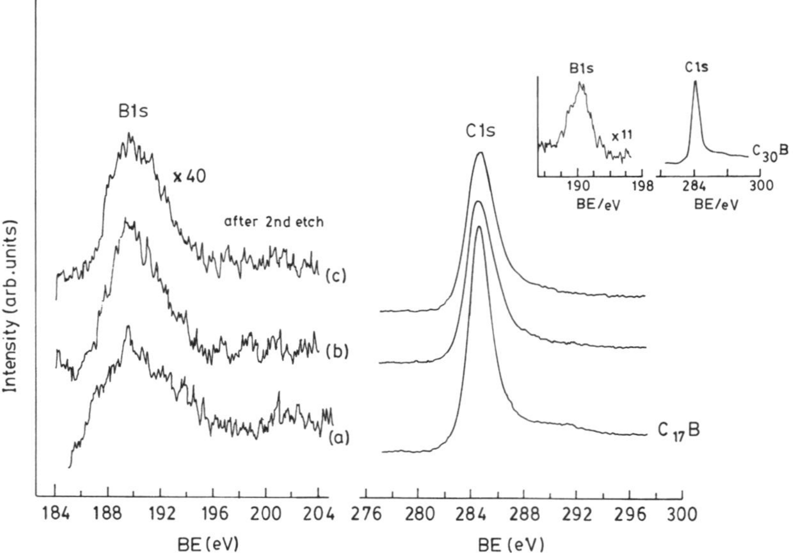

Fig. 4. B1s and C1s spectra of the nanotubes corresponding to those shown in Fig. 2, giving an average composition of $C_{35}B$. Spectra (a) are of the as-prepared nanotubes; (b) and (c) are after the first etch (20 min) and second etch (40 min) respectively. Inset shows the core-level spectra of nanotubes corresponding those shown in Fig. 2b giving an analysis of $C_{40}B$.

particles from which they grow at one end. A high resolution TEM image of the cone region is shown in Fig. 3. The growth mechanism of the B–C nanotubes appears to be similar to that of the B–C–N nanotubes [3].

Core-level X-ray photoelectron spectra of the nanotubes shown in Fig. 2 are shown in Fig. 4. The B1s feature is found as the prominant peak at 190 eV and the C1s signal at 285 eV. The shift of B1s signal towards higher binding energy compared to 188 eV in pure boron is an indication of boron substitution in the sp^2 carbon network. Based on the total B1s and C1s intensities, the boron to carbon ratios in the nanotube samples were calculated by taking the ratio of the respective integrated peak areas and dividing them by the photoionization cross-section for the 1s level. The average composition thus obtained for the nanotubes depicted in Fig. 2c works out to be $C_{17}B$. After correcting for the B1s intensity at 188 eV due to impurities, the composition works out to be close to $C_{30}B$. For the nanotubes in Fig. 2b, the core level spectra (see inset of Fig. 4) yielded a composition of $C_{30}B$ (after correction, $C_{40}B$). The B1s and C1s signals of the nanotubes shown in Fig. 2a gave the

composition of $C_{50}B$ (after correction, $C_{80}B$). The BF_3–etherate procedure of preparing B_2H_6 is not satisfactory with the experimental set-up of the type shown in Fig. 1.

In order to confirm the presence of boron in the nanotubes, it is necessary to carry out EELS analysis in the electron microscope. The EELS analysis was carried out on the different batches of nanotubes shown in Fig. 2, with the composition $C_{30}B$ from the

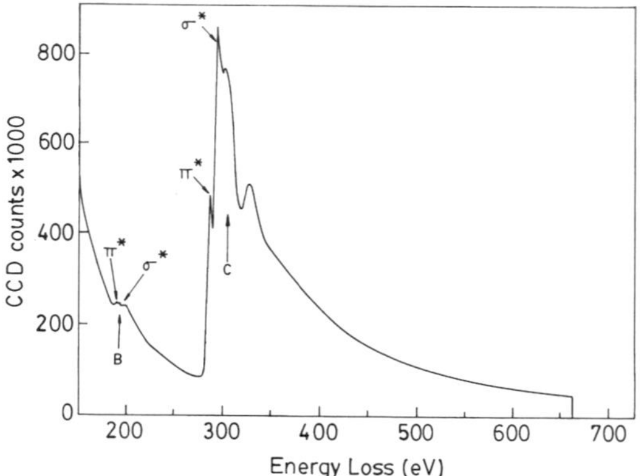

Fig. 5. EELS spectrum of the nanotubes shown in Fig. 2.

core-level spectra. These nanotubes gave characteristic edges in EELS corresponding to K-shell ionization of boron and carbon respectively. We show a typical EEL spectrum in Fig. 5. The ionization edges have sharply defined fine structures due to π^* and σ^* pre-ionizations of boron as well as carbon, characteristic of sp^2 hybridization. This is an indication of the substitution of boron at the trigonal sites in the sp^2 carbon network. The analysis of EEL spectra gave compositions to be in the range from $C_{25}B$ to $C_{50}B$. EELS analysis of the nanotubes (corresponding to Fig. 2a) with the composition of $C_{80}B$ from core-level spectra showed negligible concentration of boron.

Depth-profile experiments were carried out on the nanotubes corresponding to those in Fig. 2 with the composition of $C_{30}B$. For this purpose, B1s and C1s core-level spectra were recorded after etching with a argon ion beam of 4 keV. The core-level spectra after etching for 20 min and 40 min are shown in Fig. 4. The spectra reveal negligible variation in the boron content. It appears that boron plays an important role in the formation of these nanotubes.

X-ray diffraction studies of the different B–C nanotubes of the composition $C_{30}B$ show the (002) reflection around 3.36 Å which is lower than that found with arc-derived carbon nanotubes (3.44 Å). It may be recalled that the XRD pattern of carbon–nitrogen nanotubes of the approximate composition of $C_{35}N$ gave the reflection around 3.37 Å. Thermogravimetric studies on the B–C nanotubes show that the onset of oxidation is around 600°C, while those obtained by acetylene pyrolysis alone oxidize at 500°C. The enhanced resistance towards oxidation is an indication of improved graphitization in the boron-doped carbon nanotubes.

4. Conclusions

Boron–carbon nanotubes prepared by the pyrolysis of acetylene and diborane mixtures have an average composition of $C_{35}B$. Depth profile analysis shows that the boron content does not vary significantly with depth. The magnitude of boron doping in the B–C nanotubes is comparable to that of nitrogen in C–N nanotubes [1]. With both the p-type and and n-type nanotubes being available, one can conceive of possible applications. It may however, be necessary to prepare well-characterized B–C and C–N nanotubes with exactly known dopant concentrations before such applications become feasible.

References

[1] R. Sen, B.C. Satishkumar, A. Govindaraj, K.R. Harikumar, G. Raina, J.P. Zhang, A.K. Cheetham, C.N.R. Rao, Chem. Phys. Lett. 287 (1998) 671.

[2] R. Sen, B.C. Satishkumar, A. Govindaraj, K.R. Harikumar, M.K. Renganathan, C.N.R. Rao, J. Mater. Chem. 7 (1997) 2335.

[3] M. Terrones, A.M. Benito, C. Manteca-Diego, W.K. Hsu, O.I. Osman, J.P. Hare, D.G. Reid, H. Terrones, A.K. Cheetham, K. Prassides, H.W. Kroto, D.R.M. Walton, Chem. Phys. Lett. 257 (1996) 576.

[4] B. Ottaviani, A. Derre, E. Grivei, O.A.M. Mahmoud, M. Guimon, S. Flandrois, P. Delhaes, J. Mater. Chem. 8 (1998) 197.

[5] G.W. Kramer, A.B. Levy, M.M. Midland, Organic Synthesis via Boranes, H.C. Brown (Ed.), Wiley, New York, 1975.

[6] J.V.B. Kanth, M. Periasamy, J. Org. Chem. 56 (1991) 5964.

J. Phys. B: At. Mol. Opt. Phys. **29** (1996) 4925–4934. Printed in the UK

Novel experiments with carbon nanotubes: opening, filling, closing and functionalizing nanotubes

B C Satishkumar, A Govindaraj, J Mofokeng, G N Subbanna and C N R Rao†

Solid State and Structural Chemistry Unit and CSIR Centre of Excellence in Chemistry, Indian Institute of Science, Bangalore–560 012, India
and
Jawaharlal Nehru Center for Advanced Scientific Research, Jakkur, Bangalore–560 064, India

Abstract. Carbon nanotubes have been opened using a variety of oxidants including HF/BF_3 and OsO_4 at room temperature. The opened nanotubes have been filled with metals such as Ag, Au, Pd and Pt by simple chemical means. The opened nanotubes, containing a high concentration of acidic functional groups, get covered or closed by reaction with reagents such as ethylene glycol. The opened nanotubes can also be closed by treatment with benzene vapour in a reducing atmosphere of argon and hydrogen at high temperatures. Refluxing carbon nanotubes in a H_2SO_4–HNO_3 mixture results in a clear colourless solution, which on removal of the solvent and excess acid gives a white solid containing functionalized nanotubes. Neutralization of the acidic solution by alkali results in the precipitation of a brown solid containing nanotubes.

1. Introduction

Carbon nanotubes prepared by the arc vaporization of graphite [1–3] constitute an important new form of carbon. The nanotubes are generally closed at either end, such closure being made possible by the presence of five-membered rings. The nanotubes can be uncapped by oxidation with carbon dioxide or oxygen at elevated temperatures [4–6]. High yields of uncapped nanotubes are, however, obtained by boiling them in concentrated HNO_3 [4]. Filling such opened nanotubes by metals has also been accomplished. The best method so far reported [7, 8] involves the treatment of the nanotubes with boiling HNO_3 in the presence of metal salts such as $Ni(NO_3)_2$. The nanotubes get opened by HNO_3 and filled by the metal salt. On drying and calcination, the metal salt transforms to the metal oxide and the reduction of the encapsulated oxide in hydrogen around 673 K gives rise to the metal inside the nanotubes. We have investigated the opening of the carbon nanotubes by a variety of oxidants [9, 10] and the filling of the opened nanotubes by Ag, Au, Pd or Pt by different chemical means rather than reduction with hydrogen at high temperatures. In addition, we have examined possible ways of closing the nanotubes opened by oxidants. Besides opening, filling and closing nanotubes, we have been interested in preparing highly functionalized nanotubes. We have sought to prepare functionalized carbon nanotubes by reaction with strong acids and present rather unusual results on such nanotubes prepared from acid solutions.

† Author to whom correspondence should be addressed.

        B C Satishkumar et al

2. Experimental

We obtained carbon nanotubes from the core region of the cathode deposit produced during the arc vaporization of graphite (at 20–25 V, 100 A) in a helium atmosphere (700 Torr). The nanotubes were cleaned from the core, dispersed in methanol and sonicated for 2 hours. The portion that settled down was rich in nanotubes. The nanotubes were dried and heated at 973 K to burn the graphitic carbon [6]. The nanotubes thus prepared were used for various studies reported here. A LEICA S440i scanning electron microscope and a JEOL JEM-200CX transmission electron microscope were employed for examining the nanotubes (samples for TEM observations were dispersed on holey carbon copper grids in the presence of acetone).

In a typical experiment to open the carbon nanotubes, 100 mg of the nanotubes was refluxed in 40 ml of acid for 24 hours. In one of the methods employed to fill nanotubes with metals, 100 mg of the nanotubes was taken with 60–70 mg of the metal compound in the presence of concentrated HNO_3 and refluxed for 24–28 hours. The excess acid was distilled off and the samples were washed with acetone and dried. Acid–base titrations were carried out to determine the concentration of surface acid groups for acid-treated nanotubes. For this purpose, a known quantity of the acid-treated nanotubes was soaked in 0.005 M NaOH for 24 hours and the solution was titrated against 0.005 M potassium hydrogen phthalate in the presence of phenolphthalein indicator. In the experiments to close nanotubes (opened by nitric acid), the opened nanotubes were stirred with the organic hydroxy compounds at room temperature for 8–10 hours.

3. Results and discussion

3.1. Opening the nanotubes

The carbon nanotubes were refluxed with concentrated HNO_3, concentrated H_2SO_4, aqua regia or a $KMnO_4$ solution (acid/alkali) for 24 hours. We also treated the nanotubes with the superacid HF/BF_3, aqueous OsO_4 or OsO_4–$NaIO_4$ at room temperature for 24 hours. All these oxidants effectively opened the nanotubes, although boiling in acidified $KMnO_4$ seems to be a somewhat better procedure. The advantage with HF/BF_3 and OsO_4 is that the reaction can be carried out at room temperature. We show typical TEM images of the opened nanotubes in figures 1(a)–(d) along with schematic drawings to illustrate the nature of the openings. We see that the interaction of the nanotubes with the oxidants not only opens the tubes at the tips but also thins the tubes, the magnitude depending on the duration of the treatment. The opened nanotubes contain a large concentration of acidic carboxy and hydroxy groups [8]. Acid–base titrations of the opened nanotubes give estimates of the concentration of the surface acid groups. The concentration of the surface acid groups in the nanotubes opened by the different oxidants is in the range of 2×10^{20}–10×10^{20} acid sites/g of nanotubes, except in the case of OsO_4 which gave a concentration of 5×10^{19} sites/g (see the table below).

Table 1.

HNO_3	H_2SO_4	Aqua regia	$KMnO_4$/acid	$KMnO_4$/alkali	OsO_4/$NaIO_4$
2.5×10^{20}	6.7×10^{20}	7.6×10^{20}	8.3×10^{20}	10×10^{20}	5.2×10^{19}

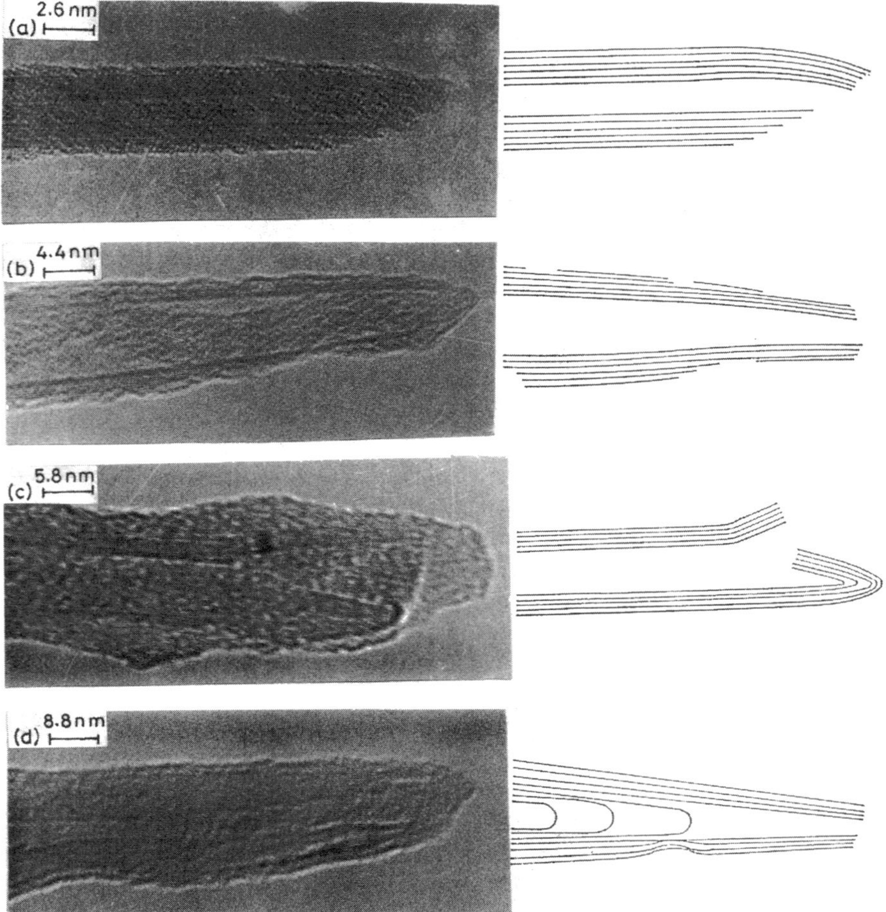

Figure 1. TEM images of carbon nanotubes opened by (*a*) boiling HNO_3, (*b*) boiling acidified $KMnO_4$, (*c*) HF/BF_3 at room temperature and (*d*) OsO_4 in H_2O at room temperature. Schematic drawings are shown to help in visualizing the opened nanotubes.

3.2. Filling the nanotubes with metals

We have examined various ways of filling the opened nanotubes with metals. The simplest way of accomplishing this in the case of metals such as Au and Pt is by merely refluxing the nanotubes with HNO_3 in the presence of $HAuCl_4$ or H_2PtCl_6 for extended periods (> 24 hours). In figures 2(*a*) and (*b*), we show the TEM images of the nanotubes filled with Au and Pt particles by such an *in situ* procedure. In figure 2(*c*), we show the TEM image of Ag particles inside the nanotube obtained by an *in situ* procedure where $AgNO_3$ was taken with HNO_3 in solution and refluxed for 24 hours. Another strategy we have employed is to first open the nanotubes with boiling HNO_3 or acidified $KMnO_4$ and then fill them with $HAuCl_4$ or H_2PtCl_6 by sonication, followed by reduction with an appropriate chemical reagent. We have employed alkaline *tetrakis* hydroxymethyl phosphonium chloride (THPC) as the reducing agent [11] in the case of Au, hydrazine in the case of Ag and hydroxylamine hydrochloride in the case of Pt. In figure 2(*d*), we show the TEM image of a nanotube filled

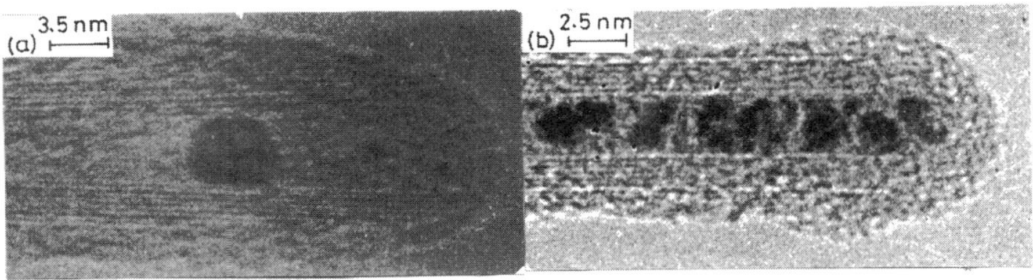

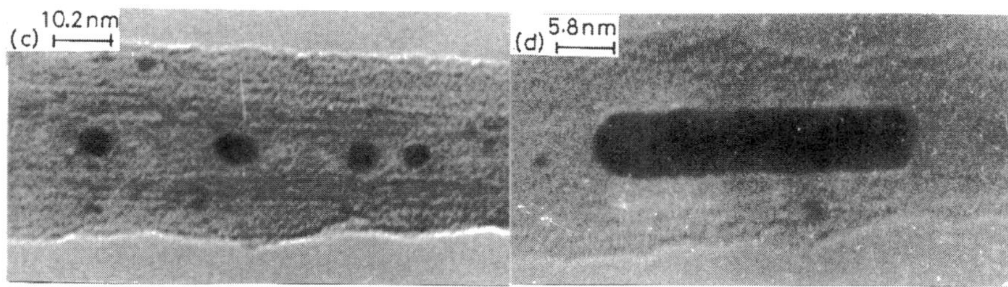

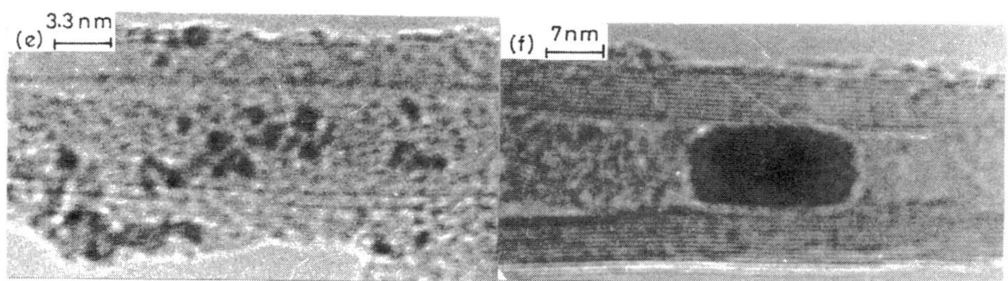

Figure 2. TEM images of carbon nanotubes filled with metals: (*a*) Au-filled nanotube obtained by refluxing in HNO_3 containing $HAuCl_4$, (*b*) Pt-filled nanotube obtained by taking H_2PtCl_6 in boiling HNO_3, (*c*) Ag-filled nanotube obtained by taking $AgNO_3$ in boiling HNO_3, (*d*) Au-filled nanotube obtained by reducing $HAuCl_4$ at room temperature with THPC, (*e*) Pd-filled nanotube obtained by reducing the metal nitrate with methanol and (*f*) Pd-filled nanotube obtained by incorporating the metal nitrate and then subjecting it to calcination and hydrogen reduction at 673 K. Some metal particles are sticking to the surface in (*e*). Electron diffraction patterns were recorded to ascertain the metallic nature of the particles inside the nanotubes.

with Au by using THPC. Metal salts incorporated into the opened nanotubes could also be reduced by using an alcohol or a glycol as the reducing agent [12]. Figure 2(*e*) shows the TEM image of a nanotube containing Pd particles, produced by the reduction of the metal nitrate by methanol. For the purposes of comparison, in figure 2(*f*) we show a nanotube filled with Pd by hydrogen reduction at 673 K. By extended sonication and optimization of the conditions, it would be possible to extensively fill the nanotubes with the metals by chemical means.

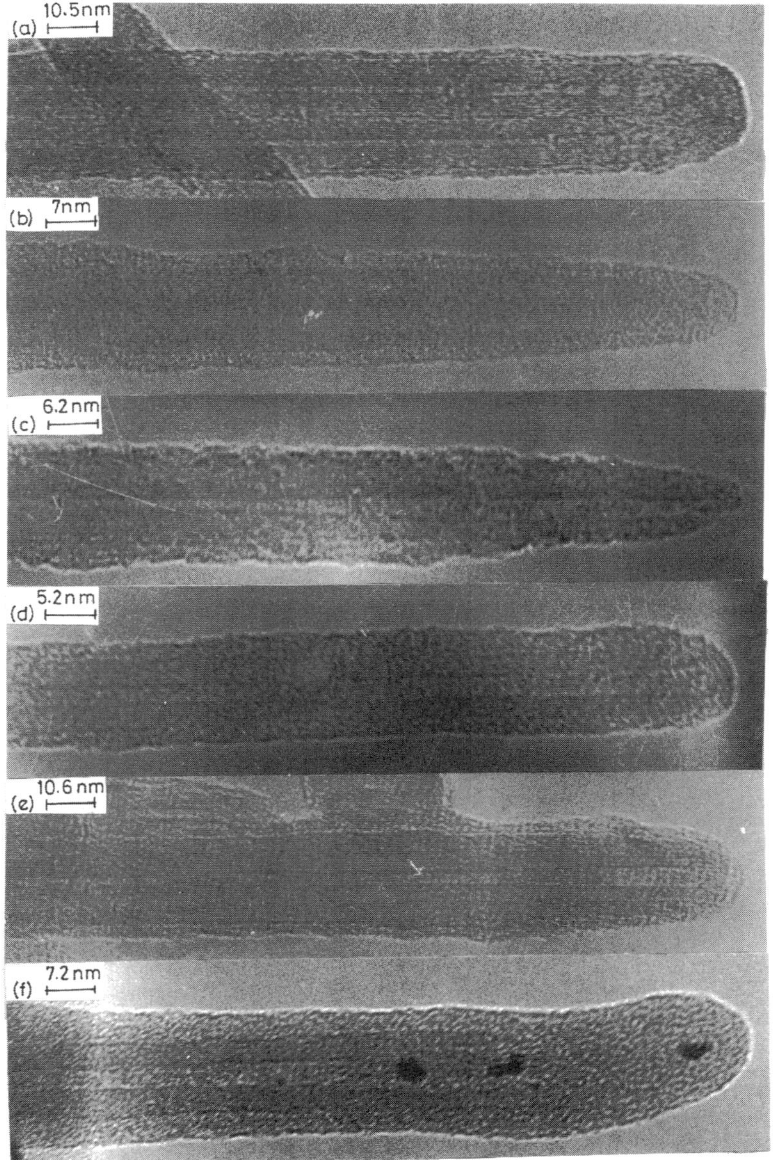

Figure 3. TEM images of opened nanotubes: (*a*) treated with methanol, (*b*) treated with methanol and heated to 673 K in air, (*c*) treated with ethylene glycol, (*d*) treated with ethylene glycol and heated to 673 K in air, (*e*) treated with propylene glycol and (*f*) filled with Pd particles and covered by interaction with ethylene glycol.

3.3. Closing the opened nanotubes

We have been making efforts to close the opened nanotubes by several means. In this process, we have tried many reagents which would react with the carboxy and/or the hydroxy groups present on the nanotubes opened by HNO_3 and have had some success with some of the organic hydroxy compounds. We show results obtained with methanol, ethylene glycol and propylene glycol in figure 3. Pristine nanotubes do not seem to interact with methanol or the glycols. The glycols sticking to the nanotubes are washed away by solvents such as diethylether and acetone. Those opened with HNO_3, however, interact to give a complete coverage. Such coverage also closes the nanotube and the coverage is not removed by solvents or by heating to 673 K as shown in figure 3. The coverage or closure of the nanotube by the hydroxy compounds is probably due to the interaction with the carboxy and the hydroxy groups present in the acid-treated nanotubes giving rise to ester and ether linkages; addition across isolated double bonds can also occur. Figure 3(f) shows the TEM image of a closed nanotube containing a few particles of Pd. The metal particles were produced by the reduction of $PdNO_3$ by ethylene glycol which also closes the tubes.

We have attempted to close the opened nanotubes by heating them in a stream of benzene vapour, argon and hydrogen at 1173 K. It is known that pyrolysis of aromatic hydrocarbons in a reducing atmosphere gives rise to interesting novel carbon species, including tubular structures. To our surprise, the opened nanotubes get closed when subjected to heating in a stream of benzene vapour, argon and hydrogen at 1173 K as can be seen from the transmission electron micrograph in figure 4(a). In figure 4(b) we show the micrograph of a nanotube containing Pt particles (prepared by the procedure outlined earlier) which was closed by the reaction with benzene vapour, Ar and H_2 at 1173 K. The fact that we can close nanotubes after opening them offers many interesting possibilities.

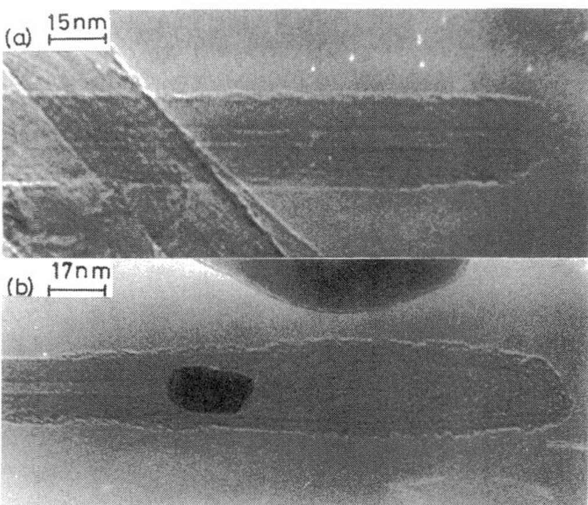

Figure 4. (a) Nanotubes closed by treatment with benzene $+$ Ar $+$ H_2 at 1173 K, (b) nanotube containing Pt particles closed by treatment with benzene $+$ Ar $+$ H_2.

3.4. *Functionalized nanotubes from acid solutions*

Carbon nanotubes which are uncapped by treatment with boiling nitric acid or other oxidants contain hydroxy and carboxylic groups [7–9], but the nanotubes retain their essential structural features during the oxidation process. In an effort to prepare carbon nanotubes containing a high proportion of acidic functional groups, we refluxed a known quantity of multi-walled carbon nanotubes with a H_2SO_4–HNO_3 mixture (3:2 by volume) for 2 hours to obtain a clear colourless solution. Removal of the solvent and the excess acid from this solution under vacuum gave a white solid residue, in contrast to the black nanotubes that

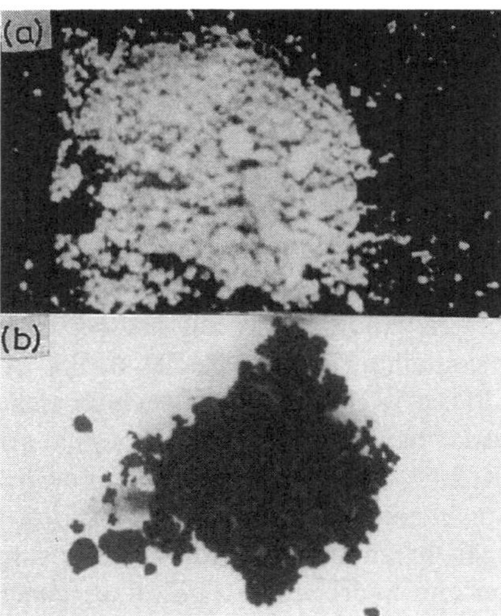

Figure 5. Photographs of (*a*) the white solid obtained from acid solution along with that of (*b*) the starting material.

Figure 6. A typical SEM image of the white solid obtained from the acidic solution of nanotubes.

we started with (figure 5). The white solid was washed extensively with distilled water and dried before subjecting it to various investigations. Scanning electron micrographs of the white solid show the presence of nanotube-like structures (figure 6). The nanotubes are clearly seen in the transmission electron microscope images. Some of the nanotubes are isolated, but most of them appear to be present as aggregates. In figure 7, we show typical TEM images of the nanotube structures found in the white solid. The images do not clearly reveal the graphitic fringes, possibly because of extensive functionalization of the nanotubes. Acid–base titrations indicate that there are 10^{21} acidic sites/g in the white solid, much higher than in nanotubes uncapped by oxidants such as HNO_3. The C(1s) core-level spectrum of the white solid (recorded with a VG ESCALAB spectrometer) does not show the $\pi-\pi^*$ shake-up satellite (figure 8(*a*)), indicating the absence of extended conjugation as in the parent nanotubes. The core-level spectrum also shows evidence of sulphur and

4932   *B C Satishkumar et al*

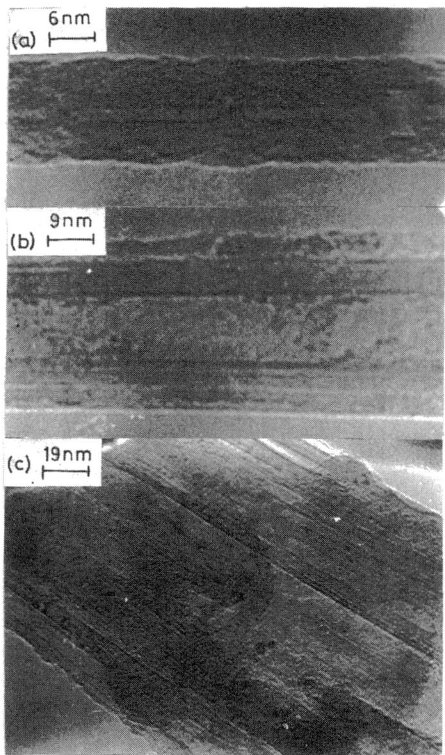

Figure 7. TEM images of the white solid obtained from the acidic solution of the nanotubes.

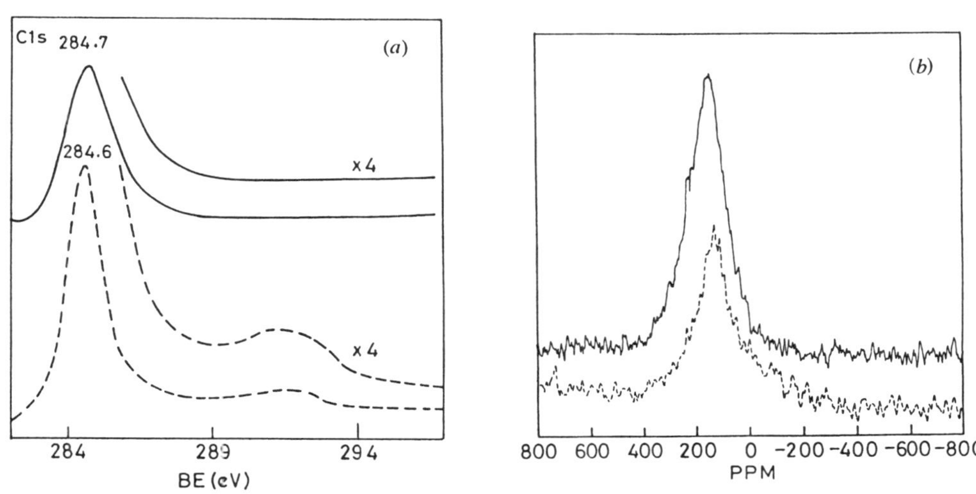

Figure 8. (*a*) C(1s) spectra of the white solid obtained from the acid solution of nanotubes (full curve) and of the parent carbon nanotubes (broken curve), (*b*) ^{13}C NMR spectra of carbon nanotubes (broken curve) and of the white solid (full curve) recorded at a magnetic field of 7.05 T with a resonance frequency of 75.467 MHz.

oxygen, the surface concentration of carbon, oxygen and sulphur bearing the ratio 2:5:1, pointing to the presence of SO_3H, OH and other groups.

The infrared spectrum of the white solid (figure 9) shows the characteristic bands due to OH (3300 cm^{-1}, very broad), SO_3H (1050, 1170, 1350 cm^{-1}) and C=C/C=O(1600 cm^{-1}, broad) groups. The ^{13}C NMR spectrum of the solid (recorded with a Brüker MSL300 spectrometer) shows a broad signal centred around 145 ppm compared to 130 ppm of the parent nanotube (figure 8(*b*)). Such a downfield shift can arise from the presence of OH and SO_3H groups. The solid, however, comprises isolated or conjugated C=C bonds, since there is no evidence for the presence of a large proportion of sp^3 carbons in the NMR spectrum. The powder x-ray diffraction pattern (recorded with Cu Kα radiation) of the white solid exhibits a relatively intense (002) reflection at a somewhat smaller angle ($2\theta = 25.22°$) than the parent nanotubes ($2\theta = 25.73°$), indicating a slight expansion of the spacing between the graphitic sheets. These characterization studies are consistent with the preponderant presence of functionalized nanotubes in the white solid.

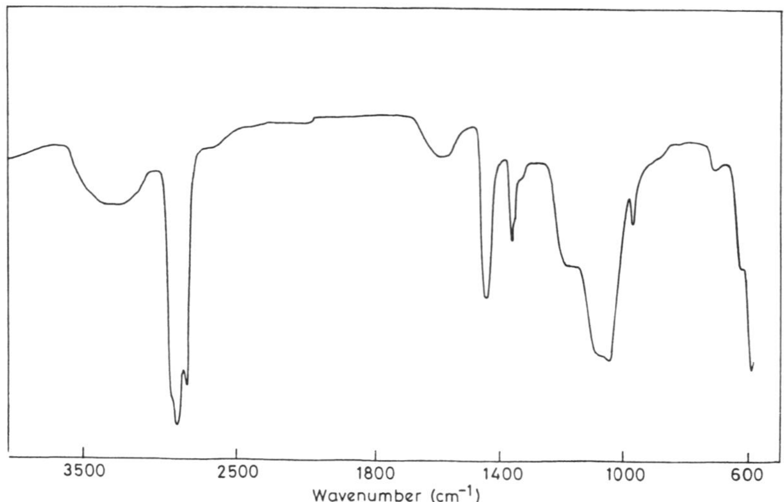

Figure 9. Infrared spectrum of the white solid recorded in nujol mull.

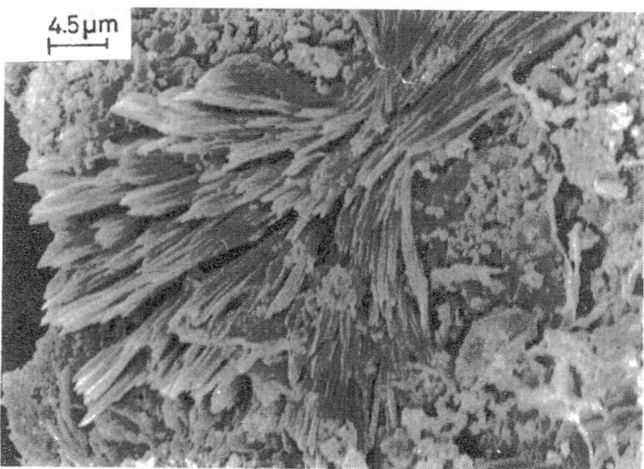

Figure 10. SEM image of the light brown solid obtained by neutralizing the acid solution of nanotubes with alkali.

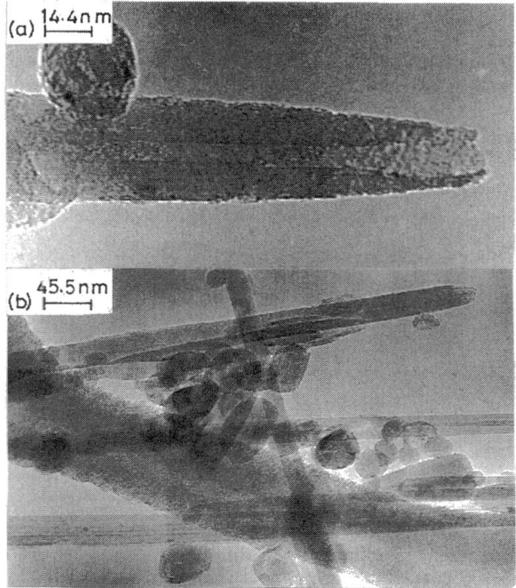

Figure 11. TEM images of the light brown solid.

Instead of distilling off the solvent from the acidic solution of the nanotubes to obtain the white solid, we have treated the solution with $BaCO_3$ to remove the excess HSO_4 and then distilled off the remaining water to get a solid. This solid also contains the nanotube structures. On neutralizing the acid solutions of nanotubes with NaOH, we find the gradual emergence of a light brown solid. The SEM image of this solid is shown in figure 10. The IR spectrum of this solid has the features due to OH ($3450\ cm^{-1}$, sharp) and SO_3^- ($1100-1230\ cm^{-1}$) and C=C/C=O ($1680\ cm^{-1}$) groups. The TEM images of the brown solid show the presence of the nanotubes with a higher proportion of well separated tubes (figure 11). It appears that the presence of SO_3H groups in the white solid gives rise to a greater association of nanotubes. The observation of the functionalized nanotubes in the solids extracted from acid solutions is of considerable interest. This could lead to a study of possible uses of functionalized nanotubes and their transformations. It is indeed remarkable that such large structures of carbon (a micron or more in length) remain in the solution phase and are regenerated in the solid state on the removal of the solvent.

References

[1] Iijima S 1991 *Nature* **354** 56

[2] Ebbesen T W and Ajayan P M 1992 *Nature* **358** 220

[3] Rao C N R, Seshadri R, Govindaraj A and Sen R 1995 *Mat. Sci. Eng.* R **15** 209

[4] Ajayan P M, Ebbesen T W, Ichihashi T, Iijima S, Tanigaki K and Hiura H 1993 *Nature* **362** 522

[5] Tsang S C, Harris P J F and Green M L H 1993 *Nature* **362** 520

[6] Seshadri R, Govindaraj A, Aiyer H N, Rahul Sen, Subbanna G N, Raju A R and Rao C N R 1994 *Current Sci. (India)* **66** 839

[7] Tsang S C, Chen Y K, Harris P J F and Green M L H 1994 *Nature* **372** 159

[8] Lago R M, Tsang S C, Lu K L, Chen Y K and Green M L H 1995 *J. Chem. Soc., Chem. Commun.* 1355

[9] Hwang K C 1995 *J. Chem. Soc., Chem. Commun.* 173

[10] Hiura H, Ebbesen T W and Tanigaki K 1995 *Adv. Mater.* **7** 275

[11] Seshadri R, Subbanna G N, Vijayakrishnan V, Kulkarni G U and Rao C N R 1995 *J. Phys. Chem.* **99** 5639

[12] Vasan H N and Rao C N R 1995 *J. Mater. Chem.* **5** 1755

J. Phys. D: Appl. Phys. **29** (1996) 3173–3176. Printed in the UK

RAPID COMMUNICATION

The decoration of carbon nanotubes by metal nanoparticles

B C Satishkumar†, Erasmus M Vogl‡, A Govindaraj§ and C N R Rao†‡§*

† Solid State and Structural Chemistry Unit, Indian Institute of Science, Bangalore-560012, India
‡ Jawaharlal Nehru Centre for Advanced Scientific Research, Indian Institute of Science, Bangalore-560012, India
§ CSIR Centre of Excellence in Chemistry, Indian Institute of Science, Bangalore-560012, India

Received 3 September 1996

Abstract. Acid-treated carbon nanotubes have been decorated with nanoscale Au, Pt and Ag clusters by employing several procedures. The nature and extent of metal coverage can be varied by changing the concentration of the metal compound or by mild sonication (treatment with ultrasound). The metal clusters seem to get deposited on the acidic surface sites of the nanotubes.

Carbon nanotubes constitute a novel class of nanomaterials with potential applications in catalysis and other areas [1–4]. Carbon nanotubes have been found to be good electron emitters and may be used in flat-screen visual displays [5]. We were interested in preparing decorated nanotubes by depositing metal clusters onto nanotube surfaces. Efforts to deposit metal clusters directly on the surface of pristine carbon nanotubes by various chemical means were not quite successful but gave rise to considerable agglomeration of the clusters. In figure 1 we show a typical transmission electron microscopy (TEM) image of agglomerated gold clusters sticking to the surface of nanotubes upon refluxing them in a solution of $HAuCl_4$ (5 mg in 10 ml distilled water) with alkaline tetrakis hydroxymethyl phosphonium chloride (THPC), the latter being an excellent reducing agent for this purpose [6]. In order to obtain nicely decorated nanotubes with well-dispersed metal clusters of nanometric dimensions adhering to the nanotubes, we have made use of acid-treated nanotubes. It is known that refluxing carbon nanotubes with nitric acid not only opens the closed tips of the tubes, but also creates acid sites on the surface [7, 8]. The number of acid sites is estimated to be around 10^{21} sites per gram. The acid sites are composed of functional groups such as COOH and OH, which can act as nucleation centres for metal ions. In this letter we present the important results related to the decoration of nanotubes by Au, Pt and Ag.

Carbon nanotubes were obtained by the arc vaporization of graphite in a helium atmosphere (650 Torr) at 20V (DC), 200 A. The nanotubes from the core region of cathode deposit were cleaned and heated at 700 °C to burn the graphite carbon. The clean nanotubes were refluxed with

* E-mail: cnrrao@sscu.iisc.ernet.in.

concentrated nitric acid (16 N) for 12 h. The acid-treated nanotubes were washed with distilled water and dried at 100 °C. The acid-treated nanotubes so prepared were used for decorating with gold, platinum or silver by the reduction of a metal compound by an appropriate reducing agent. Alternatively, we have refluxed the nanotubes with the metal compound and HNO_3 to obtain decorated nanotubes. We have examined the metal-covered nanotubes by TEM.

In figure 2(a) we show small clusters of gold deposited on the nanotube surface by refluxing 12 mg of nanotubes with HNO_3 in the presence of 5 mg $HAuCl_4$. The gold particles have a diameter of less than ~5 nm. Electron diffraction patterns showed them to be crystalline with an fcc structure. Instead of taking $HAuCl_4$ with the nanotubes during the acid treatment, we have been able to deposit gold clusters on the nanotubes by refluxing 25 mg of acid-treated nanotubes with 5 mg of $HAuCl_4.H_2O$ in 10 ml water for 4 h in the presence of alkaline THPC. The TEM image of nanotubes so decorated is shown in figure 2(b). The size of the gold clusters is in the range of 1–7 nm. The gold clusters in figure 2(c) are less than 2 nm in diameter, although they were prepared by the same procedure as those in figure 2(b), because the nanotubes were subjected to mild sonication to obtain better dispersion. Intense sonication results in some of the gold particles entering the nanotube. It may be noted that metals such as Au, Pt and Pd have indeed been incorporated into the acid opened nanotubes by various means [7, 9].

In figure 3, we present TEM images of acid-treated nanotubes decorated with platinum particles. Figure 3(a) shows a TEM image of decorated nanotubes prepared by refluxing 10 mg of H_2PtCl_6 with 20 mg nanotubes in

Rapid communication

Figure 1. A TEM image of pristine nanotubes with agglomerated gold particles.

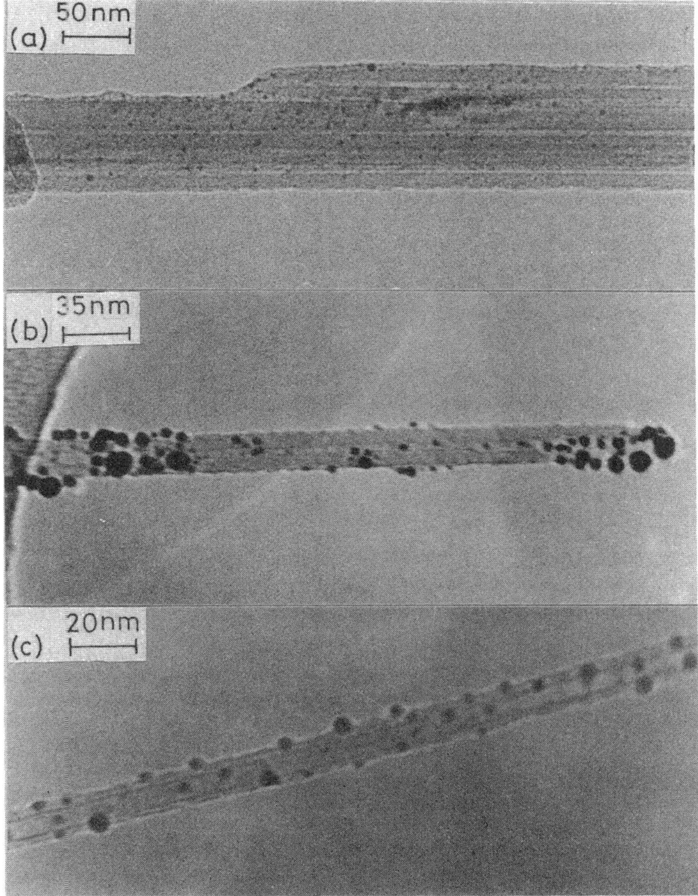

Figure 2. TEM images of nanotubes decorated by Au nanoparticles; decoration carried out (a) by refluxing with $HAuCl_4$ and HNO_3, (b) by refluxing acid treated nanotubes with $HAuCl_4$ and THPC and (c) by the same procedure as (b) except with mild sonication.

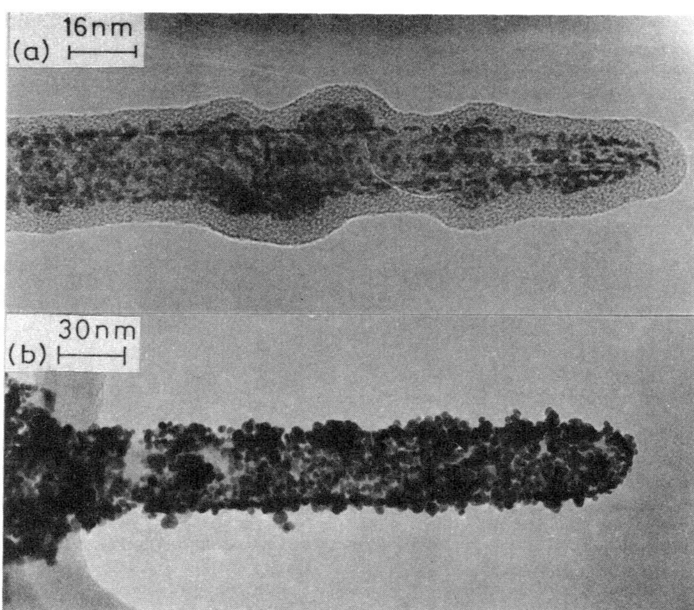

Figure 3. TEM images of nanotubes decorated by Pt nanoparticles: decoration carried out (*a*) by refluxing with H_2PtCl_6 and HNO_3 and (*b*) by refluxing with H_2PtCl_6 and ethylene glycol.

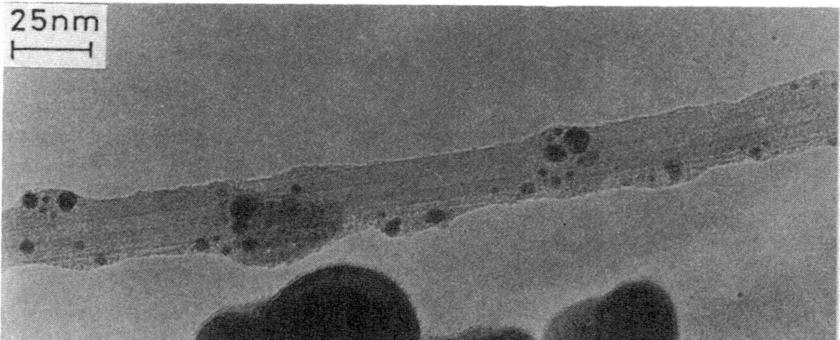

Figure 4. A TEM image of a nanotube decorated by Ag clusters.

HNO_3 for 48 h. The Pt clusters are less than 2 nm in diameter. In figure 3(*b*) we show a TEM image of nanotubes decorated with Pt, obtained by refluxing 30 mg of the acid-treated nanotubes with 20 mg H_2PtCl_6 in dilute ethylene glycol (50 ml) for 4 h. Here ethylene glycol acts as the reducing agent. Both in figure 3(*a*) and (*b*) the nanotubes are extensively covered with Pt clusters, but the particle diameter is generally less than 3 nm. One can obtain a smaller coverage of the Pt particles by using smaller proportion of the platinum chloride in the reaction.

In figure 4, we show a nanotube covered with small silver particles. Here 25 mg of carbon nanotubes were refluxed with HNO_3 along with 10 mg of $AgNO_3$ for 48 h. One can improve the coverage and dispersion by employing different reducing agents. After the completion of the present study, we noticed a recent literature report [10] on carbon nanotubes decorated by Ag.

An examination of the different metal-decorated carbon nanotubes suggests that the metal nanoparticles get deposited on the acid sites which are dispersed all over the surface of the nanotubes. This is different from the decoration of oriented graphite crystals where the metal clusters preferentially get deposited along dislocations and grain boundaries.

References

[1] Iijima S 1991 *Nature* **354** 56
[2] Ebbesen T W and Ajayan P M 1992 *Nature* **358** 220
[3] Rao C N R, Seshadri R, Govindaraj A and Sen R 1995
    Mater. Sci. Eng. **R15** 209
[4] Planeix J M, Coustel N, Coq B, Brotons V, Kumbhar P S,
    Dutartre R, Geneste P, Bernier P and Ajayan P M 1994
    J. Am. Chem. Soc. **116** 7935
[5] de Heer W A, Chatelain A and Ugarte D 1995 *Science* **270**
    1179
[6] Seshadri R, Subbanna G N, Vijayakrishnan V, Kulkarni G
    U and Rao C N R 1995 *J. Phys. Chem.* **99** 5639

Rapid communication

[7] Tsang S C, Harris P J F and Green M L H 1993 *Nature* **362** 520

[8] Satishkumar B C, Govindaraj A, Mofokeng J, Subbanna G N and Rao C N R 1996 *J. Phys. B: At. Mol. Opt. Phys.* at press

[9] Tsang S C, Chen Y K, Harris P J F and Green M L 1994 *Nature* **372** 159

[10] Ebbesen T W, Hiura H, Bisher M E, Treacy M J, Shreeve-Keyer J L and Haushalter R C 1996 *Adv. Mater.* **8** 155

30 April 1999

ELSEVIER

Chemical Physics Letters 304 (1999) 207–210

**CHEMICAL
PHYSICS
LETTERS**

A study of micropores in single-walled carbon nanotubes by the adsorption of gases and vapors

M. Eswaramoorthy, Rahul Sen, C.N.R. Rao *

Chemistry and Physics of Materials Unit, Jawaharlal Nehru Centre for Advanced Scientific Research, Jakkur P.O., Bangalore 560064, India

Received 2 February 1999; in final form 8 March 1999

Abstract

Adsorption of N_2, benzene and methanol have been studied on as-prepared single-walled carbon nanotubes (SWNT) as well as SWNTs treated with HCl and HNO_3. These nanotubes are good microporous materials with total surface areas well above 400 m^2/g and internal surface areas of 300 m^2/g or higher. Benzene molecules are shown to be adsorbed within the pores of the SWNTs. © 1999 Elsevier Science B.V. All rights reserved.

1. Introduction

Carbon nanotubes are being investigated as catalyst supports and there is considerable interest in understanding the surface and porous properties of these materials. We have been concerned with these aspects for some time and have investigated adsorption of N_2 and other molecules on single-walled carbon nanotubes (SWNT). Inoue et al. [1] have reported the mesoporous nature of multi-walled carbon nanotubes (MWNT). These workers have measured N_2 adsorption isotherms on MWNTs and used the data to derive the internal and external surface areas. We have measured the adsorption isotherms of N_2 on well-characterized SWNTs, which are more desirable as materials since we only deal with the

Fig. 1. N_2 adsorption isotherms for SWNTs at 77 K. Inset: Hysteresis in the adsorption–desorption isotherms the for as-prepared SWNT sample.

---

* Corresponding author. Fax: +91 80 846 2766; e-mail: cnrrao@jncasr.ac.in

 M. Eswaramoorthy et al. / Chemical Physics Letters 304 (1999) 207–210

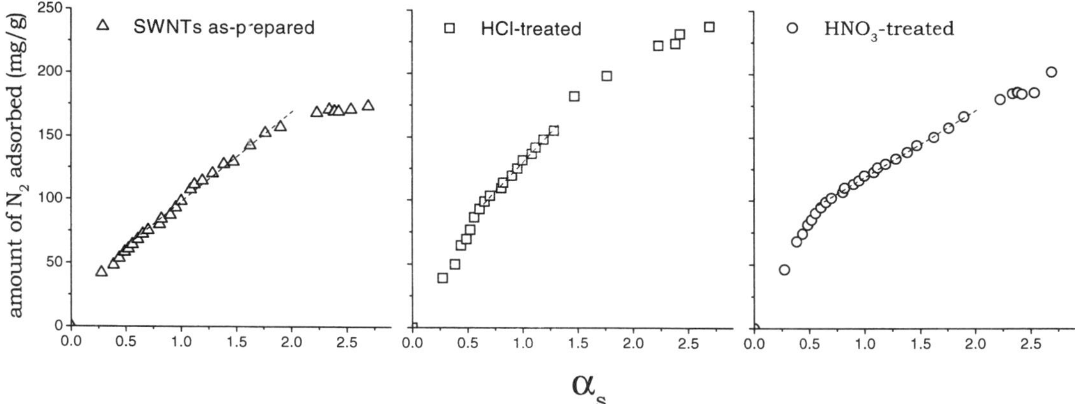

Fig. 2. α_s plots for the different SWNTs.

pore in the tubule without worrying about the surrounding graphitic layers. Furthermore, the pore size distribution is expected to be narrow with SWNTs. The present study is concerned with microporous nature of SWNTs, as distinct from the mesoporous MWNTs examined by Inoue et al. [1].

2. Experimental

SWNTs were prepared by the Y_2O_3/Ni-catalyzed arcing of graphite electrodes in a helium atmosphere [2,3]. Most of the SWNTs occur as bundles with a small fraction present as isolated nanotubes. The as-prepared nanotubes were treated with conc. HCl to open the tips [3,4]. We have also treated a batch of SWNTs with conc. HNO_3 which is known to eliminate amorphous and other carbon impurities [3,5]. We have verified this by TEM studies. The HNO_3 treatment also opens the nanotubes [6,7]. We would also expect the surfaces of the SWNTs treated by acids, especially HNO_3, to be somewhat modified by

the presence of acidic functional groups [7]. We have carried out adsorption measurements on the as-prepared SWNTs as well as the HCl and the HNO_3-treated materials.

3. Results and discussion

In Fig. 1 we show the adsorption isotherms of N_2 at 77 K on the as-prepared SWNTs and also those obtained after treatment with HCl and HNO_3. At low p/p_o, the adsorption isotherm is Type II, which also include features of a Type I isotherm due to micropores. A substantial increase in the uptake of N_2 occurs above $p/p_o = 0.6$ in all the cases, due to capillary condensation. There is also some hysteresis in the high p/p_o regime (see inset of Fig. 1). We consider the hysteresis as well as the capillary condensation at $p/p_o > 0.6$ to be due to the intertubular space. By making use of the adsorption data we have been able to obtain high-resolution α_s-plots, where $\alpha_s = V_{(p/p_o)}/V_{0.4}$; $V_{(p/p_o)}$ is the amount adsorbed at

Table 1
Surface properties of SWNTs

	Internal surface area (m^2/g)	External surface area (m^2/g)	Total surface area (m^2/g)
As-prepared SWNTs	233	143	376
HCl-treated SWNTs	297	186	483
HNO_3-treated SWNTs	323	106	429

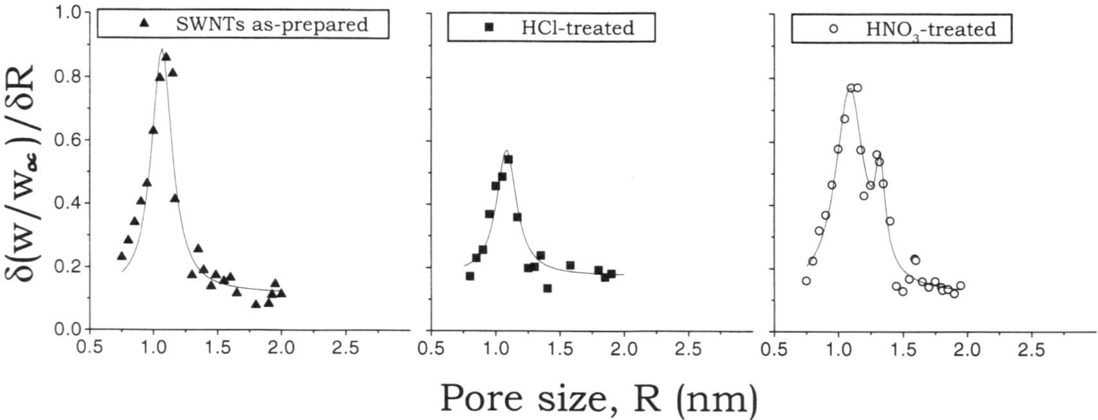

Fig. 3. Pore size distribution of the different SWNTs obtained by the Horvath–Kawazoe (HK) method.

a particular p/p_o value on non-porous carbon black used as the reference material, and $V_{0.4}$ is the amount adsorbed at $p/p_o = 0.4$ on the reference material [8]. In Fig. 2, we show the α_s-plots for the as-prepared, HCl-treated and HNO$_3$-treated SWNTs. These plots differ from the α_s-plot reported by Inoue et al. [1] who found the linear region passing through the origin due to absence of micropores. These workers employed the deviation from linearity to estimate the features of the mesopores. In the data obtained by us, we find linearity in α_s range of 0.7–2.0 in all the cases. The linear region gives an intercept, thus confirming the microporous nature of the material. The slope of the linear region provides direct information about the external surface area [9]. The α_s-plot in the region where $\alpha_s < 0.6$, shows features associated with the cooperative filling of micropores and the slope obtained from this region gives information about the internal surface area [10,11]. The values of the internal and external areas obtained for the different SWNTs samples are listed in Table 1.

We see from the data in Table 1 that acid treatment increases the internal as well as the total surface area. The high internal surface found with the HNO$_3$-treated SWNTs suggests that HNO$_3$ treatment probably opens up the nanotubes more fully, as expected [3,6,7]. Accordingly, the ratio of the internal to external surface area is highest in the HNO$_3$-treated SWNTs, being close to 3. This ratio is only 1.6 in the other two samples of SWNTs. We were able to obtain the pore size distribution by the Hor-

vath–Kawazoe (HK) method [12] and present the results in Fig. 3. All the three samples show a major peak around 1.1 nm, corresponding to the average diameter of SWNTs. The HNO$_3$-treated SWNTs show an additional peak around 1.3 nm, possibly due to the removal of some of the graphitic layers near the opened tip.

In Fig. 4 we show adsorption isotherms of benzene on the as-prepared and HNO$_3$-treated SWNTs at 298 K. The adsorption isotherms are of Type II.

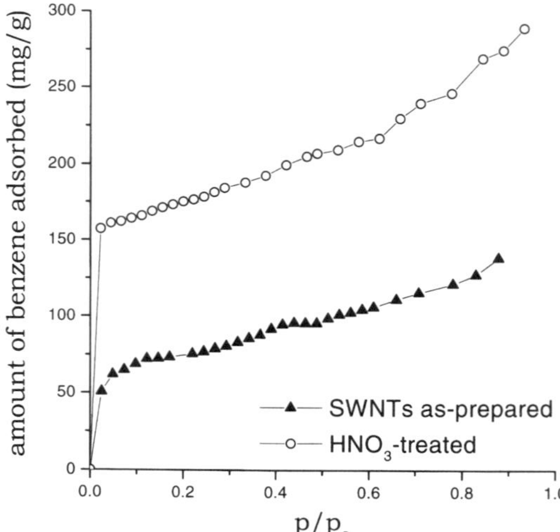

Fig. 4. Benzene adsorption isotherms for SWNTs samples at 298 K.

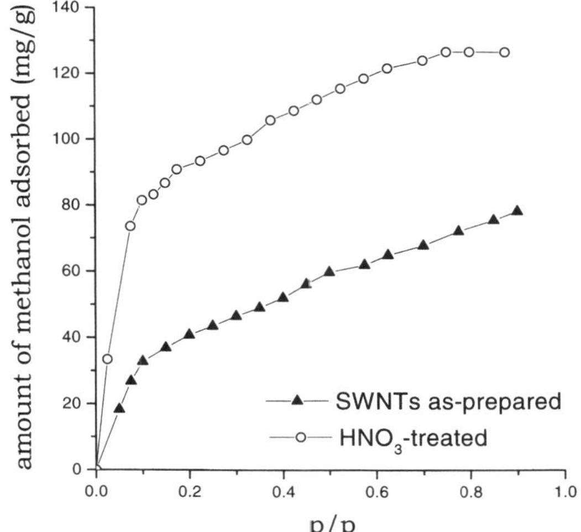

Fig. 5. Methanol adsorption isotherms for SWNTs samples at 298 K.

The HNO$_3$-treated sample shows a much higher intake of benzene than the as-prepared sample. By making use of the surface areas from N$_2$ adsorption data (Table 1) and the monolayer adsorption of benzene from Fig. 4, we have calculated the cross-sectional area, a_m, of benzene. The as-prepared SWNTs give am a_m value of 30 Å^2 when the external surface area was used and a high value of 80 Å^2 when the total surface area was used. A value of 30 Å^2 suggests that some of the benzene molecules may have standing orientation while a value of 80 Å^2 implies incomplete coverage. The HNO$_3$-treated SWNTs, however give a value of 40 Å^2 when total surface area is used and 30 Å^2 when the internal surface area is used. The value of 40 Å^2 suggests that the benzene molecules are flat on the nanotube surface [13]. It is to be noted that benzene can enter inside SWNTs in the case of the HNO$_3$-treated sam-

ple, and not in the as-prepared sample. We have measured the adsorption isotherms of methanol at 298 K and find the isotherms to be similar to those obtained with benzene (Fig. 5). We have not treated these data quantitatively.

In conclusion, the present study of the adsorption of N$_2$ on SWNTs shows that we can obtain good microporous nanotubes, particularly on acid treatment. The micropores can accommodate molecules such as benzene. Catalytic reactions carried out in the pores would be of considerable interest.

References

[1] S. Inoue, N. Ichikuni, T. Suzuki, T. Uematsu, K. Kaneko, J. Phys. Chem. 102 (1998) 4689.
[2] C. Journet, W.K. Maser, P. Bernier, A. Loiseau, M. Lamy de la Chapelle, S. Lefrant, P. Deniard, R. Lee, J.E. Fisher, Nature 388 (1997) 756.
[3] C.N.R. Rao, A. Govindaraj, R. Sen, B.C. Satishkumar, Mat. Res. Innovat. 2 (1998) 128.
[4] J. Sloan, J. Hammer, M.Z. Sibley, M.L.H. Green, Chem. Commun. (1998) 347.
[5] E. Dujardin, T.W. Ebbesen, A. Krishnan, M.M.J. Treacy, Adv. Mater. 10 (1998) 611.
[6] S.C. Tsang, Y.K. Chen, P.J.F. Harris, M.L.H. Green, Nature 372 (1994) 159.
[7] C.N.R. Rao, A. Govindaraj, B.C. Satishkumar, Chem. Commun. (1996) 1525.
[8] A. Sayari, P. Liu, M. Kruk, M. Jaroniec, Chem. Mater. 9 (1997) 2499.
[9] K.S.W. Sing, in: D.H. Everett, R.H. Ottewill (Eds.), Surface Area Determination, Butterworths, London, 1970, p. 25.
[10] K.S.W. Sing, R. Rouquerol, in: G. Ertl, H. Knözinger, J. Weitkamp (Eds.), Handbook of Heterogenous Catalysis, vol. 2, VCH, Weinheim, 1997, p. 427.
[11] K.S.W. Sing, Colloids Surf. 38 (1989) 113.
[12] G. Horvath, K.J. Kawazoe, Chem. Eng. Jpn 16 (1983) 470.
[13] A.L. McClellan, H.F. Harnsberger, J. Colloid Interface Sci. 23 (1967) 577.

PHYSICAL REVIEW B, VOLUME 63, 205417

Pressure-induced phase transformation and structural resilience of single-wall carbon nanotube bundles

Surinder M. Sharma,[1] S. Karmakar,[1] S. K. Sikka,[1] Pallavi V. Teredesai,[2] A. K. Sood,[2,3] A. Govindaraj,[3] and C. N. R. Rao[3]

[1]*High Pressure Physics Division, Bhabha Atomic Research Center, Mumbai 400 085, India*
[2]*Department of Physics, Indian Institute of Science, Bangalore 560 012, India*
[3]*Chemistry and Physics of Materials Unit, Jawaharlal Nehru Centre for Advanced Scientific Research, Jakkur Campus,*
Jakkur P.O., Bangalore 560 064, India
(Received 12 January 2001; published 2 May 2001)

We report here an *in situ* x-ray diffraction investigation of the structural changes in carbon single-wall nanotube bundles under quasihydrostatic pressures up to 13 GPa. In contrast with a recent study [Phys. Rev. Lett. **85**, 1887 (2000)] our results show that the triangular lattice of the carbon nanotube bundles continues to persist up to $\sim$10 GPa. The lattice is seen to relax just before the phase transformation that is observed at $\sim$10 GPa. Further, our results display the reversibility of the two-dimensional lattice symmetry even after compression up to 13 GPa well beyond the 5 GPa value observed recently. These experimental results explicitly validate the predicted remarkable mechanical resilience of the nanotubes.

DOI: 10.1103/PhysRevB.63.205417　　　　PACS number(s): 61.48.+c, 61.10.Nz, 61.50.Ks, 81.40.Vw

I. INTRODUCTION

Due to the quasi-one-dimensional structure, single-wall carbon nanotubes (SWNTs) have been shown to have some unique and interesting physical properties.[1] These nanotubes have also been predicted to have extraordinary mechanical properties such as enormous flexibility in terms of complete structural reversibility on bending up to 110°.[2] In addition, molecular dynamics (MD) simulations predict that SWNTs may undergo fully reversible morphological changes under extreme deformations.[3,4] The synthesis of bundles of single-wall carbon nanotubes, with a narrow size distribution, has provided tremendous impetus to the experimental investigations.[5] Several high-pressure Raman investigations have been carried out recently.[6–9] In all of these studies, Raman intensity reduces dramatically beyond a few GPa and this has been suggested to be due to the loss of the electronic resonance in the Raman scattering cross-section because of the faceting of the neighboring tubes.[6] A slight change in the slope of pressure-induced Raman shifts at $\sim$1.7 GPa has also been ascribed to a structural transformation from a triangular to a monoclinic lattice.[7] Recent Raman investigations by Teredesai *et al.*[9] indicate a structural phase transition near 10 GPa. This was conjectured to be due to faceting, as the frequency of the tangential mode approaches that of graphite. Also, all the Raman studies indicate reversibility of behavior upon unloading of the pressure. In particular, the data of Ref. 9 demonstrated the reversibility on pressure release from 25.9 GPa. However, due to the lack of information about the structural evolution of SWNTs under pressure, it was not possible to unambiguously relate these measurements to the microscopic changes in the SWNTs. A recent x-ray diffraction study[10] of SWNTs under pressure, suggests the vanishing of the triangular lattice at $\sim$1.5 GPa and its regeneration if unloaded from less than 4 GPa. Beyond 5 GPa, these x-ray results indicate an irreversible change in total contrast to the results of Ref. 9. High-pressure behavior of SWNTs has also been investigated under nonhydrostatic stresses, using a piston-cylinder device without any pressure medium. This study, restricted to $\sim$2.9 GPa, displays a reversible increase in the density of SWNTs to almost that of graphite.[11] This has been speculated to be due to the crushing or flattening of the cross section of the nanotubes from circular to elliptical shape under the nonhydrostatic stresses. So, to find a consistent interpretation of several experimental results mentioned above, it is necessary to investigate SWNTs using an intense x-ray radiation as the diffracted intensities are likely to be rather weak. In addition, it will be interesting to compare the behavior of SWNTs with that of other carbon polymorphs, e.g., C_{60} and C_{70} fullerenes that undergo irreversible and reversible pressure-induced amorphization.[12–14] We present here an *in situ* x-ray diffraction investigation of SWNT bundles and relate the results to other experimental and theoretical studies.

II. EXPERIMENTAL DETAILS

SWNT bundles were prepared by the standard arc discharge method. For this, a composite rod, made by filling powders of graphite, Y_2O_3 (1at. % Y) and Ni (4.2 at. % Ni) in a hole, was used as an anode and a simple graphite rod as a cathode. The material produced through a dc arc in He atmosphere was appropriately washed with several chemicals, decanted, filtered, dried, and characterized by transmission electron microscopy.[9,15] At ambient conditions, the x-ray diffraction peak corresponding to (1,0) plane of the two-dimensional triangular lattice shows that our SWNTs correspond to a lattice constant a_0 of 17.97 Å. As the curvature of the nanotubes reduces the contact area on which the repulsive forces act on the tubes, the intertube gap is expected to be smaller than the (002) spacing of graphite. If the intertube gap is taken as 3.12 Å,[10] present SWNTs correspond to a tube diameter of 14.85 Å. Therefore, our sample consists of either (11,11) armchair tubes or (19,0) zigzag tubes or any other appropriate combination of integers n and m.[17] Thermogravimetric analysis of the sample showed graphite abundance to be $\sim$8%. The sample, containing ran-

   PHYSICAL REVIEW B **63** 205417

domly oriented rope bundles of ~100 Å diameter and several micron length, was loaded in a steel gasket hole (~200 μm diameter) of a diamond anvil cell (DAC). A tiny ruby chip was used to measure the pressure with methanol-ethanol (4:1) as a pressure transmitter. Angle dispersive x-ray diffraction experiments on SWNTs were carried out up to a pressure of 13 GPa at the beamline BL10XU of SPring8, using a monochromatized x-ray beam of 1 Å. The diffraction patterns were recorded using the imaging plate kept at a distance of ~25 cm from the DAC. Two dimensional imaging plate records were transformed to one dimensional diffraction profiles by the radial integration of the diffraction lines.

III. RESULTS AND DISCUSSION

The triangular two-dimensional lattice of SWNTs gives the strongest feature at $\sim Q_{100}=0.402$ Å^{-1} and on the imaging plate it shows up as a diffraction circle close to the direct spot. No additional line of SWNT could be observed due to the fact that other diffraction peaks are much weaker than the first.[5] Also, the small volume of the sample in the DAC further reduces the observed intensities. Figure 1 shows variations in the intensity of the diffraction pattern corresponding to the two-dimensional triangular lattice as a function of pressure, on loading, as well as on release of pressure. The most important feature is that on increasing the pressure, the SWNT diffraction line vanishes beyond ~10 GPa [Fig. 1(c)]. On release of pressure, from ~13 to ~7 GPa, the diffracted intensity corresponding to SWNTs reappears [Fig. 1(d)] and eventually regains almost full initial intensity on complete release of the pressure [Fig. 1(e)]. The vanishing intensity of the diffraction profile of SWNTs indicates a phase transformation. However, we do not see any new diffraction lines across this pressure. This implies that either the new structure is such that it does not have any strong Bragg reflections or it has lost the translational coherence in a bundle. Alternatively, if the single-wall nanotubes graphitize at this pressure, then such a change will be irreversible as has been seen earlier.[16] Also, our calculated diffraction patterns with totally circular and ideal hexagonal tubes show that faceting does not change the diffracted intensity of the first diffraction line significantly. (Moreover, recent MD calculations suggest that SWNTs are faceted even at ambient pressures.[6,10]) Therefore, we feel that the complete loss of the low-angle diffraction line due to the triangular lattice of the SWNT at 10 GPa on compression and its retrieval on decompression clearly demonstrates a reversible loss of translational coherence. These results establish that the structural changes at ~10 GPa are not related to the uniform faceting and/or uniform flattening of the tubes as speculated in some earlier studies.[9,11] Further, the re-emergence of lattice on release of pressure provides a direct confirmation of the theoretical predictions that most structural changes in the SWNT are totally reversible.[2,3] These results are distinctly different from those of Tang *et al.*[10] on two counts. First the diffraction signal for (100) line of our (11,11) armchair or equivalent tubes vanishes at ~10 GPa, in contrast with the observed loss of diffracted intensity at ~1.5 GPa for (10,10)

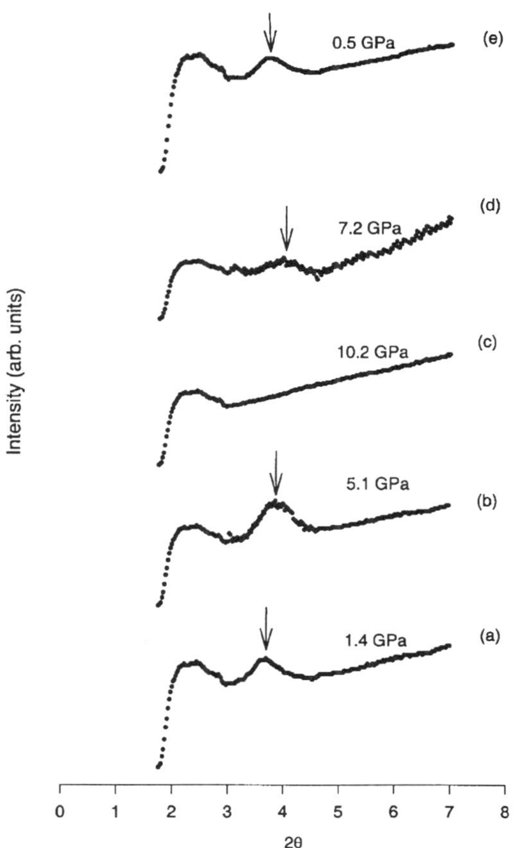

FIG. 1. One-dimensional x-ray diffraction profile of SWNTs from the (1,0) planes of the two-dimensional triangular lattice at various pressures. (a) 1.4 GPa, (b) 5.1 GPa, (c) 10.2 GPa represents the records on increasing pressure while (d) 7.2 GPa and (e) 0.5 GPa correspond to the decreasing pressure.

or equivalent tubes of Tang *et al.*[10] Second, our results establish the reversibility of transformation from unloading of pressure from ~13 GPa, in comparison with the corresponding limit of 4 GPa observed by Tang *et al.*[10] We should also point out that the reversibility observed in our experiments is consistent with the reversible behavior observed in Raman experiments carried on the same batch of samples.[9] However, as none of the theoretical work carried out so far predicts a strong dependence of the deformational behavior on the tube diameter, it is difficult to rationalize the observed different behavior for (11,11) (or equivalent) and (10,10) (or equivalent) tubes. Future work may provide some understanding of these results.

In Fig. 2, we show the variation of the d spacings of various diffraction lines with pressure as observed in this experiment. Up to ~8 GPa, we discern a systematic compression as displayed by the reduction of d_{100} of SWNT as a function of pressure. Within the experimental resolution, the absence of any additional diffraction lines or discernible

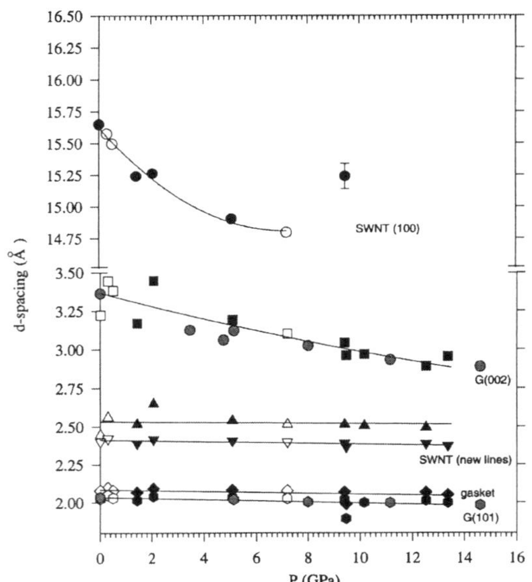

FIG. 2. Pressure variation of the d_{hkl} values. Filled symbols represent the data in an increasing pressure cycle while open symbols correspond to a decreasing pressure cycle. Hatched symbols are the published data of graphite from Ref. 23.

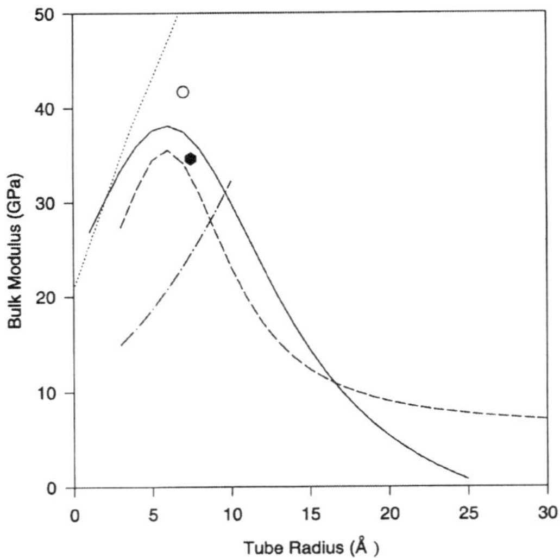

FIG. 3. Comparison between the observed bulk modulus for (11,11) or equivalent tubes (present work, represented by a solid hexagon) and for (10,10) or equivalent tubes (Ref. 10, represented by an open circle) and the theoretical results (Refs. 4,20,21). The dashed line represents the calculated results of Tersoff and Ruoff (Ref. 4), the dash dot corresponds to the model calculations of Lu (Ref. 21), the dotted line represents the computed results of Popov, Van Doren, and Balakanski (Ref. 20) in which rigid tubes interact through van der Waal forces and the solid line corresponds to their more detailed calculations in which interacting elastically deformable tubes are treated (Ref. 20).

broadening due to the splitting of the diffraction line (owing to proposed lowering of symmetry across 1.7 GPa) in our data do not support the suggestion of a structural transformation at this pressure.[7] Observed variations in the lattice spacing also highlights the differences in the behavior of SWNT under hydrostatic and non-hydrostatic stress conditions. For example, by 5 GPa, the basal plane lattice constant reduces to $\sim 0.95 a_0$. If we assume that the compression along the tube axis to be the same as that in the basal plane of graphite,[18] these results suggest that at ~ 5 GPa, $V/V_0 = 0.9$. In contrast, the compression under nonhydrostatic conditions of the piston cylinder apparatus is $V/V_0 = 0.75$ at ~ 2.5 GPa.[11] The variation of the two-dimensional triangular lattice parameter **a** with pressure can be fitted to the one-dimensional analogue[18] of the Murnaghan equation,[19] $a/a_0 = [(\beta'/\beta_0)P + 1]^{-1/\beta'}$ where β_0 is the bulk modulus and β' is the pressure derivative. Our data up to ~ 8 GPa, taken from the increasing as well as decreasing pressure runs, can be fitted with $\beta_0 = 43(\pm 4)$ GPa and $\beta' = 33(\pm 3)$. Compare this with c-axis compression of graphite, which has $\beta_0 = 35.7$ GPa and $\beta' = 10.8$.[18] This shows that up to ~ 8 GPa, SWNTs are somewhat less compressible than graphite along the c axis. Assuming the tube axial compression as that of the graphite basal plane, we find the low-pressure bulk modulus to be ~ 34 GPa.

Several authors have computed the elastic properties of SWNTs.[4,20,21] Figure 3 compares our observed bulk modulus as well as that of Tang *et al.*[10] with the results of these calculations. Our results are in excellent agreement with the theoretical estimates of Tersoff and Ruoff[4] and Popov, Van

Doren, and Balakanski[20] [which are, respectively, 34 and 37.3 GPa for (11,11) or equivalent tubes]. Observed bulk modulus of Tang *et al.*, though slightly higher (~ 41.7 GPa), is also in reasonable agreement with the theoretical calculations. Figure 3 also shows that results of the present investigation and of Tang *et al.* imply a strong decrease of the bulk modulus with the tube radii in this region. This feature contrasts with the calculated behavior of Lu,[21] which shows the bulk modulus to be a monotonically increasing function of radius. In addition, the computed bulk modulus for (11,11) or equivalent tubes is 25 GPa, which is considerably different from the observed results. These discrepancies highlight the inadequacies of the computational procedure employed by this author. Also, Fig. 3 indicates that the SWNTs behave more like coupled deformable tubes rather than the rigid tubes interacting through van der Waals forces for which calculated bulk modulus is much higher.[20] It is interesting to contrast this with the estimated bulk modulus of about 1 GPa in the piston cylinder measurements. One may speculate that if this small value of bulk modulus is not due to any remnant porosity in the sample, then the nature of SWNTs deformation under non-hydrostatic and hydrostatic pressures is very different. In addition, recent Raman scattering studies under nonhydrostatic pressures[22] show that the pressure derivative of the tangential mode frequency is almost twice as compared to the hydrostatic case. This implies that even for the intertubular compression, the mechanism may be consider-

PHYSICAL REVIEW B **63** 205417

ably different from what is observed under hydrostatic conditions.

Figure 2 shows another remarkable feature, i.e., d_{100} of the SWNT bundles relaxes to a higher value at ~ 9 GPa, just before the material undergoes the phase transformation. Reliability of our data can be inferred from the agreement of the observed pressure variation of the d (002) and d (101) diffraction lines of graphite with those published earlier.[23] The graphite diffraction lines do not show any relaxation effect at ~ 9 GPa, thereby establishing that the SWNTs undergo some structural change around this pressure, probably involving the morphology of the tubes. This is required due to the fact that total compressive strain cannot decrease under the increasing pressure. This observation of relaxation of strain on SWNTs also helps us to understand the Raman results of Ref. 9, where a discontinuous reduction was observed in the tangential mode frequency at ~ 10 GPa.[9] We attribute this to the formation of kinks and fins as described in the MD simulations of Ijima *et al.*[2] and Yakobson, Brabec, and Bernholc.[3] It should be noted that our data show that the deformation takes place at $\sim \Delta a/a_0 = 6\%$, which incidentally, is very close to the estimated value of the strain at which the morphological distortions were predicted to start under the axial compression of the isolated tubes in MD simulations.[3] However, at ~ 10 GPa, the strain along the tube axis is estimated to be $\sim 1\%$. Therefore, it is unlikely that the present deformations are initiated entirely due to the compression along the tube axis. It may be that the simultaneous presence of basal compressive strain helps trigger the mechanical deformations at a much smaller value of compression along the tube axis. This new feature should spur more theoretical activity. Also, earlier theoretical work has shown that the formation of localized deformation structures release strain on the rest of the tube.[2,3] However, it may be noted that our experimental results indicate that this relaxation of strain just precedes the loss of translational coherence. This leads us to think that despite the release in the strain, there must be a large enough number of deformed structures in each tube so that a little more compression is adequate to destabilize its two-dimensional lattice structure. This behavior is very different from what has been seen under the nonhydrostatic stresses, where deformation seems to increase monotonically.[11] In contrast, under the hydrostatic pressures, SWNTs seem to resist morphological deformations up to a critical value of the strain. However, beyond this critical value, the strain releases, probably heterogeneously, and results in a precipitous decline of any translational order at slightly higher compression. This behavior has not been predicted yet in any of the theoretical work and, thus, should prompt more realistic computations. Here it may be noted that the theoretical calculation by Yakobson, Brabec, and Bernholc[3] had predicted an inward buckling of the tubes at ~ 1 GPa under the hydrostatic pressures. Our experimental results do not support the existence of such a buckling at such low pressures.

One must note that, in principle, pressure transmitting fluid would interact with SWNTs and, hence, may influence their high-pressure behavior. Presently, not much is known to help address this issue in a detailed manner. In the context of present experiments, one may speculate that pressure transmitting fluid might enter the inter-tubular region and its chemical interaction may bring about morphological changes in the tubes at ~ 10 GPa at which alcohol mixture solidifies.[24] However, recent experimental studies indicate that just the solidification of the pressure transmitter may not be the primary driving force for the structural changes observed in SWNTs.[25] In addition to this, it should also be noted that the observed shifts of the radial modes with pressure strongly disagree with the molecular-dynamics simulations that incorporate the penetration of pressure transmitting fluid between the tubes.[6]

The presence of strong graphite diffraction lines at ambient conditions does not permit the detection of any partial irreversible graphitization of the SWNTs, as suggested in earlier static pressure,[10,16] as well as shock studies.[26] The observation of two strong diffraction peaks at $d_{hkl} \sim 2.5$ Å, is an interesting feature, not noted earlier. Though its position coincides with one of the diffraction line, (001), of the SWNTs, this line is expected to be much weaker than the first diffraction peak of the sample. Moreover, these diffraction lines continue to be present up to 13 GPa, without much change in the diffracted intensity. Further work is necessary to identify the source of this diffraction feature.

To conclude, our high-pressure x-ray diffraction investigations demonstrate that the SWNT bundles lose triangular lattice at ~ 10 GPa. These results supercede the recent observations of loss of the two dimensional lattice at ~ 1.5 GPa and reversibility of the structural changes up to ~ 4 GPa. We find a reappearance of the lattice on unloading the pressure from ~ 13 GPa. We also note that the volume compression under quasihydrostatic pressures is much less than that under nonhydrostatic stresses. The study underscores the remarkable resilience of nanotube bundles. However, despite the fact that underlying deformations may be quite different, the loss of translational order in SWNTs is similar to that of C_{70}, which shows reversible amorphization and is unlike that of C_{60}, which amorphizes irreversibly. We hope that our work will stimulate theoretical studies to understand the remarkable mechanical properties of carbon nanotube bundles.

ACKNOWLEDGMENTS

X-ray diffraction experiments were performed at SPring-8 with the approval of Japan Synchrotron Radiation Research Institute (JASRI) Proposal No. 1999B0093-ND-np. We thank the Department of Science and Technology for financial assistance. The authors are grateful to Dr. T. Watanuki for his help during the experiments.

[1] M. S. Dresselhaus, G. F. Dresselhaus, and P. C. Eklund, *Science of Fullerenes and Carbon Nanotubes* (Academic Press, New York, 1996), Chap. 19.

[2] S. Ijima, C. Brabec, A. Maiti, and J. Bernholc, J. Chem. Phys. **104**, 2089 (1996).

[3] B. I. Yakobson, C. J. Brabec, and J. Bernholc, Phys. Rev. Lett. **76**, 2511 (1996).

[4] J. Tersoff and R. S. Ruoff, Phys. Rev. Lett. **73**, 676 (1994).

[5] A. Thess, R. Lee, P. Nikolaev, H. Dai, P. Petit, J. Robert, C. Xu, Y. H. Lee, S. G. Kim, A. G. Rinzler, D. T. Colbert, G. E. Scuseria, D. Tomanek, J. E. Fischer, and R. E. Smalley, Science **273**, 483 (1996).

[6] U. D. Venkateswaran, A. M. Rao, E. Richter, M. Menon, A. Richter, R. E. Smalley, and P. C. Eklund, Phys. Rev. B **59**, 10 928 (1999).

[7] M. J. Peters, L. E. McNeil, J. P. Lu, and D. Kahn, Phys. Rev. B **61**, 5989 (2000).

[8] A. K. Sood, Pallavi V. Teredesai, D. V. S. Muthu, R. Sen, A. Govindaraj, and C. N. R. Rao, Phys. Status Solidi B **215**, 393 (1999).

[9] P. V. Teredesai, A. K. Sood, D. V. S. Muthu, R. Sen, A. Govindraj, and C. N. R. Rao, Chem. Phys. Lett. **319**, 296 (2000).

[10] J. Tang, L. C. Qin, T. Sasaki, M. Yudasaka, A. Matsushita, and S. Lijima, Phys. Rev. Lett. **85**, 1887 (2000).

[11] S. A. Chesnokov, V. A. Nalimova, A. G. Rinzler, R. E. Smalley, and J. E. Fischer, Phys. Rev. Lett. **82**, 343 (1999).

[12] D. W. Snoke, Y. S. Raptis, and K. Syassen, Phys. Rev. B **45**, 14 419 (1992).

[13] A. K. Sood, N. Chandrabhas, D. V. S. Muthu, Y. Hariharan, A. Bharathi, and C. S. Sundar, Philos. Mag. B **70**, 347 (1994).

[14] N. Chandrabhas, A. K. Sood, D. V. S. Muthu, C. S. Sundar, A. Bharathi, Y. Hariharan, and C. N. R. Rao, Phys. Rev. Lett. **73**, 3411 (1994).

[15] M. Eswaramoorthy, R. Sen, and C. N. R. Rao, Chem. Phys. Lett. **304**, 207 (1999).

[16] V. L. Kuznetsov and V. I. Zaikoviskii (unpublished), J. E. Fischer (unpublished) and cited in Ref. 10.

[17] For SWNTs the tube diameter d_t is related to integers (n,m) as $d_t = C_h / \pi = \sqrt{3} a_{C\text{-}C} (m^2 + n^2 + mn)^{1/2} / \pi$, where $a_{C\text{-}C}$ is the nearest-neighbor C-C distance ($=1.42$ Å) and C_h $=$ is the length of the chiral vector $\mathbf{C_h} = n\mathbf{a_1} + m\mathbf{a_2}$; $\mathbf{a_1}, \mathbf{a_2}$ being the primitive vectors of two-dimensional graphene sheet. For Ref. 10, the tubes diameter is determined to be 14.08 Å, therefore, these tubes will correspond to (10,10) armchair or (18,0) zigzag and equivalent tubes.

[18] M. Hanfland, H. Beister, and K. Syassen, Phys. Rev. B **39**, 12 598 (1989).

[19] F. D. Murnaghan, Proc. Natl. Acad. Sci. U.S.A. **30**, 244 (1944).

[20] V. N. Popov, V. E. Van Doren, and M. Balakanski, Solid State Commun. **114**, 395 (2000).

[21] J. P. Lu, Phys. Rev. Lett. **79**, 1297 (1997).

[22] P. V. Teredesai, A. K. Sood, S. M. Sharma, S. Karmakar, S. K. Sikka, A. Govindarajan, and C. N. R. Rao, Phys. Status Solidi B **223**, 479 (2001).

[23] T. Yagi, W. Utsumi, M. Yamakata, T. Kikegawa, and O. Shimomura, Phys. Rev. B **46**, 6031 (1992).

[24] Our recorded ruby R-lines spectra did not show any discernible line broadening up to ~11 GPa, implying that the nonhydrostatic stresses are small. This is consistent with the fact that though 4:1 methanol-ethanol mixtures solidifies at ~ 10.4 GPa, the resulting solid is a soft one to much higher pressures [Jayaraman, Rev. Mod. Phys. **55**, 65 (1983)].

[25] We rule out this possibility in light of very recent high-pressure Raman experiments (up to 6 GPa) carried out using water as a pressure-transmitting medium. These measurements do not show any phase transformations up to 6 GPa, even though water is known to solidify at ~1 GPa [Pallavi V. Teredesai and A. K. Sood (unpublished)].

[26] Y. Q. Zhu, T. Sekine, T. Kobayashi, E. Takazawa, M. Terrones, and H. Terrones, Chem. Phys. Lett. **287**, 689 (1998).

Hydrogen storage in carbon nanotubes and related materials

Gautam Gundiah,[a] A. Govindaraj,[a] N. Rajalakshmi,[b] K. S. Dhathathreyan[b] and C. N. R. Rao*[a]

[a]Chemistry and Physics of Materials Unit and CSIR Centre of Excellence in Chemistry, Jawaharlal Nehru Centre for Advanced Scientific Research, Jakkur P.O., Bangalore 560 064, India. E-mail: cnrrao@jncasr.ac.in
[b]Centre for Energy Research, SPIC Science Foundation, 64 Mount Road, Guindy, Chennai 600 032, India

Received 19th July 2002, Accepted 7th November 2002
First published as an Advance Article on the web 28th November 2002

Adsorption of hydrogen at 300 K has been investigated on well-characterized samples of carbon nanotubes, besides carbon fibres by taking care to avoid many of the pitfalls generally encountered in such measurements. The nanotube samples include single- and multi-walled nanotubes prepared by different methods, as well as aligned bundles of multi-walled nanotubes. The effect of acid treatment of the nanotubes has been examined. A maximum adsorption of *ca.* 3.7 wt% is found with aligned multi-walled nanotubes. Electrochemical hydrogen storage measurements have also been carried out on the nanotube samples and the results are similar to those found by gas adsorption measurements.

Introduction

The early report of Dillon *et al.*[1] that carbon nanotubes can be used to store considerable amounts of hydrogen created high expectations.[2] Dillon *et al.* used unpurified soot containing 0.1–0.2 wt% of single-walled carbon nanotubes (SWNTs) and obtained 0.01 wt% of H_2 storage at 0 °C. They extrapolated that SWNTs had 5 wt% H_2 storage capacity assuming that only the nanotubes were responsible for the H_2 uptake. Since then, several workers have examined H_2 adsorption with carbon fibres as well as nanotubes produced by different methods.[3–7] While Liu *et al.*[8] reported a H_2 storage capacity of 4.2 wt% in SWNTs generated by a semicontinuous hydrogen arc-discharge method, Ye *et al.*[9] found a gravimetric H_2 storage capacity as high as 8.25 wt% in high purity cut-SWNTs at 80 K under a pressure of *ca.* 7 MPa. Chen *et al.*[10] have reported a hydrogen storage capacity upto 13 wt% with aligned multi-walled carbon nanotubes (MWNTs). These workers employed quadrupole mass spectrometry and thermogravimetric analysis to study H_2 desorption properties. Interesting results on hydrogen storage have been reported with alkali-doped carbon nanotubes.[11] Li- and K-doped carbon nanotubes are reported to adsorb 20 and 14 wt% of H_2, respectively, at 1 atm. pressure, at 200–400 °C in the case of Li-doped nanotubes and near room temperature for the K-doped nanotubes. A lower, but substantial, H_2 adsorption was also reported for alkali-doped graphite. The highest capacity for H_2 storage (up to 67%) reported to date is that by Chambers *et al.*[12] for graphite nanofibres at ambient temperatures and at a pressure of 11.35 MPa. Electrochemical hydrogen storage in carbon nanotube materials was carried out by Nützenadel *et al.*[13] on as-prepared SWNTs and MWNTs. For the SWNT samples, the measured discharge capacities indicated a storage capacity of 0.39 wt% (110 mA h g^{-1}). SWNTs are also reported to show reversible charging capacities of up to 800 mA h g^{-1} corresponding to a capacity of 2.9 wt%.[14]

In spite of several of the favorable reports listed above, the scenario is not encouraging. Some of the recent papers report hardly any adsorption in the different carbon nanostructures, the maximum adsorption in some cases being lower than 1 wt%.[15,16] In view of the importance of the subject and also the wide differences amongst the various findings, we considered it important to carry out H_2 adsorption measurements systematically on well-characterized samples of carbon nanotubes and fibres prepared by different methods. We have, therefore, carried out measurements on SWNTs produced by the arc-discharge method, on MWNTs synthesized by arc-discharge and by the pyrolysis of acetylene over catalysts, on aligned multi-walled nanotubes obtained by the pyrolysis of organometallic precursors as well as carbon fibres. The measurements have been carried out on as-synthesized samples as well as samples subjected to acid treatment to remove catalyst particles. In addition to carrying out adsorption measurements, we have performed electrochemical studies of hydrogen storage on these materials.

Experimental

The carbon samples that we studied for hydrogen storage are as follows: SWNTs synthesized by the arc-discharge method (as-synthesized), **I**; SWNTs synthesized by the arc-discharge method (treated with conc. HNO$_3$), **II**; MWNTs synthesized by the pyrolysis of acetylene (as-synthesized), **III**; MWNTs synthesized by the pyrolysis of acetylene (treated with conc. HNO$_3$), **IV**; MWNTs synthesized by the arc-discharge method, **V**; aligned MWNT bundles synthesized by the pyrolysis of ferrocene (as-synthesized), **VI**; aligned MWNT bundles synthesized by the pyrolysis of ferrocene (treated with acid), **VII**; aligned MWNT bundles synthesized by the pyrolysis of ferrocene and acetylene (as-synthesized), **VIII**; and aligned MWNT bundles synthesized by the pyrolysis of ferrocene and acetylene (treated with acid), **IX**.

Scanning electron microscopy (SEM) images for all the samples were obtained on a LEICA S440i scanning electron microscope. Transmission electron microscopy (TEM) images were obtained with a JEOL JEM 3010 operating with an accelerating voltage of 300 kV. In order to find out the percentage of catalyst particles, thermogravimetric analysis (TGA) was performed on the samples.

SWNTs, **I**, were produced by the dc arc-discharge method using a composite graphite rod containing Y$_2$O$_3$ (1 at%) and Ni (4.2 at%) as anode and a graphite rod as cathode under a helium pressure of 660 Torr with a current of 100 A and 30 V as

DOI: 10.1039/b207107j

reported earlier.[17] The web-like product obtained was heated at 300 °C in air for *ca.* 24 h to remove amorphous carbonaceous materials. In order to remove the catalyst particles and open the ends of the nanotubes, the sample was stirred with concentrated nitric acid at 60 °C for *ca.* 12 h and washed with distilled water in order to remove the dissolved metal particles.[18] We denote the acid-treated SWNTs as **II**. The SWNTs in **I** had bundles of 5–50 nanotubes as revealed by high-resolution electron microscopy (HREM). The diameter of the nanotubes is in the range 1.2–1.4 nm. We show a typical HREM image of sample **II** in Fig. 1(a). There is a negligible amount of amorphous carbon and of metal particles adhering to the nanotubes.

MWNTs, **III**, were prepared by the decomposition of acetylene at a flow rate of 50 standard cubic centimeters per minute (sccm) at 730 °C over an iron/silica catalyst in flowing hydrogen (75 sccm) according to a previously reported method.[19] The silica was removed by treatment with aqueous hydrofluoric acid (HF) for 12 h. The metal catalyst particles in **III** were removed by treatment with concentrated nitric acid at 60 °C for *ca.* 12 h and washing with distilled water. The acid-treated sample was heated at 350 °C to obtain **IV**. Nanotubes **III** had diameters of between 25 and 75 nm with lengths up to tens of microns [Fig. 1(b)]. The TEM images, however, showed the presence of extraneous carbonaceous material and catalyst particles. HREM images showed that they were not very well graphitized, presumably due to the low temperatures at which they were prepared. Upon treatment with conc. HNO_3, we found that we were able to remove a majority of the catalyst particles, which was also verified by TGA. Furthermore, the nanotubes had open ends as revealed by TEM images.

MWNTs were also prepared by striking an arc between graphite electrodes in 500 Torr of helium by a previously published method.[20] A current of 60–100 A across a potential drop of *ca.* 25 V gives high yields of carbon nanotubes. The nanotubes thus obtained were opened by heating to 600 °C in air. This sample, denoted by **V**, had nanotubes with diameters in the 300 nm range with lengths of several microns. They had open ends but were well graphitized as revealed by the HREM images.

Aligned bundles of MWNTs, **VI**, were synthesized by the pyrolysis of ferrocene in an Ar atmosphere by a previously reported method.[20] The sample was scraped out from the walls of the quartz tube. We show an SEM image of the aligned MWNTs, **VI**, in Fig. 1(c). The nanotubes are well aligned with lengths extending to tens of microns. In order to remove the catalyst particles, a sample of **VI** was treated with concentrated nitric acid for 4 h, followed by washing thoroughly with distilled water and drying. This acid-treated sample, denoted

by **VII**, had open ends, but the alignment was retained. In order to synthesize dense bundles of aligned nanotubes (**VIII**), a mixture of acetylene (85 sccm) and ferrocene was sublimed. In Fig. 1(d) we show an SEM image of this sample. The catalyst particles in **VIII** were removed by treating the aligned bundles with conc. nitric acid for 4 h and then washing thoroughly with distilled water. The acid-treated sample, **IX**, was dried at 100 °C.

Carbon fibres were synthesized by the decomposition of acetylene at a flow rate of 20 sccm at 750 °C over a nickel/silica catalyst in flowing argon (80 sccm). The silica was removed by treatment with aqueous HF for 12 h. We show a low magnification SEM image of the carbon fibres prepared by us in Fig. 2(a). The yield of the fibres was good, with the fibres having diameters of *ca.* 200 nm and lengths up to 1 μm. The HREM image of this sample in Fig. 2(b) shows that the fibres are not well graphitized, with the planes inclined at an angle of *ca.* 32° with respect to the fibre axis.

In order to decorate the MNWTs with Pd and Pt nanoparticles, the nanotubes prepared by the pyrolysis route, **III**, were treated with conc. HNO_3 in order to remove the metal catalyst and then washed repeatedly with distilled water and dried at 100 °C. The decoration by Pt and Pd was carried out by reduction of the appropriate metal compound, namely H_2PtCl_4 and H_2PdCl_6, respectively, by employing the appropriate reducing agent.

The apparatus used for hydrogen adsorption experiments was custom-built and similar to that described elsewhere.[12] It consisted of a high-pressure stainless steel sample cell that was connected to a high-pressure reservoir *via* a high-pressure bellow valve. The pressure was monitored using a pressure gauge that was connected in the system. Experiments were conducted to verify that the system was leak-free. This was done by pressurizing the unit, and verifying that the pressure remained constant over a period of 10 h. Calibration of the system to account for the pressure drop brought about by the increase in volume upon opening of the valve between the reservoir and the evacuated sample cell at various pressures was also carried out. The volume of gas contained in the system at various pressures in the absence of the carbon samples was determined by allowing it to exit the system and then measuring by the displacement of water.

Hydrogen adsorption experiments were performed at 300 K. The hydrogen that was used was of ultra high purity

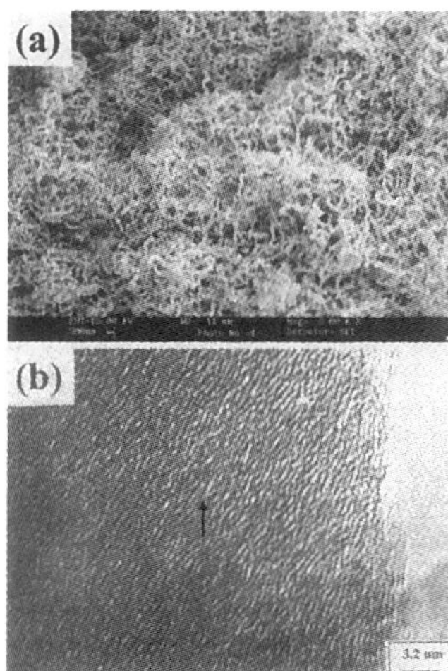

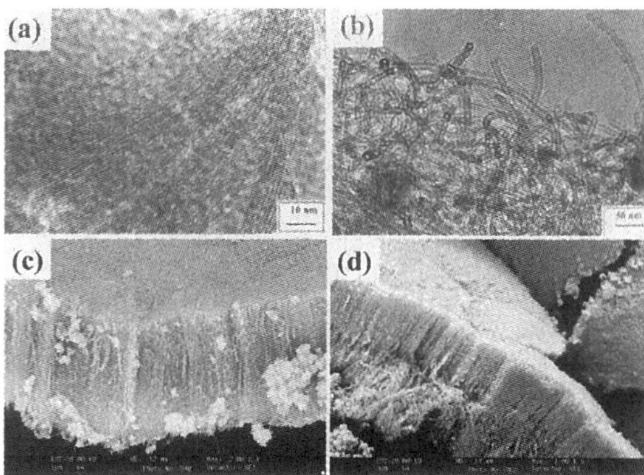

Fig. 1 (a) HREM image of **II**, (b) low magnification TEM image of **III**, (c) SEM image of **VI**, and (d) SEM image of **VIII**.

Fig. 2 (a) SEM image of as-synthesized carbon fibres; (b) HREM image of carbon fibres showing the stacking of graphitic planes. The arrow denotes the direction of growth of the fibre.

($>99.99\%$), with an impurity (*e.g.* moisture and nitrogen) content of less than 10 ppm. The carbon samples were accurately weighed and taken in the sample cell. The sample cell was evacuated to 10^{-5} Torr and heated for 12 h at 125 °C in order to degas the sample. This vacuum is sufficient as per the IUPAC recommendations on adsorption measurements.[21] Hydrogen was then introduced into the reservoir container and subsequently let into the sample cell. The drop in pressure from the initial value, of *ca.* 145 bar, was measured at regular intervals. The amount of hydrogen stored in the samples was calculated from the changes in pressure following the interaction of the material with the gas. Blank tests were repeated at regular intervals to verify that the system was free of leaks and that the pressure drop observed was only due to the uptake of hydrogen by the samples. The weight of the sample was 125 mg or more, to minimize errors in measurements. All of the adsorbed hydrogen could be desorbed, as established by adsorption–desorption experiments, the only desorbed gas being hydrogen.

Electrochemical measurements were carried out using an EG&G galvanostat/potentiostat model 273, with the help of a three electrode system using the samples as the working electrode, platinum as the counter electrode and saturated calomel as the reference electrode. Electrochemical measurements were carried out in a flooded electrolyte condition in open cells. The electrolyte, 30% KOH, which is the same as that used in alkaline batteries, was prepared from reagent grade KOH and de-ionized water. Chronopotentiometry, controlled potential coulometry, and cyclic voltametry were used. Electrodes were prepared by mixing 10 mg of the nanotube samples with Cu powder in the ratio of 1:3 with a polytetrafluoroethylene (PTFE) binder. The putty form of the mixture was mechanically pressed on to a current collector (Ni mesh) at room temperature. Then the electrode was sintered at 200 °C for about 1 h under vacuum. The geometric area of the electrode was *ca.* 2 cm^2. The electrodes were tested for their charge–discharge characteristics, initial capacity and cycle life.

Results and discussion

H$_2$ adsorption

Unlike metal alloys that adsorb hydrogen after a short period of incubation, hydrogen is adsorbed soon after it is admitted in the carbon samples studied by us. The entire adsorption was over within 100–150 min. Similar observations have been made by other workers.[8,12]

In Fig. 3, we show plots of hydrogen adsorption *versus* time

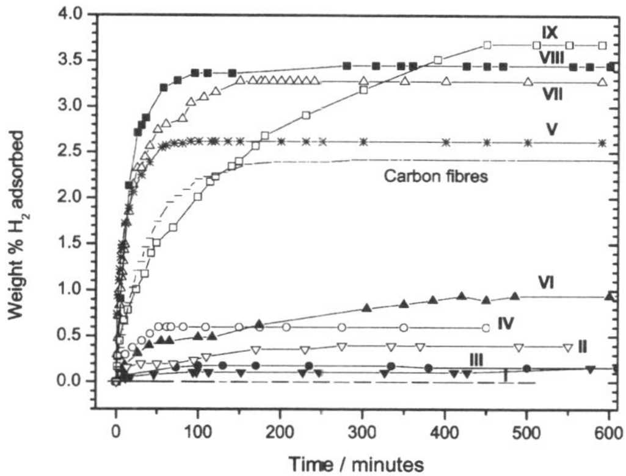

Fig. 3 Amount of hydrogen adsorbed in wt% as a function of time for the various carbon nanostructures. The broken curve represents the blank data obtained in the absence of a carbon sample.

for the various carbon nanostructured samples studied by us. The broken curve representing the blank run with an adsorption of 0.0 wt% was obtained when the system was evacuated and then pressurized with hydrogen gas in the absence of a carbon sample. This indicates that there is no adsorption by the system itself in the absence of the carbon sample. Similar blank runs, performed after each adsorption experiment with a carbon sample, showed identical curves. The SWNT sample **I** showed a storage capacity of 0.2 wt%. On removal of the catalyst particles, the storage capacity for the sample (**II**) increased only slightly to 0.4 wt%. This indicates that the catalyst particles are probably inert with regard to H$_2$ adsorption. The H$_2$ storage that we have obtained for SWNTs is considerably lower than that reported by others.[1,8] We further cleaned the surface of the tubes by treatment with a mixture of conc. HCl and H$_2$O$_2$ and were able to obtain a storage capacity of 1.2 wt%. This treatment would remove any extra carbonaceous materials that may block the pores and hinder the entry of the hydrogen molecules, as evidenced by the HREM images (not shown).

The hydrogen storage capacity for the MWNT sample **III** was 0.2 wt% as shown in Fig. 3. These results are consistent with the values obtained by Cao *et al.*[22] who obtained low H$_2$ storage for random MWNTs. Opening of the tubes (**IV**) increased the hydrogen storage capacity to 0.6 wt% (Fig. 3). The low values for adsorption may by due to the fact that the nanotubes are not well graphitized since the temperatures we used in their synthesis were low. Well-graphitized MWNTs have been reported to show higher percentages of hydrogen adsorption.[23]

We find the MWNTs, **V**, which were well graphitized, show an increased hydrogen storage capacity of 2.6 wt% as shown in Fig. 3. It appears that opened nanotubes show slightly higher storage as shown by samples **IV** and **V**.

The hydrogen storage capacity for sample **VI** is 1.0 wt%, which is considerably higher than that shown for the random MWNTs prepared by the pyrolysis route. This value is considerably less than that reported by Chen *et al.*[10] who studied the desorption of hydrogen using thermogravimetric analysis and reported a value of 5–7 wt% at around 10 atm. The present value is to be compared to that of Cao *et al.*[22] who reported a hydrogen storage of 2.4 wt% at slightly lower pressures of around 100 bar. Hydrogen is expected to be stored between the graphite layers as well as in the pores formed by the internanotube space between adjacent parallel nanotubes.[10,22] We expect that if the catalyst can be removed, hydrogen would also be stored in the hollow center of the tubes, which is the storage mechanism for SWNTs.[1,8] Accordingly, we could reproducibly achieve a H$_2$ storage capacity of 3.3 wt% for sample **VII**, as shown in Fig. 3. On increasing the packing density of the aligned nanotubes (sample **VIII**), we found only a marginal increase in the storage capacity (Fig. 3), the value being 3.5 wt%. On acid treatment and removal of metal particles, the sample, **IX**, showed even a higher storage capacity of 3.7 wt%. This is the highest storage capacity obtained by us in the present study.

In order to ensure the validity of the results in Fig. 3, we carried out independent adsorption experiments at different pressures of hydrogen. At first, measurements were performed at a high pressure followed by desorption at 300 °C in a vacuum (10^{-5} Torr) and readsorption and a lower pressure. We show the results of such studies on sample **IX** in Fig. 4, where the wt% of H$_2$ (after desorption and readsorption) is plotted against pressure. The results show the expected trend with a decrease in storage with any decrease in pressure.

The adsorption curve for the carbon fibres is also shown in Fig. 3. We see that the storage is 2.4 wt%, higher than in some of the nanotube samples, but lower than the 3.7 wt% storage found with **IX**. It appears that hydrogen is between graphite planes of the fibres.[12] The storage capacity found by us is

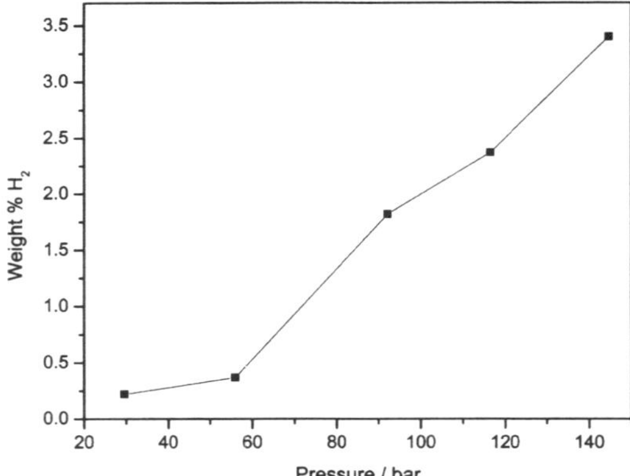

Fig. 4 Hydrogen storage in aligned MWNTs (acid-treated), sample **IX**, at different pressures.

considerably lower than that reported in the literature.[12,24] The low storage capacity may partly be due to incomplete crystallization. Furthermore, the fibres do not distinctly have a herringbone structure present in the samples employed by Chambers *et al.*[12] In order to remove the catalyst particles from the fibres, we treated the sample with conc. HNO_3 but found that the morphology was disrupted badly, resulting in a very low storage capacity (0.8 wt%).

It has been pointed out by Tibbetts *et al.*[16] that erroneous H_2 adsorption data can result from the small sample weights. We therefore performed adsorption studies using different weights of **VII**. The results of the adsorption are shown in Fig. 5. Thus, with sample weights of 138.5 and 200 mg, the shape of the adsorption curve remained the same as did the storage capacity (3.3 wt%)

Electrochemical measurements

The mechanism of electrochemical storage is different from gaseous storage in that with electrochemical storage the nanotubes behave like metal hydride electrodes in Ni–MH batteries. In metal hydride batteries, hydrogen is stored reversibly in the interstitial sites of a host metal and electric energy is produced by direct electrochemical conversion.[25,26] In the gaseous storage, however, the process is pure physisorption.

We carried out electrochemical measurements of hydrogen storage on the various nanotube samples as well as on the carbon fibres. In addition, we carried out measurements on

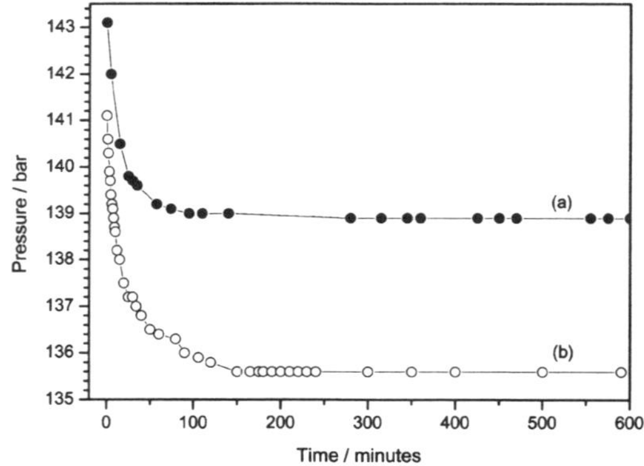

Fig. 5 Plots of pressure *versus* time for the MWNT bundles, **VII**, obtained with (a) 138.5 mg, and (b) 200 mg of the sample.

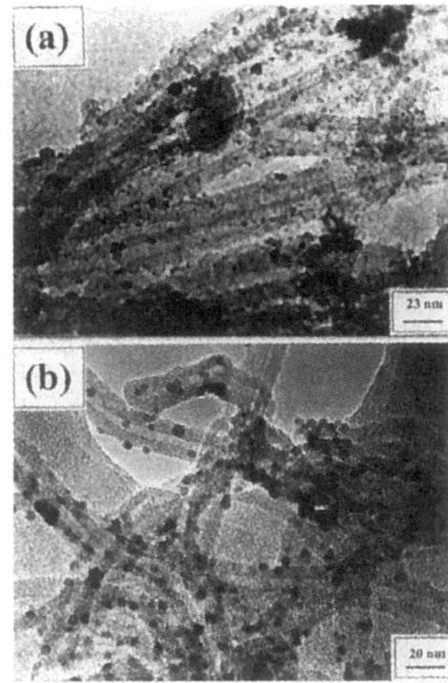

Fig. 6 Low magnification TEM images of (a) Pt-decorated MWNTs, and (b) Pd-decorated MWNTs.

MWNTs decorated with Pt and Pd nanoparticles. In Fig. 6(a), we show a low magnification TEM image of the Pt-decorated MWNTs. The size of the Pt nanoparticles is below 5 nm, while the nanotubes have diameters of *ca.* 35 nm. Thermogravimetric analysis revealed the loading of Pt in the sample to be 1.8%. We show a low magnification TEM image of the Pd-decorated nanotubes in Fig. 6(b). The Pd nanoparticles have a diameter of around 5 nm and the MWNTs have diameters of *ca.* 25 nm. The loading of Pd in the sample was 4.1%, as shown by TGA. An examination of the metal-decorated nanotubes suggests that the metal nanoparticles get deposited on the acid sites, which are dispersed all over the surface of the nanotubes.

The charging capacities of the various samples are shown in Fig. 7. Also shown are the corresponding weight percentages of hydrogen stored. The best result was obtained with sample **VIII** which showed a charging capacity of 1050 mA h g^{-1} (3.7 wt%). The hydrogen storage capacities obtained by gas adsorption and by the electrochemical method are comparable.

It is seen that the electrochemical charging capacity initially increases with the cycle. It reaches saturation after some cycles and then becomes stabilized. With most of the samples studied,

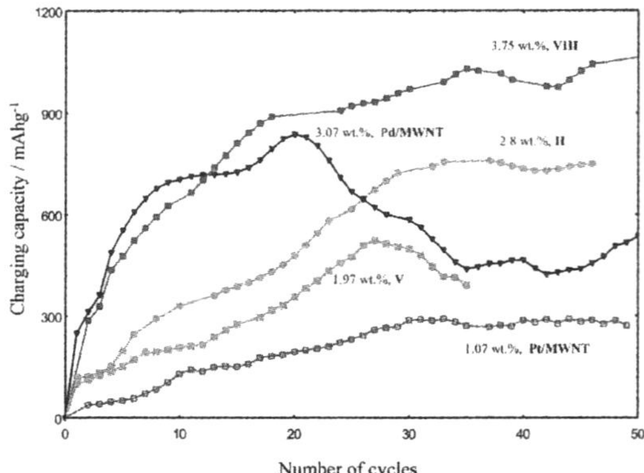

Fig. 7 Plot of the charging capacity against the number of cycles for different carbon nanostructures. Also shown are the corresponding weight percentages of H_2 stored.

the electrode was stable even after 50 cycles. In the case of Pd-loaded MNWTs, the cyclic stability was not good, as there was the degradation of the electrode after 20 cycles with the charging capacity decreasing from 800 to 200 mA h g^{-1}. Sample **V**, the arc-discharge MWNTs, also seems to have a poor cyclic stability. But the advantage of these samples is that the maximum capacity is obtained at a lower cycle number (around 20), compared with other samples, where the maximum capacity is obtained only after 30 cycles. This shows that the kinetics of hydrogen absorption are fast in these systems.

Conclusions

The maximum H_2 storage capacity achieved by us is obtained with densely aligned bundles of MNWTs that have been treated with acid. We find that carbon fibres do not show the exceptional storage capacities reported by others. Though the maximum capacity we have achieved in our studies with the nanotubes is 3.7 wt%, we are well below the 6.5 wt% benchmark set by the US Department of Energy (DOE). We have verified that the results obtained by us are reproducible as confirmed by experiments at lower pressures and with different weights of samples. The hydrogen storage in the samples has been studied electrochemically to obtain values that are similar to the gas storage values.

Though there have been a few recent reports of very low percentages of hydrogen storage in carbon nanotubes,[15,16] there have also been reports giving detailed procedures for the synthesis and purification along with good H_2 storage capacities.[22,23] What is noteworthy in the present study is that we have avoided many of the pitfalls by using different sample weights as well as by doing experiments at lower pressures. It is also significant that the hydrogen storage capacities obtained by the gas adsorption method are similar to those obtained electrochemically.

Acknowledgement

The authors thank the Department of Science and Technology, Government of India, for support of this research.

References

1 A. C. Dillon, K. M. Jones, T. A. Bekkedahl, C. H. Kiang, D. S. Bethune and M. J. Heben, *Nature*, 1997, **386**, 377.
2 C. Zandonella, *Nature*, 2001, **410**, 734.
3 R. G Ding, G. Q. Lu, Z. F. Yan and M. A. Wilson, *J. Nanosci. Nanotechnol.*, 2001, **1**, 7.
4 H. M. Cheng, Q. H. Yang and C. Liu, *Carbon*, 2001, **39**, 1447.
5 M. S. Dresselhaus, K. A. Williams and P. C. Eklund, *MRS Bull.*, 1999, **24**, 45.
6 J. E. Fischer, *Chem. Innov.*, 2000, **30**, 21.
7 A. C. Dillon and M. J. Heben, *Appl. Phys. A*, 2001, **72**, 133.
8 C. Liu, Y. Y. Fan, M. Liu, H. T. Cong, H. M. Cheng and M. S. Dresselhaus, *Science*, 1999, **286**, 1127.
9 Y. Ye, C. C. Ahn, C. Witham, B. Fultz, J. Lui, A. G. Rinzler, D. Colbert, K. A. Smith and R. E. Smalley, *Appl. Phys. Lett.*, 1999, **74**, 2307.
10 Y. Chen, D. T. Shaw, X. D. Bai, E. G. Wang, C. Lund, W. M. Lu and D. D. L. Chung, *Appl. Phys. Lett.*, 2001, **78**, 2128.
11 P. Chen, X. Wu, J. Lin and K. L. Tan, *Science*, 1999, **285**, 91.
12 A. Chambers, C. Park, R. T. K. Baker and N. M. Rodriguez, *J. Phys. Chem. B*, 1998, **102**, 4253.
13 C. Nützenadel, A. Züttel, D. Chartouni and L. Schlapbach, *Electrochem. Solid State Lett.*, 1999, **2**, 30.
14 N. Rajalakshmi, K. S. Dhathathreyan, A. Govindaraj and B. C. Satishkumar, *Electrochem. Acta*, 2000, **45**, 4511.
15 M. Ritschel, M. Uhlemann, O. Gutfleisch, A. Leonhardt, A. Graff, Ch. Täschner and J. Fink, *Appl. Phys. Lett.*, 2002, **80**, 2985.
16 G. G. Tibbetts, G. P. Meisner and C. H. Olk, *Carbon*, 2001, **39**, 2291.
17 C. Journet, W. K. Maser, P. Bernier, A. Loiseau, M. Lamy de la Chapelle, S. Lefrabt, P. Denierd, R. Lee and J. E. Fischer, *Nature*, 1997, **388**, 756.
18 C. N. R. Rao, A. Govindaraj, R. Sen and B. C. Satishkumar, *Mater. Res. Innov.*, 1998, **2**, 128.
19 K. Hernadi, A. Fonseca, J. B. Nagy, D. Bernaerts, J. Riga and A. Lucas, *Synth. Met.*, 1996, **77**, 31.
20 R. Seshadri, A. Govindaraj, H. N. Aiyer, R. Sen, G. N. Subbanna, A. R. Raju and C. N. R. Rao, *Curr. Sci. (India)*, 1994, **66**, 839; C. N. R. Rao, R. Sen, B. C. Satishkumar and A. Govindaraj, *Chem. Commun.*, 1998, 1525.
21 K. S. W. Sing, D. H. Everett, R. A. W. Haul, L. Moscou, R. A. Pierotti, J. Rouquérol and T. Siemieniewska, *Pure Appl. Chem.*, 1985, **57**, 603.
22 A. Cao, H. Zhu, X. Zhang, X. Li, D. Ruan, C. Xu, B. Wei, J. Liang and D. Wu, *Chem. Phys. Lett.*, 2001, **342**, 510.
23 P. X. Hou, Q. H. Yang, S. Bai, S. T. Xu, M. Liu and H. M. Cheng, *J. Phys. Chem. B*, 2002, **106**, 963.
24 Y. Y. Fan, B. Liao, M. Liu, Y. L. Wei, M. Q. Lu and H. M. Cheng, *Carbon*, 1999, **37**, 1649.
25 J. M. Evjen and A. J. Catotti, in *Handbook of Batteries*, D. Linden, ed., McGraw-Hill, New York, 2nd edn., 1995, ch. 27.
26 M. A. Fetcenko and S. Venkatesan, *Prog. Batteries Sol. Cells*, 1990, **9**, 259.

Inorganic nanotubes †

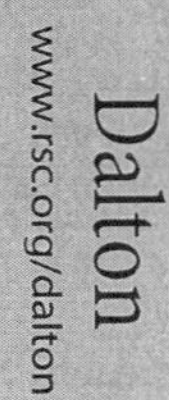

C. N. R. Rao *[a,b] and Manashi Nath [a,b]

[a] CSIR Centre for Excellence in Chemistry and Chemistry and Physics of Materials Unit, Jawaharlal Nehru Centre for Advanced Scientific Research, Jakkur P.O., Bangalore, 560 064, India. E-mail: cnrrao@jncasr.ac.in

[b] Solid State and Structural Chemistry Unit, Indian Institute of Science, Bangalore, 560 012 India

Received 13th September 2002, Accepted 14th November 2002
First published as an Advance Article on the web 2nd December 2002

Carbon nanotubes were discovered in 1991. It was soon recognized that layered metal dichalcogenides such as MoS₂ could also form fullerene and nanotube type structures, and the first synthesis was reported in 1992. Since then, a large number of layered chalcogenides and other materials have been shown to form nanotubes and their structures investigated by electron microscopy. Inorganic nanotubes constitute an important family of nanostructures with interesting properties and potential applications. In this article, we discuss the progress made in this novel class of inorganic nanomaterials.

1. Introduction

In 1991, Iijima[1] observed some unusual structures of carbon under the transmission electron microscope wherein the graphene sheets had rolled and folded onto themselves to form hollow structures. Iijima called them *nanotubes* of carbon which consisted of several concentric cylinders of graphene sheets. Graphene sheets are hexagonal networks of carbon and these layers get stacked one above the other in the c-direction to form bulk graphite. Following the initial discovery, intense research has been carried out on carbon nanotubes (CNTs).[2] The nanotubes can be open-ended or closed by caps containing five-membered rings. They can be multi- (MWNTs) or single-walled (SWNTs). We show a typical high-resolution electron

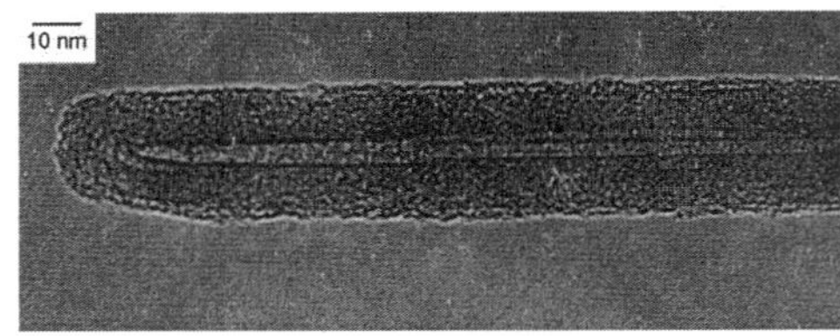

Fig. 1 A typical TEM image of a closed, multi-walled carbon nanotube. The separation between the graphite layers is 0.34 nm.

microscope (HREM) image of a multi-walled nanotube in Fig. 1. Depending on the way the graphene sheets fold, nanotubes are classified as armchair, zigzag or chiral as shown in Fig. 2. The electrical conductivity of the nanotubes depends on the nature of folding.

Several layered inorganic compounds possess structures comparable to the structure of graphite, the metal dichalcogenides being important examples. The metal dichalcogenides, MX_2 (M = Mo, W, Nb, Hf; X = S, Se) contain a metal layer sandwiched between two chalcogen layers with the metal in a trigonal pyramidal or octahedral coordination mode.[3] The MX_2 layers are stacked along the c-direction in ABAB fashion. The MX_2 layers are analogous to the single graphene sheets in the graphite structure (Fig. 3). When viewed parallel to the c-axis, the layers show the presence of dangling bonds due to the absence of an X or M atom at the edges. Such unsaturated bonds at the edges of the layers also occur in graphite. The dichalcogenide layers are unstable towards bending and have a high propensity to roll into curved structures. Folding in the

---

† The illustration of John Dalton (reproduced courtesy of the Library and Information Centre, Royal Society of Chemistry) marks the 200th anniversary of his investigations which led to the determination of atomic weights for hydrogen, nitrogen, carbon, oxygen, phosphorus and sulfur.

C. N. R. Rao was born in Bangalore, India in 1934. He received his MSc degree from the Banaras Hindu University, PhD from Purdue University and DSc from the University of Mysore. Following postdoctoral work at the University of California, Berkeley he joined the faculty at the Indian Institute of Science, Bangalore. He later moved to the Indian Institute of Technology, Kanpur where he was Professor and Head of the Department of Chemistry. He returned to the Indian Institute of Science, Bangalore in 1976 to start a new

unit devoted to Solid State and Structural Chemistry. He also founded the Jawaharlal Nehru Centre for Advanced Scientific Research, Bangalore in 1989. His main research interests are in solid state and materials chemistry. He is a Fellow of the Royal Society, London, and is a member of a several other academies. He is an honorary fellow of the Royal Society of Chemistry and is the recipient of the Hughes Medal of the Royal Society.

C. N. R. Rao

Manashi Nath was born in Calcutta in 1976 and received her BSc degree from the Presidency College, Calcutta in 1997. She is a student of the integrated PhD programme of the Indian Institute of Science, Bangalore and received her MSc degree in 2000. She has worked mainly on inorganic nanotubes.

Manashi Nath

DOI: 10.1039/b208990b

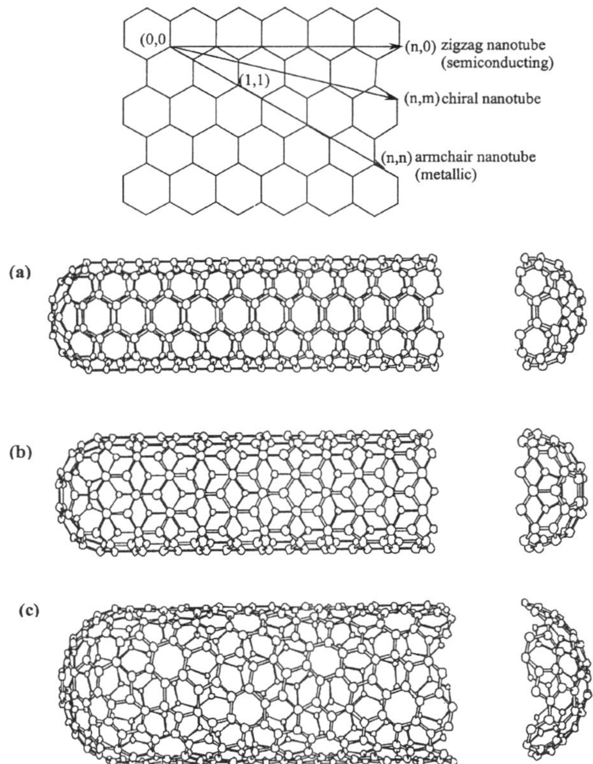

Fig. 2 Schematic representation of the folding of a graphene sheet into (a) zigzag, (b) armchair and (c) chiral nanotubes.

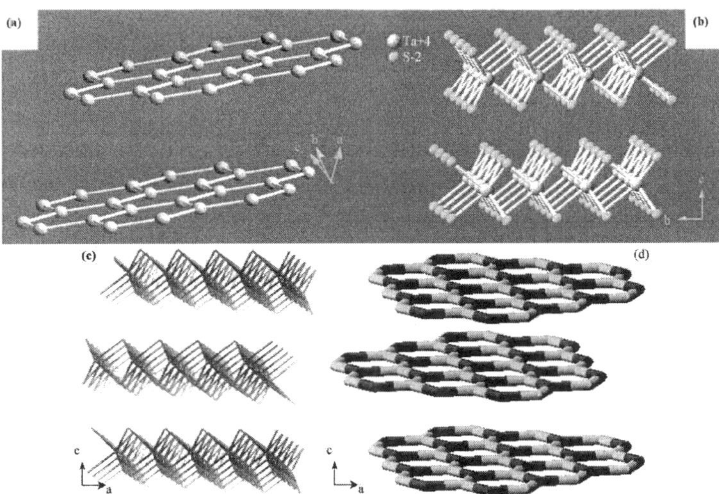

Fig. 3 Comparison of the structures of (a) graphite and inorganic layered compounds such as (b) NbS$_2$/TaS$_2$; (c) MoS$_2$; (d) BN. In the layered dichalcogenides, the metal is in trigonal prismatic (TaS$_2$) or octahedral coordination (MoS$_2$).

layered transition metal chalcogenides (LTMCs) was recognized as early as 1979, well before the discovery of the carbon nanotubes. Rag-like and tubular structures of MoS$_2$ were reported by Chianelli[4] who studied their usefulness in catalysis.

The folded sheets appear as crystalline needles in low magnification transmission electron microscope (TEM) images, and were described as layers that fold onto themselves (Fig. 4). These structures indeed represent those of nanotubes. Tenne et al.[5] first demonstrated that Mo and W dichalcogenides are capable of forming nanotubes (Fig. 5a). Closed fullerene-type structures (inorganic fullerenes) also formed along with the nanotubes (Fig. 5b). The dichalcogenide structures contain concentrically nested fullerene cylinders, with a less regular structure than in the carbon nanotubes. Accordingly, MX$_2$ nanotubes have

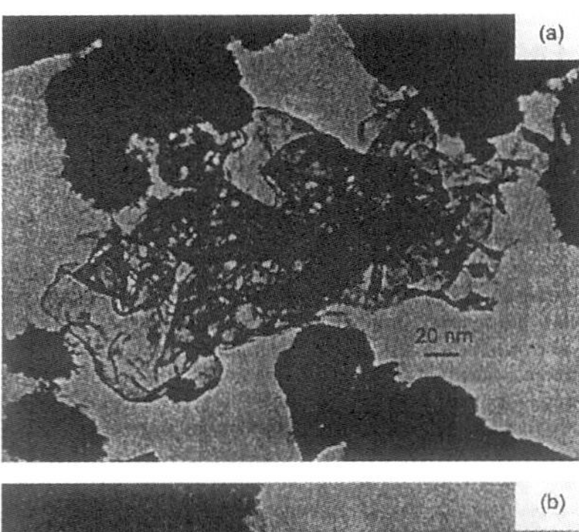

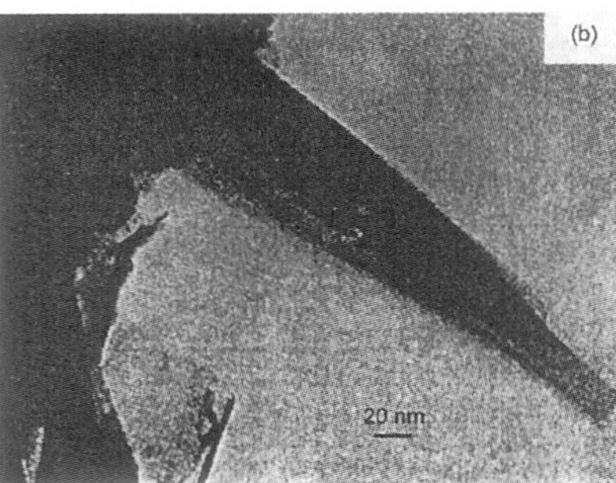

Fig. 4 Low-magnification TEM images of (a) highly folded MoS_2 needles and (b) a rolled sheet of MoS_2 folded back on itself. (Reproduced with permission from ref. 4).

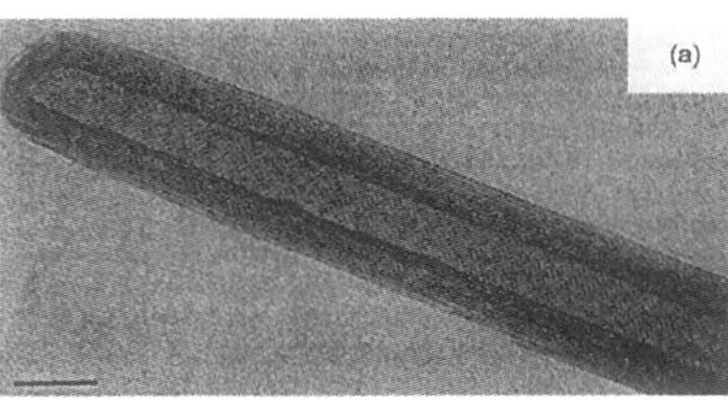

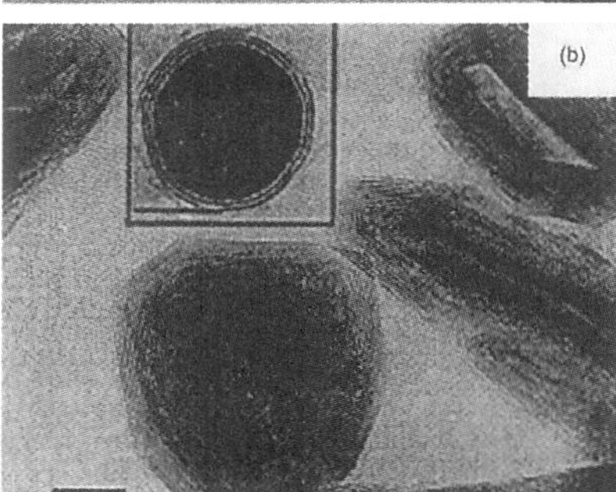

Fig. 5 TEM images of (a) a multi-walled nanotube of WS_2 and (b) hollow particles (inorganic fullerenes) of WS_2. (Reproduced with permission from refs. 5a and b, respectively).

varying wall thickness and contain some amorphous material on the exterior of the tubes. Nearly defect-free MX_2 nanotubes are rigid as a consequence of their structure and do not permit plastic deformation. The folding of a MS_2 layer in the process of forming a nanotube is shown in the schematic in Fig. 6. Considerable progress has been made in the synthesis of the nanotubes of Mo and W dichalcogenides in the last few years (Table 1). There has been some speculation on the cause of folding and curvature in the LTMCs. Stoichiometric LTMC chains and layers such as those of TiS_2 possess an inherent ability to bend and fold, as observed in intercalation reactions.

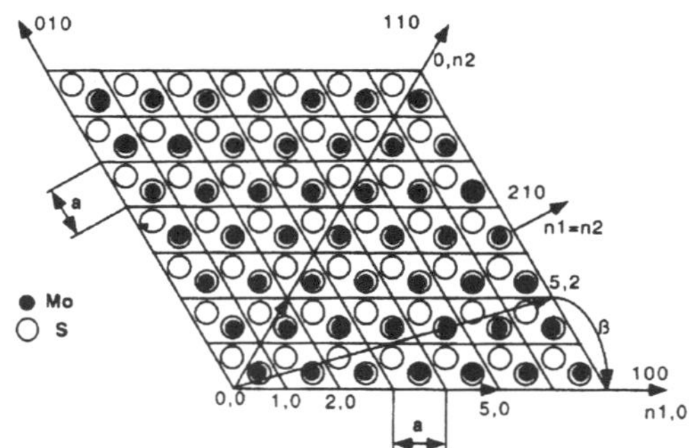

Fig. 6 Schematic illustration of the bending of a MoS_2 layer. (Reproduced with permission from ref. 19).

The existence of alternate coordination and therefore of stoichiometry in the LTMCs may also cause folding. Lastly, a change in the stoichiometry within the material would give rise to closed rings.

Transition metal chalcogenides possess a wide range of interesting physical properties. They are widely used in catalysis and as lubricants. They have both semiconducting and superconducting properties. With the synthesis and characterization of the fullerenes and nanotubes of MoS_2 and WS_2, a wide field of research has opened up enabling the successful synthesis of nanotubes of other metal chalcogenides. It may be recalled that the dichalcogenides of many of the Group 4 and 5 metals have layered structures suitable for forming nanotubes.

Curved structures are not only limited to carbon and the dichalcogenides of Mo and W. Perhaps the most well-known example of a tube-like structure with diameters in the nm range is formed by the asbestos mineral (*chrysotil*) whose fibrous characteristics are determined by the tubular structure of the fused tetrahedral and octahedral layers. The synthesis of mesoporous silica with well-defined pores in the 2–20 nm range was reported by Beck and Kresge.[6] The synthetic strategy involved the self-assembly of liquid crystalline templates. The pore size in zeolitic and other inorganic porous solids is varied by a suitable choice of the template. However, in contrast to the synthesis of porous compounds, the synthesis of nanotubes is somewhat more difficult.

Nanotubes of oxides of several transition metals as well as of other metals have been synthesized employing different methodologies.[7-12] Silica nanotubes were first produced as a spin-off product during the synthesis of spherical silica particles by the hydrolysis of tetraethylorthosilicate (TEOS) in a mixture of water, ammonia, ethanol and D,L-tartaric acid.[8] Since self-assembly reactions are not straightforward with respect to the desired product, particularly its morphology, templated reactions have been employed using carbon nanotubes to obtain nanotube structures of metal oxides.[9,10] Oxides such as V_2O_5 have good catalytic activity in the bulk phase. Redox catalytic activity is also retained in the nanotubular structure. There have been efforts to prepare V_2O_5 nanotubes by chemical methods as well.[11]

Boron nitride (BN) crystallizes in a graphite-like structure and can be simply viewed as replacing a C–C pair in the graphene sheet with the iso-electronic B–N pair. It can, therefore, be considered as an ideal precursor for the formation of BN nanotubes. Replacement of the C–C pairs partly or entirely by the B–N pairs in the hexagonal network of graphite leads to the formation of a wide array of two-dimensional phases that can form hollow cage structures and nanotubes. The possibility of replacing C–C pairs by B–N pairs in the hollow cage structure of C_{60} was predicted[13] and verified experimentally.[14] BN-doped carbon nanotubes have been prepared.[15] Pure BN nanotubes have been generated by employing several

Table 1 Synthetic strategies for various chalcogenide nanotubes

Chalcogenide	Synthetic strategy	Ref.	Chalcogenide	Synthetic strategy	Ref.
WS_2	(i) Heating MoO_3 in the presence of forming gas followed by heating in H_2S	5a	$Mo_xW_yC_zS_2$	Pyrolysis of H_2S over carbon-containing W and Mo-oxide complexes	43
	(ii) WS_3 and $(NH_4)_2WS_4$ decomposition in H_2	21	$Nb–W–S$	Heating Nb_2O_5 coated $W_{18}O_{49}$ nanorods in H_2S at 1100 °C	41
	(iii) Pyrolysis of H_2S/N_2 over WO_3-coated MWNTs	45a,b			
	(iv) Laser ablation of WS_2 target	64			
MoS_2	(i) Heating MoO_3 in the presence of forming gas followed by heating in H_2S	5b,c	$Mo–Ti–S$	Pyrolysis of H_2S/N_2 mixture over Ti–Mo alloy at elevated temperatures	42
	(ii) Decomposition of $(NH_4)_2MoS_4$ in the pores of anodic alumina	31	NbS_2	(i) NbS_3 decomposition	23
	(iii) Decomposition of $(NH_4)_2MoS_4$ MoS_3 in H_2	21		(ii) CNT templated reaction	46
	(iv) Hydrothermal treatment of ammonium thiomolybdate with ethylenediamine	62	TaS_2	TaS_3 decomposition	23
	(v) MoS_2 powder covered with Mo foil, heated to 1300 °C in H_2S	61	HfS_2	Decomposition of HfS_3	24
	(vi) Laser ablation of MoS_2 target	64	ZrS_2	Decomposition of ZrS_3	24
MoS_2I_x	MoS_2 with I_2 and C_{60} as carrier	63	$NbSe_2$	(i) $NbSe_3$ decomposition	70
				(ii) Electron irradiation of $NbSe_2$	69
$MoSe_2$	(i) MoO_3 + H_2Se	20a	$CdS, CdSe$	Surfactant-assisted synthesis	28,75
	(ii) $MoSe_3$ and $(NH_4)_2MoSe_4$ decomposition in H_2	22			
			ZnS	Sulfidization of ZnO columns by H_2S at 400 followed by etching the core	77
WSe_2	(i) WO_3 + Se vapors at 650–850 °C	20a	NiS	Treatment of $Ni(NH_3)_4^{2+}$ complex with CS_2 in aqueous ammonia	78
	(ii) Electron irradiation of WSe_2	20b			
	(iii) WSe_3 and $(NH_4)_2WSe_4$ decomposition in H_2	22	$Cu_{5.5}FeS_{6.5}$	Hydrothermal reaction between Cu and S in presence of $LiOH.H_2O$ and trace amount of Fe	79

procedures, yielding nanotubes with varying wall thickness and morphology.[16,17] It is therefore quite possible that nanotube structures of other layered materials can be prepared as well. For example, many metal halides (*e.g.*, NiCl$_2$), oxides (GeO$_2$) and nitrides (GaN) crystallize in layered structures. There is considerable interest at present to prepare exotic nanotubes and to study their properties.

In this article, we discuss the synthesis and characterization of nanotubes of chalcogenides of Mo, W and other metals, metal oxides, BN and other materials and present the current status of the subject. We briefly examine some of the important properties of the inorganic nanotubes and indicate possible future directions.

2. General synthetic strategies

Several strategies have been employed for the synthesis of carbon nanotubes.[2] They are generally made by the arc evaporation of graphite or by the pyrolysis of hydrocarbons such as acetylene or benzene over metal nanoparticles in a reducing atmosphere. Pyrolysis of organometallic precursors provides a one-step synthetic method of making carbon nanotubes.[18] In addition to the above methods, carbon nanotubes have been prepared by laser ablation of graphite or electron-beam evaporation. Electrochemical synthesis of nanotubes as well as growth inside the pores of alumina membranes have also been reported. The above methods broadly fall under two categories. Methods such as the arc evaporation of graphite employ processes which are far from equilibrium. The chemical routes are generally closer to equilibrium conditions. Nanotubes of metal chalcogenides and boron nitride are also prepared by employing techniques similar to those of carbon nanotubes, although there is an inherent difference in that the nanotubes of inorganic materials such as MoS$_2$ or BN would require reactions involving the component elements or compounds containing the elements. Decomposition of precursor compounds containing the elements is another possible route.

Nanotubes of dichalcogenides such as MoS$_2$, MoSe$_2$ and WS$_2$ are also obtained by employing processes far from equilibrium such as arc discharge and laser ablation.[19] By far the most successful routes employ appropriate chemical reactions. Thus, MoS$_2$ and WS$_2$ nanotubes are conveniently prepared starting with the stable oxides, MoO$_3$ and WO$_3$.[5] The oxides are first heated at high temperatures in a reducing atmosphere and then reacted with H$_2$S. Reaction with H$_2$Se is used to obtain the selenides.[20] Recognizing that the trisulfides MoS$_3$ and WS$_3$ are likely to be the intermediates in the formation of the disulfide nanotubes, the trisulfides have been directly decomposed to obtain the disulfide nanotubes.[21] Diselenide nanotubes have been obtained from the metal triselenides.[22] The trisulfide route is indeed found to provide a general route for the synthesis of the nanotubes of many metal disulfides such as NbS$_2$[23] and HfS$_2$.[24] In the case of Mo and W dichalcogenides, it is possible to use the decomposition of the precursor ammonium salt, such as (NH$_4$)$_2$MX$_4$ (X = S, Se; M = Mo, W) as a means of preparing the nanotubes.[21] Other methods employed for the synthesis of dichalcogenide nanotubes include hydrothermal methods where the organic amine is taken as one of the components in the reaction mixture (Table 1).

The hydrothermal route has been used for synthesizing nanotubes and related structures of a variety of other inorganic materials as well. Thus, nanotubes of several metal oxides (*e.g.*, SiO$_2$,[25] V$_2$O$_5$,[11] ZnO[26]) have been produced hydrothermally. Nanotubes of oxides such as V$_2$O$_5$ are also conveniently prepared from a suitable metal oxide precursor in the presence of an organic amine or a surfactant.[27] Surfactant-assisted synthesis of CdSe and CdS nanotubes has been reported. Here the metal oxide reacts with the sulfidizing/selenidizing agent in the presence of a surfactant such as TritonX.[28]

Sol–gel chemistry is widely used in the synthesis of metal oxide nanotubes, a good example being that of silica[8] and TiO$_2$.[29] Oxide gels in the presence of surfactants or suitable templates form nanotubes. For example, by coating carbon nanotubes (CNTs) with oxide gels and then burning off the carbon, one obtains nanotubes and nanowires of a variety of metal oxides including ZrO$_2$, SiO$_2$ and MoO$_3$.[10,30] Sol–gel synthesis of oxide nanotubes is also possible in the pores of alumina membranes. It should be noted that MoS$_2$ nanotubes are also prepared by the decomposition of a precursor in the pores of an alumina membrane.[31]

Boron nitride nanotubes have been obtained by striking an electric arc between HfB$_2$ electrodes in a N$_2$ atmosphere.[32] BCN and BC nanotubes are obtained by arcing between B/C electrodes in an appropriate atmosphere. A greater effort has gone into the synthesis of BN nanotubes starting with different precursor molecules containing B and N. Decomposition of borazine in the presence of transition metal nanoparticles and the decomposition of the 1 : 2 melamine–boric acid addition compound yield BN nanotubes.[16] Reaction of boric acid or B$_2$O$_3$ with N$_2$ or NH$_3$ at high temperature in the presence of activated carbon, carbon nanotubes or catalytic metal particles has been employed to synthesize BN nanotubes.[17]

3. Nanotubes of Mo and W dichalcogenides

Nanotubes of the disulfides and diselenides of Mo and W have been prepared by employing several strategies. The first synthesis of the MoS$_2$ and WS$_2$ nanotubes was carried out by Tenne *et al.*[5] by treating the metal oxides in an atmosphere of forming gas (95% N$_2$ + 5% H$_2$), followed by heating in a stream of H$_2$S at elevated temperatures. Initially, the oxides are reduced to the suboxides which are then converted to the sulfides. Onion-like structures analogous to nested fullerenes were also obtained in considerable yields (Fig. 5). The heating arrangement with a tubular furnace employed for the preparation of MoS$_2$ nanotubes is shown in Fig. 7. MoO$_3$ being sublimable, the

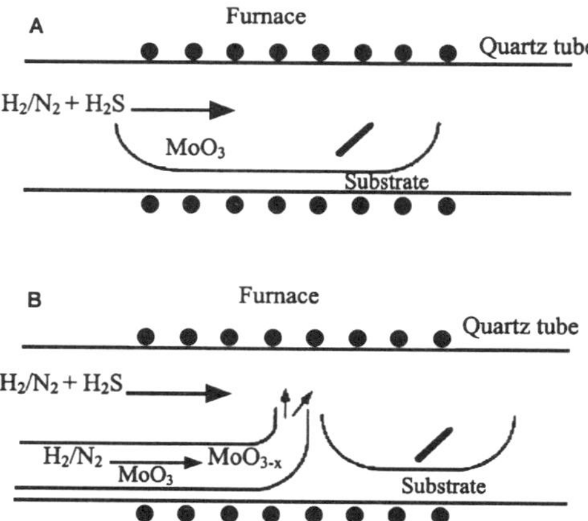

Fig. 7 (A) Schematic of the gas-phase reactor used to produce MoS$_2$ nanotubes; (B) a modification of the above where molybdenum suboxide flow was introduced through a separate tube and the collecting substrate was placed behind a high wall. (Reproduced with permission from ref. 5c).

growth of MoS$_2$ occurred from the vapor phase. On the other hand WO$_{3-x}$, not being volatile, the growth of WS$_2$ took place at the vapor–solid interface. Following the initial synthesis, various modifications have been adopted. Instead of starting with oxide particles, preformed morphologies like needles and whiskers containing the desired shapes of the nanotube products were also used as precursors. Thus, thermal treatment of the oxide needles in H$_2$S gave WS$_2$ nanotubes.[19] In these processes, the metal trisulfide is first formed on heating the oxide in an excess of H_2S. The trisulfide loses sulfur on annealing and crystallizes in the form of a disulfide nanoparticle or nanotube. MoS_2 nanotubes have also been prepared from the MoS_3 by sending a pulse through a STM tip.[33] The tip induced the crystallization of amorphous MoS_3 films deposited on a gold substrate producing inorganic fullerenes, some of them containing MoS_3 in the core, while the rest are hollow nanoparticles (Fig. 8).

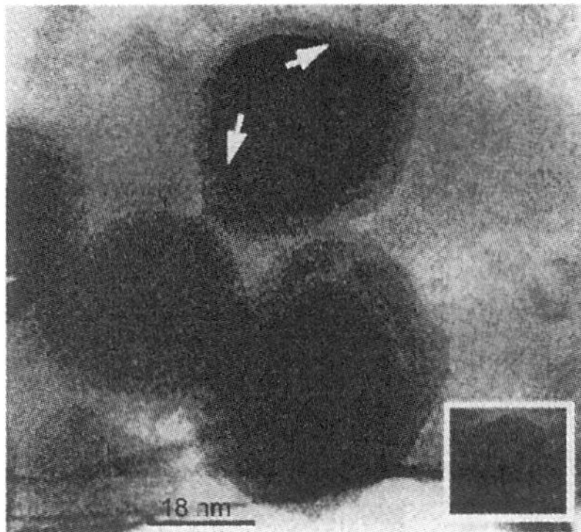

Fig. 8 Crystalline IF–MoS_2 nanoparticles with a MoS_3 core. Inset shows a fullerene-like nanoparticle of ~15 nm diameter. (Reproduced with permission from ref. 33).

On heating the nanorods or needles of the precursor oxide particles in an atmosphere of flowing H_2S or H_2Se, the outermost oxide layers are first converted to the chalcogenide layers. Further agglomeration of the oxidic particles is inhibited by the inert layer coating. Fast diffusion of H_2 into the precursor nanoparticle leads to the reduction of the core to the suboxide form. The second stage of the reaction involves the slow diffusion of the sulfur vapor into the core and the conversion of the oxide core to the sulfide. This mechanism has been substantiated by the presence of an incompletely sulfidized oxide core or even the metal in some of the WS_2 nanotubes.[34,35] The growth mechanism of the disulfide nanotubes from the suboxide particles comprises several steps. In the first step, the oxide particles react with H_2S to form one or two layers of the sulfide. This prevents further aggregation of the precursor particles and the formation of larger particles. Diffusion of H_2 into the particle and the out-diffusion of O_2 leads to the reduction of the oxide particles and the creation of crystallographic shear planes. In the next step of the reaction, sulfur vapor diffuses slowly into the precursor particles converting the suboxide core to the sulfide, which becomes hollow at the end of the process. The growth front is near the core of the precursor particle.

The knowledge that MoS_3 and WS_3 are the stable phases at low temperatures in the excess sulfur regime,[36] and that these amorphous trisulfides are first formed when the metal trioxide reacts with H_2S,[37] underscores the role of the trisulfide as a likely intermediate in the formation of the disulfide. A study of the binary metal–sulfur (Mo(W)–S) phase diagram reveals that the inorganic fullerene phase is obtained at the phase boundary between the amorphous MS_3 and the crystalline MS_2 phases. The loss of sulfur from the MS_3 phase triggers the nucleation of the MS_2 nanoparticles. By starting with a trisulfide, it should therefore be possible to obtain the disulfide nanostructure by thermal decomposition or by reacting with hydrogen.[38,39]

$$MS_3 \longrightarrow MS_2 + S \quad (M = W \text{ or } Mo)$$
$$MS_3 + H_2 \longrightarrow MS_2 + H_2S$$

Thermal decomposition of amorphous Mo and W trisulfides has been investigated under a steady gas flow, to explore the

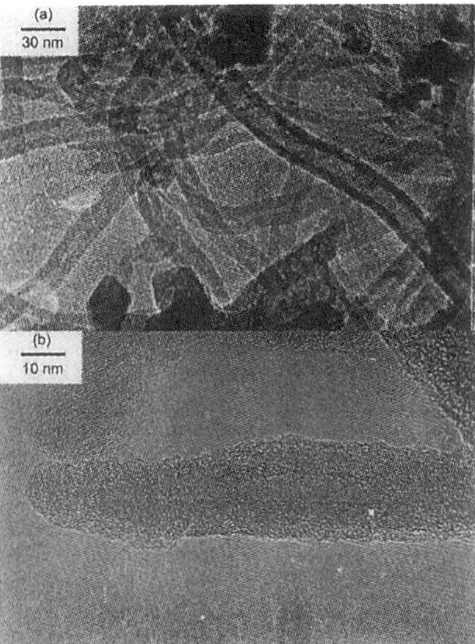

Fig. 9 TEM images of MoS_2 nanotubes grown by the decomposition of MoS_3. (Reproduced with permission from ref. 21).

formation of the disulfide nanotubes under a steady flow of hydrogen (100 sccm).[21] The decomposition products consisted of a high proportion of disulfide nanotubes (Fig. 9).

Mo and W trisulfides themselves are prepared by thermal decomposition of the ammonium thiometallates, $(NH_4)_2MS_4$ (M = Mo, W).[38] Accordingly, direct decomposition of the ammonium thiometallates in H_2 yields the MoS_2 and WS_2 nanotubes (Fig. 10).[21] In addition to providing a direct method

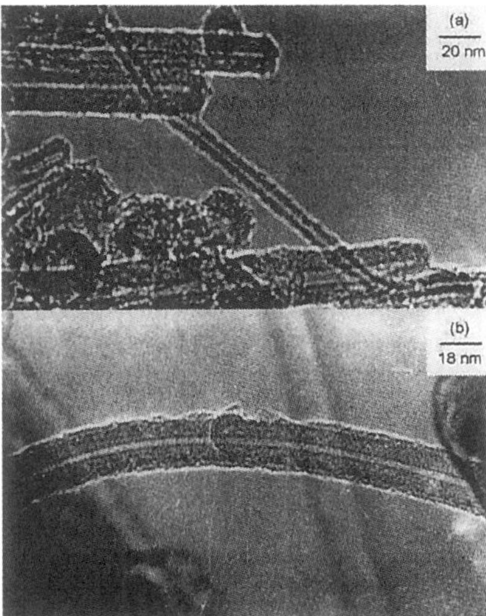

Fig. 10 (a) Low-resolution TEM images of MoS_2 nanotubes grown by the decomposition of ammonium thiomolybdate; (b) HREM image of the MoS_2 nanotube. (Reproduced with permission from ref. 21).

for the preparation of dichalcogenide nanotubes,[21] the trichalcogenide or the ammonium chalcometallate route enables the easy synthesis of nanotubes of the other layered dichalcogenides as well.[22–24]

$$(NH_4)_2MX_4 + H_2S \longrightarrow MX_2 + H_2X + NH_3$$
$$(X = S, Se; M = Mo, W)$$
$$MX_3 + H_2 \longrightarrow MX_2 + H_2X$$
$$MX_3 \longrightarrow MX_2 + X$$

Thus, nanotubes of Mo and W diselenides have been prepared by the decomposition of the triselenide or the ammonium selenometallate at elevated temperatures under a flow of H_2.[22] Apart from the nanotubes, nanorods of WSe_2 were also obtained. Some of the nanorods were attached to an amorphous particle at the tip (Fig. 11a). Several single-walled WSe_2

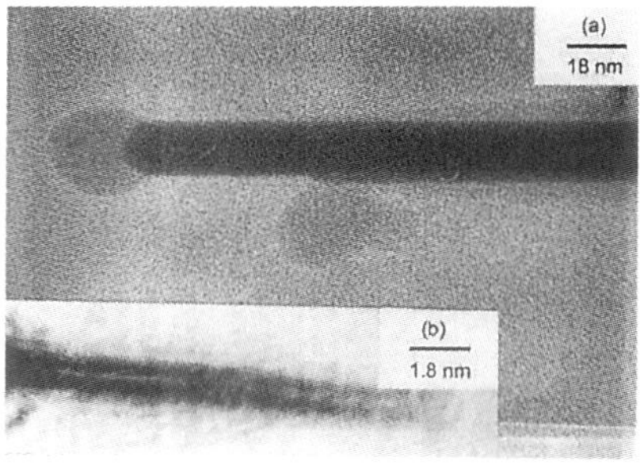

Fig. 11 (a) TEM image of a WSe_2 nanorod with a particle attached at the tip, grown by the decomposition of WSe_3; (b) single-walled WSe_2 nanotube. (Reproduced with permission from ref. 22).

nanotubes were also observed (Fig. 11b). This strategy has been employed to obtain W doped MoS_2 nanotubes.[40] Thus, solid solutions of the ammonium thiometallates, $(NH_4)_2Mo_{1-x}W_xS_4$ with varying ratios of Mo : W were used as precursors, to yield $Mo_{1-x}W_xS_2$ nanotubes (x = 0.15–0.5) on thermal decomposition under a gas flow (H_2, He or Ar) at ~770 °C (Fig. 12).

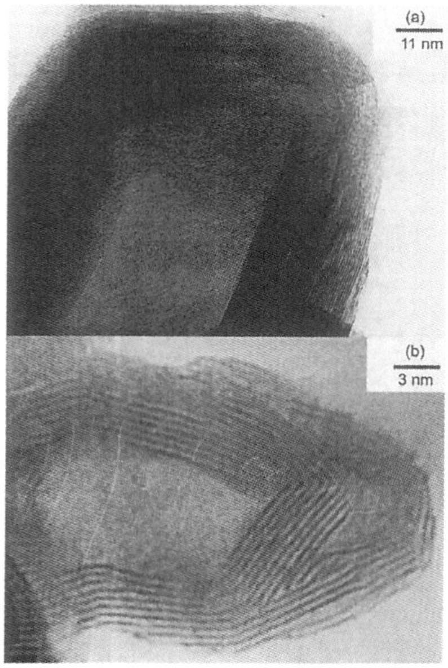

Fig. 12 HREM image of $Mo_{1-x}W_xS_2$ nanotubes showing a rectangular tip in (a) and (b) bamboo-like stacking. The layer separation in the walls is ~0.62 nm. (Reproduced with permission from ref. 40).

SWNTs of the doped sulfide structures were also observed. In general, the yield of nanotubes decreased with the increase in W content in the host MoS_2 lattice. Increasing the W content in MoS_2 also increases the layer mismatch in the tube walls and defects.

Other metals have been doped in the host disulfide layers of the nanotubes by using methodologies similar to those employed for the parent disulfides. Thus, composite nanotubes of W–Nb–S,[41] Ti–Mo–S[42] and Mo–W–C–S[43] have been prepared where the precursor was a mixed oxide of the two metals. In the case of Ti–Mo–S, the Ti–Mo alloy heated in oxygen to yield the mixed oxide, which was further heated in forming gas followed by H_2S to produce Ti doped MoS_2 nanotubes (Fig. 13).

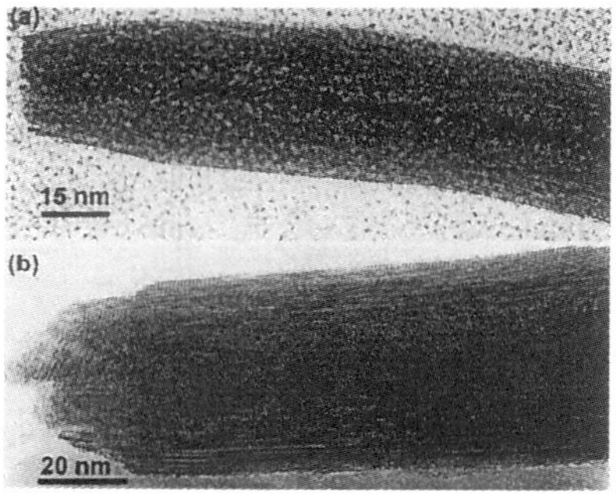

Fig. 13 TEM images of hollow Ti–Mo–S nanotubes. (Reproduced with permission from ref. 42).

The W–Nb mixed metal oxide was prepared by sonicating the solution containing the precursors. Treatment of the mixed metal oxide precursors in H_2S produced the Nb doped WS_2 nanotubes (Fig. 14). The mechanism of formation of the com-

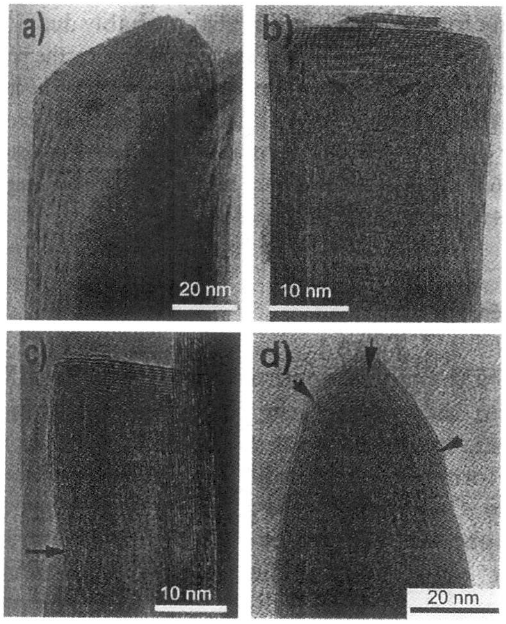

Fig. 14 HREM images of Nb–W–S nanotubes showing various tube closures: (a) an irregular tube closure; (b) a 90° wall–tip junction; (c) another near 90° closure (the arrow points to the buckling defect in the tube wall); (d) an irregular closure with severe bending defects. (Reproduced with permission from ref. 41).

posite nanotubes is similar to that of the binary nanotubes of MoS_2 and WS_2 and involves a layer by layer conversion of the mixed metal oxide to the sulfide. Some of the layers in the nanotube walls terminate abruptly, probably because of the exhaustion of the growing materials at that edge. The metal particles present at the terminated edge also inhibit the growth,[44] as in the case of metal oxide promoted growth of the BN nanotubes.

The methods of preparation discussed above do not involve any template, and the nanoparticles of the oxide or the tri-sulfide act as nucleation centers for tube growth. Recently, CNTs have been used as templates to grow MoS_2, WS_2 and NbS_2 coated carbon nanotubes, some of which contain 1–2 layers of the chalcogenide at the exterior.[45,46] The CNTs were coated with the metal oxide or its precursor and treated in a $H_2S/H_2/N_2$ atmosphere at elevated temperatures to convert the oxide to the sulfide. However, the CNT core was not removed in the nanostructures (Fig. 15).

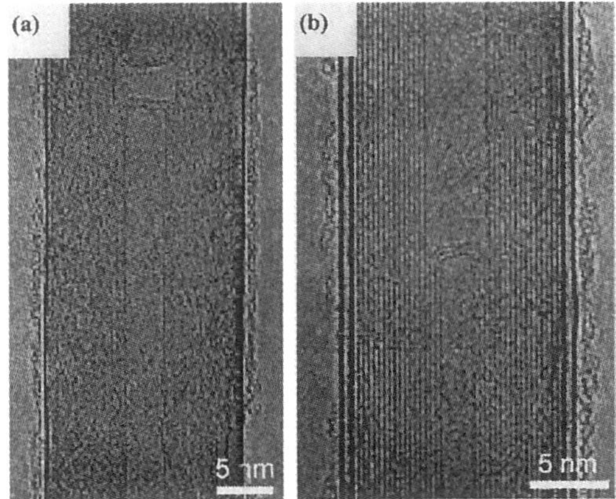

Fig. 15   HREM image of a CNT coated with (a) a single layer and (b) a double layer of WS_2. (Reproduced with permission from ref. 45a).

Confined reactions have been used to control the size of MoS_2 nanotubes. For example, MoS_2 nanotubes have been grown in the voids of an anodic alumina membrane, by decomposing $(NH_4)_2MoS_4$ inside the pores of the membrane.[31] The membrane is dissolved by treatment with alkali to yield free tubes. This method yields large diameter MoS_2 nanotubes.

As mentioned earlier, the structure of MoS_2 consists of the disulfide layers stacked along the c-direction.[3] This implies that the S–S interaction between the MoS_2 slabs is weaker compared to the intralayer interactions. The S–S interlayer distances are therefore susceptible to distortions during the folding of the layers. This is exemplified by the slight expansion of the c-axis (2%) in the MoS_2 nanotubes.[19] High resolution (HREM) images of the disulfide nanotubes show stacking of the (002) planes parallel to the tube axis. The distance between the layer fringes corresponds to the d(002) spacing.

Nanotubes of the disulfides are open-ended or capped. However, the closed end of the nanotube is not exactly spherical. Polygonal caps and rectangular tips are frequently observed in the disulfide nanotubes[47] (Fig. 16). Various types of open ends of nanotubes have been observed. They include flat open ends, conical open ends and also open-ended tubes where the layers at the tip arrange in peculiar ways giving the appearance of a pseudo closed tip.[47,48] The outer layer of the disulfide nanotubes and inorganic fullerenes are almost always complete, but the inner layers show defects, dislocations and terminated growth. This results in differences in the wall thickness on the two sides and in the varying diameter of the inner core along the length of the tube wall. This type of terminated growth is not observed in the CNTs, but has been found in metal-filled CNTs.[49] The terminated layer may be a manifestation of the defective structure of the starting precursor, or can be due to the absence of growing material at the edges. Layer defects have been observed in WS_2 nanotubes. They can be traced back to the crystallographic shear in the starting precursor phase.[50] Rémskar et al.[51] showed that each WS_2 layer has to satisfy the stacking order and orientation relationship with respect to the previous layer

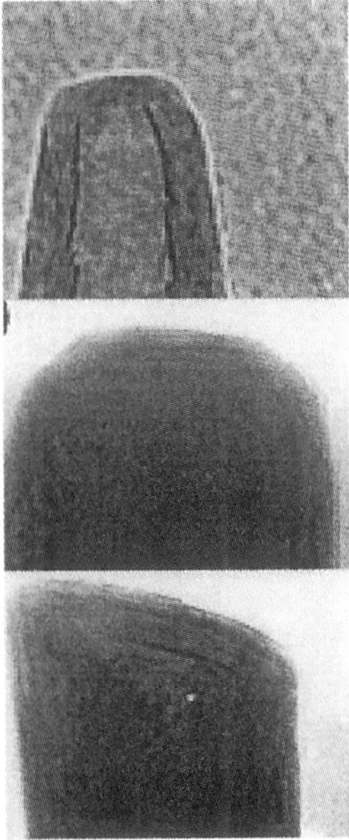

Fig. 16   Various closed tips observed in WS_2 tubes possibly containing square or octagonal defects. (Reproduced with permission from ref. 47).

and the strain involved can be relaxed by the formation of edge dislocations. The outer cylinders are subjected to less stress than the inner cylinders, and this internal stress stabilizes the inner shells, while the outer layers become easy to deform resulting in a greater number of defects near the edges.

The dichalcogenide nanotubes exhibit positive and negative curvatures similar to those in the CNTs.[52] In the CNTs, the curvature is believed to be due to the presence of a pentagonal ring (positive curvature) and a heptagonal ring (negative curvature) in the otherwise hexagonal network of the graphene sheet.[53] In the MX_2 layers, the absence of a central M or X atom can give rise to triangular or rhombohedral point defects, which can cause curvature in the MX_2 layers similar to those obtained in the graphene sheets (Fig. 17).[5,19,54] These point defects lead to topological defects and a combination of topological defects can cause tube closure.[55] Typical topological defects are square-like and octagonal-like defects. Based on the TEM images and electron diffraction (ED) patterns, it has been shown that a combination of these can close the armchair and zigzag tubes.[55] The presence of a rhombohedral point defect gives sharp edges at the corners of a rectangular inorganic fullerene. Similar point defects provide nearly 90° bends at the corners of the disulfide nanotubes[47] as well as the doped disulfide nanotubes (Fig. 18).[41] Analogous to the CNTs, the MX_2 tubes exhibit the energy minimized zigzag and armchair type tube morphologies.[47,56,57]

In $Mo_{1-x}W_xS_2$ nanotubes, defects in the layers, such as edge mismatch, growth termination of some of the layers and disordered layer stacking, increase with increasing tungsten doping.[40] Layer mismatch such as Nb–WS_2,[41] and Ti–MoS_2,[42] due to the simultaneous growth of the disulfide layers from different points on the precursor particle. In the $Mo_{1-x}W_xS_2$ nanotubes, W occurs in the MoS_2 layers, rather than having a structure with alternate layers of MoS_2 and WS_2.[58] The intra-layer doping of W in the MoS_2 layer does not cause a major structural deformation as the lattice constants of the MoS_2 and WS_2 are comparable. However, a slight expansion in the c-axis occurs in composite nanotubes similar to that in the MoS_2 nanotubes, due to the increased strain in curving the doped layers. With the

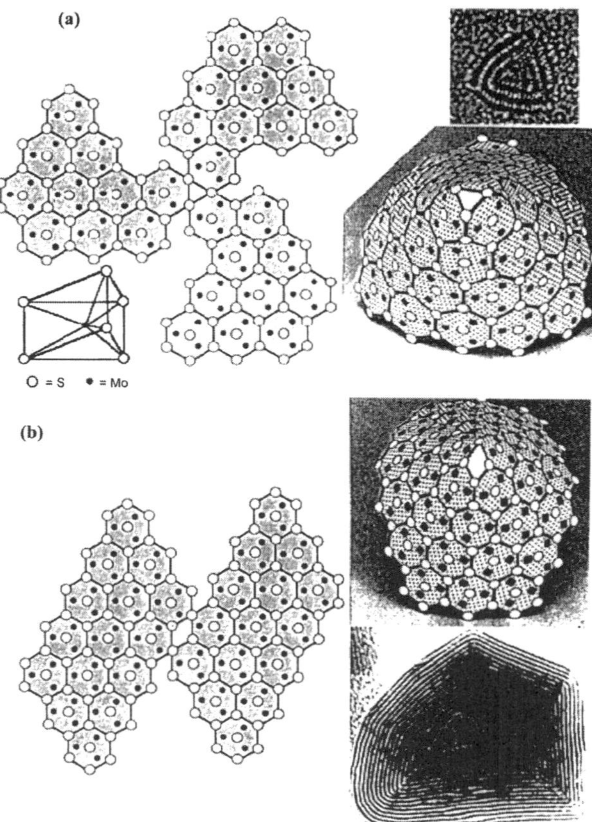

Fig. 17 Illustration of various point defects which exists in the vertices of IF MoS_2: (a) a triangular point defect; (b) a rhombohedral point defect. Insets show IF structures that are likely to contain such point defects. (Reproduced with permission from ref. 19).

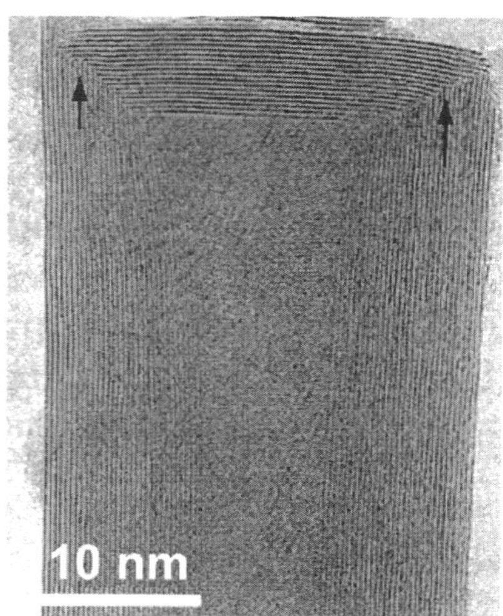

Fig. 18 TEM image showing the sharp 90° bend observed in the W–Nb–S nanotubes. (Reproduced with permission from ref. 41a).

increasing W doping, the yield of the nanotubes decreases considerably, while other nanostructures like nanorods and nanowires are obtained in moderate yield.

Disulfide nanotubes formed from the trisulfide precursors are structurally similar to those obtained by other means. Some of the nanotubes obtained by this method are open-ended, while others are closed. The nanotube tips are mostly non-spherical, polygonal with faceted edges. In the TEM images of MoS_2, WS_2, and $Mo_xW_{1-x}S_2$ nanotubes, the resolution at the tip of some of the nanotubes is lost. The ED patterns of the MoS_2 and WS_2 nanotubes prepared by this method match those reported for the armchair and zigzag tubes. It has been noticed that placing a small amount of sulfur powder near the inlet of the reaction zone, increases the yield of the nanotubes in some of the reactions. This observation suggests that the diffusion of vapors inside the particle may play a role in the formation of the nanostructures. Tenne *et al.*[59] have shown that the yield and diameter of the nanostructures depend on the diffusion length of the reactant vapors. In this study, MoO_3 was reduced to the suboxide under N_2 which was then carried into another reaction zone where it was converted to the sulfide. A quartz rod was used to precipitate the reaction products in order to quantify and hence study the influences of temperature, gas flow rates and diffusion rates of the reactant gases on the morphology and formation of the nanostructures.

The growth of the disulfide nanotubes from the trisulfide precursors starts with the reduction of the trisulfide precursors to a nanoparticulate form on heating in a gas flow. As the temperature reaches the decomposition temperature of the trisulfide, a few layers of the disulfide are formed at the periphery of the trisulfide particle. The growth then proceeds from "inside outwards" as the trisulfide at the core decomposes to form the disulfide layers that starts projecting outwards. This mechanism is supported by the presence of the precursor particle at the tip of some of the nanostructures, the diameter of such particles being greater than the outer diameter of the nanostructures (Fig. 11a).[22] There is a simultaneous growth of the disulfide layers from different points in the precursor particle, thus resulting in a layer mismatch at the edges where the growing layers meet. In the case of WSe_2, while the nanostructures show lattice fringes with the layer separation matching that reported for the bulk, the particle at the tip appears to be due to the amorphous triselenide.

Metal-catalyzed CNT growth is diffusion-controlled, and the carbon vapor adsorbed onto the metal catalyst particles, diffuses to the rear end, from where the graphite layer grows thus encapsulating the particle inside the growing nanotubes in some cases.[60] The dichalcogenide tube growth is also diffusion-controlled and involves the diffusion of H_2S into the inner core of the precursor particle, which is slowly converted to the disulfide. However, in the nucleation-mediated growth of the dichalcogenide nanotubes where the trisulfide precursor acts as the nucleation center, the disulfide layer grows from the inner core of the particle and proceeds outwards.

MoS_2 nanotubes are also obtained by heating MoS_2 powder covered with a Mo foil in the presence of H_2S.[61] MoS_2 evaporates on heating and is deposited on the Mo foil. The hollow tubes formed exhibit a zigzag arrangement of the layers in the tube walls. In a separate hydrothermal synthesis, molydenum polysulfide microtubules and hollow fibers were grown at room temperature from a solution containing condensed ammonium thiomolybdate and ethylene diamine. Initially, a condensed phase of the formula $(NH_3OH)_{3.9}MoS_{4.8}$ was formed. Thermal decomposition of these phases led to the formation of highly dispersed tubular MoS_2.[62] The amine plays an important role in nanotube formation, the absence of which yields bulk MoS_2. Ethylene diamine dihydrate appears to form chain-like structures, providing a support for formation of the nanotube. The amine is washed off to separate the hollow tubes. Bundles containing MoS_2I_x single-walled nanotubes have been prepared in the presence of C_{60} as a carrier.[63] The presence of C_{60} was crucial in the growth process as the nanotubes failed to grow in their absence. The single-walled tubes are hexagonally packed in bundles with a cell constant of 4.0 Å along the bundle axis and 9.6 Å perpendicular to the axis.

Electron beam irradiation and laser ablation has been successful in producing the metal dichalcogenide nanotubes. Laser ablation of MoS_2 and WS_2 targets produces substantial amounts of inorganic fullerenes and nanotubes.[64] Electron beam irradiation of bulk WS_2 powder yields various nanostructures of WS_2 including inorganic fullerenes, nanotubes and nanorods.[65]

The disulfide nanotubes have been characterized by Raman spectroscopy. The Raman bands for the MoS_2 nanotubes are similar to those of the bulk 2H-phase, except that the bands of the nanostructures show slight broadening. The composite nanotubes like those of Ti–Mo–S and Nb–W–S were also characterized by Raman spectroscopy.[41,42] The bands were similar to those of the undoped nanotubes, except for the appearance of a new band at 313 cm^{-1} which has been assigned to the increase in disorder owing to the presence of Nb.[41]

4. Nanotubes of dichalcogenides of Group 4 and 5 metals

With the success in synthesising nanotubes of Mo and W dichalcogenides starting from the trichalcogenides,[21,22] it was realised that this methodology could prove useful to prepare the nanostructures of other metal dichalcogenides, even though the trichalcogenides may be crystalline (rather than amorphous as in the case of Mo and W), since the dichalcogenide phase appears at the boundary of the trichalcogenide phase, and the thermal decomposition of the bulk samples of the latter is known to yield dichalcogenides.[39]

Nath and Rao[23,24] have investigated the thermal decomposition of the trisulfides of Group 4 and 5 metals prepared by the conventional solid state synthesis route. The decomposition of the trisulfides of Zr, Hf, Nb and Ta in a reducing atmosphere at elevated temperatures has indeed produced good yields of nanostructures, including nanotubes and nanorods.

4.1. HfS₂ nanotubes

Nanotubes of HfS_2 are obtained by the decomposition of HfS_3 in an atmosphere of H_2 + Ar (1 : 9), at 1170 K.[24] The product contained a good yield of the nanostructures as can be seen from the SEM image in Fig. 19a. EDX analysis revealed the

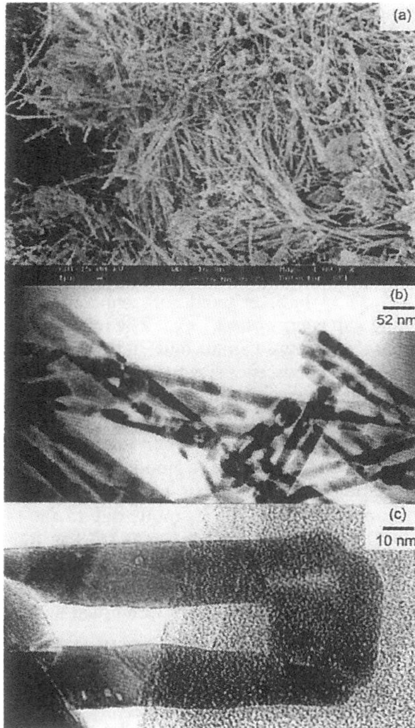

Fig. 19 (a) SEM image of the HfS_2 nanostructures; (b) and (c) low-resolution TEM images showing hollow nanotubes. The tube in (c) has a flat tip.

chemical composition to be HfS_2, while the XRD pattern showed it to be hexagonal. There is a slight expansion of the lattice in the c-direction (~1%) of the nanotubes as compared to bulk HfS_2. The lattice expansion is less than that observed with the MoS_2 and WS_2 nanotubes (2–3%).[19] This can be attributed to the fact that the mean compressibility factor of the c-axis in HfS_2 is higher than that observed for MoS_2.[66] The nanostructures as can be seen from the SEM image on Fig. 19a are quite lengthy, some being more than a micron long. Interestingly, a large proportion of these nanostructures is nanotubes. The low-resolution TEM image in Fig. 19b shows several nanotubes, some of which are closed with a non-spherical, nearly rectangular tip. The image in Fig. 19c shows a single nanotube with a rectangular tip, showing ripple-like undulations near the bend, which can arise from the strain involved in bending the layers. On close inspection, layer fringes are visible along the tube walls. Interrupted layer growth is observed in the inner edge of the tube wall, causing terminated layers and has non-uniformity in the wall thickness. This type of rectangular tip and terminated layer growth has also been observed in MoS_2, WS_2 and $Mo_{1-x}W_xS_2$ nanotubes.[40,47]

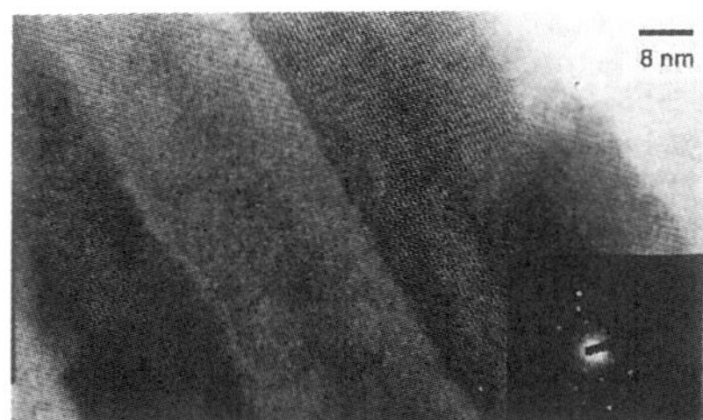

Fig. 20 HREM image of the HfS₂ nanotubes, showing a layer separation of ~0.6 nm in the walls. Inset shows a typical ED pattern.

A high-resolution image of a HfS$_2$ nanotube is shown in Fig. 20. The layers are separated by ~5.8 Å corresponding to the spacing of the (001) planes. Several terminated layers are observed at the outer edge of the tube wall possibly due to the absence of the growing material at these edges. A considerable number of defects and edge dislocations are also present along the length of the tube wall. The inset in Fig. 20 shows a ED pattern of the nanotube, characteristic of the hexagonal arrangement of the layers. Bragg spots corresponding to d(002) plane (2.923 Å) are seen. The ED pattern together with the high-resolution image indicate that the growth axis of the nanotube is perpendicular to the c-direction.

Bulk HfS$_2$ is an indirect band gap semiconductor with an indirect band gap energy of ~2.1 eV.[67] The reflectance spectrum of the nanotubes shows a small blue shift compared with the bulk. The photoluminescence spectrum of the nanotubes shows a band at 676 nm due to trapped states, and the band is blue-shifted with respect to that of bulk HfS$_2$ powder (see Fig. 21a). The Raman spectrum of the HfS$_2$ nanotubes is shown in Fig. 21b. It shows a band due to the A$_{1g}$ mode, corresponding to the S atom vibration along the c-axis perpendicular to the basal plane, and another due to the E$_g$ mode due to the movement of the S and Hf atoms in the basal plane.[68] The full-width at half maximum (FWHM) of the A$_{1g}$ band is 11 cm^{-1} in the nanotubes compared to 8 cm^{-1} for the bulk sample. Such broadening of the Raman band has been noted with MoS$_2$ and WS$_2$ nanotubes.[19]

4.2. ZrS₂ nanotubes

ZrS$_2$ nanotubes admixed with nanorods have been prepared by the thermal decomposition of ZrS$_3$ under H$_2$ + Ar at 1170 K.[24] Many of the nanotubes exhibit rectangular tips. The inner wall of some of the ZrS$_2$ nanotubes show non-uniformity near the tip resulting from the discontinuous growth of the ZrS$_2$ layers.

4.3. NbS₂ nanotubes

NbS$_3$ when heated in a stream of H$_2$ (100 sccm) at 1000 °C for 30–60 min, produces good yield of nanostructures.[23] A c-axis expansion of ~3% is observed in the nanotubes. The SEM image in Fig. 22a reveals a high yield of the nanostructures from the thermal decomposition of the trisulfide. These nanostructures contained a considerable amount of nanotubes as seen from the TEM images. The TEM image in Fig. 22b shows a couple of nanotubes, having closed rectangular tips. While some of the nanotubes are closed with flat non-spherical polygonal tips, most of the tubes are open at one or both ends. The image in Fig. 22c shows the high-resolution image of an open nanotube. The layer separation in the walls is ~6 Å, corresponding to the (002) plane of bulk NbS$_2$. The ED pattern in the inset shows the nanotube to be single-crystalline with Bragg spots corresponding to the known d values. Some of the diffraction spots show diffuse scattering or streaking due to bent layers or disorder. The NbS$_2$ nanostructures were not superconducting.

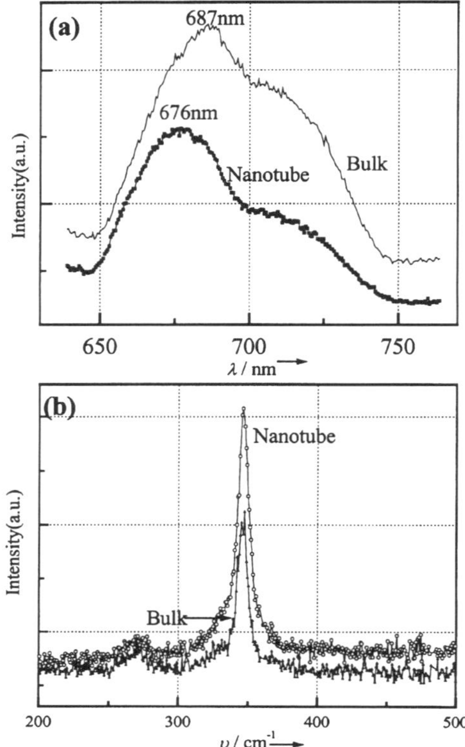

Fig. 21 (a) Photoluminescence spectra of bulk HfS₂ and HfS₂ nanotubes. (b) Raman spectrum of bulk HfS₂ and HfS₂ nanotubes.

4.4. TaS₂ nanotubes

TaS$_2$ nanotubes could be prepared by decomposing TaS$_3$ at 1270 K under H$_2$ (100 sccm). TEM observation of the product showed the presence of hollow core nanotubes in the product. The inner wall of some of the TaS$_2$ nanotubes showed non-uniformity along the length of the tube. Some of the tubes are closed with flat rectangular tips. In both NbS$_2$ and TaS$_2$ nanotubes, most of the tube tips are nearly flat or rectangular.[23] This is in sharp contrast to the CNTs.

4.5. NbSe₂ nanotubes

Galván *et al.*[69] have reported that NbSe$_2$ nanotubes can be prepared by the use of intense electron irradiation. The nanotubes and nanorods of NbSe$_2$ have been prepared by the decomposition of the triselenide, at ~970 K under a gas flow of Ar.[70] The product contained a mixture of nanotubes and nanorods. The powder X-ray diffraction was characteristic of 6aH–NbSe$_2$, with no appreciable shift in the d(002) line position relative to the bulk sample. Most of the NbSe$_2$ nanotubes are open-tipped. The NbSe$_2$ nanorods generally possess smaller diameters as well as shorter lengths compared to the nanotubes. Under the most favourable conditions, the ratio of the nanotubes to nanorods was around 2 : 1. The rate of argon flow appears to have an effect on the ratio of nanotubes and nanorods in the final product. At a smaller flow rate, there was a higher percentage of nanotubes in the final product, while higher flow rates gave a greater percentage of nanorods. High argon flow also gave rise to platelets, which also occurred, on increasing the temperature.

Fig. 23a shows the HREM image of a NbSe$_2$ nanotube in the tip region. The structure of the tip deviates from the hemi-

Fig. 22 (a) SEM image of NbS$_2$ nanotubes; (b) low-resolution TEM image and (c) HREM image of NbS$_2$ nanotubes. Inset shows the typical ED pattern. The layer separation is 0.62 nm in the walls. (Reproduced with permission from ref. 23).

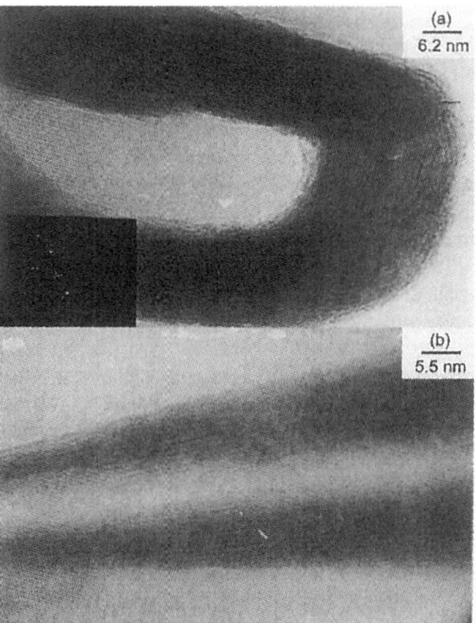

Fig. 23 HREM images of the NbSe$_2$ nanotubes. The tube in (a) has a closed tip with a 90° bend in one corner, while the tube in (b) is near-conical in shape due to several terminated layers in the walls. The inset in (a) shows a typical ED pattern. (Reproduced with permission from ref. 70).

spherical nature, and has a 90° bend at one corner while the other corner is comparatively smooth. The outer diameter of the tube is ~57 nm and the wall contains ~16 layers. The inner wall of the tube shows evidence for terminated layer growth. Sharp facets of the type reported by Tenne *et al.*[71] in the fuller-

ene structures of NbS$_2$ are however not present in these nanotubes. The layers near the tip are somewhat less regularly ordered, but the number of defects, dislocations and stacking faults in the NbSe$_2$ nanotubes is generally much smaller than that in the other chalcogenide nanotubes.[23,24] Fig. 23b shows a HREM image of a nanotube with a near-conical closed tip showing an interlayer separation of ~6.2 Å. This tube has a tapering shape due to the disrupted growth of the layers as it proceeds towards the tip. As a result, the tube walls near the tip are thinner than that along the body. There are approximately 10 layers near the tip and ~14 layers along the body of the tube, ~77 nm away from the tip.

It is interesting that most of the NbSe$_2$ nanotubes contain more than 10 layers with one or two containing a smaller number of layers. Some of the nanotubes exhibit a different type of stacking due to the presence of different polytypes just as in BN nanotubes where the local rhombohedral stacking occurs within the hexagonal phase.[44]

The Raman spectrum of NbSe$_2$ single crystals exhibit three first order lines at 29.6, 230.9 and 238.3 cm^{-1}.[72] The low frequency line at 29.6 cm^{-1} is due to the rigid layer vibration mode (E_g^2) which accounts for the weak interlayer bonding. The high-frequency lines at 230.9 and 238.3 cm^{-1} are due to the A_{1g} and E_{2g}^1 modes, respectively. The high frequency Raman modes in the NbSe$_2$ nanostructures were found to be identical to those of the bulk crystal. Bulk NbSe$_2$ shows a photoluminescence band of very weak intensity band at around 825 nm possibly due to trapped states. The band is shifted to 820 nm in the nanostructures.

NbSe$_2$ is a metallic conductor, becoming superconducting at low temperatures.[3b] The NbSe$_2$ nanostructures show metallic conductivity from 300 K down to lower temperatures (Fig. 24),

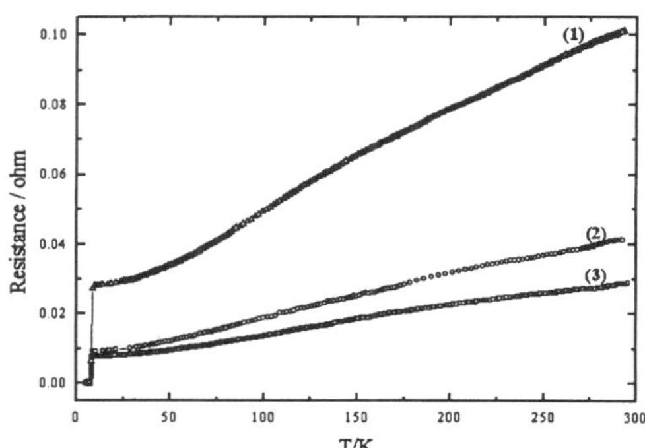

Fig. 24 Resistance *vs.* temperature curves of the NbSe$_2$ nanostructures, (2) and (3), are compared with the behavior of the bulk sample in (1).

in agreement with theoretical predictions.[57] The resistance–temperature curve for the bulk sample has a much higher slope than for the nanostructures, possibly because of the dominance of temperature-independent scattering in the latter. The NbSe$_2$ nanostructures become superconducting at 8.3 K (Fig. 24). The superconducting transition temperature of the bulk sample was found at 8.6 K, indicating a very small decrease in T_C, if at all, in the nanostructures. It has been suggested that the superconducting T_C of NbSe$_2$ is dependent on factors such as stoichiometry and the number of layers of NbSe$_2$ when the number of layers goes below 6, the T_C is expected to shift to 3 K.[73] The NbSe$_2$ nanostructures studied had diameters in excess of 30 nm, with the number of layers greater than 10. It would be worthwhile to prepare NbSe$_2$ nanostructures containing a smaller number of layers to study whether the T_C is substantially affected compared to the bulk material. In this context, it must be noted that MgB$_2$ nanowires of 50–200 nm diameter have been shown recently to have a superconducting T_C identical to that of the bulk material.[74]

5. Nanotubes of other metal chalcogenides

Nanotubes and nanowires of II–VI semiconductor compounds such as CdS and CdSe have been obtained by a soft chemical route involving surfactant-assisted synthesis.[28,75] For CdSe nanotubes, the metal oxide was reacted with the selenidizing reagent in the presence of a surfactant such as Triton 100X. Substantial amounts of nanotubes were obtained by this method (Fig. 25a and b). Annealing of the as-prepared nano-

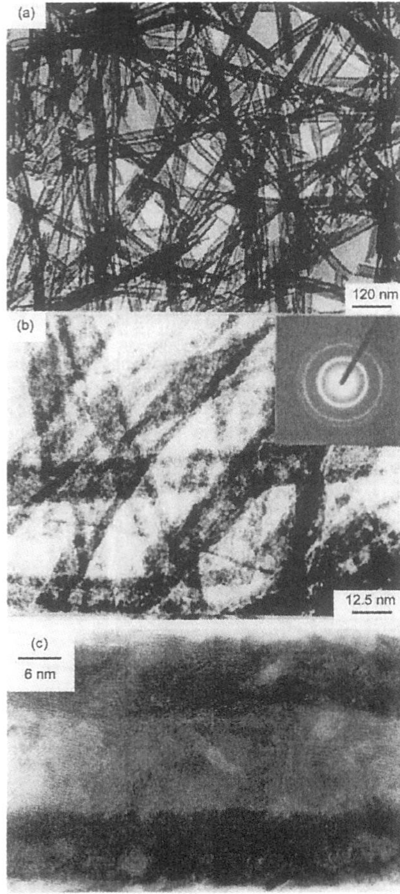

Fig. 25 (a) and (b) low-resolution TEM images of CdSe nanotubes. Inset shows a typical ED pattern; (c) HREM image of the CdSe nanotube showing walls containing several nano-crystallites. (Reproduced with permission from ref. 28).

tubes appears to improve the crystallinity of the nanotubes (Fig. 25c). To obtain nanotubes of CdS, a similar procedure was followed, except that thioacetamide was used as the sulfidizing agent in place of NaHSe. The formation of the nanotubes and nanowires is controlled by varying the surfactant concentration which in turn changes the morphology and shape of the micellar cavity. Thus, in both CdS and CdSe, a higher concentration of the surfactants produce nanotubes, while a lower concentration of the surfactants preferentially produce nanowires. The surfactant is readily removed from the nanotubes by washing with a solvent. Both the CdSe and CdS nanotubes seem to be polycrystalline formed by aggregates of

nanoparticles.[76] The nanotubes of CdSe, though extended in one direction show quantum confinement and the absorption band is blue-shifted to 550 nm from 650 nm of the bulk sample. A quantum confinement effect is also observed in the PL spectrum of the nanotubes, which shows a band at 560 nm compared to 750 nm for the bulk sample (Fig. 26). The CdS nano-

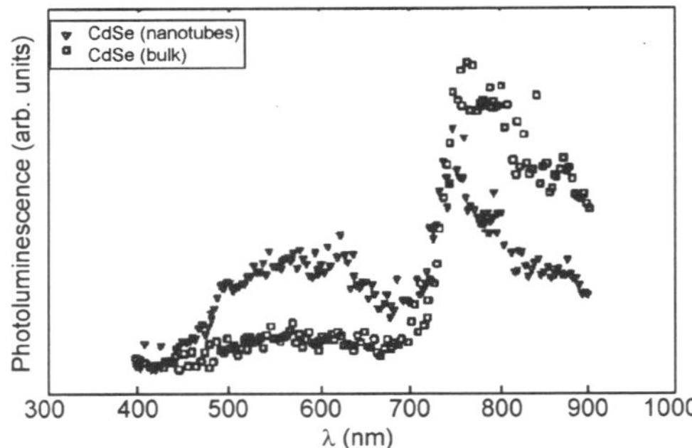

Fig. 26 Photoluminescence spectra of nanotubes of CdSe and bulk CdSe. (Reproduced with permission from ref. 28).

tubes also show a blue-shift of the absorption maximum due to confinement effects. The Raman spectrum of CdSe nanotubes show modes at 208 and 198 cm^{-1}, the former arising from the longitudinal optic phonon, red-shifted due to phonon confinement.[76]

ZnS nanotubes have been prepared by sulfidizing ZnO templates obtained in columnar form by electrochemical deposition.[77] Heating the ZnO column in H_2S above 400 °C gave the ZnS coated ZnO columns. The ZnO cores were then etched out giving hollow ZnS tubes.

Hydrothermal methods have been successfully used to synthesize a variety of nanotubes and nanorods. Layer-rolled structures of nickel sulfide have been obtained by sulfidizing the $Ni(NH_3)_4^{2+}$ complex by CS_2 in the presence of aqueous ammonia[78] under hydrothermal conditions. Nanotubes of $Cu_{5.5}FeS_{6.5}$[79] have been prepared hydrothermally. These nanotubes are large in diameter, have thick walls and are mostly open-ended. The crystallinity of the thin-walled nanotubes is not generally satisfactory. $Cu_{5.5}FeS_{6.5}$ has a layered structure characterized by S–S pairs and MS_4 tetrahedra. The nanotubes are thought to be formed by rolling the basal planes. The ED patterns collected from the nanotubes indicates that they are of the armchair and zigzag types. In the hydrothermal synthesis of the nanotubes and related structures, organic amines seem to act as the templating agents. The amine probably intercalates into the lattice of the precursor which is subsequently decomposes to yield the chalcogenide nantoubes around the templating amine.

6. Metal oxide nanotubes

Metal oxide nanotubes have been prepared by using a variety of techniques including templating reactions, sol–gel chemistry and hydrothermal methods.[7] Besides CNTs, organic surfactants have been employed to grow oxidic nanotubes. Pores of anodic alumina have been used as confining templates for the growth of TiO_2 and Al_2O_3 nanotubes, as well as coaxial nanotubes of TiO_2 sheathed SiO_2.[80,92]

6.1. VO_x nanotubes

Vanadium oxides deserve special attention owing to their structural flexibility and interesting catalytic, electrochemical and other properties. V_2O_5 can be regarded as a layer structure in which the VO_5 square pyramids are connected by sharing corners and edges and thereby forms the layer.[81] The interlayer interaction is weak as shown by the long V–O distances.[82] This structural characteristic permits the intercalation of various cationic species in the interlamellar space.[83] The cations include alkylammonium ions which are readily intercalated in between the layers under hydrothermal conditions.[84] Thus, this system is analogous to graphite and the layered dichalcogenides. In 1998 Nesper and co-workers[11,85] synthesized nanotubules of alkylammonium intercalated VO_x by hydrothermal means. The vanadium alkoxide precursor was hydrolyzed in the presence of hexadecylamine and the hydrolysis product (lamellar structured composite of the surfactant and the vanadium oxide) yielded VO_x nanotubes along with the intercalated amine under hydrothermal conditions (Fig. 27). The interesting feature of this

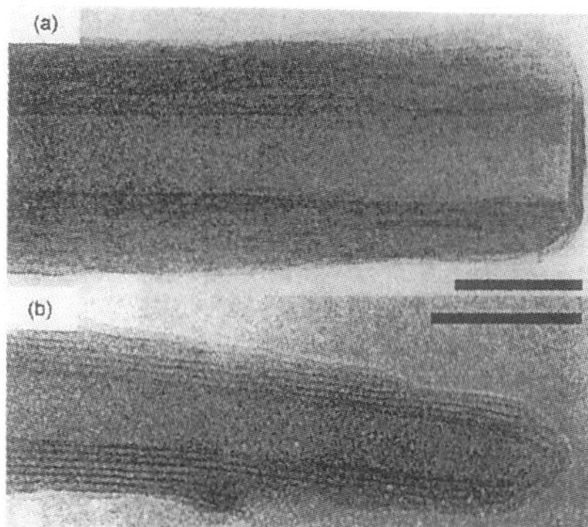

Fig. 27 TEM images of VO_x nanotubes with intercalated amine having varying chain lengths; (a) C_4VO_x–NT; (b) C_{16}–VO_x–NT. The length of the bar is 50 nm. (Reproduced with permission from ref. 85).

vanadium oxide nanotube is the presence of vanadium in the mixed valent state, thereby rendering it redox-active. The template could not be removed by calcination as the structural stability was lost above 523 K. Nevertheless, it was possible to partially extract the surfactant under mildly acidic conditions. These workers have later shown that the alkylamine intercalated in the intertubular space could be exchanged with other alkylamines of varying chain lengths as well as α,ω-diamines.[85] The distance between the layers in the VO_x nanotubes can be controlled by the length of the $-CH_2-$ chain in the amine template. The tubes obtained with the diamine had thicker walls compared to those obtained with the monoamines. A change in the interlayer distance observed in the XRD pattern as a function of the alkyl chain length is rendered to be due to the monolayer arrangement of the amine molecules in between the layers. In the VO_x nanotubes, the amine may be acting as a structure-directing agent and is incorporated in the tube walls.

Most of the VO_x nanotubes obtained by the hydrothermal method are open-ended. Very few closed tubes had flat or pointed conical tips. Cross-sectional TEM images of the nanotubular phases show that instead of concentric cylinders, (*i.e.* layers that fold and close within themselves), the tubes are made up of single or double layer scrolls providing a serpentine-like morphology.[85,86] The scrolls are seen as circles that do not close in the images (Fig. 28). Non-symmetric fringe patterns in the tube walls exemplify that most of the nanotubes are not rotationally symmetric and carry depressions and holes in the walls. Diamine-intercalated VO_x nanotubes are multilayer scrolls with narrow cores and thick walls, composed of packs of several vanadium oxide layers (Fig. 29). The diamine-containing VO_x nanotubes also show a smaller number of holes in the wall structure and the tubes are well ordered with uniform distances throughout the tube length.[85] The scroll-like structure of the nanotubes may be the most obvious cause for their high structural flexibility. This is also why facile exchange reactions occur.

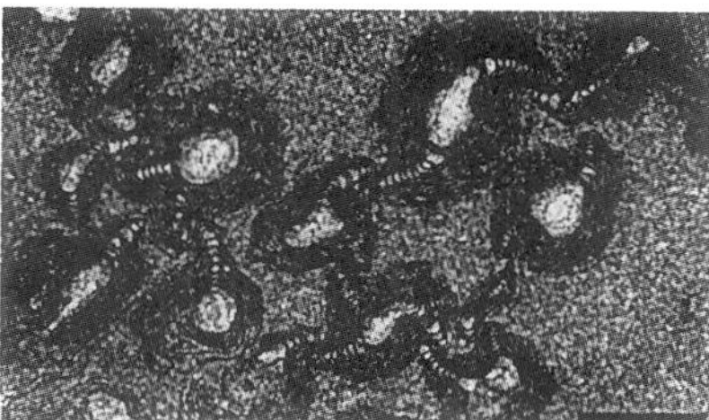

Fig. 28 Cross-sectional TEM images of monoamine-intercalated VO_x nanotubes showing serpentine-like scrolls.

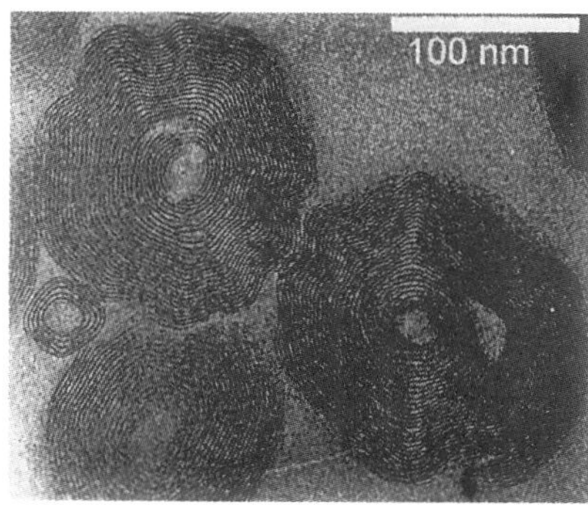

Fig. 29 Cross-sectional TEM images of diamine-intercalated VO_x nanotubes showing a larger thickness of the tube walls and a smaller inner core. (Reproduced with permission from ref. 85).

It must be pointed out that the VO_x scroll-like structures are not genuine nanotubes of the type formed by carbon or metal dichalcogenides.

The alkylammonium intercalated VO_x nanotubes are paramagnetic and show semimetallic conductivity probably due to the mixed valent V centers. The percentage content of V(IV) calculated on the basis of the effective magnetic moment is ~45–50%.[85] The structure contains VO_4 tetrahedra and VO_5 square-pyramids simultaneously.

The as-synthesized amine intercalated VO_x nanotubes could be aligned on glass substrates by using micromolding in capillaries (MIMIC).[86] An elastomeric polydimethylsiloxane (PDMS) stamp where parallel capillaries of 5 μm were patterned was used as the mould.[86] The capillaries were filled with a VO_x nanotube suspension in octanol. After evaporation of the solvent and removal of the mould, long lines of assemblies of well aligned nanotubes were obtained.

Vanadium oxide nanotubes containing primary monoamines with long alkyl chains have been prepared by employing nonalkoxide vanadium precursors such as $VOCl_3$ and V_2O_5. The amine complexes of the vanadium precursors are then hydrolyzed. Hydrothermal treatment of the precursors gives good yields of VO_x nanotubes incorporating the amines.[27] The distance between the layers is proportional to the length of the alkyl amine chain which acts as the structure-directing template. Cross-sectional TEM images demonstrate the predominance of serpentine-like scrolls rather than of concentric tubes.

Mn–V-oxides have been prepared starting from vanadium oxide–dodecylamine composite nanotubes.[87] The composite nanotubes were prepared by mixing V_2O_5 with dodecylamine in the presence of ethanol and water. The amine templates can be easily substituted or even ion-exchanged with ions like Mn^{2+} in an aqueous alcohol solution to obtain the Mn–V–O nanotubes. Most of the nanotubes had open ends, while some of them had closed ends, with the side of the tubes wrapped around the end to close it. The Mn^{2+} ions replace the organic cations in the structures and hence are intercalated in between the layers.

6.2. Use of carbon nanotubes as templates for the growth of oxide nanotubes

Carbon nanotubes have been successfully used as removable templates for the synthesis of a variety of oxide nanotubes. Ajayan *et al.*[9] reported the preparation of V_2O_5 nanotubes by using partially oxidized carbon nanotubes as templates. Apart from coating of CNTs by the oxide phase, metal oxide fillings in the internal cavities and thin oxide layers between the concentric shells of the tubes were also obtained (Fig. 30). A mix-

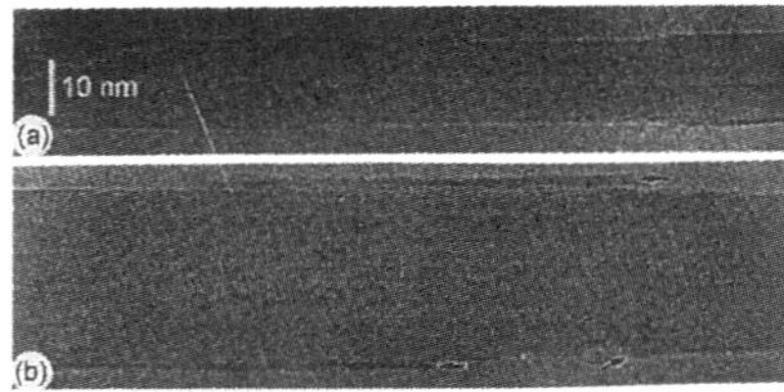

Fig. 30 TEM images of the VO_x coated carbon nanotubes. (a) Low-magnification image showing partial coating; (b) image at higher magnification showing the 0.34 nm fringes of the CNT and oxide coatings of uniform thickness on the upper and lower parts of the tubes. (Reproduced with permission from ref. 9).

ture of partially oxidized CNTs and V_2O_5 powder was annealed in air to produce a nanotube–oxide composite. The external coating of the tubes consists of crystalline V_2O_5 layers, which grow with the c-axis parallel to that of the nanotube layers. Intercalation of the oxide occurs where there are missing shells of the nanotubes. The coating on the isolated nanotubes is of uniform thickness, but does not always cover the entire length of the nanotube surface. Continuous cylindrical oxidic films on the nanotube surface are generally formed over a length of a few hundred nanometers. Attempts have been made to remove the carbon nanotube templates by oxidation at 650 °C. These samples show a dramatic increase in the ratio of the coated to the uncoated tubes. On observing the tips of the remaining coated tubes, oxidation was found to have occurred from the inner wall of the tube and progressed outwards leaving few outer layers and the outer coating of the oxide. This is in contrast to the oxidation of pure nanotubes where the etching of the layers starts from outside and proceeds inwards leaving chisel-shaped tips. The complete removal of the nanotube template resulted in larger hollow oxidic structures with thin skins.

Rao *et al.*[10] have prepared a variety of oxide nanotubes including SiO_2, Al_2O_3, V_2O_5, MoO_3 and RuO_2 employing carbon nanotube templates. By using multi-walled carbon nanotubes as the template they also made nanotubes of zirconia and yttrium-stabilized zirconia.[30] In these preparations, acid-treated MWNT bundles were coated with a suitable precursor of the metal oxide, and the coated composites heated to high temperatures to remove the carbon template (Fig. 31). TEM images show that the carbon nanotubes get fully coated with the oxidic material on reaction with the organometallic compound followed by calcination at 500 °C. The nanotube features are essentially retained in the oxide–nanotube composites even after calcination at 500 °C (Fig. 31). Acid-treated tubes give a better coating than the pristine ones. The nanotube templates could be removed successfully by heating at 750 °C for several hours. The oxidation proceeded from the hollow regions inside the tubes. Silica nanotubes prepared by this method could be modified by doping with transition metal ions for possible use in catalysis. Thus Cu, Cr or Ni containing SiO_2 tubes have been prepared by this method by taking the relevant transition metal ions in the starting reaction mixture.[10] The transition metals are present as oxides in the final structure, and the oxide appeared as islands on the surface of the hollow silica tubes in the case of Cu.

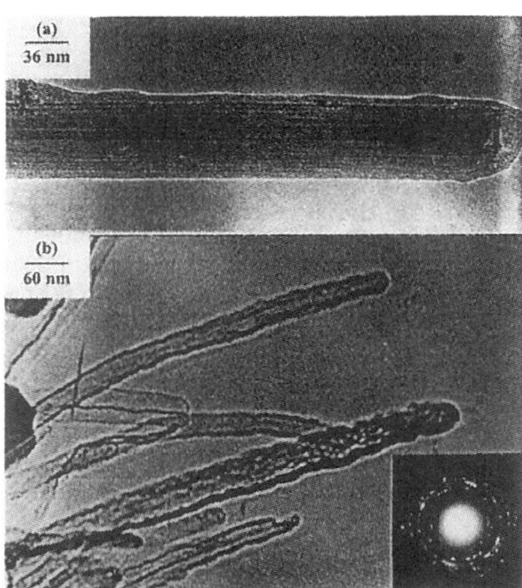

Fig. 31 (a) TEM image of a ZrO_2 coated CNT. (b) Hollow ZrO_2 nanotubes after removal of the CNT template. (Reproduced with permission from ref. 30).

6.3. TiO_2

Nanotubes of transition metal oxides that do not possess layered structures, (*e.g.*, TiO_2, ZrO_2) have been prepared by employing the sol–gel technique, membrane-confined growth and templated growth. Titania tubes were first prepared electrochemically by Hoyer.[88] A polymer mould with a negative (replicated) structure of an anodic alumina porous membrane was used for electrodeposition of the titania nanotubes. The titania phase obtained was amorphous and calcination to induce crystallization led to the deformation of the tube structure. The diameters of the tubes were in the ~70–100 nm range, being controlled by the pore size of the membrane. Needle-shaped anatase nanotubes could be precipitated from a gel containing a mixture of SiO_2 and TiO_2.[29] A mixture of titanium isopropoxide and tetraethylorthosilicate (TEOS) was hydrolyzed and gelled in an incubator, and the gel further heated to 870 K resulting in the precipitation of fine TiO_2 (anatase) crystals. This was further treated with NaOH at 380 K for 20 h to yield the TiO_2 nanotubular phase. An amorphous SiO_2-related phase present in the product could be removed by chemical treatment. The nanotubes formed by this method had a diameter of ~8 nm and lengths upto 100 nm (Fig. 32a).

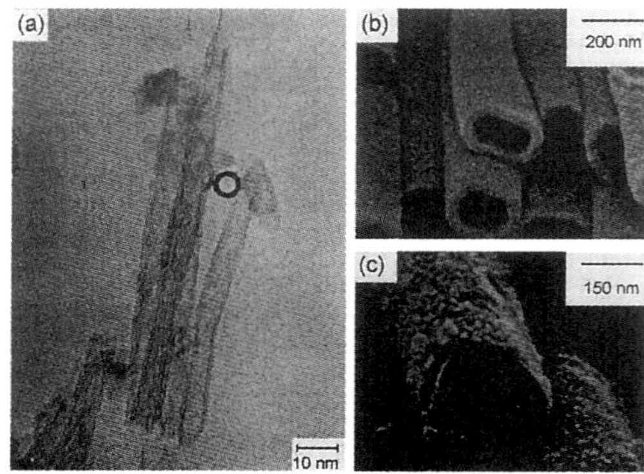

Fig. 32 (a) TEM image of TiO_2 nanotubes (Reproduced with permission from ref. 29); (b) and (c) FE-SEM images of TiO_2 nanotubes deposited from a TiF_4 solution. (Reproduced with permission from ref. 80).

Titania nanotubes are obtained by the direct deposition of titanium tetraflouride (TiF_4) in the pores of an alumina membrane.[80] The porous alumina membrane prepared electrochemically was immersed in an aqueous solution of TiF_4 and ammonia, and maintained at 60 °C. The walls of the nanotubes contained small nanoparticles of anatase (Fig. 32b). Titania rods were obtained at longer deposition times. The initial solution supersaturated with titania produces nuclei of anatase which deposit on the inner walls of the porous membranes, and give rise to the nanotubes. Photocatalytic activity of the as-deposited anatase nanotubes shows a performance similar to that of the conventionally prepared anatase powder.[80] Titania nanotubes are also prepared by using electrospun polymer fibers as templates.[89] The polymer fibers are coated with titanium oxide, by dipping in a titanium isopropoxide solution followed by dipping in ethanol–water for hydrolysis and condensation. Thermal treatment of the metal oxide coated polymer fibers results in the loss of the polymeric core, producing hollow nanotubes of titania. HREM images show that the individual particles that make the nanotube walls are crystalline (anatase phase). The sol–gel coating was able to mimic the finer details of the polymeric fiber, thereby forming nodules in the inner walls of the tubes. Well aligned, uniform arrays of titania nanotubes could be fabricated by the anodic oxidation of a pure Ti sheet in an aqueous solution containing 0.5 to 3.5 wt% of HF.[90] The upper ends of these tubes were mostly open. Recently, single crystalline nanotubes of TiO_2 (anatase) have been prepared by an unconstrained solution growth, by hydrolyzing TiF_4 under acidic conditions at 60 °C (Fig. 33).[91] A majority of these TiO_2 nanotubes were closed, with spherical or polyhedral caps.

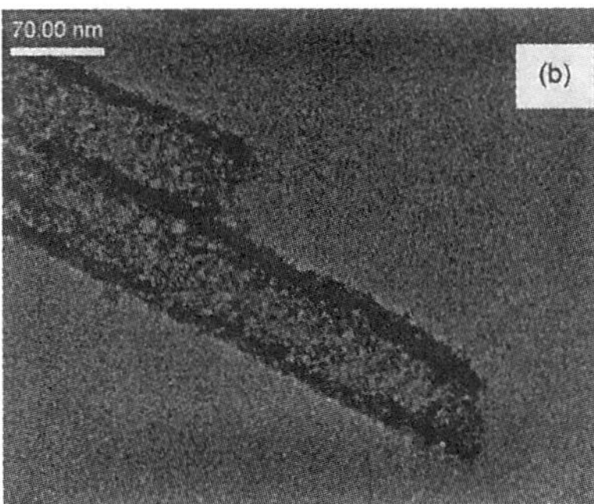

Fig. 33 (a) SEM image of unidirectionally aligned TiO_2 nanotubes; (b) TEM image of a TiO_2 nanotube formed with nanocrystalline TiO_2 particles. (Reproduced with permission from ref. 91).

Pores of anodic alumina have been used for the growth of coaxial nanotubes of TiO_2 sheathed SiO_2.[92] Electrochemical deposition in the pores of a polymeric alumina membrane generates arrays of TiO_2 nanotubes. For the coaxial nanotubes, the SiO_2 nanotubes are first grown inside the pores of the anodic alumina membrane. The TiO_2 nanotubes are then grown inside the SiO_2 nanotubes. The existence of Ti–O–Si bonds in the amorphous sheaths is believed to play a role in the formation of these composite nanotubes.

6.4. SiO_2 and Al_2O_3 nanotubes

6.4.1. SiO_2.
Sol–gel techniques have been extensively used to form silica gel nanotubes[93] as well as mesoporous silica nanotubes.[25] Nakamura and Matsui[93] prepared silica nanotubes as a spin-off product of sol–gel synthesis wherein, tetraethylorthosilicate (TEOS) was hydrolyzed in the presence of ammonia and D,L-tartaric acid. Mesoporous silica nanotubes were also prepared by a special post-synthesis-ammonia-hydrothermal treatment of the mesoporous silica material.[25] It was possible to simultaneously restructure the pore size, nanochannel regularity and morphology of the mesoporous material by this method. The restructured products were highly ordered and exhibited nanotubular forms of silica. This methodology gained momentum with the successful synthesis of individual nanotubes or small bundles containing a few nanotubes. Adachi et al.[94] were the first to report the synthesis of very long silica nanotubes by employing surfactant-assisted growth. Laurylamine hydrochloride was used as the surfactant template around which TEOS was hydrolyzed. Tube formation was followed by trisilylation treatment. Trimethylsilylation inactivated the silanol groups on the surface of the tube, thus inhibiting the condensation of silanol groups between the different bundles, and yielding long individual silica tubes. Another advantage of the trisilylation treatment was that the surfactant was removed without calcination. Individual silica nanotubes have also been prepared by the sol–gel method in the presence of citric acid as the structure modifier.[95] The silica nanotubes obtained were, however, mostly amorphous as determined by XRD. Electron microscopy revealed that the individual silica nanotubes were long (~0.5–20 µm), with wide diameters.

The sol–gel technique has been used to dope silica in TiO_2 nanotubes.[96] Different loadings of Si in the TiO_2 nanotubes was achieved by varying the amount of TEOS in the starting solution containing tetrapropylorthotitanate in butanediol. After several days of aging the dry gel was calcined to induce the crystallization and formation of the nanotubes. The pore size of the nanotubes decreased with increasing Si content. EDAX mapping images revealed that the Ti and Si were homogeneously distributed in the nanotubes. It is believed that Si doping influences the sintering process and suppresses the grain growth of titania nanoparticles, thus increasing the surface area of the doped tube.[96]

Mesoporous silica materials with hierarchical tubule-within-tubule structures have been prepared and characterized by spectroscopic methods.[97] Strong photoluminescence of these materials is explained as due to the presence of Si–OH complexes located on the nanotube surface, which also explains the persistence of the signal for some time after the pumping laser is turned off.

6.4.2. Al_2O_3.
Alumina nanotubes have been prepared by electrochemical means.[12] Two different preparation methods designated as normal stepwise anodization (NSA) and lateral stepwise anodization (LSA) have been employed for the purpose. The major difference between the two methods is the way the potential was applied. An aluminium film deposited on a p-type Si substrate was anodized in dilute H_2SO_4. In NSA, the potential was applied to the bottom surface of the Si substrate, while in LSA, it was applied to the top surface of the alumina film. The nanotubes formed were attached to the anodic porous alumina film (Fig. 34a). The NSA tubes were smaller than the LSA tubes. In addition to straight alumina nanotubes, branched alumina nanotubes were also obtained in the same synthesis (Fig. 34b).[12,98]

6.5. Other oxide nanotubes

Recently, ZnO nanotubes have been prepared hydrothermally.[99] In this procedure, ethanol was added to an aqueous solution containing the $Zn(NH_3)_4^{2+}$ complex, and the mixture heated to

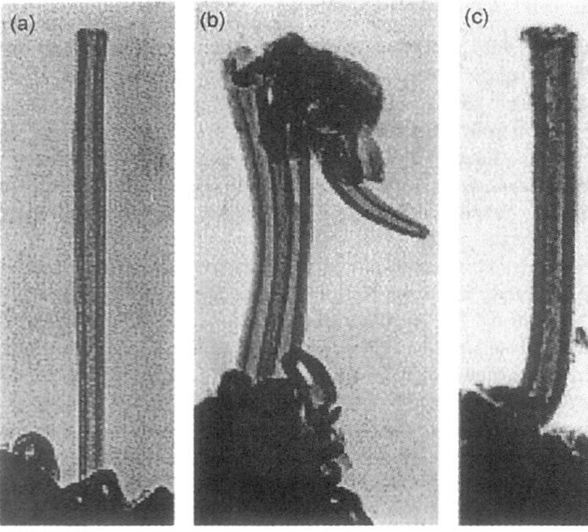

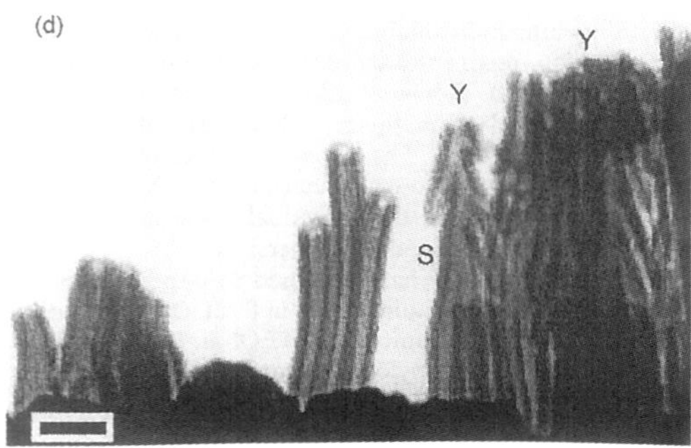

Fig. 34 TEM images of alumina nanotubes: (a and b) NSA tubes; (c) LSA tubes; (d) side-view of the branched alumina tubes attached to the APA film, (Y indicates Y-shaped branched cells, S indicates the stem of such branched cells). (Reproduced with permission from refs. 12 and 98).

450 K for 13 h in a Teflon-lined autoclave. There was, however, deviation in the peak intensities compared to the standard diffraction pattern, indicating the preferred orientation of the tubular ZnO (wurtzite). TEM images showed that the walls of the ZnO tubes were not very smooth due to the build-up by polycrystalline nanoparticles. The $Zn(NH_3)_4^{2+}$ complex is hydrolyzed in the presence of ammonia and nanoparticles of ZnO are produced along with the nanotubes as seen in the TEM images. As the temperature and duration of the reaction are increased, there is increased evolution of ammonia and the ZnO nanoparticles assemble along certain orientations and aggregate to the hollow tubular structures.

Oxides of Er, Tm, Yb and Lu have been prepared in the nanotubular form employing template-mediated reactions using dodecylsulfate assemblies.[100] These oxidic nanotubes were synthesized by the homogeneous precipitation method using urea. The oxidic phase was precipitated from the reaction mixture containing the salt of the rare earth element, sodium dodecylsulfate and water, the pH of the medium being varied by the progressive addition of urea. Hydrolysis of the rare earth salt occurs on heating to 60 °C and the oxidic phase forms a precipitate. The rare earth oxide nanotubes so obtained have small inner diameters and thin walls. Only Yb- and Lu-oxide nanotubes have been obtained reproducibly by this method.

Nanotubes of In_2O_3 and Ga_2O_3 have been synthesized by employing sol–gel chemistry and porous alumina templates.[101] Aqueous ammonia is added to In (or Ga) containing sols to obtain precipitates which are then peptized with nitric acid to produce stable sols. The alumina membrane is immersed in the sol and then air-dried followed by annealing in air at elevated temperatures for 12 h to obtain the oxides. The alumina template is dissolved in alkali solution to give the free tubes. The hollow nanotubes so obtained have lengths of upto 10 μm. The inner diameters could be varied by selecting the template dimensions and the immersion time. The positively charged sol particles adhere to the negatively charged pore walls of the templating membrane leading to the formation of alumina–semiconductor composite nanotubes after annealing.

Nanotubes of perovskite oxides have been prepared by employing a template-mediated growth mechanism.[102] Nanotubes of the ceramic materials such as $BaTiO_3$ and $PbTiO_3$ are obtained by heating the metal acetate sol with Ti-isopropoxide in ethanol. Masked Whatman anodic membranes (200 nm pores) were used as the templates and were immersed in the mixed sol, and then air-dried. The templates could be removed after calcination followed by treatment with 6 M NaOH. TEM images revealed 50 μm long $BaTiO_3$ tubes bundled together after removal of the template. Most of the tubes obtained were open-ended (Fig. 35). Nanotubes of both $PbTiO_3$ and $BaTiO_3$

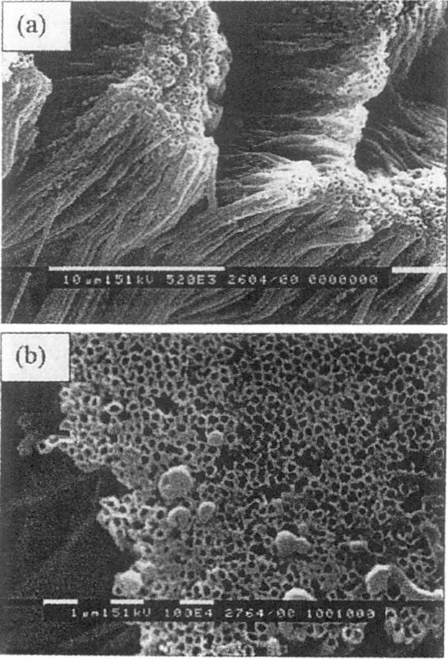

Fig. 35 SEM images of the perovskite nanotubes: (a) side-view showing the bundle formation after removal of the template; (b) top-view of $PbTiO_3$ bundles showing the open ends. (Reproduced with permission from ref. 102).

were found to be hollow throughout their length. XRD, Raman and electron diffraction measurements indicated that $BaTiO_3$ was in the cubic (paraelectric) phase, while $PbTiO_3$ was in a tetragonal ferroelectric phase after calcination. TEM images also showed that the tubes comprise small polycrystalline grains.

7. Nanotubes of BN and other nitrides

Based on theoretical calculations, the existence of nanotube structures of BN was predicted in 1994[13,103] which was soon verified by the first synthesis of BN nanotubes in 1995.[104] Based on theoretical calculations, it was predicted that unlike the carbon nanotubes whose electronic properties (metallic or semiconducting nature) are controlled by the tube diameter, wrapping, twisting and topological defects, the electronic properties of the BN nanotubes were independent of the tube diameter and chirality. In contrast to the CNTs, the BN nanotubes were predicted to be constant band-gap materials, with a large band gap of ~5.5 eV.[13]

Thin BN tubes of less than 200 nm diameter were obtained by arc discharge with hollow tungsten electrodes filled with h-BN powder. Following this initial report, a variety of methods have been employed to prepare BN nanotubes. In some cases, BN nanocages and fullerenes were also obtained.[105] The methods of synthesis of BN nanotubes include those which are far from the equilibrium, such as the electrical arc method,[32,104] arcing between h-BN and Ta rods in a N_2 atmosphere,[106] laser ablation of h-BN,[107] and continuous laser heating of h-BN surfaces.[108] The last method is particularly useful in providing long ropes of BN nanotubes with thin walls. Faceted BN polyhedra filled with metallic boron are found in a few cases. Single-walled BN nanotubes are also observed in some cases.[32,109] Recently, single-walled nanotubes of BN deposited on polycrystalline W substrates were obtained by using electron-cyclotron resonance nitrogen and electron beam boron sources.[110] Fig. 36 shows a

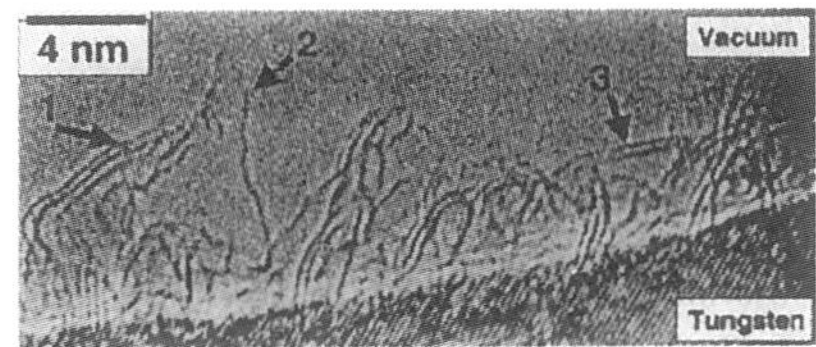

Fig. 36   HREM image of single-walled BN nanotubes adhering on a W substrate. (Reproduced with permission from ref. 110).

HREM image of the single-walled BN nanostructures adhering to the W substrate. *In-situ* HREM images show that the single-walled BN nanotubes contain both four- and eight-fold rings and the tubes are closed with fullerene-like structures. This is in contrast to the rectangular terminations observed earlier[109] which were interpreted in terms of the closure *via* four-fold rings and by nitrogen-rich pentagons.

Boron nitride nanotubes co-existing with a web and an amorphous phase were obtained by a plasma jet method wherein a sintered BN disk was subjected to a direct current arc plasma jet with Ar–H_2 as the plasma gas.[111] As revealed by the HREM image (Fig. 37), the tubes grow from the amorphous phase present at the root of the tubes. The tubes were closed by parallel bases of co-axial cylindrical shape. Solid state processes have also been employed to obtain BN nanotubes. Hexagonal BN powder was first ball-milled to generate highly disordered and amorphous nanostructures which were then annealed under N_2 to 1570 K for about 10 h.[112] Several tubes with bamboo-like morphology were observed some of which contained metal particles at the tip as a contaminant from the stainless steel reaction chamber. The nanostructure of the ball-milled powder which acts as the catalyst was crucial for the tube growth.

Apart from the above methods, some of which employ drastic conditions, processes close to equilibrium conditions such as pyrolysis and chemical vapor deposition (CVD) have also been employed to prepare BN nanotubes. The CVD growth of hollow, crystalline BN nanotubules by the pyrolysis of borazine on nickel boride catalyst particles maintained at 1270–1370 K, produced nanotubes with bulbous or flag-like caps (Fig. 38).[16b] The reaction is given by,

$$B_3N_3H_6 - Ni\ boride \longrightarrow 3BN + 3H_2$$

A root-growth mechanism has been proposed for the growth of BN nanotubes, wherein the nanotubes nucleate on the nickel boride catalyst particle often with irregular initiation caps and

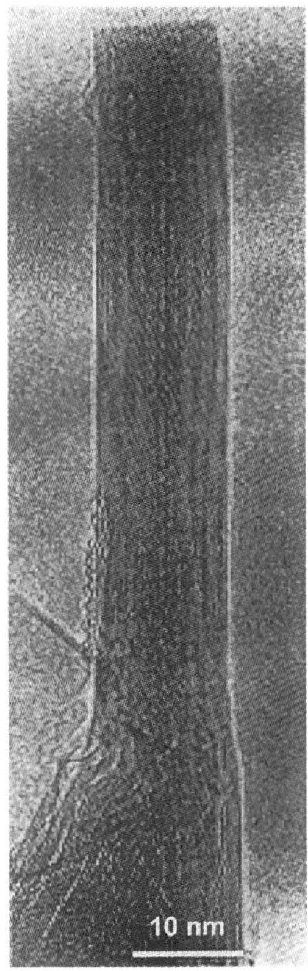

Fig. 37   TEM image of a BN nanotube having a phase boundary between the amorphous phase and the nanotube structure at the roots. (Reproduced with permission from ref. 111).

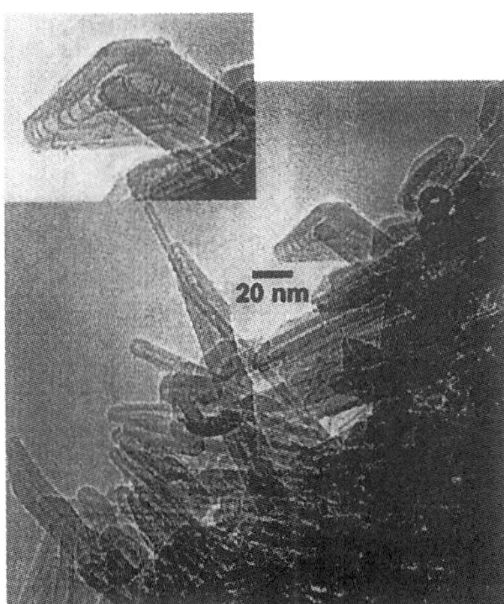

Fig. 38   TEM image showing different BN nanotube tip closures. The club-shaped tip is magnified in the inset. (Reproduced with permission from ref. 16b).

grow out by the incorporation of additional BN at the catalyst–nanotube junction. An efficient CVD method involving the thermal decomposition of the 1 : 2 adduct of melamine and boric acid in a N_2 atmosphere at 1970 K was developed for the synthesis of BN nanotubes. This method does not use any metal catalyst.[16c] The nanotubes have the stoichiometric composition and the HREM images (Fig. 39a) reveal the regular

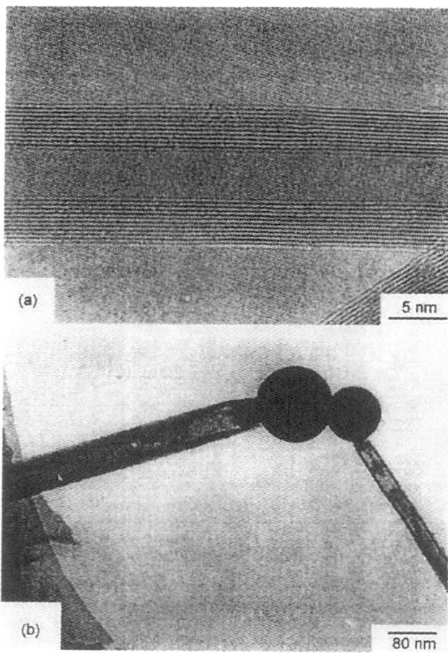

Fig. 39 (a) HREM image of a multi-walled BN tube (layer separation = 5.2 nm); (b) low-magnification image of the tubes showing that they may grow out from the bulbous tips. (Reproduced with permission from ref. 113).

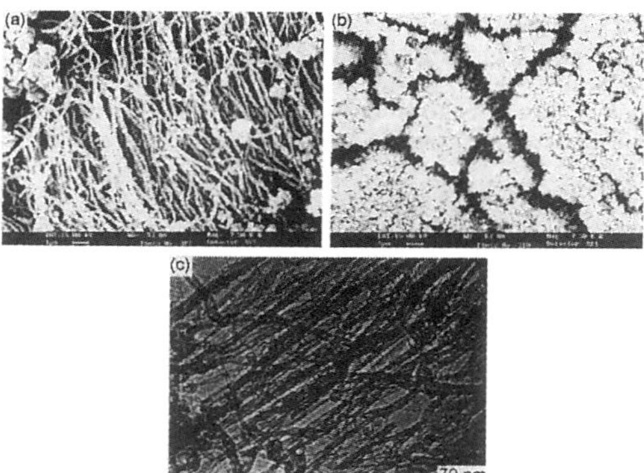

Fig. 40 SEM and TEM images of aligned BN nanotubes: (a) and (b) give side and top view SEM images, respectively; (c) TEM image of pure BN nanotube. (Reproduced with permission from ref. 17).

spacing of the BN layers (d = 5.2 nm). The nanotubes had bulbous tips filled with amorphous material (Fig. 39b) and the tip-growth mechanism seems to be valid. The amorphous material contains mainly $B_4N_3O_2H$.[113] It is believed that at the reaction temperature (1700 °C), some species such as B_2O_3 form along with the amorphous B–N–O cluster originating from decomposition of the precursor. The BN nanotube growth from the oxide phase follows the reaction

$$B_2O_3 + 4C + H_2O + N_2 \longrightarrow 2BN + 4CO + H_2$$

Boron powder has also been used in the CVD method for growing BN nanotubes. By heating a mixture of B and iron oxide in flowing ammonia gas, nanotubes and nano-bamboo structures of BN were obtained.[114] The *in-situ* generated Fe particles act as a catalyst and the growth of the BN nanotubes and nano-bamboo structures is ascribed to the vapor–liquid–solid (VLS) catalytic growth mechanism. Recently, B_2O_2 (obtained by heating B and MgO at 1300 °C) was heated in the presence of NH_3 in a BN-made reaction tube, to obtain BN nanotubes as the product with ~40% yield.[115] Most of the nanotubes obtained were open-tipped showing irregular fracture surfaces.

An exhaustive study has been carried out recently on the synthesis of BN nanotubes and nanowires by various CVD techniques.[17] The methods examined include heating boric acid with activated carbon, multi-walled carbon nanotubes, catalytic iron particles or a mixture of activated carbon and iron particles, in the presence of ammonia. With activated carbon, BN nanowires are obtained as the primary product. However, with multi-walled carbon tubes, high yields of pure BN nanotubes are obtained as the major product. BN nanotubes with different structures were obtained on heating boric acid and iron particles in the presence of NH_3. Aligned BN nanotubes are obtained when aligned multi-walled nanotubes are used as the templates (Fig. 40). Prior to this report, alignment of BN nanotubes was achieved by the synthesis of the BN nanotubule composites in the pores of the anodic alumina oxide, by the decomposition of 2,4,6-trichloroborazine at 750 °C.[116] Attempts had been made earlier to align BN nanotubes by

depositing them on the surfaces of carbon fibers.[113] The tubes grow almost vertically from the surface and are nearly aligned.

Carbon nanotubes have been used successfully as templates for the growth of BN nanotubular structures.[15,17,117] MWNTs were heated to 1773 K with B_2O_3 in a N_2 atmosphere. Several metal oxides like MoO_3, V_2O_5, Ag_2O or PbO[118] were used as growth promoters, and added separately into the starting powder mixtures. Pyrolysis of borazine in the presence of acetylene over metal catalyst particles yielded B–C–N nanotubes.[15a] Previously $B_xC_yN_z$ nanotubules were prepared by the arc-discharge method.[14a] B–C–N composite rods were used as the anode and arced against pure graphite cathodes in a He gas environment to yield the nanotubules.

Boron nitride nanotubes have been used in a wide variety of ways to generate nanocables where the BN nanotubes play host to 1D nanowires or nanoclusters occupying the hollow cavity. Zhang and co-workers[119] have reported the synthesis of $(BN)_xC_y$ nanotubes filled with a SiC and SiO_2 core by laser ablation. Carbon nanotube-confined reactions involving substitution reactions are employed to synthesize SiC nanowires encapsulated in BN nanotubes.[120] The CNTs react with boron oxide vapor in the presence of N_2 to yield BN nanotubes. The SiO vapor then penetrates into the cavity of the nanotubes and reacts with the internal wall of the CNTs to give SiC nanowires. In some cases, the filling occurs through the entire length of the nanotube. In a slightly modified method, BN and $(BN)_xC_y$ nanotubes filled with boron carbide nanowires were prepared using CNTs as templates.[121] The $(BN)_xC_y$ nanotubes are formed by capillary filling of boron oxide vapor in the inner cavity of the CNTs, followed by the substitution of the inner layers of the CNTs with B_2O_3 in the presence of N_2 gas. Inside the CNTs, boron oxide reacts simultaneously with the gaseous carbon monoxide or the interior layers of the CNTs to produce the boron carbide filling. The final product contained boron carbide filled $(BN)_xC_y$ nanotubes with an outer layer of pure C and inner layers of pure BN. Fig. 41 shows an illustration of the boron carbide nanowires inside the B_xC_yN composite tubes. Pure BN tubes filled with boron carbide were also formed in the product.

Nanocables of BN nanotubes filled with Mo clusters are reported by Golberg *et al.*[122] The Mo cluster-filled BN nanotubes are prepared by the treatment of CVD grown CNTs with B_2O_3, CuO and MoO_3 in a N_2 atmosphere. It has been proposed that the filling of CNTs with MoO_3 precedes the formation of BN tubes on the CNT template. The filling of MoO_3 is then further reduced to metallic Mo by the carbon of the CNTs. Continuous filling of Mo could not be obtained by this method.

BN nanotubes have been used as hosts to oxide materials such as α-Al_2O_3 nanorods.[123] The $B_4N_3O_2H$ precursor (obtained

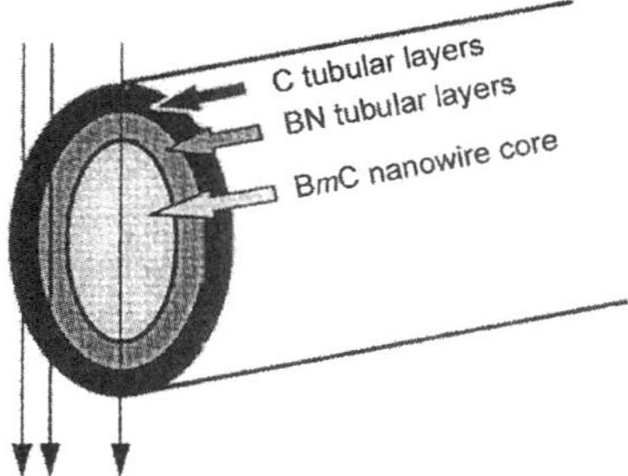

Fig. 41 Schematic illustration of a multi-phase filled B_xC_yN nanotube. (Reproduced with permission from ref. 121).

by the thermal decomposition of melamine diborate) when pyrolyzed on $Si–SiO_2–Al_2O_3$ substrate, yielded $\alpha-Al_2O_3$ nanorod-filled BN nanocables in abundance.

According to Menon and Srivastava,[124] the chirality of a nanotube directly depends on the tip-end morphology, *i.e.*, for a flat tip end, a zigzag arrangement of the layers in the tube walls is energetically favorable. Rhombohedral stacking in relatively thick BN tubular fibers has been observed,[125] in contrast to the belief that the BN nanotubes are exact analogues of CNTs which usually show random stacking between the layers and no preferential tube helicity. HREM studies of the BN nanotubes reveal layered structures somewhat similar to those in the CNTs (Fig. 42a). However, there are some distinct characteristics. A

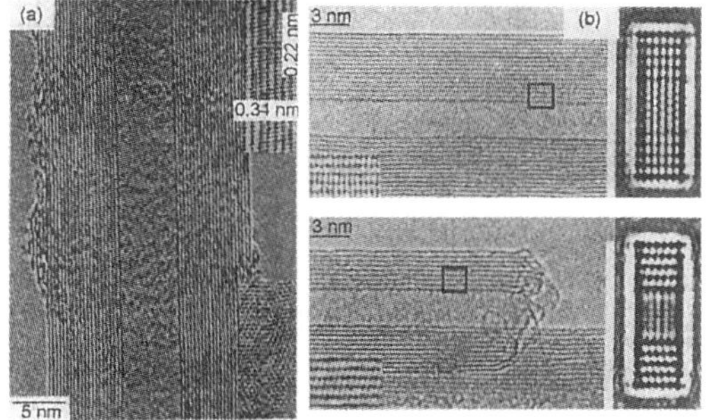

Fig. 42 (a) A BN nanotube with a uniform layer separation of 0.34 nm. The atomic columns in the wall fragments exhibit layer fringes separated by 0.22 nm and makes an angle of 12.5° with respect to the tube axis, exhibiting *r*-BN stacking; (b) hexagonal stacking and *r*-BN stacking confirmed by the computer-simulated HREM images (right hand side panels). (Reproduced with permission from ref. 120).

field emission high-resolution analytical TEM study has revealed the following characteristic features of the multi-walled BN nanotubes: (i) Hexagonal and rhombohedral (3R) stackings co-exist in nanotube shell assembly, a feature readily seen in HREM images where the layers due to 3R stacking occur in the walls and sometimes in the core besides the regular hexagonal stacking (Fig. 42a). In Fig. 42b the observed HREM images of BN nanotubes are compared with the computer simulated images for the *r*-BN and *h*-BN stacking in the walls. (ii) Flattening of the nanotube cross-section makes clear atomic resolution of the pore structure possible in a three-shelled nanotube. (iii) There is a change in chirality of tubular layers from armchair to zigzag arrangement in a 30° double-walled nanotube kink. BN nanotubes with open tips exhibit local *r*-BN stacking in the walls, while those with close tips exhibit *h*-BN stacking[120] (Fig. 42b).

BN nanotubes obtained by laser ablation and arc-discharge seem to have very few layers and the laser grown tubes self-assemble into long ropes. It is proposed that the growth of BN

nanotubes is due to surface diffusion along the external surface, which also ensures morphological stability of the open end during growth. The nanotube heights are limited by the corresponding diffusion lengths. Recombination of B and N, both in plasma and on the surface, may serve as a possible nucleation center. In the pyrolytically grown BN nanotubes, it is commonly observed that the nanotubes have bulbous tips.[113] The amorphous clusters present in the tip region may play a catalytic role in the nanotube tip-growth process similar to the metal catalyst in the CVD process of CNT growth. Formation of open/flat-tip ends are also observed in some of the BN tubes. There appears to be a preference for the growth of open BN nanotubes in metal oxide-promoted CVD synthesis. One of the reasons behind this may be that the metal atoms occurring at the edges of the growing nanotube may prevent tube closure.[44] Fig. 43 shows the presence of a dark contrast spot on the inner

Fig. 43 The innermost terminated layer of the BN nanotube shows a spot assigned to the Pb atom (or cluster), which may prevent tube closure. (Reproduced with permission from ref. 44).

layer of the tube at the terminated edge. The spot is likely to be related to the presence of the metal atom (Pb) or the metal cluster at that edge, which prevents closure of the layer.

Recently, in their effort to prepare Si_3N_4 nanowires, Gundiah *et al.*[126] have found occasional nanotubes in the TEM images. Similarly, in the preparations of GaN nanowires, GaN nanotubes have been observed by Deepak *et al.*[127]

8. Nanotubes of other materials

Transition metal halides such as $NiCl_2$ crystallize in the $CdCl_2$ structure, with the metal halide layers held together by weak van der Waals forces. $NiCl_2$ has been shown to form closed cage structures and nanotubes.[128] These were prepared by heating $NiCl_2·6H_2O$ initially in air to lose the water of crystallization, and then heated further at 450 °C under N_2 (Fig. 44).

Very few metallic nanotubes have been synthesized to date. Martin and co-workers[129] have prepared Au nanotubules with lengths upto 6 μm and inner diameters of 1 nm by using a pore-wall modified alumina membrane. Co and Fe nanotubules have been synthesized using polycarbonate membranes as templates.[130] Cu and Ni microtubules have also been prepared by the pyrolysis of composite fibers consisting of a poly(ethyleneterephthalate) (PET) core fiber and electroless-plated metal skin at the exterior.[131] While Ni microtubules prepared by this method were single-crystalline, the Cu microtubules were polycrystalline. Ordered arrays of Ni nanotubules have been prepared by electrodeposition in the pores of an alumina membrane, the pore walls being modified with an organic amine.[132] Nickel when electrodeposited in the pores binds preferentially to the pore walls because of its strong affinity towards the

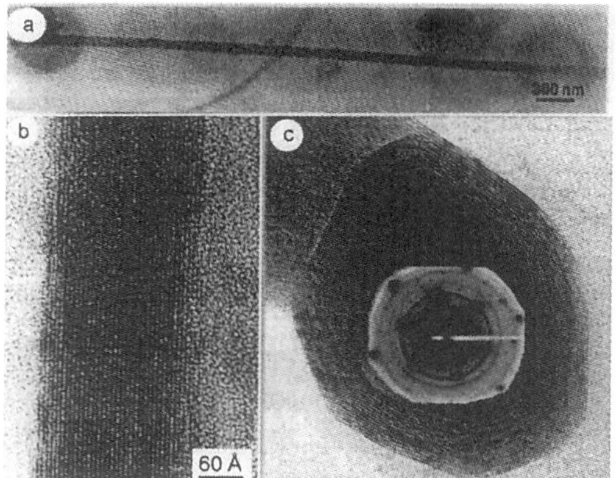

Fig. 44 (a) Low-magnification view of the NiCl₂ nanotube; (b) HREM image of the nanotube wall; (c) many-layered cage structure of NiCl₂, with the hexagonal ED pattern superimposed. (Reproduced with permission from ref. 128).

amine. In the absence of the amine in the pore walls, solid Ni nanowires were obtained. The alumina membrane could be removed by treatment with NaOH giving highly ordered arrays of Ni nanotubules (Fig. 45a). The Ni nanotubules were ferromagnetic with enhanced coercivity compared to bulk Ni (Fig. 45b).

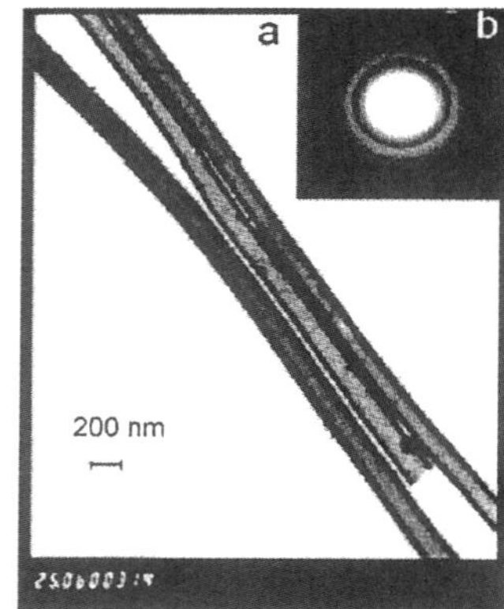

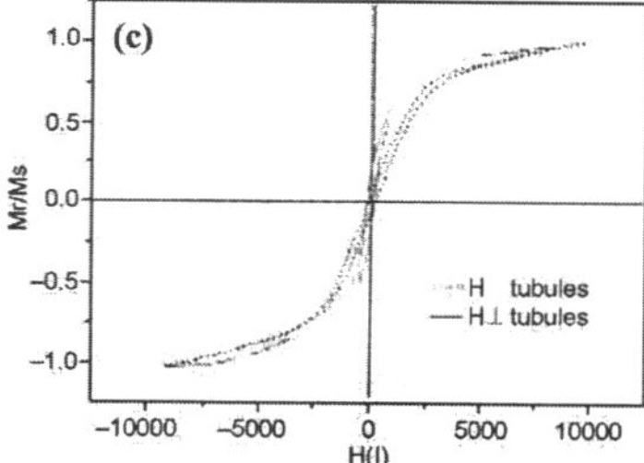

Fig. 45 (a) TEM image showing the Ni nanotubules; (b) ED pattern of the Ni nanotubules; (c) magnetization *vs.* applied field (*M–H*) curve showing hysteresis. (Reproduced with permission from ref. 132).

Tellurium nanotubes have been prepared using the polyol method. Orthotelluric acid in ethylene glycol was added to a refluxing solution of ethylene glycol.[133] TEM images taken after stopping the reaction at different stages showed the formation of cylindrical seeds and the subsequent growth of nanotubules along the peripheral edge of the seeds. The walls of the nano-

tubes have a uniform thickness, but appear to be fragile compared to CNTs. Decreasing the amount of the Te precursor in the initial reaction mixture led to the preferential formation of solid nanorods rather than hollow tubes.

9. Useful properties of inorganic nanotubes

Various properties of carbon nanotubes of potential technological value are known.[2,134] The properties and applications of the inorganic nanotubes, however, have not been investigated as extensively as would be desirable. The electronic structures of MoS₂ and WS₂ have been examined briefly and the semiconducting nature of the nanotubes confirmed.[55,57a] It is necessary to investigate the optical, electrical and other properties of the various chalcogenide nanotubes. This is especially true of nanotubes of NbS₂ and such materials which are predicted to be metallic.[57b] NbSe₂ nanotubes have been found to be metallic at ordinary temperatures, becoming superconducting at lower temperatures.[70] Electronic and optical properties of the BN nanotubes have not yet been investigated in detail. Theoretical calculations suggest BN tubes to be insulating with a wide band gap of 5.5 eV.[13]

Carbon nanotubes have been investigated for H₂ storage properties.[135] It would be worthwhile to look into the H₂ storage ability of some of the inorganic nanotubes. The chalcogenide nanotubes with an ~6 Å van der Waals gap between the layers, are potential candidates for showing storage capacity. It has been shown recently that BN nanotubes can store a reasonable quantity of H₂.[136] Multi-walled BN nanotubes have been shown to possess a capacity of 1.8–2.6 wt% of H₂ uptake under ~10 MPa at room temperature. This value, though smaller than that reported for CNTs, nevertheless suggests the possible use of BN nanotubes as a hydrogen storage system. MoS₂ nanotubes could be electrochemically charged and discharged with a capacity of 260 mA h g⁻¹ at 20 °C, corresponding to a formula of H₁.₂₄MoS₂.[137] The high storage capacity is believed to be due to the enhanced electrochemical-catalytic activity of the highly nanoporous structure. This may find wide applications in high energy batteries.

Single-walled carbon nanotubes are known to possess extraordinary strength.[138] Mechanical properties of BN nanotubes would be worthy of exploration. Unlike carbon nanotubes, BN nanotubes are predicted to have stable insulating properties independent of their structure and morphology. Thus, BN tubes can be used as nano-insulating devices for encapsulating conducting materials like metallic wires. Filled BN nanotubes are expected to be useful in nanoscale electronic devices and for the preparation of nano-structured ceramics.

Electrochemical studies have been performed with the alkylammonium intercalated VOₓ nanotubes[139] as well as Mn intercalated VOₓ nanotubes.[87] Cyclic voltammetry studies of alkylammonium-VOₓ nanotubes showed a single reduction peak, which broadened on replacing the amine with Na with an additional peak. Li ion reactivity has also been tested with Mn–VOₓ nanotubes by reacting with *n*-butyllithium, and found that ~2 lithiums per V ion are consumed. Electrochemical Li intercalation of Mn–VOₓ nanotubes show that 0.5 Li ions per V atom were intercalated above 2 V.[87] This observation may be relevant to battery applications.

Conventional microfabricated AFM tips are of limited use for investigating high aspect ratio features (*i.e.* deep and narrow features), mainly because, without special treatment typical aspect ratios of such tips would be around 3 : 1 or lower. Thus, the width of such a tip at a certain height from the apex is much larger than that of the nanotubes of uniform thickness adhered to the tip. The nanotubes therefore would be more suitable for the analysis of deep and narrow structures than the commonly available tips. CNTs have been used as AFM tips and there appears to be every likelihood that extremely narrow structures can be probed.[140] WS₂ could be mounted on the ultrasharp Si tip following a similar methodology. These tips were tested in an AFM microscope by imaging a replica of high aspect ratio, and it was observed that these WS_2 nanotube tips provide a considerable improvement in the image quality compared to the conventional ultrasharp Si tips.[141]

The most likely application of the chalcogenide nanotubes is as solid lubricants. Mo and W chalcogenides are widely used as solid lubricants. It has been observed that the hollow nanoparticles of WS_2 show better tribological properties and act as a better lubricant compared to the bulk phase in every respect (friction, wear and life-time of the lubricant).[142] Tribological properties of $2H-MoS_2$ and WS_2 powder can be attributed to the weak van der Waals forces between the layers which allow easy shear of the films with respect to each other. The mechanism in the WS_2 nanostructures is somewhat different and the better tribological properties may arise from the rolling friction allowed by the round shape of the nanostructures.

Recently, open-tipped MoS_2 nanotubes were prepared by the decomposition of ball-milled ammonium thiomolybdate powder under a H_2 + thiophene atmosphere, and used as a catalyst for the methanation of CO with H_2.[143] The conversion of CO to CH_4 was achieved at a much lower temperature compared to polycrystalline MoS_2 particles, and there was no deterioration even after 50 h of consecutive catalyzing cycles. This observation is of importance in the context of energy conversion of global CO_2.

10. Concluding remarks

Inorganic nanotubes have emerged to become a group of novel materials. Although this area of research started with the layered metal chalcogenides, recent results suggest that other inorganic materials can also be prepared in the form of nanotubes, as typified by the metal oxides. It is likely that many new types of inorganic nanotubes will be made in the near future. These would include metal nanotubes as well as nanotubes of inorganic compounds such as Mg_3B_2, GeO_2 and GaSe. Theoretical calculations indeed predict a stable nanotubular structure for GaSe.[144] Various layered materials could be explored for this purpose. It is noteworthy that the nanotubes of metal chalcogenides have been made by employing several methods ranging from soft chemical routes to techniques such as arc evaporation and laser ablation (Table 1).

Nanotubes of MoS_2, WS_2 and a few other layered materials are single-crystalline in the sense that the layers run through the entire structure. Some of the chalcogenide nanotubes are, however, polycrystalline, the nanotubular form being produced by an aggregation of nanoparticles, just as in some of the metal oxide nanotubes.

Properties of inorganic nanotubes such as those of MoS_2 have been investigated to some extent. However, by and large, there is much to be studied with respect to the electronic, optical and other properties of most of the inorganic nanotubes. Properties such as sorption, hydrogen storage and catalytic activity are worthy of exploration. Mechanical properties of BN, B–N–C and related nanotubes are also worthy of study.

Acknowledgements

The authors thank the Department of Science & Technology and DRDO (India) for support of this research.

References

1 S. Iijima, *Nature*, 1991, **354**, 56.

2 C. N. R. Rao, B. C. Satishkumar, A. Govindaraj and M. Nath, *Chem. Phys. Chem.*, 2001, **2**, 78.

3 (*a*) P. Ratnasamy, L. Rodrigues and A. J. Leonard, *J. Phys. Chem.*, 1973, **77**, 2242; (*b*) J. Wilson and A. D. Yoffe, *Adv. Phys.*, 1969, **269**, 193.

4 R. R. Chianelli, E. Prestridge, T. Pecorano and J. P. DeNeufville, *Science*, 1979, **203**, 1105.

5 (*a*) R. Tenne, L. Margulis, M. Genut and G. Hodes, *Nature*, 1992, **360**, 444; (*b*) L. Margulis, G. Salitra and R. Tenne, *Nature*, 1993, **365**, 113; (*c*) Y. Feldman, E. Wasserman, D. J. Srolovitch and R. Tenne, *Science*, 1995, **267**, 222.

6 (*a*) C. T. Kresge, M. E. Leonowicz, W. J. Roth, J. C. Vartulli and J. S. Beck, *Nature*, 1992, **259**, 710; (*b*) J. S. Beck, J. C. Vartulli, W. J. Roth, M. E. Leonowicz, C. T. Kressge, K. D. Schmitt, C. T. W. Chu, D. H. Olson, E. W. Sheppard, S. B. McCullen, J. B. Higgins and J. C. Scwenker, *J. Am. Chem. Soc.*, 1992, **114**, 10834.

7 G. R. Pratzke, F. Krumeich and R. Nesper, *Angew. Chem., Int. Ed.*, 2002, **41**, 2446.

8 (*a*) W. Stöber, A. Fink and E. Bohn, *J. Colloid Interface Sci.*, 1968, **26**, 62; (*b*) M. Nakamura and Y. Matsui, *J. Am. Chem. Soc.*, 1995, **117**, 2651.

9 P. M. Ajayan, O. Stephane, Ph. Redlich and C. Colliex, *Nature*, 1995, **375**, 564.

10 (*a*) B. C. Satishkumar, A. G. Govindaraj, E. M. Vogl, L. Basumallick and C. N. R. Rao, *J. Mater. Res.*, 1997, **12**, 604; (*b*) B. C. Satishkumar, A. Govindaraj, M. Nath and C. N. R. Rao, *J. Mater. Chem.*, 2000, **10**, 2115.

11 M. E. Spahr, P. Bitterli, R. Nesper, M. Müller, F. Krumeich and H. U. Nissen, *Angew. Chem., Int. Ed.*, 1998, **37**, 1263 (*Angew. Chem.*, 1998, **110**, 1339).

12 L. Pu, X. Bao, J. Zou and D. Feng, *Angew. Chem., Int. Ed.*, 2001, **40**, 1490.

13 (*a*) M. L. Cohen, *Solid State Commun.*, 1994, **92**, 45; (*b*) J. L. Corkill and M. L. Cohen, *Phys. Rev. B*, 1994, **49**, 5081; (*c*) Y. Miyamoto, A. Rubio, S. G. Louie and M. L. Cohen, *Phys. Rev. B*, 1994, **50**, 18360.

14 (*a*) K. Kobayashi and N. Kurita, *Phys. Rev. Lett.*, 1993, **70**, 3542; (*b*) Z. W. -Sieh, K. Cherrey, N. G. Chopra, X. Blasé, Y. Miyamoto, A. Rubio, M. L. Cohen, S. G. Louie, A. Zettl and P. Gronsky, *Phys. Rev. B*, 1994, **51**, 11229.

15 (*a*) R. Sen, B. C. Satishkumar, A. Govindaraj, K. R. Harikumar, G. Raina, J. P. Zhang, A. K. Cheetham and C. N. R. Rao, *Chem. Phys. Lett.*, 1998, **287**, 671; (*b*) O. Stephan, P. M. Ajayan, C. Colliex, Ph. Redlich, J. M. Lambert, P. Bernier and P. Lefing, *Science*, 1994, **266**, 1683.

16 (*a*) P. Gleize, M. C. Schouler, P. Gadelle and M. Caillet, *J. Mater. Sci.*, 1994, **29**, 1575; (*b*) O. R. Lourie, C. R. Jones, B. M. Bertlett, P. C. Gibbons, R. S. Ruoff and W. E. Buhro, *Chem. Mater.*, 2000, **12**, 1808; (*c*) R. Ma, Y. Bando and T. Sato, *Chem. Phys. Lett.*, 2001, **337**, 61.

17 F. L. Deepak, C. P. Vinod, K. Mukhopadhyay, A. Govindaraj and C. N. R. Rao, *Chem. Phys. Lett.*, 2002, **353**, 345.

18 C. N. R. Rao and A. G. Govindaraj, *Acc. Chem. Res.*, 2002 and references therein.

19 R. Tenne, M. Homyonfer and Y. Feldman, *Chem. Mater.*, 1998, **10**, 3225 and references therein.

20 (*a*) M. Hershfinkel, L. A. Gheber, V. Volterra, J. L. Hutchison, L. Margulis and R. Tenne, *J. Am. Chem. Soc.*, 1994, **116**, 1914; (*b*) T. Tsirlina, Y. Feldman, M. Homyonfer, J. Sloan, J. L. Hutchison and R. Tenne, *Fullerene Sci. Technol.*, 1998, **6**, 157.

21 M. Nath, A. Govindaraj and C. N. R. Rao, *Adv. Mater.*, 2001, **13**, 283.

22 M. Nath and C. N. R. Rao, *Chem. Commun.*, 2001, 2236.

23 M. Nath and C. N. R. Rao, *J. Am. Chem. Soc.*, 2001, **123**, 4841.

24 M. Nath and C. N. R. Rao, *Angew. Chem., Int. Ed.*, 2002, **41**, 3451.

25 A. P. Lin, C. Y. Mou and S. D. Liu, *Adv. Mater.*, 2000, **12**, 103.

26 J. Zhang, L. Sun, C. Liao and C. Yan, *Chem. Commun.*, 2002, 262.

27 M. Niederberger, H.-J. Muhr, F. Krumeich, F. Bieri, D. Günther and R. Nesper, *Chem. Mater.*, 2000, **12**, 1995.

28 C. N. R. Rao, A. G. Govindaraj, F. L. Deepak, N. A. Gunari and M. Nath, *Appl. Phys. Lett.*, 2001, **78**, 1853.

29 T. Kasuga, M. Hiramatsu, A. Hason, T. Sekino and K. Niihara, *Langmuir*, 1998, **14**, 3160.

30 C. N. R. Rao, B. C. Satishkumar and A. Govindaraj, *Chem. Commun.*, 1997, 1581.

31 C. M. Zelenski and P. K. Dorhout, *J. Am. Chem. Soc.*, 1998, **120**, 734.

32 A. Loiseau, F. Williame, N. Demonecy, G. Hug and H. Pascard, *Phys. Rev. Lett.*, 1996, **76**, 4737.

33 M. Homyonfer, Y. Mastai, M. Hershfinkel, V. Volterra, J. L. Hutchison and R. Tenne, *J. Am. Chem. Soc.*, 1996, **118**, 7804.

34 Y. Feldman, G. L. Frey, M. Homyonfer, V. Lyakhovitskaya, L. Margulis, H. Cohen, G. Hodes, J. L. Hutchison and R. Tenne, *J. Am. Chem. Soc.*, 1996, **118**, 5362.

35 M. Homyonfer, B. Alperson, Y. Rosenberg, L. Sapir, S. R. Cohen, G. Hodes and R. Tenne, *J. Am. Chem. Soc.*, 1997, **119**, 2693.

36 S. P. Cramer, K. S. Liang, A. J Jacobson, C. H. Chang and R. R. Chianelli, *Inorg. Chem.*, 1984, **23**, 1215.

37 G. U. Kulkarni and C. N. R. Rao, *Catal. Lett.*, 1991, **11**, 63.

38 (a) E. Diemann and A. Müller, *Coord. Chem. Rev.*, 1973, **10**, 79; (b) R. R. Chianelli and M. B. Dines, *Inorg. Chem.*, 1978, **17**, 2758; (c) A. Wildervanck and F. Jellinek, *Z. Anorg. Allg. Chem*, 1964, **328**, 309.

39 (a) C. N. R. Rao and K. P. R. Pisharody, *Prog. Solid State Chem.*, 1976, **10**, 207; (b) N. Allali, V. Gaborit, E. Prouzet, C. Geantet, M. Danot and A. Nadiri, *J. Phys. IV (France)*, 1997, **7**, C2–927; (c) E. Bjerkelund and A. Kjekhus, *Z. Anorg. Allg. Chem.*, 1964, **328**, 235; (d) F. K. McTaggart and A. D. Wadsley, *Aust. J. Chem*, 1958, **11**, 445.

40 M. Nath, K. Mukhopadhyay and C. N. R. Rao, *Chem. Phys. Lett.*, 2002, **352**, 163.

41 (a) Y. Q. Zhu, W. K. Hsu, M. Terrones, S. Firth, N. Grobert, R. J. H. Clark, H. W. Kroto and D. R. M. Walton, *Chem. Commun.*, 2001, 121; (b) Y. Q. Zhu, W. K. Hsu, M. Terrones, S. Firth, N. Grobert, R. J. H. Clark, H. W. Kroto and D. R. M. Walton, *Chem. Phys. Lett.*, 2001, **342**, 15.

42 W. K. Hsu, Y. Q. Zhu, N. Yao, S. Firth, R. J. H. Clark, H. W. Kroto and D. R. M. Walton, *Adv. Funct. Mater.*, 2001, **11**, 69.

43 W. K. Hsu, Y. Q. Zhu, C. B. Bothroyd, I. Kinlöch, S. Trasobares, H. Terrones, N. Grobert, M. Terrones, R. Escudero, G. Z. Chen, C. Colliex, A. H. Windle, D. H. Fray, H. W. Kroto and D. R. M. Walton, *Chem. Mater.*, 2000, **12**, 3541.

44 D. Golberg and Y. Bando, *Appl. Phys. Lett.*, 2001, **79**, 415.

45 (a) R. L. D. Whitby, W. K. Hsu, C. B. Bothroyd, P. K. Fearon, H. W. Kroto and D. R. M. Walton, *Chem. Phys. Chem.*, 2001, **10**, 620; (b) R. L. D. Whitby, W. K. Hsu, P. K. Fearon, N. C. Billingham, I. Maurin, H. W. Kroto, D. R. M. Walton, C. B. Bothroyd, S. Firth, R. J. H. Clark and D. Collison, *Chem. Mater.*, 2002, **14**, 2209; (c) R. L. D. Whitby, W. K. Hsu, P. C. P. Watts, H. W. Kroto and D. R. M. Walton, *Appl. Phys. Lett.*, 2001, **79**, 4574; (d) W. K. Hsu, Y. Q. Zhu, H. W. Kroto, D. R. M. Walton, R. Kamalakaran and M. Terrones, *Appl. Phys. Lett.*, 2000, **77**, 4130.

46 Y. Q. Zhu, W. K. Hsu, H. W. Kroto and D. R. M. Walton, *Chem. Commun*, 2001, 2184.

47 Y. Q. Zhu, W. K. Hsu, H. Terrones, N. Grobert, B. H. Chang, M. Terrones, B. Q. Wei, H. W. Kroto, D. R. M. Walton, C. B. Bothroyd, I. Kinlöch, G. Z. Chen, A. H. Windle and D. J. Fray, *J. Mater. Chem.*, 2000, **10**, 2570.

48 Y. Q. Zhu, W. K. Hsu, N. Grobert, B. H. Chang, M. Terrones, H. Terrones, H. W. Kroto and D. R. M. Walton, *Chem. Mater.*, 2000, **12**, 1190.

49 N. Grobert, M. Terrones, A. J. Osborne, H. Terrones, W. K. Hsu, S. Trasobares, Y. Q. Zhu, J. P. Hare, H. W. Kroto and D. R. M. Walton, *Appl. Phys. A*, 1998, **69**, 595.

50 J. Sloan, J. L. Hutchison, R. Tenne, Y. Feldman, T. Srilina and M. Homyonfer, *J. Solid State Chem.*, 1999, **144**, 100.

51 M. Rémskar, Z. Skraba, C. Ballif, R. Sanjines and F. Levy, *Surf. Sci.*, 1999, **435**, 637.

52 M. Homyonfer, Y. Feldman, L. Margulis and R. Tenne, *Fullerene Sci. Technol.*, 1998, **6**, 59.

53 (a) S. Iijima, S. Ichihashi and Y. Ando, *Nature*, 1992, **356**, 776; (b) D. E. H. Jones, *Nature*, 1991, **351**, 526; (c) S. Iijima, P. M. Ajayan and T. Ichihashi, *Phys. Rev. Lett.*, 1991, **69**, 3900.

54 R. Tenne, *Adv. Mater.*, 1995, **7**, 965.

55 G. Seifert, H. Terrones, M. Terrones, G. Jungnickel and T. Frauenheim, *Phys. Rev. Lett.*, 2000, **85**, 146.

56 G. Seifert, T. Köhler and R. Tenne, *J. Phys. Chem. B*, 2002, **106**, 2497.

57 (a) G. Seifert, H. Terrones, M. Terrones, G. Jungnickel and T. Frauenheim, *Solid State Commun.*, 2000, **114**, 245; (b) G. Seifert, H. Terrones, M. Terrones, G. Jungnickel and T. Frauenheim, *Solid State Commun.*, 2000, **115**, 635.

58 C. Thomazeau, C. Geantet, M. Lacroix, V. Harlé, S. Benazeth, C. Marhic and M. Danot, *J. Solid State Chem.*, 2001, **160**, 147.

59 A. Zak, Y. Feldman, V. Alperovich, R. Rosentsveig and R. Tenne, *J. Am. Chem. Soc.*, 2000, **122**, 11108.

60 (a) P. J. F. Harris, in *Carbon Nanotube and Related Structures*, Cambridge University Press, UK, 1999 pp. 62–107; (b) S. Amelinckx, D. Bernaerts, X. B. Zhang, G. Van Tendeloo and J. Van. Landuyt, *Science*, 1995, **267**, 1334; (c) A. Oberlin, M. Endo and T. Kayama, *J. Cryst. Growth*, 1992, **32**, 6941; (d) H. Dai, A. Z. Rinzler, P. Nikolaev, A. Thess, D. T. Colbert and R. E. Smalley, *Chem. Phys. Lett.*, 1996, **260**, 471.

61 W. K. Hsu, B. H. Chang, Y. Q. Zhu, W. Q. Han, H. Terrones, M. Terrones, N. Grobert, A. K. Cheetham, H. W. Kroto and D. R. M. Walton, *J. Am. Chem. Soc.*, 2000, **122**, 10155.

62 P. Afanasiev, C. Geantet, C. Thomazeau and B. Jouget, *Chem. Commun.*, 2000, 1001.

63 M. Remškar, A. Mrzel, Z. Skraba, A. Jesih, M. Ceh, J. Demšar, P. Stadelmann, F. Lévy and D. Mihailovic, *Science*, 2001, **292**, 479.

64 R. Sen, A. Govindaraj, K. Suenaga, S. Suzuki, H. Kataura, S. Iijima and Y. Achiba, *Chem. Phys. Lett.*, 2001, **340**, 242.

65 D. H. Galván, R. Rangel and G. Alonso, *Fullerene Sci. Technol.*, 1998, **6**, 1025.

66 H. D. Flack, *J. Appl. Crystallogr.*, 1972, **5**, 138.

67 D. L. Greenaway and R. Nitsche, *J. Phys. Chem. Solids*, 1965, **26**, 1445.

68 M. I. Nathan, M. W. Shafer and J. E. Smith, *Bull. Am. Phys. Soc.*, 1972, **17**, 336.

69 D. H. Galván, J.-H. Kim, M. B. Maple, M. Avalos-Berja and E. Adem, *Fullerene Sci. Technol.*, 2000, **8**, 143.

70 M. Nath S. Kar A. K. Raychaudhuri C. N. R. Rao, *Chem. Phys. Lett.*, in press.

71 C. Schuffenhauer, R. P-Biro and R. Tenne, *J. Mater. Chem.*, 2002, **12**, 1587.

72 (a) C. S. Wang and J. M. Chen, *Solid State Commun.*, 1974, **14**, 1145; (b) A. LeBlanc-Soreau, P. Molinié and E. Faulques, *Physica C*, 1997, **282–287**, 1937.

73 R. F. Frindt, *Phys. Rev. Lett.*, 1972, **28**, 299.

74 Y. Wu, B. Messer and P. Yang, *Adv. Mater.*, 2001, **13**, 1487.

75 A. Govindaraj, F. L. Deepak, N. A. Gunari and C. N. R. Rao, *Israel J. Chem.*, 2001, **41**, 23.

76 P. V. Teredesai, F. L. Deepak, A. Govindaraj, C. N. R. Rao and A. K. Sood, *J. Nanosci. Nanotechnol.*, 2002, **2**, 495.

77 L. Dloczik, R. Engelhardt, K. Ernst, S. Fiechter, I. Seiber and R. Könenkamp, *Appl. Phys. Lett.*, 2001, **78**, 3687.

78 X. Ziang, Y. Xie, L. Zhu, W. He and Y. Qian, *Adv. Mater.*, 2001, **13**, 1278.

79 Y. Peng, Z. Meng, C. Zhong, J. Lu, L. Xu, S. Zhang and Y. Qian, *New J. Chem.*, 2001, **25**, 1359.

80 H. Imai, Y. Takei, K. Shimizu, M. Matsuda and H. Hirashima, *J. Mater. Chem.*, 1999, **9**, 2971.

81 H. G. Bachmann, F. R. Ahmed and W. H. Barnes, *Z. Kristallogr.*, 1961, **115**, 110.

82 R. Enjalbert and J. Galy, *Acta Crystallogr., Sect. C*, 1986, **42**, 1467.

83 J. Galy, *J. Solid State Chem.*, 1992, **100**, 229.

84 M. S. Whittingham, J. Guo, R. Chen, T. Chirayil, G. Janauer and P.Y. Zavalji, *Solid State Ionics*, 1995, **75**, 257.

85 F. Krumeich, H.-J. Muhr, M. Niederberger, F. Bieri, B. Schnyder and R. Nesper, *J. Am. Chem. Soc.*, 1999, **121**, 8324.

86 H.-J. Muhr, F. Krumeich, U. P. Schönholzer, F. Bieri, M. Niederberger, L. J. Gauckler and R. Nesper, *Adv. Mater.*, 2000, **12**, 231.

87 A. Dobley, K. Ngala, S. Yang, P. Y. Zavalji and M. S. Whittingham, *Chem. Mater.*, 2001, **13**, 4382.

88 P. Hoyer, *Langmuir*, 1996, **12**, 1411.

89 R. A. Caruso, J. H. Schattka and A. Greiner, *Adv. Mater.*, 2001, **13**, 1577.

90 D. Gong, C. A. Grimes, O. K. Varghese, W. Hu, R. S. Singh, Z. Chen and E. C. Dickey, *J. Mater. Res.*, 2001, **16**, 3331.

91 S. M. Liu, L. M. Gan, L. H. Liu, W. D. Zhang and H. C. Zeng, *Chem. Mater.*, 2002, **14**, 1391.

92 M. Zhang, Y. Bando and K. Wada, *J. Mater. Res.*, 2001, **16**, 1408.

93 (a) H. Nakamura and Y. Matsui, *J. Am. Chem. Soc.*, 1995, **117**, 2651; (b) H. Nakamura and Y. Matsui, *Adv. Mater.*, 1995, **7**, 871.

94 (a) M. Harada and M. Adachi, *Adv. Mater.*, 2000, **12**, 839; (b) M. Adachi, T. Harada and M. Harada, *Langmuir*, 1999, **15**, 7079.

95 L. Wang, S. Tomura, F. Ohashi, M. Maeda, M. Suzuki and K. Inukai, *J. Mater. Chem.*, 2001, **11**, 1465.

96 Y. Zhang and A. Reller, *Chem. Commun.*, 2002, 606.

97 H. J. Chang, Y. F. Chen, H. P. Lin and C. Y. Mou, *Appl. Phys. Lett.*, 2001, **78**, 3791.

98 J. Zou, L. Pu, X. Bao and D. Feng, *Appl. Phys. Lett.*, 2002, **80**, 1079.

99 J. Zhang, L. Sun, C. Liao and C. Yan, *Chem. Commun*, 2002, 262.

100 M. Yada, M. Mihara, S. Mouri, M. Kuroki and T. Kijima, *Adv. Mater.*, 2002, **14**, 309.

101 B. Cheng and E. T. Samulski, *J. Mater. Chem.*, 2001, **11**, 2901.

102 B. A. Hernandez, K.-S. Chang, E. R. Fisher and P. K. Dorhout, *Chem. Mater.*, 2002, **14**, 480.

103 X. Blase, A. Rubio, S. G. Louie and M. L. Cohen, *Europhys. Lett.*, 1994, **28**, 335.

104 N. G. Chopra, R. G. Luyken, K. Cherrey, V. H. Crespi, M. L. Cohen, S. G. Louie and A. Zettl, *Science*, 1995, **269**, 966.

105 O. Stephan, Y. Bando, A. Loiseau, F. Williame, N. Schramechenko, T. Tamiya and T. Sato, *Appl. Phys. A: Mater. Sci. Process.*, 1998, **67**, 107.

106 M. Terrones, W. K. Hsu, H. Terrones, J. P. Zhang, S. Ramos, J. P. Hare, R. Castillo, K. Prassides, A. K. Cheetham. H. W. Kroto and D. R. M. Walton, *Chem. Phys. Lett.*, 1996, **259**, 568.
107 (*a*) D. P. Yu, X. S. Sun, C. S. Lee, I. Bello, S. T. Lee, H. D. Gu, K. M. Leung, G. W. Zhou, Z. F. Dong and Z. Zhang, *Appl. Phys. Lett*, 1998, **72**, 1966; (*b*) G. W. Zhou, Z. Zhang, Z. G. Bai and D. P. Yu, *Solid State Commun.*, 1999, **109**, 555; (*c*) M. Terauchi, M. Tanaka, H. Matsuda, M. Takeda and K. Kimura, *J. Electron Microsc.*, 1997, **1**, 75.
108 T. Laude, Y. Matsui, A. Marraud and B. Jouffrey, *Appl. Phys. Lett.*, 2000, **76**, 3239.
109 (*a*) D. Golberg, Y. Bando, W. Han, K. Kurashima and T. Sato, *Chem. Phys. Lett.*, 1999, **308**, 337; (*b*) P. W. Fowler, K. M. Rogers, G. Seifert, M. Terrones and H. Terrones, *Chem. Phys. Lett.*, 1999, **299**, 359.
110 E. Bengu and L. D. Marks, *Phys. Rev. Lett.*, 2001, **86**, 2385.
111 Y. Shimizu, Y. Moriyoshi, H. Tanaka and S. Komatsu, *Appl. Phys. Lett.*, 1999, **75**, 929.
112 Y. Chen, L. T. Chadderton, J. F. Gerals and J. S. Williams, *Appl. Phys. Lett.*, 1999, **74**, 2960.
113 R. Ma, Y. Bando, T. Sato and K. Kurashima, *Chem. Mater.*, 2001, **13**, 2965.
114 C. C. Tang, M. L. de la Chapelle, P. Li, Y. M. Liu, H. Y. Dang and S. S. Fan, *Chem. Phys. Lett.*, 2001, **342**, 492.
115 C. C. Tang, Y. Bando, T. Sato and K. Kurashima, *Chem. Commun.*, 2002, 1290.
116 K. B. Shelimov and M. Moskovits, *Chem. Mater.*, 2000, **12**, 250.
117 (*a*) W. Han, Y. Bando, K. Kurashima and T. Sato, *Appl. Phys. Lett*, 1998, **73**, 3085; (*b*) D. Golberg, W. Han, Y. Bando, L. Burgeois, K. Kurashima and T. Sato, *J. Appl. Phys.*, 1999, **86**, 2364; (*c*) D. Golberg, Y. Bando, K. Kurashima and T. Sato, *Chem. Phys. Lett.*, 2000, **323**, 185.
118 D. Golberg, Y. Bando, K. Kurashima and T. Sato, *Solid State Commun.*, 2000, **116**, .
119 Y. Zhang, K. Suenaga, C. Colliex and S. Iijima, *Science*, 1998, **281**, 973.
120 W. Han, Ph. K.- Redlich, F. Ernst and M. Rühle, *Appl. Phys. Lett.*, 1999, **75**, 1875.
121 W. Han, P. K. -Redlich, F. Ernst and M. Rühle, *Chem. Mater.*, 1999, **11**, 3620.
122 D. Golberg, Y. Bando, K. Kurashima and T. Sato, *J. Nanosci. Nanotechnol.*, 2000, **1**, 49.
123 R. Ma, Y. Bando and T. Sato, *Adv. Mater.*, 2002, **14**, 366.
124 M. Menon and D. Srivastava, *Chem. Phys. Lett.*, 1999, **307**, 407.
125 L. Burgeois, Y. Bando and T. Sato, *J. Phys. D: Appl. Phys.*, 2000, **33**, 1902.
126 G. Gundiah, G. V. Madhav, A. Govindaraj, Md. Motin Seikh and C. N. R. Rao, *J. Mater. Chem.*, 2002, **12**, 1606.
127 F. L. Deepak, A. Govindaraj and C. N. R. Rao, *J. Nanosci. Nanotechnol.*, 2001, **1**, 303.
128 Y. R. Hacohen, E. Grunbaum, R. Tenne, J. Sloan and J. L. Hutchison, *Nature*, 1998, **395**, 337.
129 (*a*) C. R. Martin, M. Nishizawa, K. Jirage and M. Kang, *J. Phys. Chem. B*, 2001, **105**, 1925; (*b*) J. C. Hutleen, K. B. Jirage and C. R. Martin, *J. Am. Chem. Soc.*, 2000, **120**, 6603; (*c*) C. J. Brumlik and C. R. Martin, *J. Am. Chem. Soc*, 1991, **113**, 3174.
130 G. Tourillon, L. Pontonnier, J. P. Levy and V. Langlais, *Electrochem. Solid-State Lett.*, 2000, **3**, 20.
131 C. C. Han, M. Y. Bai and J. T. Lee, *Chem. Mater.*, 2001, **13**, 4260.
132 J. Bao, C. Tie, Z. Xu, Q. Zhou, D. Shen and Q. Ma, *Adv. Mater.*, 2001, **13**, 1631.
133 B. Mayers and Y. Xia, *Adv. Mater.*, 2002, **14**, 279.
134 (*a*) S. Saito, S. I. Sawada, M. Hawada and A. Oshima, *Mater. Sci. Eng.*, 1993, **B19**, 105; (*b*) R. Saito, M. Fujita, G. Dresslhaus and M. S. Dresslhaus, *Mater. Sci. Eng.*, 1993, **B19**, 185; (*c*) R. Dagani, *Chem. Eng. News*, Jan 11 1999, 31.
135 (*a*) H. M. Cheng, Q. H. Yang and C. Liu, *Carbon*, 2001, **39**, 1447; (*b*) P. Hou, Q. Yang, S. Bai, S. Xu, M. Liu and H. Cheng, *J. Phys. Chem. B*, 2002, **106**, 963 and references therein.
136 R. Ma, Y. Bando, H. Zhu, T. Sato, C. Xu and D. Wu, *J. Am. Chem. Soc.*, 2002, **124**, 7672.
137 J. Chen, N. Kuriyama, H. Yuan, H. T. Takeshita and T. Sakai, *J. Am. Chem. Soc.*, 2001, **123**, 11813.
138 (*a*) M. Treacy and T. Ebbesen, *Nature*, 1996, **381**, 678; (*b*) C. F. Cornwell and L. T. Wille, *Solid State Commun.*, 1997, **101**, 505.
139 M. E. Spahr, P. S. Bitterli, R. Nesper, O. Haas and P. Novak, *J. Electrochem. Soc.*, 1999, **146**, 2780.
140 H. Dai, J. Hafner, A. G. Rinzler, D. T. Colbert and R. E. Smalley, *Nature*, 1996, **384**, 147.
141 A. Rothschild, S. R. Cohen and R. Tenne, *Appl. Phys. Lett.*, 1999, **75**, 4025.
142 L. Rapoport, Y. Bilik, Y. Feldman, M. Homyonfer, S. R. Cohen and R. Tenne, *Nature*, 1997, **387**, 791.
143 J. Chen, S. L. Li, Q. Xu and K. Tanaka, *Chem. Commun.*, 2002, 1722.
144 M. Cote and M. L. Cohen, *Phys. Rev. B*, 1998, **58**, R2477.

202   *Chem. Mater.* **2000**, *12*, 202–205

Metal Nanowires and Intercalated Metal Layers in Single-Walled Carbon Nanotube Bundles

A. Govindaraj,[†,‡] B. C. Satishkumar,[‡] Manashi Nath,[‡] and C. N. R. Rao*[,†,‡]

CSIR Centre of Excellence In Chemistry, Jawaharlal Nehru Centre for Advanced Scientific Research, Jakkur P.O., Bangalore 560 064, India, and Solid State and Structural Chemistry Unit, Indian Institute of Science, Bangalore 560 012, India

Received August 23, 1999

Nanowires of Au, Ag, Pt, and Pd (1.0–1.4 nm diam) have been produced in the capillaries of single-walled carbon nanotubes (SWNTs). The nanowire is single-crystalline in some cases. Dispersions of the nanowires in alcohol show longitudinal plasmon absorption bands at different wavelengths, suggesting the presence of a distribution of aspect ratios. A novel phenomenon involving the intercalation of metal layers (~0.5 nm thick) in the intertubular space of SWNT bundles has been observed. SWNTs decorated by metal nanoparticles are formed in some of the preparations.

Introduction

Electron transport in low-dimensional materials is not only of academic interest but also holds promise in nanoelectronics. The synthesis of metal nanowires therefore assumes considerable significance. Carbon nanotubes[1] provide an excellent host matrix for the preparation of metal nanowires. Quantum wire behavior has been reported in single-walled carbon nanotubes (SWNTs) at very low temperatures.[2] Preparation of metal nanorods covered by carbon and other materials has been reported in the recent literature.[3,4] Hsu et al.[5] have reported the electrolytic formation of carbon-sheathed Sn–Pb nanowires with diameters in the 40–90 nm range. We have explored the preparation of thin metal nanowires by filling SWNTs by metals. It may be recalled that multiwalled carbon nanotubes have been opened by acids and filled with various metals.[6,7] Single-walled carbon nanotubes (SWNTs) are also readily opened by acids and filled with metals.[8,9] Sloan et al.[10]

have just reported that SWNTs can be filled up to 50% by silver, by employing the KCl–UCl$_4$ and AgCl–AgBr eutectic systems, to produce nanowires. We have found that a variety of metal nanowires of 1.0–1.4 nm diameter can be readily prepared by employing SWNTs opened by acid treatment prior to the filling with metals. In addition, we have also found evidence of the incorporation of thin layers of metals in the intertubular space of the SWNT bundles.

Experimental Section

SWNTs were produced by the direct current arc-discharge method using a composite graphite rod containing Y$_2$O$_3$ (1 atom %) and Ni (4.2 atom %) as the anode and a graphite rod as the cathode,[11] under a helium pressure of 660 Torr with a current of 100 A and 30 V. The web produced from the arc discharge predominantly contained SWNT bundles and amorphous carbon along with metal-encapsulated carbon particles. It was heat-treated at 573 K in air for 24 h to remove the amorphous carbonaceous materials. The heat-treated material was stirred with concentrated nitric acid at 333 K for about 12 h and washed with distilled water to remove the dissolved metal particles. The SWNT material so obtained was suspended in ethanol by using a ultrasonicator and filtered through a micropore filter paper (0.3 μm) from Millipore to remove the polyhedral carbon particles present. The product was then dried at 423 K for about 12 h. Purified SWNT samples so obtained were heat-treated at 623 K for 30 min to remove the acid sites on the surface of the tubes. This treatment is known to open the nanotubes.[8,9] High-resolution electron microscopic (HREM) observations showed that the SWNTs with an average diameter of 1.4 nm were present in bundles of 5–50 nanotubes. The pore size (inner tube diameter) of these nanotubes has been established to be ~1.4 nm by gas adsorption measurements.[12]

To prepare gold nanowires, 5 mg of purified SWNTs was mixed with ~10 mg of HAuCl$_4$·xH$_2$O, and the mixture was dried in a 10 mm diameter quartz tube at 373 K for 2 h under

* E-mail: cnrrao@jncasr.ac.in.
† Jawaharlal Nehru Centre for Advanced Scientific Research.
‡ Indian Institute Of Science.
(1) Ajayan, P. M.; Iijima, S. *Nature* **1993**, *361*, 333–334.
(2) Tans, S. J.; Devoret, M. H.; Dai, H.; Thess, A.; Smalley, R. E.; Geerligs, L. J.; Dekker, C. *Nature* **1997**, *386*, 474–476.
(3) Dai, H.; Wong, E. W.; Lu, Y. J.; Fan, S. S.; Leiber, C. M. *Nature* **1995**, *375*, 769–771. Morales, A. M.; Leiber, C. M. *Science* **1998**, *279*, 208–210.
(4) Zhang, Y.; Suenaga, K.; Colliex, C.; Iijima, S. *Science* **1998**, *281*, 973–975. Rao, C. N. R.; Sen, R.; Satishkumar, B. C.; Govindaraj, A. *Chem. Commun.* **1998**, 1525–1526.
(5) Hsu, W. K.; Trasobares, S.; Terrones, H.; Terrones, M.; Grobert, N.; Zhu, Y. Q.; Li, W. Z.; Escudero, R.; Hare, J. P.; Kroto, H. W.; Walton, D. R. M. *Chem. Mater.* **1999**, *11*, 1747–1751.
(6) Tsang, S. C.; Chen, Y. K.; Harris, P. J. F.; Green, M. L. H. *Nature* **1994**, *372*, 159–161.
(7) Satishkumar, B. C.; Govindaraj, A.; Subbanna, G. N.; Mofokeng, J.; Rao, C. N. R. *J. Phys. B: Atom. Mol. Opt. Phys.* **1996**, *29*, 4925–4934.
(8) Sloan, J.; Hammer, J.; Zwiefka-Sibley, M.; Green, M. L. H. *Chem. Commun.* **1998**, 347–348
(9) Rao, C. N. R.; Govindaraj, A.; Sen, R.; Satishkumar, B. C. *Mater. Res. Innov.* **1998**, *2*, 128–141.
(10) Sloan, J.; Wright, D. M.; Woo, H. G.; Bailey, S.; Brown, G.; York, A. P. E.;, Coleman K. S.; Hutchison J. H.; Green, M. L. H. *Chem. Commun.* **1999**, 699–700

(11) Journet, C.; Maser, W. K.; Bernier, P.; Loiseau, A.; Lamy de la Chapelle, M.; Lefrant, S.; Deniard, P.; Lee, R.; Fischer, J. E. *Nature* **1997**, *388*, 756–758.
(12) Eswaramoorthy, M.; Sen, R.; Rao, C. N. R. *Chem. Phys. Lett.* **1999**, *304*, 207–210.

10.1021/cm990546o CCC: $19.00   © 2000 American Chemical Society
Published on Web 12/23/1999

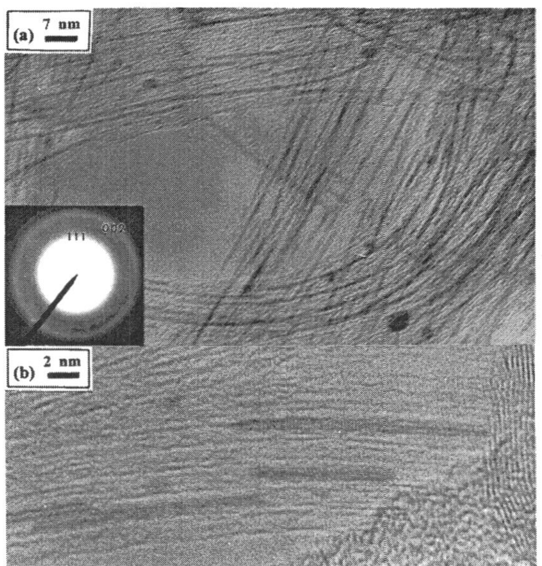

Figure 1. (a) TEM image of Au nanowires inside SWNTs obtained by the sealed-tube reaction. Inset shows an SAED pattern of the nanowires. (b) HREM image showing the single-crystalline nature of some of the Au nanowires.

vacuum (10^{-3} Torr). The quartz tube was sealed under vacuum and heated at 643 K (decomposition temperature of $AuCl_3$ is 527 K) for 20 h. To prepare Pt nanowires, a mixture of SWNTs (5 mg) with $H_2PtCl_6 \cdot 6H_2O$ (10 mg) was dried at 393 K for 2 h under vacuum and the sealed quartz tube containing the mixture was heated in a furnace at 773 K (melting point of $H_2PtCl_6 \cdot 6H_2O$ is 333 K) for 20 h. In a similar manner, $PdCl_2$ (10 mg) was mixed with the SWNTs (5 mg), heated at 373 K for 2 h, and sealed under vacuum. The sealed tube was maintained at 873 K (decomposition temperature of $PdCl_2$ is 773 K) for 20 h. In all of the sealed-tube reactions, the temperature was maintained at ~100 K above the decomposition (or melting point) temperature of the respective metal salt. After treatment with metal salts, the sealed tubes were broken open, and the product was treated with hydrogen at 723 K for 1 h to reduce any metal salt present. The hydrogen treatment was not found to be essential in the case of gold nanowires. To prepare Pt nanowires, we have employed a simple solution method as well. The mixture of SWNTs (5 mg) and $H_2PtCl_6 \cdot 6H_2O$ (10 mg) was refluxed in concentrated HCl or HNO_3 (2 mL) for 1 h. The product was washed with distilled water, dried, and treated with hydrogen at 723 K for 1 h. We have also attempted to prepare metal nanowires by sonochemical means,[13] wherein SWNTs in $AgNO_3$ solution were sonicated (35 MHz ultrasound treatment) for a few hours in a hydrogen atmosphere. It should be noted that SWNTs generally get filled by the metal salt/metal through the open ends.

The products obtained from the various processes discussed above were dispersed in carbon tetrachloride and deposited onto holey carbon copper grids for transmission electron microscope (TEM) observations with a JEOL JEM-3010 operating at 300 kV. Plasmon absorption bands of the dispersions of the nanowires in alcohol were recorded using a Hitachi U3400 spectrometer.

Results and Discussion

In Figure 1a, we show a TEM image that reveals the presence of a large number of gold nanowires obtained

(13) Salkar, R. A.; Jeevandam, P.; Aruna, S. T.; Koltpin, Y.; Gedanken, A. J. *J. Mater. Chem.* **1999**, *9*, 1333–1335.

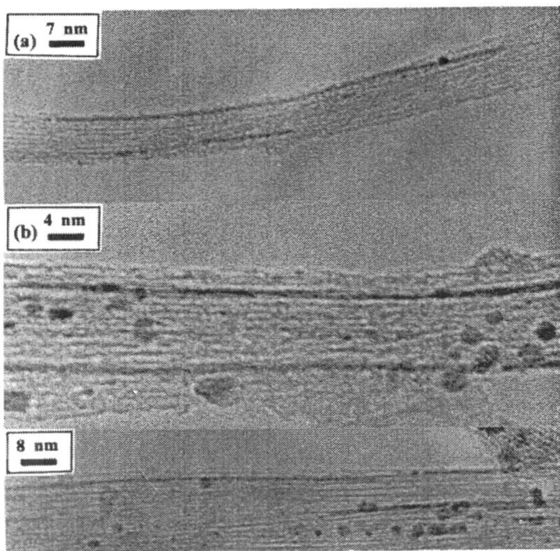

Figure 2. TEM image of Pt nanowires inside SWNTs obtained (a) by the sealed-tube reaction and (b) by the solution method.

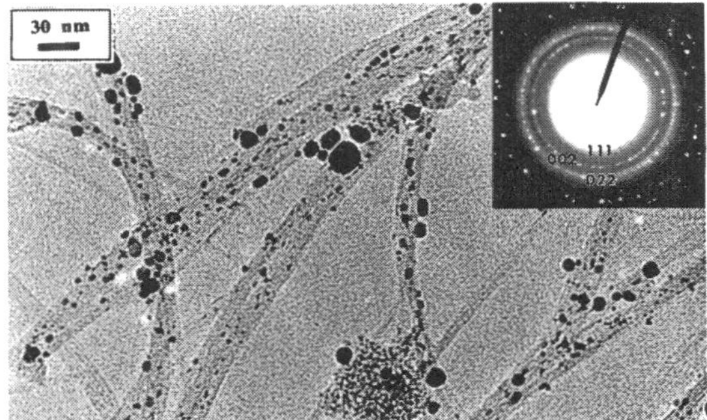

Figure 3. TEM image of SWNT bundles decorated by Pt nanoparticles. Inset shows an SAED pattern of the metal nanoparticles.

from the sealed-tube reaction. The image clearly shows extensive filling of SWNT tubes. The length of the gold nanowires is in the range 15–70 nm, and the diameters are in the range 1.0–1.4 nm. Detailed TEM observations show that the extent of filling of Au nanowires inside the SWNTs is in excess of 50%. Selected area electron diffraction (SAED) patterns taken from the region of nanowires (see inset of Figure 1a) show diffuse rings due to small polycrystalline metal particles. The diffraction rings correspond to the (111) and (002) reflections of gold. In some of the nanowires, however, we have found the metal to be single crystalline, as shown in the HREM image of Figure 1b. This image reveals the resolved lattice of gold with a spacing of ~0.23 nm, corresponding to the (111) planes. Polycrystalline Au aggregates could be transformed into the single-crystalline form by suitable annealing.

We have obtained platinum nanowires by the sealed-tube reaction, as shown in the TEM image in Figure 2a. The image shows a nanowire, 90 nm long and ~1 nm in diameter. Long nanowires of Pt were obtained by the solution method as well. In Figure 2b, we show a TEM image of Pt nanowires obtained on refluxing the

204   *Chem. Mater., Vol. 12, No. 1, 2000*

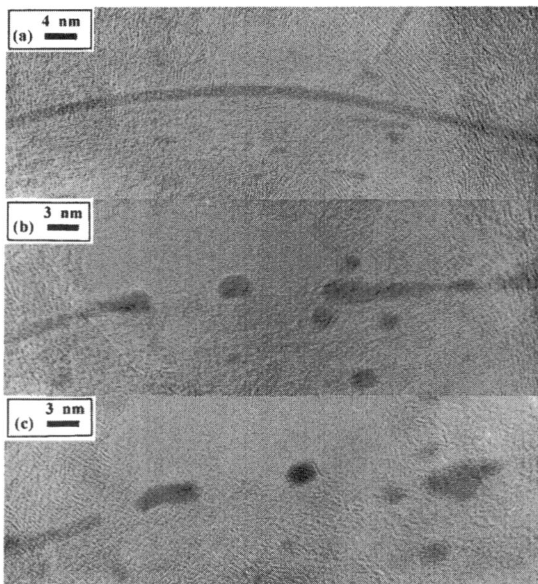

Figure 4. TEM images showing the breakup of a Au nanowire to nanoparticles on continuous exposure to an electron beam at 300 kV (at magnifications >600 000×). SWNTs are also damaged under these conditions.

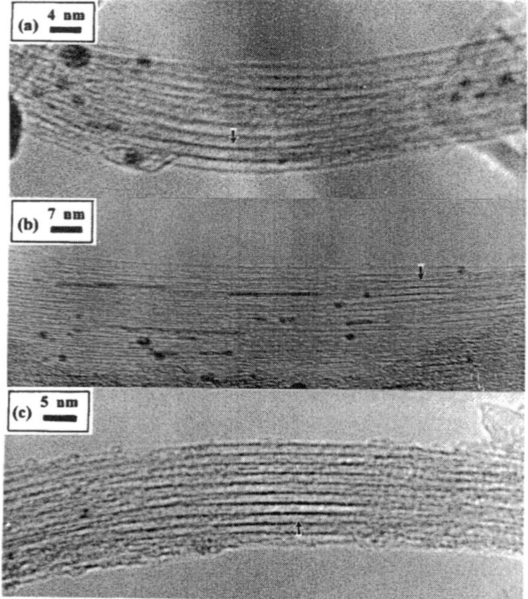

Figure 5. HREM images of metals intercalated in the intertubular space of SWNT bundles: (a) intercalated Pt obtained by the sealed-tube reaction, (b) intercalated Pt (coexisting with nanowires) obtained by the solution method, and (c) intercalated Pd obtained by sealed-tube reaction. Arrows point to the intercalated metal layers.

SWNTs with the metal salt and concentrated HCl. The length of the nanowires here is ~70 nm, with the diameter remaining at ~1 nm. We could obtain platinum nanowires by refluxing SWNTs with H_2PtCl_6 and HNO_3. Palladium nanowires could also be obtained by

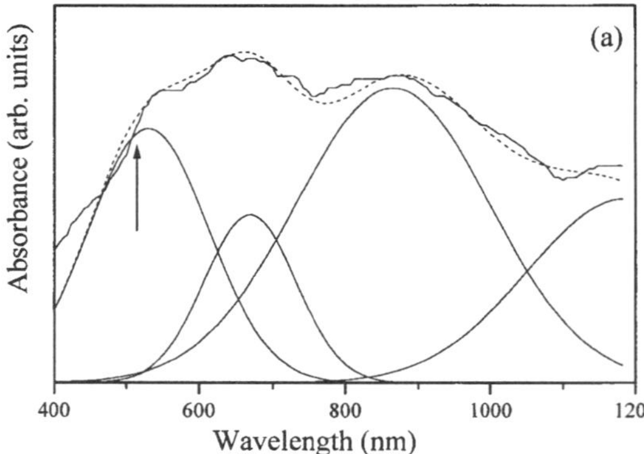

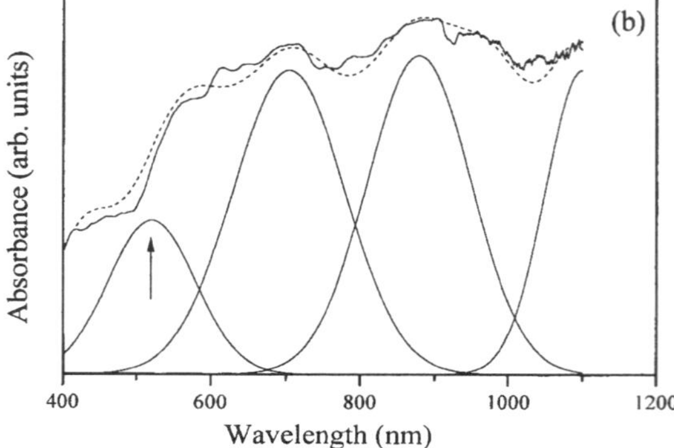

Figure 6. Optical absorption spectra of two alcohol dispersions of Au nanowires in SWNTs. The band around 520 nm corresponds to the transverse plasmon absorption. The other bands are longitudinal plasmon absorption bands.

the sealed-tube reaction, but these were not as long as those of gold or platinum. We could obtain silver nanowires by heating the SWNTs obtained after sonication in $AgNO_3$ solution at 623 K for 1 h, in a $Ar-H_2$ atmosphere.

The products of the sealed-tube reaction occasionally contained metal-decorated SWNTs along with the encapsulated-metal nanowires. The metal nanoparticles (3–20 nm diam) were uniformly dispersed on the nanotube surface (Figure 3). The nanotube surface, having been acid treated, contains active centers[14,15] that can bind the nanoparticles. The SAED patterns showed spotty rings corresponding to (111), (002), and (022) reflections of platinum (Figure 3), arising from the fairly large metal nanoparticles.

During the course of our investigations of gold nanowires by transmission electron microscopy, we have found the nanowires to disintegrate into particles on continuous exposure to an electron beam, when we attempted to examine them under high magnification. We have been able to photograph the sequence of changes with exposure and show a typical set of images in Figure 4. The nanowire in Figure 4a breaks up into smaller fragments, as shown in parts b and c of Figure

(14) Lago, R. M.; Tsang, S. C.; Lu, K. L.; Chen, Y. K.; Green, M. L. H. *J. Chem Soc., Chem. Commun.* **1995**, 1355–1358.

(15) Rao, C. N. R.; Govindaraj, A.; Satishkumar, B. C. *J. Chem. Soc., Chem. Commun.* **1996**, 1525–1526.

4, after exposure to the beam for 30 and 60 s, respectively. While the microscope is operated at high magnifications (>600 000×), the electron beam energy gets concentrated on a very small region of the sample (in this case, the Au nanowire), causing breakup and melting as a result of the beam-induced heating. The nanoparticles thus produced often exhibit single-crystalline character, as revealed in parts b and c of Figure 4. These images show lattice fringes corresponding to the (111) planes of gold.

An interesting phenomenon that we have observed in the process of filling of the SWNTs with metals is the incorporation of metal layers in the intertubular space of the SWNT bundles. Such intertubular intercalation in the SWNT bundles is demonstrated by the TEM images in Figure 5. Figure 5a shows platinum-intercalated SWNTs obtained in a sealed-tube reaction. The solution method also gives rise to such intercalated SWNTs. Figure 5b shows a TEM image of intercalated platinum coexisting with the nanowires. A HREM image of a SWNT bundle intercalated with palladium is shown in Figure 5c. In all of the images, we notice an increase in the intertubular spacing or the diameter of the SWNT bundle due to intercalation. The thickness of the intercalated metal layer present in the intertubular spacing is generally between 0.4 and 0.5 nm. This is slightly larger than the van der Waals spacing (~0.3 nm) between the single-walled nanotubes.

We have recorded the electronic absorption spectra of dispersions of gold nanowires in ethanol. The spectra show transverse and longitudinal plasmon absorption bands.[16] Although the transverse absorption band (~520 nm) does not vary with the aspect ratio of Au nanowires as pointed out by Link et al.,[16] the position of the longitudinal absorption band varies markedly with the aspect ratio and the medium dielectric constant. Decomposition of the observed band envelopes indicates that the dispersions of Au nanowires are characterized by the bands around 670, 860, and 1150 nm due to the longitudinal absorption (Figure 6). These bands indicate that there is distribution of aspect ratios in the nanowire dispersion, with the aspect ratio corresponding to the longest wavelength being rather large. We estimate the range of aspect ratios in the nanowire dispersions examined by us to be 3−6. Clearly, more detailed studies of the plasmon bands of metal nanowires would be worthwhile.

Conclusions

Nanowires of metals such as gold, silver, platinum, and palladium with diameters in the range of 1.0−1.4 nm have been prepared in the capillaries of single-walled carbon nanotubes by simple chemical or sonochemical methods. The nanowires can be polycrystalline or single crystalline. The nanowires exhibit transverse and longitudinal plasmon absorption bands in the electronic spectra, with the latter showing the presence of nanowires with a distribution of aspect ratios. An interesting phenomenon found along with the formation of the nanowires is that of metal intercalation in the intertubular space of the SWNT bundles causing an increase in the diameter of the bundles. Metal-decorated nanotubes have been found to occur in some of the preparations.

CM9905460

(16) Link, S.; Mohammed, M. B.; El-Sayed, M. A. *J. Phys. Chem.* **1999**, *103*, 3073−3077 and the references listed therein.

Synthesis of metal oxide nanorods using carbon nanotubes as templates

B. C. Satishkumar,[a] A. Govindaraj,[a,b] Manashi Nath[a] and C. N. R. Rao*[a,b]

[a]Chemistry and Physics of Materials Unit and CSIR Centre of Excellence in Chemistry, Jawaharlal Nehru Centre for Advanced Scientific Research, Jakkur P. O., Bangalore 560 064, India. E-mail: cnrrao@jncasr.ac.in
[b]Solid State and Structural Chemistry Unit, Indian Institute of Science, Bangalore 560 012, India

Received 10th April 2000, Accepted 14th June 2000
Published on the web 19th July 2000

Nanorods of several oxides, with diameters in the range of 10–200 nm and lengths upto a few microns, have been prepared by templating against carbon nanotubes. The oxides include V_2O_5, WO_3, MoO_3 and Sb_2O_5 as well as metallic MoO_2, RuO_2 and IrO_2. The nanorods tend to be single-crystalline structures. Nanotube structures have also been obtained in MoO_3 and RuO_2.

1. Introduction

The preparation of metal oxide nanocomposites and oxide fibres by using carbon nanotubes as removable templates was first reported by Ajayan et al.,[1] who prepared such materials based on V_2O_5. Carbon nanotubes have been converted to nanorods of metal carbides by reaction with volatile oxide or halide species.[2] The reaction of gallium oxide vapor with carbon nanotubes in the presence of ammonia gives rise to GaN nanorods,[3] while the reaction of silicon oxide vapor in a nitrogen atmosphere gives silicon nitride nanorods.[4] Metal oxide nanotubes have been prepared by using carbon nanotubes as templates, with particular success in the case of zirconia.[5,6] Metal oxide nanostructures have been prepared by the use of sol–gel chemistry within the pores of alumina and polymer membranes.[7] We have found that carbon nanotubes can be effectively used as templates to obtain nanostructures of metal oxides. We have prepared nanorods of V_2O_5, WO_3, MoO_3, Sb_2O_5, MoO_2, RuO_2 and IrO_2 in good yields, the last three being metallic. Furthermore, most of the nanorods are single-crystalline structures. Nanotube structures of the oxides have also been obtained in some instances.

2. Experimental

The method of preparation of metal oxide nanorods and nanotubes makes use of multi-walled carbon nanotubes (MWNTs). The MWNTs were prepared by the arc vaporization of graphite rods in a helium atmosphere (550 Torr) at 30 V, 100 A direct current.[8] The nanotubes formed in the core region of the cathode deposit were dispersed in methanol and sonicated for 2 hours in a separating funnel. The bulky graphitic carbon was removed from the suspension. The remaining solid suspension was allowed to settle, dried and heated in air at 700 °C for 20 minutes to oxidize the graphitic carbon.[9] The pure carbon nanotubes thus obtained were closed at either end or open at one end in some instances. They were treated with boiling HNO_3 for 24 h. Such nanotubes are almost always open and contain a considerable number of acidic sites on the surface.[10,11] The concentration of surface acidic sites is around 6×10^{20} sites g^{-1} of nanotubes. The acid-treated nanotubes were washed with water and dried in an oven at 60 °C for 12 h. The nanotubes so obtained had an inner diameter in the range of 2–8 nm and an outer diameter in the

range of 10–30 nm (length up to 1 μm). The acid-treated carbon nanotubes were coated with an appropriate oxide precursor such as an alkoxide and dried at 100 °C, followed by calcination at 450 °C. The calcined sample was heated at 700 °C in air to burn off the carbon. While generally we got nanorods by this method, in some instances we got nanotubes as well.

The experimental procedure for preparing the oxide nanostructures is as follows: In order to prepare V_2O_5 coated carbon nanotubes, aqueous sodium meta-vanadate was first passed through a cation exchange column (DOWAX) to get vanadic acid (HVO_3). 100 mg of acid-treated carbon nanotubes were stirred with 2 ml of HVO_3 for 48 h. The excess gel was removed by washing with water and acetone. The nanotubes so obtained were dried in an oven at 100 °C for 6 h.

In the case of WO_3 and MoO_3, an aqueous solution of Na_2WO_4 or Na_2MoO_4 was passed through a cation exchange column. The resulting tungstic or molybdic acid (H_2WO_4 or H_2MoO_4) was used for coating the acid-treated carbon nanotubes (100 mg) by magnetic stirring for 48 h. The excess tungstic or molybdic acid was washed off with water. The nanotube sample was dried at 100 °C for 6 h. To coat the nanotubes with Sb_2O_5, the acid-treated nanotubes (100 mg) were mixed with 2 ml of $SbCl_5$ and stirred for 48 h. The sample was filtered and washed with methanol and dried at 100 °C for 6 h.

In the case of RuO_2, the gel was obtained by the reaction of aqueous $RuCl_3$ (Aldrich) with NaOH. The gel was washed repeatedly with distilled water to remove free Na^+ and Cl^- ions. 100 mg of the acid-treated carbon nanotubes were stirred with the gel for 48 h. The sample was dried at 100 °C for 6 h and treated with H_2O_2 to oxidize Ru^{3+} to Ru^{4+}. The resulting nanotubes were dried at 100 °C for 6 h and calcined at 450 °C for 12 h.

Anhydrous $IrCl_3$ (Aldrich) was fused with NaOH to obtain Na_2IrCl_6. This was diluted with distilled water to get an intensely blue-colored solution. 100 mg of acid-treated carbon nanotubes were stirred with this solution for 48 h. The sample was washed with distilled water and dried at 100 °C for 6 h.

After the oxide-coated nanotubes were heated in air at 700 °C (except in the case of V_2O_5 where it was 500 °C), the oxide nanostructures obtained were subjected to X-ray diffraction (XRD), scanning electron microscopy (SEM) and transmission electron microscopy (TEM).

DOI: 10.1039/b002868l

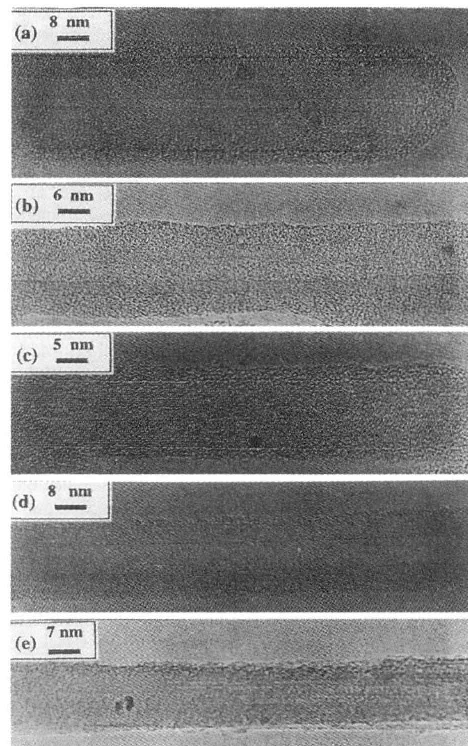

Fig. 1 HREM images of carbon nanotubes coated with (a) V_2O_5, (b) WO_3, (c) MoO_3, (d) RuO_2 and (e) IrO_2 obtained on calcination of the oxide-coated carbon nanotubes at 450 °C for 12 h.

3. Results and discussion

The calcined samples of different oxide-coated carbon nanotubes revealed the presence of satisfactory coverage of the oxidic material as shown in the typical TEM images in Fig. 1. Thus, the high resolution electron microscopic (HREM) image of the nanotubes coated with V_2O_5 in Fig. 1(a) shows that the oxidic coating is uniform throughout the length of the nanotube. The situation is similar with the other oxide coatings as can be seen from the images in Fig. 1(b)–(e). The strong interaction of the oxidic material with the surface of the carbon nanotube is due to the presence of surface acidic sites, resulting from the acid treatment.[10,11]

On the removal of the nanotube template, the resulting oxidic species showed the presence of interesting nanostructures. In Fig. 2(a) we show the SEM image of V_2O_5 obtained after the removal of the template. The image shows the presence of copious quantities of nanorods. A TEM bright-field image of the V_2O_5 nanorods is shown in Fig. 2(b). The nanorods have diameters in the range of 20–55 nm and are 2–3 μm long. The selected area electron diffraction (SAED) pattern on a nanorod shows the presence of regular spots due to (110) and (200) planes as shown in the inset of Fig. 2(b), signifying the single crystalline nature of the V_2O_5 nanorods. The zone axis projection is along [10$\bar{2}$] of the orthorhombic V_2O_5. The XRD pattern of the oxide powder showed it to be orthorhombic, with unit cell parameters of $a = 11.44$ Å, $b = 3.54$ Å and $c = 4.35$ Å, in agreement with the literature values (JCPDS file: 9-387). The template could be removed at a relatively lower temperature in the case of V_2O_5, since it catalyzes the oxidation of the carbon nanotubes.

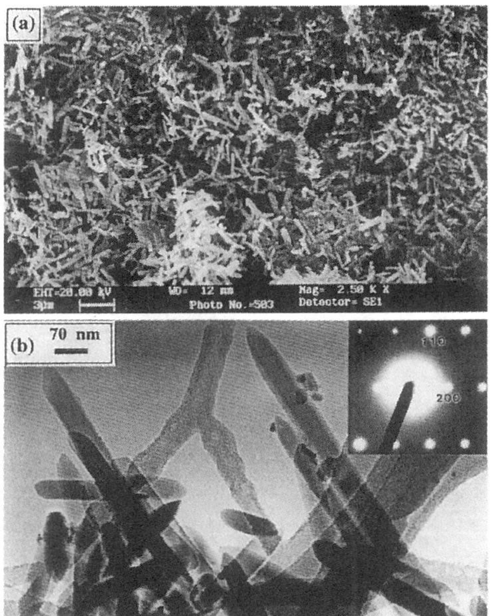

Fig. 2 (a) SEM image of V_2O_5 nanorods and (b) TEM image of a V_2O_5 nanorod. The inset shows the SAED pattern taken on a nanorod.

In the case of WO_3, SEM observations indicated a relatively low yield of the nanorods. A typical TEM image of the WO_3 nanorods is shown in Fig. 3(a). The diameter of the nanorods is in between 15 and 60 nm. The TEM image in Fig. 3(b) shows a relatively long nanorod, with a length of 5 μm. The SAED pattern shown in the inset of Fig. 3(b) indicates a zone axis projection along [010], with the Bragg spots corresponding to the (001) and (200) planes of monoclinic WO_3. The XRD pattern of the oxide powder gave unit cell parameters of the monoclinic phase ($a = 7.331$ Å, $b = 7.52$ Å and $c = 7.71$ Å with $\beta = 90.1°$, JCPDS file: 24-747).

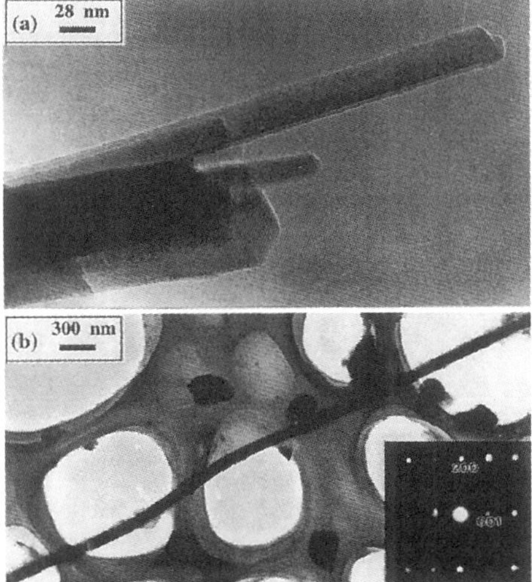

Fig. 3 (a) TEM image of nanorods of WO_3 after removal of carbon template at 700 °C/12 h. (b) TEM image of a relatively long nanorod. Inset shows the SAED pattern of a nanorod showing a zone axis projection along [010].

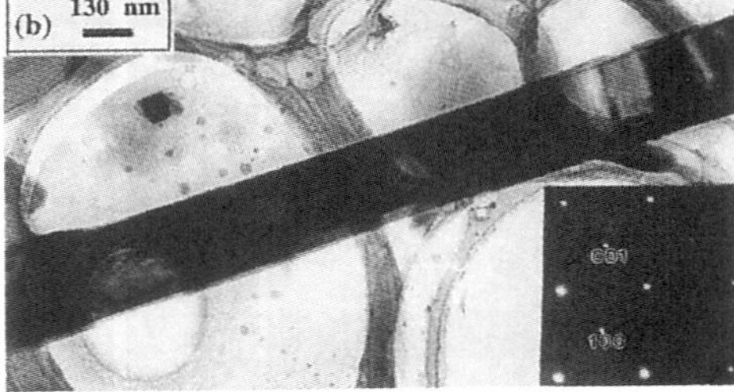

Fig. 4 (a) SEM image of MoO₃ nanorods and (b) TEM image of a nanorod. The inset in (b) shows the SAED pattern along the [010] direction.

We obtained nanorods of MoO_3 in high yields (after template removal at 600 °C for 12 h), as revealed by the SEM image in Fig. 4(a). The diameter of the nanorods is in the 80–150 nm range, with the length in the 5–15 µm range. By subjecting the MoO_3-coated carbon nanotubes to repeated washing (before calcination), we were able to obtain thinner nanorods of the oxide, after removal of the template. The TEM image of a MoO_3 nanorod is shown in Fig. 4(b). The SAED pattern revealing the single crystalline nature of the nanorod is shown in the inset of Fig. 4(b). The XRD pattern gave the orthorhombic unit cell parameters, $a = 3.96$ Å, $b = 13.85$ Å and $c = 3.7$ Å (JCPDS file: 35-609).

Sb_2O_5 gave short nanorods in relatively good yields. The

TEM image of the nanorods is shown in Fig. 5(a). The image in Fig. 5(b) shows the TEM image of an isolated nanorod of Sb_2O_5. The nanorods have diameters in the range of 10–60 nm and lengths in the range 100–400 nm. The SAED pattern shown as an inset in Fig. 5(a) reveals bright spots due to the (222) planes. The XRD pattern showed it to have a cubic structure with the cell parameter, $a = 10.98$ Å (JCPDS file: 5-534).

While the nanostructures discussed above are all of nonmetallic oxides, we sought to prepare nanorods of metallic oxides such as MoO_2, RuO_2 and IrO_2. In order to obtain MoO_2 nanorods, nanorods of MoO_3 were reacted with hydrogen at 500 °C for 48 h. The SEM and TEM images of MoO_2 nanorods are shown in Fig. 6(a) and (b) respectively. The nanorods are 150–300 nm in diameter and 5–10 µm long. The SAED pattern of the nanorod (see inset of Fig. 6(b)) shows the projection along the [010] direction. The XRD pattern showed the presence of a monoclinic phase with the unit cell parameters, $a = 5.60$ Å, $b = 4.85$ Å and $c = 5.54$ Å with $\beta = 119.4°$ (JCPDS file: 32-671).

A TEM image of the RuO_2 nanorods is shown in Fig. 7(a). The nanorods are single-crystalline in nature as indicated by the SAED pattern given as an inset in Fig. 7(a). It shows the projection along the [001] direction. The nanorods have diameters in the 5–20 nm range and are 50–200 nm long. An HREM image of a RuO_2 nanorod is shown in Fig. 7(b). The image shows the lattice planes with $d = 3.1$ Å corresponding to (110) planes of tetragonal RuO_2. The XRD pattern confirmed the tetragonal structure with the tetragonal unit cell parameters, $a = 4.48$ Å and $c = 3.12$ Å (JCPDS file: 40-1290).

IrO_2 also gave a relatively good yield of nanorods. In this case, we also observed an intermediate oxide structure, after heat-treating the oxide-coated carbon nanotubes at 500 °C. This temperature is slightly lower than that required for the complete removal of the carbon template. We show a TEM image of this structure in Fig. 8(a). HREM observations showed the polycrystalline nature for the intermediate nanostructure. It appears that during the template removal,

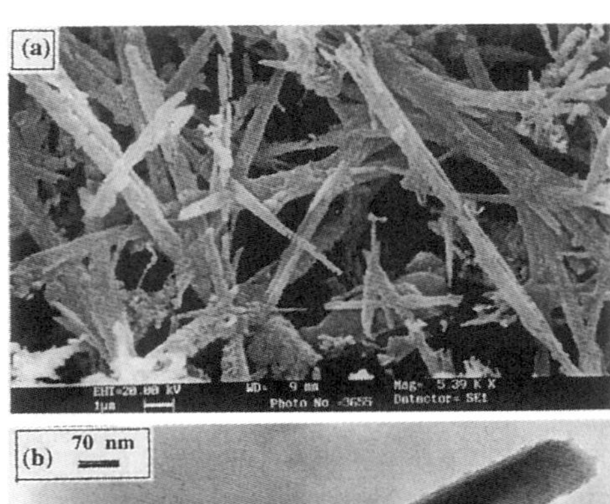

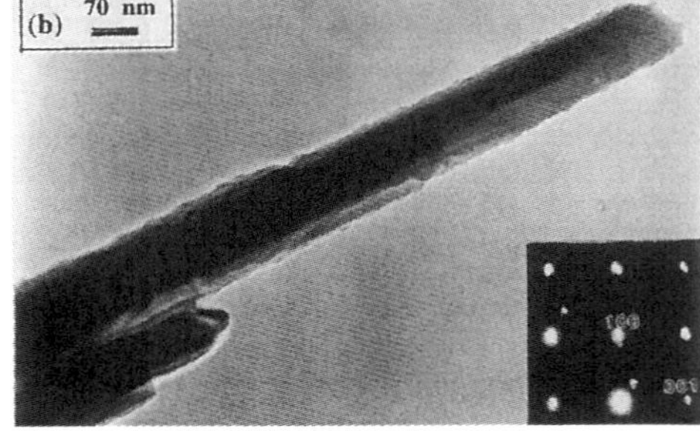

Fig. 6 (a) SEM image and (b) TEM image of nanorods of MoO_2 obtained by the treatment of MoO_3 nanorods with H_2 at 500 °C for 48 h. Inset in (b) shows the zone axis projection along [010].

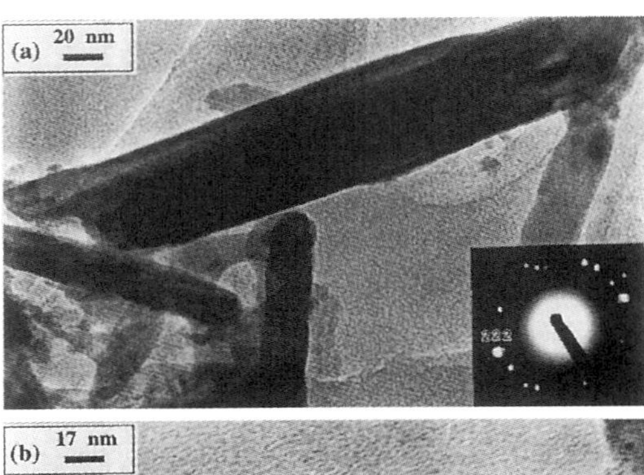

Fig. 5 (a) and (b) TEM images of Sb_2O_5 nanorods obtained on removal of the template at 600 °C/12 h. The SAED pattern of the nanorods is shown in the inset of (a).

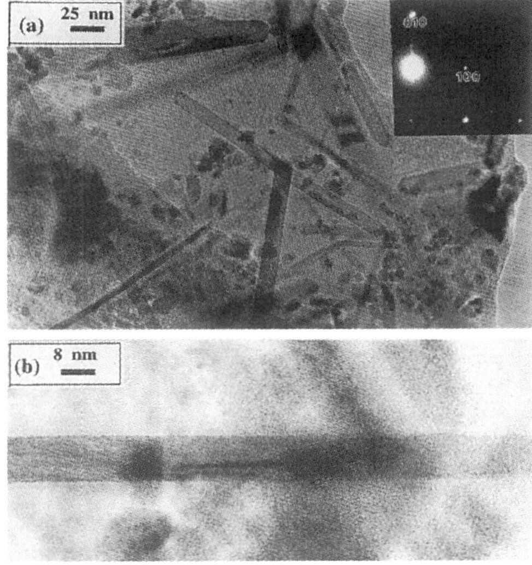

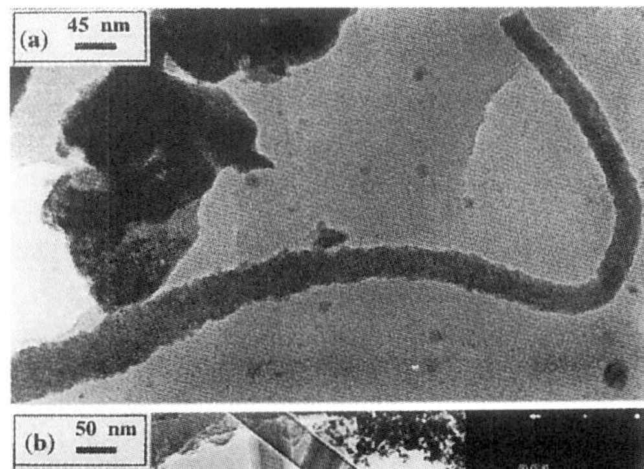

Fig. 7 (a) TEM image of RuO_2 nanorods and (b) HREM image of a RuO_2 nanorod showing the (110) planes. The inset in (a) shows the SAED pattern of a nanorod. The small particles seen in the image are also due to the oxide.

the oxide coatings on neighboring carbon nanotubes coalesce to form the rod-like nanostructures. Upto a certain temperature the nanorod formed is polycrystalline, and becomes single-crystalline around 600 °C. This is seen from the TEM image of the IrO_2 nanorods in Fig. 8(b). The diameter of the nanorods is in the range of 30–60 nm and the length goes up to 1 μm. The

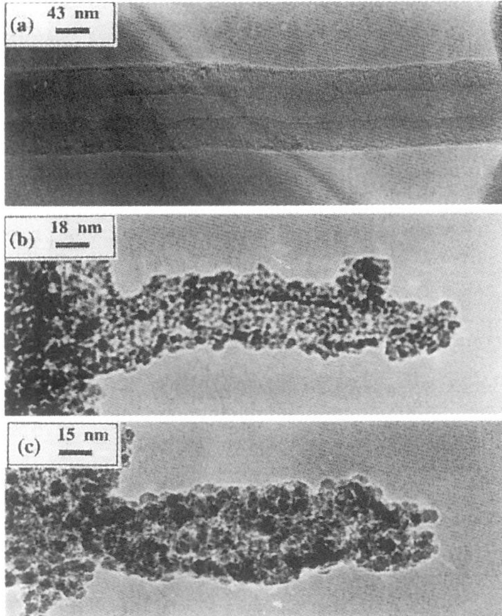

Fig. 8 TEM images of (a) intermediate nanostructure of IrO_2 obtained on heat-treatment of oxide-coated carbon nanotubes at 500 °C for 12 h and (b) nanorods of IrO_2 obtained after the complete removal of carbon template at 600 °C for 12 h. The SAED pattern of a nanorod is shown as the inset.

Fig. 9 (a) TEM image of a MoO_3 nanotube and (b), (c) TEM images of nanotube-like structures of RuO_2.

inset in Fig. 8(b) shows the SAED pattern of a single nanorod, with the spots arising from the (100) and (001) planes of tetragonal IrO_2. The XRD pattern revealed the rutile phase with the unit cell parameters, $a = 4.50$ Å and $c = 3.15$ Å (JCPDS file: 15-870). It is noteworthy that both RuO_2 and IrO_2 are metallic.

The formation of single-crystalline nanorods in the present study is noteworthy. A possible mechanism of formation of such nanorods may be as follows. CO or/and CO_2 is produced when the oxide-coated carbon nanotubes are heated, the oxygen being (at least partly) derived from the coated oxide. Subsequently, the remaining metal or sub-oxide may get reoxidized and undergoes recrystallization. Another possibility is that the decomposition of the oxide precursor in the hot combustion zone of the nanotubes gives rise to the crystals *in situ*. The crystals could get elongated because of the evolution of gases during the transformation. The precursor decomposition also gives H_2O or/and CO_2.

In some of the preparations, we obtained nanotube structures of the oxide, in addition to the nanorod structures. In Fig. 9(a) we show the TEM image of a MoO_3 nanotube. The inner diameter of the oxidic nanotube is 23 nm. Nanotube-like structures of RuO_2 are also obtained in some preparations. TEM images of such structures are shown in Fig. 9(b) and (c). The nanotube in RuO_2 is not as well formed and we barely see a nanotube shape in the TEM images. The nanotube comprises the stable rutile phase.

4. Conclusions

In conclusion, the present study establishes that nanorods of a variety of metal oxides can be readily prepared by using carbon nanotubes as templates. The method also offers certain advantages. Thus, most of the preparations yield single-crystalline nanorods of fairly large dimensions. Furthermore, the nanorods are generally much longer than the starting nanotube template. The observation of oxide nanotubes in some cases is also significant.

References

1   P. M. Ajayan, O. Stephan, Ph. Redlich and C. Colliex, *Nature*, 1995, **375**, 564.
2   H. Dai, E. W. Wong, Y. Z. Lu, S. Fan and C. M. Lieber, *Nature*, 1995, **375**, 769.
3   W. Han, S. Fan, Q. Li and Y. Hu, *Science*, 1997, **277**, 1287.
4   W. Han, S. Fan, Q. Li, B. Gu, X. Zhang and D. Yu, *Appl. Phys. Lett.*, 1997, **71**, 2271
5   B. C. Satishkumar, A. Govindaraj, E. M. Vogl, L. Basumallick and C. N. R. Rao, *J. Mater. Res.*, 1997, **12**, 604.
6   C. N. R. Rao, B. C. Satishkumar and A. Govindaraj, *Chem. Commun.*, 1997, 158 .
7   B. B. Lakshmi, C. J. Patrissi and C. R. Martin, *Chem. Mater.*, 1997, **9**, 2544.
8   C. N. R. Rao, R. Seshadri, A. Govindaraj and R. Sen, *Mater. Sci. Eng.*, 1995, **R15**, 209.
9   R. Seshadri, A. Govindaraj, H. N. Aiyer, R. Sen, G. N. Subbanna, A. R. Raju and C. N. R. Rao, *Curr. Sci. (India)*, 1994, **66**, 839.
10 R. M. Lago, S. C. Tsang, K. L. Lu, Y. K. Chen and M. L. H. Green, *J. Chem. Soc., Chem. Commun.*, 1995, 1355.
11 B. C. Satishkumar, A. Govindaraj, J. Mofokeng, G. N. Subbanna and C. N. R. Rao, *J. Phys. B, Atm. Mol. Opt. Phys.*, 1996, **29**, 4925.

10 January 2002

Chemical Physics Letters 351 (2002) 189–194

CHEMICAL PHYSICS LETTERS

www.elsevier.com/locate/cplett

Nanowires, nanobelts and related nanostructures of Ga_2O_3

Gautam Gundiah, A. Govindaraj, C.N.R. Rao [*]

Chemistry and Physics of Materials Unit and CSIR Centre of Excellence in Chemistry, Jawaharlal Nehru Centre for Advanced Scientific Research, Jakkur P.O., Bangalore 560 064, India

Received 6 November 2001; in final form 14 November 2001

Abstract

By carrying out the reaction of Ga_2O_3 powder with carbon nanotubes around 1100 °C, nanowires, nanobelts and nanosheets of Ga_2O_3 have been obtained, the diameter and proportion of the nanowires depending on the flow rate of argon through the reaction zone. Reaction of Ga_2O_3 powder with activated carbon mainly gives rise to nanosheets and nanorods. The procedures employed in this study are attractive since they give high yields of nanowires and nanobelts. The nanowires are single crystalline with the growth direction perpendicular to the $(\bar{1}\,0\,2)$ planes. The Ga_2O_3 nanowires exhibit good photoluminescence characteristics. © 2002 Elsevier Science B.V. All rights reserved.

1. Introduction

Nanotubes, nanorods and nanowires of various inorganic materials are likely to have potential technological applications. In the last two to three years, nanowires of GaN [1,2], SiC [3] and other inorganic materials have been prepared and characterized. We have been interested in preparing nanowires and nanorods of β-Ga_2O_3, which is a wide band gap semiconductor ($E_g = 4.9$ eV) with intense luminescent properties [4]. Zhang et al. [5] prepared Ga_2O_3 nanowires by the evaporation of bulk Ga at 300 °C under a pressure of 100 torr in a mixture of 90% Ar and 10% H_2 gases at a flow rate of 30 standard cubic centimeter per minute (sccm). β-Ga_2O_3 nanowires

have been synthesized by Choi et al. [6] using direct current arc discharge of GaN powders in a mixture of Ar and O_2 gases in the presence of a small amount of a transition metal catalyst. Wu et al. [7], on the other hand, have prepared Ga_2O_3 nanowires by the carbothermal reduction starting with a mixture of gallium oxide powder and graphite. The reactions were carried out at 980 °C for 2 h in flowing N_2 atmosphere (20 sccm). Liang et al. [8] have recently obtained β-Ga_2O_3 nanowires by heating a composite material of GaAs and pre-evaporated Au at 1240 °C in dry O_2 atmosphere. We have developed a simple method to synthesize nanostructures of β-Ga_2O_3, starting with Ga_2O_3 powder mixed with activated carbon or carbon nanotubes. The nanostructures obtained include not only nanowires, but also nanobelts. The width of the nanowires could be controlled by adjusting the flow rate of the argon gas going through the

[*] Corresponding author. Fax: +91-80-846-2766.
E-mail address: cnrrao@jncasr.ac.in (C.N.R. Rao).

190   *G. Gundiah et al. / Chemical Physics Letters 351 (2002) 189–194*

furnace. Besides determining the direction of growth of the nanowires, we have measured the luminescent properties.

2. Experimental

For the preparation of β-Ga_2O_3 nanowires, we have employed several procedures. In procedure (i), gallium oxide was mixed with activated charcoal (Sarabhai Chemicals, India) according to the weight ratio of 1:1.5 and the mixture ground to a fine powder. The mixture was taken in a quartz tube (10 cm in length) with an outer diameter of 8 mm. This tube was kept in the centre of a quartz tube (10 mm outer diameter) placed horizontally in a tubular furnace. The mixture was heated at 1000 °C for 4 h in a flow of Ar gas at 40 sccm. The flow rate of Ar gas was controlled by a unit mass flow controller. After a period of 4 h, air was let into the reaction tube and the product allowed to reach ambient temperature. This helped to burn out the unreacted carbon and to convert any suboxide of gallium into Ga_2O_3. A white, wool-like product was obtained at the outlet of the inner quartz tube along with a white powder at the outlet of the outer tube. These products were collected separately and analyzed. Procedure (ii) was the same as procedure (i) except that activated carbon prepared by the thermal decomposition of polyethylene glycol (600 units) at 700 °C in a N_2 atmosphere was used in place of activated charcoal.

In procedure (iii), we employed multi-walled carbon nanotubes as the source of carbon. The multi-walled carbon nanotubes were prepared by the arc-discharge method outlined in the literature [9]. The rest of the procedure was the same as in (i). In order to see the effect that the flow rate plays on the final product, we performed the reaction at different flow rates of Ar (40, 60 and 80 sccm). As we increased the flow rate, the amount of product carried to the outlet increased. Thus, at a flow rate of 80 sccm, the amount of product obtained at the inner tube was negligible and the entire product was present at the outlet.

Powder X-ray diffraction (XRD) patterns were recorded using CuKα radiation on a Rich-Siefert, XRD-3000-TT diffractometer. Scanning electron

microscopy (SEM) images were obtained on a LEICA S440i SEM. Transmission electron microscopic (TEM) images were obtained with a JEOL JEM 3010 operating with an accelerating voltage of 300 kV. Photoluminescence (PL) measurements were carried out at room temperature using 265 nm wavelength as the excitation wavelength with a Perkin–Elmer model LS50B luminescence spectrometer.

3. Results and discussion

The reaction of activated charcoal with gallium oxide powder by procedure (i) yields a mixture of nanosheets and nanowires of Ga_2O_3 at the outlet of the inner tube as revealed in Fig. 1a. The diameter of the nanowires is between 300 and 400 nm and the length extends to tens of microns. The nanosheets are rectangular with a width of around 5 μm and the length going upto tens of microns. XRD patterns of these nanostructures showed them to be β-Ga_2O_3 with a monoclinic structure (JCPDS card: No. 11-0370).

When the reaction was carried out with active carbon by procedure (ii), we obtain nanorods of β-Ga_2O_3 at the outlet of the outer tube, as shown in Fig. 1b. These have diameters in the range 500–1000 nm and lengths between 10 and 15 μm. XRD patterns confirmed the nanorods to be of β-Ga_2O_3. On increasing the flow rate of Ar from 40 to 60 sccm, there was a distinct change in the morphology. We obtained nanosheets and nanowires at a flow rate of 40 sccm with activated charcoal while at 60 sccm with activated carbon, we obtained nanorods.

Activated carbon has a large surface area and is oxidized around 500 °C, and is more reactive than multi-walled carbon nanotubes prepared by the arc-discharge method. The nanotubes have a considerably lower surface area and are oxidized around 700 °C. The reaction of Ga_2O_3 powder with multi-walled nanotubes by procedure (iii) at an Ar flow rate of 40 sccm yielded a mixture of nanosheets and nanobelts at the inner tube. The XRD pattern confirmed that the product obtained was β-Ga_2O_3. An SEM image of the sample collected at the inner tube (Fig. 2a) reveals the morphology of the belts and sheets. The nanosheets

G. Gundiah et al. / Chemical Physics Letters 351 (2002) 189–194

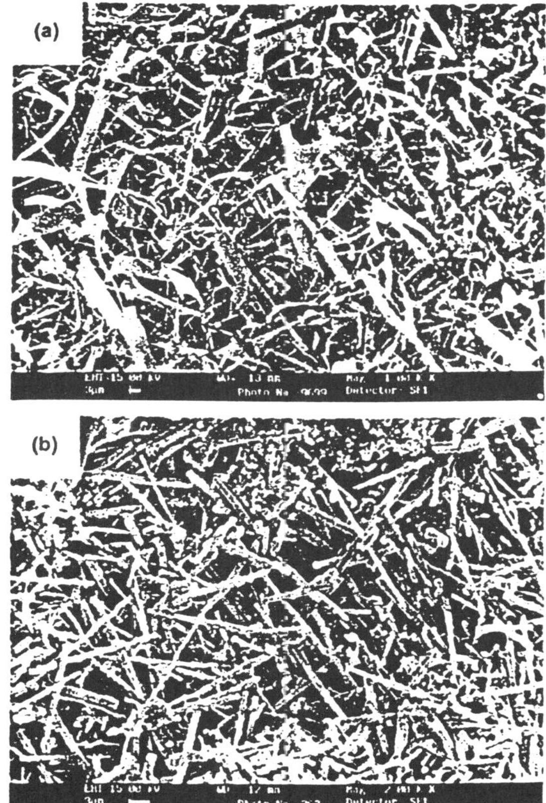

Fig. 1. (a) SEM image of Ga_2O_3 nanosheets and nanowires prepared by procedure (i). (b) SEM image of Ga_2O_3 nanorods prepared by procedure (ii).

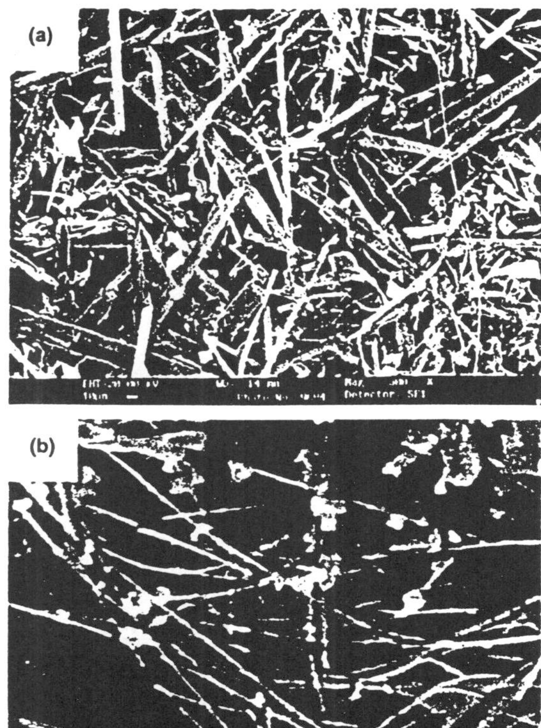

Fig. 2. SEM images of nanostructures obtained by procedure (iii) when the Ar flow rate was maintained at 40 sccm and the product was collected at the (a) inner tube and (b) outlet.

have widths going upto 10 μm and lengths of tens of microns. The nanobelts, however, have a much smaller width as can be seen from the low magnification TEM image of a nanobelt in Fig. 3a. Nanobelts of semiconducting oxides such as SnO_2 and Ga_2O_3 have been prepared by Pan et al. [10] by the thermal evaporation of oxide powders under controlled conditions. The nanobelts obtained by us typically have widths of 150–200 nm, with lengths extending to tens of microns. The selected area electron diffraction (SAED) pattern of a nanobelt is shown in the inset in Fig. 3a. The reflections correspond to the $(1\,0\,4)$, $(\bar{2}\,1\,1)$ and $(\bar{2}\,0\,2)$ planes of β-Ga_2O_3. The SEM image of the product obtained at the outlet by procedure (iii) is shown in Fig. 2b. The product mainly consists of nanowires with a diameter of around 1 μm and a length of

several microns. It appears that the lighter products such as the nanowires are carried to the outlet whereas the heavier nanobelts and nanosheets are deposited on the inner tube. When the reaction was carried out in the absence of any carbon source under similar conditions, we did not obtain any nanowires.

On increasing the flow rate of Ar to 60 sccm in procedure (iii) and maintaining the rest of the parameters the same, there was a marked change in the morphology and in the dimensions of the nanostructures. In Fig. 4a we show a SEM image of the product obtained at the outlet of the inner tube. At this flow rate, the yield of nanowires was high, compared to that of the nanosheets or the nanobelts, which are predominantly formed at lower flow rates of Ar. The nanowires have a diameter of

192   *G. Gundiah et al. / Chemical Physics Letters 351 (2002) 189–194*

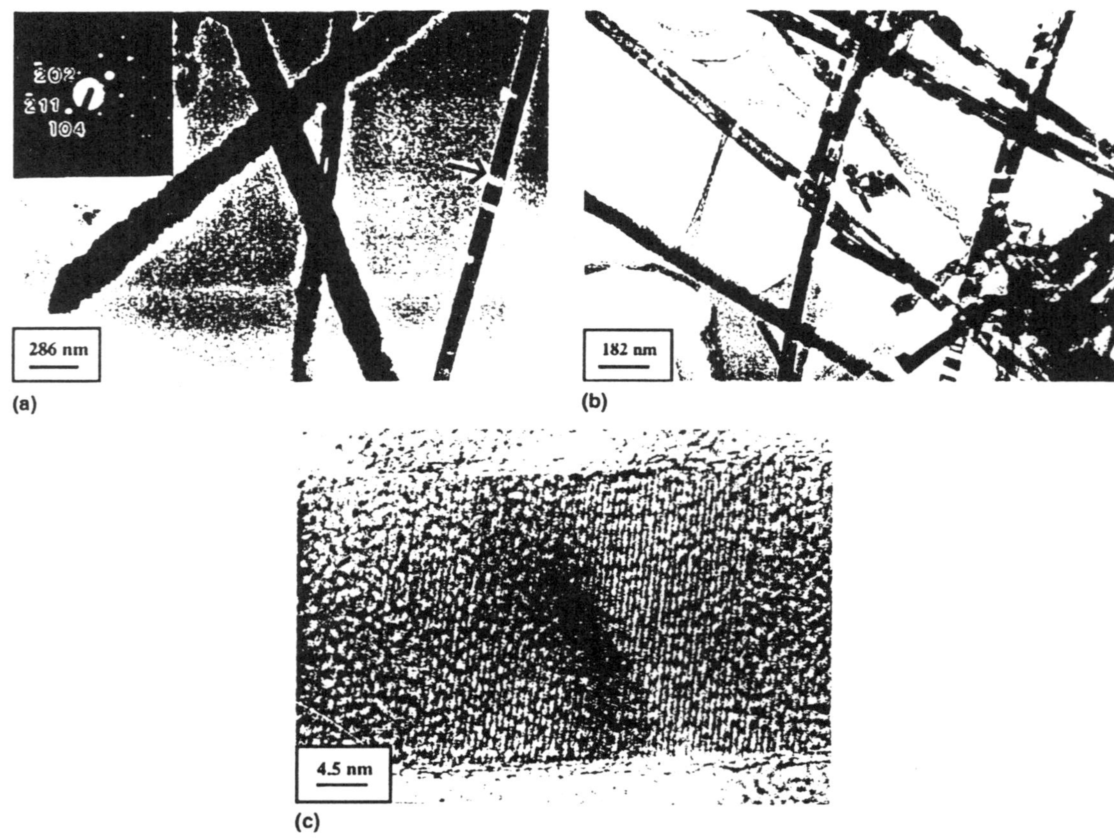

Fig. 3. Low magnification TEM images of the nanowires obtained by the procedure (iii) (a) at a flow rate of Ar maintained at 40 sccm and the product collected at the inner tube. The arrow shows a nanobelt. Inset is the SAED pattern of the sample. (b) Flow rate of Ar maintained at 80 sccm and the product collected at the outlet. (c) HREM of a gallium oxide nanowire obtained on the inner tube by procedure (iii) on maintaining the flow of the Ar gas at 60 sccm. The arrow indicates the growth direction that makes an angle of ∼6° with the normal to the ($\bar{1}$02) planes.

around 500 nm with lengths of several microns. Thinner nanowires were formed at the outlet as shown in the SEM image in Fig. 4b. These nanowires have a diameter of around 300 nm. On further increasing the flow rate of Ar to 80 sccm, no product accumulated at the inner tube. The entire product was present at the outlet and SEM images of this sample showed them to entirely comprising nanowires of a considerably smaller diameter than that obtained at lower flow-rates. The low magnification TEM image in Fig. 3b shows nanowires with diameters of around 70 nm.

A high-resolution electron microscopic (HREM) image of a nanowire synthesized by procedure (iii) obtained at Ar flow-rate of 60 sccm is shown in Fig. 3c. The image clearly shows a lattice spacing of 0.47 nm, corresponding to the ($\bar{1}$02) planes of β-Ga_2O_3. The nanowire is clearly single crystalline with the growth direction being nearly perpendicular to the ($\bar{1}$02) planes. In actuality, the normal to the ($\bar{1}$02) planes make a small angle of 6° with the growth direction.

The growth mechanism of the nanowires can be explained on the basis of a vapor–solid mechanism since there are no droplets present at the ends of the nanowires. The reactions that may be involved in the formation of the nanowires are as follows:

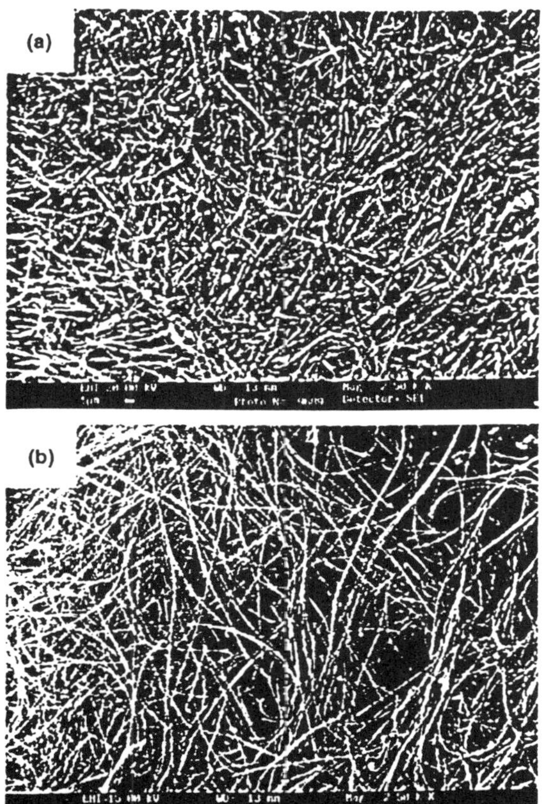

Fig. 4. SEM image of gallium oxide nanowires obtained by procedure (iii) when the reaction was carried out under a flow rate of 60 sccm of Ar gas at (a) the inlet and (b) the outlet.

$$Ga_2O_3(s) + 2C(s) \rightarrow Ga_2O(v) + 2CO(v) \qquad (1)$$

$$Ga_2O_3(s) + 3C(s) \rightarrow 2Ga(v) + 3CO(v) \qquad (2)$$

$$Ga_2O(v) + O_2(g) \rightarrow Ga_2O_3(s) \qquad (3)$$

$$4Ga(v) + 3O_2(g) \rightarrow 2Ga_2O_3(s) \qquad (4)$$

The first step involves the reduction of gallium oxide into its suboxide or metal in vapor form by reactions (1) and (2). The suboxide or metal particles get deposited on the walls of the inner tube or at the outlet, depending on the flow rate of Ar, and get oxidized to Ga_2O_3 in the presence of air.

Gallium oxide has been extensively studied for its luminescent properties [4]. We carried out photoluminescence measurements at room temperature at an excitation wavelength of 265 nm. In Fig. 5, we show the PL spectrum of the nanowires prepared by procedure (iii). Two broad peaks are seen at 324 and 405 nm. The intensity of the peak at 324 nm is considerably smaller than that of the peak at 405 nm. The peak at 405 nm is slightly shifted from the peak (436 nm) of β-Ga_2O_3 single crystal [11]. The blue-shift may be due to the small grain size in the nanowires compared to the bulk sample. The mechanism of PL in Ga_2O_3 is likely to involve the recombination of an electron on the donor and a hole on the acceptor formed by the gallium vacancies [4]. Vasil'tsiv et al. [12] propose that the acceptor would be formed by a gallium–oxygen vacancy pair. According to Binet and Gourier [11], after excitation of the acceptor, a hole on the acceptor and an electron on a donor would be created. These combine radiatively to emit a blue photon. By increasing the temperature, the blue emission can be quenched either by electron detrapping from a donor to the conduction band or by hole detrapping from an acceptor to the valence band. The holes and electrons recombine via a self-trapped exciton to emit an UV photon. In the β-Ga_2O_3 nanowires prepared by us, it is likely that oxygen vacancies and gallium–oxygen vacancy pair are produced, accounting for

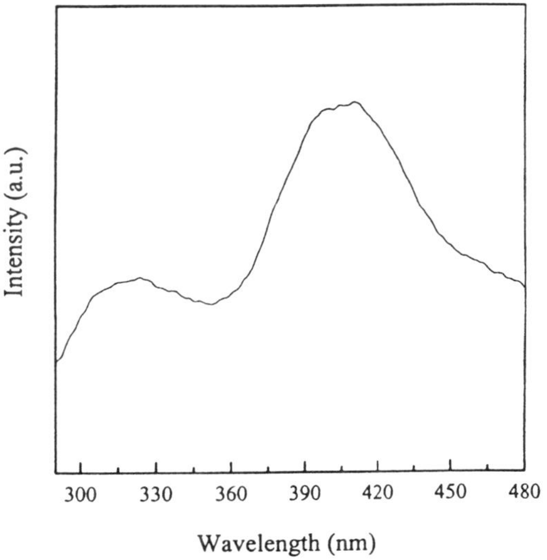

Fig. 5. PL spectrum of gallium oxide nanosheets obtained by procedure (iii) on maintaining the flow of the Ar gas at 40 sccm.

the presence of one peak in the UV region and another peak slightly shifted from the blue region in the PL spectrum.

4. Conclusions

In summary, we have been able to prepare different nanostructures of β-Ga_2O_3 by the reaction of gallium oxide with activated carbon and carbon nanotubes. These include nanosheets and nanobelts besides nanowires. The flow rate of the Ar gas determines the morphology of the final nanostructures, thin nanowires being favored by a high flow rate. The formation of nanobelts is significant since they would be ideal for understanding dimensionally confined transport phenomena in functional oxides and for building functional devices along individual nanobelts.

Acknowledgements

The authors thank DRDO (India) for support.

References

[1] X. Duan, C.M. Lieber, J. Am. Chem. Soc. 122 (2000) 188.
[2] F.L. Deepak, A. Govindaraj, C.N.R. Rao, J. Nanosci. Nanotechnol. 1 (2000) 1, and references therein.
[3] H. Dai, E.W. Wong, Y.Z. Lu, S. Fan, C.M. Lieber, Nature (London) 375 (1995) 769.
[4] T. Harwig, F. Kellendonk, J. Solid State Chem. 24 (1978) 255.
[5] H.Z. Zhang, Y.C. Kong, Y.Z. Wang, X. Du, Z.G. Bai, J.J. Wang, D.P. Yu, Y. Ding, Q.L. Hang, S.Q. Feng, Solid State Commun. 109 (1999) 677.
[6] Y.C. Choi, W.S. Kim, Y.S. Park, S.M. Lee, D.J. Bae, Y.H. Lee, G.S. Park, W.B. Choi, N.S. Lee, J.M. Kim, Adv. Mater. 12 (2000) 746.
[7] X.C. Wu, W.H. Song, W.D. Huang, M.H. Pu, B. Zhao, Y.P. Sun, J.J. Du, Chem. Phys. Lett. 328 (2000) 5.
[8] C.H. Liang, G.W. Meng, G.Z. Wang, Y.W. Wang, L.D. Zhang, S.Y. Zhang, Appl. Phys. Lett 78 (2001) 3202.
[9] R. Seshadri, A. Govindaraj, H.N. Aiyer, R. Sen, G.N. Subbanna, A.R. Raju, C.N.R. Rao, Current Sci. 66 (1994) 839.
[10] Z.W. Pan, Z.R. Dai, Z.L. Wang, Science 291 (2001) 1947.
[11] L. Binet, D. Gourier, J. Phys. Chem. Solids 59 (1998) 1241.
[12] V.I. Vasil'tsiv, Ya.M. Zakharko, Ya.I. Prim, Ukr. Fiz. Zh 33 (1988) 1320.

Synthesis and characterization of silicon carbide, silicon oxynitride and silicon nitride nanowires

Gautam Gundiah, G. V. Madhav, A. Govindaraj, Md. Motin Seikh and C. N. R. Rao*

Chemistry and Physics of Materials Unit and CSIR Centre of Excellence in Chemistry, Jawaharlal Nehru Centre for Advanced Scientific Research, Jakkur P.O., Bangalore 560 064, India. E-mail: cnrrao@jncasr.ac.in

Received 4th January 2002, Accepted 27th February 2002
First published as an Advance Article on the web 3rd April 2002

Several methods have been employed to synthesize SiC nanowires. The methods include heating silica gel or fumed silica with activated carbon in a reducing atmosphere, the carbon particles being produced *in situ* in one of the methods. The simplest method to obtain β-SiC nanowires involves heating silica gel with activated carbon at 1360 °C in H_2 or NH_3. The same reaction, if carried out in the presence of catalytic iron particles, at 1200 °C gives α-Si_3N_4 nanowires and Si_2N_2O nanowires at 1100 °C. Another method to obtain Si_3N_4 nanowires is to heat multi-walled carbon nanotubes with silica gel at 1360 °C in an atmosphere of NH_3. In the presence of catalytic Fe particles, this method yields α-Si_3N_4 nanowires in pure form.

Introduction

The discovery of carbon nanotubes[1] has triggered exploration for various types of one-dimensional nanomaterials over the last few years. Several inorganic nanotubes and nanowires have been reported, those of the metal chalcogenides being noteworthy.[2] These nanomaterials are associated with interesting mechanical, electronic, optical and other properties. Thus, silicon carbide (SiC) nanowires possess high elasticity and strength and are good candidates for making various types of composites. Nanowires of silicon nitride (Si_3N_4) may have potential applications in nanodevices and in the fabrication of composites. There are several reports on the preparation of SiC nanowires in the literature, but fewer on the preparation of Si_3N_4 nanowires. The methods employed for the synthesis of SiC nanowires have been varied. SiC nanowires were prepared by Dai *et al.*[3] by the reaction between carbon nanotubes and SiO or SiI_2 in a sealed tube under vacuum at 1300–1400 °C and 1100–1200 °C, respectively. Han *et al.*[4] employed a two-step reaction in which SiO vapour was first generated *via* the reduction of silica and then reacted with carbon nanotubes at 1400 °C in an Ar atmosphere to form SiC nanowires. β-SiC nanowires with and without amorphous silica (SiO_2) wrapping layers have also been obtained by the carbothermal reduction of sol–gel-derived silica xerogels containing carbon nanoparticles, at 1800 °C and 1650 °C, respectively, in an Ar atmosphere.[5] SiC nanorods have also been prepared from solid sources of carbon and silicon by hot filament chemical vapour deposition.[6] Liang *et al.*[7] used the reaction between activated carbon and sol–gel derived silica embedded with Fe nanoparticles at 1400 °C in an H_2 atmosphere to produce β-SiC nanowires, while Hu *et al.*[8] prepared β-SiC nanowires by the reaction of silicon with carbon tetrachloride (CCl_4) and metallic sodium at 700 °C. Zhang *et al.*,[9] on the other hand, used a floating catalyst method wherein $SiCl_4$ was reacted with benzene and floating Fe catalyst particles derived from ferrocene in the presence of H_2 and Ar at around 1150 °C. In the various procedures listed above, the diameters of the nanowires varied between 10 and 100 nm, while the lengths were in the micrometer (µm) range. The reaction of aligned carbon nanotubes with SiO has however been considered to be advantageous, the diameter and the length of the nanowires depending on the nature of the starting carbon nanotubes.[10]

In spite of many of the studies mentioned above, there is a need for a simple procedure for the synthesis of SiC nanowires which uses common chemicals as starting materials and avoids extreme conditions. Since both SiC and Si_3N_4 are products of the carbothermal reduction of SiO_2, it should be possible to establish conditions wherein one set of specific conditions favours one over the other. We have carried out detailed investigations on the preparation of SiC nanowires starting with silica gel and activated carbon and by employing NH_3 or H_2 as the reducing agent, the former enabling us to obtain silicon nitride nanowires as well, under slightly modified conditions. The procedure employed by us for SiC nanowires has the advantage that it does not require carbon nanotubes or the use of SiO as the starting material. In addition to the synthesis of SiC nanowires, we have carried out studies to establish a procedure for the synthesis of pure silicon nitride nanowires as well. Han *et al.*[11] obtained nanorods of a mixture of α- and β-forms of silicon nitride along with Si_2N_2O, by heating a mixture of Si and SiO_2 powders with carbon nanotubes in a nitrogen atmosphere at 1400 °C. Silicon nitride whiskers have been prepared by the gas phase reaction between SiO, CO and N_2 at 1350 °C[12] while α-Si_3N_4 nanowires sheathed with Si and SiO_2 have been obtained by heating silicon oxide nanoparticles with carbon in a flowing N_2 atmosphere at 1450 °C.[13] We have found that the reaction of SiO_2 gel with carbon nanotubes or activated carbon with Fe catalyst in the presence of NH_3 yields pure α-Si_3N_4 nanowires.

Experimental

For the preparation of SiC nanowires, we have employed several methods. In procedure (i), silica gel prepared in admixture with activated carbon, was dried, and heated to 1360 °C (4–7 h) in an NH_3 or a H_2 atmosphere. Activated carbon was prepared by the thermal decomposition of polyethylene glycol (600 units) at 700 °C for 3 h in an Ar atmosphere. In a typical experiment, 2 ml of tetraethylorthosilicate (TEOS) was mixed with 10 ml of ethanol under stirring for 10–15 min. To this solution, 0.424 g of activated carbon was added (giving a C : Si molar ratio of 4 : 1), followed by 1 ml of aqueous HF (48% A.R.) and the stirring continued. After gelation of the above solution, the gel was dried at 125 °C for 12–15 h. The gel containing finely distributed activated carbon was powdered, taken in an alumina boat and placed in a

DOI: 10.1039/b200161f

tubular furnace, purged earlier with NH_3 gas for 15 min. The gel powder was heated at an appropriate temperature in the 1100–1360 °C range for several hours with the flow of NH_3 maintained at 10 ml min^{-1}. Instead of NH_3, H_2 was also used in this procedure. A grey, wool-like product was deposited on the walls of the alumina boat. This was collected and analyzed.

Procedure (ii) for the preparation of SiC nanowires involved a solid state synthesis in which fumed silica (Grade M-5, surface area 210 m^2 g^{-1}, Cabot Corporation) was finely ground with activated carbon, keeping the molar ratio of C : Si at 4 : 1. The mixture was reduced under conditions similar to those in procedure (i). In procedure (iii), a homogenous gel was prepared by the reaction of ethylene glycol with citric acid in the presence of TEOS at elevated temperatures, by the following procedure. In a typical synthesis, 1.5 ml of ethylene glycol was mixed with 3.5 g of citric acid followed by the addition of 2 ml of TEOS. The sample was heated at 80 °C for 12 h, at 120 °C for 6 h and finally at 180 °C for 12 h. The dried gel so obtained was reduced as in procedure (i).

In order to prepare Si_3N_4 nanowires, we have used multi-walled carbon nanotubes as the carbon source instead of activated carbon. The nanotubes have higher thermal stability than activated carbon. The multi-walled nanotubes were prepared by the arc discharge method as well as by the pyrolysis technique outlined in the literature.[14,15] The procedure was similar to procedure (i) used for the synthesis of SiC nanowires. NH_3 was used to provide a reducing atmosphere as well as to carry out nitridation. The reaction was also carried out in the presence of an Fe catalyst prepared *in situ* by taking ferric nitrate along with the other reactants. The proportion of Fe was varied between 0.1 and 0.5 mol%. Instead of using multi-walled carbon nanotubes prepared by the arc-discharge method, we have also used aligned multi-walled carbon nanotubes with or without catalytic Fe particles. Aligned multi-walled nanotubes were prepared by the pyrolysis of ferrocene along with acetylene in an Ar atmosphere.[16] Most of the reactions were carried out at 1360 °C and a few of them at 1100 °C. The grey coloured, wool-like product in each case was collected and analyzed.

In order to prepare Si_3N_4 nanowires, we have also used the reaction of silica gel and NH_3 in the presence of activated carbon (as in procedure (i) for SiC nanowires) and catalytic iron particles. The Fe particles were incorporated by taking ferric nitrate along with TEOS and activated carbon during the preparation of the silica gel.

Graphite powder was also used in place of the other carbon sources in procedure (i). The reaction of graphite powder at 1360 °C did not give us the carbide or the nitride due to the stability of graphite as compared with the other carbon sources.

Powder X-ray diffraction (XRD) patterns were recorded using Cu-Kα radiation on a Rich-Siefert, XRD–3000-TT diffractometer. Scanning electron microscopy (SEM) images were obtained on a LEICA S440i scanning electron microscope. Transmission electron microscopy (TEM) images were obtained with a JEOL JEM 3010 operating with an accelerating voltage of 300 kV. Photoluminescence (PL) measurements were carried out at room temperature using 260 nm wavelength as the excitation wavelength with a PerkinElmer model LS50B luminescence spectrometer. Infrared spectra were recorded with a Bruker FT-IR spectrometer. The Raman experiments were performed in a quasi-backscattering geometry. The Raman spectra were measured at room temperature using a Nd-YAG laser at 532 nm, and an Ar-ion laser at 488 and 514.5 nm. The Raman spectra show no remarkable resonance enhancement. The scattered light was collected using an optical fiber through a Super Notch filter, dispersed by a single monochromator with $f = 0.55$ m (Spex 550) and detected by a cooled CCD.

Fig. 1 SEM images of the β-SiC nanowires obtained by procedure (i) (heating the gel containing the activated carbon and silica at 1360 °C in NH_3) for (a) 4 h and (b) 7 h.

Results and discussion

SiC nanowires

The reaction of activated carbon with silica gel at 1360 °C by procedure (i) gave a good yield of SiC nanowires after a period of 4 h. In Fig. 1a we show the SEM image of the nanowires obtained by this procedure. The diameter of the nanowires is around 350 nm and the length extends to several tens of microns. The XRD pattern of this sample shown in Fig. 2a

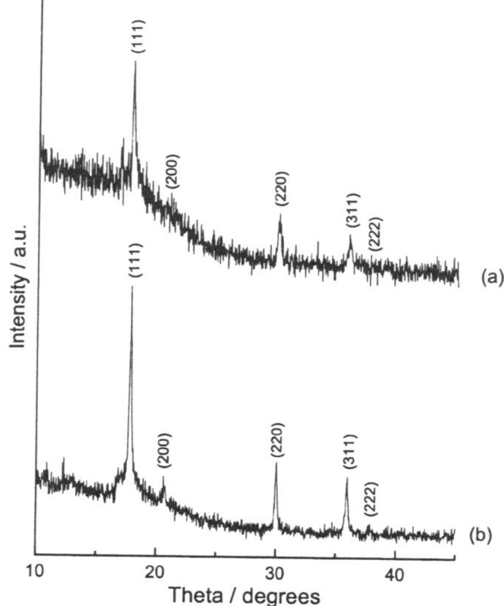

Fig. 2 XRD patterns of the SiC nanowires obtained by procedure (i) (heating the gel containing the activated carbon and silica at 1360 °C in NH_3) for (a) 4 h and (b) 7 h.

matches with that of cubic β-SiC with the unit cell parameters $a = 4.3589$ Å (JCPDS file: 29-1129). There are no reflections corresponding to α-SiC (hexagonal form) or SiO_2 in the diffraction pattern. We have carried out the same reaction for a period of 7 h (instead of 4 h) to improve the crystallinity of the SiC nanowires. The SEM image shown in Fig. 1b indicates that the nanowires so obtained have a diameter of ~40 nm with lengths of several tens of microns. The XRD pattern of this sample (Fig. 2b) is characteristic of the β-SiC phase. The pattern also shows that the crystallinity of the sample is considerably improved. The SiC nanowires did not have a silica coating as evidenced from the XRD pattern and transmission electron microscope images. Low magnification electron microscope images of the nanowires did not reveal the presence of any droplets at the ends of the nanowires. A bead-necklace morphology, similar to that reported by Wu et al.[17] was, however, present in the nanowires, suggesting that the growth of the nanowires probably occurs by the VLS mechanism.[18,19] In this mechanism, the SiO_2 nanoparticles form droplets that are in contact with the activated carbon to give SiO vapour. The SiO vapour further reacts with carbon to form β-SiC particles, which act as the nuclei for the growth of the nanowires. Once the droplets are saturated with β-SiC, crystallization of the β-SiC occurs, followed by growth in one direction to yield nanowires.

We have carried out the reaction of silica gel with activated carbon in the presence of H_2 gas instead of NH_3 gas. This also gives excellent yields of nanowires of β-SiC. In Fig. 3a we show the SEM image of β-SiC obtained by this method. The nanowires are thick with a uniform diameter of ~325 nm and have lengths of several tens of microns.

The HREM image of a SiC nanowire presented in Fig. 4, shows a spacing of 2.5 Å between the (111) planes. The normal to the (111) planes forms an angle of 35° with the growth direction of the nanowires. The selected area electron

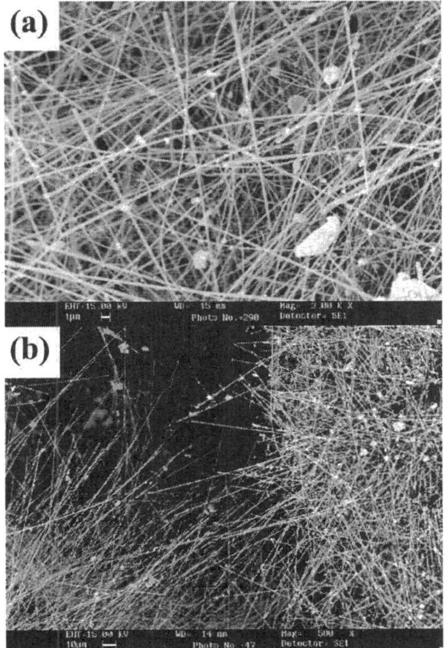

Fig. 3 SEM images of SiC nanowires obtained by (a) procedure (i) (heating the gel containing the activated carbon and silica for 7 h at 1360 °C) in the presence of H_2 gas and (b) procedure (iii) (heating the gel prepared by the reaction of ethylene glycol with citric acid in the presence of TEOS at 1360 °C for 7 h in NH_3).

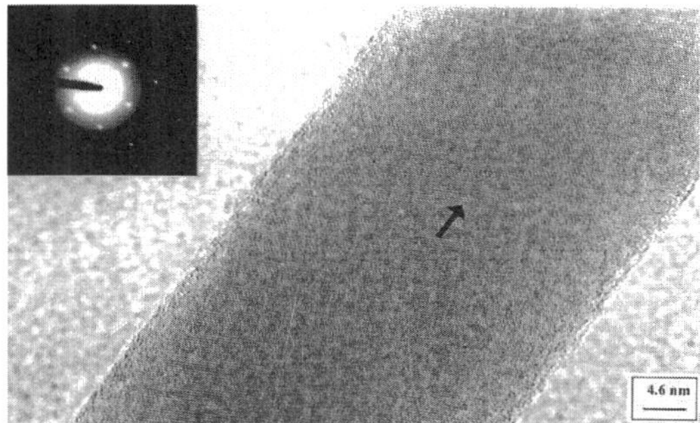

Fig. 4 HREM image of a SiC nanowire prepared by procedure (i) (heating the gel containing activated carbon and silica for 7 h at 1360 °C in NH_3). The inset shows the SAED pattern. The arrow denotes the normal to the (111) planes and the direction of growth of the nanowire.

diffraction pattern (SAED) pattern showed Bragg spots corresponding to the (111) planes, with some streaking due to the presence of stacking faults.

The basic reactions involved in the formation of the SiC nanowires are given by,

$$SiO_2 + C \rightarrow SiO + CO$$

$$SiO + 2C \rightarrow SiC + CO$$

The NH_3 and H_2 in the gas stream favour the formation of SiO from SiO_2. The SiO reacts readily with carbon to form SiC.

In procedure (i), we have employed TEOS as the starting material since it gives fine particles of silica at low pH values. We have carried out the reaction of fumed silica (instead of a silica gel) with activated carbon by grinding the reactants in order to see if this procedure also gives us a good yield of SiC nanowires as in the case of the sol–gel route. Procedure (ii) yielded a grey, wool-like product, which had β-SiC as the major component. The diameter of the nanowires was ~500 nm and with the lengths extending to tens of microns. The XRD pattern, however, showed the presence of some silica.

We could obtain β-SiC nanowires by procedure (iii) wherein the gel formed by ethylene glycol, citric acid and TEOS was heated in a reducing atmosphere at 1360 °C. The ethylene glycol initially reacts with citric acid (in the presence of silica gel) to form silica particles embedded in carbon precursors. On treatment with NH_3 or H_2, this composite generates active carbon *in situ* which reacts with the silica nanoparticles to yield SiC. The SEM image (Fig. 3b) shows that the nanowires have large diameters (~750 nm) with lengths of up to tens of microns. The nanowires appear to be straight, unlike the ones obtained earlier, and seem to grow out from the carbon–silica composite particles.

The photoluminescence (PL) spectrum of the β-SiC nanowires synthesized by procedure (i) is shown in Fig. 5. The sample exhibits a single broad peak at 400 nm. Such a PL feature is characteristic of SiC, though there is a considerable blue shift as compared to the bulk sample.[20] The 3C form SiC shows a PL band maximum around 540 nm due to the presence of defect centres within the band gap.

Infrared spectra of the β-SiC nanowires show a strong band at 810 cm^{-1} due to the Si–C stretching band. The Raman spectrum of the nanowires is more interesting. Fig. 6 shows the room temperature Raman spectrum of the SiC nanowires with a quasi-backscattering geometry. We observe a broad band around 780 cm^{-1} corresponding to the TO phonon at the Γ point of cubic SiC. The absence of the folded modes that are related to the SiC polytypes (natural superlattices due to various stacking orders of Si–C bilayers) in the Raman

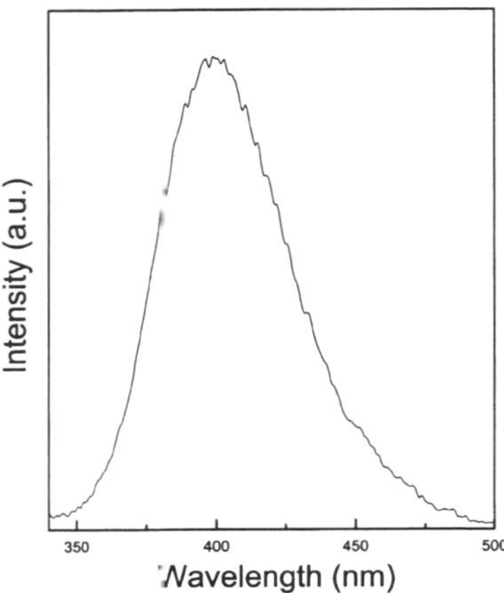

Fig. 5 Photoluminescence spectrum of SiC nanowires synthesized by procedure (i) (heating the gel containing activated carbon and silica for 7 h at 1360 °C in NH_3).

spectrum is consistent with the X-ray data, and suggests the nanowires are only cubic SiC.[21] The TO phonon line is shifted by around 16 cm^{-1} from the bulk SiC[22] towards lower energy. This is due to the small grain size of SiC and the presence of growth faults of the nanowires.

In the zinc-blende structure (as in cubic SiC) the LO phonon has a higher energy than the TO phonon near the zone center. LO phonon is around 972 cm^{-1} in bulk at ambient conditions in the bulk 3C-SiC.[22] In the Raman spectrum recorded by us, we do not see the LO phonon. As shown in Fig. 6, the LO phonon is absent even when different excitation wavelengths are used. Raman selection rules for the zinc-blende structure governs this observation as explained below. While preparing the SiC sample for Raman experiments, the nanowires are compressed to form a continuous surface. This orients the nanowires, such that the long cylindrical axes of the nanowires are parallel to the sample plane. From the TEM images it is clear that the cylindrical axis of the nanowires is [011] for the zinc-blende 3C-SiC. Raman experiments are carried out in the quasi-backscattering geometry on planes perpendicular to the (011) planes, such as (0$\bar{1}$1). In the case of such a backscattering geometry, the incident wave vector (k_i) and scattered wave vector (k_s) are along the [0$\bar{1}$1] direction and the polarization of the incident (e_i) and scattered (e_s) wave vectors

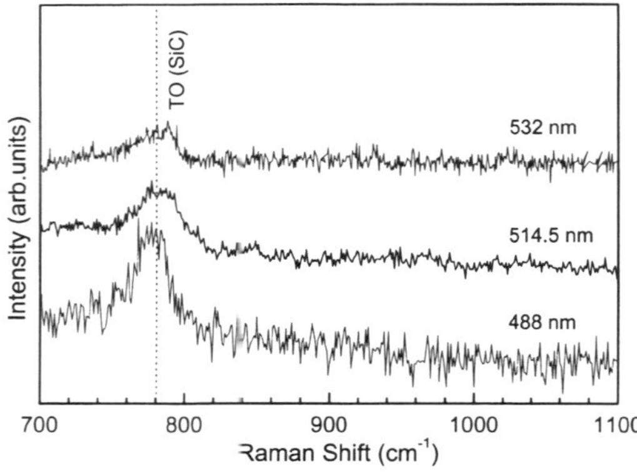

Fig. 6 Raman spectra of SiC nanowires synthesized by procedure (i) (heating the gel containing activated carbon and silica for 7 h at 1360 °C in NH_3).

are along the [011] direction. According to the Raman selection rules, the TO phonon is allowed whereas the LO phonon is forbidden for a zinc-blende structure.[23] This is the reason for the absence of the LO mode in the present experiment.

Si_3N_4 nanowires

The reaction of silica gel with ammonia alone or in the presence of activated carbon in the temperature range 1100–1360 °C did not yield Si_3N_4 nanowires. However, when the reaction of silica gel was carried out with arc-generated multi-walled carbon nanotubes in an NH_3 atmosphere at 1360 °C for 4 h, we obtained a good yield of Si_2N_2O nanowires. The XRD pattern of the nanowires, shown in Fig. 7a, is characteristic of Si_2N_2O with unit cell parameters a = 8.868 Å, b = 5.497 Å, c = 4.854 Å, $\alpha = \beta = \gamma = 90°$ (JCPDS file: 47-1627). The SEM image of the Si_2N_2O nanowires shown in Fig. 8a reveals the diameter of the nanowires to be ~ 300 nm with lengths of up to tens of microns.

When the reaction of arc-generated multi-walled carbon nanotubes with silica gel was carried out at 1360 °C for longer periods ($\geqslant 7$ h), the product contained mainly Si_3N_4 nanowires. The XRD pattern of the sample, shown in Fig. 7b, establishes the nanowires to be predominantly of the hexagonal α-Si_3N_4

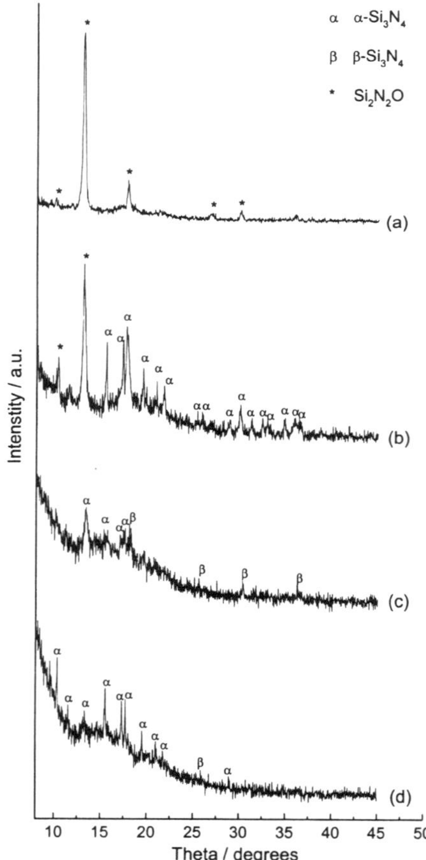

Fig. 7 XRD patterns of the products obtained by the reaction of multi-walled carbon nanotubes with silica gel at 1360 °C in an NH_3 atmosphere: (a) for 4 h, (b) for 7 h, (c) for 4 h in the presence of 0.1 mol% Fe, and (d) for 4 h in the presence of 0.5 mol% Fe. Relative intensities of the α-Si_3N_4 peaks show variations due to the orientational effects.

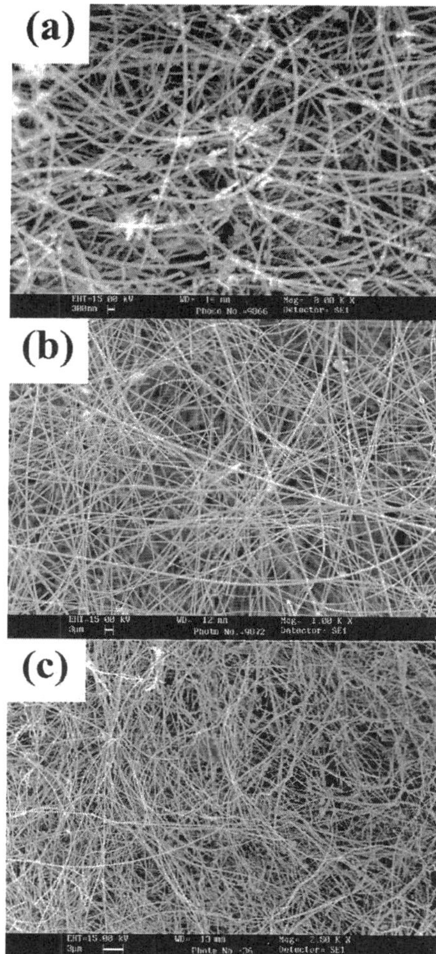

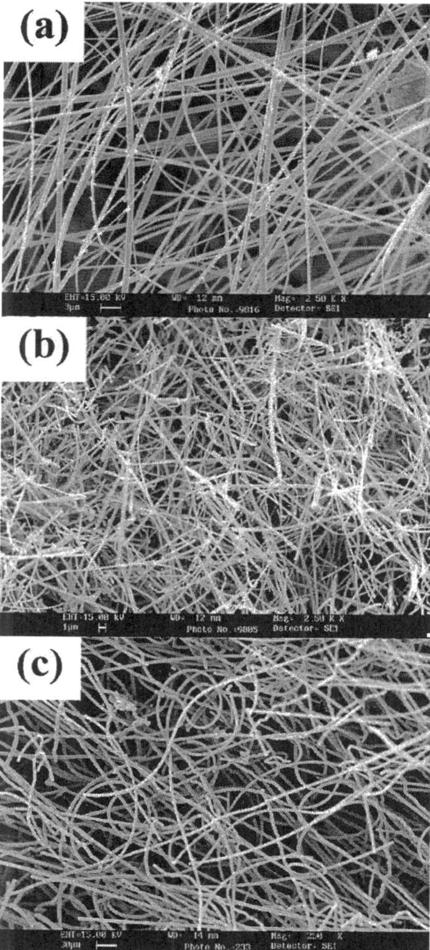

Fig. 8 SEM images of (a) Si$_2$N$_2$O nanowires obtained by the reaction of silica with NH$_3$ in the presence of multi-walled carbon nanotubes with silica at 1360 °C for 4 h, (b) Si$_3$N$_4$ nanowires prepared by the reaction of silica with NH$_3$ in the presence of multi-walled carbon nanotubes with at 1360 °C for 7 h and (c) Si$_2$N$_2$O nanowires obtained by the reaction of silica with NH$_3$ at 1100 °C in the presence of activated carbon and 0.5% Fe catalyst.

Fig. 9 SEM images of the Si$_3$N$_4$ nanowires prepared by the reaction of multi-walled carbon nanotubes with silica at 1360 °C for 4 h in the presence of (a) 0.1% Fe and (b) 0.5% Fe catalysts, respectively. (c) SEM image of the Si$_3$N$_4$ nanowires prepared by the reaction of aligned multi-walled carbon nanotubes with silica at 1360 °C for 4 h.

phase with unit cell parameters $a = 7.7541$ Å, $c = 5.6217$ Å (JCPDS file: 41-0360), with a minor component of Si$_2$N$_2$O. The composition of the final product was dependent on the duration over which the reaction was carried out. Fig. 8b shows an SEM image of these nanowires, which have diameters of between 150 and 175 nm. In the presence of catalytic Fe particles, the reaction of silica gel with carbon nanotubes occurs within 4 h giving a good yield of nanowires of pure Si$_3$N$_4$ with no Si$_2$N$_2$O. Fig. 7c and d show the XRD patterns of the products obtained in the presence of 0.1% and 0.5% Fe catalysts. While the 0.1% Fe catalyst gives a mixture of the α-phase with the hexagonal β-phase with unit cell parameters $a = 7.6044$ Å, $c = 2.9075$ Å (JCPDS file: 33-1160), the 0.5% Fe catalyst yields almost monophasic α-Si$_3$N$_4$ nanowires, as can be seen from Fig. 7d. Fig. 9a shows an SEM image of the Si$_3$N$_4$ nanowires obtained with 0.1% catalyst. The diameter is ~400 nm and the length extends to hundreds of microns. The nanowires prepared with the 0.5% Fe catalyst however have larger diameters (~700 nm) as can be seen from Fig. 9b.

It should be noted that both the length and diameter of the Si$_3$N$_4$ nanowires are generally larger than those of the carbon nanotubes.

The reaction involved in the formation of Si$_3$N$_4$ nanowires in the presence of carbon nanotubes can be represented as,

$$3SiO_2 + 6C + 4NH_3 \rightarrow Si_3N_4 + 6H_2 + 6CO$$

The role of the catalytic iron particles is likely to be in facilitating the removal of oxygen from the silica. The iron oxide formed in such a reaction would readily get reduced back to metal particles in the reducing atmosphere. The formation of Si$_2$N$_2$O is given by the reaction

$$2SiO_2 + 3C + 2NH_3 \rightarrow Si_2N_2O + 3CO + 3H_2$$

Carbon nanotubes prepared by precursor pyrolysis get oxidized at a relatively low temperature (~550 °C) compared to the nanotubes prepared by arc evaporation (which oxidize at around 700 °C). We have carried out the reaction of silica and NH$_3$ with multi-walled carbon nanotubes prepared by the

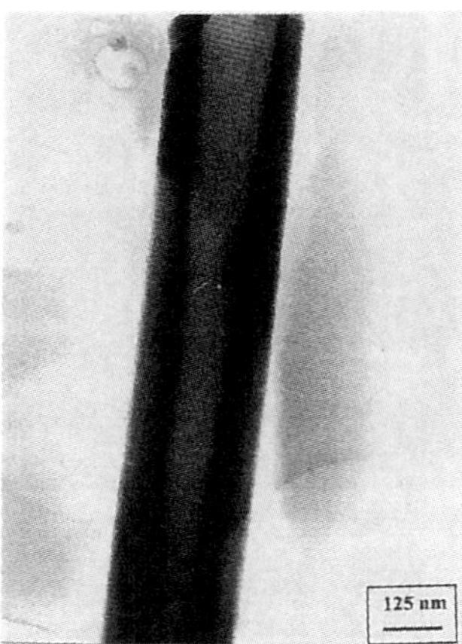

Fig. 10 TEM image of Si₃N₄ nanotubes obtained by the reaction of aligned multi-walled nanotubes with silica gel at 1360 °C.

decomposition of ferrocene. The reaction of these nanotubes with silica gel and NH_3 at 1360 °C yields a mixture of α- and β-Si_3N_4. Fig. 9c shows the SEM image of the product obtained by this reaction. The nanowires have a large diameter (5–7 µm), with lengths of the order of hundreds of microns. Activated carbon has a larger surface area than the nanotubes produced from hydrocarbon pyrolysis and gets oxidized at temperatures as low as 500 °C (compared to 800 °C for graphite). We would therefore expect activated carbon to yield Si_3N_4 by reacting with silica gel in the presence of NH_3 under milder conditions. However, as mentioned earlier, we did not obtain Si_3N_4 nanowires with activated carbon. In the presence of 0.5% Fe, however, the reaction of silica gel with activated carbon at 1200 °C for 8 h yielded α-Si_3N_4. The same reaction at 1360 °C for 4–7 h yields β-SiC nanowires. When the reaction was carried out at 1100 °C for 4 h, it gave Si_2N_2O nanowires. We show an SEM image of these nanowires in Fig. 8c.

Infrared spectra of the Si_3N_4 nanowires show a prominent band around 825 cm⁻¹ due to the Si–N stretching vibration. There were no bands due to Si–O stretching.

In the synthesis of Si_3N_4 nanowires using aligned multi-walled nanotubes as the carbon source, we occasionally obtained nanotubes of α-Si_3N_4 along with the nanowires. The outer diameter of the nanotubes is ∼200 nm as seen in the low resolution TEM image in Fig. 10.

Conclusions

The present study establishes that pure β-SiC (cubic) nanowires are easily obtained by reacting silica gel with activated carbon in a reducing atmosphere at 1360 °C. This procedure eliminates the need to use SiO as the starting material. Furthermore, carbon nanotubes are not necessary to produce SiC nanowires. The reaction of activated carbon with silica gel and NH_3 in the presence of catalytic iron particles yields α-Si_3N_4 nanowires at 1200 °C, and Si_2N_2O nanowires at 1100 °C. The reaction of multi-walled carbon nanotubes with silica at 1360 °C in an atmosphere of NH_3 also yields pure Si_3N_4 nanowires, while the same reaction at 1100 °C gives Si_2N_2O nanowires. The presence of catalytic iron particles favours the formation of nanowires of α-Si_3N_4.

Acknowledgement

The authors thank DRDO (India) for support and Dr. Chandrabhas Narayana for help with the Raman spectra.

References

1   S. Iijima, *Nature (London)*, 1991, **345**, 56.
2   (a) C. N. R. Rao, B. C. Satishkumar, A. Govindaraj and M. Nath, *ChemPhysChem*, 2001, **2**, 78; (b) Y. Feldman, E. Wasserman, D. J. Srolovitch and R. Tenne, *Science*, 1995, **267**, 222; (c) M. Nath, A. Govindaraj and C. N. R. Rao, *Adv. Mater.*, 2001, **13**, 283; (d) C. N. R. Rao, A. Govindaraj, F. L. Deepak, N. A. Gunari and M. Nath, *Appl. Phys. Lett.*, 2001, **78**, 1853.
3   H. Dai, E. W. Wong, Y. Z. Lu, S. Fan and C. M. Lieber, *Nature (London)*, 1995, **375**, 769.
4   W. Han, S. Fan, Q. Li, W. Liang, B. Gu and D. Yu, *Chem. Phys. Lett.*, 1997, **265**, 374.
5   G. W. Meng, L. D. Zhang, C. M. Mo, S. Y. Zhang, Y. Qin, S. P. Feng and H. J. Li, *J. Mater. Res.*, 1998, **13**, 2533.
6   X. T. Zhou, N. Wang, H. L. Lai, H. Y. Peng, I. Bello, N. B. Wong, C. S. Lee and S. T. Lee, *Appl. Phys. Lett.*, 1999, **74**, 3942.
7   C. H. Liang, G. W. Meng, L. D. Zhang, Y. C. Wu and Z. Cui, *Chem. Phys. Lett.*, 2000, **329**, 323.
8   J. Q. Hu, Q. Y. Lu, K. B. Tang, B. Deng, R. R. Jiang, Y. T. Qian, W. C. Yu, G. E. Zhou, X. M. Liu and J. X. Wu, *J. Phys. Chem. B*, 2000, **104**, 5251.
9   Y. Zhang, N. Wang, R. He, X. Chen and J. Zhu, *Solid State Commun.*, 2001, **118**, 595.
10 Z. Pan, H. L. Lai, F. C. K. Au, X. Duan, W. Zhou, W. Shi, N. Wang, C. S. Lee, N. B. Wong, S. T. Lee and S. Xie, *Adv. Mater.*, 2000, **12**, 1186.
11 W. Han, S. Fan, Q. Li, B. Gu, X. Zhang and D. Yu, *Appl. Phys. Lett.*, 1997, **71**, 2271.
12 M. J. Wang and H. Wada, *J. Mater. Sci.*, 1990, **25**, 1690.
13 X. C. Wu, W. H. Song, B. Zhao, W. D. Huang, M. H. Pu, Y. P. Sun and J. J. Du, *Solid State Commun.*, 2000, **115**, 683.
14 R. Seshadri, A. Govindaraj, H. N. Aiyer, R. Sen, G. N. Subbanna, A. R. Raju and C. N. R. Rao, *Curr. Sci.*, 1994, **66**, 839.
15 M. Nath, B. C. Satishkumar, A. Govindaraj, C. P. Vinod and C. N. R. Rao, *Chem. Phys. Lett.*, 2000, **322**, 333.
16 C. N. R. Rao, R. Sen, B. C. Satishkumar and A. Govindaraj, *Chem. Commun.*, 1998, 1525.
17 X. C. Wu, W. H. Song, W. D. Huang, M. H. Pu, B. Zhao, Y. P. Sun and J. J. Du, *Mater. Res. Bull.*, 2001, **36**, 847 and references therein.
18 E. I. Givargizov, *J. Cryst. Growth*, 1975, **31**, 20.
19 R. S. Wagner and W. C. Ellis, *Appl. Phys. Lett.*, 1964, **4**, 89.
20 A. O. Konstantinov, A. Henry, C. I. Harris and E. Jansén, *Appl. Phys. Lett.*, 1995, **66**, 2250.
21 (a) T. Tomita, S. Saito, M. Baba, M. Hundhausen, T. Suemoto and S. Nakashima, *Phys. Rev. B*, 2000, **62**, 12896; (b) S. Nakashima, H. Katashama, Y. Nakamura and A. Mitsuishi, *Phys. Rev. B*, 1986, **33**, 5721.
22 D. Olego and M. Cardona, *Phys. Rev. B*, 1982, **25**, 3889.
23 P. Y. Yu and M. Cardona, in *Fundamentals of Semiconductors*, Springer-Verlag, Berlin, 1996, p. 366.

Available online at www.sciencedirect.com

SCIENCE @DIRECT®

Chemical Physics Letters 374 (2003) 314–318

CHEMICAL PHYSICS LETTERS

www.elsevier.com/locate/cplett

Photoluminescence spectra and ferromagnetic properties of GaMnN nanowires

F.L. Deepak, P.V. Vanitha, A. Govindaraj, C.N.R. Rao *

Chemistry and Physics of Materials Unit and CSIR Centre of Excellence in Chemistry, Jawaharlal Nehru Centre for Advanced Scientific Research, Jakkur P.O., Bangalore 560 064, India

Received 7 January 2003; in final form 7 April 2003

Abstract

Mn-doped GaN nanowires have been prepared by reacting a mixture of acetylacetonates with NH_3 at 950 °C in the presence of multi-walled (MWNTs) and single-walled (SWNTs) carbon nanotubes, the nanowires prepared with SWNTs being considerably smaller in diameter. GaMnN nanowires with 1%, 3% and 5% Mn so obtained have been characterized by X-ray diffraction, EDAX analysis and photoluminescence (PL) spectra. The GaMnN nanowires are all ferromagnetic even at 300 K, exhibiting magnetic hysteresis. The PL spectra of the GaMnN nanowires prepared with SWNTs show a large blue-shift of the Mn^{2+} emission.
© 2003 Elsevier Science B.V. All rights reserved.

1. Introduction

Dilute magnetic semiconductors are being investigated extensively over the last several years, typical of these materials being Mn doped GaAs which is ferromagnetic with a T_c of 110 K. There is great current interest in the study of Mn doped GaN, with the main objective of examining whether ferromagnetism can be induced in this material. Zajac et al. [1] reported Mn doped GaN to be paramagnetic, but Theodoropoulou et al. [2] find that p-type GaN films, on implantation with high doses of Mn (3–5%), exhibit ferromagnetic behavior upto ~250 K. Reed et al. [3] have recently reported ferromagnetism in $Ga_{1-x}Mn_xN$ films in the temperature range 228–370 K. We were interested in exploring whether single-crystalline Mn doped GaN nanowires could be produced and also in examining their magnetic and optical properties of the nanowires. Pure single-crystalline GaN nanowires themselves have been synthesized by a variety of techniques. The methods of synthesis include the reaction of Ga_2O vapor with NH_3 in the presence of multi-walled carbon nanotubes (MWNTs) [4], gas-phase reaction of Ga_2O vapor with NH_3 in an anodic alumina membrane at 1273 K [5], and the reaction of Ga and NH_3 over catalytic Ni nanoparticles or In powder [6,7]. The other methods employed for the synthesis of GaN nanowires include laser-assisted catalytic growth [8], and hot-filament chemical vapor deposition

* Corresponding author. Fax: +91-80-846-2760.
E-mail address: cnrrao@jncasr.ac.in (C.N.R. Rao).

0009-2614/03/$ - see front matter © 2003 Elsevier Science B.V. All rights reserved.
doi:10.1016/S0009-2614(03)00635-7

using a solid source consisting of a mixture of Ga_2O_3 and C powders [9]. We have synthesized single-crystal GaN nanowires by a variety of procedures [10], of which the use of carbon nanotubes gave good yields of nanowires devoid of carbon or of catalyst particles. We have modified the procedure based on carbon nanotubes to prepare Mn doped GaN nanowires and investigated their photoluminescence and magnetic properties. We have employed both MWNTs and single-walled carbon nanotubes (SWNTs) to prepare GaMnN nanowires in order to vary the diameter of the nanowires.

2. Experimental

Mn-doped GaN nanowires were prepared starting with a mixture of the acetylacetonates of gallium and manganese taken in the required mole ratios corresponding to 1%, 3% and 5% doping. MWNTs were prepared by arc discharge [11] and were of high purity. Since no catalyst was employed there was no contamination from the metal particles. Accordingly, thermogravimetry showed 100% weight loss. The nanotubes were used as removable templates which burn off during the course of the reaction. The diameter of the nanotubes was in the 40–50 nm range. A thoroughly ground mixture of the acetylacetonates was mixed with MWNTs and the mixture was taken in a quartz tube (12 cm length, 6 mm dia.) closed at one end. The quartz tube was placed inside another quartz tube of a larger diameter and connected to a dynamic NH_3 atmosphere (flow rate: 150 ml min^{-1}). This set-up was placed vertically in a tubular furnace, heated slowly (1 °C/min) to reach the melting point of $Ga(acac)_3$ (195 °C), and maintained at that temperature for 1 h. It was then heated to the melting point of $Mn(acac)_2$ (260 °C) and maintained at that temperature for 1 h. The temperature was then raised to 950 °C (3 °C/min) and held at that temperature for 4 h. This procedure was crucial in obtaining high yields of GaMnN nanowires of good quality. The same procedure was employed to prepare Mn-doped GaN nanowires with SWNTs as templates, having prepared the SWNTs themselves by the literature

procedure [12]. The SWNTs individually had diameters of ∼1.2 nm and occurred in bundles.

Scanning electron microscope (SEM) images of the nanowires were obtained with a Leica S-440i microscope and transmission electron microscope (TEM) images with a JEOL JEM 3010 instrument operating at an accelerating voltage of 300 kV. EDAX analysis was carried out with the SEM instrument. Photoluminescence (PL) measurements were carried out with a Perkin–Elmer Model LS50B spectrometer.

Magnetization measurements were carried out with a vibrating sample magnetometer (VSM 7300).

3. Results and discussion

Powder X-ray diffraction patterns showed the materials to have the hexagonal structure, with the c - parameter increasing from 5.178 to 5.190 Å with the increase in Mn dopant concentration from 1 to 5% and the a - parameter remaining essentially constant around 3.190 Å [13,14]. This observation showed that Mn was present substitutionally in GaN. Energy dispersive X-ray analysis (EDAX) carried out on several samples and on different regions of the nanowires, showed the Ga/Mn ratios to be close to those expected from the starting compositions (Fig. 1).

SEM images of the GaN nanowires doped with 1%, 3% and 5% Mn prepared with MWNTs showed that the diameter of the nanowires to be generally in the 50–75 nm range, with the lengths going upto 1–2 μm. In Figs. 2a and b, we show typical TEM images of Mn doped GaN nanowires prepared by the use of SWNTs as templates. It can be seen that the diameter of these nanowires are much smaller than those obtained previously by the use of MWNTs. The average diameter of the GaMnN nanowires obtained from SWNTs is in the 20–25 nm range. The inset in the figure shows an electron diffraction pattern (Fig. 2b) with the Bragg spots corresponding to the [1 0 0] reflection of the wurtzite structure of GaN. The electron diffraction patterns showed occasional streaks, but no extra spots due to other phases, and confirmed that the GaMnN samples had the same structure of GaN although with slightly different lattice

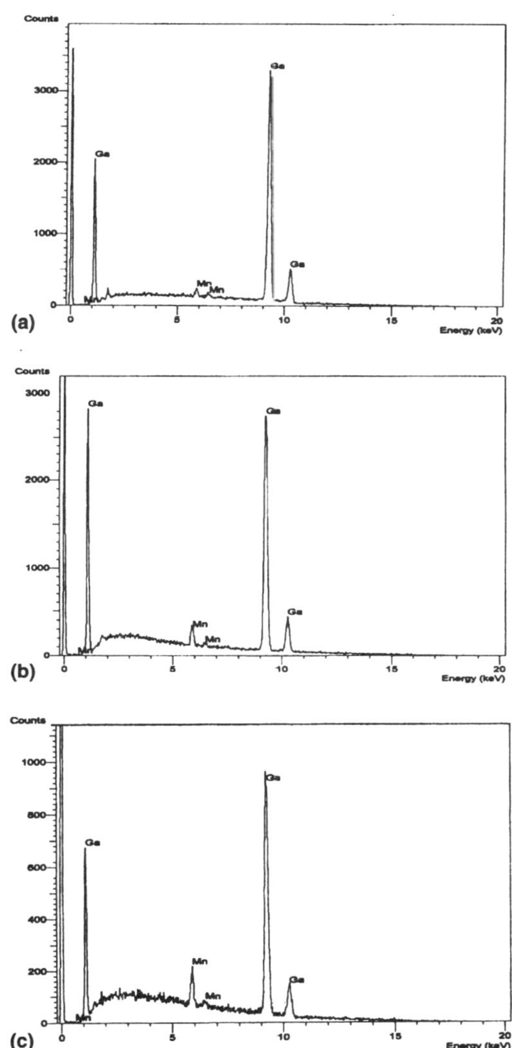

Fig. 1. EDAX spectra of GaMnn nanowires: (a) 1% Mn, (b) 3% Mn, (c) 5% Mn.

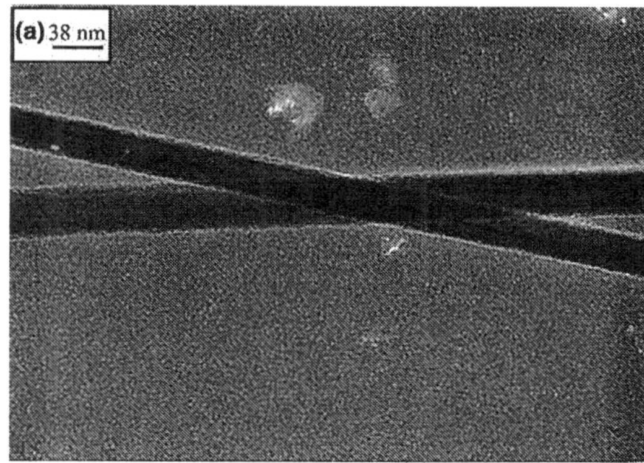
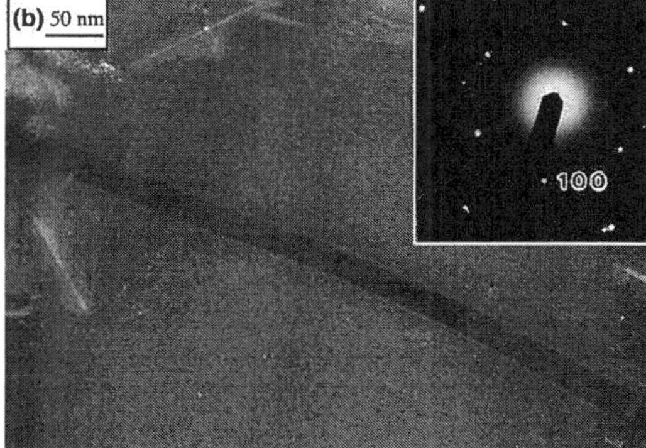
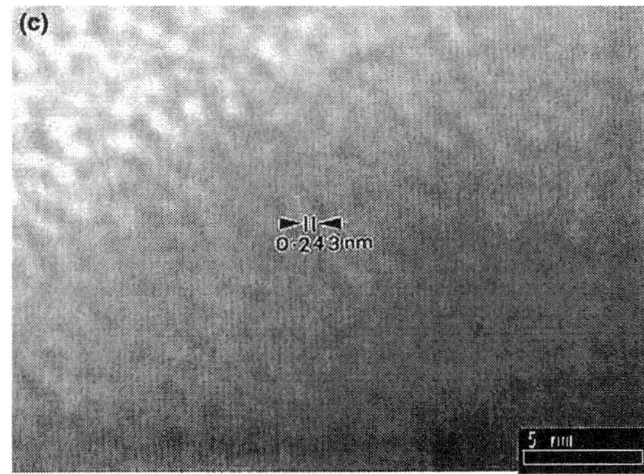

Fig. 2. TEM images of the Mn doped GaN nanowires prepared with SWNTs: (a) 1% Mn, (b) 5% Mn, (c) HREM image of 5% Mn doped nanowire. Inset in (b) shows the electron diffraction pattern.

constants [15]. In Fig. 2c we show a high resolution electron microscope image which not only reveals the single-crystalline nature of the nanowire, but also that the growth direction is at an acute angle with respect to the $(1\,0\,1)$ plane. No defects or other imperfections are apparent.

Magnetization measurements show that the nanowires prepared by using MWNTs with 1%, 3% and 5% Mn are all ferromagnetic with a T_c of around 325 K. The samples exhibit magnetic hysteresis at room temperature. We show typical hysteresis curves at 300 and 50 K in Fig. 3. The M vs H curves show a smooth S-shaped loop with a remnant magnetization (M_R) of 0.026 emu/g at 297 K for the 1% Mn doped nanowires and 0.02 emu/g for the 5% Mn doped GaN nanowires. The coercivity (H_c) in the case of the 1% Mn doped GaN nanowires is 970 Oe and for the 5% doped nanowires 630 Oe. The value of remnant magnetization for the 5% doped sample at 50 K is 0.03 emu/g, whereas the coercivity is 620 Oe. The thinner diameter (25 nm) nanowires obtained from the use of SWNTs show a coercivity of 1400 Oe in the case of the 5% Mn doped samples. The 25 nm nanowires also show a tendency for saturation at 50 K. The remnant magnetization of this sample is 0.06 emu/g at 50 K. The values of coercivity in the GaMnN nanowires found here are higher than those reported for thin films [16].

PL spectra of the GaMnN nanowires prepared with MWNTs are shown in Fig. 4. At an excitation wavelength of 325 nm, the undoped nanowires show a broad emission band with a well-defined peak around 380 nm and a weak feature around 530 nm. These PL features are comparable to those of bulk GaN [4,6,7,9,10], the 380 nm feature arising from defect states. The Mn doped samples show a PL band centred around 420 nm, the red shift arising from hole-doping [17]. We also see the characteristic Mn^{2+} emission around 610 nm, due

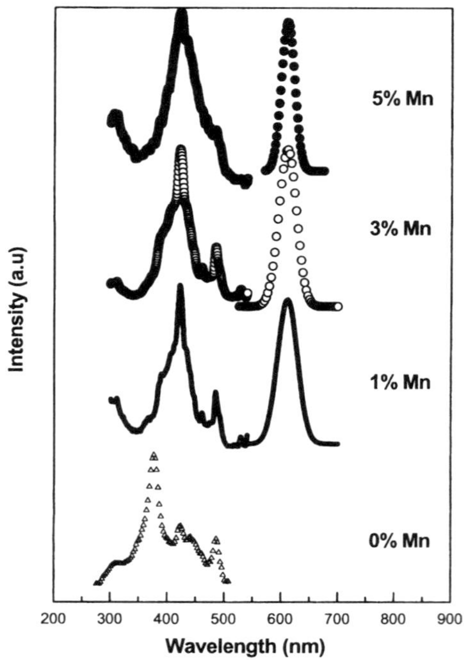

Fig. 4. Photoluminescence spectra of the GaMnN nanowires prepared with MWNTs. The PL spectrum of the undoped nanowires is also shown. Mn^{2+} emission from the GaMnN nanowires is seen around 610 nm.

to the $^4T_1 \rightarrow {}^6A_1$ internal Mn^{2+} ($3d^5$) transition [18]. The FWHM of the Mn^{2+} emission band decreases with increase in Mn concentration.

We have also examined the PL spectra of the 25 nm GaMnN nanowires prepared with SWNTs. The spectra show interesting differences from the spectra of the nanowires prepared with MWNTs as evident. The PL spectrum of the 25 nm nanowires of GaN in Fig. 5 shows the main feature at ~410 nm, slightly blue-shifted compared to that of the larger diameter nanowires (Fig. 4). The Mn emission from the smaller nanowires is significantly different. The characteristic Mn^{2+} emission is strongly blue-shifted in the spectra of the smaller diameter nanowires as can be seen from Fig. 5, compared to the spectra in Fig. 1. Thus, the emission is shifted to around 574 nm. A similar blue-shift of the Mn^{2+} emission has been reported in doped ZnS nanoparticles, where the shift is essentially due to the decrease in the cluster size

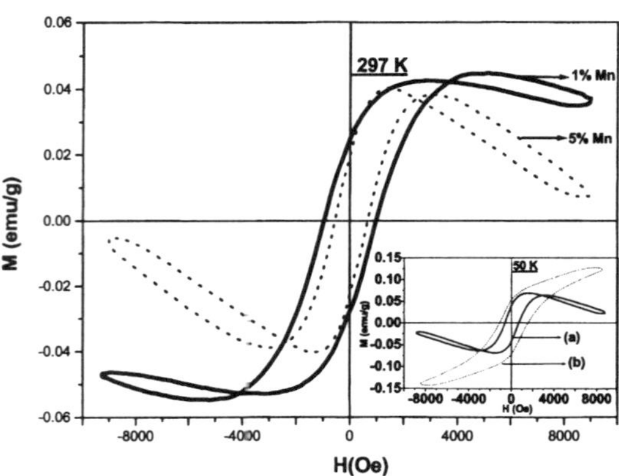

Fig. 3. Typical magnetic hysteresis curves of GaMnN nanowires at 297 K. Inset shows the hysteresis curves at 50 K for the 5% Mn doped nanowires. (a) MWNTs, (b) SWNTs.

        F.L. Deepak et al. / Chemical Physics Letters 374 (2003) 314–318

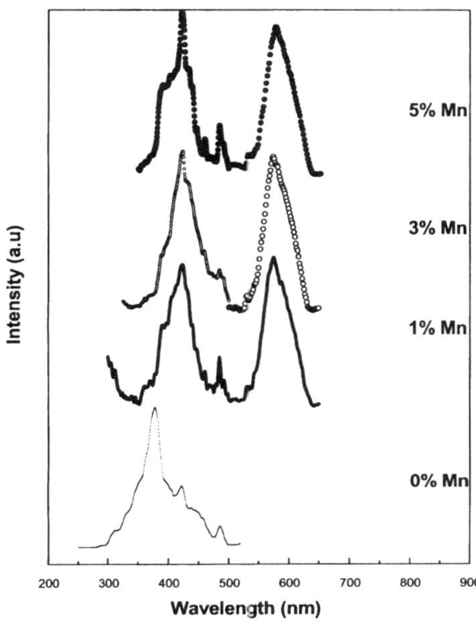

Fig. 5. Photoluminescence spectra of the GaMnN nanowires prepared with SWNTs. The spectrum of the undoped nanowires is also shown. Mn^{2+} emission exhibits a blue-shift compared to the bands of the larger diameter nanowires in Fig. 4.

[19–21]. The intensity of the Mn emission also increases with the Mn content in the case of the 25 nm nanowires.

4. Conclusions

In conclusion, Mn-doped GaN nanowires have been prepared by heating appropriate mixtures of Ga and Mn acetylacetonates along with carbon nanotubes in the presence of NH_3 at 950 °C. The nanowires prepared with single-walled carbon nanotubes are much thinner ($\sim$25 nm) than those prepared with multi-walled carbon nanotubes ($\sim$70 nm). The nanowires are ferromagnetic at room temperature, a feature that may be of value in device technologies. Photoluminescence spectra show a significant blue-shift of the Mn^{2+} emission in the 25 nm nanowires.

References

[1] M. Zajac, R. Doradzinski, J. Gosk, J. Szczytko, M. Lefeld-Sosnowska, M. Kaminska, A. Twardowski, M. Palczewska, E. Grzanka, W. Gebicki, Appl. Phys. Lett. 78 (2001) 1276.

[2] N. Theodoropoulou, A.F. Hebard, M.E. Overberg, C.R. Abernathy, S.J. Pearton, S.N.G. Chu, R.G. Wilson, Appl. Phys. Lett. 78 (2001) 3475.

[3] M.L. Reed, N.A. El-Masry, H.H. Stadelmaier, M.K. Ritums, M.J. Reed, C.A. Parker, J.C. Roberts, S.M. Bedair, Appl. Phys. Lett. 79 (2001) 3473.

[4] W. Han, S. Fan, Q. Li, Y. Hu, Science 277 (1997) 1287.

[5] G.S. Cheng, L.D. Zhang, Y. Zhu, G.T. Fei, L. Li, C.M. Mo, Y.Q. Mao, Appl. Phys. Lett. 75 (1999) 2455.

[6] X. Chen, J. Li, Y. Cao, Y. Lan, H. Li, M. He, C. Wang, Z. Zhang, Z. Qiao, Adv. Mater. 12 (2000) 1432.

[7] C.-C. Chen, C.-C. Yeh, Adv. Mater. 12 (2000) 738.

[8] X. Duan, C.M. Lieber, J. Am. Chem. Soc. 122 (2000) 188.

[9] H.Y. Peng, X.T. Zhou, N. Wang, Y.F. Zheng, L.S. Liao, W.S. Shi, C.S. Lee, S.T. Lee, Chem. Phys. Lett. 327 (2000) 263.

[10] F.L. Deepak, A. Govindaraj, C.N.R. Rao, J. Nanosci. Nanotechnol. 1 (2001) 303.

[11] C.N.R. Rao, R. Seshadri, A. Govindaraj, R. Sen, Mat. Sci. Eng. R. 15 (1995) 209.

[12] C. Journet, W.K. Maser, P. Bernier, A. Loiseau, M. Lamy de la Chapelle, S. Lefrant, P. Denierd, R. Lee, J.E. Fischer, Nature 388 (1997) 756.

[13] G.S. Sudhir, Y. Peyrot, J. Krüger, Y. Kim, R. Klockenbrink, C. Kisielowski, M.D. Rubin, E.R. Weber, W. Kriegseis, B.K. Meyer, Mat. Res. Soc. Symp. 482 (1998) 525.

[14] M.C. Park, K.S. Huh, J.M. Myoung, J.M. Lee, J.Y. Chang, K.I. Lee, S.H. Han, W.Y. Lee, Solid. State. Commun. 124 (2002) 11.

[15] N. Theodoropoulou, K.P. Lee, M.E. Overberg, S.N.G. Chu, A.F. Hebard, C.R. Abernathy, S.J. Pearton, R.G. Wilson, J. Nanosci. Nanotechnol. 1 (2001) 101.

[16] M.E. Overberg, C.R. Abernathy, S.J. Pearton, N. Theodoropoulou, K.T. McCarthy, A.F. Hebard, Appl. Phys. Lett. 79 (2001) 1312.

[17] Y. Shon, Y.H. Kwon, Sh.U. Yuldashev, J.H. Leem, C.S. Park, D.J. Fu, H.J. Kim, T.W. Kang, X.J. Fan, Appl. Phys. Lett. 81 (2002) 1845.

[18] L. Chen, P.J. Klar, W. Heimbrodt, F. Brieler, M. Froba, Appl. Phys. Lett. 76 (2000) 3531.

[19] W. Chen, R. Sammynaiken, Y. Huang, J.-O. Malm, R. Wallenberg, J.-O. Bovin, V. Zwiller, N.A. Kotov, J. Appl. Phys. 11 (2001) 1120.

[20] M.A. Malik, P.O. Brien, N. Revaprasadu, J. Mater. Chem. 11 (2001) 2382.

[21] G. Yi, B. Sun, F. Yang, D. Chen, J. Mater. Chem. 11 (2001) 2928.

VI. Molecular Solids

C.N.R. Rao

CSIR Centre of Excellence in Chemistry,
Chemistry & Physics of Materials Unit
Jawaharlal Nehru Centre for Advanced Scientific Research
Jakkur P.O., Bangalore-560 064, INDIA
cnrrao@jncasr.ac.in

A few decades ago, research on molecular solids was limited to a few specialists and properties of these solids were not considered that attractive. The situation changed when molecular solids with fascinating electronic and magnetic properties, including metallicity and superconductivity began to be discovered. The discovery of fullerenes and the increasing attention being paid to supramolecular architectures have brought about a sea change in this area.[1,2]

The articles in this section mainly deal with supramolecular hydrogen-bonded structures besides hybrid structures and solids with nonlinear optical properties. Carbon structures involving diamond-graphite hybrids and amorphization of fullerenes are also discussed. Among the hydrogen bonded structures, the adduct between trithiocyanuric acid and bipyridyl to form shape-selective channels is noteworthy, just as the rosette structure between cyanuric acid and melamine.

Experimental electronic charge densities of organic crystals obtained from X-ray crystallography are of value in understanding chemical bonds. This is specifically illustrated in the case of organic crystals showing nonlinear optical properties.

References

1. C.N.R. Rao and J. Gopalakrishnan, *New Directions in Solid State Chemistry*, Cambridge University Press, Second Edition, 1997.
2. W. Jones and C.N.R. Rao (Eds.), *Supramolecular Organization and Materials Design*, Cambridge University Press, 2002.

ELSEVIER

Journal of Molecular Structure 436–437 (1997) 11–18

Journal of
MOLECULAR
STRUCTURE

Investigations of diamond–graphite hybrids and fullerenes with seven-membered rings[1]

Rahul Sen, R. Sumathy, B.C. Satishkumar, C.N.R. Rao*

Solid State and Structural Chemistry Unit and Materials Research Centre, Indian Institute of Science, Bangalore 560012, India, and Jawaharlal Nehru Centre for Advanced Scientific Research, Bangalore 560064, India

Received 21 April 1997; accepted 12 May 1997

Abstract

The nature of diamond–graphite hybrids has been studied by molecular mechanics, by examining the structures of species such as $C_{84}H_x$ wherein the sp^3 to sp^2 carbon ratio is varied progressively. The dependence of the average coordination number on the atom fraction of hydrogen has been examined in the light of the random covalent network model. The HOMO–LUMO gap has been estimated in a graphite-like $C_{110}H_x$ and in a diamond-like $C_{120}H_x$ as a function of the sp^3/sp^2 carbon atom ratio. The gap increases exponentially with the fraction of sp^3 carbon. Shapes of fullerene-like structures with 7-membered rings, in addition to 6- and 5-membered rings, have been investigated along with structures of bent nanotubes having similar ring systems. © 1997 Elsevier Science B.V.

Keywords: Diamond–graphite hybrids; Fullerenes; Molecular mechanics

1. Introduction

There has been increasing interest in carbon structures in recent years, particularly because of the discovery of fullerenes and related forms of carbon containing five- and six-membered rings, with the hybridization of carbon between sp^2 and sp^3 and with carbon–carbon bond distances between those of single and double bonds [1]. Another important aspect of carbon relates to hybrid structures between graphite and diamond, involving mixtures of sp^2 and sp^3 hybridized carbon atoms [2–4]. In addition, several forms of carbon structures have been proposed recently on the basis of theoretical considerations. Thus, on the basis of topological considerations, schwarzites involving 5-, 6-, 7-, 8- and 9-membered rings have been suggested by Townsend et al. [5]. Hoffmann et al. [6] have proposed a possible new form of conducting carbon. We have studied the stability of structures of diamond–graphite hybrids with different sp^2/sp^3 atom ratios by employing the force field method. In addition to visualizing the hybrid structures, we have tried to understand their structures in terms of the dependence of the average coordination number as well as the sp^3/sp^2 atom ratio on the hydrogen atom fraction, x_H, the hydrogen atoms being present on the surface and the edges of diamond-type structures. Some of the amorphous carbons are known to contain 20–60% hydrogen, the percentage varying with the sp^3/sp^2 ratio [7,8].

---

* Corresponding author.

[1] Dedicated to Professor Henryk Ratajczak, a dear friend and an outstanding structural chemist.

R. Sen et al./Journal of Molecular Structure 436–437 (1997) 11–18

We have made use of the random covalent network model [3] to examine the structural changes with x_H. For this purpose, we have started with graphitic-type sheets containing only sp^2 carbons and examined the effect of progressively introducing sp^3 carbons. We have also studied the effect of introducing sp^2 carbons in diamond-like sp^3 networks.

It is of importance to know whether diamond–graphite hybrids exhibit a range of band gaps between those of graphite and diamond. Balaban et al. [9] have theoretically examined diamond layers connected by graphite strips and have predicted that the band gap should decrease with increase in the strip width. Band gap estimates for diamond–graphite hybrid structures with different fractions of sp^3 carbons which take into account several possible alternatives are, however, not available. We have carried out a systematic study of the HOMO–LUMO gaps in a variety of diamond–graphite hybrids, with varying sp^3/sp^2 ratios, starting with a pure sp^2 layer at the graphite end and a three dimensional sp^3 network at the diamond end.

Besides diamond–graphite hybrid structures, we have investigated fullerenes and nanotubes containing seven-membered rings by employing force-field calculations. Fullerenes containing five-, six- and seven-membered rings provide synthetic challenges and we describe some of the simplest fullerenes of this category. The presence of seven-membered rings in nanotubes, giving rise to bent nanotubes, has indeed been observed [10]. We have compared force-field minimized structures of such namotubes with observed transmission electron microscopic (TEM) and atomic force microscopic (AFM) images.

2. Method of study

In order to visualize the alternative structures of different hydrocarbons of the general formula $C_{84}H_x$, $C_{110}H_x$ and $C_{120}H_x$, we have carried out energy minimization using the molecular simulation program DISCOVER (Biosym Technologies, San Diego) and obtained the actual structures by using Insight II. This molecular mechanics program solves for the electronic wavefunction by using the force field approach and focuses on obtaining the geometries and their static energies. The quasi-Newton–Raphson (VA09A) algorithm coupled with

the Broyden–Fletcher–Goldfarb and Shanno (BFGS) method for Hessian updating is used for the minimization. The force field calculations are much faster than ab initio or semiempirical quantum chemical methods and describe the structures and energies of non-polar systems fairly satisfactorily [11,12]. We have not given much credence to the actual energy values of the different hydrocarbon structures, but have used the relative energies of alternative structures of the same system to identify the likely stable structure.

We have calculated the average coordination number, m, in structures such as those in Fig. 1, from Eq. (1) by employing the relation

$$m = (3 - 2x_H) + (4 - 3x_H)(N_{sp^3}/N_{sp^2})/\left(1 + (N_{sp^3}/N_{sp^2})\right) \tag{1}$$

where (N_{sp^3}/N_{sp^2}) is the ratio of sp^3 to sp^2 hybridized carbon atoms. We compare the m values so obtained with those predicted by Eq. (2).

$$m = (20 - 8x_H)/7 \tag{2}$$

While Eq. (1) gives the observed m value by simple counting, Eq. (2) of Angus and Jansen [3] gives the m value obtained by applying constraint theory to the random covalent network. A random covalent network is completely constraint when the number of constraints per atom is equal to the number of mechanical degrees of freedom per atom [13,14], a completely constraint network being one where the stabilization due to the bond energy is just balanced

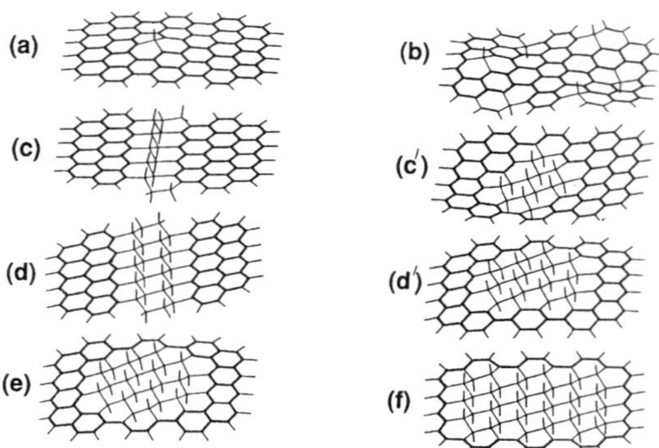

Fig. 1. Some possible structures of the general composition $C_{84}H_x$: (a) $C_{84}H_{25}$ (1.2% sp^3 carbons); (b) $C_{84}H_{30}$ (7.1% sp^3 carbons); (c) and (c') $C_{84}H_{37}$ (15.5% sp^3 carbons); (d) and (d') $C_{84}H_{46}$ (26.2% sp^3 carbons).

R. Sen et al./Journal of Molecular Structure 436–437 (1997) 11–18

by the destabilization due to the strain energy. The number of constraints of an atom is the number of directed covalent bonds with its nearest neighbours, in other words its coordination number. Increasing the coordination number (covalent bonds) of an atom contributes to stabilization due to bond energy, but also causes destabilization due to strain energy. The optimal coordination number is one that just uses up the allowed degrees of freedom to balance these effects. The number of degrees of freedom is equal to the dimensionality (three). Thus, the coordination number of an atom or the average coordination number of the system can be derived using the constraint theory.

In order to study the effect of incorporating sp^3 (or sp^2) carbons in an sp^2 (or sp^3) network, on the band (HOMO–LUMO) gap, we have started with large hydrocarbons containing either only sp^2 or sp^3 carbons and estimated the gaps by using extended Huckel theory (EHT) [15]. It may be noted that, for a series of related systems, the EHT method provides estimates of HOMO–LUMO gaps which follow the trend in experimental values [15,16]. The atomic coordinates for the structures were obtained from force field minimization by using the molecular simulation program DISCOVER described earlier. Extended Huckel calculations were performed by using the ICON version 8 (by R. Hoffmann) and were run in a VAX-88 machine. The program uses parametrized basis sets consisting of H 1s and C 2s and 2p orbitals for hydrocarbon molecules.

Fullerene-type structures having pentagons, hexagons and heptagons were built following Euler's theorem:

$$v - e + f = k \tag{3}$$

where v, e and f are the number of vertices, edges and faces of a polyhedral object, and k is the Euler characteristic and is related to the number of holes, g, according to the relation $k = 2(1 - g)$. g equals zero for closed-cage structures like fullerenes having no holes. In fullerenes three edges extend from each vertex, hence

$$3v = 2e \tag{4}$$

If there are f_n faces of n-membered rings, then

$$f = \Sigma f_n \tag{5}$$

and

$$2e = \sum_{nf_n} \tag{6}$$

Eqs. (3)–(6) give rise to

$$\Sigma (6 - n)f_n = 6k$$

for fullerenes containing pentagons, hexagons and heptagons,

$$f_5 - f_7 = 12$$

We can therefore build closed-cage fullerene structures when the number of pentagons minus the number of heptagons equals twelve. All nanotube and fullerene-like structures were built and energy minimized using the force-field approach in the molecular simulation program DISCOVER.

3. Results and discussion

3.1. Structural studies of diamond–graphite hybrids

We have first studied the effect of increasing the number of sp^3 carbons in a graphite-like system with the general formula $C^{84}H_x$. In Fig. 1 we show typical structures of hydrogenated carbons with the composition varying anywhere between $C_{84}H_{25}$ and $C_{84}H_{66}$ wherein, with increasing proportion of sp^3 carbons, the fraction of hydrogen atoms, x_H, also increases. Thus, $C_{84}H_{25}$ has one sp^3 carbon, while $C_{84}H_{66}$ has 42 sp^3 carbons. In some of the compositions, especially those with small x_H, we have considered alternative structures. For example, in $C_{84}H_{30}$ with six sp^3 carbons we have considered three alternative structures, the structure shown in Fig. 1b being the most stable. The structures (c) and (c′) as well as (d) and (d′) in Fig. 1 (containing 13 and 22 sp^3 carbons respectively) are alternative structures. With the increase in x_H or the number of sp^3 carbons, the number of alternative structures with random arrangements decreases significantly.

Diamond-like carbons (DLCs) generally contain 10–25% sp^2 carbons [2]. Although there has been some speculation on the structures of DLCs, very few studies have been carried out on them. We have

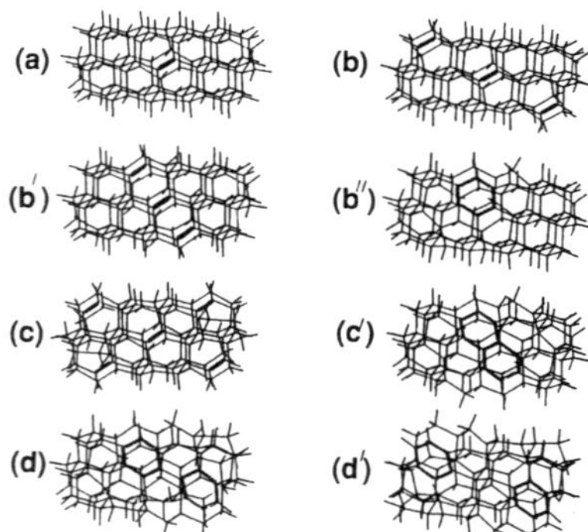

Fig. 2. Some possible structures of diamond-like carbons of composition $C_{84}H_x$: (a) $C_{84}H_{76}$ (2.4% sp^2 carbons); (b), (b′) and (b″) $C_{84}H_{80}$ (7.1% sp^2 carbons); (c) and (c′) $C_{84}H_{84}$ (11.9% sp^2 carbons); (d) and (d′) $C_{84}H_{86}$ (14.3% sp^2 carbons).

considered a large number of structures with the general formula $C_{84}H_x$, with the percentage of sp^2 carbons varying in the range of 2–30%. In Fig. 2, we show a few typical structures with varying numbers of C=C bonds. Among the structures containing three double bonds (Fig. 2(b), (b′) and (b″)), 2(b″) with a benzene ring is considerably more stable (by 440–560 kJ) than the structures with isolated double bonds. Similarly, the structure with a naphthalene ring (Fig. 2(c′)) is more stable (by 775 kJ) than the structure with five isolated double bonds (Fig. 2(c)). Between the structures containing two benzene rings, the structure 2(d′), where the two benzene rings are apart, is more stable (by 548 kJ) than where they form a biphenyl unit (Fig. 2(d)). On the basis of band structure calculations, Robertson [2] has proposed that sp^2 carbons in DLCs tend to cluster together to form single aromatic rings or very small aromatic clusters. Structures (b)–(d) with a sufficient proportion of sp^2 carbons could be models for these amorphous carbons.

In Fig. 3 we have plotted the values of the average coordination number, m, obtained from Eq. (1) for the various $C_{84}H_x$ structures. We have also shown the theoretical line from Eq. (2) in the figure. We see that the m values from Eq. (1) for graphite-like structures shown in Fig. 1 (represented by circles in Fig. 3) vary linearly with x_H, but the slope is smaller than that of the theoretical line predicted by Eq. (2). In the structures considered here, x_H is increased by replacing one 3-coordinate carbon atom (sp^2 carbon)

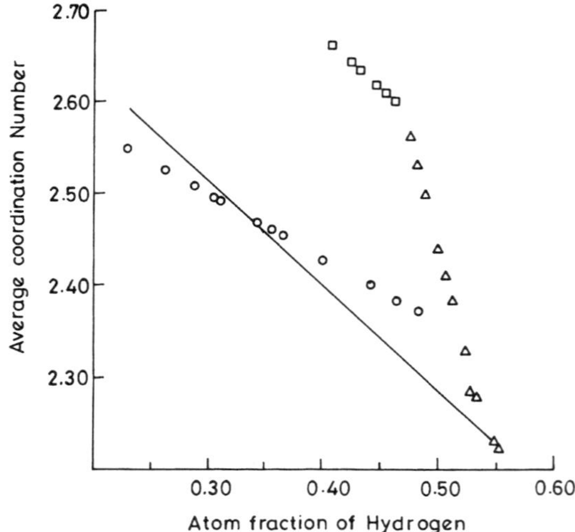

Fig. 3. A plot of the average coordination number, m, against the hydrogen atom fraction, x_H, for diamond–graphite hybrids: circles, $C_{84}H_x$ structures of Fig. 1; triangles, $C_{84}H_x$ structures of Fig. 2; squares, $C_{84}H_x$ structures obtained by scheme (b) in Fig. 4. The straight line is from Eq. (2).

by one 4-coordinate carbon (sp^3 carbon) and one 1-coordinate hydrogen. While this decreases m with increase in x_H, the decrease is not as much as is expected from Eq. (2). The m values from Eq. (1), for these structures, fall below the theoretical line up to $x_H = 0.35$ and above the theoretical line when $x_H > 0.35$. The composition with $x_H \simeq 0.35$ with about 30% sp^3 carbons is close to that of amorphous carbon, whose structure could indeed be similar to that of $C_{84}H_{46}$ (Fig. 1(d′)).

We have also plotted the m values from Eq. (1) against x_H for various diamond-like structures. The m values of the structures shown in Fig. 2 (represented by triangles in Fig. 3) are all higher than the values from Eq. (2). In the structures of Fig. 2, sp^2 carbons are created from sp^3 carbons by scheme (a) of Fig. 4, wherein the creation of sp^2 carbons is accompanied by the formation of C–H bonds. The sharp decrease in m with increase in x_H happens because we are replacing one 4-coordinate atom with one 2-coordinate atom and one 1-coordinate atom.

We have considered other ways of generating sp^2 carbons from sp^3 carbons, as shown in schemes (b) and (c) of Fig. 4. Scheme (b) involves a decrease in x_H with increase in the number of sp^2 carbons. The m values of such structures (shown by squares in

R. Sen et al./Journal of Molecular Structure 436–437 (1997) 11–18

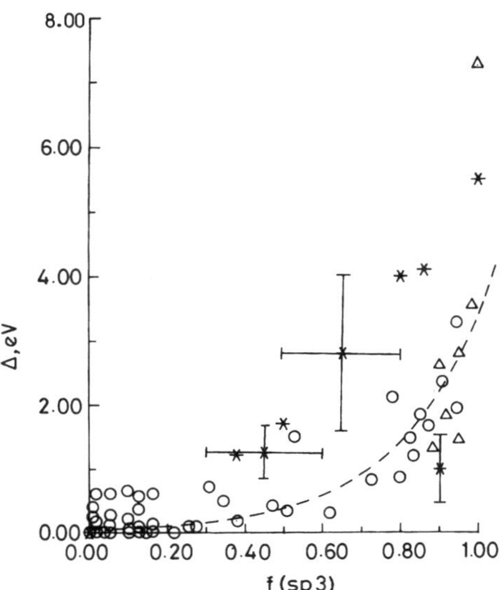

Fig. 5. Plot of the HOMO–LUMO gap, Δ, in $C_{110}H_x$ compositions (circles) against the fraction of sp^3 carbons, $f(sp^3)$. Triangles represent data on $C_{120}H_x$ compositions. Experimental values are shown by stars.

Fig. 4. Possible schemes for the transformation of sp^3 carbons to sp^2 carbons.

Fig. 3) fall well above the line from Eq. (2), but the slope is about the same. Thus, if we start with an sp^3 network and introduce hydrogens by breaking the C–C bonds, so that the initial m value falls somewhere near the theoretical, and then introduce sp^2 carbons by scheme (b), we may obtain m values close to the line predicted by Eq. (2). This is not possible with 84 carbon atoms because the connectivity of the network breaks down at high hydrogen content. Scheme (c) in Fig. 4 does not allow a change in x_H and is not relevant. Clearly, Eq. (2) fails in diamond-like structures with high proportions of sp^3 carbons, unless some of the C–C bonds are broken to introduce hydrogens. Eq. (2) appears to be more applicable to carbon networks with high proportions of sp^2 carbons.

3.2. Band gaps in diamond–graphite hybrids

Initial calculations on several hydrocarbon structures showed that the aromatic hydrocarbon

$C_{110}H_{30}$ containing only sp^2 carbons has a gap of 0.0009 eV and provides a good model for a graphitic sheet. We have progressively increased the number of sp^3 carbons in this network, taking into account the different possible structures of such hybrids. As we progressively increase the number of sp^3 carbons, we finally obtain the structure with only sp^3 carbons and with a high value of the gap ($\sim$9 eV). In Fig. 5 we have plotted the HOMO–LUMO gap, Δ, against the fraction of sp^3 carbons in the $C_{110}H_x$ system. (Note that the hydrogen content will also increase with the increase in the fraction of sp^3 carbons.) We see from Fig. 5 that the gap is in the 0.0–0.8 eV range up to an sp^3 fraction of $\sim$0.4 and increases significantly when the sp^3 fraction is increased further. Structures with a high sp^3 fraction have gaps in the range 1.0–4.0 eV. When the sp^3 fraction is less than 0.4, we find a range of gaps (0.0–0.6 eV) for different structures with the same fraction of the sp^3 carbons because of the different ways of locating the sp^3 carbons in the sp^2 network. Such differences give rise to changes in the gap, but the changes are relatively small. When the sp^3 carbons cluster together to form cyclohexane, decalin or the saturated forms of pyrene and coronene in the sp^2 layer, the gap becomes very small

16   *R. Sen et al./Journal of Molecular Structure 436–437 (1997) 11–18*

($\sim 10^{-3}$ eV). Structures with a high sp^3 fraction (0.7–0.8) but low band gap (0.8 eV) are those where the sp^2 carbons form strips of fused benzene rings. Significant changes in the gap come about when we increase the number of sp^3 carbons not merely at the edges but over the entire network.

The HOMO–LUMO gaps, Δ, in the hybrids of the general formula $C_{110}H_x$ are described satisfactorily by the expression

$$\Delta \text{ (in eV)} = 0.04 \exp(4.34f)$$

where f is the fraction of sp^3 carbons. The broken curve in Fig. 5 is that predicted by the above expression. Experimental values of the optical gaps reported in the literature [8] also show an exponential increase with the fraction of sp^3 carbons, although they tend to be somewhat higher than those predicted by the above expression.

The variation in the HOMO–LUMO gap in a diamond network with progressive increase in the number of sp^2 carbons is also of interest. The parent diamond network, $C_{120}H_{92}$, has a gap of 7.22 eV. Although the experimental optical gap for diamond is 5.5 eV, theoretical values of 7–10 eV are considered to be good estimates [17]. We have calculated the gap of the $C_{120}H_x$ system by increasing the number of sp^2 carbons in the form of double bonds or aromatic rings as in diamond-like carbon. Thus $C_{120}H_{98}$, which has a benzene ring in the sp^3 network, has a gap of 2.81 eV. The gap in pure benzene is, however, calculated to be 4.5 eV. Similarly, the gaps found in the structures containing isolated naphthalene and anthracene units are 1.84 eV and 1.34 eV respectively, which are much lower than the gaps calculated for the pure hydrocarbons (2.8 eV and 1.8 eV respectively). The band gaps of diamond-like $C_{120}H_x$ structures with varying sp^2 content are shown in the Δ vs $f(sp^3)$ plot in Fig. 5 (represented by triangles) to demonstrate that they are described reasonably well by the same expression. The experimental values of the gap, with their uncertainties, are also shown in Fig. 5 to illustrate that they are not far from the predicted trend.

The present study demonstrates how the HOMO–LUMO gap in diamond–graphite hybrids is a sensitive function of the sp^3/sp^2 ratio as well as the structure of the hybrid. The marked variation in the gap when $f(sp^3)$ is between 0.4 and 1.0 is noteworthy.

3.3. Fullerene-like structures with 7-membered rings

Electron microscopic investigation of the cathodic deposits obtained during arc vaporization of graphite has revealed bent nanotubes [10] and that such bent nanotubes can occur due to the presence of heptagons (seven-membered rings).

Seven-membered rings give rise to negative curvature on a flat sheet of hexagons, whereas five-membered rings give rise to positive curvature. A heptagon/pentagon pair on opposite sides of a nanotube can thus cause the tube to bend. We have simulated such a bent tube with seven- and five-membered rings in Fig. 6a. In Fig. 6b we show a TEM image which may correspond to such bent

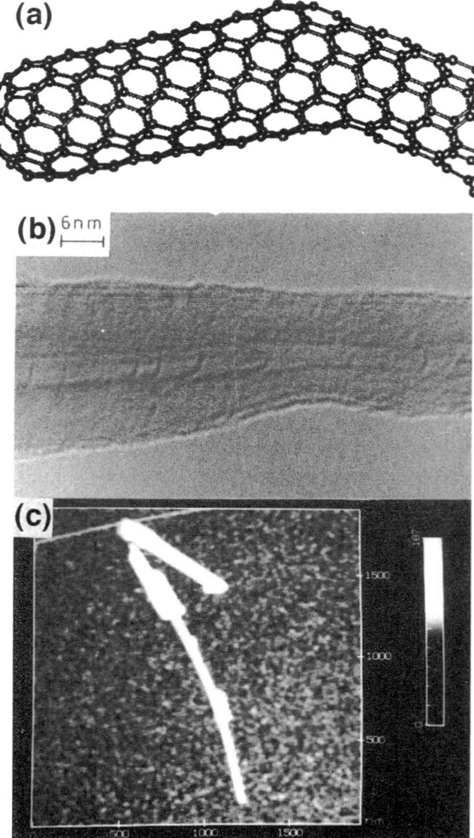

Fig. 6. (a) Structure of a bent nanotube containing seven- and five-membered rings; (b) TEM image of a nanotube with a negative curvature; (c) AFM image of a bent nanotube.

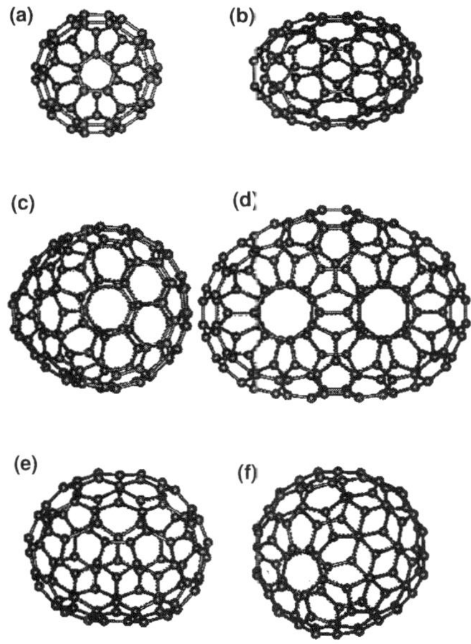

Fig. 7. Small fullerene-like structures having varying numbers of pentagons and heptagons.

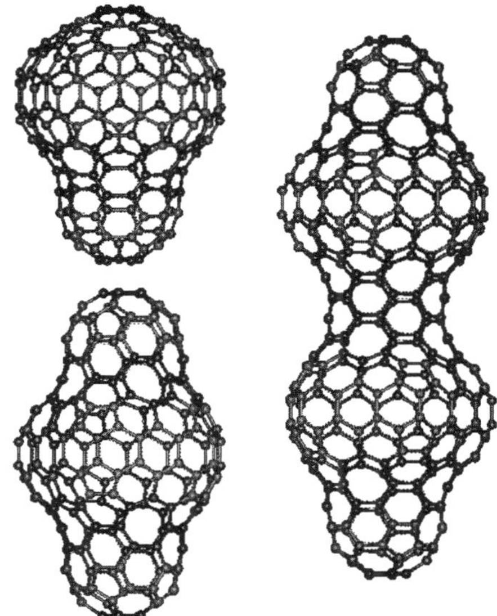

Fig. 8. Fullerene-like and tube-like structures with 5- and 7-membered rings showing unusual shapes.

tubes. Fig. 6c gives an AFM image of a bent nanotube. Seven-membered rings can, in principle, occur in closed-cage structures which are fullerene-like.

In Fig. 7 we show a few of the small fullerene-like structures containing varying numbers of pentagons and heptagons, obtained by energy minimization. Fig. 7(a) represents C_{60} with twelve pentagons and zero heptagons. Fig. 7(b) shows C_{84} containing two heptagons and fourteen pentagons, this structure being highly symmetrical with the two heptagons present in the opposite ends. This would be one of the simplest of such fullerene-like molecules with 7-membered rings that chemists may one day synthesize. Fig. 7(c) shows C_{120} with one heptagon and thirteen pentagons. Although the numbers of heptagons and pentagons are small in this case, the odd numbers make the structure unsymmetrical, requiring a larger number of hexagons to close the structure. Fig. 7(d) is that of C_{144} with four heptagons and sixteen pentagons. In the structures shown in Figs. 7(b)–(d), the heptagons and pentagons are isolated, but structures can also be built where pentagon and heptagon occur in pairs. A structure of C_{102} with a pentagon–heptagon pair is shown in Fig. 7(e), while a structure of C_{98} with a

pentagon–heptagon–pentagon triplet is given in Fig. 7(f). We can have unusual shapes with various combinations of pentagons, hexagons and heptagons. In Fig. 8 we show fullerene-like or tube-like structures with unusual shapes.

References

[1] H.W. Kroto, C.N.R. Rao, E. Osawa (Eds.), MRS Bull., special issue (1994) 21.

[2] J. Robertson, Pure Appl. Chem. 66 (1994) 1789.

[3] J.C. Angus, F. Jansen, J. Vac. Sci. Technol. A6 (1988) 1776.

[4] R. Sen, R. Sumathy, C.N.R. Rao, J. Mater. Res. 10 (1995) 2531.

[5] S.J. Townsend, T.J. Lenosky, D.A. Muller, C.S. Nichols, V. Elser, Phys. Rev. Lett. 69 (1992) 921.

[6] R. Hoffmann, T. Hughbanks, M. Kertez, J. Am. Chem. Soc. 105 (1983) 4831.

[7] J.C. Angus, P. Koidl, S. Domitz, in J. Mort and F. Jansen (Eds.), Plasma Deposited Thin Films, CRC, Boca Raton, FL, 1984.

[8] S. Kaplan, S. Jansen, M. Machonkin, Appl. Phys. Lett. 47 (1985) 750.

[9] A.T. Balaban, D.J. Klein, C.A. Folden, Chem. Phys. Lett. 217 (1994) 266.

[10] S. Iijima, T. Ichihashi, Y. Ando, Nature 356 (1992) 776.

[11] E.M. Engler, J.D. Andose, P.v.R. Schleyer, J. Am. Chem. Soc. 95 (1973) 8005.

[12] T. Clark, in: A Handbook of Computational Chemistry, Wiley Interscience, 1985.

[13] J.C. Phillips, J. Non-Cryst. Solids 34 (1979) 153.

[14] M.F. Thorpe, J. Non-Cryst. Solids 57 (1983) 355.

[15] R. Hoffmann, J. Chem. Phys. 39 (1963) 1397.

[16] R. Hoffmann, C. Janiak, C. Kollmar, Macromolecules 24 (1991) 3725.

[17] H.R. Karfunkel, T. Dressler, J. Am. Chem. Soc. 114 (1992) 2285.

16814 *J. Phys. Chem.* **1995**, *99*, 16814–16816

Polymerization and Pressure-Induced Amorphization of C₆₀ and C₇₀†

C. N. R. Rao,* A. Govindaraj, Hemantkumar N. Aiyer, and Ram Seshadri

CSIR Centre of Excellence in Chemistry, Solid State & Structural Chemistry Unit and Materials Research Centre, Indian Institute of Science, Bangalore 560 012, India

Received: July 13, 1995; In Final Form: September 18, 1995⊗

Scanning tunneling microscopy of solid films of C_{60} and C_{70} clearly demonstrate the occurrence of photochemical polymerization of these fullerenes in the solid state. X-ray diffraction studies show that such a polymerization is accompanied by contraction of the unit-cell volume in the case of C_{60} and expansion in the case of C_{70}. This is also evidenced from the STM images. These observations help to understand the differences in the amorphization behavior of C_{60} and C_{70} under pressure. Amorphization of C_{60} under pressure is irreversible because it is accompanied by polymerization associated with a contraction of the unit cell volume. Monte Carlo simulations show how pressure-induced polymerization is favored in C_{60} because of proper orientation as well as the required proximity of the molecules. Amorphization of C_{70}, on the other hand, is reversible because C_{70} is less compressible and polymerization is not favored under pressure.

It has been reported recently that C_{60} undergoes polymerization in the solid state involving 2+2 cycloaddition induced photochemically[1] or at high pressures and temperatures.[2] C_{60} is also known to undergo irreversible amorphization at high pressures, and the irreversibility has been attributed to the formation of cycloaddition products.[3] C_{70} seems to undergo photopolymerization in the solid state involving 2+2 cycloaddition,[4] but unlike C_{60}, it undergoes reversible amorphization at high pressures.[5] We have been interested in understanding the polymerization of C_{60} and C_{70} in the solid state[6] and its relation to the nature of amorphization of these fullerenes. For this purpose, we have investigated photopolymerization of solid films of the two fullerenes by scanning tunneling microscopy (STM) and X-ray diffraction and carried out certain simulation studies on the effects of pressure on these solids. The study has indeed revealed the subtle differences between C_{60} and C_{70} with regard to features of their polymerization and amorphization.

Photopolymerization of solid films of C_{60} and C_{70} was studied by recording the STM images before and after UV irradiation. Typical STM images of C_{60} and C_{70} shown in Figure 1 bear clear evidence for the occurrence of photopolymerization. Photopolymerization of C_{60} (Figure 1a) does not change the symmetry of the top surface. The internal features of the molecules become visible after photopolymerization because of the complete freezing of molecular motion.[7] Internal features become visible after photopolymerization of C_{70} as well (Figure 1b), but the lattice symmetry of the top layer appears to change from fcc(111)-like to fcc(100)-like. STM results indicate that in C_{60}, a lattice contraction of $\approx13\%$ occurs on polymerization while in C_{70}, an expansion of $\approx10\%$ is observed. X-ray diffraction measurements on C_{60} and C_{70} films confirm similar changes in the unit-cell volume. Accordingly, we see from Figure 2 that the 2θ values decrease on photopolymerization of C_{70} while the opposite holds in the case of C_{60}. The changes in the unit cell volume can be understood by means of the model for the dimerization of C_{70} and C_{60} shown in Figure 2. The volume expansion in C_{70} arises due to the ellipsoidal shape of

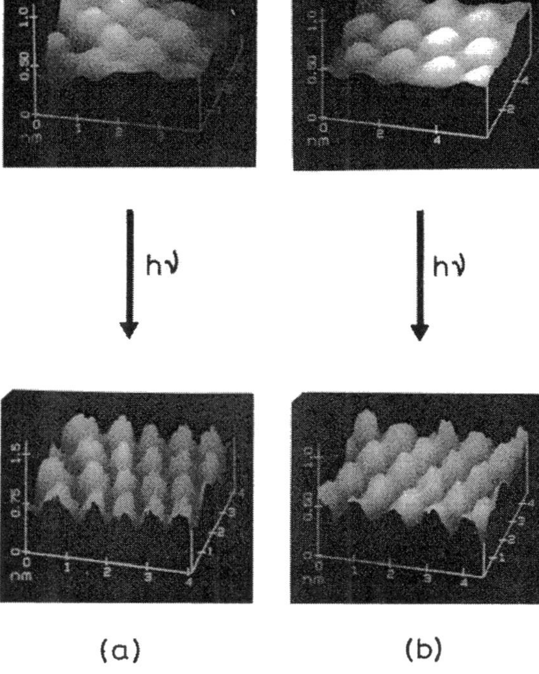

Figure 1. (a) STM images of C_{60} films on highly oriented pyrolytic graphite (HOPG) before and after UV irradiation. (b) STM images of C_{70} films on HOPG before and after UV irradiation. The conditions were 254 nm UV, 12 h, 10^{-5} Torr vacuum.

the molecule as well as the high reactivity of the bonds radial to the capping pentagons toward 2+2 addition, i.e., the pentagons defining the 5-fold axis.[8] Since the molecules do not have to reorient for the polymerization of C_{60} to occur, the only effect is due to intermolecular bonding resulting in the observed contraction. The observation of unit-cell expansion in the case of C_{70} and contraction in the case of C_{60} is indeed noteworthy and has direct bearing on the amorphization of these fullerenes.

† Supported by the Jawaharlal Nehru Centre for Advanced Scientific Research.
* Correspondence to be addressed at the CSIR Centre of Excellence in Chemistry, Indian Institute of Science, Bangalore 560 012, India.
⊗ Abstract published in *Advance ACS Abstracts*, October 15, 1995.

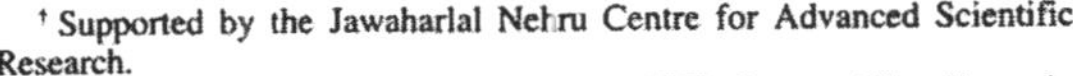

Letters

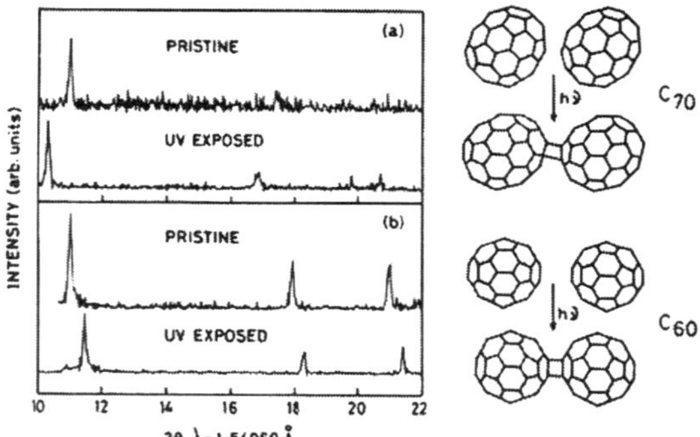

Figure 2. XRD patterns of (a) C_{70} and (b) C_{60} on glass substrates before and after UV irradiation. Whereas the C_{60} lattice contracts on photopolymerization, for C_{70}, the lattice is seen to expand. The 2+2 addition of C_{60} and C_{70} (resulting in photopolymerized products) is shown schematically.

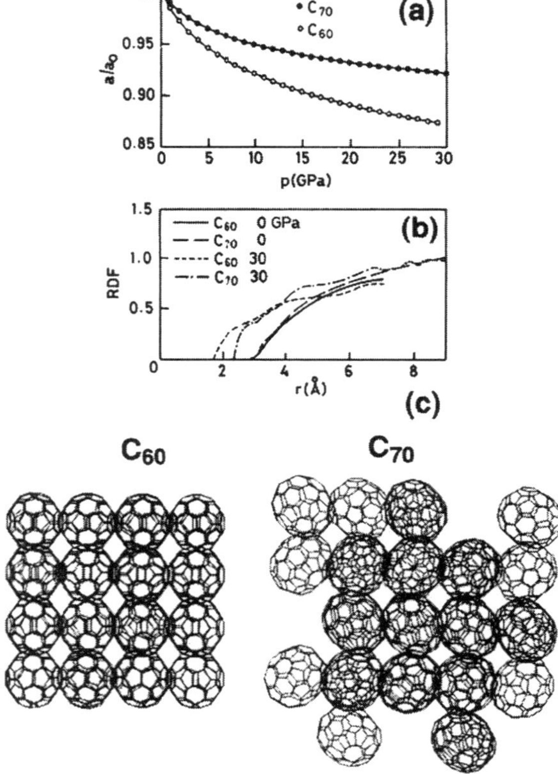

Figure 3. (a) Pressure dependence of the ratio of the pseudocubic lattice parameter a of C_{60} and C_{70} to the parameter a_0 at ambient pressure as given by the Murnaghan equation of state. Experimental compressibilities were employed to produce these plots. It is assumed that there are no intervening phase changes and that C_{60} remains cubic and C_{70}, rhombohedral throughout. (b) Atom–atom radial distribution functions of C_{60} and C_{70} molecules at 0 and 30 GPa and 300 K from NVT Monte Carlo studies. The molecules of C_{60} approach closer at a given pressure than do C_{70}. (c) Snapshots of the C_{60} lattice at 20 GPa and C_{70} lattice at 25 GPa obtained from NVT-MC simulations, looking down the $\langle 100 \rangle$ direction.

What is also interesting about C_{60} and C_{70} is that their compressibilities are quite different.[9] The ratios of the pseudocubic lattice parameters of C_{60} and C_{70} with respect to their corresponding room-pressure values are shown as a function

of pressure in Figure 3a. These plots were made using the Murnaghan equation of state and experimental compressibilities. We see that solid C_{60} is more compressible than C_{70}. Our NVT Monte Carlo studies[10] also throw light on the nature of changes brought about on application of pressure. Among the various results obtained from such studies, the one most relevant to the context of polymerization under pressure is the following: On application of pressure, the C_{60} molecules approach one another much more closely than C_{70} molecules. The calculated closest approach between two C_{60} molecules at 300 K and 30 GPa is ≈1.8 Å, while it is ≈2.2 Å in C_{70} (Figure 3b). Although the C_{60} molecules are randomly oriented in the solid up to moderate pressures, the molecules align themselves at higher pressures providing so as to be favorable for the 2+2 addition to occur. In the case of C_{70} on the other hand, the alignment of the molecules under pressure occurs with their long axes along the face and body diagonals of the unit cell so as to render the reactive end caps of the molecule to be relatively farther apart. Snapshots of the simulated C_{60} and C_{70} lattices under pressure are shown in Figure 3c.

Clearly, polymerization under pressure is favored in the case of C_{60} but not in C_{70}. Since the polymerization is also accompanied by volume contraction in the case of C_{60}, pressure would indeed favor such a process. This is not so in C_{70}, where polymerization is accompanied by volume expansion. On the basis of these observations, we are able to understand why the amorphization of C_{60} under pressure is irreversible because of polymerization.

References and Notes

(1) (a) Rao, A. M.; Zhou, P.; Wang, K. A.; Hager, G. T.; Holden, J. M.; Wang, Y.; Lee, W. T.; Bi, X. X.; Ecklund, P. C.; Cornett, D. S.; Duncan, M. A.; Amster, I. J. *Science* **1993**, *259*, 955. (b) Weaver, J. H.; Poirier, D. M. *Solid State Phys.* **1994**, *48*, 1. (c) Zhou, P.; Rao, A. M.; Wang, K. A.; Robertson, J. D.; Eloi, C.; Meier, M. S.; Rin, S. L.; Bi, X. X.; Ecklund, P. C.; Dresselhaus, M. S. *Appl. Phys. Lett.* **1992**, *60*, 2871. (d) Bacsa, W. S.; Lannin, J. S. *Phys. Rev.* **1994**, *B49* 14750. (e) Menon, M.; Subbaswamy, K. R. *Phys. Rev.* **1994**, *B49* 13966.

(2) (a) Nunez-Reguero, M.; Marques, L.; Hodeau, J. L.; Bethoux, O.; Perroux, M. *Phys. Rev. Lett.* **1995**, *74*, 278. (b) Xu, C. H.; Scuseria, G. E. *Phys. Rev. Lett* **1995**, *74*, 274.

(3) Rao, A. M.; Menon, M.; Wang, K. A.; Ecklund, P. C.; Subbaswamy, K. R.; Cornett, D. S.; Duncan, M. A.; Amster, I. J. *Chem. Phys. Lett.* **1994**, *224*, 106.

(4) (a) Yoo, C. S.; Nellis, W. J. *Chem. Phys. Lett.* **1992**, *198*, 379. (b) Moshary, F.; Chen, N. H.; Silvera, I. F.; Brown, C. A.; Dorn, H. C.; deVries, M. S.; Bethune, D. S. *Phys. Rev. Lett.* **1992**, *69*, 466. (c) Snoke, D. W.; Syassen, K.; Mittelbach, A. *Phys. Rev.* **1993**, *B47*, 4146.

(5) Chandrabhas, N.; Sood, A. K.; Muthu, D. V. S.; Sundar, C. S.; Bharathi, A.; Hariharan, Y.; Rao, C. N. R. *Phys. Rev. Lett.* **1994**, *73*, 3411.

(6) Since 2+2 cycloaddition is symmetry allowed under photochemical conditions but not under thermal conditions, under the application of pressure, it must take place by routes other than concerted, for example, via polar intermediates.

(7) Both C_{60} and C_{70} are orientationally disordered in the solid state at ordinary temperatures. Moreover, the disorder is dynamic. This prevents the probing of intramolecular features by techniques such as STM. However, when the fullerenes are deposited as films on surfaces such as single crystal metals, ordered structures can be obtained because the molecules are attached to the surface relatively strongly because of charge transfer from the metal to the fullerenes. When the molecules polymerize, they cannot rotate freely and the internal structure becomes visible. For details see: (a) Wang, X. D.; Hashizume, T.; Sakurai, T. *Mod. Phys. Lett.* **1994**, *8*, 1397. (b) Aiyer, H. N.; Govindaraj, A.; Rao, C. N. R. *Bull. Mater. Sci.* **1994**, *17*, 563. (c) Joachim, C.; Gimzewski, J.; Schittler, R.; Chavy, C. *Phys. Rev. Lett.* **1995**, *74*, 2102. (d) Aiyer, H. N.; Govindaraj, A.; Rao, C. N. R. *Philos. Mag. Lett.*, in print.

(8) (a) Henderson, C. C.; Cahill, P. A. *Science* **1994**, *263*, 397. (b) Karfunkel, H. R.; Hrisch, A. *Angew. Chem., Int. Ed. Engl.* **1992**, *31*, 1468. (c) Rathna, A.; Chandrasekhar, J. *Curr. Sci. (India)* **1993**, *65*, 768. (d) Govindaraj, A.; Rathna, A.; Chandrasekhar, J.; Rao, C. N. R. *Proc. Indian Acad. Sci. (Chem. Sci.)* **1993**, *105*, 303.

(9) (a) Duclos, S. J.; Brister, K.; Haddon, R. C.; Kortan, A. R.; Thiel, F. A. *Nature* **1991**, *351*, 380. (b) Christides, C.; Thomas, I. M.; Dennis, T. J. S.; Prassides, K. *Europhys. Lett.* **1993**, *22*, 545.

16816 *J. Phys. Chem., Vol. 99, No. 46, 1995*

Letters

(10) (a) NVT Monte Carlo simulations were performed on cells containing 32 rigid C_{60} or C_{70} molecules. The intermolecular potential was of the Lennard-Jones form. The unit cells for the simulation were taken from experiments referred to in the text. Visualization was done using the InsightII software from BIOSYM Technologies, San Diego on Indigoll workstations. For details of the simulation as well as the potential parameters used, see: (a) Cheng, A.; Klein, M. L. *J. Phys. Chem.* **1991**, *95*, 6750. (b) Girifalco, L. A. *J. Phys. Chem.* **1992**, *98*, 858. (c) Chakrabarti, A.; Yashonath, S.; Rao, C. N. R. *Chem. Phys. Lett.* **1993**, *215*, 519.

JP951970P

MOLECULAR PHYSICS, 1996, VOL. 89, No. 1, 267–277

A combined experimental and theoretical study of the charge-transfer compound between $C_{60}Br_8$ and tetrathiafulvalene

By C. N. R. RAO, A. GOVINDARAJ, R. SUMATHY and A. K. SOOD

CSIR Centre of Excellence in Chemistry and Department of Physics, Indian Institute of Science, Bangalore 560 012, India and Jawaharlal Nehru Centre for Advanced Scientific Research, Jakkur, Bangalore 560 064, India

(*Received 31 January 1996; accepted 12 February 1996*)

$C_{60}Br_8$, unlike $C_{60}Br_6$ and $C_{60}Cl_6$, forms a solid charge-transfer compound with tetrathiafulvalene (TTF), the composition being $C_{60}Br_8(TTF)_8$. The unique complex-forming property of $C_{60}Br_8$ can be understood on the basis of the electronic structures of the halogenated derivatives of C_{60}. Molecular orbital calculations show that the low LUMO energy of $C_{60}Br_8$ compared with the other halogen derivatives renders the formation of the complex with TTF favourable, the four virtual LUMOs being able to accept 8 electrons. The Raman spectrum of $C_{60}Br_8(TTF)_8$ shows a marked softening of the bands ($-46\ cm^{-1}$ on average) with respect to $C_{60}Br_8$, suggesting that indeed 8 electrons are transferred per $C_{60}Br_8$ molecule, one from each TTF molecule. The complex is weakly paramagnetic and shows a magnetic transition around 80 K.

1. Introduction

Co-crystals of C_{60} with several organic molecules have been reported in recent years, but in most of these systems there is no extensive charge transfer between the fullerene and the other molecule. The complex between C_{60} and tetrakis(dimethyl-amino)ethylene (TDAE) is exceptional [1] in that C_{60} acquires a negative charge because of electron donation from the powerful electron donor molecule TDAE. The compound is a soft ferromagnet with a T_c of ~ 16 K. On the other hand, even the complex between C_{60} and hexamethylenetetratellurafulvalene is only weakly para-magnetic, involving negligible donor–acceptor charge transfer [2]. We were interested in exploring the formation of charge-transfer compounds involving donors such as tetrathiafulvalene (TTF) with C_{60} derivatives with increased electron affinity of the fullerene due to the presence of electron attracting substituents such as the halogens. Earlier efforts [3] in this direction have not yielded pure compounds with well defined properties. With this objective in mind, we prepared $C_{60}Cl_6$, $C_{60}Br_6$ and $C_{60}Br_8$ by employing the literature procedures [4, 5] and studied their interaction with tetrathiafulvalene. To our surprise, we found that only $C_{60}Br_8$ forms a solid adduct with TTF. We have characterized the $C_{60}Br_8$–TTF complex by various physical methods and explored why the complexation with TTF is specific to $C_{60}Br_8$ on the basis of detailed molecular orbital calculations. For this purpose, we have examined the electronic structures of $C_{60}X_6$ (X = H, Cl and Br) derivatives as well as $C_{60}Br_{60-6n}$ (n = 4–8). We have investigated the nature of the $C_{60}Br_8$–TTF complex, in particular the magnitude of electron transfer from the TTF donor to $C_{60}Br_8$, by means of Raman studies of the complex, plus ESR and magnetic measurements. It is to be noted that the magnitude of softening of the Raman modes of doped fullerenes provides a direct

268   C. N. R. Rao *et al.*

means of titrating the number of electrons transferred to the fullerene by a metal or a electron donor molecule [6].

2. Experimental

$C_{60}Br_6$ was prepared by the reaction of excess bromine (15 ml, 0·29 mol) with C_{60} (100 mg, 0·14 mmol) in dry benzene (100 ml) solution. The reaction mixture was stirred for two days under an argon atmosphere. Slow evaporation of the solvent and excess bromine gave magenta coloured platelets of $C_{60}Br_6$ [4]. $C_{60}Br_8$ was prepared by the reaction of bromine (15 ml, 0·29 mol) with C_{60} (100 mg, 0·14 mmol) in dry CS_2 (30 ml) solution. Stirring the reaction mixture under argon at 303 K for 48 h, followed by the slow evaporation of the solvent and the bromine, yielded brown crystals of $C_{60}Br_8$ [4]. In order to prepare $C_{60}Cl_6$, a freshly prepared solution of ICl (300 mg, 1·9 mmol) in dry benzene (10 ml) was added to a solution of C_{60} (100 mg, 0·14 mmol) in dry benzene (100 ml). The mixture was left to stand at room temperature for three days under an argon atmosphere. The solvent and evolved iodine were removed under reduced pressure at 323 K, yielding the orange coloured $C_{60}Cl_6$ [5]. $C_{60}Br_8$ crystals contain some free Br_2. From these crystals bromine was removed by reacting with excess toluene followed by vacuum evaporation of the solvent and drying at 323 K for 5 h.

A saturated solution of each of the halogenated fullerene derivatives ($C_{60}Cl_6$, $C_{60}Br_6$ or $C_{60}Br_8$) in CS_2 was mixed with a CS_2 solution of TTF (TTF was in slight excess compared with the halogenated fullerene derivative). $C_{60}Br_8$ gave a deep brown precipitate of a solid complex. No solid product was obtained with the other two. The solid $C_{60}Br_8$–TTF complex was subjected to chemical and spectroscopic analysis, taking care not to expose the solid to the atmosphere or to direct light. EDX analysis was carried out using a Leica scanning electron microscope.

A sample of $(TTF)Br_y$ ($y > 1$) was prepared for use as a reference material to ensure that this impurity was not present in the complex of $C_{60}Br_8$ with TTF. This was accomplished by the addition of bromine (5 ml) to a TTF (50 mg) solution in CS_2 (50 ml). The orange red precipitate of $(TTF)Br_y$ was filtered, washed with CS_2 and dried.

Powder X-ray diffraction (XRD) patterns were recorded with $Cu\,K_\alpha$ radiation using STOE STADI diffractometer using graphite-monochromatized $Cu\,K_\alpha$ radiation ($\lambda = 1·54056$ Å) and a linear position sensitive detector. The transmission geometry was employed with the samples being held between Mylar sheets. The diffraction pattern of the $C_{60}Br_8$–TTF complex was distinctly different from that of $(TTF)Br_y$ or $C_{60}Br_8$, as can be seen from figure 1.

A Brüker IFS 113 FTIR spectrometer was used for infrared studies, the samples being in the form of KBr pellets. The spectrum of the $C_{60}Br_8$–TTF complex was significantly different from that of $C_{60}Br_8$ (figure 2). Raman spectra were recorded from the pellets of solid samples at room temperature in back-scattering geometry using a DILOR-XY spectrophotometer equipped with a liquid nitrogen cooled CCD detector. The excitation source was the 5145 Å line of an argon ion laser coherent model (Innova 300) at a power level of ~ 1 mW (focused spot ~ 40 μm).

ESR measurements on the $C_{60}Br_8$–TTF complex were performed with a Brüker X-band spectrometer operating at 9·45 GHz. Low temperatures were obtained (6·5 K) using a continuous flow helium cryostat (Oxford Instruments) with a temperature stability of $\pm 0·1$ K. In the discussion that follows, the signal intensity I refers to the

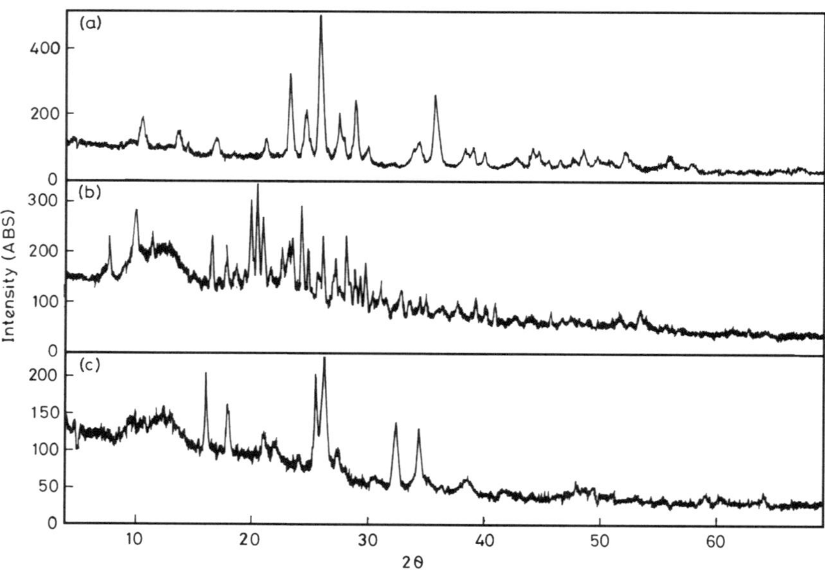

Figure 1.   XRD patterns of (*a*) TTFBr$_y$, (*b*) C$_{60}$Br$_8$, and (*c*) C$_{60}$Br$_8$(TTF)$_8$.

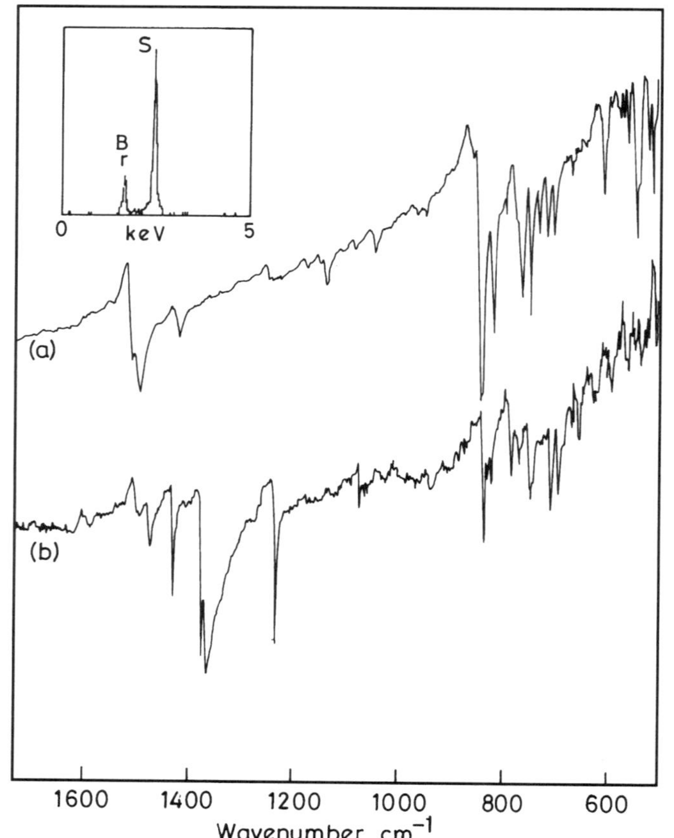

Figure 2.   FTIR spectra of (*a*) C$_{60}$Br$_8$, and (*b*) C$_{60}$Br$_8$(TTF)$_8$. Inset shows the EDX profile of C$_{60}$Br$_8$(TTF)$_8$ giving a Br:S ratio of 1:4.

peak height, and the line width Γ refers to the peak-to-peak width. The product $I\Gamma^2$ of the peak height and the square of the line-width is proportional to the magnetic susceptibility. Dc magnetic susceptibility was measured on powdered samples using a Lewis coil force magnetometer (George Associates, USA, model 200). The temperature was varied in the range 300–20 K using a closed cycle helium cryostat (APD Cryogenics, USA). The Lewis coil magnetometer has the advantage of measuring magnetic susceptibility at low magnetic fields. The maximum field and gradient that can be achieved are 8 kOe and 180 kOe cm^{-1}, respectively.

3.   Computational details

All the calculations were carried out at the AM1 Hamiltonian level built in the MOPAC [7] package with standard parameters. Molecular geometries were completely optimized within a given point group by employing the BFGS algorithm [8–11]. A force constant analysis was performed for $C_{60}Br_6$ and $C_{60}Br_8$ derivatives at the AM1-SCF level to characterize the structures as stationary points.

The Schlegel representation of the hexa and octo derivatives examined in the study are shown in figure 3. Although a large number of isomeric structures are possible for $C_{60}Br_6$ and $C_{60}Br_8$, we have considered only the C_s structures proposed by Birkett *et al.* [4]. For higher brominated derivatives, full geometry optimization was carried out with the following symmetry constraints: $C_{60}Br_{12}(T_h4)$, $C_{60}Br_{18}(C_3)$, $C_{60}Br_{24}(T_h)$, $C_{60}Br_{30}(C_3)$ and $C_{60}Br_{36}(T)$. The point groups correspond to the highest possible symmetry in each case.

4.   Results and discussion

4.1. *Characterization of the $C_{60}Br_8$–TTF complex*

EDX analysis of the solid product obtained by the reaction of $C_{60}Br_8$ with TTF showed the Br:S ratio to be 1:4 (see inset of figure 2) suggesting the formula to be $C_{60}Br_8(TTF)_8$. Chemical analysis also confirmed the formula. The XRD patterns of $C_{60}Br_8(TTF)_8$ (figure 1) shows low angle peaks which are rather broad. Making use of a peak fit program, we obtained unit cell dimensions of $a = 23 \cdot 03$ Å, $b = 11 \cdot 24$ Å and $c = 18 \cdot 93$ Å (volume ~ 4903 Å^3). It may be noted that the unit cell dimensions of $C_{60}Br_8$ are $a = 24 \cdot 48$ Å, $b = 16 \cdot 895$ Å and $c = 11 \cdot 343$ Å (volume ~ 4692 Å^3). The sample of $(TTF)Br_y$ ($y > 1$) prepared by us has a monoclinic unit cell with a cell volume of 1418 Å^3 ($a = 11 \cdot 2795$ Å, $b = 12 \cdot 5684$ Å and $c = 10 \cdot 5512$ Å, and $\beta = 108 \cdot 45°$). The FTIR spectrum of $C_{60}Br_8(TTF)_8$ (figure 2) shows the C—Br stretching band at 837 cm^{-1} compared to 847 cm^{-1} in $C_{60}Br_8$. The bands at 1236 cm^{-1} and 1065 cm^{-1} are due to the TTF moiety. The strong doublet at 1377 cm^{-1} and 1366 cm^{-1} in the complex corresponds to the 1420 cm^{-1} band of the parent bromo compound. The Raman spectrum of $C_{60}Br_8(TTF)_8$ throws greater light on the nature of the compound, as discussed later.

4.2. *Results from calculations*

The calculated heats of formation and heats of bromination of the various C_{60} derivatives are listed in table 1. The heats of formation of $C_{60}Br_6$ as well as of $C_{60}H_6$ and $C_{60}Cl_6$ are higher than that of C_{60}, suggesting some relative instability of the hexasubstituted compounds. These derivatives contain 15 benzenoid units, which are not isolated and have 10 non-conjugated carbons. $C_{60}Br_8$ has 9 benzenoid units and 42

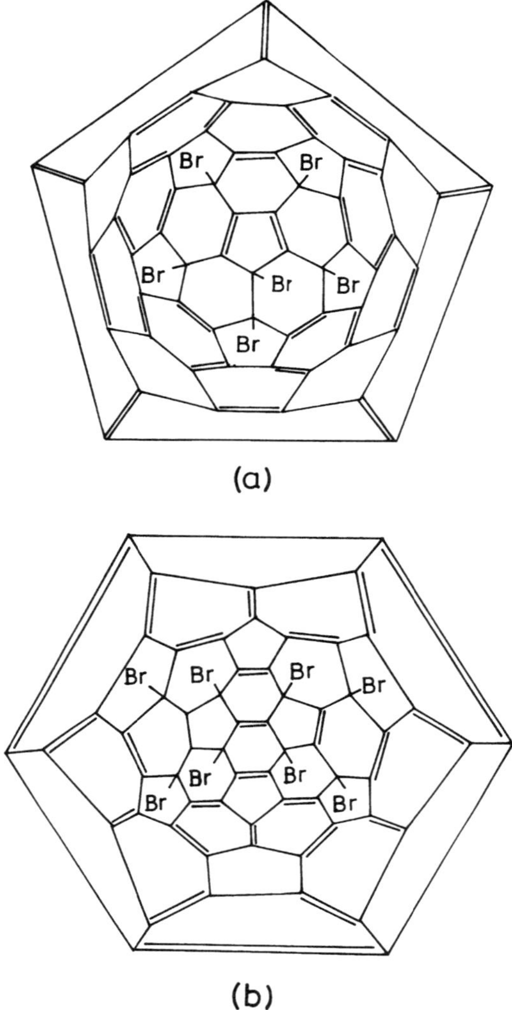

Figure 3.   Schlegel diagrams of $C_{60}Br_6$ and $C_{60}Br_8$.

carbons are conjugated. The heats of bromination (table 1) suggest that $C_{60}Br_6$ is easier to form than $C_{60}Br_8$, but neither is very stable relative to C_{60}. $C_{60}Br_{12}$ and $C_{60}Br_{18}$ are also associated with large negative heats of bromination. There is, however, a large reduction in this energy on going from $C_{60}Br_{18}$ to $C_{60}Br_{24}$ with isolated double bonds. Both $C_{60}Br_{24}$ and $C_{60}Br_{30}$ are predicted to be stable, although further bromination to $C_{60}Br_{36}$ and beyond is unfavourable. Indeed, $C_{60}Br_{24}$ is known to be the most stable brominated derivative of C_{60} [4].

In order to determine the nature of the stationary points by geometry optimization with symmetry constraints, force constraints were computed at the AM1 level for $C_{60}Br_6$ and $C_{60}Br_8$. Diagonalization of the force constant matrix revealed no negative eigenvalues. Correspondingly, a vibrational analysis assuming the harmonic approximation yielded ($3N$-6) real frequencies and no imaginary frequency. $C_{60}Br_6$ and $C_{60}Br_8$ considered in the present study (with the C_s symmetry) are minima on the potential energy surface.

C. N. R. Rao *et al.*

Table 1. Heats of formation and bromination of the bromo derivatives of C_{60}.

Molecule	Symmetry	Heat of formation[a] kcal mol^{-1}	Heat of bromination kcal mol^{-1}
C_{60}	I_h	973·3	—
$C_{60}Br_6$	C_s	1068·0	−94·7
$C_{60}Br_8$	C_s	1148·6	−175·3
$C_{60}Br_{12}$	T_h	1195·3	−127·3
$C_{60}Br_{18}$	C_3	1882·4	−687·1
$C_{60}Br_{24}$	T_h	1148·6	743·8
$C_{60}Br_{30}$	C_3	944·7	204·0
$C_{60}Br_{36}$	T	1224·9	−280·3

[a] The heats of formation of $C_{60}H_6(C_s)$ and $C_{60}Cl_6(C_s)$ are 993·1 kcal mol^{-1} and 1021·1 kcal mol^{-1}, respectively, and heats of hydrogenation and chlorination are −19·7 kcal mol^{-1} and −47·8 kcal mol^{-1}, respectively.

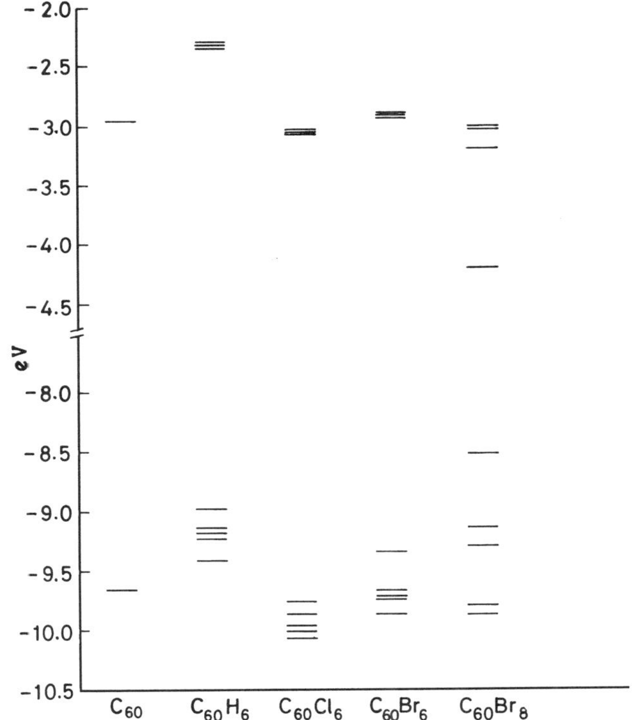

Figure 4.   Electron energy levels of halogenated C_{60} derivatives.

In figure 4 we show the energetic ordering of the HOMO and the LUMO of the various derivatives studied by us. In C_{60}, the HOMO is fivefold degenerate and the LUMO is threefold degenerate in the I_h point group. The hexaderivatives (figure 3) have a σ plane with a drastically reduced symmetry and the degeneracies associated with the MOs are therefore lifted (figure 4). Hydrogenation of C_{60} to $C_{60}H_6$ results in the destabilization of both the HOMO and the LUMO, and the optical gap remains nearly the same. On the other hand, chlorination to $C_{60}Cl_6$ causes stabilization of both the HOMO and the LUMO, while bromination to $C_{60}Br_6$ gives rise to a small destabilization of the frontier orbitals. The nature of the HOMO remains essentially of the π-type in all the derivatives. On examining the trends in the computed HOMO energies (figure 4), it becomes evident that there is only a small variation in this energy on going from C_{60} to $C_{60}Br_{36}$. The ionization energy of all the derivatives would therefore be comparable.

The LUMO energies can, in principle, be related to the electron affinities through Koopman's theorem. The LUMO of C_{60} is computed to be bound clearly by over 2.9 eV at the AM1 level. This result is consistent with the well known electron deficiency of C_{60} and its readiness to undergo sequential electron addition. The LUMO orbitals of $C_{60}H_6$ constitute a less bound state (-2.34 eV) compared with $C_{60}Cl_6$ (-3.061 eV) and $C_{60}Br_6$ (-2.93 eV). This result suggests that halogenated derivatives are more electron deficient than the hydrogenated counterparts, and should undergo sequential electron addition. It is interesting that the LUMO of $C_{60}Br_8$ is bound by 4.2 eV indicating its tendency to accept electrons. When we compare C_{60} and $C_{60}Br_8$, the latter is more electron deficient. The tendency to form electron donor–acceptor complexes should therefore be greater in $C_{60}Br_8$ than C_{60} or any other derivative considered in the present study. Furthermore, figure 4 shows that there are four virtual orbitals of $C_{60}Br_8$ below the LUMO of C_{60}. If we assume C_{60} to be the limiting case, then $C_{60}Br_8$ can accept 8 electrons in the four virtual orbitals. According to the present molecular orbital calculations, the formation of the $C_{60}Br_8(TTF)_8$ complex would be accompanied by the filling of the four LUMOs.

In table 2, we have tabulated the HOMO–LUMO gaps as well as the $(I_D - E_A)$ values for all the derivatives. Here, I_D represents the energy of the TTF donor (7.818 eV) and E_A is the electron affinity of the acceptor. The smaller the magnitude of $(I_D - E_A)$, the greater would be the tendency towards complex formation, since it represents the activation barrier for the process. $C_{60}Br_8$ has the lowest $(I_D - E_A)$ value amongst all the C_{60} derivatives studied by us, again showing why it alone forms a complex with TTF.

4.3. *Electronic structure of $C_{60}Br_8(TTF)_8$ as revealed by Raman, ESR and other studies*

The Raman spectrum of $C_{60}Br_8(TTF)_8$ is compared with the spectra of $C_{60}Br_8$ and C_{60} in figure 5. The Raman spectrum of $C_{60}Br_8$ (curve b) shows 15 clearly resolved lines as compared with 35 lines observed in the Fourier transform (FT) Raman spectrum recorded using red laser light [12]. The main peaks in our spectrum at 284, 493, 708, 1427, 1464 and 1586 cm^{-1} correspond to the peaks at 289, 494, 708, 1429, 1460 and 1581 cm^{-1} in the FT Raman spectrum. The number of observed modes is much less than the 198 Raman-active modes predicted by the normal mode analysis ($52A_1 + 48A_2 + 49B_1 + 49B_2$) of the molecular vibrations of the point group symmetry C_s of $C_{60}Br_8$. The main peaks at 1464 cm^{-1} and 493 cm^{-1} in $C_{60}Br_8$ are close to the two A_g modes of C_{60} (symmetry I_h) at 1469 and 469 cm^{-1}, implying that the cage structure of C_{60} is preserved in $C_{60}Br_8$. The highest frequency mode in $C_{60}Br_8$ is at 1586 cm^{-1} compared with 1573 cm^{-1} in C_{60}. The reason for the higher frequency in brominated C_{60} is attributed to the localization of the C=C double bonds. The geometry of the molecules localizes all the 18 double bonds in $C_{60}Br_{24}$, 3 in $C_{60}Br_8$ and none in $C_{60}Br_6$.

C. N. R. Rao *et al.*

Table 2. Values of the HOMO–LUMO gaps and $(I_D - E_A)$.

Molecule	HOMO–LUMO gap/eV	$(I_D - E_A)$/eV[a]
C_{60}	6·7	4·87
$C_{60}H_6$	6·6	5·48
$C_{60}Cl_6$	6·7	4·76
$C_{60}Br_6$	6·4	4·89
$C_{60}Br_8$	4·3	3·64
$C_{60}Br_{12}$	7·1	5·26
$C_{60}Br_{13}$	5·1	4·46
$C_{60}Br_{24}$	5·2	4·82

[a] I_D is the ionization energy of TTF and E_A is the electron affinity of the fullerene given by the lowest LUMO energy.

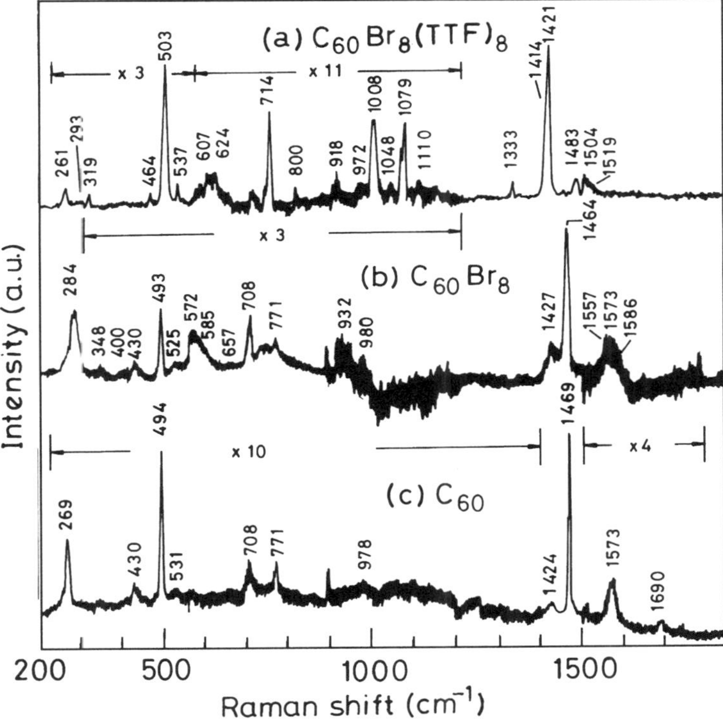

Figure 5. Raman spectra of (*a*) $C_{60}Br_8(TTF)_8$, (*b*) $C_{60}Br_8$ and (*c*) C_{60} in the solid state. The magnification factors for different spectral ranges are shown for each spectrum.

The modes at 525, 572 and 585 cm^{-1} in $C_{60}Br_8$ are attributed to the stretching motions of carbon–bromine bonds which belong to irreducible representations $2A_1 + 2A_2 + 1B_2 + 2B_2$.

The strongest band of $C_{60}Br_8(TTF)_8$ (figure 5, curve a) is a doublet with peaks at 1414 cm^{-1} and 1421 cm^{-1}. Comparing with $C_{60}Br_8$, where the strongest mode occurs at 1464 cm^{-1}, the phonon softening $\Delta\omega$ for the modes are -50 cm^{-1} and -43 cm^{-1}. The softening of the corresponding $A_g(2)$ mode in A_xC_{60} (A = K, Rb and Cs) is -17 cm^{-1} to -20 cm^{-1} for $x = 3$ and -36 cm^{-1} to -38 cm^{-1} for $x = 6$. In C_{60}·TDAE, where

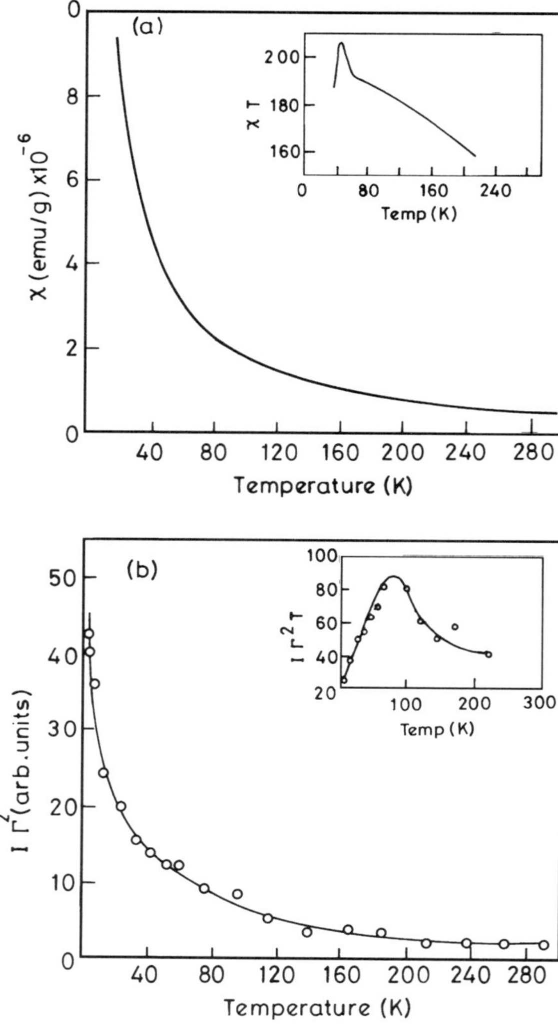

Figure 6.   Temperature variation of (*a*) the magnetic susceptibility and (*b*) $I\Gamma^2$ from ESR spectra of $C_{60}Br_8(TTF)_8$. Insets show the temperature variation of χT and $I\Gamma^2 T$, respectively.

one electron is transferred per C_{60} molecule, $\Delta\omega = -14\ cm^{-1}$ [13]. This phonon softening is attributed to C—C bond expansion due to charge transfer in the antibonding t_{1u} band. The average softening of $-46\ cm^{-1}$ observed in $C_{60}Br_8(TTF)_8$ is higher than the values for A_6C_{60} and a linear extrapolation (i.e., $\Delta\omega = -6\ cm^{-1}$ per electron transfer to the C_{60} molecule) would imply that ~ 8 electrons are donated per C_{60} molecule, consistent with the complex having the formula $C_{60}Br_8(TTF)_8$. The modes at 1483, 1504 and 1519 cm^{-1} in the complex correspond to the modes at 1557, 1573 and 1586 cm^{-1} in $C_{60}Br_8$, implying a softening of $\sim 70\ cm^{-1}$. The mode at 1333 cm^{-1} in the complex corresponds to the 1427 cm^{-1} mode in $C_{60}Br_8$, implying a rather large phonon softening of $-94\ cm^{-1}$. By comparison, the $H_g(7)$ mode at 1424 cm^{-1} in C_{60} experiences $\Delta\omega \approx -45\ cm^{-1}$ in A_6C_{60}.

In contrast to the high frequency tangential modes, the $A_g(1)$ radial mode in C_{60} at

276   C. N. R. Rao *et al.*

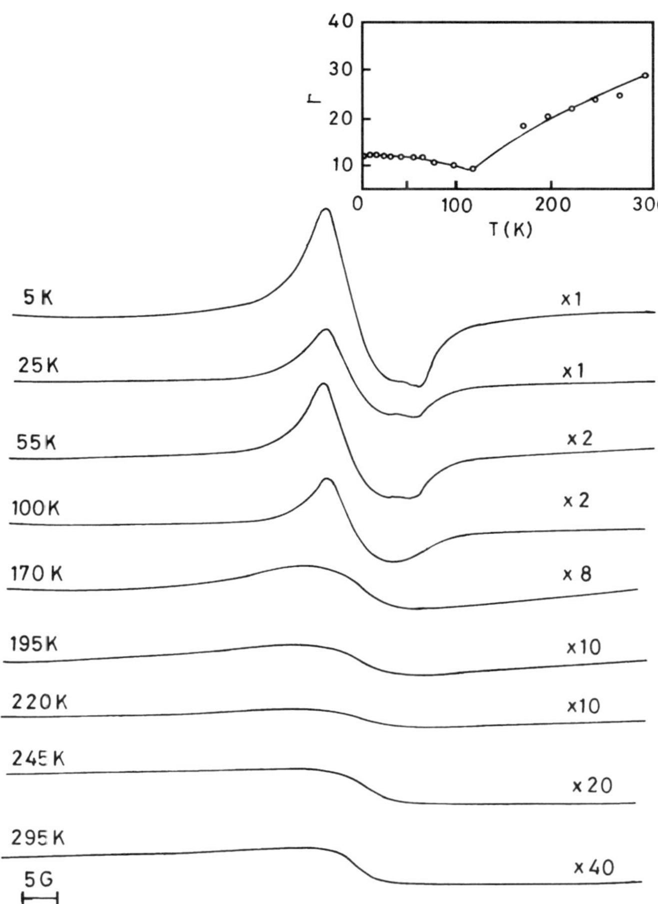

Figure 7.   ESR spectra of $C_{60}Br_8(TTF)_8$ at different temperatures. Inset shows the variation of line width with temperature.

$493\ cm^{-1}$ is shifted upwards to $\sim 500\ cm^{-1}$ in A_6C_{60}. This upward shift is attributed partly to an electrostatic interaction arising from the charged C_{60} ball which compensates the effect of C—C bond elongation [14]. Similar behaviour is seen in the present case. The mode at $493\ cm^{-1}$ in $C_{60}Br_8$ is shifted to $503\ cm^{-1}$ in $C_{60}Br_8(TTF)_8$, giving a $\Delta\omega$ of $+10\ cm^{-1}$. The squashing mode at $284\ cm^{-1}$ in $C_{60}Br_8$ is shifted to $261\ cm^{-1}$ in $C_{60}Br_8(TTF)_8$, showing a shift of $-23\ cm^{-1}$. We also observe that the relative intensities of the modes in $C_{60}Br_8$ and $C_{60}Br_8(TTF)_8$ are different, just as in the charge transfer salt, $C_{60}\cdot TDAE$.

Clearly, there is a considerable softening of phonons in the charge transfer compound with respect to $C_{60}Br_8$. As mentioned earlier, a comparison with the observed phonon softening in alkali-doped C_{60} suggests that about 8 electrons are transferred per $C_{60}Br_8$ molecule, one electron by each TTF molecule. This conclusion is supported fully by the results of our AM1 calculations discussed earlier. It should be noted that C_{60} itself can accept up to 12 electrons [15].

Four-probe electrical conductivity measurements of compressed pellets of $C_{60}Br_8(TTF)_8$ show the conductivity to be $0\cdot05\ S\ cm^{-1}$ at 300 K. This is slightly higher than the conductivity of $C_{60}\cdot TDAE$.

In figure 6(*a*) we show the magnetic susceptibility data of $C_{60}Br_8(TTF)_8$. The material is weakly paramagnetic, but the susceptibility increases markedly around 80 K. The plot of χT versus temperature shows a peak around this temperature (figure 6(*a*), inset). This peak could correspond to a ferromagnetic transition, being similar to the feature exhibited by the soft ferromagnet C_{60}.TDAE at a lower temperature [1]. The ESR spectrum of $C_{60}Br_8(TTF)_8$ shows a signal which varies in intensity and width with temperature (figure 7). The magnetic susceptibility behaviour from ESR measurements (the product of peak height and the square of the line width) is comparable with that from DC magnetic susceptibility measurements (figure 6). The ESR linewidth shows a marked change around 100 K and the $I\Gamma^2T$ versus temperature plot showing peak around 80 K. The paramagnetic nature and the possibility of ferromagnetism at low temperatures suggest that the LUMOs of $C_{60}Br_8$ are not completely filled. The magnitude of electron transfer is therefore expected to be slightly less than 8.

The authors thank Professors T. N. Guru Row and S. V. Bhat for assistance with XRD and ESR measurements.

The authors dedicate this paper to David Buckingham, a dear friend and a fine theoretical chemist.

References

[1] ALLEMAND, P. M., KHEMANI, K. C., KOCH, A., HOLCZER, K., DONOVAN, S., GRUNER, G., and THOMPSON, J. D., 1991, *Science*, **253**, 301; SESHADRI, R., RASTOGI, A., BHAT, S. V., RAMASESHA, S., and RAO, C. N. R., 1993, *Solid State Commun.*, **85**, 97; EHRENREICH, H., and SAEPEN, F., 1994, Vol. 48 *Solid State Physics* (San Diego: Academic Press).
[2] PRADEEP T., SINGH, K. K., SINHA, A. P. B., and MORRIS, D. E., 1992, *J. chem. Soc. chem. Commun.*, 1747.
[3] LI, Y., ZHANG, D., BAI, F., ZHU, D., YIN, B., LI, J. W., and ZHAO, Z., 1993, *Solid State Commun.*, **86**, 475.
[4] BIRKETT, P. R., HITCHCOCK, P. B., KROTO, H. W., and WALTON, D. R. M., 1992, *Nature*, **357**, 479.
[5] BIRKETT, P. R., AVENT, A. G., DARWISH, A. D., KROTO, H. W., TAYLOR, R. and WALTON, D. R. M., 1993, *J. chem. Soc. chem. Commun.*, 1230.
[6] PICHLER, T., MATUS, M., KURTI, J., and KUZMANY, H., 1992, *Phys. Rev.* B, **45**, 13841.
[7] DEWAR, M. J. S., ZOEBISCH, E. G., HEALY, E. F., and STEWART, J. J. P., 1985, *J. Amer. chem. Soc.*, **107**, 3902; DEWAR, M. J. S., ZOEBISCH, E. G., HEALY, E. F., and STEWART, J. J. P., 1985, *QCPE Bull.*, Package No. 455.
[8] BROYBEN, C. G., 1970, *J. Inst. Math. Appl.*, **6**, 76.
[9] FLETCHER, R., 1970, *Computer J.*, **13**, 317.
[10] GOLDFARB, D., 1970, *Math. Comput.*, **24**, 23.
[11] SHANNO, D. F., 1970, *Math Comput.*, **24**, 647.
[12] BIRKETT, P. R., KROTO, H. W., TAYLOR, R., WALTON, D. R. M., GROSE, R. I., HENDRA, P. J., and FOWLER, P. W., 1993, *Chem. Phys. Lett.*, **205**, 399.
[13] MUTHU, D. V. S., SHASHIKALA, M. N., SOOD, A. K., SESHADRI, R., and RAO, C. N. R., 1994, *Chem. Phys. Lett.*, **217**, 146.
[14] CHEN, C. T., HO, K. M., and KAMITAKAHARA, W. A., 1987, *Phys. Rev.* B, **36**, 3499.
[15] MAXWELL, A. J., BRUHWILER, P. A., ANDERSSON, S., MARTENSSON, N., and RUDOLF, F., 1995, *Chem. Phys. Lett.*, **247**, 257.

J. Am. Chem. Soc. **1999**, *121*, 1752–1753

1752

Hydrothermal Synthesis of Organic Channel Structures: 1:1 Hydrogen-Bonded Adducts of Melamine with Cyanuric and Trithiocyanuric Acids

Anupama Ranganathan, V. R. Pedireddi, and C. N. R. Rao*

*Chemistry & Physics of Materials Unit
Jawaharlal Nehru Centre for Advanced Scientific Research
Jakkur P.O. Box 6436, Bangalore 560 064, India*

Received November 12, 1998

Whitesides and co-workers[1] have extensively cited the 1:1 hydrogen-bonded adduct between cyanuric acid (**CA**) and melamine (**M**)-forming rosettes, as an outstanding case of noncovalent synthesis. The hydrogen-bonded **CA·M** lattice is the template on which these workers based the design of a family of self-assembled hydrogen-bonded aggregates, stable in solution.[2,3] The **CA·M** adduct is expected to form a hexagonal network through the formation of hydrogen bonds between the two molecules, but the structure has not been established by single-crystal X-ray crystallography.[3] The crystal structure of **CA·M·3HCl** reported by Wang et al.[4] reveals one-dimensional tapes rather than the rosette structure. The difficulty in growing crystals of the **CA·M** adduct is partly because **CA** and **M** are both highly hydrogen-bonded solids melting at high temperatures, with limited solubility in most organic solvents. Mixtures of **CA** and **M** in water or alcohol solution give precipitates containing the adduct, but do not yield crystals suitable for X-ray crystallography.[1,3] We felt that to obtain single crystals of the **CA·M** adduct, it may be necessary to employ more drastic conditions than normally employed. We have, therefore, made use of hydrothermal conditions,[5] commonly employed in the synthesis of quartz, zeolites, and inorganic open-frame structures, to obtain the crystals of the 1:1 adduct. We have indeed been able to obtain good crystals of the 1:1 **CA·M** adduct by this means. Crystals of the 1:1 adduct of trithiocyanuric acid (**TCA**) and **M** could also be

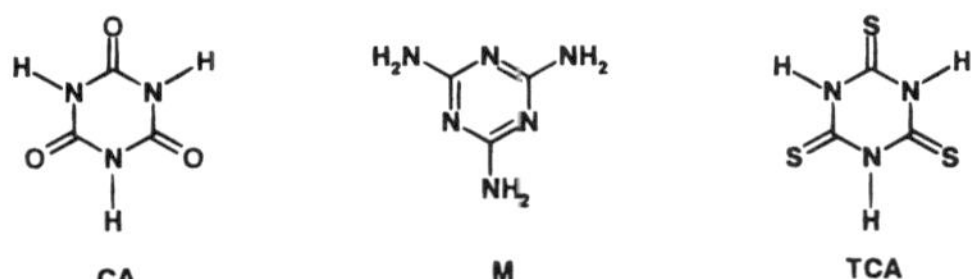

prepared hydrothermally. We report the structures of both of these fascinating adducts of **M** in this paper.

The procedure for preparing the crystals of the **CA·M** adduct was as follows. Aqueous solutions of cyanuric acid (0.1 mmol) and melamine, (0.1 mmol) were mixed in a Teflon flask, and the mixture (15 mL) was kept in a stainless steel bomb. The bomb was sealed and maintained in a furnace at 180 °C. Rectangular platelike crystals of good quality separated from the solution, upon cooling the bomb to room temperature over a period of 4 h. These crystals were used for collecting intensity data on the single-crystal

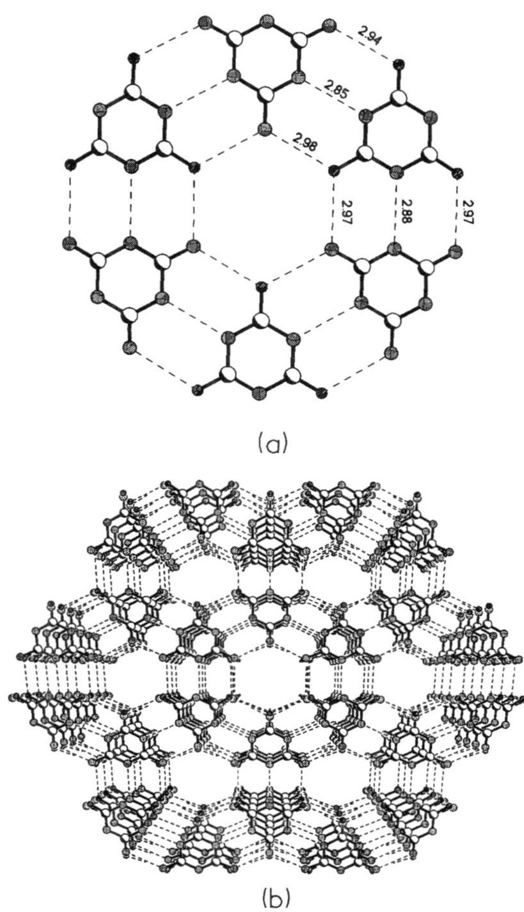

(a)

(b)

Figure 1. (a) A rosette formed between cyanuric acid (**CA**) and Melamine (**M**) in the adduct of **CA·M** with a cavity diameter of approximately 4 Å. Dashed lines represent intermolecular hydrogen bonds. (b) Three-dimensional arrangement of the **CA·M** adduct forming channels along the crystallographic *c*-axis.

X-ray diffractometer. Crystals of the 1:1 **TCA·M** adduct were obtained by a similar procedure except that the temperature was maintained at 100 °C.[6]

The 1:1 **CA·M** adduct[7] has an asymmetric unit consisting of superimposed **CA** and **M** molecules.[8] Packing analysis shows that

(1) Whitesides, G. M.; Simanek, E. E.; Mathias, J. P.; Seto, C. T.; Chin, D. N.; Mammen, M.; Gordon, D. M. *Acc. Chem. Res.* **1995**, *28*, 37. Whitesides, G. M.; Mathias J. P.; Seto, C. T. *Science* **1991**, *254*, 1312.

(2) Seto C. T.; Whitesides, G. M. *J. Am. Chem. Soc.* **1993**, *115*, 905.

(3) Mathias, J. P.; Simanek, E. E.; Zerkowski, J. A.; Seto, C. T.; Whitesides, G. M. *J. Am. Chem. Soc.* **1994**, *116*, 4316. (b) Zerkowski, J. A.; Seto, C. T.; Wierda, D. A.; Whitesides, G. M. *J. Am. Chem. Soc.* **1990**, *112*, 9025.

(4) Wang, Y.; Wei, B.; Wang, Q. *J. Crystallogr. Spectrosc. Res.* **1990**, *20*, 79.

(5) Rao, C. N. R. *Chemical Approaches to the Synthesis of Inorganic Materials*; John Wiley: New York, 1993.

(6) **TCA** decomposed when the temperature was maintained at 180 °C as in the case of **CA**.

(7) Crystal data for the **CA·M** adduct: (C₃H₃N₃O₃):(C₃H₆N₆), $M = 255.22$, crystal dimensions, $0.3 \times 0.2 \times 0.2$ mm, monoclinic, space group, *C2/m*, $a = 14.853(2)$ Å, $b = 9.641(2)$ Å, $c = 3.581(1)$ Å, $\beta = 92.26(1)°$, $V = 512.4(2)$ Å³, $Z = 2$, $\rho_{calcd} = 1.654$, μ(Mo Kα) $= 0.136$ mm⁻¹, $F(000) = 264$, $\lambda = 0.710\,73$ Å, Smart CCD area detector, Siemens, $\omega{-}2\theta$ scan, $2° < \theta < 24°$ ($-16 \leq h \leq 14$, $-8 \leq k \leq 10$, $-3 \leq l \leq 3$), 1095 total reflections, 395 independent reflections which were used for refinement. The structure was solved by direct methods (SHELXTL-PLUS) and refined by full-matrix least squares on F^2 (SHELX-93; G. M. Sheldrick, Gottingen, 1993) to R1 = 0.056 and wR2 = 0.141. Residual density, min/max $-0.495/0.393$ e·Å⁻³. Hydrogen atoms were not included in the refinement, but including the hydrogen atoms in the calculated positions improves the *R*-factor by 1%.

(8) The best stacking arrangement was one with alternating **CA** and **M** molecules (with 50% occupation each), as established by the low *R*-factor and also by energy calculations using the MOPAC program. The least stable arrangement is the one with **CA** on top of another **CA**, as one would expect.

CA and **M** are held together by N—H···O and N—H···N hydrogen bonds yielding the hexamer (rosette) shown in Figure 1a. The interatomic N···O and N···N distances corresponding to N—H··· O and N—H···O bonds are in the range of 2.94—2.98 and 2.85—2.88 Å, respectively. The hexamers are arranged in two dimensions to form planar sheets, the structure being exactly as predicted by Whitesides.[1] A significant feature is that the planar sheets are stacked in three dimensions[8] to give channels with a diameter of 4 Å (Figure 1b). These channels are comparable to the cavities in cryptands.[9]

Crystal structure determination of the **TCA·M** adduct[10] reveals that it has features similar to those of the **CA·M** adduct, except that the N—H···O hydrogen bonds are replaced by the N—H···S hydrogen bonds as shown in Figure 2a. The N···S and N···N distances corresponding to N—H···S and N—H···N bonds are in the range of 2.96—2.99 and 2.86—2.88 Å, respectively. The sheets are stacked in a three-dimensional channel arrangement as depicted in Figure 2b. The diameter of the channel is approximately 4 Å here as well.

The present study illustrates the efficacy of the hydrothermal method to prepare crystals of organic materials in certain special situations, as in the present case where the compounds are highly hydrogen-bonded (with limited solubility in common solvents) and give precipitates when mixed in water.[11]

It is noteworthy that the adducts of **M** with **CA** and **TCA** constitute members of a rare class of organic solids containing channels.[12] Trimesic acid is a good example of a system forming hexagonal channels.[13] The other example of such a solid is the adduct of **TCA** with 4,4′-bipyridyl.[14] Experiments to explore whether the channels in the **CA·M** adduct can be filled by cations and other species are in progress.

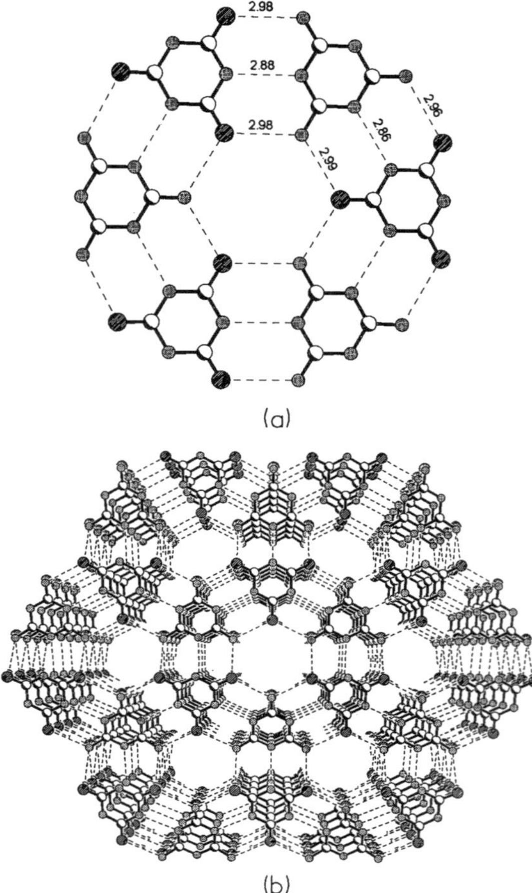

(a)

(b)

Figure 2. (a) Hexagonal network (rosette) formed between trithiocyanuric acid (**TCA**) and Melamine (**M**) in the adduct of **TCA·M** with a cavity diameter of approximately 4 Å. (b) Three-dimensional arrangement of the **TCA·M** adduct forming channels along crystallographic *c*-axis.

Supporting Information Available: Two figures (ORTEP drawings of the adducts **CA·M** and **TCA·M** with labeling scheme) and 6 tables containing general crystallographic information, fractional coordinates of non-hydrogen atoms, list of bond lengths and angles, and anisotropic displacement parameters (PDF). This material is available free of charge via the Internet at http://pubs.acs.org.

JA983928O

CA on top of **M** is more stable by 37 kcal mol^{-1} relative to **CA** on **CA**. The packing coefficient of the **CA·M** adduct calculated by using Cerius is 67.5%.

(9) Vogtle, F. *Supramolecular Chemistry*; Wiley: Chichester, 1971.

(10) Crystal data for the **TCA·M** adduct: (C$_3$H$_3$N$_3$S$_3$):(C$_3$H$_6$N$_6$) M = 303.39, crystal dimensions, 0.3 × 0.25 × 0.35 mm, monoclinic, space group, *C2/m*, *a* = 14.862(3) Å, *b* = 9.650(2) Å, *c* = 3.588(1) Å, β = 92.32(1)°, *V* = 514.2(2) Å^3, *Z* = 2, ρ_{calcd} = 1.901, μ(Mo Kα) = 0.716 mm^{-1}, *F*(000) = 294, λ = 0.710 73 Å, Smart CCD area detector, Siemens, ω−2θ scan, 2° < θ < 24° (−13 ≤ *h* ≤ 16, −10 ≤ *k* ≤ 10, −3 ≤ *l* ≤ 3), 1091 total reflections, 397 independent reflections which were used for refinement. The structure was solved by direct methods (SHELXTL-PLUS) and refined by full-matrix least squares on F^2 (SHELX-93; G. M. Sheldrick, Gottingen, 1993) to R1 = 0.062 and wR2 = 0.134. Residual density, min/max −0.396/0.487e·Å^{-3}. Hydrogen atoms were not included in the refinement, but including the hydrogen atoms in the calculated positions improves the *R*-factor by 1%. The best stacking arrangement is **TCA** on top of **M** (see (8) for details). The packing coefficient of the adduct is 74.6% as calculated from Cerius.

(11) The powder X-ray diffraction pattern of the **CA·M** adduct simulated on the basis of structure found by us agrees with the experimental powder pattern of the precipitate or of the polycrystalline sample.

(12) Zimmerman, S. C. *Science* **1997**, *276*, 544.

(13) Kolotuchin, S. V.; Fenlon, E. E.; Wilson, S. R.; Loweth, C. J.; Zimmerman, S. C. *Angew. Chem., Int. Ed. Engl.* **1995**, *34*, 2654.

(14) Pedireddi, V. R.; Chatterjee, S.; Ranganathan, A.; Rao, C. N. R. *J. Am. Chem. Soc.* **1997**, *119*, 10867.

An organic channel structure formed by the supramolecular assembly of trithiocyanuric acid and 4,4′-bipyridyl

Anupama Ranganathan, V. R. Pedireddi, Swati Chatterjee and C. N. R. Rao*

Chemistry & Physics of Materials Unit, Jawaharlal Nehru Centre for Advanced Scientific Research, Jakkur P.O., Bangalore 560 064, India

Received 18th February 1999, Accepted 5th July 1999

Trithiocyanuric acid (TCA) and 4,4′-bipyridyl (BP) form hydrogen-bonded co-crystals with aromatic compounds such as benzene, toluene, *p*-xylene and anthracene. The TCA–BP co-crystal is composed of cavities formed by the N–H···N hydrogen bonds between the two molecules, and the three-dimensional structure contains channels of approximately 10 Å where aromatic molecules are accommodated. The molar ratios of TCA, BP and the aromatic compound in the co-crystals are 2:1:1 or 2:1:0.5. Benzene, toluene and *p*-xylene are removed from the channels around 190, 183 and 170 °C respectively, and these aromatic guests can be reintroduced into the empty channels of the apo-hosts. The apo-hosts with empty channels have reasonable thermal stability and exhibit shape selectivity in that the empty channels accommodate *p*-xylene but not *m*- or *o*-xylene or mesitylene.

1 Introduction

A variety of inorganic microporous solids have been synthesized in recent years because of their use in catalysis and separation technology.[1–3] These structures are generally prepared by using structure-directing template molecules such as organic amines. Organic amines are also employed extensively to synthesize open-framework metal phosphates and carboxylates.[3] A few organic porous structures containing transition metal ions have been synthesized.[4,5] The supramolecular assembly formed by trimesic acid (benzene-1,3,5-tricarboxylic acid) with cobalt nitrate and pyridine reported by Yaghi *et al.*[4] is a good example of such a metal–organic microporous material. The recently reported nanoporous material formed between trimesic acid and cupric nitrate reported by Chui *et al.*[6] highlights the role of metal–organic interactions in supramolecular assemblies. There is also considerable interest in designing organic microporous solids by making use of supramolecular assemblies. Thus, hexagonal channel structures formed by trimesic acid and its anion have been described.[7] There are also other organic channel structures comprising large pores in three-dimensional structures reported in the literature.[8,9] The nanoporous molecular sandwiches formed by guanidium ions and disulfonate ions serve as representative examples.[8] It has been shown recently that the hexagonal rosette structure formed through hydrogen bonding between cyanuric acid and melamine gives rise to channels in the three-dimensional structure.[10] Self-assembly of three-dimensional networks with large chambers has been reported by using a pyridone as the tecton.[11] Inclusion compounds of urea and thiourea as well as the quinol clathrates also belong to this class of organic solids, but they are not engineered supramolecular designs. We have been exploring several possible designs of organic porous solids for some time and have found that the hydrogen bonded co-crystal between trithiocyanuric acid (TCA) and 4,4′-bipyridyl (BP) has channels in the three-dimensional structure that can accommodate benzene.[12] We have carried out detailed investigations to explore whether other aromatic compounds can be introduced in these channels and if so, whether there is shape selectivity. The study has shown that the supramolecular hydrogen-bonded structure formed by TCA and BP accommodates several aromatic molecules, with some shape selectivity and thermal stability.

TCA   **BP**

2 Experimental

The 2:1 co-crystal of TCA with BP was first prepared by the co-crystallization of the two compounds in methanol solution, but the crystals were not very stable, because of the loss of methanol. After the loss of methanol, the crystallinity was not sufficiently good to carry out a single crystal study. In the presence of benzene, toluene, *o*-xylene, *m*-xylene, *p*-xylene or anthracene, however, the crystals of the 2:1 co-crystal of TCA and BP were stable. In a typical preparation, 106 mg of TCA and 47 mg of BP were taken up in 15 mL of methanol in the presence of an aromatic compound (5 mL). In the case of anthracene, the same quantities of TCA and BP as above were taken up in 15 mL of methanol along with 55 mg of anthracene. After slow evaporation, crystals containing the 2:1 co-crystal of TCA with BP, containing different proportions of the aromatic compounds were obtained. We designate the co-crystals containing benzene, toluene, *o*-xylene, *m*-xylene, *p*-xylene, and anthracene by **1**, **2**, **3**, **4**, **5** and **6** respectively. We could not prepare the 2:1 TCA–BP cocrystal with mesitylene. The crystals were generally needle-shaped and yellow in colour. The as-prepared crystals were dried at room temperature over a period of two days prior to the characterization studies.

Crystals of **1–6** prepared as above were employed for the determination of molecular structure by single crystal X-ray diffraction. The relevant details of the crystal structures of the co-crystals are listed in Table 1.† The intensity data were collected on a SMART system, Siemens, equipped with a CCD area detector,[13] using Mo-Kα radiation. The structures were solved and refined using SHELXTL[14] software. The refinements were uncomplicated and converged to good *R*-factors as

---

†CCDC reference number 1145/171. See http://www.rsc.org/suppdata/jm/1999/2407 for crystallographic files in .cif format.

Table 1 Crystal data for the complexes, **1–6**, formed between trithiocyanuric acid and 4,4′-bipyridyl along with various guest molecules

	1	**2**	**3**	**4**	**5**	**6**
Formulae	$2(C_3H_3N_3S_3)$: $C_{10}H_8N_2$:C_6H_6	$2(C_3H_3N_3S_3)$: $C_{10}H_8N_2$:C_7H_8	$2(C_3H_3N_3S_3)$: $C_{10}H_8N_2$:C_8H_{10}	$2(C_3H_3N_3S_3)$: $C_{10}H_8N_2$:C_8H_{10}	$2(C_3H_3N_3S_3)$: $(C_{10}H_8N_2)$: $0.5(C_8H_{10})$	$2(C_3H_3N_3S_3)$: $C_{10}H_8N_2$: $0.5(C_{14}H_{10})$
Mol. wt.	588.82	602.85	616.87	616.87	872.23	599.82
Crystal system	triclinic	triclinic	triclinic	triclinic	triclinic	triclinic
Space group	$P\bar{1}$	$P\bar{1}$	$P\bar{1}$	$P\bar{1}$	$P\bar{1}$	$P\bar{1}$
$a/\text{Å}$	7.018(1)	7.112(1)	7.282(1)	10.324(1)	10.359(1)	10.472(1)
$b/\text{Å}$	10.344(1)	10.489(1)	10.621(1)	11.615(1)	11.467(1)	11.419(1)
$c/\text{Å}$	10.757(1)	10.681(1)	10.648(1)	12.319(1)	18.412(1)	12.932(1)
$\alpha/°$	63.75(1)	65.40(1)	66.30(1)	102.16(1)	83.65(1)	100.46(1)
$\beta/°$	75.78(1)	76.58(1)	76.53(1)	93.46(1)	75.00(1)	107.69(2)
$\gamma/°$	73.82(1)	73.05(1)	70.42(1)	90.68(1)	67.94(1)	111.28(2)
Cell vol./Å^3	665.8(1)	687.4(1)	705.9(1)	1441.0(2)	1957.7(3)	1296.5(2)
Z	1	1	1	2	1	2
μ/mm^{-1}	0.54	0.53	0.52	0.51	0.55	0.56
T/K	293	293	293	293	293	293
Total reflectn.	2581	2887	2781	6134	7548	4996
Non-zero reflectn.	1838	1928	1970	4109	5489	3556
R	0.032	0.063	0.067	0.082	0.058	0.049
R_w	0.077	0.183	0.184	0.219	0.132	0.098

shown in Table 1. Except for the atoms of the guest molecules (toluene, *o*-xylene and *m*-xylene) in the complexes **2**, **3** and **4**, all the other atoms are oriented in well resolved positions. Non-hydrogen atoms in all the structures except guest molecules in **2**, **3** and **4** were refined anisotropically while the hydrogen atoms were refined isotropically, except in the structures where the hydrogen atoms were placed in the calculated positions using AFIX routines of SHELXTL. The guest molecules in **2** and **5** were refined isotropically only owing to the disorder. The crystal structures reveal that the ratio of TCA, BP and the aromatic molecules in the co-crystals **1–4** was 2:1:1. The ratio was 2:1:0.5 in **5** and **6**. We see from Table 1 that the unit cells of **1–3** are similar and the unit cells of **4–6** form another set. As an illustrative example, we list the structural parameters of **2** in Table 2. These data typify the structures of the TCA–BP co-crystals examined here. We have not given the data for the other co-crystals for the purpose of brevity. Intermolecular interactions were calculated using the PLATON[15] programme.

Thermogravimetric analysis (TGA) of the co-crystals was carried out over the temperature range 25 °C to 200 °C by employing a Mettler Toledo instrument. These curves showed mass loss due to the removal of the aromatic guest molecules. The apo-hosts thus obtained did not dissolve in benzene and other aromatic solvents. The apo-hosts were immersed in the respective aromatic liquid for several hours. The crystals were then taken out, and the TGA repeated. This procedure was repeated more than once to find out whether the inclusion of the guest molecule was reversible and also whether there was any change in the temperature of decomposition or the proportion of the aromatic compound in the co-crystal, with such cycling.

3 Results and discussion

Co-crystallization of TCA with BP from methanol solution gives a 2:1 hydrogen bonded co-crystal containing the solvent of crystallization. The crystal structure of the co-crystal reveals the presence of intermolecular N–H$\cdots$N hydrogen bonds between TCA and BP as shown in Fig. 1. This hydrogen bonded structure has a cavity of 10 Å (calculated using CERIUS, version 3.0), occupied by methanol molecules. These crystals were unstable under ambient conditions because of the evaporation of methanol. However the 2:1:1 co-crystals of TCA with BP incorporating benzene, **1**, or toluene, **2**, were highly stable. These crystals gave the same two-dimensional hydrogen bonded structure with a cavity as in Fig. 1. Besides

Table 2 Bond lengths and angles of **2**

Distances/Å		Angles/°	
S(23)–C(23)	1.654(6)	C(21)–N(23)–C(23)	123.3(5)
S(22)–C(22)	1.651(6)	C(16)–C(11)–C(12)	116.5(6)
S(21)–C(21)	1.662(6)	N(14)–C(15)–C(16)	122.4(7)
N(23)–C(21)	1.357(8)	C(22)–N(22)–C(23)	125.4(5)
N(23)–C(23)	1.369(8)	C(13)–C(12)–C(11)	119.6(6)
C(11)–C(16)	1.387(9)	N(23)–C(23)–N(22)	115.8(5)
C(11)–C(12)	1.393(9)	N(23)–C(23)–S(23)	123.4(4)
C(15)–N(14)	1.339(8)	N(22)–C(23)–S(23)	120.9(5)
C(15)–C(16)	1.393(9)	C(11)–C(16)–C(15)	120.5(6)
N(22)–C(22)	1.363(7)	C(22)–N(21)–C(21)	126.1(5)
N(22)–C(23)	1.378(7)	N(21)–C(22)–N(22)	113.4(5)
C(12)–C(13)	1.392(9)	N(21)–C(22)–S(22)	122.8(5)
N(21)–C(22)	1.361(8)	N(22)–C(22)–S(22)	123.7(5)
N(21)–C(21)	1.369(7)	C(13)–N(14)–C(15)	117.5(6)
N(14)–C(13)	1.327(9)	N(23)–C(21)–N(21)	115.9(6)
C(4)–C(7)	1.26(3)	N(23)–C(21)–S(21)	122.6(4)
C(4)–C(5)	1.47(3)	N(21)–C(21)–S(21)	121.5(5)
C(4)–C(6)	1.78(3)	N(14)–C(13)–C(12)	123.5(7)
C(3)–C(5)	1.35(4)	C(7)–C(4)–C(5)	71(2)
C(3)–C(6)	1.53(4)	C(7)–C(4)–C(6)	115(2)
C(3)–C(2)	1.97(5)	C(5)–C(3)–C(6)	50(2)
C(5)–C(6)	1.23(3)	C(5)–C(3)–C(2)	88(3)
C(5)–C(7)	1.60(3)	C(6)–C(3)–C(2)	38(2)
C(2)–C(6)	1.20(3)	C(6)–C(5)–C(3)	72(3)
		C(6)–C(5)–C(4)	82(2)
		C(3)–C(5)–C(4)	154(4)
		C(6)–C(5)–C(7)	130(3)
		C(3)–C(5)–C(7)	158(4)
		C(6)–C(2)–C(3)	51(2)
		C(2)–C(6)–C(5)	148(3)
		C(2)–C(6)–C(3)	91(3)
		C(5)–C(6)–C(3)	57(2)
		C(2)–C(6)–C(4)	157(3)
		C(5)–C(6)–C(4)	55(2)
		C(3)–C(6)–C(4)	112(3)
		C(4)–C(7)–C(5)	61(2)

the strong N–H$\cdots$N bonds (H$\cdots$N, 1.82 Å) responsible for the formation of the cavity, there are weak C–H$\cdots$S bonds (H$\cdots$S, 2.91, 2.93 Å) between the TCA and BP molecules. In addition, there are N–H$\cdots$S hydrogen bonds between TCA molecules (H$\cdots$S, 2.5 Å). The three-dimensional structure of **2** is shown in Fig. 2. The structure clearly reveals the presence of channels formed by the stacking of the layers with cavities. The channels accommodate aromatic molecules as can be seen in Fig. 2(a). Fig. 2(b) shows another projection of **2** where we see layers of TCA·BP with toluene molecules residing in the regions of the

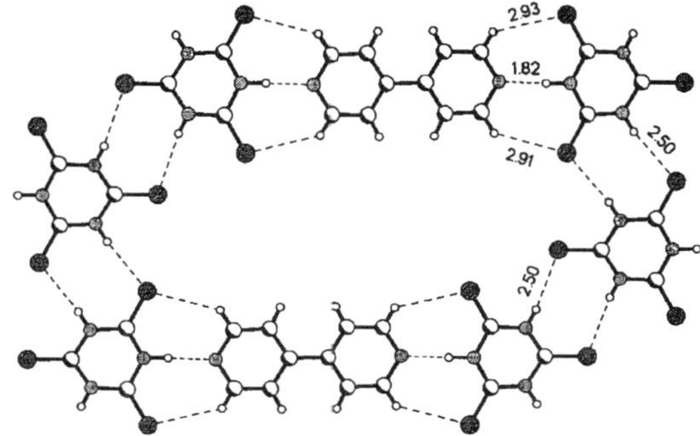

Fig. 1 Self assembly of TCA and BP forming a layered network with a cavity.

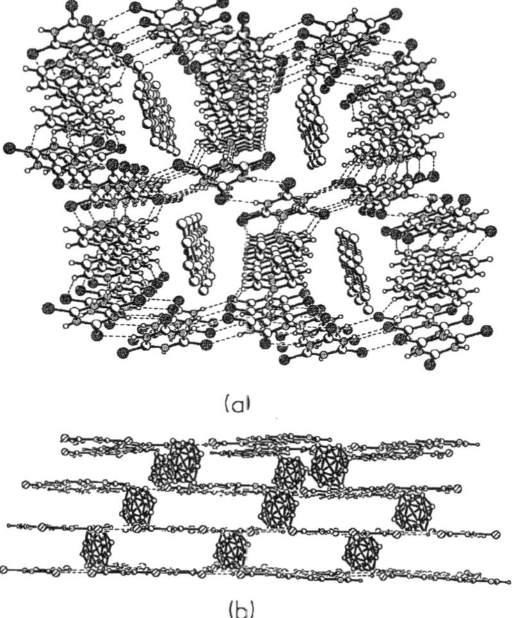

(a)

(b)

Fig. 2 (a) Three-dimensional structure of the 2:1 TCA–BP co-crystal containing toluene, 2, in the channels. (b) Stacking arrangement of molecular sheets perpendicular to the channels.

cavity. The crystals were stable and heating them up to 180 °C or slightly higher, did not destroy them.

The hydrogen-bonding pattern observed in 1 is interesting in terms of the multi-point recognition.[16] The hydrogen-bonding pattern with one N–H···N and two C–H···S bonds corresponds to the highly robust three-point recognition pattern. Different possible three-point recognition patterns of the type are shown below. While the patterns represented in (a) and (b) correspond to the motifs comprising strong hydrogen bonds, pattern (c) is a result of weak C–H···O hydrogen bonds.[17] A search of the Cambridge Structural Database[18] (CSD) suggests that the pattern present in 1 corresponding to (d) is indeed unique. No other such system appears to be reported in the literature.

Crystal structures of 3–6 show that the cavities in these co-crystals are similar to those in 2, a minor exception being 4 containing m-xylene, where the size and shape of the cavity are somewhat different. In Fig. 3, we show the two-dimensional

structure of the TCA–BP in 4 to illustrate the small differences between the cavities in this co-crystal and 1. Here, the intermolecular N–H···N as well as the C–H···S bonds are slightly longer but the N–H···S bonds between the TCA molecules are similar to those in 2 (Fig. 1). A noticeable difference is the slight asymmetry in the cavity arising from the different intermolecular hydrogen bond distances on the two sides. In Fig. 4a, we show the three-dimensional structure of 6 where the guest molecule is anthracene. In Fig. 4b, we show another projection of 6 showing TCA·BP layers wherein anthracene molecules protrude through the cavity. Fig. 2 and 4 serve to demonstrate the similarity of the three-dimensional channels in the various co-crystals.

In Fig. 5, we show the TGA curve of the benzene co-crystal 1, recorded at a heating rate of $2\,°C\,min^{-1}$. All the benzene is removed (mass loss, 13.9%) around 190 °C which is well above the boiling point of benzene (see curve 1). After the removal of benzene, the apo-host crystals were soaked in benzene for 36 hours and again subjected to TGA. (These crystals are insoluble in benzene as evidenced by the retention of the crystal morphology and absence of the relevant spots on the TLC plates.) The crystals showed the loss of benzene at a slightly lower temperature (158 °C), but the quantity of benzene in the channels is less by about 4% (see curve 2 in Fig. 5), compared to the as-prepared co-crystal. A repetition of this procedure showed that the removal of benzene continued to occur around 158 °C with the same mass loss (curve 3 in Fig. 5). It appears that after the benzene in the channels of the initial co-crystal is removed, there are some definitive changes in the channel structure, with the empty channels accommodating benzene to a smaller extent. The lower temperature at which benzene is removed also suggests that the guest molecules do not interact as strongly with the host channels. One difference that we notice is that the weak C–H···S interactions present in the as-formed co-crystal is absent after the removal of the guest molecules. The crystal structure of the TCA–BP co-crystal 1 heated to 200 °C (to remove all the benzene) gave a single crystal X-ray diffraction pattern, but the data were not sufficiently good to obtain the structure. We could not determine the structure after reintroducing benzene. Powder

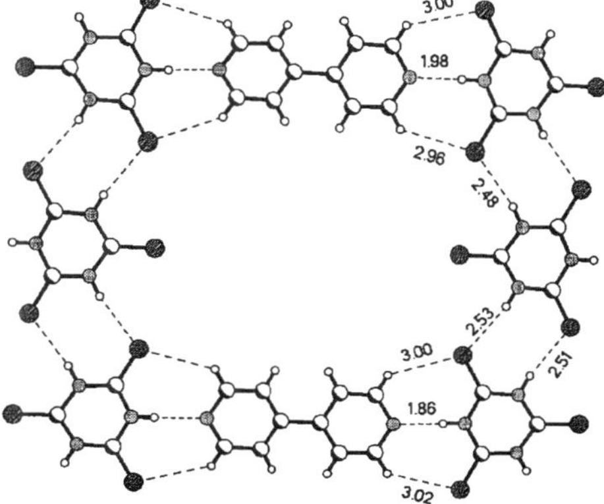

Fig. 3 The two-dimensional layered network in the TCA–BP co-crystal containing m-xylene, 4.

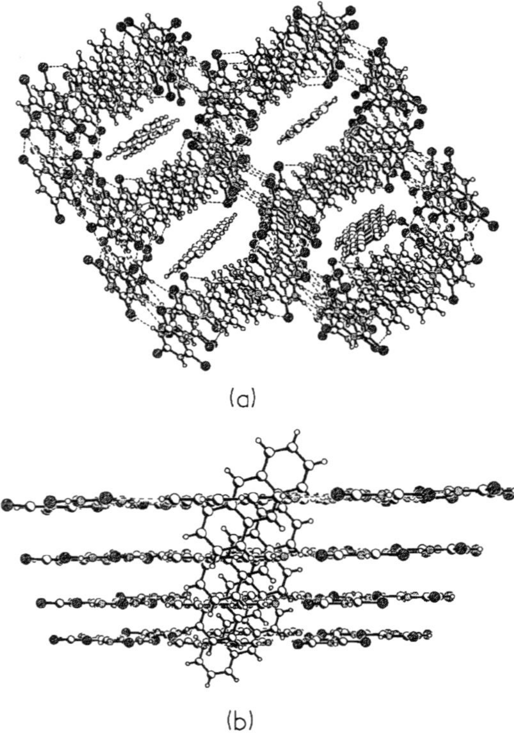

(a)

(b)

Fig. 4 (a) Three-dimensional structure of the 2:1 TCA–BP co-crystal containing anthracene, **6**. (b) Stacking arrangement of planar molecular sheets perpendicular to the channels.

XRD patterns were identical after two cycles of guest removal, but the patterns of the apo-host were generally poorer in quality.

TGA of the toluene co-crystal **2** showed a mass loss of 14.8% at 183 °C. The host crystals with the empty channels were soaked in toluene for 36 hours and then subjected to TGA. These crystals showed a mass loss of 9.2% at 166 °C, a behaviour similar to that of the benzene co-crystal, **1**. TGA of the p-xylene cocrystal, **5**, showed a mass loss (12.2%) due to the removal of the aromatic guest around 167 °C. After reintroduction of p-xylene in the channels, the mass loss occurred around 139 °C, but the magnitude of mass loss was smaller

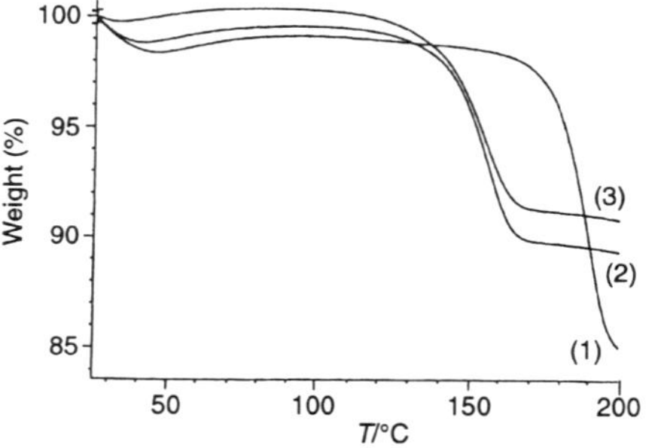

Fig. 5 Thermogravimetic analysis (heating rate 2 °C min^{-1}) of the 2:1 TCA–BP co-crystal containing benzene. Curve 1 is that of the as-prepared sample. Curve 2 was obtained with a sample obtained by immersing the host crystal with empty channels in the aromatic liquid for several hours. Curve 3 was obtained after repeating this procedure for a second time.

(8%). The temperature of decomposition and the mass loss did not change appreciably after further cycling in these cases as well. The co-crystals of o-, m-xylene, **3** and **4**, however, behaved entirely differently. Once the guest molecules were removed, they could not be reintroduced in the channels even after soaking the host crystals with empty channels for long periods. This observation suggests that there is some selectivity in the channels of the TCA–BP co-crystal. This can be understood from the differences in the shapes of p-, o- and m-xylenes. In p-xylene, the two methyl groups are symmetrically positioned and favor the ready re-incorporation into the channels of the host crystal. This is not the case in the other two isomers. Accordingly, the empty channel of the TCA–BP co-crystal only takes in p-xylene from a mixture of the three xylene isomers. It should be recalled that the channel does not accommodate mesitylene as well. In the apo-host obtained by removing benzene from co-crystal **1**, toluene could be introduced readily, but not p-xylene. We have attempted to understand the selectivity of the apo-host channels by carrying out powder XRD measurements before and after the removal of the guest molecules. Although the apo-hosts are still crystalline, we have not yet been able to come to any meaningful conclusions from the XRD patterns since the data are not entirely sufficient.

4 Conclusions

The stable organic solid containing channels formed by the supramolecular hydrogen-bonded assembly of trithiocyanuric acid and 4,4′-bipyridyl can accommodate aromatic molecules such as benzene, toluene and p-xylene. The apo-host is thermally stable up to 200 °C and exhibits shape selectivity with respect to the xylene isomers. The channels do not accommodate mesitylene. It would be interesting to carry out chemical reactions in these channels.

5 References

1 (a) G. Ozin, *Adv. Mater.*, 1992, **4**, 612; (b) J. M. Thomas, *Angew. Chem., Int. Ed. Engl.*, 1994, **33**, 913.
2 (a) S. Zone and M. E. Davis, *Curr. Opin. Solid State Mater. Sci.*, 1996, **1**, 107; (b) A. K. Cheetham, *Science*, 1994, **264**, 794; (c) X. Bu, P. Feng and G. D. Stucky, *J. Chem. Soc., Chem. Commun.*, 1995, 1337.
3 (a) S. Ayyappan, A. K. Cheetham, S. Natarajan and C. N. R. Rao, *Chem. Mater.*, 1998, **10**, 3746 and references cited therein; (b) S. Natarajan, S. Ayyappan, A. K. Cheetham and C. N. R. Rao, *Chem. Mater.*, 1998, **10**, 1627 and references cited therein.
4 (a) O. M. Yaghi, G. Li and H. Li, *Nature*, 1995, **378**, 703; (b) O. M. Yaghi, H. Li and T. L. Groy, *J. Am. Chem. Soc.*, 1996, **118**, 9096; (c) O. M. Yaghi and H. Li, *J. Am. Chem. Soc.*, 1995, **117**, 10401; (d) H. Li, M. Eddaoudi, T. L Groy and O. M. Yaghi, *J. Am. Chem. Soc.*, 1998, **120**, 8571.
5 (a) D. Venkataraman, G. B. Gardner, S. Lee and J. S. Moore, *J. Am. Chem. Soc.*, 1995, **117**, 11600; (b) B. F. Abrahams, J. Coleiro, B. Hoskins and R. Robson, *Chem. Commun.*, 1996, 603.
6 S. S. Y. Chui, S. M. F. Lo, J. P. H. Charmant, A. G. Orpen and I. D. Williams, *Science*, 1999, **283**, 1148.
7 (a) S. V. Kolotuchin, E. E. Fenlon, S. R. Wilson, C. J. Loweth and S. C. Zimmerman, *Angew. Chem., Int. Ed. Engl.*, 1995, **34**, 2654; (b) R. Melendez, C. V. K. Sharma, M. J. Zaworotko, C. Baner and R. D. Rogers, *Angew. Chem., Int. Ed. Engl.*, 1996, **35**, 2213; (c) S. C. Zimmerman, *Science*, 1997, **276**, 543.
8 (a) V. A. Russel, C. C. Evans, W. Li and M. D. Ward, *Science*, 1997, **276**, 575; (b) V. A. Russel and M. D. Ward, *Chem. Mater.*, 1996, **8**, 1654.
9 J. S. Moore and S. Lee, *Nature*, 1995, **374**, 792.
10 A. Ranganathan, V. R. Pedireddi and C. N. R. Rao, *J. Am. Chem. Soc.*, 1999, **121**, 1752.
11 M. Simard, D. Su and J. D. Wuest, *J. Am. Chem. Soc.*, 1991, **113**, 4696.
12 V. R. Pedireddi, S. Chatterjee, A. Ranganathan and C. N. R. Rao, *J. Am. Chem. Soc.*, 1997, **119**, 10867.
13 Siemens, *SMART System*, Siemens Analytical X-ray Instruments Inc., Madison, WI, U.S.A., 1995.

14 G. M. Sheldrick, SHELXTL, *Users Manual*, Siemens Analytical X-ray Instruments Inc., Madison, WI, U.S.A., 1993.

15 A. L. Spek, PLATON, *Molecular Geometry Program*, University of Utrecht, The Netherlands, 1995.

16 (*a*) J. C. MacDonald and G. M. Whitesides, *Chem. Rev.*, 1994, **94**, 2383; (*b*) G. M. Whitesides, E. E. Simanek, J. P. Mathias, C. T. Seto, D. N. Chin, M. Mammen and D. M. Gordon, *Acc. Chem. Res.*, 1995, **28**, 37.

17 (*a*) G. R. Desiraju, *Angew. Chem., Int. Ed. Engl.*, 1995, **34**, 2311; (*b*) G. R. Desiraju, *Crystal Engineering: The Design of Organic Solids*, Elsevier, New York, 1989; (*c*) G. R. Desiraju, *Science*, 1997, **278**, 404; (*d*) G. R. Desiraju, *Curr. Opin. Solid State Mater. Sci.*, 1997, 451.

18 F. H. Allen and O. Kennard, *Chemical Design Automation News*, 1993, **8(1)**, 31.

Paper 9/05344A

ELSEVIER

Journal of Molecular Structure 520 (2000) 107–115

Journal of
**MOLECULAR
STRUCTURE**

www.elsevier.nl/locate/molstruc

Self-assembled four-membered networks of trimesic acid forming organic channel structures

S. Chatterjee, V.R. Pedireddi, A. Ranganathan, C.N.R. Rao[*]

*Chemistry and Physics of Materials Unit and the CSIR Centre of Excellence in Chemistry,
Jawaharlal Nehru Centre for Advanced Scientific Research, Jakkur P.O., Bangalore 560 064, India*

Received 27 April 1999; accepted 21 June 1999

Abstract

By the co-crystallization of trimesic acid, TMA, with molecules such as dimethylamine, N,N,N',N'-tetramethylethylenediamine and methanol, it has been possible to generate hydrogen-bonded four-membered networks of TMA. The three-dimensional arrangement of the four-membered networks gives rise to channels occupied by the guest molecules. It has also been possible to generate a four-membered network by co-ordination of TMA with Co(II). © 2000 Elsevier Science B.V. All rights reserved.

Keywords: Trimesic acid; Hydrogen-bonding; Supramolecular chemistry; Crystal engineering; Organic channel structures

1. Introduction

Hydrogen-bonded assemblies of trimesic acid (TMA) have attracted attention because of their interesting structures, which provide the basis for the design of organic porous solids. Thus, the two-dimensional hexagonal network, formed by hydrogen bonding between TMA molecules, has been investigated by several workers. Kolotuchin et al. [1] have described the hexagonal channel structures of TMA containing ethanol, pyrene and tetrahydrofuran. Hexagonal networks have also been formed between TMA anions and ammonium cations [2]. The hexagonal networks have been enlarged by inserting 4,4'-bipyridyl between the TMA units [3]. Yaghi et al. [4] have obtained a stable layered network by using transition metal salts of TMA, with the aromatic guest molecules residing in the interlayer spacing. We want to explore other types of networks giving rise to channels, especially the four-membered networks formed by the self-assembly of TMA molecules. One such structure formed by trimesic acid hydrate, TMA·H$_2$O, giving rise to channels wherein picric acid molecules reside, has been described in the literature [5]. We can visualize at least four different types of hydrogen-bonded four-membered networks (**1–4**) formed by TMA, as shown in Fig. 1, of which **1** corresponds to the network formed by TMA·H$_2$O [5]. Such a four-membered network could also result from metal ion co-ordination as shown in **5** in Fig. 1. In this article, we describe different types of four-membered networks formed by TMA through supramolecular organization.

2. Experimental

All the chemicals used for the preparation of the

---

* Corresponding author. Tel.: + 91-80-8462-762; fax: + 91-80-846-2766.

E-mail address: cnrrao@jncasr.ac.in (C.N.R. Rao).

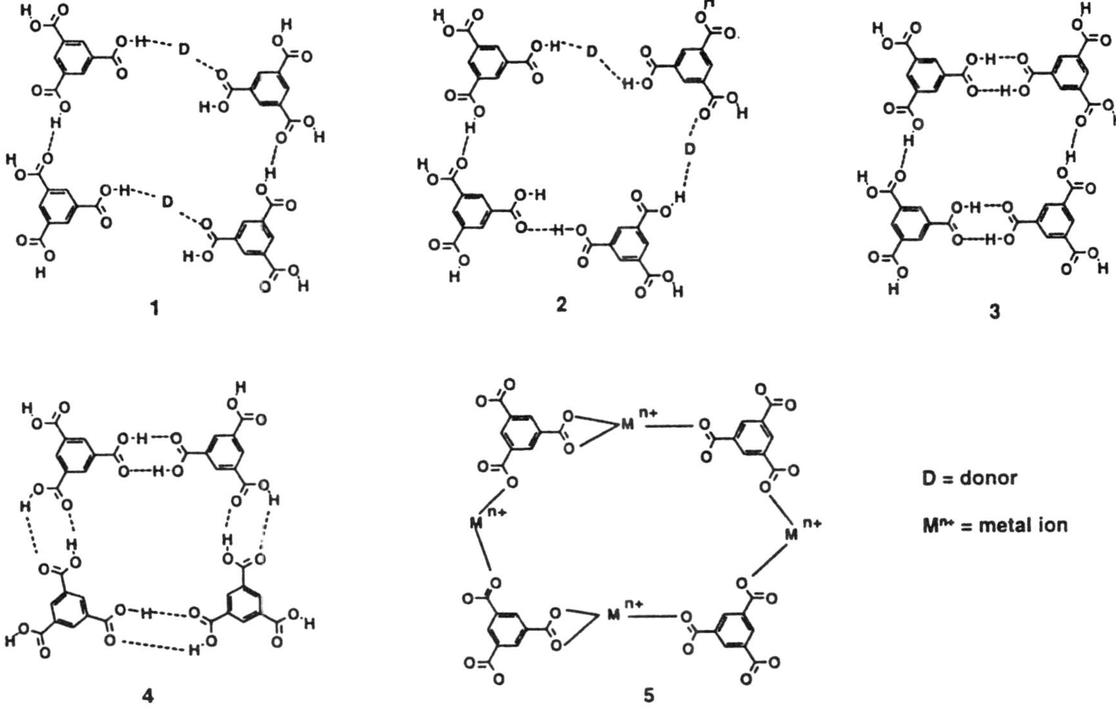

Fig. 1. Self-assembled four-membered networks formed by TM based on hydrogen bonding (**1–4**) and metal coordination (**5**).

co-crystals discussed in this article were obtained commercially from Sigma–Aldrich. Crystallization of TMA from dimethylformamide (DMF) gave a 1:1 adduct of TMA and dimethylamine (the dimethylamine resulting from the decomposition of DMF). However, a 1:2 adduct, between TMA and DMF could be obtained by the crystallization of TMA from a DMF–benzene (1:10) mixture. Co-crystallization of TMA with N,N,N',N'-tetramethylethylenediamine (TMEDA) from a methanol solution gave a 2:1 adduct incorporating water molecules. Adducts of TMA with methanol and acetone were prepared by crystallization from methanol/benzene and acetone solutions, respectively. Co-crystallization of TMA and Co(NO$_3$)$_2$·6H$_2$O from a dimethylsulfoxide (DMSO) solution gave a complex which incorporated DMSO molecules.

We designate the various adducts of TMA studied by us as follows:

Label	Adduct
A	TMA·2DMF
B	TMA:DMA
C	2TMA:TMEDA·2H$_2$O
D	TMA·CH$_3$OH·H$_2$O
E	TMA:1.5Co(NO$_3$)$_2$·6H$_2$O·2DMSO

The single crystals obtained as above were used for the elucidation of three-dimensional crystal structures by X-ray diffraction. The intensity data were collected on a Siemens X-ray diffractometer equipped with a CCD-area detector [6]. The crystal structures were determined and refined using the Siemens SHELXTL-PLUS package [7]. The analyses were uncomplicated and the non-hydrogen atoms were refined anisotropically. The hydrogen atoms were refined isotropically

Table 1
Crystal structures of some of the supramolecular assemblies of trimesic acid, TMA

	A	B	C	D	E
Formulae	$C_9H_6O_6$:$2(C_3H_7N_1)$	$C_9H_6O_6$:$C_2H_7N_1$	$2(C_9H_6O_6)$:$C_6H_{16}N_2$:$2(H_2O)$	$C_9H_6O_6$:CH_3OH:H_2O	$C_9H_3O_6$:$1.5Co(II)$:$2(C_2H_6S_2O_1)$
Mol. wt.	356.33	254.22	572.52	260.2	451.77
Crystal system	Monoclinic	Triclinic	Monoclinic	Triclinic	Monoclinic
Space group	$P2_1/c$	$P\bar{1}$	$P2_1/c$	$P\bar{1}$	$P2_1/c$
a	16.912(6)	7.207(1)	12.587(1)	3.679(1)	10.124(1)
b	14.475(5)	9.246(1)	15.127(1)	8.971(1)	10.281(1)
c	7.390(2)	9.285(1)	15.426(1)	18.038(3)	16.330(1)
α	90	69.34(1)	90	77.76(1)	90
β	90.55(3)	86.50(1)	113.40(1)	86.86(1)	95.34(1)
γ	90	86.60(1)	90	88.04(1)	90
Cell vol. (Å^3)	1809.0(10)	577.41(1)	2695.6(3)	580.8(2)	1692.3(3)
Z	4	2	4	2	4
$F(000)$	752	266	1208	272	918
D_{calcd} (Mg m^{-3})	1.308	1.462	1.411	1.488	1.773
λ (Å)	0.7107	0.7107	0.7107	0.7107	0.7107
μ (mm^{-1})	0.107	0.121	0.117	0.132	1.772
Crystal size (mm)	$0.35 \times 0.20 \times 0.25$	$0.30 \times 0.25 \times 0.25$	$0.40 \times 0.35 \times 0.25$	$0.30 \times 0.30 \times 0.30$	$0.40 \times 0.20 \times 0.20$
Diffractometer	Smart, CCD area detector	Smart, CCD area detector	Smart, CCD area detector	Smart, CCD area detector	Smart, CCD area detector
T (K)	293	293	293	293	293
X-radiation	Mo-Kα	Mo-Kα	Mo-Kα	Mo-Kα	Mo-Kα
θ range (°)	1–24	2–24	1–24	1–24	2–24
h	$-$ 12 to 15	$-$ 4 to 8	$-$ 13 to 13	$-$ 4 to 4	$-$ 9 to 11
k	$-$ 16 to 6	$-$ 10 to 10	$-$ 14 to 16	$-$ 8 to 9	$-$ 11 to 10
l	$-$ 8 to 8	$-$ 10 to 10	$-$ 10 to 17	$-$ 19 to 18	$-$ 18 to 17
Total reflection	3513	2524	10946	2426	6799
Non-zero reflection	2100	1647	3855	1630	2429
σ-level	3	3	3	3	3
R	0.059	0.049	0.073	0.133	0.039
R_w	0.116	0.110	0.084	0.308	0.075
Max. (e Å^{-3})	0.239	0.440	0.332	0.967	0.334

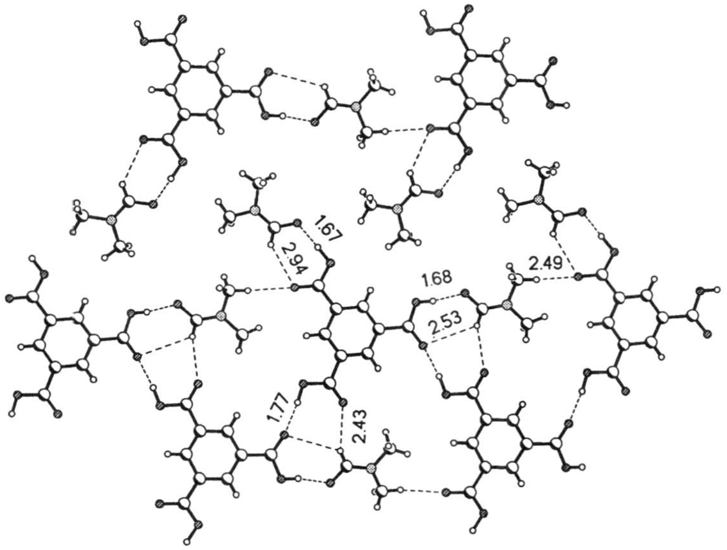

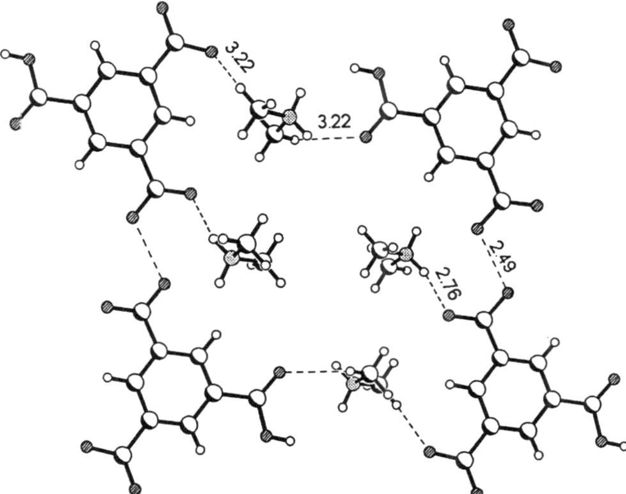

Fig. 2. Arrangement of molecules of TMA and dimethylforamide in a two-dimensional planar sheet.

in all the adducts except **A** and **D**. In the former, the atoms were placed in calculated positions while in the latter, the hydrogen atoms were not included in the refinement owing to the presence of disorder. The non-covalent bond distances were calculated using the program PLATON [8]. We have listed the crystal structure data of all the adducts in Table 1.

3. Results and discussion

Crystallization of TMA from DMF in the presence of benzene yields the 1:2 adduct, **A** (Table 1). In the two-dimensional arrangement, this adduct consists of molecular tapes formed between the molecules of TMA and DMF through hydrogen bonds as shown

Fig. 3. Four-membered networks formed by the hydrogen-bonded self-assembly of TMA and dimethylamine.

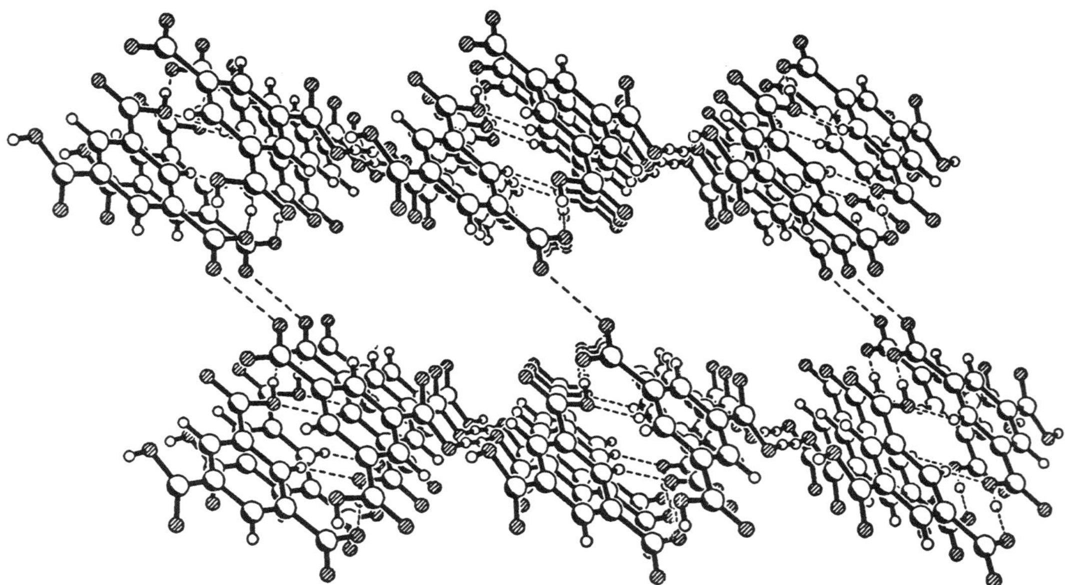

Fig. 4. Three-dimensional arrangement of the four-membered network forming channels. The amine molecules in the channels are not shown.

in Fig. 2. Out of the three available carboxyl groups in each TMA molecule, only two interact with the DMF molecules, forming O–H⋯O and C–H⋯O hydrogen-bond couplings with H⋯O distances in

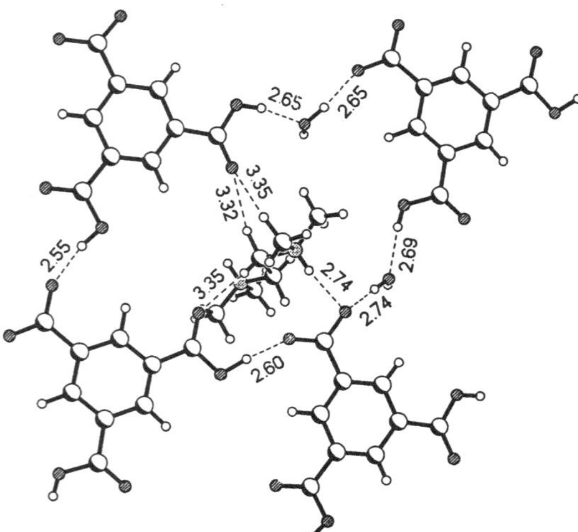

Fig. 5. Four-membered network formed by the hydrogen-bonded self-assembly of TMA and H_2O molecules with N,N,N',N'-tetramethylethylenediamine.

the range of 1.67–1.68 and 2.53–2.94 Å, respectively. The third carboxyl group also forms similar hydrogen-bond couplings, by interacting with an adjacent TMA·DMF supermolecule as shown in Fig. 2. In this case, the H⋯O distances corresponding to the O–H⋯O and C–H⋯O bonds are 1.77 and 2.43 Å, respectively.

The 1:1 adduct of TMA with dimethylamine, **B**, forms a four-membered network consisting of hydrogen bonds between the TMA molecules as well as between TMA and the amine molecules as shown in Fig. 3. In this hydrogen-bonded structure, corresponding to **1** in Fig. 1, two TMA molecules are held together by a single O–H⋯O hydrogen bond (O⋯O, 2.49 Å) between them, and each TMA is also bonded to the amine by a N–H⋯O bond (N⋯O, 2.76 Å). Two such pairs of TMA molecules are connected to each other through the amine molecules by C–H⋯O bonds (C⋯O, 3.22 Å). The four-membered networks form two-dimensional sheets which are stacked to give rise to channels in the three-dimensional arrangement as shown in Fig. 4.

Crystallization of TMA with N,N,N',N'-tetramethylethylenediamine from an aqueous methanol solution gives the 2:1 adduct, **C**, which incorporates water

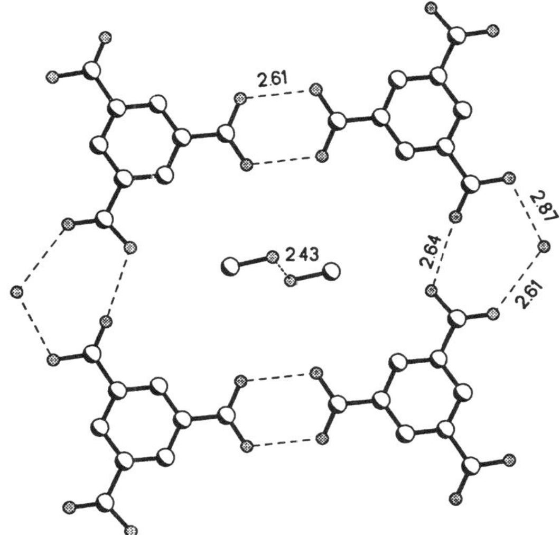

Fig. 6. Four-membered network formed by the hydrogen-bonded self-assembly of TMA and H_2O molecules with CH_3OH in the cavity.

molecules. This hydrogen-bonded adduct also has a four-membered network (Fig. 5), wherein three TMA molecules are held together by single O–H···O hydrogen bonds between the carboxylic groups (O···O, 2.55 and 2.60 Å), and the unit of three TMA molecules is connected to the fourth molecule, through two H_2O molecules, by means of O–H···O hydrogen bonds (O···O, 2.65–2.74 Å). The hydrogen-bonding pattern here is similar to **2** in Fig. 1. The cavity is occupied by the amine forming N–H···O (N···O, 2.74 Å) and C–H···O (C···O, 3.32–3.35 Å) hydrogen bonds with the TMA molecules. In the three-dimensional arrangement, the cavities get aligned to yield a channel structure.

The more interesting four-membered hydrogen-bonded network of TMA is the one wherein dimeric hydrogen bonds are formed between the carboxyl groups. We obtain such a network involving dimeric hydrogen bonds when TMA is crystallized from a 1:1 mixture of CH_3OH and benzene. The 1:1 adduct with methanol, **D**, includes a water molecule as well. Fig. 6 shows the hydrogen-bonded square network in the

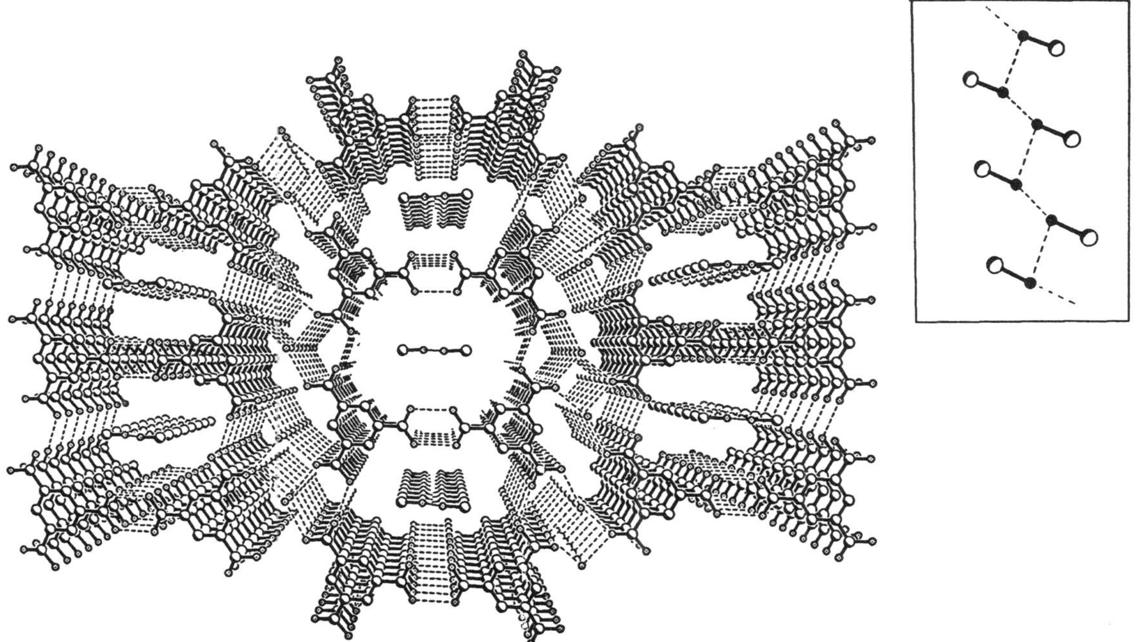

Fig. 7. The three-dimensional channel structure with CH_3OH molecules in the channels. The linear hydrogen-bonded polymer of CH_3OH is shown in the inset.

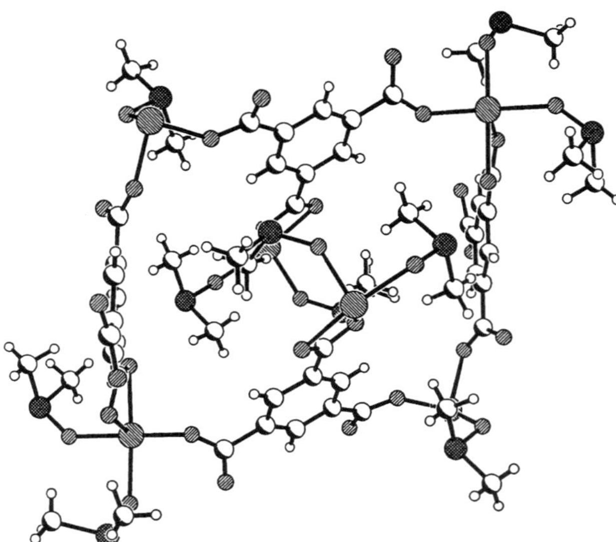

Fig. 8. Four-membered network formed by the coordination of TMA molecules with Co(II) and DMSO. Molecules of DMSO in the cavity are also shown.

TMA–methanol adduct. Here, two sides of the network are made up of dimeric hydrogen bonds between the TMA molecules (O···O, 2.61 and 2.64 Å) and these sides are connected to each other through H_2O molecules by O–H···O bonds (O···O, 2.61 and 2.87 Å), to yield planar sheets as in **3** in Fig. 1. This layered structure of TMA·H_2O is similar to that reported in the literature [5]. These sheets are stacked in a three-dimensional arrangement yielding channels occupied by CH_3OH molecules as shown in Fig. 7. The CH_3OH molecules are hydrogen-bonded to one another (as shown in the inset of Fig. 7). Such a polymeric chain of methanol molecules with O–H···O hydrogen bonds (O···O, 2.43 Å) has not been characterized hitherto. This has been possible in the present study since the CH_3OH molecules present in the channel have no option but to form a linear polymer. The 1:1 adduct of TMA with acetone has a structure identical to that with methanol in Fig. 7 except that acetone molecules are present in the channel.

We have not been able to obtain a four-membered network with dimeric hydrogen-bonds on all the four sides as in **4**, Fig. 1. We have, however, obtained a structure wherein all the four sides are connected by metal ions as in **5**. Such a structure was obtained by the co-crystallization of TMA and Co(NO$_3$)$_2$·6H$_2$O in dimethylsulfoxide (DMSO). We show this network in Fig. 8, where the molecules of TMA are connected to each other through Co(II) by different types of dative bonds. The coordination of Co(II) is depicted in Fig. 9. One of the carboxylate groups bonds to Co(II) as a

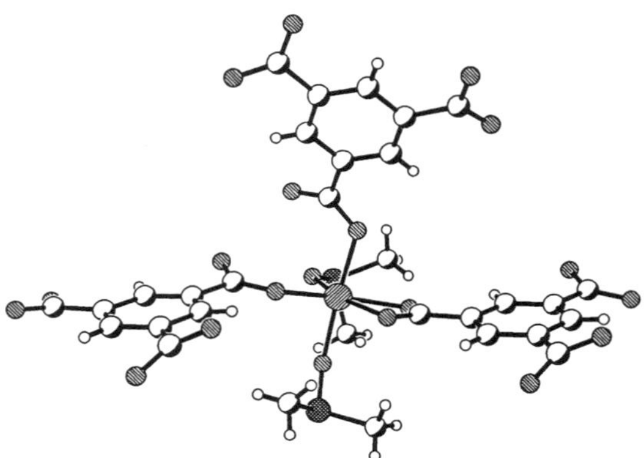

Fig. 9. Hexa-coordination around Co(II) with three molecules of TMA and two DMSO molecules. One of the TMA molecules forms two dative bonds.

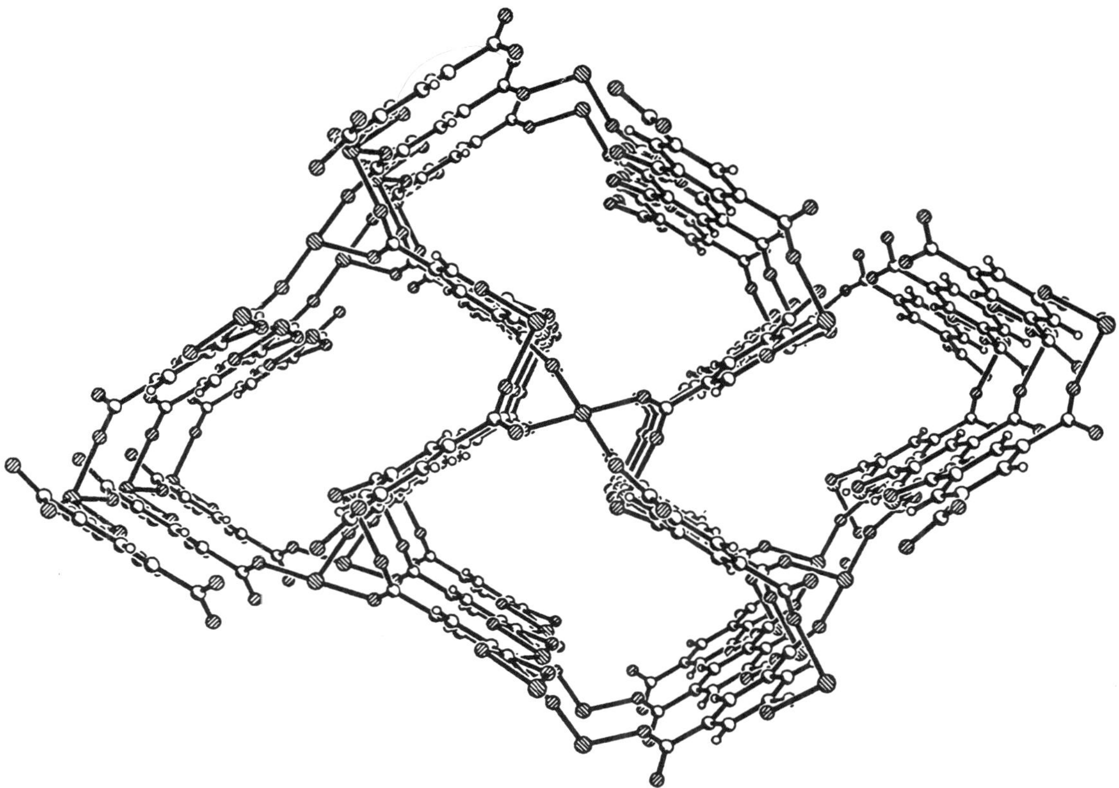

Fig. 10. Three-dimensional channel structure of the four-membered network of TMA involving metal coordination. DMSO molecules in the channels are not shown.

bidentate ligand, with Co–O distances of 2.24 and 2.12 Å. The other two carboxylate groups of TMA interact with the metal ion as monodentate ligands with Co–O distances of ∽2.05 Å. The hexa-coordination of Co(II) is satisfied by bonding with DMSO molecules. The four-membered rectangular units get aligned in the three-dimensional structure to form channels as shown in Fig. 10. The DMSO molecules residing in the channels get connected to the walls through Co–O bonds of 2.09 and 2.14 Å (Fig. 8).

4. Conclusions

The present study demonstrates how TMA molecules can self-assemble to form various types of four-membered networks involving hydrogen-bonding or metal ion coordination. The four-membered networks form three-dimensional channels that accommodate various guest molecules. The packing coefficients of these structures are all generally around 0.7 as expected of such solids, the channel structure with CH_3OH exhibiting the lowest value.

References

[1] S.V. Kolotuchin, E.E. Fenlon, S.R. Wilson, C.J. Loweth, S.C. Zimmerman, Angew. Chem. Int. Ed. Engl. 34 (1995) 2654.

[2] R.E. Melendez, C.V.K. Sharma, M.J. Zaworotko, C. Bauer, R.D. Rogers, Angew. Chem. Int. Ed. Engl. 35 (1996) 2213.

[3] C.V.K. Sharma, M.J. Zaworotko, Chem. Commun. (1996) 2655.

[4] O.M. Yaghi, G. Li, H. Li, Nature 378 (1995) 703.

[5] F.H. Herbstein, R.E. Marsh, Acta Cryst. B34 (1977) 2358.

[6] Siemens, SMART System, Siemens Analytical X-ray Instruments Inc., Madison, WI, USA, 1995.

[7] G.M. Sheldrick, SHELXTL, Users Manual, Siemens Analytical X-ray Instruments Inc., Madison, WI, USA, 1993.

[8] A.L. Spek, PLATON, Molecular Geometry Program, University of Utrecht, The Netherlands, 1995.

A novel hybrid layer compound containing silver sheets and an organic spacer

C. N. R. Rao,* Anupama Ranganathan, V. R. Pedireddi and A. R. Raju

Chemistry & Physics of Materials Unit and CSIR Centre for Excellence in Chemistry, Jawaharlal Nehru Centre for Advanced Scientific Research, Jakkur P.O., Bangalore 560 064, India. E-mail: cnrrao@jncasr.ac.in

Received (in Cambridge, UK) 12th October 1999, Accepted 23rd November 1999

A novel compound of the formula $Ag_2\cdot CA$ (CA = cyanuric acid) possessing Ag sheets and hydrogen-bonded CA chains, exhibits anisotropic conductivity and acts as an infinite parallel plate capacitor with a high dielectric constant.

Design of infinite two- and three-dimensional arrays of metal–ligand networks has attracted considerable attention in the last few years not only because of the structural and topological novelty of such engineered solids but also due to the potentially interesting electrical, magnetic and other properties.[1,2] Recently, interesting structures containing polymeric Ag(I) species and heterocyclic as well as aromatic compounds have been described.[2–4] For instance, Ag(I)–benzenesulfonate has a layered structure containing a planar hexagonal array of Ag(I) ions incorporating the anion.[4] A coordination network of dicyanodiphenylacetylene comprising Ag(I) sheets with an Ag⋯Ag separation of 3.39 Å has also been reported.[5] Equally interesting are the supramolecular Ag(I) complexes constructed with several aromatic compounds involving novel stacking of the aromatics such as the herringbone packing pattern.[6,7] Many of these compounds have Ag⋯Ag separation significantly shorter than the van der Waals contact distance, with Ag(I) having linear, trigonal, tetrahedral or hexagonal coordination,[8] however, the materials are generally either insulators or semiconductors with no unusual properties.

During the course of our investigations of supramolecular assemblies of cyanuric acid, $C_3H_3N_3O_3$ (CA), involving both hydrogen bonding and metal-ion coordination, we have isolated a novel silver compound possessing two-dimensional Ag sheets with the CA molecules in the interlayer space, forming linear hydrogen-bonded chains. This compound of composition $Ag_2\cdot CA$, is a unique organic–inorganic hybrid with novel electrical properties, and is entirely different from the supramolecular assemblies described above, and from other Ag complexes and salts with short Ag⋯Ag distances.[9] Here, we describe the fascinating structure and properties of this Ag(I) compound.

Reaction of $AgNO_3$ with CA under hydrothermal conditions† gave single crystals of composition $Ag_2\cdot CA$ suitable for X-ray diffraction studies. The structure was determined‡ using the SHELXTL package,[10] with the intensity data collected on a Siemens smart diffractometer equipped with CCD area detector. The asymmetric unit of the compound is shown in Fig. 1. The structure viewed down the *b*-axis (Fig. 2) reveals the presence of two-dimensional sheets of Ag atoms separated by CA molecules, the inter-sheet separation being *ca.* 6 Å. The average Ag⋯Ag distance in the sheets is 2.95 Å, slightly longer than the Ag–Ag distance in metallic silver (2.89 Å). The dative Ag–O and Ag–N bond distances are in the range 2.22–2.76 and 2.09–2.12 Å, respectively, and the CA molecules are linked by relatively short N–H⋯O hydrogen bonds (H⋯O 1.90 Å, N⋯O, 2.75 Å), giving rise to a linear chain (Fig. 2). The arrangement of the CA molecules in the layers perpendicular to the Ag sheets is illustrated in Fig. 3. Another noteable feature of $Ag_2\cdot CA$ is that the organic spacer itself is the anion. $Ag_2\cdot CA$ can also be compared with Ag_3O with an *anti*-BiI_3 structure with the O atoms occupying 2/3 of the octahedral holes.[11] The Ag⋯Ag and Ag–O distances in $Ag_2\cdot CA$ are slightly longer than in Ag_3O, except for one Ag–O bond of 2.22 Å.

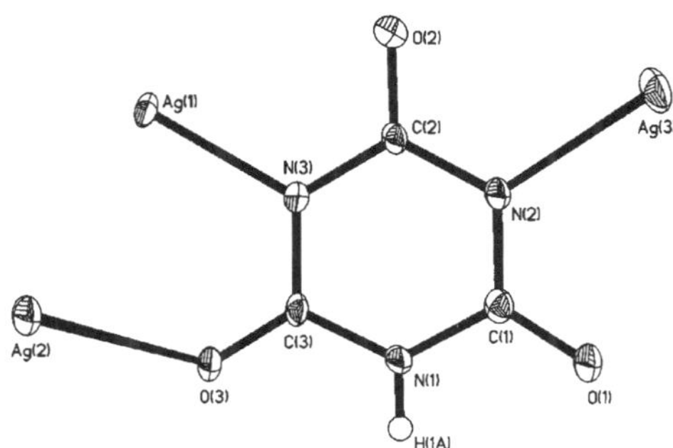

Fig. 1 ORTEP drawing showing the asymmetric unit of $Ag_2\cdot CA$.

The presence of multipoint recognition patterns between CA and Ag in $Ag_2\cdot CA$ is of interest. For instance, the hydrogen bonding motif (a) in Scheme 1, found in the structure of CA,[12] is replaced by the motif (b) in $Ag_2\cdot CA$ by the substitution of Ag for H. Motif (b) is similar to that present in Ag carboxylates, except that two of the O atoms are replaced by N atoms. In addition, there are three-point recognition patterns (c) and (d) in $Ag_2\cdot CA$, comparable to the hydrogen bonding pattern (e) found in the adduct of CA with melamine.[13]

$Ag_2\cdot CA$ crystals are mica-like and are readily cleaved because of the layer structure and the presence of weakly bound Ag sheets. The presence of two-dimensional Ag sheets is expected to give rise to anisotropic conductivity. Accordingly, the values of the dc conductivity parallel and perpendicular to the Ag sheets (*bc* plane) are *ca.* 5×10^{-3} and *ca.* 2×10^{-5} S cm^{-1}, respectively, at 300 K. The conductivity along the sheets is temperature-independent down to 15 K. $Ag_2\cdot CA$, in which the conducting Ag sheets are separated by the organic spacer molecules, can be considered as an infinite parallel plate capacitor. In accord with this, the crystals possess a high static dielectric constant of *ca.* 22 000 at 300 K, a phenomenally high value which promises potential applications. This value of the

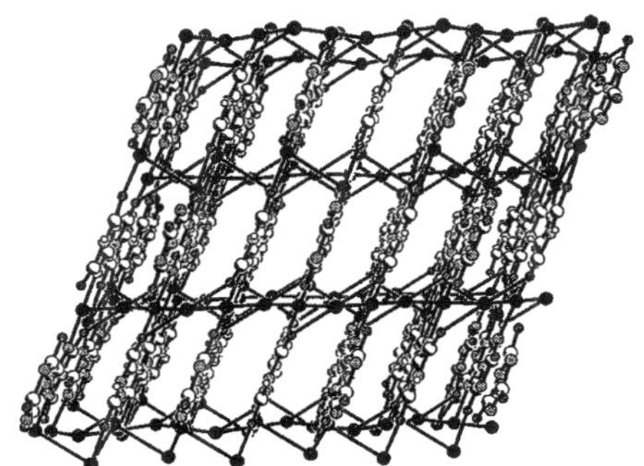

Fig. 2 Structure of $Ag_2\cdot CA$ showing Ag sheets and linear CA chains.

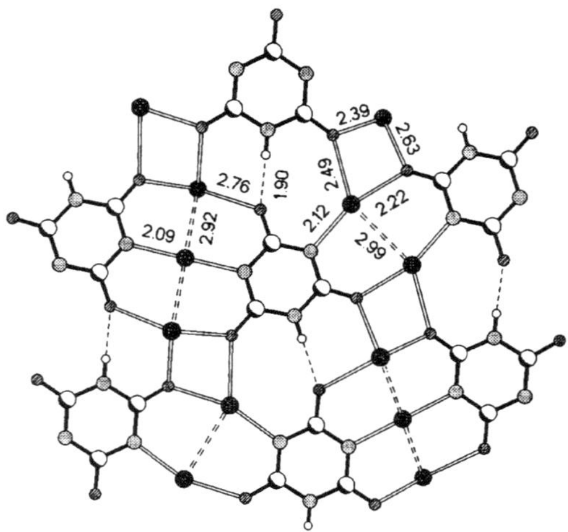

Fig. 3 Structure of a layer (*ab* plane) perpendicular to the Ag sheets: Solid lines, covalent bonds; double lines, dative bonds; dashed lines, hydrogen bonds; double dashed lines, Ag–Ag bonds.

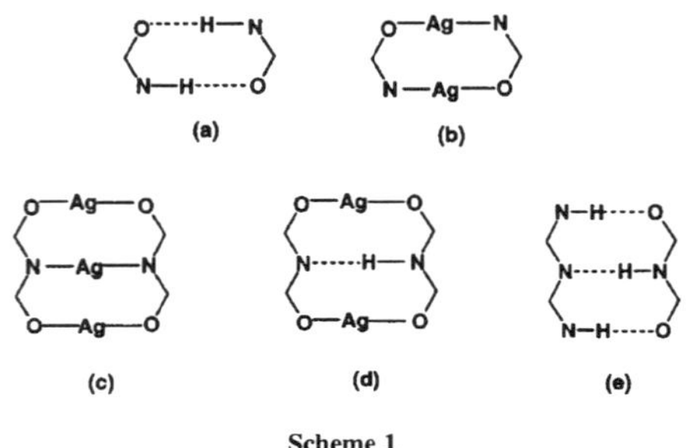

Scheme 1

dielectric constant is comparable with that of barium titanate (*ca.* 12000 at 300 K).

Notes and references

† A solution (10 mL) consisting of a mixture of AgNO₃ (0.170 g) and CA (0.129 g) in water in a Teflon flask was placed in a steel bomb. The bomb was placed in an oven maintained at 180 °C for 24 h and then cooled to room temperature (25 °C) over a period of 3 h. Good quality off-white plate-like single crystals, were obtained and the composition of the product was established as Ag₂·CA, consistent with that derived from X-ray crystallography. There were no other products in the reaction.

‡ *Crystal data* for Ag₂·CA(Ag₂C₃N₃HO₃): crystal dimensions, 0.35 × 0.25 × 0.20 mm, monoclinic, space group, *C*2/*c*, $a = 12.726(1)$, $b = 13.064(1)$, $c = 6.623(1)$ Å, $\beta = 97.35(1)°$, $V = 1092.0(2)$ Å³, $Z = 8$, $D_c = 4.170$ Mg m⁻³, μ(Mo-Kα) = 7.11 mm⁻¹, F(000) = 1264, $\lambda = 0.71073$ Å, ω–2θ scan, $2 < \theta < 24°$ ($-13 \leqslant h \leqslant 14$, $-11 \leqslant k \leqslant 14$, $-6 \leqslant l \leqslant 7$), 2266 total reflections, 785 independent reflections which were used in the refinement. The structure was solved to $R1 = 0.035$ and $wR2 = 0.083$. Hydrogen atoms were placed in calculated positions.

CCDC 182/1491. See http://www.rsc.org/suppdata/cc/a9/a908171b/ for crystallographic files in .cif format.

1 R. Robson, B. F. Abrahams, S. R. Batten, R. W. Gable, B. F. Hoskins and J. Liu, *Supramolecular Architecutre*, ACS Symp. Ser., Washington, D.C., 1992, vol. 449, 1992, ch. 19.
2 L. Carlucci, G. Ciani, D. M. Proserpio and A. Sironi, *Angew. Chem., Int. Ed. Engl.*, 1995, **34**, 1895 and references therein.
3 G. K. H. Shimizu, G. D. Enright, C. I. Ratcliffe, J. A. Ripmeester and D. D. M. Wayner, *Angew. Chem., Int. Ed.*, 1998, **37**, 1407.
4 G. K. H. Shimizu, G. D. Enright, C. I. Ratcliffe, K. F. Preston, J. L. Reid and J. A. Ripmeester, *Chem. Commun.*, 1999, 1485.
5 K. A. Hirsch, S. R. Wilson and J. S. Moore, *Inorg. Chem.*, 1997, **36**, 2960.
6 M. Munakata, L. P. Wu, T. Kuroda-Sowa, M. Maekawa, Y. Suenaga, G. L. Ning and T. Kojima, *J. Am. Chem. Soc.*, 1998, **120**, 8610.
7 G. L. Ning, L. P. Wu, K. Sugimoto, M. Munakata, T. Kuroda-Sowa and M. Maekawa, *J. Chem. Soc., Dalton Trans.*, 1999, 2529.
8 D. Venkataraman, Y. Du, S. R. Wilson, P. Zhang, K. Hirsch and J. S. Moore, *J. Chem. Educ.*, 1997, **74**, 915.
9 A. F. Wells, *Structural Inorganic Chemistry*, Oxford University Press, Oxford, 5th edn., 1995, p. 1099.
10 G. M. Sheldrick, SHELXTL, Users Manual, Siemens Analytical X-ray Instruments Inc., Madison, WI, 1993.
11 W. Beesk, P. G. Jones, H. Rumpel, E. Schwarzmann and G. M. Sheldrick, *J. Chem. Soc., Chem. Commun.*, 1981, 664.
12 P. Coppens and A. Vos, *Acta Crystallogr., Sect. B*, 1971, **27**, 146.
13 A. Ranganathan, V. R. Pedireddi and C. N. R. Rao, *J. Am. Chem. Soc.*, 1999, **121**, 1752.

Communication a908171b

ELSEVIER

Journal of Molecular Structure (Theochem) 500 (2000) 339–362

THEO CHEM

www.elsevier.nl/locate/theochem

Experimental and theoretical electronic charge densities in molecular crystals

G.U. Kulkarni, R.S. Gopalan, C.N.R. Rao*

Chemistry and Physics of Materials Unit, Jawaharlal Nehru Centre for Advanced Scientific Research, Jakkur P.O., Bangalore 560 064, India

Abstract

Electronic charge density distribution in molecular systems has been described in terms of the topological properties. After briefly reviewing methods of obtaining charge densities from X-ray diffraction and theory, typical case studies are discussed. These studies include rings and cage systems, hydrogen bonded solids, polymorphic solids and molecular NLO materials. It is shown how combined experimental and theoretical investigations of charge densities in molecular crystals can provide useful insights into electronic structure and reactivity. © 2000 Elsevier Science B.V. All rights reserved.

Keywords: Electronic charge densities; Molecular crystals; Topological properties

1. Introduction

The description of charge distribution in crystalline lattices has come a long way since the first quantum model of the atom. It was known from early days that a quantitative account of the chemical bonds in molecules and crystals would require the calculation of the probability density of the electron cloud between atoms. The experimental possibility itself was considered soon after the discovery of X-ray diffraction. As early as 1915, Debye [1] stated "that experimental study of scattered radiation, in particular from light atoms, should get more attention, since along this way it should be possible to determine the arrangement of electrons in the atoms". Calculation of charge densities in molecules has been a preoccupation of theoretical chemists for sometime [2] and several charge density investigations of crystalline solids by both

experiment and theory have been reported in the literature [3–11]. These comprise molecular crystals including non-linear materials, metallo-organic complexes and inorganic compounds. Koritsanszky [12] has provided a summary of charge density studies with reference to the topological properties and the electrostatic potential. The more recent literature has been surveyed by Spackman [13,14] and the charge density analysis in relation to metal–ligand and intermolecular interactions, has been discussed by Coppens [15].

In this article, we discuss electronic charge density in molecular systems as obtained from both experiment and theory. Besides introducing the topological analysis of charge density, we examine the experimental methods based on X-ray crystallography. In addition to dealing with the multipolar formalism for treating the experimental data, the program packages for orbital calculations in free molecules and crystals are mentioned. In particular, we discuss ring and cage systems, intermolecular hydrogen bonds, polymorphism and non-linear optical crystals.

---

* Corresponding author. Tel.: + 91-80-846-2762; fax: + 91-80-846-2766.

E-mail address: cnrrao@jncasr.ac.in (C.N.R. Rao).

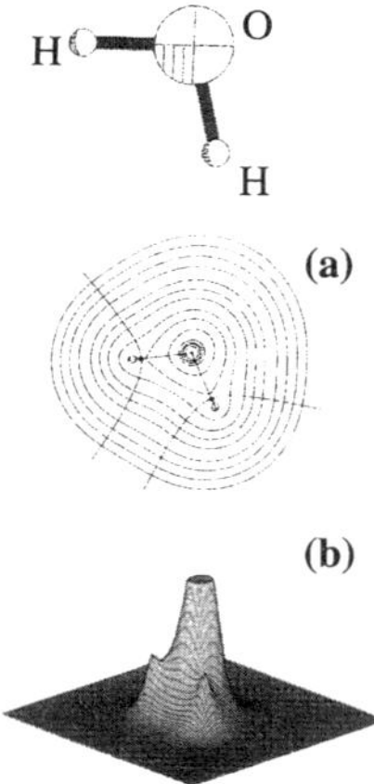

Fig. 1. Water: charge density in the molecular plane, (a) the contour map. The outermost contour has the value $0.0067\ e\text{Å}^{-3}$. The density increases almost exponentially for inner contours. The bond paths, the interatomic surfaces and the bond critical points are also indicated. (b) The relief map where the atom-cores are seen as peaks (reproduced with permission from Bader [2]).

2. Description of electronic charge density

The electronic charge density in an N-electron system is the probability per unit volume of finding any of the electrons in the phase space τ,

$$\rho(r) = N \int \psi^* \psi \, d\tau \tag{1}$$

where ψ is the stationary state function; τ denotes the spin coordinates of all the electrons and the Cartesian coordinates of all N electrons but one [2]. It is expressed in $e\text{Å}^{-3}$ or au (1 au $= 6.7483\ e\text{Å}^{-3}$). The description of electronic structure of a molecule in real space therefore relates to the charge density distribution around the constituent atoms. The density

Table 1
Critical points in molecular systems

Function	CP (rank, signature)	Chemical entity
$\rho(r)$	$(3, -3)$	Atom
$\rho(r)$, $V(r)$	$(3, -1)$	Bond
$\rho(r)$	$(3, +1)$	Ring
$\rho(r)$	$(3, +3)$	Cage
$\nabla^2\rho(r)$	$(3, -3)$	Lone-pairs
$V(r)$	$(3, +3)$	Lone-pairs

in a molecule can be conveniently modeled by partitioning into core, spherical valence and deformation valence around each atom [16],

$$\rho_{\text{atom}}(r) = \rho_{\text{core}}(r) + \rho_{\text{valence}}(r) + \rho_{\text{deformation}}(r, \theta, \phi) \tag{2}$$

The topology of a charge distribution has many rich features—maxima, minima, saddles and nodes which help characterize intuitional elements such as atom cores, bonds and lone-pair electrons. As an example, the charge density distribution in water molecule [2] is depicted in Fig. 1 in the form of contour as well as relief maps. The density is maximum at the oxygen core position and decreases steeply towards the mid-region between oxygen and hydrogen reaching the minimum value at the 'critical point' ($\nabla\rho = 0$). This point carries maximum densities from the other two perpendicular directions. A quantitative description of charge density thus boils down to examining the number and the nature of such critical points in and around a molecule. A critical point (CP) is characterized not only by its density and location but also by the curvatures and the associated signs. The curvature of charge density at a point ($\nabla^2\rho$) well known as the Laplacian, is a measure of the charge concentration ($\nabla^2\rho < 0$) or depletion ($\nabla^2\rho > 0$) at that point. It is obtained as the sum of eigenvalues—λ_1, λ_2 and λ_3 of the Hessian matrix diagonalized against principal axes of curvature with λ_3 set along the internuclear vector. The rank of the matrix (3 for a stable molecular system) and the signature (sum of the signs of eigenvalues) imply the nature of the critical point. At a $(3, -3)$ CP, for example, all the curvatures are negative and ρ is locally the maximum. All the atom-cores exhibit $(3, -3)$ CPs in ρ (see Table 1). The relative magnitudes of the curvatures perpendicular to the bond direction (λ_3) determine the ellipticity [17] associated with a bond, $\epsilon = (\lambda_1/\lambda_2) - 1$.

An estimate of bond polarization [17] can be obtained from the location of the bond CP with respect to the internuclear vector,

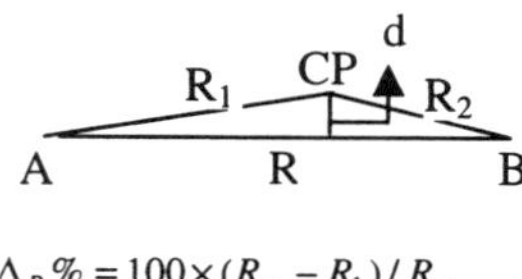

$$\Delta_B\% = 100 \times (R_m - R_1)/R_m$$

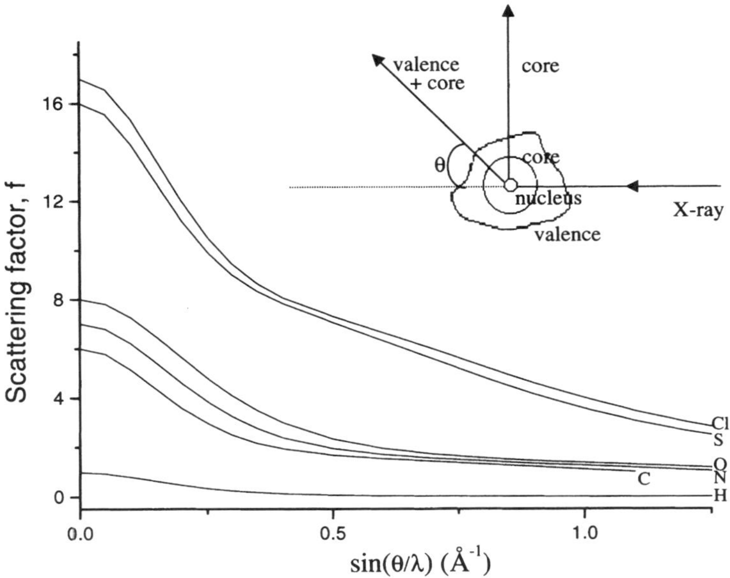

Fig. 2. Variations of the scattering factor with sin θ/λ for different atoms. Above 0.5 Å^{-1}, the scattering due to valence electrons decreases gradually and the atom-core becomes visible. This is shown schematically in the inset. The scattering factors were obtained from the International Tables for Crystallography, Vol. IV, (1974) p. 71.

$$\Delta_B\% = 100 \times (R_m - R_1)/R_m \tag{3}$$

where $R_m = (R_1 + R_2)/2$. The Δ value is used to describe relative electronegativities of the atoms involved. The strain involved in a bond can be estimated [17] in terms of the vertical displacement, d of the bond path from the internuclear vector

$$d = 2 \times \sqrt{s \times (s - R_1) \times (s - R_2) \times (s - R)}/R \tag{4}$$

where $s = (R_1 + R_2 + R)/2$.

The electrostatic potential [18,19] generated by a molecule containing nuclear charges, z_i, placed at R_i, with a charge distribution $\rho(r)$, is given by

$$V(r) = \sum_i \frac{z_i}{|r - R_i|} - \int \frac{\rho(r')}{|r' - r|} d^3r' \tag{5}$$

It is more useful in describing attractive and repulsive interactions and also in determining the electrophilic and nucleophilic sites in molecules. The other quantity of interest is the kinetic energy density at the critical point, G_{CP} [20] which has been obtained by

$$G_{CP} = \frac{3}{10}(3\pi)^{2/3}\rho_{CP}^{5/3} + \frac{\nabla^2\rho_{CP}}{6} \tag{6}$$

This quantity has been used in some cases for the calculation of hydrogen bond energies [21] and their classification [22].

3. Charge density from X-ray diffraction

Experimental determination of charge density relies mostly on X-ray diffraction although other techniques have been applied in some instances. X-ray diffraction arises from scattering by electrons and therefore carries information on the distribution of electronic charge in real space [10]. The intensity of a Bragg reflection, $I(\mathbf{h})$, at a given temperature, is proportional to the square of its structure factor,

$$I(\mathbf{h}) \propto |F(\mathbf{h})|^2 \propto \left|\sum_i f_i(\mathbf{h})\, e^{2\pi i \mathbf{h} \cdot \mathbf{r}}\right|^2 \tag{7}$$

where $f_i(\mathbf{h})$ is the scattering factor of the ith atom in the unit cell of volume, V. The charge density is obtained by the Fourier summation of the experimentally measured reflections

$$\rho(r) = \frac{1}{V} \sum_{\mathbf{h}} F(\mathbf{h}) \, e^{-2\pi i \mathbf{h} \mathbf{r}} \tag{8}$$

In conventional structure determination, $f_i(\mathbf{h})$ is approximated to the scattering factor from spherical electronic density, while for a complete description of bonding, accurate modeling of $f_i(\mathbf{h})$ becomes necessary. In parallel to $\rho(r)$ (see Eq. (2)),

$$f(\mathbf{h}) = f_{core}(\mathbf{h}) + f_{valence}(\mathbf{h}) + f_{deformation}(\mathbf{h}) \tag{9}$$

Such a partitioning of $f(\mathbf{h})$ is justifiable in X-ray diffraction since one can select regions of reciprocal space where core scattering is predominant. In Fig. 2, we show variation of $f(\mathbf{h})$ with scattering angle, θ, for various elements. Each curve is composed of two regions. At low angles or Bragg vector ($\mathbf{h} = 2\sin\theta/\lambda$), $f(\mathbf{h})$ decreases steeply and above $\sim 0.5\ \text{Å}^{-1}$, the fall is gradual. The first part has contributions from both the atom-core and the valence density while the second arises mainly due to the core. Thus, X-ray diffraction facilitates extraction of the bonding or the deformation density,

$$\rho_{deformation} = \rho_{total} - \rho_{promolecule} \tag{10}$$

where promolecule is obtained by the superposition of atoms without any interaction between them. This is called the (X–X) method. In the (X–N) method, the core positions along with the thermal parameters are obtained from a neutron diffraction experiment. The latter is particularly useful while dealing with hydrogen atom positions though it requires two data sets, which can be expensive besides having to grow bigger crystals. In recent years, the (X–X) method has become more popular.

Eq. (8) above necessitates data collection covering a wide range of the reciprocal space. For small unit cells with dimensions $\sim 30\ \text{Å}$, data collection up to moderately high resolution ($1.25\ \text{Å}^{-1}$) can be achieved using short wavelength radiations such as MoK_{α} ($0.71\ \text{Å}$). Moreover, data collection strategy critically depends on the type of the diffractometer. In the past, point detectors mounted on a four circle diffractometers were used for charge density measurements with the data collection extending in some

cases to a period of few weeks. Of late, area detectors or image plates are preferred over the conventional ones, as the experiments can be carried out faster with greater redundancy [23]. Area detectors in combination with synchrotron radiation are becoming increasingly popular [24,25]. Koritsanszky et al. [25] demonstrated that the charge density data can be collected within a day.

Thermal smearing of the charge density caused by atomic vibrations can hamper the extraction of subtle features of bonding,

$$f(T)|_{(h,k,l)} = f(0) \exp\left(-(b_{11}h^2 + b_{12}hk + b_{13}hl\right.$$
$$\left. + b_{22}k^2 + b_{23}kl + b_{33}l^2)\right) \tag{11}$$

where

$$2\pi^2 a^{*2} U_{11} = b_{11} \qquad 2 \times 2\pi^2 a^* b^* U_{12} = b_{12} \qquad \text{etc.}$$

where U_{ij}s are the anisotropic displacement parameters and a^*, b^* and c^* are the reciprocal lattice vectors. Low temperature experiments at ~ 100 K, are carried out by allowing a stream of liquid nitrogen to fall on the crystal. In some cases however, much lower temperature (~ 20 K) has been achieved using one or two stage He-closed-cycle cryostats [9]. It is also necessary to choose a good quality, least mosaic crystal for charge density work.

4. Data refinement and computer codes

A preliminary knowledge of the crystal structure is important prior to a detailed charge density analysis. Direct methods are commonly used to solve structures in the spherical atom approximation. The most popular code is the SHELX from Sheldrick [26] which provides excellent graphical tools for visualization. The refinement of the atom positional parameters and anisotropic temperature factors are carried out by applying the full-matrix least-squares method on a data corrected if found necessary, for absorption and diffuse scattering. Hydrogen atoms are either fixed at idealized positions or located using the difference Fourier technique.

In the absence of inputs from neutron diffraction, a higher-order refinement of X-ray data ($>0.6\ \text{Å}^{-1}$) becomes essential to obtain accurate core positions and the associated thermal parameters (the X–X

G.U. Kulkarni et al. / Journal of Molecular Structure (Theochem) 500 (2000) 339–362

method). In this case, the hydrogen atom positions are often adjusted to the average neutron diffraction values [27] and are held there during the refinement. Often, the rigid bond test [28] is carried out and the parameters are corrected for translation–libration motions of the molecule [29]. The aspherical atom electron density is obtained in a local coordination system using the Hansen–Coppens formalism [16],

$$\rho(\mathbf{r}) = \rho_c(r) + P_v \rho_v(\kappa r) + \sum_l R_l(\kappa' r) \sum_{m=-l}^{l} P_{lm} y_{lm}\left(\frac{\mathbf{r}}{r}\right) \tag{12}$$

Here, ρ_c and ρ_v are the spherically averaged Hartree–Fock core and valence densities, respectively, with ρ_v normalized to one electron. The Slater type radial functions $R_1 = N_1 r^n \exp(-\kappa' \xi r)$, modulated by the multipolar spherical harmonic angular functions y_{lm} define the deformation density. The population parameters, P_v and P_{lm}, are floated along with κ, κ' during the refinement. The kappa parameters control the expansion or the contraction of the radial part of the electron cloud with respect to the free atom. The multipoles on the first row atoms are generally refined up to octapole moments, while for the heavier ones, moments up to hexadecapole are used. Hydrogen atoms are restricted to dipole, although occasionally quadrupole moments are included in the refinement. The above formalism is well adopted in the recently developed user-friendly program package, XD [30]. Older codes such as MOLLY [16], VALRAY [31], LSEXP [32], POP [33] are also still in use. The quality of a refined model can be monitored based on the residuals and the goodness-of-fit apart from closely inspecting the deformation density maps.

The valence population coefficients P^i can be used to estimate the pseudo-atomic charges on the different atoms according to the equation,

$$q_i = n_i - P^i_{\text{core}} - P^i_{\text{valence}} \tag{13}$$

where n_i is the total number of electrons of atom i. The molecular dipole moment is given by

$$\mathbf{p}_i = \sum_i z_i \mathbf{R}_i + \int_v \mathbf{r}\rho_i(\mathbf{r}_i)\, d\mathbf{r} \tag{14}$$

5. Theoretical methods

Electronic charge density distribution in a molecule or a crystal may be obtained by Hartree–Fock calculations [34]. It involves calculation of antisymmetrized many-electron wavefunction and minimizing the energy with respect to the coefficients of the one-electron wave function. When the energy is minimized, the wavefunction is said to have achieved self-consistency (SCF). Slater type atomic orbitals were used in the past which posed lot of difficulty in analytically integrating the polynomial functions. Use of gaussian functions in the radial part of the wavefunction has made HF method more applicable. Pople [35] has described the current status of ab initio quantum chemical models in his recent Nobel lecture. The accuracy that can be reached with these models critically depends on how many gaussians, polarization and diffuse functions make the basis set. In principle, using a full configuration interaction (FCI) with a large number of gaussians on each orbital should give the best results. For example, calculations performed using $6\text{-}311G^{**}++$ basis sets at FCI level of theory is common with small molecules. Computer program packages like GAUSSIAN [36] and GAMESS [37] are available for ab initio calculations on molecules. MOPAC [38] is used at a semi-empirical level. Periodic Hartree–Fock calculations suitable for crystalline substances has been incorporated in CRYSTAL algorithm by Dovesi et al. [39]. It uses one-particle basis function made up of Bloch functions,

$$\chi_i(r, k) = (1/\sqrt{N}) \sum_t \chi_a^t\, e^{ikt} \tag{15}$$

Here χ_a^t refers to the ath atomic orbital in the unit cell of the crystal described by the lattice translation vector, t. The CRYSTAL code is also capable of calculating charge density in a solid using the density functional theory (DFT) at local density approximation (LDA) or at generalized gradient approximation (GGA).

The charge density obtained using the theoretical procedures can be usefully compared with that from X-ray diffraction by several means. Charge density maps either in total or in deformation provide the obvious tools to evaluate how well the two models agree. CRYSTAL95 [39] offers routines to calculate

G.U. Kulkarni et al. / Journal of Molecular Structure (Theochem) 500 (2000) 339–362

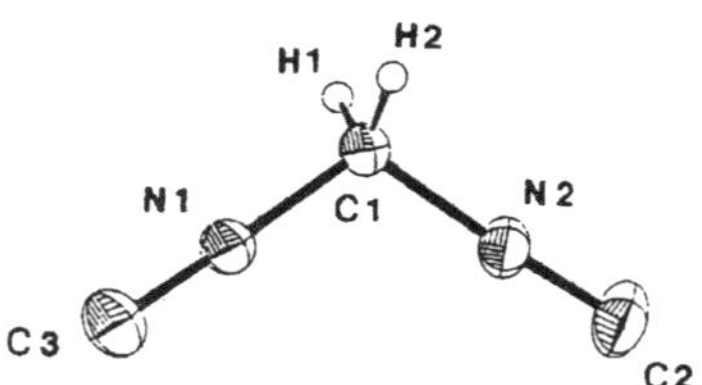

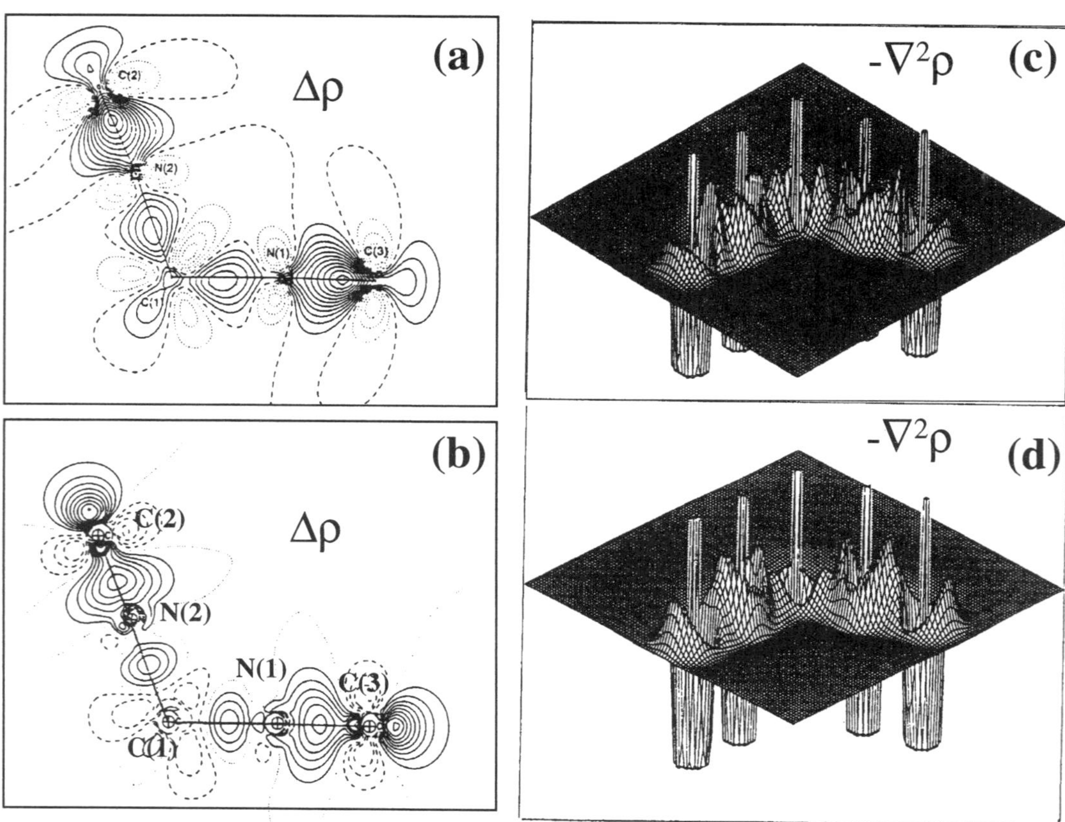

Fig. 3. Charge density in diisocyanomethane: deformation density maps in the molecular plane (a) experimental (b) theoretical (contours at $0.1\ e\text{Å}^{-3}$). The non-bonded regions of C(2) and C(3) are more depleted in (a) with the density migrating to the inside of the molecule. The corresponding Laplacians (range -20 to $250\ e\text{Å}^{-5}$) are shown in (c) and (d), respectively (reproduced with permission from Koritsanszky et al. [40]).

X-ray structure factors from the theoretical density. Using the theoretical structure factors, one can carry out multipolar refinement in parallel to experimental X-ray data and make in depth comparison of the topological properties from the two sets.

6. Charge density in bonds, rings and cages

Covalent bonds are associated with high charge densities $(1.5–3\ e\text{Å}^{-3})$ and negative Laplacians, while ionic bonds are characterized by small densities and positive Laplacians. Hydrogen bonds are

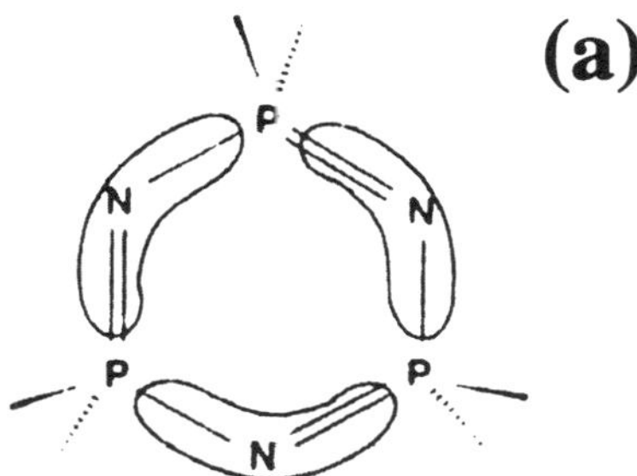

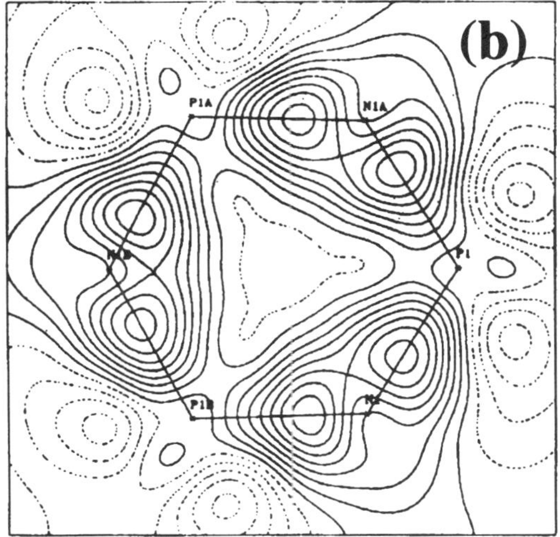

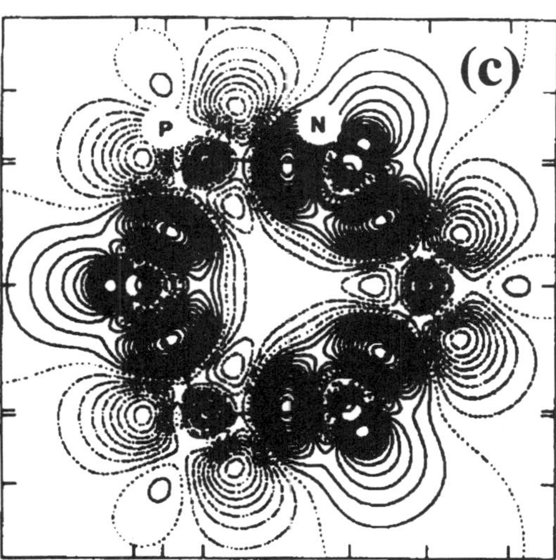

associated with even smaller densities and Laplacians. Following Cramer and Kraka [17], the charge density at critical point, ρ_{CP}, is a measure of the bond strength in covalent bonds. Thus, a typical C–C bond carries a density of $\sim 1.7\ e\text{Å}^{-3}$ at the CP while a C=C bond exhibits much higher density $\sim 2.5\ e\text{Å}^{-3}$. Similarly, the density associated with a C≡C bond is $\sim 2.8\ e\text{Å}^{-3}$. The ellipticity of a bond, ϵ, is a measure of its extent of double bond character. For cylindrically symmetric bonds, the ellipticity is therefore zero, while in the case of ideal carbon double bonds, the theoretical estimate gives $\epsilon \sim 0.74$. These quantities, in combination with bond polarity (Δ), pseudo-atomic charges and the bent bond character (d) describe a bond quantitatively.

Let us examine the case of diisocyanomethane as a typical example. Isocyanides possess a formally divalent carbon atom with a coordination number of only one. Though they are divalent, their bond lengths are only slightly longer than ideal C≡N bond and therefore a resonance is expected. Koritsanszky et al. [40] analyzed topographs of electron density of diisocyanomethane (Fig. 3) derived from both experiment and theory. Refinements of the experimental data were carried out with different levels of constraints. The most restricted model contained C_{2v} symmetry on the tetrahedral carbon (C1), rotational symmetry on isocyano groups and thermal motion correction on all non-hydrogen atoms. This model gave the best convergence for the data. They also carried out ab initio calculations at the Hartree–Fock and MP2 levels, optimizing the molecule with 6-311++G(3d,3p) basis sets starting from X-ray structural data. Fig. 3 shows the deformation density and the Laplacians in the mean molecular plane from both theory and experiment. There is a striking difference between the two deformation density maps in that the interatomic regions of the experimental map are richer in electron density at the expense of charge in the non-bonded regions of the terminal carbon atoms

Fig. 4. The phosphazene ring: (a) island delocalization model predicting nodes in π-density at the phosphorus atoms (b) dynamic deformation density (at $0.1\ e\text{Å}^{-3}$) in the plane of the ring (c) theoretical deformation density (at $0.05\ e\text{Å}^{-3}$) of cyclic phosphazene was used as a model (reproduced with permission from Cameron et al. [43]).

(compare Fig. 3a and b). The isocyano bonds are also more polarized. The same is reflected in the Laplacians shown in Fig. 3c and d, where bonded charge concentration appears as a sharp peak in the experimental map. This has been taken to indicate a greater swing of the NC bond resonance towards the $N{\equiv}C$ bond.

Intramolecular bonding in benzene, triazine, phosphazene and such cyclic systems are of interest because of the varying degree of superposition of the s, p and d orbitals as the case may be. The Hückel rule predicts benzene to be aromatic with the π-electrons delocalized over the ring. According to the island delocalization model [41] for the phosphazene ring, the overlap of the d orbitals on the phosphorus atoms and the p orbitals on the nitrogen atoms in the ring would produce a π-system above and below the plane of the ring with nodes at each phosphorous atom (Fig. 4a). An extension of the π-system within and in the plane of the ring is also expected based on the π/π' model due to Craig and Paddock [42]. Cameron et al. [43] applied charge density methods to study small cyclic systems. For this purpose, they carried out high resolution X-ray diffraction at 200 K with MoK_{α} radiation on hexaazirdinylcyclo-triphosphazene crystallized from benzene. This system is particularly interesting in that, the solvent benzene gets securely trapped between two phosphazene molecules and offers one to use it as an internal standard. They observed that the benzene ring in plane density was symmetric with respect to the carbon–carbon bonds while the density in a plane perpendicular to the ring across the center of a bond showed elongation in the direction of the π-system as expected. In contrast, the in-plane density in phosphazene was highly polarized as shown in Fig. 4b. Nodes in density at phosphorus atoms can be clearly seen from the figure. The electron density appears to spread from one P–N bond through the nitrogen to the second N–P bond as predicted by the island delocalization model. Further, there is a considerable spread of electron density inside the ring validating the π'-bonding model. Theoretical electron densities determined from ab initio calculations using GAMESS [37] with 6-31G* basis set depict similar features in the deformation density (see Fig. 4c).

Charge density distribution in cage rings has been investigated in few cases. An adduct of C_{60} was

studied by Irngartinger et al. [44] who discussed the degree of aromaticity. These workers have also carried out charge density measurements on a cubane derivative, methyl 3,4-difluorocubane-1-carboxylate [45]. This compound with the fluorine substituents fixed in a *cis*-like orientation to the rigid cubane cage (see Fig. 5) was expected to serve as a model for the *cis* isomer of 1,2-difluoroethylene. The latter along with its cousins such as *gauche*-1,2-difluoroethane is known to be energetically favorable compared to the *trans* isomer (*anti* conformer). Wilberg and co-workers [46] explained this '*cis/gauche* stability effect' and predicted the C–C bond to be bent due the strongly electronegative property of the fluorine substituents. Difference density maps in various diagonal planes of the cubane indeed show bending of the C–C bonds (Fig. 5). The CF–CF bond and the CH–CF bonds were found to be more bent $(d \sim 0.16\,\text{Å})$ than the CH–CH bonds $(d_{\text{average}} \sim 0.12\,\text{Å})$. This study clearly provides an evidence that strong electronegative substituents increase the bending of the bonds to which they are attached.

7. Charge density of hydrogen bonds

Hydrogen bonds can be classified on the basis of charge density. Alkorta and co-workers studied various types of hydrogen bonds including dihydrogen bonds [47], bifurcated hydrogen bonds [48], $H{\cdots}\pi$ interactions [49], inverse H bonds [50] and hydrogen bonds involving carbenes and silylenes as acceptors [51]. In general, the hydrogen bond CPs are associated with small densities and positive Laplacians characteristic of closed-shell interactions. Zhang et al. [52] carried out theoretical studies on hydrogen bonded complexes, with strained organic systems like tetrahedrane acting as pseudo-π-acceptors while Larsen and co-workers [53,54] have studied the strong hydrogen bond in methylammonium hydrogen maleate and benzoylacetone. Such strong bonds arise due to shared interactions exhibiting relatively large density $(\sim 1\,e\text{Å}^{-3})$ and negative Laplacians $(-7\,e\text{Å}^{-5})$ like typical intramolecular bonds. They also find a ring critical point $(3, +1)$ near the center of the keto–enol dimer in benzoylacetone. Experimental charge density study of *cis*-HMn(CO)$_4$PPh$_3$

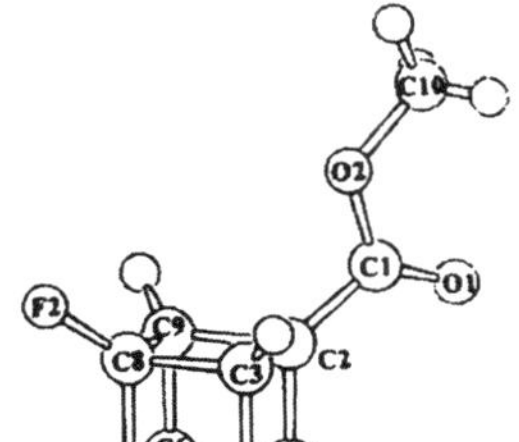

Methyl 3,4-difluorocubane-1-carboxylate

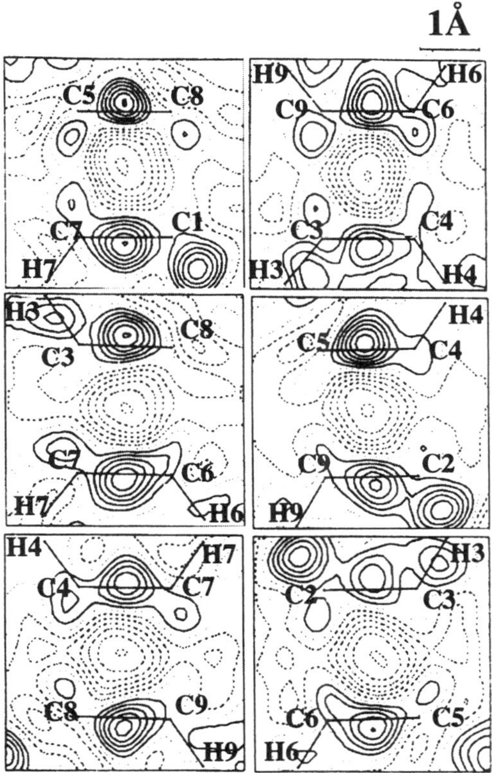

Fig. 5. Methyl 3,4-difluorocubane-1-carboxylate: difference density maps in various diagonal planes of the cubane cage. The maxima of the bond densities lie outside, implying that the cage bonds are bent (reproduced with permission from Irngartinger et al. [45]).

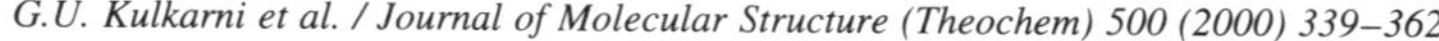

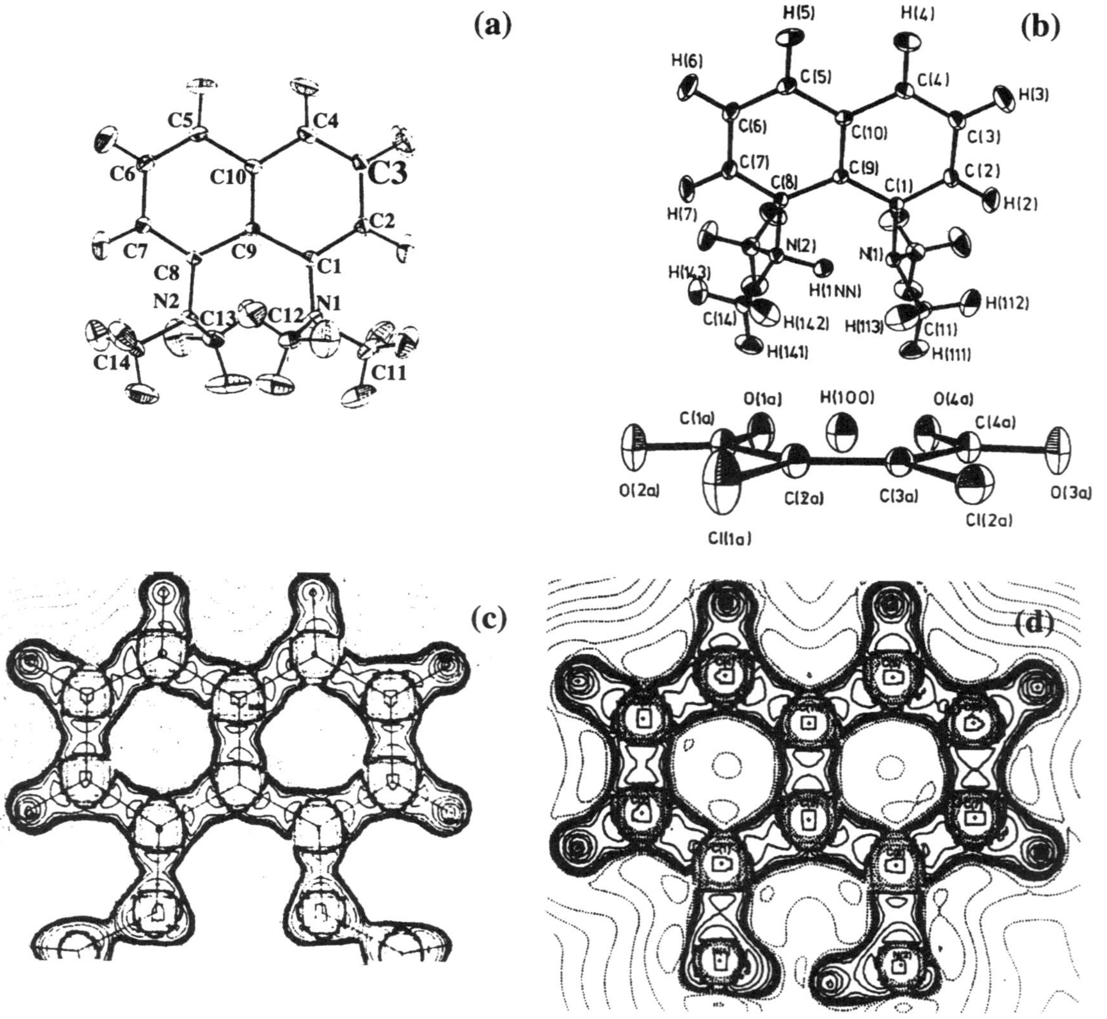

Fig. 6. A proton-sponge: (a) 1,8-Bis(dimethylamino)napthalene in the pristine form; (b) after protonation using 1,2-dichloro maleic acid. The contour maps of the corresponding Laplacians are shown at logarithmic intervals in (c) and (d), respectively (reproduced with permission from Mallinson et al. [56,57]).

has provided evidence for the C–H···H–Mn bond [55]. They found that the hydrogen atom in the Mn–H bond is nucleophilic carrying a charge of $-0.4e$ while that in the C–H is electrophilic $(0.3e)$. The electrostatic part of the H···H interaction energy was estimated to be 5.7 kcal/mol, which is in the range of H-bond interactions. It is characterized by a critical point carrying a small density $(0.066\,e\text{Å}^{-3})$ and a positive Laplacian

$(0.79\,e\text{Å}^{-5})$, somewhat higher compared to a typical C–H···O interaction.

Mallinson et al. [56,57] have carried out experimental and theoretical charge density determinations on a proton sponge compound, bis(dimethyamino)-napthalene (DMAN), in both pristine and protonated forms (Fig. 6). They used positional and thermal parameters of hydrogen atoms from an independent neutron diffraction experiment and treated the thermal

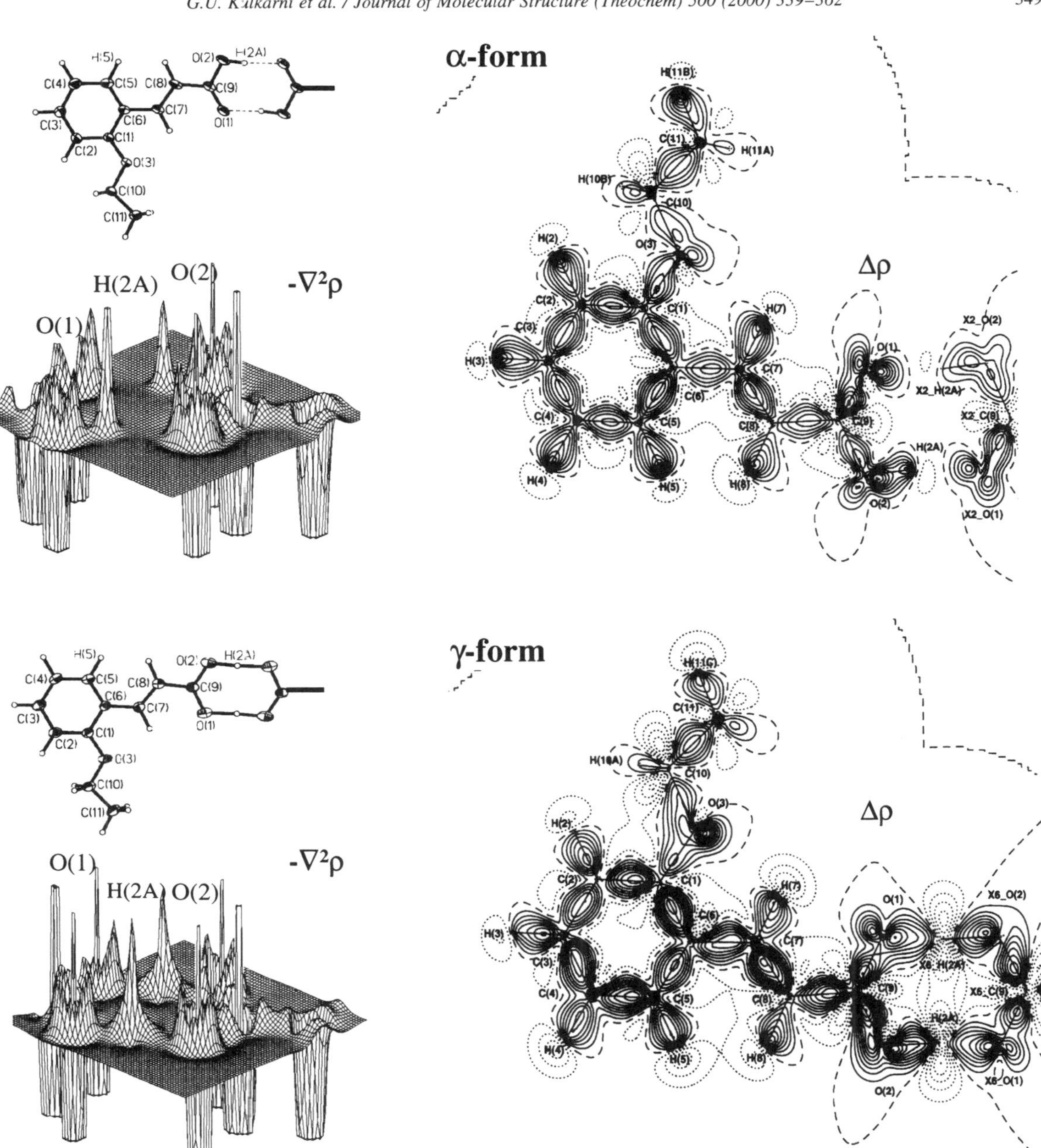

Fig. 7. Polymorphic forms of *o*-ethoxy cinnamic acid: molecular diagrams and deformation density maps close to the mean plane of the molecules in the α- and the γ-forms (contours at 0.12 $e\text{Å}^{-3}$). Subtle differences in the cinnamoyl bond and the hydrogen bond region are noticeable. The Laplacians of the intermolecular hydrogen bonds in the acid dimer are shown in the relief maps along side (range −250 to 250 $e\text{Å}^{-5}$).

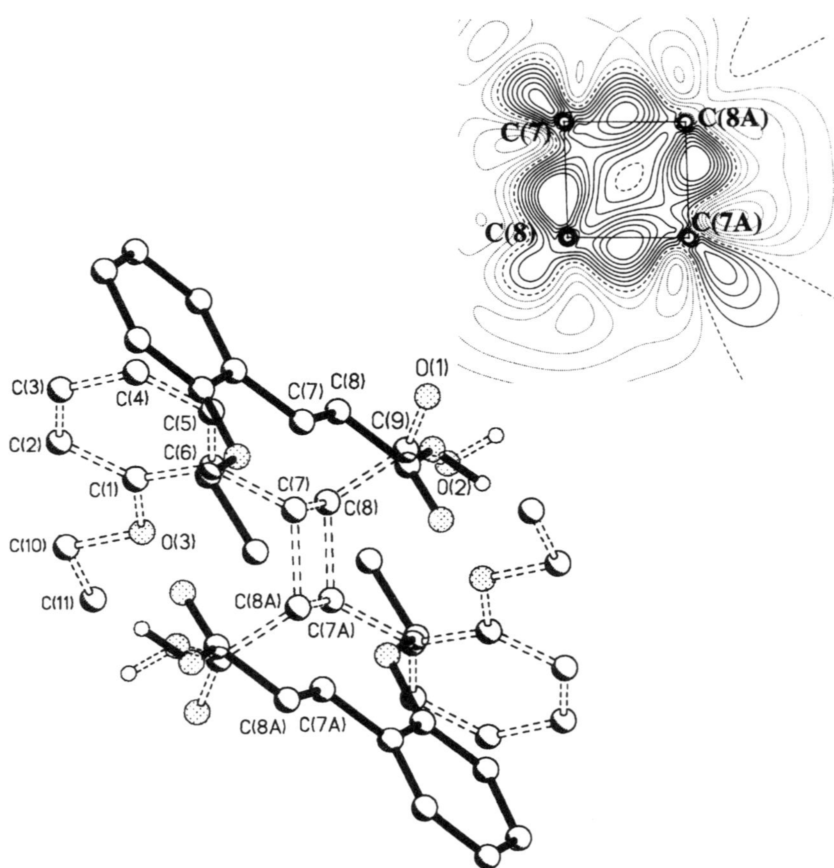

Fig. 8. Photodimerization in *o*-ethoxy cinnamic acid: Two centrosymmetric molecules of the α-form before (full lines) and after (dashed lines) the reaction. Hydrogen atoms other than the hydroxyl are omitted for the sake of clarity. The cinnamoyl double bonds are spaced at ~4.3 Å and following the cycloaddition, a new pair of bonds is formed which are slightly longer (~1.57 Å) than typical C–C bonds. The inset shows the deformation density (at 0.075 $e\text{Å}^{-3}$) in the plane of the cyclobutane ring in the α-dimer.

motions of all atoms anisotropically. The molecule in the pristine form (Fig. 6a) which is expected to have a two-fold symmetry like naphthalene is unsymmetrical, the asymmetry being reflected only as a small difference in the Laplacian between the two rings. From Fig. 6c, we see that the shape of the contours of the corresponding regions of the two halves of the molecule are somewhat different. Accordingly, the Laplacian values of the N2–C8 and N1–C1 bonds are -15.9 and $-13.3\,e\text{Å}^{-5}$, respectively, while those of C6–C5 and C4–C3 are -21.2 and $-18.7\,e\text{Å}^{-5}$, respectively. Upon protonation (Fig. 6b), noticeable changes occur in bond lengths in the

molecule. The C–C bonds are shortened by ~0.01 Å, while the C–N and C_{aromatic}–H bonds are lengthened by 0.04 and 0.02 Å, respectively. The $C_{\text{aliphatic}}$–H bonds however, do not change considerably. The atomic charges which are negative on the outer carbons decrease as a result of the migration of charge towards proton. These authors also carried out a ^{13}C NMR study in the solid state to show that the outer carbons are deshielded in the complexed proton sponge compared to the uncomplexed one and obtained useful correlations among ρ, $\nabla^2\rho$ and bond lengths. Another interesting aspect of this study is the curved interaction bond path joining two stacked

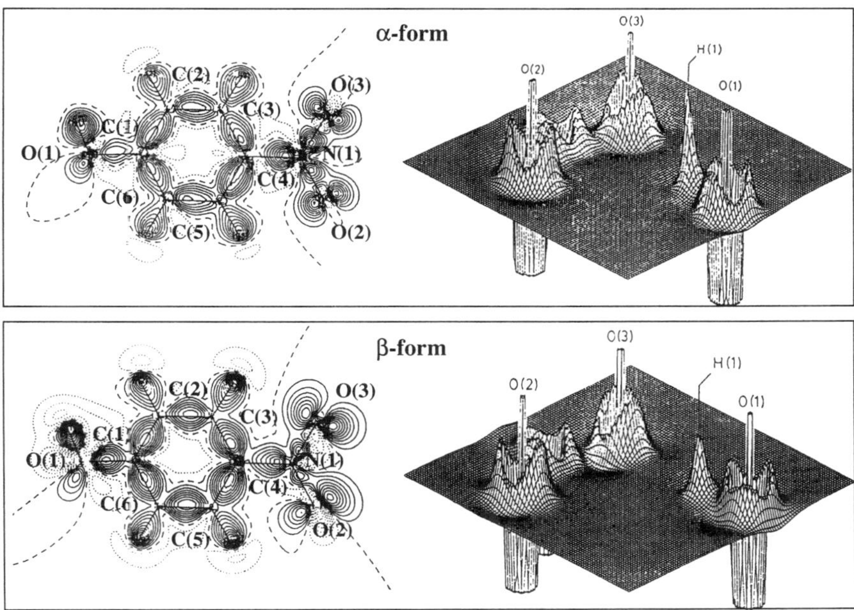

Fig. 9. Polymorphism in *p*-nitrophenol: static deformation density in the plane of the phenyl rings for the α- and the β-forms (contours at $0.1\ e\text{Å}^{-3}$). Intramolecular and lone-pair regions exhibit many differences. Relief maps of the Laplacians in the intermolecular hydrogen bond region are also shown (range -250 to $250\ e\text{Å}^{-5}$). In the α-form, H(1) bonds not only with O(3) but also with O(2) and N(1) of the neighboring nitro group (reproduced with permission from Kulkarni et al. [61]).

DMAN molecules attributed to the C–H···π interaction.

Intermolecular hydrogen bonds play a major role in deciding the properties of molecules in the solid state. Cinnamic acids for example, crystallize in two or three polymorphic forms [58], some being photoreactive forming cycloadditives while one polymorph may be photostable. We have carried out a charge density study using a CCD detector on the α- and γ-forms of *o*-ethoxy cinnamic acid (Fig. 7) obtained, respectively, from an ethyl acetate solution and slow cooling of an aqueous ethanol solution [59]. In the reactive α-form, the molecule is found to be quite planar, while in the photostable γ-form, the ethoxy and the cinnamoyl groups make angles of 6.5 and 3.5°, respectively, with the phenyl ring. Interestingly, the latter exhibits near-symmetric hydrogen bonds in the intermolecular region. The hydrogen H(2A) was located midway between the oxygens with the O–H and H···O distances of 1.23 and 1.39 Å, respectively. Accordingly, the ρ_{CP} values associated with the two bonds are comparable, ~ 1.6 and $0.8\ e\text{Å}^{-3}$,

respectively. Moreover, the intermolecular bond carries a negative Laplacian ($\sim -12.4\ e\text{Å}^{-5}$) like a hydroxy bond which is indicative of a highly shared interaction. In the α-form, on the other hand, the O–H and H···O distances are usual (0.96 and 1.68 Å, respectively) with ρ_{CP} of 2.2 and $0.32\ e\text{Å}^{-3}$, respectively. The H(2A)···O(1) hydrogen bond shows a small positive Laplacian of $4.81\ e\text{Å}^{-5}$ as is generally expected for a closed shell interaction. The other interesting aspect of this study is the inference on delocalization of the π-density in the α-molecule, from the cinnamoyl double bonds to the neighboring single bonds and across the phenyl group. This is also in compliance with the molecule being planar in this polymorph. A complete absence of such an effect in the γ-form was interpreted as due to the ionic nature induced by the symmetric hydrogen bonds. These factors influence the molecular geometry and the packing which in turn decide reactivity of a polymorphic form. The cinnamoyl bond undergoes (2 + 2) cycloaddition in the α-polymorph. Structural analysis indicated favorable approach of the

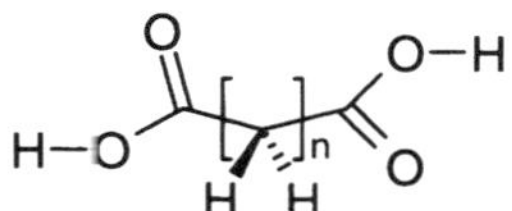

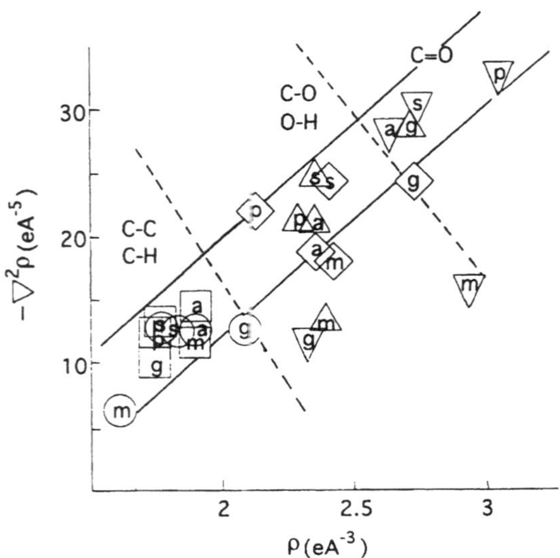

Fig. 10. Alkanedioic acids: variation of Laplacian with charge density at the critical point for various bonds, C–C, circle; C–H, square; C–O, up-triangle; C=O, down-triangle and O–H, rhombus. The letters m, s, g, a and p marked inside the data symbols represent bonds belonging to malonic, succinic, glutaric, adipic and pimelic acids, respectively. The different bond regions are delineated (reproduced with permission from Gopalan et al. [63]).

centrosymmetrically related cinnamoyl bonds in the photoactive α-form (Fig. 8). The deformation density of the cyclobutyl ring resulting from the photodimerization of α-form is shown in the inset of Fig. 8. It exhibits a $(3, +1)$ CP at the inversion center of the ring associated with small density $(0.62\ e\text{Å}^{-3})$ and Laplacian $(6.6\ e\text{Å}^{-5})$. The densities of the ring bonds are small $(\sim 1.51\ e\text{Å}^{-3})$ implying that the bonds are weak. They are also associated with high ellipticity (~ 0.2) and polarization (6%). The contours of these bonds lie outside the interatomic vector, the vertical displacement being ~ 0.04 Å, characteristic of bent bonds in a strained ring [17,45].

p-Nitrophenol is known to show interesting

photochemical activity only in one of its polymorphic forms. The α-form which crystallizes from benzene undergoes a topochemical transformation up on irradiation, changing its color from yellow to red, though structural changes associated with the transformation have been found to be insignificant [60]. On the other hand, the β-form obtained from aqueous solution, is light-stable. This has been the subject of a charge density study [61] (Fig. 9). It is found that the phenyl ring bonds in the α-form exhibit less density $(\sim 2.05\ e\text{Å}^{-3})$ compared to those in the β-form $(\sim 2.21\ e\text{Å}^{-3})$ as though charge had migrated outwardly in the former. Accordingly, the nitro and the hydroxyl bonds in the latter were found to carry relatively higher densities. The authors found many differences in the hydrogen bonding as well. In the β-form, there are four hydrogen bonds compared to six in the α-form. The striking difference between the two polymorphs is that in the α-form, the entire nitro group participates in hydrogen bonding with the neighboring hydroxyl hydrogen while in the β-form only the nitro oxygens involve in the hydrogen bonding. Interestingly upon photoirradiation of the α-form, charge density redistribution seems to occur with the reacted product assuming an intramolecular density similar to that found in the light stable β-form [62]. The authors have estimated the molecular dipole moments to be 18.0 and 21.5 Debye for the α- and β-forms, respectively and a somewhat lower moment in the case of the irradiated α-form, ~ 9.9 Debye.

The lattice cohesion of the first few members of the homologous series of aliphatic dicarboxylic acids has been investigated in terms of charge density [63]. These acids form an interesting class of organic compounds in that the structure and properties in the solid state exhibit undulatory behavior with the number of methylene groups being odd or even [64]. For example, the melting point alternates in the series malonic, succinic, glutaric, adipic, pimelic and so on. The study yielded interesting systematics in the charge densities and the Laplacians along the series. The ρ_{CP} values of C–C, C–O and O–H bonds increase from malonic (1.61, 2.27 and $2.45\ e\text{Å}^{-3}$, respectively) to glutaric (2.18, 2.72 and $2.7\ e\text{Å}^{-3}$, respectively) and decrease thereafter. The C=O and C–H bonds exhibit the opposite trend with glutaric acid carrying the minimum charge density ~ 2.33 and $1.66\ e\text{Å}^{-3}$, respectively. An overall

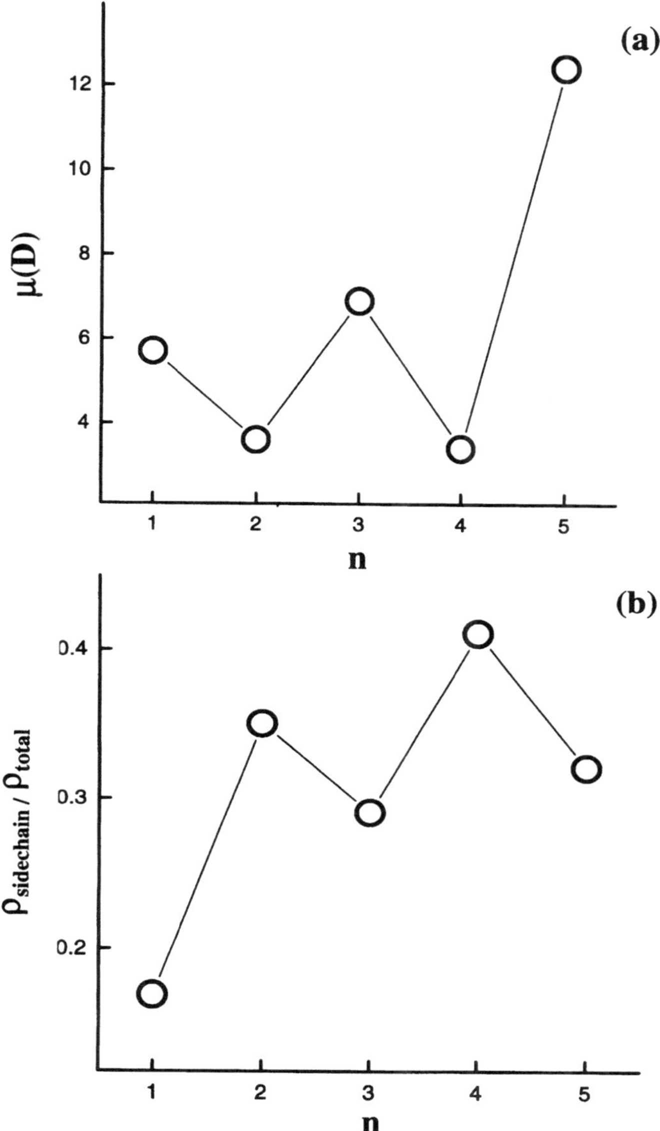

Fig. 11. (a) Molecular dipole moments of the dicarboxylic acids. Values obtained for the asymmetric units are shown since net dipole moment for an even acid vanishes due to center of symmetry. Here, n denotes the number of methylene groups in the acid. (b) A plot of the sum of ρ_{CP} obtained for the side-chain interactions normalized with the total ρ_{CP} due to intermolecular interactions, against the number of methylene groups, n, in the acid (reproduced with permission from Gopalan et al. [63]).

assessment of the charge distribution among various bonds can be made by plotting the Laplacian against the density for various bonds as shown in Fig. 10. We see that most bonds lie in a region where the Laplacian is roughly proportional to the bond density, as one would normally expect. Thus, the C–C and the C–H bonds fall in the first region of the plot while the C–O and the O–H bonds group fall in the second

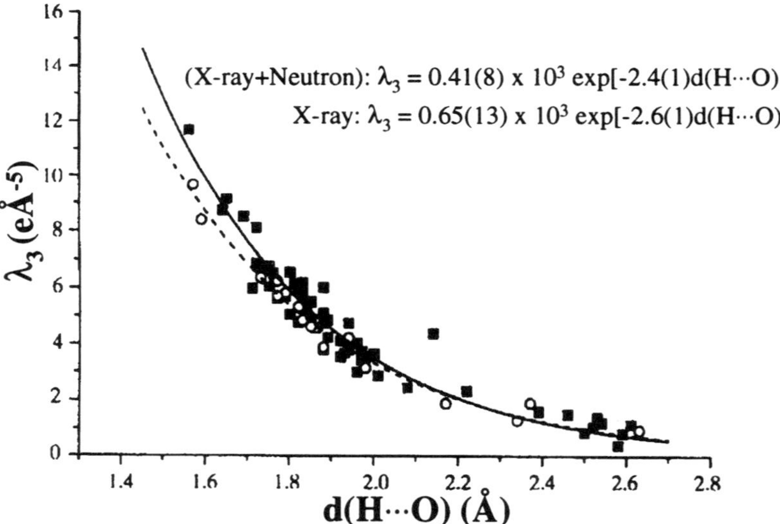

Fig. 12. Variation of the positive curvature, λ_3 at bond CP, with hydrogen contact distance, $d(H\cdots O)$. The data points obtained from a number of charge density investigations reported in the literature and were fitted using exponential equations. Circles and dashed line; joint X-ray and neutron, filled squares and solid line; X-ray data only. Among the three eigenvalues, λ_3 was found to be giving the best representation for hydrogen bonds having closed-shell interactions (reproduced with permission from Espinosa et al. [65]).

region. The C=O bonds are found in the third region. The C–O and C=O bonds of the malonic and the glutaric acids are somewhat unusual. In the case of the malonic acid, the Laplacians of the two bonds are noticeably lower. In glutaric acid, on the other hand, the Laplacian of the C=O bond is lower than that of the C–O bond, a trend which is reflected in the charge densities as well.

An important outcome of the study of the dicarboxylic acids is the analysis of the dipole moments of the asymmetric units in relation to intermolecular interactions in the crystal lattice (Fig. 11a). It is interesting that the dipole moments showed alternation along the series with the odd acids having higher moments than the even neighbors with pimelic acid exhibiting the highest moment, 12.7 Debye. The strength of molecular packing in the lattice was guided by the value of the sum of the ρ_{CP}s associated with side chain C–H$\cdots$O interactions expressed as a fraction of the total intermolecular density due to C–H$\cdots$O and O–H$\cdots$O interactions (Fig. 11b). Interestingly, there is an alternation in this value along the series with the even acids exhibiting higher values compared to their odd neighbors. Thus, it appears

that increased side-chain interactions in the even acids lead to distributed bond dipoles thereby decreasing the net dipole moment in the asymmetric unit. This may also be the reason for the relatively higher melting points in the even acids, since side-chain interactions as compared to the dimeric bonds play a decisive role in the cohesion of acid molecules in the solid state.

Hydrogen bonds have been classified by many workers in terms of varying degrees of shared and closed-shell interactions. Correlations between $\nabla^2\rho$ and ρ, as shown above in Fig. 10 find good agreement only within a group of like-hydrogen bonds. A more general approach has been suggested by Espinosa et al. [65] who used the variation of positive curvature along the interaction axis (λ_3) at CP with the hydrogen bond distance and obtained a well-defined behavior as shown in Fig. 12. The value of λ_3 was found to increase exponentially with decreasing hydrogen bond distances. It represents the overlap between the electron cloud of both H and O atoms at the critical point and is proportional to the kinetic energy density, G_{CP}.

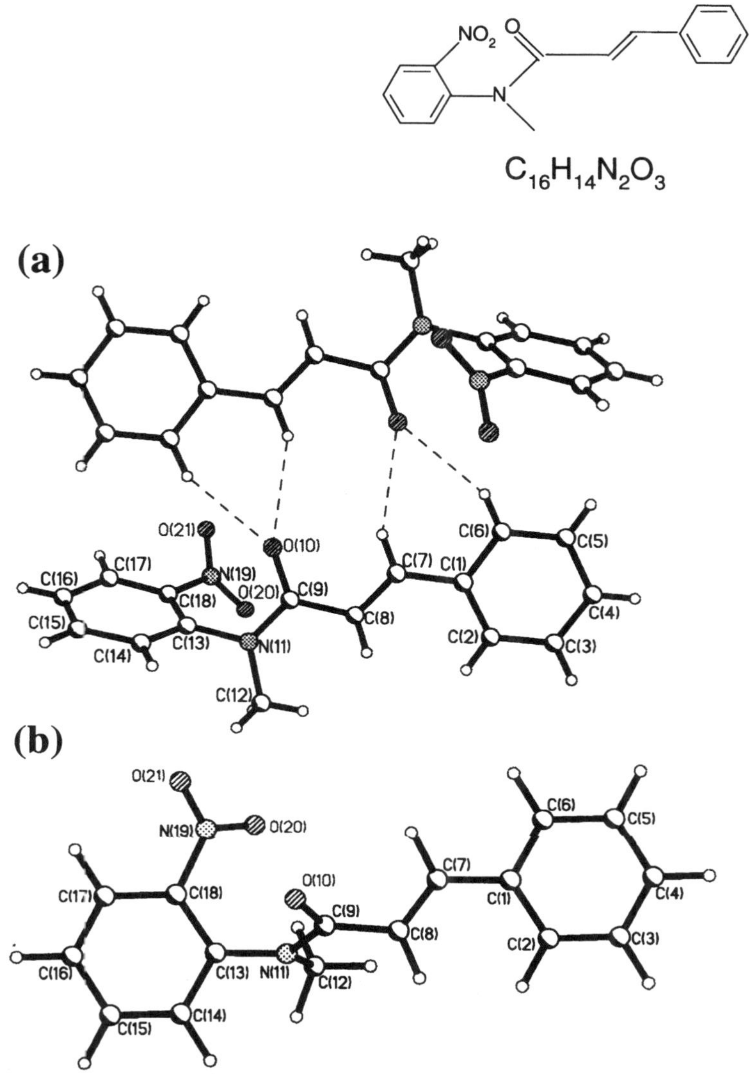

Fig. 13. Molecular structure of *N*-methyl-*N*-(2-nitrophenyl)cinnamanilide from (a) X-ray crystallography (bifurcated intermolecular hydrogen bonds are also shown) (b) AM1 calculation. Formula diagram is given in the inset (reproduced with permission from Gopalan et al. [68]).

8. Molecular packing in crystals

The geometry of a molecule in the solid state may deviate significantly from that in the free state, an effect which is mainly due to the constraints of packing in a lattice. For instance, the nitrobenzene molecule which is known to be planar in the free state is twisted by about 2° in the solid state across the nitro–benzene link. On the contrary, biphenyl [66] is non-planar in both gaseous and liquid states while it is planar [67] in the crystal at room temperature, exhibiting D_{2h} symmetry compared to D_2 in the former. Polymorphism in molecular crystals discussed above, is another subject of interest in this context. It is considered to be important to understand the symmetries prevailing in a molecular solid in terms of

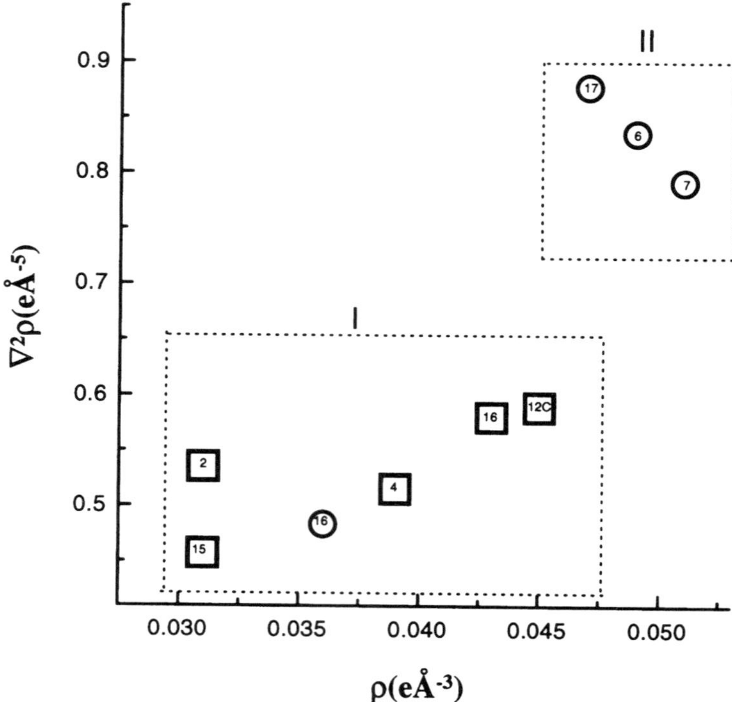

Fig. 14. Variation of the Laplacian with density at the critical points for various hydrogen bonds, $O_{amido}\cdots H$, circles; $O_{nitro}\cdots H$, squares. The numbers inside the symbols refer to the hydrogens involved in bonding. Regions I and II emphasize different trends in the plot (reproduced with permission from Gopalan et al. [68]).

the distortions of the molecules and the nature of the intermolecular interactions.

Gopalan et al. [68] have carried out a charge density investigation on N-methyl-N-(2-nitrophenyl)cinnamanilide. The molecule exhibits a highly favored bifurcated $C–H\cdots O$ hydrogen bond ring structure in the lattice and is also considerably distorted (Fig. 13a). It is twisted at the amide-phenyl link with the cinnamide portion being nearly planar (torsion angle, $C(2)–C(1)–C(7)–C(8)$, $-6.24°$). The nitrobenzene ring is highly non-planar, the nitro group being rotated with respect to the phenyl ring by $\sim 43.5°$. The nitrobenzene group as a whole is twisted away from the mean cinnamide plane making an angle of $63.3°$. The molecular structure in the lattice as discussed above was compared with that of a free molecule shown in Fig. 13b. The latter was obtained using AM1-PRECISE calculation of MOPAC [38] after optimizing the bond lengths, angles and torsion angles. The

experimental coordinates served as the initial input. An important finding from the calculation is that the benzene rings are parallel within $8°$ while the intervening bonds are buckled with high torsion angles (see Fig. 13b). This calculation provides a reference state of the molecule using which constraints imposed by packing in a lattice could be examined.

Several intermolecular hydrogen bond contacts were examined. Four $C–H\cdots O$ contacts were found to originate from the amidic oxygen (two involved in bifurcated bond) while six from the nitro-oxygens. The results of the charge density analysis is shown in Fig. 14. The $\nabla^2\rho–\rho$ plot consists of two regions of interactions as shown—one region, where the hydrogen bonds exhibit low values of both the density and the Laplacian and the other, where both quantities are disproportionately higher. What is interesting is that all the bonds involving the nitro-oxygens belong to the first region and those from the amidic oxygen

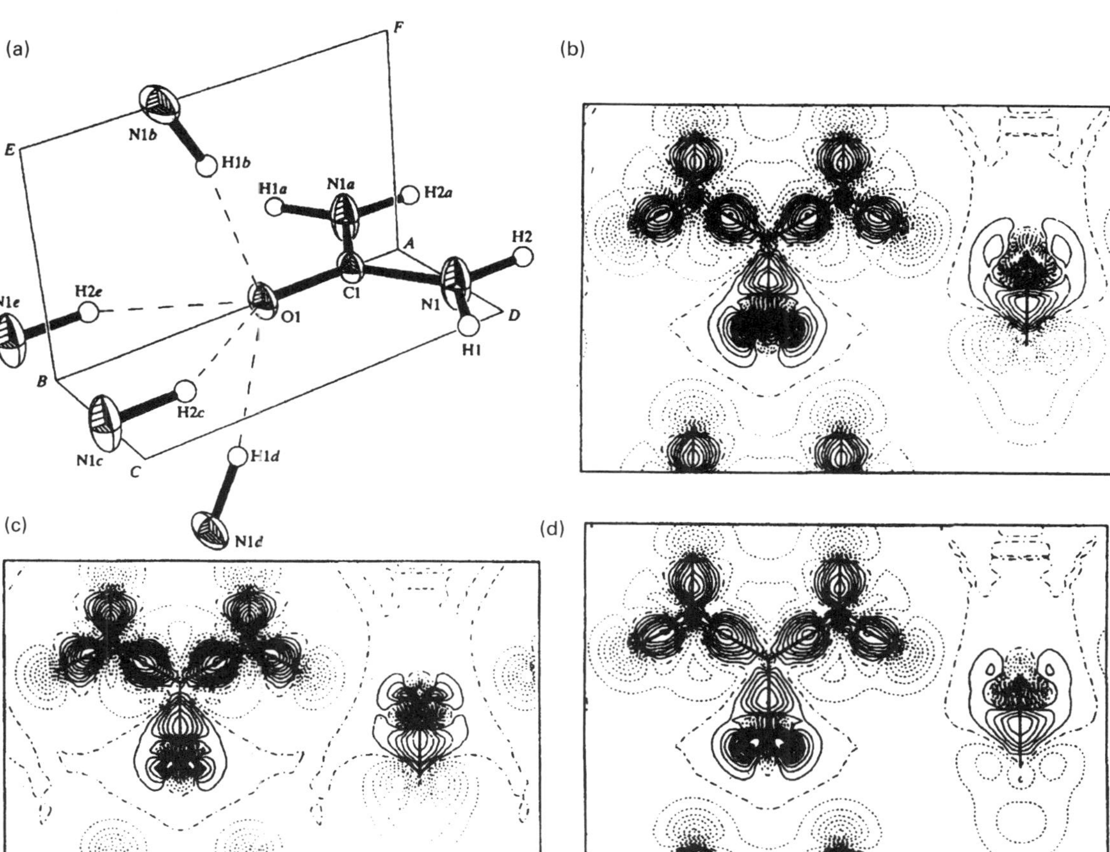

Fig. 15. (a) Intermolecular hydrogen bonds in urea crystal with displacement ellipsoids at 50% probability. (b) Static deformation density obtained from the multipolar analysis of the experimental data corrected for the thermal diffuse scattering. Theoretical deformation density obtained using (c) the Hartree–Fock method; (d) the DFT method by generalized gradient approximation (contours at 0.0675 $e\text{Å}^{-3}$) (reproduced with permission from Zavodnik et al. [69]).

fill the second region. Clearly, the amidic oxygen forms stronger hydrogen bonds compared to the more ionic nitro-oxygens. It is as though to favor bifurcated hydrogen bonding from the amidic oxygen, the molecule is highly twisted at the amide–nitrobenzene link.

Urea, despite being simple and highly symmetric (C_2), is known to crystallize in a non-centric crystal system ($P\text{-}42_1m$). The molecules are linked to each other through hydrogen bonds forming infinite tapes, adjacent tapes being held by orthogonal hydrogen bonds. This is perhaps the only example containing

a carbonyl group involved in four hydrogen bonds (Fig. 15a). Feil and co-workers [69] performed X-ray diffraction at 148 K and carried out charge density analysis on the data corrected for thermal diffuse scattering. They have also done orbital calculations in lattice using CRYSTAL95 [39] by both Hartee–Fock and DFT methods, the latter in both local density and generalized gradient approximations. The authors compared deformation densities and structure factors obtained from the experimental charge density analysis with those obtained from theoretical procedures. The deformation density maps from experiment and

358   *G.U. Kulkarni et al. / Journal of Molecular Structure (Theochem) 500 (2000) 339–362*

theory (compare Fig. 15b–d) were found to be in good agreement. The density functional theory seemed to yield slightly better results compared to the Hartree–Fock calculations.

Gatti et al. [70] have described the unusual hydrogen bonding in urea based on computations. They considered three cases of the urea molecule. A molecule as in the bulk structure with inputs from neutron diffraction, was computed using CRYSTAL92 [39] with 6-31 G** basis sets. In the second case, the free molecule was held at the crystal geometry and in the third, at more optimized geometry with only the C_{2v} constraint. Geometry optimization were done using GAUSSIAN92 with spherical harmonic gaussian functions. Compared to the free molecule, they found that the heavy atoms of the molecule in the bulk gain electrons at the expense of the hydrogens, the flux being $\sim 0.06e$. The fields created by such charge transfer polarizes the molecule in the bulk enhancing dipole moments. They observed that the ellipticity and ρ increase for the C–N bonds on passing from gas to bulk while for all the other bonds, both ellipticity and ρ decrease. These changes were found to comply with strengthening of the C–N bonds in terms of increased covalency and π-character. Other bonds in the molecule become more ionic on passing from gas to bulk. The ellipticity associated with the intermolecular hydrogen bonds along infinite planar tapes was found to be similar to that in the π-plane of the molecule. Based on this, the authors have argued that the π-conjugation propagates through these hydrogen bonds.

The above study has also shown that the carbonyl oxygen actually forms an active center for hydrogen bonding. The oxygen lone-pairs being electron-rich regions in the base, are characterized by high negative Laplacians in the free molecule (145.82 $e\text{Å}^{-5}$) while those associated with nitrogen carry much smaller Laplacians (54.85 $e\text{Å}^{-5}$). In the bulk, when oxygens get involved in hydrogen bonding, the lone-pair Laplacians were found to decrease to ~ 136.62 $e\text{Å}^{-5}$. The two saddle points in between the lone-pairs which lie above and below the molecular plane are the second most electron-rich regions (89.26 $e\text{Å}^{-5}$) which are seen as maxima by hydrogens approaching perpendicularly. Accordingly, the Laplacian associated with the saddle point increases to 94.3 $e\text{Å}^{-5}$ in the bulk. The valence shell charge

concentration of oxygen therefore changes in such a way so as to form a torus of nearly uniform charge concentration in the non-bonded region. The authors conclude that the lengthening of the C=O bond in bulk urea by 0.13 Å accompanied by decreasing ellipticity and associated changes in the oxygen non-bonded regions is a fundamental step for the creation of a three-dimensional network of hydrogen bonds. This may be a more general mechanism in the formation of hydrogen bonded molecular crystals as was shown in the case of the cinnamanilide molecule.

9. Molecular NLO materials

In the last few years, several workers have analyzed charge density distribution in molecular crystals with non-linear optical (NLO) properties [71–74]. The NLO response can, in principle, be explained by an anharmonic distortion of the electron density distribution due to the electric field of an applied optical pulse. The polarization P induced in a molecule is

$$P = \mu + \alpha E + \beta E^2 + \cdots \tag{16}$$

where μ is the ground state dipole moment and α and β are the linear and the quadratic polarizabilities of the molecules, respectively. However, determination of polarizability from X-ray diffraction requires careful handling of data. Many of the studies on NLO crystals using charge density concentrate merely on the accurate extraction of phases associated with the structure factors [75–78],

$$\Delta\rho(\mathbf{r}) = V^{-1}\sum_{\mathbf{h}}[|F_{\mathrm{m}}(\mathbf{h})|\exp i\varphi_{\mathrm{m}}(\mathbf{h})$$
$$- |F_{\mathrm{s}}(\mathbf{h})|\exp i\varphi_{\mathrm{s}}(\mathbf{h})]\exp(-2\pi i\mathbf{hr}) \tag{17}$$

where the subscripts m designate the atom-centered multipolar density model, and s, the spherically averaged free-atom superposition model. The deformation density therefore, can be written as the sum of an amplitude deformation density and a phase deformation density,

$$\Delta\rho = \Delta\rho(\Delta|F|) + \Delta\rho(\Delta\varphi) \tag{18}$$

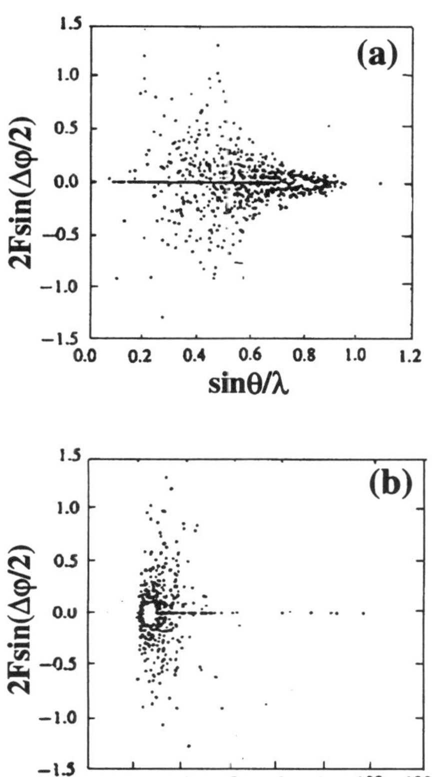

Fig. 16. Distribution of $2F \sin(\Delta\varphi/2)$ against (a) (sin θ/λ; and (b) $|F|$ (reproduced with permission from Hamazaoui et al. [76]).

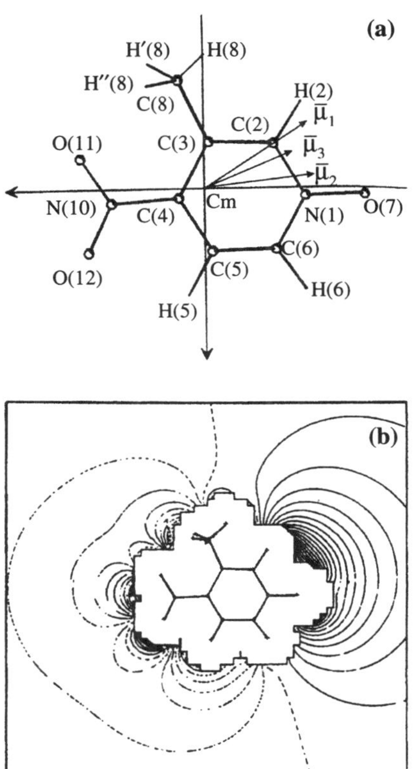

Fig. 17. (a) Orientation of molecular dipole moments in 3-methyl-4-nitropyridine *N*-oxide: μ_1, molecular dipole moment from direct integration methods; μ_2, from multipolar model; and μ_3 from semi-empirical calculation. (b) Electrostatic potential around the molecule in the plane of ring atoms. Contours at 0.2 kcal/mol (reproduced with permission from Hamazaoui et al. [79]).

where

$$\Delta\rho(\Delta|F|) = V^{-1} \sum (|F_{\mathrm{m}}| - |F_{\mathrm{s}}|) \exp(i\varphi_{\mathrm{m}})$$
$$\times \exp(-2\pi i\mathbf{hr}) \tag{19}$$

and

$$\Delta\rho(\Delta\varphi) = V^{-1} \sum 2|F_{\mathrm{s}}| \sin(\Delta\varphi/2) \exp[i(\varphi_{\mathrm{s}} + \varphi_{\mathrm{m}}$$
$$+ \pi)/2] \exp(-2\pi i\mathbf{hr}) \tag{20}$$

For a given data set, it is instructive to examine the magnitude of $2F \sin(\Delta\varphi/2)$ in order to understand the effect of phase correction on the final solution. Fig. 16 shows the distribution of $2F \sin(\Delta\varphi/2)$ as a function of sin θ/λ and $|F|$,

from a data set due to Hamazaoui et al. [76]. It is apparent that the magnitude of phase correction is relatively higher for the low-order reflections (<0.6 Å^{-1}) as well as for weak- and medium-intensity reflections. Consequently, the measurement and the processing of the low angle weak reflections deserve special care. The authors have also shown that the overall charge density increases by $\sim 25\%$ after phase correction. Besides phase correction, the authors carried out thermal analysis involving rigid bond tests. Such corrections are highly recommendable while determining charge density in non-centric crystals.

There has been some attempt to calculate of

hyperpolarizability from higher moments, but a more acceptable quantitative connection is yet to be established. Fkyerat et al. using octupole moments of the charge distribution in N-(4-nitrophenyl)-L-prolinol [73,74] estimated β to be 42.9 $e\text{Å}^{-3}$ and compared with the experimental result $\sim$39.14 $e\text{Å}^{-3}$. They concluded that NPP is a one-dimensional non-linear compound with β oriented almost along the main axis of the molecule making only $19.8°$ with the dipole moment vector.

Other studies on NLO crystals report topological analysis of deformation density and electrostatic potential and also on the calculation of molecular dipole moment. Hamazaoui et al. have carried out an experimental charge density study on 3-methyl-4-nitro-pyridine-N-oxide [79] and have computed the molecular dipole moments by three different procedures namely semi-empirical calculation, multipolar model and direct integration method. In all the methods, the dipole moment points nearly towards N-oxide group as shown in Fig. 17a. They found a close agreement between the semi-empirical and multipolar models while that from direct integration differed considerably in the direction of the dipole moment. The authors attributed this difference to the deconvolutability of thermal parameters from electron density in the case of multipolar refinement which is not possible in direct integration. The dipole moment of this molecule (1 Debye) was compared with that of nitropyridine N-oxide, a closely related molecule in which the methyl group in the *meta*-position is absent. The latter was found to have a much smaller dipole moment (0.4 Debye) [80]. On this basis, the authors concluded that the introduction of a methyl group in the *meta*-position favors intramolecular charge transfer. They have also determined the electrostatic potential in the plane of the ring which showed positive electrostatic potential around the N-oxide group and a negative electrostatic potential around the nitro group (Fig. 17b) thereby confirming the nature of charge transfer as found by the orientation of the molecular dipole moment.

10. Conclusions

In this article, we have dealt with the salient features of the electronic charge density distribution in molecular solids obtained by both theory and experiment. The importance of comparative experimental and theoretical studies of electron density is pointed out. The topological analysis of the density is illustrated taking representative examples. The chemical aspects of the charge distribution is described in terms of the deformation density, the Laplacian of the total density as well as the electrostatic potential. The polarization of the electron density in cyano bonds of diisocyanomethane, π/π' delocalization in the phosphazene ring and the bent cage bonds in cubane are some of the examples discussed. In addition, cases revealing the interplay between the intermolecular hydrogen bonding and properties are examined. These include proton sponges, polymorphic forms of cinnamic acid and *p*-nitrophenol where one of the forms exhibits photochemical reactivity. The role of intermolecular hydrogen bonding in lattice cohesion is discussed for a series of dicarboxylic acids. Molecular distortions in urea and cinnamanilide molecules are explained on the basis of the strength of the surrounding hydrogen bonds. Molecules exhibiting NLO properties in the solid state such as N-(4-nitrophenyl)-L-prolinol have been discussed, focusing on the intramolecular charge transfer leading to enhanced dipole moments.

References

[1] P. Debye, Dispersion of Roentgen rays, Ann. Phys. 46 (1915) 809.
[2] R.F.W. Bader, Atoms in Molecules—A Quantum Theory, Clanderon Press, Oxford, 1990.
[3] P. Coppens, Annu. Rev. Phys. Chem. 43 (1992) 663.
[4] D. Feil, J. Mol. Struct. 255 (1992) 221.
[5] F.L. Hirshfeld, in: A. Domenicano, I. Hargittai (Eds.), Accurate Molecular Structures. Their Determination and Importance, IUCr/Oxford University Press, Oxford, 1992, p. 237.
[6] F.L. Hirshfeld, Crystallogr. Rev. 2 (1991) 169.
[7] M.A. Spackman, Chem. Rev. 92 (1992) 1769.
[8] V.G. Tsirelson, R.P. Ozerov, J. Mol. Struct. 255 (1992) 335.
[9] G.A. Jeffrey, J.F. Piniella (Eds.), The Application of Charge Density Research to Chemistry and Drug Design Plenum Press, New York, 1991.
[10] P. Coppens, X-ray Charge Densities and Chemical Bonding, Oxford University Press, Oxford, 1997.
[11] R. Blessing (Ed.), Studies of electron distributions in molecules and crystals, Trans. Am. Crystallogr. Assoc., 26 (1990).
[12] T. Koritsanszky, in: W. Gans, A. Amann, J.C.A. Boeyens (Eds.), Fundamental Principles of Molecular Modelling, Plenum Press, New York, 1996, p. 143.

[13] M.A. Spackman, A.S. Brown, Annu. Rep. Prog. Chem. Sect. C: Phys. Chem. 91 (1994) 175.

[14] M.A. Spackman, Annu. Rep. Prog. Chem. Sect. C: Phys. Chem. 94 (1998) 177.

[15] P. Coppens, Acta Crystallogr. A54 (1998) 779.

[16] N.K. Hansen, P. Coppens, Acta Crystallogr. A34 (1978) 909.

[17] D. Cremer, E. Kraka, Croat. Chem. Acta 57 (1984) 1259.

[18] P. Politzer, D.G. Truhlar (Eds.), Chemical Applications of Atomic and Molecular Electrostatic Potentials Plenum Press, New York, 1981.

[19] J.S. Murray, K.D. Sen (Eds.), Molecular Electrostatic Potentials: Concepts and Applications Elsevier, Amsterdam, 1996.

[20] Y. Abramov, Acta Crystallogr. A53 (1997) 264.

[21] P. Coppens, Y. Abramov, M. Carducci, B. Korjov, I. Novozhilova, C. Alhambra, M.R. Pressprich, J. Am. Chem. Soc. 121 (1999) 2585.

[22] E. Espinosa, E. Mollins, C. Lecomte, Chem. Phys. Lett. 285 (1998) 170.

[23] P. Macchi, D.M. Prosperpio, A. Sironi, R. Soave, R. Destro, J. Appl. Crystallogr. 31 (1998) 583.

[24] A. Volkov, G. Wu, P. Coppens, J. Synchroton. Rad. 6 (1997) 1007.

[25] T. Koritsansky, R. Flaig, D. Zobel, H.-G. Crane, W. Morgenroth, P. Luger, Science 279 (1998) 356.

[26] G.M. Sheldrick, SHELX-76, Program for crystal structure determination, University of Göttingen, Germany.

[27] F.H. Allen, O. Kennard, D.G. Watson, L. Brammer, A.G. Orpen, R. Taylor, J. Chem. Soc. Perkin Trans. II (1987) S1.

[28] F.L. Hirshfeld, Acta Crystallogr. A32 (1976) 239.

[29] V. Schoemaker, K.N. Trueblood, Acta Cryst. B54 (1998) 507.

[30] T. Koritsansky, S.T. Howard, T. Richter, P.R. Mallinson, Z. Su, N.K. Hansen, XD, A computer program package for multipole refinement and analysis of charge densities from diffraction data, Cardiff, Glasgow, Buffalo, Nancy, Berlin.

[31] R.F. Stewart, M.A. Spackman, VALRAY User's Manual, Carnegie Mellon University, Pittsburgh, PA.

[32] H.L. Hirshfeld, Acta Crystallogr. A32 (1976) 239.

[33] B.M. Craven, H.P. Weber, X. He, Technical Report TR-87-2, Department of Crystallography, University of Pittsburgh, PA, 15260, 1987.

[34] W.J. Hehre, L. Radom, P.v.R. Schleyer, J.A. Pople, Ab initio Molecular Orbital Theory, Wiley, New York, 1986.

[35] J.A. Pople, Angew. Chem. Int. Ed. Engl. 38 (1999) 1894.

[36] M.J. Frisch, G.W. Trucks, E.B. Schlegel, P.M.W. Wong, J.B. Foresman, M.A. Robb, M. Head-Gordon, E.S. Replogle, R. Gomperts, J.L. Andres, K. Raghavachari, J.S. Binkley, C. Gonzalez, R.L. Martin, D.J. Fox, D.J. Defrees, J. Baker, J.J.P. Stewart, J.A. Pople, GAUSSIAN 92/DFT, Revision G.4, Gaussian Inc., Pittsburgh, PA, 1993.

[37] M.W. Schmidt, K.K. Baldridge, J.A. Boatz, S.T. Elbert, M.S. Gordon, J.H. Jensen, S. Koseki, N. Matsunaga, K.A. Nguyen, S.J. Su, T.L. Windus, M. Dupuis, J.A. Montgomery, Gamess, J. Comput. Chem. 14 (1993) 1347.

[38] J.J.P. Stewart, J. Comput.-Aided Mol. Des. 4 (1990) 1.

[39] R. Dovesi, V.R. Saunders, C. Roetti, M. Causa, N.M. Harrison, R. Orlando, R. Apra, CRYSTAL95 User's Manual, University of Turin, Turin, Italy, 1996.

[40] T. Koritsanszky, J. Buschmann, D. Lentz, P. Luger, G. Peretuo, M. Rottger, Chem. Eur. J. 5 (1999) 3413.

[41] M.J.S. Dewar, E.A.C. Lucken, M.A. Whitehead, J. Chem. Soc. (1960) 243.

[42] D.P. Craig, N.L. Paddock, J. Chem. Soc. (1962) 4118.

[43] T.S. Cameron, B. Boreka, W. Kwiatkowski, J. Am. Chem. Soc. 116 (1994) 1211.

[44] H. Irngartinger, A. Wesler, T. Oeser, Angew. Chem. Int. Ed. Engl. 38 (1999) 1279.

[45] H. Irngartinger, S. Strack, J. Am. Chem. Soc. 120 (1998) 5818.

[46] K.B. Wilberg, C.M. Hadad, C.M. Breneman, K.E. Laidig, M.A. Murcko, T.J. LePage, Science 252 (1991) 1266.

[47] I. Alkorta, J. Elguero, C. Foces-Foces, Chem. Commun. (1996) 1633.

[48] I. Rozas, I. Alkorta, J. Elguero, J. Phys. Chem. A 102 (1998) 9925.

[49] I. Rozas, I. Alkorta, J. Elguero, J. Phys. Chem. A 101 (1997) 9457.

[50] I. Rozas, I. Alkorta, J. Elguero, J. Phys. Chem. A 101 (1997) 4236.

[51] I. Alkorta, J. Elguero, J. Phys. Chem. (1996) 100.

[52] Y.H. Zhang, J.-K. Hao, X. Wang, W. Zhou, T.-H. Tang, J. Mol. Struct. (Theochem) 455 (1998) 85.

[53] D. Madsen, C. Flensburg, S. Larsen, J. Phys. Chem. A 102 (1998) 2177.

[54] G.K.H. Madsen, B.B. Iversen, F.K. Larsen, M. Kapon, G.M. Reisner, F.H. Herbstein, J. Am. Chem. Soc 120 (1998) 10040.

[55] Y.A. Abramov, L. Brammer, W.T. Klooster, R.M. Bullock, Inorg. Chem. 37 (1998) 6317.

[56] P.R. Mallinson, K. Wozniak, T. Garry, K.L. McCormak, J. Am. Chem. Soc. 119 (1997) 11502.

[57] P.R. Mallinson, K. Wozniak, C.C. Wilson, K.L. McCormak, D.M. Yufit, J. Am. Chem. Soc. 121 (1999) 4640.

[58] M.D. Cohen, G.M.J. Schmidt, F.I. Sonntag, J. Chem. Soc. (1964) 2000.

[59] R.S. Gopalan, G.U. Kulkarni, C.N.R. Rao, Acta Cryst. B (2000) in press.

[60] P. Coppens, G.M.J. Schmidt, Acta Crystallogr. 17 (1964) 222.

[61] G.U. Kulkarni, P. Kumaradhas, C.N.R. Rao, Chem. Mater. 10 (1998) 3498.

[62] P. Kumaradhas, R.S. Gopalan, G.U. Kulkarni, Proc. Indian Acad. Sci. (Chem. Sci.) 111 (1999) 569.

[63] R.S. Gopalan, P. Kumaradhas, G.U. Kulkarni, C.N.R. Rao, J. Mol. Struct. (2000) (in press).

[64] R.T. Morrison, R.N. Boyd, Organic Chemistry, 5th ed., Prentice-Hall, London, 1987 (p. 846).

[65] E. Espinosa, M. Souhassou, H. Lachekar, C. Lecomte, Acta Crystallogr. B55 (1999) 563.

[66] O. Bastiansen, Acta Chem. Scand. 3 (1949) 408.

[67] A. Chakrabarti, S. Yashonath, C.N.R. Rao, Mol. Phys. 84 (1995) 49.

[68] R.S. Gopalan, G.U. Kulkarni, E. Subramanian, S. Renganayaki, J. Mol. Struct. (2000) (in press).

[69] V. Zavodnik, A. Stash, V. Tsirelson, R. De Vries, D. Feil, Acta Crystallogr. B55 (1999) 45.

362   *G.U. Kulkarni et al. / Journal of Molecular Structure (Theochem) 500 (2000) 339–362*

[70] C. Gatti, V.R. Saunders, C. Roetti, J. Chem. Phys. 101 (1994) 10686.

[71] S.T. Howard, M.B. Hursthouse, C.W. Lehmann, P.R. Mallinson, C.S. Frampton, J. Chem. Phys. 97 (1992) 5616.

[72] E. Espinosa, C. Lecomte, E. Molins, S. Veintemillas, A. Cousson, W. Paulus, Acta Crystallogr. B52 (1996) 519.

[73] A. Fkyerat, A. Guelzim, F. Baert, W. Paulus, G. Heger, J. Zyss, A. Perigaud, Acta Crystallogr. B51 (1995) 197.

[74] A. Fkyerat, A. Guelzim, F. Baert, J. Zyss, A. Perigaud, Phys. Rev. B 53 (1996) 16236.

[75] M. Souhassou, C. Lecomte, R.H. Blessing, A. Aubry, M.-M. Rohmer, R. Wiest, M. Benard, M. Merraud, Acta Crystallogr. B47 (1991) 253.

[76] F. Hamazaoui, F. Baert, C. Wojcik, Acta Crystallogr. B52 (1996) 159.

[77] A. El-Haouzi, N.K. Hansen, C. Le Henaff, J. Protas, Acta Crystallogr. A52 (1996) 291.

[78] M.A. Spackman, P.G. Byrom, Acta Crystallogr. B53 (1997) 553.

[79] F. Hamazaoui, F. Baert, J. Zyss, J. Mater. Chem. 6 (1996) 1123.

[80] P. Coppens, Phys. Rev. Lett. 34 (1975) 98.

An Experimental Charge Density Study of the Effect of the Noncentric Crystal Field on the Molecular Properties of Organic NLO Materials**

R. Srinivasa Gopalan,[a] Giridhar U. Kulkarni,[a] and C. N. R. Rao*[a]

The structure, packing, and charge distribution in molecules of nonlinear optical materials have been analysed with reference to their counterparts in centrosymmetric structures based on low temperature X-ray measurements. The systems studied are the centric and noncentric polymorphs of 5-nitrouracil as well as the diamino, dithio, and thioamino derivatives of 1,1-ethylenedicarbonitrile; the latter possesses a noncentric structure. The molecular structure of 5-nitrouracil is invariant between the two forms, while the crystal packing is considerably different, leading to dimeric N—H···O rings in the centric polymorph and linear chains in noncentric one. There is an additional C—H···O contact in the centric form with a significant overlap of the electrostatic potentials between the alkenyl hydrogen atom and an oxygen atom of the nitro group. The dipole moment of 5-nitrouracil in the noncentric form is much higher ($\mu = 9\,D$) than in the centric form ($\approx 6\,D$). Among the 1,1-ethylenedicarbonitriles, there is an increased charge separation in the noncentric thioamino derivative,

leading to an enhanced dipole of 15 D compared to the centric diamino (5 D) and dithio (6 D) derivatives. The effect of the crystal field is borne out by semiempirical AM1 calculations on the two systems. Dipole moments calculated for the molecules in the frozen geometries match closely with those obtained for centric crystals from the experimental charge densities. The calculated values of the dipole moment in the frozen or optimized geometries in the noncentric structures are, however, considerably lower than the observed value. Furthermore, the conformation of the S—CH$_3$ group in the noncentric crystal is anti with respect to the central C=C bond while the syn conformation is predicted for the free molecule in the optimized geometry.

KEYWORDS:

charge density · dipole moment · materials science · nonlinear optics · semiempirical calculations

Several organic materials with nonlinear optical (NLO) properties have been described in the recent literature.[1–5] Nitroaniline and push–pull ethylenes are typical examples of NLO materials involving intramolecular charge transfer. Besides having a polarizable π-cloud with electron-donating and -accepting groups, an essential feature of such compounds is that they crystallize in noncentrosymmetric space groups. Espinosa et al.[6] have reported a charge density study of L-arginine phosphate. Puig-Molina et al.[7] have compared the topological properties of charge density of 2-amino-5-nitropyridinium dihydrogen phosphate from experiment and theory. Charge density studies of 2-methyl-4-nitroaniline[8] and urea[9] show that there is a considerable enhancement of the molecular dipole moment in the solid state. Based on a charge density study on N-(4-nitrophenyl)-L-prolinol, Fkyerat et al.[10, 11] extracted hyperpolarizability from the octapole moments. Such a possibility has been discussed in connection with inorganic NLO materials as well.[12, 13] While there is some speculation as to whether charge densities can provide hyperpolarizability,[14] it is generally accepted that reliable molecular dipole moments can be extracted from charge densities.[15] Thus, Hamazaoui et al.[16] obtained the dipole moment of 3-methyl-4-nitropyridine-N-oxide from its charge density, which agrees well with that from semiempirical calculations. Recently, Madsen et al.[17] have evaluated the dipole moment of phosphangulene from its charge

density to be 42% higher than that measured in a chloroform solution. They also obtained the pyroelectric coefficient by combining the derived dipole moment with temperature-dependent measurements of the unit cell volume.

We have carried out an experimental charge density investigation of organic NLO systems in order to understand the manner in which the noncentric nature of the crystal field affects the molecular dipole moment and other properties in the solid. For this purpose, we have examined two systems—one where the center of symmetry is destroyed by crystallizing the molecule in a related polymorphic structure and the other where the noncentric nature is introduced by changing the pattern of substitution. For the first system, we have chosen the two polymorphic forms of 5-nitrouracil, **1**, which crystallizes in orthorhombic $P2_12_12_1$ and *Pbca* space groups, the former being noncentrosymmetric. The NLO activity of the noncentric form at $1.06\,\mu m$ is ~ 160 times that of potassium dihydrogen phos-

[a] *Prof. Dr. C. N. R. Rao, R. S. Gopalan, Dr. G. U. Kulkarni*
Chemistry and Physics of Materials Unit
Jawaharlal Nehru Centre for Advanced Scientific Research
Jakkur P.O., Bangalore, 560064 (India)
E-mail: cnrrao@jncasr.ac.in

[**] *The authors thank Dr. S. N. Ghanshyam Acharya for the synthesis of dicarbonitrile derivatives 2 – 4.*

CHEMPHYSCHEM

phate.[18] In order to explore the effect of substitution, we have studied 1,1-ethylenedicarbonitriles (EDCN), in which the positions of the dimethylamino and methylthio substituents have been varied to give the symmetric and unsymmetric molecules bis(*N,N*-dimethylamino)-EDCN (**2**), 2,2-bis(methylthio)-EDCN (**3**), and 2-(*N,N*-dimethylamino)-2-methylthio-EDCN (**4**). The diamino and the dithio derivatives, **2** and **3**, crystallize in centric

structures whereas the thioamino, **4**, occurs in a noncentrosymmetric structure exhibiting NLO properties. In both the systems, we have examined the intramolecular bonding using the Laplacian of the total electron density. Where necessary, we have analyzed the intermolecular interactions in terms of the electrostatic potential. We have computed the dipole moments from the experimental charge densities and compared the results with those from semiempirical AM1 calculations.

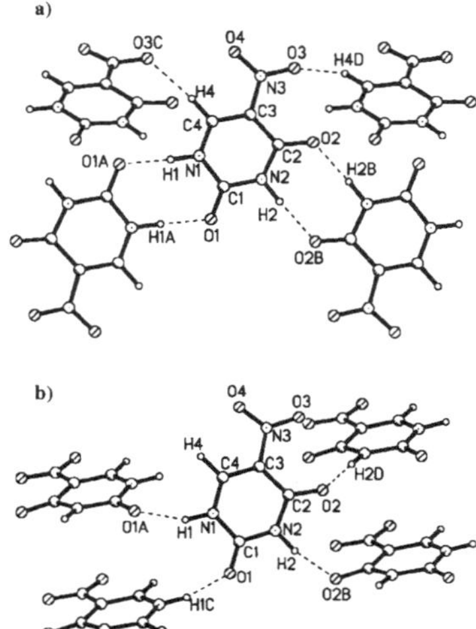

[*] *Members of the Editorial Advisory Board will be introduced to the readers with their first manuscript.*

5-Nitrouracil

The centric polymorph of **1** crystallizes with eight molecules per unit cell (*Pbca*; $a = 8.308(3)$, $b = 10.426(3)$, $c = 13.363(4)$ Å) while the unit cell of the noncentric form contains four molecules ($P2_12_12_1$; $a = 5.4342(1)$, $b = 9.8406(1)$, $c = 10.3659(1)$ Å). The densities of both forms are similar (Table 1).

The space groups *Pbca* and $P2_12_12_1$ contain the same set of symmetry elements but for the inversion center, which makes the problem interesting. In both polymorphs, the asymmetric unit comprises a whole molecule. The intramolecular geometry is essentially the same in both cases and the molecule is nearly planar (Table 2): The nitro group forms an angle of 1.5° with the plane of the uracil ring in the centric form and 2.0° in the noncentric polymorph.

In Figure 1, the packing diagram along with the atom labels are shown. In the centric form, the molecule is held by the dimeric N—H···O contacts of $R_2^2(8)$ type[19] with the two neighboring coplanar molecules (Figure 1 a). The N—H···O contacts in the noncentric form are of D(2) type arising from four different molecules (Figure 1 b). In both the cases, the N—H···O bonds are ~1.8 Å long, with ∢N—H···O angles of 155–174°. The centric polymorph, in addition, exhibits an intermolecular C—H···O interaction with the nitro group at a relatively short distance of 2.15 Å. The nitro group of the noncentric polymorph is in an unfavorable geometry to establish such a contact (Figure 1 b).

The results of the multipolar refinement (see Experimental Section) are described using the Laplacian of the total charge density. The topology of the Laplacian field allows one to obtain a chemical model of the bonded and nonbonded pairs and to characterize the local concentration ($\nabla^2\rho < 0$) and depletion ($\nabla^2\rho > 0$) of the molecular charge distribution.[20, 21] Figure 2 shows the contour maps of the negative Laplacian obtained for the two polymorphs. The lone electron pairs on the oxygen

Figure 1. *Molecular packing in a) centric and b) noncentric **1** showing the intermolecular hydrogen bonds. Atom labels are also shown.*

CHEMPHYSCHEM

Crystal	1		2	3	4
	centric	noncentric			
Chemical formula (g mol^{-1})	$C_4H_3N_3O_4$ (157.09)		$C_8H_{12}N_4$ (164.22)	$C_6H_6N_2S_2$ (170.25)	$C_7H_9N_3S$ (167.23)
Crystal system	orthorhombic	orthorhombic	orthorhombic	monoclinic	orthorhombic
Space group	*Pbca* (No. 61)	$P2_12_12_1$ (No. 19)	*Pcab* (No. 6)	$P2_1/n$ (No. 14)	$Pna2_1$ (No. 33)
a [Å]	8.308(3)	5.4342(1)	7.6280(1)	4.0261(1)	7.904(2)
b [Å]	10.426(3)	9.8406(1)	14.455	13.227	8.630(2)
c [Å]	13.363(4)	10.3659(1)	16.3489(2)	14.3906(1)	12.856(3)
β [°]	90	90	90	95.445(1)	90
V [cm^3]	1157.5(6)	554.32(1)	1802.70(3)	762.89(2)	876.9(3)
Z	8	4	8	4	4
F_{000}	640	320	704	352	352
ρ [g cm^{-3}]	1.803	1.882	1.210	1.482	1.267
No. of reflections for cell parameters	45	37	60	60	60
μ [mm^{-1}]	0.16	0.17	0.079	0.617	0.309
Crystal form	cuboidal	hexagonal	cuboidal	cuboidal	cuboidal
Crystal size [mm]	$0.3 \times 0.2 \times 0.2$	$0.15 \times 0.15 \times 0.1$	$0.2 \times 0.1 \times 0.1$	$0.2 \times 0.15 \times 0.1$	$0.2 \times 0.1 \times 0.1$
Crystal color	colorless	colorless	colorless	yellow	colorless
No. of measured reflections	16262	16515	29153	12914	24387
No. of independent reflections	5026	8371	8589	6451	8576
No. of observed reflections	3879	3239	4894	4863	2874
R_{merge}	0.0644	0.0434			
R_{int}	0.0763	0.0454	0.0361	0.0301	0.0692
θ_{min} [°]	3.0	2.8	2.49	2.10	2.84
θ_{max} [°]	49.4	49.4	49.48	49.47	49.89
Range of h, k, l	$-12 \leq h \leq 16$	$-10 \leq h \leq 11$	$-15 \leq h \leq 15$	$-8 \leq h \leq 7$	$-16 \leq h \leq 15$
	$-22 \leq k \leq 11$	$-19 \leq k \leq 20$	$-15 \leq k \leq 30$	$-21 \leq k \leq 27$	$-18 \leq k \leq 18$
	$-28 \leq l \leq 27$	$-21 \leq l \leq 21$	$-34 \leq l \leq 34$	$-29 \leq l \leq 28$	$-27 \leq l \leq 27$
After multipole refinement					
Weighting scheme	0.02, 0.3	0.02, 0.1	0.05, 0.5	0.03, 0.3	0.04, 0.3
$R1$	0.0345	0.0391	0.0501	0.0397	0.0332
$wR2$	0.0416	0.0454	0.0775	0.0475	0.0513
S	1.05	1.00	0.94	1.06	1.27
No. of variables	399	404	161	178	197
N_{ref}/N_v	20.5	17.2	30.4	27.3	14.6
Cambridge Crystallographic Database	145990	145991	145987	145988	145989

Table 1. *Crystal data for 5-nitrouracil (1) and the 1,1-ethylenedicarbonitriles (2 – 4).*[a]

[a] Experimental details: Siemens CCD diffractometer, crystal – detector distance 5.0 cm, Mo$_{K\alpha}$ radiation ($\lambda = 0.71073$ Å, 50 kV, 40 mA), 130 ± 1 K

atoms occur as (3, − 3) critical points. There are some differences in the Laplacian maps of the two polymorphs. The atomic basins are generally linked in the centrosymmetric form but they appear to be disjoint in the noncentric case. The N3—O3 bond region in the centrosymmetric form contains nonoverlapping atomic lobes, which indicate a closed shell interaction. This is reflected in the properties of the bond critical points, listed in Table 3. This bond carries a much smaller Laplacian (− 3.0 e Å^{-5}) compared to the other N—O bond of the molecule as well as the N—O bonds of the noncentric form, though the electron density itself is quite comparable. We also notice from Table 3 that the densities and the associated Laplacians of the non-hydrogen bonds are somewhat higher in the noncentric form while those of the N—H bonds are higher in the centric form. Moreover, the pseudo-atomic charges are generally higher in the noncentric polymorph. This implies that there is an increased charge separation in the noncentric case. The nitro group charge is − 0.19 and − 0.38 e respectively, in the centric and the non-centric forms, which compare well with − 0.25 e obtained for 3-methyl-4-nitropyridine-*N*-oxide.[16] We also observe that the properties of the various bonds of the uracil moiety match closely with those reported in the case of 1-methyluracil.[22]

The nature of the intermolecular bonding is described in terms of the electrostatic potential. In Figure 3, contour maps of the electrostatic potential maps in the N—H $\cdots$ O and the C—H $\cdots$ O bond regions are shown. The contours originating from the N—H region overlap with those from the oxygen atom (Figures 3 a, b). The penetration is significant and indicates that these are strong contacts.[23] They are associated with (3, − 1) critical points in ρ as well as in electrostatic potential. The Laplacian at the CPs are small and positive (Table 4), as generally found in hydrogen bonds.[24] The C—H $\cdots$ O contact of the centric form also exhibits a noticeable overlap of the potentials ($\rho = 0.08(2)$ e Å^{-3}, $V = 0.28(1)$ e Å^{-1}), while in the noncentric form a zero potential surface passes between the proton and the oxygen atom to indicate no interaction between them (Figure 3 c). This is understandable, since in the presence of a carbonyl group, a nitro group acts as a secondary hydrogen bond acceptor.[25] The C—H $\cdots$ O contact of the centric form is special in that the oxygen O3, which participates in intermolecular bonding, tends to have nearly ionic interaction with the nitro group (Figure 2). Such bonding would be favored by the centrosymmetric packing of the molecules. For the same reason, this contact is absent in the noncentric form.

CHEMPHYSCHEM

Table 2. Bond lengths and angles of **1**.

Bond length [Å]	centric **1**	noncentric **1**
O1-C1	1.2284(13)	1.2310(13)
O2-C2	1.2252(13)	1.2290(12)
O4-N3	1.2364(12)	1.235(2)
O3-N3	1.2294(12)	1.2306(14)
N2-C1	1.3667(13)	1.3663(12)
N2-C2	1.3945(13)	1.3900(14)
N2-H2	1.01	1.01
N1-C4	1.3427(14)	1.341(2)
N1-C1	1.3807(13)	1.3786(14)
N1-H1	1.01	1.01
N3-C3	1.4432(13)	1.4412(14)
C2-C3	1.4568(14)	1.454(2)
C3-C4	1.3575(14)	1.3621(14)
C4-H4	1.08	1.08

Bond angle [°]	centric **1**	noncentric **1**
C1-N2-C2	127.18(9)	127.47(9)
C1-N2-H2	116.41(6)	116.27(6)
C2-N2-H2	116.41(6)	116.27(6)
C4-N1-C1	122.51(9)	122.90(8)
C4-N1-H1	118.74(6)	118.55(6)
C1-N1-H1	118.74(6)	118.55(5)
O3-N3-O4	123.09(9)	123.46(11)
O3-N3-C3	119.36(9)	119.20(11)
O4-N3-C3	117.56(9)	117.33(10)
O1-C1-N2	123.08(9)	122.60(10)
O1-C1-N1	121.61(9)	122.54(9)
N2-C1-N1	115.31(9)	114.85(9)
O2-C2-N2	119.86(9)	118.82(10)
O2-C2-C3	127.55(10)	128.32(10)
N2-C2-C3	112.59(8)	112.85(8)
C4-C3-N3	116.49(9)	116.94(10)
C4-C3-C2	120.79(9)	120.32(10)
N3-C3-C2	122.62(8)	122.73(9)
N1-C4-C3	121.39(9)	121.40(10)
N1-C4-H4	119.30(6)	119.30(6)
C3-C4-H4	119.30(6)	119.30(7)

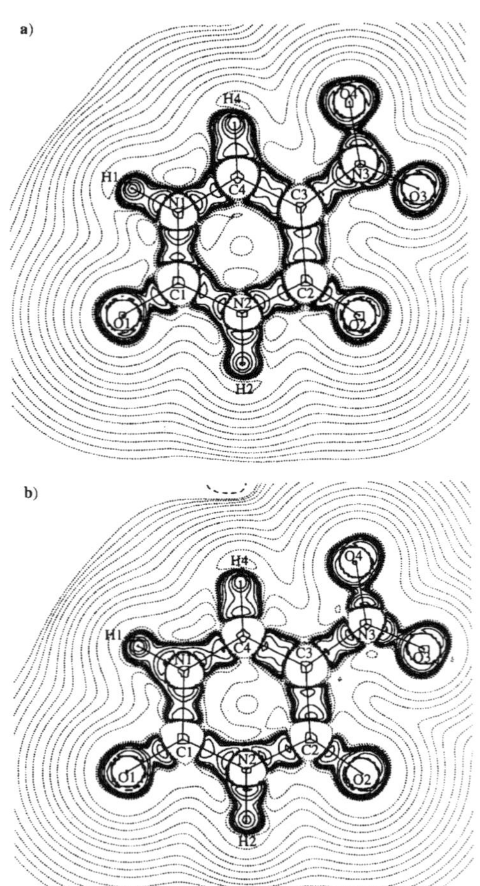

Figure 2. Contour maps of the total density Laplacian in the C1-C2-C4 plane for a) centric and b) noncentric **1**.

In situ dipole moments of molecules in crystals can be obtained from charge densities by using Equation (1).

$$p_i = \sum_i z_i R_i + \int r \rho_i(r_i)\,dr \tag{1}$$

The XDPROP routine (the XD program) was used to calculate dipole moments and the dipole moment vectors are shown in Figure 4. The vectors lie close to the N(1)—H(1) bond direction in the plane of the molecule, with magnitudes of $\mu = 5.5(6)$ and 9(1) D in the centric and the noncentric polymorphs, respectively.[26] As the intramolecular geometry remains the same in the two cases (Table 2), the enhancement of the dipole moment in the noncentric structure is likely to arise mainly due to packing. The effect of crystal packing can be understood by calculating the dipole moment of the molecule ex situ in the frozen geometry. For this purpose, we carried out AM1, PRECISE calculations by freezing the molecular geometries obtained from X-ray diffraction. The calculated dipole moments are very similar in the two polymorphs ($\mu = 5.5$ and 5.6 D for centric and noncentric cases, respectively) and point in the same direction (Figure 4). Interestingly, the vectors obtained from the calculation lie close (in magnitude as well as in direction) to that from

Table 3. Analysis of the bond critical points in **1**.

Bond	centric $\rho^{[a]}$	centric $\nabla^2\rho^{[b]}$	centric $\varepsilon^{[c]}$	noncentric $\rho^{[a]}$	noncentric $\nabla^2\rho^{[b]}$	noncentric $\varepsilon^{[c]}$
O1-C1	2.91(5)	−38.2(3)	0.21	3.31(9)	−51.9(5)	0.27
O2-C2	2.77(6)	−26.8(4)	0.18	3.21(97)	−51.3(4)	0.17
N1-C1	2.16(5)	−26.8(2)	0.06	2.36(8)	−27.9(3)	0.29
N1-C4	2.29(5)	−27.3(2)	0.25	2.50(9)	−31.5(4)	0.28
N1-H1	1.98(6)	−28.8(4)	0.06	1.9(1)	−30.1(8)	0.00
N2-C1	2.12(4)	−23.9(2)	0.21	2.34(8)	−30.1(3)	0.38
N2-C2	2.09(4)	−23.5(2)	0.12	2.22(7)	−23.7(3)	0.24
N2-H2	1.92(6)	−29.8(4)	0.04	1.8(1)	−36(1)	0.04
N3-C3	1.87(4)	−15.4(2)	0.37	1.96(8)	−17.1(4)	0.39
C2-C3	1.93(3)	−16.0(1)	0.27	1.98(7)	−7.0(2)	0.46
C3-C4	2.21(3)	−22.1(1)	0.34	2.31(7)	−24.7(2)	0.43
C4-H4	1.79(5)	−20.1(2)	0.05	1.9(1)	−29.5(7)	0.07
O3-N3	3.19(5)	−3.0(2)	0.09	3.48(9)	−14.5(3)	0.05
O4-N3	3.28(5)	−11.9(2)	0.08	3.61(9)	−18.6(4)	0.08

[a] electron density, in e Å^{-3}. [b] Laplacian, in e Å^{-5}. [c] Ellipticity.

Table 4. Hydrogen bond critical points in **1**.

Polymorph	D-H ··· A	ρ [e Å^{-3}]	$\nabla^2\rho$ [e Å^{-5}]	V [e Å^{-1}]
centric	N(1)-H(1) ··· O(1)[a]	0.18(3)	3.32(2)	0.55(2)
	N(2)-H(2) ··· O(2)[a]	0.17(3)	2.88(2)	0.47(2)
	C(4)-H(4) ··· O(3)[b]	0.08(2)	1.39(1)	0.28(1)
noncentric	N(1)-H(1) ··· O(1)[c]	0.16(5)	2.99(4)	0.52(3)
	N(2)-H(2) ··· O(2)[d]	0.15(6)	3.05(5)	0.61(4)
	O(1) ··· H(1)-N(1)[e]	0.14(5)	2.64(5)	0.53(3)
	O(2) ··· H(2)-N(2)[f]	0.13(5)	2.69(5)	0.60(3)

Symmetries: [a] $-x$, $-y$, $1-z$. [b] $^1/_2-x$, $^1/_2+y$, z. [c] $^1/_2+x$, $^1/_2-y$, $1-z$. [d] $-^1/_2+x$, $^1/_2-y$, $-z$; [e] $-^1/_2+x$, $^1/_2-y$, $1-z$. [f] $^1/_2+x$, $^1/_2-y$, $-z$.

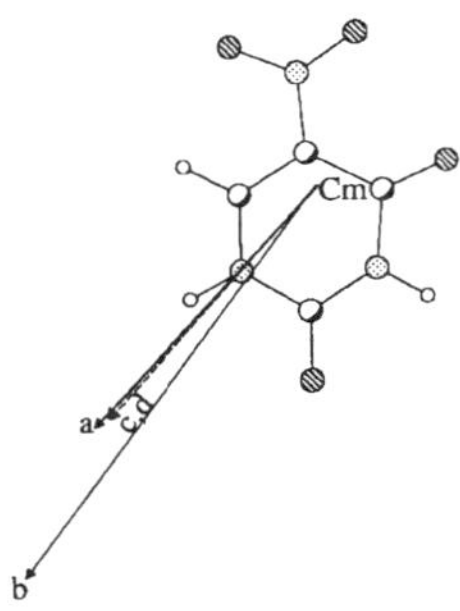

Figure 4. Orientation of the molecular dipole moment of **1**: a) centric, multipole; b) noncentric, multipole; c) centric, AM1; d) noncentric, AM1. Cm refers to the center of mass of the molecule.

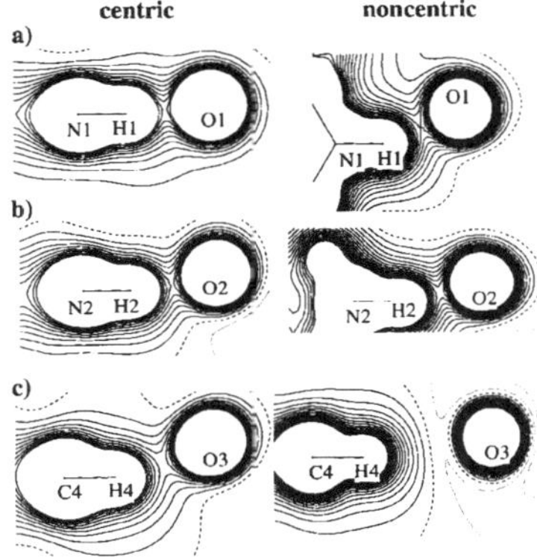

Figure 3. Contour maps of the electrostatic potential in the intermolecular regions (contour level at 0.05 e Å^{-1}): a) N1–H1 ··· O1, b) N2–H2 ··· O2, c) C4–H4 ··· O3. The centric and noncentric forms are given on the left and right sides, respectively.

experiment for the centric polymorph. This seems to suggest that the packing in the centric form of **1** has a minimal effect on the intramolecular charge distribution. This observation also demonstrates that an asymmetric crystal field in a noncentric structure can significantly enhance the dipole moment of a molecule. To our knowledge, this is the first instance of charge density study of a compound crystallizing in closely related space groups exhibiting differing molecular properties.

1,1-Ethylenedicarbonitriles

The two nitriles, **2** and **3**, crystallize in the centrosymmetric *Pcab* and *P2₁/n* space groups, respectively, whereas **4** crystallizes in the noncentric *Pna*2₁ space group (Table 1). We show the molecular diagrams along with the thermal ellipsoids in Figure 5, where the bond lengths and angles associated with the non-hydrogen atoms are also indicated. Diamine **2** possesses a pseudo two-fold symmetry along C3–C4 while the dithio **3** exhibits a *syn – anti* methylthio conformation. Thioamine **4** has methylthio group in the *anti* conformation. The 1,1-ethylenedicarbonitrile moiety, which contains a formal C3=C4 bond and two single C1–C3 and C2–C3 bonds, along with the cyano group, is essentially planar in all three molecules. The C≡N bond lengths are also similar (1.14(3) Å) and close to those reported in the literature. There are however, some differences between the formal C–C single and double bonds. In **2**, the single and the

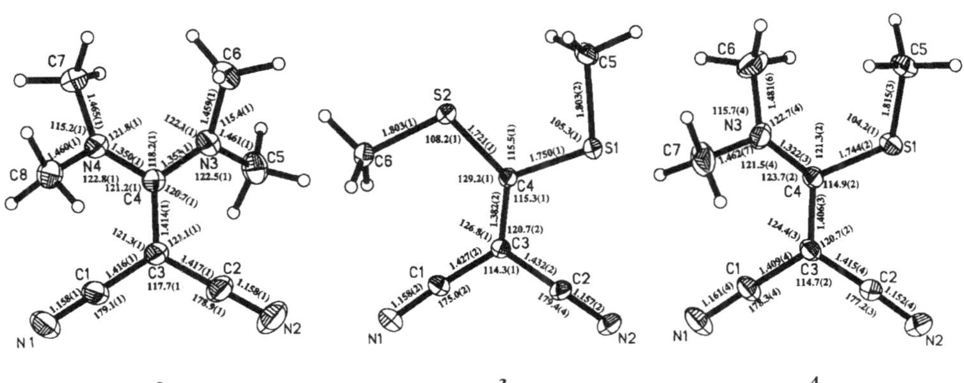

Figure 5. Molecular diagrams of the diamino **2**, dithio **3**, and thioamino **4** EDCN derivatives, showing thermal ellipsoids. The bond lengths and angles of the non-hydrogen bonds are shown.

double bonds are of equal length (~ 1.416 Å), while they differ in **3** (1.432(2) and 1.427(2), versus 1.382(2) Å). In **4**, the differences are moderate (see Figure 5). A survey of formally uncharged molecules containing the ethylenedicarbonitrile moiety (466 compounds with $r < 10\%$) listed in the Cambridge Crystallographic Database have mean C—C and C=C bond lengths of 1.43(2) and 1.37(3) Å, respectively. The shortening of the C—C single bonds in this moiety is noteworthy. The C_{sp^2}—N (or C_{sp^2}—S) bonds are shorter than C_{sp^3}—N (or C_{sp^3}—S). In **3**, there exist considerable differences between the bond lengths and the angles associated with the *syn* and *anti* conformations of the methylthio groups. The C—C≡N bond angle associated with the *syn*-methythio group is 175° instead of $\sim 180°$. An earlier structural report on this molecule suggested steric interaction to be responsible for the discrepancy.[27] However, 3,3-bis(methylthio)-2-nitro-2-propene-1-nitrile does not show significant differences in the bond lengths and angles between the *syn*- and the *anti*-configurations of the methylthio groups.[28]

Figure 6 shows the contour maps of the negative Laplacian of the total density in the plane defined by the N1-N2-C3 atoms. The lone pairs on the nitrogen and sulfur atoms exhibit (3, − 3) as critical points. The lobes near sulfur are similar to those seen previously by Mallinson and co-workers in the case of 3,3,6,6-tetramethyl-*S*-tetrathiane.[29] In the diamine **2**, the contours associated with the amino groups are small, since this group is out of plane with the rest of the molecule. The C—S bonds of both **3** and **4** exhibit disjoint lobes.

Figure 7 depicts the critical points in the total density for the various bonds in the molecules. The values of total electron density, the Laplacian, and the ellipticity at the critical points are also shown. The C≡N group exhibits similar densities ($\sim 3.2\,e\,\text{Å}^{-3}$) in the three molecules although the Laplacian values are somewhat different, − 36, − 20, and − 16 $e\,\text{Å}^{-5}$ in **1**, **2**, and **3**, respectively. Interestingly, the densities of the C—C single and double bonds are not significantly different in the three compounds. The Laplacian values are also similar, the differences are less than $\sim 5\,e\,\text{Å}^{-5}$. This result is in contrast with what is expected of normal single and double bonds. The electron densities associated with the isolated single and double bonds are generally 1.71 and 2.5 $e\,\text{Å}^{-3}$ respectively.[30] In the present

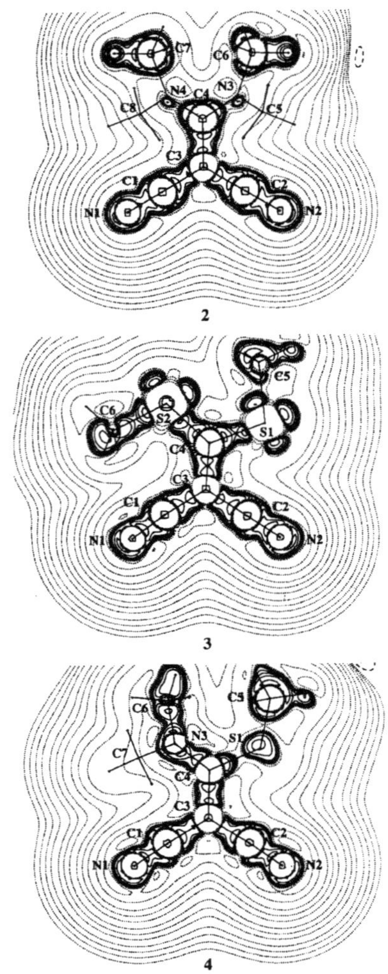

Figure 6. Contour maps of the Laplacian of the total electron density of the diamino **2**, dithio **3**, and thioamino **4** EDCN derivatives in the N1-C3-C4 plane.

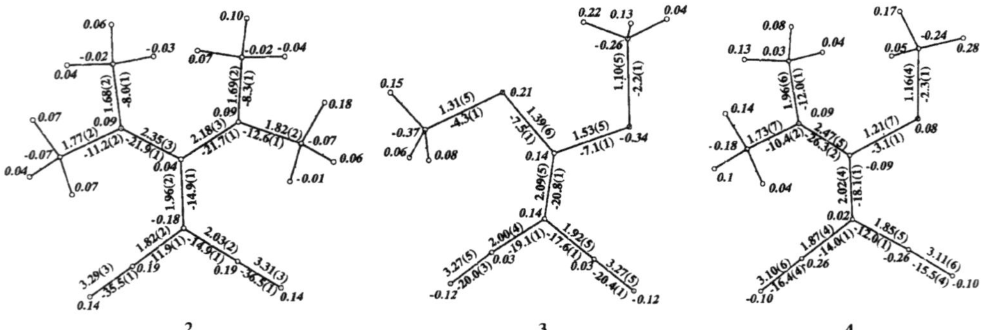

Figure 7. Stick diagrams showing the critical points for the various bonds of the diamino **2**, dithio **3**, and thioamino **4** EDCN derivatives. The electron density and the Laplacians associated with each bond are shown nearby. The pseudo-atomic charges are also indicated in italics. The atoms are labeled as in Figure 5.

case, there is probably some delocalization of electrons in the bonding region. A similar observation was made earlier by Espinosa et al.[31] in the case of the BTDMTTF-TCNQ complex. The low values of the densities and the Laplacians in the C–S bonds are characteristic of hypervalent species.[29] A recent study on tetrasulfurtetranitride[32] has reported low values of densities and Laplacians in the N–S bonds. The charge density analysis reflects the differences between the C_{sp^2} and C_{sp^3} bonding regions, as expected from the structural features discussed above. Accordingly, the C_{sp^2} bonding regions carry higher densities and Laplacians. Based on the pseudoatomic charges shown in Figure 7, there appears to be an increased charge separation in the noncentric thioamine **4**.

The intermolecular contacts of the structures **2 – 4** are listed in Table 5. Both **2** and (**3**) exhibit five C–H···N contacts while the thioamine (**4**) makes only three such contacts. These are typical C–H···N contacts in terms of both the geometry and the charge density. In addition, there are intermolecular S···S and S···N contacts in the dithio compound and S···N contacts in the thioamine (Table 5). The distances as well as the charge densities associated with these contacts are similar to the contacts found in other molecules.[29, 32]

The in situ dipole moments of the molecules in the three crystals are projected in Figure 8. The dipole moment vectors lie

Figure 8. *Molecular diagrams showing the dipole moment vectors computed from charge density in **2 – 4**. Cm refers to the center of mass of the molecule.*

close to the mean planes of the molecules. An important observation from Figure 8 is that the dipole moment of the thioamine **4**, in the noncentric structure, is much higher ($\mu =$ 15(2) D) than the values of the diamino (**2**) and the dithio (**3**) derivatives in the centric structures ($\mu = 5(2)$ and 6(2) D).[26] The enhancement of the dipole moment in the thioamine could arise from the asymmetric substitution of the electron donating groups or due to the noncentric crystal field. In order to shed more light, we have computed dipole moments of the three molecules in the free state in both the frozen and the optimized (AM1, PRECISE) geometries using MOPAC. In Figure 9, we depict

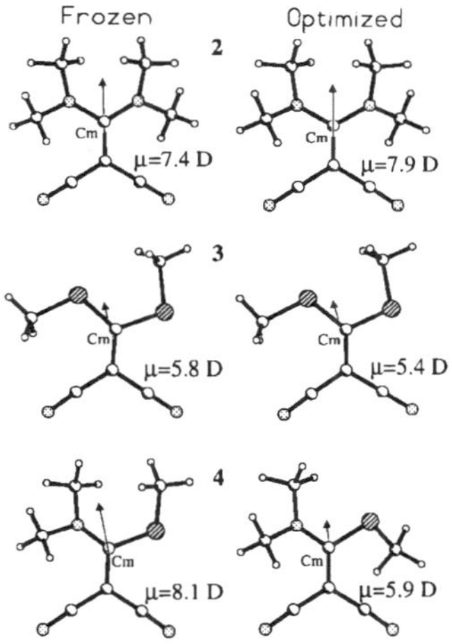

Figure 9. *Molecular dipole moments calculated using AM1 (MOPAC) in the frozen (left) and the optimized (right) geometries of the diamino **2**, dithio **3**, and thioamino **4** EDCN derivatives. Cm refers to the center of mass of the molecule. Atomic labeling is given in Figure 8.*

the frozen and the optimized dipole moments projected along with the molecule. The calculated dipole moment remains essentially unchanged between the frozen and the optimized geometries in the case of **2** and **3**. Compared to the in-crystal dipole moment, these values are slightly higher in the diamino and are comparable in the case of the dithio compound (see Figures 8 and 9). The noncentric thioamine **4**, on the other hand, exhibits much smaller dipole moments outside the lattice. Both the optimized ($\mu = 5.9$ D) and the frozen ($\mu = 8.1$ D) moments are considerably lower than the in-crystal dipole moment of $\mu = 15$ D. It is interesting that in the optimized geometry, the

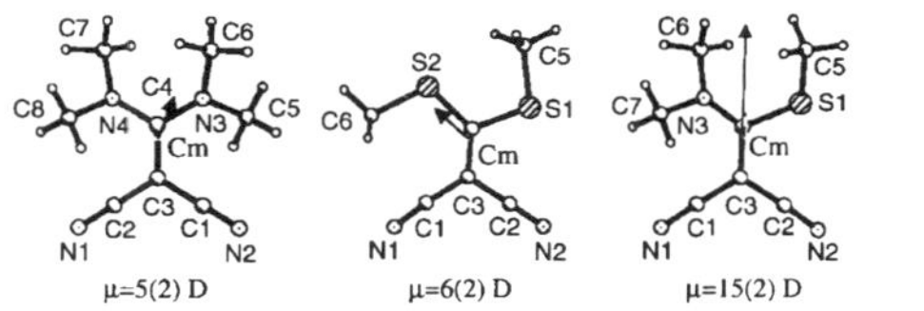

Table 5. *Intermolecular contacts in **2 – 4**.*

Hydrogen contacts between donor (D) and acceptor (A)

D–H···A	H···A [Å]	D···A [Å]	∢D-H···A [°]	ρ [e Å⁻³]	$\nabla^2\rho$ [e Å⁻⁵]
2					
C5-H5A···N1[a]	2.745	3.531	130.9	0.061(3)	0.601(2)
C7-H7A···N1[b]	2.510	3.467	149.6	0.103(5)	0.967(2)
C8-H8C···N1[c]	2.872	3.649	130.4	0.041(2)	0.41(1)
C8-H8B···N2[d]	2.720	3.353	137.2	0.059(3)	0.616(1)
C7-H7B···N2[e]	2.493	3.504	136.8	0.103(5)	0.951(2)
3					
C6-H6C···N1[f]	2.564	3.541	152.8	0.035(6)	0.653(5)
C6-H6A···N1[g]	2.840	3.500	120.6	0.042(5)	0.523(3)
C5-H5B···N1[h]	2.586	3.631	168.6	0.04(2)	0.619(5)
C5-H5A···N2[i]	2.560	3.603	167.8	0.040(2)	0.522(2)
C6-H6A···N2[j]	2.886	3.554	121.3	0.048(4)	0.566(3)
4					
C5-H5B···N1[k]	2.594	3.617	162.0	0.032(8)	0.572(9)
C6-H6B···N1[k]	2.780	3.633	137.5	0.025(8)	0.420(4)
C5-H5A···N2[l]	2.443	3.463	161.2	0.05(1)	0.691(4)

Sulfur contacts	Distance [Å]	ρ [e Å⁻³]	$\nabla^2\rho$ [e Å⁻⁵]
3			
S1···S1[m]	3.515	0.067(1)	0.654(2)
S1···S1[n]	3.952	0.030(1)	0.305(1)
N2···S2[o]	3.307	0.055(1)	0.603(1)
N2···S2[p]	3.549	0.055(1)	0.603(1)
4			
N1···S1[q]	3.194	0.055(2)	0.586(3)

Symmetries: [a] $x, -1/2+y, 1/2+z$. [b] $3/2-x, -1/2+y, -z$. [c] $1/2+x, 3/2-y, z$. [d] $2-x, 3/2-y, -1/2+z$. [e] $3/2-x, y, -1/2+z$. [f] $1+x, y, z$. [g] $-x, 1-y, -z$. [h] $-1/2-x, 1/2+y, 1/2-z$. [i] $1-x, 1-y, 1-z$. [j] $1/2-x, 1/2+y, 1/2-z$. [k] $2-x, -y, 1/2+z$. [l] $2-x, 1-y, 1/2+z$. [m] $1-x, 1-y, 1-z$. [n] $-x, 1-y, 1-z$. [o] $1/2-x, -1/2+y, 1/2-z$. [p] $-1/2-x, -1/2+y, 1/2-z$. [q] $3/2-x, -1/2+y, -1/2+z$.

CHEMPHYSCHEM

thioamine exhibits a *syn*-methylthio conformation with a smaller dipole moment compared to the *anti*-methylthio conformation found in the frozen geometry of the crystal. It appears that the noncentric structure induces significant distortions in the molecule and enhances the dipole moment.

In conclusion, the present charge density study on the centric and noncentric molecular systems has provided valuable insight into the effect of the crystal field on molecular properties. In the case of 5-nitrouracil, the dipole moment is 5.5(6) D in the centric structure and increases to 9(1) D in the noncentric lattice, while the molecular geometry essentially remains the same; the difference in the dipole moments clearly arises from the crystal packing. In the centric structure, there are $N-H\cdots O$ hydrogen bonded dimers, whereas in the noncentric the molecules form $N-H\cdots O$ linear chains. The centric crystal exhibits an additional contact, $C-H\cdots O_{nitro}$, absent in the noncentric structure. The molecular dipole moment in the noncentric crystal is considerably greater than the value calculated for a free molecule of outside the crystal (frozen geometry). In centric polymorph, the dipole moment is close to that obtained from the free molecule.

In the ethylenedicarbonitriles, the crystal is centrosymmetric when the two substituents are the same, as in the diamino (**2**) and dithio (**3**) derivatives and they have low dipole moments ($\mu \sim 6$ D). The dipole moment values are close to those obtained by AM1 calculations on the frozen molecular geometries. In contrast, the thioamine (**4**), with the *anti*-methylthio conformation of the $S-CH_3$ group, occurs in a noncentric structure where there is an increased charge separation within the molecule and, hence, a considerably higher dipole moment ($\mu = 15(2)$ D) than that calculated for the molecule of frozen geometry. When the molecule is subjected to optimization, it adopts a *syn*-methylthio conformation and the dipole moment falls to below $\mu = 5.9$ D. The enhancement of the dipole moment in the thioamino system is partly due to the *anti*-conformation forced on the molecule by the noncentric crystal field and partly due to the field itself.

Experimental Section

On cooling an aqueous solution of 5-nitrouracil from 70 °C, crystals of the centrosymmetric polymorph were obtained. The noncentric crystals were obtained from an acetonitrile solution at room temperature. The dinitriles **2–4** were synthesized by employing reported procedures[33–35] and the crystals were grown from toluene, ethyl acetate, and benzene solutions, respectively. High quality crystals were chosen after examination under an optical microscope. X-ray diffraction intensities were measured by ω scans using a Siemens three-circle diffractometer.

The unit cell parameters and the orientation matrix of the crystal were initially determined using ~ 45 reflections from 25 frames collected over a small ω scan of 7.5° sliced at 0.3° intervals. A hemisphere of reciprocal space was then collected in two shells using the SMART software[35] with 2θ at 28° and 70°. Data reduction was performed using the SAINT program[36] and the orientation matrix along with the detector and the cell parameters were refined for every 40 frames. Further experimental details are listed in Table 1. The crystal structures were first determined with the low-resolution

data up to $(\sin\theta)/\lambda = 0.56$ Å^{-1}. The phase problem was solved by direct methods and the non-hydrogen atoms were refined anisotropically, by means of the full-matrix least-squares procedures using the SHELXTL program.[37] All the hydrogen atoms were located using the difference Fourier method and the temperature factors of these atoms were refined isotropically. The crystal structures agreed well with those reported previously.[27, 38, 39] The structure of the centrosymmetric 5-nitrouracil polymorph (*Pbca*) is reported for the first time and deposited in the Cambridge Crystallographic Database.

Charge density analysis was carried out based on multipole expansion of the electron density centered at the nucleus of the atom.[40] Accordingly, the aspherical atomic density can be described in terms of spherical harmonics [Eq. (2)]. Thus, for each atom, in which the origin at the atomic nucleus, [Eq. (3)] holds. The population coefficients, P_{lmp}, are to be refined along with the κ and κ' parameters which control the radial dependence of the valence shell density. The analysis was carried out in several steps.

$$\rho_{atom}(r) = \rho_{core}(r) + \rho_{valence}(r) + \rho_{def}(r) \tag{2}$$

$$\rho_{atom}(r) = \rho_{core}(r) + P_v \kappa^3 \rho_{valence}(\kappa r) + \sum_{l=0} \kappa'^3 R_l(\kappa' \zeta r) \sum_{m=0} \sum_{p=\pm l} P_{lmp} Y_{lmp}(\theta,\varphi) \tag{3}$$

The hydrogen atomic positions were found using the difference Fourier method and were adjusted to average neutron values,[41] as is usually done during the multipole refinement ($C_{sp^3}-H$ 1.06, $C_{sp^2}-H$ 1.083, $N-H$ 1.01 Å). A high order refinement of the data was performed using reflections with $(\sin\theta)/\lambda \geq 0.6$ Å^{-1} and $F_0 \geq 4\sigma$. All the hydrogen atoms were held constant throughout the refinement along with their isotropic temperature factors. Multipolar refinement for the charge density analysis was carried out using the XDLSM routine of the XD package;[42] details are given in Table 1. The atomic coordinates and the thermal parameters obtained from the high order refinement were used as input to XD refinement. Sulphur atoms were refined up to hexadecapole moments, carbon, nitrogen, and oxygen atoms up to octapole moments, while hydrogen atoms were restricted to dipole moments. During the refinement, the charge neutrality constraint was applied in all the cases. Kappa refinement was carried out on both the spherical and deformation valence shells of the non-hydrogen atoms, while those for the hydrogen atoms were restricted to the spherical valence. The multipolar refinement strategy was the following: a) scale factor, b) P_v, c) P_{lm}, d) repeat (b, c) until convergence, e) κ, f) repeat (b, c) until convergence, g) κ', h) repeat (b, c) until convergence, i) positional and thermal parameters of all nonhydrogen atoms, and finally j) P_v and P_{lm} together. The quality of the refinement was monitored using the residual density; the magnitude of the highest residual peak in the final refinement was 0.19, 0.2, 0.2, 0.2, 0.15 e Å^{-3} for the centric **1**, noncentric **1**, **2**, **3**, and **4**, respectively. Difference mean-square displacement amplitudes for the various bonds obtained from XD were found to follow closely the Hirshfeld criterion.[43]

The XDPROP routine was used to calculate the total electron density $\rho(r)$, the Laplacian $\nabla^2\rho$, and the ellipticity ε, at the bond critical points (CPs). Electrostatic potential (*V*) and dipole moments were also obtained using this routine. The Laplacian maps have been plotted using the XDGRAPH routine. The MOPAC program[44] was used for the calculation of the dipole moments at the frozen and optimized molecular geometries. This calculation, although semiempirical, is known to provide reliable estimates of molecular dipole moments.[16]

[1] P. N. Prasad, J. D. Williams, *Introduction to Nonlinear Optical Effects in Molecules and Polymers*, Wiley, New York, **1990**.

[2] J. Zyss, *Molecular Nonlinear Optics: Materials, Physics, Devices*, Academic Press, Boston, **1994**.

[3] C. Bosshard, K. Sutter, P. Pretre, J. Hullinger, M. Florsheimer, P. Kaatz, P. Gunter, *Organic Nonlinear Optical Materials*, Gordon & Breach, Amsterdam, **1995**.

[4] D. S. Chemla, J. Zyss, *Nonlinear Optical Properties of Organic Molecules and Crystals*, Academic Press, Orlando, FL, **1987**.

[5] D. R. Kanis, M. A. Ratner, T. J. Marks, *Chem. Rev.* **1994**, *94*, 195.

[6] E. Espinosa, C. Lecomte, E. Molins, S. Veintemillas, A. Cousson, W. Paulus, *Acta Crystallogr. Sect. B* **1996**, *52*, 519.

[7] A. Puig-Molina, A. Alvarez-Larena, J. F. Piniella, S. T. Howard, F. Baert, *Struct. Chem.* **1998**, *9*, 395.

[8] S. T. Howard, M. B. Hursthouse, C. W. Lehmann, P. R. Mallinson, C. S. Frampton, *J. Chem. Phys.* **1992**, *97*, 5616.

[9] C. Gatti, V. R. Saunders, C. Roetti, *J. Chem. Phys.* **1994**, *101*, 10686.

[10] A. Fkyerat, A. Guelzim, F. Baert, W. Paulus, G. Heger, J. Zyss, A. Perigaud, *Acta Crystallogr. Sect. B* **1995**, *51*, 197.

[11] A. Fkyerat, A. Guelzim, F. Baert, J. Zyss, A. Perigaud, *Phys. Rev. B* **1996**, *53*, 16236.

[12] N. K. Hansen, J. Protas, G. Marrier, *C. R. Acad. Sci. Ser. B* **1988**, *307*, 475.

[13] N. K. Hansen, J. Protas, G. Marnier, *Acta Crystallogr. Sect. B*, **1991**, *47*, 660.

[14] M. A. Spackman, *Annu. Rep. Prog. Chem. Sect. C* **1998**, *94*, 177.

[15] P. Coppens in *X-ray Charge Densities and Chemical Bonding*, Oxford University Press, **1997**.

[16] F. Hamazaoui, F. Baert, J. Zyss, *J. Mater. Chem.* **1996**, *6*, 1123.

[17] G. K. H. Madsen, F. C. Krebs, B. Lebech, F. K. Larsen, *Chem. Eur. J.* **2000**, *6*, 1797.

[18] H. Youping, S. Genbo, W. Bochang, J. Rihong, *J. Cryst. Growth* **1992**, *119*, 393.

[19] M. C. Etter, *Acc. Chem. Res.* **1990**, *23*, 120.

[20] P. R. Mallinson, K. Wozniak, T. Garry, K. L. McCormak, *J. Am. Chem. Soc.* **1997**, *119*, 11502.

[21] P. R. Mallinson, K. Wozniak, C. C. Wilson, K. L. McCormak, D. M. Yufit, *J. Am. Chem. Soc.* **1999**, *121*, 4640.

[22] W. T. Klooster, S. Swaminathan, R. Nanni, B. M. Craven, *Acta Crystallogr. Sect. B* **1992**, *48*, 217.

[23] R. F. Stewart in *The Application of Charge Density Research to Chemistry and Drug Design* (Eds.: G. A. Jeffrey, J. F. Piniella), Plenum, New York, **1991**.

[24] R. F. W. Bader, *Atoms in Molecules—A Quantum Theory*, Clarendon Press, Oxford, **1990**.

[25] R. Srinivasa Gopalan, G. U. Kulkarni, S. Renganayaki, E. Subramanian, *J. Mol. Struct.* **2000**, *524*, 169.

[26] Following referee comments, R_{int} for centric **1** and **4** were improved to ~ 0.04 by eliminating reflections above 0.9 Å^{-1}. The dipole moments did not vary significantly.

[27] H. U. Hummel, H. Procher, *Acta Crystallogr. Sect. C* **1986**, *42*, 1602.

[28] N. U. Kamath, K. Venkatesan, *Acta Crystallogr. Sect. C* **1984**, *40*, 1211.

[29] K. L. McCormack, P. R. Mallinson, B. C. Webster, D. S. Yufit, *J. Chem. Soc. Faraday Trans.* **1996**, *92*, 1709.

[30] D. Cremer, E. Kraka, *Croat. Chem. Acta.* **1984**, *57*, 1259.

[31] E. Espinosa, E. Molins, C. Lecomte, *Phys. Rev. B* **1997**, *57*, 1820.

[32] W. Scherer, M. Spiegler, B. Pedersen, M. Tafipolsky, W. Hieringer, B. Reinhard, A. J. Downs, G. S. McGrady, *Chem. Commun.* **2000**, 635.

[33] E. Ericsson, J. Sandström, I. Wennerbeck, *Acta Chem. Scand.* **1970**, *24*, 3102.

[34] L. Dalgaard, H.-K. Andersen, S.-O. Lawesson, *Tetrahedron* **1973**, *29*, 2077.

[35] N. H. Nilsson, *Synthesis* **1974**, 433.

[36] Siemens Analytical X-ray Instruments, **1995**, Madison, WI, USA.

[37] SHELXTL (SGI version), Siemens Analytical X-ray Instruments, **1995**, Madison, WI, USA.

[38] J. G. Bergman, G. R. Crane, B. F. Levine, C. G. Bethea, *Appl. Phys. Lett.* **1972**, *20*, 21.

[39] D. Adhikesavulu, K. Venkatesan, *Acta Crystallogr. Sect. C* **1983**, *39*, 589.

[40] N. K. Hansen, P. Coppens, *Acta Crystallogr. Sect. A* **1978**, *34*, 909.

[41] F. H. Allen, O. Kennard, D. G. Watson, L. Brammer, A. G. Orpen, R. Taylor, *J. Chem. Soc. Perkin Trans. 2* **1987**, S1.

[42] T. Koritsansky, S. T. Howard, T. Richter, P. R. Mallinson, Z. Su, N. K. Hansen, *XD, A Computer Program Package for Multipole Refinement and Analysis of Charge Densities from Diffraction Data*, Cardiff University (UK), **1995**.

[43] H. L. Hirshfeld, *Acta Crystallogr. Sect. A* **1976**, *32*, 239.

[44] J. J. P. Stewart, *J. Comput. Aided Mol. Des.* **1990**, *4*, 1.

Received: June 26, 2000 [Z 57]

VII. Porous Solids

C.N.R. Rao

CSIR Centre of Excellence in Chemistry,
Chemistry & Physics of Materials Unit
Jawaharlal Nehru Centre for Advanced Scientific Research
Jakkur P.O., Bangalore-560 064, INDIA
cnrrao@jncasr.ac.in

Porous solids are of great importance in catalysis, adsorption and separation processes.[1] Microporous solids containing channels or pores of 5 to 20 Å diameter are well known, zeolites being typical examples. The discovery of mesoporous solids with pore diameters of 20 to 200 Å by Mobil scientists added a new dimension to the area of porous solids. This section includes articles on the design and synthesis of mesoporous solids, including new ones such as the aluminoborates. Mesoporous solids can have lamellar, hexagonal or cubic structures and transitions amongst these phases is an important aspect discussed in one of the papers. Enhanced catalytic activity of metal complexes incorporated in mesoporous materials forms the subject matter of one of the articles. Another article deals with solids with pore diameters intermediate between those of conventional microporous and mesoporous solids.

Macroporous materials with pore diameters in the 200–10,000 Å range are of interest in various areas, in particular as optical gap materials. Two articles in this section deal with macroporous oxides and carbons.

Reference

1. C.N.R. Rao, *J. Mater. Chem.* **9**, 1 (1999).

Evidence for supramolecular organization of alkane and surfactant molecules in the process of forming mesoporous silica

N. Ulagappan and C. N. R. Rao*

CSIR Centre of Excellence in Chemistry, Indian Institute of Science, Bangalore 560012, India
Jawaharlal Nehru Centre for Advanced Scientific Research, Jakkur 560064, Bangalore, India

Investigations of the pore expansion in mesoporous silica in the presence of *n*-alkanes suggest a cooperative organization of the surfactant and alkane molecules, involving additivity of chain lengths.

The discovery of mesoporous silica by Beck *et al.*[1] involving an unusual combination of supramolecular cationic surfactant aggregates and anionic silicate species is a fine example of cooperative organization. The formation of supramolecular aggregates here is favoured by the hydrophobic interactions between the surfactant chains. Such an organization of ionic organic and inorganic molecular species to produce periodic arrays can be understood on the basis of the model of Stucky and coworkers.[2,3] The pore size in mesoporous silica can be expanded by the addition of various organic molecules, in particular 1,3,5-trimethylbenzene (TMB), the hydrophobic solvation interactions of the aromatic molecules playing the key role.[1,2] We have investigated the systematics in the pore size expansion when normal alkanes, C_nH_{2n+2}, are employed as solubilizing agents, with the expectation that the study may throw light on the nature of the supramolecular organization of the different species involved in the process.

By employing hexadecyltrimethylammonium bromide as the surfactant (SA), we have prepared mesoporous silicas starting with gels of the composition $SiO_2 : Na_2O : SA : alkane : H_2O$ ($1 : 0.08 : 0.16 : 0.16 : 63$) prepared from the appropriate quantities of tetraethylorthosilicate, NaOH, SA, alkane and distilled water. The molar ratio of the surfactant and the alkane was kept at $1 : 1$, since this ratio was optimal for obtaining the maximum possible pore size with a given alkane just as in the case of TMB.[2] The gels were subjected to hydrothermal treatment at

363 K for 24 h. The solids obtained were examined by powder X-ray diffraction (XRD) and transmission electron microscopy.

Fig. 1 shows the XRD patterns of some mesoporous silica samples prepared with the *n*-alkanes as the solubilizing agents along with the pattern of the sample prepared in the absence of any solubilizing agent. The d_{100} value of 39 Å found in the absence of a solubilizing agent suggests that the surfactant molecule in the micelle is fully extended as in Fig. 2(*a*). The d_{100} spacing (corresponding to the reflection peak) shows a progressive increase with the increase in the chain length of the alkane, reaching a value of *ca.* 74 Å with $C_{14}H_{30}$. In Fig. 3 the observed d_{100} values are plotted against the number of carbon atoms, *n*, of the alkane. There is some increase in the d_{100} value up to $n = 8$, but a significant increase from $n = 9$ to 15. The X-ray lines are somewhat broad suggesting the presence of

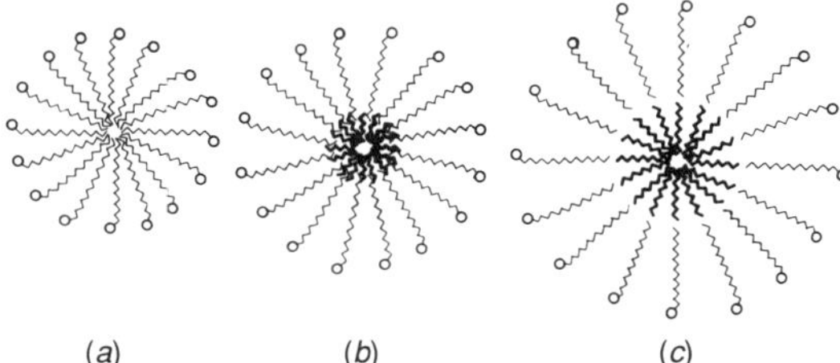

Fig. 2 Schematic drawings of a micelle of surfactant molecules (*a*) in the absence of a solubilizing agent and (*b*), (*c*) in the presence of normal alkanes as solubilizing agents. The alkane molecules at the centre of the micelle are represented by thicker lines.

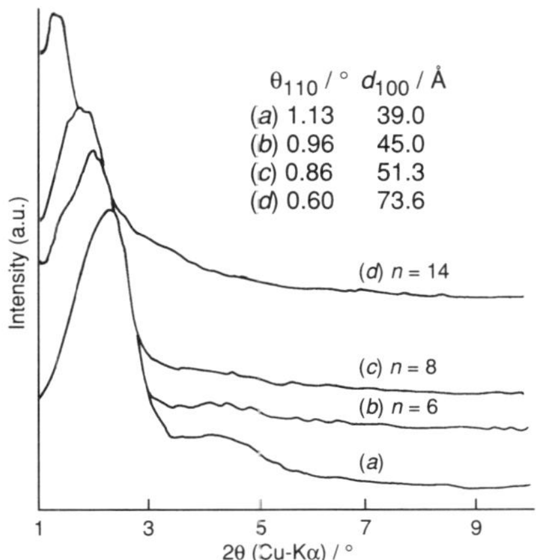

Fig. 1 X-Ray diffraction patterns of mesoporous silica formed with a cationic surfactant (*a*) in the absence of a solubilizing agent and (*b*)–(*d*) in the presence of normal alkanes, C_nH_{2n+2} (with $1 : 1$ molar ratio of the surfactant and the alkane). The *n* values are indicated beside the diffraction profiles.

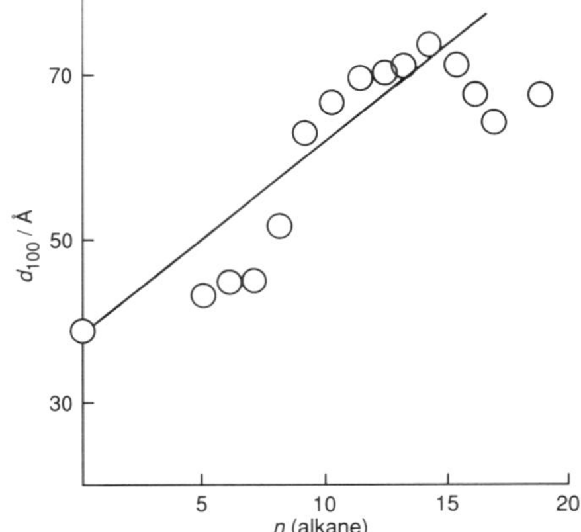

Fig. 3 Plot of the observed d_{100} values of mesoporous silica against the number of carbon atoms, *n*, of the alkane. The straight line is that expected if the lengths of the alkane chain and surfactant tail were to be additive in forming the micelle template for the formation of mesoporous silica.

disorder or variability in structure, but the TEM images confirm the systematic nature of the increase in pore size with chain length of the alkane. Fig. 4 shows the TEM image of mesoporous silica obtained by using n-$C_{10}H_{22}$. The variation of d_{100} with n can be understood with reference to the theoretical line (shown in Fig. 3) representing the d_{100} values expected if there were additivity of the chain lengths of the alkane and the surfactant molecule in forming the initial micelle.

The d_{100} values found for alkanes with n = 5–8 fall well below the theoretical line in Fig. 2, suggesting that the entire chain length of the alkane is not involved in increasing the size of the micelle. Thus, hexane with a chain length of 6.4 Å would be expected to increase the d_{100} by about 12.8 Å if the entire chain length were to contribute to the formation of the micelle, giving d_{100} of 39 + 12.8 = 51.8 Å, whereas the observed d_{100}

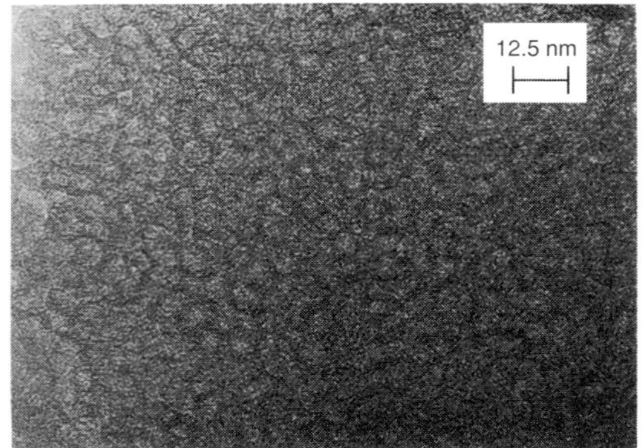

Fig. 4 TEM image of mesoporous silica obtained by using n-$C_{10}H_{22}$ as the solubilizing agent. The pore size is *ca.* 66 Å.

value is 45.0 Å. The micelles formed with the n = 5–8 alkanes and the surfactant molecules can be described as molecular dispersions of the solubilizing agent between the tails of the surfactant molecules as depicted in Fig. 2(*b*). With the higher alkanes (n = 9–15), on the other hand, the increase in d_{100} is close to that expected on the basis of the chain length of the alkanes and accordingly the observed d_{100} values are close to the theoretical line in Fig. 3. For example, with $C_{14}H_{30}$ (chain length of 16.6 Å), the expected increase in d_{100} would be 33.2 Å, predicting a d_{100} value of 39 + 33.2 = 72.2 Å; the observed d_{100} of 73.6 Å is quite close to this value. Alkanes with n > 15 do not show such additivity (*e.g.* d_{110} with n = 16 is 66.9 Å) suggesting that the chains may not be rigid. The behaviour of the n = 9–15 alkanes is best described by Fig. 2(*c*) where the alkane molecules form a core which is then surrounded by a layer of the cationic surfactant molecules, with a one-to-one alignment of the alkane chain and the surfactant tail. Such a cooperative organization arising from hydrophobic interaction is noteworthy, lending credence to the expectation that the structure-directing effect of organic molecules may make it possible to design inorganic solids.[4]

References

1 J. S. Beck, J. C. Vartuli, W. J. Roth, M. E. Leonowicz, C. T. Kresge, K. D. Schmitt, C. T. W. Chu, D. H. Olson, E. W. Sheppard, S. B. McCullen, J. B. Higgins and J. L. Schlenker, *J. Am. Chem. Soc.*, 1992, **114**, 10834.
2 Q. Huo, D. I. Margolese, U. Ciesla, D. G. Demuth, P. Feng, T. E. Gier, P. Sieger, A. Firouzi, B. F. Chmelka, F. Schuth and G. D. Stucky, *Chem. Mater.*, 1994, **6**, 1176.
3 A. Firouzi, D. Kumar, L. M. Bull, T. Besier, P. Sieger, Q. Huo, S. A. Walker, J. A. Zasadzinski, C. Glinka, J. Nicol, D. Margolese, G. D. Stucky and B. F. Chmelka, *Science*, 1995, **267**, 1138.
4 M. E. Davies, *CHEMTECH*, 1994, 22.

Received, 28th August 1996; Com. 6/05926K

Phase transformations of mesoporous zirconia

Neeraj and C. N. R. Rao*

Chemistry and Physics of Materials Unit, Jawaharlal Nehru Center for Advanced Scientific Research, Jakkur, Bangalore 560064, India

The kinetics of the lamellar→hexagonal→cubic transformations of mesoporous zirconia prepared by using a neutral organic amine as the amphiphile have been studied in phosphoric acid medium. The lamellar→hexagonal transformation is preceded by a loss of the template molecules and the hexagonal→cubic transformation proceeds only when the lamellar form has entirely transformed to the hexagonal phase. The kinetics of the thermal transformation of lamellar zirconia to the hexagonal form have also been examined; this transformation also occurs and is accompanied by a loss of template molecules. Accordingly, the activation energy of the transformation is comparable to the hydrogen bond energy between the amine and the oxo–zirconium species. The phase transformations of mesoporous zirconia can be understood in terms of minimum energy surfaces.

Mesoporous solids are generally prepared by making use of self-assembled ordered aggregates of surfactants as templates, thus rendering the structures exhibited by the mesoporous solids to be similar to those of the self-assembled[1–3] surfactants. Accordingly, mesoporous solids occur in lamellar, hexagonal and cubic forms, just as the surfactant aggregates. Phase transitions among these forms occur in mesoporous materials in the solution phase. For example, changing the pH of the medium or ageing, leads to transformation of the lamellar phase of silica to the hexagonal phase.[4] Lamellar, hexagonal and cubic forms of aluminoborates have been prepared by changing the pH.[5] Other factors such as temperature, presence of counter ions and concentration of the surfactant also affect the phase transitions.[4–10] The phase transitions in mesoporous solids bear some similarity to those in surfactant assemblies.

In surfactant systems, decreasing the concentration of the surfactant instantaneously transforms the lamellar phase to the hexagonal phase.[11,12] The transitions in the surfactant self-assemblies are commonly understood in terms of the surface/interface energies of the ordered aggregates.[13] The effect of ageing on the lamellar to hexagonal phase transformation in mesoporous silica has been related to the extent of polymerization of the silica framework and the balancing of charges.[4] High pH and a low degree of polymerization favour the lamellar phase whereas low pH and highly condensed silica favour the hexagonal phase of silica.[4] The transformation from the hexagonal to the cubic phase in silica has been explained in terms of the formation of a periodic minimal surface governed by a competition between the curvature and packing and the transformation is associated with kinetic barriers.[14] Since there is limited quantitative information on the nature of the phase transitions in mesoporous solids in the literature, we considered it important to investigate the transitions in some detail. For this purpose, we have chosen to investigate mesoporous zirconia which exhibits interesting phase transitions both in solution and in the solid state. Here, we report the results of a kinetic study of the lamellar→hexagonal→cubic transitions of mesoporous zirconia in the presence of phosphoric acid and of the thermal lamellar → hexagonal transition of zirconia in the solid state.

Experimental

The lamellar form of mesoporous zirconium oxide was prepared using dodecylamine (DA) as the surfactant. In a typical synthesis zirconium isopropoxide (0.01 mol) was added to a solution of the dodecylamine (DA) (0.03 mol) in propan-1-ol (0.1 mol) to which ammonium sulfate (0.12 mol) was added under stirring. The pH of the gel was adjusted to 1.5–2.0 by using dilute hydrochloric acid (HCl). The gel was subjected to hydrothermal treatment at 373 K for 20 h, filtered, washed with acetone and dried at 373 K for 2 h. The lamellar nature of the product was verified by X-ray diffraction (XRD) and transmission electron microscopy (TEM). To study the transformation of lamellar ZrO_2 in solution phase, *ca.* 100 mg of the as-synthesized sample was added to 50 ml of 0.87 M phosphoric acid and stirred for different times, filtered, washed with water, acetone and dried at ambient temperature. XRD patterns and TEM images of the samples were recorded for each time.

The kinetics of the lamellar→hexagonal→cubic transition of zirconia in phosphoric acid solution was studied as follows. The (100) reflection of the lamellar and hexagonal phases differ both in intensity and position. Thus, as the lamellar phase ($d_{100} = 3.34$ nm) transforms to the hexagonal phase the intensity of the (100) reflection decreases until it attains a much smaller value, characteristic of that of a disordered hexagonal mesophase (Fig. 1). In addition, the *d*-spacing of the (100) reflection decreases until it reaches a minimum value after the transformation to the hexagonal phase is complete. The intensities of the (200) and (300) reflections of the lamellar phase also decrease continuously with time, as the lamellar→hexagonal transformation proceeds. We have employed both the intensity and position of the (100) reflection to follow the kinetics of the lamellar–hexagonal transformation. These two measurements give slightly different estimates of the phase compositions and we have taken the average value in the kinetic study. The hexagonal to cubic transformation of mesoporous zirconia was followed by the increase in the intensity of the (220) reflection of the cubic phase with time (Fig. 1).

The lamellar→hexagonal thermal transformation of mesoporous zirconia in the solid state was studied by heating the lamellar form at a fixed temperature for different periods of time. The phase composition of the sample subjected to heat treatment was estimated on the basis of the intensity of the (100) reflection, the *d*-spacing showing only small changes in the thermal transformation (Fig. 2). The kinetics of the transformation was studied as a function of time at 373, 403 and 413 K. The transformation was also studied by heating the lamellar sample for a fixed period of 2 h at different temperatures in the range 360–430 K.

Results and Discussion

The lamellar form of mesoporous zirconia, on contact with 0.87 M phosphoric acid, first transforms to the hexagonal form.

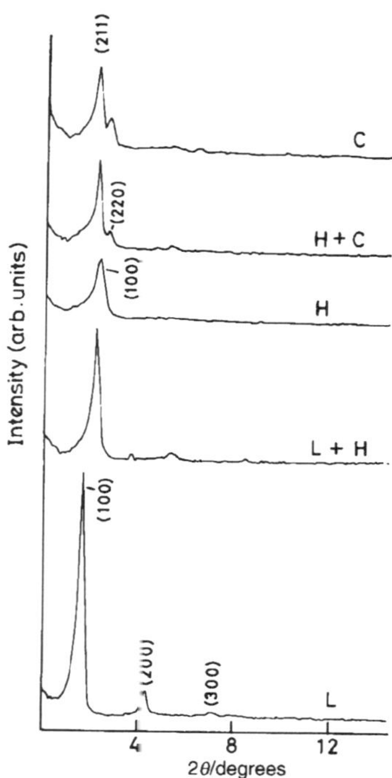

Fig. 1 XRD patterns showing the transformation of lamellar (L) zirconia to hexagonal (H) and then cubic (C) phases in phosphoric acid solution. Intermediate stages during the L–H and H–C transformations are shown.

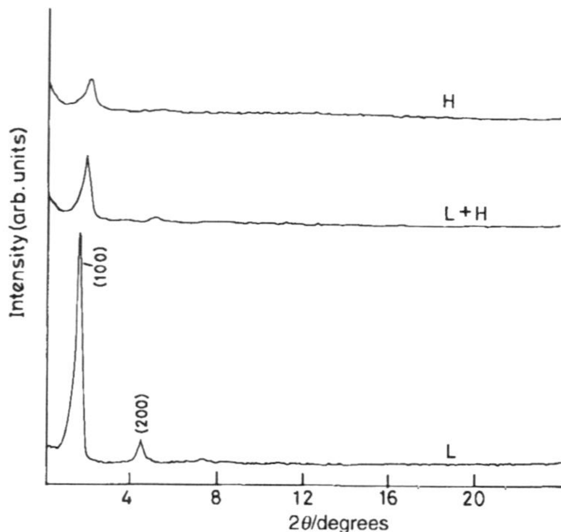

Fig. 2 XRD patterns showing the thermal transformation of lamellar (L) zirconia to the hexagonal (H) form. An intermediate stage during the transformation is shown.

This transformation is complete in 7 h. We have followed the lamellar→hexagonal transformation as a function of time. Fig. 3 shows the progress of the lamellar→hexagonal transformation by plotting the percentages of the lamellar and hexagonal phases *vs.* time. We see that the proportion of the lamellar form decreases while that of the hexagonal form increases. It is noteworthy that the transformation of the hexagonal form to the cubic form starts only after the lamellar phase has completely

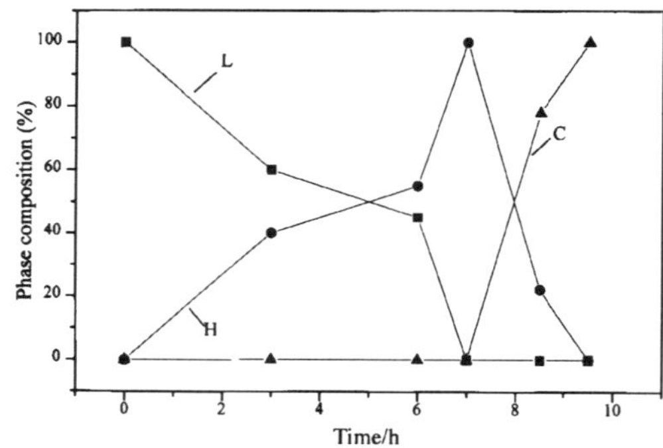

Fig. 3 Time variation in the phase composition of zirconia in phosphoric acid solution; L, lamellar; H, hexagonal; C, cubic

transformed to the hexagonal phase. The hexagonal to cubic transformation occurs over a short time (<3 h).

The lamellar→hexagonal transformation of ZrO_2 is likely to be initiated first by the removal of some of the surfactant species, followed by the curling of the surfactant bilayer in order to minimize the surface/interface energy as shown in Fig. 4(a) and (b).[13,15] The curled bilayers transform to cylindrical rods to further minimize the surface energy as shown in Fig. 4(c) and the cylindrical rods assemble to give the ordered hexagonal structure shown in Fig. 4(d). In order to examine

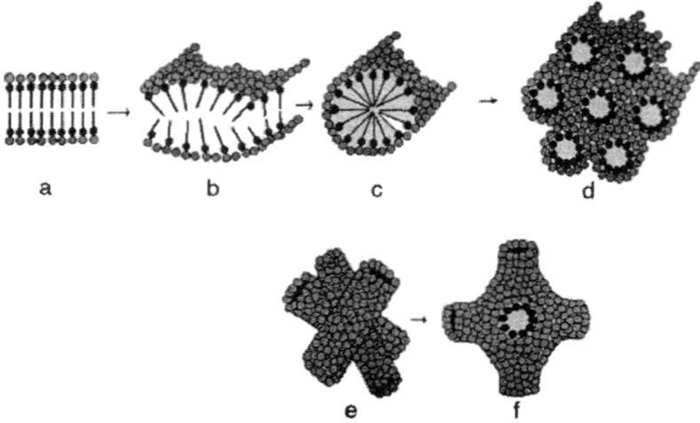

Fig. 4 Schematic representation of lamellar→hexagonal phase transformation (a through d) and the hexagonal→cubic transformation (e and f). The shaded circles around the surfactant aggregates represent the inorganic species (generally metal alkoxides or other metal–oxo species).

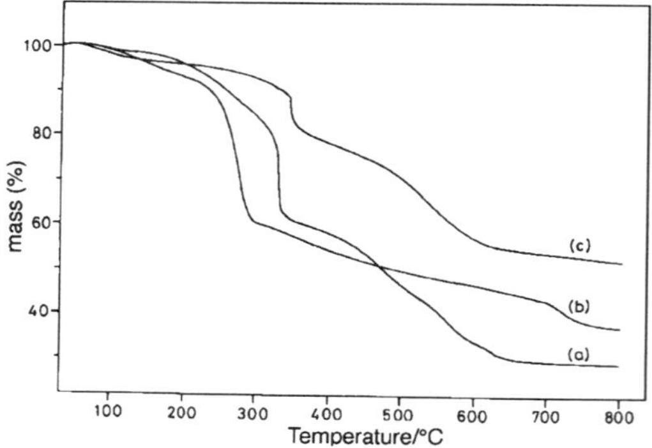

Fig. 5 Thermogravimetric curves of lamellar zirconia maintained for different periods in phosphoric acid solution: (a) as-prepared lamellar zirconia, (b) after 3 h and (c) after 7 h

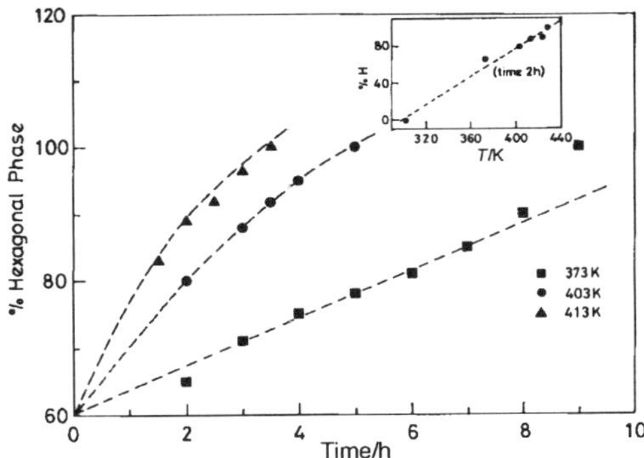

Fig. 6 Kinetics of the lamellar to hexagonal transformation of zirconia at different temperatures. Inset shows the temperature variation of the percentage of the hexagonal (H) phase in a fixed period of 2 h.

whether the loss of the surfactant species precedes the lamellar→hexagonal transformation, we have carried out thermogravimetric analysis (TGA) studies. We find that there is a significant loss of the surfactant in the lamellar→hexagonal→cubic transformation in phosphoric acid solution.

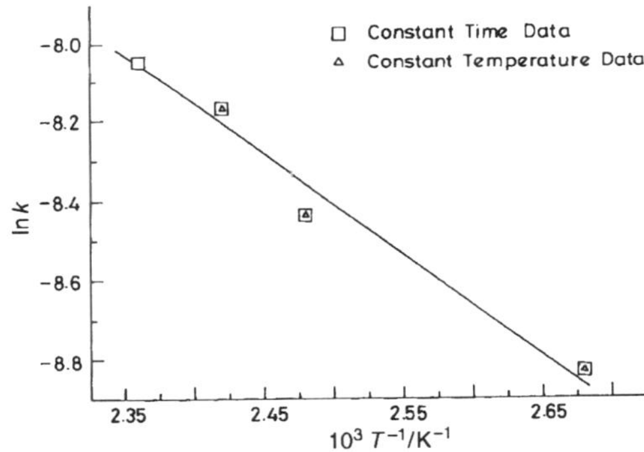

Fig. 7 Arrhenius plots of the kinetic data of the lamellar→hexagonal transformation of zirconia in the solid state

TGA of the lamellar zirconia [Fig. 5(a)] shows a marked mass loss at around 593 K and gradual loss of mass from 593 to 973 K. There is negligible loss of mass below 430 K due to water and other species. The mass loss at 593 K and above appears to be entirely due to the loss of the surfactant. The lamellar ZrO_2 sample treated for 3 h in phosphoric acid [Fig. 5(b)] showed a sharp mass loss around 543 K, followed by a gradual loss up to 973 K. Since the entire mass loss is only due to the surfactant, we can take the difference in the mass loss, say at 973 K, between the curves Fig. 5(a) and (b) as equal to the amount of the surfactant lost by phosphoric acid treatment. We see that around 8% of the surfactant is lost after 3 h of treatment. The ZrO_2 sample treated in phosphoric acid for 7 h was entirely hexagonal and showed a sharp mass loss around 613 K and a gradual loss thereafter, up to 973 K [Fig. 5(c)]. The difference between curves Fig. 5(a) and (c) shows that 23% of the surfactant has been removed after 7 h of treatment. Phosphoric acid treatment for 9.5 h gave the cubic phase, and *ca.* 30% of the template had been removed at this stage. The TGA studies suggest that the loss of the surfactant is a necessary initial step in the lamellar→hexagonal transformation.

The transformation from the hexagonal to the cubic phase is driven by the tendency of the cylindrical rods of the hexagonal phase to minimize their energy by forming a three-dimensional network of rods, forming bicontinuous cubic phases as shown in Fig. 4(e) and (f). The cubic phase can be described using the concept of periodic minimal surface,[15,16] with the space groups *Pn3m, Pm3n, P4₃32, Im3m, Ia3d* or *Fd3m*.[16] The XRD pattern of the cubic phase obtained after the complete transformation of the hexagonal ZrO_2 phase in phosphoric acid, appears to be consistent with the space group *Ia3d*.

The thermal transformation of the lamellar form of mesoporous ZrO_2 to the hexagonal form occurs in the solid state. Thus, on heating at 428 K for 2 h, the lamellar phase completely transforms to the hexagonal phase. We have followed the kinetics of the lamellar→hexagonal transformation at three fixed temperatures. Fig. 6 shows the kinetics of the lamellar→hexagonal transformation at different temperatures. We have also examined the kinetics of this transformation by heating lamellar zirconia at different temperatures for a fixed time of 2 h. The inset of Fig. 6 shows how the proportion of the hexagonal form increases with temperature at a given time (2 h). We were able to fit these kinetic data to a first order

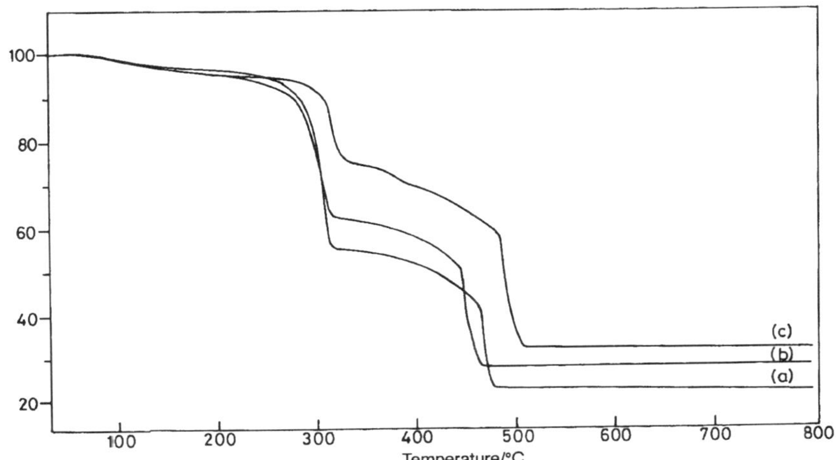

Fig. 8 Thermogravimetric curves of lamellar form heated at different temperatures: (a) as-prepared lamellar zirconia, (b) 403 K, 2 h and (c) 428 K, 2 h.

rate equation. The rate data follow the Arrhenius equation as shown in Fig. 7. The data give an activation energy of *ca.* 22 kJ mol^{-1} for the lamellar→hexagonal transformation in the solid state. This value of the activation energy is comparable to the energy of a medium strength hydrogen bond. This is understandable since the removal of surfactant molecules from the lamellar phase is necessary for the transformation to occur. The surfactants interact with the oxo–zirconium species primarily through hydrogen bonding.

TGA studies show that on heating the lamellar phase to 403 K for 2 h, around 6% of the surfactant is lost (Fig. 8). On heating to 428 K the sample loses 11% of the surfactant. These data demonstrate that during the thermal transformation of the lamellar phase to the hexagonal phase, the amine template is partially removed, leading to the reorganization of the self-assembled surfactant aggregate. The magnitude of the loss of the surfactant in the thermal transformation is somewhat smaller than that accompanying in the transformation in phosphoric acid solution.

In conclusion, the present study of the kinetics of the lamellar to hexagonal transformation of mesoporous zirconia shows that a loss of surfactant molecules accompanies the transformation. Transformation to the cubic form seems to require that all the starting material be in the hexagonal form. The thermally induced lamellar→hexagonal transformation is associated with an activation energy comparable to the hydrogen bond energy.

References

1 P. Mariani, V. Luzatti and H. Delacroix, *J. Mol. Biol.*, 1988, **204**, 165.
2 P. O. Eriksson, G. Lindblom and G. Arvidson, *J. Phys. Chem.*, 1985, **89**, 1050.
3 J. C. Vartuli, K. D. Schmitt, C. T. Kresge, W. J. Roth, M. E. Leonowicz, S. B. McCullen, S. D. Hellring, J. S. Beck, J. L. Schlenker, D. H. Olson and E. W. Sheppard, *Chem. Mater.*, 1994, **6**, 2317.
4 A. Monnier, F. Schuth, Q. Huo, D. Kumar, D. Margolese, R. S. Maxwell, G. D. Stucky, M. Krishnamurthy, P. Petroff, A. Firouzi, M. Janicke and B. F. Chmelka, *Science*, 1993, **261**, 1299.
5 S. Ayyappan and C. N. R. Rao, *Chem. Commun.*, 1997, 575.
6 J. Luo and S. L. Suib, *Chem. Commun.*, 1997, 1031.
7 M. Ogawa, *J. Am. Chem. Soc.*, 1994, **116**, 7941.
8 S. Oliver, A. Kuperman, N. Coombs, A. Laugh and G. A. Ozin, *Nature (London)*, 1995, **378**, 47.
9 C. A. Fyfe and G. Fu, *J. Am. Chem. Soc.*, 1995, **117**, 9709.
10 H. P. Lin and C. Y. Mou, *Science*, 1996, **273**, 765.
11 E. S. Blackmore and G. J. T. Tiddy, *J. Chem. Soc., Faraday Trans. 2*, 1988, **84**, 1115.
12 A. Sein and J. B. F. N. Engberts, *Langmuir*, 1995, **11**, 455.
13 J. M. Seddon, *Biochim. Biophys. Acta.*, 1990, **1031**, 1.
14 S. M. Gruner, *J. Phys. Chem.*, 1989, **93**, 7562.
15 G. Porte in *Micelles, Membranes, Microemulsions and Monolayers*, ed. W. M. Gelbart, A. Ben-Shaul and D. Roux, Springer-Verlag, New York, 1994, p. 105.
16 S. T. Hyde, *Curr. Opin. Solid State Mater. Sci.*, 1996, **1**, 653.

Paper 8/01419A; Received 19th February, 1998

Mesoporous phases based on SnO_2 and TiO_2

N. Ulagappan and C. N. R. Rao*

CSIR Centre of Excellence in Chemistry, Indian Institute of Science, Bangalore 56012, India
Jawaharlal Nehru Centre for Advanced Scientific Research, Jakkur, Bangalore 560 064, India

A hexagonal mesoporous phase based on SnO_2 is synthesized for the first time by using an anionic surfactant; hexagonal phases of TiO_2 are prepared with neutral amine surfactants.

There has been intense research activity in the synthesis and characterization of non-siliceous mesoporous solids based on metal oxides, since the discovery of mesoporous silica of the MCM-41 type.[1-3] Thus, Huo *et al.*[4] and Ciesla *et al.*[5] attempted to synthesize mesostructured materials of tungsten, molybdenum, iron and lead oxides, and had definitive success in obtaining a hexagonal mesoporous phase in the case of tungsten oxide. The surfactant (alkyltrimethylammonium salt), however, could not be removed thermally or by the use of different solvents. Stein *et al.*,[6] on the other hand, obtained a nonporous CTMA–tungsten oxide composite (CTMA = cetyltrimethylammonium) from hydrothermal synthesis. Abe *et al.*[7] have reported vanadium phosphate-based hexagonal mesoporous phases by employing alkyltrimethylammonium chlorides as surfactants, but were unable to remove the surfactants from the pores. A hexagonal mesoporous phase based on vanadium oxide has also been prepared by using CTMA, again with the surfactant remaining in the pores.[8] We have explored ways of synthesizing mesoporous solids based on tin and titanium oxides for some time. We have been able to prepare hexagonal mesoporous phases of TiO_2 by using $Ti(OPr^i)_4$ and neutral amine surfactants, but have since noticed the work of Antonelli and Ying[9] reporting a hexagonal phase with alkyl phosphate surfactants. We have however succeeded in preparing a hexagonal phase of SnO_2, for the first time, by using an anionic surfactant.

Having failed to obtain a mesoporous phase of SnO_2 with cationic and neutral surfactants, we employed anionic surfactants such as sodium lauryl sulfate, sodium stearate and aerosol OT (sodium dioctylsulfosuccinate, AOT). It is indeed known that Sn^{II} and Sn^{IV} occur as cationic species in solution and polymerize to a tin–oxygen network between pH 4 and 8.5.[10] In order to prepare the mesophase of SnO_2, compositions corresponding to molar ratios of $1\ SnO_2 : 1–2\ AOT : 1000–2000\ H_2O$ were prepared by adding the AOT solution to $SnCl_4$. This resulted in a white thick mass which settled quickly on standing. The supernatant water was periodically decanted and the solid allowed to age in water for 24 h. The resulting material was filtered off and washed with acetone, and air dried. The X-ray diffraction pattern of the solid showed a single intense peak at a low angle (*d ca.* 3.2 nm) characteristic of mesoporous materials [Fig. 1(*a*)]. We could not obtain a mesoporous phase when we used sodium lauryl sulfate or sodium stearate as the surfactant [Fig. 1(*b*)].

The IR and ^{13}C NMR spectra of the SnO_2–AOT composite showed the known features of AOT confirming the presence of the surfactant in the pores. Thermogravimetric analysis of the composite showed the loss of the water below 420 K and surfactant loss starts at *ca.* 513 K and was completely removed at *ca.* 670 K. Based on the mass loss, the composition of the starting material was $(SnO_2 \cdot 2H_2O)_4 \cdot AOT$. Calcining the composite at *ca.* 700 K in air for 12 h results in the collapse of the mesoporous structure [Fig. 1(*c*)], giving the dense cassiterite

form of SnO_2. Attempts to remove the surfactant by extraction with solvents including acetonitrile or dimethyl sulfoxide were unsuccessful, resulting in the collapse of the mesophase structure. Further efforts to remove the surfactant are in progress.

Transmission electron microscope images of the SnO_2–AOT composite (using a JEOL JEM 3010 instrument) confirm the

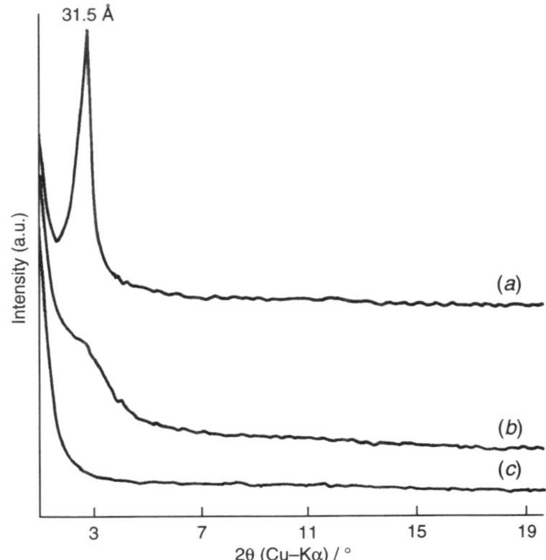

Fig. 1 X-Ray diffraction patterns of (*a*) the mesoporous SnO_2–AOT, (*b*) material obtained using lauryl sulfate and (*c*) of SnO_2–AOT calcined at 700 K

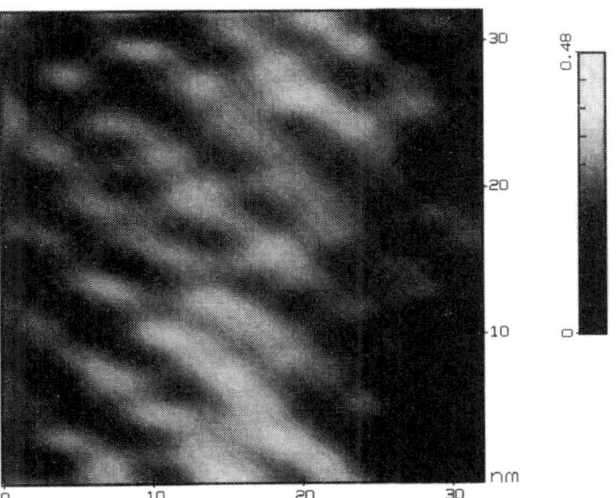

Fig. 2 Atomic force microscope image of SnO_2–AOT showing pores of *ca.* 3 nm diameter

presence of hexagonal channels with a pore size of *ca.* 3 nm. Atomic force microscope images (obtained with a Nanoscope II instrument) also show pores of *ca.* 3 nm (Fig. 2). The ^{119}Sn MAS NMR spectrum of SnO_2–AOT shows a signal at δ −629, close to the chemical shift of SnO_2 (δ −604), suggesting that the coordination around tin in the mesoporous material is octahedral, as in SnO_2. The resonance from the mesoporous material is however broader than that of SnO_2, indicating the presence of some local disorder around the tin site.

Since we were unsuccessful in obtaining the hexagonal phase of TiO_2 by employing cationic or anionic surfactants, we used titanium alkoxide in combination with decylamine or hexadecylamine. In a typical preparation, 0.6 mmol of the amine was dispersed in 20 mmol of isopropyl alcohol and 2 mmol of $Ti(OPr^i)_4$ was added to the mixture. Addition of 60 mmol of water to this solution resulted in a pale yellow slurry which was

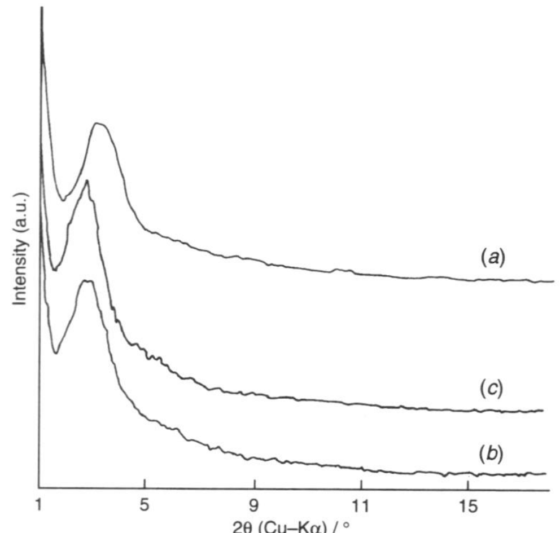

Fig. 3 X-Ray diffraction patterns of mesoporous TiO_2 prepared with amine surfactants: (*a*) decylamine at room temperature, (*b*) hexadecylamine at room temperature and (*c*) hexadecylamine under hydrothermal conditions

aged at room temperature for 18 h in one experiment and subjected to hydrothermal treatment at 363 K for 18 h in another. The product obtained from either procedure after washing and drying at 373 K was mesoporous as revealed by X-ray diffraction. The pore sizes were 2.9 and 3.2 nm, with decylamine and hexadecylamine as surfactants, respectively (Fig. 3). Thermogravimetric analysis showed the loss of the amine at *ca.* 700 K. The observed mass loss suggests the composition of the mesoporous amine adduct to be $(TiO_2)_{3.5}\cdot2H_2O\cdot$amine. The oxide obtained after the removal of the surfactant by calcination has the anatase structure. Removal of the amine from the pores thermally or by the use of solvents generally destroys the mesoporous structure. Leaching of the amine in an acidified dilute alcohol medium, however, appears more promising. It should be noted that Antonelli and Ying[9] could partially remove alkyl phosphate from the pores with the phosphate units being retained in the oxide phase.

We thank Ms Gargi Raina for assistance with AFM studies and Drs A. K. Cheetham and Lucy Bull for advice regarding NMR measurements.

References

1 J. S. Beck, J. C. Vartuli, W. J. Roth, M. E. Leonowicz, C. T. Kresge, K. D. Schmitt, C. T. W. Chu, D. H. Olson, E. W. Sheppard, S. B. McCullen, J. B. Higgins and J. L. Schlenker, *J. Am. Chem. Soc.*, 1992, **114**, 10 834.
2 P. Behrens, *Angew. Chem., Int. Ed. Engl.*, 1996, **35**, 515.
3 J. S. Beck and J. C. Vartuli, *Current Opinion in Solid State Mater. Sci.*, 1996, **1**, 76.
4 Q. Huo, D. J. Margolese, U. Ciesla, D. G. Demuth, P. Feng, T. E. Gier, P. Sieger, A. Firouzi, B. F. Chmelka, F. Schuth and F. D. Stucky, *Nature*, 1994, **368**, 317.
5 U. Ciesla, D. Demuth, R. Leon, P. Petroff, G. Stucky, K. Unger and F. Schuth, *J. Chem. Soc., Chem. Commun.*, 1994, 1387.
6 A. Stein, M. Fendorf, T. P. Jarvie, K. T. Mueller, A. J. Benesi and T. E. Mallouk, *Chem. Mater.*, 1995, **7**, 304.
7 T. Abe, A. Taguchi and M. Iwamoto, *Chem. Mater.*, 1995, **7**, 1429.
8 V. Luca, D. J. Maclachlan, J. M. Hook and R. Withers, *Chem. Mater.*, 1995, **7**, 2220.
9 D. M. Antonelli and J. Y. Ying, *Angew. Chem., Int. Ed. Engl.*, 1995, **34**, 2014.
10 F. R. G. Gimblett, *Inorganic Polymer Chemistry*, Butterworths, London, 1963.

Received, 23rd April 1996; Com. 6/02837C

Mesoporous aluminoborates

S. Ayyappan and C. N. R. Rao*

CSIR Centre of Excellence in Chemistry and Materials Research Centre, Indian Institute of Science, Bangalore 560 012, India and Jawaharlal Nehru Centre for Advanced Scientific Research, Jakkur, Bangalore 560 064, India

Hexagonal, cubic and lamellar aluminoborate mesophases containing octahedral aluminium and tetrahedral boron are prepared and characterized for the first time.

Since the first synthesis of mesoporous silica,[1] there has been intense activity in the design and synthesis of a variety of mesoporous solids.[2–4] These studies have also unravelled the mechanism of formation of these materials by means of supramolecular templating involving an unusual combination of surfactant aggregates and anionic silicate species.[4,5] By varying the pH or the surfactant to silica ratio, hexagonal, cubic and lamellar forms of silica mesophases have been prepared.[6] Besides ionic surfactants, neutral amines[7] and polyalkene oxides[8] have been employed to synthesise the mesophases. The strategy to prepare silica-based mesophases has been extended to prepare a variety of mesoporous metal oxides such as Al_2O_3,[8] TiO_2[9,10] and ZrO_2[11] and Nb_2O_5.[12] $AlPO_4$ which is isoelectronic with SiO_2 has provided the basis for a distinct family of microporous materials.[13] However, efforts to make hexagonal mesoporous $AlPO_4$ have not been successful; instead, only lamellar $AlPO_4$ has been obtained.[14] We considered it most worthwhile to explore the synthesis of the mesoporous phases of the analogous aluminoborates containing the $AlBO_4{}^{2-}$ species. There has been some effort[15] to make microporous aluminoborates with different B/Al ratios, but the products do not seem to have been adequately characterized. Here, we report the first successful synthesis and characterization of mesoporous aluminoborate phases in hexagonal, cubic and lamellar forms. It has been possible to remove the surfactant from the hexagonal phase and expand the pore size of the hexagonal phase by using aliphatic hydrocarbons as solubilizing agents.

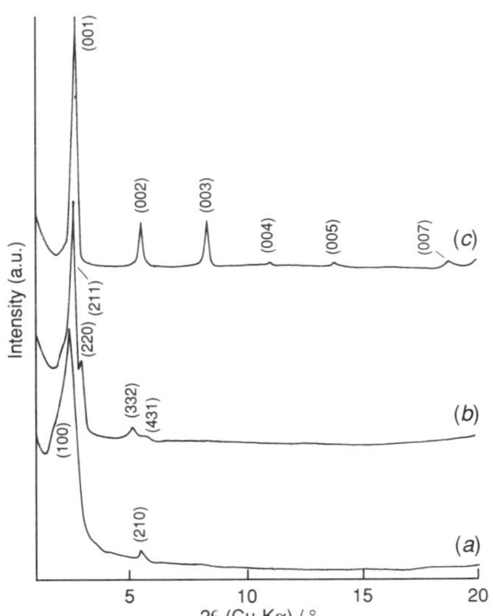

Fig. 1 X-Ray diffraction patterns of the aluminoborate mesophases: (*a*) hexagonal, (*b*) cubic and (*c*) lamellar

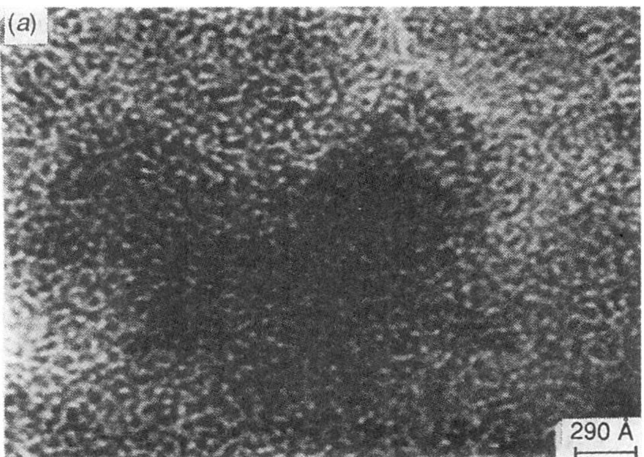

Fig. 2 TEM images of (*a*) hexagonal and (*b*) lamellar aluminoborate mesophases

Mesoporous aluminoborate phases could not be obtained when cationic surfactants or long-chain amines were employed for the synthesis. They could however be prepared by employing sodium dodecyl sulfate (SDS) as the surfactant. In a typical preparation, to a solution containing Ca, Al and B in 1:1:1 molar ratio, prepared by dissolving appropriate quantities of CaO and $Al(NO_3)_3$ and boric acid in aqueous nitric acid, sodium dodecyl sulfate (1.0 mol with respect to Ca or Al) was added under stirring. Ammonia was added to the resulting turbid solution under constant stirring to reach the desired value of pH. The resultant gel was maintained at room temperature for 48 h. The product so obtained was washed several times with distilled water and dried at 100 °C in a hot air oven. X-Ray diffraction (XRD) patterns of the dried solids were recorded with a Rich–Seifert (model XRD 3000TT) instrument. Fig. 1 shows the XRD patterns of the products obtained at different pH values. The XRD pattern of the product obtained at pH 2 [Fig. 1(*a*)] is characteristic of a hexagonal phase with a d_{100} value of 36.1 Å (a_0 = 41.7 Å). The XRD pattern of the product obtained at a pH of 3.5 [Fig. 1(*b*)] corresponds to that of a cubic phase. This pattern could be indexed with $a = 83.5$ Å. The XRD pattern of the product obtained at pH *ca.* 5.5 [Fig. 1(*c*)] is characteristic of a lamellar phase and accordingly exhibits the various 00*l* reflections ($d_{001} = 32.5$ Å). The scanning electron micrographs of the hexagonal and lamellar phases exhibit characteristic differences in morphology, the lamellar form showing the flaky structure. The transmission electron microscope (TEM) image (recorded with a JEOL JEM 3010 microscope) reveals the presence of disordered hexagonal pores with an opening of *ca.* 32 Å with a wall thickness of *ca.* 11 Å [Fig. 2(*a*)]. The TEM image of the lamellar phase shows an interlayer separation of *ca.* 32 Å [Fig. 2(*b*)].

EDAX analysis of the aluminoborate mesophases showed the Ca : Al molar ratio to be 1 : 1, while an ICP analysis gave a Al : B molar ratio of 1 : 1. Thermogravimetry showed the loss of water by 150 °C and of the surfactant by 600 °C. Based on the mass loss data and assuming the Ca : Al : B molar ratio to be 1 : 1 : 1, the composition of the aluminoborate–surfactant adduct works out to be $CaO \cdot 0.5Al_2O_3 \cdot 0.5B_2O_3 \cdot 0.6SDS \cdot 1.5H_2O$. The %mass of boron from the ICP analysis ($3.15 \pm 0.25\%$) is also consistent with the theoretical value of 3.2% based on the empirical formula. The IR spectrum of this adduct showed a boron–oxygen stretching band in the region 900–1100 cm^{-1} due to BO_4 tetrahedra. The IR band characteristic of BO_3 units in the region 1200–1400 cm^{-1} was absent. Significant structural information was obtained by means of ^{27}Al and ^{11}B MAS NMR spectra (recorded with a Bruker MSL-300 spectrometer) of the hexagonal aluminoborate mesophase. The ^{27}Al NMR signal is centred around δ 0.0 [Fig. 3(*a*)], characteristic of an octahedral environment of Al.[16] The ^{11}B NMR signal is also centred around δ 0.0 [Fig. 3(*b*)] suggesting a tetrahedral environment for boron.[17]

By heating the as-prepared hexagonal mesophase to 300 °C, we could remove some of the surfactant without destroying the structure. Thus, the XRD pattern of the resulting product shows the mesoporous structure with a d_{100} value of 35.2 Å. We could

however remove the surfactant by refluxing the hexagonal mesophase with acidified ethanol for 3 h. The removal of the template was confirmed by the absence of mass loss over the temperature range where the template is normally removed as well as by the XRD pattern. The removal of the surfactant from the hexagonal phase was confirmed by the IR spectra,which showed the near-absence of bands arising from CH_2 groups. Accordingly, the surface area of the hexagonal aluminoborate obtained by calcination was *ca.* 100 m^2 g^{-1} and that of the sample obtained by alcohol extraction was 470 m^2 g^{-1}; their XRD patterns were not identical.

We have been able to expand the pore size of the hexagonal mesoporous aluminoborate by the use of *n*-alkanes as solubilizing agents. For this purpose, we added the alkane along with the surfactant solution to the acidic solution containing Ca, Al and B, maintaining the molar ratio of the alkane to the surfactant to be 1 : 1. This molar ratio was found to be optimal for expanding the pore size. The XRD patterns gave d_{100} values of 49.0 and 56.6 Å respectively with *n*-octane and *n*-hexadecane, compared to the value of 36.1 Å obtained in the absence of the alkanes.

We have also varied the Al : B ratio and obtained hexagonal mesoporous phases with Al : B ratios down to 0.5 : 1.5. By using Na$^+$ as the cation instead of Ca^{2+}, a hexagonal mesoporous phase with a Na : Al : B ratio of 2 : 1 : 1 was obtained with a d_{100} value of 35.5 Å.

References

1 J. S. Beck, J. C. Vartuli, W. J. Roth, M. E. Leonowicz, C. T. Kresge, K. D. Schmitt, C. T. W. Chu, D. H. Olson, E. W. Sheppard, S. B. McCullen, J. B. Higgins and J. L. Schlenker, *J. Am. Chem. Soc.*, 1992, **114**, 10 834.
2 P. Behrens and G. D. Stucky, *Angew. Chem., Int. Ed. Engl.*, 1993, **32**, 696.
3 P. Behrens, *Angew. Chem., Int. Ed. Engl.*, 1996, **35**, 515.
4 J. S. Beck and J. C. Vartuli, *Curr. Opinion Solid State Mater. Sci.*, 1996, **1**, 76.
5 Q. Huo, D. I. Margolese, U. Ciesla, D. G. Demuth, P. Feng, T. E. Gier, P. Sieger, A. Firouzi, B. F. Chmelka, F. Schüth and G. D. Stucky, *Chem. Mater.*, 1994, **6**, 1176; Q. Huo, D. Margolese and G. D. Stucky, *Chem. Mater.*, 1996, **8**, 1147.
6 A. Monnier, F. Schüth, Q. Huo, D. Kumar, D. Margolese, R. S. Maxwell, G. D. Stucky, M. Krishnamurty, P. Petroff, A. Firouzi, M. Janicke and B. F. Chmelka, *Science*, 1993, **261**, 1299; J. C. Vartuli, K. D. Schmitt, C. T. Kresge, W. J. Roth, M. E. Leonowicz, S. B. McCullen, S. D. Helbring, J. S. Beck, J. S. Schlenker, D. H. Olson and E. W. Sheppard, *Chem. Mater.*, 1994, **6**, 2317.
7 P. T. Tanev and T. J. Pinnavaia, *Science*, 1995, **267**, 365.
8 S. A. Bagshaw and T. J. Pinnavaia, *Angew. Chem., Int. Ed. Engl.*, 1996, **35**, 1102.
9 D. M. Antonelli and J. Y. Ying, *Angew. Chem., Int. Ed. Engl.*, 1995, **34**, 2014.
10 N. Ulagappan and C. N. R. Rao, *Chem. Commun.*, 1996, 1685.
11 U. Ciesla, S. Schacht, G. D. Stucky, K. K. Unger and F. Schüth, *Angew. Chem., Int. Ed. Engl.*, 1996, **35**, 426; N. Ulagappan, Neeraj, B. V. N. Raju and C. N. R. Rao, *Chem. Commun.*, 1996, 2243.
12 D. M. Antonelli and J. Y. Ying, *Angew. Chem., Int. Ed. Engl.*, 1996, **35**, 426.
13 S. T. Wilson, B. M. Lok, C. A. Messina, T. R. Connan and E. M. Flanigan, *J. Am. Chem. Soc.*, 1982, **104**, 1176.
14 A. Sayari, V. R. Karra, J. S. Reddy and I. L. Moudrakovski, *Chem. Commun.*, 1996, 411.
15 J. Wang, S. Feng and R. Xu, *J. Chem. Soc., Chem. Commun.*, 1989, 265; J. Yu, K. Tu and R. Xu, in *Zeolites and Related Microporous Materials: State of the Art*, ed. J. Weitkamp, H. G. Karge, H. Pfeifer and W. Hölderich, Elsevier, Amsterdam, 1994.
16 C. A. Fyfe, G. C. Gobbi, J. Klinowski, J. M. Thomas and S. Ramdas, *Nature*, 1982, **96**, 530.
17 C. A. Fyfe, L. Berni, H. G. Clark, J. A. Davies, G. C. Cobbi, J. S. Hartman, P. J. Hayes and R. E. Wasilyshen, *Adv. Chem. Ser.*, 1983, **211**, 405.

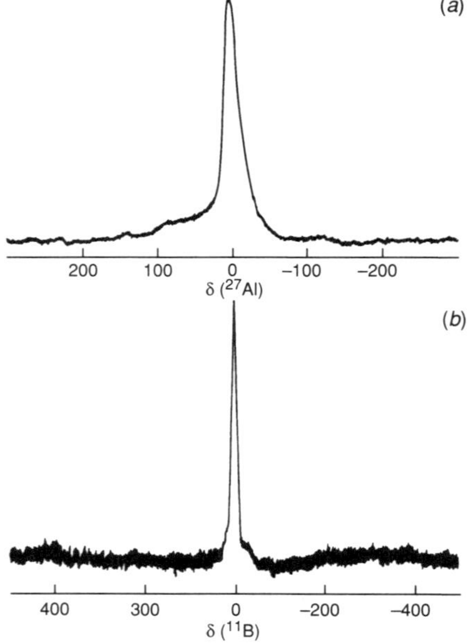

Fig. 3 MAS NMR spectra of mesoporous aluminoborate: (*a*) ^{27}Al spectrum and (*b*) ^{11}B spectrum

Received, 18th November 1996; Com. 6/07782J

High catalytic efficiency of transition metal complexes encapsulated in a cubic mesoporous phase

M. Eswaramoorthy, Neeraj and C. N. R. Rao,*

Chemistry and Physics of Materials Unit, Jawaharlal Nehru Centre for Advanced Scientific Research, Jakkur Post, Bangalore 560 064, India

Copper(II) acetate dimer and $[Mn^{II}(bipy)_2]^{2+}$ encapsulated in cubic Al-MCM-48 show high catalytic activity in the oxidation of phenol to catechol by oxygen activation, and of styrene to styrene oxide by singlet oxygen, respectively.

Transition metal complexes encapsulated in the cavities of zeolites are known to exhibit high catalytic activity in certain oxidation reactions, suggesting that these catalytic systems are good enzyme mimics.[1] Oxidation of phenols with O_2 by copper acetate dimer incorporated in MCM-22 or VPI-5 is a case in instance wherein the copper(II) complex mimics the phenolase activity of tyrosinase.[2,3] Besides the activation of O_2, there have been studies of the oxidation of organic compounds with singlet oxygen sources such as H_2O_2, by metal complexes encapsulated in molecular sieves. Selective oxidation of alkenes by bis(2,2′-bipyridyl)manganese, $[Mn(bipy)_2]^{2+}$, encapsulated in zeolites X and Y is one such example.[4] The $[Mn^{II}(bipy)_2]^{2+}$ complex immobilized in mesoporous Al-MCM-41 has been recently shown to exhibit high catalytic activity for styrene oxidation.[5] Based on the geometry of the pore structures of mesoporous solids, it was our view that the cubic phase should be an excellent host for enhancing the catalytic activity of metal complexes. We have therefore investigated the catalytic activity of two transition metal complexes incorporated in mesoporous Al-MCM-48, in oxidation reactions. The metal complexes examined are copper(II) acetate dimer which has the structural features of Cu-containing monooxygenase enzymes[2,3] and $[Mn^{II}(bipy)_2]^{2+}$. While copper(II) acetate incorporated in Al-MCM-48 was primarily meant to examine the catalytic activity for oxygen activation at ambient conditions, the manganese(II) complex system was intended to study the oxidation of styrene by singlet oxygen.

Al-MCM-48 was prepared by employing a modified procedure. To a solution of 2.9 g of cetyltrimethylammonium bromide in 40 ml of deionised water, 2.8 ml of 5 M NaOH was added and the solution stirred for 30 min. To this, a solution containing 0.09 g of aluminium sulfate dissolved in 10 ml of water was added followed by the dropwise addition of 6.2 ml of tetraethylorthosilicate. The mixture was stirred for 1 h, transferred to a stainless steel autoclave and kept at 383 K for 5 days. The final product was filtered, washed several times with deionised water, dried at 353 K for 3 h and calcined at 773 K for 10 h in air to remove the template. The Si/Al ratio in the product was *ca.* 50. The cubic nature of the mesoporous phase was confirmed by X-ray diffraction (Fig. 1). Distinct (211), (220), (321), (400) and (332) reflections were seen in the pattern. Cu-acetate-Al-MCM-48 was prepared by stirring 0.4 g of Al-MCM-48 with 0.2 g of copper acetate monohydrate in distilled, deionised water for 12 h. The product was filtered, washed with water and dried at 383 K for 24 h in vacuum. The Cu/Al ratio in the final product was 0.15. The XRD pattern given in Fig. 1 shows the cubic mesoporous nature of the catalyst. The surface area of the catalyst was 600 m^2 g^{-1}. Al-MCM-48-$[Mn(bipy)_2]^{2+}$ was prepared by the treatment of Al-MCM-48 (0.3 g) with a solution of 0.3 g of $[Mn(bipy)_2][NO_3]_2$ in 20 ml of 1:9 (by volume) DMF–acetonitrile mixture at room temperature for 48 h. The sample was filtered, washed with

acetonitrile and dried at 353 K under vacuum for 1 h. Chemical analysis showed that the Mn/Al ratio in the product was 0.14. The XRD pattern (Fig. 1) confirmed that the cubic mesoporous structure was retained. The surface area of the catalyst was 770 m^2 g^{-1}.

The second derivative EPR spectrum of Cu-acetate-Al-MCM-48 ($g_\perp$ = 2.06, $g_\parallel$ = 2.18) exhibited the expected

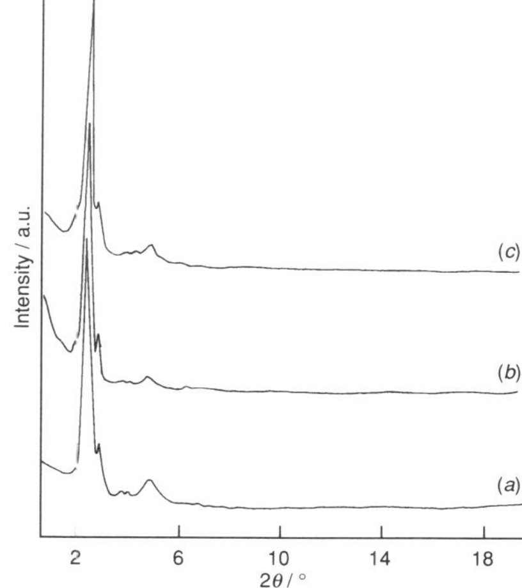

Fig. 1 X-Ray diffraction patterns of (*a*) calcined Al-MCM-48, (*b*) copper acetate dimer encapsulated Al-MCM-48 and (*c*) $[Mn(bipy)_2]^{2+}$ encapsulated Al-MCM-48

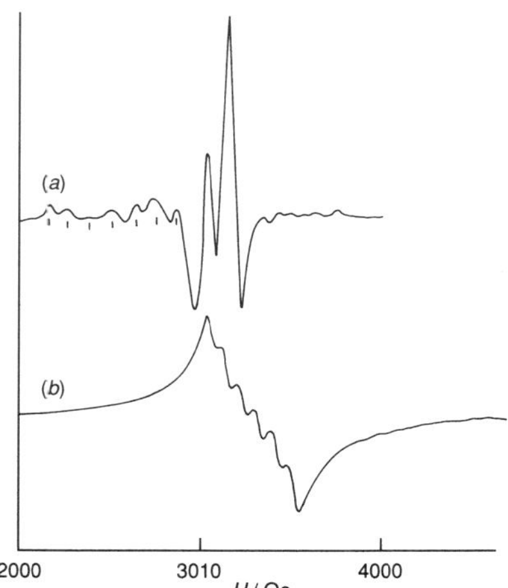

Fig. 2 EPR spectra of (*a*) copper acetate dimer and (*b*) $[Mn(bipy)]^{2+}$ encapsulated in Al-MCM-48

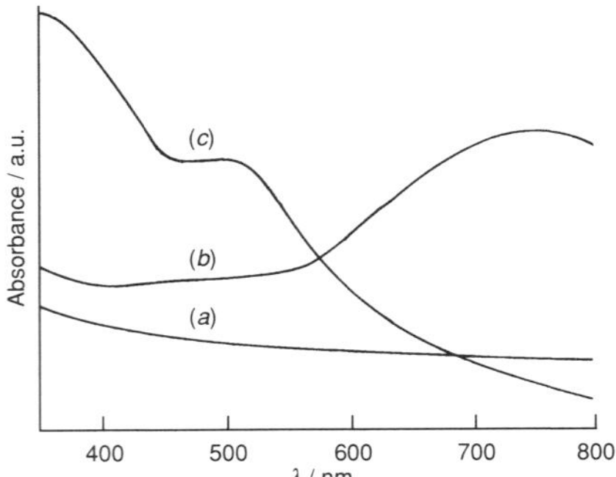

Fig. 3 Electronic absorption spectra of (*a*) Al-MCM-48, (*b*) copper acetate dimer encapsulated in Al-MCM-48

hyperfine structure [Fig. 2(*a*)], establishing the presence of the copper(II) acetate dimer.[3] The IR spectrum showed the carboxylate absorption at 1629 cm^{-1}. The diffuse reflectance spectrum [Fig. 3(*b*)] gave a band around 740 nm, just as in the case of the copper acetate encapsulated zeolites. Oxidation of phenol by the Cu-Al-MCM-48 catalyst was studied by stirring 100 mg of the catalyst in a phosphate buffer solution with 0.57 mmol of phenol in an oxygen atmosphere at 303 K. Gas chromatographic analysis of the product indicated 36% conversion, with catechol as the primary product. The turnover number was 37, a value considerably higher than that found (*ca.* 4) with copper(II) acetate alone. This result demonstrates the high catalytic activity of Cu-Al-MCM-48 in the orthohydroxylation of phenol by oxygen activation.

The Al-MCM-48-[Mn(bipy)$_2$]$^{2+}$ gave an EPR spectrum with the expected hyperfine structure due to Mn^{2+} [Fig. 2(*b*)]. It was pink with a broad band around 490 nm in its DRS, due to the metal–ligand charge transfer transition [Fig. 3(*c*)]. The IR

spectrum showed bands at 760 and 773 cm^{-1} due to the out-of-plane C–H deformation of the bipy. Oxidation of styrene was studied by taking 100 mg of the catalyst in a solution of 0.87 mmol of styrene in 5 ml of acetonitrile, to which 3.5 mmol of H$_2$O$_2$ was added. Gas chromatographic analysis showed the conversion to be *ca.* 40% with styrene oxide as the primary product. The turnover number was 82 compared with *ca.* 7 for the [Mn(bipy)$_2$]$^{2+}$ complex alone. In the hexagonal mesoporous host, Al-MCM-41, the maximum turnover number was 58.[5] This result establishes Al-MCM-48-[Mn(bipy)$_2$]$^{2+}$ to be an excellent catalyst for such oxidation reactions. It is to be noted that the turnover number for styrene oxidation with Cu-acetate-Al-MCM-48 catalyst was 46, but there was hardly 5% conversion to styrene oxide, the main product being benzaldehyde. Similarly, the turnover number of the manganese catalyst for the oxidation of phenol through the activation of molecular oxygen was 14 and the product contained almost no catechol. These results reveal the specificity of the metal complexes encapsulated in the cubic mesoporous phase.

The present study not only demonstrates the high catalytic potential of transition metal complexes encapsulated in cubic mesoporous phases in oxidation reactions, but also the need to explore other reactions as well as mesophase compositions with different Si/Al ratios.

Notes and References

* E-mail: cnrrao@jncasr.ac.in

1 D. R. Corbin and N. Herron, *J. Mol. Catal.*, 1994, **86**, 343.
2 L. M. Sayre and D. V. Nadkarni, *J. Am. Chem. Soc.*, 1994, **116**, 3157.
3 R. Robert and P. Ratnasamy, *J. Mol. Catal.*, 1995, **100**, 93.
4 P. P. Knops-Gerrits, D. D. Vos, F. Thibault-Starzyk and P. A. Jacobs, *Nature*, 1994, **369**, 543.
5 S. S. Kim, W. Zhang and T. J. Pinnavaia, *Catal. Lett.*, 1997, **43**, 149.

Received in Cambridge, UK, 18th December 1997; 7/09095A

Metal chalcogenide–organic nanostructured composites from self-assembled organic amine templates

Neeraj and C. N. R. Rao*

Chemistry and Physics of Materials Unit, Jawaharlal Nehru Center for Advanced Scientific Research, Jakkur Post, Bangalore 560 064, India

Hexagonal and lamellar nanostructured organic–metal chalcogenide composites have been prepared by the reaction of metal salt aliphatic-amine nanostructured adducts with Na_2S or Na_2Se solution; nanostructured composites of CdS, SnS_2, Sb_2S_3 and CdSe with long-chain aliphatic amines obtained in this manner have been characterized.

Braun *et al.*[1] have recently described semiconductor–organic nanostructured composites of hexagonal symmetry based on cadmium sulfide obtained by using non-ionic amphiphiles such as poly(ethylene oxide). Such nanocomposites have been prepared by starting with different cadmium salts.[2] By employing hydrated polyol amphiphiles, Osenar *et al.*[3] have obtained lamellar, nanostructured cadmium sulfide. In all these preparations, the nanostructured adduct of a cadmium salt with the amphiphiles was treated with H_2S gas. Since the preparation of the chalcogenide nanocomposites using amphiphiles involves methods akin to those employed in the synthesis of mesoporous metal oxides,[4–6] we considered it important to evolve a general method for the synthesis of mesostructured semiconductor chalcogenide–organic nanostructures and characterize the materials suitably. By employing long-chain amines as the amphiphiles,[7] we have prepared both hexagonal and lamellar nanostructures of CdS, SnS_2, Sb_2S_3 and CdSe.

The general procedure for the synthesis employed by us is as follows: to an aqueous solution of cadmium acetate (5 mmol) was added an alcoholic solution of the amphiphilic amine (5 mmol) and the mixture was stirred to obtain a gel. The gel was aged at ambient temperature for 18 h and dried. X-Ray diffraction (XRD) patterns of the gel indicated that nanostructured mesophases of the amine and $Cd(CH_3CO_2)_2$ had indeed formed. To the gel, a concentrated aqueous solution of sodium sulfide was slowly added and the pH adjusted to 9.0–9.5. The resulting product was aged at 333 K for 18 h. The product thus obtained was washed first with water, followed by a ethanol–diethyl ether (50:50) mixture and dried at 333 K. The X-ray diffraction pattern of the product was then recorded. Fig. 1(a) and (b) show the XRD patterns of the mesophases obtained with dodecylamine (DA) and stearylamine (SA), respectively. The diffraction patterns are characteristic of a hexagonal mesophase with d_{100} values of 4.1 and 5.5 nm, respectively, for DA and SA. EDX analysis of these products gave a Cd:S ratio of 1:1 (see inset of Fig. 1) showing that the sulfide had the expected composition. Thermogravimetry (TG) showed that the amine template was completely removed below 573 K while the water of hydration, if any, was removed at 393 K. TG gave the compositions of the chalcogenide amine adducts as 3CdS·DA and 7CdS·6SA·9H$_2$O for DA and SA, respectively. The hexagonal nature of the CdS–amine adducts was also confirmed by recording transmission electron microscope (TEM) images. The TEM image of the adduct of CdS with DA shown in Fig. 2(a) suggests that the mesophase has a

hexagonal structure consistent with the XRD pattern in Fig. 1(a). The image shows the wall thickness to be *ca.* 2.0 nm, however, there is considerable disorder.

When we employed a $Cd(CH_3CO_2)_2$:amine ratio of 1:2 instead of 1:1, we obtained a lamellar structure as evident from the XRD pattern of an adduct with SA shown in Fig. 1(c) with *d*-values of 5.0, 2.5 and 1.64 nm corresponding to the (001), (002) and (003) reflections, respectively. TG showed that the amine was completely removed at 623 K and the water removed at 393 K. The composition of the chalcogenide–SA adduct from TG gave the composition 9CdS·8SA·3.5H$_2$O. A typical TEM image of the lamellar mesophase is shown in Fig. 2(b) which shows a well defined striped pattern with a periodicity of *ca.* 5 nm. No change was observed on tilting the particle perpendicular to the stripes, confirming the lamellar morphology. When we employed thiourea instead of Na_2S as the sulfiding agent, we obtained a lamellar nanostructure of CdS with DA of composition 4CdS·3DA. The XRD pattern of this adduct is shown in Fig. 1(d), with *d* values of 3.53, 1.74, 1.17, 0.88 and 0.7 nm, respectively, due to (001), (002), (003), (004) and (005) reflections. We also obtained excellent lamellar mesophases by using long-chain thiols with $Cd(CH_3CO_2)_2$. For example, the adduct with dodecanethiol (DT) had the composition $3Cd(CH_3CO_2)_2$·4DT. However, on heating this adduct we could not obtain pure CdS.

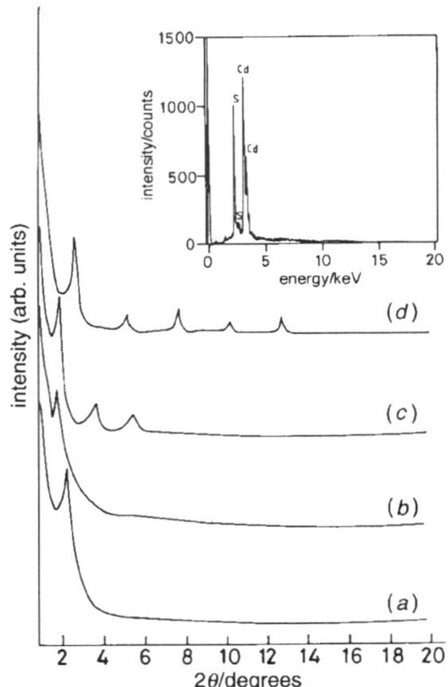

Fig. 1 X-Ray diffraction patterns of (Cu-Kα radiation) CdS–amine nanostructures. Hexagonal phases obtained with (*a*) dodecylamine and (*b*) stearylamine. Lamellar phases obtained with (*c*) stearylamine and (*d*) using thiourea as the sulfiding agent. Inset shows EDX of an adduct with dodecylamine.

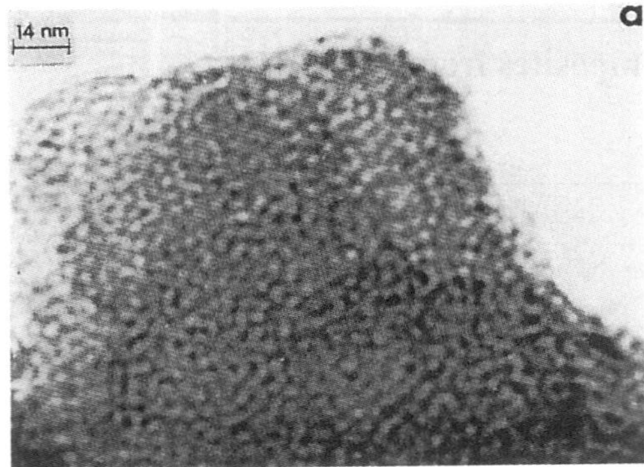

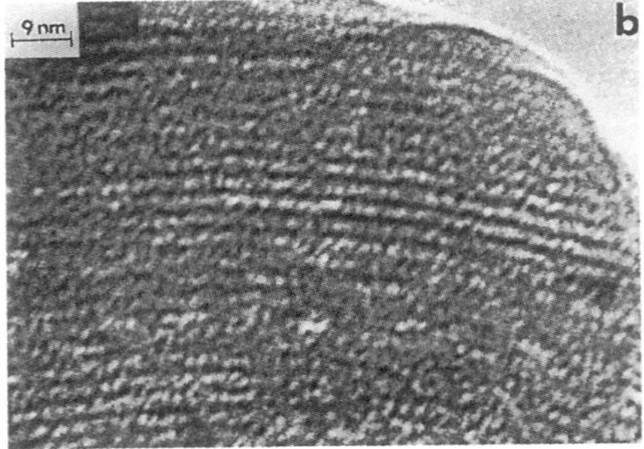

Fig. 2 TEM images of CdS–amine nanostructures: (a) hexagonal phase with dodecylamine, (b) lamellar phase with stearylamine

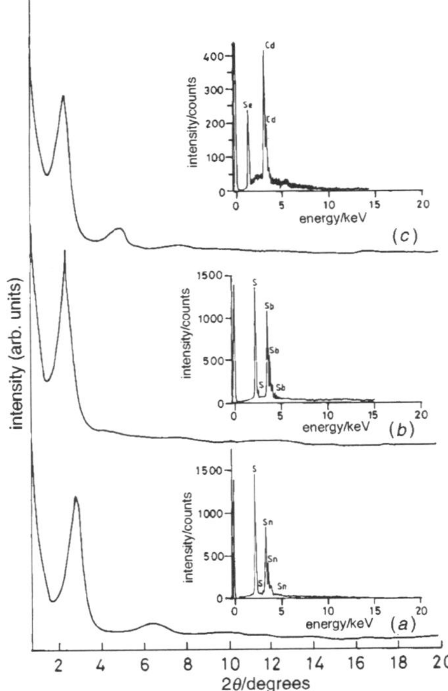

Fig. 3 X-Ray diffraction patterns of metal chalcogenide–amine nanostructures: (a) SnS$_2$–dodecylamine, (b) Sb$_2$S$_3$–dodecylamine, (c) CdSe–dodecylamine. The EDX results are shown alongside the XRD patterns.

On heating the hexagonal CdS adduct with DA at 473 K for 2 h, the mesostructure collapsed, but the resulting sulfide exhibited high surface area (90 m^2 g^{-1}). We have been able to remove amine partly from the hexagonal phase of the CdS adduct with SA by heating it slowly at 448 K for 2 h. The XRD pattern of the product showed a feature corresponding to a d_{100} value of *ca.* 5.5 nm, although somewhat weaker in intensity compared to the adduct.

We have also been able to synthesize metal sulfide–organic amphiphile nanostructures of tin and antimony by starting from SnCl$_4$·2H$_2$O and SbCl$_3$·5H$_2$O, respectively, and keeping the metal salt:amine ratio at 1:1. XRD patterns of the hexagonal mesophases of the adducts of SnS$_2$ and Sb$_2$S$_3$ with DA are shown in Fig. 3(a) and (b), respectively, with d_{100} values of 3.12 and 3.57 nm. The d_{110} and d_{200} reflections are also observed at larger angles. The composition of the sulfides was confirmed by EDX analysis (see insets of Fig. 3). TG indicated the adduct compositions to be 2SnS$_2$·3DA·H$_2$O and 5Sb$_2$S$_3$·7DA·H$_2$O. TEM images showed that the adducts possessed disordered hexagonal structures.

In order to prepare CdSe–amine nanostructures, we employed a procedure similar to that with CdS, except that an aqueous solution of Na$_2$Se was reacted with the Cd(CH$_3$CO$_2$)$_2$–amine gel. Fig. 3(c) shows the XRD pattern of the hexagonal mesophase of the CdSe–DA adduct with d values of 3.66, 1.9 and 1.75 nm for the (100), (110) and (200) reflections, respectively. The adduct had the composition 9CdSe·4DA and the amine could be removed completely at 573 K. We have also been able to obtain CdSe–amine nanostructures by employing sodium selenosulfate as a seleniding agent instead of Na$_2$Se; Na$_2$Se however appears to be a better seleniding agent.

References

1 P. V. Braun, P. Osenar and S. I. Stupp, *Nature (London)*, 1996, **380**, 325.
2 V. Tohver, P. V. Braun, M. U. Pralle and S. I. Stupp, *Chem. Mater.*, 1997, **9**, 1495.
3 P. Osenar, P. V. Braun and S. I. Stupp, *Adv. Mater.*, 1996, **8**, 1022.
4 J. S. Beck and J. C. Vartuli, *Curr. Opinion Solid State Mater. Sci.*, 1996, **1**, 76.
5 P. Behrens, *Angew. Chem., Int. Ed. Engl.*, 1996, **35**, 515.
6 S. Ayyappan and C. N. R. Rao, *Chem. Commun.*, 1997, 575.
7 P. T. Tanev and T. J. Pinnavaia, *Science*, 1995, **267**, 365.
8 N. Ulagappan, Neeraj, B. V. N. Raju and C. N. R. Rao, *Chem. Commun.*, 1996, 2243.

Communication 7/07690H; Received 24th October, 1997

MICROPOROUS AND
MESOPOROUS MATERIALS

Microporous and Mesoporous Materials 28 (1999) 205–210

Synthesis of hexagonal microporous silica and aluminophosphate by supramolecular templating of a short-chain amine

M. Eswaramoorthy, S. Neeraj, C.N.R. Rao *

Chemistry and Physics of Materials Unit, Jawaharlal Nehru Centre for Advanced Scientific Research, Jakkur P.O., Bangalore 560064, India

Received 30 July 1998; accepted 20 October 1998

Abstract

Hexagonal microporous phases of SiO_2 and $AlPO_4$ with pore sizes in the range intermediate between traditional microporous and mesoporous materials have been prepared, by making use of supramolecular organization of short-chain amine ($n\text{-}C_6H_{12}NH_2$) template molecules. The pore diameters are around 1.4 nm in both cases. The hexagonal microporous phase of $AlPO_4$ is formed only in the presence of fluoride ions. The surface area of SiO_2 after removal of the template is 800 $m^2\,g^{-1}$. In the case of $AlPO_4$, the surface area is 190 $m^2\,g^{-1}$ after removal of 60% of the template. © 1999 Elsevier Science B.V. All rights reserved.

Keywords: Hexagonal microporous solids; Mesoporous solids; Microporous aluminophosphate ; Microporous silica

1. Introduction

A variety of crystalline microporous oxides with pore diameters in the 0.4–1.5 nm range have been prepared over the last few years, by using organic template molecules [1–3]. By making use of self-assembled aggregates of long-chain surfactant molecules mesoporous oxides (pore diameter of 2–10 nm) have been prepared, the very first instance being that of silica by Mobil chemists [4,5]. There is, however, a lack in the spectrum of hexagonal porous solids corresponding to those with pore diameters in the range of 1–2 nm, i.e., in the intermediate range of pore sizes between those of the traditional microporous and meso-

porous solids. Mesoporous MCM-41 silicates prepared with C_8–C_{10} surfactants are known to have pore sizes in the 1.6–2.0 nm range [6]. Silica with a bimodal pore-size distribution containing both micro- and mesopores with diameters of 1 nm and 3 nm, respectively, has been obtained using a cationic surfactant, on post-synthesis hydrothermal treatment [7]. Hexagonal, mesoporous aluminophosphate with a pore diameter of 3.5 nm has been prepared using a cationic surfactant by Zhao et al. [8]. The use of neutral amine surfactants, however, does not yield the hexagonal mesoporous aluminophosphate. Recently, Sun and Ying [9] have reported that hexagonal microporous niobia with a pore diameter of less than 2 nm can be prepared by supramolecular templating using short-chain C_4–C_7 n-alkylamines. We have investigated the synthesis of hexagonal microporous silica

---

* Corresponding author. Fax: +91 80 846 2766;
E-mail: cnrrao@jncasr.ac.in

and aluminophosphate employing a short-chain amine as the template molecule and have succeeded in obtaining these materials with well-defined pore distributions between 1 and 2 nm.

2. Experimental

In a typical preparation of hexagonal microporous silica, to a solution of 0.019 mol of hexylamine in 50 ml of H_2O, 0.01 mol of tetraethylorthosilicate (TEOS) was added and stirred for 30 min. The mixture was aged at ambient temperature for 5 days. The product was filtered, washed with water and dried at 343 K for 24 h. The as-synthesized sample was calcined at different temperatures in a nitrogen atmosphere.

To prepare hexagonal microporous aluminophosphate, 1 g of $Al(OH)_3 \cdot xH_2O$ was added to a solution containing 40 ml of H_2O and 10 ml of methanol and stirred for 10 min. To this slurry, 0.5 ml of 80% H_3PO_4 was added, followed by 1.5 ml of 48% hydrofluoric acid to obtain a clear solution. The final pH was adjusted to 7–8 using ~4 ml of hexylamine. The gel formed was allowed to stir for 3 h and aged at ambient temperature for 24 h. The product was filtered, washed with water and dried at 343 K for 24 h. The presence of a fluoride-containing medium was essential, in the absence of which only a lamellar phase was obtained.

X-ray diffraction (XRD) patterns of the samples were obtained using a Rich-Siefert instrument (model XRD-3000-TT) and Cu Kα radiation. Thermogravimetric analyses (TGA) were carried out with a Mettler Toledo-TG-850 instrument at a heating rate of 10 K min^{-1} in flowing oxygen atmosphere. Energy-dispersive X-ray (EDX) analysis was carried out with a Leica scanning electron microscope fitted with an EDX spectrometer and a Link ISIS detector. Transmission electron microscopy (TEM) images were obtained with JEOL JEM-3010 (300 kV) instrument operating at 200 kV. The samples for TEM were prepared by dispersing them in acetone on a holey carbon–copper grid. Surface areas were determined by the Brunauer–Emmet–Teller (BET) method on a Cahn-2000 electrobalance. Pore-size distribution

was calculated by means of the Barrett–Joyner–Halenda [10] or the Horvath–Kawazoe [11] method.

3. Results and discussion

Thermogravimetric analysis of the as-synthesized silica sample showed complete removal of water around 393 K and complete loss of the amine around 573 K. Dehydroxylation of the silanol groups occurs from 573 K up to 900 K. Based on the TGA results, the composition of the as-synthesized sample was $SiO_2:0.12\ C_6H_{13}NH_2: 0.33\ H_2O$. The XRD pattern of the as-synthesized silica is characteristic of a hexagonal phase with a d_{100} value of 2.51 nm [Fig. 1(a)]. After calcina-

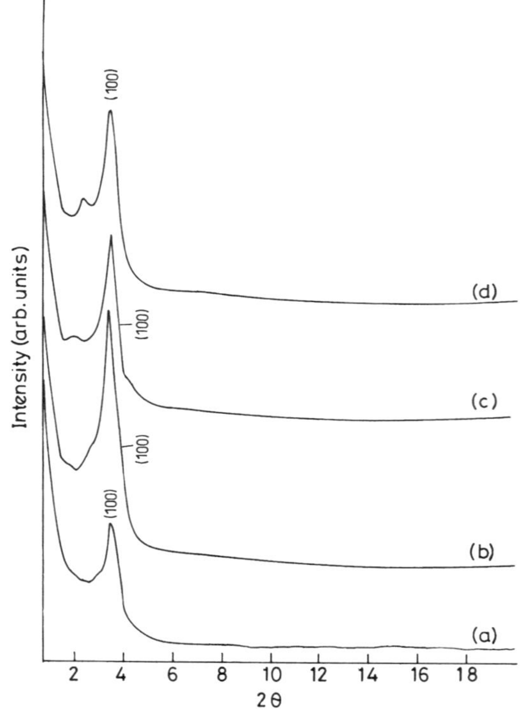

Fig. 1. X-ray diffraction patterns of microporous silica prepared with n-hexylamine as the templating agent: (a) as-synthesized; (b) calcined at 523 K for 3 h in N_2; (c) calcined at 673 K for 3 h in N_2; and (d) calcined at 723 K for 3 h in N_2. The small feature at low angles before the (100) reflection may be due to the impurity of a mesoporous phase.

tion at 523 K for 3 h in N_2 atmosphere, the diffraction pattern showed little change [Fig. 1(b)] although $\sim$80% of the amine template as well as water were removed. Calcination at 673 K for 3 h in N_2 atmosphere also did not cause any significant change in the d_{100} spacing in the XRD pattern [Fig. 1(c)], even though all the template had been eliminated. Such an absence of any appreciable change in the d_{100} value on calcination has to be contrasted with that of the mesoporous silica, where there was a considerable change in the d-spacing during calcination. Calcination at 723 K for 3 h in N_2 atmosphere, however, gives rise to a small additional peak at a smaller angle [Fig. 1(d)].

In Fig. 2(a) we show the N_2-adsorption isotherm of the porous silica sample calcined at 523 K (corresponding to 80% loss of template). The isotherm is of type I, similar to that found in zeolites [12] and microporous niobia [9]. The BET

surface area of the sample was 700 $m^2\,g^{-1}$. The pore-size distribution calculated by the Horvath–Kawazoe method shows two distinct pore diameters of 1.3 and 1.5 nm [Fig. 2(b)]. It is noteworthy that the pore diameters are well below the mesoporous range. The wall thickness for the calcined sample is around 1.5 nm. The TEM image of the sample calcined at 523 K (Fig. 3) shows the presence of disordered hexagonal channels with a pore diameter of $\sim$1.42 nm.

The N_2-adsorption isotherm of the porous silica calcined at 673 K for 3 h in a N_2 atmosphere was also examined. The isotherm is of type I as can be seen from Fig. 4(a), indicating the microporous nature of the sample. The BET surface area of this sample was 800 $m^2\,g^{-1}$. The increase in surface area is not surprising since all the template is removed at this temperature. When the sample calcined at 673 K for 3 h in N_2 atmosphere was exposed to the atmosphere for 2 days, it lost part of the crystallinity and showed a type IV adsorption isotherm [Fig. 4(b)]. Accordingly, the pore-size distribution obtained by the Barrett–Joyner–Halenda method showed the existence of both micropores (pore diameter of $\sim$1.5 nm) and mesopores (pore diameter of $\sim$3.5 nm) as shown in the inset of Fig. 4.

Thermogravimetric analysis of the as-synthesized sample of aluminophosphate (in the presence of F^- ions) showed weight losses at three different temperatures: $\sim$423 K due to water,

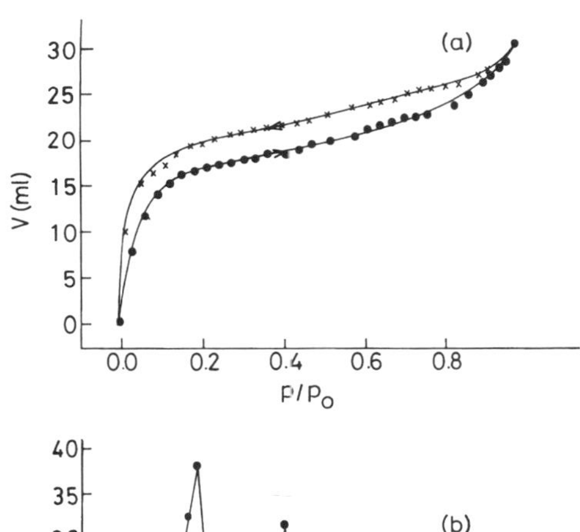

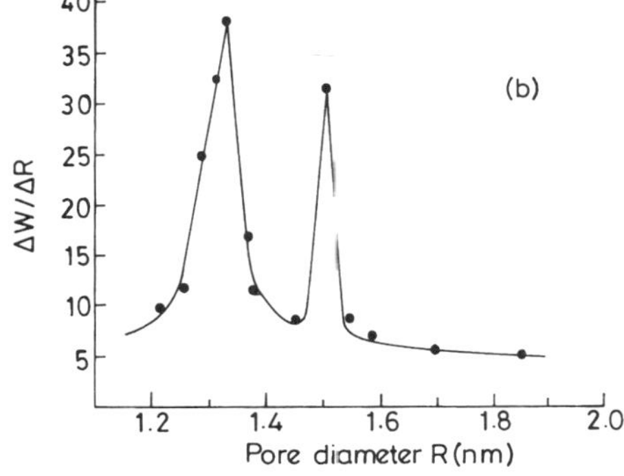

Fig. 2. (a) N_2 adsorption–desorption isotherm (at 77 K) of microporous silica calcined at 523 K for 3 h. (b) Pore-size distribution of the same sample obtained by the Horvath–Kawazoe method.

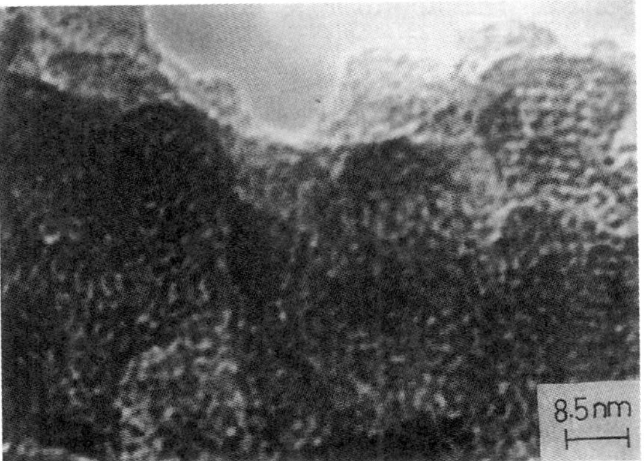

Fig. 3. TEM image of microporous silica calcined at 523 K for 3 h.

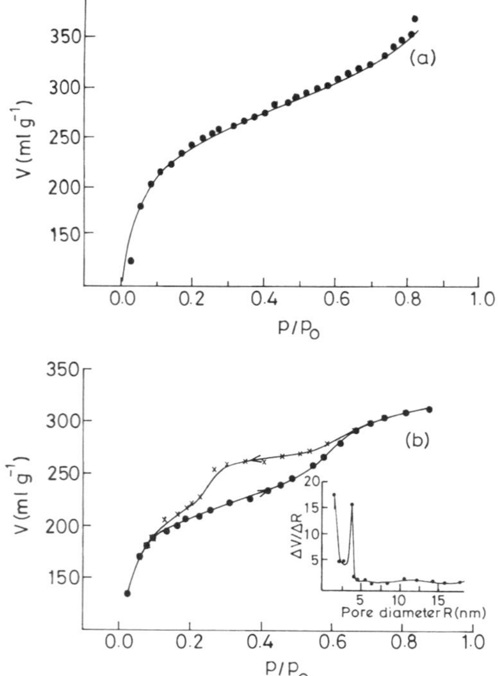

Fig. 4. (a) N_2-adsorption isotherm (at 77 K) of microporous silica calcined at 673 K for 3 h. (b) N_2 adsorption–desorption isotherm (at 77 K) of the sample in (a) exposed to the atmosphere for 2 days. Inset shows the pore-size distribution of the sample in (b).

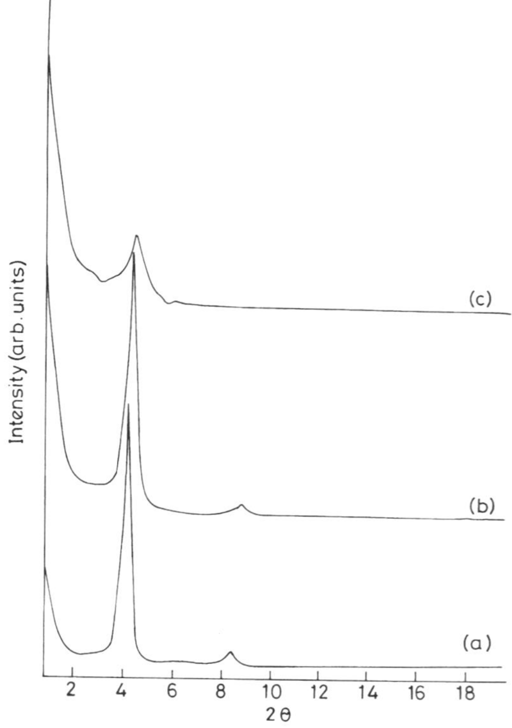

Fig. 5. X-ray diffraction patterns of microporous aluminophosphate prepared with n-hexylamine as template (in the presence of F^- ions): (a) as-synthesized; (b) calcined at 503 K for 2 h in N_2; and (c) calcined at 573 K for 2 h in N_2.

$\sim$423–623 K due to the loss of template and above $\sim$623 K due to dehydroxylation of the surface. The approximate composition of the as-synthesized material calculated from TGA (assuming the absence of flururide) was Al_2O_3:P_2O_5:2.6 $C_6H_{13}NH_2$:H_2O. The Al/P ratio of unity was also verified by energy-dispersive X-ray analysis. The presence of fluoride ions in the reaction mixture was found to be necessary to obtain the Al/P ratio of unity. In the absence of fluoride ions, the Al/P ratio was around 0.5. Furthermore, the $AlPO_4$ so obtained had a lamellar structure.

The XRD pattern of the as-synthesized aluminophosphate sample (prepared in the presence of fluoride ions) is shown in Fig. 5(a). The presence of a low-angle peak in the pattern with d_{100} value of 2.1 nm is attributed to the hexagonal micro-

porous material, rather than to a mesoporous material of the M41S family. Calcination of the as-synthesized $AlPO_4$ sample at 503 K for 2 h in N_2 atmosphere does not affect the crystallinity, as can be seen from Fig. 5(b). On calcination at 573 K for 2 h in a N_2 atmosphere, we observe a significant decrease in the intensity of the d_{100} reflection [Fig. 5(c)]. This is likely to be due to the formation of disordered pores resulting from removal of the template. It is to be noted that calcination around 573 K removes $\sim$60% of the template. The BET surface area of this sample from N_2 adsorption was 190 m^2 g^{-1}. The adsorption isotherm was of type I as shown in Fig. 6(a) and the pore-size distribution calculated by the Horvath–Kawazoe method shows pores of two sizes, with diameters of 1.3 and 1.38 nm, as illustrated in Fig. 6(b). The wall thickness of the

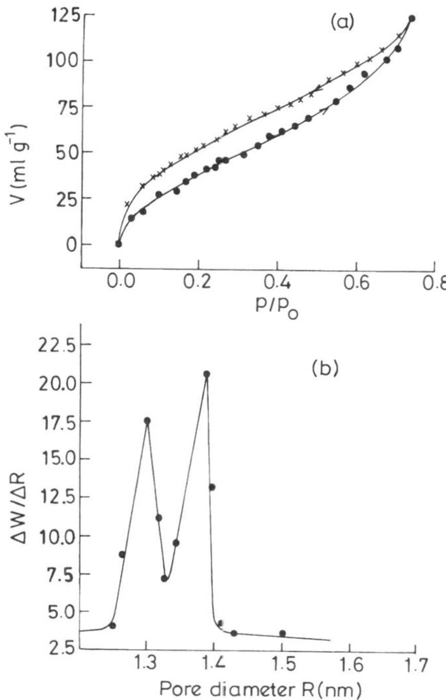

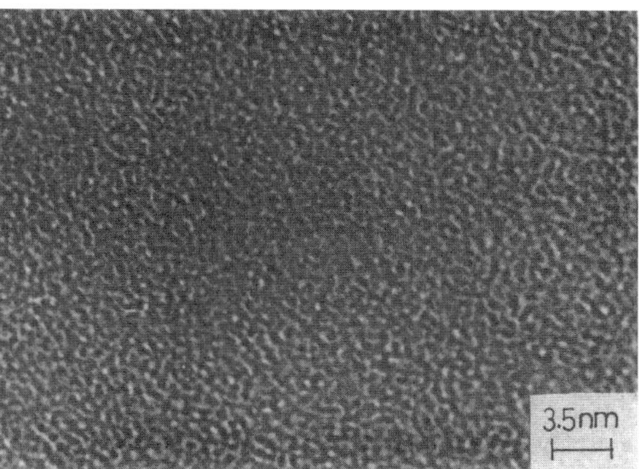

Fig. 7. TEM image of microporous aluminophosphate calcined at 573 K for 3 h.

Fig. 6. N_2 adsorption–desorption isotherm (at 77 K) of microporous aluminophosphate calcined at 573 K for 2 h in N_2: (b) pore-size distribution of the sample in (a) calculated by the Horvath–Kawazoe method.

sample obtained by subtracting the Horvath–Kawazoe pore size from the a_0 value was ~ 1.0 nm. The TEM image of the sample calcined at 503 K (Fig. 7) reveals disordered hexagonal micropores, with a pore size comparable to that obtained from the N_2 adsorption–desorption data.

The XRD pattern of the aluminophosphate prepared in the absence of fluoride ions using n-hexylamine as the template, is characteristic of a lamellar phase [Fig. 8(c)], with a d_{100} spacing of 2.05 nm. It was possible to exchange the hexylamine with other amines, by soaking the lamellar aluminophosphate sample in the corresponding amine in methanol solution. The XRD patterns of the lamellar phases with different amines thus obtained are shown in Fig. 8. The d-spacing of the lamellar material varies in proportion to the chain length of the amine, as expected.

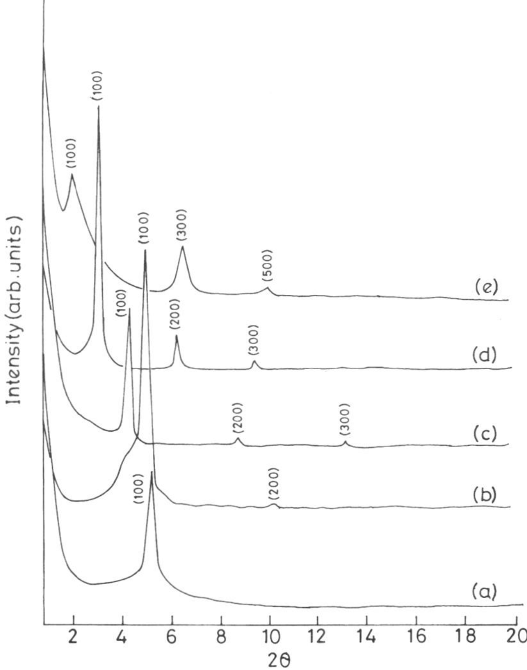

Fig. 8. X-ray diffraction patterns of lamellar aluminophosphate samples prepared with different amines in the absence of F^- ions: (a) 1,3 diaminopropane; (b) butylamine; (c) hexylamine; (d) octylamine; and (e) dodecylamine.

4. Conclusions

The present study shows that it is possible to prepare hexagonal microporous silica with a pore diameter between 1 and 2 nm by using a short-chain amine such as n-hexylamine as the template molecule. The pores are disordered. Hexagonal microporous aluminophosphate of similar pore size can also be prepared by using n-hexylamine as the surfactant. The presence of F^- ions is necessary to obtain the hexagonal phase of $AlPO_4$, the lamellar phase being obtained in the absence of F^- ions. The pore diameters in the hexagonal microporous silica and aluminophosphate samples prepared in the present work are both in the range of 1.3–1.5 nm. These values of pore diameters are in the range intermediate between those in microporous zeolites and mesoporous solids.

References

[1] M.E. Davis, Chem. Eur. J. 3 (1997) 1745.
[2] J.M. Thomas, Angew. Chem. Intl. Ed. Engl. 33 (1994) 913.
[3] C.J. Brinker, Curr. Opinion Solid State Mater. Sci. 1 (1996) 795.
[4] C.T. Kresge, M.E. Leonowicz, W.J. Roth, J.C. Vartuli, J.S. Beck, Nature 359 (1992) 710.
[5] S. Ayyappan, C.N.R. Rao, J. Chem. Soc., Chem. Commun. (1997) 575 (and references listed therein).
[6] P.T. Tanev, T.J. Pinnavaia, Chem. Mater. 8 (1996) 2068.
[7] A. Sayari, M. Kruk, M. Jaroniec, Catal. Lett. 49 (1997) 147.
[8] D. Zhao, Z. Luan, L. Kevan, J. Chem. Soc., Chem. Commun. (1997) 1009.
[9] T. Sun, J.Y. Ying, Nature 389 (1998) 704.
[10] E.P. Barrett, L.G. Joyner, P.P. Halender, J. Am. Chem. Soc. 73 (1951) 373.
[11] G. Horvath, J. Kawazoe, J. Chem. Eng. Jpn 16 (1983) 470.
[12] R. Szostak, Molecular Sieves; Principles of Synthesis and Identification, Van Nostrand, New York, 1989.

Solid State Sciences 2 (2000) 877–882

Solid State
Sciences

www.elsevier.com/locate/ssscie

Macroporous oxide materials with three-dimensionally interconnected pores

Gautam Gundiah, C.N.R. Rao *

*Chemistry and Physics of Materials Unit, Jawaharlal Nehru Centre for Advanced Scientific Research, Jakkur P.O.,
Bangalore 560 064, India*

Received 25 July 2000; accepted 21 September 2000

Dedicated to Professor J.M. Hönig on the occasion of his 75th birthday

Abstract

Ordered mesoscale hollow spheres (1000 nm diameter) of binary oxides such as TiO_2 and ZrO_2 as well as of ternary oxides such as ferroelectric $PbTiO_3$ and $Pb(ZrTi)O_3$ have been prepared by templating against colloidal crystals of polystyrene, by adopting different procedures. © 2000 Éditions scientifiques et médicales Elsevier SAS. All rights reserved.

Keywords: Ordered mesoscale hollow spheres; Binary and ternary oxides; Colloidal crystals of polystyrene

1. Introduction

Research on uniform, hollow capsules of nanometer to micrometer dimensions is being pursued with great interest because of several possible technical applications in drug delivery systems, catalysis, separation techniques, photonics as well as piezoelectric and other dielectric devices [1,2]. Velev et al. [3] obtained ordered macroporous silica by templating colloidal crystals of polystyrene while Imhof and Pine [4] obtained macroporous titania by using surfactants. The preparation of a few ceramic oxides with periodic three-dimensional arrays of macropores by using polystyrene spheres as templates has been reported [5–8]. Macroporous TiO_2 with diameters between 240 and 2000 nm has also been prepared by filling voids in opals by a liquid phase chemical reaction [9]. While some of the workers have removed the organic template by dissolution in

an appropriate solvent [7,8], others have burnt off the polymer template [4] or extracted it by refluxing in a solvent [5,6].

We have prepared ordered mesoscale hollow spheres of ~ 1000 nm diameter of binary as well as ternary oxides by employing crystalline arrays of polystyrene beads as templates. Typical of the ordered macroporous ternary oxides obtained by us is ferroelectric $PbTiO_3$ and $Pb(ZrTi)O_3$. In preparing these ordered porous networks, we have employed calcination as well as washing with a solvent as the means of removing the template.

2. Experimental

Polystyrene beads of 1000 nm diameter, prepared by emulsifier-free polymerization of styrene, were centrifuged at 1100 rpm for 6 h and dried in air. These beads yielded satisfactory colloidal crystals, as can be seen from a scanning electron microscopy (SEM) image (obtained with a Leica microscope) of

* Correspondence and reprints: Fax: +91-80-8462776.
E-mail address: cnrrao@jncasr.ac.in (C.N.R. Rao).

a dispersion shown in Fig. 1a. We prepared macroporous networks of SiO_2, TiO_2 and ZrO_2 by using tetraethylorthosilicate (TEOS), tetrabutylorthotitanate (TBOT) and zirconium i-propoxide (ZIP), respectively as the precursors. Macroporous SiO_2, TiO_2 and ZrO_2 networks were prepared from using

alkoxide precursors by the following procedure — polystyrene spheres were taken on a Büchner funnel connected to a vacuum pump. The ratio of the weights of the metal precursors (TEOS, TBOT, ZIP) to the template was 10.0, 5.0 and 2.5 for SiO_2, TiO_2 and ZrO_2, respectively. The pure precursors were

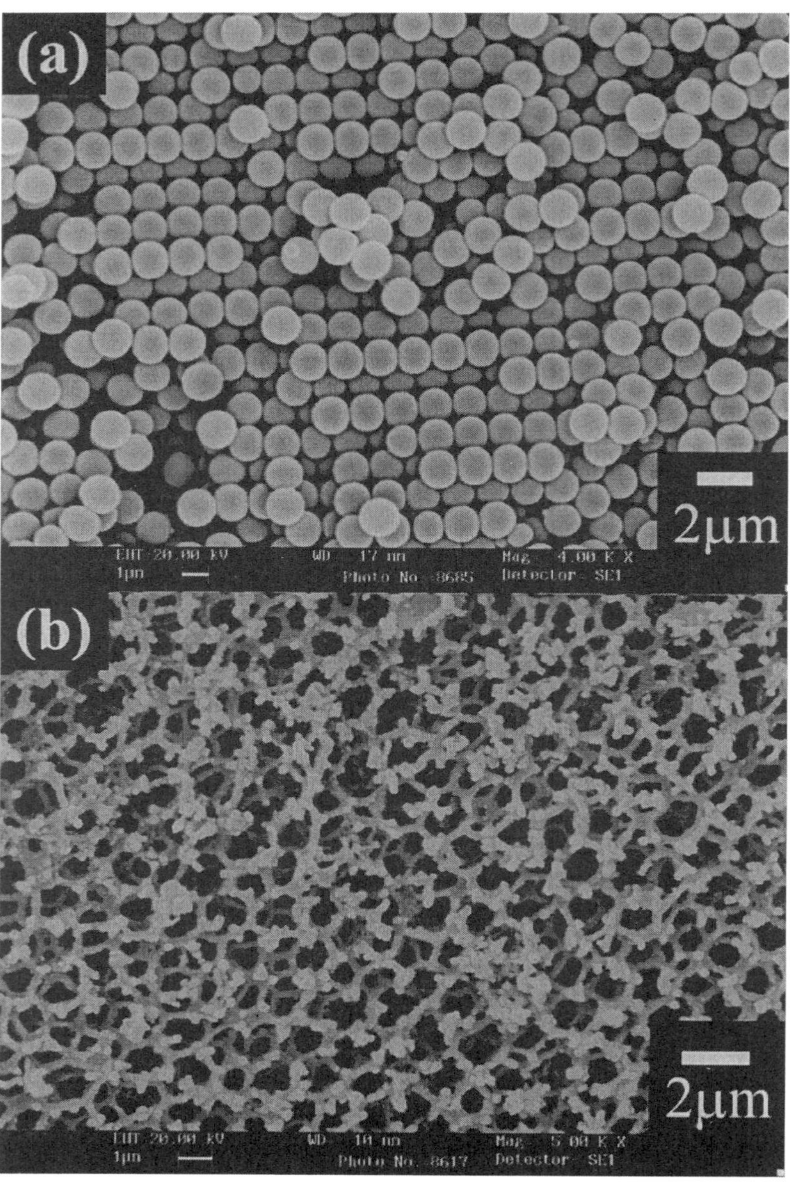

Fig. 1. (a) SEM image of an array of polystyrene beads. (b) SEM image of ordered macroporous SiO_2.

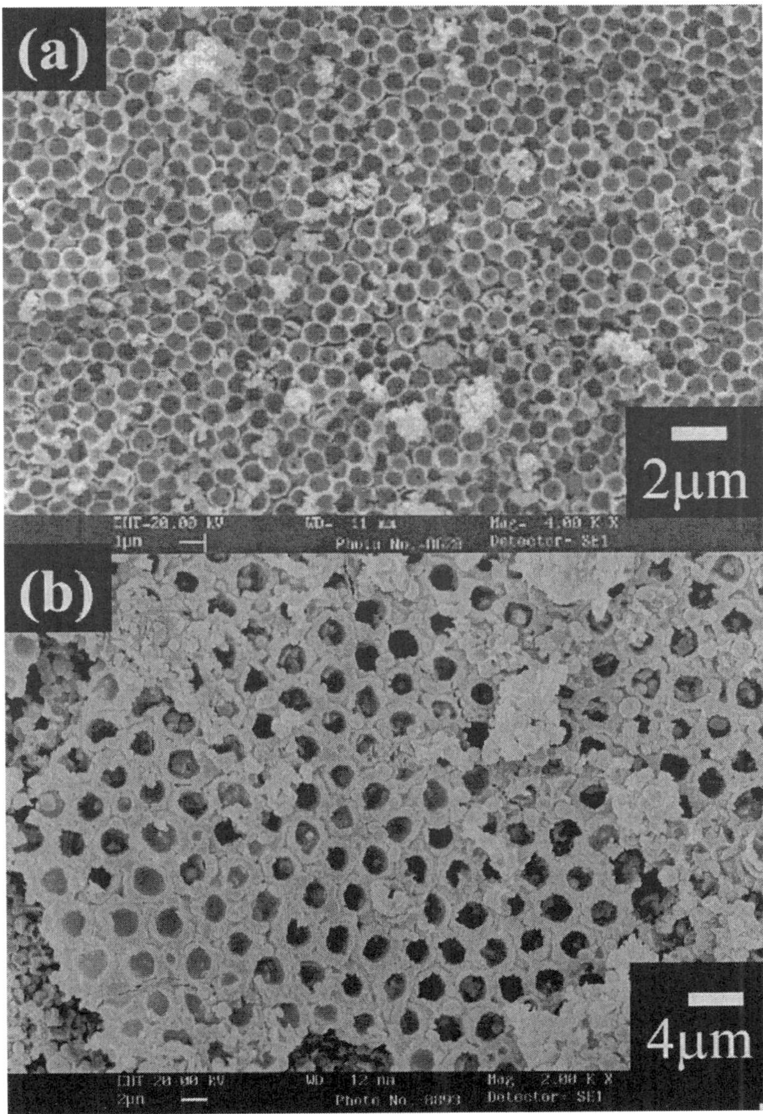

Fig. 2. SEM images of ordered macroporous TiO_2 obtained by two different routes.

poured on the template to wet it completely. The precursors hydrolyzed on exposure to air. The composites were left to dry for 12 h at room temperature. The template was removed by calcination at 575°C for 7 h.

Macroporous TiO_2 was prepared by an alternative route as well. $TiCl_4$ was taken in a three-necked round bottom flask and kept in salt–ice mixture. Distilled water ice pieces were added to it slowly to form a yellow coloured solution. Distilled water was finally added to get a transparent solution. This was poured on polystyrene beads kept in a glass vial and allowed to dry at room temperature. The template was removed by treatment with toluene.

Macroporous $PbTiO_3$ and $Pb(ZrTi)O_3$ (PZT) were prepared by the following procedure — 0.1 M solu-

G. Gundiah, C.N.R. Rao / Solid State Sciences 2 (2000) 877–882

tion of lead acetate, 0.1 M TBOT and 0.1M ZIP (all in methanol) were prepared. In order to prepare PbTiO$_3$, stoichiometric amounts of the Pb and Ti precursor solutions were taken in a beaker. The solution was stirred for 15 min after which the temperature was increased to 60°C. A few drops of water were added to the solution, which resulted in gel. This was poured on the polystyrene spheres and allowed to dry at room temperature for 12 h. The template was removed by treatment with toluene. The sample was crystallized by heating to 400°C for 60 min. Pb(Zr$_{0.5}$,Ti$_{0.5}$)O$_3$ was obtained by a similar procedure (starting with Pb, Zr, Ti precursor solutions).

X-ray diffraction of all the oxide products were recorded with a Siefert XRD 3000TT. SEM images of these oxides were recorded. Surface areas of the

various macroporous oxide networks tend to be small [6] and therefore are not elaborated on.

3. Results and discussion

In Fig. 1b we show the SEM image of the ordered macroporous SiO$_2$. The pore diameter is slightly smaller ($\sim$ 850 nm) than that of the template beads because of shrinkage occurring during calcination. What is interesting is that we see lower layers in the SEM image due to the three-dimensional ordering in the material, although it is X-ray amorphous.

The SEM image of macroporous TiO$_2$ prepared by the TBOT route is shown in Fig. 2a. The pores are highly ordered and there is some shrinkage due to calcination. The pore diameter is approximately 800 nm. The material is crystalline as revealed by the X-ray diffraction (XRD) pattern (Fig. 3). TiO$_2$ in this porous structure is in the anatase phase with $a = 14.3858$; $c = 4.8429$ Å. Macroporous TiO$_2$ obtained by the TiCl$_4$ route gave an ordered structure as well, as can be seen from the SEM image in Fig. 2b, but the material was X-ray amorphous. There was little shrinkage of the pores in this case. By carrying out the hydrolysis around 65°C, one obtains the rutile phase [10].

A SEM image of ordered macroporous zirconia is shown in Fig. 3. The material is crystalline as evidenced by the XRD pattern (Fig. 3). ZrO$_2$ has the expanded monoclinic structure ($a = 6.0522$; $b = 7.6176$; $c = 3.388$ Å, $\beta = 111.539°$). The image in Fig. 4 shows small circular windows approximately one-quarter the diameter of the big pores. The windows connect the spherical voids and are formed when the close packed spheres are in contact.

The as-prepared macroporous network of PbTiO$_3$ was amorphous. However, on heating at 400°C for 2 h it became crystalline as evidenced by the XRD pattern shown in Fig. 3. (tetragonal: $a = 7.8151$; $c = 4.0984$ Å). The stoichiometry was verified by EDX analysis. There was no shrinkage of the pores in the network and we show the porous structure in Fig. 5.

The SEM image of the macroporous structure of Pb(ZrTi)O$_3$ obtained by us is shown in Fig. 6. The material was however X-ray amorphous. Heating the material for purpose of crystallization destroyed the porous structure.

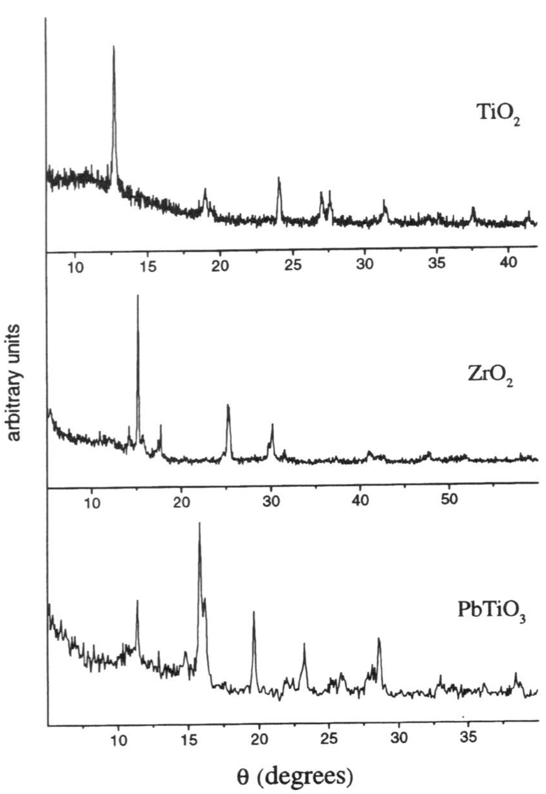

Fig. 3. XRD patterns of macroporous TiO$_2$, ZrO$_2$ and PbTiO$_3$.

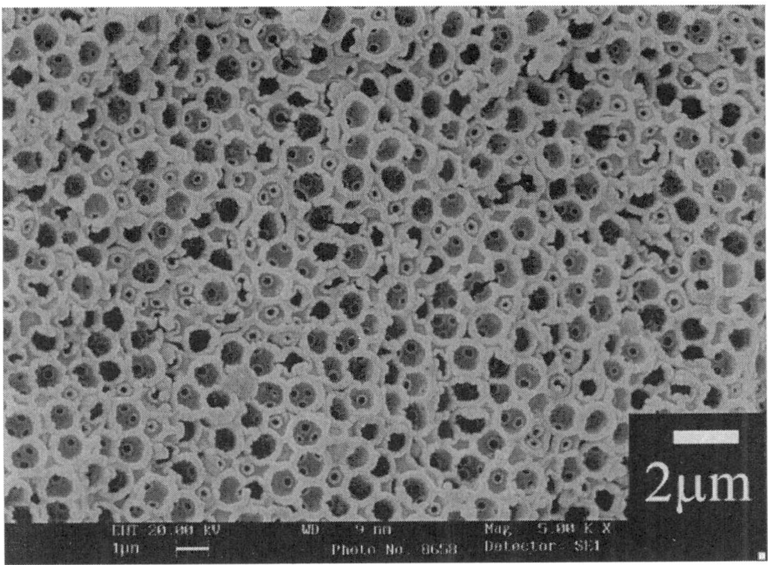

Fig. 4. SEM image of ordered macroporous ZrO$_2$.

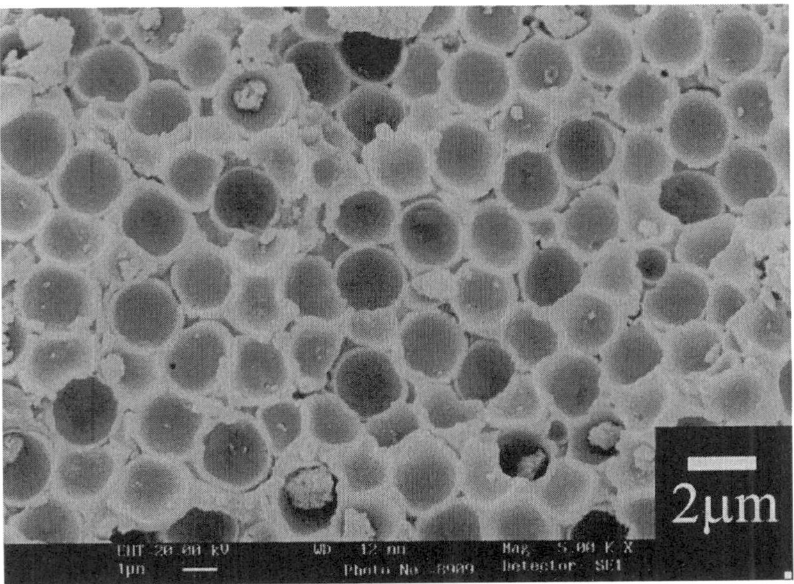

Fig. 5. SEM image of ordered macroporous PbTiO$_3$.

4. Conclusions

In summary, we have prepared three-dimensional ordered macroporous structures, with pore diameters of 1000 nm, of not only binary oxides, but also of ternary oxides such as ferroelectric PbTiO$_3$. Since gels give good macroporous solids, the scope of this technique is immense. Thus, we are in the process of

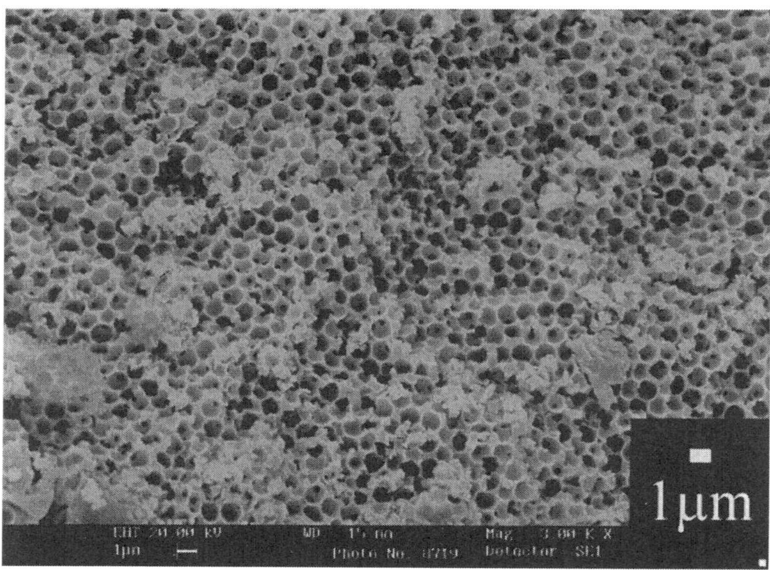

Fig. 6. SEM image of ordered macroporous $Pb(ZrTi)O_3$.

completing studies of macroporous aluminosilicates and other complex oxidic materials. It appears that washing with a solvent is a better way of removing the template from these structures. The macroporous structures of ferroelectric materials may indeed find useful applications.

References

[1] S.H. Park, Y. Xia, Adv. Mater. 10 (13) (1998) 1045.

[2] F. Caruso, Chem. Eur. J. 6 (3) (2000) 413.

[3] O.D. Velev, T.A. Jede, R.F. Lobo, A.M. Lenhoff, Nature 389 (1997) 447.

[4] A. Imhof, D.J. Pine, Nature 389 (1997) 948.

[5] B.T. Holland, C.F. Blanford, A. Stein, Science 281 (1998) 538.

[6] B.T. Holland, C.F. Blanford, T. Do, A. Stein, Chem. Mater. 11 (1999) 795.

[7] B. Gates, Y. Yin, Y. Xia, Chem. Mater. 11 (1999) 2827.

[8] Z. Zhong, Y. Yin, B. Gates, Y. Xia, Adv.Mater. 12 (2000) 206.

[9] J.E.G.J. Wijnhoven, W.L. Vos, Science 281 (1998) 802.

[10] S.D. Park, Y.H. Cho, W.W. Kim, S.J. Kim, J. Solid State Chem. 146 (1999) 230.

Pergamon

Materials Research Bulletin 36 (2001) 1751–1757

Materials Research Bulletin

Macroporous carbons prepared by templating silica spheres

Gautam Gundiah, A. Govindaraj, C.N.R. Rao*

Chemistry and Physics of Materials Unit, Jawaharlal Nehru Centre for Advanced Scientific Research, Jakkur P.O., Bangalore 560 064, India

(Refereed)
Received 16 February 2001; accepted 16 March 2001

Abstract

Macroporous carbons of different pore sizes, containing three-dimensionally connected voids, have been prepared by an elegant method. The method involves the coating of ordered silica spheres with sucrose, followed by carbonization using sulfuric acid, and the removal of silica with aqueous hydrofluoric acid. The carbon samples show the expected optical properties. The surface area of the macroporous carbon samples varies between 120 to 550 m^2g^{-1}, depending on whether nonporous or mesoporous silica spheres were used as templates. © 2001 Elsevier Science Ltd. All rights reserved.

Keywords: Macroporous carbon, photonic gap materials

1. Introduction

Macroporous materials with three-dimensionally connected voids are of great current interest because of possible applications such as photonic gap and dielectric materials and catalyst supports [1,2]. Recently, crystals of silicon inverse opal have been produced by growing silicon inside the voids of an opal template of close-packed silica spheres, followed by the removal of the silica template [3]. Porous carbons with similar features have been prepared by infiltering silica opal plates with a phenolic resin or by chemical vapor deposition using propylene, and then dissolving the silica [4]. We have prepared macroporous carbon with three-dimensionally interconnected voids by a novel method, involving the coating of close-packed monodisperse silica spheres with sucrose, converting sucrose

---

* Corresponding author.
E-mail address: cnrrao@jncasr.ac.in (C.N.R. Rao).

 G. Gundiah et al. / Materials Research Bulletin 36 (2001) 1751–1757

Fig. 1. SEM image of nonporous, ordered silica spheres of ~625 nm diameter.

into carbon by mild carbonization using sulfuric acid as the catalyst [5], and then dissolving the silica with aqueous hydrofluoric acid. In this article, we wish to draw attention to the excellent results obtained by this technique in preparing three-dimensionally ordered macroporous carbon networks. We have also prepared carbon samples with macropores by using mesoporous silica spheres instead of the nonporous silica spheres. We also present the adsorptive properties of the various macroporous carbon samples prepared by us.

2. Experimental

The procedure employed by us for the synthesis of macroporous carbon is as follows. Monodisperse silica spheres were synthesized following the method of Stöber et al. [6]. Spheres of different sizes could be prepared by the hydrolysis of tetraethylorthosilicate (TEOS) in an alcohol medium with aqueous ammonia (25% solution) as the base. Evaporation of the alcohol solvent from between the spheres packs them into arrays. In Fig. 1 we show a scanning electron microscope (SEM) image (obtained with a Leica Microscope) of ordered silica beads of diameter ~625 nm. The spheres have a narrow size distribution, and are reasonably ordered. To template the silica spheres with carbon, 0.245 g of the spheres was treated with a solution containing 0.2 g sucrose along with 4 ml H_2O and 0.5 ml H_2SO_4. The resulting mixture was kept in a drying oven at 60°C for 6 h, and the temperature gradually raised to 100°C to allow complete drying. The temperature was then increased to 150°C at a rate of 2°/min. Carbonization was carried out by heating the sample to 800°C under vacuum (10^{-6} Torr). The composite thus formed was treated with an aqueous solution of HF (48%) for 24 h to dissolve the templating silica spheres. Aqueous HF was found to be better than aqueous NaOH for the dissolution. The product was washed several times with water and dried at 60°C. Macroporous carbon samples were prepared by this procedure by employing nonporous silica spheres of diameters ~625 and ~200 nm.

We also prepared macroporous carbon by templating mesoporous silica spheres with sucrose. For this purpose, monodisperse silica spheres containing mesopores (diameter $\sim$450 nm) were prepared by a method similar to Schumacher et al. [7]. In a typical synthesis, 0.667 g of *N-cetyl-N,N,N*-tetramethylammonium bromide was dissolved in a solution containing 14 ml H_2O and 14 ml C_2H_5OH. To the above solution, 3.5 ml ammonia (25% solution) was added under stirring. After 10 min, 1 ml of TEOS was added dropwise and stirred for a period of 2 h. The mixture was aged at room temperature (25°C) for 2 h, filtered, washed with deionized water, and dried at ambient temperature. The dried product was calcined at 400°C for 2 h. Powder X-ray diffraction (XRD) patterns recorded using CuKα radiation (Rich-Siefert, XRD-3000-TT) established the mesoporous nature of the sample. Thermogravimetric analysis (Mettler-Toledo-TG-850) showed that $\sim$95% of the surfactant could be removed by calcination at 400°C. SEM images showed the particles to be spherical, but there was a distribution of sizes. These spheres were taken along with water and sonicated for 30 min to disperse the spheres. This was left undisturbed at room temperature until a major part of the water had evaporated. The rest of the water was then evaporated at 60°C. Due to the variation in the sizes of the spheres, it was difficult to order them into arrays. Using these spheres, macroporous carbon networks were prepared by a procedure similar to the one used for templating the nonporous silica spheres.

We also employed an array of silica spheres containing 25% mesoporous spheres ($\sim$450 nm diameter) and 75% nonporous spheres ($\sim$625 nm diameter) for preparing porous carbon samples. This was done by taking the spheres in water, followed by sonicating for 30 min. The spheres settled down under the influence of gravity. The ordering is considerably improved due to the presence of a larger percentage of uniform sized nonporous spheres. Employing these arrays, we repeated the synthesis of macroporous carbon.

The application of porous carbons as catalyst supports is widely recognized, and we, therefore, carried out nitrogen adsorption studies at liquid nitrogen temperature on these samples using a Cahn-2000 Electrobalance. Reflectance spectra of the macroporous carbons were recorded in the visible region for the samples prepared with nonporous silica spheres ($\sim$200 nm diameter).

3. Results and discussion

In Fig. 2a and b we show the SEM images of the macroporous carbon obtained by templating nonporous silica spheres of $\sim$625 nm diameter. Fig. 2a reveals long-range ordering extending to several tens of microns. There is considerable shrinkage on removal of the template, as seen from Fig. 2b, which shows pores of $\sim$400 nm diameter. Similarly, the pores of the carbon sample prepared from silica spheres of $\sim$200 nm diameter, have a diameter of $\sim$125 nm, as shown in Fig. 2c. Three-dimensional (3D) ordering is indicated by the circular windows connecting the spheroidal voids, formed when the spheres in the array are in contact with one another. Due to extensive 3D ordering and interconnected voids, these carbon materials are ideal for photonic bandgap applications.

In Fig. 3 we show the reflectance spectra of the silica spheres ($\sim$200 nm) coated with carbon as well as the macroporous carbon sample obtained after removal of the silica

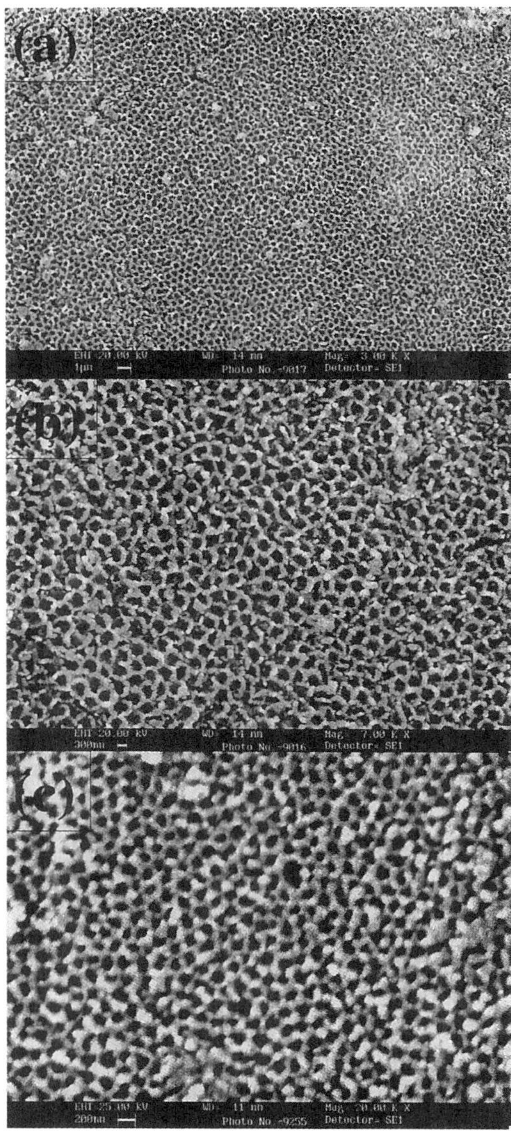

Fig. 2. (a,b) SEM images of macroporous carbon obtained by templating nonporous silica spheres of ~625 nm diameter. (c) SEM image of macroporous carbon obtained by templating nonporous silica spheres of ~200 nm diameter.

template. We see a clear maximum at 434 nm for the silica coated with carbon and a maximum at 465 nm for the macroporous carbon sample. There is a red shift in the wavelength of absorption for the carbon sample due to the increase in the effective refractive index after the silica spheres are removed. The color of the sample was bluish green.

The N_2 adsorption isotherm for the macroporous carbon with ~400 nm diameter pores (prepared by starting with ~625 nm nonporous silica spheres) is shown in Fig. 4a. It is a

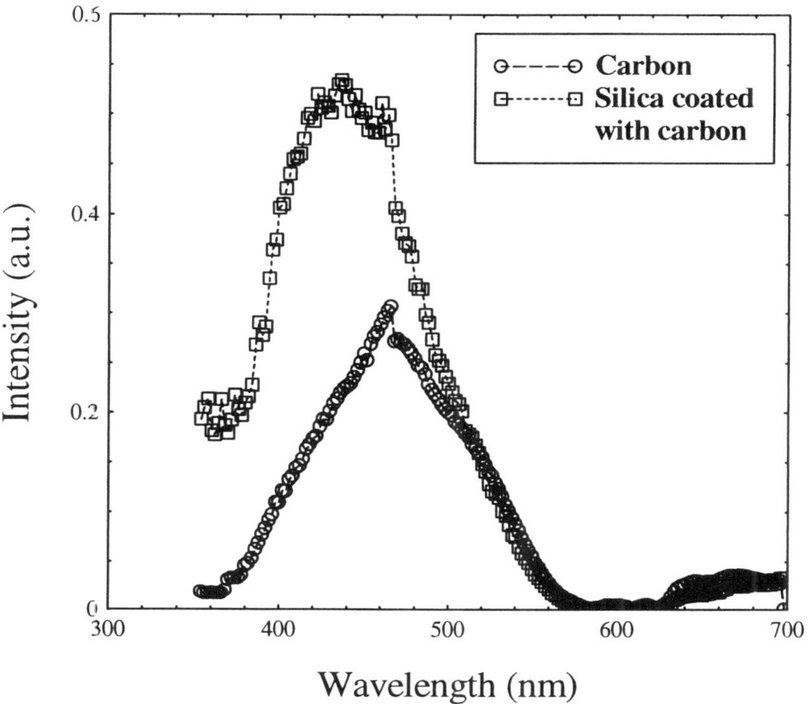

Fig. 3. Reflectance spectra of silica spheres (~200 nm diameter) coated with carbon and of the macroporous carbon obtained after removal of silica.

typical Type I isotherm, similar to that exhibited by microporous carbons [8]. The slope is high initially, indicating that the micropores get filled at low partial pressures. At higher partial pressures, there is saturation. The surface area calculated by the BET method is 119 m^2g^{-1}. The surface area of the macroporous carbon sample with ~125 nm diameter pores (prepared by starting with ~200 nm nonporous silica spheres) was 226 m^2g^{-1}. Although these values of the surface area are somewhat lower that those of other porous carbons (500–3000 m^2/g) [8], the presence of ordered macropores may provide certain advantage to macroporous carbons as potential catalyst supports.

The network prepared by templating mesoporous silica spheres (~450 nm diameter) did not show as good an ordering as the networks obtained with nonporous silica spheres (Fig. 2). This is mainly because the initial silica arrays themselves are not well ordered. The N_2 adsorption isotherm of the carbon network prepared by templating mesoporous silica spheres was of type II, and gave a surface area of 230 m^2g^{-1}. Because we had started with mesoporous silica spheres, it was expected that the macroporous carbon would have mesoporous walls. However, the mesopores seemed to have been blocked. To get a better porous structure, we reduced the sucrose content of the initial reaction mixture to 0.1 g. By this means we obtained a macroporous carbon sample exhibiting a N_2 adsorption isotherm of type I as shown in Fig. 4b. The surface area calculated by the BET method was 551 m^2g^{-1}. This value of the surface area is the largest that we have obtained amongst the macroporous carbon samples prepared by us.

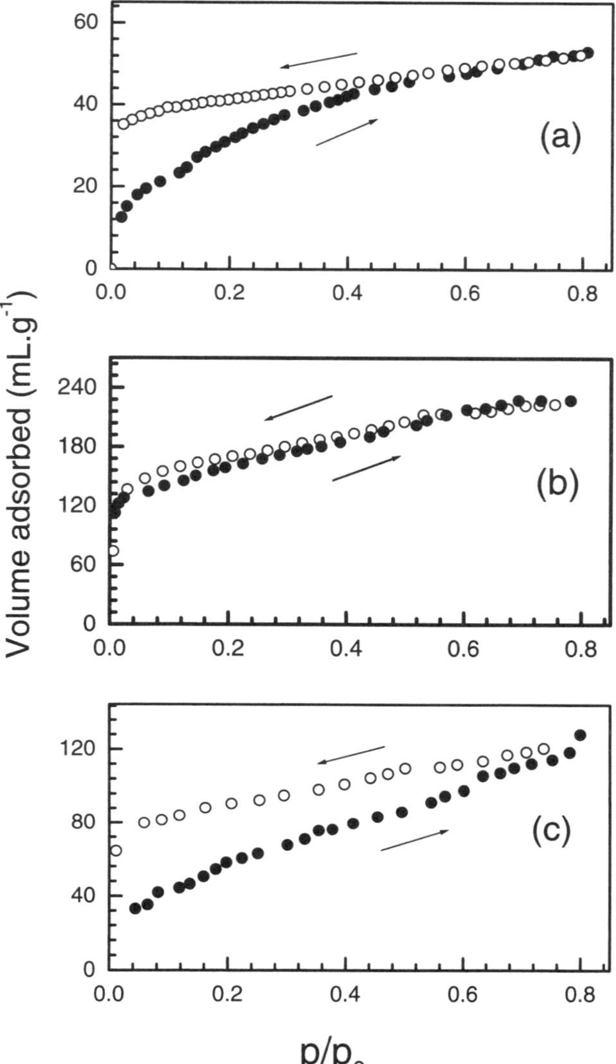

Fig. 4. Nitrogen adsorption isotherms for macroporous carbons: (a) prepared by templating nonporous silica spheres (~625 nm diameter); (b) prepared by templating mesoporous silica spheres (~450 nm diameter); and (c) prepared by templating a mixture of 25% mesoporous (~450 nm diameter) and 75% nonporous (~625 nm) silica spheres.

In Fig. 5, we show the SEM image of the macroporous carbon obtained by templating a mixture of 25% mesoporous (~450 nm diameter) and 75% nonporous (~625 nm diameter) silica spheres. Accordingly, we see smaller pores in the SEM image along with the large pores. There is reasonable ordering in this sample, although not as good as in the images shown in Fig. 2. The N_2 adsorption isotherm of this sample shown in Fig. 4c is of Type II, with a large hysterisis. The BET surface area was 207 m^2g^{-1}, which is in between the values obtained with the samples prepared using nonporous and mesoporous silica spheres.

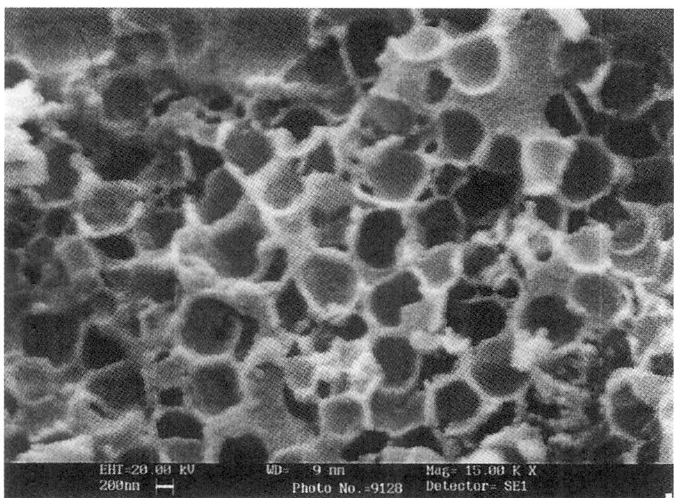

Fig. 5. SEM image of macroporous carbon obtained by templating a mixture of mesoporous (25%) and nonporous (75%) silica spheres.

4. Conclusions

We have demonstrated a method for fabricating three-dimensionally ordered macroporous carbon networks. The method can be used to template silica spheres of different sizes with good results, but the surface areas of these carbon samples is relatively small. We have templated mesoporous silica spheres by the same method to obtain different porous carbons with somewhat higher surface areas. Although the surface areas of all these macroporous carbons are in the $120-550\ \mathrm{m^2 g^{-1}}$ range, they could be useful as supports for metal catalysts. The presence of macropores may aid the diffusion of the reacting species to the active sites or of the product to leave the catalyst surface. The 3D ordered macroporous carbons may also be useful as photonic bandgap materials.

References

[1] S.H. Park, Y. Xia, Adv. Mater. 10 (1998) 1045.
[2] F. Caruso, Chem. Eur. J. 6 (2000) 413.
[3] A. Blanco, E. Chomski, S. Grabtchak, M. Ibisate, S. John, S.W. Leonard, F. Meseguer, J.P. Mondia, G.A. Ozin, O. Toader, H.M. van Driel, Nature 405 (2000) 437.
[4] A.A. Zakhidov, R.H. Baughman, Z.Iqbal, C. Cui, I. Khayrullin, S.O. Dantas, J. Marti, V.G. Ralchenko, Science 282 (1998) 897.
[5] R. Ryoo, S.H. Joo, S. Jun, J. Phys. Chem. B 103 (1999) 77430.
[6] W. Stöber, A. Fink, E. Bohn, J. Colloid Interface Sci. 26 (1968) 62.
[7] K. Schumacher, M. Grün, K.K. Unger, Microporous Mesoporous Mater. 27 (1999) 201.
[8] J.F. Bryne, H. Marsh, in: J.W. Partick, (Ed.), Porosity in Carbons, Edward Arnold, Great Britain, 1995.

VIII. Open Framework Materials

C.N.R. Rao

CSIR Centre of Excellence in Chemistry,
Chemistry & Physics of Materials Unit
Jawaharlal Nehru Centre for Advanced Scientific Research
Jakkur P.O., Bangalore-560 064, INDIA
cnrrao@jncasr.ac.in

Zeolitic aluminosilicates and aluminophosphates are the most well-known examples of inorganic open-framework solids. In the last few years, a variety of metal phosphates with open architectures have been synthesized and characterized, some of them exhibit interesting magnetic properties. These include one-, two- and three-dimensional structures. An important aspect of these materials concerns their mode of formation.[1] Are the complex 3D structures formed by making use of secondary structural building units? Is there a progressive building up of structures from 1D to 3D? An attempt to answer such questions is presented in the first two papers of this section. Besides a few representative papers on open-framework metal phosphates, the section contains papers on open-framework metal carboxylates. Organically templated open-framework carboxylates constitute an important new family.[2] Interestingly, even simple metal carboxylates crystallized under appropriate conditions exhibit novel channel structures.

It has been shown recently that oxyanions such as sulfate and selenite can also be used to design open architectures. Typical examples of such inorganic compounds are provided in this section. More importantly, novel inorganic-hybrid materials are also discussed in a few papers. The incorporation of alkali halide structures of different dimensionalities in metal oxalate hosts is noteworthy. Equally fascinating is the formation of sodalite structures by the four-membered ring of metal squarates which is probably the first example of such an assembly process.

References

1. C.N.R. Rao, in *Frontiers of Solid State Chemistry*, Eds. S.H. Feng and J.S. Chen, World Scientific, Singapore, 2002.
2. C.N.R. Rao, S. Natarajan and R. Vaidhyanathan, *Angew. Chem.* (2003).

Acc. Chem. Res. **2001**, 34, 80–87

Aufbau Principle of Complex Open-Framework Structures of Metal Phosphates with Different Dimensionalities

C. N. R. RAO,*,† SRINIVASAN NATARAJAN, AMITAVA CHOUDHURY,† S. NEERAJ, AND A. A. AYI

Chemistry and Physics of Materials Unit and CSIR Centre of Excellence in Chemistry, Jawaharlal Nehru Centre for Advanced Scientific Research, Jakkur P.O., Bangalore, 560 064, India

Received July 25, 2000

ABSTRACT

Open-framework metal phosphates occur as one-dimensional (1D) chains or ladders, two-dimensional (2D) layers, and complex three-dimensional (3D) structures. Zero-dimensional monomers have also been isolated recently. These materials are traditionally prepared by hydrothermal means, in the presence of organic amines, but the reactions of amine phosphates with metal ions provide a facile route for the synthesis, and also throw some light on the mode of formation of these fascinating architectures. Careful studies of the transformations of monophasic zinc phosphates of well-characterized structures show that the 1D structures transform to 2D and 3D structures, while the 2D structures transform to 3D structures. The zero-dimensional monomers transform to 1D, 2D, and 3D structures. There is reason to believe that the 0D monomers, comprising four-membered rings, are the most basic structural units of the open-framework phosphates and that after an optimal precursor state, such as the ladder structure, is formed, further building may occur spontaneously. Evidence for the occurrence of self-assembly in the formation of complex structures is provided by the presence of the structural features of the one-dimensional starting material in the final products. These observations constitute the beginning of our understanding of the building-up principle of such complex structures.

1. Introduction

While supramolecular chemistry of organic compounds has developed to maturity in the past few years,[1] supramolecular inorganic chemistry is somewhat at a nascent stage. Supramolecular design provides a means to generate a variety of novel inorganic materials with complex, unusual structural features in areas such as host–guest chemistry, open-framework structures, and the like. In these compounds, one can visualize structures of different dimensionalities, the dimensionality varying from zero to three. Such complex structures would be expected to have subunits, which not only are structural motifs but also act as building blocks in the building up of the complex structures. Such building units can be considered to be synthons of complex structures. The synthons would be simple geometrical figures such as squares, cubes, or polyhedra, their corners acting as linkage points. The challenge that one faces in supramolecular inorganic materials chemistry is to establish whether such synthons exist in reality and, if so, whether one can demonstrate how they are involved in the formation of the larger structures. Müller et al.[2] have made use of the concepts of supramolecular inorganic chemistry to provide a beautiful description of large assemblies based on the chemistry of polyoxometalates. Recently, Férey[3] has described the concept of building units to understand inorganic solid-state structures, to visualize new topologies, and to conceive the formation of new solids with novel designs. On the basis of such building units, Férey also defines the notion of scale chemistry, which is concerned with the edification of the solids with building units and the consequences it has on the structure of the framework and the voids in them. Of particular interest is the fact that not only are the structures of open-framework compounds topologically interesting, but also the cavities present in them have potential applications. When one looks at the complexity as well as the beauty of the myriad of structures of both open-framework and host–guest compounds, one cannot escape the feeling that the building-up process cannot occur by conventional chemical means alone, involving simple making and breaking of chemical bonds. The formation of such organized structures would be expected to involve self-assembly or some such spontaneous process at a certain stage. In this Account, we demonstrate how the formation of complex open-framework metal phosphates with open architectures involves a building-up process from simple building units, possibly leading to an ultimate step where spontaneous self-assembly occurs.

Several classes of inorganic open-framework structures have been synthesized and characterized in the past several years. While zeolitic aluminosilicates constitute the best-known class of open-framework structures,[4] metal phosphates have been gaining considerable importance, and a variety of metal phosphates with open architectures have been reported in the past decade.[5] The open-framework phosphates are generally synthesized under hydrothermal conditions in the presence of organic

C. N. R. Rao obtained his Ph.D. degree from Purdue University and a D.Sc. degree from the Mysore University. He is Linus Pauling Research Professor at the Centre and Honorary Professor at the Indian Institute of Science. He is a member of several academies, including the Royal Society, London, the U.S. National Academy of Sciences, and the Pontifical Academy of Sciences.

Srinivasan Natarajan obtained his M.Sc. degree from the Madurai Kamaraj University and his Ph.D. from the Indian Institute of Technology, Madras. He did postdoctoral work at the Royal Institution London and at the University of California, Santa Barbara, and is a Faculty Fellow at the Centre.

Amitava Choudhury obtained his M.Sc. degree from North Bengal University and is now a Ph.D. scholar at the Indian Institute of Science.

S. Neeraj obtained his M.S. degree at the Centre and has just completed his Ph.D. work.

A. A. Ayi has an M.Sc. degree from the University of Calabar, Nigeria, where he is a Lecturer. He is on a visiting fellowship of the Third World Academy of Sciences.

* To whom correspondence should be addressed. E-mail: cnrrao@jncasr.ac.in. Fax: 91-80-8462766.
† Also at the Solid State & Structural Chemistry Unit, Indian Institute of Science, Bangalore 560012, India.

10.1021/ar000135+ CCC: $20.00© 2001 American Chemical Society
Published on Web 11/15/2000

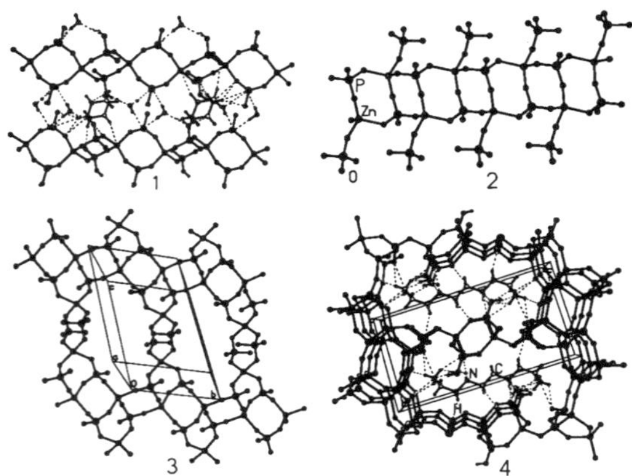

FIGURE 1. Examples of open-framework metal phosphates of different dimensionalities: **1**, one-dimensional (1D) linear chain phosphate, $[C_4N_2H_{10}][Zn(HPO_4)_2]\cdot H_2O$ with piperazine (PIP); **2**, one-dimensional (1D) ladder phosphate obtained with triethylenetetramine (TETA), $[C_6N_4H_{22}]_{0.5}[Zn(HPO_4)_2]$; **3**, two-dimensional (2D) layer phosphate $[C_6N_4H_{22}]_{0.5}[Zn_2(HPO_4)_3]$ with TETA; **4**, three-dimensional (3D) structure with 16-membered channels obtained with TETA. When counting the number of atoms forming a feature such as a ring or a channel, only metal and phosphorus atoms (T atoms) are taken into account.

amines. A noteworthy aspect of the metal phosphates is the occurrence of a hierarchy of structures with different dimensionalities. These include one-dimensional linear chains possessing corner-shared four-membered rings, one-dimensional ladders with edge-shared four-membered rings, two-dimensional layers, and three-dimensional structures with channels.[5,6] In Figure 1, we show typical examples of these structures (**1−4**) from the zinc phosphate family. Zero-dimensional metal phosphate monomers comprising four-membered rings have also been isolated recently.[7,8] Four-membered rings are generally the simplest units in the open-framework metal phosphates, and they seem to readily transform to six-membered, eight-membered, and higher membered rings.[9] Among the hierarchy of open-framework structures, the three-dimensional ones are most commonly observed, being associated with greater stability, and the lower dimensional architectures such as the linear chain and the ladder are somewhat rare. There has been considerable effort to understand the processes involved in the formation of open-framework structures.[10] Our knowledge of the mechanism(s) of formation of the beautiful architectures of metal phosphates with varying degrees of complexity, however, remains limited. It is difficult to readily obtain information related to the mechanism(s), partly because the materials are generally prepared under hydrothermal conditions. We know little about the nature of the species in solution or the exact role of the organic amine. The hydrothermal reaction vessel is the proverbial black box. The processes involved are kinetically controlled, and the energies associated with the different structures are likely to be comparable. Quite often, with a single organic amine, one obtains several open-framework phosphates with different structures. There

have been some suggestions with regard to the role of the amine in the formation of these structures.[11] The amine could act as a structure-directing agent or merely fill the available voids and stabilize the structure through hydrogen-bonding and other interactions. Since the amine generally gets protonated in the reaction, it also helps in charge compensation with respect to the framework. Recently, it has been suggested that phosphates of the organic amines may act as intermediates in the formation of the metal phosphates.[12]

In situ synchrotron X-ray diffraction as well as NMR studies of gallium phosphates under hydrothermal conditions have revealed the spontaneous nature of the transformation of preformed precursor units into the open-framework structure, and also the initial formation of a four-membered ring phosphate.[13] It is possible to conceive of basic building units involved in the formation of open-framework metal phosphates, since a variety of complex three-dimensional structures with varying cell sizes are often produced by the arrangement of similar building units.[3] What is of vital importance, however, is to understand the relationship between the structures of different dimensionalities and complexity. For example, it is of value to explore whether it is possible to transform one-dimensional chain or ladder structures to two-dimensional layer or three-dimensional structures under certain conditions. If such transformations do occur, one can rationalize the formation of structures of different dimensionality in terms of a "building-up" (Aufbau) principle. In the case of aluminum phosphates, it has been proposed that a one-dimensional chain, on hydrolysis and rotation of bonds, may transform to higher dimensional structures.[6] In Figure 2, we show a few such transformations schematically, but there has not been sufficient experimental evidence in support of such a hypothesis.

We have been trying to understand the formation of the hierarchy of open-framework structures in metal phosphates by several methodologies. The amine phosphate route[12] has provided some insight into the mode of formation of these materials but does not adequately reveal the nature of interconvertibility of the structures. In this context, our recent finding that a zero-dimensional monomeric zinc phosphate transforms to a layered structure on heating[7] was encouraging and prompted us to explore the transformations of zero-, one-, and two-dimensional structures to higher dimensional ones under different conditions. In this Account, we shall briefly present some of the results obtained by the study of the reactions of organic amine phosphates with metal ions, to show how they throw some light on the formation of open-framework metal phosphates. We shall then discuss the results of our investigations of the transformations of well-characterized zinc phosphates of different dimensionalities and demonstrate how these studies provide valuable insight into the building-up (Aufbau) principle of these complex architectures. In particular, we discuss the transformation of the one-dimensional (1D) ladder structures to two-dimensional (2D) layer and three-dimensional (3D) structures with channels, and the trans-

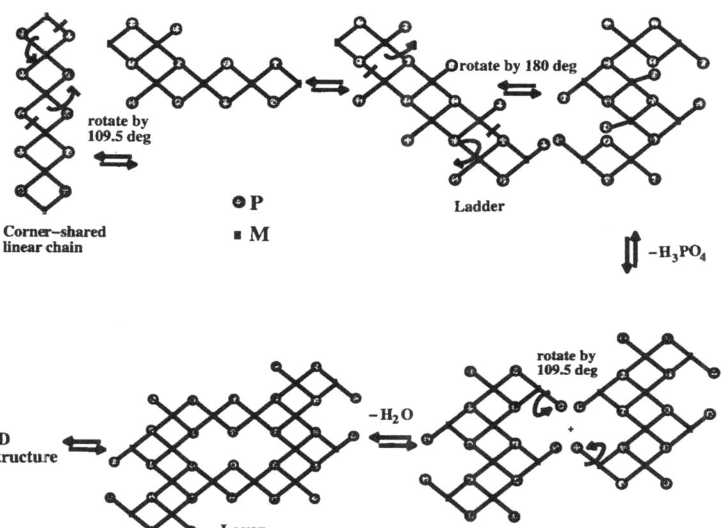

FIGURE 2. Schematic representation of possible types of transformations in open-framework phosphates.

formations of the 2D layer structures to 3D structures. We also examine the transformations of the zero-dimensional monomers to higher dimensional structures. We believe that these studies are of great significance not only to the area of open-framework structures, but also to our understanding of the nature of the processes involved in the building up of complex structures. Studies employing in situ methods and better facilities are necessary to make further progress. As pointed out by Müller et al.,[2] unraveling the Aufbau principle of complex inorganic systems and understanding such systems in terms of the topological classifications based on dimensionality constitute an important direction in the chemistry of materials.

2. Information Revealed by the Reactions of Organic Amine Phosphates

Phosphates of organic amines are often found as byproducts during the hydrothermal synthesis of open-framework metal phosphates. The role of the amine phosphates in the formation of metal phosphates was, however, not clear. Our recent studies of the reactions of well-characterized amine phosphates with metal ions have thrown some light in this regard.[12,14,15] Thus, amine phosphates react with Zn^{2+} ions to yield open-framework metal phosphates of different dimensionalities. What is noteworthy is that many of these reactions can be carried out under ambient conditions, thereby avoiding the hydrothermal route. For example, piperazine phosphate (PIPP) on reaction with Zn^{2+} ions gives the linear-chain phosphate $[C_4N_2H_{10}][Zn(HPO_4)_2]$ (1), formed by corner-shared four-membered rings (Figure 1) at temperatures as low as 85 °C. 1,3-Diaminopropane phosphate (DAPP) on reaction with Zn^{2+} ions gives a ladder phosphate, $[C_3N_2H_{12}][Zn(HPO_4)_2]$ (5), comprising edge-shared four-membered rings (Figure 3) and on prolonged reaction yields a layered structure, $[C_3N_2H_{12}][Zn_2(HPO_4)_3]$ (6). These

(a)

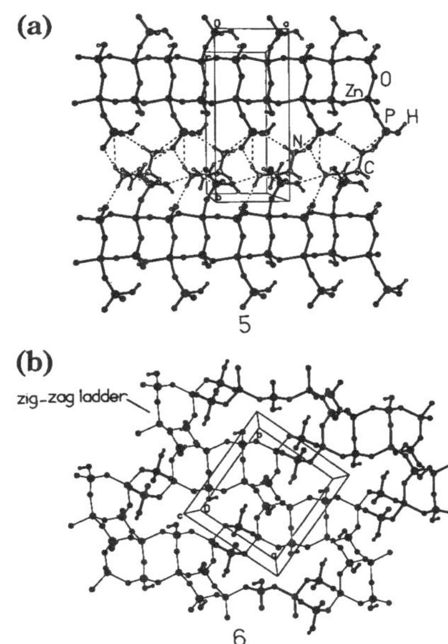

FIGURE 3. (a) One-dimensional ladder phosphate with 1,3-diaminopropane (DAP), $[C_3N_2H_{12}][Zn(HPO_4)_2]$ (5), and (b) the two-dimensional layer phosphate, $[C_3N_2H_{12}][Zn_2(HPO_4)_3]$ (6), obtained by the reaction of Zn^{2+} ions with 1,3-diaminopropane phosphate (DAPP). We can see the features of a zigzag ladder in **6**.

reactions could be carried out in the 30–50 °C temperature range. The layers in **6** are formed from a zigzag chain of four-membered rings, constructed from two Zn and P atoms ($Zn_2P_2O_4$ units), that are connected to each other via two PO_4 units, creating a bifurcation within the layer.

Reaction of DAPP with Zn^{2+} ions in aqueous solution at 30 °C for 24 h gives a product whose XRD pattern shows lines due to ladder structure 5 ($d_{011} = 9.76$ Å), while the XRD pattern of the product obtained from the reaction

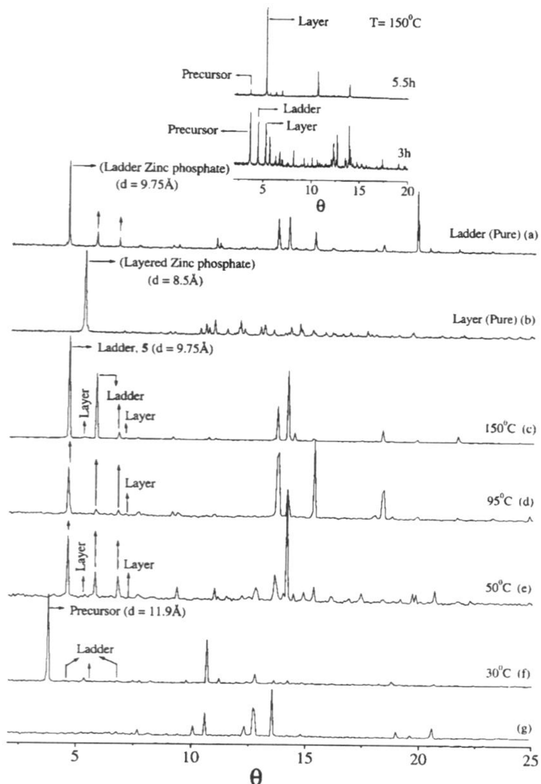

FIGURE 4. X-ray diffraction patterns of the products of the reaction of DAPP with Zn^{2+} ions. (a) XRD pattern of the monophasic zinc phosphate with a ladder structure, and (b) XRD pattern of the monophasic layered zinc phosphate. The diffraction patterns (c)−(f) are those of the products obtained from the reaction of DAPP with Zn^{2+} ions at different temperatures, as indicated (duration of reaction, ∼24 h). Notice the presence of reflections due to the ladder and the layer structures in the patterns and the time evolution of phases. The XRD pattern (f), in addition, shows a unique reflection due to an unidentified precursor. The XRD pattern (g) is that of the amine phosphate (DAPP). The inset at the top of the figure shows the formation of the ladder and the layer structures and their time evolution at 150 °C.

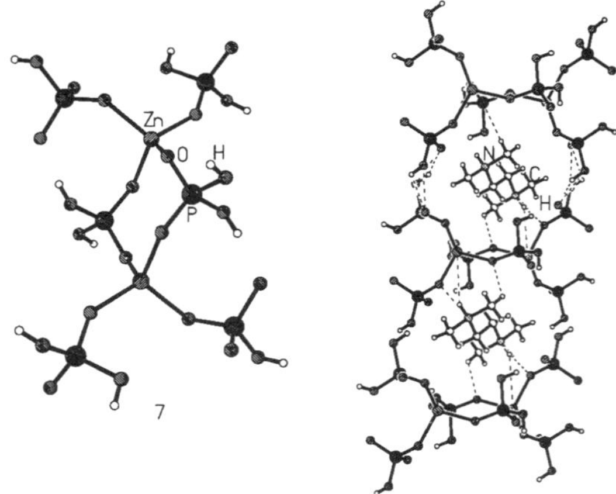

FIGURE 5. Monomeric zinc phosphate possessing a four-membered ring $[C_6N_2H_{18}][Zn(HPO_4)(H_2PO_4)_2]$ (**7**), formed by the reaction of *N,N,N′,N′*-tetramethylethyelenediamine (TMED) with Zn^{2+} ions in the presence of phosphoric acid.

at 50 °C (24 h) shows a reflection due to the ladder structure, **5**, as well as the layered structure, **6** (d_{002} = 8.5 Å), as can be seen from Figure 4. Furthermore, immediately after mixing DAPP with Zn^{2+} ions at room temperature, in addition to the weak line due to the ladder phase, one finds another reflection (d = 11.9 Å) due to an uncharacterized precursor phase (Figure 4). This precursor phase subsequently transforms to the ladder, and finally to the layered structure (see Figure 4), suggesting that the transformation to the layer structure probably occurs through the ladder phase. These results are also supported by in situ ^{31}P NMR studies carried out at 85 °C, which showed the disappearance of the amine phosphate signal followed by the immediate appearance of a signal due to the precursor phase, before the ladder phase is formed. We have followed the time evolution of

the products of the reaction of DAPP with Zn^{2+} ions by powder X-ray diffraction (XRD). The reaction at 150 °C initially gives a mixture of the ladder structure, **5**, and the layered structure, **6**. After a period of 3 h (see inset Figure 4), reflections due to the various phases can be seen. After 5.5 h, we mainly see reflections due to the layered phase, **6**, suggesting that the ladder, **5**, transforms to the layer, **6**.

We have been able to prepare a four-membered-ring zinc phosphate monomer, $[C_6N_2H_{18}][Zn(HPO_4)(H_2PO_4)_2]$ (**7**) by the reaction of *N,N,N′,N′*-tetramethylethylenediamine phosphate (TMEDP) with Zn^{2+} ions under ambient conditions (Figure 5). The structure consists of four-membered rings formed by ZnO_4 and $PO_2(OH)_2$ tetrahedra, with the $PO_3(OH)$ and $PO_2(OH)_2$ moieties hanging from the Zn center. Having discovered that the amine phosphates react with Zn^{2+} ions in a facile manner, we have employed the amine phosphate route as an effective method for the synthesis of a variety of open-framework metal phosphates.[14−17] Thus, starting with PIPP, we have isolated a large number of open-framework structures with various metal ions.[14]

3. Transformations of One-Dimensional Zinc Phosphates

As mentioned earlier, one-dimensional metal phosphates can possess either linear chain or ladder structures. The linear chain zinc phosphates do not have pendant phosphate groups, and we have found them to be quite unreactive, unlike the one-dimensional ladder structures with pendant phosphate groups (see Figures 1 and 3). We have carried out transformations of the ladder structures **2** and **5** under different conditions.[18] On heating the ladder structure, $[C_6N_4H_{22}]_{0.5}[Zn(HPO_4)_2]$ (**2**), in water (2:H₂O = 1:100) at 150 °C for 100 h, we obtained the 3D structure $[C_6N_4H_{22}]_{0.5}[Zn_3(PO_4)_2(HPO_4)]$ (**4**). The structure of **4** is built up of ZnO_4 and PO_4 tetrahedra sharing vertexes, forming a 3D structure (Figure 1), the connectivity between the ZnO_4 and PO_4 units giving rise to 16-membered 1D

(a)

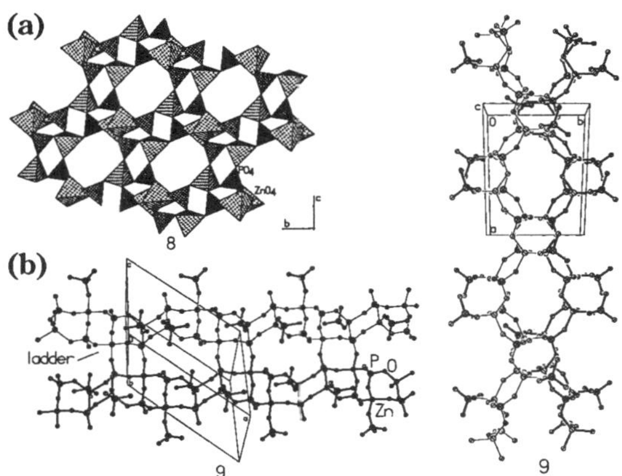

(b)

FIGURE 6. 3D zinc phosphates with eight-membered rings obtained from the transformation of the ladder compound **2**: (a) $[C_6N_4H_{22}]_{0.5}$-$[Zn_2(PO_4)_2]$ (**8**) and (b) $[C_2N_2H_{10}][Zn_2(PO_4)_2]$ (**9**). The features of the ladder structure can be clearly seen in the 3D structure of **9**. Such features are also present in **4** and **8**.

channels along the *bc* plane. However, by heating **2** in the presence of a base such as piperazine (PIP) or triethylenetetramine (TETA) at 150 °C in water (2:amine:H_2O = 1:5:100) for relatively short periods (24 h), we obtained another 3D structure, $[C_6N_4H_{22}]_{0.5}[Zn_2(PO_4)_2]$ (**8**) (Figure 6a). The structure of **8** is also built up of ZnO_4 and PO_4 tetrahedra, but the connectivity between these units creates eight-membered channels along all the crystallographic directions. The same reaction carried out at a slightly higher temperature (165 °C) gave a new 3D structure, $[C_2N_2H_{10}][Zn_2(PO_4)_2]$ (**9**) (Figure 6b). The connectivity between ZnO_4 and PO_4 tetrahedra in **9** results in four-membered rings, which are connected to each other via oxygens. The connectivity between the four-membered rings is such that they form an edge-shared four-ring ladder. The ladders are then connected together within and out of the plane, forming eight-membered channels along the *a* axis. Along the *c* axis, the connectivity between the tetrahedra gives rise to another eight-membered channel. While forming **9**, the amine (TETA) decomposed to ethylenediamine. The 3D structure **4** with 16-membered rings appears to be the more stable phase, since we could transform **8** into **4** by heating it in water at 150 °C under acidic conditions. It is noteworthy that all the three 3D structures, **4**, **8**, and **9**, obtained by the transformation of **2** contain ladder-like features of the starting material.

By heating the ladder structure **2** in the presence of PIP at 165 °C, we were able to obtain a two-dimensional layer compound, $[C_4N_2H_{10}][Zn_2(PO_4)_2]$ (**10**), shown in Figure 7a. The ladder phosphate, **5**, obtained with DAP (Figure 3a) could be transformed to the layered structure, **6** (Figure 3b), by heating it in water (5:H_2O = 1:100) at different temperatures (50–150 °C). We have followed this transformation by recording the X-ray powder diffraction patterns at different times. By heating **5** with zinc acetate in water at 85 °C (5:ZnAce H_2O = 1:5:100), we obtained the layered phosphate, $[C_3N_2H_{12}][Zn_4(PO_4)_2(HPO_4)_2]$ (**11**), shown in Figure 7b. Both of the layer structures, **10** and

(a)

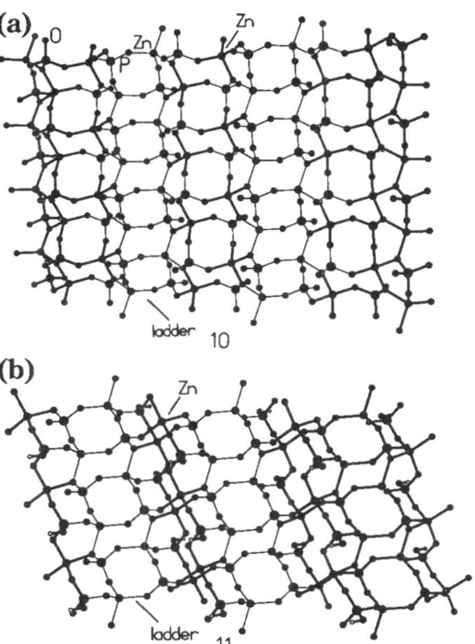

(b)

FIGURE 7. (a) 2D layer phosphate, $[C_4N_2H_{10}]$ $[Zn_2(PO_4)_2]$ (**10**), obtained from the transformation of the ladder compound **2**. (b) A 2D layer compound, $[C_3N_2H_{12}][Zn_4(PO_4)_2(HPO_4)_2]$ (**11**), obtained from the transformation of the ladder compound **5**. Both **10** and **11** exhibit features of the ladder structure, from which they are formed.

11, contain ladder-like features, as can be seen from Figure 7. The layers in **11** can be considered to be formed from the ladders in **2**, wherein the pendant groups connect the ladders, mediated by zinc ions, in an "*in-* and *out-of-plane*" fashion to result in chains with alternating three- and four-membered rings running in opposite directions. The connectivity between the chains and the parent ladder results in the formation of a tubular structure with six-membered rings capped by two or three three-membered rings on top and bottom and by two four-membered rings along the direction of the tubule. Thus, we have been able to transform the ladder structures **2** and **5** to 2D layers and 3D structures, all of which possess the structural features of the ladder, indicative of the likely occurrence of self-assembly in the formation of the complex structures.

We attempted to transform linear chain zinc phosphates of type **1** by carrying out reactions under different conditions but were generally unsuccessful. The only case of transformation of a linear chain phosphate was found with $[C_{10}N_4H_{26}][Zn(HPO_4)_2]$ (**12**), prepared with 1,4-bis(3-aminopropyl)piperazine, which on heating in the presence of phosphoric acid at 150 °C (12:H_3PO_4:H_2O = 1:5:100) yielded $[C_{10}N_4H_{26}]$ $[Zn_3(PO_4)_2(HPO_4)]$ (**13**) with tubular layers (Figure 8). The tubules themselves are made of "strips" (formed by fusion of two linear chains via a three-coordinated oxygen) in which the three-membered rings get capped by Zn atoms alternately on top and bottom. We have also been able to transform the ladder phosphate **2** to the linear chain phosphate **1** (Figure 1a) by heating it with PIP in water. It appears that the transformation of

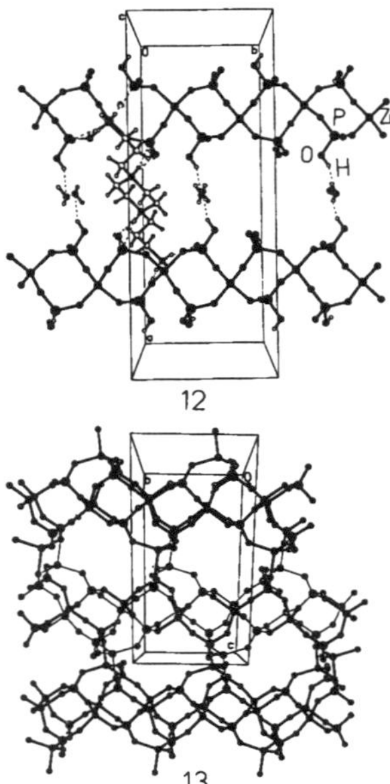

FIGURE 8. Linear chain phosphate, [C_{10}N_4H_{26}][Zn(HPO_4)_2] (**12**), which transforms to a tubular layered phosphate, [C_{10}N_4H_{26}] [Zn_3(PO_4)_2-(HPO_4)] (**13**). The features of the linear chain can be delineated in the tubular structure of **13**.

the ladder to other structures involves the deprotonation of the pendant HPO_4 groups when heated with an organic amine and/or water. In such a reaction, the ladder can go to a higher dimensional or to a one-dimensional linear chain structure where the pendant HPO_4 groups are absent.

4. Transformations of Two-Dimensional Zinc Phosphates

We have carried out the reactions of 2D layered zinc phosphates to see whether they transform to 3D structures. Thus, the layer structure [C_6N_4H_{22}]_{0.5}[Zn_2(HPO_4)_3] (**3**), on heating in water at 150 °C (**3**:H_2O = 1:200), gave the 3D structure **4** with 16-membered channels. It must be recalled that we could obtain this 3D structure from the ladder structure, **2**, as well. Heating the tubular layer phosphate obtained with TETA, [C_6N_4H_{22}]_{0.5}[Zn_3(PO_4)_2-(HPO_4)] (**14**), at 150 °C in water (**14**:H_2O = 1:100), produced the 3D structure, **8**.

5. Transformations of Zero-Dimensional Monomers

The zero-dimensional monomeric zinc phosphate, **7**, in Figure 5, transforms to the layered compound, [C_6N_2H_{18}]-[Zn_3(H_2O)_4(HPO_4)_4] (**15**), shown in Figure 9, on heating in water (**7**:H_2O = 1:100) at 50 °C. On heating **7** with piperazine in water at 60 °C (**7**:PIP:H_2O = 1:1:500), we obtained a 3D structure [C_4N_2H_{10}][Zn(H_2O)Zn(HPO_4)(PO_4)]_2, which has been described in the literature.[19] An in situ ^{31}P NMR study at 85 °C showed that the intensity of the signal due to the monomer decreases with time, accompanied by an

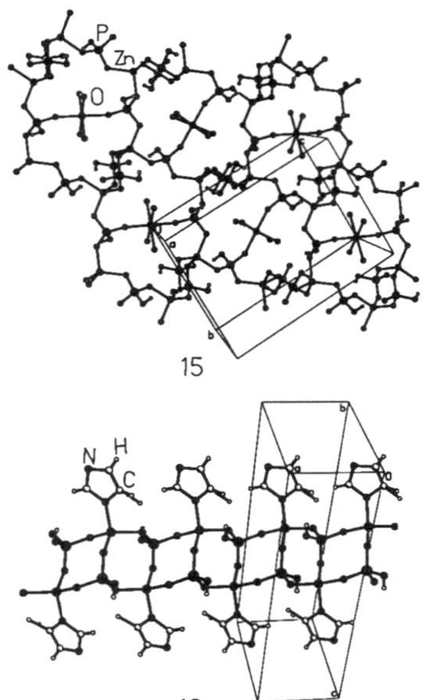

FIGURE 9. Layer phosphate, **15**, and ladder phosphate, **16**, obtained from the transformation of the zero-dimensional monomer, **7**.

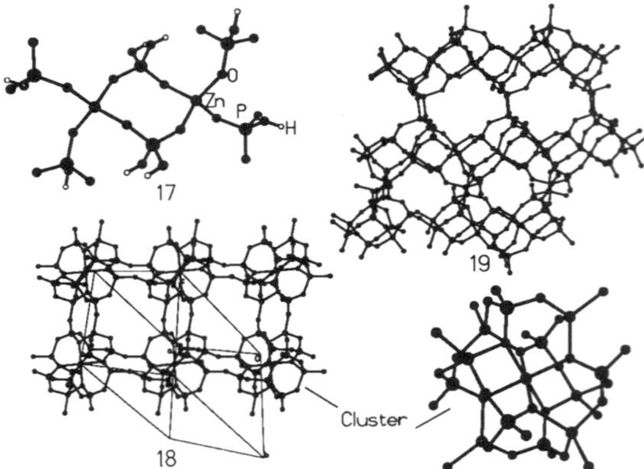

FIGURE 10. Monomeric zinc phosphate, **17**, which transforms to the 3D structure, **18**, and the tubular layer structure, **19**.

increase in the signal due to the ladder structure. The monomer structure disappears after ~4 h, and the intensity of the signal due to the ladder starts decreasing thereafter due to subsequent transformation. Heating the monomer **7** with imidazole in water (**7**:imidazole:H_2O = 1:1:500) at 60 °C gave a ladder compound of the formula [C_3N_2H_5][Zn(HPO_4)], **16**, where the pendant HPO_4 groups are replaced by imidazole molecules (Figure 9).

The monomeric zinc phosphate obtained with tris(2-aminoethyl)amine (TREN), [C_6N_4H_{21}][Zn(HPO_4)_2(H_2PO_4)] (**17**), on heating in water (**17**:H_2O = 1:100) at 180 °C, transformed to a 3D structure, [C_6N_4H_{21}]_{1.33}[Zn_7(PO_4)_6] (**18**), shown in Figure 10. The structure of **18** is made of ZnO_6, ZnO_4, and PO_4 polyhedra, which connect in such a way as to give rise to a Zn_7 cluster. The clusters are arranged

Scheme 1. Various Types of Transformations in Open-Framework Zinc Phosphates: (a) Transformations of One-Dimensional Ladder and 2D Layer Compounds and (b) Transformations of 0D Monomer[a]

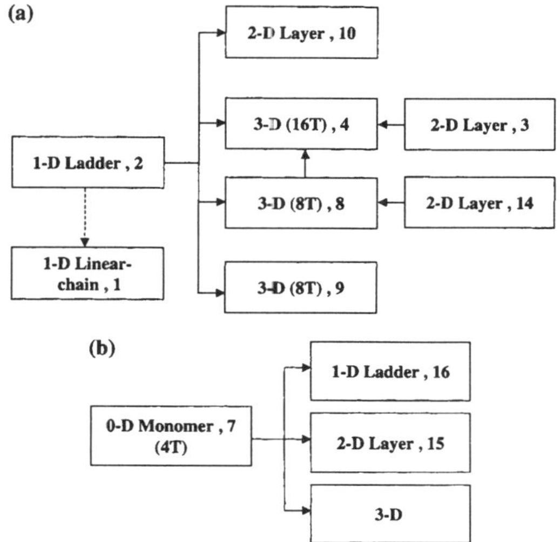

[a] T refers to tetrahedral framework atoms (Zn or P in the present case).

in such a manner that each cluster is displaced by half the length of c-axis from its neighbor, forming a honeycomb network. The next layer of clusters is identical to the first one but is displaced along the a-axis by half the unit cell, so that the honeycomb channels get capped. This type of three-dimensional AB-type stacking results in the formation of eight-membered channels along the b-axis between two such layers. On reaction of the monomer **17** with zinc acetate in water (**17**:ZnAce:H_2O = 1:1:100), a tubular layer structure of the composition $[C_6N_4H_{21}][NH_4][Zn_6(PO_4)_4(HPO_4)_2] \cdot H_2O$ (**19**) was obtained (Figure 10).

6. Conclusions and Outlook

The transformations of lower dimensional structures to higher dimensional ones in the open-framework zinc phosphates are truly fascinating. In Scheme 1, we summarize the important observations, to highlight the changes in dimensionality and the interconvertibility of structures. Among the zinc phosphates investigated, the one-dimensional ladder structure is the most reactive, rather than the linear chain phosphate. The exact nature of the reactive low-dimensional structure may vary from one system to another, as exemplified by the linear chain gallium fluorophosphate described recently.[20] This one-dimensional gallium fluorophosphate, formed at room temperature, transforms to a 3D structure under hydrothermal conditions. This is comparable to the transformations of the ladder structures **2** and **5**, described in Figures 2 and 6 as well as Scheme 1. The transformations of the monomeric zero-dimensional zinc phosphates comprising four-membered rings to one-, two-, and three-dimensional structures are especially noteworthy, since the four-membered ring is the common structural unit in open-framework metal phosphates. It is possible that the four-membered ring is related to the building units described

by Férey.[3] Thus, the hexameric unit of Férey has two four-membered rings.

Since the four-membered ring appears to be the first unit formed in the process of building of these complex open-framework structures,[13] it is possible that the four-membered rings initially form a one-dimensional chain or a ladder structure, which then transforms to 2D and 3D structures. In principle, we could consider the four-membered rings or/and the one-dimensional structures as synthons of the more complex structures. Self-assembly of ladder structures is suggested by the present study. The spontaneous assembly observed in in situ X-ray diffraction studies[13] could also correspond to the transformation of a 1D or 2D structure to a 3D structure by self-assembly. The formation of six-membered, eight-membered, and higher membered rings, commonly present in the open-framework phosphates, may follow that of the 0D/1D structures, the four-membered rings themselves transforming to the higher rings as suggested in the literature.[9] Although we are not yet in position to provide a definitive mechanism for the formation of complex 3D structures, it seems reasonable to state that the 3D and the 2D structures are likely to be formed through the 0D and 1D precursors. Clearly, there is a need for detailed in situ X-ray diffraction and NMR studies for fully characterizing the precursor states and the transformations. It would be of value to establish the occurrence of progressive 0D−1D−2D−3D transformations. It is therefore necessary to pursue the study of the transformations of well-characterized monophasic compounds of lower (zero, one, and two) dimensionalities.

In our studies of the open-framework zinc oxalates, we have recently isolated monomeric, dimeric, 1D linear chain, 2D layer, and 3D structures by the reaction of amine oxalates with Zn^{2+} ions,[21] suggesting thereby that the presence of a hierarchy of structures is not unique to the phosphates alone. We believe that the evidence provided by our studies for the existence of an Aufbau principle of open-framework complex structures is of considerable significance. Many other complex inorganic structures are also likely to be formed by similar building-up processes, involving basic building units and self-assembly.

References

(1) *Supramolecular Chemistry, Concepts and Perspectives*; Lehn, J.-M., Ed.; VCH: Weinheim, 1995.

(2) Müller, A.; Reuter, H.; Dillinger, S. Supramolecular inorganic chemistry: small guests in small and large hosts. *Angew. Chem., Int. Ed. Engl.* **1995**, *34*, 2328.

(3) Férey, G. Building units, design and scale chemistry. *J. Solid State Chem.* **2000**, *152*, 37.

(4) *Hydrothermal Chemistry of Zeolites*; Barrer, R. M., Ed.; Academic Press: London, 1982. *An Introduction to molecular sieves*; Dyer, A., Ed.; Wiley: Chichester, 1988.

(5) Cheetham, A. K.; Ferey, G.; Loiseau, T. Open-framework inorganic materials. *Angew. Chem., Int. Ed.* **1999**, *38*, 3268.

(6) Oliver, S.; Kuperman, A.; Ozin, G. A. A new model for aluminophosphate formation: Transformation of a linear chain aluminophosphate to chain, layer and framework structures. *Angew. Chem., Int. Ed.* **1998**, *37*, 46.

(7) Neeraj, S.; Natarajan, S.; Rao, C. N. R. Isolation of a zinc phosphate primary building unit [C$_6$N$_2$H$_{18}$]$^{2+}$[Zn(HPO$_4$)(H$_2$PO$_4$)$_2$]$^{2-}$ and its transformation to open-framework phosphate [C$_6$N$_2$H$_{18}$]$^{2+}$[Zn$_3$(H$_2$O)$_4$(HPO$_4$)$_4$]$^{2-}$. *J. Solid State Chem.* **2000**, *150*, 417.

(8) Ayyappan, S.; Cheetham, A. K.; Natarajan. S.; Rao, C. N. R. A novel monomeric Tin(II) Phosphate: [N(C $_2$H$_5$NH$_3$)$_3$]$^{3+}$[Sn(PO$_4$)(HPO$_4$)]$^{3-}$· 4H$_2$O, connected through hydrogen bonding. *J. Solid State Chem.* **1998**, *139*, 207.

(9) Chidambaram, D.; Neeraj, S.; Natarajan, S.; Rao, C. N. R. Open-framework zinc phosphates synthesized in the presence of structure-directing organic amines. *J. Solid State Chem.* **1999**, *147*, 154. Ayyappan, S.; Bu, X.; Cheetham, A. K.; Natarajan, S.; Rao, C. N. R. A simple ladder-tin phosphate and its layered relative. *Chem. Commun.* **1998**, 218.

(10) Thomas, J. M. New Microcrystalline Catalysts. *Philos. Trans. R. Soc. London A*. **1990**, *333*, 173. Davis, M. E. The quest for extra-large pore, crystalline molecular sieves *Chem. Eur. J.* **1997**, *3*, 1745. Francis, R. J.; O'Hare, D. The kinetics and mechanisms of the crystallization of microporous materials. *J. Chem. Soc., Dalton Trans.* **1998**, 3133.

(11) Davis, M. E.; Lobo, R. F. Zeolite and Molecular sieve synthesis. *Chem. Mater.* **1992**, *4*, 756. Ferey, G. The new microporous compounds and their design. *C. R. Acad. Ser. Paris Ser. II* **1998**, 1.

(12) Neeraj, S.; Natarajan, S.; Rao, C. N. R. Amine phosphates as intermediates in the formation of open-framework structures. *Angew. Chem., Int. Ed.* **1999**, *38*, 3480.

(13) Francis, R. J.; O'Brien, S.; Fogg, A. M.; Halasyamani, P. S.; O'Hare, D.; Loiseau, T.; Férey, G. Time-resolved in-situ energy and angular dispersive X-ray diffraction studies of the formation of the microporous gallophosphates ULM-5 under hydrothermal conditions. *J. Am. Chem. Soc.* **1999**, *121*, 1002. Taulelle, F.; Haoqs, M.; Gerardin, C.; Estournes, C.; Loiseau, T.; Férey G. NMR of microporous compounds: From in-situ reactions to solid paving. *Colloids Interfaces* **1999**, *158*, 229.

(14) Rao, C. N. R.; Natarajan, S.; Neeraj, S. Exploration of a simple universal route to the myriad of open-framework metal phosphates. *J. Am. Chem. Soc.* **2000**, *122*, 2810.

(15) Rao, C. N. R.; Natarajan, S.; Neeraj, S. Building open-framework metal phosphates from amine phosphates and a monomeric four-membered ring phosphate. *J. Solid State Chem.* **2000**, *152*, 302.

(16) Natarajan, S.; Neeraj, S.; Choudhury, A.; Rao, C. N. R. Three-dimensional open-framework cobalt(II) phosphates by novel routes. *Inorg. Chem.* **2000**, *39*, 1426.

(17) Choudhury, A.; Natarajan, S.; Rao, C. N. R. A layered aluminum phosphate by the amine phosphate by the amine phosphate route. *Int. J. Inorg. Mater.* **2000**, *2*, 87. Natarajan, S.; Neeraj, S.; Rao, C. N. R. Three-dimensional open-framework Co ^{II} and Zn^{II} phosphates synthesized via the amine phosphate route. *Solid State Sci.* **2000**, *2*, 87.

(18) The general procedure employed in these studies was to isolate a specific low-dimensional structure (monomer, ladder, or layer), solve its structure by single-crystal X-ray crystallography, and then study its transformation in an appropriate medium at a desired temperature. The medium could be just water or water with additives such as an amine, phosphoric acid, or zinc acetate. After the reaction, the product was analyzed by single-crystal X-ray crystallography.

(19) Feng, P.; Bu, X.; Stucky, G. D. Designed assemblies in open-framework materials synthesis: An interrupted sodalite and an expanded sodalite. *Angew. Chem., Int. Ed. Engl.* **1995**, *34*, 1745.

(20) Walton, R. I.; Millange, F.; Le Bail, A.; Loiseau, T.; Serre, C.; O'Hare, D.; Férey G. The room-temperature crystallization of a one-dimensional gallium fluorophosphate, a precursor to three-dimensional microporous gallium flurophosphates. *Chem. Commun.* **2000**, 203.

(21) Vaidhyanathan, R.; Natarajan, S.; Rao, C. N. R. Synthesis of a hierarchy of open-framework zinc oxalates from amine oxalates, communicated.

AR000135 +

COMMUNICATION

Understanding the building-up process of three dimensional open-framework metal phosphates: Acid degradation of the 3D structures to lower dimensional structures†

Amitava Choudhury[a] and C. N. R. Rao*[ab]

[a] Chemistry and Physics of Materials Unit, Jawaharlal Nehru Centre for Advanced Scientific Research, Jakkur P.O, Bangalore 560 064, India
[b] Solid State and Structural Chemistry Unit, Indian Institute of Science, Bangalore 560 012, India.
E-mail: cnrrao@jncasr.ac.in

Received (in Cambridge, UK) 11th October 2002, Accepted 13th December 2002
First published as an Advance Article on the web 6th January 2003

Acid degradation of 3D zinc phosphates primarily yields a one-dimensional ladder compound, an observation that is significant considering that the latter forms 3D structures on heating in water.

A large variety of open-framework inorganic materials with fascinating architectures has been synthesized and characterized in recent years. These include aluminosilicates,[1] phosphates,[2] carboxylates,[3] selenites[4] and sulfates.[5] One of the intriguing aspects of these materials is concerned with their mode of formation.[5] It is not clear as to how such complex three-dimensional (3D) structures are formed and whether the formation of such structures involves a building-up process starting from distinct secondary building units (SBUs). It is important to unravel these mechanistic issues to be able to rationally design these materials. The possible role of SBUs in the building-up process has been recognised by Férey.[7] In the case of open-framework metal phosphates, it has been shown that low-dimensional structures do indeed transform to 3D structures.[6,8] Thus, zero-,[8a] one-,[8b] and two-dimensional[8c] zinc phosphates transform to 3D structures under relatively simple reaction conditions, specially in the presence of amines. A linear chain gallium phosphate has been found to transform to a 3D structure as well.[9] A relevant question in this connection pertains to the reversibility of such transformations. Does a 3D open-framework structure undergo degradative transformations to lower dimensional structures? If so, can one isolate 1D or 2D structures on acid treatment of 3D metal phosphates? In order to answer this important question, we have carried out a systematic study of the acid-induced degradation of 3D zinc phosphates. The study has revealed that the 3D structures do indeed transform to the 1D ladder structure under acidic conditions, thereby demonstrating the reversibility of the 1D–3D transformation.

Acid-induced degradation studies were carried out‡ on the 3D zinc phosphate, $[C_6N_4H_{22}]_{0.5}[Zn_2(PO_4)_2]$,[10] I, analogous to gismondine,[1] built up of a double-crankshaft chain and possessing an 8-membered ring channel in all three crystallographic directions (Fig. 1a). On treatment with H_3PO_4 under hydrothermal conditions at 150 °C for 24 h, I transforms to another 3D structure of composition $[C_6N_4H_{22}]_{0.5}[Zn_3(PO_4)_2(HPO_4)]$,[10] II, even at relatively low acid concentrations (molar ratio of $I:H_3PO_4 = 1:0.25-0.5$). II, with a 16-membered channel along the a-axis (Fig. 1b), is a slightly lower density structure (2.62 g cm^{-3} compared to 2.63 g cm^{-3} of I). As the acid concentration is increased, however, a one-dimensional ladder structure, $[C_6N_4H_{22}]_{0.5}[Zn(HPO_4)_2]$,[10] III (Fig. 1c), and a two-dimensional layer structure, $[C_6N_4H_{22}]_{0.5}[Zn_2(HPO_4)_3]$,[10] IV (Fig. 1d), are obtained along with II, the proportion of IV increasing with the acid concentration and becoming 100% when the $I:H_3PO_4$ ratio is 1:1.25 or greater. The evolution of the

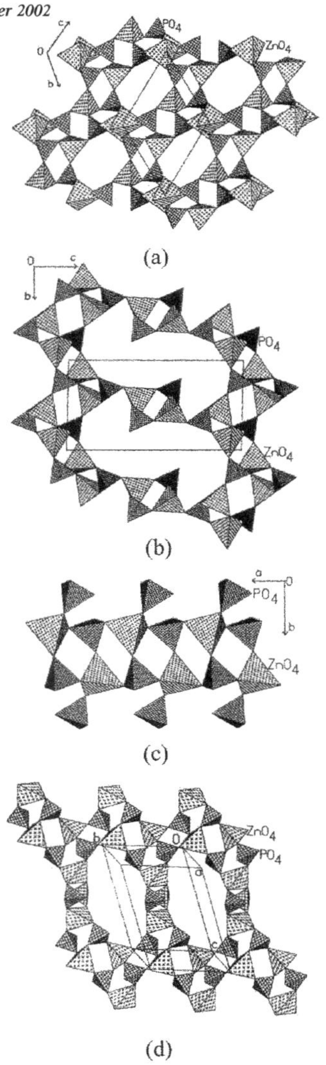

Fig. 1 (a) Polyhedral view of I along the a-axis showing the 8-membered channel. Amine molecules are omitted for clarity. (b) Polyhedral view of II along the a-axis showing the 16-membered channel. Amine molecules are omitted for clarity. (c) Structure of III showing the 4-membered ladder like chain propagating along the a-axis. (d) The layered structure of IV, with the protonated amine molecule sitting in the 12-membered aperture. (Compounds I, II, III and IV in this communication correspond to the compounds III, IV, I and II respectively in ref. 10).

† Electronic supplementary information (ESI) available: typical experimental parameters. See http://www.rsc.org/suppdata/cc/b2/b210037c/

DOI: 10.1039/b210037c

Table 1 Conditions for the degradation of compound **I** on treatment with 1M H_3PO_4

Compositions	Conditions				
	$T/°C$	t/h	pH	end pH	Product(s)
I:1M H_3PO_4:H_2O					
1:0.25:200	150	24	<2	<2	**II**
1:0.5:200	150	24	<2	<2	**II**
1:0.75:200	150	24	<2	<2	**II, III, IV**
1:1:200	150	24	<2	<2	**II, III, IV**
1:1.25:200	150	24	<2	<2	**IV**
1:1.5:200	150	24	<2	<2	**IV**
1:2–5:200	150	24	<2	<2	**IV**
1:>5:200	150	24	<2	<2	Dissolution

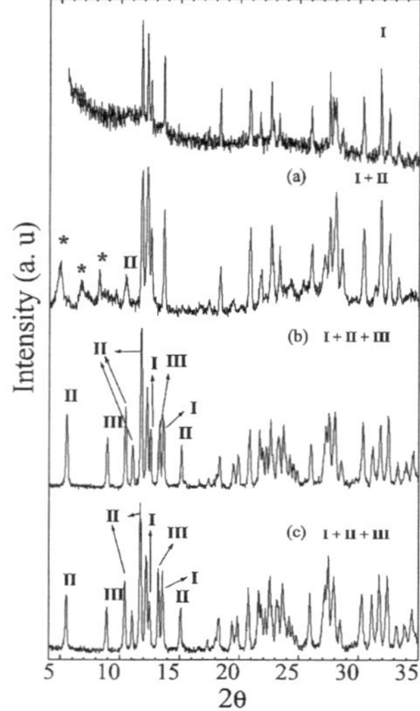

Fig. 2 Time dependent changes observed in the reaction of **I** with 1M H_3PO_4 under hydrothermal conditions (**I**:H_3PO_4 = 1:0.25). The powder XRD patterns show the evolution of the various phases as a function of reaction time. (a) 10 min, (b) 25 min and (c) 50 min. Asterisk denotes an unidentified phase.

different phases with the increasing acid concentration is summarised in Table 1. Using H_2SO_4 or HCl in place of H_3PO_4 also transforms **I** to **II** at low acid concentrations, and to **IV** at high acid concentrations.

In order to understand the evolution of different phases in the acid degradation of **I**, we carried out a time-dependent study, keeping the **I**:H_3PO_4 ratio fixed at 1:0.25. Some of the results are shown in Fig. 2 in the form of XRD data. Within 10 min of the reaction, **I** transforms to **II**, admixed with an unidentified phase with low angle lines at d = 17.5, 12.1 and 10.2 Å. After 25 min, however, the ladder phase, **III**, is formed along with **II**, but the unidentified phase is no longer present. A mixture of **II** and **III** along with the unreacted **I** is obtained after 30, 40 and 50 min of the reaction. A mixture of **II** and **III** (without any **I**) is found after 12 h. In order to examine whether **I** transforms to **II** or **III** under nonhydrothermal conditions, we reacted **I** with a H_3PO_4 solution in an open beaker. While no transformation occurred at 25 °C, **I** transformed to **II** or **III** at 100 °C, depending on the **I**:H_3PO_4 ratio.

Since the 3D zinc phosphate **II** is the initial product of the acid treatment of **I**, we have examined the transformation of **II** in acid media under hydrothermal conditions‡ **II** transforms to

III at low acid concentrations (**II**:H_3PO_4 = 1:0.25–0.75) and to the layered structure, **IV**, at higher acid concentrations, hopeite being the other product.

The present study clearly shows that the 3D zinc phosphate, **I**, transforms in a facile manner to the lower dimensional ladder (**III**) and/or layer (**IV**) structure through a lower density 3D structure, **II**.

The primary degradation of the 3D structures is the 1D ladder compound **III**, which undergoes a transformation to the layered compound **IV** on acid treatment at 150 °C.‡

It is known that the ladder compound **III** transforms to **II** on heating with water and to **I** on heating with water in the presence of an amine.[6,8b] The present finding that the 3D zinc phosphates, **I** and **II**, transform primarily to the 1D ladder compound, **III**, under acidic conditions demonstrates the reversibility of the 3D–1D transformation.

Notes and references

‡ *Synthesis*: The syntheses of **I–IV** (formed in the presence of triethylenetetramine, TETA) are reported in the literature.[10] In order to obtain a good yield of pure materials, we had to modify the synthetic conditions in some cases. The detailed synthetic conditions are listed in Table S1 and supplied as supplementary information.† All the four starting materials were characterized using powder XRD. The powder XRD patterns were in good agreement with their simulated patterns generated from known single crystal data.

In order to study the behavior of **I** under acidic conditions, **I** was hydrothermally treated in the presence of, 1M H_3PO_4, 1M H_2SO_4 or 1M HCl, with varying **I**:acid ratios. The acid concentration was progressively increased till the reactants dissolved fully to give a clear solution and no solid product resulted after hydrothermal treatment. Typically, 0.100 g of **I** was taken in a 7 ml PTFE-lined acid digestion bomb and treated under hydrothermal conditions at 150 °C for a period of 24 h. Similar studies were performed with **II** and **III**. In some cases, time dependent reactions were carried out in which hydrothermal treatment of the relevant composition of reactants was carried out for different periods, and the final products analyzed by powder XRD. The resulting products were generally single crystalline, although mixtures of polycrystalline and single crystalline phases were obtained in certain cases. Typical experimental parameters along with the products obtained in the acid-degradation reactions of **I** are listed in Tables 1 and S2 (electronic supplementary information).

1 Ch. Baerlocher, W. M. Meier and D. H. Olson, *Atlas of zeolite framework types*, Elsevier, Amsterdam, 2001.
2 A. K. Cheetham, G. Férey and T. Loiseau, *Angew Chem., Int. Ed.*, 1999, **38**, 3268.
3 (*a*) G. Férey, *Chem. Mater.*, 2001, **13**, 3084 and reference therein (*b*) M. Eddaouddi, D. B. Moler, H. Li, B. Chen, T. M. Reineke, M. O'Keeffe and O. M. Yaghi, *Acc. Chem. Res.*, 2001, **34**, 319 and references therein (*c*) R. Vaidhyanathan, S. Natarajan and C. N. R. Rao, *Inorg Chem.*, 2002, **41**, 1537 and references therein.
4 (*a*) W. T. A. Harrison, M. L. F. Phillips, J. Stanchfield and T. M. Nenoff, *Angew Chem., Int. Ed.*, 2000, **39**, 3808; (*b*) A. Choudhury, U. Kumar D and C. N. R. Rao, *Angew. Chem., Int. Ed.*, 2002, **41**, 158.
5 (*a*) A. Choudhury, J. Krishnamoorthy and C. N. R. Rao, *Chem. Commun.*, 2001, 2610; (*b*) Geo Paul, A. Choudhury and C. N. R. Rao, *Chem. Commun.*, 2002, 1904; (*c*) Geo Paul, A. Choudhury and C. N. R. Rao, *J. Chem. Soc., Dalton Trans.*, 2002, 3859.
6 C. N. R. Rao, S. Natarajan, A. Choudhury, S. Neeraj and A. A. Ayi, *Acc. Chem. Res.*, 2001, **34**, 80.
7 (*a*) G. Férey, *J. Fluorine Chem.*, 1995, **72**, 187; (*b*) G. Férey, *C. R. Acad. Sci. Ser. IIc*, 1998, **1**, 1.
8 (*a*) A. A. Ayi, A. Choudhury, S. Natarajan, S. Neeraj and C. N. R. Rao, *J. Mater. Chem.*, 2001, **11**, 1181; (*b*) A. Choudhury, S. Neeraj, S. Natarajan and C. N. R. Rao, *J. Mater. Chem.*, 2001, **11**, 1537; (*c*) A. Choudhury, S. Neeraj, S. Natarajan and C. N. R. Rao, *J. Mater. Chem.*, 2002, **12**, 1044.
9 (*a*) R. I. Walton, F. Millange, D. O'Hare, C. Paulet, T. Loiseau and G. Férey, *Chem. Mater.*, 2000, **12**, 1977; (*b*) R. I. Walton, F. Millange, T. Le Bail, T. Loiseau, C. Serre, D. O'Hare and G. Férey, *Chem. Commun.*, 2000, 203.
10 A. Choudhury, S. Natarajan and C. N. R. Rao, *Inorg. Chem.*, 2000, **39**, 4295. Compound **III** in this reference is the same as **I** in this communication, possessing the gismondine structure.

www.rsc.org/chemcomm

CHEMCOMM
Communication

A new route for the synthesis of open-framework metal phosphates using organophosphates

S. Neeraj,[a] P. M. Forster,[a] C. N. R. Rao*[ab] and A. K. Cheetham*[a]

[a] Materials Research Laboratory, University of California, Santa Barbara, CA 93106, USA.
  E-mail: cheetham@mrl.ucsb.edu
[b] Jawaharlal Nehru Center for Advanced Scientific Research, Jakkur Bangalore 560064, India

Received (in Columbia, MO, USA) 3rd August 2001, Accepted 9th October 2001
First published as an Advance Article on the web 6th December 2001

Use of tributylphosphate, an organophosphate, as the phosphorus source in place of phosphoric acid, has enabled the synthesis of several new open-framework zinc(II) and cobalt(II) phosphates, under solvothermal conditions.

Amongst the many inorganic open-framework structures known to-date, the metal phosphates constitute a large family.[1,2] The synthesis of these compounds is generally carried out under hydrothermal conditions by taking a metal salt and phosphoric acid in the presence of an organic amine, which may act as a template or structure-directing agent. Other strategies include the use of amine phosphates.[3] We have examined whether the use of an organic phosphorus source in place of H_3PO_4 would provide a new way to synthesize these fascinating compounds. This is of interest because the formation of open-framework structures is kinetically controlled and can be highly sensitive to the reaction conditions. In addition, the use of an organic source of phosphorus enables one to employ non-aqueous media for the synthesis. We have carried out several reactions of metal ions with tributylphosphate, both in alcohol and aqueous media, and obtained several open-framework cobalt, zinc and iron phosphates possessing different architectures. In this communication we report this new route for the synthesis of open-framework phosphates.

In a typical synthesis, zinc(II) chloride was dissolved in a butan-2-ol–water mixture. Tributylphosphate was added to the solution followed by an organic amine under constant stirring. The homogenized gel was sealed in a Parr autoclave and heated at 180 °C for 60 h. With different amines (N-(2-aminoethyl)-1,3-propanediamine in I, 4-(aminomethyl)piperidine in II and piperazine in III) the reactions yielded three-dimensional phosphates of the following compositions: $[C_5N_3H_{18}]$-$[Zn_3(HPO_4)_3(PO_4)]$ I, $[C_4N_2H_{14}]_4[Zn_5(PO_4)_4]_4 \cdot 5H_2O$, II and $[C_4N_2H_{12}][Zn_2(PO_4)(H_2PO_4)_2]$ III. While I and II, which possess channels are new, III has been reported in the literature.[3]

The structure of I is based on a three-dimensional network involving ZnO_4, PO_4 and HPO_4 tetrahedra with all the zinc atoms being linked to P atoms via oxygen. There are no Zn–O–Zn linkages present in the structure, hence no three-ring features are observed. The structure of I is built up of a stack of parallel layers linked by the corner-shared (CS) 4-ring chains embedded in the interlayer space (Fig. 1a). The layer in turn is made up of zigzag ladders consisting of 8-ring apertures as shown in Fig. 1b. The oxygens of the phosphate of the CS chains link on either side to the zinc tetrahedra of the layers. Such linkages between layers and CS chains result in 16-ring apertures along the c-axis, into which the terminal hydroxy groups project.

The framework of II is built up from layers and 4-ring ladders. The layers are made up of three types of chains (A, B and C) as can be visualized in Fig. 1c. The phosphate tetrahedra on the chains of type A and B add on to another set of ZnO_4 tetrahedra to form the 'tubule' like feature, whereas the phosphate tetrahedra common to these chains and the oxygens of the newly added ZnO_4 tetrahedra connect to the oxygens of the 4-ring ladder to form the three-dimensional architecture

with 12- and 6-ring channels along the a-axis (Fig. 1d). The protonated amine molecule (4-(aminomethyl)piperidine decomposes to give 1,4-diaminobutane) sits in the cavities and interacts with the framework via hydrogen bonding.

For the synthesis of cobalt(II) phosphates, a known amount of $CoCl_2 \cdot 6H_2O$ was dissolved in butan-2-ol and tributyl phosphate added to the solution under stirring. The amine (piperazine in IV, V and diethylentriamine in VI) was added after a few minutes and the mixture stirred until it became homogeneous. The gels were sealed in Parr autoclaves at 180 °C for 60 h. By

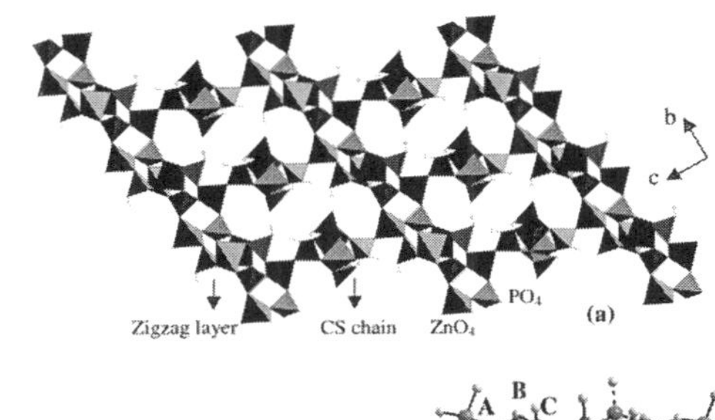

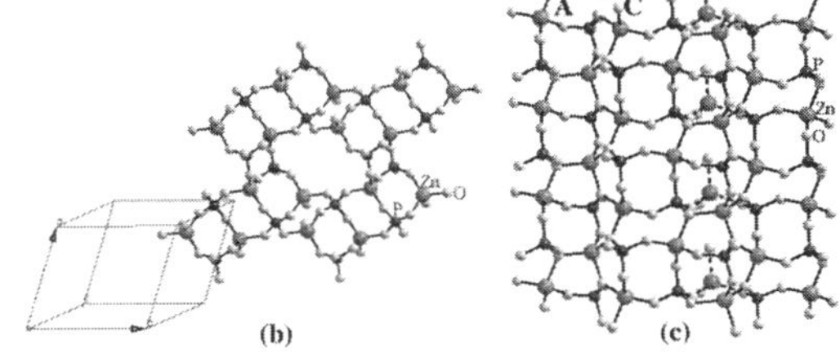

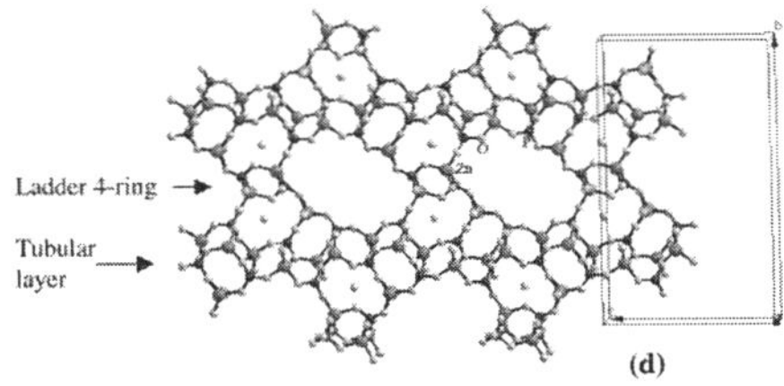

Fig. 1 (a) Polyhedral view of I, showing the 16-membered channels along the a-axis. Also marked are the CS chain and the layer forming the 16-ring channels and the pendant hydroxy groups protruding into the cavity. (b) A section of I showing the zigzag layer containing 8-ring apertures. (c) A section of II showing the layer formed of chains A, B and C. Also shown by dotted lines is the ZnO_4 tetrahedron connecting to the oxygen atoms of the phosphate groups of chains (type A and B) to form the tubule. (d) Ball and stick view of 6- and 12-ring cavities of II along the a-axis.

DOI: 10.1039/b107898b

486

such reactions, we obtained three new open-framework cobalt phosphates $[C_4N_2H_{11}][Co_2(PO_4)(H_2PO_4)_2]$ **IV**, $[C_4N_2H_{12}][Co_2(PO_4)_2]$ **V** and $[C_6N_4H_{22}][Co_7(PO_4)_6]$ **VI**. Of these, **IV** and **VI** possess three-dimensional frameworks and **V** has a layered architecture. Although these cobalt phosphate are new, their zinc phosphate analogues are known in the literature.[3–5] We have also synthesized a known mixed valence iron phosphate[6] by this route.

Framework **IV** is isostructural with the zinc phosphate **III** and has a three-dimensional structure built up of infinite CS chains containing 4-rings running perpendicular to each other and connected at various junctions.[3] Four such junctions form a 16-membered clover-shaped aperture, with the pendant hydroxyl groups from the phosphorus projecting in the cavities. The protonated piperazinium cations occupy the 4-ring windows (Fig. 2a). Framework **V** has a layered topology made up of 3- and 4-membered rings formed by vertex linkage between

CoO$_4$ and PO$_4$ tetrahedral units. The anionic layer is built up of two kinds of chains, **D** and **E**, made from 3- and 4-rings (Fig. 2b). The presence of a 3-ring chain results in a step-like feature in the layer due to the strain involved in accommodating it and also gives rise to an infinite one-dimensional Co–O–Co chain.

Compound **VI** possesses a three-dimensional architecture and is made of CoO$_6$, CoO$_4$ and PO$_4$ polyhedra, which connect to give rise to a Co$_7$O$_6$ cluster. The cobalt octahedron is surrounded by six cobalt tetrahedra to form the Co$_7$ cluster. Such clusters are linked *via* six PO$_4$ tetrahedra to other clusters to form the structure. The clusters are arranged in such a manner that each cluster is displaced by half the length of the *c*-axis from its neighbour, forming a honeycomb layer. The next layer of clusters is identical to the first but is displaced along the *a*-axis by half the unit cell so that the honeycomb channels are capped. This type of three-dimensional AB type stacking results in the formation of 8-membered channels at an angle to the *c*-axis (Fig. 2c).

The various structures obtained by using tributylphosphate as the phosphorylating agent confirm that we have found a new route for the formation of open-framework phosphates. The use of the phosphate ester enables us to carry out the reactions under non-aqueous conditions and to isolate new framework structures. It may be noted that pure open-framework Co(II) phosphates have been considered difficult to prepare. The use of the phosphate ester influences the release of phosphate ions in the solution, thereby affecting the course of the reaction and the products formed. The method can be readily extended to phosphonates, carboxylates and other systems.

We thank Unilever plc for their generous support of this work.

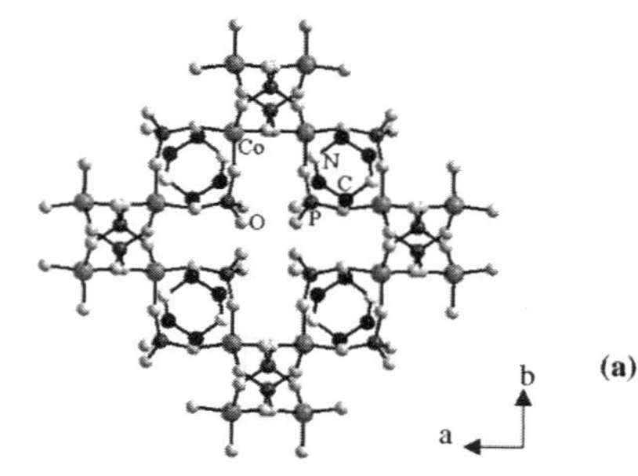

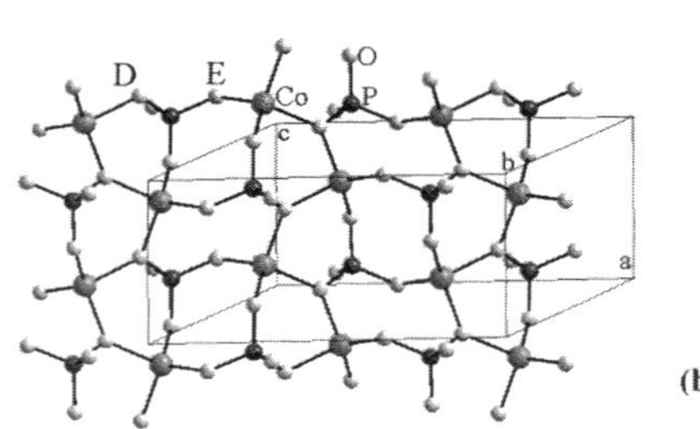

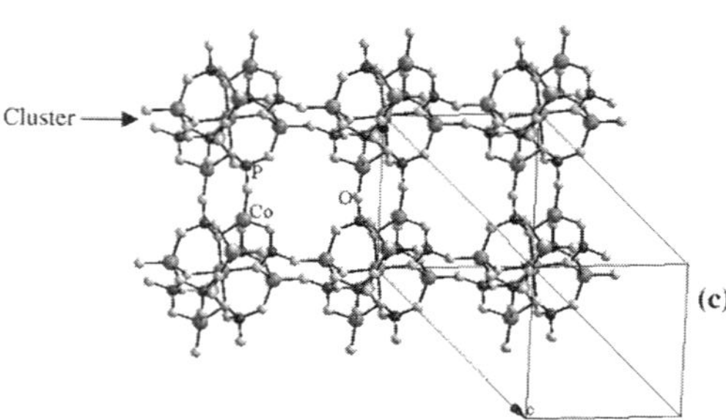

Fig. 2 (a) Structure of **IV** showing clover-shaped 16-ring channels. Also seen are the piperazinium molecules in the 4-ring windows. (b) Structure of **V** showing a single layer. Note the presence of two types of chains **D** and **E** and the infinite Co–O–Co chains. (d) Structure of **VI** showing the 8-membered channels at an angle to the *c*-axis formed by the linkage of Co$_7$O$_6$ clusters.

Notes and references

1 A. K. Cheetham, T. Loiseau and G. Férey, *Angew. Chem., Int. Ed. Engl.*, 1999, **38**, 3268.
2 W. H. Meier and D. H. Olson, *Atlas of Zeolite Structure Types*, Butterworth-Heinemann, London, 1992.
3 C. N. R. Rao, S. Natarajan and S. Neeraj, *J. Am. Chem. Soc.*, 2000, **122**, 2810.
4 A. Choudhury, S. Natarajan, S. Neeraj and C. N. R. Rao, *J. Mater. Chem.*, 2001, **11**, 1537.
5 A. A. Ayi, A. Choudhury, S. Natarajan, S. Neeraj and C. N. R. Rao, *J. Mater. Chem.*, 2001, **11**, 1181.
6 V. Zima, K. H Lii, N. Nguyen and A. Ducouret, *Chem. Mater.*, 1998, **10**, 1914.
7 Single crystal X-ray data was collected using a Siemens SMART-CCD diffractometer equipped with a normal focus, 2.4 kW sealed tube X-ray source (Mo-Kα radiation, λ = 0.71073 Å) operating at 45 kV and 40 mA. A hemisphere of intensity data was collected at room temperature in 1321 frames with ω scans (width 0.30° and exposure time of 10–20 s per frame). The structure was solved by direct methods using SHELX-97 and difference Fourier syntheses. Full matrix least-squares refinement against $|F^2|$ was carried out using the SHELXTL-PLUS package of programs. *Unit cell parameters*: for **I**: space group $P\bar{1}$ (no. 2), a = 8.4408(13), b = 9.2630(16), c = 13.8427(20) Å, α = 81.119(15), β = 83.643(12), γ = 72.647(11)°, R = 0.037. For **II**: space group $P2_1/n$ (no. 14), a = 5.1601(6), b = 25.0748(31), c = 14.9860(19) Å, β = 92.610(2)°, R = 0.046. For **IV**: space group $C2/c$ (no. 15), a = 13.4441(49), b = 12.8745(41), c = 8.2243(23) Å, β = 94.640(22)°, R = 0.047. For **V**: space group $P\bar{1}$ (no. 2), a = 5.1534(7), b = 10.7578(13), c = 10.8329(14) Å, α = 66.381(2), β = 89.058(2), γ = 81.674(2)°, R = 0.054. For **VI**: space group $R\bar{3}$ (no. 148), a = 13.4892(18), b = 13.4892(18), c = 14.9839(30) Å, γ = 120°, R = 0.066.
CCDC reference numbers 168260–168264 for **I**, **II**, **IV**, **V** and **VI** respectively.
See http://www.rsc.org/suppdata/cc/b1/b107898b/ for crystallographic data in CIF or other electronic format.

An open-framework iron phosphate with large voids, exhibiting spin-crossover

ChemComm

Amitava Choudhury,[ab] Srinivasan Natarajan[a] and C. N. R. Rao*[ab]

[a] Chemistry and Physics of Materials Unit, Jawaharlal Nehru Centre for Advanced Scientific Research, Jakkur P.O., Bangalore 560 064, India. E-mail: cnrrao@jncasr.ac.in
[b] Solid State and Structural Chemistry Unit, Indian Institute of Science, Bangalore 560 012, India

Received (in Cambridge, UK) 7th May 1999, Accepted 8th June 1999

A novel iron phosphate, $[(C_4N_3H_{16})(C_4N_3H_{15})]^{5+}$-$[Fe_5F_4(H_2PO_4)(HPO_4)_3(PO_4)_3]^{5-}\cdot H_2O$, consisting of a Fe–O/F–Fe network crosslinked by PO_4 groups is shown to possess unusually large elliptical voids of 24 T atoms (T = Fe, P) and to exhibit a gradual low- to high-spin transformation.

Amongst the variety of open-framework metal phosphates investigated in recent years, those of transition metals are of particular interest not only because of the novelty in their structures, but also because of their properties of potential value. One of the important objectives in synthesizing open-framework transition metal phosphates is the possibility of obtaining new solids with novel magnetic properties. Several open-framework iron phosphates have been synthesized and characterized in recent years,[1–3] but most of them order antiferromagnetically with the exception of one, which shows ferrimagnetic behavior at low temperatures.[4] Open-framework cobalt phosphates also order antiferromagnetically.[5] We have synthesized a new iron phosphate, exhibiting spin-crossover, which is rather unusual in an oxidic material. The material also has several noteworthy structural features which include the presence of infinite Fe–O/F–Fe chains, large voids formed by 24 T atoms (T = Fe, P), crosslinking by PO_4 tetrahedra.

The new iron phosphate $[(C_4N_3H_{16})(C_4N_3H_{15})]^{5+}[Fe_5F_4(H_2PO_4)(HPO_4)_3(PO_4)_3]^{5-}\cdot H_2O$ **I**, was prepared by employing hydrothermal methods in the presence of diethylenetriamine (DETA).† The light green colored crystals of **I** belonged to the monoclinic space group $P2_1/n$ and since the asymmetric unit of **I** contains 59 non-hydrogen atoms and it is convenient to describe the structure in terms of small building blocks. The three-dimensional structure of **I** can be considered as made from layers along the [001] direction, consisting of a network of FeO_6, FeO_5F and FeO_4F_2 octahedra and PO_4 tetrahedra (Fig. 1). The framework has the formula $[Fe_5F_4(H_2PO_4)(HPO_4)_3(PO_4)_3]^{5-}$. Charge neutrality is achieved by the presence of the organic structure-directing amine (DETA) in its proto-

nated form. There are two molecules of DETA in the unit cell, one triply protonated and the other doubly protonated. The layers are connected to each other *via* phosphate groups, completing the three-dimensional architecture (Fig. 2). This connectivity creates large elliptical voids bound by 24 T atoms (T = Fe, P) forming one-dimensional channels along the [010] direction, within which the DETA and water molecules reside. The width of the channels is 15.3×4.5 Å (longest and shortest atom—atom contact distances, not including van der Waals radii). To our knowledge, this is the first open-framework iron phosphate material with such large voids. The layers themselves contain pores bound by 8 T atoms (T = Fe, P) and the H_2PO_4 groups protrude into this opening rendering the pores inaccessible. Along the [100] direction, the structure has another narrow channel bound by 16 T atoms and the protruding H_2PO_4 and HPO_4 moieties occupy this channel.

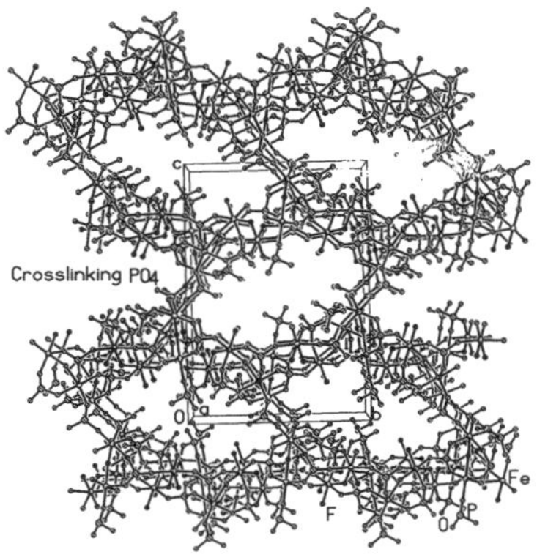

Fig. 2 Structures of **I**, showing large elliptical tunnels along the [010] direction. Note that the walls of the tunnels are composed of PO_4 tetrahedra. Amine and water molecules are not shown for clarity.

The most striking aspect of the structure of **I** is the cationic sub-network of Fe atoms. The connectivity between the various iron–oxygen(fluorine) octahedra is such that infinite Fe–O/F–Fe polymeric chains are formed as shown in Fig. 3(a). This network results from the presence of three-coordinate oxygen atoms and fluorine bridges in the structure. The Fe–O/F–Fe polymeric chains and the phosphate tetrahedra are so connected as to give rise to channels which appear as though they are decorated by PO_4 tetrahedra. This material may, thus, be considered to be an infinite iron–oxy fluoride network, $Fe_2O_{3-x}F_x(x \leqslant 0.4)$, crosslinked by the phosphate groups. Pillaring of layers in framework solids is not common, the only such material being an indium phosphate,[6] where the pillaring is by InO_6 octahedra. The crosslinking by PO_4 tetrahedra found in **I** is analogous to the pillaring in the indium phosphate.[6]

Besides its novel structural features, the iron phosphate **I** also exhibits unusual magnetic properties. Magnetic susceptibility measurements show that the magnetic moment increases gradually from 2.0 μ_B around 20 K to *ca.* 6.0 μ_B around 250 K [Fig. 3(b)]. The low-temperature moment corresponds to that of low-spin Fe^{III} ($^2T_{2g}$) and the high-temperature moment to that of

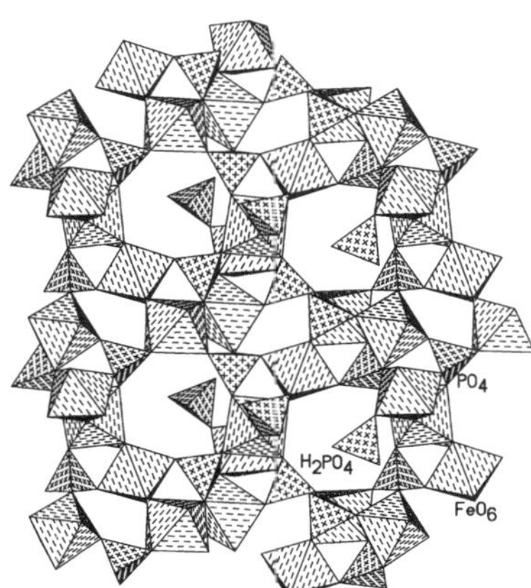

Fig. 1 Layer type arrangement in **I** along the [001] direction. Note that the H_2PO_4 moieties project into the eight-membered pores within the layers.

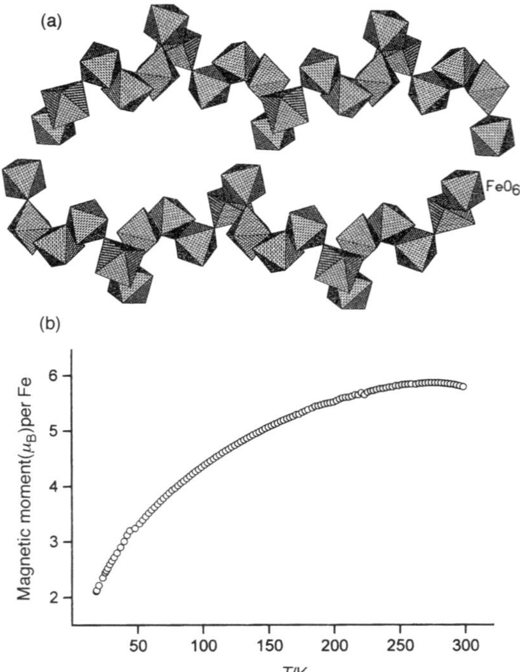

Fig. 3 Network of Fe–O/F octahedra along the [100] direction. The network shows strong anisotropy and is crosslinked by PO$_4$ tetrahedra (not shown). (b) Temperature variation of the magnetic moment.

high-spin Fe^{III} ($^6A_{1g}$). The variation of the magnetic moment [Fig. 3(b)] is characteristic of a gradual spin-crossover similar to that found in some iron(III) complexes.[7] Mössbauer spectra recorded at 298 and 50 K show the presence of the high- and low-spin Fe^{III} ions, respectively, with characteristic differences in the isomer shift and the quadruple splitting (δ = 0.32, 0.1 and ΔE_Q = 0.34, 1.03 mm s^{-1}). The occurrence of spin-crossover in Fe^{III} ions in an oxide material is rather unusual and suggests that the possibility of internal pressure being exerted on the Fe–O polyhedra.

In summary, the synthesis of a novel open-framework iron phosphate with interesting structure and magnetic properties has been accomplished. The presence of Fe–O/F–Fe chains and large one-dimensional channels decorated by phosphate groups are two of the novel features of the material. The occurrence of a smooth magnetic spin-crossover as a function of temperature is noteworthy in that iron(III) oxides generally have Fe^{III} in the high-spin state.

A. C. thanks the Council of Scientific and Industrial Research (CSIR), Government of India for the award of a research fellowship.

Notes and references

† The title compound was synthesized from an iron phosphate gel containing DETA as a structure-directing agent. iron(III) chloride, phosphoric acid (85 wt%), DETA, HF and water in a ratio 1:6:3:1:200 were mixed and stirred until homogeneous. The mixture was sealed in a 23 ml Teflon-lined, stainless steel autoclave (Parr, USA) and heated at 180 °C for 72 h. The resulting product, containing predominantly light green crystals, was filtered off and washed thoroughly with deionized water. A suitable single crystal (0.06 × 0.06 × 0.08) was carefully selected under a polarizing microscope. The crystal structure determination was performed on a Siemens SMART CCD diffractometer equipped with a normal focus, 2.4 kW sealed-tube X-ray source (Mo-Kα radiation, λ = 0.71073 Å) operating at 50 kV and 40 mA. A hemisphere of intensity data were collected in 1321 frames with ω scans (width of 0.30° and exposure time of 30 s per frame).

Crystal data for [(C$_4$N$_3$H$_{16}$)(C$_4$N$_3$H$_{15}$)]$^{5+}$[Fe$_5$F$_4$(H$_2$PO$_4$)(HPO$_4$)$_3$(PO$_4$)$_3$]$^{5-}$·H$_2$O, **I**: monoclinic, space group $P2_1/n$ (no. 14), a = 9.670(1), b = 15.618(1), c = 22.563(1), β = 90.82(1)°, V = 3407.1(1) Å^3, Z = 4, M = 1254.1, μ = 1.915 mm^{-1} and D_c = 1.836(1) g cm^{-3}. A total of 13848 reflections were collected at 298 K in the θ range 1.59–23.19° and merged to give 4844 unique data (R_{int} = 0.065) of which 3635 with $I > 2\sigma(I)$ were considered to be observed. The structure was solved by direct methods with SHELXS-86[8] and difference Fourier synthesis. Final R = 0.06, R_w = 0.13 and S = 1.13 were obtained for 541 parameters. Part of the hydrogen atoms were located initially in the difference Fourier maps and for the final refinements, hydrogen atoms for both the framework as well as the amine molecules were placed geometrically and held in the riding mode. Final Fourier map minimum and maximum: −1.009 and 0.908. Full-matrix least-squares structure refinement against |F^2| were carried out with SHELXTL-PLUS program package.[9]

CCDC 182/1280. See: http://www.rsc.org/suppdata/cc/1999/1305/ for crystallographic files in .cif format.

1 K.-H. Lii, Y.-F. Huang, V. Zima, C.-Y. Huang, H.-M. Lin, Y.-C. Jiang, F.-L. Liao and S.-L. Wang, *Chem. Mater.*, 1998, **10**, 2599 and references therein; Z. A. D. Lethbridge, P. Lightfoot, R. E. Morris, D. S. Wragg, P. A. Wright, Å. Kvick and G. Vaughan, *J. Solid State Chem.*, 1999, **142**, 455.
2 J. R. D. DeBord, W. M. Reiff, R. C. Haushalter and J. Zubieta, *J. Solid State Chem.*, 1996, **125**, 186; J. R. D. DeBord, W. M. Reiff, C. J. Warran, R. C. Haushalter and J. Zubieta, *Chem. Mater.*, 1997, **9**, 1994.
3 M. Cavellec, D. Riou and G. Ferey, *J. Solid State Chem.*, 1994, **112**, 441; M. Cavellec, D. Riou, J.-M. Greneche and G. Ferey, *Zeolites*, 1996, **17**, 252; M. Cavellec, C. Egger, J. Linares, M. Nogues, F. Varret and G. Ferey, *J. Solid State Chem.*, 1997, **134**, 349.
4 M. R. Cavellec, J.-M. Greneche, D. Riou and G. Ferey, *Chem. Mater.*, 1998, **10**, 2434.
5 P. Feng, X. Bu and G. D. Stucky, *Nature*, 1997, **388**, 735; X. Bu, P. Feng and G. D. Stucky, *Science*, 1997, **278**, 2080.
6 A. M. Chippindale, S. J. Brech, A. R. Cowley and W. M. Simpson, *Chem. Mater.*, 1996, **8**, 2259.
7 M. S. Haddad, M. W. Lunch, W. D. Ferderer and D. N. Hendrickson, *Inorg. Chem.*, 1981, **20**, 123.
8 G. M. Sheldrick, SHELXS-86 Program for Crystal Structure Determination, Universität Göttingen, 1986, *Acta Crystallogr., Sect. A*, 1990, **46**, 467.
9 G. M. Sheldrick, SHELXTL-PLUS Program for Crystal Structure Refinement, Universität Göttingen, 1993.

Communication 9/03683K

An Unusual Open-Framework Cobalt(II) Phosphate with a Channel Structure That Exhibits Structural and Magnetic Transitions**

Amitava Choudhury, S. Neeraj, Srinivasan Natarajan, and C. N. R. Rao*

Among the wide variety of open-framework materials, metal phosphates constitute an important family,[1–4] and interest in these materials is mainly because of potential uses associated with their micoporosity. Open-framework transition metal phosphates are of additional interest because of possible magnetism.[5–12] Magnetic channel structures could have many useful applications, including the separation of oxygen and nitrogen from air. The most common transition metal phosphates are those of iron[5–8], and only a few derivatives of cobalt[9–11] and nickel[12] are known. However, very few open-framework structures with magnetic channels are known to date. Ferri- and ferromagnetic behavior was observed in certain iron phosphates[5–7] and Ni[II] diphosphonates[13] at low temperatures (less than 10 K). Magnetism in the Ni[II] diphosphonate arises from a change in the dimensionality and the coordination of the Ni[II] center on dehydration. Ferrimagnetism was reported in a three-dimensional cobalt succinate[14, 15] at about 10 K, and a metamagnetlike behavior in an organically pillared Co[II] sulfate.[16] In our pursuit of open-framework metal phosphates with magnetic channels we employed novel synthetic routes to obtain an unusual cobalt(II) phosphate possessing a 12-membered channel with novel structural features and magnetic properties. Interestingly, the cobalt phosphate undergoes a structural phase transition around 170 K followed by the emergence of ferrimagnetism around 16 K. We believe that this is the first observation of such a temperature-induced change in the channel structure of an open-framework metal phosphate.

Open-framework cobalt phosphates are difficult to prepare by the conventional hydrothermal procedure.[17, 18] We therefore employed two routes: with an amine phosphate[19] or a metal amine complex as starting material. Thus, the reaction of Co[II] ions with diethylenetriamine phosphate (detap) at 180 °C gave [enH$_2$][Co$_{3.5}$(PO$_4$)$_3$] (**I**) along with a mixture of a few other open-framework cobalt phosphates; the triamine deta decomposed to ethylenediamine (en). The reaction of a mixture with the composition [Co(en)$_3$Cl$_3$]·2H$_2$O/2H$_3$PO$_4$/piperazine/150H$_2$O at 165 °C for 3 d gave **I** in pure form.[20a] The asymmetric unit of **I** (Figure 1 a) contains four crystallographically distinct Co and three P atoms. The final atomic

[*] Prof. Dr. C. N. R. Rao, A. Choudhury, S. Neeraj, Dr. S. Natarajan
Chemistry and Physics of Materials Unit and CSIR Centre of Excellence in Chemistry
Jawaharlal Nehru Centre for Advanced Scientific Research
Jakkur P.O., Bangalore 560 064 (India)
Fax: (+91)80-846-2766
E-mail: cnrrao@jncasr.ac.in

Prof. Dr. C. N. R. Rao, A. Choudhury
Solid State and Structural Chemistry Unit
Indian Institute of Science
Bangalore 560 012 (India)

[**] The authors thank Professor S. K. Malik of the TIFR for help with the SQUID measurements.

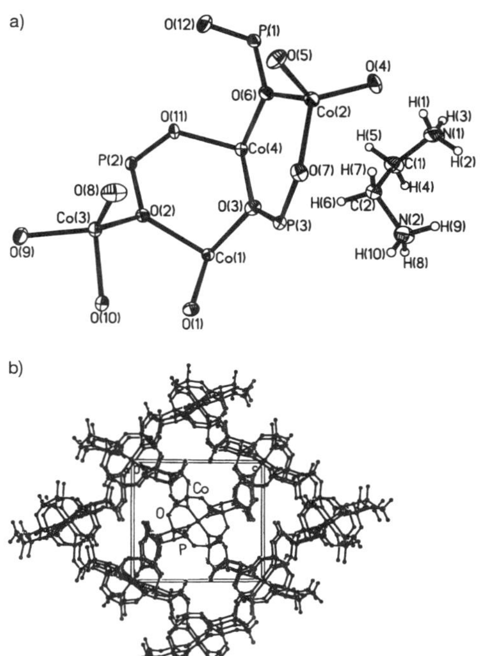

Figure 1. a) Asymmetric unit of **I** at room temperature (50 % probability thermal ellipsoids). b) Structure of **I** along the [100] direction at room temperature showing the 12-membered elliptical channels. The enH$_2$ cations are omitted for clarity.

coordinates of **I** at 298 K are given in Table 1. The room-temperature structure consists of a network of CoO$_4$ [Co(2)], CoO$_5$ [Co(3), Co(4)], and CoO$_6$ [Co(1)] polyhedra that are interconnected by PO$_4$ tetrahedra to form a three-dimensional architecture (Figure 1 b). The Co–O and P–O distances are in the ranges 1.931–2.427 and 1.499–1.573 Å, respectively. The polyhedral connectivity between the various cobalt species in **I** (Figure 2) is such that Co(1) is linked to Co(3) and Co(4) while Co(4) is linked to Co(2) through oxygen atoms to form a secondary building unit (SBU) of composition Co$_4$O$_8$. The SBUs are interconnected by phosphate tetrahedra to form a one-dimensional 12-membered elliptical channel of width 6.5 × 14.2 Å (the longest atom–atom distance not including the van der Waals radii), as shown in Figures 1 b and 2 a.

A preliminary magnetic susceptibility investigation of **I** in the range 30–300 K showed a Curie–Weiss behavior, wherein the slope exhibited a definite change around 200 K. Differential scanning calorimetry (DSC) indicated a phase transition with a small enthalpy change ($\Delta H \approx 20$ cal mol^{-1}) around this temperature. A crystallographic study of **I** at 140 K showed no change in the space group ($P2_1/n$), but changes in the lattice parameters.[20b] The atomic coordinates at 140 K are presented in Table 1. The unit-cell volume at 140 K was significantly larger (ca. 1 %) than at 298 K—a behavior reminiscent of materials with negative coefficients of thermal expansion.[21, 22] The main changes in the structure of **I**

COMMUNICATIONS

Table 1. Atomic coordinates [× 10⁴] and equivalent isotropic displacement parameters [Å × 10³] for **I** at 298 and 140 K.

Atom	298 K				140 K			
	x	y	z	U(eq)[a]	x	y	z	U(eq)[a]
Co(1)	− 10000	0	0	14(1)	− 10000	0	0	7(1)
Co(2)	− 4655(2)	1771(1)	2836(1)	16(1)	− 4654(2)	1771(1)	2835(1)	9(1)
Co(3)	− 6543(2)	− 1969(1)	815(1)	15(1)	− 6567(2)	− 1974(1)	822(1)	9(1)
Co(4)	− 10085(2)	710(1)	1802(1)	15(1)	− 10104(2)	715(1)	1796(1)	8(1)
P(1)	− 9154(4)	1008(1)	3699(1)	14(1)	− 9151(4)	999(1)	3695(1)	7(1)
P(2)	− 10754(4)	− 1306(1)	1622(1)	14(1)	− 10756(4)	− 1305(1)	1622(1)	7(1)
P(3)	− 4748(4)	1140(1)	882(1)	18(1)	− 4761(4)	1141(1)	875(1)	7(1)
O(1)	− 7102(10)	− 402(3)	− 663(3)	17(1)	− 7095(9)	− 403(3)	− 668(3)	9(1)
O(2)	− 9702(10)	− 1148(3)	773(3)	17(1)	− 9706(10)	− 1148(3)	771(3)	9(1)
O(3)	− 7543(10)	786(3)	950(3)	23(1)	− 7572(10)	794(3)	940(3)	12(1)
O(4)	− 6000(11)	2942(3)	2988(3)	22(1)	− 6042(10)	2949(3)	2989(3)	13(1)
O(5)	− 2164(10)	1105(3)	3572(3)	17(1)	− 2178(10)	1095(3)	3572(3)	11(1)
O(6)	− 8059(10)	1141(3)	2841(3)	18(1)	− 8065(9)	1141(3)	2837(3)	9(1)
O(7)	− 3468(10)	1451(3)	1746(3)	27(1)	− 3483(10)	1461(3)	1744(3)	11(1)
O(8)	− 3630(11)	− 1588(4)	1581(3)	22(1)	− 3628(10)	− 1603(3)	1585(3)	13(1)
O(9)	− 7057(11)	− 3258(3)	738(3)	17(1)	− 7071(10)	− 3273(3)	735(3)	11(1)
O(10)	− 5229(11)	− 1912(3)	− 288(3)	17(1)	− 5210(10)	− 1911(3)	− 283(3)	13(1)
O(11)	− 10326(11)	− 493(3)	2171(3)	20(1)	− 10342(10)	− 489(3)	2176(3)	13(1)
O(12)	− 8330(11)	121(3)	4035(3)	23(1)	− 8341(10)	100(3)	4026(3)	13(1)
N(1)	− 10214(14)	4305(4)	1880(4)	28(12)	− 10167(13)	4286(4)	1896(4)	18(2)
N(2)	− 8043(13)	3522(5)	297(4)	28(2)	− 8023(12)	3507(4)	286(4)	12(1)
C(1)	− 11296(17)	3628(6)	1311(5)	28(2)	− 11348(16)	3623(5)	1304(5)	16(2)
C(2)	− 9187(17)	3077(5)	984(5)	27(2)	− 9211(16)	3065(5)	974(5)	16(2)

[a] U(eq) is defined as one third of the trace of the orthogonalized U_{ij} tensor.

a)

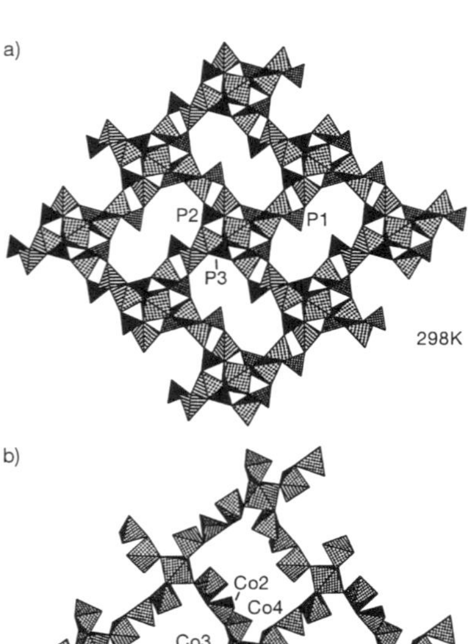

b)

Figure 2. a) The polyhedral connectivity in **I** showing the channels along the a axis at 298 K. b) The cobalt–oxygen polyhedral connectivity showing the channels at 140 K.

Table 2. Some interatomic distances of **I** at 298 and 140 K

Moiety	Bond length [Å]	
	298 K	140 K
Co(3)−O(8)	1.930(5) [0.526][a]	1.940(5) [0.512][a]
Co(3)−O(9)	1.980(5) [0.460]	1.999(5) [0.436]
Co(3)−O(10)	1.987(5) [0.452]	2.007(5) [0.427]
Co(3)−O(2)	2.029(5) [0.402]	2.031(5) [0.400]
Co(3)−O(4)	2.427(5) [0.137]	2.405(5) [0.146]
Co(3)−P(2)	2.809(2)	2.807(2)
Co(2)−P(2)	3.056(2)	3.065(2)
Co(3)−Co(2)	3.620(2)	3.616(2)
Co(2)−O(4)	1.932(5)	1.955(5)
P(2)−O(4)	1.548(6)	1.556(5)

[a] Sum of the bond valences.

on cooling occurs around Co(3), especially for the bonds to O(10) and O(4) (Table 2). The Co(3)−O(4) distance is rather long in the room-temperature phase. If we ignore the Co(3)−O(4) bond in the room-temperature phase, the coordination number of Co(3) would change from four to five. If the Co(3)−O(4) bond is taken into account, Co(3) would essentially be unaffected in that it is a pentacoordinate in both the phases, possibly in a trigonal-bipyramidal environment. However, a slight change in the orientation of the enH₂ cations could occur. The Co−Co and Co−P distances in the two phases bear testimony to the structural transition (Table 2). These changes are consistent with the fact that the transition is not truly first order, but is likely to be displacive. Note that the single crystal remains intact across the transition. The face-sharing of the Co(3) and P(2) polyhedra in **I** is rather unusual and is noteworthy. Furthermore, if the Co(3)−O(4) bond is taken into consideration, the structure is three-dimensional. If it is not, the Co network would consist of isolated ribbons shared by PO₄ groups, a situation intermediate between that of magnetic clusters commonly found in porous transition metal phosphates and the Ni phosphates with a three-dimensional Ni network.[12]

The change in the slope of the Curie – Weiss plot of **I** around the structural transition temperature was mentioned above.

© WILEY-VCH Verlag GmbH, D-69451 Weinheim, 2000   1433-7851/00/3917-3092 $ 17.50+.50/0   *Angew. Chem. Int. Ed.* **2000**, *39*, No. 17

The paramagnetic Curie temperature was calculated as −79 K from susceptibility data in the range 200–300 K, and as −45 K for the range 40–150 K. These values indicate an increase in the ferromagnetic interaction after the structural transition. We examined the low-temperature magnetic properties of the material with a SQUID magnetometer. Figure 3 shows the zero-field-cooled (ZFC) and field-cooled

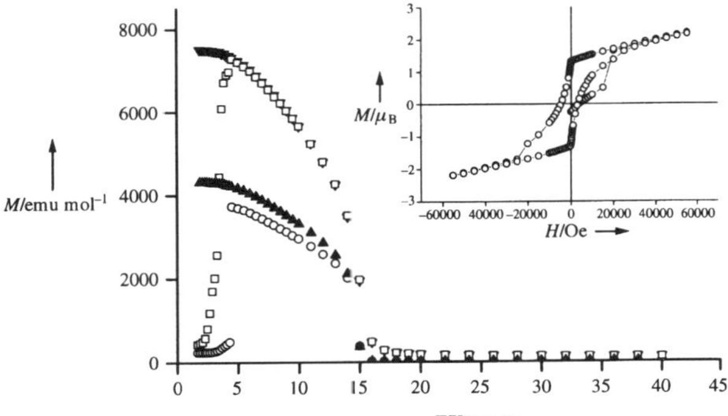

Figure 3. Variation of the magnetic susceptibility of I with temperature. The field-cooled (FC) data show a saturation behavior. ○ ZFC, ▲ FC, 50 Oe; □ ZFC, ▼ FC, 1000 Oe. The inset shows the variation of magnetization with the applied field with hysteresis at 2 K. Note the sudden rise in magnetization around 15 kOe.

(FC) magnetization data of the sample at 50 and 1000 Oe. The ZFC data showed a low magnetization at 1.2 K, an abrupt increase between 2 and 4 K on warming, and attainment of a maximum at 4.2 K. It then shows a sharp drop at 16 K where the magnetism vanishes. The drop in magnetization is not found in the FC samples, which exhibit saturation. Such a ferrimagnetic behavior was found recently in a layered Co^{II} hydroxide.[23] The Co-O-Co angles in I are in the range of 94–122.4°, which favors ferromagnetic interaction. The angle of 122.4° corresponds to the shared corner of the Co(1) octahedron and the Co(3) trigonal bipyramid. The magnetization versus field plot of I shows hysteresis (inset of Figure 3). The magnetization initially rises with increasing field, followed by a sharp jump around 15 kOe. After the jump, the hysteresis loop is that of a typical ferromagnet. Apparently, a transition from one ferrimagnetic to another ferrimagnetic state occurs in a magnetic field; the magnetic moment is too low for a ferromagnetic state.

The isolation of an open-framework Co^{II} phosphate that exhibits a structural phase transition constitutes a novel finding. It is noteworthy that the change in the structure is accompanied by the evolution of ferrimagnetism at low temperatures. We expect that novel routes such as those employed here will enable the synthesis of open-framework structures which are magnetic at considerably higher temperatures.

Received: March 3, 2000 [Z 14801]

[1] J. M. Thomas, *Nature* **1994**, *368*, 289.

[2] J. M. Thomas, *Angew. Chem.* **1999**, *111*, 3800; *Angew. Chem. Int. Ed.* **1999**, *38*, 3589.

[3] J. M. Thomas, *Chem. Eur. J.* **1997**, *3*, 1557.

[4] A. K. Cheetham, G. Ferey, T. Loiseau, *Angew. Chem.* **1999**, *111*, 3466; *Angew. Chem. Int. Ed.* **1999**, *38*, 3268.

[5] M. Cavellec, D. Riou, C. Ninclaus, J.-M. Greneche, G. Ferey, *Zeolites* **1996**, *17*, 250.

[6] M. Riou-Cavellec, J.-M. Greneche, G. Ferey, *J. Solid State Chem.* **1999**, *148*, 150.

[7] K.-H. Lii, Y.-F. Huang, V. Zima, C.-Y. Huang, H.-M. Lin, Y.-C. Jiang, F.-L. Liao, S.-L. Wang, *Chem. Mater.* **1998**, *10*, 2599.

[8] A. Choudhury, S. Natarajan, C. N. R. Rao, *Chem. Commun.* **1999**, 1305.

[9] P. Feng, X. Bu, S. H. Tolbert, G. D. Stucky, *J. Am. Chem. Soc.* **1997**, *119*, 2497.

[10] J.-S. Chen, R. H. Jones, S. Natarajan, M. B. Hursthouse, J. M. Thomas, *Angew. Chem.* **1994**, *106*, 667; *Angew. Chem. Int. Ed. Engl.* **1994**, *33*, 639.

[11] J. R. D. DeBord, R. C. Haushalter, J. Zubieta, *J. Solid State Chem.* **1996**, *125*, 270.

[12] N. Guillou, Q. Gao, M. Nogues, R. E. Morris, M. Hervieu, G. Ferey, A. K. Cheetham, *C. R. Acad. Sci. Paris II* **1999**, 387.

[13] Q. Gao, N. Guillou, M. Nogues, A. K. Cheetham, G. Ferey, *Chem. Mater.* **1999**, *11*, 2937.

[14] C. Livage, C. Egger, M. Nogues, G. Ferey, *J. Mater. Chem.* **1998**, *8*, 2743.

[15] C. Livage, C. Egger, G. Ferey, *Chem. Mater.* **1999**, *11*, 1546.

[16] A. Rujiwatra, C. J. Kepert, M. J. Rosseinsky, *Chem. Commun.* **1999**, 2307.

[17] P. Feng, X. Bu, G. D. Stucky, *Nature* **1997**, *388*, 735.

[18] X. Bu, P. Feng, G. D. Stucky, *Science* **1997**, *278*, 2080.

[19] S. Neeraj, S. Natarajan, C. N. R. Rao, *Angew. Chem.* **1999**, *111*, 3688; *Angew. Chem. Int. Ed.* **1999**, *38*, 3480.

[20] a) In a typical synthesis $[Co(en)_3Cl_3] \cdot 2H_2O$ (0.191 g) was dispersed in water (1.35 mL), and H_3PO_4 (85 wt %, 0.07 mL) was added to solution. Then, piperazine (0.043 g) was added with constant stirring to attain a pH of 3.0. The resulting mixture was homogenized for 30 min, transferred into a 23-mL PTFE-lined stainless autoclave, and heated at 165 °C for 72 h. After the reaction, the mixture contained purple-blue needle-shaped crystals of I, which were collected by filtration, washed with deionized water, and dried under ambient conditions. A suitable single crystal was subjected to X-ray diffraction studies on a Siemens SMART-CCD diffractometer. Crystal data for I (298 K): monoclinic, space group $P2_1/n$ (no. 14), $a = 5.0798(1)$, $b = 15.2030(4)$, $c = 16.3535(1)$ Å, $\beta = 95.69(1)°$, $V = 1256.7(3)$ Å^3, $Z = 4$, $M_r = 553.3$, $\rho_{calcd} = 2.895$ g cm^{-3}, $\mu(Mo_{K\alpha}) = 4.955$ mm^{-1}; 5228 reflections were collected at $3.5 \leq 2\theta \leq 46.5°$ and merged to give 1808 unique reflections ($R_{int} = 0.08$), of which 1422 with $I > 2\sigma(I)$ were considered to be observed. The structure was solved and refined by using the SHELXTL-PLUS suite of programs to $R_1 = 0.04$, $wR_2 = 0.1$, and $S = 1.13$ for 206 parameters. b) Crystal data for I (140 K): monoclinic, space group $P2_1/n$ (no. 14), $a = 5.0979(1)$, $b = 15.2338(1)$, $c = 16.4247(4)$ Å, $\beta = 95.67(1)°$, $V = 1269.3(3)$ Å^3, $Z = 4$, $M_r = 553.3$, $\rho_{calcd} = 2.895$ g cm^{-3}, $\mu(Mo_{K\alpha}) = 4.955$ mm^{-1}; 5193 reflections were collected at $3.5 \leq 2\theta \leq 46.5°$ and merged to give 1818 unique reflections ($R_{int} = 0.08$), of which 1434 with $I > 2\sigma(I)$ were considered to be observed. The structure was solved and refined by using the SHELXTL-PLUS suite of programs to $R_1 = 0.04$, $wR_2 = 0.08$, and $S = 1.11$ for 206 parameters. Crystallographic data (excluding structure factors) for the structures reported in this paper have been deposited with the Cambridge Crystallographic Data Centre as supplementary publication no. CCDC-141075 (298 K) and -141076 (140 K). Copies of the data can be obtained free of charge on application to CCDC, 12 Union Road, Cambridge CB21EZ, UK (fax: (+44) 1223-336-033; e-mail: deposit@ccdc.cam.ac.uk).

[21] T. A. Mary, J. S. O. Evans, T. Vogt, A. W. Sleight, *Science* **1996**, *272*, 90.

[22] M. P. Attfield, A. W. Sleight, *Chem. Commun.* **1998**, 601.

[23] M. Kurmoo, *Chem. Mater.* **1999**, *11*, 3370.

Organically Templated Mixed-Valent Iron Sulfates Possessing Kagomé and Other Types of Layered Networks

Geo Paul, Amitava Choudhury,
E. V. Sampathkumaran, and C. N. R. Rao*

We have been interested in developing strategies for designing new open-framework architectures. One such strategy is the case of the sulfate tetrahedron and its utilization as a primary building unit, in place of the silicate or the phosphate tetrahedron commonly employed for the purpose.[1,2] Our efforts in this direction have enabled us to obtain organically templated open-framework metal sulfates under hydrothermal conditions with one-dimensional chain structures as well as two-dimensional layered structures.[3] In the case of iron sulfates, Fe is present generally in the $+2$ or the $+3$ oxidation state; they all show the presence of antiferromagnetic interactions. In this communication, we report the successful synthesis of organically templated mixed-valent iron sulfates of the compositions, $[H_2N(CH_2)_4NH_2]_2[Fe_2^{III}Fe_3^{II}F_{12}(SO_4)_2(H_2O)_2]$ **(1)** and $[HN(CH_2)_6NH][Fe^{III}Fe_2^{II}F_6(SO_4)_2] \cdot [H_3O]$ **(2)**, with novel structural features and magnetic properties. In particular, **2** is an unusual example of an iron compound with a perfect Kagomé structure. The discovery of such an unusual Kagomé lattice is significant considering the current interest in these compounds[4] and the fact that iron compounds possessing the Kagomé lattice generally contain Fe in a single oxidation state.

The asymmetric unit of **1** contains four crystallographically distinct iron atoms and one sulfur atom. The fluorine and oxygen neighbors are coordinated octahedrally to the Fe atoms ($Fe(1)F_5O$, $Fe(2)F_4O_2$, $Fe(3)F_2O_2(H_2O)_2$ and $Fe(4)$-F_4O_2). The structure of **1** is constructed from the anionic framework layers of $[Fe_2^{III}Fe_3^{II}F_{12}(SO_4)_2(H_2O)_2]^{4-}$. The $Fe(2)$-F_4O_2 and $Fe(4)F_4O_2$ octahedra are vertex-shared through *trans*-Fe-F-Fe linkages to form an infinite chain along the a axis. Such chains are connected by an edge-shared trimer of two $Fe(1)F_5O$ and one $Fe(3)F_2O_4$ octahedra to yield a layered network in the ac plane. In the trimer, the $Fe(3)F_2O_4$ octahedra are flanked by two $Fe(1)F_5O$ octahedra. Such connectivity creates a triangular lattice formed by $Fe(1)F_5O$, $Fe(2)F_4O_2$, and $Fe(4)F_4O_2$ octahedra. The SO_4 tetrahedron caps the triangular lattice, creating a 10-membered aperture

within the layer, as shown in Figure 1 a. The layers are stacked one over the other along the b axis in *AAAA* fashion and are held together by hydrogen bonding interactions with the diprotonated piperazine (PIP) molecules residing in the interlamellar space (see Figure 1 b). The bond lengths and

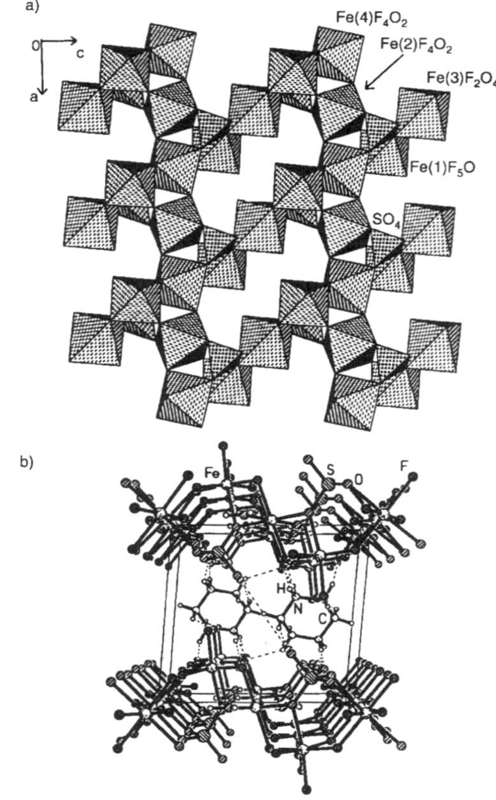

Figure 1. a) Polyhedral view of the $[Fe_2^{III}Fe_3^{II}F_{12}(SO_4)_2(H_2O)_2]^{4-}$ layer in **1**, viewed perpendicular to the ac plane. Here, Fe(1) is in the $+3$ oxidation state and the remaining iron atoms are in the $+2$ state. Note the symmetrical capping of the sulfate tetrahedra in the triangular lattice and the 10-membered aperture within the layer; b) Structure of **1** showing the presence of the amine group located in the interlayer space.

angles indicate perfect octahedral geometry around Fe^{II}-centers and a slightly distorted geometry for the Fe^{III} centers. Bond-valence sum (BVS) calculations[5] ($Fe(1) = 3.03$, $Fe(2) = 2.02$, $Fe(3) = 1.98$, and $Fe(4) = 2.05$) and the average Fe—O/F bond lengths indicate the oxidation state of Fe(1) to be $+3$, and the remaining Fe centers to be $+2$. The positions of the fluorine atoms find indirect support from BVS calculations ($F(1) = 0.616$, $F(2) = 0.563$, $F(3) = 0.896$, $F(4) = 0.804$, $F(5) = 0.786$ and $F(6) = 0.74$). The mixed-valent nature of **1** was also confirmed by Mössbauer spectroscopy. DC magnetic susceptibility data (at 100 Oe) show a sharp magnetic transition around 15 K and what appears to be a ferrimagnetic transition at around 3 K, the field-cooled (FC) and the zero-field-cooled (ZFC) data being identical (Figure 2 a). The inverse susceptibility data recorded at 5 kOe (Figure 2 b) show a linear

[*] Prof. Dr. C. N. R. Rao, G. Paul, A. Choudhury
Chemistry and Physics of Materials Unit and
CSIR Center of Excellence in Chemistry
Jawaharlal Nehru Center for Advanced Scientific Research
Jakkur, P.O., Bangalore 560 064 (India)
Fax: (+91) 80-846-2766
E-mail: cnrrao@jncasr.ac.in

Prof. Dr. C. N. R. Rao, A. Choudhury
Solid State and Structural Chemistry Unit
Indian Institute of Science
Bangalore 560 012 (India)

Prof. Dr. E. V. Sampathkumaran
Department of Condensed Matter Physics and Materials Science
Tata Institute of Fundamental Research
Mumbai 400005 (India)

COMMUNICATIONS

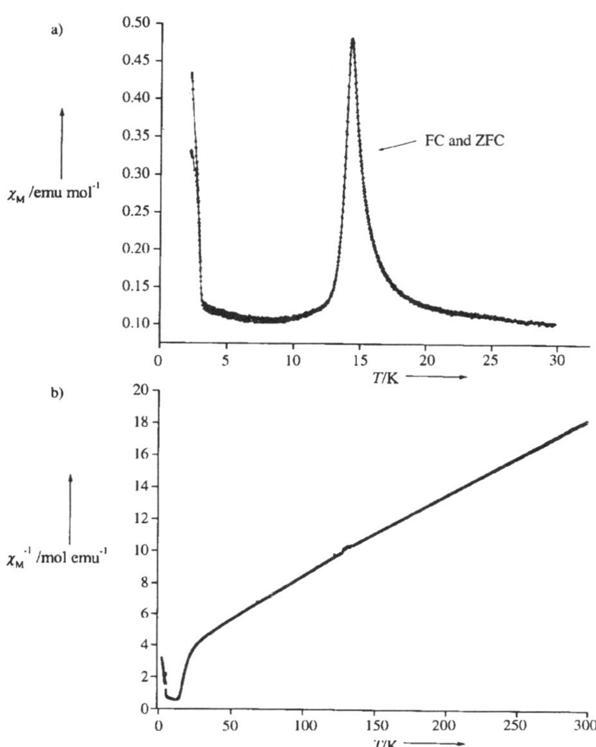

Figure 2. a) Temperature dependence of the DC magnetic susceptibility of **1** measured at 100 Oe under field-cooled (FC) and zero-field-cooled (ZFC) conditions; b) Temperature variation of the inverse DC susceptibility measured at 5 kOe.

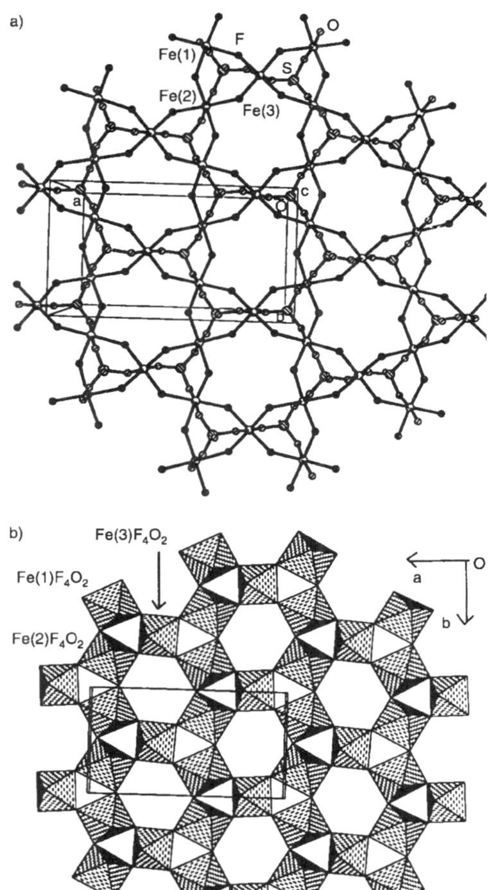

Figure 3. a) Ball-and-stick representation of a section of $[Fe^{III}Fe_2^{II}F_6(SO_4)_2]^{3-}$ in **2**, which shows the presence of the 3- and 6-rings; the 3-rings are capped by the sulfate tetrahedra; b) Polyhedral view of the 2D network of corner-sharing iron octahedra in **2**, viewed perpendicular to the *ab* plane. Here, Fe(1) and Fe(2) are in the +2 oxidation state and Fe(3) is in the +3 state.

behavior at high temperatures, which yields a negative Curie temperature ($T_c = -86$ K). These data also reveal a transition to a canted antiferromagnetic state at low temperatures.

The asymmetric unit of **2** contains 28 non-hydrogen atoms, of which 21 belong to the inorganic framework and seven belong to the extra-framework guest molecules; one extra-framework oxygen atom belongs to the hydronium ion. The structure of **2** consists of anionic layers of vertex-sharing $Fe^{III}F_4O_2$ and $Fe^{II}F_4O_2$ octahedra and tetrahedral SO_4 units, which are fused together by $Fe^{III/II}$-F-$Fe^{III/II}$ and $Fe^{III/II}$-O-S moieties. Each $Fe^{III/II}F_4O_2$ unit shares four of its $Fe^{III/II}$-F vertices with similar neighbors, with the $Fe^{III/II}$-F-$Fe^{III/II}$ bonds roughly aligned in the *ab* plane. The $Fe^{III/II}$-O bond is canted from the *ab* plane and this $Fe^{III/II}$-O vertex effectively forces a three-ring trio of apical $Fe^{III/II}$-O bonds closer together to allow them to be capped by the SO_4 tetrahedra. The three- and six-rings of octahedra result from the in-plane connectivity, as shown in Figure 3a. The layers in **2** are typical of a Kagomé lattice with hexagonal tungsten bronze-type sheets [6] of vertex-sharing $Fe^{III/II}F_4O_2$ octahedra, as shown in Figure 3b. The amine molecules are present between the layers just as in **1**. The structure of **2** is akin to that of the mineral jarosite[7] where all of the Fe atoms are in the +3 oxidation state. The Fe-O bond lengths in **2** are in the range 2.003(3)–2.149(3) Å ((Fe(1)-O)$_{av}$ 2.140(3), (Fe(2)-O)$_{av}$ 2.146(3), and (Fe(3)-O)$_{av}$

2.011(3) Å), and the Fe-F bond lengths are in the range 1.906(2)–2.157(2) Å ((Fe(1)-F)$_{av}$ 2.081(2), (Fe(2)-F)$_{av}$ 2.080(2), and (Fe(3)-F)$_{av}$ 1.936(2) Å). BVS calculations (Fe(1) = 1.93, Fe(2) = 1.92, and Fe(3) = 3.01) as well as the average bond distances indicate that the oxidation state of Fe(1) and Fe(2) is +2 and that of Fe(3) is +3. The positions of the six F atoms are also supported by the bond valence calculations (F(1) = 0.734, F(2) = 0.669, F(3) = 0.797, F(4) = 0.70, F(5) = 0.76, and F(6) = 0.71). Thus, the framework stoichiometry of the $[Fe^{III}Fe_2^{II}F_6(SO_4)_2]$ unit with a −3 charge requires the amine to be doubly protonated, in addition to the presence of the hydronium ion. Mössbauer spectra also confirm the presence of the Fe^{II} and Fe^{III} states in the ratio 2:1.

The Kagomé lattice in **2** is unusual in several respects. There is 100% occupation of the iron sites[8] and this is the first example of a Kagomé lattice where the Fe centers are found in two oxidation states. Extensive protonation of the OH sites in jarosite results in vacancies in the Fe lattice, but we seem to be able to overcome this problem by replacing OH groups with fluorine atoms. The magnetic behavior of **2** is quite different from that of **1** and is even more complex. Compound **2** shows magnetic frustration, as typified by the divergence in the FC and ZFC magnetization data shown in Figure 4a. The spin-freezing temperature (T_f) is around 12 K.[9] AC suscept-

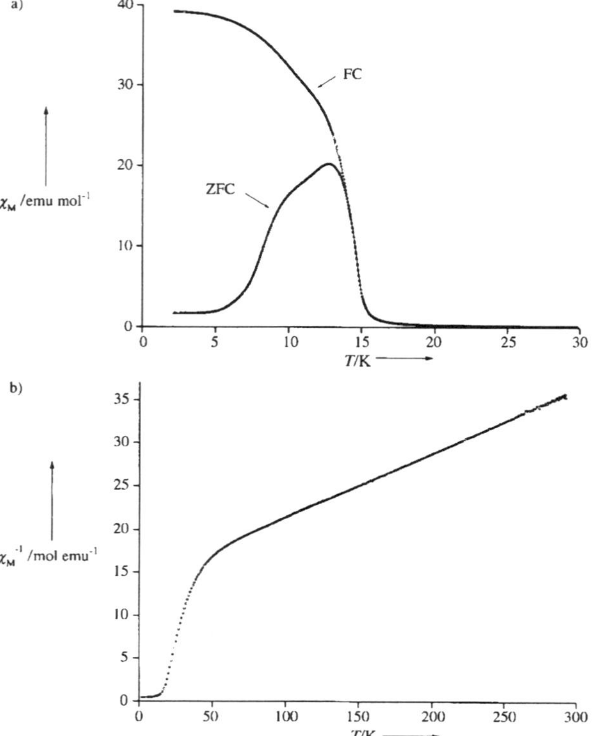

Figure 4. a) DC susceptibility data of **2** measured at 100 Oe which displays the divergence between the ZFC and FC measurements; b) Temperature variation of the inverse DC susceptibility measured at 5 kOe.

ibility measurements confirm a typical spin-glass transition behavior. The inverse magnetic susceptibility data recorded at 5 kOe are linear at high temperatures giving a negative Curie temperature ($T_c = -180$ K). The data show a ferrimagnetic-type transition, as can be seen from Figure 4b. This unusual behavior of **2** requires further study. It may be noted that most of the Fe compounds with the Kagomé lattice show either spin-glass behavior or long-range antiferromagnetic ordering.[4,10,11] The only other Kagomé-type structures exhibiting ferromagnetic features are the vanadium compound reported by Grohol et al.[12] and the dimeric copper compound reported by Zaworotko and co-workers[13] The magnetic moment obtained by extrapolating the high-field [12 T] portion of M versus H to $H = 0$ is 2.95 μ_B mol^{-1} at 10 K for **2** (calcd 9 μ_B mol^{-1}), compared to 4.8 μ_B mol^{-1} for **1** (calcd

16 μ_B mol^{-1}). It should be noted that other mixed-valent iron compounds comparable to **2** exhibit entirely different magnetic properties. Thus [Fe$_3$F$_8$(H$_2$O)$_2$], with hexagonal bronze-type layers, is antiferromagnetic (Néel temperature, $T_N = 157$ K) and shows weak ferromagnetism at low temperatures.[14] [Fe$_2$F$_5$(H$_2$O)$_2$], with an inverse weberite structure, is ferrimagnetic ($T_C = 48$ K).[15] Na$_2$NiFeF$_7$, with the weberite structure, is also ferrimagnetic ($T_C = 88$ K)[16] while in BaMn-FeF$_7$, ferromagnetic sublattices of Mn and Fe are coupled antiferromagnetically giving a T_N value of 85 K.[17] In contrast, compound **2** exhibits a clear spin-glass behavior, as well as ferrimagnetism.

Besides demonstrating that the sulfate tetrahedron can be conveniently employed to generate novel organically templated transition metal sulfates, the present study reveals that layered mixed-valent iron sulfates have unusual magnetic properties. The mixed-valent iron Kagomé compound deserves to be further explored both experimentally and theoretically.

Experimental Section

The synthesis of **1** and **2** was carried out in teflon-lined acid digestion bombs with an internal volume of 23 cm³ under constant pressure by heating the starting mixture at 180 °C for 2 days. Compound **1** was prepared from the following reagents: ferric citrate (1 mmol), piperazine sulfate[18] (4 mmol), HF (48 %, 8 mmol), H$_2$O (50 mmol), and ethylene glycol (50 mmol). The product (yield 50 %), which consisted of red rod-shaped crystals, was monophasic. Compound **2** was prepared from the following reagents: ferric citrate (1 mmol), H$_2$SO$_4$ (98 %, 4 mmol), 1,4-diazabicy-clo[222]octane (DABCO, 6 mmol), HF (48 %, 4 mmol), H$_2$O (50 mmol), and *n*-butanol (30 mmol). The product (yield = 40 %), consisting of red hexagonal plate-shaped crystals was monophasic. The identities of **1** and **2** were established by various techniques, which included X-ray crystallog-raphy. X-ray powder diffraction patterns of **1** and **2** were in good agreement with the simulated patterns based on single-crystal data, indicative of their phase purity. The Fe:S ratios in **1** and **2** were determined by energy-dispersive X-ray analysis (EDAX) to be 5:2 and 3:2, respectively, which is commensurate with their molecular formulas. Furthermore, thermogravi-metric analysis gave the exact mass losses expected for the removal of water, amine, and SO$_3$ in both **1** and **2**. The final product of the decomposition of **1** was Fe$_2$O$_3$, which was also verified from its powder XRD pattern. Complex **2** gave FeO as the final product. The fluorine analysis was also satisfactory, which indicated that the OH groups are, generally, replaced by fluorine atoms.

Structure determination: Single-crystal data were collected on a Sie-mens SMART-CCD diffractometer (graphite-monochromated Mo$_{K\alpha}$ radi-ation, $\lambda = 0.71073$ Å, $T = 298$ K). An absorption correction based on symmetry-equivalent reflections was applied using SADABS.[19] The structures were solved by direct methods using SHELXS-86[20] and difference Fourier synthesis. Full-matrix least-squares structure refinement against $|F^2|$ was carried out using the SHELXTL-PLUS[21] package of programs. Hydrogen positions for the bonded water molecule (O(3) in **1**) as well as the extra-framework hydronium ion (O(100) in **2**) were located from the difference Fourier map and placed in the observed position and refined isotropically. All the remaining hydrogen positions for **1** and **2** were initially located in the difference Fourier maps, and for the final refinement, the hydrogen atoms were placed geometrically and held in the riding mode. The non-hydrogen atoms were refined anisotropically.

Crystal data. **1**: $M_r = 911.72$, triclinic, space group $= P\bar{1}$ (no. 2), $a = 7.222(2)$, $b = 9.043(2)$, $c = 9.463(4)$ Å, $\alpha = 94.58(3)$, $\beta = 91.71(2)$, $\gamma = 95.08(2)°$, $V = 613.2(3)$ Å³, $Z = 2$, $\mu = 3.204$ mm^{-1}, $\rho_{calcd} = 2.469$ mg m^{-3}, 1724 unique re-flections ($R_{int} = 0.0208$), final $R = 0.0359$, $R_w = 0.0905$, GOF = 1.043. **2**: $M_r = 606.89$, orthorhombic, space group $= Pna2_1$ (no. 33), $a = 12.9269(6)$, $b = 7.4091(3)$, $c = 17.5970(8)$ Å, $\alpha = \beta = \gamma = 90°$, $V = 1685.38(13)$ Å³, $Z = 4$, $\mu = 2.908$ mm^{-1}, $\rho_{calcd} = 2.392$ mg m^{-3}, 2362 unique reflections ($R_{int} = 0.0333$), final $R = 0.0243$, $R_w = 0.0562$, GOF = 1.000.

COMMUNICATIONS

CCDC-185168 (**1**) and CCDC-185167 (**2**) contains the supplementary crystallographic data for this paper. These data can be obtained free of charge via www.ccdc.cam.ac.uk/conts/retrieving.html (or from the Cambridge Crystallographic Data Centre, 12, Union Road, Cambridge CB21EZ, UK; fax: (+ 44)1223-336-033; or deposit@ccdc.cam.ac.uk).

Received: May 14, 2002
Revised: August 26, 2002 [Z19300]

[1] a) D. W. Breck, *Zeolite Molecular Sieves*, Wiley, New York, **1974**; b) W. M. Meier, D. H. Oslen, C. Baerlocher, *Atlas of Zeolite Structure Types*, Elsevier, London, **1996**.

[2] A. K. Cheetham, G. Férey, T. Loiseau, *Angew. Chem.* **1999**, *111*, 3466 – 3492; *Angew. Chem. Int. Ed.* **1999**, *38*, 3268 – 3292.

[3] a) A. Choudhury, J. Krishnamoorthy, C. N. R. Rao, *Chem. Commun.* **2001**, 2610 – 2611; b) G. Paul, A. Choudhury, C. N. R. Rao, *Chem. Commun.* **2002**, 1904 – 1905.

[4] a) A. P. Ramirez, *Annu. Rev. Mater. Sci.* **1994**, *24*, 453 – 480; b) J. E. Greedan, *J. Mater. Chem.* **2001**, *11*, 37 – 53; c) A. S. Wills, A. Harrison, *J. Chem. Soc. Faraday Trans.* **1996**, *92*, 2161 – 2166; d) T. Inami, M. Nishiyama, S. Maegawa, Y. Oka, *Phys. Rev. B* **2000**, *61*, 12181 – 12186; e) J. N. Reimers, A. J. Berlinsky, *Phys. Rev. B* **1993**, *48*, 9539 – 9554; f) J. Frunzke, T. Hansen, A. Harrison, J. S. Lord, G. S. Oakley, D. Visser, A. S. Wills, *J. Mater. Chem.* **2001**, *11*, 179 – 185.

[5] a) N. E. Brese, M. O'Keeffe, *Acta Crystallogr. Sect. B* **1991**, *47*, 192 – 197; b) I. D. Brown, D. Altermatt, *Acta Crystallogr. Sect. B* **1985**, *41*, 244 – 247.

[6] A. Magneli, *Acta Chem. Scand.* **1953**, *7*, 315.

[7] a) J. E. Dutrizac, S. Kaiman, *Can. Mineral.* **1976**, *14*, 151 – 158; b) F. C. Hawthrone, S. V. Krivovichev, P. C. Burns, *Rev. Mineral. Geochem.* **2000**, *40*, 1 – 101.

[8] A. S. Wills, A. Harrison, S. A. M. Mentink, T. E. Mason, Z. Tun, *Europhys. Lett.* **1998**, *42*, 325 – 330.

[9] a) G. S. Oakley, D. Visser, J. Frunzke, K. H. Andersen, A. S. Wills, A. Harrison, *Phys. B* **1999**, *267–268*, 142 – 144; b) A. Harrison, A. S. Wills, C. Ritter, *Phys. B* **1998**, *241–243*, 722 – 723.

[10] H. Serrano-Gonzalez, S. T. Bramwell, K. D. M. Harris, B. M. Kariuki, L. Nixon, I. P. Parkin, C. Ritter, *Phys. Rev. B* **1999**, *59*, 14451 – 14460.

[11] a) G. S. Oakley, S. Pouget, A. Harrison, J. Frunzke, D. Visser, *Phys. B* **1999**, *267–268*, 145 – 148; b) A. S. Wills, A. Harrison, C. Ritter, R. I. Smith, *Phys. Rev. B* **2000**, *61*, 6156 – 6169; c) E. A. Earle, A. P. Ramirez, R. J. Cava, *Phys. B* **1999**, *262*, 199 – 204; d) S.-H. Lee, C. Broholm, M. F. Collins, L. Heller, A. P. Ramirez, Ch. Kloc, E. Bucher, R. W. Erwin, N. Lacevic, *Phys. Rev. B* **1997**, *56*, 8091 – 8097.

[12] a) D. Grohol, D. Papoutsakis, D. G. Nocera, *Angew. Chem.* **2001**, *113*, 1567 – 1569; *Angew. Chem. Int. Ed.* **2001**, *40*, 1519 – 1521; b) D. Papoutsakis, D. Grohol, D. G. Nocera, *J. Am. Chem. Soc.* **2002**, *124*, 2647 – 2656; c) D. Grohol, D. G. Nocera, *J. Am. Chem. Soc.* **2002**, *124*, 2640 – 2646.

[13] B. Moulton, J. Lu, R. Hajndl, S. Hariharan, M. J. Zaworotko, *Angew. Chem.* **2002**, *114*, 2945 – 2948; *Angew. Chem. Int. Ed.* **2002**, *41*, 2821 – 2824.

[14] a) M. Leblanc, G. Férey, Y. Calage, R. De Pape, *J. Solid State Chem.* **1984**, *53*, 360 – 368; b) M. Leblanc, G. Férey, P. Lacone, J. Punnetier, *J. Magn. Magn. Mater.* **1991**, *92*, 359 – 365.

[15] Y. Laligant, M. Leblanc, J. Pannetier, G. Férey, *J. Phys. C* **1986**, *19*, 1081 – 1095.

[16] Y. Laligant, Y. Calage, G. Heger, J. Pannetier, G. Férey, *J. Solid State Chem.* **1989**, *78*, 66 – 77.

[17] P. Lacorre, J. Pannetier, J. Pebler, S. Nagel, D. Babel, A. De Kozak, M. Samouel, G. Férey, *J. Solid State Chem.* **1992**, *101*, 296 – 308.

[18] K. Jayaraman, A. Choudhury, C. N. R. Rao, *Solid State Sci.* **2002**, *4*, 413 – 422.

[19] G. M. Sheldrick, SADABS Siemens Area Detector Absorption Correction Program, University of Göttingen, Göttingen (Germany), **1994**.

[20] a) G. M. Sheldrick, SHELXS-86 Program for crystal structure determination, University of Göttingen, **1986**; b) G. M. Sheldrick, *Acta Crystallogr. Sect. A* **1990**, *35*, 467.

[21] G. M. Sheldrick, SHELXTL-PLUS Program for Crystal Structure Solution and Refinement, University of Göttingen, Göttingen (Germany).

COMMUNICATIONS

Three-Dimensional Organically Templated Open-Framework Transition Metal Selenites**

Amitava Choudhury, Udaya Kumar D, and
C. N. R. Rao*

A large variety of open-framework metal phosphates has been synthesized and characterized in the last few years.[1] Several open-framework metal carboxylates have also been reported.[2] Besides the phosphates and the carboxylates, open-framework phosphites,[3] arsenates,[4] and germanates[5] are also known. An organically templated layered zinc selenite was described recently by Harrison et al.,[6] but no three-dimensional (3D) open-framework structure with the oxo anions of sulfur or selenium has been isolated to date. We have synthesized for the first time a 3D open-framework iron fluoroselenite templated by various organic amines. Herein we report the synthesis and structure of $[A]^+[Fe_2F_3(SeO_3)_2]$ ($[A] = [C_4N_2H_{12}]_{0.5}$, $[C_4N_3H_{14}]_{0.5}$ or $[NH_4]$) possessing an 8-ring channel. This selenite is an interesting geometrically, frustrated magnetic material by virtue of possessing near-tetrahedral $Fe_4F_6O_{12}$ clusters, comprising equilateral triangles of Fe atoms. In addition to the Fe selenite, we have obtained a new 3D zinc selenite.

Synthesis of open-framework transition metal selenites by amine templating is difficult because of the low reduction potential of the Se^{IV}/Se^0 couple, which under hydrothermal conditions may cause reduction to metallic selenium. We have therefore employed the fluoride route (addition of HF to the reaction medium) of Guth et al.,[7] as elaborated by Férey and co-workers[8] in the case of metal phosphates. The fluoride ion stabilizes the higher oxidation states of the metal and acts as a good mineralizer besides getting incorporated into the frame-work. Reaction of a mixture of $FeCl_3 \cdot 6H_2O$, HF, SeO_2, and piperazine (PIP) in water under hydrothermal conditions yields pure **I** (see Experimental Section). This inorganic

$$[C_4N_2H_{12}]_{0.5}[Fe_2F_3(SeO_3)_2] \quad \textbf{I}$$

framework was obtained with diethylenetetramine (DETA), 1,3-diaminopropane (DAP), and ethylenediamine (en) as well. In the case of DAP and en, the amine decomposes to give the ammonium ion, but DETA remains intact with its two end nitrogen atoms being protonated. In the absence of fluorine in the reaction medium, only dense iron selenites were obtained.

[*] Prof. Dr. C. N. R. Rao,[+] A. Choudhury,[+] U. Kumar D
    Chemistry and Physics of Materials Unit and CSIR Center of
    Excellence in Chemistry
    Jawaharlal Nehru Center for Advanced Scientific Research
    Jakkur, P.O., Bangalore 560 064 (India)
    Fax: (+91) 80-8462766
    E-mail: cnrrao@jncasr.ac.in

[+] Solid State and Structural Chemistry Unit
    Indian Institute of Science
    Bangalore 560 012 (India)

[**] The authors thank Dr. E. V. Sampathkumaran for help with the
    magnetic measurements. A.C. thanks the CSIR, Government of India
    for the award of Senior Research Fellowship.

    Supporting information for this article is available on the WWW under
    http://www.angewandte.com or from the author.

The 3D inorganic framework of $[A]^+[Fe_2F_3(SeO_3)_2]$ is built up from the vertex-sharing FeF_3O_3 octahedra and SeO_3 pseudo-tetrahedra, and the structure can be described in terms of a unique tetrameric cluster, $Fe_4F_6O_{12}$ (Figure 1a). Since the 3D structure obtained with all the amines was the

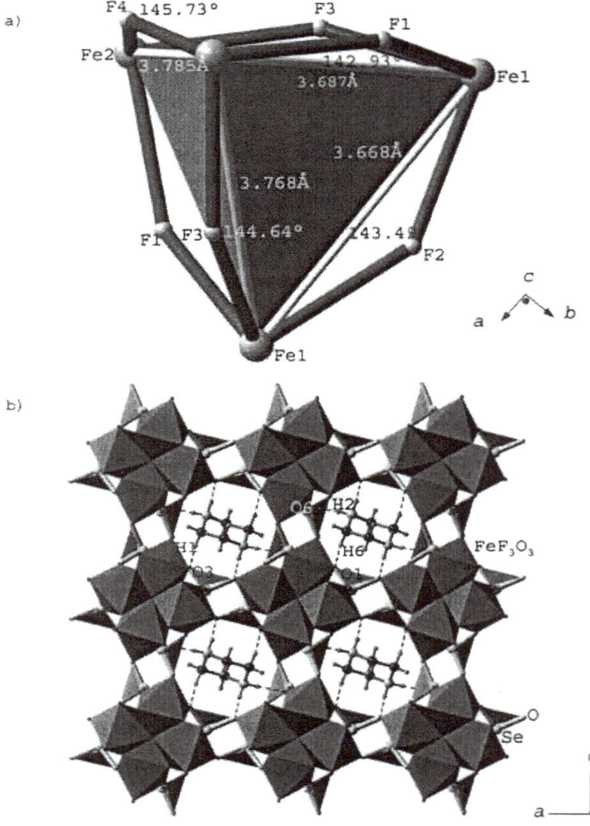

Figure 1. a) The tetrameric cluster of $Fe_4F_6O_{12}$ in **I** formed by the corner-sharing of FeO_3F_3 through six Fe-μF-Fe linkages. Note near perfect tetrahedron formed by connecting the Fe atoms. b) Polyhedral view of the inorganic framework of **I** along the b axis showing one-dimensional 8-ring channel. The amine molecules are located in the middle of the channel and interact with the framework through N—H$\cdots$O hydrogen bonds (dotted lines).

same, we shall only discuss **I** obtained with piperazine. The asymmetric unit of **I** contains 14 non-hydrogen atoms with two crystallographically distinct Fe and Se atoms. The Fe atoms are six-coordinate, surrounded by three oxygen atoms and three fluorine atoms. The average Fe—O bond lengths are 1.964 and 1.953 Å for Fe1 and Fe2, respectively; the corresponding Fe—F bond lengths of 1.987 and 1.982 Å are longer than in other oxyfluorinated Fe^{III} compounds.[9] The two Se atoms, each bound to three O atoms form a pyramid characteristic of the $Se^{III}O_3$ group, with the stereochemically active lone pair of electrons occupying the remaining tetrahedral site. The average Se—O bond lengths are 1.696 and 1.702 Å for Se1 and Se2 respectively, in agreement with

      1433-7851/02/4101-0158 $ 17.50+.50/0   *Angew. Chem. Int. Ed.* **2002**, *41*, No. 1

the literature values.[6] Both the Fe atoms form three Fe-F-Fe linkages and three Fe-O-Se linkages, the former being responsible for the $Fe_4F_6O_{12}$ cluster, wherein four FeF_3O_4 octahedra are corner-shared through six Fe-F-Fe linkages (Figure 1a). This connectivity between the FeF_3O_3 octahedra creates four 3-rings in the cluster. The cluster, which can be considered to be the secondary building unit (SBU), is connected by Fe-O-Se linkages to form the 3D structure, possessing 4-, 5-, and 6-rings as well as an 8-ring 1D channel (with the dimensions 5.847×7.593 Å, the longest atom–atom distance, not including the van der Waal radii) along the b axis (Figure 1b). A section of the intercluster connectivity along the b axis is shown in Figure 2. The protonated PIP molecules reside in the channel and interact with the framework through strong N–H $\cdots$ O hydrogen bonding (d(H1$\cdots$O3) 1.913(8), d(N1$\cdots$O3) 2.764(8) Å, θ(N1-H1$\cdots$O3) 156.9(7)°; d(H2$\cdots$O6) 1.841(7), d(N1$\cdots$O6) 2.740(7) Å, θ(N1-H2$\cdots$O6) 175.2(7)°). There is evidence for C-H$\cdots$O (d(C$\cdots$O)$_{av}$ 3.320 Å) and C-H $\cdots$ F (d(C$\cdots$F) 3.076 Å) interactions as well.

Of the many noteworthy features in the structure of **I**, the tetrameric SBU is rather unusual. The cluster in $[A][Fe_2F_3(SeO_3)_2]$ has a near-tetrahedral geometry with four nearly equilateral triangles, which do not share a common vertex; they are linked by Fe-μ_2-F-Fe and not capped by SeO_3 pyramids (Figure 1a). The literature examples of SBUs formed by Fe-X-Fe (X = O, F, OH) linkages in open-framework phosphates or oxyfluorinated phosphates are generally constructed from a dimer, trimer, or a tetramer of edge- or corner-shared $Fe-F_xO_{6-x}$ ($x = 0-3$) octahedra and capped by PO_4 units; the few known iron–oxo tetranuclear clusters are also different.[10] A somewhat related tetranuclear cluster is that formed by joining of four trigonal-bipyramidal iron centers in $[enH_2]_2[Fe_4O(PO_4)_4] \cdot H_2O$.[11]

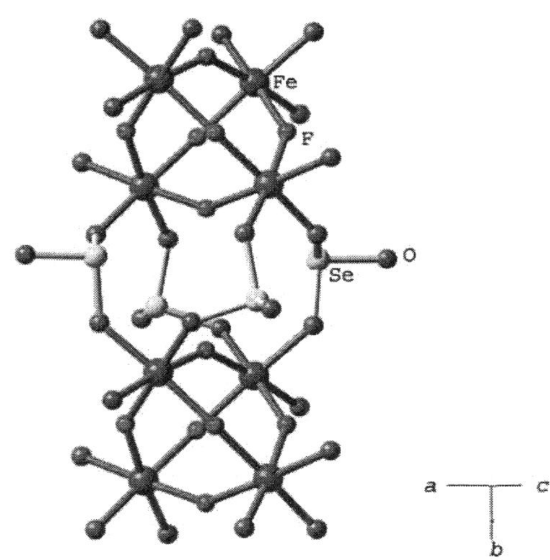

Figure 2. Section of the intercluster connectivity along the b axis showing the presence of 5- and 6-rings.

Variable-temperature magnetic susceptibility χ measurements on **I** in the 2–300 K range show a continuous increase in the susceptibility with the decrease in temperature, with a sharp jump occurring around 20 K (T_t) (Figure 3). The plot of $1/\chi$ versus T ($T > 150$ K) is linear with a μ_{eff} value of 6.1 characteristic of the high-spin Fe^{III} center and a θ_P of -366 K showing predominantly antiferromagnetic interaction. We have measured the isothermal magnetization (M versus H) and the zero-field cooled (ZFC) susceptibility. Careful measurements showed that a small remnant field (~ 1 Oe) was sufficient to force the orientation of the moments below

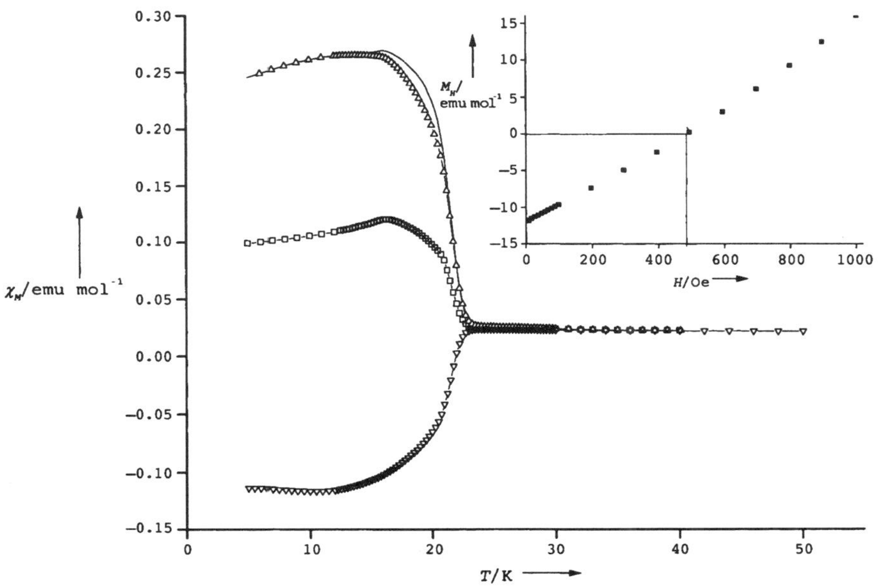

Figure 3. Temperature dependence of the magnetic susceptibility for **I** measured in the presence of a magnetic field of 100 Oe under ZFC and FC conditions. Inset: Section of the M versus H curve at 5 K with a -1 Oe field-cooled state that demonstrates that a field as high as 500 Oe is required to make M positive; (—— FC while decreasing T, $\triangle$ FC while warming, $\square$ ZFC, $\triangledown$ ZFC (in the presence of a remnant field of ~ -1 Oe)).

COMMUNICATIONS

T_t. The ZFC data, recorded after making sure that the remnant field was much smaller than 1 Oe but on the positive side of zero, reveal the divergence shown in Figure 3. We have measured the M versus H curve as well as the ZFC curve after cooling at a small remnant field of -1 Oe. The ZFC curve becomes negative under these conditions as shown in Figure 3, a behavior reported earlier for Ca_3LiRuO_6[12] and for low-dimensional $AFe^{II}Fe^{III}(C_2O_4)_3$.[13] A field of about 500 Oe is required to make M positive in the M versus H plot (Figure 3). The M versus H curve at 5 K is practically linear and does not exhibit any spontaneous magnetization. While the magnetic behavior of **I** is clearly due to magnetic frustration, it is not due to a spin glass as corroborated by AC susceptibility measurements. The frustrated behavior may arise from the tetrahedral geometry[14] of the Fe^{3+} subnetwork of the cluster (Figure 1a) involving four near-equilateral triangles. Such tetrahedral frustrating topology has been observed earlier in the pyrochlore form of FeF_3 (Pyr-FeF_3).[15, 16] The transition temperature in Pyr-FeF_3 is about 18 K compared to $T_t = 23$ K in **I**. The higher T_t value in **I** indicates a relaxation of the frustrated character compared to FeF_3, due to the absence of a long-range frustration by super exchange via the SeO_3 group.

We have, however, obtained an unusual open-framework 3D zinc selenite **II** by hydrothermal synthesis in the presence

$$[C_2N_2H_8]_{0.5}ZnSeO_3 \quad \textbf{II}$$

of en. The polyhedral network of this selenite is built up from ZnO_3N tetrahedra and SeO_3 pyramids connected to form a layer in the ac plane. The nitrogen atom in ZnO_3N tetrahedra comes from the amine (en), the other nitrogen atom of the amine coordinates to the Zn atom of the next nearest layer along the b axis to form a 3D structure. The en molecules thus act as pillars through end-to-end (Zn-N-C-C-N-Zn) connectivity and create channels along the a and c axis (Figure 4).

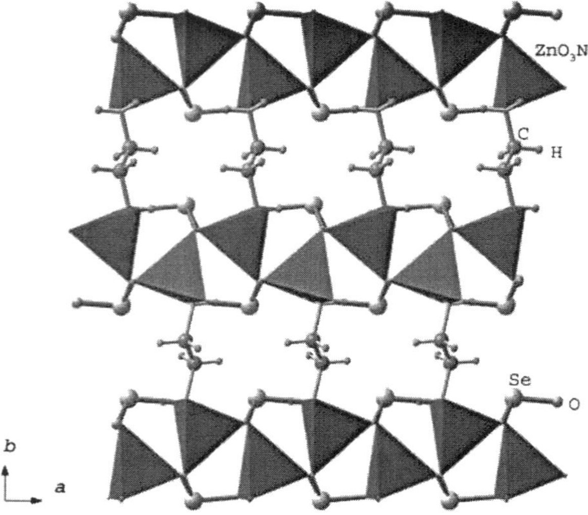

Figure 4. Structure of $[H_2N(CH_2)_2NH_2]_{0.5} \cdot ZnSeO_3$ along the c axis, showing the zinc selenite layers pillared by the en molecule through coordinate bonds.

The structure of **II** is similar to a recently reported zinc phosphite, $[C_2N_2H_8]_{0.5}Zn(HPO_3)$, in which there are two such completely independent interpenetrating networks.[3b]

Experimental Section

I: $FeCl_3 \cdot 6H_2O$ (0.8153 g) was dissolved in water (5.4 mL) in a teflon beaker. To this solution HF (0.43 mL) was added with stirring followed by SeO_2 (0.6692 g). Finally, PIP (0.2598 g) was added to this mixture and stirring continued for another 30 min to obtain a homogeneous gel. The resulting gel with a molar ratio $FeCl_3 \cdot 6H_2O : 4HF : 2SeO_2 : PIP : 100H_2O$ was transferred into a 23 mL PTFE-lined stainless steel autoclave and heated at 150 °C for 5 days. The pH of the starting reaction mixture was 2, and there was little change in pH after completion of the reaction. Rod-shaped faint brown crystals (yield 70 %) of **I** were collected, washed with deionized water, and dried under ambient conditions. Similar reactions were carried out with the amines DETA, DAP, and en.

II: Compound **II** was synthesized starting with the following mixture: ZnO (0.0814 g), H_2SeO_4 (0.515 mL), en (0.067 mL), 36% HCl (0.062 mL), and H_2O (3.6 mL) at 100 °C for three days.

Structure determination: Single-crystal data were collected on a Siemens SMART-CCD diffractometer (graphite-monochromated $Mo_{K\alpha}$ radiation, $\lambda = 0.71073$ Å $(T = 298$ K)). An absorption correction based on symmetry-equivalent reflections was applied using SADABS.[17] The structures were solved by direct methods using SHELXS-86[18] and difference fourier synthesis. The hydrogen atoms associated with piperazine molecules in **I** were placed geometrically and refined in the riding mode. Full-matrix least-squares structure refinement against $|F^2|$ was carried out using SHELXTL-PLUS[19] package of programs. The non-hydrogen atoms were refined anisotropically.

Characterization: The X-ray powder data for all the compounds were in excellent agreement with their simulated patterns based on single-crystal data, indicating the phase purity. The thermogravimetric analysis of **I** reveals that the selenites are thermally stable up to 300 °C, above which there is a sharp one-step weight loss corresponding to the loss of amine molecules, SeO_2, and HF. The calcined products diffracted poorly and showed a few reflections of Fe_2O_3.

Crystal data: $[C_4N_2H_{12}]_{0.5}[Fe_2F_3(SeO_3)_2]$, **I**, $M_r = 466.70$, monoclinic, space group = $C2/c$ (no. 15), $a = 16.1683(14)$, $b = 7.3973(6)$, $c = 16.1275(14)$ Å, $\beta = 90.044(2)°$, $V = 1928.9(3)$ Å^3, $Z = 8$, $\mu = 10.616$ mm^{-1}, $\rho_{calcd} = 3.14$ gcm^{-3}. $[C_2N_2H_8]_{0.5}ZnSeO_3$, **II**: $M_r = 222.38$, monoclinic, space group = $P2_1/n$ (no. 14), $a = 4.8180(7)$, $b = 15.197(2)$, $c = 6.4863(9)$ Å, $\beta = 91.724(2)°$, $V = 474.71(12)$ Å^3, $Z = 4$, $\mu = 12.724$ mm^{-1}, $\rho_{calcd} = 3.112$ gcm^{-3}. Unique reflections: **I**, $1391(R_{int} = 0.0352)$; **II**, $686(R_{int} = 0.0432)$. Final R, R_w: **I**, 0.03, 0.07; **II**, 0.04, 0.09. GOF: **I**, 1.07; **II**, 1.05.

Crystallographic data (excluding structure factors) for the structures reported in this paper have been deposited with the Cambridge Crystallographic Data Centre as supplementary publication nos. CCDC-170619 (**I**) and CCDC-170620 (**II**). Copies of the data can be obtained free of charge on application to CCDC, 12 Union Road, Cambridge CB21EZ, UK (fax: (+44) 1223-336-033; e-mail: deposit@ccdc.cam.ac.uk).

Received: September 3, 2001 [Z17846]

[1] A. K. Cheetham, G. Férey, T. Loiseau, *Angew. Chem.* **1999**, *111*, 3466; *Angew. Chem. Int. Ed.* **1999**, *38*, 3268–3292.

[2] a) C. Livage, C. Egger, G. Ferey, *Chem. Mater.* **1999**, *11*, 1546–1550; b) T. M. Reineke, M. Eddaoudi, M. O'Keeffe, O. M. Yaghi, *Angew. Chem.* **1999**, *111*, 2712–2716; *Angew. Chem. Int. Ed.* **1999**, *38*, 2590–2594; c) H. Li, M. Eddaoudi, T. L. Groy, O. L. Yaghi, *J. Am. Chem. Soc.* **1998**, *120*, 8571–8572; d) R. Vaidhyanathan, S. Natarajan, C. N. R. Rao, *Chem. Mater.* **2001**, *13*, 185–191.

[3] a) S. Fernandez, J. L. Mesa, J. L. Pizarro, L. Lezama, M. I. Arriortua, R. Olazcuaga, T. Rojo, *Chem. Mater.* **2000**, *12*, 2092–2098; b) J. A. Rodgers, W. T. A. Harrison, *Chem. Commun.* **2000**, 2385–2386; c) W. T. A. Harrison, M. L. F. Phillips, T. M. Nenoff, *J. Chem. Soc. Dalton Trans.* **2001**, 2459–2461.

© WILEY-VCH Verlag GmbH, 69451 Weinheim, Germany, 2002   1433-7851/02/4101-0160 $ 17.50+.50/0   *Angew. Chem. Int. Ed.* **2002**, *41*, No. 1

[4] a) T. E. Gier, X. Bu, P. Feng, G. D. Stucky, *Nature* **1998**, *395*, 154–157; b) S. Ekambaram, S. C. Sevov, *Inorg. Chem.* **2000**, *39*, 2405–2410.

[5] a) X. Bu, P. Feng, G. D. Stucky, *J. Am. Chem. Soc.* **1998**, *120*, 11204–11205; b) X. Bu, P. Feng, T. E. Gier, D. Zhao, G. D. Stucky, *J. Am. Chem. Soc.* **1998**, *120*, 13389–13397.

[6] W. T. A. Harrison, M. L. F. Phillips, J. Stanchfield, T. M. Nenoff, *Angew. Chem.* **2000**, *112*, 3966–3968; *Angew. Chem. Int. Ed.* **2000**, *39*, 3808–3810.

[7] J. L. Guth, H. Kessler, R. Wey, *Stud. Surf. Sci. Catal.* **1986**, *28*, 121.

[8] a) G. Férey, *J. Fluorine Chem.* **1995**, *72*, 187–193; b) G. Férey, *C. R. Acad. Sci. Ser. C* **1998**, *1*, 1–13; c) M. Cavellec, D. Riou, G. Férey, *Inorg. Chim. Acta* **1999**, *291*, 317–325.

[9] M. Riou-Cavellec, J.-M. Grenèche, D. Riou, G. Férey, *Chem. Mater.* **1998**, *10*, 2434–2439.

[10] a) V. Zima, K. H. Lii, *J. Chem. Soc. Dalton Trans.* **1998**, 4109–4112; b) K. H. Lii, Y. F. Huang *Chem. Commun.* **1997**, 1311–1312.

[11] a) J. R. D. DeBord, W. M. Reiff, C. J. Warren, R. C. Houshalter, J. Zubieta, *Chem. Mater.* **1997**, *9*, 1994–1998; b) C. Y. Huans, S. L. Wang, K. H. Lii, *J. Porous Mater.* **1998**, *5*, 147–152.

[12] M. Mahesh Kumar, E. V. Sampathkumaran, *Solid State Commun.* **2000**, *144*, 643–647.

[13] C. J. Nutall, P. Day, *Chem. Mater.* **1990**, *10*, 3050–3057.

[14] J. E. Greedan, *J. Mater Chem.* **2001**, *11*, 37–53.

[15] G. Férey, R. De Pape, M. Leblanc, J. Pannetier, *Rev. Chim. Miner.* **1986**, *23*, 474–484.

[16] R. De Pape, G. Férey, *Mater. Res. Bull.* **1986**, *21*, 971–978.

[17] G. M. Sheldrick, SADABS Siemens Area Detector Absorption Correction Program, University of Götingen, Götingen, Germany, **1994**.

[18] a) G. M. Sheldrick, SHELXS-85 Program for crystal structure determination, University of Götingen, **1986**; b) G. M. Sheldrick, *Acta Crystallogr. Sect A.* **1990**, *35* 467.

[19] G. M. Sheldrick, SHELXTL-PLUS Program for Crystal Structure Solution and Refinement, University of Götingen, Götingen, Germany.

Sequential Catalytic Asymmetric Allylic Transfer Reaction: Enantioselective and Diastereoselective Construction of Tetrahydropyran Units**

Chan-Mo Yu,* Jae-Young Lee, Byungran So, and Junghyun Hong

There is considerable interest in the enhancement of catalysts for Lewis acid promoted reactions so that practical and useful levels of asymmetric synthesis can be achieved.[1] Other research groups have used a ligand-accelerated strategy[2] to activate the chiral Lewis acid through the internal regulation of the chiral catalyst by a modification of the ligand, thus leading to impressive advances in carbonyl addition reactions.[3] In previous studies we demonstrated

[*] Prof. Dr. C.-M. Yu, J.-Y. Lee, B So, J. Hong
Department of Chemistry and BK-21 School of Molecular Science
Sungkyunkwan University
Suwon 440-746 (Korea)
Fax: (+82) 31-290-7075
E-mail: cmyu@chem.skku.ac.kr

[**] Generous financial support from the Center for Molecular Design and Synthesis (CMDS: KOSEF SRC) at KAIST and the Korea Ministry of Science and Technology through the National Research Laboratory program is gratefully acknowledged.

that the use of molecular synergistic reagents in catalytic asymmetric allylic transfer reactions resulted in a significant increased catalytic ability as the chiral catalyst could be regenerated expediently. Our strategy involves the use of BINOL–Ti^{IV} complex (BINOL = 2,2'-binaphthol) as a chiral promoter along with iPrSBEt$_2$ or iPrSSiMe$_3$ as an accelerating synergistic reagent. Recently, this approach provided highly catalytic versions of enantioselective allylic transfer reactions of achiral aldehydes, for example, allylation,[4] propargylation,[5] allenylation,[6] and dienylation.[7] The efficiency of this protocol in terms of enantioselectivity and catalytic ability has encouraged us to apply the extension of this method to more versatile systems, which would expand the scope and utility of allylic transfer reactions. Described herein is an extension of our strategy aimed at finding new reagents and realizing practical ways to advance new levels of asymmetric synthesis. In this study we focus on the sequential addition of a bifunctional reagent to two aldehydes to form an asymmetric tetrahydropyran system (Scheme 1). An efficient method for

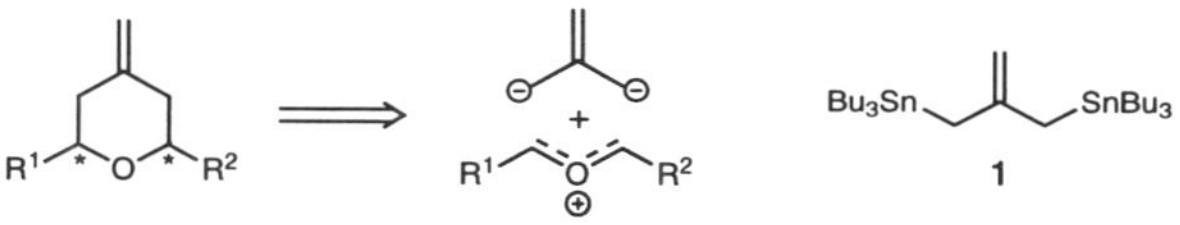

Scheme 1. General protocol for the synthesis of a tetrahydropyran by means of sequential allylic transfer reactions.

this reaction would be useful in the synthesis of biologically active substances that contain a tetrahydropyran unit.[8] Several crucial points emerged from this investigation, including the development of a new catalyst and reagents for the sequential allylic transfer reaction, the introduction of highly efficient promoters for the formation of the tetrahydropyran ring, and the highly stereoselective synthesis of four different tetrahydropyran systems that contain an exocyclic double bond.

Our initial studies began with bis-stannane **1** as a dianion equivalent which was prepared from Bu$_3$SnLi (two equivalents) and methallyl dichloride at $-78\,^{\circ}$C in THF. Initial attempts to add **1** to benzaldehyde in the presence of BINOL–Ti^{IV} [a 2:1 mixture of BINOL and Ti(OiPr)$_4$][9] were not successful. The reactivity was improved by introducing a synergistic reagent.[4–7] Treatment of **1** with benzaldehyde in the presence of the catalyst (10 mol%), followed by the dropwise addition of iPrSBEt$_2$ at $-20\,^{\circ}$C for 20 h in CH$_2$Cl$_2$ afforded undesired **3** and **2** in a combined yield of 44% (2:1) [Eq. (1)]. The formation of alcohol **2** was attributed to a

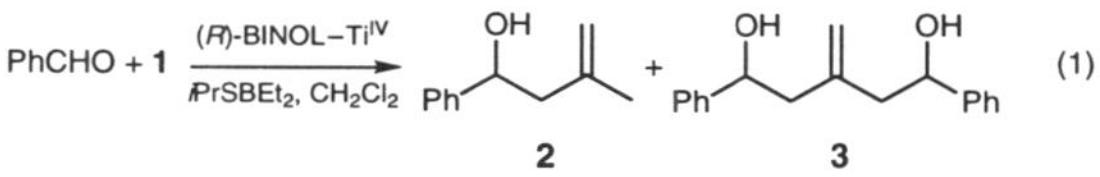

proton source in the reaction mixture (iPrOH with Ti^{IV} species). We speculated that the limited scope and side reaction with **1** could be solved by introducing a relatively less bulky and more stable silyl substituent instead of the stannyl

DALTON
FULL PAPER

Synthesis of a hierarchy of zinc oxalate structures from amine oxalates

R. Vaidhyanathan, Srinivasan Natarajan and C. N. R. Rao *

Chemistry and Physics of Materials Unit, Jawaharlal Nehru Centre for Advanced Scientific Research, Jakkur P.O., Bangalore 560 064, India. E-mail: cnrrao@jncasr.ac.in

Received 24th October 2000, Accepted 18th December 2000
First published as an Advance Article on the web 12th February 2001

A hierarchy of novel zinc oxalates including monomers and dimers has been prepared by reaction of amine oxalates with zinc ions, the amine oxalates having been characterized for the first time. In most of the amine oxalates one of the carboxyl groups transfers a proton to the amino nitrogen, leaving the other carboxyl group free to form hydrogen bonds. The zinc oxalates obtained are composed of a network of ZnO_6 octahedra and oxalate units, and possess zero-, one-, two- and three-dimensional structures. The monomer, dimer, and chain zinc oxalates are the first members of the hierarchy of structures. Relationships amongst these various oxalate structures are noteworthy and give indications as to the manner in which these structures are formed. Thus, the three-dimensional structure can be formed by the linking of layers, and the layer structure by condensation of linear chains. The isolation and characterization of a hierarchy of zinc oxalates of differing dimensionalities assumes significance in the light of the recent finding that low-dimensional structures transform to higher, more complex structures in the phosphate family.

Introduction

A variety of open-framework metal phosphates possessing one-, two- and three-dimensional structures are known today.[1] There is reason to believe that the low-dimensional structures, those of zero and one dimension in particular, may act as starting building units in the formation of the more complex open-framework phosphates.[2] Among the novel open-framework materials not based on phosphates, those of phosphonates[3,4] and carboxylates[5,6] are noteworthy. Open-framework zinc and tin(II) oxalates have been synthesized hydrothermally in the presence of organic amines.[7] It has been shown recently that amine phosphates may act as intermediates in the formation of open-framework metal phosphates, and provide a convenient route for the synthesis of these materials.[8] Thus, reactions of amine phosphates with metal ions yield a variety of metal phosphates with different structures. Furthermore, this route also yields metal phosphates of different dimensionalities. It occurred to us that amine carboxylates could play a similar role in the formation of metal carboxylates with open architectures. We have, therefore, investigated the reaction of oxalates of organic amines with zinc ions. For this purpose, amine oxalates had to be synthesized and characterized for the first time, as there is hardly any report on these compounds in the literature. Interestingly, we have obtained zinc oxalates with one-, two- and three-dimensional architectures, in addition to the zero-dimensional monomeric and dimeric compounds, by the reaction of zinc ions with amine oxalates. These different structures exhibit interesting relationships, suggesting the possible

presence of an *aufbau* principle in the formation of the hierarchy of oxalate structures.

Experimental

Synthesis of amine oxalates

In order to study the reaction of the amine oxalates with zinc ions, we first prepared the oxalates of organic amines such as propylamine (PRO), guanidine (GUO), piperazine (PIPO) and 1,4-diazabicyclo[2.2.2]octane (DABCO-O). In a typical experiment, the amine was added dropwise to an aqueous solution of oxalic acid with continuous stirring. The resulting solution was heated at 85 °C for 12 h and left at room temperature to obtain single crystals of the amine oxalates. The compositions of the amine oxalates are: $[CN_3H_6][HC_2O_4]\cdot H_2O$ (GUO), $[C_4N_2H_{12}][HC_2O_4]_2$ (PIPO), $[C_6N_2H_{14}][HC_2O_4]_2$ (DABCO-O) and $[C_3NH_{10}][HC_2O_4]\cdot H_2O$ (PRO). Elemental analysis using ICP-MS (inductively coupled plasma mass spectroscopy) of the amine oxalates confirmed the above compositions. The synthesis conditions for the amine oxalates and the analysis of the products are given in Table 1. The amine oxalates, thus obtained, were characterized using single crystal X-ray diffraction and elemental analysis.

Synthesis of zinc oxalates

The amine oxalates were treated with zinc ions under hydrothermal conditions. In a typical experiment, 0.137 g of ZnO

Table 1 Compositions of the amine oxalates[a]

Starting composition	Elemental analysis[b] (%)	Product composition
$2\ H_2C_2O_4 : CN_3H_6 : 25\ H_2O$	N 24.95 (25.15), C 21.17 (21.55), H 5.39 (5.39)	$[CN_3H_6][HC_2O_4]\cdot H_2O$ (GUO)
$2\ H_2C_2O_4 : C_4H_{10}N_2 : 25\ H_2O$	N 10.95 (10.45), C 36.48 (36.09), H 5.56 (5.26)	$[C_4N_2H_{12}]_{0.5}[HC_2O_4]$ (PIPO)
$2\ H_2C_2O_4 : C_6N_2H_{12} : 25\ H_2O$	N 9.84 (9.59), C 41.86 (41.09), H 5.41 (5.49)	$[C_6N_2H_{14}][HC_2O_4]_2$ (DABCO-O)
$H_2C_2O_4 : C_3H_9N : 25\ H_2O$	N 8.58 (8.81), C 37.80 (37.74), H 7.71 (8.17)	$[C_3NH_{10}][HC_2O_4]\cdot H_2O$ (PRO)

[a] The synthesis was carried out at 85 °C for 12 h. [b] The values in parenthesis are the calculated values based on the single crystal structure.

DOI: 10.1039/b008571p

Table 2 Composition and synthesis conditions for the zinc oxalates **1**–**5**

Starting composition	$T/^{\circ}C$	t/h	Product composition
$ZnO : 5GUO : 100H_2O$	85	360	$[CN_3H_6]_2[Zn(H_2O)_2(C_2O_4)_2]$ **1**
$ZnO : 2PIPO : 100H_2O$	50	120	$[C_4N_2H_{12}]_3[Zn_2(C_2O_4)_5]\cdot 8H_2O$ **2**
$ZnO : 2PIPO : 100H_2O$	85	96	**2**
$ZnO : 5PIPO : 100H_2O$	70	120	$[C_4N_2H_{12}][Zn_2(C_2O_4)_3]\cdot 4H_2O$ **4**
$ZnO : 5PIPO : 12DMF$	110	84	**4**
$ZnO : DABCO\text{-}O : 200H_2O$	150	72	$[C_6N_2H_{14}][Zn(C_2O_4)_2]\cdot 3H_2O$ **3**
$ZnO : 5DABCO\text{-}O : 100H_2O$	110	192	**3**
$ZnO : 5DABCO\text{-}O : 100H_2O$	150	120	**3**
$1.5ZnO : 0.5HCl : 2H_2C_2O_4 : 1DABCO\text{-}O : 200H_2O$	130	72	**3**
$ZnO : 1HCl : 2H_2C_2O_4 : 1.5DABCO\text{-}O : 200H_2O$	130	156	**3**
$ZnO : 2.2PRO : 250H_2O$	150	48	$[C_3NH_{10}]_2[Zn_2(C_2O_4)_3]\cdot 3H_2O$ **5** $+ Zn(C_2O_4)\cdot 2H_2O$
$ZnO : 2.2PRO : 250H_2O$	110	84	**5** + PRO
$ZnO : 2.2PRO : 250H_2O$	150	72	**5**

Table 3 Crystal data and structure refinement parameters for the amine oxalates

	GUO	PIPO	DABCO-O	PRO
Chemical formula	$C_3H_9N_3O_5$	$C_8H_{14}N_2O_8$	$C_{10}H_{16}N_2O_8$	$C_{10}H_{24}N_2O_9$
Formula mass	167.13	266.21	292.25	316.31
Crystal symmetry	Monoclinic	Monoclinic	Triclinic	Monoclinic
Space group	$P2_1/c$ (no. 14)	$C2/c$ (no. 15)	$P\bar{1}$ (no. 2)	$C2/c$ (no. 15)
T/K	293(2)	293(2)	293(2)	293(2)
$a/Å$	6.699(1)	15.963(11)	6.971(11)	18.004(11)
$b/Å$	10.552(6)	5.716(2)	9.607(2)	5.679(2)
$c/Å$	10.224(4)	12.354(4)	10.633(4)	13.986(4)
$a/^{\circ}$			91.25(7)	
$\beta/^{\circ}$	103.77(3)	108.07(3)	107.5(3)	126.85 (3)
$\gamma/^{\circ}$			109.3(6)	
$V/Å^3$	702.0(9)	1071.5(9)	634.8(9)	1144.4(9)
Z	4	4	2	4
μ/mm^{-1}	0.149	0.49	0.134	0.117
No. of measured/observed reflections	1012/663	760/686	1781/1436	1134/970
$R1, wR2 [I > 2\sigma(I)]$	0.08, 0.21	0.029, 0.076	0.049, 0.139	0.043, 0.12

Table 4 Crystal data and structure refinement parameters for the zinc oxalates **1**–**4**

	1	2	3	4
Chemical formula	$C_6H_{16}N_6O_{10}Zn$	$C_{11}H_{18}N_3O_{14}Zn$	$C_{10}H_{20}N_2O_{11}Zn$	$C_{10}H_{20}N_2O_{16}Zn_2$
Formula mass	397.62	481.65	409.65	555.1
Crystal symmetry	Monoclinic	Monoclinic	Monoclinic	Monoclinic
Space group	$C2/c$ (no. 15)	$C2/c$ (no. 15)	$P2_1/n$ (no. 14)	$C2/c$ (no. 15)
T/K	293(2)	293(2)	293(2)	293(2)
$a/Å$	14.170(2)	13.799(1)	9.433(1)	16.725(3)
$b/Å$	10.129(1)	11.524(2)	16.860(2)	9.254(1)
$c/Å$	11.324(3)	25.412(2)	9.788(1)	31.298(2)
$\beta/^{\circ}$	115.4(2)	105.2(1)	91.4(1)	98.59(1)
$V/Å^3$	1468.2(3)	3897.2(2)	1556.1(2)	4789.9(3)
Z	4	8	4	4
μ/mm^{-1}	1.739	1.337	1.641	2.086
No. of measured/observed reflections	1053/916	2758/2421	2246/1366	3439/2403
$R1, wR2 [I > 2\sigma(I)]$	0.044, 0.098	0.042, 0.116	0.048, 0.104	0.054, 0.135

[a] For structural details for compound **5**, please see ref. 12.

was dispersed in 3 ml of water and 2 g of GUO added with continuous stirring. The contents were homogenized, transferred to a PTFE-lined stainless steel acid digestion bomb (Parr, USA) and heated at 85 °C for 15 days. The final composition of the mixture was $ZnO : 5GUO : 100H_2O$. The above reaction resulted in the formation of colorless rod-shaped crystals of composition $[CN_3H_6]_2[Zn(H_2O)_2(C_2O_4)_2]$ **1**. A similar procedure was adopted for the preparation of compounds **2**–**5** by employing different amine oxalates. In the majority of cases good quality single crystals, stable under laboratory conditions, were obtained. The single crystals were employed for all subsequent characterization purposes. Powder X-ray diffraction (XRD) patterns on the powdered crystals indicated that the products were consistent with the structure determined by single crystal X-ray diffraction. A summary of all the synthesis conditions and compositions of the products obtained is presented in Table 2.

Single crystal structure determination

A suitable single crystal of each compound including the amine oxalates was carefully selected under a polarizing microscope and glued to a thin glass fiber with cyanoacrylate (super glue) adhesive. Single crystal structure determination by X-ray diffraction was performed on a Siemens Smart-CCD diffractometer (Mo-Kα radiation, $\lambda = 0.71073$ Å). A hemisphere of intensity data was collected at room temperature. Pertinent experimental details for the amine oxalates and zinc oxalates are presented in Tables 3 and 4.

Table 5 Important hydrogen bond distances and angles for the amine oxalates

Compound	N⋯O/Å	N–H⋯O/°	O⋯O/Å	O–H⋯O/°	C⋯O/Å	C–H⋯O/°
GUO[a]	2.881(5)	177.0(4)	—	—	—	—
	2.923(5)	171.0(4)	—	—	—	—
PIPO[b]	2.816(19)	150.0(2)	2.563(16)	172.0(2)	3.431(2)	163.0(17)
	2.784(2)	158.0(2)	—	—	3.328(2)	159.0(18)
	—	—	—	—	3.373(2)	158.0(19)
DABCO-O[c]	2.619(3)	152.0(2)	2.619(3)	158.0(3)	3.393(3)	165.0(2)
	2.619(3)	148.8(19)	2.741(3)	157.0(3)	3.359(3)	141.0(18)
	—	—	—	—	3.419(3)	153.1(18)
PRO[b]	2.868(4)	142	2.580(2)	176	3.427(8)	147
	2.862(6)	172	—	—	—	—

[a] Amine and oxalate units alternate in the layer resulting in only N–H⋯O type hydrogen bonds. [b] Linear chains are formed by oxalate moieties. [c] Dimer units are formed by oxalate moieties as in oxalic acid.

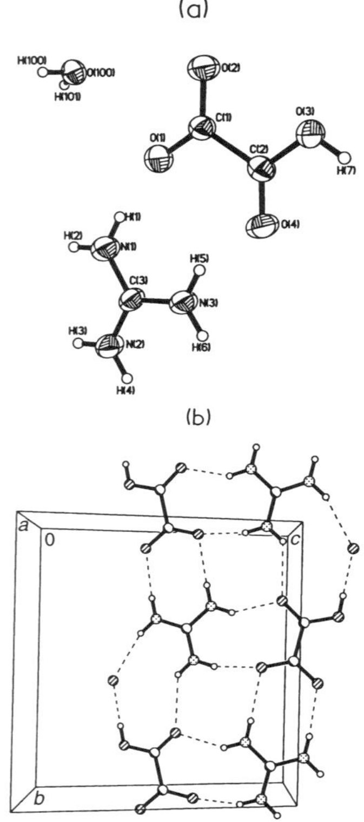

Fig. 1 (a) An ORTEP[12] plot of guanidinium oxalate, $[CN_3H_6]$-$[HC_2O_4]\cdot H_2O$ (GUO). Thermal ellipsoids (in all the structures) are given at 50% probability. (b) Packing diagram of GUO along the bc plane showing the hydrogen bond interactions (dashed lines).

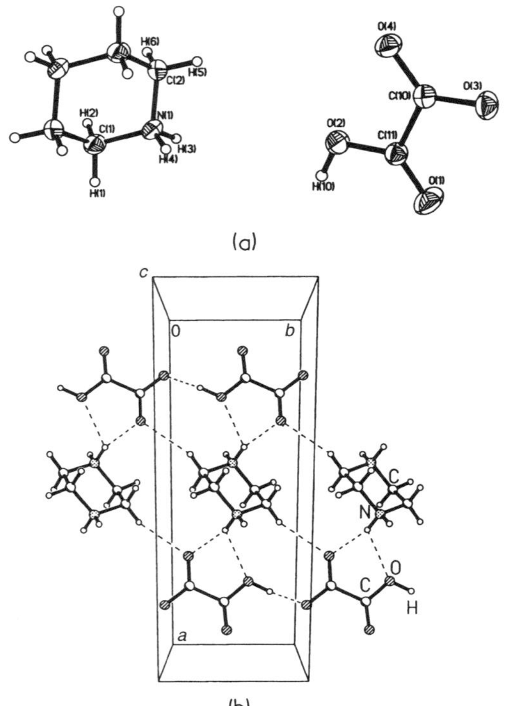

Fig. 2 (a) ORTEP plot of piperazinium oxalate, $[C_4N_2H_{12}][HC_2O_4]_2$ (PIPO). Asymmetric unit is labeled. (b) Packing diagram of PIPO along the ab plane showing the arrangement of the oxalate and the amine. Note that the oxalate units form a chain-like arrangement with the amine in between.

An empirical absorption correction based on symmetry equivalent reflections was applied for all five compounds using the SADABS[9] program and the structures of the amine oxalates as well as of the zinc oxalates were solved by direct methods using SHELXS 86[10] and Fourier difference syntheses. All the hydrogen positions for the amine oxalates and compounds **1–4** were initially located in the Fourier difference maps, and for the final refinement the hydrogen atoms were placed geometrically and held in the riding mode. The last cycles of refinement included positions for all the atoms, anisotropic thermal parameters for all non-hydrogen atoms and isotropic thermal parameters for all the hydrogen atoms. Full-matrix least-squares structure refinement against $|F^2|$ was carried out using the SHELXTL PLUS[11] package of programs. Important bond distances and angles related to the hydrogen bonds in the

amine oxalates are listed in Table 5, selected bond distances and angles of **1**, **2** and **3** in Tables 6, 7 and 8 and for **4** in Tables 9 and 10 respectively.

CCDC reference numbers 151822–151825, 151827–151830.

See http://www.rsc.org/suppdata/dt/b0/b008571p/ for crystallographic data in CIF or other electronic format.

Results

Amine oxalates

The structures of the amine oxalates synthesized and characterized in the present study are given in Figs. 1–4. They show strong hydrogen bond interactions typical of non-covalent solids involving the oxalate and amine moieties. The structurally significant hydrogen bonds arise mainly from N–H⋯O and O–H⋯O interactions. In most of the amine oxalates one

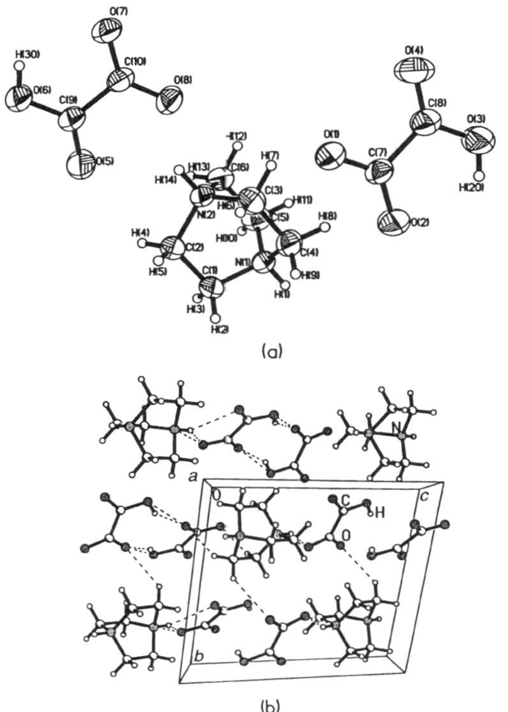

(a)

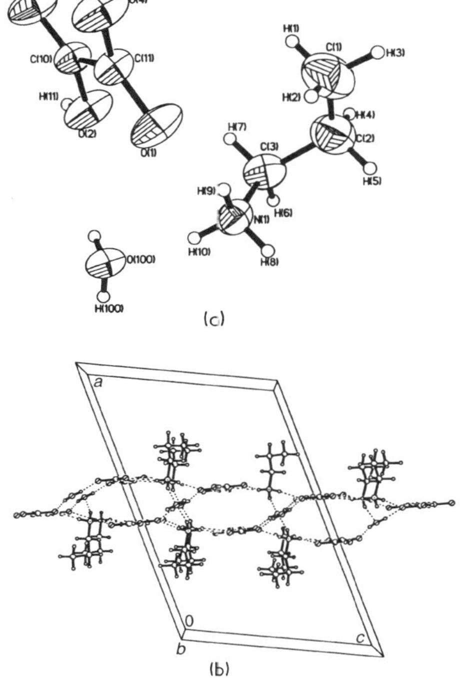

(b)

Fig. 3 (a) ORTEP plot of DABCO oxalate, [C₆N₂H₁₄][HC₂O₄]₂ (DABCO-O). (b) Structure of DABCO-O along the *bc* direction. Note the formation of oxalic acid dimers.

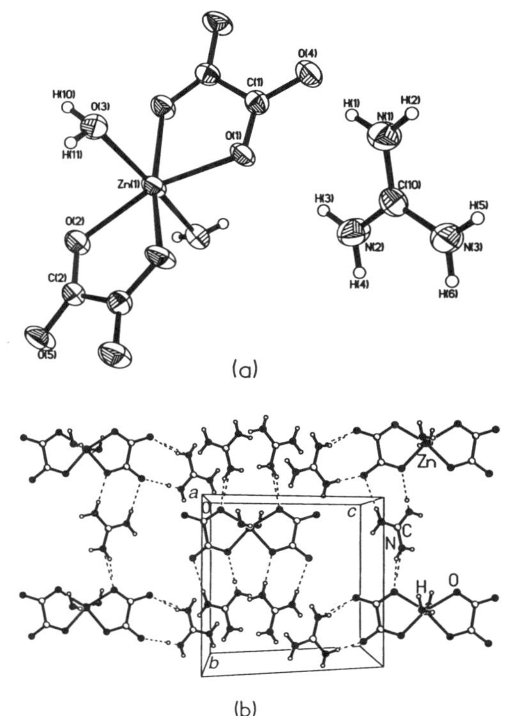

(a)

(b)

Fig. 5 (a) ORTEP plot of the zinc oxalate monomer, [CN₃H₆]₂[Zn-(H₂O)₂(C₂O₄)₂] **1**. Asymmetric unit is labeled. (b) Structure of the monomer along the [100] direction. Dashed lines represent hydrogen bond interactions.

(A)

(B)

of the carboxyl groups transfers a proton to the amino nitrogen, leaving the other carboxyl group free to form a linear hydrogen bonded chain as shown schematically in structure **A**. In DABCO-O the amine forms two N–H · · · O bonds giving a distorted cyclic carboxylic dimer in the *anti* conformation as shown in structure **B**. The N · · · O and O · · · O distances in the amine oxalates are all around 2.5 Å and the N–H · · · O and O–H · · · O bond angles are larger than 150° (Table 5).

Zinc oxalates

Zero-dimensional [CN₃H₆]₂[Zn(H₂O)₂(C₂O₄)₂] 1 and [C₄N₂H₁₂]₃[Zn₂(C₂O₄)₅]·8H₂O 2. The asymmetric unit of compound **1**, presented in Fig. 5a, contains 12 non-hydrogen atoms. The structure of **I** is that of a monomer consisting of two oxalate units directly linked to Zn atoms, which are also bonded to two water molecules. The monomeric zinc oxalate units are held by strong hydrogen bonds with the monoprotonated amine (Fig. 5b). The Zn atoms are octahedrally coordinated with respect to oxygen atoms with Zn–O distances in the range 2.082(3)–2.154(4) Å (av. 2.113 Å) and C–O bond distances in the range 1.238(5)–1.265(5) Å (av. 1.254 Å) (Table 6). The O–Zn–O angles are in the range 83.08(14)–171.0(2)° (av. 108.6°) and the O–C–O bond angles have an average value of 125.7°. These values are typical for this type of bonding and similar bond distances and angles have been observed earlier in similar compounds.[13,14]

Fig. 4 (a) ORTEP plot of propylammonium oxalate, [C₃NH₁₀]-[HC₂O₄]·H₂O (PRO). (b) Structure of PRO along the *ac* plane. Note the strong hydrogen bond interaction between the oxalate and water molecule.

Table 6 Selected bond distances (Å) and angles (°) for [CN₃H₆]₂[Zn(H₂O)₂(C₂O₄)₂] 1

Zn(1)–O(1)	2.082(3)	O(1)–C(1)	1.256(5)
Zn(1)–O(1)[#1]	2.082(3)	O(2)–C(2)	1.265(5)
Zn(1)–O(2)[#1]	2.102(3)	O(4)–C(1)	1.257(5)
Zn(1)–O(2)	2.102(3)	O(5)–C(2)	1.238(5)
Zn(1)–O(3)	2.154(4)	C(1)–C(2)[#1]	1.563(6)
Zn(1)–O(3)[#1]	2.154(4)		
O(1)–Zn(1)–O(1)[#1]	85.9(2)	O(1)[#1]–Zn(1)–O(3)	90.2(2)
O(1)–Zn(1)–O(2)[#1]	79.28(11)	O(2)–Zn(1)–O(3)	92.19(13)
O(1)–Zn(1)–O(2)	162.87(12)	O(2)[#1]–Zn(1)–O(3)	83.08(14)
O(2)[#1]–Zn(1)–O(2)	116.6(2)	O(3)–Zn(1)–O(3)[#1]	171.0(2)
O(1)–Zn(1)–O(3)	96.39(14)	O(4)–C(1)–C(2)[#1]	119.2(4)
C(1)–O(1)–Zn(1)	114.3(3)	O(5)–C(2)–C(1)[#1]	117.6(4)
C(2)–O(2)–Zn(1)	113.5(3)	O(1)–C(1)–O(4)	124.3(4)
O(1)–C(1)–C(2)[#1]	116.5(4)	O(5)–C(2)–O(2)	127.0(4)
O(2)–C(2)–C(1)[#1]	115.4(4)		

Symmetry transformation used to generate equivalent atoms: #1 $-x + 1, y, -z + \frac{3}{2}$.

Table 7 Selected bond distances (Å) and angles (°) in [C₄N₂H₁₂]₃[Zn₂(C₂O₄)₅]·8H₂O 2

Zn(1)–O(1)	2.042(3)	O(1)–C(1)	1.279(5)
Zn(1)–O(2)	2.069(3)	O(7)–C(1)	1.233(5)
Zn(1)–O(4)	2.093(3)	O(2)–C(2)	1.257(5)
Zn(1)–O(5)	2.097(3)	O(3)–C(2)	1.242(6)
Zn(1)–O(9)	2.160(3)	O(4)–C(3)	1.268(5)
Zn(1)–O(10)	2.164(3)	O(6)–C(3)	1.242(5)
C(1)–C(3)	1.555(6)	O(5)–C(4)	1.269(5)
C(2)–C(4)	1.569(6)	O(8)–C(4)	1.226(6)
C(5)–C(5)[#1]	1.544(8)	O(9)–C(5)	1.268(5)
		O(10)–C(5)[#1]	1.252(5)
O(1)–Zn(1)–O(2)	171.33(13)	O(5)–Zn(1)–O(10)	91.35(12)
O(1)–Zn(1)–O(4)	79.98(11)	O(9)–Zn(1)–O(10)	77.14(10)
O(2)–Zn(1)–O(4)	96.35(12)	C(1)–O(1)–Zn(1)	114.9(3)
O(1)–Zn(1)–O(5)	93.59(12)	C(2)–O(2)–Zn(1)	114.5(3)
O(2)–Zn(1)–O(5)	79.64(13)	C(3)–O(4)–Zn(1)	112.9(3)
O(4)–Zn(1)–O(5)	104.41(12)	C(4)–O(5)–Zn(1)	114.0(3)
O(1)–Zn(1)–O(9)	101.35(13)	C(5)–O(9)–Zn(1)	111.6(2)
O(2)–Zn(1)–O(9)	86.41(12)	C(5)[#1]–O(10)–Zn(1)	111.3(3)
O(4)–Zn(1)–O(9)	89.57(12)	O(7)–C(1)–O(1)	125.9(4)
O(5)–Zn(1)–O(9)	161.16(11)	O(3)–C(2)–O(2)	125.2(4)
O(1)–Zn(1)–O(10)	90.85(12)	O(6)–C(3)–O(4)	126.2(4)
O(2)–Zn(1)–O(10)	94.69(13)	O(8)–C(4)–O(5)	126.0(4)
O(4)–Zn(1)–O(10)	162.13(12)	O(10)[#1]–C(5)–O(9)	125.5(4)

Symmetry transformation used to generate equivalent atoms: #1 $-x, y, -z + \frac{3}{2}$.

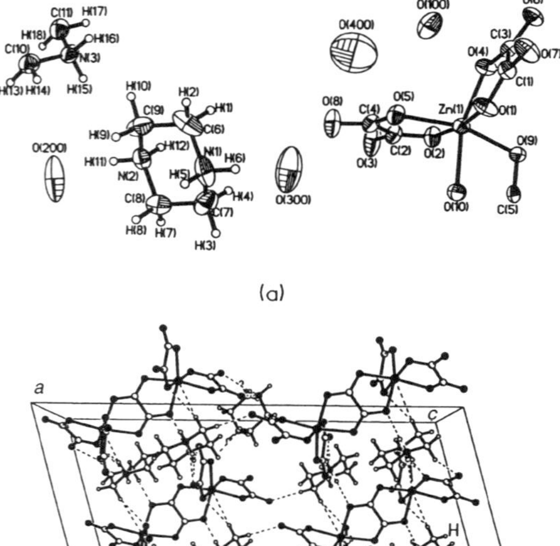

(a)

(b)

Fig. 6 (a) ORTEP plot of the zinc oxalate, dimer [C₄N₂H₁₂]₃[Zn₂(C₂O₄)₅]·8H₂O 2. (b) Structure of the dimer, along the [010] direction. The dimer and the amine alternate in a plane. Water molecules are omitted for clarity. Dashed lines are hydrogen bond interactions.

The asymmetric unit of compound **2** contains 29 non-hydrogen atoms (Fig. 6a). Unlike in **1**, in **2** there are two Zn atoms, which are connected by one oxalate unit. The Zn atoms also possess two terminal oxalates forming a dimeric unit. Just as in **1**, the dimeric zinc oxalate is strongly hydrogen bonded to the diprotonated amine and water molecules (Fig. 6b). The octahedrally coordinated Zn atoms have Zn–O distances in the range 2.042(3)–2.164(3) Å (av. 2.104 Å). Of the six oxygens bound to zinc, three are associated with Zn–O distances in the range 2.042(3)–2.093(3) and the remaining three have distances in the range 2.097(3)–2.164(5) (Table 7). Formally these correspond to singly and doubly bonded oxygens bound to the carbon atoms respectively. The C–O bond distances are

in the range 1.226(6)–1.279(5) Å (av. 1.255 Å) and the variations in them are also reflected in this bonding as well. The O–Zn–O angles are in the range 77.14(10)–171.33(13)° (av. 105.33°). The O–C–O bond angles are in the range 125.2(4)–126.0(4)° (av. 125.8°). The observed geometric parameters are as expected.

One-dimensional [C₆N₂H₁₄][Zn(C₂O₄)₂]·3H₂O 3. The asymmetric unit of compound **3** contains 24 non-hydrogen atoms, of which 13 belong to the framework (Fig. 7a). Of the remaining 11 atoms, 3 belong to the water molecules and 8 to the amine. The Zn atoms are six-coordinated with respect to the oxygens with Zn–O distances in the range 2.074(4)–2.134(4) Å (av. 2.099 Å). Of the six oxygens bound to zinc, three are associated with Zn–O distances in the range 2.074(4)–2.089(4) Å and the remaining three have distances in the range 2.103(4)–2.134(4) Å (Table 8). These correspond to the singly and doubly bonded oxygens bound to the carbon atoms respectively. The C–O bond distances are in the range 1.238(7)–1.257(7) Å (av. 1.249 Å). The O–Zn–O angles are in the range 79.2(2)–169.9(2)° (av. 105.33°) and the O–C–O bond angles are in the range 125.6(6)–126.2(6)° (av. 125.9°). These geometric parameters are in agreement with those reported earlier for similar compounds. Unlike the zero-dimensional monomeric and dimeric zinc oxalates, in **3**, the Zn atoms and the oxalate units are so connected as to form a one-dimensional chain with each Zn possessing a terminal oxalate unit. The diprotonated amine is situated in between these chains and interacts with the terminal oxalate *via* hydrogen bonds (Fig. 7b). Zinc oxalate dihydrate, [Zn(C₂O₄)(H₂O)₂], with a chain architecture is known,[14] but **3** is the first example of a chain zinc oxalate synthesized in the presence of an organic amine.

Two-dimensional [C₄N₂H₁₂]₂[Zn₂(C₂O₄)₃]₂·4H₂O 4. The asymmetric unit of compound **4** contains 36 non-hydrogen atoms (Fig. 8a). It consists of macroanionic sheets of formula [Zn₂(C₂O₄)₃]²⁻ with interlamellar [C₄N₂H₁₂]²⁺ ions. The asymmetric unit contains two crystallographically independent Zn atoms. The Zn atoms are octahedrally coordinated with respect to oxygen atoms with Zn–O distances in the range 2.074(5)–2.126(5) Å (av. Zn(1)–O 2.1007; Zn(2)–O 2.1007 Å). Of the six oxygens bound to zinc, three are associated with Zn–O distances in the range 2.079(5)–2.100(5) Å for Zn(1) and 2.074(5)–

Table 8 Selected bond distances (Å) and angles (°) for $[C_6N_2H_{14}][Zn(C_2O_4)_2]\cdot 3H_2O$ 3

Zn(1)–O(1)	2.074(4)	O(1)–C(1)	1.253(7)
Zn(1)–O(2)	2.085(4)	O(2)–C(4)[#1]	1.245(6)
Zn(1)–O(3)	2.089(4)	O(3)–C(2)	1.238(7)
Zn(1)–O(4)	2.103(4)	O(4)–C(3)	1.252(6)
Zn(1)–O(5)	2.112(4)	O(5)–C(4)	1.252(7)
Zn(1)–O(6)	2.134(4)	O(6)–C(1)[#2]	1.257(7)
C(1)–C(1)[#2]	1.546(11)	O(7)–C(3)	1.248(7)
C(2)–C(3)	1.549(9)	O(8)–C(2)	1.248(7)
C(4)–C(4)[#1]	1.550(12)		
O(1)–Zn(1)–O(2)	100.1(2)	O(4)–Zn(1)–O(6)	100.6(2)
O(1)–Zn(1)–O(3)	159.9(2)	O(5)–Zn(1)–O(6)	169.9(2)
O(2)–Zn(1)–O(3)	93.1(2)	C(1)–O(1)–Zn(1)	114.2(4)
O(1)–Zn(1)–O(4)	90.1(2)	C(4)[#1]–O(2)–Zn(1)	113.4(4)
O(2)–Zn(1)–O(4)	166.3(2)	C(2)–O(3)–Zn(1)	112.9(4)
O(3)–Zn(1)–O(4)	79.7(2)	C(3)–O(4)–Zn(1)	112.3(4)
O(1)–Zn(1)–O(5)	102.2(2)	C(4)–O(5)–Zn(1)	112.7(3)
O(2)–Zn(1)–O(5)	79.6(2)	C(1)[#2]–O(6)–Zn(1)	112.6(4)
O(3)–Zn(1)–O(5)	94.9(2)	O(1)–C(1)–O(6)[#2]	126.1(5)
O(4)–Zn(1)–O(5)	89.4(2)	O(3)–C(2)–O(8)	125.6(6)
O(1)–Zn(1)–O(6)	79.2(2)	O(7)–C(3)–O(4)	126.2(6)
O(2)–Zn(1)–O(6)	90.3(2)	O(2)[#1]–C(4)–O(5)	125.9(5)
O(3)–Zn(1)–O(6)	85.7(2)		

Symmetry transformations used to generate equivalent atoms: #1 $-x, -y + 1, -z$; #2 $-x, -y + 1, -z + 1$.

Table 9 Selected bond distances (Å) for $[C_4N_2H_{12}][Zn_2(C_2O_4)_3]\cdot 4H_2O$ 4

Zn(1)–O(1)	2.079(5)	Zn(2)–O(7)	2.074(5)
Zn(1)–O(2)	2.092(5)	Zn(2)–O(8)	2.082(5)
Zn(1)–O(3)	2.100(5)	Zn(2)–O(9)	2.098(5)
Zn(1)–O(4)	2.103(5)	Zn(2)–O(10)	2.101(5)
Zn(1)–O(5)	2.107(5)	Zn(2)–O(11)	2.123(5)
Zn(1)–O(6)	2.123(5)	Zn(2)–O(12)	2.126(5)
O(1)–C(1)[#1]	1.257(8)	O(7)–C(3)	1.248(9)
O(2)–C(2)[#1]	1.264(8)	O(8)–C(1)	1.257(8)
O(3)–C(3)	1.256(9)	O(9)–C(5)[#2]	1.250(9)
O(4)–C(4)	1.244(8)	O(10)–C(2)	1.251(8)
O(5)–C(5)	1.259(9)	O(11)–C(6)	1.260(8)
O(6)–C(6)	1.251(8)	O(12)–C(4)[#2]	1.253(8)
C(1)–C(2)	1.540(11)	C(3)–C(6)	1.554(10)
C(4)–C(5)	1.548(10)		

Symmetry transformations used to generate equivalent atoms: #1 $x - \frac{1}{2}, y + \frac{1}{2}, z$; #2 $x, y - 1, z$.

O–Zn–O angles are in the range 78.9(2)–169.7(2)° (av. O–Zn(1)–O 105.7; O–Zn(2)–O 105.8°) (Table 10). The O–C–O bond angles are in the range 125.4(7)–126.9(7)° (av. 126.0°). The geometrical parameters associated with the Zn and oxalate units are as expected.

The individual layers in compound 4 involve a network of ZnO$_6$ octahedra and C$_2$O$_4$ units. Each Zn is cross-linked to six oxalate oxygens, forming a two-dimensional layer structure involving the honeycomb motif as in many of the divalent metal oxalates (Fig. 8b). The honeycomb layers stack one over the other, along the c axis. Such perforated sheets have been observed in layered aluminophosphates[15] and the recently discovered iron oxalate phosphates.[16] The diprotonated amine sits in the middle of a 12-membered ring (6 Zn and 6 oxalate units) as shown in Fig. 8(b). The four water molecules, present in 4, are present in between the layers. The amine and water molecules participate in extensive hydrogen bonding, lending structural stability.

Three-dimensional $[C_3NH_{10}]_2[Zn_2(C_2O_4)_3]\cdot 3H_2O$ 5. The asymmetric unit of compound 5 contains 19 non-hydrogen atoms and is shown in Fig. 9(a). The structure also consists of a network of ZnO$_6$ octahedra and oxalate units and three oxalate units connect with the Zn atoms. Of the three oxalate units, two connect *via* an in-plane linkage and the third is cross-linked to the Zn atom in an out-of-plane manner as shown in Fig. 9(b). The connectivity between Zn and the oxalates units gives rise to a 20-membered aperture, which on projection forms 12-membered square channels. The amine sits in the middle of the channels. 5 was recently synthesized by employing the conventional hydrothermal procedure and the structure reported.[13] We will not, therefore, discuss its structure in detail.

Discussion

Five zinc oxalates, $[CN_3H_6]_2[Zn(H_2O)_2(C_2O_4)_2]$ 1, $[C_4N_2H_{12}]_3[Zn_2(C_2O_4)_5]\cdot 8H_2O$ 2, $[C_6N_2H_{14}][Zn(C_2O_4)_2]\cdot 3H_2O$ 3, $[C_4N_2H_{12}][Zn_2(C_2O_4)_3]\cdot 4H_2O$ 4 and $[C_3NH_{10}]_2[Zn_2(C_2O_4)_3]\cdot 3H_2O$ 5, possessing novel structures have been synthesized by the reaction of amine oxalates with zinc ions, the amine oxalates themselves having been synthesized and characterized for the first time. The amine oxalates exhibit strong hydrogen bond interactions *via* O–H$\cdots$O and N–H$\cdots$O bonding. While there is no simple relation between the starting amine oxalate and the product zinc oxalates, what is significant is that zinc oxalates with one-, two- and three-dimensional architectures, in addition to the monomeric and dimeric oxalates, could be synthesized by using amine oxalates under mild conditions.

Compounds 1–3 are new members of the zinc oxalate family. Monomeric 1 and dimeric 2 can be considered to be zero-dimensional oxalate structures, in contrast to 3 which has a

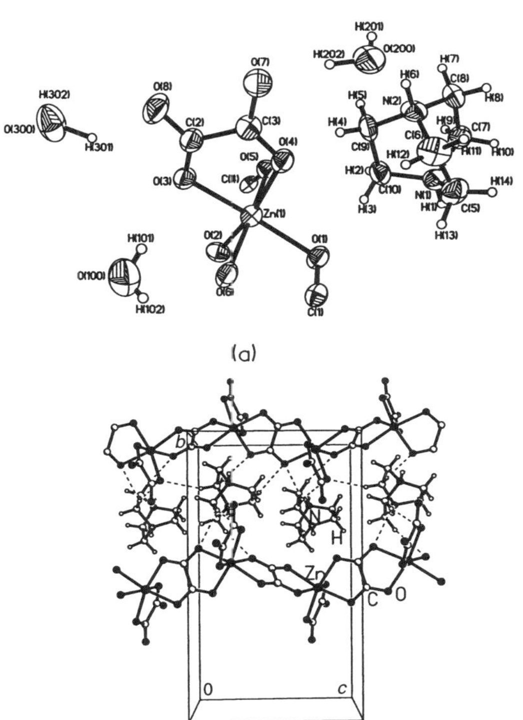

(a)

(b)

Fig. 7 (a) ORTEP plot of the zinc oxalate chain $[C_6N_2H_{14}][Zn(C_2O_4)_2]\cdot 3H_2O$ 3. (b) Structure of the linear chain along the [100] direction. The oxalate chains are separated by the amine.

2.098(5) Å for Zn(2), and the remaining three have distances in the range 2.103(5)–2.123(5) for Zn(1) and 2.110(5)–2.126(5) Å for Zn(2) (Table 9). These distances formally correspond to the singly and doubly bonded oxygens bound to the carbon atoms respectively. The C–O bond distances are in the range 1.244(8)–1.264(8) Å (av. 1.254 Å) and the variations in their distances are also reflected in this bonding as well. The

Table 10 Selected bond angles (°) for $[C_4N_2H_{12}][Zn_2(C_2O_4)_3]\cdot4H_2O$ **4**

O(1)–Zn(1)–O(2)	79.8(2)	O(7)–Zn(2)–O(8)	92.6(2)
O(1)–Zn(1)–O(3)	95.1(2)	O(7)–Zn(2)–O(9)	165.8(2)
O(2)–Zn(1)–O(3)	166.5(2)	O(8)–Zn(2)–O(9)	97.9(2)
O(1)–Zn(1)–O(4)	92.0(2)	O(7)–Zn(2)–O(10)	98.5(2)
O(2)–Zn(1)–O(4)	99.5(2)	O(8)–Zn(2)–O(10)	80.1(2)
O(3)–Zn(1)–O(4)	93.1(2)	O(9)–Zn(2)–O(10)	92.8(2)
O(1)–Zn(1)–O(5)	167.8(2)	O(7)–Zn(2)–O(11)	79.6(2)
O(2)–Zn(1)–O(5)	93.2(2)	O(8)–Zn(2)–O(11)	167.4(2)
O(3)–Zn(1)–O(5)	93.8(2)	O(9)–Zn(2)–O(11)	91.6(2)
O(4)–Zn(1)–O(5)	79.1(2)	O(10)–Zn(2)–O(11)	91.3(2)
O(1)–Zn(1)–O(6)	97.8(2)	O(7)–Zn(2)–O(12)	90.6(2)
O(2)–Zn(1)–O(6)	88.9(2)	O(8)–Zn(2)–O(12)	94.9(2)
O(3)–Zn(1)–O(6)	79.4(2)	O(9)–Zn(2)–O(12)	78.9(2)
O(4)–Zn(1)–O(6)	168.1(2)	O(10)–Zn(2)–O(12)	169.7(2)
O(5)–Zn(1)–O(6)	92.1(2)	O(11)–Zn(2)–O(12)	94.9(2)
C(1)[#1]–O(1)–Zn(1)	113.3(5)	C(3)–O(7)–Zn(2)	114.3(5)
C(2)[#1]–O(2)–Zn(1)	113.1(4)	C(1)–O(8)–Zn(2)	113.1(4)
C(3)–O(3)–Zn(1)	113.4(5)	C(5)[#2]–O(9)–Zn(2)	113.6(5)
C(4)–O(4)–Zn(1)	113.8(5)	C(2)–O(10)–Zn(2)	112.2(4)
C(5)–O(5)–Zn(1)	113.2(5)	C(6)–O(11)–Zn(2)	111.9(4)
C(6)–O(6)–Zn(1)	112.5(4)	C(4)[#2]–O(12)–Zn(2)	113.6(5)
O(1)[#3]–C(1)–O(8)	126.0(7)	O(4)–C(4)–O(12)[#4]	126.9(7)
O(10)–C(2)–O(2)[#3]	125.8(7)	O(9)[#4]–C(5)–O(5)	125.4(7)
O(7)–C(3)–O(3)	126.4(7)	O(6)–C(6)–O(11)	125.7(7)

Symmetry transformations used to generate equivalent atoms: #1 $x - \frac{1}{2}$, $y + \frac{1}{2}$, z; #2 x, $y - 1$, z; #3 $x + \frac{1}{2}$, $y - \frac{1}{2}$, z; #4 x, $y + 1$, z.

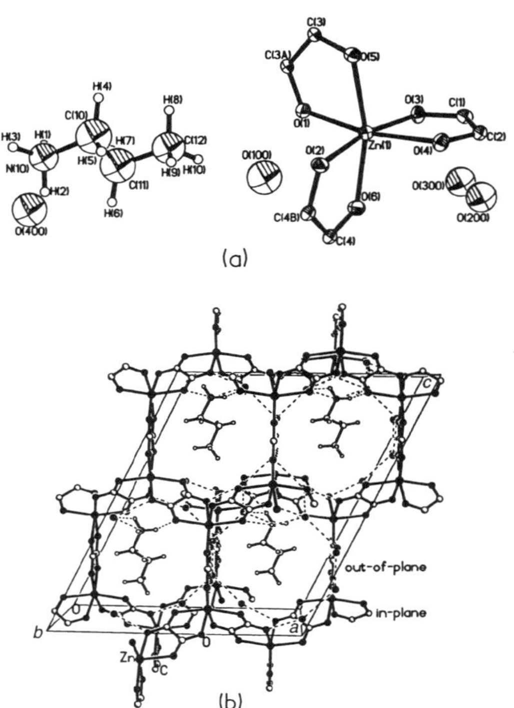

Fig. 9 (a) ORTEP plot of the zinc oxalate $[C_3NH_{10}]_2[Zn_2(C_2O_4)_3]\cdot3H_2O$ **5** possessing three-dimensional structure. (b) Three-dimensional structure along the [010] direction showing 12-membered square channels. Dashed lines represent the various hydrogen bond interactions.

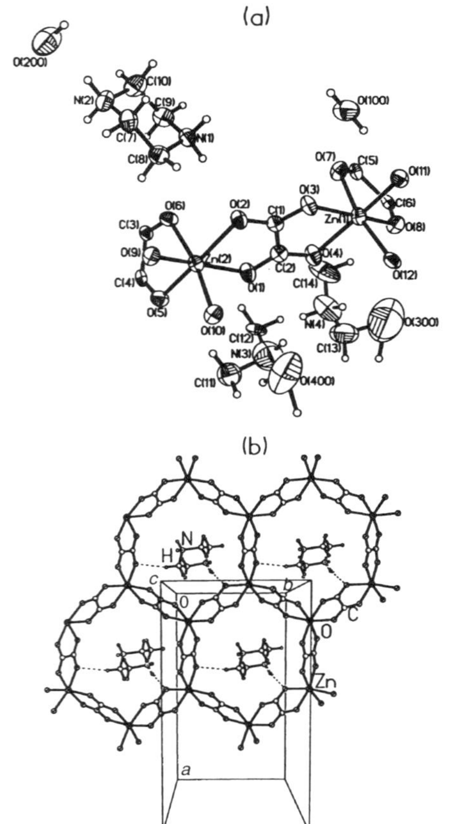

Fig. 8 (a) ORTEP plot of the zinc oxalate layer $[C_4N_2H_{12}][Zn_2(C_2O_4)_3]\cdot4H_2O$ **4**. (b) The layer structure along the [001] direction showing the honeycomb architecture. The amine molecules sit in the middle of the 12-membered aperture. The water molecules are not shown for clarity.

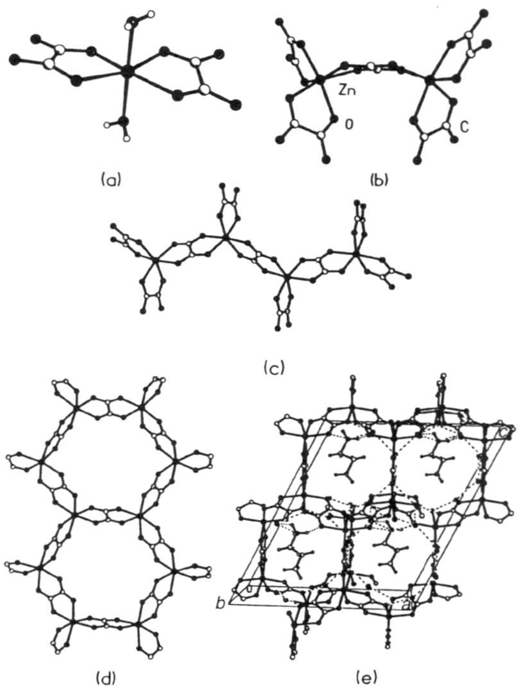

Fig. 10 Various types of zinc oxalate structures obtained in the present study: (a) monomer, (b) dimer, (c) one-dimensional chain, (d) two-dimensional layer and (e) three-dimensional structure. Note the close relationships amongst them.

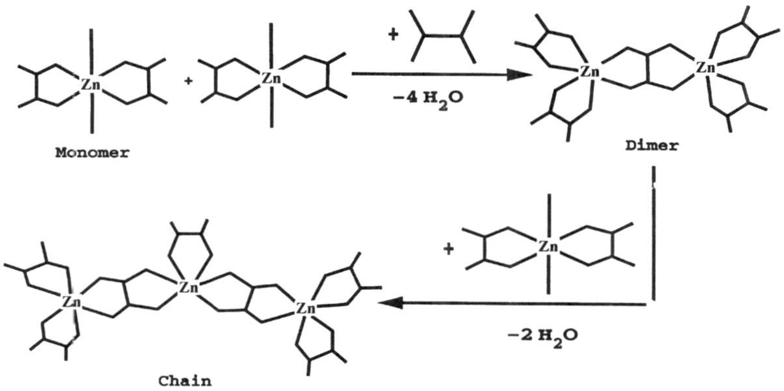

Fig. 11 Schematic showing the formation of chain architecture from the monomer *via* hydrolysis and, condensation.

one-dimensional chain structure. In **1**, the connectivity between the Zn and the oxalate moieties forms a monomer, *i.e.* the Zn atoms have two free oxalate units and possess two terminal water molecules, and in **2** one oxalate joins two zinc centers with two free oxalates. In the case of **3** the connectivity forms a one-dimensional zinc oxalate chain and the Zn has one free oxalate group. These lower dimensional structures are stabilized by extensive hydrogen bonding. Unlike in **3**, the oxalates link up with Zn to form a honeycomb-like layer in **4**. The Zn atoms do not have any free oxalate groups in the layer structure. The number of free oxalate groups attached to the Zn probably governs the reactivity of the zinc oxalates. Thus, the monomer must be most reactive followed by the dimer, chain and the layer.

In Fig. 10 we present the various types of structures obtained by us to demonstrate the similarities and relationships. We can derive the structure of the dimer from that of the monomer, the chain from the dimer, and the layer from the chain. It is easy to see how the layers in **4** get connected by the oxalate units to form the three-dimensional structure of **5**. Just as the four-membered ring monomeric phosphate unit plays a crucial role in the building of framework phosphates,[2,17] it is possible that the monomeric and dimeric oxalates are involved in the construction of the extended oxalate framework structures.

Conclusion

The present study shows that new types of zinc oxalates can be obtained under relatively mild conditions by the reaction of amine oxalates with zinc ions. Open-framework zinc phosphates are known to occur in four structure classes: monomers, chains, sheets and 3-D structures. We have obtained the oxalate analogues of all these four structures by the amine oxalate route, along with a dimer structure, which is unique to oxalates. Hydrothermal synthesis of the metal oxalates in the presence of amines is known to yield both two- and three-dimensional structures.[13] While it is possible that the zero- and one-dimensional structures are also formed under hydrothermal conditions, their isolation might be rendered difficult. This is because once the zero-dimensional structures are formed they may readily transform to higher dimensional structures. Such transformations from the zero-dimensional monomer to the linear chain structure are shown schematically in Fig. 11. The chains can readily condense to give the two-dimensional honeycomb structures. The reaction of amine oxalates with metal ions is likely to provide a better control of the reactant concentrations and the kinetics of the reaction, thereby enabling isolation of the initial building units such as the zero- and one-dimensional structures. The variety of structures obtained by us by the amine oxalate route suggests that it would be most worthwhile to investigate the reactions of various amine dicarb-

oxylates with metal ions. The hierarchy of oxalate structures formed here also suggests that there is likely to be a building up principle in the formation of open-framework oxalates just as in the phosphates,[2a] wherein low-dimensional structures transform to higher dimensional ones.

References

1 A. K. Cheetham, G. Ferey and T. Loiseau, *Angew. Chem., Int. Ed.*, 1999, **38**, 3268.

2 (*a*) C. N. R. Rao, S. Natarajan, A. Choudhury, S. Neeraj and A. A. Ayi, *Acc. Chem. Res.*, 2000, **34**, 80; (*b*) S. Neeraj, S. Natarajan and C. N. R. Rao, *J. Solid State Chem.*, 2000, **150**, 417; (*c*) R. I. Walton, F. Millange, A. Le Bail, T. Loiseau, C. Serre, D. O'Hare and G. Ferey, *Chem. Commun.*, 2000, 203.

3 Q. Gao, N. Guillou, M. Nogues, A. K. Cheetham and G. Ferey, *Chem. Mater.*, 1999, **11**, 2937; D. Riou and G. Ferey, *J. Mater. Chem.*, 1998, **8**, 2733; D. Riou, C. Serre and G. Ferey, *Microporous Mesoporous Mater.*, 1998, **23**, 23.

4 G. B. Hix, D. S. Wragg, I. Bull, R. E. Morris and P. A. Wright, *Chem. Commun.*, 1999, 2421; B. Zhang, D. M. Poojary and A. Clearfield, *Inorg. Chem.*, 1998, **37**, 1844; A. Clearfield, *Chem. Mater.*, 1998, **10**, 2733 and references therein.

5 F. Serpaggi and G. Ferey, *J. Mater. Chem.*, 1998, **8**, 2737; F. Serpaggi and G. Ferey, *J. Mater. Chem.*, 1998, **8**, 2749; C. Livage, C. Egger, M. Nogues and G. Ferey, *J. Mater. Chem.*, 1998, **8**, 2743; C. Livage, C. Egger and G. Ferey, *Chem. Mater.*, 1999, **11**, 1546 and references therein.

6 T. M. Reineke, M. Eddaoudi, M. O'Keeffe and O. M. Yaghi, *Angew. Chem., Int. Ed.*, 1999, **38**, 2590; H. Li, M. Eddaoudi, T. L. Groy and O. M. Yaghi, *J. Am. Chem. Soc.*, 1998, **120**, 8571.

7 S. Ayyappan, A. K. Cheetham, S. Natarajan and C. N. R. Rao, *Chem. Mater.*, 1998, **10**, 3746; S. Natarajan, R. Vaidhyanathan, C. N. R. Rao, S. Ayyappan and A. K. Cheetham, *Chem. Mater.*, 1999, **11**, 1633.

8 S. Neeraj, S. Natarajan and C. N. R. Rao, *Angew. Chem., Int. Ed.*, 1999, **38**, 3480; C. N. R. Rao, S. Natarajan and S. Neeraj, *J. Am. Chem. Soc.*, 2000, **122**, 2810.

9 G. M. Sheldrick, SADABS, Siemens Area Detector Absorption Correction Program, University of Göttingen, 1994.

10 SHELXS 86, Program for Crystal Structure Determination, University of Göttingen, 1986; G. M. Sheldrick, *Acta Crystallogr., Sect. A*, 1990, **35**, 467.

11 G. M. Sheldrick, SHELXTL-PLUS, Program for Crystal Structure Solution and Refinement, University of Göttingen, 1993.

12 C. K. Johnson, ORTEP II, Report ORNL-5138, Oak Ridge National Laboratory, Oak Ridge, TN, 1976.

13 R. Vaidhyanathan, S. Natarajan, A. K. Cheetham and C. N. R. Rao, *Chem. Mater.*, 1999, **11**, 3636.

14 R. Deyrieux, C. Bero and A. Penelous, *Bull. Soc. Chim. Fr.*, 1973, 25.

15 J. M. Thomas, R. H. Jones, J. Chen, R. Xu, A. M. Chippindale, S. Natarajan and A. K. Cheetham, *J. Chem. Soc., Chem. Commun.*, 1992, 929.

16 A. Choudhury, S. Natarajan and C. N. R. Rao, *Chem. Mater.*, 1999, **11**, 2316.

17 S. Oliver, A. Kuperman and G. A. Ozin, *Angew. Chem., Int. Ed.*, 1998, **37**, 46; S. Oliver, A. Kuperman, A. Lough and G. A. Ozin, *Chem. Mater.*, 1996, **8**, 2391.

FULL PAPER

Aliphatic dicarboxylates with three-dimensional metal–organic frameworks possessing hydrophobic channels

Ramanathan Vaidhyanathan, Srinivasan Natarajan and C. N. R. Rao*

Chemistry and Physics of Materials Unit, Jawaharlal Nehru Centre for Advanced Scientific Research, Jakkur P. O., Bangalore, 560 064, India. E-mail: cnrrao@jncasr.ac.in

Received 11th December 2002, Accepted 21st February 2003
First published as an Advance Article on the web 12th March 2003

www.rsc.org/dalton

Dalton

Two three-dimensional open-framework metal dicarboxylates possessing channels with a hydrophobic environment have been synthesized. One is a malonate of the formula, $[Cd(O_2C\text{-}CH_2\text{-}CO_2)(H_2O)]\cdot H_2O$, **I**, and the other a glutarate of the formula $[Mn(O_2C\text{-}(CH_2)_3\text{-}CO_2)]$, **II**. The three-dimensional structure of **I** is attained through infinite Cd–O–Cd corner-linkages. On the other hand, **II**, contains Mn–O–Mn layers cross-linked by MnO_6 octahedra and glutarate moieties.

Introduction

There is intense interest today in the synthesis of open-framework hybrid compounds, specially those employing rigid organic linkers such as carboxylates, bipyridyl and other multifunctional ligands.[1–14] Several materials with large channels and high porosity have been synthesized and characterized.[15,16] Important contributors to this family are Forster and Cheetham,[17] who have described an open-framework nickel succinate containing an infinite Ni–O–Ni framework, possessing uni-dimensional channels. The nickel succinate is quite distinct from the open-framework dicarboxylates described by Kim et al.[18] and Férey and co-workers[19] in that the channels have a hydrophobic environment because of the protruding methylene groups. Férey and co-workers have described cobalt succinates based on two-dimensional sheets of edge-sharing CoO_6 octahedra. The Mn^{II} and Fe^{II} carboxylates, on the other hand, contain one-dimensional MnO_6 ribbons and two-dimensional metal carboxylate sheets. We have synthesized two interesting three-dimensional open-framework metal dicarboxylates, a malonate $[Cd(O_2C\text{-}CH_2\text{-}CO_2)(H_2O)]\cdot H_2O$, **I**, and a glutarate $[Mn(O_2C\text{-}(CH_2)_3\text{-}CO_2)]$, **II**, possessing channels with hydrophobic environments. While the three-dimensional structure of **I** arises from the infinite Cd–O–Cd connectivity, that of **II** is formed by two-dimensional Mn–O–Mn layers cross-linked by the glutarate and the MnO_6 octahedra.

Experimental

Compound **I** was prepared at room temperature by reacting $CdCl_2\cdot 2H_2O$ with malonic acid. Typically, 0.1 g of $CdCl_2\cdot 2H_2O$ was dissolved in a n-hexanol + water (1 ml + 2 ml) mixture and 0.16 g of malonic acid and 0.1 ml of 1,2-diaminopropane added to the above mixture. The mixture was stirred to render it homogeneous. The final mixture of composition $CdCl_2\cdot 2H_2O$: $3.1C_3O_4H_4$: $2.3C_3H_{10}N_2$: $16.2n\text{-}C_6H_{13}OH$: $223.6H_2O$, was left in a closed polypropylene bottle for three days at room temperature. The product, colorless hexagonal plate-like crystals, was filtered off, washed with deionized water and dried under ambient conditions. The yield was about 40%. Anal. calcd.: C, 14.4; H, 2.4. Anal. obsd.: C, 14.8; H, 1.8%.

Compound **II** was synthesized by reacting manganese(II) chloride with glutaric acid under solvothermal conditions. Typically, 0.2 g of $MnCl_2\cdot 4H_2O$ was dissolved in 2 ml of n-hexanol and 0.33 g of glutaric acid and 0.22 g of piperazine added to the above. The contents were stirred to homogeneity and the final mixture of composition $MnCl_2\cdot 4H_2O$: $2.47C_5$-H_8O_4 : $2.53C_4N_2H_{10}$: 15.58 n-$C_6H_{13}OH$, was sealed in a PTFE-lined stainless steel autoclave and heated at 180 °C for 72 hours. The product, colorless hexagonal plate-like crystals, thus

obtained, was filtered off, washed with ethanol and n-butanol, and dried under ambient conditions. The yield was about 55%. Anal. calcd.: C, 32.5; H, 3.3. Anal. obsd.: C, 34.3; H, 3.2%.

Initial characterization of the products was carried out by powder X-ray diffraction (XRD), chemical analysis, thermogravimetric analysis (TGA) and IR spectroscopy. The powder XRD pattern exclusively exhibited reflections of a hitherto unknown material; the pattern was entirely consistent with the structure determined by single crystal XRD. A least squares fit (Cu-Kα) of the XRD patterns, using the *hkl* indices generated from the single crystal structure gave the following lattice parameters: $a = 17.047$ (4); $b = 17.047$ (4); $c = 12.389$ (1) Å; $\alpha = \beta = 90°$ and $\gamma = 120°$ for **I**; $a = 11.287$ (8); $b = 11.287$ (8); $c = 29.437$ (6) Å; $\alpha = \beta = 90°$ and $\gamma = 120°$ for **II**.

Infrared spectroscopic studies of KBr pellets showed the characteristic features of the dicarboxylate.[20] The various observed bands were: $v_{as}(C=O)$ at 1546(s) cm^{-1} and $v_s(C=O)$ at 1427(s) cm^{-1} (**I**) and 1561(s) cm^{-1} and 1401(s) cm^{-1} (**II**); $v_s(O\text{-}C\text{-}O)$ at 1261(s) cm^{-1} (**I**) and 1245(s) cm^{-1} (**II**); $\delta(O\text{-}C\text{-}O)$ at 829(s) cm^{-1} (**I**) and 833(s) cm^{-1} (**II**). The presence of bound water molecules in **I** gives rise to $\rho_w(H_2O)$ in the 550–600 cm^{-1} region and $\rho_r(H_2O)$ at 850–900 cm^{-1}. In both of the compounds multiple bands in the region 2400–3000 cm^{-1} appear due to the $v_s(CH_2)$ and $v_a(CH_2)$ vibrations. The $\rho_w(CH_2)$ and $\rho_r(CH_2)$ bands occur in the 950–1200 cm^{-1} region. In **I**, a broad band appears at 3604 cm^{-1} due to the presence of water molecules in the structure.

Single crystal structure determination

A suitable single crystal of each compound was carefully selected under a polarizing microscope and glued at the tip of a thin glass fiber with cyano-acrylate (super glue) adhesive. Single crystal structure determination by X-ray diffraction was performed on a Siemens Smart-CCD diffractometer equipped with a normal focus, 2.4 kW sealed tube X-ray source (Mo-K$_\alpha$ radiation, $\lambda = 0.71073$ Å) operating at 50 kV and 40 mA. A hemisphere of intensity data was collected at room temperature in 1321 frames with ω scans (width of 0.30° and exposure time of 20 s per frame) in the θ range 2.07 to 23.29°. Pertinent experimental details for the structure determinations are presented in Table 1.

The structure was solved and refined using the SHELXTL-PLUS[21] suite of programs. An absorption correction based on symmetry equivalent reflections was applied using the SADABS program.[22] The hydrogen positions for both of the compounds were initially located in the difference Fourier maps, and for the final refinement the hydrogen atoms were placed geometrically and held in the riding mode. The last cycles of refinement included atomic positions for all the atoms, aniso-

DOI: 10.1039/b212314m

Table 1 Crystal data and structure refinement parameters for $[Cd(O_2C\text{-}CH_2\text{-}CO_2)(H_2O)]\cdot H_2O$, **I**, and, $[Mn(O_2C\text{-}(CH_2)_3\text{-}CO_2)]$, **II**

Structural parameter	I	II
Chemical formula	$CdC_3H_6O_6$	$MnC_5H_6O_4$
Formula mass	250.48	185.04
Crystal symmetry	Rhombohedral	Rhombohedral
Space group	$R\bar{3}$ (No. 148)	$R\bar{3}$ (No. 148)
T/K	293 (2)	293 (2)
a/Å	17.0685 (2)	11.2944 (10)
c/Å	12.4302 (3)	29.4830 (4)
V/Å^3	3136.17 (9)	3257.08 (6)
Z	18	18
ρ_{calc}/g cm^{-3}	2.387	1.698
μ/mm^{-1}	3.107	1.775
No. of measured/observed reflections	4504/990	4523/957
R1, wR2 [$I > 2\sigma(I)$]	$0.0336,^a\ 0.0898^b$	$0.0205,^a\ 0.0500^b$

^{a} $R_1 = ||F_o| - |F_c||/|F_o|$. ^{b} $wR_2 = \{[w(F_o^2 - F_c^2)^2]/[w(F_o^2)^2]\}^{1/2}$. $w = 1/[\sigma^2(F_o)^2 + (aP)^2 + bP]$, $P = [\max(F_o^2,0) + 2(F_c)^2]/3$, where $a = 0.0000$ and $b = 72.3680$ for **I** and $a = 0.0189$ and $b = 2.2708$ for **II**.

Table 2 Selected bond distances (Å) and angles (°) for $[Cd(O_2C\text{-}CH_2\text{-}CO_2)(H_2O)]\cdot H_2O$, **I**

Cd(1)–O(10)	2.288(7)	Cd(1)–O(2)#1	2.535(6)
Cd(1)–O(1)	2.288(5)	Cd(1)–O(1)#2	2.545(5)
Cd(1)–O(2)	2.293(6)	O(1)–Cd(1)#5	2.545(5)
Cd(1)–O(3)	2.312(6)	O(2)–Cd(1)#6	2.535(6)
Cd(1)–O(4)	2.319(6)		
O(10)–Cd(1)–O(1)	165.5(2)	O(4)–Cd(1)–O(2)#1	53.6(2)
O(10)–Cd(1)–O(2)	83.6(2)	O(10)–Cd(1)–O(1)#2	85.1(2)
O(1)–Cd(1)–O(2)	82.0(2)	O(1)–Cd(1)–O(1)#2	93.7(3)
O(10)–Cd(1)–O(3)	106.7(2)	O(2)–Cd(1)–O(1)#2	89.3(2)
O(1)–Cd(1)–O(3)	83.8(2)	O(3)–Cd(1)–O(1)#2	53.2(2)
O(2)–Cd(1)–O(3)	138.6(2)	O(4)–Cd(1)–O(1)#2	174.2(2)
O(10)–Cd(1)–O(4)	89.2(2)	O(2)#1–Cd(1)–O(1)#2	124.7(2)
O(1)–Cd(1)–O(4)	92.1(2)	C(1)#5–O(1)–Cd(1)	126.7(5)
O(2)–Cd(1)–O(4)	91.3(2)	C(1)#5–O(1)–Cd(1)#5	87.1(4)
O(3)–Cd(1)–O(4)	127.9(2)	Cd(1)–O(1)–Cd(1)#5	143.1(2)
O(10)–Cd(1)–O(2)#1	86.5(2)	C(2)#6–O(2)–Cd(1)	127.7(5)
O(1)–Cd(1)–O(2)#1	105.8(2)	C(2)#6–O(2)–Cd(1)#6	87.3(5)
O(2)–Cd(1)–O(2)#1	143.6(2)	C(1)–O(3)–Cd(1)	98.3(5)
O(3)–Cd(1)–O(2)#1	77.7(2)	C(2)–O(4)–Cd(1)	97.9(5)

Symmetry transformations used to generate equivalent atoms in **I**: #1 $-x + y + 2/3, -x + 1/3, z + 1/3$. #2 $y + 1/3, -x + y + 2/3, -z + 5/3$. #3 $x - y, x, -z + 2$. #4 $y, -x + y, -z + 2$. #5 $x - y + 1/3, x - 1/3, -z + 5/3$. #6 $-y + 1/3, x - y - 1/3, z - 1/3$.

tropic thermal parameters for all the non-hydrogen atoms and isotropic thermal parameters for all hydrogen atoms. Full-matrix-least-squares structure refinement against $|F^2|$ was carried out using the SHELXTL-PLUS package of programs.

CCDC reference numbers 199941 and 199942.

See http://www.rsc.org/suppdata/dt/b2/b212314m/ for crystallographic data in CIF or other electronic format.

Results

The asymmetric unit of the malonate, $[Cd(O_2C\text{-}CH_2\text{-}CO_2)\text{-}(H_2O)]\cdot H_2O$, **I**, contains 10 non-hydrogen atoms, of which 9 belong to the cadmium malonate framework and one is the oxygen of the free water molecule. The Cd atom is seven-coordinated and has a distorted pentagonal bipyramidal arrangement. The equatorial positions are occupied by the malonate oxygen atoms and the apical positions are occupied by a water molecule and a malonate oxygen. The Cd–O distances are in the range 2.288(5)–2.545(5) Å (av. 2.369 Å) and the average O–Cd–O angle is 102.3°. The malonate unit has the expected bond distances and angles (Table 2). A similar coordination for cadmium has been observed in other cadmium dicarboxylates.[24] There are two types of water molecules in **I**, one bound to the Cd atom and the other present in the channels. Bond valence sum calculations[23] indicated the oxidation numbers of Cd, C and O to be 2+, 4+ and 2−, as expected. Ancillary ligation by water molecules has been observed in other metal carboxylates.[25,26]

The structure of **I** can be described as a one-dimensional chain formed by corner-sharing CdO_7 pentagonal bipyramids, with the adjacent pentagonal bipyramids shifted and rotated with respect to each other, rendering the chain sinusoidal (see top of Fig. 1). Six such chains, arranged in an hexagonal array through corner-sharing, form the three-dimensional framework. Thus, the entire structure is built up of an infinite array of corner-sharing CdO_7 pentagonal bipyramids, with uni-directional hexagonal channels (6.95 × 6.95 Å, shortest C–C contact not including the van der Waals radii) along the c-axis (Fig. 1). The inner-walls of the channels are lined by the methylene groups of the malonate moiety, making them hydrophobic. The water molecule bound to Cd protrudes into the 12-membered channels along the a-axis, whilst the free water molecule resides in a smaller aperture formed due to corner-sharing between two adjacent chains.

The asymmetric unit of the glutarate, $[Mn(O_2C\text{-}(CH_2)_3\text{-}CO_2)]$, **II**, consists of 12 non-hydrogen atoms, of which three are crystallographically independent Mn atoms. Mn(1) has 1/3 occupancy, Mn(2) has 1/6 occupancy and Mn(3) has 1/2 occupancy. All three Mn atoms are octahedrally coordinated by the glutarate oxygens, with average Mn–O distances of 2.229 Å, 2.148 Å and 2.172 Å for Mn(1)–O, Mn(2)–O and Mn(3)–O respectively and average O–Mn–O angles of 105.1°, 108.0° and 108.07° for Mn(1), Mn(2) and Mn(3) respectively. Open-framework manganese dicarboxylates involving octahedrally coordinated Mn atoms have been reported recently.[27] Bond valence sum calculations[23] indicated that the valence states of the Mn, C and O in **II** were +2, +4 and −2 respectively. The glutarate moiety has the expected bond distances and angles (Table 3). The edge-sharing Mn(1)O$_6$ and Mn(3)O$_6$ octahedra form an infinite two-dimensional array possessing honeycomb-like apertures in the ab-plane (6.642 × 6.642 Å, shortest C–C contact not including the van der Waals radii) (Fig. 2). These layers are cross-linked by isolated Mn(2)O$_6$ octahedra and *out-of-plane* glutarate moieties to give rise to the three-dimensional architecture (Fig. 3). Two-dimensional structures with open architectures are known to achieve efficient packing by shifting of adjacent layers.[28] Accordingly, in **II** the layers formed by the edge-sharing octahedra get shifted by 1/2 the unit cell length along the b- and a-axes and thereby contribute negatively to the openness of the structure.

Thermogravimetric analysis (TGA) of **I** and **II** was carried out in a N_2 atmosphere (50 ml min^{-1}) in the 25 to 700 °C range (Fig. 4). **I** showed a mass loss of 14.8% in the range 100–180 °C due to the loss of the lattice and the bonded water (calc.14.3%), followed by a sharp mass loss of 35.4% in the range 220–350 °C due to the loss of the malonate moiety (calc. 34.6%). The

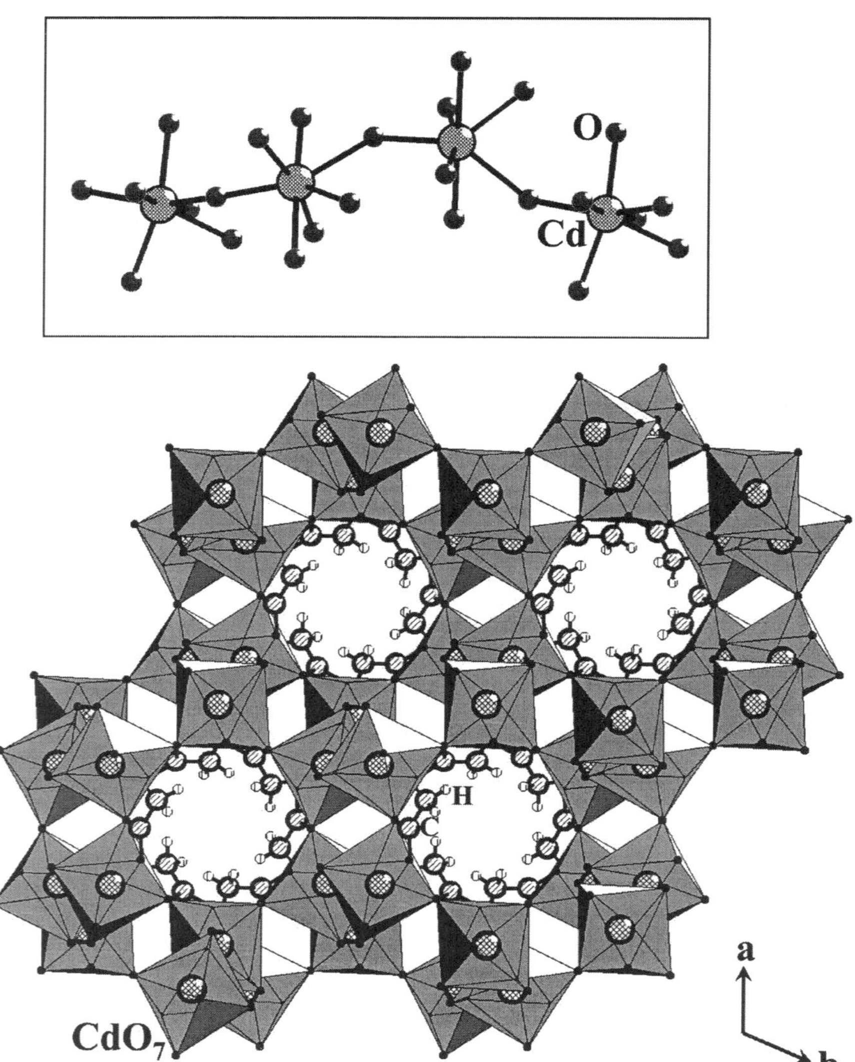

Fig. 1 Structure of **I** showing the 12-membered, methylene-lined hydrophobic channels. Note the presence of the free water molecules in the small apertures.

powder XRD pattern of the product of decomposition of **I**, showed characteristic peaks of the mineral monteporite, CdO (JCPDS: 05–0640). **II** showed a gradual mass loss of 53.2% in the range 210–450 °C due to the loss of the glutarate moiety (calc. 53.0%). A very small weight loss is observed in the range 50–100 °C due to the loss of some adsorbed species. The product of decomposition of **II**, as revealed by the powder XRD pattern, was MnO_2 (JCPDS: 44–0992).

Discussion

Both **I** and **II** have extended inorganic structures formed by corner- and edge-sharing of metal–oxygen polyhedra. **I** has a three-dimensional structure arising entirely from the infinite Cd–O–Cd connectivities while **II** exhibits infinite Mn–O–Mn connectivities in two dimensions. The three-dimensional nature of **II** is derived through the *out-of-plane* connectivities between the Mn–O–Mn sheets and the glutarate and the isolated MnO_6

octahedral units. The M–O–M connectivity in **I** bears some resemblance to that in the Ni succinate reported by Foster and Cheetham.[17a] The M–O–M linkage in **I** occurs through corner-sharing of CdO_7 pentagonal bipyramids, whilst in the Ni succinate, it is through the edge-shared NiO_6 octahedra. The M–O–M linkages in **II**, are comparable to those observed in two- and three-dimensional Co succinates,[19] but the interlayer connectivity between the two structures is different. The Co–O–Co layers in the cobalt succinate are connected by carboxylate pillars and the Mn–O–Mn layers in **II** are linked through isolated MnO_6 octahedra.

Several two-dimensional metal–dicarboxylate frameworks with apertures have been reported in the literature.[18,19a,27b,28,29] there are also examples where organic linkers to the metal ions give rise to three-dimensional structures.[15,19b] Some of the possible ways of building up three-dimensional metal–dicarboxylate frameworks are illustrated in Scheme 1. Many such three-dimensional structures are known. Motif (e) in the

Table 3 Selected bond distances (Å) and angles (°) for [Mn(O$_2$C-(CH$_2$)$_3$-CO$_2$)], **II**

Mn(1)–O(1)#1	2.1812(13)	Mn(2)–O(3)#4	2.1483(13)
Mn(1)–O(1)	2.1812(13)	Mn(2)–O(3)#2	2.1483(13)
Mn(1)–O(1)#2	2.1813(13)	Mn(2)–O(3)#5	2.1483(13)
Mn(1)–O(2)	2.2772(13)	Mn(3)–O(4)#6	2.1118(14)
Mn(1)–O(2)#1	2.2772(13)	Mn(3)–O(4)	2.1117(14)
Mn(1)–O(2)#2	2.2773(13)	Mn(3)–O(1)#6	2.1927(13)
Mn(2)–O(3)	2.1482(13)	Mn(3)–O(1)	2.1926(13)
Mn(2)–O(3)#1	2.1483(13)	Mn(3)–O(2)	2.2122(13)
Mn(2)–O(3)#3	2.1483(13)	Mn(3)–O(2)#6	2.2122(13)
O(1)#1–Mn(1)–O(1)	105.12(4)	O(3)#1–Mn(2)–O(3)#5	180
O(1)#1–Mn(1)–O(1)#2	105.12(4)	O(3)#3–Mn(2)–O(3)#5	88.60(5)
O(1)–Mn(1)–O(1)#2	105.12(4)	O(3)#4–Mn(2)–O(3)#5	88.60(5)
O(1)#1–Mn(1)–O(2)	86.89(5)	O(3)#2–Mn(2)–O(3)#5	91.39(5)
O(1)–Mn(1)–O(2)	76.64(5)	O(4)#6–Mn(3)–O(4)	180
O(1)#2–Mn(1)–O(2)	166.64(5)	O(4)#6–Mn(3)–O(1)#6	91.17(5)
O(1)#1–Mn(1)–O(2)#1	76.64(5)	O(4)–Mn(3)–O(1)#6	88.83(5)
O(1)–Mn(1)–O(2)#1	166.64(5)	O(4)#6–Mn(3)–O(1)	88.82(5)
O(1)#2–Mn(1)–O(2)#1	86.89(5)	O(4)–Mn(3)–O(1)	91.18(5)
O(2)–Mn(1)–O(2)#1	90.33(5)	O(1)#6–Mn(3)–O(1)	180
O(1)#1–Mn(1)–O(2)#2	166.64(5)	O(4)#6–Mn(3)–O(2)	89.52(6)
O(1)–Mn(1)–O(2)#2	86.89(5)	O(4)–Mn(3)–O(2)	90.48(6)
O(1)#2–Mn(1)–O(2)#2	76.64(5)	O(1)#6–Mn(3)–O(2)	102.22(5)
O(2)–Mn(1)–O(2)#2	90.33(5)	O(1)–Mn(3)–O(2)	77.78(5)
O(2)#1–Mn(1)–O(2)#2	90.32(5)	O(4)#6–Mn(3)–O(2)#6	90.48(6)
O(3)–Mn(2)–O(3)#1	88.61(5)	O(4)–Mn(3)–O(2)#6	89.52(6)
O(3)–Mn(2)–O(3)#3	91.39(5)	O(1)#6–Mn(3)–O(2)#6	77.78(5)
O(3)#1–Mn(2)–O(3)#3	91.40(5)	O(1)–Mn(3)–O(2)#6	102.22(5)
O(3)–Mn(2)–O(3)#4	180	O(2)–Mn(3)–O(2)#6	180
O(3)#1–Mn(2)–O(3)#4	91.40(5)	C(1)–O(1)–Mn(1)	125.50(12)
O(3)#3–Mn(2)–O(3)#4	88.61(5)	C(1)–O(1)–Mn(3)	133.99(12)
O(3)–Mn(2)–O(3)#2	88.61(5)	C(2)#7–O(2)–Mn(3)	132.14(13)
O(3)#1–Mn(2)–O(3)#2	88.60(5)	C(2)#7–O(2)–Mn(1)	128.82(12)
O(3)#3–Mn(2)–O(3)#2	180	C(1)–O(3)–Mn(2)	129.01(13)
O(3)#4–Mn(2)–O(3)#2	91.39(5)	C(2)–O(4)–Mn(3)	136.86(13)
O(3)–Mn(2)–O(3)#5	91.39(5)		

Symmetry transformations used to generate equivalent atoms in **II**: #1 $-y, x - y, z$. #2 $-x + y, -x, z$. #3 $x - y, x, -z$. #4 $-x, -y, -z$. #5 $y, -x + y, -z$. #6 $-x - 1/3, -y - 2/3, -z + 1/3$. #7 $y + 2/3, -x + y + 1/3, -z + 1/3$. #8 $x - y - 1/3, x - 2/3, -z + 1/3$. #9 $-x + y, -x - 1, z$. #10 $-y - 1, x - y - 1, z$.

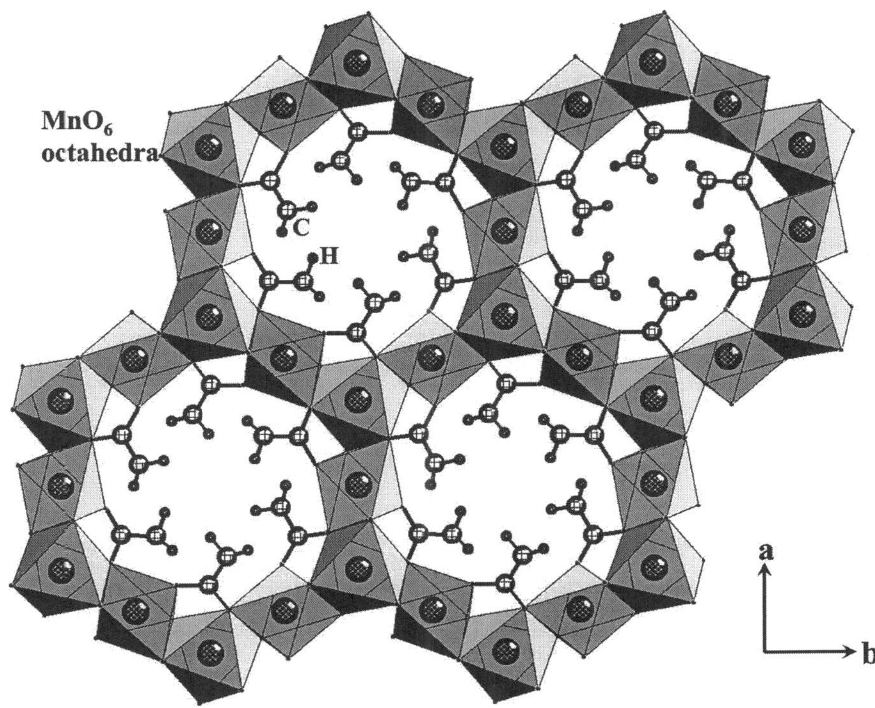

Fig. 2 Structure of **II** showing the 12-membered C-backbone of glutarate-lined hydrophobic channels, formed by the edge-sharing MnO$_6$ octahedra.

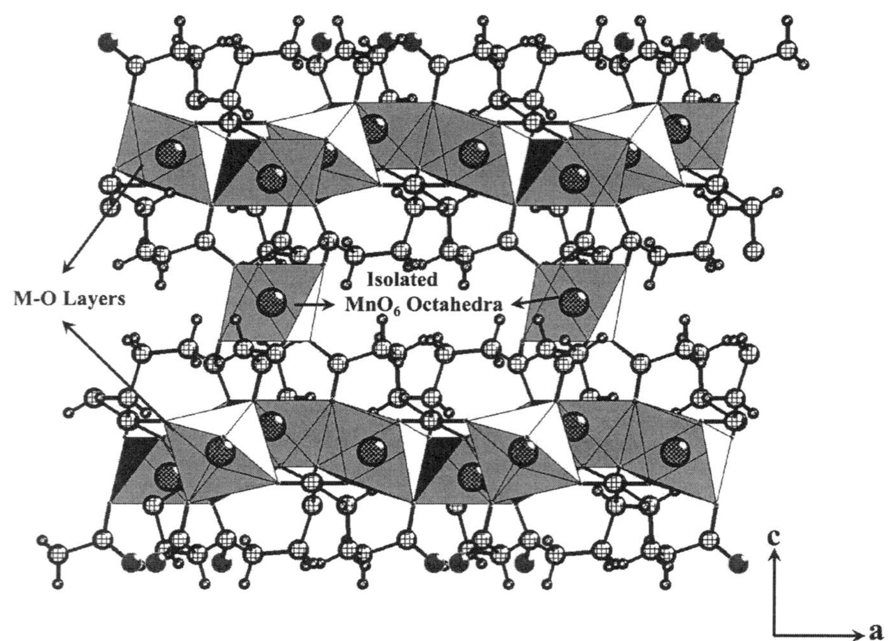

Fig. 3 The connectivity between adjacent MnO layers *via* isolated Mn(3)O$_6$ octahedra. Note that the isolated octahedra are covalently linked by the glutarate units to the inorganic layers.

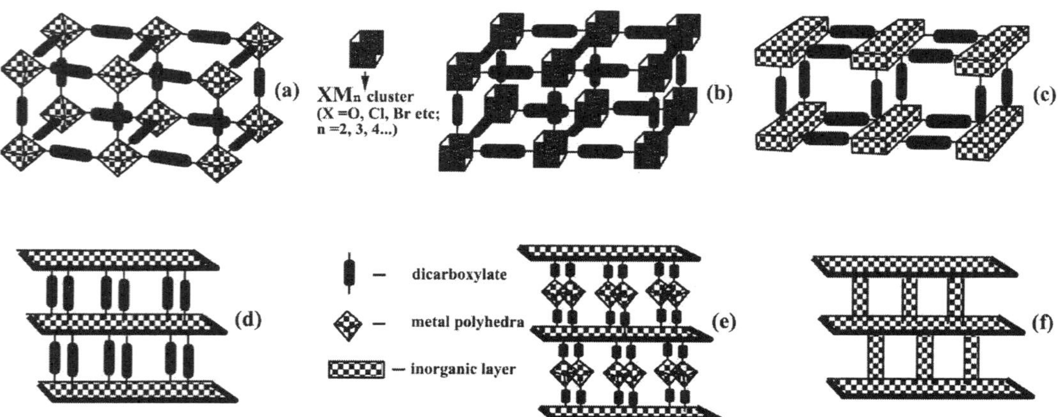

Scheme 1 A schematic illustrating how different inorganic units can be linked through dicarboxylate units to generate three-dimensional open-framework architectures. Motif (a) represents isolated polyhedra, which could be a dimer or a trimer, connected in all directions by dicarboxylate linkers;[17b,30] (b) represents the metal clusters linked *via* dicarboxylates;[15] (c) shows the aligning of one-dimensional inorganic metal–oxygen chains by dicarboxylate linkers;[16c,31] (d) shows the classic pillaring of inorganic MO layers by the dicarboxylate units;[19b,32] (e) represents inorganic layers pillared by isolated polyhedra through coordinating dicarboxylate linkers (*e.g.* compound **II**); (f) shows three-dimensionality arising from the connectivities among the polyhedra (*e.g.* compound **I**).[17a,33] It is important to note that the dicarboxylate oxygens are directly involved in the formation of the inorganic units.

scheme is similar to compound **II** and motif (f) is similar to compound **I**.

Considering the structures of nickel[17a] and cobalt succinates[19b] along with the structure of **I**, we notice that the common feature in the compounds under discussion is the presence of channels lined by the methylene groups of the dicarboxylate moieties, giving rise to a highly hydrophobic environment within them. It is worth mentioning that a three-dimensional vanadyl carboxylate framework possessing hydrophobic channels, exhibiting interesting adsorption properties has been reported recently.[16b] This is indeed a noteworthy feature and it is possible that such channels may show significant selectivity for non-polar species. The formation of compound **I** at room temperature also opens up the possibility of forming other open-framework structures with infinite M–O–M connectivities, under mild reaction conditions (*chimie douce*), instead of hydrothermal conditions.[19c] It is likely that many such metal–carboxylates with open-architectures can be made by a suitable choice of the metal center and the ligand, in the presence of organic amines. We are currently pursuing such experiments.

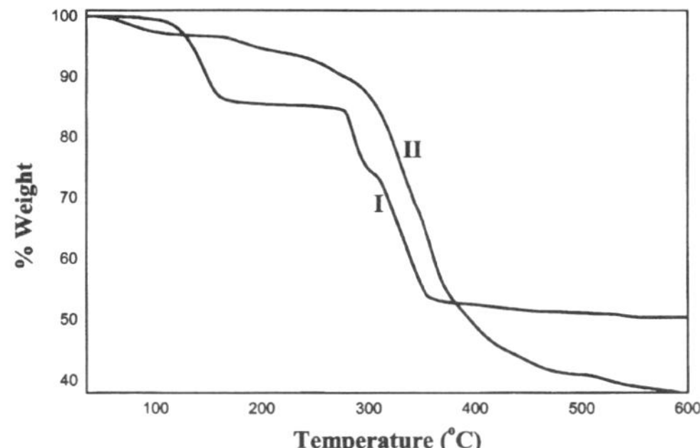

Fig. 4 TGA curves of **I** and **II**.

References

1 (*a*) E. J. Cussen, J. B. Claridge, M. J. Rosseinsky and C. J. Kepert, *J. Am. Chem. Soc.*, 2002, **124**, 9574; (*b*) G. S. Papaefstathiou and L. R. MacGillivray, *Angew. Chem., Int. Ed*, 2002, **41**, 2070.
2 B. F. Hoskins and R. Robson, *J. Am. Chem. Soc.*, 1990, **112**, 1546 and references therein.
3 (*a*) O. M. Yaghi, H. Li, C. Davis, D. Richardson and T. Groy, *Acc. Chem. Res.*, 1998, **31**, 474; (*b*) G. Férey, *J. Solid State Chem.*, 2000, **152**, 37.
4 (*a*) G. Férey, *Chem. Mater.*, 2001, **13**, 3084; (*b*) P. J. Hagraman, D. Hagraman and J Zubieta, *Angew. Chem., Int. Ed.*, 1999, **38**, 2638.
5 (*a*) *Crystal Engineering: The Design and Application of Functional Solids*, eds. K. R. Seddon and M. Zaworotko, Kluwer Academic Publishers, NATO ASI series, 1999; (*b*) M. J. Zaworotko, *Angew. Chem., Int. Ed.*, 2000, **39**, 3052.
6 (*a*) S. R. Batten and R. Robson, *Angew. Chem., Int. Ed.*, 1998, **37**, 1460 and references therein.; (*b*) S. R. Batten, *CrystEngComm*, 2001, **18**, 1.
7 A. Dimos, D. Tsaousis, A. Michaelides, S. Skoulika, S. Golhen, L. Quahab, C. Didierjean and A. Aubry, *Chem. Mater.*, 2002, **14**, 2616.
8 (*a*) J. Y. Lu, A. M. Babb and O. R. Evans, *Inorg. Chem.*, 2002, **41**, 1339; (*b*) P. Ayyappan, O. R. Evans and W. Lin, *Inorg. Chem.*, 2002, **41**, 3328.
9 (*a*) O. R. Evans and W. Lin, *Cryst. Growth Des.*, 2001, **1**, 9; (*b*) C. Janiak, *Angew. Chem., Int. Ed. Engl.*, 1997, **36**, 1431.
10 (*a*) T. Kitazawa, S. Nishikiori, R. Kuroda and T. Iwamoto, *J. Chem. Soc., Dalton Trans.* 1994, 1029.
11 O. R. Evans, R.-G. Xiong, Z. Wang, G. K. Wong and W. Lin, *Angew. Chem., Int. Ed.*, 1999, **38**, 536.
12 (*a*) R. Vaidhyanathan, S. Natarajan and C. N. R. Rao, *Inorg. Chem.*, 2002, **41**, 4496 and references therein; (*b*) R. Vaidhyanathan, S. Natarajan and C. N. R. Rao, *Chem. Mater.*, 2001, **13**, 3524.
13 (*a*) L. Falvello, R. Garde and M. Tomas, *Inorg. Chem.*, 2002, **41**, 4599; (*b*) R. C. Finn, R. S. Rarig, Jr. and J. Zubeita, *Inorg. Chem.*, 2002, **41**, 2109; (*c*) J.-G. Mao and A. Clearfield, *Inorg. Chem.*, 2002, **41**, 2319.
14 (*a*) J.-C. Dai, X.-T. Wu, Z.-Y. Fu, C.-P. Cui, S.-M. Hu, W.-X. Du, L.-M. Wu, H.-H. Zhang and R.-Q. Sun, *Inorg. Chem.*, 2002, **41**, 1391; (*b*) Y.-H. Liu, Y.-L. Lu, H.-C. Wu, J.-C. Wang and K.-H. Lu, *Inorg. Chem.*, 2002, **41**, 2592.
15 (*a*) M. Eddaoudi, D. B. Moler, H. Li, B. Chen, T. M. Reineke, M. O'Keeffe and O. M. Yaghi, *Acc. Chem. Res.*, 2001, **34**, 319; (*b*) H. Li, M. Eddaoudi, M. O'Keeffe and O. M. Yaghi, *Nature (London)*, 1999, **402**, 276; (*c*) T. M. Reineke, M. O'Keeffe and O. M. Yaghi, *Angew. Chem., Int. Ed.*, 1999, **38**, 2590; (*d*) B. Chen, M. Eddaoudi, T. M. Reineke, J. W. Kampf, M. O'Keeffe and O. M. Yaghi, *J. Am. Chem. Soc.*, 2000, **122**, 11559; (*e*) N. L. Rosi, M. Eddaoudi, J. Kim, M. O'Keeffe and O. M. Yaghi, *CrystEngComm*, 2002, **4**, 401.
16 (*a*) K. Biradha, Y. Hongo and M. Fujita, *Angew. Chem., Int. Ed.*, 2000, **39**, 3843; (*b*) K. Barthelet, J. Marrot, D. Riou and G. Férey, *Angew. Chem., Int. Ed.*, 2002, **41**, 281; (*c*) F. Millange, C. Serre and G. Férey, *Chem. Commun.*, 2002, 822; (*d*) S. S. Chui, S. M.-F. Lo, J. P. H. Charmant, A. G. Orpen and I. D. Williams, *Science*, 1999, **283**, 1148; (*e*) S.-I. Noro, S. Kitagawa, M. Kondo and K. Seki, *Angew. Chem., Int. Ed.*, 2000, **39**, 2082.
17 (*a*) P. M. Forster and A. K. Cheetham, *Angew. Chem., Int. Ed.*, 2002, **41**, 457; (*b*) P. M. Forster and A. K. Cheetham, *Chem. Mater.*, 2002, **14**, 17.
18 (*a*) Y. Kim and D.-Y. Jung, *Bull. Korean Chem. Soc.*, 1999, **20**, 827; (*b*) Y. Kim and D.-Y. Jung, *Bull. Korean Chem. Soc.*, 2000, **21**, 656; (*c*) Y. Kim, E. W. Lee and D.-Y. Jung, *Chem. Mater.*, 2001, **13**, 2684.
19 (*a*) C. Livage, C. Egger and G. Férey, *Chem. Mater.*, 1999, **13**, 1546; (*b*) C. Livage, C. Egger, M. Nogues and G. Férey, *J. Mater. Chem.*, 1998, **8**, 2743; (*c*) C. Livage, N. Guillou, J. Marrot and G. Férey, *Chem. Mater.*, 2001, **13**, 4387.
20 (*a*) *Infrared and Raman Spectra of Inorganic and Coordination Compounds*, ed. K. Nakamoto, John Wiley, New York, 5th edn., 1997; (*b*) *Spectrometric Identification of Organic Compounds*, ed. R. M. Silverstein and F. X. Webster, John Wiley and Sons, Inc., New York, 6th edn., 1997.
21 G. M. Sheldrick, SHELXTL-PLUS, Program for Crystal Structure Solution And Refinement, University of Göttingen, Germany 1993.
22 G. M. Sheldrick, SADABS, Siemens Area Detector Absorption Correction Program, University of Göttingen, Germany 1994.
23 I. D. Brown and D. Aldermatt, *Acta Crystallogr., Sect. B*, 1984, **41**, 244.
24 (*a*) E. V. Brusau, J. C. Pedregosa, G. E. Narda, G. Echeverria and G. Punte, *J. Solid State Chem.*, 2000, **153**, 1; (*b*) E. A. H. Griffith, N. G. Charles and E. L. Amma, *Acta Crystallogr., Sect. B*, 1982, **38**, 262.
25 (*a*) E. V. Brusau, J. C. Pedregosa, G. E. Narda, G. Echeverria and G. Punte, *J. Solid State Chem.*, 1999, **143**, 174; (*b*) K. H. Chung, E. Hong, Y. Do and C. H. Moon, *Chem. Commun.*, 1995, 2333.
26 (*a*) R. Vaidhyanathan, S. Natarajan and C. N. R. Rao, *Inorg. Chem.*, 2002, **41**, 5226 and references therein.; (*b*) R. Vaidhyanathan, S. Natarajan and C. N. R. Rao, *Cryst. Growth Des.*, 2003, **3**, 47.
27 (*a*) Y.-Q. Zheng and Z.-P. Kong, *J. Solid State Chem.*, 2002, **166**, 279; (*b*) Y. J. Kim and D.-Y. Jung, *Inorg. Chem.*, 2000, **39**, 1470.
28 (*a*) E. Coronado, M. C.-L. León, J. R. Galán-Mascaros, C. Giménez-Saiz, C. J. Gómez-García and E. Martinez-Ferrero, *J. Chem. Soc., Dalton Trans.*, 2000, 3955; (*b*) K. Biradha, H. Hongo and M. Fujita, *Angew. Chem., Int. Ed.*, 2002, **41**, 3395.
29 A. Michaelides, S. Skoulika, V. Kiritsis and A. Aubry, *J. Chem. Soc., Chem. Commun.*, 1995, 1415.
30 K. Barthelet, D. Riou and G. Férey, *Chem. Commun.*, 2002, 1492.
31 N. Guillou, S. Pastre, C. Livage and G. Férey, *Chem. Commun.*, 2002, 2358.
32 L.-S. Long, X.-M. Chen, M.-L. Tong, Z.-G. Sun, Y.-P. Ren, R.-B. Huang and L.-S. Zheng, *J. Chem. Soc., Dalton Trans.*, 2001, 2888.
33 R. Kuhlman, G. L. Schimek and J. W. Kolis, *Inorg. Chem.*, 1999, **38**, 194.

FULL PAPER

Hybrid Open-Framework Iron Phosphate – Oxalates Demonstrating a Dual Role of the Oxalate Unit

Amitava Choudhury,[a, b] Srinivasan Natarajan,[a] and C.N.R. Rao*[a, b]

Abstract: New inorganic – organic hybrid open-framework materials of the phosphate – oxalate family, $[Fe_2(H_2O)_2(HPO_4)_2(C_2O_4)]\cdot H_2O$ (**I**), $[Fe_2(H_2O)_2(HPO_4)_2(C_2O_4)]\cdot 2H_2O$ (**II**), $[C_3N_2H_{12}][Fe_2(HPO_4)_2(C_2O_4)_{1.5}]_2$ (**III**), and $[C_3N_2OH_{12}][Fe_2(HPO_4)_2(C_2O_4)_{1.5}]_2$ (**IV**) have been synthesized hydrothermally in the presence of structure-directing amines. The amine molecules are incorporated in **III** and **IV**, whereas **I** and **II** are devoid of them. The oxalate units act as a bridge between the layers in all the compounds. The layers in **I** and **II** are entirely inorganic, being formed by FeO_6 and PO_4 units, whereas in **III** and **IV** oxalate units constitute the inorganic

Keywords: bridging ligands · crystal growth · iron · phosphorus · X-ray scattering

layers and act as the bridge between these layers. Such a dual role of the oxalate unit is unique and noteworthy. The formation of two types of inorganic layers in **I** and **II** consisting of four-, six-, and eight-membered rings, indicates the interconversions between the various rings in the phosphate – oxalates to be facile. All the phosphate – oxalates show antiferromagnetic ordering at low temperatures.

Introduction

Of the various inorganic open-framework materials discovered in the last few years, the metal phosphates occupy a prime position. The family of open-framework metal phosphates includes one-dimensional chains or ladders, two-dimensional layers, and three-dimensional structures.[1] An interesting variant of the metal phosphate is that obtained by incorporating the oxalate unit in the materials.[2] A few metal phosphate – oxalates, especially those of iron, constituting a novel kind of inorganic – organic hybrid materials, have been reported during the last year.[3–7] The important structural features of these compounds is the presence of iron phosphate sheets, crosslinked by the oxalate units giving rise to the three-dimensional architecture. We have been investigating phosphate – oxalates not only with a view to producing new materials but also to unravelling the structural features of these novel hybrid material, and we have succeeded in isolating iron phosphate – oxalates with a variety of structural features. Two of the iron phosphate – oxalates discovered by us possess iron phosphate layers that are crosslinked by

oxalate bridges. The layers in these materials consist of different ring systems reminiscent of the open-framework metal phosphates. More importantly, we have obtained iron phosphate – oxalates wherein the oxalate unit, besides acting as a bridge between the metal phosphate layers, is part of the layer system. The presence of oxalates performing two functions in these hybrid materials is noteworthy. Such a dual role of the oxalate unit has also been found by us in a metal oxalate structure. Thus, the oxalate unit in iron phosphate – oxalates acts like the phosphate unit in metal phosphates and as the oxalate unit in metal oxalates.

Results and Discussion

$[Fe_2(H_2O)_2(HPO_4)_2(C_2O_4)]\cdot H_2O$ (**I**) and $[Fe_2(H_2O)_2(HPO_4)_2(C_2O_4)]\cdot 2H_2O$ (**II**): The asymmetric units of **I** and **II** are identical and each consist of 11 non-hydrogen atoms (Figure 1). Both **I** and **II**, have three-dimensional structures built from a networking of FeO_6 octahedra, PO_4 tetrahedra, and oxalate units. The FeO_6 octahedra and PO_4 tetrahedra form layers that are held together and apart by the crosslinking oxalate groups. Of the six oxygen atoms of a Fe octahedron, three belong to the phosphate groups and two to the oxalate units, while the remaining oxygen atom is a terminal water molecule.

In **I**, the FeO_6 octahedra and PO_4 tetrahedra strictly alternate along the b axis forming a layer possessing six-membered apertures (made up of 6 T atoms, T = Fe, P, Figure 2). Layers exclusively possessing six-membered aper-

[a] Prof. Dr. C.N.R. Rao, A. Choudhury, Dr. S. Natarajan
Chemistry and Physics of Materials Unit and
CSIR Centre of Excellence in Chemistry
Jawaharlal Nehru Centre for Advanced Scientific Research
Jakkur P.O., Bangalore 560 064 (India)
Fax: (+91) 80-846-2766
E-mail: cnrrao@jncasr.ac.in

[b] Prof. Dr. C.N.R. Rao, A. Choudhury
Solid State and Structural Chemistry Unit
Indian Institute of Science, Bangalore 560 012 (India)

© WILEY-VCH Verlag GmbH, D-69451 Weinheim, 2000   0947-6539/00/0607-1168 $ 17.50+.50/0   *Chem. Eur. J.* **2000**, *6*, No. 7

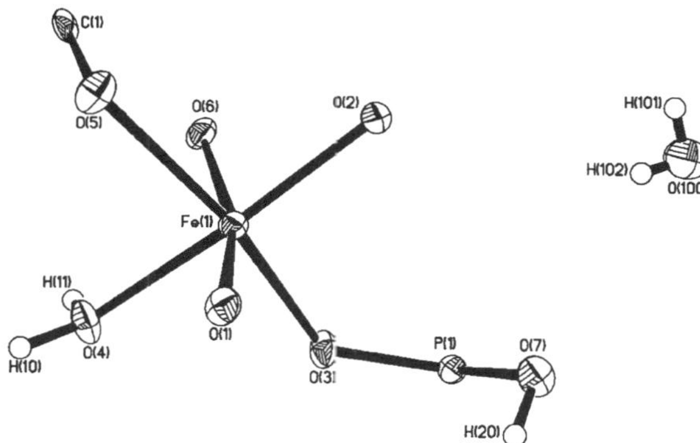

Figure 1. ORTEP plot of the asymmetric unit in **I** and **II**. Thermal ellipsoids are given at 50 % probability.

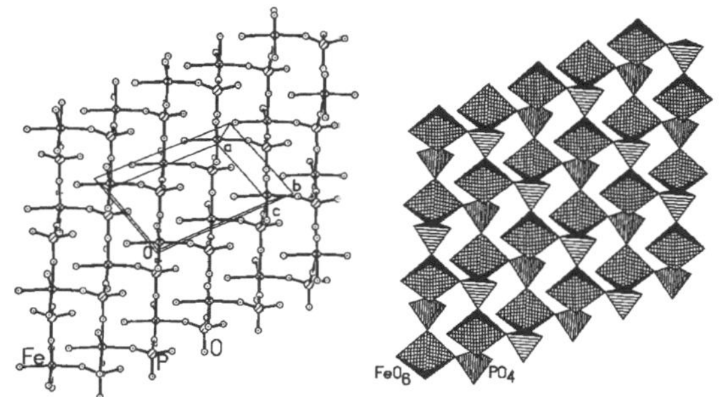

Figure 2. Structure of **I** showing the layers made by the networking of FeO$_6$ and PO$_4$ units. Note that the layers are made of six-membered rings. Left: Ball-and-stick view; right: polyhedral view.

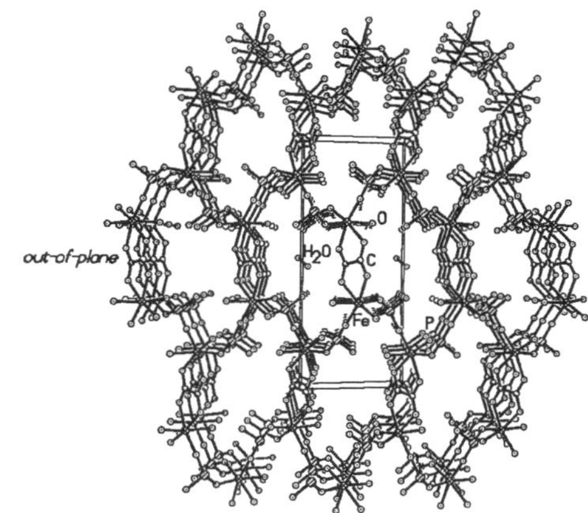

Figure 3. Structure of **I** showing the channels. The *out-of-plane* oxalate unit bridges the layers, and the water molecules are present in the channels. Hydrogen atoms are omitted for clarity.

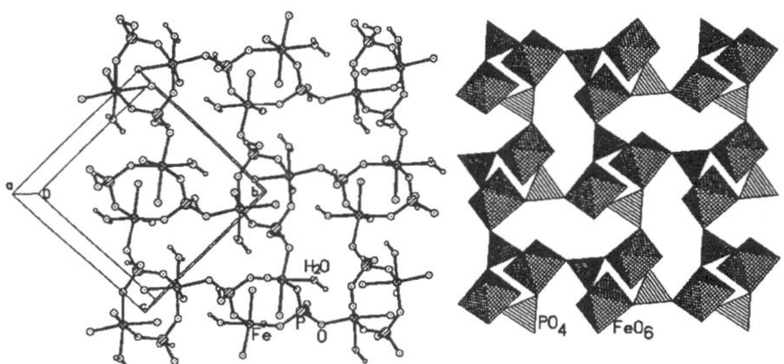

Figure 4. Structure of **II** along the *bc* plane showing the layers. The layers are made by four- and eight-membered rings, and the bound water molecule from the Fe center protrude into the eight-membered aperture. Left: Ball-and-stick view; right: polyhedral view.

tures are rare, the only other example being a tin(II) phosphate – oxalate.[2] The layers are connected through the anionic oxalate bridges (*out-of-plane* oxalate units) completing the neutral three-dimensional architecture. Along the *a* axis, the connectivity produces eight-membered one-dimensional channels wherein the bonded water and the free water molecules reside (Figure 3).

In **II**, the strictly alternating FeO$_6$ and PO$_4$ moieties form four- and eight-membered apertures (made up of 4 and 8 T atoms, respectively) along the *bc* plane (Figure 4). The water molecules attached to the iron centers project into the eight-membered apertures. The oxalate units link these layers completing the three-dimensional connectivity. Along the *a* axis, the eight-membered apertures form one-dimensional channels (Figure 5).

Selected bond lengths and angles of **I** and **II** are given in Tables 1 and 2 respectively. The three different types of Fe—O bonds (Fe—O—P; Fe—O—C; Fe—O$_{Water}$) show the expected differences in average bond lengths and angles. The phosphorus atoms are connected to three neighboring iron atoms through the oxygens, and have a terminal OH group. Bond valence sum calculations[8] indicate that one of the oxygen atoms [O(4)] is a water molecule and that the one bound to

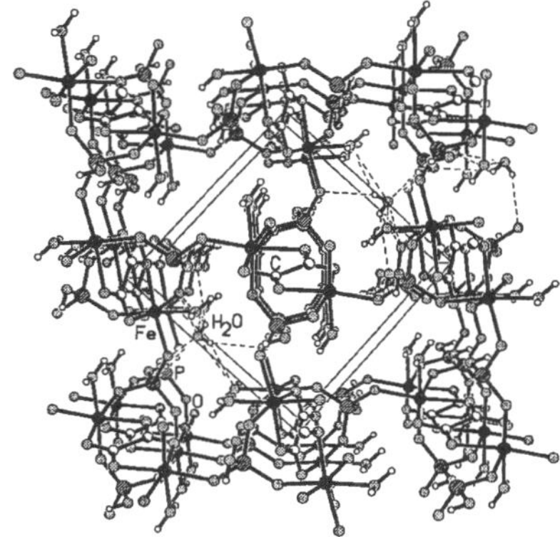

Figure 5. Structure of **II** along *a* axis, showing channels.

P [O(7)] is a terminal hydroxy group. The bonding of water molecules directly to the iron center is not uncommon, and similar bonding is known in open-framework iron phosphate materials.[9] The observed bond lengths and angles observed in **I** and **II** are in general agreement with those reported in the literature.

Table 1. Selected bond lengths [Å] and angles [°] for **I**.

Fe(1)–O(1)	1.893(3)	Fe(1)–O(2)	1.936(3)
Fe(1)–O(3)	1.948(2)	Fe(1)–O(4)	2.082(3)
Fe(1)–O(5)	2.102(2)	Fe(1)–O(6)	2.125(2)
P(1)–O(1)[#1]	1.508(3)	P(1)–O(2)	1.517(3)
P(1)–O(3)[#2]	1.528(2)	P(1)–O(7)	1.564(3)
C(1)–O(5)	1.249(4)	C(1)[#3]–O(6)	1.255(4)
C(1)–C(1)[#3]	1.536(6)		
O(3)-Fe(1)-O(4)	93.55(11)	O(1)-Fe(1)-O(5)	167.86(10)
O(2)-Fe(1)-O(5)	90.43(10)	O(3)-Fe(1)-O(5)	90.47(10)
O(4)-Fe(1)-O(5)	85.16(13)	O(1)-Fe(1)-O(6)	92.27(11)
O(2)-Fe(1)-O(6)	86.58(10)	O(3)-Fe(1)-O(6)	168.20(9)
O(4)-Fe(1)-O(6)	84.39(11)	O(5)-Fe(1)-O(6)	77.79(9)
O(1)[#1]-P(1)-O(2)	111.9(2)	O(1)[#1]-P(1)-O(3)[#2]	109.10(14)
O(2)-P(1)-O(3)[#2]	113.08(14)	O(1)[#1]-P(1)-O(7)	108.7(2)
O(1)-Fe(1)-O(2)	95.98(11)	O(2)-P(1)-O(7)	107.7(2)
O(1)-Fe(1)-O(3)	99.24(11)	O(1)[#2]-P(1)-O(7)	106.10(14)
O(2)-Fe(1)-O(3)	94.77(11)	O(5)-C(1)-O(6)[#3]	127.0(3)
O(1)-Fe(1)-O(4)	86.96(13)	O(5)-C(1)-C(1)[#3]	116.9(4)
O(2)-Fe(1)-O(4)	170.62(12)	O(6)[#3]-C(1)-C(1)[#3]	116.1(4)

Symmetry transformations used to generate equivalent atoms: #1: $x, -y + 3/2, z + 1/2$; #2: $x - 1, y, z$; #3: $-x, -y + 1, -z + 1$.

Table 2. Selected bond lengths [Å] and angles [°] for **II**.

Fe(1)–O(1)	1.935(3)	Fe(1)–O(2)	1.935(3)
Fe(1)–O(3)	1.942(3)	Fe(1)–O(4)	2.088(4)
Fe(1)–O(5)	2.104(3)	Fe(1)–O(6)	2.110(3)
P(1)–O(2)[#1]	1.521(3)	P(1)–O(1)[#2]	1.527(3)
P(1)–O(3)	1.530(3)	P(1)–O(7)	1.578(3)
C(1)–O(5)	1.244(5)	C(1)–O(6)[#3]	1.259(5)
C(1)–C(1)[#3]	1.535(8)		
O(3)-Fe(1)-O(4)	83.03(13)	O(1)-Fe(1)-O(5)	87.69(12)
O(2)-Fe(1)-O(5)	94.42(13)	O(3)-Fe(1)-O(5)	167.84(13)
O(4)-Fe(1)-O(5)	85.52(13)	O(1)-Fe(1)-O(6)	164.92(12)
O(2)-Fe(1)-O(6)	88.43(12)	O(3)-Fe(1)-O(6)	96.02(12)
O(4)-Fe(1)-O(6)	81.0(2)	O(5)-Fe(1)-O(6)	78.10(12)
O(2)[#1]-P(1)-O(1)[#2]	112.5(2)	O(2)[#1]-P(1)-O(3)	110.4(2)
O(1)[#2]-P(1)-O(3)	111.7(2)	O(2)[#1]-P(1)-O(7)	109.1(2)
O(1)-Fe(1)-O(2)	97.78(13)	O(1)[#2]-P(1)-O(7)	104.0(2)
O(1)-Fe(1)-O(3)	96.96(13)	O(3)-P(1)-O(7)	108.8(2)
O(2)-Fe(1)-O(3)	96.07(12)	O(5)-C(1)-O(6)[#3]	126.9(4)
O(1)-Fe(1)-O(4)	93.0(2)	O(5)-C(1)-C(1)[#3]	116.9(5)
O(2)-Fe(1)-O(4)	169.2(2)	O(6)[#3]-C(1)-C(1)[#3]	116.3(5)

Symmetry transformations used to generate equivalent atoms: #1: $-x + 1, y + 1/2, -z + 3/2$; #2: $x, -y + 1/2, z - 1/2$; #3: $-x + 2, -y, -z + 2$.

$[C_3N_2H_{12}][Fe_2(HPO_4)_2(C_2O_4)_{1.5}]_2$ (**III**) and $[C_3N_2OH_{12}]$-$[Fe_2(HPO_4)_2(C_2O_4)_{1.5}]_2$ (**IV**): The asymmetric units of **III** and **IV** contain 24 and 25 non-hydrogen atoms, respectively, with two crystallographically distinct iron and phosphorus atoms (Figure 6). The final atomic coordinates for **III** are presented in Table 3. The structure consists of layers of formula $[Fe_2(HPO_4)_2(C_2O_4)]$, linked through another oxalate unit to complete the anionic framework. Charge neutrality is achieved by the incorporation of organic amines in their diprotonated form. Thus, there are $0.5[C_3N_2H_{12}]^{2+}$ molecules in **III** and $0.5[C_3N_2OH_{12}]^{2+}$ molecules in **IV**, respectively, per framework formula. The structure comprises a network of FeO_6, PO_4, and C_2O_4 moieties with each iron bound to six oxygens, which are, in turn bound to carbon and phosphorus atoms completing the network.

The structures of **III** and **IV** are made from a networking of FeO_4, PO_4, and C_2O_4 forming layers. Of the two iron atoms in the asymmetric unit, Fe(1) is connected to two oxalate and

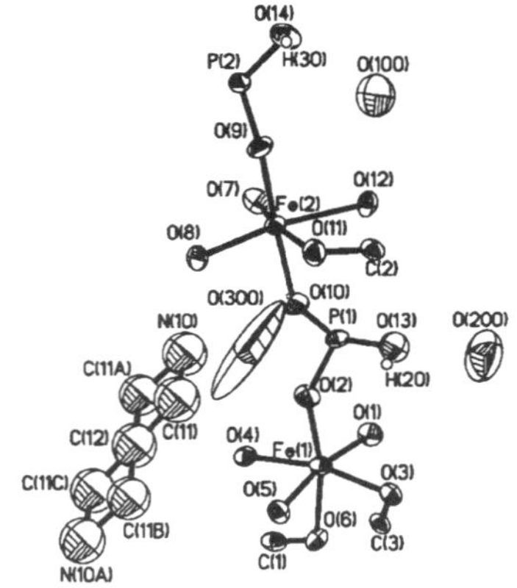

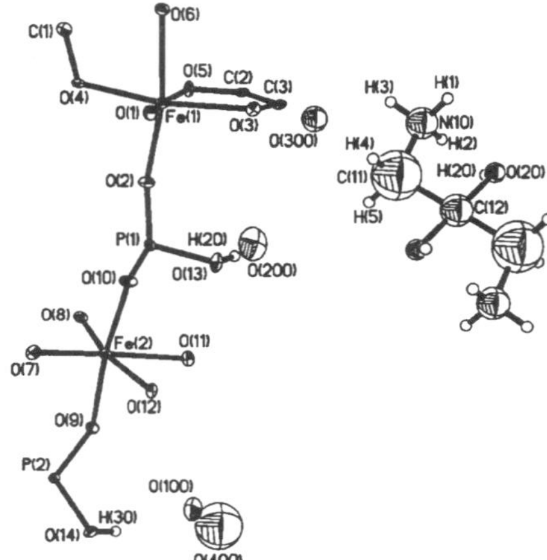

Figure 6. Top: ORTEP plot of **III**. Thermal ellipsoids are given at 50 % probability. Bottom: ORTEP plot of **IV**. Thermal ellipsoids are given at 50 % probability.

phosphate units, and Fe(2) is bonded with one oxalate and four phosphate units. The FeO_6 and PO_4 units are connected to each other in such a way that they form zigzag one-dimensional ladderlike chains that are linked through an oxalate unit to form an oxalate–phosphate layer (Figure 7). This is indeed a unique inorganic–organic hybrid layer network. The hybrid layers are linked through another oxalate unit acting like a bridge (Figure 8). Thus, two different types of oxalate units occur in **III** and **IV**. One *in-plane* (with respect to the layer) and the other *out-of-plane*. To our knowledge, this is the first example of the existence of two different types of oxalate moieties in such a material.

The linkages between the oxalates and the oxalate–phosphate layers in **III** and **IV** create channels of width 5.2 × 4.5 Å (shortest atom–atom contact distance not including the van der Waals radii) along a axis (Figure 8). Figure 9 shows the connectivity along the c axis. Thus, **III** and **IV** are members of three-dimensional solids possessing different channels. All the iron atoms are hexacoordinated with respect to oxygen. Of the two crystallographically independent iron atoms in **III**

Table 3. Atomic coordinates [× 10⁴] and equivalent isotropic displacement parameters [$Å^2 \times 10^3$] for **III**.

Atom	x	y	z	U(eq)[a]
Fe(1)	6348(1)	1125(1)	6645(1)	24(1)
Fe(2)	12478(1)	619(1)	10020(1)	22(1)
P(1)	8945(2)	1073(2)	8908(2)	24(1)
P(2)	15512(2)	− 30(2)	11833(2)	23(1)
O(1)	5195(7)	829(4)	7747(5)	33(2)
O(2)	8243(6)	1005(4)	7643(5)	34(2)
O(3)	6001(7)	2432(4)	6906(5)	30(2)
O(4)	6675(6)	− 74(4)	5978(5)	28(2)
O(5)	7383(6)	1693(4)	5436(5)	28(2)
O(6)	4444(1)	1046(4)	5347(5)	29(2)
O(7)	11847(7)	− 521(4)	10366(5)	35(2)
O(8)	13183(6)	318(4)	8670(5)	35(2)
O(9)	14422(6)	532(4)	11019(5)	30(2)
O(10)	10540(6)	849(4)	8996(5)	40(2)
O(11)	12936(7)	1969(4)	9834(5)	31(2)
O(12)	11571(6)	1214(4)	11337(5)	26(2)
O(13)	8828(8)	2041(4)	9319(6)	51(2)
O(14)	16170(7)	540(4)	12900(5)	39(2)
C(1)	4352(10)	324(6)	4817(8)	29(2)
C(2)	12386(9)	2480(6)	10448(7)	25(2)
C(3)	6578(9)	2957(6)	6311(7)	24(2)
O(100)	15052(13)	2098(6)	12736(8)	114(4)
O(200)	6432(11)	2231(8)	10239(10)	128(4)
O(300)	13122(42)	2458(24)	7633(26)	738(42)
N(10)[b]	10318(15)	− 363(9)	7045(11)	109(4)
C(12)[b]	10000	0	5000	107(8)
C(11A)[b]	9926(33)	− 744(10)	5874(12)	95(14)
C(11)[b]	10521(35)	352(13)	6199(9)	110(15)

[a] U(eq) is defined as one third of the trace of the orthogonalized U_{ij} tensor. [b] Refined isotropically.

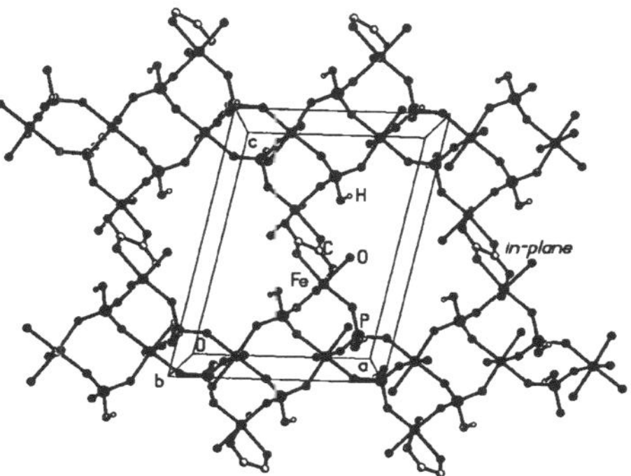

Figure 7. Structure of **III** along the *a* axis, showing the ladder like chains and the connecting oxalate unit (*in-plane* oxalate) forming a hybrid layer.

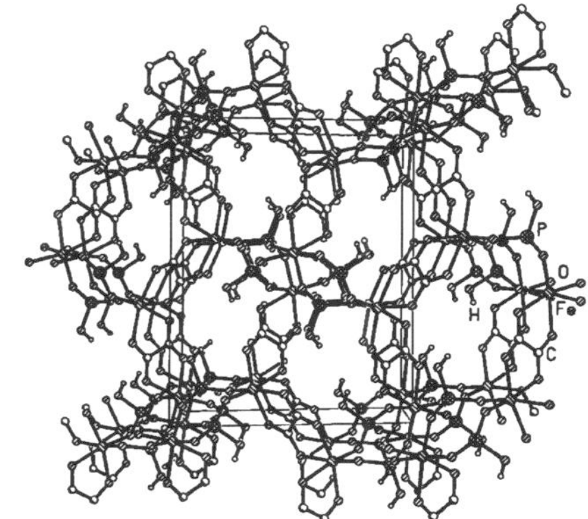

Figure 8. Structure of **III** showing channels along the *a* axis formed by the bridging oxalates with the hybrid layer.

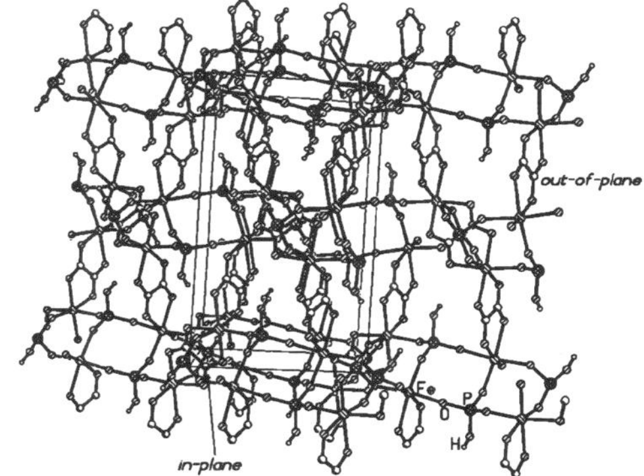

Figure 9. Structure of **III** showing channels along the *c* axis. Note the two different types of oxalate units (*in-plane* and *out-of-plane*).

and **IV**, Fe(1) is coordinated with four oxygen atoms with distances in the range 2.041–2.101 Å (av 2.068 Å, for **III**, 2.074 Å for **IV**) and two oxygen atoms have distances in the range 1.893–1.916 Å (av 1.915 Å for **III**, 1.90 Å for **IV**). Fe(2), on the other hand, is coordinated with two oxygen atoms in the range 2.116–2.148 Å (av 2.128 Å for **III**, 2.133 Å for **IV**) and four oxygen atoms in the range 1.902–1.981 Å (av 1.939 Å for **III**, 1.946 Å for **IV**). These are the oxygen atoms linked to phosphorus and carbon atoms, respectively. Similar bond lengths have been observed before. The O-Fe-O bond angles are as expected. The P–O and C–O distances and O-P-O and O-C-O bond angles are in the range expected for this type of bonding. Similar structural parameters have been observed for other phosphate-oxalates. Bond valence sum calculations[8] indicate that two of the oxygen atoms attached to the phosphorus atom have a hydrogen atom attached to it making it a terminal OH group. The selected bond lengths and angles for **III** are presented in Table 4. We have not given the various structural details of **IV** since they are very similar to those of **III**.

The iron phosphate – oxalates **I – IV** are members of a new family of inorganic – organic hybrid framework structures and have been obtained as good quality single crystals by employing hydrothermal methods. Although all these materials involve bonding between the iron atoms and the phosphate, oxalate units, they exhibit interesting differences. While **III** and **IV** are formed with the amine in the final framework solid, **I** and **II** are formed devoid of them. Such elimination of amines during the hydrothermal synthesis is known.[2, 10] It is noteworthy that although **I** and **II** are formed under identical conditions of composition and temperature with only a slight difference in the duration of reaction, the resultant products possess perceptible differences in structural features. In the few known iron phosphate – oxalates, the iron phosphate layers are neutral and the anionicity of the framework is

Table 4. Selected bond lengths [Å] and angles [°] in **III**.

Fe(1)–O(1)	1.916(6)	Fe(2)–O(7)	1.902(6)
Fe(1)–O(2)	1.914(6)	Fe(2)–O(8)	1.927(6)
Fe(1)–O(3)	2.046(6)	Fe(2)–O(9)	1.945(6)
Fe(1)–O(4)	2.041(6)	Fe(2)–O(10)	1.981(6)
Fe(1)–O(5)	2.085(6)	Fe(2)–O(11)	2.116(6)
Fe(1)–O(6)	2.101(6)	Fe(2)–O(12)	2.140(6)
P(1)–O(10)	1.491(6)	P(2)–O(9)	1.514(6)
P(1)–O(2)	1.531(6)	P(2)–O(1)[#1]	1.514(6)
P(1)–O(7)[#1]	1.505(9)	P(2)–O(8)[#2]	1.519(6)
P(1)–O(13)	1.562(7)	P(2)–O(14)	1.565(6)
C(1)–O(4)[#3]	1.259(10)	C(2)–O(5)[#5]	1.256(10)
C(3)–O(12)[#4]	1.260(10)	C(1)–C(1)[#3]	1.54(2)
C(2)–C(3)[#5]	1.548(12)	N(10)–C(11)	1.528(10)
N(10)–C(11A)	1.499(10)	C(11)–C(12)	1.522(10)
C(11A)–C(12)	1.556(10)		
O(2)-Fe(1)-O(1)	96.1(3)	O(7)-Fe(2)-O(8)	98.0(3)
O(2)-Fe(1)-O(4)	88.4(2)	O(7)-Fe(2)-O(9)	94.9(3)
O(1)-Fe(1)-O(4)	102.2(2)	O(8)-Fe(2)-O(9)	94.5(2)
O(2)-Fe(1)-O(3)	98.5(3)	O(7)-Fe(2)-O(10)	91.2(3)
O(1)-Fe(1)-O(3)	89.7(3)	O(8)-Fe(2)-O(10)	86.2(3)
O(4)-Fe(1)-O(3)	165.6(2)	O(9)-Fe(2)-O(10)	173.7(3)
O(2)-Fe(1)-O(5)	89.4(2)	O(7)-Fe(2)-O(11)	169.7(2)
O(1)-Fe(1)-O(5)	168.3(2)	O(8)-Fe(2)-O(11)	91.8(3)
O(4)-Fe(1)-O(5)	88.2(2)	O(9)-Fe(2)-O(11)	87.3(2)
O(3)-Fe(1)-O(5)	79.3(2)	O(10)-Fe(2)-O(11)	86.4(3)
O(2)-Fe(1)-O(6)	167.6(2)	O(7)-Fe(2)-O(12)	92.2(2)
O(1)-Fe(1)-O(6)	90.2(3)	O(8)-Fe(2)-O(12)	167.8(2)
O(4)-Fe(1)-O(6)	79.9(2)	O(9)-Fe(2)-O(12)	91.2(2)
O(3)-Fe(1)-O(6)	92.1(2)	O(10)-Fe(2)-O(12)	87.0(2)
O(5)-F1(1)-O(6)	86.4(2)	O(11)-Fe(2)-O(12)	77.7(2)
O(10)-P(1)-O(7)[#1]	114.8(4)	O(10)-P(1)-O(2)	105.3(4)
O(7)[#1]-P(1)-O(2)	112.6(4)	O(10)-P(1)-O(13)	108.9(4)
O(7)[#1]-P(1)-O(13)	105.5(4)	O(2)-P(1)-O(13)	109.7(4)
O(9)-P(2)-O(1)[#1]	113.3(3)	O(9)-P(2)-O(8)[#2]	112.2(3)
O(1)[#1]-P(2)-O(8)[#2]	109.4(4)	O(9)-P(2)-O(14)	108.2(3)
O(1)[#1]-P(2)-O(14)	106.9(3)	O(8)[#2]-P(2)-O(14)	106.4(4)
O(4)[#3]-C(1)-O(6)	127.4(8)	O(11)-C(2)-O(5)[#5]	128.0(8)
O(12)[#4]-C(3)-O(3)	127.8(8)	N(10)-C(11A)-C(12)	108.7(11)
N(10)-C(11)-C(12)	109.0(11)	C(11)[#6]-C(12)-C(11)	109.7(10)

Symmetry transformations used to generate equivalent atoms: #1: $-x+2$, $-y$, $-z+2$; #2: $-x+3$, $-y$, $-z+2$; #3: $-x+1$, $-y$, $-z+1$; #4: $x+1/2$, $-y+1/2$, $z-1/2$; #5: $x+1/2$, $-y+1/2$, $z+1/2$; #6: $-x+2$, $-y$, $-z+1$.

derived from the oxalate bridges. In **I** and **II**, however, cationic iron phosphate layers are neutralized by the oxalate bridges to render the framework neutral.

The amine molecules in **III** and **IV** (1,3-diaminopropane (1,3-DAP) and 1,3-diamino-2-hydroxy propane (1,3-DAHP), respectively) are disordered along with the water molecules within the channels. The disorder in the case of 1,3-DAP occurs on the carbon atom attached to the nitrogen atom and in 1,3-DAHP, it is the oxygen atom on the central carbon atom that is disordered. Disorder of the amine molecules in open-framework solids is not uncommon, and most of the earlier observations relate to the terminal atoms[11] and only recently disorder of the amine molecules involving nonterminal atoms has been observed.[12]

The coordination environment and connectivity of Fe atoms in **I**–**IV** presents an interesting comparison. Though all the Fe atoms are octahedrally coordinated, the manner they link up with the phosphate and oxalates units show distinct differences. In **I** and **II** the iron atoms do not show any differences and in **III** and **IV** the iron atoms presents two different coordination environments. In **III** and **IV**, Fe(1) is

bonded to two oxalate and two phosphate units, and Fe(2) is bonded to one oxalate and four phosphate units. This is possibly due to the ambidendate coordinating ability of the oxalate units requiring two phosphate units to make similar connections as that of one oxalate unit.

The framework composition of **I** and **II** is similar, but there are differences in the networking of the polyhedra. Compound **I** forms layers made up of six-membered apertures within the layers, **II** possess four- and eight-membered apertures. It is proposed that the four-memberd rings can readily transform to six-, eight-, and other higher membered rings.[13] In the case of **I** and **II**, ring conversions could occur as shown in Scheme 1. The six-membered ring within the layers

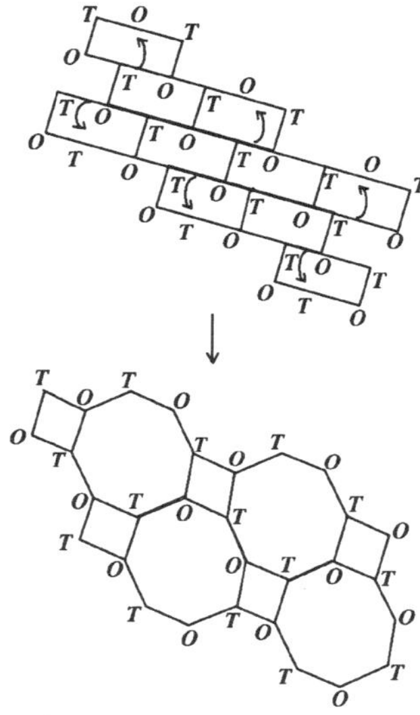

Scheme 1. Schematic representation of a possible conversion between the six-membered ring aperture of **I** into four- and eight-membered apertures in **II**. Note that oxygen atoms are not shown and only bond shifting is indicated. O and T represent octahedral (Fe) and tetrahedral (P) atoms respectively.

of **I** can readily transformed to four- and eight-membered rings of **II** by a simple shifting of bonds. In general, n-edge-sharing four-membered rings can give rise to a ring with $4n - 2(n-1)$ atoms. Or, if we add n four-membered rings to a m-membered ring, we get a ring with $m+2n$ atoms (m and n represent T atoms; T = Fe, P in the present case). The facile transformations of these ring structures within the layers of **I** and **II** suggests that one is dealing with structures of comparable energies that render it difficult to exactly pin-down the stepwise mechanism involved in these transformations.

The layers in **III** and **IV** are different from those in **I** and **II**. Whereas the layers in **I** and **II** are formed by the linkages between the FeO_6 octahedra and PO_4 tetrahedra, the layers in **III** and **IV** are made by linkages involving the (*in-plane*) C_2O_4 units, in addition to the FeO_6 and PO_4. The layers are crosslinked by the (*out-of-plane*) oxalate units in all the phosphate–oxalates. Such a dual role of the oxalate unit has

© WILEY-VCH Verlag GmbH, D-69451 Weinheim, 2000   0947-6539/00/0607-1172 $ 17.50+.50/0   *Chem. Eur. J.* **2000**, *6*, No. 7

not been observed previously. Significantly, we have just discovered a zinc oxalate containing both the *in-plane* and *out-of-plane* oxalate linkages with three-dimensional connectivity.[14] In Figure 10, we show the structure of this material to illustrate the presence of the oxalates within the layers as well as a bridge between the layers. This dual functionality of the oxalate units, in the Zn oxalate, gives rise to an elliptical aperture made by the linkages between 10 Zn and 10 oxalate units within the same plane, with the other oxalate unit connecting the elliptical pores such that two such rings are perpendicular to each other (Figure 10 bottom).

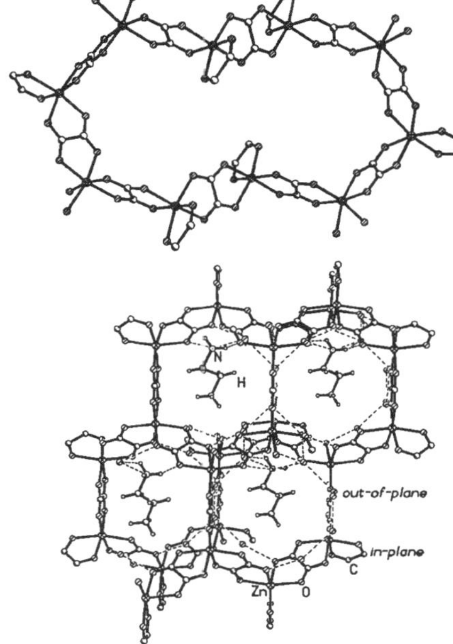

Figure 10. Top: Structure showing the two types of oxalate units in a zinc oxalate. Note that one of zinc atoms is coordinated to two oxalate units going out of plane leading to three-dimensional connectivity. Bottom: Three-dimensional zinc oxalate structure showing the channels formed by the oxalate units. Note the presence of two types of oxalate units (*in-plane* and *out-of-plane*).

The coordination environment of iron atoms in phosphates and oxalates also presents an interesting comparison. While in most of the phosphate-based open-framework structures iron is essentially either five- or six-coordinate, forming a trigonal bipyramidal FeO_5 or octahedral FeO_6 as the building units, in oxalates[15] and phosphate – oxalates[4–7] including the present solids, iron is present exclusively in an octahedral environment. We believe that this is because the average charge per oxygen atom on the oxalate (0.5) is less than that on the phosphate (0.75), so that more oxalate oxygen atoms are needed to satisfy the valence of iron.

Magnetic susceptibility measurements indicate strong antiferromagnetic interactions in **I – IV**, with the Neel temperature in the range 25 – 40 K (Figure 11). The magnetic behavior above the ordering temperature obeys Curie – Weiss behavior. The various magnetic parameters are given in Table 5. The μ_{eff} calculated from the Curie – Weiss law

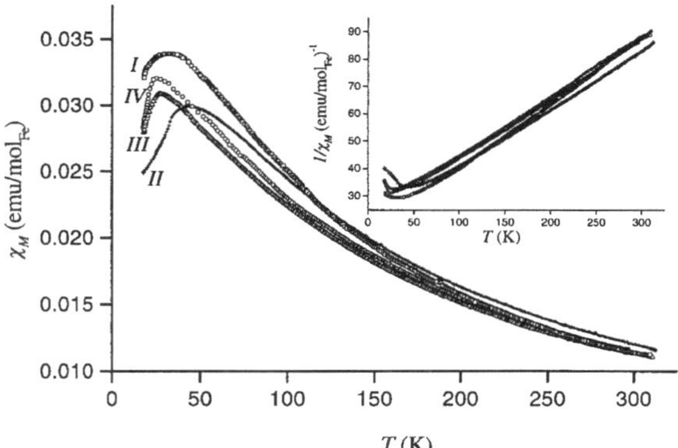

Figure 11. Temperature dependence of magnetic susceptibility for compounds **I – IV**. Inset shows inverse of the susceptibility.

Table 5. Magnetic parameters for compounds **I – IV**.

Compound	θp	T_N [K]	μ_{eff}
I	− 63.1	30	5.73
II	− 90.9	40	6.15
III	− 88.8	25	6.13
IV	− 92.3	25	6.01

shows that iron is present in the high-spin state in all the compounds.

Thermogravimetric analysis (TGA) of **I – IV** was carried out in flowing nitrogen in the range 30 – 700 °C. Compounds **I** and **II** showed three steps of decomposition. The first mass loss occurring at 150 °C (4 % for **I** and 8.3 % for **II**) corresponds to the loss of free water molecule (calcd 4.04 % for **I** and 7.8 % for **II**), the second mass loss of about 23 % in the range 220 – 290 °C corresponds to the loss of bound water and oxalate molecules (calcd 24.3 % for **I** and 23.3 % for **II**), and the final mass loss of 4 % in the region 300 – 350 °C corresponds to the loss of the OH group (calcd 3.9 % for **I** and 3.88 % for **II**). In **III** and **IV**, the mass loss occurs in two steps with the second step having a large tail. For **III**, the mass loss of 9.8 % occurring at 170 °C corresponds to the loss of free water (calcd 10.2 %), and the second total mass loss including the tail of 33.3 % occurring in the temperature range 250 – 550 °C corresponds to the loss of oxalate and amine (calcd 31 %). In the case of **IV**, the mass loss of 8.7 % occurring at 140 °C corresponds to the loss of free water (calcd 10.1 %), and the second mass loss including the tail of 34.4 % in the temperature range 300 – 600 °C corresponds to the loss of oxalate and the amine (calcd 32.1 %). In all the cases, the final decomposed products were poorly crystalline and showed broad X-ray diffraction lines that correspond to the condensed iron phosphate $FePO_4$ [JCPDS: 29-0715].

Conclusion

Four new iron phosphate – oxalates belonging to the hybrid inorganic – organic family of open-framework structures have been synthesized hydrothermally in the presence of structure-directing amines. Compounds **III** and **IV** are formed with the amine molecules within the framework, but **I** and **II** are devoid of the amines. All the compounds have the oxalate unit acting as the bridge between the layers. The layers in **I** and **II** are purely inorganic (formed only FeO_6 and PO_4 units), whereas **III** and **IV** have hybrid layers containing oxalate units in addition to FeO_6 and PO_4 moieties. Such a dual role of the oxalate unit is noteworthy. The formation of two types of inorganic layers in **I** and **II** suggest that the interconversions between the various sizes are facilitated by small energy changes. The observation of antiferromagnetic interactions in all the compounds shows that superexchange is facilitated through the phosphate and oxalate moieties. It would be worthwhile exploring the possibility of introducing other magnetic elements such as cobalt, and manganese as part of the iron phosphate – oxalate framework, to obtain materials with novel magnetic properties. Work on this theme is currently in progress and our initial results indicate that one can indeed form such hetero-substituted frameworks.

Experimental Section

Synthesis and initial characterization: The iron oxalate – phosphates **I**–**IV** were prepared under mild hydro/solvothermal conditions starting from a mixture of $FeCl_3$, H_3PO_4, and oxalic acid with an organic amine. The synthetic conditions are presented in Table 6. The starting mixtures were stirred to attain homogeneity and then sealed in 30 mL polypropylene bottles. All the chemicals were purchased from Aldrich and were used without any further purification. The initial pH of the mixture was 2.0 and the fill-factor was $\approx 40\%$. The mixtures after the heat treatments did not show any appreciable change in the pH (2.0). The reaction led to the formation of large quantities of transparent colorless rodlike (**I**), pale pink octahedral (**II**), and platelike (**III** and **IV**) crystals. The yields of the products were 20 (**I**), 60 (**II**), 50 (**III**), and 45% (**IV**), respectively. These crystals were used for all further studies. Initial characterization was carried out by subjecting the materials to powder X-ray diffraction (XRD), as well as to different forms of analysis, including thermogravimetric analysis (TGA), energy dispersive X-ray analysis (EDAX), and IR spectroscopy. XRD patterns on the powdered crystals indicated that the products were new; the patterns were entirely consistent with the structures determined by single-crystal X-ray diffraction. TGA studies revealed the content of water and oxalate in **I** and **II** and that of the oxalate and amine in **III** and **IV**. EDAX analysis gave the metal:phosphorus ratios. Infrared spectra showed

Table 6. Synthesis conditions and analysis for compounds **I–IV**.

	Synthetic conditions			Analysis			Composition
				EDAX	TGA		
Mole ratio		Temp [°C]	Time [h]	Fe/P	Free H_2O [%]	Bound H_2O, oxalate + amine [%]	
$FeCl_3:2H_3PO_4:H_2C_2O_4\cdot 2H_2O:CHA:122H_2O$		110	120	1:1	4	27	$[Fe_2(H_2O)_2(HPO_4)_2(C_2O_4)]\cdot H_2O$ (**I**)
$FeCl_3:2H_3PO_4:H_2C_2O_4\cdot 2H_2O:CHA:122H_2O$		110	220	1:1	8.3	27	$[Fe_2(H_2O)_2(HPO_4)_2(C_2O_4)]\cdot H_2O$ (**II**)
$FeCl_3:6H_3PO_4:H_2C_2O_4\cdot 2H_2O:21,3\text{-}DAP:200H_2O$		110	96	1:1	9.8	33.3	$[C_3N_2H_{12}][Fe_2(HPO_4)_2(C_2O_4)_{1.5}]_2$ (**III**)
$FeCl_3:2H_3PO_4:H_2C_2O_4\cdot 2H_2O:1,3\text{-}DAHP:200H_2O$		110	144	1:1	8.7	34.4	$[C_3N_2OH_{12}][Fe_2(HPO_4)_2(C_2O_4)_{1.5}]_2$ (**IV**)

Table 7. Crystal data and structure refinement parameters for compounds **I** – **IV**.

	I	II	III	IV
empirical formula	$Fe_2P_2O_{15}C_2H_8$	$Fe_2P_2O_{16}C_2H_{10}$	$Fe_4P_4O_{34}C_9N_2H_{28}$	$Fe_4P_4O_{35}C_9N_2H_{28}$
crystal system	monoclinic	monoclinic	monoclinic	monoclinic
space group	$P2_1/c$	$P2_1/c$	$P2_1/n$	$P2_1/n$
crystal size [mm]	$0.08 \times 0.1 \times 0.16$	$0.08 \times 0.12 \times 0.16$	$0.08 \times 0.12 \times 0.16$	$0.08 \times 0.12 \times 0.16$
a [Å]	4.840(2)	7.211(1)	9.214(1)	9.266(1)
b [Å]	17.571(6)	9.294(1)	15.201(2)	15.204(1)
c [Å]	7.342(4)	9.567(1)	12.046(1)	11.984(1)
β [°]	106.6(1)	103.4(1)	101.9(1)	102.2(1)
volume [Å³]	598.5(4)	623.8(1)	1650.9(3)	1650.5(2)
Z	4	4	4	4
formula mass	445.7	463.7	527.7	535.7
ρ_{calcd} [g cm^{-3}]	1.930	1.852	2.152	2.152
λ (Mo$_{K\alpha}$) [Å]	0.71073	0.71073	0.71073	0.71073
μ [mm^{-1}]	2.093	2.008	2.045	2.044
θ range [°]	2.32 – 23.25	2.90 – 23.33	2.19 – 23.31	2.19 – 23.29
total data collected	2410	2532	6765	6801
index ranges	$-5 \leq h \leq 5, -18 \leq k \leq 19, -4 \leq l \leq 8$	$-8 \leq h \leq 7, -10 \leq k \leq 6, -10 \leq l \leq 10$	$-10 \leq h \leq 10, -16 \leq k \leq 13, -11 \leq l \leq 13$	$-10 \leq h \leq 9, -16 \leq k \leq 16, -7 \leq l \leq 13$
unique data	857	894	2387	2370
observed data ($\sigma > 2\sigma(I)$)	781	747	1533	1617
refinement method	full-matrix least-squares on $|F^2|$	full-matrix least-squares on $|F^2|$	full-matrix least-squares on $|F^2|$	full-matrix least-squares on $|F^2|$
R_{merg}	0.04	0.03	0.08	0.08
R indexes [$I > 2\sigma(I)$]	$R = 0.03$,[a] $wR_2 = 0.06$[b]	$R = 0.03, wR_2 = 0.08$	$R = 0.06, wR_2 = 0.14$	$R = 0.08, wR_2 = 0.18$
goodness of fit (S)	1.11	1.13	1.04	1.14
no. of variables	118	117	232	236
largest difference map peak and hole [e Å^{-3}]	0.455/ – 0.402	0.360/ – 0.422	1.408/ – 0.711	1.547/ – 0.734

[a] $R_1 = \Sigma||F_o| - |F_c||/\Sigma|F_o|$. [b] $wR_2 = \{\Sigma[w(F_o^2 - F_c^2)^2]/\Sigma[w(F_o^2)^2]\}^{1/2}$, $w = 1/[\sigma^2(F_0)^2 + (aP)^2 + bP]$, $P = [\max.(F_o^2,0) + 2(F_c)^2]/3$, where $a = 0.0239$ and $b = 1.1632$ for **I**, $a = 0.049$ and $b = 0.0$ for **II**, $a = 0.9$ and $b = 0.0$ for **III** and $a = 0.0423$ and $b = 33.461$ for **IV**.

Iron Phosphate – Oxalates Open-Frameworks

the absence of the amine in **I** and **II**. The results of the analysis on the synthesized products are presented in Table 6. The composition of the various iron phosphate – oxalates were calculated to be as follows: **I**, $Fe_2(H_2O)_2(HPO_4)_2(C_2O_4) \cdot H_2O$, **II**, $Fe_2(H_2O)_2(HPO_4)_2(C_2O_4) \cdot 2H_2O$, **III**, $[C_3N_2H_{12}][(Fe_2(HPO_4)_2(C_2O_4)_{1.5}]_2$, and **IV**, $[C_3N_2OH_{12}][(Fe_2(HPO_4)_2-(C_2O_4)_{1.5}]_2$. These compositions exactly agree with those derived from X-ray crystallography. Compositionally **I** and **II** are similar except for the water of hydration. Structures of **III** and **IV** differ with respect to the composition of the amine. Magnetic susceptibility measurements on all the four compounds were carried out in the 30 – 300 K range using a Lewis coil magnetometer.

Single-crystal structure determination: A suitable single crystal of each compound was selected carefully under a polarizing microscope and mounted at the tip of a glass fiber using cyanoacrylate (superglue) adhesive. Crystal structure determination by X-ray diffraction was performed on a Siemens Smart-CCD diffractometer equipped with a normal focus, 2.4 kW sealed tube X-ray source ($Mo_{K\alpha}$ radiation, $\lambda = 0.71073$ Å) operating at 50 kV and 40 mA. A hemisphere of intensity data were collected at room temperature in 1321 frames with ω scans (width of $0.30°$ and exposure time of 20 s per frame). The final unit cell constants were determined by a least-squares fit of 1822 reflections for I, 1681 reflections for **II**, 2562 reflections for **III**, and 2549 reflections for **IV** in the range $4 \leq 2\theta \leq 46.5$ and are presented in Table 7.

The structure was solved by direct methods using SHELXS-86[16] and difference Fourier syntheses. For compounds **I** and **II**, the hydrogen positions were easily located in the difference Fourier maps and for the final refinement the hydrogen atom on one of the oxygen atoms [O(7)] was placed geometrically and held in the riding mode. In the case of **III** and **IV**, both the amine and water molecules were disordered, and the location of the hydrogen positions for these molecules was not possible. The hydrogen atoms on the oxygen atoms [O(13), O(14)] were, however, located in the difference Fourier maps and were placed geometrically and held in the riding mode. The disorder of the amine molecules in **III** and **IV** made it difficult to refine anisotropically. The last cycles of refinement included atomic positions, anisotropic thermal parameters for all the non-hydrogen framework atoms, and isotropic thermal parameters for all the hydrogen atoms. Full-matrix least-squares structure refinement against $|F^2|$ was carried out using the SHELXTL-PLUS[17] package of programs. Pertinent refinement parameters are presented in Table 6. Further details of crystal structure investigations may be obtained from the Fachinformationszentrum, Karlsruhe, D-76344, Eggenstein-Leopoldshagen, Germany (fax: ($+49$) 7247-808-666; e-mail: crysdata@fiz-karlsruhe.de) on quoting the depository numbers CSD-391074 (**I**), CSD-391075 (**II**), CSD-391076 (**III**), and CSD-391077 (**IV**).

[1] *Atlas of Zeolite Structure Types* (Eds.: W. H. Meier, D. H. Olson), Butterworth-Heinemann, London, **1992**; *Handbook of Heterogeneous Catalysis* (Eds.: G. Ertl, H. Knözinger, J. Weitkamp), VCH, Berlin **1997**; A. K. Cheetham, G. Ferey, T. Loiseau, *Angew. Chem.* **1999**, *111*, 3466; *Angew. Chem. Int. Ed.* **1999**, *38*, 3268.

[2] S. Natarajan, *J. Solid State Chem.* **1998**, *139*, 200.

[3] P. Lightfoot, Z. A. D. Lethbridge, R. E. Morris, D. S. Wragg, P. A. Wright, Å. Kvick, G. B. M. Vaughan, *J. Solid State Chem.* **1999**, *143*, 74.

[4] Y.-F. Huang, K.-H. Lii, *J. Chem. Soc. Dalton Trans.* **1998**, 4085; H.-M. Lin, K.-H. Lii, Y.-C. Jiang, S.-L. Wang, *Chem. Mater.* **1999**, *11*, 519.

[5] Z. A. D. Lethbridge, P. Lightfoot, *J. Solid State Chem.* **1999**, *143*, 58.

[6] A. Choudhury, S. Natarajan, C. N. R. Rao, *J. Solid State Chem.* **1999**, *146*, 538.

[7] A. Choudhury, S. Natarajan, C. N. R. Rao, *Chem. Mater.* **1999**, *11*, 2316.

[8] I. D. Brown, D. Altermatt, *Acta Crystallogr. B* **1985**, *41*, 244.

[9] M. Cavellec, D. Riou, C. Ninclaus, J.-M. Gerenèche G. Férey, *Zeolites* **1996**, *17*, 250; K.-H. Lii, Y.-F. Huang, V. Zima, C.-Y. Huang, H.-M. Lin, Y.-C. Jiang, F.-L. Liao S.-L. Wang, *Chem. Mater.* **1998**, *10*, 2599.

[10] W. T. A. Harrison, L. L. Dussack, A. J. Jacobson, *J. Solid State Chem.* **1996**, *125*, 234.

[11] S. Natarajan, J.-C. P. Gabriel, A. K. Cheetham, *Chem. Commun.* **1996**, 1415, and references therein.

[12] X. Bu, P. Feng, T. E. Gier, G. D. Stucky, *J. Solid State Chem.* **1998**, *136*, 210.

[13] S. Ayyappan, X. Bu, A. K. Cheetham, S. Natarajan, C. N. R. Rao, *Chem. Commun.* **1998**, 2181.

[14] R. Vaidhyanathan, S. Natarajan, A. K. Cheetham, C. N. R. Rao, *Chem. Mater.* **1999**, *11*, 3636.

[15] C. Mathoniere, S. G. Carling, P. Day, *J. Chem. Soc. Chem. Commun.* **1994**, 1551, and references therein.

[16] G. M. Sheldrick, SHELXS-86 Program for Crystal Structure Determination, University of Göttingen **1986**; G. M. Sheldrick, *Acta Crystallogr.* **1990**, *A35*, 467.

[17] G. M. Sheldrick, SHELXTL-PLUS Program for Crystal Structure Solution and Refinement, University of Göttingen, **1993**.

Received: August 23, 1999 [F1993]

3524 *Chem. Mater.* **2001**, *13*, 3524–3533

Articles

Hybrid Inorganic–Organic Host–Guest Compounds: Open-Framework Cadmium Oxalates Incorporating Novel Extended Structures of Alkali Halides

R. Vaidhyanathan, Srinivasan Natarajan, and C. N. R. Rao*

Chemistry and Physics of Materials Unit, Jawaharlal Nehru Centre for Advanced Scientific Research, Jakkur P.O., Bangalore 560 064, India

Received January 18, 2001

Four unusual inorganic–organic nanocomposites containing extended alkali halide structures, present as layers or as three-dimensional units, have been synthesized by a metathetic reaction carried out under hydrothermal conditions. These compounds, with cadmium oxalate host lattices and the alkali halide structures as guests, have the compositions [RbCl][Cd$_6$(C$_2$O$_4$)$_6$]·2H$_2$O, **I**; 2[CsBr][Cd(C$_2$O$_4$)]·H$_2$O, **II**; 2[CsBr][Cd$_2$(C$_2$O$_4$)(Br)$_2$]· 2H$_2$O, **III**; and 3[RbCl][Cd$_2$(C$_2$O$_4$)(Cl$_2$)]·H$_2$O, **IV**. Crystal data for **I**, rhombohedral, space group = R-3 (no. 148), a = 9.3859(3), c = 23.9086(8) Å; V = 1824.05(1) Å^3, Z = 18, M = 1359.51, ρ_{calcd} = 3.702 g cm^{-3}, μ(Mo Kα) = 7.375 mm^{-1}, R_1 = 0.04, wR_2 = 0.0904 [537 observed reflections with I > 2σ(I)]; for **II**, orthorhombic, space group = $Pbcm$ (no. 57), a = 6.0854(6), b = 11.0793(11), c = 16.889(2) Å; V = 1138.7(2) Å^3, Z = 8, M = 644.06, ρ_{calcd} = 3.745 g cm^{-3}, μ(Mo Kα) = 15.219 mm^{-1}, R_1 = 0.04, wR_2 = 0.08 [408 observed reflections with I > 2σ(I)]; for **III**, orthorhombic, space group = $Cmcm$ (no.63), a = 23.6151(14), b = 10.2528(6), c = 7.8199(6) Å; V = 1894.2(2) Å^3, Z = 16, M = 970.33, ρ_{calcd} = 3.374 g cm^{-3}, μ(Mo Kα) = 14.487 mm^{-1}, R_1 = 0.03, wR_2 = 0.0744 [651 observed reflections with I > 2σ(I)]; for **IV**, monoclinic, space group = $P2_1/c$ (no. 14), a = 8.0648(2)(3), b = 22.9026(4), c = 9.3967(3) Å; V = 1681.10(7) Å^3, Z = 4, M = 782.53, ρ_{calcd} = 3.076 g cm^{-3}, μ(Mo Kα) = 11.961 mm^{-1}, R_1 = 0.04, wR_2 = 0.093 [2071 observed reflections with I > 2σ(I)]. Compounds **I–IV** possess many unusual structural features. While **I** has interpenetrating lattices of the three-dimensional cadmium oxalate and RbCl with an expanded lattice, giving rise to a super rock-salt cell, **II** possesses alternating cadmium oxalate and CsBr layers, the latter comprising six-membered CsBr rings. Compounds **III** and **IV** have isolated cadmium oxalate units (zero-dimensional) linked and stabilized by the alkali halides.

Introduction

A large variety of supramolecular organic and inorganic structures have been described in the recent literature.[1-4] These include several types of host–guest compounds,[2] as well as metal–organic and –inorganic open-framework structures.[4-7] In many of the host–guest compounds, ions, ion pairs, or molecules have been accommodated in the cages or cavities.[3,8] A novel departure from the traditional host–guest chemistry is the recent discovery of cadmium oxalates incorporating extended alkali halide structures of different dimensionalities.[9] These materials can be considered to be

* Corresponding author. E-mail: cnrrao@jncasr.ac.in.

(1) (a) Lehn, J.-M. *Supramolecular Chemistry: Concepts and Perspectives*; VCH: New York, 1995.

(2) (a) Dalton Discussion No. 3, Inorganic Crystal Engineering. *J. Chem. Soc., Dalton Trans.* **2000** (21). (b) Braga, D.; Grepioni, F. *Acc. Chem. Res.* **2000**, *33*, 601.

(3) Muller, A.; Reuterm, H.; Dillinger, S. *Angew. Chem., Int. Ed. Engl.* **1995**, *34*, 2328.

(4) (a) Cheetham, A. K.; Ferey, G.; Loiseau, T. *Angew. Chem., Int. Ed. Engl.* **1999**, *38*, 3268. (b) Serpaggi, F.; Ferey, G. *J. Mater. Chem.* **1998**, *8*, 2737. (c) Serpaggi, F.; Ferey, G. *Microporous and Mesoporous Mater.* **1999**, *32*, 311. (d) Riou, D.; Ferey, F. *J. Mater. Chem.* **1998**, *8*, 2733. (e) Livage, C.; Egger, C.; Nogues, M.; Ferey, G. *J. Mater. Chem.* **1998**, *8*, 2743. (f) Livage, C.; Egger, C.; Ferey, G. *Chem. Mater.* **1999**, *11*, 1546. (g) Kuhlman, R.; Schemek, G. L.; Kolis, J. W. *Inorg. Chem.* **1999**, *38*, 194.

(5) Chui, S. S.-Y.; Lo, S. M.-F.; Charmant, J. P. H.; Orpen, A. G.; Williams, I. A. *Science* **1999**, *283*, 1148.

(6) (a) Reineke, T. M.; Eddaoudi, M.; Fehr, M.; Kelly, D.; Yaghi, O. M. *J. Am. Chem. Soc.* **1999**, *121*, 1651. (b) Reineke, T. M.; Eddaoudi, M.; O'Keeffe, M.; Yaghi, O. M. *Angew. Chem., Int. Ed. Engl.* **1999**, *38*, 2590. (c) Li, H.; Eddaoudi, M.; O'Keeffe, M.; Yaghi, O. M. *Nature* **1999**, *402*, 276. (d) Reineke, T. M.; Eddaoudi, M.; Moler, D.; O_Keeffe, M.; Yaghi, O. M. *J. Am. Chem. Soc.* **2000**, *122*, 4843.

(7) (a) Ayyappan, S.; Cheetham, A. K.; Natarajan, S.; Rao, C. N. R. *Chem. Mater.* **1998**, *10*, 3746. (b) Natarajan, S.; Vaidhyanathan, R.; Rao, C. N. R.; Ayyappan, S.; Cheetham, A. K. *Chem. Mater.* **1999**, *11*, 1633. (c) Vaidhyanathan, R.; Natarajan, S.; Cheetham, A. K.; Rao, C. N. R. *Chem. Mater.* **1999**, *11*, 3636. (d) Vaidhyanathan, R.; Natarajan, S.; Rao, C. N. R. *Chem. Mater.* **2001**, *13*, 185.

(8) (a) Wragg, D. S.; Morris, R. E. *J. Am. Chem. Soc.* **2000**, *122*, 11 246. (b) Taulelle, F.; Poblet, J.-M.; Ferey, G.; Benard, M. *J. Am. Chem. Soc.* **2001**, *123*, 111.

(9) Vaidhyanathan, R.; Neeraj, S.; Prasad, P. A.; Natarajan, S.; Rao, C. N. R. *Angew. Chem., Int. Ed. Engl.* **2000**, *39*, 3470.

10.1021/cm011011+ CCC: $20.00 © 2001 American Chemical Society
Published on Web 09/07/2001

Table 1. Synthesis Conditions and Analysis for Compounds I–IV

reactant mole ratios	temp, °C	time, h	EDAX ratios	product
$CdCl_2$:$0.5Rb_2CO_3$:$5H_2C_2O_4$: $1.5CH_3COOH$:$88n\text{-}C_4H_9OH$: $220H_2O$	150	78	Cd:Rb:Cl 6.5:1.0:1.1	$[RbCl][Cd_6(C_2O_4)_6]\cdot 2H_2O$ (**I**)
$CdBr_2$:Cs_2CO_3:$1.47H_6C_4O_4$:$70n\text{-}C_4H_9OH$:$150H_2O$	150	108	Cd:Cs:Br 1.1:1.5:1.5	$2[CsBr][Cd(C_2O_4)]\cdot H_2O$ (**II**)
$CdBr_2$:$(COOCs)_2$:$1.95CH_3COC$:$40n\text{-}C_4H_9OH$:$62H_2O$	150	80	Cd:Cs:Br 1.1:1.1:1.8	$2[CsBr][Cd_2(C_2O_4)(Br)_2]\cdot 4H_2O$ (**III**)
$CdCl_2$:Rb_2CO_3:$H_2C_2O_4$:$2CH_3COOH$:$33n\text{-}C_4H_9OH$:$33H_2O$	150	80	Cd:Rb:Cl 1.0:1.5:2.5	$3[RbCl][Cd_2(H_2O)C_2O_4)(Cl)_2]\cdot H_2O$ (**IV**)

organic–inorganic nanocomposites or hetero superlattices wherein the oxalate and alkali halide units are either intricately connected in three-dimensions or occur discretely in alternate layers. In view of the novelty of these hybrid structures, we considered it highly desirable to explore newer systems incorporating such alkali halide structures. In the present study, we describe a three-dimensional cadmium oxalate structure, $[RbCl]$-$[Cd_6(C_2O_4)_6]\cdot 2H_2O$, **I**, incorporating RbCl in the rock-salt ($Fm3m$) structure, with a unit-cell parameter double that of the normal RbCl. The structure of **I** is even more fascinating in that the oppositely charged cluster ions form a rock-salt supercell, by using the RbCl lattice as a template. We then have a layered cadmium oxalate, $2[CsBr][Cd(C_2O_4)]\cdot H_2O$, **II**, having CsBr in a graphitic layered architecture formed by six-membered rings, and two other cadmium oxalates, $2[CsBr][Cd_2\text{-}(C_2O_4)(Br)_2]\cdot 2H_2O$, **III**, and $3[RbCl][Cd_2(C_2O_4)(Cl_2)]\cdot$ H_2O, **IV**, where the alkali halide layers comprising different types of rings act as bridges between the oxalate units. The alkali halide units in **III** form a layered structure with eight-ring apertures. Compounds **I–IV** constitute members of a new class of hybrid host–guest materials and are likely to possess novel dielectric, optical, and other properties in view of the periodic arrangement of the organic and inorganic components.

Experimental Section

Compounds **I–IV** were prepared by metathetic reaction under hydrothermal conditions. Typically for the synthesis of **I**, 0.057 g of Rb_2CO_3 was dissolved in 4 mL of *n*-butanol and 2 mL of water mixture. Oxalic acid, 0.313 g, and $CdCl_2$, 0.1 g, were added, and the contents were stirred for 20 min. Finally 0.04 mL of glacial acetic acid was added and the mixture was stirred to homogeneity. The final mixture with the composition $CdCl_2$:$0.5Rb_2CO_3$:$5H_2C_2O_4$:$1.5CH_3COOH$:$90C_4H_9OH$:$220H_2O$ was sealed in a PTFE-lined stainless steel autoclave and heated at 150 °C for 72 h. The resulting product, a crop of truncated cubelike crystals suitable for single-crystal X-ray study, was filtered, washed, and dried at room temperature. A similar synthesis procedure was employed for the preparation of compounds **II–IV** and is presented in Table 1.

The initial characterization was carried out using powder X-ray diffraction (XRD), EDAX, and chemical analysis. The powder XRD patterns indicated that the products were new materials; the patterns were entirely consistent with the structures determined by using single-crystal X-ray diffraction. As a representative example, powder data for compound **I** is presented in Table 2.

Thermogravimetric analysis of compounds **I** and **II** were carried out in O_2 atmosphere (50 mL min^{-1}) in the range room temperature to 700 °C. While **I** showed two distinct mass losses, **II** had three distinct mass losses. For **I**, a gradual mass loss around 100 °C followed by a sharp mass loss in the range 350–400 °C (42.0%) was found due to the loss of the extra framework water and oxalate (calcd 41.5%). A mass loss above 500 °C was found due to the slow evaporation of CdO. For compound **II**, a sharp mass loss of 2.3% around 150 °C corresponds to the loss of extra framework water (calcd 2.8) and the second mass loss of 16% around 350 °C corresponds

Table 2. X-ray Powder Data for I,
$[RbCl][Cd_6(C_2O_4)_6]\cdot 2H_2O$

H	K	L	$2\theta_{obsd}$	$\Delta(2\theta)^a$	d_{calcd}	$\Delta(d)^b$	I_{rel}^c
0	0	3	11.102	0.000	7.970	0.000	11
1	0	1	11.498	0.000	7.696	0.000	9
1	0	2	13.163	0.008	6.722	0.004	52
1	0	4	18.485	0.060	4.815	0.015	8
1	1	0	19.007	0.097	4.693	0.024	20
1	1	3	21.996	0.017	4.044	0.003	14
0	0	6	22.369	0.059	3.985	0.010	11
2	0	2	23.179	0.065	3.848	0.011	8
2	0	4	26.560	0.040	3.361	0.005	100
1	0	7	28.480	0.137	3.149	0.015	11
2	1	1	29.384	0.075	3.047	0.008	17
1	1	6	29.444	0.040	3.038	0.004	13
2	1	2	30.135	0.105	2.976	0.010	9
2	1	4	32.804	0.030	2.732	0.002	7
0	0	9	33.739	0.000	2.657	0.000	15
2	0	7	34.395	0.101	2.615	0.007	6
2	1	5	34.751	0.046	2.585	0.003	18
3	0	3	35.045	0.068	2.565	0.005	7
2	0	8	37.439	0.091	2.408	0.006	11
1	1	9	39.086	0.128	2.312	0.007	6
3	1	1	40.211	0.034	2.244	0.002	48
3	0	6	40.248	0.001	2.241	0.000	27
3	1	2	40.703	0.024	2.215	0.001	11
3	1	5	44.424	0.002	2.039	0.000	4

^{a} $2\theta_{obsd} - 2\theta_{calcd}$. ^{b} $d_{obsd} - d_{calcd}$. ^{c} $100I/I_{max}$. LSQ fitted lattice parameter (Cu Kα): $a = 9.3773(5)$, $b = 9.3773(5)$, $c = 23.8786(9)$ Å, $V = 1818.4(3)$ Å^3.

to the loss of oxalate units. Above 600 °C there was loss due to the evaporation of CdO. In both **I** and **II**, the powder XRD pattern of the product of decomposition was characteristic of CdO, mineral monteporite (JCPDS: 05-0640).

Infrared spectroscopic studies as a KBr pellet showed characteristic features of the bischelating oxalate.[10] The various bands for compounds **I–IV** are as follows: $\nu_{as}(C=O)$ 1612–1672 cm^{-1}; $\nu_s(O-C-O)$ 1372 and 1311 (s) cm^{-1} (**I**), 1374 and 1310 (s) cm^{-1} (**II**), 1372 (m) and 1310 (s) cm^{-1} (**III**), 1373 and 1312 (s) cm^{-1} (**IV**); $\delta(O-C=O)$ 803 (m), 790 (w), and 774 (w) cm^{-1} (**I**), 800 (w) and 773 (s) cm^{-1} (**II**), 773 (s) and 731 (w) cm^{-1} (**III**), 802 and 776 (m) cm^{-1} (**IV**). The extraframework water molecules showed characteristic bands around 3550 (s) cm^{-1} in all the cases. The M–O (M = Cd) stretching vibrations $\nu_s(M-O)$ and $\nu_s(C-C)$ are also observed at 512 and 441 (s) cm^{-1} (**I**), 513 (m) cm^{-1} (**II**), 514 (w) cm^{-1} (**III**), 512 and 441 (s) cm^{-1} (**IV**). The various infrared bands are consistent with the structure determined by single-crystal studies.

Single-Crystal Structure Determination. A suitable single crystal of each compound was carefully selected under a polarizing microscope and glued to a thin glass fiber with cyano-acrylate (super glue) adhesive. Single-crystal structure determination by X-ray diffraction was performed on a Siemens Smart-CCD diffractometer equipped with a normal focus, 2.4 kW sealed tube X-ray source (Mo Kα radiation, $\lambda = 0.710\ 73$ Å) operating at 50 kV and 40 mA. A hemisphere of intensity data was collected at room temperature in 1321 frames with ω scans (width of 0.30° and exposure time of 20s per frame) in the 2θ range 3–46.5°. Pertinent experimental details for the structure determinations are presented in Table 3.

(10) *Infrared and Raman Spectra of Inorganic and Coordination Compounds*, 5th ed.; Nakamoto, K., Ed.; John Wiley: New York, 1997.

Table 3. Crystal Data and Structure Refinement Parameters for $[RbCl][Cd_6(C_2O_4)_6]\cdot2H_2O$, I; $2[CsBr][Cd(C_2O_4)]\cdot H_2O$, II; $2[CsBr][Cd_2(C_2O_4)(Br)_2]\cdot4H_2O$, III; and $3[RbCl][Cd_2(H_2O)(C_2O_4)(Cl)_2]\cdot H_2O$, IV

parameters	I	II	III	IV
empirical formula	$H_4C_{12}Cd_6Cl_1O_{26}Rb_1$	$H_2C_2Cd_1Br_2O_5Cs_2$	$H_8C_2Cd_1Br_4O_5Cs_2$	$H_4C_2Cd_2Cl_5O_6Rb_3$
crystal system	rhombohedral	orthorhombic	orthorhombic	monoclinic
space group	R-3 (no. 148)	$Pbcm$ (no. 57)	$Cmcm$ (no. 63)	$P2_1/c$ (no. 14)
crystal size, mm	0.09, 0.09, 0.09	0.08, 0.06, 0.07	0.08, 0.10, 0.12	0.08, 0.06, 0.09
a, Å	9.3859(3)	6.0854(6)	23.6251(14)	8.0648(2)
b, Å	9.3859(3)	11.0793(11)	10.2528(6)	22.9026(4)
c, Å	23.9086(8)	16.889(2)	7.8199 (58)	9.3967(3)
α	90°	90°	90°	90°
β	90°	90°	90°	104.399(**10**)°
γ	120°	90°	90°	90°
vol, Å^3	1824.05(10)	1138.7(2)	1894.2(2)	1681.10(7)
Z	18	8	16	4
formula mass	1359.51	644.06	970.33	782.53
r_{calcd}, g cm^{-3}	3.702	3.745	3.374	3.076
λ (Mo Kα), Å	0.71073	0.71073	0.71073	0.71073
μ, mm^{-1}	7.375	15.219	14.487	11.961
θ range	2.56−23.26°	2.41−23.27°	1.72−23.24°	1.78−23.25°
total data collected	2598	4407	3894	6939
index ranges	$-10 \leq h \leq 6, -7 \leq k \leq 10,$ $-26 \leq l \leq 24$	$-6 \leq h \leq 6, -12 \leq k \leq 11,$ $-18 \leq l \leq 14$	$-21 \leq h \leq 26, -11 \leq k\ 11,$ $-8 \leq l \leq 8$	$-7 \leq h \leq 8, -21 \leq k \leq 25,$ $-10 \leq l \leq 8$
unique data	589	855	770	2405
obsd data ($I > 2\sigma(I)$)	537	408	651	2071
refinement method	full-matrix least-squares on $|F^2|$	full-matrix least-squares on $|F^2|$	full-matrix least-squares on $|F^2|$	full-matrix least-squares on $|F^2|$
R_{int}	0.0887	0.0647	0.0535	0.0950
R indexes [$I > 2\sigma(I)$]	$R_1 = 0.0355,^a\ wR_2 = 0.0904^b$	$R_1 = 0.0377,^a\ wR_2 = 0.0827^b$	$R_1 = 0.0334,^a\ wR_2 = 0.0744^b$	$R_1 = 0.0409,^a\ wR_2 = 0.0931^b$
R (all data)	$R_1 = 0.0414, wR_2 = 0.0980$	$R_1 = 0.0485, wR_2 = 0.0879$	$R_1 = 0.0428, wR_2 = 0.0784$	$R_1 = 0.0516, wR_2 = 0.1008$
goodness of fit	1.155	1.005	1.073	1.133
no. of variables	71	66	55	163
largest difference map peak and hole eÅ^{-3}	0.806 and 1.595	1.129 and 1.625	1.338 and 0.618	0.775 and 1.652

^{a} $R_1 = \Sigma||F_o| - |F_c||/\Sigma|F_o|$. ^{b} $wR_2 = \{\Sigma[w(F_o^2 - F_c^2)^2]\ /\Sigma[w(F_o^2)^2]\}^{1/2}$, $w = 1/[\sigma^2(F_o)^2 + (aP)^2 + bP]$, $P = [\max. (F_o^2, 0) + 2(F_c)^2]/3$; where a = 0.0565, b = 9.1718 for I; a = 0.0493, b = 0.0000 for **II**; a = 0.0431, b = 14.3797 for **III**; and a = 0.0352, b = 10.3425 for **IV** .

The structure was solved by direct methods using SHELXS-86[11] and difference Fourier syntheses. An absorption correction based on symmetry equivalent reflections was applied using SADABS program.[12] The hydrogen positions associated with the water molecules in the case of compounds **I−IV** could not be located from the difference Fourier map. The last cycles of refinement included atomic positions for all the atoms, anisotropic thermal parameters for all non-hydrogen atoms, and isotropic thermal parameters for all the hydrogen atoms. Full-matrix least-squares structure refinement against $|F^2|$ was carried out using SHELXTL-PLUS[13] package of programs. Details of the final refinements are given in Table 3. The final atomic coordinates, selected bond distances, and angles for compounds **I** are presented in Tables 4 and 5, for **II** in Tables 6 and 7, for **III** in Tables 8 and 9, and for **IV** in Tables 10 and 11.

Results

$[RbCl][Cd_6(C_2O_4)_6]\cdot2H_2O$, I. The asymmetric unit of **I** contains 10 non-hydrogen atoms, of which seven atoms make up cadmium oxalate host framework, and the remaining three are the guest molecules (Rb, Cl, and water). The Cd is six coordinated with respect to the oxygen atoms forming a triangular prism with Cd−O distances in the range 2.292(5)−2.384(6) Å (av 2.339 Å). The oxygens are linked to the carbon atoms forming the oxalate unit, with average C−O distances of 1.248 Å (Table 5). The O−Cd−O bond angles are in

Table 4. Final Atomic Coordinates [x10^4] and Equivalent Isotropic Displacement Parameters [Å^2 x 10^3] for I, $[RbCl][Cd_6(C_2O_4)_6]\cdot2H_2O$

atom	x	y	z	U(eq)^a
Cd(1)	2567(1)	3862(1)	96(1)	15(1)
Rb(1)	0	0	0	18(1)
Cl(1)	3333	6667	1667	15(1)
O(1)	1121(7)	5001(7)	587(2)	19(1)
O(2)	3151(8)	3164(8)	96(2)	24(2)
O(3)	2194(8)	1212(8)	1048(2)	25(2)
O(4)	4191(8)	3795(7)	1697(2)	19(1)
C(1)	342(11)	4467(11)	136(4)	19(2)
C(2)	2770(11)	934(10)	1481(4)	15(2)
O(100)	6667	3333	767(20)	232(18)

^{a} U(eq) is defined as one-third of the trace of the orthogonalized U_{ij} tensor.

the range 69.8(2)−148.5(2)° (av 102.9°) (Table 5). The Rb ions are coordinated by 12 oxygen atoms at a distance of ∼3 Å and the Cl$^-$ ions are surrounded by six cadmium atoms at ∼2.9 Å (Table 5). These geometrical parameters are in good agreement with those reported for similar compounds.

The framework structure of **I** consists of a network of cadmium and oxalate units. The connectivity between these units is such that they form an unusual cadmium oxide cluster of the composition $[Cd_6O_{24}]$. The clusters are connected into three-dimensions by the oxalate units. The Cl$^-$ ions are located inside these clusters (Figure 1) with the charge compensating Rb$^+$ ions located in the cavities formed between these clusters (Figure 2). The Rb$^+$ and Cl$^-$ ions are perfectly ordered in three dimensions forming an independent rock-salt ($Fm3m$) structure with interpenetrating fcc lattices (Figure 3). The cubic unit cell parameter of RbCl is 13.452 Å, which is roughly double that in ordinary RbCl

(11) Sheldrick, G. M. *SHELXS-86 Program for Crystal Structure Determination*; University of Gottingen: Gottingen, Germany 1986; *Acta Crystallogr.* **1990**, *A35*, 467.

(12) Sheldrick, G. M. *SADABS Siemens Area Detector Absorption Correction Program*; University of Gottingen: Gottingen, Germany, 1994.

(13) Sheldrick, G. M. *SHELXS-93 Program for Crystal Structure Solution and Refinement*; University of Gottingen: Gottingen, Germany, 1993.

Table 5. Selected Bond Distances and Bond Angles for I, [RbCl][Cd₆(C₂O₄)₆]·2H₂O

moiety[a]	distance, Å	moiety[a]	distance, Å
Cd(1)O(1)	2.292(5)	Rb(1)O(3)#5	3.078(6)
Cd(1)O(2)	2.327(6)	Rb(1)O(3)	3.078(6)
Cd(1)O(3)	2.341(6)	Cl(1)Cd(1)#7	2.8919(6)
Cd(1)O(4)	2.342(6)	Cl(1)Cd(1)#1	2.8918(6)
Cd(1)O(4)#1	2.348(6)	Cl(1)Cd(1)#8	2.8918(6)
Cd(1)O(1)#2	2.384(6)	Cl(1)Cd(1)#7	2.8918(6)
Cd(1)Cl(1)	2.8917(6)	Cl(1)Cd(1)#2	2.8918(6)
Rb(1)O(2)	2.973(6)	O(1)C(1)	1.259(11)
Rb(1)O(2)#3	2.973(6)	O(1)Cd(1)#7	2.384(6)
Rb(1)O(2)#4	2.973(6)	O(2)C(1)#6	1.237(11)
Rb(1)O(2)#5	2.973(6)	O(3)C(2)	1.252(10)
Rb(1)O(2)#6	2.973(6)	O(4)C(2)#9	1.253(10)
Rb(1)O(2)#4	2.973(6)	O(4)Cd(1)#8	2.348(5)
Rb(1)O(3)#3	3.078(6)	C(1)O(2)#4	1.237(11)
Rb(1)O(3)#6	3.078(6)	C(1)C(1)#10	1.57(2)
Rb(1)O(3)#4	3.078(6)	C(2)O(4)#9	1.253(10)
Rb(1)O(3)#4	3.078(6)	C(2)C(2)#9	1.53(2)

moiety[a]	angle, deg	moiety[a]	angle, deg
O(1)Cd(1)−O(2)	93.2(2)	O(2)Cd(1)−O(1)#2	69.8(2)
O(1)Cd(1)−O(3)	136.1(2)	O(3)Cd(1)−O(1)#2	123.0(2)
O(2)Cd(1)−O(3)	73.0(2)	O(4)Cd(1)−O(1)#2	87.7(2)
O(1)Cd(1)−O(4)	148.5(2)	O(4)#1−Cd(1)O(1)#2	147.4(2)
O(2)Cd(1)−O(4)	114.3(2)	O(2)#4−C(1)O(1)	126.4(8)
O(3)Cd(1)−O(4)	70.2(2)	O(2)#4−C(1)C(1)#10	117.3(10)
O(1)Cd(1)−O(4)#1	81.8(2)	O(1)C(1)−C(1)#10	116.2(10)
O(2)Cd(1)−O(4)#1	141.2(2)	O(3)C(2)−O(4)#9	124.8(8)
O(3)Cd(1)−O(4)#1	84.2(2)	O(3)C(2)−C(2)#9	118.0(9)
O(4)Cd(1)−O(4)#1	85.35(7)	O(4)#9−C(2)C(2)#9	117.3(9)
O(1)Cd(1)−O(1)#2	87.9(3)		

[a] Symmetry transformations used to generate equivalent atoms: #1, $y - \frac{1}{3}, -x + y + \frac{1}{3}, -z + \frac{1}{3}$; #2, $-y + 1, x - y + 1, z$; #3, $-x, -y, -z$; #4, $x - y, x, -z$; #5, $-x + y, -x, z$; #6, $y, -x + y, -z$; #7, $-x + \frac{2}{3}, -y + \frac{4}{3}, -z + \frac{1}{3}$; #8, $x - y + \frac{2}{3}, x + \frac{1}{3}, -z + \frac{1}{3}$; #9, $-x + \frac{2}{3}, -y + \frac{1}{3}, -z + \frac{1}{3}$; #10, $-x, -y + 1, -z$.

Table 6. Final Atomic Coordinates [x10⁴] and Equivalent Isotropic Displacement Parameters [Å² x 10³] for II, 2[CsBr][Cd(C₂O₄)]·H₂O

atom	x	y	z	U(eq)[a]
Cd(1)	8496(2)	871(1)	2500	21(1)
Cs(1)	6536(1)	3910(1)	3847(1)	32(1)
Br(1)	8483(2)	809(1)	896(1)	40(1)
C(1)	8168(20)	1994(10)	2500	15(3)
C(2)	3271(21)	1789(12)	2500	23(3)
O(1)	7119(15)	1053(8)	2500	28(2)
O(2)	5319(13)	1983(8)	2500	30(2)
O(3)	2291(15)	816(7)	2500	29(2)
O(4)	9807(13)	2882(8)	2500	29(2)
O(100)	2733(18)	2500	5000	93(5)

[a] U(eq) is defined as one-third of the trace of the orthogonalized U_{ij} tensor.

Table 7. Selected Bond Distances and Bond Angles for II, 2[CsBr][Cd(C₂O₄)]·H₂O

moiety[a]	distance, Å	moiety[a]	distance, Å
Cd(1)O(1)	2.290(8)	Br(1)Cs(1)#10	3.6706(13)
Cd(1)O(2)	2.293(8)	Br(1)Cs(1)#3	3.7153(12)
Cd(1)O(3)#1	2.310(9)	Br(1)Cs(1)#5	3.7336(13)
Cd(1)O(4)	2.367(9)	C(1)O(1)	1.223(13)
O(3)Cd(1)#11	2.310(9)	C(1)O(4)#3	1.240(13)
Cd(1)Br(1)	2.7099(11)	C(1)C(2)#5	1.61(2)
Cd(1)Br(1)#2	2.7100(11)	C(2)O(3)	1.233(14)
Cs(1)Br(1)#2	3.6603(13)	C(2)O(2)	1.264(14)
Cs(1)Br(1)#8	3.6706(13)	C(2)C(1)#7	1.61(2)
Cs(1)Br(1)#9	3.7153(12)	O(3)Cs(1)#5	3.185(6)
Cs(1)Br(1)#7	3.7336(13)	O(3)Cs(1)#6	3.185(6)
Br(1)Cs(1)#2	3.6602(13)	O(4)C(1)#9	1.240(13)

moiety[a]	angle, deg	moiety[a]	angle, deg
O(1)Cd(1)−O(2)	101.0(3)	O(3)#1−Cd(1)Br(1)	90.14(3)
O(1)Cd(1)−O(3)#1	110.0(3)	O(4)Cd(1)−Br(1)	91.43(3)
O(2)Cd(1)−O(3)#1	149.0(3)	O(1)C(1)−O(4)#3	127.8(11)
O(1)Cd(1)−O(4)	178.2(3)	O(1)C(1)−C(2)#5	115.5(11)
O(2)Cd(1)−O(4)	77.2(3)	O(4)#3−C(1)C(2)#5	116.7(10)
O(3)#1−Cd(1)O(4)	71.8(3)	O(3)C(2)−O(2)	128.7(12)
O(1)Cd(1)−Br(1)	88.58(3)	O(3)C(2)−C(1)#7	118.0(11)
O(2)Cd(1)−Br(1)	90.63(3)	O(2)C(2)−C(1)#7	113.3(11)

[a] Symmetry transformations used to generate equivalent atoms: #1, $x + 1, y, z$; #2, $x, y, -z + \frac{1}{2}$; #3, $-x + 2, y - \frac{1}{2}, -z + \frac{1}{2}$; #4, $-x + 2, y - \frac{1}{2}, z$; #5, $-x + 1, y - \frac{1}{2}, -z + \frac{1}{2}$; #6, $-x + 1, y - \frac{1}{2}, z$; #7, $-x + 1, y + \frac{1}{2}, -z + \frac{1}{2}$; #8, $x, -y + \frac{1}{2}, z + \frac{1}{2}$; #9, $-x + 2, y + \frac{1}{2}, -z + \frac{1}{2}$; #10, $x, -y + \frac{1}{2}, z - \frac{1}{2}$; #11, $x - 1, y, z$; #12, $x, -y + \frac{1}{2}, -z + 1$.

Table 8. Final Atomic Coordinates [x10⁴] and Equivalent Isotropic Displacement Parameters [Å² x 10³] for III, 2[CsBr][Cd₂(C₂O₄)(Br)₂]·2H₂O

atom	x	y	z	U(eq)[a]
Cd(1)	1243(1)	4005(1)	2500	28(1)
Cs(1)	3116(1)	3170(1)	2500	47(1)
Br(1)	1692(1)	1663(1)	2500	52(1)
Br(2)	3074(1)	0	0	33(1)
C(1)	0	3522(9)	1499(13)	22(2)
O(1)	468(2)	3522(5)	772(6)	32(1)
O(200)	4173(3)	1145(7)	2500	33(2)
O(100)	101(69)	118(72)	6151(186)	739(227)

[a] U(eq) is defined as one-third of the trace of the orthogonalized U_{ij} tensor.

Table 9. Selected Bond Distances and Bond Angles for III, 2[CsBr][Cd₂(C₂O₄)(Br)₂]·2H₂O

moiety[a]	distance, Å	moiety[a]	distance, Å
Cd(1)O(1)#1	2.329(5)	Cs(1)Br(1)	3.703(2)
Cd(1)O(1)	2.329(5)	Br(1)Cs(1)#6	3.610(2)
Cd(1)O(200)#2	2.405(7)	Br(2)Cd(1)#6	2.7332(8)
Cd(1)Br(1)	2.6248(14)	Br(2)Cd(1)#4	2.7332(8)
Cd(1)Br(2)#3	2.7332(8)	C(1)O(1)	1.243(7)
Cd(1)Br(2)#4	2.7332(8)	C(1)O(1)#8	1.243(7)
Cs(1)Br(1)#2	3.610(2)	C(1)C(1)#1	1.57(2)

moiety[a]	angle, deg	moiety[a]	angle, deg
O(1)#1−Cd(1)O(1)	70.9(2)	O(1)C(1)−C(1)#1	117.2(5)
O(1)#1−Cd(1)O(200)#2	82.7(2)	O(1)#8−C(1)C(1)#1	117.2(5)
O(1)Cd(1)−O(200)#2	82.7(2)	C(1)O(1)−Cd(1)	115.7(5)
O(1)C(1)−O(1)#8	125.6(9)	Cd(1)#6−O(200)Cs(1)	105.6(2)

[a] Symmetry transformations used to generate equivalent atoms: #1, $x, y, -z + \frac{1}{2}$; #2, $-x + \frac{1}{2}, y + \frac{1}{2}, -z + \frac{1}{2}$; #3, $-x + \frac{1}{2}, -y + \frac{1}{2}, z + \frac{1}{2}$; #4, $-x + \frac{1}{2}, -y + \frac{1}{2}, -z$; #5, $-x + \frac{1}{2}, -y + \frac{1}{2}, -z + 1$; #6, $-x + \frac{1}{2}, y - \frac{1}{2}, -z + \frac{1}{2}$; #7, $-x, -y, z - \frac{1}{2}$; #8, $-x, y, z$; #9, $x, -y, -z + 1$.

(6.570 Å) as shown in Figure 4. The [100] RbCl planes run along the [111] direction of I (Figure 3). An alternate way to describe the structure of **I** would be to consider the Cl⁻ ion as being at the center of a cluster of six Cd atoms arranged in such a manner that they form the six vertexes of an octahedron, with the next nearest oxygen neighbors forming the trigonal prisms with Cd atoms. Accordingly, **I** could be considered as a super rock-salt arrangement of two types of clusters, [ClCd₆(C₂O₄)₃]⁵⁺ and [Rb(H₂O)₂(C₂O₄)₃]⁵⁻. This type of arrangement is reminiscent of Chevrel phases.[14]

2[CsBr][Cd(C₂O₄)]·H₂O, II. The asymmetric unit of **II** contains 10 non-hydrogen atoms, of which eight atoms make up cadmium bromo-oxalate host framework

and the remaining two are the guest molecules (Rb and water). The Cd atom is in an octahedral coordination with four oxygen and two bromine atoms. The Cd–O distances are in the range 2.290(8)−2.367(9) Å (av 2.315 Å), and the Cd–Br distances have an average value of 2.710 Å. The oxygens, in turn, are linked with carbon

(14) (a) Subba Rao, G. V.; Balakrishnan, G. *Bull. Mater. Sci. (India)* **1984**, *6*, 283 and the references therein. (b) Subba Rao, G. V. *Proc. Ind. Natl. Sci. Acad.* **1986**, 52A, 292.

Table 10. Final Atomic Coordinates [x10⁴] and Equivalent Isotropic Displacement Parameters [Å² x 10³] for IV, 3[RbCl][Cd₂(C₂O₄)(Cl₂)]·H₂O

atom	x	y	z	U(eq)[a]
Cd(1)	3176(1)	4425(1)	4647(1)	24(1)
Cd(2)	62(1)	3703(1)	9969(1)	22(1)
Rb(1)	393(1)	3103(1)	5741(1)	33(1)
Rb(2)	4892(1)	2793(1)	2861(1)	36(1)
Rb(3)	4799(1)	4150(1)	9398(1)	36(1)
Cl(1)	4605(3)	3526(1)	6111(3)	30(1)
Cl(2)	1497(3)	3657(1)	2765(3)	35(1)
Cl(3)	4832(3)	5177(1)	6845(3)	33(1)
Cl(4)	2435(3)	3006(1)	10294(3)	32(1)
Cl(5)	2212(3)	3009(1)	9392(3)	28(1)
C(1)	902(11)	5002(4)	10186(10)	22(2)
C(2)	47(12)	5220(4)	4394(10)	23(2)
O(1)	1282(8)	5181(3)	3804(7)	30(2)
O(2)	1511(8)	4507(3)	9666(7)	28(2)
O(3)	1605(8)	4528(3)	10268(7)	29(2)
O(4)	1083(8)	5612(3)	4100(7)	29(2)
O(10)	1496(8)	3782(3)	7500(7)	28(2)
O(100)	7572(10)	3677(4)	3508(9)	51(2)

[a] U(eq) is defined as one-third of the trace of the orthogonalized U_{ij} tensor.

Table 11. Selected Bond Distances and Bond Angles for IV, 3[RbCl][Cd₂(C₂O₄)(Cl₂)]·H₂O

moiety[a]	distance, Å	moiety[a]	distance, Å
Cd(1)O(4)#1	2.286(6)	Rb(2)Cl(4)#9	3.646(3)
Cd(1)O(1)	2.315(6)	Rb(3)Cl(5)	3.343(3)
Cd(1)Cl(3)#2	2.549(2)	Rb(3)Cl(3)	3.365(3)
Cd(1)Cl(1)	2.583(2)	Rb(3)Cl(1)	3.371(3)
Cd(1)Cl(2)	2.620(3)	Rb(3)Cl(4)#10	3.408(3)
Cd(1)Cl(3)	2.765(3)	Rb(3)Cl(3)#11	3.796(3)
Cd(2)O(2)	2.293(6)	Cl(1)Rb(2)#5	3.418(3)
Cd(2)O(3)	2.316(6)	Cl(2)Cd(2)#8	2.618(2)
Cd(2)O(10)	2.327(6)	Cl(3)Cd(1)#2	2.549(2)
Cd(2)Cl(4)	2.569(2)	Cl(3)Rb(3)#11	3.796(3)
Cd(2)Cl(5)	2.585(2)	Cl(4)Rb(2)#12	3.281(3)
Cd(2)Cl(2)#3	2.618(2)	Cl(4)Rb(1)#5	3.369(3)
Rb(1)Cl(5)#6	3.342(3)	Cl(4)Rb(3)#4	3.408(3)
Rb(1)Cl(4)#6	3.369(3)	Cl(4)Rb(2)#13	3.646(3)
Rb(1)Cl(2)	3.386(3)	Cl(5)Rb(1)#5	3.342(3)
Rb(1)Cl(5)	3.386(3)	Cl(5)Rb(2)#5	3.414(2)
Rb(1)Cl(1)	3.465(3)	Cl(5)Rb(2)#3	3.470(3)
Rb(2)Cl(4)#7	3.281(3)	C(1)O(3)	1.237(11)
Rb(2)Cl(2)	3.362(3)	C(1)O(2)#14	1.249(11)
Rb(2)Cl(5)#6	3.414(2)	C(1)C(1)#14	1.58(2)
Rb(2)Cl(1)#6	3.418(3)	C(2)O(1)	1.258(11)
Rb(2)Cl(5)#8	3.470(3)	C(2)O(4)	1.259(11)
Rb(2)Cl(1)	3.543(3)	C(2)C(2)#1	1.54(2)

moiety[a]	angle, deg	moiety[a]	angle, deg
O(4)#1−Cd(1)O(1)	72.3(2)	O(3)C(1)−C(1)#14	118.1(9)
O(2)Cd(2)−O(3)	72.0(2)	O(2)#14−C(1)C(1)#14	116.1(10)
O(2)Cd(2)−O(10)	88.2(2)	O(1)C(2)−O(4)	124.3(8)
O(3)Cd(2)−O(10)	84.8(2)	O(1)C(2)−C(2)#1	118.0(10)
O(3)C(1)−O(2)#14	125.8(8)	O(4)C(2)−C(2)#1	117.7(10)

[a] Symmetry transformations used to generate equivalent atoms: #1, $-x, -y + 1, -z + 1$; #2, $-x + 1, -y + 1, -z + 1$; #3, $x, y, z + 1$; #4, $x - 1, y, z$; #5, $x, -y + 1/2, z + 1/2$; #6, $x, -y + 1/2, z - 1/2$; #7, $x + 1, -y + 1/2, z - 1/2$; #8, $x, y, z - 1$; #9, $x + 1, y, z - 1$; #10, $x + 1, y, z$; #11, $-x + 1, -y + 1 -z + 2$; #12, $x - 1, -y + 1/2, z + 1/2$; #13, $x - 1, y, z + 1$; #14, $-x, -y + 1, -z + 2$.

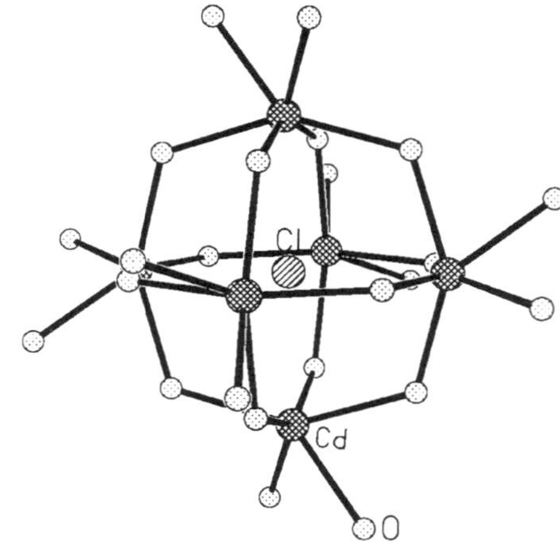

Figure 1. Figure showing a single [Cd₆O₂₄] cluster with the Cl⁻ ions at the center.

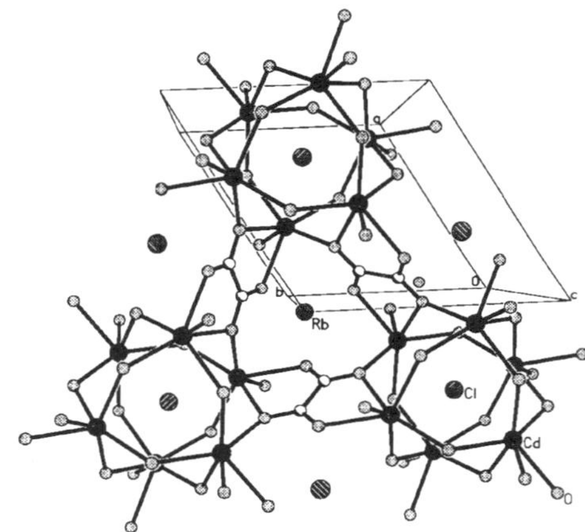

Figure 2. Structure of **I** in the *ab* plane. Note that the Rb⁺ ions are positioned outside the cluster unit.

forming the oxalate unit, with average C−O distances of 1.240 Å (Table 7). The various bond angles, viz., O−Cd−O, O−Cd−Br, and Br−Cd−Br have values in the range 71.8(3)−178.2(3)° (av 114.5°), 88.58(3)−91.43(3)° (av 90.2°), and 177.06(6)° (Table 7). The Cs⁺ ions are eight-coordinated by four oxygen and bromine atoms at average distances of ~3.2 and 3.7 Å (Table 7). The various geometrical parameters found in **II** are in good agreement with those reported for similar compounds.

The framework structure of **II** consists of a network of cadmium and oxalate units forming layers (Figure 5). The Br⁻ ions connected to the cadmium protrude into the interlamellar space. The Cs⁺ ions along with a water molecule occupy the interlamellar space (Figure 5). The relative positions between the Cs⁺ and Br⁻ ions are such that they form a graphite-like layer (Figure 6). Such an arrangement creates two closely spaced hexagonal CsBr layers separated by 3.67 Å. The CsBr layers are made of neutral hexagonal units of Cs₃Br₃, with chair conformation as in cyclohexane. The unusual layered alkali halide structures in **II** are stabilized by the cadmium oxalate layers, present after every two CsBr layers, by providing additional coordination for the Cs⁺ ions through the oxalate oxygens.

2[CsBr][Cd₂(C₂O₄)(Br)₂]·2H₂O, III, and 3[RbCl]-[Cd₂(C₂O₄)(Cl₂)]·H₂O, IV. The structures of **III** and **IV** have some common features and can be described together. The asymmetric units of **III** and **IV** contain 8 and 18 non-hydrogen atoms, respectively. The crystallographically distinct cadmium atom in **III** is octahedrally coordinated to oxygen and bromine atoms. There

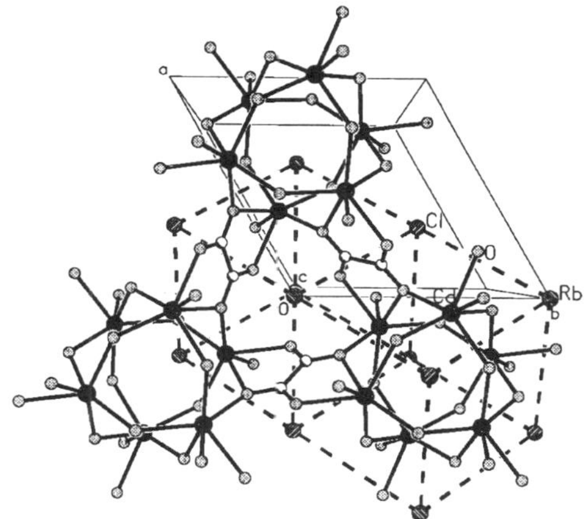

Figure 3. Structure showing the $Fm3m$ lattice of RbCl within the parent cadmium oxalate structure.

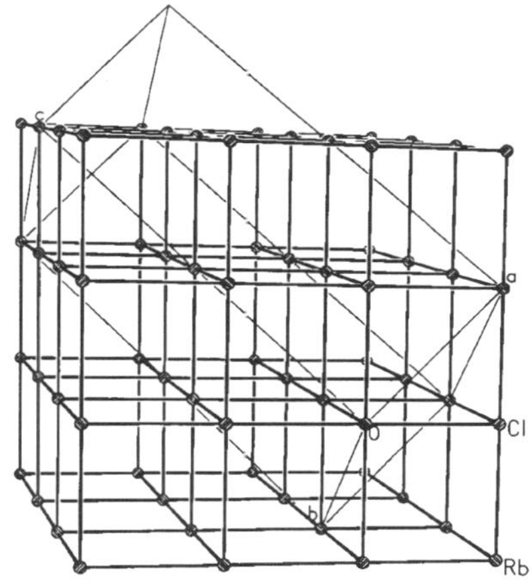

Figure 4. Structure of RbCl in **I**. Note that RbCl is along the [*111*] lattice.

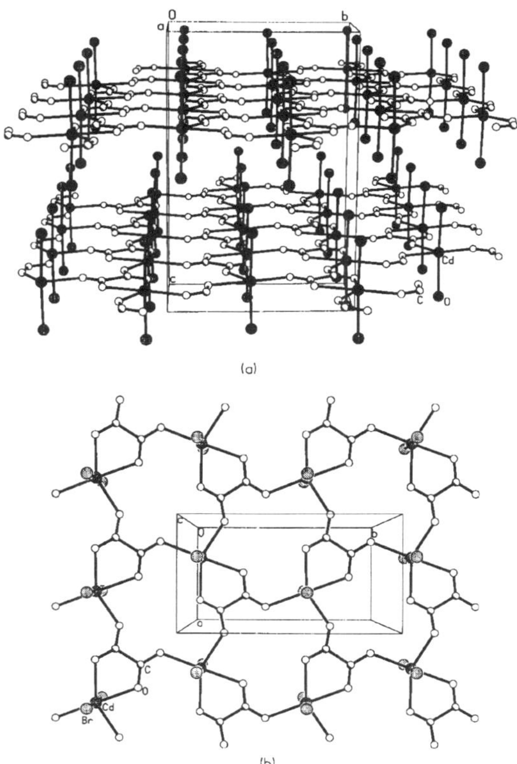

Figure 5. (a) Structure of **II** in the bc plane showing the layer arrangement. Note that the Br atoms point into the interlamellar space. Cs and water molecules are not shown for purposes of clarity. (b) Structure of **II** showing a single layer of the cadmium oxalate.

are three Cd–O and three Cd–Br linkages with average distances of 2.354 and 2.697 Å. The longer Cd–O linkage with a distance of 2.405(7) Å belongs to the terminal water molecule. The oxygen atoms are connected to a carbon atom forming the oxalate unit, with average C–O distances of 1.243 Å (Table 9). The Cs$^+$ ions are eight coordinated by bromine atoms and possess an additional coordination through the terminal water molecule. The average Cs–Br distance is 3.824 Å, while the O–Cd–O, O–Cd–Br, and Br–Cd–Br bond angles have values in the ranges 70.9(2)–82.7(2)° (av 78.8°), 84.30(12)–179.7(2) (av 118.3°), and 91.33(4)–95.91(3)° (av 94.4°) (Table 9). The O–C–O bond angles and the other observed geometrical parameters are in satisfactory agreement with those reported for similar compounds (Table 9).

There are two crystallographically independent cadmium atoms in **IV**, both of which are octahedrally coordinated by oxygen and chlorine atoms. While Cd(1) possesses two Cd–O (av 2.3005 Å) and four Cd–Cl linkages (av 2.629 Å), Cd(2) has three Cd–O (av 2.312 Å) and Cd–Cl (av 2.591 Å) connections. The oxygen

atoms are connected to carbon atoms and form the oxalate unit with average C–O distances of 1.251 Å (Table 11). There are three unique Rb atoms with Rb-(1) and Rb(8), being eight coordinated, and Rb(3) is seven coordinated with chlorine and oxygen atoms. Thus, Rb(1) has five Rb–Cl (av 3.390 Å) and three Rb–O (av 2.978 Å) linkages, Rb(2) has seven Rb–Cl (av 3.448 Å) and one Rb–O (2.914(9) Å) bondings, and Rb-(3) has five Rb–Cl (av 3.457 Å) and two Rb–O (av 2.893 Å) connectivities (Table 11). The average O–Cd–O, O–Cd–Cl, and Cl–Cd–Cl bond angles have values of 72.3, 107.24, and 107.5° for Cd(1) and 81.7, 117.1, and 95.18° for Cd(2). These are typical values and are in agreement with those observed earlier for similar compounds.

The framework structures of **III** and **IV** consist of linkages involving cadmium and oxalate units. Connectivity between the two moieties gives rise to interesting structural features. Thus, both **III** and **IV** possess isolated monomeric cadmium oxalate units connected together by the alkali halides, viz., CsBr and RbCl. In **III**, the isolated cadmium oxalate units are connected with the alkali halide in such a manner as to form a three-dimensionally extended channel structure containing water molecules (Figure 7). The alkali halide units are present as eight-membered rings with the chair conformation as in cyclooctane (Figure 8). The

3530   *Chem. Mater., Vol. 13, No. 10, 2001*   *Vaidhyanathan et al.*

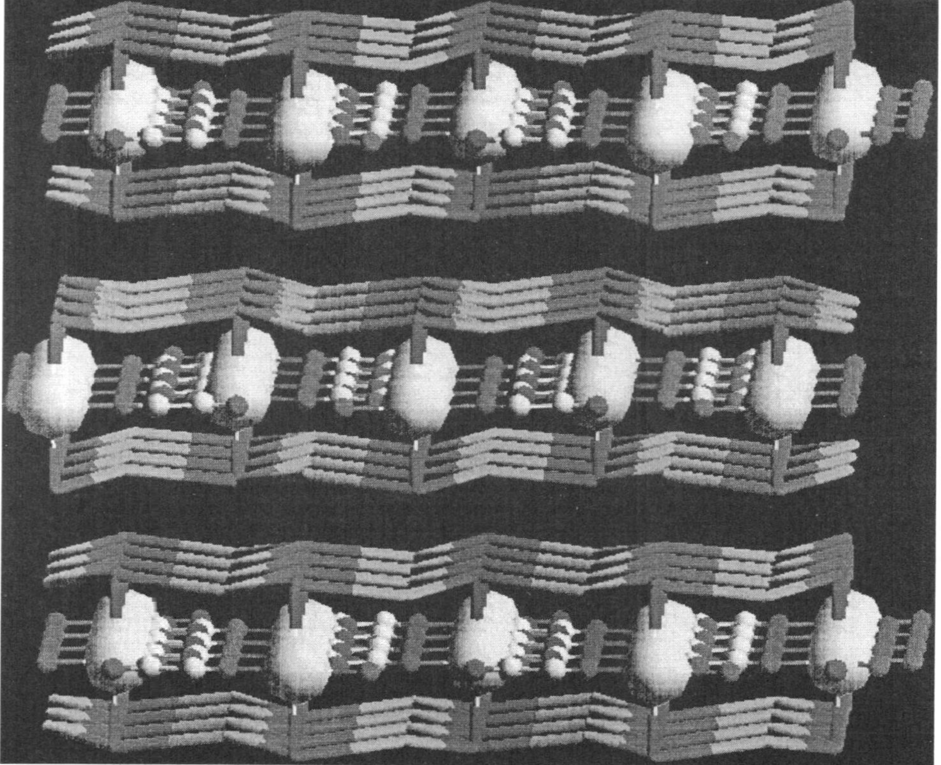

Figure 6. Structure of **II** showing the CsBr lattice within the layers. Inset shows the CsBr lattice alone. Note that alternate CsBr layers have larger separation. (Color scheme: yellow, Cd; white, C; red, O; purple, Br; green, Cs.)

connectivity between the isolated oxalate units along with the chloride ions in **IV** results in an interpenetrating layerlike arrangement (Figure 9). The layers are so positioned in a helical fashion, with the immediate next layer arranged in the opposite direction, thus canceling total helicity of the structure (Figure 10). Accordingly, the Rb^+ ions also reflect the helical nature of the layers and form a helical strand between the cadmium chlorooxalate layers (Figure 10). The alkali halide, however, does not form an independent lattice in **IV**.

Discussion

Four novel cadmium oxalates incorporating extended alkali halide structures, $[RbCl][Cd_6(C_2O_4)_6]\cdot 2H_2O$, **I**, $2[CsBr][Cd(C_2O_4)]\cdot H_2O$, **II**, $2[CsBr][Cd_2(C_2O_4)(Br)_2]\cdot 2H_2O$, **III**, and $3[RbCl][Cd_2(C_2O_4)(Cl_2)]\cdot H_2O$, **IV**, have

been prepared hydrothermally as good quality single crystals. The reaction itself involves simple metathetic exchange between a cadmium halide and an alkali oxalate. The cadmium oxalates **I** and **IV** are formed with RbCl; **II** and **III** are formed with CsBr. Though RbCl is part of the structure in both **I** and **IV**, the way the alkali halide units are arranged within the host cadmium oxalate is very different. In **II** and **III**, the CsBr has a layered structure with different apertures, but the host lattice possesses completely different structures. These fine differences in the structures are likely to result from the differences in experimental conditions such as the composition, time, and temperature (Table 1), but we are not yet in a position to relate the two.

The host oxalate structures have different architectures in **I**–**IV**, ranging from isolated cadmium oxalate

Figure 7. Structure of **III** showing the channels formed by the connectivity between the isolated cadmium oxalate and CsBr. Note that the water molecules occupy these channels. (Color scheme: yellow, Cd; white, C; red, O; purple, Br; green, Cs.)

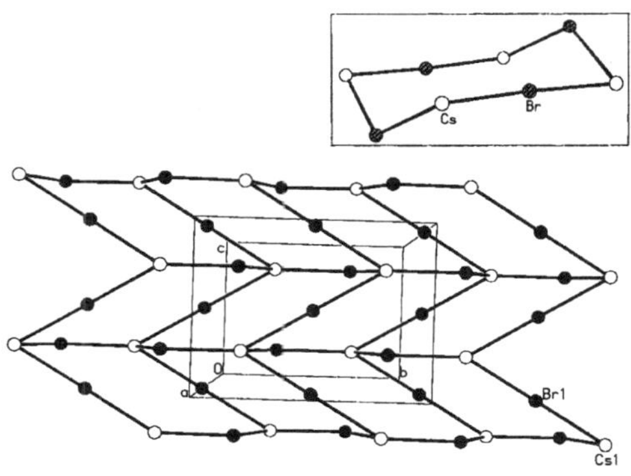

Figure 8. Layered structure of CsBr in **III**. Inset shows the CsBr in chair conformation.

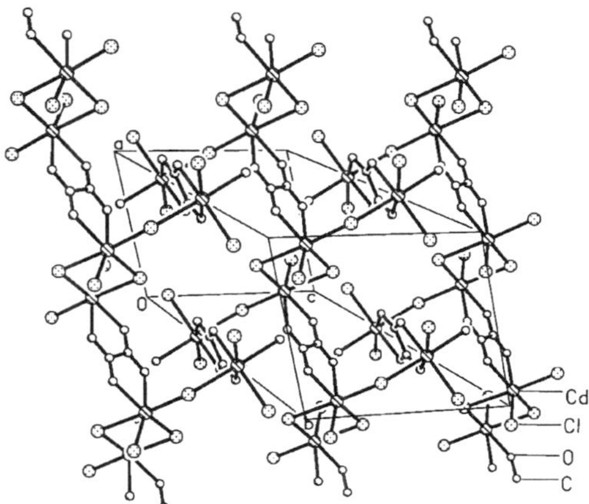

Figure 9. Structure of **IV** showing a single layer resulting from the connectivity between the isolated cadmium oxalate and chlorine atoms.

units to three-dimensionally extended architectures. While **I** possesses two interpenetrating lattices of three-dimensional cadmium oxalate and a $Fm3m$ RbCl lattice, **II** has a layered cadmium oxalate architecture (two-dimensional) with a layered CsBr structure. Compounds **III** and **IV** have isolated cadmium oxalate units (zero-dimensional) linked and stabilized by the alkali halides. The isolated cadmium oxalate units in **IV** are linked via a chlorine atom to form a sheet of cadmium chlorooxalate that is arranged in a helical fashion. These are indeed unique architectures. In addition to the uniqueness of the host structures, the alkali halide guest species also possesses remarkable architectures.

In **I**, RbCl forms a face-centered lattice with unit-cell parameters which are approximately double that of the normal RbCl. In **II**, CsBr forms a layered architecture with the layers closely resembling that of graphite. The six-membered Cs_3Br_3 ring, in **II**, is in a chair conformation of the cyclohexane. The two closely spaced CsBr layers have a thickness of 3.67 Å. The thickness of the oxalate layers is 10.62 Å. Accordingly, **II** can be considered to be a genuine hetero-lattice or nanocomposite

3532   *Chem. Mater., Vol. 13, No. 10, 2001*   *Vaidhyanathan et al.*

Figure 10. Structure of **IV** along the *a* axis showing the cadmium chloro-oxalate layers and the Rb atoms. Note that both the cadmium chloro-oxalate and the Rb atoms form a helical structure. (Color scheme: yellow, Cd; white, C; red, O; purple, Cl; green, Rb)

with alternating organic and inorganic layers (Figure 6). In **III**, CsBr again forms a layered structure but is unusual in that it consists exclusively of eight-membered rings that resemble the chair conformation of cyclooctane (Figure 8). What is interesting in **III** is that the isolated cadmium oxalate units are situated between these CsBr layers and act as a pillar through Cd–Br connectivity. The thickness of the CsBr layers in **III** is ~3.7 Å, which is similar to the closely spaced CsBr layers in **II** (Figure 7). The oxalate layer has a thickness of 5.87 Å, which is roughly half that observed in **II**. In **IV**, the chlorine atoms connect the Cd ions of the isolated cadmium oxalate unit and form a two-dimensional layer with a right-handed helical strand with the immediate next layer having the opposite handedness, canceling the total helicity of the structure. The Rb atoms situated between these sheets reflect the helical nature of the cadmium chloro-oxalate layer (host), with the Rb atoms from the neighboring strand so positioned as to cancel the helicity of the structure. Thus, there is no independent alkali halide lattice in **IV** (Figure 10).

The distances between the alkali metal and the halogen ions in compounds **I**–**III** provide an interesting comparison. In **I**, the Rb–Cl distance is 6.726 Å, while in RbCl with rock-salt structure, it is 3.285 Å (Figure 4). In the case of compounds **II** and **III**, the average Cs–Br distance is ~3.7 Å, which is comparable to the Cs–Br distance (3.72 Å) in body-centered CsBr. It may be noted that in **I**, the RbCl forms a lattice identical to that of normal RbCl, but with double the unit-cell dimensions, whereas in **II** and **III** the CSBr forms unusual layered architectures despite having nearly similar distances between Cs and Br as in normal CsBr. These observations demonstrate the role of host–guest interactions in these structures.

Host–guest structures of the type observed in **I**–**IV** possessing expanded RbCl and layered CsBr are indeed most unusual. These structures are formed because the host structures provide the additional coordination necessary for the alkali halide ions. Thus, the expanded rock-salt structure of RbCl is made possible by the stability provided by the extra coordination due to the

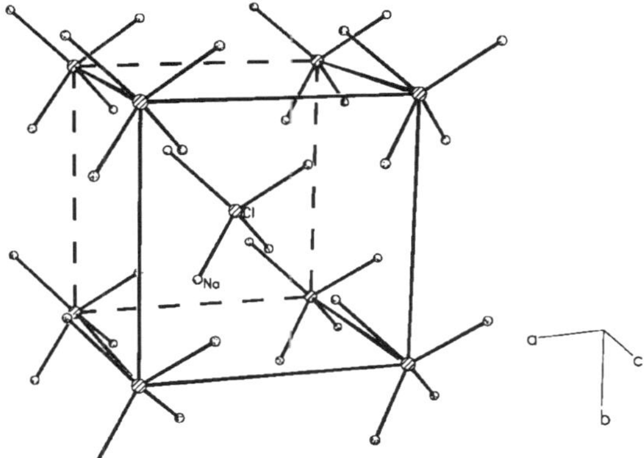

Figure 11. Structure of **V**, showing the body-centered cubic arrangement of the [ClNa$_4$] units.

oxalate units. What is truly interesting, however, is that such beautiful structures are formed by a simple metathetic reaction carried out under hydrothermal conditions. Such a reaction under ambient conditions would clearly give a mixture of the Cd oxalate and the alkali halide. In this connection, it may be noted that the structures reported herein are different from those resulting from ion-exchange in layered solids, such as metal—anion arrays within the oxide hosts.[15] Thus, in the layered RbLaNb$_2$O$_7$, the Rb$^+$ ions, situated in the interlamellar region, are exchanged topotactically by copper halide, CuX$_2$ (X = Cl and Br), in evacuated and sealed quartz tubes at 325 °C.[15] This gives rise to a unique situation wherein the ability to prepare hosts at high temperatures, with varying interatomic spacings and topologies, is combined with the low-temperature ion-exchange methods to construct metal—anion arrays. The host—guest structures of **I**—**IV** are somewhat

comparable to the KCl/CsCl structures reported in a phosphate framework.[16] Accordingly, in A$_2$[M$_3$(X$_2$O$_7$)$_2$]-[salt] (A = Rb, Cs; M = Mn, Cu; X = P, As), synthesized under molten salt conditions (>600 °C), the phosphate network is formed around the alkali metal chloride salts, the salt probably acting as a structure-directing template. The super rock-salt structure formed by the clusters in **I** can be considered to be a result of templating by the underlying RbCl lattice. This is reminiscent of organic molecules employed in conventional low-temperature hydrothermal methods for the synthesis of extended framework architectures.

It is likely that many more materials containing alkali and alkaline earth halides in unusual coordinations and structures may be formed by a suitable choice of the framework. In this direction, we have just isolated a new cadmium succinate, [NaCl][Cd$_5$(H$_4$C$_4$O$_4$)$_2$Na$_6$], **V**, incorporating NaCl. In **V**, however, we have extra Na$^+$ ions in addition to the alkali halide, similar to the metal—anion arrays within the oxide hosts established in Dion–Jacobson structures.[15] The alkali halide does not form any independent lattice in **V**. The relative positions of Na$^+$ and Cl$^-$ ions in **V** are such that four Na$^+$ ions coordinate one Cl$^-$ ion in a tetrahedral fashion. The tetrahedral [ClNa$_4$] units form a body-centered cube within the cadmium succinate structure (Figure 11).

The new types of hybrid host—guest compounds incorporating novel alkali halide structures are likely to possess novel properties. In particular, they would be expected to exhibit interesting optical and dielectric properties. The three-dimensional organic—inorganic periodic superlattice would also bestow certain unusual properties. It would be of interest to analyze the Infrarad and Raman spectra of the alkali halide guests in these compounds. Such studies will be pursued, although at this stage we primarily report the structures, which are unique in their own merit.

CM011011+

(15) Kodenkandath, T. A.; Lalena, J. N.; Zhou, W. I.; Carpenter, E. E.; Sangregorio, Falster, A. U.; Simmons, J. R.; O'Connor, C. J.; Wiley, J. B. *J. Am. Chem. Soc.* **1999**, *121*, 10 743.

(16) Huang, Q.; Ulutagy, M.; Michener, P. A.; Hwu, S.-J. *J. Am. Chem. Soc.* **1999**, *121*, 10323.

Available online at www.sciencedirect.com

SCIENCE @ DIRECT®

Solid State Sciences 4 (2002) 1231–1236

www.elsevier.com/locate/ssscie

Sodalite networks formed by metal squarates

S. Neeraj [a], M.L. Noy [a], C.N.R. Rao [a,b], A.K. Cheetham [a,*]

[a] *Materials Research Laboratory, University of California, Santa Barbara, CA 93106, USA*
[b] *Jawaharlal Nehru Center for Advanced Scientific Research, Jakkur, Bangalore 560064, India*

Received 27 August 2002; accepted 29 August 2002

Abstract

Sodalite structures have not hitherto been obtained by rational design based on linking of square building units. We have, for the first time, succeeded in synthesizing sodalite networks by making use of squaric acid as the 4-membered square unit. Cobalt, manganese and zinc squarates with this architecture have been obtained (Crystal data for **1**, $Co(H_2O)_2(C_4O_4)$: cubic, space group $Pn\text{-}3n$ (No. 222), $a = 16.280(5)$ Å, $\alpha = \beta = \gamma = 90°$, $V = 4315(2)$ Å^3, $Z = 48$, $\rho_{calc} = 1.912$ mg m^{-3}, $R_1 = 0.041$ for Mo $K\alpha$, $\lambda = 0.71073$ Å. Crystal data for **2**, $Mn(H_2O)_2(C_4O_4)$: trigonal, space group $R\text{-}3$ (No. 148), $a = b = 11.607(5)$, $c = 14.66(3)$ Å, $\alpha = \beta = 90°$, $\gamma = 120°$, $V = 1711(4)$ Å^3, $Z = 18$, $\rho_{calc} = 1.773$ mg m^{-3}, $R_1 = 0.029$. Crystal data for **3**, $Zn(H_2O)_2(C_4O_4)$: cubic, space group $Pn\text{-}3n$ (No. 222), $a = 16.256(3)$ Å, $\alpha = \beta = \gamma = 90°$, $V = 4295.6(15)$ Å^3, $Z = 48$, $\rho_{calc} = 1.980$ mg m^{-3}, final $R_1 = 0.033$. In the sodalite structures obtained by us, twelve metal$(H_2O)_2O_4$ units link the squares, instead of oxygen atoms as in the normal aluminosilicate sodalite.
© 2002 Éditions scientifiques et médicales Elsevier SAS. All rights reserved.

1. Introduction

There is an implicit conviction today that extended solids with novel structures based on common topologies can be designed and synthesized by assembling simple structural building units [1,2]. Thus, several diamondoid networks have been obtained by using tetrahedral building units. Metal carboxylate clusters have been employed for the design of porous metal-organic frameworks containing square grids wherein paddle wheel metal–O–C clusters act as secondary building units [3]. For example, NbO type networks have been synthesized by this means. Structures required for linking square units to create polyhedra and networks have been adequately described [4], but in spite of the progress in such rational design, it is not always possible to obtain all the desired structures. For example, sodalite networks have not been found in structures based on metal-organic frameworks [4]. In an attempt to form sodalite networks based on metal-organic frameworks, we have carried out reactions of squaric acid with metal ions under hydrothermal conditions in the presence of 4,4′-bipyridine. Several workers have prepared metal squarates under hydrothermal conditions and reported some interesting compounds with porous or polymeric structures, and useful properties [5–7]. In this communication, we report the successful synthesis of sodalite structures based on metal squarates. Though the structures themselves are topologically not new and analogous compounds with the same stoichiometry have been reported in literature [8–10], their relationship with sodalite has not hitherto been realized.

2. Experimental

2.1. Synthesis

1: A mixture of Co_3O_4 (0.1204 g), squaric acid (0.1743 g), 4,4′-bipyridine (0.151 g), H_3PO_4 (85% wt) (0.18 g), CsOH (0.3 ml, w/w 50%) and H_2O (4 g) was sealed in a Teflon-lined stainless steel autoclave (23 ml) and heated at 180°C for 3 days. The pH of the reaction mixture prior to heating was ~ 3. The orange-red cubic crystals of **1**, along with uncharacterized orange needles, were filtered off, and washed with ethanol and dried. Thermogravimetric analysis of the compound **1** in air shows the loss of water molecules around 180–250°C whereas the squarate units decompose around 350°C.

2: A mixture of Mn_3O_4 (0.113 g), squaric acid (0.1743 g), 4,4′-bipyridine (0.151 g), CsOH (0.3 ml, w/w 50%) and H_2O (4 g) was sealed in a Teflon-lined stainless steel autoclave (23 ml) and heated at 180°C for 3 days (pH $\sim$ 3–4). Very

---

* Correspondence and reprints.
  E-mail address: cheetham@mrl.ucsb.edu (A.K. Cheetham).

1293-2558/02/$ – see front matter © 2002 Éditions scientifiques et médicales Elsevier SAS. All rights reserved.
PII: S 1 2 9 3 - 2 5 5 8 (0 2) 0 0 0 0 3 - 1

 S. Neeraj et al. / Solid State Sciences 4 (2002) 1231–1236

Table 1
Crystal data and structure refinement parameters for **1–3**

Structural parameter	1	2	3
Chemical formula	$Co(H_2O)_2(C_4O_4)$	$Mn(H_2O)_2(C_4O_4)$	$Zn(H_2O)_2(C_4O_4)$
Formula mass	206.93	202.93	213.4
Crystal system	Cubic	Trigonal	Cubic
Space group	Pn-$3n$	R-3	Pn-$3n$
T (K)	293	293	293
a (Å)	16.280(5)	11.607(11)	16.256(3)
b (Å)	16.280(5)	11.607(11)	16.256(3)
c (Å)	16.280(5)	14.66(3)	16.256(3)
γ (°)	90	120	90
V (Å^3)	4315(2)	1711(4)	4295.6(15)
Z	48	18	48
μ (mm^{-1})	2.366	1.717	3.410
2θ range data collected	3.54, 56.48	4.92, 56.40	3.54, 56.84
Total data collected	23211	3517	23333
Unique data	909	898	909
Observed data ($\sigma > 2\sigma(I)$)	603	838	603
R_1, wR_2 [$I > 2\sigma(I)$]	0.041, 0.121	0.029, 0.072	0.033, 0.093
R (all data)	0.063, 0.153	0.030, 0.073	0.053, 0.107

Table 2
Atomic coordinates [$\times 10^4$] and equivalent isotropic displacement parameters [Å$^2 \times 10^3$] for $Mn(H_2O)_2(C_4O_4)$, **2** and $Zn(H_2O)_2(C_4O_4)$, **3**

Atom	x	y	z	U_{eq}
Mn(1)	3333	1667	1667	16(1)
O(1)	1739(1)	84(1)	914(1)	29(1)
O(2)	4833(1)	1257(2)	1137(1)	31(1)
O(3)	3497(1)	2895(1)	480(1)	31(1)
C(1)	4914(2)	565(2)	513(1)	24(1)
C(2)	4324(2)	4052(2)	222(1)	24(1)
Zn(1)	4981(1)	4981(1)	2500	13(1)
O(1)	4615(2)	5370(1)	3632(1)	21(1)
O(2)	3803(1)	4871(2)	1977(1)	24(1)
O(3)	5039(2)	6190(1)	1998(1)	26(1)
C(1)	3092(2)	4847(2)	2271(2)	18(1)
C(2)	5009(2)	6906(2)	2281(2)	20(1)

pale pink cubic crystals of **2** were filtered off, and washed with ethanol and dried.

3: A mixture of ZnO (0.121 g), squaric acid (0.1743 g), H_3PO_4 (85% wt) (0.18 g), CsOH (0.37 ml, w/w 50%) and H_2O (4 g) was sealed in a Teflon-lined stainless steel autoclave (23 ml) and heated at 180°C for 3 days (pH $\sim$ 3–4). Colorless cubic crystals of **3** were filtered off, and washed with ethanol and dried. Thermogravimetric analysis of compound **3** in air shows the loss of water molecules around 200°C whereas the squarate units decompose around 400°C.

All the compounds are stable in air and insoluble in water. These compounds can also be synthesized in the absence of 4,4′-bipyridine by addition of slight excess of cesium hydroxide.

2.2. Single-crystal X-ray crystallography

A suitable single crystal of each of the title compound was carefully selected under a polarizing microscope and glued to a thin glass fiber with cyanoacrylate (Superglue) adhesive. Crystal structure determination by X-ray diffraction was performed on a Siemens SMART-CCD diffractometer equipped with a normal focus, 2.4 kW sealed tube X-ray source (Mo $K\alpha$ radiation, $\lambda = 0.71073$ Å) operating at 50 kV and 40 mA. A hemisphere of intensity data was collected at room temperature in 1321 frames with ω scans (width 0.30° and exposure time of 30 sec per frame). An empirical correction based on symmetry equivalent reflections was applied using the SADABS program [11]. The structure was solved by direct methods using SHELX-97 and difference Fourier syntheses [12]. The relevant details of structure determination are presented in Table 1. Full matrix least-squares refinement against $|F^2|$ was carried out using the SHELXTL-PLUS package of programs [13]. The last cycles of refinement included atomic positions and anisotropic thermal parameters for all atoms. The final atomic coordinates and thermal parameters for **2** and **3** are presented in Table 2.

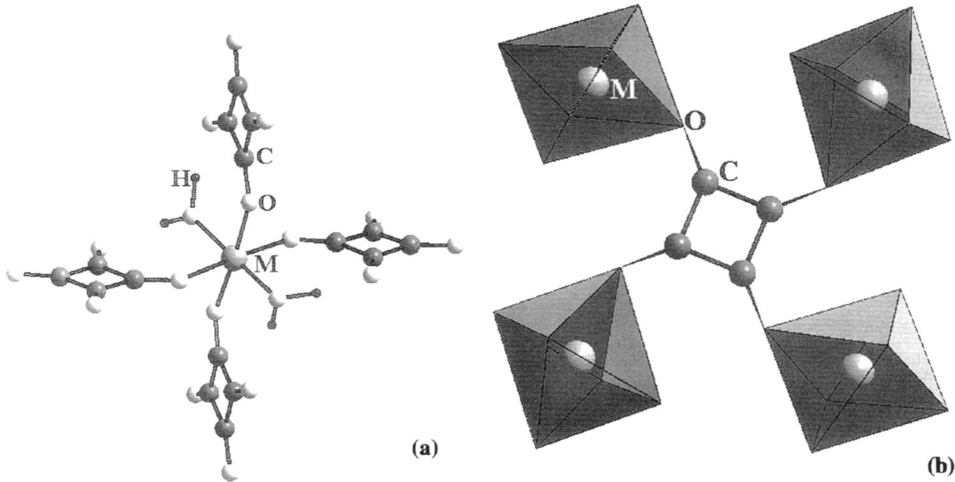

Fig. 1. (a) Coordination environment around the metal atom (M = Co, Mn, Zn) in **1–3**. (b) Figure showing linking of squarate unit to four $M(H_2O)_2O_4$ octahedra in **1–3**.

2.3. Thermal characterization

Thermogravimetric analysis for the compounds was carried out on a Mettler-Toledo TGA/SDTA851 instrument. The TGA of the compound **1** in air shows the loss of water molecules around 180–250°C whereas the squarate units decompose around 350°C.

The thermal stability of **1** was also studied by variable temperature powder X-ray diffraction on a Bruker-D8 Advance powder X-ray diffractometer fitted with a PSD detector.

3. Results

We have obtained three metal squarates of the general composition $M(H_2O)_2(C_4O_4)$ with M = Co, Mn and Zn, designated **1**, **2** and **3**, respectively, by the hydrothermal method. The three-dimensional framework connectivity is identical in **1**, **2** and **3**, although **2** crystallizes in a trigonal spacegroup (*R-3*) whereas **1** and **3** crystallize in the cubic spacegroup (*Pn-3n*). The asymmetric units of **1**, **2** and **3** contain 6 non-hydrogen atoms: the metal, three oxygen (one of the oxygens belonging to a water molecule) and two carbon atoms. The structures of **1–3** are formed by the vertex sharing of the squarate dianions with the $M(H_2O)_2O_4$ octahedra. Each metal octahedron is linked to four different squarate units and two water molecules, which are *trans* to each other. The adjacent (*cis*) squarate units lie in planes perpendicular to each other, as shown in Fig. 1a. Each squarate unit is linked to four metal polyhedra (Fig. 1b). The μ-4 type linking of the squarate units and the metal polyhedra leads to a three-dimensional metal squarate framework, with chan-

nels along the [111] direction in **1** and **3**, and along the *c*-axis in **2**, as can be seen from Fig. 2.

The fascinating aspect of **1**, **2** and **3** is that they all possess a structure similar to that of the aluminosilicate sodalite. The relationship between the sodalite structure and that of the metal squarates with the sodalite topology is as follows. The sodalite cage is formed of six 4-membered rings with four 4-rings lying in the equatorial plane, one 4-ring above and one 4-ring below (Fig. 3a). The six 4-rings are linked to each other by 12 oxygen atoms (not shown in the T-atom connectivity model) to form the sodalite cage. The sodalite topology in **1–3** is obtained by replacing the six 4-membered Al_2Si_2 rings of the sodalite by the 4-membered squarate (C_4) rings, and linking the squarate rings by twelve $M(H_2O)_2O_4$ (M = Co, Mn and Zn in **1**, **2** and **3** respectively) octahedra as shown in Fig. 3b and c. The opposite pairs of the squarate 4-rings in the sodalite cage in **1** and **3** (stacked in a ABAB.. fashion) appear staggered, when viewed along the *a*- *b*- or *c*-axis, whereas the corresponding 4-rings in **2** are stacked in an eclipsed manner (AAAA.) (Fig. 4). The M–O bond distances formed by the squarate units are longer than those formed by the water molecules ($Co-O_{sq} = 2.100(2)$, $2.118(2)$ Å, $Co-O_w = 2.045(2)$ Å; $Mn-O_{sq} = 2.165(2)$, $2.197(3)$ Å, $Mn-O_w = 2.149(2)$ Å; $Zn-O_{sq} = 2.104(2)$, $2.129(2)$ Å, $Zn-O_w = 2.035(2)$ Å). The O–M–O bond angles lie in the range $83.94(8)$–$179.26(13)°$ for **1**, $84.61(10)$–$180.0°$ for **2**, and $83.81(9)$–$178.57(14)°$ for **3**. The bond valence sums calculated using program VALIST [14] are in accordance with the assignment of all the metal atoms in **1**, **2**, and **3** as divalent.

The channels in all the metal squarate frameworks are lined with the water molecules bonded to the metal center, which may render the interior hydrophilic. The channel dimensions are $\sim 7.8 \times 7.8$ Å when taking into account

S. Neeraj et al. / Solid State Sciences 4 (2002) 1231–1236

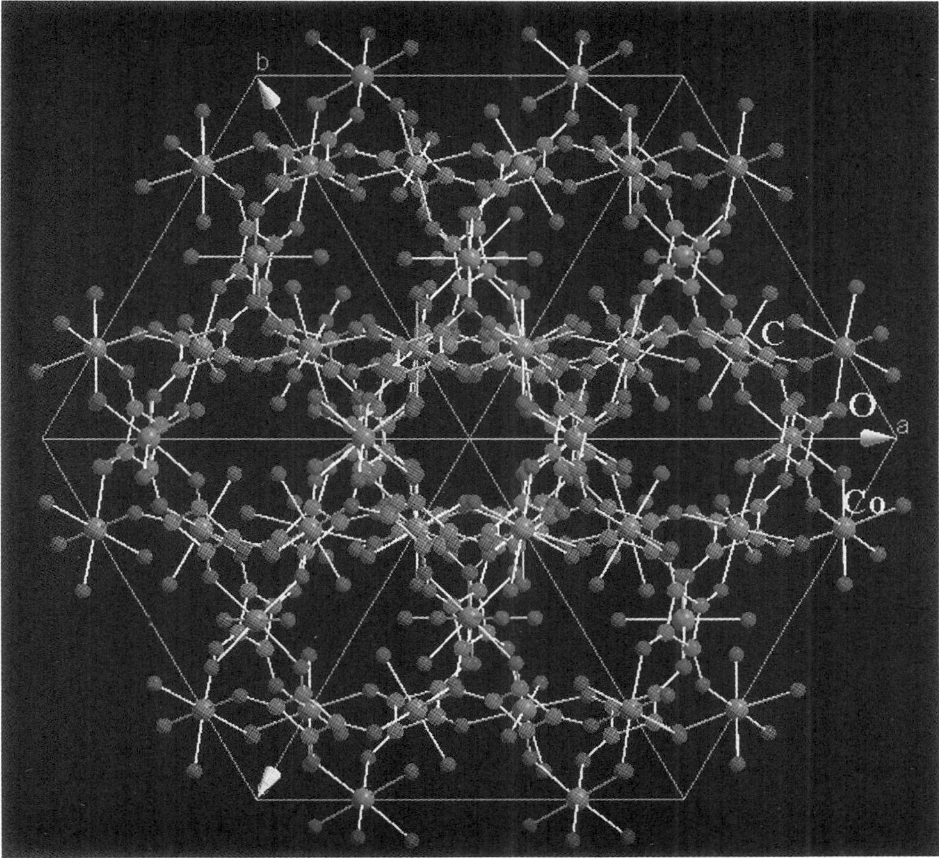

Fig. 2. (a) Ball and stick view of **1** showing the channels along the [111] direction. The hydrogen atoms have been omitted for the sake of clarity. Note that the channels are lined with water molecules linked to the metal atoms.

the water molecules lining the channel (oxygen-to-oxygen contact distance excluding the Van der Waals radii). The effective channel dimension decreases further when considering the hydrogen atoms on the water molecule to $\sim 6.75 \times 6.75$ Å. The hydrogen atoms on the water molecule interact strongly with the framework oxygens via O–H$\cdots$O hydrogen bonds. The hydrogen bond distances and angles (O(1)–H(1)$\cdots$O(3) = 1.6580 Å, 165.86°, O(1)–H(1)$\cdots$O(2) = 1.9729 Å, 166.99° for **1**; O(1)–H(1)$\cdots$O(3) = 1.9437 Å, 161.00°, O(1)–H(1)$\cdots$O(2) = 1.9162 Å, 161.92° for **2**; and O(1)–H(1)$\cdots$O(3) = 1.6075 Å, 161.30°, O(1)–H(1)$\cdots$O(2) = 1.9893 Å, 168.47° for **3**) indicate strong hydrogen bonding in all the framework structures.

Thermogravimetric analyses of **1** and **3** carried out in air shows the loss of water molecules around 180–250°C and the decomposition of the squarate units around 350–400°C. Variable temperature powder X-ray diffraction carried on **1**

shows that the framework structure is unstable following the loss of water.

In conclusion, we have been able to synthesize sodalite networks based on the metal-organic framework of metal squarates. Although the cobalt phase, **1**, has been reported previously [10], and closely related solvates have been described for several divalent transition metals [8,9] their relationship to the sodalite has not been recognized previously. Clearly, the 4-membered square of the squarate acts as the basic building unit in generating the sodalite framework. The sodalite network obtained here is highly expanded since the linkers of the squares are the $M(H_2O)_2O_4$ units rather than oxygen atoms. The role of the amine employed in the synthesis is far from clear, although we believe that it is involved in the formation of a precursor species in addition to controlling the pH of the reaction mixture.

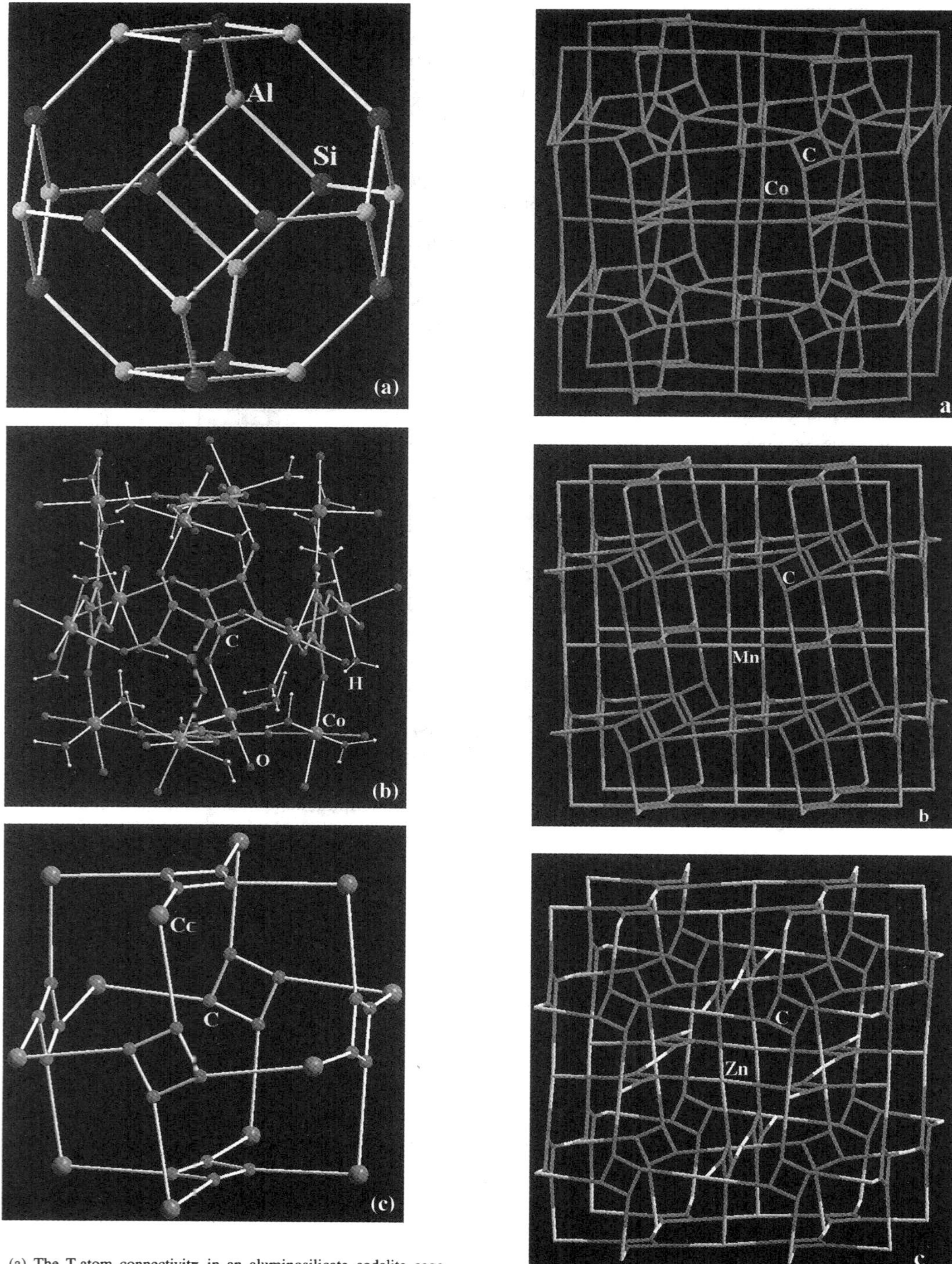

Fig. 3. (a) The T-atom connectivity in an aluminosilicate sodalite cage. (b) The ball and stick view of analogous sodalite cage in **1**. (c) A different depiction of the sodalite cage in **1**. The oxygen atoms have been omitted and the metal atom is linked to the carbon atoms of the squarate ring to show the analogy with the sodalite cage.

Fig. 4. (a) The three-dimensional sodalite networks in **1**, **2**, and **3** are shown (a), (b) and (c) respectively.

Acknowledgements

This work was supported by Unilever plc and the MRSEC Program of the National Science Foundation under Award No. DMR00-80034.

References

[1] M. O'Keeffe, M. Eddaoudi H. Li, T. Reineke, O.M. Yaghi, J. Solid State Chem. 152 (2000) 3.

[2] G. Férey, J. Solid State Chem. 152 (2000) 37.

[3] M. Eddaoudi, J. Kim, M. O'Keeffe, O.M. Yaghi, J. Am. Chem. Soc. 124 (2002) 376, and the references cited therein.

[4] M. Eddaoudi, J. Kim, D. Vodak, A. Sudik, J. Wachter, M. O'Keeffe, O.M. Yaghi, Proc. Natl. Acad. Sci. USA 99 (2002) 4900.

[5] S.O.H. Gutschke, M. Molinier, A.K. Powell, P.T. Wood, Angew. Chem. Int. Ed. Engl. 36 (1997) 991.

[6] D.S. Yufit, D.J. Price, J.A.K. Howard, S.O.H. Gutschke, A.K. Powell, P.T. Wood, Chem. Commun. (1999) 1561.

[7] K.J. Lin, K.H. Lii, Angew. Chem. Int. Ed. Engl. 36 (1997) 2076.

[8] A. Weiss, E. Riegler, C. Robl, Z. Naturforsch. 41b (1986) 1329.

[9] A. Weiss, E. Riegler, C. Robl, Z. Naturforsch. 41b (1986) 1333.

[10] C.R. Lee, C.C. Wang, Y. Wang, Acta. Crystallogr., Sect. B 52 (1996) 966.

[11] G.M. Sheldrick, SADABS User Guide, University of Göttingen, Göttingen, 1995.

[12] G.M. Sheldrick, SHELXL-97, A program for crystal structure determination, University of Göttingen, Göttingen, 1997.

[13] G.M. Sheldrick, SHELXTL-PLUS Program for Crystal Structure Solution and Refinement, University of Göttingen, Göttingen, 1993.

[14] A.S. Wills, I.D. Brown, VaList, CEA, France, 1999.

Brief Biodata of Professor C.N.R. Rao

Professor C.N.R. Rao is the Linus Pauling Research Professor and Honorary President of the Jawaharlal Nehru Centre for Advanced Scientific Research, Bangalore. His main research interests are in solid state and materials chemistry, surface phenomena, spectroscopy and molecular structure. He is an author of over 1000 research papers and 36 books. He received the M.Sc. degree from Banaras, Ph.D. from Purdue, D.Sc. from Mysore universities and has received honoris causa doctorate degrees from 32 universities including Purdue, Bordeaux, Banaras, Mysore, IIT Bombay, IIT Kharagpur, Notre Dame, Novosibirsk, Uppsala, Wales, Wroclaw, Caen, Khartoum and Sri Venkateswara University.

Besides being Fellow of the Indian National Science Academy and the Indian Academy of Sciences, Professor Rao is a Fellow of the Royal Society, London, Foreign Associate of the National Academy of Sciences, U.S.A., Foreign Member of the Russian Academy of Sciences, French Academy of Sciences, Japan Academy as well as the Polish, Czechoslovakian, Serbian, Slovenian, Brazil, Spanish, Korean and African Academies and the American Philosophical Society. He is a Member of the Pontifical Academy of Sciences, Foreign Member of Academia Europaea and Foreign Fellow of the Royal Society of Canada. He is on the editorial boards of 15 leading professional journals.

Among the various medals, honours and awards received by him, mention must be made of the Marlow Medal of the Faraday Society (1967), Bhatnagar Prize (1968), Jawaharlal Nehru Fellowship (1973), Padma Shri (1974), Sir C.V. Raman Award (1975), Centennial Foreign Fellowship of the American Chemical Society (1976), S.N. Bose Medal of the Indian National Science Academy (1980), Royal Society of Chemistry (London) Medal (1981), Padma Vibhushan (1985), Honorary Fellowship of the Royal Society of Chemistry, London (1989), Hevrovsky Gold Medal of the Czechoslovak Academy (1989), Meghnad Saha Medal of the Indian National Science Academy (1990), Blackett Lectureship of the Royal Society (1991), CSIR Golden Jubilee Prize in physical sciences (1991), TWAS Medal in Chemistry (1995), Einstein Gold Medal of UNESCO (1996), Linnett Professorship of the University of Cambridge (1998), Centenary Lectureship and Medal of the Royal Society of Chemistry, London (2000), the Hughes Medal of the Royal Society, London for original discovery in physical sciences (2000), Karnataka Ratna (2001) by the Karnataka Government, the Order of Scientific Merit (Grand-Cross) from the President of Brazil (2002),

Commander of the Order of Rio Branco from Brazil (2002) and the Gauss Professorship of Germany (2003).

Professor Rao is the President of the Third World Academy of Sciences, Member of the Atomic Energy Commission of India and of the Executive Board of the Science Institutes Group, Princeton, and Chairman, Indo-Japan Science Council. Prof. Rao was President of the Indian National Science Academy (1985-86), the Indian Academy of Sciences (1989-91), the International Union of Pure and Applied Chemistry (1985-87), the Indian Science Congress Association (1987-88), the Materials Research Society of India (1989-91) and Chairman, Advisory Board of the Council of Scientific and Industrial Research (India). He was the Director of the Indian Institute of Science (1984-94), Chairman of the Science Advisory Council to Prime Minister Rajiv Gandhi (1985-89) and Chairman, Scientific Advisory Committee to the Union Cabinet (1997-98), and Albert Einstein Research Professor (1995-99).

* * *

Scientific Colleagues

Graduate Students

Agarwal D.C.
Agarwal U.
Aiyer H.N.
Arulraj A.
Anupama T.R.
Arunarkavalli T.
Ayyappan S.*
Bahadur D.
Balasubramanian A.
Basu P.K.
Bhaskar K.R.
Bhat S.N.
Chakrabarti A.
Chaturvedi G.C.
Choudhury A.
Dan M.
Deepak F.L.
Dwivedi P.C.
Ganapathi L.
Ganguli A.K
Ganguly S.*
Gosavi R.K.
Govindaraj A.*
Gundiah G.
Gupta A.
Harikumar K.R.
Jayaram V.
Jayaraman K.
Kalaiselvam M.
Kamath P.V.
Kannan K.R.
Kulkarni G.U.*
Madhusudan W.H.

Mahesh R.
Manivannan V.
Mehrotra P.N.
Mohan Ram R.A.
Murthy A.S.N.*
Murugavel P.
Murugesan T.
Nagarajan R.
Natarajan M.
Nath M.
Neeraj S.*
Om Parkash
Parashar S.
Parthasarathy R.
Patil K.C.
Prabhakar S.
Prabhakaran K.
Pradeep T.*
Ch. Pulla Rao
Raina G.
Raju A.R.*
Rajumon M.K.
Rama Rao G.
Ramasesha S.
Ramasesha S.K.
Ramdas S.*
Ranga Rao G.*
Rangavittal N.
Rao A.M.K.
Rao K.G.
Rao K.J.*
Sankar G.*
Santra A.K.

Sardar K.
Sarma D.D.
Sastry R.L.N.
Satish Kumar B.C.
Sen P.
Sen R.
Seshadri R.
Shukla A.K. *
Singh S.
Sivashankar K.
Somasundaram P.
Srinivasan A.
Srinivasa Gopalan R.
Subbanna G.N.
Subba Rao G.V.*
Sudheendra L.
Swamy H.R.
Tantry K.N.
Thomas P. J.
Uppal M.K.
Vanitha P.V.
Varma V.
Vasudevan S.
Vaidhyanathan R
Venkataraghavan R.
Vidyasagar K.
Vinod C.P.
Vijayakrishnan V.
Vijayaraghavan R.
Vijayasarathy K.
Yashonath S.
Yoganarasimhan S.R.

* also worked as postdoctoral associates

Collaborators during their Graduate Studies

Ayi A.A.
Bita B.
Bouet L.
Buttrey D.
Chakraborti P.
Chandrabhas N.
Ebenso E.E.
Ghosh A.
Guha A.

Jadhao V.
Kalyanaraman V.
Ch. Laurent
Loehman R.
Mahendiran R.
Nanjundaswamy K.S.
Ramachandran J.
Ramanan A.
Singh K.K.

de Santos (Antonio)
Singh K.P.
Somasundaram T.
Sreedhar K.*
Thornton G.
Vasan H.N.
Vasanthacharya N.Y.

* also worked as postdoctoral associates

Postdoctoral Associates

Bandekar N.
Bharati S.
Bhat V.
Bhattacharya R.
Chandrashekhar G.V.
Chandrashekhar M.
Desiraju G.R.
Durrant M.C.
Fernández J.R.
Ganguly P.

Grantscharova E.
Hegde M.S.
Jagannathan K.
Jagannathan N.R.
Kudchadkar S.A.
Kumardhas P.
Prakash B.
Ponnambalam V.
Randhawa H.S.
Reller A.

Santosh P.N.
Sarode P.R
Singh R.S.
Sreekanth C.S.
Sugandha
Tomar M.S.
Viswanathan B.
Verelst M.
Yadav R.T.

Senior Collaborators

Anderson J.S.
Attfield P.
Bhat S.V.
Bhide V.G.
Cheetham A.K.
Edwards P.P.
Ferraro J.R.
Feuer H
Ferey G.

Goodenough J.B
Gopalakrishnan J.
Honig J.M.
Jefferson D.
Jones W.
Kumar N.
Mahanty J.
Mott N.F.
Nagarajan R.

Ramakrishnan T.V.
Raychaudhuri A.K.
Raveau B.
Roberts M.W.
Rousset A.
Sampathkumaran E.V.
Singru R.M.
Sood A.K.
Thomas J.M.

* * *